ENVIRONMENTAL
Science

TOWARD A SUSTAINABLE FUTURE 13E

Richard T. Wright | Dorothy F. Boorse

Gordon College

Boston   Columbus   Indianapolis   New York   San Francisco   Amsterdam
Cape Town   Dubai   London   Madrid   Milan   Munich   Paris   Montréal   Toronto
Delhi   Mexico City   São Paulo   Sydney   Hong Kong   Seoul   Singapore   Taipei   Tokyo

VP/Editor in Chief: Beth Wilbur
Acquisitions Editor: Alison Rodal
Project Manager: Arielle Grant
Program Manager: Anna Amato
Development Editor: Julia Osborne
Editorial Assistant: Alison Cagle
Executive Editorial Manager: Ginnie Simione Jutson
Program Management Team Lead: Mike Early
Project Management Team Lead: David Zielonka

Production Management, Compositor: Integra
Production Project Managers: Angel Chavez, Charles Fisher
Design Manager: Marilyn Perry
Interior and Cover Designer: tani hasegawa
Rights & Permissions Project Manager: Donna Kalal
Rights & Permissions Management,
   Photo Researcher: Amanda Larkin, QBS
Manufacturing Buyer: Maura Zaldivar-Garcia
Executive Marketing Manager: Lauren Harp

ABOUT OUR SUSTAINABILITY INITIATIVES

Pearson recognizes the environmental challenges facing this planet, as well as acknowledges our responsibility in making a difference. This book is carefully crafted to minimize environmental impact. The paper's chain of custody is certified by the Forest Stewardship Council® (FSC®) . The binding, cover, and paper come from facilities that minimize waste, energy consumption, and the use of harmful chemicals. Pearson closes the loop by recycling every out-of-date text returned to our warehouse.

Along with developing and exploring digital solutions to our market's needs, Pearson has a strong commitment to achieving carbon neutrality. In 2009, Pearson became the first carbon- and climate-neutral publishing company. Since then, Pearson remains strongly committed to measuring, reducing, and offsetting our carbon footprint.

The future holds great promise for reducing our impact on Earth's environment, and Pearson is proud to be leading the way. We strive to publish the best books with the most up-to-date and accurate content, and to do so in ways that minimize our impact on Earth. To learn more about our initiatives, please visit **www.pearson.com/responsibility**.

MIX
Paper from
responsible sources
FSC® C101537

PEARSON

Library of Congress Cataloging-in-Publication Data available upon request from Publisher.

2 16

www.pearsonhighered.com   ISBN-10: 0-134-01127-9; ISBN-13: 978-0-134-01127-1

About the Authors

Richard T. Wright is Professor Emeritus of Biology at Gordon College in Massachusetts, where he taught environmental science for 28 years. He earned a B.A. from Rutgers University and an M.A. and a Ph.D. in biology from Harvard University. For many years, Wright received grant support from the National Science Foundation for his work in marine microbiology, and in 1981, he was a founding faculty member of Au Sable Institute of Environmental Studies in Michigan, where he also served as Academic Chairman for 11 years. He is a Fellow of the American Association for the Advancement of Science, Au Sable Institute, and the American Scientific Affiliation. In 1996, Wright was appointed a Fulbright Scholar to Daystar University in Kenya, where he taught for two months. He is a member of many environmental organizations, including the Nature Conservancy, Habitat for Humanity, the Union of Concerned Scientists, and the Audubon Society, and is a supporting member of the Trustees of Reservations. He volunteers his services at the Parker River National Wildlife Refuge in Newbury, Massachusetts, and is an elder in First Presbyterian Church of the North Shore. Wright and his wife, Ann, live in Byfield, Massachusetts, and they drive a Toyota Camry hybrid vehicle as a means of reducing their environmental impact. Wright spends his spare time birding, fishing, hiking, and enjoying his three children and seven grandchildren.

Dorothy F. Boorse is a professor of biology at Gordon College in Wenham, Massachusetts. Her research interest is in drying wetlands, such as vernal pools and prairie potholes, and in salt marshes. Her research with undergraduates has included wetland and invasive species projects. She earned a B.S. in biology from Gordon College, an M.S. in entomology from Cornell University, and a Ph.D. in oceanography and limnology from the University of Wisconsin–Madison. Boorse teaches, writes, and speaks about biology, the environment, ecological justice, and care of creation. She was recently an author on a report on poverty and climate change. In 2005, Boorse provided expert testimony on wildlife corridors and environmental ethics for a Congressional House subcommittee hearing. Boorse is a member of a number of ecological and environmental societies, including the Ecological Society of America, the Society of Wetland Scientists, the Nature Conservancy, the Audubon Society, the New England Wildflower Society, and the Trustees of Reservations (the oldest land conservancy group in the United States). She and her family live in Beverly, Massachusetts. They belong to Appleton Farms, a CSA (community-supported agriculture) farm. At home, Boorse has a native plant garden and has planted two disease-resistant elm trees.

Dedication

This edition is dedicated to Sylvia Earle (1935–), marine scientist and tireless advocate for the environment. An oceanographer, explorer, author, company founder, and lecturer, Earle has lived out a fantastic dream to study the oceans, and fearlessly pursued that goal when opportunities for women were limited. After receiving her Ph.D. in 1966, Earle was a research fellow at Harvard and then moved to Florida, where she took underwater research dives, setting records for women's depth diving, and leading an all-female team of aquanauts in an underwater research project. While she has many other accomplishments as well, Earle is particularly noted for being Chief Scientist at the National Oceanic and Atmospheric Administration from 1990 to 1992, and a National Geographic Explorer-in-Residence since 1998. Earle has founded three companies, which produce robotics and other ocean exploration equipment.

In 1998, Earle was named the first "Hero for the Planet" by *Time* magazine. In 2009, she won a TED prize, which come with money to carry out a vision for global change. She used that opportunity to launch a nonprofit, Mission Blue, which aims to establish what Earle calls "hope spots," or marine protected areas around the globe. It is an honor to dedicate this book to someone who is such a good scientist and has done so much to help the environment. Earle represents the themes of sound science, stewardship, and sustainability in a way few people do. She is a real hero for our time. In her own words, she calls upon us to act to protect the ocean:

"People ask: Why should I care about the ocean? Because the ocean is the cornerstone of Earth's life support system, it shapes climate and weather. It holds most of life on Earth. Ninety-seven percent of Earth's water is there. It's the blue heart of the planet—we should take care of our heart. It's what makes life possible for us. We still have a really good chance to make things better than they are. They won't get better unless we take the action and inspire others to do the same thing. No one is without power. Everybody has the capacity to do something."

– Sylvia Earle in the film Bag It: Is Your Life Too Plastic?
(Paramount Classics 2010)

Contents

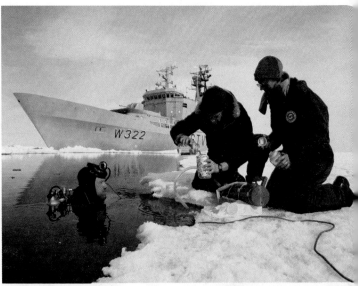

PART THREE THE HUMAN
POPULATION AND
ESSENTIAL RESOURCES **183**

PART SIX STEWARDSHIP FOR A SUSTAINABLE FUTURE   567

23 Sustainable Communities and Lifestyles   568

Essay List

Preface

We are now well into the 21st century and are at critical junctures in the relationship between humans and the rest of the environment. Globally, major changes are taking place in the atmosphere and climate, the human population and its well-being, and the Earth's natural resources. We are still recovering from a major global economic recession, the scientific evidence for climate change continues to accumulate, terrorism and conflict continue to grip the Middle East, and an extended drought in the western half of the United States and elsewhere is affecting food production and water availability.

In contrast to these trends, there are some changes that point to a brighter future. Renewable energy is ramping up swiftly in its share of the world's energy portfolio; many of the UN Millennium Development Goals (MDGs) were achieved by their target date of 2015; the Sustainable Development Goals that follow them have been crafted and are close to launching.

Even though international accord on climate change is slow in coming, many countries are achieving major reductions in greenhouse gas emissions; death by tobacco use is being addressed in a global campaign; AIDS, tuberculosis, and malaria are on the defensive as public-health agencies expand treatment options and research; and population growth rates in many regions are continuing to decline.

The most profound change that must happen, and soon, is the transition to a sustainable civilization—one in which a stable human population recognizes the finite limits of Earth's systems to produce resources and absorb wastes, and acts accordingly. This is hard to picture at present, but it is the only future that makes any sense. If we fail to achieve it by our deliberate actions, the natural world will impose it on us in highly undesirable ways.

Core Values

Environmental science stands at the interface between humans and Earth and explores the interactions and relations between them. This relationship will need to be considered in virtually all future decision making. This text considers a full spectrum of views and information in an effort to establish a solid base of understanding and a sustainable formula for the future. What you have in your hands is a readable guide and up-to-date source of information that will help you to explore the issues in more depth. It will also help you to connect them to a framework of ideas and values that will equip you to become part of the solution to many of the environmental problems confronting us.

In this new edition, we hope to continue to reflect accurately the field of environmental science; in so doing, we have constantly attempted to accomplish each of the following objectives:

- To write in a style that makes learning about environmental science both interesting to read and easy to understand, without overwhelming students with details;
- To present well-established scientific principles and concepts that form the knowledge base for an understanding of our interactions with the natural environment;
- To organize the text in a way that promotes sequential learning yet allows individual chapters to stand on their own;
- To address all of the major environmental issues that confront our society and help to define the subject matter of environmental science;
- To present the latest information available by making full use of resources such as the Internet, books, and journals, every possible statistic has been brought up to date;
- To assess options or progress made in solving environmental problems; and
- To support the text with excellent supplements for teachers and students that strongly enhance the teaching and learning processes.

Because we believe that learning how to live in the environment is one of the most important subjects in every student's educational experience, we have made every effort to put in their hands a book that will help the study of environmental science come alive.

New to This Edition

Building on its core values, *Environmental Science* has several new features that help make the content more approachable to students.

- **Concept Check** questions appear at the end of sections to help students check their understanding. Each concept check aligns with the learning objectives at the start of the chapter.
- **Understanding the Data** questions appear alongside figures, such as graphs or maps, to help students further engage with the data and build data analysis skills.
- Content has been thoroughly updated throughout; environmental science and the related issues are complex and constantly evolving. The thirteenth edition features the most current research presented in a balanced manner.

- Transitioning beyond the Millennium Development Goals of 2000–2015, discussions of environmental issues and solutions have been reframed to reflect the new Sustainable Development Goals.

- New Everyday Environmental Science videos and coaching activities in MasteringEnvironmentalScience accompany the thirteenth edition and will help students to see the application of the key concepts presented throughout the text to the real world.

Content Updates to the Thirteenth Edition of Environmental Science

Part One—Framework for a Sustainable Future

- The Part One opener focuses on the vision of sustainability and the challenges facing us that are inconsistent with that vision.

- Chapter 1 (Science and the Environment) The chapter has been restructured so that each of our three unifying themes (sustainability, sound science, and stewardship) now has its own major section. A new essay (Stewardship: "Protecting Forests") introduces the concept of stewardship. Hypothesis formation is illustrated with a revised essay, "Oysters Sound the Alarm," while a new graphic better explains the scientific method, which now includes more emphasis on the community of science. A new table (1–2) lists previous and ongoing ecosystem and biodiversity assessments.

- Chapter 2 (Economics, Politics, and Public Policy) A new chapter opener shows the impact of China's rapid economic growth on air pollution there. The chapter was reorganized into six sections rather than five. A figure illustrating the global environmental footprint and human development index shows the narrowing safe space for humanity. The chapter's end was significantly rewritten. It includes new information on international politics that informs discussion on global issues in later chapters. It also has a stronger tie between ethics and indigenous rights.

Part Two—Ecology: The Science of Organisms and Their Environment

- Part two begins with a part opener about the importance of science in understanding the ecosystems around us. Chapters 3–5 include topics that flow from basic to more complex, small to large, and species to ecosystems and humans.

- Chapter 3 (Basic Needs of Living Things) has a new chapter opener on the ecology of the Emperor penguin, which is followed as an illustration throughout the chapter. The chapter covers the science of ecology, needs of organisms, matter and energy, respiration and photosynthesis, and four material cycles. The carbon cycle has been updated to include the impact of volcanoes.

- Chapter 4 (Populations and Communities) opens with a new chapter opener (savanna elephants controlling a non-native plant by their herbivory). The examples of population growth have been simplified somewhat. While mathematical equations are not included, descriptions of the meanings of those equations are. A new Sustainability box ("Elephants in Kruger National Park: How Many Is Too Many?") illustrates concepts covered in sections on growth and limits of populations.

- Chapter 5 (Ecosystems: Energy, Patterns, and Disturbance) includes a new Sound Science essay ("What Are the Effects of Reintroducing a Predator?") and a new Stewardship essay ("Local Control Helps Restore Woodlands"). There is more information on the oceans, including the importance of picophytoplankton. New biomes or ecosystems (chaparral, tropical dry forest, coral reefs, and deep sea vents) are described and a new biome map is included. A new figure depicts ecosystem services. The concept of Human Appropriation of Net Primary Production (HANPP) has been added in the discussion of human effects on ecosystems.

- Chapter 6 (Wild Species and Biodiversity) features expanded discussion on the Aichi Biodiversity Targets. The section on the Endangered Species Act has been rewritten and a new table (6–4, Important U.S. Federal and International Conservation Regulations) streamlines discussion on regulation. The newest *Global Biodiversity Outlook 4* report is also included.

- Chapter 7 (The Value, Use, and Restoration of Ecosystems) has been reorganized. A central section on ecosystems under pressure has been divided into two sections—one on forests and grasslands and another on oceans—with restoration and conservation moved to the end of the chapter. A section on maximum sustainable yield is explained more thoroughly. Land protection has been expanded to include international protection as well. The chapter has an even greater focus on ecosystem goods and services and a completely rewritten final section on restoration, with examples of restoration from a number of biomes.

Part Three—The Human Population and Essential Resources

- Part 3 begins with a new part opener about community-supported agriculture and the resources we need to support 7.2 billion people.

- Chapter 8 (The Human Population) still covers population growth, the various revolutions that have increased growth, and concepts such as the IPAT and ImPACT equations and the Gini index of inequality. Sections 8.1 and 8.3 have been streamlined substantially to place more emphasis on humans as populations.

- Chapter 9 (Population and Development) has a new Sustainability box ("Dealing with Graying Populations"), which covers the needs of increasingly elderly populations, using Japan as an example. The examples of China and India have been made into case studies of population growth. Later in the chapter, the Millennium Development Goals are discussed with a new table (9–2) that describes their levels of success. The new Sustainable Development Goals are rolled out in a new table (9–3).

- Chapter 10 (Water: Hydrologic Cycle and Human Use) has a new chapter opener on drought in California. More has been added on climate change and the concept of peak water. The section on dams has been made clearer and shortened. There is an added figure on the over-pumping of ground water. Additional information on water loss from aging infrastructure, water in agriculture and fracking, and on the UN's efforts to promote water planning is included.

- Chapter 11 (Soil: Foundation for Land Ecosystems) begins with a new picture of a dust storm in Arizona and a story about the Dust Bowl. A new story of land creation in China illustrates a different type of mountaintop removal. Throughout the chapter, there is more on mining, soil pollution, and reclamation. A new concept (the land-degradation neutral world) is introduced. And a new section on international efforts to protect soil is added. Soil is connected to the new Sustainable Development Goals as the chapter wraps up.

- Chapter 12 (The Production and Distribution of Food) has been altered significantly. The chapter is organized around three ideas—production, environmental sustainability, and effective distribution. The section on GMO crops has been substantially rewritten and updated. The last section, "Feeding the World as We Approach 2030–2050," has been rewritten and broken into four subsections: increasing food production, using current production more efficiently, sustainable agriculture, and policy changes. The term *food justice* is introduced and described.

- Chapter 13 (Pests and Pest Control) maintains many of the changes made in the twelfth edition, including a chapter opener on bedbugs and the updated essays. Throughout the chapter, as in all chapters, data and graphics have been updated. A new example of APHIS stopping a coconut rhinoceros beetle infestation in Hawaii is included.

Part Four—Harnessing Energy for Human Societies

- Part Four begins with an updated part opener with bad news and good news about energy.

- Chapter 14 (Energy from Fossil Fuels) includes an expanded discussion on oil sands, and the Keystone pipeline proposal is presented and illustrated. Discussions of natural gas, coal, and hydraulic fracturing (fracking) are expanded, with a new figure illustrating fracking. Policy is better described with two new tables dealing with demand-side and supply-side policy actions to lower U.S. dependence on foreign oil. More analysis of three major laws makes it clear they have both positive and negative aspects. The chapter also includes two new rules by the EPA, one about power plants and emissions and the other about coal ash.

- Chapter 15 (Nuclear Power) opens with a shortened and updated chapter opener about the earthquake and tsunami that rocked northern Japan in 2011, leading to the nuclear disaster at the Fukushima Daiichi power plant. New information is presented on the recent ecology of the area around Chernobyl, the current status of U.S. nuclear waste disposal, and on issues with extending power plant life spans.

- Chapter 16 (Renewable Energy) opens with a new opener about Germany's extraordinary effort to change its energy economy. An initial section ("Strategic Issues") examines the issues surrounding calls for great changes in renewable energy. A new figure (16-4) shows relative amounts of energy availability from different sources. The chapter has been changed to better reflect the pros and cons of each type of renewable energy and updates to their levels of use. Sections on dams and on energy laws have been shortened, as these concepts are covered in other chapters. A Sustainability box ("Transfer of Energy Technology to the Developing World") has been rewritten to reflect new dilemmas in that transfer.

Part Five—Pollution and Prevention

- Part Five begins with a part opener that describes the problems of pollution and lays out the coming chapters.

- Chapter 17 (Environmental Hazards and Human Health) opens with a new opener on Ebola covering the 2014–2015 international outbreak. The chapter has been restructured to begin with definitions of environmental health, focusing more on environmental harms and less on cultural hazards. Pollution is moved later and set in the context of other hazards. A new figure on malaria simplifies the discussion on that disease. A new Stewardship box ("A Cultural Hazard Worsens Other Risks") connects tobacco use to a range of other public health issues. New information on heavy metal, mining and industry as a pathway to risk, unintended poisonings, urban air pollution, radiation, climate change and public health, and exposure to animals as a pathway to infectious risk broadens the discussion on risks and pathways. A new Sound Science box ("Water Pollution Drives Malnutrition in India") describes cutting-edge research on the connection between water-borne disease and childhood stunting.

- Chapter 18 (Global Climate Change) is significantly revised. The chapter has been reframed around the newest (2013–2014) IPCC report (AR5) and includes information from the AAAS report *What We Know*

(2014) and *National Climate Assessment 3* (2014). The sections are reorganized so that the physical science basis of the atmosphere and how climate change occurs is in the first section, followed by evidence that climate change is occurring. The third section covers the effects of climate change, including new information on effects on glaciers. The term "risk multiplier" is introduced. Adaptation, mitigation, and geo-engineering are expanded. The major international climate conferences are rewritten and summarized in a new table. Information on talks between the United States and India, and the United States and China climate talks in 2014, is included, and the idea of "contract and converge" is added.

- Chapter 19 (Atmospheric Pollution) begins with an updated chapter opener on air pollution in Donora, Pennsylvania. The chapter now more clearly emphasizes two trends: global air quality is declining, while U.S. air quality is improving. There is a great deal more on international air pollution. A new Sound Science feature ("Complex Clouds and Co-Benefits of Solutions") explores the complexity of the environmental and health effects of atmospheric brown clouds and the positive benefits of solving multiple problems at the same time.

- Chapter 20 (Water Pollution and Its Prevention) begins with an updated chapter opener. A new Sound Science box ("Can Salt Marshes Absorb Our Nutrients?") explores new long-term research on the effects of nutrient pollution on salt marshes. International water issues are emphasized more, particularly in sections on sewage treatment and on policy. A new graphic makes the workings of composting toilets more clear. Water pollution is connected to themes from other chapters on global N and P cycles and planetary boundaries, as well as to both Millennium Development Goals and Sustainable Development Goals.

- Chapter 21 (Municipal Solid Waste: Disposal and Recovery) has a new chapter opener on positive waste management examples in both the country of Sweden and in Lagos, Nigeria. The chapter is revised to introduce waste disposal globally first, followed by waste disposal solutions in the United States such as recycling, and public policy. There is more on the ocean and new art depicting garbage gyres and the effects of trash on wildlife. More types of recycling are showcased, especially the recycling of tires, batteries, and e-waste. The chapter is more international, with more about informal recycling and reuse, people who live as waste pickers, and international regulations. There are more examples of positive change such as Big Belly Solar trash compactors, new materials replacing plastics, bag laws, and student volunteerism. New box features (Stewardship: "Citizen Power," and Sustainability: "Taking the Waste Out of Take-Out") highlight positive stories.

- Chapter 22 (Hazardous Chemicals: Pollution and Prevention) has a new opener on the contamination of water and soil from the mining of rare earth minerals in Inner Mongolia. Information on Bisphenol A, an endocrine disruptor, has been moved to a new box feature (Sound Science: "Disrupting Hormones with Pollution"). The historic protests in Warren County, North Carolina, in 1982 are used to introduce the environmental justice movement, a new topic in the final section.

Part Six—Stewardship for a Sustainable Future

- Chapter 23 (Sustainable Communities and Lifestyles) has been significantly reorganized to begin with a section on definitions and megatrends in communities (urbanization and the collapse of rural communities), followed by trends in U.S. communities. The urban heat-island effect and new art illustrating it have been added. The development of sustainable communities follows. A new box feature (Sustainability: "Curitiba, Brazil—City Planning Meets Growth") describes some of the successes and struggles of one of the most sustainable cities in Latin America. Discussions of urban homesteading, the revitalization of Detroit, city climate adaptation, smart growth (planning to avoid sprawl), smart cities (using big data), green buildings, and networks of big cities have been added. A new feature (Stewardship: "Living in Tiny Houses") highlights decisions some are making to lower their resource footprint.

Reviewers

Lisa Doner, *Plymouth State University*

Chris Cogan, *Ventura College*

Melinda Huff, *NEO A&M College*

Clark Adams, *Texas A&M*

Eric Atkinson, *Northwest College*

Chris D'Elia, *Louisiana State University*

Edward Laws, *Louisiana State University*

Matthew Eick, *Virginia Tech*

Nisse Goldberg, *Jacksonville University*

John S. Campbell, *Northwest College*

Meg Rawls, *Carteret Community College*

Carole Neidich-Ryder, *Nassau Community College*

Kim Bjorgo-Thorne, *West Virginia Wesleyan College*

Kathy McCann Evans, *Reading Area Community College*

Sarah Havens, *East Central*

Gregory S. Keller, *Gordon College*

Karen McReynolds, *Hope International University*

Ken Peterson, *Bethel University*

Aldemaro Romero Jr., *Southern Illinois University, Edwardsville*

Dr. David Unander, *Eastern University*

Alison Varty, *College of the Siskiyous*

Reviewers of Previous Editions

Clark Adams, *Texas A&M University*

Anthony Akubue, *St. Cloud State University*

Mary Allen, *Hartwick College*

Walter Arenstein, *San Jose State University*

Eric C. Atkinson, *Northwest College*

Abbed Babaei, *Cleveland State University*

Kenneth Banks, *University of North Texas*

Raymond Beiersdorfer, *Youngstown State University*

Lisa Bonneau, *Metropolitan Community College*

William Brown, *State University of New York, Fredonia*

John S. Campbell, *Northwest College*

Geralyn Caplan, *Owensboro Community and Technical College*

Kelly Cartwright, *College of Lake County*

Kathy Carvalho-Knighton, *University of South Florida*

Jason Cashmore, *College of Lake County*

Christopher G. Coy, *Indiana Wesleyan University*

Christopher F. D'Elia, *Louisiana State University*

Lisa Doner, *Plymouth State University*

Matthew J. Eick, *Virginia Tech*

Anne Ehrlich, *Stanford University*

Cory Etchberger, *Johnson County Community College*

Carri Gerber, *Ohio State University Agricultural Technical Institute*

Marcia Gillette, *Indiana University–Kokomo*

Nisse Goldberg, *Jacksonville University*

Sue Habeck, *Tacoma Community College*

Crista Haney, *Mississippi State University*

Sherri Hitz, *Florida Keys Community College*

Bryan Hopkins, *Brigham Young University*

David W. Hoferer, *Eastern University*

Aixin Hou, *Louisiana State University*

Melinda Huff, *Northeastern Oklahoma A&E College*

Ali Ishaque, *University of Maryland, Eastern Shore*

Les Kanat, *Johnson State College*

Barry King, *College of Santa Fe*

Cindy Klevickis, *James Madison University*

Steven Kolmes, *University of Portland*

Peter Kyem, *Central Connecticut State University*

Owen Lawlor, *Northeastern University*

Marcie Lehman, *Shippensburg University*

Kurt Leuschner, *College of the Desert*

Gary Li, *California State University, East Bay*

Heidi Marcum, *Baylor University*

Kenneth Mantai, *SUNY Fredonia*

Allan Matthias, *University of Arizona*

Charles A. McClaugherty, *University of Mount Union*

Blodwyn McIntyre, *University of the Redlands*

Dan McNally, *Bryant University*

Karen E. McReynolds, *Hope International University*

Glynda Mercier, *Austin Community College*

Grace Ju Miller, *Indiana Wesleyan University*

Kiran Misra, *Edinboro University of Pennsylvania*

Michael Moore, *Delaware State University*

Natalie Moore, *Lone Star College*

Carrie Morjan, *Aurora University*

Carole Neidich-Ryder, *Nassau Community College*

Robert Andrew Nichols, *City College–Gainesville Campus*

Tim Nuttle, *Indiana University of Pennsylvania*

Bruce Olszewski, *San Jose State University*

John Pleasants, *Iowa State University*

John Reuter, *Portland State University*

Virginia Rivers, *Truckee Meadows Community College*

Kim Schulte, *Georgia Perimeter College*

Brian Shmaefsky, *Lone Star College–Kingwood*

Annelle Soponis, *Reading Area Community College*

Shamili Stanford, *College of DuPage*

Keith Summerville, *Drake University*

Todd Tarrant, *Michigan State University*

Bradley Turner, *McLennan Community College*

Alison Kate Varty, *College of the Siskiyous*

Dave Wartell, *Harrisburg Area Community College*

John Weishampel, *University of Central Florida*

Arlene Westhoven, *Ferris State University*

Robert S. Whyte, *California University of Pennsylvania*

Danielle Wirth, *Des Moines Area Community College*

Todd Christian Yetter, *University of the Cumberlands*

Acknowledgments

I am deeply indebted to both Bernard Nebel and Richard Wright for their diligent work in developing the text and for producing successive editions. They laid a solid foundation and strong legacy to build upon. I joined Richard Wright as a co-author on the eleventh and twelfth editions, and am thankful to have had such a strong writing partner. Starting with this thirteenth edition, I am now lead author and responsible for all chapters of the book. However, revising the thirteenth edition has been a collaborative effort, and I would like to express my heartfelt thanks to all those at Pearson Education who have contributed to the book in so many ways and helped make this revision possible. My editor, Alison Rodal, encouraged me and helped me reorganize and write this new edition. Developmental Editor Julia Osborne worked closely with me on every aspect of the book; thank you, Julia, for your work on every page and for the wonderful times we had working together. Project Managers Arielle Grant and Lori Newman kept me on schedule from start to finish. Program Manager Anna Amato kept all aspects of the process running smoothly. Editorial Assistant Alison Cagle skillfully managed all reviews and text supplements. Supervising Project Manager for Instructional Media Eddie Lee and Associate Content Producer Chloe Veylit managed the production of our Instructor Resources and the updates to MasteringEnvironmentalScience. Thanks also to Executive Marketing Manager Lauren Harp and her team, and the Pearson sales team, for their hard work on campus every day.

In addition, I want to thank Nisse Goldberg, Matt Eick, Kayla Rihani, Thomas Pliske, and Heidi Marcum for their work on MasteringEnvironmentalScience and the Instructor Resources that accompany the thirteenth edition.

I would like to thank my husband, Gary, my biggest supporter. I am also very thankful to my mentors—particularly Richard Wright, who has known me for years and helped me inestimably—and Calvin DeWitt of the University of Wisconsin–Madison, who has been one of the foremost figures in motivating young people to care about the environment and ethics.

Finally, it is our hope that this book can inspire a new generation to work toward bringing healing to a Creation suffering from human misuse.

Dorothy F. Boorse

BE EQUIPPED TO UNDERSTAND THE ROLES OF
SCIENCE, SUSTAINABILITY, AND STEWARDSHIP

The thirteenth edition builds on its student friendly approach with new built-in study tools, and retains its focus on science, sustainability and stewardship, equipping students with a current understanding of environmental science issues and research.

Chapter 4

Populations and Communities

African savanna elephants (*Loxodonta africana africana*) saunter through the glaring sun of the East African savanna. Trunks swaying side to side, they seek out shrubs and grasses. Nearby, the indigenous people graze their herds of rangy cattle on the sparse, tough grasses. These pastoralists view the elephants as competitors that reduce the amount of food available to their cattle. The herders resent the elephants' use of the land, while the villagers fear that the elephants will trample their crops.

Current Threats. Like many large, slow-growing animals, these elephants are increasingly rare. Numbering from 3 to 5 million at the start of the 20th century, today there are only 450,000 to 700,000. Loss of habitat is one reason for their decline. In addition, poachers kill them for their ivory, even though the ivory trade is illegal. As climate change increases drought, elephants have to migrate farther and face more conflicts with humans than before. In 2014, a team of researchers from Elephants Without Borders began using light aircraft to document every group of elephants in the 13 sub-Saharan countries where 90% of savanna elephants live. Their initial findings documented the heavy poaching of elephants in numerous places.

While the number of elephants is declining, another native species, the aggressive Sodom apple (*Solanum campylacanthum*), is increasing. Unfortunately for the herders, this weedy shrub is toxic to grazing animals such as cattle, sheep, and zebras. In Kenya, the eradication of the noxious Sodom apple is costing the government millions.

A Win-Win Solution? In spite of the problems caused by the ongoing changes in East Africa, there is a small amount of good news. Unlike grazing animals, browsers such as elephants and impalas can eat the Sodom apple. In fact, elephants have a voracious appetite for the shrub, which may provide a

▲ African savanna elephants can eat plants that are unpalatable to sheep and cattle.

Savanna elephant range / Kenya

Learning Objectives

4.1 Dynamics of Natural Populations: Describe three models of the way populations grow and the graph that would illustrate each.

4.2 Limits on Populations: Identify factors that limit populations, including those that increase as populations become more dense (such as predation and resource limitation) and factors that are unrelated to population density.

4.3 Community Interactions: Define the types of interactions that can occur between species in a community and the effect of those interactions on each species.

4.4 Evolution as a Force for Change: Describe the major ideas in the theory of evolution, such as inheritance and natural selection, and list examples of adaptations that allow organisms to survive. Explain how major changes in the Earth facilitate evolutionary change.

4.5 Implications for Management by Humans: Describe at least three ways in which human actions alter populations and communities.

◀ **Numbered Learning Objectives** open each chapter and introduce you to key concepts that you will understand at the conclusion of the chapter.

CONCEPT CHECK Which kind of population growth curve would you expect to see in a population of squirrels in a forest? Which would you expect to see in ragweed growing on a construction site? Why? ☑

CONCEPT CHECK How might the removal of a predator harm a community of a species? How might the removal of a competitor affect the community? ☑

CONCEPT CHECK Describe a human action that increases the size of a species' population and an action that decreases the size of a wild population. Give an example of each. ☑

▲

NEW! Concept Check Questions align with each learning objective and appear at the end of each chapter section to provide you with opportunities to check and deepen understanding as you read each chapter. You can quickly check your knowledge by consulting the answer key at the back of the text.

RESOURCE NEEDED FOR A CONSUMED MEGACALORIE (1,000 kcal)

The Environmental Costs of Producing Meat. The graphs compare the environmental impact of the production of five types of animal-based foods. The resources needed to produce each food are listed per Megacalorie (1,000 kilocalories), about half of the daily energy requirement for an adult. (The nutritional unit 1 Calorie is equal to 1 kilocalorie). Arrows at the top show the environmental costs of growing wheat, rice, and potatoes. Pastureland is marked with hatch marks. For some factors, the value for beef is off the scale and the correct number is marked at the end of the bar.

UNDERSTANDING THE DATA

1. What is the ratio between the amount of nitrogen used to grow beef and that used to produce eggs?
2. Which is greater, the pastureland or the other land used for beef production?
3. Is there a difference in the amount of irrigation water needed for poultry and pork?
4. Which emits more greenhouse gases, producing a Megacalorie (1,000 kilocalories) of rice or of eggs?

NEW! Understanding the Data questions prompt you to practice data interpretation skills and build understanding of environmental issues presented in select graphs, maps and tables in each chapter.

Chapter 10

Water: Hydrologic Cycle and Human Use

Learning Objectives

10.1 Water, a Vital Resource: Describe the unique properties that make water so vital, the differences in water availability in different societies, and conflicts over availability of clean water.

10.2 Hydrologic Cycle and Human Impacts: Explain the movement of water through the hydrologic cycle and human impacts on the cycle.

10.3 Water: Getting Enough, Controlling Excess: Describe the ways humans try to provide clean freshwater and some of their outcomes.

10.4 Water Stewardship, Economics, and Policy: Describe options for meeting rising demands for water, new innovations in water science and technology, and public policies for water in a water-scarce world.

Drought in the American Southwest. It's a tough time in America's almond-growing region—a section of California's Central Valley that produces more than 80% of the world's almonds. For several years this region has languished in deep drought, threatening the survival of the almond orchards. Almonds are a water-needy crop—it takes about 1.1 gallons of water to produce each nut. Indeed, almond trees require a great investment of water during all the years they are growing, including the first five years when they produce no nuts. The millions of gallons of water used to grow almonds cannot simply be turned off in a drought year. If farmers do not irrigate the almond trees, they will die, and the farmer will lose an investment of years.

California contains one sixth of the irrigated land in the country. Much of the water used for this irrigation has been diverted from far-away rivers. Those diversions endanger the salmon in the rivers, which require more water than is left for them. Some of the water used for irrigation is pumped from new wells that tap into ancient reserves of water that percolated underground millions of years ago. That groundwater is being withdrawn faster than any rain could replace it, depleting an underground resource owned by all.

Mega-Drought. California is not the only state in drought—the drought encompasses much of the Southwest. The drought, which has been going on since 2000, is now being called a "mega-drought." It is estimated to be the fifth most severe drought in that region since AD 1000. Stressed by the lack of water, trees in the forests are dying. The dry forests sit like tinder, waiting for a spark to set off a wildfire. The risk of wildfire is increased by the heavy load of unburned brush lying on the forest floor, the product of years of fire suppression. Large,

▲ These California almond trees are being removed because there is not enough water to irrigate them.

233

U.S. Drought Monitor September 2, 2014

Intensity:
- Abnormally Dry
- Moderate Drought
- Severe Drought
- Extreme Drought
- Exceptional Drought

UPDATED! Engaging stories open each chapter and draw students into reading the chapter topic at hand. New topics include the drought in the western United States, the recent Ebola outbreak, and more.

Three unifying themes of science, sustainability, and stewardship help students conceptualize the task of forging a sustainable future. Essays appear at appropriate points within chapters and provide a current perspective on the topic.

STEWARDSHIP
A Cultural Hazard Worsens Other Risks

Lifestyle choices such as driving fast, imbibing alcohol, and climbing mountains involve a significant risk of accident and death. But one cultural hazard—tobacco use—is particularly problematic. In the United States, tobacco use is the leading preventable cause of death, linked to about 450,000 deaths annually. Every year U.S. adults who smoke cigarettes cost the United States $170 billion in health care and another $151 billion in lost productivity. In some developed countries, tobacco use is declining, but it remains high in the former socialist countries of Eastern Europe and is on the increase in most developing countries. The World Health Organization (WHO) puts the number of smokers worldwide at more than 1.1 billion, with 6 million deaths per year due to cigarette smoking.

Cigarette smoking has been clearly correlated with cancer and other lung diseases; it is also related to cardiovas-

known human carcinogen; this move meant that exposure to smoking by others was clearly an environmental hazard.

In the United States, several measures have been taken to reduce the use of tobacco. Raising cigarette taxes has been shown to be one of the most effective ways to reduce smoking, especially in young people, so tobacco taxes are high. Warning labels are required on smoking materials, advertising has been curtailed, and smoking has been banned in most public places. The 2009 Tobacco

Activists in Seoul, South Korea, march to celebrate World No Tobacco Day on March 31, 2014.

SOUND SCIENCE
Complex Clouds and Co-Benefits of Solutions

One of the striking features of air pollution is its complexity. Atmospheric brown clouds (ABCs) provide an excellent example of this complexity. These huge clouds contain both sulfate aerosols and black carbon particles, such as those released by diesel fumes. The black carbon particles absorb energy, causing a warming effect. At the same time, the sulfate aerosols reflect light, which causes cooling. The sulfate aerosols also affect water droplets, causing the clouds to produce smaller droplets, so rain is less likely to form. These aerosols and particles are linked together in tiny droplets, so the processes of cooling and warming occur simultaneously throughout the clouds. The haze of the ABCs reduces the amount of light reaching the surface of the land and ocean (global dimming), thus decreasing temperatures and evaporation near the surface,

from cleaner air, but also an improvement in the quality of water, as seen in the Chesapeake Bay.

On the global scale, efforts to solve the problem of climate change include reducing the burning of wood and fossil fuels. Such changes will improve smog as well.

In many developing nations, forests are burned to clear land for agriculture. In Indonesia, palm oil companies have been accused of setting illegal fires to clear vast tracts of forest for planta-

SUSTAINABILITY
Taking the Waste Out of Take-Out

In today's busy world, we often have to choose between convenience and sustainable actions. For example, take-out meals arrive in pizza boxes or Styrofoam containers, which typically end up in the trash. The Take Out Without campaign, started by Canadian designer Lisa Borden, states, "The American population tosses out enough paper bags, plastic cups, forks, and spoons every year to circle the equator 300 times." To cut down on this waste, the campaign recommends bringing reusable containers to stores and restaurants and refusing unneeded items, such as extra straws, napkins, and ketchup. But sometimes, disposable packaging is necessary. In those cases, it would be helpful to have packaging that can be com-

plastic-like materials made from cellulose, starch, and even certain leaves.

Recently researchers have found that chitosan, a material made from shrimp shells, can be used to make biodegradable packaging. Chitosan has qualities that are similar to plastic, but unlike plastic, chitosan can decompose quickly in landfills. Chitosan is also antimicrobial, an excellent characteristic for food packaging. Furthermore, it is made from a by-product of shrimp processing, so its use would reduce waste in two ways. Chitosan is not yet ready for scaled-up production

This filament is made of a plastic-like substance produced from the shells of shrimp and crabs.

Source: Take Out Without, http://takeoutwithout.org/behind-the-campaign/; J. Fernandez and D. Ingber, "Manufacturing of Large-Scale

PERSONALIZE LEARNING WITH
MasteringEnvironmentalScience®

The text and MasteringEnvironmentalScience work together to help you understand the science behind environmental issues.

NEW! Everyday Environmental Science videos connect concepts with current stories in the news. Produced by the BBC, these high-quality videos can be assigned for pre- and post-lecture homework, or can be shown in class to engage students in the topic at hand.

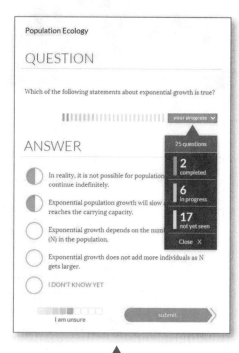

Population Ecology

QUESTION

Which of the following statements about exponential growth is true?

your progress

ANSWER

- In reality, it is not possible for population continue indefinitely.
- Exponential population growth will slow reaches the carrying capacity.
- Exponential growth depends on the num (N) in the population.
- Exponential growth does not add more individuals as N gets larger.
- I DON'T KNOW YET

I am unsure   submit

25 questions
2 completed
6 in progress
17 not yet seen
Close X

EXPANDED! Interpreting Graphs and Data coaching activities help students practice basic quantitative analysis skills. Each assignable activity includes personalized feedback for wrong answers.

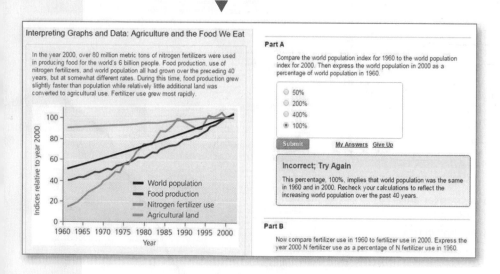

Interpreting Graphs and Data: Agriculture and the Food We Eat

In the year 2000, over 80 million metric tons of nitrogen fertilizers were used in producing food for the world's 6 billion people. Food production, use of nitrogen fertilizers, and world population all had grown over the preceding 40 years, but at somewhat different rates. During this time, food production grew slightly faster than population while relatively little additional land was converted to agricultural use. Fertilizer use grew most rapidly.

Part A

Compare the world population index for 1960 to the world population index for 2000. Then express the world population in 2000 as a percentage of world population in 1960.

- 50%
- 200%
- 400%
- 100%

Submit   My Answers   Give Up

Incorrect; Try Again

This percentage, 100%, implies that world population was the same in 1960 and in 2000. Recheck your calculations to reflect the increasing world population over the past 40 years.

Part B

Now compare fertilizer use in 1960 to fertilizer use in 2000. Express the year 2000 N fertilizer use as a percentage of N fertilizer use in 1960.

NEW! Dynamic Study Modules help students study effectively on their own by continuously assessing their activity and performance in real time.

These are available as graded assignments prior to class, and accessible on smartphones, tablets, and computers.

PART ONE

FRAMEWORK FOR A SUSTAINABLE FUTURE

Nights spent pouring over blueprints, days looking at budgets, time spent deciding what parts need to be purchased and who will maintain a structure after it is built: you might imagine these activities being performed by professionals building a road or school, but they aren't. This is the work of a team of dedicated engineering undergraduates from Purdue University and their advisors. They are designing a small hydropower facility in Banyang, Cameroon, which will provide water and irrigation for a village.

The project is designed to promote sustainability—the ongoing thriving of people in a context of the natural world, living in a way that doesn't use up resources, harm other creatures, or degrade our environment in the long term. Right now, there is evidence that we are not living sustainably: Our global economy relies on the use of fossil fuels and nuclear power, but continued emissions of carbon dioxide into the atmosphere and oceans are bringing major changes in the climate as Earth warms up and the oceans acidify. The human population will likely exceed 9 billion by 2050, but the populations of many wild plant and animal species are declining. For the poorest people, the availability of food, clean water, and basic health care remains low. Natural disasters such as the Japanese tsunami of 2011 can interact with the built environment to cause radiation leaks and other problems. Human activities such as coal mining can interact with the natural environment to cause events like the 2014 spill of thousands of gallons of an industrial chemical used to clean coal; when the chemical leaked into the Elk River in West Virginia, thousands were without drinkable water for weeks. We are all bound together: humanity, the other creatures, and the world around us.

We begin our framework for a sustainable future in Chapter 1, where we introduce environmental science and what it might mean for you. In Chapter 2, we look at three features of human societies that interact with science—economics, politics, and public policy—as we come to understand the environment and address the challenges we face.

Chapter 1

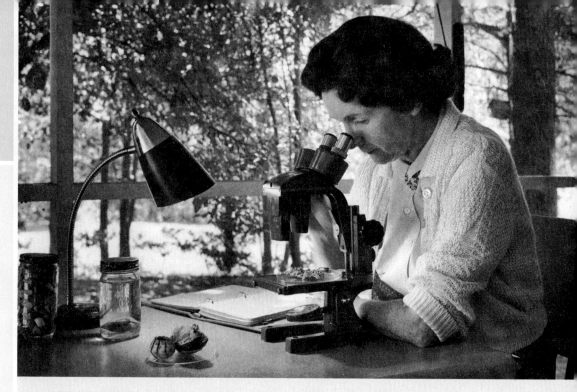

Learning Objectives

1.1 The State of the Planet: Explain the main reasons for concern about the health of our planet today. Describe what the environmental movement has achieved in recent years, and explain how environmental science has greatly contributed to the environmental movement.

1.2 Sustainability: Define sustainability and explain ways in which our relationship with the environment needs to be more sustainable.

1.3 Sound Science: Explain the process of science, how the scientific community tests new ideas, and contrast sound science with junk science, with examples.

1.4 Stewardship: Define the principle of stewardship and give examples.

1.5 Moving Toward a Sustainable Future: Identify trends that must be overcome in order to pursue a sustainable future and trends that promote sustainability.

Science and the Environment

"There was once a town in the heart of America where all life seemed to live in harmony with its surroundings. The town lay in the midst of a checkerboard of prosperous farms, with fields of grain and hillsides of orchards where, in spring, white clouds of blossom drifted above the green fields. . . . The countryside was, in fact, famous for the abundance and variety of its bird life, and when the flood of migrants was pouring through in spring and fall people traveled from great distances to observe them. . . . So it had been from the days many years ago when the first settlers raised their houses, sank their wells, and built their barns.[1]"

These are words from the classic *Silent Spring*, written by biologist Rachel Carson to open her first chapter, titled "A Fable for Tomorrow." After painting this idyllic picture, the chapter goes on to describe "a strange blight" that began to afflict the town and its surrounding area. Fish died in streams, farm animals sickened and died, families were plagued with illnesses and occasional deaths. The birds had disappeared, their songs no longer heard—it was a "silent spring." And on the roofs and lawns and fields remnants of a white powder could still be seen, having fallen from the skies a few weeks before.

Rachel Carson explained that no such town existed, but that all of the problems she described had already happened somewhere, and that there was the very real danger that ". . . this imagined tragedy may easily become a stark reality we all shall know."[2] She published her book in 1962, during an era when pesticides and herbicides were sprayed widely on the landscape to control pests in agricultural crops, forests, towns, and cities. In *Silent Spring*, Carson was particularly critical

---

[1]Rachel Carson, *Silent Spring* (Boston: Houghton Mifflin Company, 1962), 1, 2.
[2]Ibid., 3.

▲ Rachel Carson, the author of *Silent Spring*.

of the widespread spraying of DDT. This pesticide was used to control Dutch elm disease, a fungus that invades trees and eventually kills them. The fungus is spread by elm bark beetles and DDT was used to kill the beetles. In towns that employed DDT spraying, birds began dying off, until in some areas people reported their yards were empty of birds. Thousands of dead songbirds were recovered and analyzed in laboratories for DDT content; all had toxic levels in their tissues. DDT was also employed in spraying salt marshes for mosquito control, and the result was a drastic reduction in the fish-eating bald eagle and osprey.

Fallout. Rachel Carson brought two important qualities to her work: she was very careful to document every finding reported in the book, and she had a high degree of personal courage. She was sure of her scientific claims, and she was willing to take on the establishment and defend her work. In spite of the fact that her work was thoroughly documented, her book ignited a firestorm of criticism from the chemical and agricultural establishment. Even respected institutions such as the American Medical Association joined in the attack against her.

Despite this criticism, *Silent Spring* caught the public's eye, and it quickly made its way to the President's Science Advisory Committee when John F. Kennedy read a serialized version of it in the *New Yorker*. Kennedy charged the committee with studying the pesticide problem and recommending changes in public policy. In 1963, Kennedy's committee made recommendations that fully supported Carson's thesis. Congress began holding hearings, public debate followed, and Carson's voice was joined by others who called for new policies to deal

not only with pesticides, but also with air and water pollution and more protection for wild areas. Finally, in 1969, Congress passed a bill known as the Environmental Policy Act, the first legislation to recognize the interconnectedness of ecological systems and human enterprises. Shortly after that, a commission appointed by President Richard Nixon to study environmental policy recommended the creation of a new agency that would be responsible for dealing with air, water, solid waste, the use of pesticides, and radiation standards. The new agency, called the Environmental Protection Agency (EPA), was given a mandate to protect the environment, on behalf of the public, against pressures from other governmental agencies and from industry. The year was 1970, the same year that 20 million Americans celebrated the first Earth Day.

In what must be seen as a triumph of Rachel Carson's work, DDT was banned in the United States and most other industrialized countries in the early 1970s. (The DDT story is more fully documented in Chapter 13.) Unfortunately, Rachel Carson did not live long after her world-shaking book was published; she died of breast cancer in 1964. Her legacy, however, is a lasting one: she is credited with initiating major reforms in pesticide policy as well as an environmental awareness that eventually led to the modern environmental movement and the creation of the EPA.

Moving On. This is a story of science and the environment, but it is more than that; it is a story of a courageous woman who changed the course of history. In this chapter, we briefly explore the current condition of our planet and then introduce three themes that provide structure to the primary goal of this text: *to promote a sustainable future.*

1.1 A Paradox: What Is the Real State of the Planet?

Paradox (n.): *A statement exhibiting contradictory or inexplicable aspects or qualities.*[3] A group of scientists from McGill University recently published a paper in which they identified a so-called **environmentalist's paradox.**[4] The paradox, they said, is this: over the past 40 years, human well-being has been steadily improving, while natural ecosystems (from which we derive many goods and services) have been declining.

To explain this paradox, the authors advanced four hypotheses:

1. The measurements of human well-being are flawed; it is actually declining.

2. Food production, a crucial ecosystem service that has been enhanced, outweighs the effects of declines in other ecosystem services.

3. Human technology, such as irrigation and synthetic fertilizers, makes us less dependent on ecosystem services.

4. There is a time lag between ecosystem decline and human well-being; the worst is yet to come.

We will take a brief look at four important global trends and keep in mind these hypotheses (the scientific method and hypotheses are explained later in the chapter) as we engage in our initial examination of the state of our planet: (1) human population and well-being, (2) the status of vital ecosystem services, (3) global climate change, and (4) the loss of biodiversity. Each of these topics is explored in greater depth in later chapters.

Population Growth and Human Well-Being

The world's human population, more than 7.3 billion in 2014, has grown by 2 billion in just the past 25 years. It is continuing to grow, at the rate of about 80 million persons per year. Even though the growth rate (now 1.1%/year) is gradually slowing, the world population in 2050 is likely to exceed 9.3 billion, according to the most recent projections from

---

[3]*Webster's II New College Dictionary* (Boston: Houghton Mifflin Company, 1995), s.v. "paradox."

[4]Ciara Raudsepp-Hearne et al., "Untangling the Environmentalist's Paradox: Why Is Human Well-Being Increasing as Ecosystem Services Degrade?" *Bioscience* 60 (September 2010): 576–589.

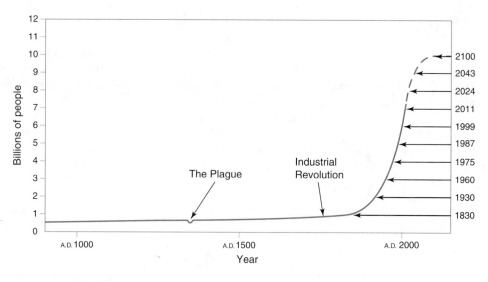

Figure 1–1 World population explosion. World population started a rapid growth phase in the early 1800s and has increased sixfold in the past 200 years. At present it is growing by 80 million people per year. Future projections are based on assumptions that birthrates will continue to decline.

(*Source:* Data from UN Population Division, 2012 revision, and from Population Reference Bureau 2014 report.)

the United Nations (UN) Population Division (Figure 1–1). The 2.2 billion persons added to the human population by 2050 will all have to be fed, clothed, housed, and, hopefully, supported by gainful employment. Virtually all of the increase will be in developing countries.

Human Development Index. Each year since 1990, the United Nations Development Program (UNDP) has published a Human Development Report.[5] A key part of the report is the *Human Development Index* (HDI), a comprehensive assessment of human well-being in most countries of the world. With this index, well-being is measured in health, education, and basic living standards. The 2014 report highlighted the importance of resiliency and the vulnerability of the poor, suggesting that poverty is not simply a function of the amount of money people have, but is also a function of factors such as literacy and stability.

The 2010 report included a unique four-decade comparison in which worldwide trends in HDI were plotted over the 40 years since 1970 (Figure 1–2). Only three of the 135 countries analyzed declined in HDI, while most of the countries showed marked improvement. During those 40 years, life expectancy rose from 59 years to 70, school enrollment climbed from 55% to 70%, and per capita gross domestic product (GDP) doubled to more than $10,000. It is this overall progress that has provided one side of the environmentalist's paradox. As a result of these facts, the McGill team concluded that its first hypothesis is not supported; there are too many indications that human well-being has indeed improved markedly.

Is It All Good? However, the overall progress can, and does, mask serious inequalities. Economic growth has been extremely unequal, both between and within countries. And there are huge gaps in human development across the world. For example, in developing countries, an estimated 1.1 billion people still experience extreme poverty, existing on an income of $1.25 a day. More than 800 million people, about 13% of

the people living in developing countries, remain undernourished. Some 6.9 million children per year do not live to see their fifth birthday.

Addressing these tragic outcomes of severe poverty has been a major concern of the UNDP, and in 2000, all UN member countries adopted a set of goals—the **Millennium Development Goals (MDGs)**—to reduce extreme poverty and its effects on human well-being by 2015 (see Table 9–2 for a list of the eight goals). Several of the MDGs were met ahead of schedule, while others were not met. The world has now moved to a post-2015 development agenda, driven by a set of seventeen **Sustainable Development Goals (SDGs)** discussed at length later (Chapter 9). The SDGs are a set of goals for world development and poverty alleviation, described by the UN, for 2015–2030.

Ecosystem Goods and Services

Natural and managed ecosystems support human life and economies with a range of goods and services. As crucial as they are, there is evidence that these vital resources are not being managed well. Around the world, human societies are depleting groundwater supplies, degrading agricultural soils, overfishing the oceans, and cutting forests faster than they can regrow. The world economy depends heavily on many renewable resources, as we exploit these systems for goods—water, all of our food, much of our fuel, wood for lumber and paper, leather, furs, raw materials for fabrics, oils and alcohols, and much more.

These same ecosystems also provide a flow of **services** that support human life and economic well-being, such as the breakdown of waste, regulation of the climate, erosion control, pest management, and maintenance of crucial nutrient cycles. In a very real sense, these goods and services can be thought of as capital—ecosystem capital. Human well-being and economic development are absolutely dependent on the products of this capital—its income, so to speak. As a result, the stock of ecosystem capital in a nation and its income-generating capacity represent a major form of the wealth of the nation (see Chapter 2). These goods and services are provided year after year, as long as the ecosystems producing them are protected.

[5]United Nations Development Program, *Human Development Report 2014: Sustaining Human Progress: Reducing Vulnerabilities and Building Resilience.* (New York: UNDP, July 24, 2014), http://hdr.undp.org/en/2014-report.

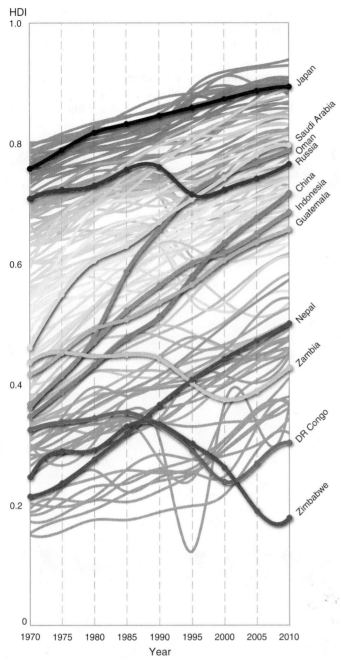

Figure 1–2 Human Development Index, 1970–2010. This complex graph shows the Human Development Index (HDI) of more than 100 countries over a period of 40 years. Countries represented by lines with similar colors began the time period with similar HDI values. Highlighted countries include top and bottom performers (in terms of increasing HDI) and selected others. (This four-decade graph accompanied a 2010 special report. Newer reports highlight these trends in other ways.)

(*Source:* United Nations Development Program, *Human Development Report 2010.* New York: UNDP, p. 27.)

UNDERSTANDING THE DATA

1. All but three of the 135 countries have a higher level of human development today than in 1970. What explains the general upward trend for most countries?
2. Which two labeled countries appear to have improved the most? Which of the labeled countries decreased?
3. What historical event might explain the pattern for the line representing Russia?

Patterns of Resource Consumption. As the human population grows, each person requires food, water, shelter, clothing, and other resources. Many of the goods and services that people need are derived from ecosystems. However, not all people use the same amount of resources. Any discussion of population growth, development, and the preservation of ecosystem services needs to include the idea that some people consume more resources than is necessary. One way to imagine individual consumption patterns is to picture what it would require to have everyone consume resources at the same level. For example, if all humans alive today replicated the resource consumption patterns of the average American, we would need more than four Earths to accommodate all of their needs. The concept of individual consumption will be covered extensively later (Chapters 2, 8, and 23).

Measuring Ecosystem Health: A Huge Undertaking. To protect ecosystem goods and services for future generations, we need to know what they are, how they are being used, and what is happening to them. To find out, scientists have carried out a number of large-scale assessments.

The most prominent, the *Millennium Ecosystem Assessment*, compiled available information on the state of ecosystems across the globe. During a period of four years, some 1,360 scientists from 95 countries gathered, analyzed, and synthesized information from published, peer-reviewed research. The project focused especially on the linkages between ecosystem services and human well-being on global, regional, and local scales. Ecosystem goods and services were grouped into *provisioning* services (goods such as food and fuel), *regulating* services (processes such as flood protection), and *cultural* services (nonmaterial benefits such as recreation) (see **Table 1–1** on the following page). *Supporting* services (not included in the table), such as primary productivity and habitat, are necessary to the other three.

In a summary report, the most prominent finding of the scientists was the widespread degradation and overexploitation of ecosystem resources. More than 60% of the classes of ecosystem goods and services assessed by the team were being degraded or used unsustainably (Table 1–1). The scientists concluded that if this trend is not reversed, the next half century could see deadly consequences for humans as the ecosystem services that sustain life are further degraded. Since the Millennium Ecosystem Assessment, a number of other assessments have been conducted; several are listed in **Table 1–2** (on the following page). Some of these assessments considered global patterns and others focused on regional ecosystems, but all found similar trends.

One set of provisioning ecosystem services has actually been enhanced over recent years: the production of crops, livestock, and aquaculture. As a result, the production of food has kept pace with population growth, improving human health and increasing life expectancy. However, many ecosystem services and resources such as groundwater, soil, wild fish, and forestry products have declined, in part because of the way we use land and other resources to provide food, shelter, and consumer goods for humans.

Paradox Resolved? The McGill team set out to explain the environmentalist's paradox—the fact that human well-being has been improving while natural ecosystems have been declining. It rejected hypothesis 1, which stated that human well-being is actually declining. The team concluded that *hypothesis 2 was confirmed*: enhanced food production outweighs the effects of declines in other ecosystem services. Two further hypotheses remain: (3) our use of technology makes us less dependent on ecosystem services, and (4) the existence of a time lag between the loss of goods and services and the impact on human well-being, with the possibility of exceeding limits and bringing on ecosystem collapse. The McGill University team concluded that these last two hypotheses help explain the environmentalist's paradox, although not as

Table 1–1 The global status of ecosystem services. Human use has degraded almost two-thirds of the identified services; 20% are mixed, meaning they are degraded in some areas and enhanced in others; and 17% have been enhanced by human use.

Ecosystem Services	Degraded	Mixed	Enhanced
Provisioning (goods obtained from ecosystems)	Capture fisheries Wild foods Wood fuel Genetic resources Biochemicals Fresh water	Timber Fiber	Crops Livestock Aquaculture
Regulating (services obtained from the regulation of ecosystem processes)	Air quality regulation Climate regulation Erosion regulation Water purification Pest regulation Pollination Natural hazard regulation	Water regulation (flood protection, aquifer recharge) Disease regulation	Carbon sequestration (trapping atmospheric carbon in trees, etc.)
Cultural (nonmaterial benefits from ecosystems)	Spiritual and religious values Aesthetic values	Recreation and ecotourism	

Source: Millennium Ecosystem Assessment, *Ecosystems and Human Well-Being: Synthesis.* Washington, DC: Island Press, 2005.

Table 1–2 Examples of previous and ongoing ecosystem and biodiversity assessments.

Title	Organization	Year(s)
Global Biodiversity Assessment	UN Environmental Programme (UNEP)	1995
Millennium Ecosystem Assessment	UNEP; World Resources Institute	2005
State of the Nation's Ecosystems	The Heinz Center	2002, 2008
National Ecosystem Assessment	United Kingdom	2011
Sustaining Environmental Capital: Protecting Society and the Economy	President's Council of Advisors on Science and Technology	2011
Intergovernmental Platform on Biodiversity and Ecosystem Services (IPBES)	UNEP	2012
Global Environmental Outlook (GEO)	UNEP	1997, 1999, 2002, 2007, 2012
IUCN Red List of Threatened Species	International Union for Conservation of Nature (IUCN)	2000, 2004, 2008, 2012
Biodiversity Ecosystems and Ecosystem Services, Technical Input	Part of National Climate Assessment	2014

Source: Adapted from Grimm, N., M. Staudinger, A. Staudt, S. Carter, F. S. Chapin III, P. Kareiva, M. Ruckelhaus, and B. Stein. "Climate-Change Impacts on Ecological Systems: Introduction to a US Assessment." *Frontiers in Ecology and the Environment* 9, no. 11 (2013): 456–464.

clearly as hypothesis 2. The team concluded that the paradox is not fully explained by any of the hypotheses, although hypothesis 1 was rejected. It also concluded that ecosystem conditions are indeed continuing to decline, with unknown and perhaps severe impacts on human well-being in the future. The most serious concern is **global climate change**, the worldwide alteration of patterns of temperature, precipitation, and the intensity of storms.

Global Climate Change

The global economy runs on fossil fuel. Every day in 2013 we burned some 91.3 million barrels of oil, 324 billion cubic feet of natural gas, and 11.6 million tons of coal. All of this combustion generates carbon dioxide (CO_2), which is released into the atmosphere at a rate of 80 million tons a day. Because of past and present burning of fossil fuels, the CO_2 content of the atmosphere increased from 280 parts per million (ppm) in 1900 to 400 ppm in 2014. For the past decade, the level of atmospheric CO_2 has increased by 2 ppm per year, and given our dependency on fossil fuels, there is no end in sight.

Monitoring Carbon Dioxide and Its Effects. Carbon dioxide is a natural component of the lower atmosphere, along with nitrogen and oxygen. It is required by plants for photosynthesis and is important to the Earth-atmosphere energy system. Carbon dioxide gas absorbs infrared (heat) energy radiated from Earth's surface, thus slowing the loss of this energy to space. The absorption of infrared energy by CO_2 and other gases warms the lower atmosphere in a phenomenon known as the **greenhouse effect**. As the greenhouse gases trap heat, they keep Earth at hospitable temperatures.

Although the concentration of CO_2 is a small percentage of the atmospheric gases, increases in the volume of this gas affect temperatures. **Figure 1–3** graphs changes in the concentration of CO_2 in the atmosphere from 1958 to the present. (The yearly peaks and valleys on the graph result from seasonal changes in the uptake of CO_2. In summer, plants take in more CO_2 for photosynthesis than they do in winter, causing the

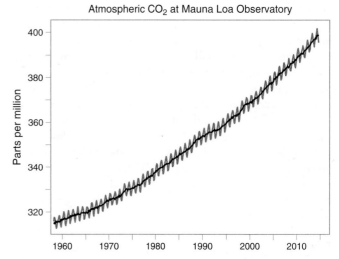

Atmospheric CO_2 at Mauna Loa Observatory

Figure 1–3 Atmospheric carbon dioxide concentrations. This record of CO_2 has been measured at the Mauna Loa Observatory since 1958. The atmospheric content of CO_2 has risen by 45% since the Industrial Revolution began around 1750.

(*Source:* Mauna Loa Observatory, Hawaii, NOAA Research Laboratory, Scripps Institution of Oceanography.)

total concentration of CO_2 in the atmosphere to decrease. The pattern of peaks and valleys reflects the seasons of the Northern Hemisphere, where most of Earth's landmasses and plants are located.) **Figure 1–4** shows changes in global temperature since 1880. Both of these parameters are increasing. This is not *proof* that increases in CO_2 caused the increase in global temperature, but the argument based on the well-known greenhouse effect is quite convincing.

The Intergovernmental Panel on Climate Change (IPCC) was established by the UN in 1988 and given the responsibility to report its assessment of climate change at five-year intervals. The latest of these assessments, the Fifth Assessment Report (AR5), was released during 2013 and 2014. The work of thousands of scientific experts, this assessment produced convincing evidence of human-induced global warming that is

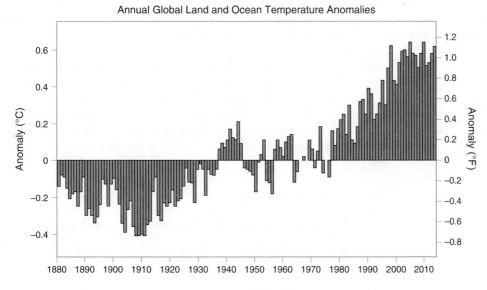

Annual Global Land and Ocean Temperature Anomalies

Figure 1–4 Global temperatures since 1880. This graph shows the course of global temperatures as recorded by thousands of stations around the world. The baseline, or zero point, is the 20th century average temperature. A temperature anomaly is the amount the global mean temperature for a particular year differs from that baseline.

(*Source:* National Climatic Data Center, NOAA, 2014.)

already severely affecting the global climate. Polar ice is melting at an unprecedented rate, glaciers are retreating, storms are increasing in intensity, and sea level is rising. Because the oceans are absorbing half of the CO_2 produced by burning fossil fuels and producing cement, the pH of seawater is declining, making the oceans more acidic. The Sound Science essay, *Oysters Sound the Alarm* (see p. 13) explores one consequence of ocean acidification. The IPCC concluded that future climate change could be catastrophic if something is not done to bring the rapidly rising emissions of CO_2 and other greenhouse gases under control. With few exceptions, the scientists who have studied these phenomena have reached a clear consensus: Climate change is a huge global problem, and it must be addressed on a global scale.

Responses. The solutions to this problem are not easy. The most obvious need is to reduce global CO_2 emissions; the reduction of emissions is called **mitigation**. At issue for many countries is the conflict between the short-term economic impacts of reducing the use of fossil fuels and the long-term consequences of climate change for the planet and all its inhabitants. Future climate changes are likely to disrupt the provision of ecosystem goods and services essential to human well-being, and because the extremely poor depend especially on natural ecosystems, they will suffer disproportionately. International agreements to reduce greenhouse gas emissions have been forged, and some limited mitigation has been achieved.

Most observers believe that the best course forward is to aim toward an effective, binding international treaty to reduce emissions, but for countries also to act independently on a more immediate scale. The United States is moving forward to regulate greenhouse gas emissions under existing air pollution laws and to encourage renewable energy development. Without doubt, this is one of the defining environmental issues of the 21st century. (Chapter 18 explores the many dimensions of global climate change.)

Loss of Biodiversity

Biodiversity is the variability among living organisms, both terrestrial and aquatic. It includes the variety within species, among species, and within ecosystems. The rapidly growing human population, with its increasing appetite for food, water, timber, fiber, and fuel, is accelerating the conversion of forests, grasslands, and wetlands to agriculture and urban development. The inevitable result is the loss of many of the wild plants and animals that occupy those natural habitats. Pollution also degrades habitats—particularly aquatic and marine habitats—eliminating the species they support. Further, hundreds of species of mammals, reptiles, amphibians, fish, birds, and butterflies, as well as innumerable plants, are exploited for their commercial value. Even when species are protected by law, many are hunted, killed, and marketed illegally. According to the Global Biodiversity Outlook 4 (GBO 4) assessment, the majority of wild plant and animal species are declining in their range and/or population size.[6]

As a result of these human activities, Earth is rapidly losing many of its species, although no one knows exactly how many. About 2 million species have been described and classified, but scientists estimate that 5 to 30 million species may exist on Earth. Because so many species remain unidentified, the exact number of species becoming extinct can only be estimated. Recently the World Wildlife Fund reported that since 1970, more than 10,000 populations of vertebrates, representing more than 3,000 species, have been reduced on average by half.[7] The most dramatic declines occurred in freshwater species; regionally, more declines occurred in the tropics.

Why is the loss of biodiversity so critical? Biodiversity is the mainstay of agricultural crops and of many medicines. It is a key factor in maintaining the stability of natural systems and enabling them to recover after disturbances such as fires or volcanic eruptions. Many of the essential goods and services provided by natural systems are derived directly from various living organisms. These goods and services are especially important in sustaining the poor in developing countries. There are also aesthetic and moral arguments for maintaining biodiversity: Once a species is gone, it is gone forever. Finding ways to protect the planet's biodiversity is one of the major challenges of environmental science. (Chapter 6 covers loss of biodiversity.)

Environmental Science and the Environmental Movement

As you read this book, you will encounter descriptions of the natural world and how it works. You will also encounter a great diversity of issues and problems that have arisen because human societies live in this natural world. We use materials from it (goods taken from natural ecosystems); we convert parts of it into the built environment of our towns, cities, factories, and highways; and we transform many natural ecosystems into food-producing agricultural systems. We also use the environment as a place to dump our wastes—solids, liquids, gases—which affects the rest of the world. The **environment**, then, includes the natural world, human societies, and the human-built world; it is an extremely inclusive concept.

Environmental Science. Things go wrong in the environment, sometimes badly. We have already considered four trends—human population growth, ecosystem decline, global climate change, and loss of biodiversity—that signal to us that we are creating problems for ourselves that we ignore at our peril. Lest you think that all we do is create problems, however, consider some of the great successes human societies have achieved. We have learned how to domesticate landscapes and ecosystems, converting them into highly productive food-producing systems that provide sustenance for more than 7 billion people. We have learned how to convert natural materials into an endless number of manufactured goods and structures, all useful for the successful building of cities, roadways, vehicles, and all that makes up a 21st-century human society. All of these successes, however, carry with them hazards.

[6]Secretariat of the Convention on Biological Diversity, *Global Biodiversity Outlook 4* (Montreal, 2014), accessed September 5, 2014, www.cbd.int/gbo3.

[7]World Wildlife Fund, *Living Planet Report 2014*.

Human actions have negative effects on the environment in two broad categories: cumulative impacts and unintended consequences. Sometimes we simply do too much of any one activity—too much burning, too much tree cutting, too much mining on steep slopes. Activities that would not pose a problem if a few people engaged in them are big problems if millions of people do. Sometimes it isn't the accumulation of an activity; it is that we are not paying attention to how the world works. There are unintended consequences of using chemical pesticides, as we've seen, or of dumping trash in wetlands. These two concepts, cumulative impacts and unintended consequences, will come up in the chapters ahead. This is where environmental science comes in.

Simply put, **environmental science** is the study of how the world works. Scientists help figure out ways to lower the negative impacts of our actions, to find alternative ways to meet the same needs, and to better anticipate the likely effects of what we are doing. All sorts of disciplines contribute to environmental science: history, engineering, geology, physics, medicine, biology, and sociology, to name a few. It is perhaps the most multidisciplinary of all sciences. As you study

environmental science, you have an opportunity to engage in something that can change your life and that will certainly equip you to better understand the world you live in now and will encounter in the future.

The Early Environmental Movement. To understand how the world works today, we need some sense of history. Figure 1–5 is a timeline of some of the events and scientific findings, people, and policies of the American environmental movement as well as several international environmental events that will come up throughout the book.

In the United States, the modern environmental movement began less than 60 years ago. The roots of this movement were in the late 19th century, when some people realized that the unique, wild areas of the United States were disappearing. Environmental degradation, resource misuse, and disastrous events sparked scientific study and sometimes grassroots action. Scientific study yielded information on how the world works. Individuals and groups worked as stewards to make changes. Policies were put into place to better protect resources and people from environmental degradation.

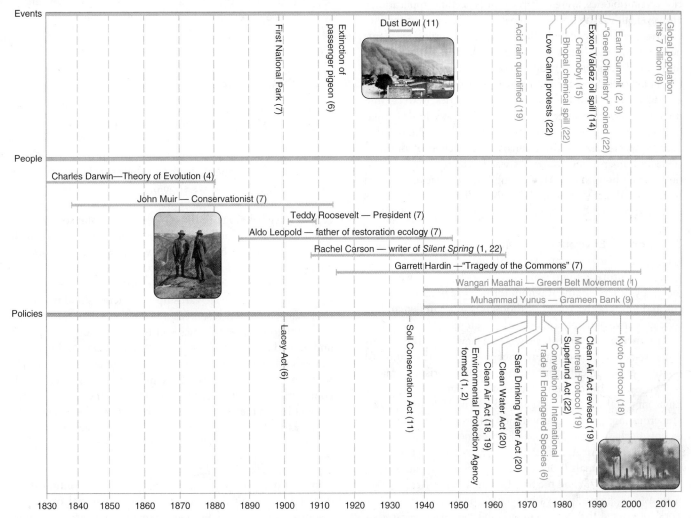

Figure 1–5 Timeline of selected American and international environmental events, people, and policies. These critical pieces of environmental history will be covered throughout the book. Numbers next to each of the events and scientific findings, people, and policies are the chapters in which they are discussed. International subjects are labeled in green.

In the late 19th century, the indiscriminate killing of birds and other animals and the closing of the western frontier sparked a reaction. Around that time, several groups devoted to conservation formed: the National Audubon Society, the National Wildlife Federation, and the Sierra Club, which was founded in California by naturalist John Muir, who helped popularize the idea of wilderness. President Theodore Roosevelt promoted the conservation of public lands, and the national parks were formed.

An increasing awareness of the environment marked the first half of the 20th century. Unwise agricultural practices following World War I eventually helped create an environmental crisis—the Dust Bowl of the 1930s (enormous soil erosion in the American and Canadian prairielands). During the Great Depression (1930–1936), conservation provided a means of both restoring the land and providing work for the unemployed, such as workers in the Civilian Conservation Corps (CCC), who built trails and did erosion control in national parks and forests.

The two decades following World War II (1945–1965) were full of technological optimism. New developments, ranging from rocket science to computers and from pesticides to antibiotics, were redirected to peacetime applications. However, although economic expansion enabled most families to have a home, a car, and other possessions, certain problems became obvious. The air in and around cities was becoming murky and irritating to people's eyes and respiratory systems. Rivers and beaches were increasingly fouled with raw sewage, garbage, and chemical wastes from industries, sewers, and dumps. Conspicuous declines occurred in many bird populations (including that of our national symbol, the bald eagle), aquatic species, and other animals. The decline of the bald eagle and other bird populations was traced to the accumulation in their bodies of DDT, the long-lasting pesticide that had been used in large amounts since the 1940s. In short, it was clear that we were seriously contaminating our environment.

The Modern Environmental Movement. In 1962, biologist Rachel Carson wrote *Silent Spring*, presenting her scenario of a future with no songbirds. Carson's voice was soon joined by others, many of whom formed organizations to focus and amplify the voices of thousands more in demanding a cleaner environment. This was the beginning of the modern **environmental movement**, in which a newly motivated citizenry demanded the curtailment of pollution, the cleanup of polluted environments, and the protection of pristine areas.

Pressured by concerned citizens, Congress created the Environmental Protection Agency (EPA) in 1970 and passed numerous laws promoting pollution control and wildlife protection, including the National Environmental Policy Act of 1969, the Clean Air Act of 1970, the Clean Water Act of 1972, the Marine Mammals Protection Act of 1972, the Endangered Species Act of 1973, and the Safe Drinking Water Act of 1974. A disastrous hazardous waste problem in Love Canal, New York, prompted the Superfund Act of 1980. Scientific studies on acid rain, the decline of the ozone layer, and global climate change caused more environmental groups to form and more policies to be put into place. Concerns about hazardous waste

disposal triggered the rise of the Green Chemistry Movement. You will read more about these and more recent laws and their effects later in this book.

As a result of these efforts, the air in our cities and the water in our lakes and rivers are far cleaner than they were in the late 1960s. Without a doubt, absent the environmental movement, our air and water would now be a toxic brew. By almost any measure, the environmental movement has been successful, at least in solving some pollution problems.

In the early stages of the environmental movement, sources of problems were specific and visible, and the solutions seemed relatively straightforward: for example, install waste treatment and pollution control equipment and ban the use of very toxic pesticides, substituting safer pesticides. Today, there are political battles surrounding almost every environmental issue. Bitter conflicts have emerged over issues involving access to publicly owned resources such as water, grazing land, and timber. As environmental public policies are developed and enforced, the controversies that accompany those policies remind us that it is often difficult to get people with different priorities to agree on solutions.

CONCEPT CHECK How is the environmental movement of today different from that of the early 20th century? From that of 1970? ☑

1.2 Sustainability

What will it take to move our civilization in the direction of a long-term sustainable relationship with the natural world? The answer to this question is not simple, but we would like to next present three unifying themes that provide coherence to the issues and topics covered in this text. In reality, these themes deal with how we should conceptualize our task of forging a sustainable future (**Figure 1–6**). Each theme is a concept that we believe is essential to environmental science. The first theme is **sustainability**—the practical goal toward which our interactions with the natural world should be working. The second is **sound science**—the basis for our understanding of how the world works and how human systems interact with it. The third theme is **stewardship**—the actions and programs that manage natural resources and human well-being for the common good. These themes will be applied to public policy and individual responsibility throughout the text and will also be explored in brief essays within each chapter. As each chapter of this text develops, the concepts behind these themes will come into play at different points. At the end of each chapter, we will revisit the themes, summarizing their relevance to the chapter topics and often adding our own perspective.

Sustainable Yields

A system or process is sustainable if it can be continued indefinitely, without depleting any of the material or energy resources required to keep it running. The term was first applied to the idea of **sustainable yields** in human endeavors such as forestry and fisheries. Trees, fish, and other biological species normally grow and reproduce at rates faster than

(a) Sustainability   **(b)** Stewardship   **(c)** Sound Science

Figure 1–6 Three unifying themes. Sustainability, stewardship, and sound science are three vital concepts that move societies toward a sustainable future if they are applied to public policies and private environmental actions. **(a)** Sustainability includes the actions we take to protect ecosystems and their services, here represented by wind turbines. **(b)** These people represent stewardship by caring for the place they are in. **(c)** Sound science, demonstrated by these polar studies, is critical to understanding the effects of human activities and making good policy.

those required just to keep their populations stable. Thus it is possible to harvest a certain percentage of trees or fish every year without depleting the forest or reducing the fish population below a certain base number. As long as the size of the harvest stays within the capacity of the population to grow and replace itself, the practice can be continued indefinitely. The harvest then represents a sustainable yield. The concept of a sustainable yield can also be applied to freshwater supplies, soils, and the ability of natural systems to absorb pollutants without being damaged.

The notion of sustainability can be extended to include ecosystems. Sustainable ecosystems are entire natural systems that persist and thrive over time by recycling nutrients, maintaining a diversity of species, and using the Sun as a source of sustainable energy. As we'll see, ecosystems are enormously successful at being sustainable (Chapters 3, 4, and 5).

Sustainable Societies

Applying the concept of sustainability to human systems, we say that a **sustainable society** is a society in balance with the natural world, continuing generation after generation, neither depleting its resource base by exceeding sustainable yields nor producing pollutants in excess of nature's capacity to absorb them. Many primitive societies were sustainable in this sense for thousands of years.

Many of our current interactions with the environment are *not* sustainable, however. This is demonstrated by such global trends as the decline of biodiversity and essential ecosystems and the increased emissions of greenhouse gases. Although population growth in industrialized countries has almost halted, these countries are using energy and other resources at unsustainable rates, producing pollutants that are accumulating in the atmosphere, water, and land. In contrast, developing countries are experiencing continued population growth, yet are often unable to meet the needs of many of their people in spite of heavy exploitation of natural resources. As our modern societies pursue continued economic growth and consumption, we

are apparently locked into an unsustainable mode. How do we resolve this dilemma? One answer is the concept of sustainable development.

Sustainable Development

Sustainable development is a term that was first brought into common use by the World Commission on Environment and Development, a group appointed by the United Nations. The commission made sustainable development the theme of its final report, *Our Common Future*, published in 1987. The report defined sustainable development as a form of development or progress that "meets the needs of the present without compromising the ability of future generations to meet their own needs." The concept arose in the context of a debate between the respective *environmental* and *developmental* concerns of groups of developed and developing countries. **Development** refers to the continued improvement of human well-being, usually in the lower and middle income countries. Both developed and developing countries have embraced the concept of sustainable development, although industrialized countries are usually more concerned about environmental sustainability, while developing countries are more concerned about economic development. The basic idea, however, is to maintain and improve the well-being of both humans and ecosystems.

The concept of sustainable development sounds comforting, so people want to believe that it is possible, and it appears to incorporate some ideals that are sorely needed, such as **equity**—whereby the needs of the present are actually met and future generations are seen as equally deserving as those living now. Sustainable development means different things to different people, however. Economists, for example, are concerned mainly with growth, efficiency, and the optimum use of resources. Sociologists focus on human needs and concepts such as equity, empowerment, social cohesion, and cultural identity. Ecologists show their greatest concern for preserving the integrity of natural systems, for living within the carrying capacity of the environment, and for dealing effectively with

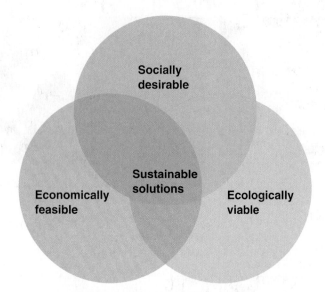

Figure 1–7 Sustainable solutions. Sustainable solutions are much more achievable when the concerns of sociologists, economists, and ecologists intersect.

pollution. It can be argued that sustainable solutions will be found mainly where the concerns of these three groups intersect, as illustrated in Figure 1–7.

There are many dimensions to sustainable development—environmental, social, economic, and political—and no societies today have achieved it. Nevertheless, as with justice, equality, and freedom, it is important to uphold sustainable development as an ideal—a goal toward which all human societies need to be moving, even if we have not achieved it completely anywhere. For example, policies that reduce infant mortality, improve air quality, and restore coastal fisheries move societies toward a sustainable future. In addition, the outcomes of such policies are measurable, so it is possible to assess our progress in achieving sustainable development. Communities and organizations are forging sustainable development indicators and goals to track their progress, such as the Environmental Sustainability Index (produced by Yale University, 1999–2005), which evaluated a society's natural resources and its stewardship of those resources and the people they support.

An Essential Transition

What will it take to make the transition to a sustainable future? There is broad agreement on the following major points:

- A population transition from a continually increasing human population to one that is stable or even declining.

- A resource transition to an economy that is not dedicated to growth and consumption, but instead protects ecosystem capital from depletion.

- A technology transition from pollution-intensive economic production to environmentally friendly processes—in particular, moving from fossil fuel energy to renewable energy sources.

- A political/sociological transition to public policies that embrace a careful and just approach to people's needs and that eliminate large-scale poverty.

- A community transition from car-dominated urban sprawl to the "smart growth" concepts of smaller, functional settlements and more livable cities.

This is not an exhaustive list, but it does give a glimpse of what a sustainable future might look like. One thing is certain: we cannot continue on our present trajectory and hope that everything will turn out well. The natural world is being degraded, its ecosystem capital eroded. One essential tool for achieving sustainability is sound science.

CONCEPT CHECK If a society is not using up resources rapidly but continues to have high inequality between rich and poor, what aspects of sustainability are missing? How does equity promote sustainability? ☑

1.3 Sound Science

Some environmental issues are embroiled in controversies so polarized that no middle ground seems possible. On one side are those who argue using seemingly sound scientific facts to support their position. On the other side are those who present opposing explanations of the same facts. Both groups may have motives for arguing their case that are not apparent to the public. In the face of such controversy, many people are understandably left confused. It is our objective to give a brief overview of the nature of science and the scientific method and then to show how science forms the foundation for understanding controversial issues.

The Scientific Method

In its essence, science is simply a way of gaining knowledge; that way is called the **scientific method**. The term **science** further refers to all the knowledge gained through that method. We employ the term **sound science** to distinguish legitimate science from what can be called **junk science**, information that is presented as valid science but that does not conform to the rigors of the methods and practice of legitimate science. Sound science involves a disciplined approach to understanding how the natural world works—the scientific method.

Assumptions. To begin, the scientific method rests on four basic assumptions that most of us accept without argument.

1. What we perceive with our five senses represents an objective reality, not some kind of dream or mirage.

2. This objective reality functions according to certain basic principles and natural laws that remain consistent through time and space.

3. Every result has a cause, and every event in turn will cause other events.

4. Through our powers of observation, manipulation, and reason, we can discover and understand the basic principles and natural laws by which the universe functions.

SOUND SCIENCE
Oysters Sound the Alarm

Oysters are big business on the U.S. West Coast, netting some $85 million a year and employing 3,000 workers. The oyster of choice on that coast is *Crassostrea gigas,* a native of Japanese waters imported many years ago and planted in many coastal waters. Because this oyster species needs warm water in order to spawn and West Coast waters are cold, oyster farms purchase "seed" oysters, young oysters 2 to 3 mm in diameter that are raised in hatcheries (see photograph). They then "plant" the seed on selected sandy banks or in special trays that are suspended in coastal bays. There the oysters grow to maturity in a year or two as they feed on natural food in the saline waters.

On a good day, Whiskey Creek Shellfish Hatchery on Netarts Bay, Oregon, ships out millions of seed oysters to oyster farmers. At the hatchery, oysters spawn under laboratory conditions. A female oyster produces millions of eggs, which are then fertilized by male oysters. The fertile eggs are held in tanks, where they develop into swimming larvae and start to form shells. The larvae settle down as they grow into seed oysters; all this time they have been bathed in filtered coastal seawater and provided with microscopic algae for food.

Unfortunately, the good days at Whiskey Creek came to a halt in 2007. During the summer, oyster larvae and seed oysters began dying off, sometimes simply disappearing from the incubation tanks. Co-owners Sue Cudd and her husband Mark Wiegardt immediately started looking for the culprit; they quickly identified a naturally occurring pathogenic bacterium, *Vibrio tubiashii,* that had become established in their tanks. They installed an expensive filtration system that removed the bacteria from incoming water, but the larvae continued to die. So it wasn't a pathogen killing the young oysters.

Then, in July 2008, all of the remaining larvae in their tanks suddenly died. At that time, the owners noticed that the bay water brought into the tanks was colder than normal. This colder water had recently been brought up to the surface as offshore winds pushed the warmer surface water away, a process called upwelling. The owners also observed that the pH of the water coming into the tanks was lower than normal. Surface seawater normally has a pH around 8.1, which is slightly basic; the pH of the upwelled water was in the range

of 7.5. (A pH above 7.0 is basic, a pH of 7.0 is neutral, and a pH below 7.0 is acidic.) A change that moves a liquid closer to the acid range of the pH scale is called *acidification,* even if the water is not actually acidic. The reason for the change in pH was high concentrations of CO_2, released as plankton died and sank to the deeper waters and decomposed.

New studies by oceanographers showed that waters rising from depths of the Pacific Ocean hold the CO_2 from normal decomposition plus CO_2 absorbed from the atmosphere of past decades. Over the past 250 years, the oceans have taken up about 30% of the total emissions of CO_2, or about half of what comes from burning fossil fuels and manufacturing cement. By absorbing this atmospheric CO_2, the oceans have been acidified. It appears that ocean acidification is well under way and is now beginning to affect shellfish.

The solution for the hatchery owners was now clear: They had to avoid ocean water with a lower pH. They installed chemical monitors to measure the CO_2 and pH of incoming water and avoided withdrawing water that was recently upwelled and had a lower pH. They also limited their withdrawals to daytime hours when algae were removing CO_2 from the water for photosynthesis and avoided nighttime

hours when the water had higher CO_2 levels and a lower pH. Many shellfish hatcheries now buffer their intake water by adding limestone or another ingredient to bring the pH back up. Of course, wild shellfish have to live under whatever conditions the ocean offers.

Today Whiskey Creek Hatchery has recovered, and seed oysters are being shipped out once again. The death of oyster larvae was an early sign of a process that is now under way and looks to become increasingly serious—ocean acidification.

This story shows parts of the scientific process in action. The hatchery owners identified several different possible explanations for the death of the oyster larvae and tested them by simple experiments. The results of their experiments caused the owners to reject some explanations, but eventually one explanation was better supported. In addition, scientists used data from the Whisky Creek Hatchery in further experiments studying the effects of acidification on oysters.

Source: Barton, A., B. Hales, G. G. Waldbusser, C. Langdon, and R. A. Feely. "The Pacific Oyster, *Crassostrea gigas,* Shows Negative Correlation to Naturally Elevated Carbon Dioxide Levels: Implications for Near-Term Ocean Acidification Effects." *Limnology and Oceanography* 57, no. 3 (2012): 698. doi:10.4319/lo.2012.57.3.0698.

Although assumptions, by definition, are premises that are simply accepted, the fact is that the assumptions underlying science have served us well and are borne out by everyday experience. For example, we suffer severe consequences if we do not accept our perception of fire as real. Similarly, our experience confirms that gravity is a predictable force acting throughout the universe and that it is not subject to unpredictable change. Thus, whether we are conscious of the fact or not, we all accept the basic assumptions of science in the conduct and understanding of our everyday lives.

Observation. The foundation of all science and scientific discovery is **observation**. Indeed, many branches of science, such as natural history (the study of where and how various plants and animals live, mate, reproduce, etc.), astronomy, anthropology, and paleontology, are based almost entirely on observation because experimentation is either inappropriate or impossible. For example, it simply is impossible to conduct experiments on stars or past events. Even experimentation, as we will discuss shortly, is conducted to gain another window of observation.

How can we be sure that observations are accurate? One answer is that scientists use a variety of tools and technologies to help them make accurate observations. Our current understanding of climate, for example, depends on remarkably precise maps and measurements from satellites. Even so, not every reported observation is accurate, for reasons ranging from honest misperceptions or instrument error to calculated mischief. Therefore, an important aspect of science, and a trait of scientists, is to be *sceptical* of any new findings until they are replicated.

Constructing Models. Various observations, like the pieces of a puzzle, are often put together into a larger picture—a **model** of how a system works. Models may be mathematical formulas, computer simulations, diagrams, conceptual descriptions, or other representations. What all models have in common is an attempt to portray the important parts of a real system; although a model is simpler than the system it portrays, it can be used to predict outcomes.

To give a simple example, we observe that water evaporates, making the air moist, and that water from moist air condenses on a cool surface. We also observe clouds and precipitation. Putting these observations together logically, we derive a conceptual description of the hydrologic cycle, described later (Chapter 10). Water evaporates and then condenses as air is cooled, condensation forms clouds, and precipitation follows. Water thus makes a cycle from the surface of Earth into the atmosphere and back to Earth (Figure 10–3). Note how this simple example incorporates the four assumptions described previously: there is an objective reality, it operates according to principles, every result has a cause, and we can discover the principles according to which reality operates.

So, the essence of science and the scientific method may be seen as a process of making observations and logically integrating those observations into a model of how some natural system works. This is where the sequence consisting of observation, hypothesis, test (experiment), and explanation comes in. Figure 1–8 shows the typical sequence of these step.

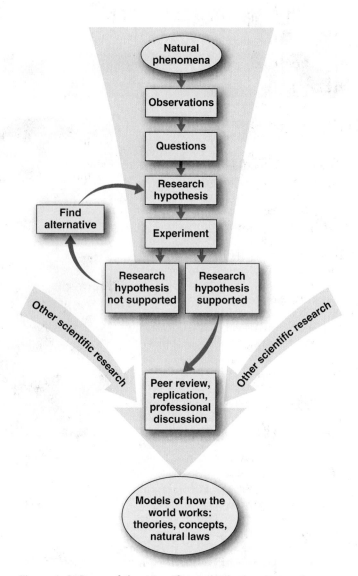

Figure 1–8 Steps of the scientific method. The process of science is a continual interplay among observations, hypotheses, tests (experiments), explanations, and eventually theories and natural laws, as well as further refinement. The process involves the work of individual scientists and peer review by the scientific community.

Experimentation. **Experimentation** is simply setting up situations to make more systematic observations regarding causes and effects. For example, biologists put organisms into specific situations in which they can carefully observe and measure their responses to particular conditions or treatments. Figure 1–9 depicts scientists exposing marine organisms to different pH levels in order to determine the effects of acidification. To solve a particular problem (e.g., What is the cause of this event?), a systematic line of experimentation is used. One of the reasons experimentation is so powerful is because it includes a control—a point against which you compare what you find under experimental conditions. Experiments need to include replication—that is, more than a one-time measurement of a factor. After the measurements are taken, scientists use statistics to determine whether or not their findings show an effect big enough to conclude that the results probably do not result from random chance.

Hypotheses. Let's consider the problem of the death of oyster larvae at the Whiskey Creek hatchery (see Sound Science, p. 13). The owners' question was What is killing the oyster larvae in the incubation tanks? The first step in solving the problem was to make educated guesses as to the cause. Each such educated guess is a **hypothesis**. Each hypothesis is then tested by making further field observations or by conducting experiments to determine whether or not the hypothesis really accounts for the observed effect.

In the case of the oysters, the owners performed what were essentially a series of informal experiments. The owners found high numbers of pathogenic bacteria in the tanks (hypothesis: the bacteria were killing the larvae). They installed a new filtration system that removed bacteria before they could multiply, essentially conducting an experiment. The larvae still died. They rejected that hypothesis. Then a crucial observation was made about the timing of the bay water brought into the tanks, water that had come from recently upwelled coastal surface water. This water had a lower pH (closer to the acid range) than usual (hypothesis: water with a lower pH was killing the larvae). By monitoring the pH of incoming water, the owners were able to avoid the water with lower pH and time water withdrawals to daytime hours when the pH was higher. The results were dramatic: the larvae were able to thrive and grow. In this case, the hypothesis was supported because the larval death ended. Figure 1–8 shows how such a process moves from observations to hypotheses, experiments, and finally to an answer that fits the data better than other answers and can be used to make predictions—that is, a better model of how the world works.

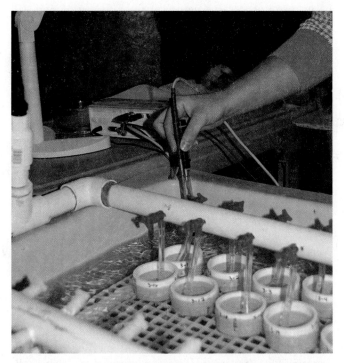

Figure 1–9 Experiment demonstrating the effect of ocean acidification on marine organisms. Researchers from the Alaska Fisheries Science Center conduct experiments to test effects of ocean acidification on fish, shellfish, and cold water corals in the North Pacific ocean, a region considered to be especially vulnerable to acidification.

Uncertainty. Testing hypotheses helps scientists learn how the world works, but scientists never confirm a hypothesis. Instead, they *reject* a **null hypothesis**, which is what would be true if there were no relationship between the variables they are investigating. Scientists compare their findings to what they would expect to see from random chance. Then they can say something like, "With 95% confidence, we reject that this result came from random chance, that is, we reject the null hypothesis." Notice that the statement includes a measure of confidence or its opposite, uncertainty. Scientific uncertainty can result from errors in measurement or understanding; it can also result from accurate (correct) but imprecise (not finely divided) measurements. Because there is always a limit to what we can think of, measure, or know, we can never be 100% certain that our model is correct. There is always a possibility that some other explanation fits the data better than the explanation we have. However, that possibility can be so small that we can use our model to predict outcomes with a high degree of success. That is the sign of a good model.

Theories. We have already noted how various specific observations (e.g., evaporation, precipitation) may fit together to give a logically coherent conceptual framework (e.g., the hydrologic cycle). A hypothesis is a tentative explanation that answers a question. An overarching conceptual framework that explains many observations and is well supported by evidence is called a **theory**.

Through logical reasoning, theories can be used to suggest or predict certain events or outcomes. Predictions require experiments, testing, further data gathering, and more observation. When theories reach a state of providing a logically consistent framework for relevant observations and when they can be used to reliably predict outcomes, they are accepted. For example, we have never seen atoms directly (or, until recently, even indirectly), but innumerable observations and experiments support and are explained by the atomic theory of matter. Of course, because of the nature of science, theories are always less than absolutely certain.

Natural Laws. The second assumption underlying science—that the universe functions according to certain basic principles and natural laws that remain consistent through time and space—cannot be established with absolute certainty. Still, all our observations, whether direct or through experimentation, demonstrate that matter and energy behave neither randomly nor even inconsistently, but in precise and predictable ways. We refer to these principles by which we can define and precisely predict the behavior of matter and energy as **natural laws**. Examples are the *Law of Gravity,* the *Law of Conservation of Matter,* and the various *Laws of Thermodynamics.* Our technological success in space exploration and many other fields is in no small part due to our recognition of these principles and our precise calculations based on them. Conversely, trying to make something work in a manner contrary to a natural law invariably results in failure.

In many situations, the outcome of scientific work results in well-established explanations that must be expressed in the mathematical language of probability and statistics. This is especially true of biological phenomena such as predator–prey relationships and the effects of pesticides. In such cases, it is appropriate to speak of concepts rather than laws. **Concepts** are perfectly valid explanations of data gathered from the natural world, but they don't reach the status of laws. Concepts model the way we believe the natural world works and enable us to make qualified predictions of future outcomes. Thus, on the basis of our understanding of the effects of DDT (a potent pesticide) on mosquitoes and on the basis of our observation that mosquitoes can develop resistance to DDT, we might predict that regular spraying of the local salt marshes with DDT to eradicate unwelcome mosquitoes is likely to result in the development of a more resistant mosquito population in the future.

The Scientific Community

The process of science and its outcomes take place in the context of a scientific community and a larger society. There is no single authoritative source that makes judgments on the validity of scientific explanations. Instead, it is the collective body of scientists working in a given field who, because of their competence and experience, establish what is sound science and what is not. Scientists communicate their findings to each other and to the public by publishing their work in peer-reviewed journals. In the process of peer review, experts in a given field scrutinize the results of their colleagues' work, with the objective of rooting out poor or sloppy science and affirming work that is of good quality.

Professional Societies. Scientists also operate through professional societies, which may put out statements, white papers, or reports that represent the best scientific understanding of their members. For example, the report *What We Know*, published in 2014 by the American Academy for the Advancement of Science (AAAS), describes the current state of scientific understanding on climate change.[8] Members of the U.S. **National Academies of Science** (NAS), the top scientific society in the United States, are preeminent American scientists chosen by their peers. The NAS provides leadership for the country by describing the current state of scientific knowledge in different fields.

Disagreements. Even with the checks and balances that work to establish scientific knowledge and with all the procedures of the scientific method, disagreements arise that pit one group against another, with both groups claiming that they have science on their side. For example, we will make the claim that the global climate is changing because of human activities (Chapter 18). A few scientists disagree with this claim, although for most the evidence is convincing. There are at least four reasons why such disagreements occur.

1. **New information.** We are continually confronted by new observations—the hole in the ozone layer, for instance (see Chapter 19). It takes some time before all the hypotheses regarding the cause of the new observation can be adequately tested. During this time, there is often much debate as to which hypothesis is most likely.

2. **Complex phenomena.** Certain phenomena, such as the hole in the ozone layer, do not lend themselves to simple tests or experiments. Therefore, it is difficult and time consuming to confirm the causative role of one factor or to rule out the involvement of another. Gradually, different lines of evidence come together to support one hypothesis and exclude another, enabling the issue to be resolved.

3. **Multiple perspectives.** The third reason for disagreement is that multiple perspectives may be involved. This is particularly true in environmental science because many issues have social, economic, or political ramifications. For example, there is no controversy regarding nuclear power, as long as it is considered at the purely scientific level of physics. However, when it comes to regulating the use of nuclear power to generate electrical power, disagreement arises because different groups have different perspectives about the relative costs, risks, and benefits involved.

4. **Bias.** The fourth reason for disagreement is that there are many vested interests that wish to maintain and promote disagreement because they stand to profit by doing so. For example, tobacco interests argued for years that the connection between smoking and illness had not been established and that more studies were necessary. In spite of the overwhelming body of evidence supporting the connection between smoking and illness, the tobacco lobby succeeded in keeping the issue controversial and thereby delayed regulatory restrictions on smoking.

Some disagreement is the inevitable outcome of the scientific process itself, but much of it is attributable to less-noble causes. Unfortunately, the media and the public may be unaware of the true nature of the information and will often give equal credibility to opposing views on an issue. This leads us to the topic of junk science.

Junk Science. Junk science refers to information that is presented as scientifically valid but that does not conform to the rigors imposed by the scientific community. Junk science can take many forms: the presentation of selective results (picking and choosing only those results that agree with one's preconceived ideas), politically motivated distortions of scientifically sound information, and the publication of poor work in quasi-scientific (unreviewed) journals and books. Quite often, junk science is generated by special interest groups trying to influence the public debate about science-related concerns. As a result, the public arena may consider the junk science to be equally as valid as the sound science. Information presented by the Tobacco Institute, for example, is clearly suspect, due to its corporate-related funding. Junk science has been heavily employed in promoting the (scientifically debunked) idea that vaccines cause autism, a common misconception in the United States.

In recent years, another use of the term *junk science* has appeared that has created some confusion. In some cases,

---

[8]Molina, M., et al., *What We Know: The Reality, Risks and Response to Climate Change* (AAAS, 2014), accessed November 7, 2014, http://whatweknow.aaas.org/wp-content/uploads/2014/07/whatweknow_website.pdf.

people with a bias have used the term to refer to anything that threatens their preferred viewpoint; conversely, *sound science* becomes anything that supports their viewpoint. One example of this usage has come from those who do not support the Endangered Species Act. They label as "junk science" the legitimate use of computer models to make predictions about the survival of endangered species. This limits scientific input to a much smaller number of studies and could mean that decisions are made without the full range of available information.

Evaluating Issues. Whether we are scientists or simply concerned citizens, we can use facets of the scientific method to deal with disagreements about science and to develop our own capacity for logical reasoning. Here are some basic questions to ask:

- What are the data underlying a particular claim? Can they be satisfactorily confirmed?
- Do the explanations and theories follow logically from the data?
- Are there reasons that a particular explanation is favored? Who profits from having that explanation accepted broadly?
- Is the conclusion supported by the community of scientists with the greatest competence to judge the work? If not, it is highly suspect.

In sum, sound science is absolutely essential to forging a sustainable relationship with the natural world. Our planet is dominated by human actions. Our activities have reached such an intensity and such a scale that we are now one of the major forces affecting the rest of nature. Our influence on the natural ecosystems that support most of the world economy and process our waste is strong and widespread, and we need to know how to manage the planet so as to maintain a sustainable relationship with it. As a result, the information gathered by scientists needs to be accurate, credible, and communicated clearly to policy makers and the public; the policy makers, in turn, need to handle that information responsibly (the interaction between science and politics is explored in Chapter 2).

CONCEPT CHECK If an advertisement claims that magnets can cure cancer, what information would you want to have before you considered the claim to be sound science rather than junk science? ✓

1.4 Stewardship

The third of our themes is **stewardship**—the actions and programs that manage natural resources and human well-being for the common good. Stewardship is a concept that can be traced back to ancient civilizations. A steward was put in charge of his master's household, responsible for maintaining the welfare of the people and the property of the owner. Because a steward did not own the property himself, the steward was responsible for caring for something on behalf of someone else. Applying this concept to the world today, stewards are those who care for something—from the natural world or from human culture—that they do not own and that they will pass on to

STEWARDSHIP
Protecting Forests

Many organizations have developed stewardship principles to guide their policies and actions. Sometimes these principles can serve as broader guidelines for society, although they often address specific areas of concern and are often focused on human needs. Stewardship of forests is a good example.

Several groups working to protect forests have stewardship as a foundational principle. One such group is the Forest Service Employees for Environmental Ethics, which has the mission of protecting national forests and bringing reform to the U.S. Forest Service by advocating environmental ethics. The group is active in halting unauthorized logging and protecting Forest Service employees from unjust disciplinary actions, as well as lobbying Congress to maintain stewardship of the National Forests. Another group is the Forest Stewardship Council (FSC),

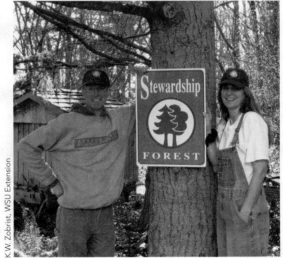

K.W. Zobrist, WSU Extension

Forest Stewardship. Students at Washington State University can take courses in forest stewardship to learn how to manage forests for sustainability.

created in 1993 to "promote environmentally sound, socially beneficial and economically prosperous management of the world's forests." FSC has a forest products certification program used around the world and works on other sustainable forest issues, such as lowering the amount of pesticides used on forests.

Many Native American tribes, such as the Snoqualmie in Washington State, have a strong stewardship ethic. Some members of the Snoqualmie tribe have taken forest stewardship courses at Washington State University, along with other forest managers. Groups such as the Nature Conservancy also are stewards of forestlands. All these groups work with the rest of society to balance the desires and needs of different stakeholders, protect resources for the future, preserve species, encourage care of others, and limit waste.

the next generation. Stewardship is compatible with the goal of sustainability, but it is different from it, too, because stewardship deals more directly with how sustainability is to be achieved—what actions are taken, and what values and ethical considerations are behind those actions.

Ethics is the study of moral principles; it is concerned with what is good—those values and virtues we should encourage—and what is right—our moral duties as we face practical problems. A stewardship ethic is concerned with right and wrong as they apply to actions taken to care for the natural world and the people in it. It recognizes that even our ownership of land is temporary; the land will be there after we die, and others will own it in turn. Because a steward is someone who cares for the natural world on behalf of others, we might ask, "To whom is the steward responsible?" Many would answer, "The steward is responsible to present and future generations of people who depend on the natural world as their life-support system."

There are several underlying reasons for a stewardship ethic, some stronger than others. For people with religious convictions, stewardship stems from a belief that they are stewards on behalf of a higher power. For others, stewardship becomes a matter of concern that stems from a deep understanding and love of the natural world and the necessary limitations on our use of that world.

Is there a well-established stewardship ethic? Insofar as human interests are concerned, ethical principles and rules are fairly well established, even if they are often violated by public and private acts. However, there is no firmly established ethic that deals with care for natural lands and creatures for their own sake. Most of our ethic concerning natural things deals with how those things serve human purposes; that is, our current ethic is highly anthropocentric. For example, a basic ethical principle from the UN Declaration on Human Rights and the Environment is, "All persons have the right to a secure, healthy, and ecologically sound environment." Here, the point of the healthy environment is for humanity. Ethics for protection of other species will be covered later (Chapter 6).

Who Are the Stewards?

How is stewardship achieved? Sometimes stewardship leads people to try to stop the destruction of the environment or to stop the pollution that is degrading human neighborhoods and health. For example, Rachel Carson was a steward who alerted the public about the dangers of pesticides and their role in decimating bird populations.

The late Wangari Maathai (the first Kenyan woman to earn a Ph.D.) founded the Green Belt Movement to help rural Kenyan women. The grassroots movement eventually involved more than 6,000 groups of women who planted 47 million trees in Kenya. Maathai was beaten and jailed when she went on to protest government corruption. Eventually, she was appointed deputy environment minister under a new Kenyan president and received the Nobel Peace Prize in 2004 (Figure 1–10)—the first environmental activist to receive that honor. Ramesh Agrawal (Figure 1–11) is an Indian activist who opposed the powerful Jindal Steel and Power Company and led a grassroots campaign of India's

Figure 1–10 Stewardship at work. Wangari Maathai receiving the 2004 Nobel Peace Prize from Ole Danbolt Mjøs, chairman of the Norwegian Nobel Committee. She is an example of unusual environmental stewardship at work.

poor villagers against environmentally damaging mining operations. Agrawal was imprisoned and later shot in the leg for his activism. Even so, his efforts have succeeded in helping the poor in India who did not know their legal rights. In recognition of this work, Agrawal was awarded the 2014 Goldman Environmental Prize.

More often, stewardship is a matter of everyday people caring enough for each other and for the natural world that

Figure 1–11 Goldman Prize winner, 2014. Ramesh Agrawal was shot in the leg as a result of his efforts to help poor Indian villagers demand their right to information about the environmental impact of industrial projects.

(*Source:* Goldman Environmental Prize.)

they do the things that are compatible with that care. These include participating fully in recycling efforts, purchasing cars that pollute less and use less energy, turning off the lights in an empty room, refusing to engage in the conspicuous consumption constantly being urged on them by commercial advertising, supporting organizations that promote sustainable practices, staying informed on environmentally sensitive issues, and expressing their citizenship by voting for candidates who are sympathetic to environmental concerns and the need for sustainable development.

Justice and Equity

The stewardship ethic is concerned not only with caring for the natural world, but also with establishing just relationships among humans. This concern for justice is seen in the United States in the environmental justice movement. One major problem addressed by the movement is environmental racism—the placement of waste sites and other hazardous industries in towns and neighborhoods where most of the residents are nonwhite. The flip side of this problem is seen when wealthier, more politically active, and often predominantly white communities receive a disproportionately greater share of facility improvements, such as new roads, public buildings, and water and sewer projects.

People of color are seizing the initiative to correct these wrongs, creating citizen groups and watchdog agencies to bring effective action and monitor progress. For example, the city of Chattanooga, Tennessee, recently underwent a huge renewal that changed its reputation from being "the dirtiest city in America" to the "Renaissance City of the South." Yet most of this renewal bypassed Chattanooga's Southside, an area that is populated largely by African Americans. In 2010, Southside community members formed Chattanooga Organized for Action (COA), a group that has quickly become a voice for social justice. Among other actions, COA has organized the restoration of funding for family-planning programs for low-income and minority women and forced a more thorough cleanup of a former chemical plant site in Southside.

Justice is not only about racial concerns. Injustices can occur wherever people lack equal protection from environmental degradation. In some countries, extreme poverty is brought on by injustices within societies where wealthy elites maintain political power and, through corruption and nepotism, steal money and give corporations preferential treatment. It takes significant government reform to turn this situation around so that the poor can achieve justice in securing property rights and gain greater access to ecosystem goods and services.

Some of the poverty of the developing countries can be attributed to unjust economic practices of the wealthy industrialized countries. The current pattern of international trade is an example. The industrialized countries have maintained inequities that discriminate against the developing countries by taxing and restricting imports from the developing countries and by flooding the world markets with agricultural products that are subsidized (and, therefore, priced below real cost). This is the number one issue at every meeting of the World Trade Organization, where global trade agreements are negotiated (see Chapter 2). Wealth inequality and development are discussed at length later (Chapters 8 and 9). Operating under a stewardship ethic ought to cause us to care for the natural world in ways that protect ecosystem resources for all of us—in a just way.

CONCEPT CHECK How does a stewardship ethic relate to climate change and biodiversity loss, described in Section 1.1? ✓

1.5 Moving Toward a Sustainable Future

Without doubt, one of the dominant trends of our times is globalization—the accelerating interconnectedness of human activities, ideas, and cultures. For centuries, global connections have brought people together through migration, trade, and the exchange of ideas. What is new about the current scene is the increasing intensity, speed, extent, and impact of connectedness, to the point where its consequences are profoundly changing many human enterprises. These changes are most evident in the globalization of economies, cultural patterns, political arrangements, environmental resources, and pollution.

For many in the world, globalization has brought undeniable improvements in well-being and health: economic exchanges have connected people to global markets, information exchanges have improved public health practices, and agricultural research has improved crop yields. For others, however, globalization has brought the dilution or even destruction of cultural norms and religious ideals while doing little to improve economic well-being. Whether helpful or harmful, globalization is taking place, so it is important to recognize its impacts and to identify where those impacts are harmful or undesirable. Moving toward a sustainable future will occur in the context of globalization.

Social Changes

The major elements of globalization may be the economic reorganization of the world and the changes wrought by social media. These transformations have been facilitated by the globalization of communication, whereby people are instantly linked through the Internet, satellite, and cable. Other components of the global economic reorganization are the relative ease of transportation and financial transactions and the dominance of transnational corporations, such as Nokia, Roche, Philips Electronics, and Nestlé. All of these changes occur within the context of the demographic patterns of urbanization, mass migration, and aging populations in some countries. Moreover, economic changes bring cultural, environmental, and technological changes in their wake. Western diet, styles, and culture are marketed throughout the world, to the detriment of local customs and diversity.

Environmental Changes

On the one hand, the increased dissemination of information has made it possible for environmental organizations and government agencies to connect with the public and has enabled people to become politically involved with current issues. It has also enabled consumers to find more environmentally friendly consumer goods and services, such as shade-grown coffee and sustainably harvested timber. On the other hand, globalization has contributed to some notoriously harmful outcomes, such as the worldwide spread of emerging diseases such as the SARS virus, H1N1 swine influenza, and HIV/AIDS; the global dispersion of exotic species; the trade in hazardous wastes; the spread of persistent organic pollutants; the radioactive fallout from nuclear accidents; the exploitation of oceanic fisheries; the destruction of the ozone layer; and global climate change. These outcomes must be addressed in the move toward a sustainable future.

A New Commitment

People in all walks of life—scientists, sociologists, workers and executives, economists, government leaders, and clergy, as well as traditional environmentalists—are recognizing that "business as usual" is not sustainable. Many global trends are on a collision course with the fundamental systems that maintain our planet as a tolerable place to live. A finite planet cannot continue to grow by 80 million persons annually without significant detrimental effects. The current degradation of ecosystems, atmospheric changes, losses of species, and depletion of water resources inevitably lead to a point where resources are no longer adequate to support the human population without significant changes in the way we live.

The news is not all bad, however. Food production has improved the nutrition of millions in the developing world, life expectancy continues to rise, and the percentage of individuals who are desperately poor continues to decline. Population growth rates continue to fall in many countries. A rising tide of environmental awareness in the industrialized countries has led to the establishment of policies, laws, and treaties that have improved environmental protection. Professional groups such as ecologists, climatologists, and engineers are proposing scientific and technological solutions to environmental problems. **Figure 1–12** shows one such technological solution—one of the many types of fuel-efficient stoves developed for use in places where wood is used for cooking.

Numerous caring people are beginning to play an important role in changing society's treatment of Earth. For example, in January 2011, a group of 70 engineers produced a report proposing engineering development goals and stated, ". . . the engineers of today, and the future, will need to be innovative in the application of sustainable solutions and increasingly engaged with the human factors that influence their decisions."[9] Likewise, people in business have formed the Business Council for Sustainable Development, economists have formed the International Society for Ecological Economics, religious leaders have formed the National Religious Partnership for the

Environment, and environmental philosophers are calling for a new ethic of "caring for creation." Higher education is involved in transforming society as well. The Association for the Advancement of Sustainability in Higher Education (AASHE) promotes sustainability and its teaching in thousands of institutions.

The movement toward sustainability is growing as people, businesses, and governments slowly but surely begin to change direction. Some examples include: wind energy is the fastest-growing source of electricity; cap-and-trade markets are being established to curb carbon emissions; car-sharing companies are springing up in our cities; chemists have embraced the initiatives of "green chemistry"; and microfinance is opening up opportunities for small-scale businesses, thereby helping very poor people. In addition, "green investments" are attracting increasing capital, and the nations of the world are working on agreements to bring down greenhouse gas emissions. These developments show that people's behavior toward the environment *can* be transformed from exploitation to conservation.

CONCEPT CHECK Name two patterns of globalization that make environmental sustainability more difficult and two modern developments that make it easier. ☑

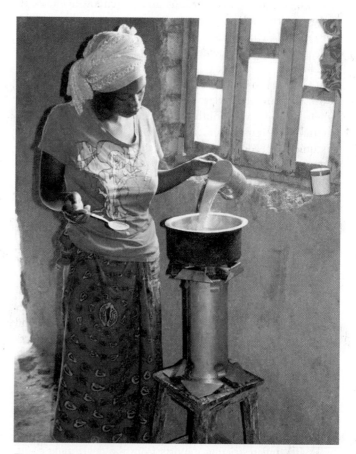

Figure 1–12 Innovative technologies helping sustainability.
WorldStove is one organization that is producing innovative technologies that meet real-world needs. This Rwandan woman is cooking with one of their stoves; the stove is designed to use local waste products for fuel and to produce smokeless heat.

[9]Institution of Mechanical Engineers, "Population: One Planet, Too Many People?" (2011), accessed September 21, 2011, http://www.imeche.org/Libraries/2011_Press_Releases/Population_report.sflb.ashx.

REVISITING THE THEMES

It will be our practice in each chapter to revisit the three themes that we have introduced in the chapter, summarizing the ways in which the chapter topics have connected to the themes and sometimes adding editorial comments.

SUSTAINABILITY

Many unsustainable practices result from a combination of cumulative impacts and the unintended consequences of human actions. The story of DDT and other chemicals broadcast over vast areas of land is a great example of an unsustainable action. At first, spraying the land with DDT seemed a wise way to protect food supplies and eliminate malaria; we did not expect that DDT would affect the food chain. As we gathered more information, we discovered the unintended consequences of those actions. Likewise, as the human population expands, the effects of individuals are multiplied. Both natural and managed ecosystems are being pressed to provide increasing goods and services. We may have reached the limit of some resources, and most of the ways we use the other resources are currently unsustainable. For example, we extract and burn fossil fuels much faster than they could ever be replenished, bringing on global climate change. In addition, we tolerate the continued loss of biodiversity. Many individuals suffer from hunger and other effects of poverty. The Millennium Development Goals moved us in a better direction, but there is still a great deal to do, thus the need for the post-2015 development agenda and the Sustainable Development Goals. A movement toward sustainability is under way, and it will occur within the context of globalization.

SOUND SCIENCE

Our brief look at the global environment was based largely on the information uncovered by sound science. The Millennium Ecosystem Assessment and other assessments demonstrate the crucial value of scientific study of the status of and trends in global ecosystems and the ways in which human systems affect them. Environmental science itself is a multidisciplinary enterprise, one that can readily lead to disputes and misuse. To understand these, it is essential to understand the scientific method as a disciplined way to investigate the natural world. Rachel Carson is a good example of a scientist who communicated sound science to popular audiences. The research done at the Whiskey Creek Hatchery is another example of the way science can be used to answer a question. The issue of "junk science" points to the importance of a functioning scientific community and the need for an educated public able to evaluate opposing claims of scientific support for a given issue.

STEWARDSHIP

Although current cultures and religions place some value on stewardship, it is not always a fundamental part of the way we think about caring for the natural world and for our fellow humans. If it were, there would not be the widespread degradation of essential ecosystems and the devastating impacts of poverty and injustice. However, many people care deeply about the natural world and its human inhabitants. Rachel Carson is an example of a person using her expertise to promote stewardship and to advocate for more sustainable policies. Ramesh Agrawal and Wangari Maathai also exemplify stewardship, as do the students studying forestry (see Stewardship, *Protecting Forests* p. 17). As we will see throughout the book, it is no easy matter to establish a valid stewardship ethic to guide the actions and programs that manage natural resources and human well-being for the common good.

REVIEW QUESTIONS

1. What information made Rachel Carson concerned about chemical pollution?

2. What is the paradox of human population and well-being?

3. Cite four global trends that indicate that the health of planet Earth is suffering.

4. Define *environment*, and describe the general development and successes of modern environmentalism.

5. What are the concepts behind each of the three unifying themes—sound science, sustainability, and stewardship?

6. Explain the role of assumptions, observation, experimentation, and theory formation in the operation of scientific research and thinking.

7. Cite some reasons for the existence of controversies within science, and carefully distinguish between sound science and junk science.

8. Define *sustainability* and *sustainable development*. What is a sustainable society?

9. List five transitions that are necessary for a future sustainable civilization.

10. Describe the origins of stewardship and its modern applications.

11. Give an example of an environmental injustice.

12. How does justice become an issue between the industrialized countries and the developing countries?

13. What is globalization? What are its most significant elements?

14. Cite some of the recent developments in the movement toward sustainability.

THINKING ENVIRONMENTALLY

1. Imagine a class debate between people representing the developing countries and people representing the developed countries. Characterize the arguments of the two sides in terms of the issues surrounding population growth, energy use, resource use, and sustainable development. Then describe what should be the common interests between the two.

2. Some people say that the concept of sustainable development is an oxymoron or that it represents going back to some kind of primitive living. Develop an argument contradicting this opinion, and give your own opinion on sustainable development.

3. Study Table 1–1, pick one of the ecosystem services listed there, and investigate the reasons why it is placed where it is by the Millennium Ecosystem Assessment.

4. Look for three articles about a current scientific issue where one reflects sound science, one reflects junk science, and one is hard to distinguish as either. Identify the features of the studies described that qualify as sound science or junk science.

MAKING A DIFFERENCE

1. Find out whether there is an environmental organization or club on your campus or in your community. Join it if there is, or help to create one if there isn't. Once in the organization, join efforts to maintain awareness of important environmental issues and problems.

2. Write a letter to a local newspaper about an issue of sustainability, such as a bottle bill, curbside recycling, public transportation, or something else. Figure out what you recommend and briefly state your case.

3. Encourage sound science on a national level by researching and voting in favor of political candidates who support your environmental viewpoint. Share your views with your friends and relatives.

4. Investigate the placement of waste-treatment facilities and industrial plants in your city or community, especially if they are in the planning stage. If they are located in low-income or largely nonwhite communities, join the local population in protesting these situations.

MasteringEnvironmentalScience®

Students Go to MasteringEnvironmentalScience for assignments, the eText, and the Study Area with animations, practice tests, and activities.

Professors Go to MasteringEnvironmentalScience for automatically graded tutorials and questions that you can assign to your students, plus Instructor Resources.

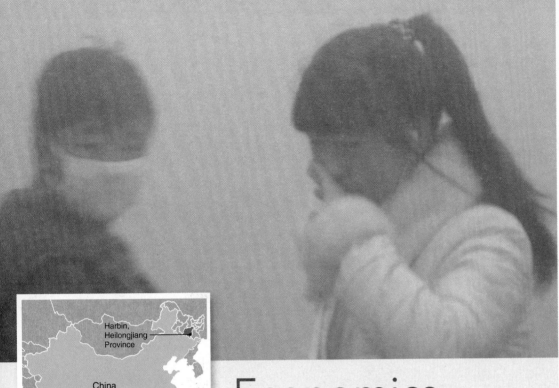

Economics, Politics, and Public Policy

By all measures, China has become the economic giant of the early 21st century. Take your pick: umbrellas, toys, shoes, appliances, electronics, souvenirs, clothes—chances are that they were made in China. In 2012, the Chinese economy was growing at 7.8 percent, which represented a slowing from the massive growth from the late 1970s to the early 2000s, when it averaged 10% per year. In contrast, the U.S. economy grew only 2.5% in 2012 and averaged only about 3% per year in the latter decades of the 20th century. A new Chinese middle class numbers between 100 million and 150 million. Middle-class Chinese citizens commonly own a car and an apartment and have opportunities to travel and purchase pricey consumer goods. With a population of 1.35 billion, China has a huge pool of labor. Each year millions of rural Chinese move to urban centers, attracted by jobs created by factories that spring up almost overnight. Chinese schools have created a workforce with a literacy rate that is over 95%. These changes have not come without stress. Traditions are being overthrown by the Internet, by the fierce drive to learn English, and by the social mobility that accompanies the economic boom.

Toxic Air. There are clear signs that China's race to achieve economic superpower status carries with it some costs that go well beyond a stretching of the social fabric. Several Chinese cities are among the most polluted cities in the world. The air is loaded with sulfur and nitrogen oxides and particulates from burning fuels, especially coal. In the last 10 years, the number of lung cancer patients has increased by 60%, although smoking rates have not increased. According to one estimate, 750,000 premature deaths occur annually because of air pollution.

▲ Chinese citizens walk through heavy smog in Harbin, northeast China's Heilongjiang province, in October 2013.

The city of Harbin, known for its winter festival, is an example of the complexity of modernization. A city of 10 million, Harbin is located in China's northeast Heilongjiang province. In October 2013, Harbin had an extended period when air pollution levels were so severe that residents could not see 10 meters (33 ft.) in front of themselves. A lack of wind, the burning of stubble in farm fields, and the burning of coal contributed to the crisis, which resulted in officials closing schools, the airport, and other city functions. Air quality index readings hit 500, the maximum possible level; levels above 300 are considered dangerous to human health.

Environmental Policy. The smog in Harbin is only one example of the pollution problems China faces as a result of rapid economic growth. The Chinese economy has flourished in large degree because of the decentralization of power in the 1980s, when local governments and private industries were encouraged to develop their resources relatively free of the restraints of central planning. The booming economy has also put greater power in the hands of the local governments and industries and has enabled them to flout existing environmental laws and policies. China has a Ministry of Environmental Protection, similar to the U.S. Environmental Protection Agency (EPA), but the country is so large that the ministry has been unable to keep up with the need for enforcement of its environmental regulations. Recently, the Chinese government has been turning to international nongovernmental organizations (NGOs) and has encouraged concerned citizens to form home-grown NGOs that focus attention on pollution and other environmental issues.

Bottom Line. China's example demonstrates that economic growth can lift millions out of poverty and establish a country's strong place in the world market. But when this growth is at the cost of natural resources and the health of many of its people, it is clearly unsustainable. China is now in the process of developing smart growth that is less environmentally damaging. As a new study suggests, sustainable growth will be better for the economy. The Chinese example shows that economic growth should be accompanied by the development and enforcement of just and effective environmental public policies.

In this chapter, we look first at economics and its connections to the natural environment.

2.1 Economics and the Environment

Economics is the social science that deals with the production, distribution, and consumption of goods and services and with the theory and management of economies or economic systems. Because economic goods and services are usually connected to—and indeed dependent on—environmental goods and services, economics needs to be grounded in an understanding of the biological and physical world. Simply put, no environmental goods and services, no economy.

Relationships Between Economic Development and the Environment

An **economy** is the system of exchanges of goods and services worked out by the members of a society. Goods and services are produced, distributed, and consumed as people make economic decisions about what they need and want and what they will do to become players in the system—what they might provide that others would need and want. As societies develop, economic activity assumes increasingly broader dimensions, with increasingly pervasive impacts on the whole society and the environment. For example, in future chapters we will see numerous instances in which pollution from economic activity has damaged the environment and human health. Furthermore, an unregulated economy can make intolerable inroads into natural resources. These changes result from development that is unsustainable.

Fortunately, economic activity in a nation can also provide the resources needed to solve the problems that economic activity creates. A strong relationship exists between the level of development of a nation and the effectiveness of its environmental public policies. **Figure 2–1** shows three patterns that describe relationships between various environmental problems and per capita (per individual in the population) income levels.

- Many problems decline (for example, sickness due to inadequate sanitation or water treatment) as income levels rise because the society decides to apply the resources available—usually from taxes—to address the problems with effective technologies.

- Some problems increase and then decline (for example, urban air pollutants such as sulfur dioxide, particulate matter, and carbon monoxide) when the serious consequences of the problem are recognized and public policies are developed to address them.

- Increased economic activity causes some problems to increase without any end in sight (for example, suburban sprawl, municipal wastes, and CO_2 emissions).

The key to solving problems brought on by economic activity is the development of effective public policies and institutions. To understand these relationships better, let us look at how economic systems work.

Economic Systems

Economic systems are social and legal arrangements that people construct in order to satisfy their needs and wants and to improve their well-being. Economic systems range from communal barter systems to the current global economy, which is so complex that it defies description. In any kind of economy, there are basic components that determine the economic flow

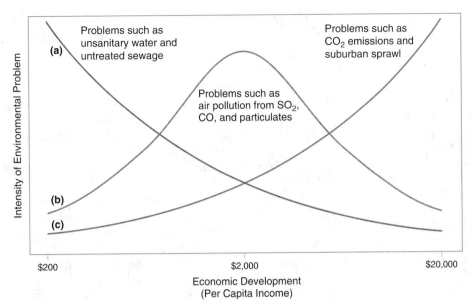

Figure 2–1 **Environmental problems and per capita income. (a)** Some of the most serious environmental problems can be improved with income growth, **(b)** other problems get worse and then improve, **(c)** while some problems just worsen.

of goods and services. Classical economic theory considers **land** (natural resources), **labor**, and **capital** to be the three elements that constitute the "factors of production." *Economic activity* involves the circular flow of money and products, as shown in **Figure 2–2**. Money flows from households to businesses as people pay for goods and services, and from businesses to households as people are paid for their labor. Labor, land, and capital are invested in the production of goods and services by businesses, and the products provided by businesses are *consumed*, or used, by households.

Two Economic Systems. In the organized societies of today's world, two broad kinds of economic systems have emerged: **centrally planned economies**, characteristic of dictatorships, and **free-market** (or **capitalist**) **economies**, characteristic of democracies. The cycles shown in Figure 2–2 apply to both kinds of economic systems; the two systems differ mostly in how economic decisions are made. Economies and the policies that drive them determine many of the effects of human activity on environmental systems, in somewhat different ways.

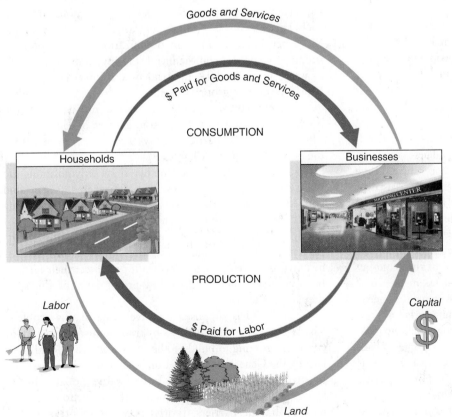

Figure 2–2 **Classical view of economic activity.** Land (natural resources), labor, and capital are the three elements constituting the "factors of production" (bottom green arrow). Economic activity involves the circular flow of money (blue arrows) and products (top green arrow).

In a centrally planned economy, the rulers make the basic decisions regarding what and how much will be produced, where, and by whom. The government owns or manages most industries, and the workers often lack the freedom to choose or change jobs. This system was most prominent during the 20th century in the communist regimes of the Soviet Union and its satellites and in mainland China and several Asian countries. Government control of economic activity has largely failed, both economically and environmentally. The failure of the centrally planned economy in the former Soviet Union led to economic chaos in the countries that it once comprised or dominated; the countries of the former Soviet Union are some of the most polluted on Earth. China also has many pollution problems. North Korea, Cuba, and a few African countries are the last to retain centrally planned economies. In fact, Cuba has some of the world's most extensive support of urban agriculture (Chapter 12).

In a pure free-market economy, the market itself determines what will be exchanged. Goods and services are offered in a market that is free from government interference. The whole system is in private hands and is driven by the desire of people and businesses to acquire goods, services, and wealth as they act in their self-interest. The system is completely open to competition and the interplay of supply and demand. If supply is limited relative to demand, prices rise. If there is an oversupply, prices fall because people and corporations economize and pay the lowest price possible. Competition spurs efficiency as inferior products and services are forced out of the market. In an ideal world, all "players" have free access to the market. People can make informed decisions about their purchases because there is sufficient information about the benefits and harms associated with economic goods. However, these theoretical conditions of a free-market economy rarely, if ever, occur. For example, powerful business interests may manipulate facets of a market economy, and unscrupulous people may exploit the freedom of the market in order to defraud people.

The Role of Governments. In reality, no country has a pure form of either kind of economy. Recently, a hybrid economic system has emerged, called **state capitalism**, most fully developed in China, Russia, and the Gulf oil states. These countries, rich in natural resources or in abundant labor (China), have used their exports to amass capital, which they have used to build up the country's infrastructure and to form state-owned corporations. Although major players in the global economy, these are countries where democracy is restricted. As they develop, their people are increasingly joining a global middle class and the tensions between state control of the economy and political freedom are becoming evident (as in the 2011 "Arab spring" upheaval in the Middle East, which triggered ongoing conflicts in several countries, including Syria).

Even in democracies with market economies, government involvement occurs at many levels. Governments can, for example, own and operate different facets of the country's infrastructure, such as postal services, power plants, and public transport systems. Governments may control interest rates, determine the amount of money in circulation, and adopt policies that stimulate economic growth in times of slowdown. Governments also maintain surveillance over financial processes.

Limits to the Free Market? Many people believe that the conventional capitalist economic system lacks a "conscience." They argue that many workings of the market economy create hardships for people (and countries) at the bottom of the economic ladder. For example, if population growth is rapid and there are not enough jobs to go around, the result often is exploitation of workers. Furthermore, a market economy only offers people *access* to goods and services; if people lack the means to pay, access alone will not meet their needs. In this situation, a society could employ a "safety net" to provide food security for impoverished people—something the market does not do (see Chapter 12). One of the chief problems with the free market is the pressure to *externalize costs,* that is, to put part of the cost of producing a good or service someplace else in the economy, rather than in the price of the good or service itself. (We will discuss this problem further in Section 2.2.)

International Economies and Trade

Trade is a fundamental economic activity. The exchange of goods drives the engine of the free-market economy. In today's world, trade has become increasingly globalized; in the process, world trade has become one of the most dominant forces in society.

On the global level, there are a number of organizations that work to foster cooperation in economic activity and development. The **World Trade Organization** (WTO) was created to enforce trade rules between nations, although such an outcome is hard to guarantee. The **International Monetary Fund** (IMF) is an organization of 188 countries supporting cooperation, financial stability, and trade, as well as working to promote employment and lower poverty. The **World Bank** is a group of agencies with the goal of ending extreme poverty and promoting shared prosperity by improving the incomes of the bottom earners in each country. The World Bank in particular plays a role in development for the poorest countries (Chapter 9).

People disagree about whether these organizations are effective, and in some cases, just. Governments can find ways to manipulate the market in their own favor, imposing protective barriers and subsidies that prevent the free flow of goods and services. The developed nations are major exporters of agricultural products, and they subsidize their farmers heavily; the developing nations want to protect their farmers from competition with these low-priced farm products, so they impose tariffs.

The WTO is supposed to solve some of this problem. The WTO was created in 1993 by trade ministers from many countries meeting to build on the foundation for global trade laid by the General Agreement on Tariffs and Trade (GATT). In effect, the WTO became a force for the globalization of trade, facilitating it by encouraging the reduction or elimination of trade barriers (tariffs) between countries and of subsidies to domestic industries that tended to discriminate against foreign-produced goods.

Unfortunately, the WTO has established a reputation for elevating free trade over concerns about human rights and environmental resources. In Seattle, Washington, in December 1999, more than 40,000 protesters took to the streets to block the delegates from assembling—and were subjected to tear gas, pepper spray, and rubber bullets as things began to turn ugly **(Figure 2–3)**. The protesters were an unusual alliance of environmental activists and organized labor, groups that are often at odds over timber harvests and fishing. They joined forces to protest what they saw as unfair disadvantages to the poorest countries.

Talks at the 1999 WTO meetings in Seattle eventually broke down because the key players—the United States, Japan, and Europe—refused to move from their positions. Subsequent WTO meetings have failed to reach agreement on the issues of farming subsidies and trade barriers.

Recently, there have been small steps forward. During a 2013 meeting in Bali, Indonesia, progress was made as WTO members adopted the "Bali Package," an agreement with components designed to facilitate trade, cut port delays, and improve development. There was some commitment to reducing subsidies for agricultural exports, although it remains to be seen how that will play out over time.

The Need for a Sustainable Economy

The recent economic growth of China illustrates the power of the global economy in mobilizing people to engage in economic activity. As a result, the well-being of many people around the globe has increased; life expectancy has never been higher; labor-saving devices remove much of the toil in wealthy countries; and aircraft, cars, computers, and cell phones open up lifestyles and work opportunities unimaginable a few decades ago. However, the global economy has also helped to create environmental problems that harm human well-being. For example, according to a recent study, the recent decline in fish stocks in much of Asia has made it harder and harder to catch fish, so more labor is needed. This need for more labor has led to an increase in the use of child slavery in fisheries.

The continued poverty and hunger in developing countries, decline of ecosystem goods and services, rising global temperatures and sea levels, and loss of biodiversity result, in part, from an unsustainable economy. In contrast to what can be called the **brown economy**—all the current economic activities that are contributing to environmental degradation and resource depletion—there is a clear need for a different kind of economy, a sustainable **green economy**.

CONCEPT CHECK A country has national energy companies and nationally owned industries such as steel and various manufacturing sectors, but does not have workers' rights or competition for goods and services. What term best describes this type of economy? ☑

2.2 Resources in a Sustainable Economy

There is a movement afoot to change the way we look at economies. Classical economics says that land (which represents the environment), labor, and capital are the components needed for a country to be able to mount its economy. In recent years, however, a new breed of economists, often called **ecological economists**, has taken issue with this view. Ecological economists point out that the classical view (Figure 2–2) sees the environment as just one set of resources within the larger sphere of the human economy. They argue that the classical approach looks at things backward, because the natural environment actually *encompasses*

Figure 2–3 Protestors at the WTO meeting.
The World Trade Organization meeting in Seattle, Washington, in 1999, drew thousands of protestors who highlighted the problems globalization can cause. Some protestors were repelled by police using tear gas and water cannons.

the economy, which is constrained by the limits of resources in the environment. The ecosystems and natural resources found in a given country provide the context for that country's economy. Similarly, the total global ecosphere is the context for the global economy (Figure 2–4).

Ecological economists describe **economic production** as the process of converting the natural world to the manufactured world. Renewable and nonrenewable resources and the ecosystems containing them are turned into cars, toys, books, food, buildings, highways, computers, and so forth. Wastes are returned to the ecosystem.[1]

The ecological economists' view emphasizes the role of natural ecosystems as essential life-support elements—*ecosystem capital* (Chapter 1). The more inclusive term **natural capital** refers to the combination of ecosystem capital and nonrenewable mineral resources, such as fossil fuels and metal ores. Ecological economics brings sustainability

sharply into focus. In time, economic production and consumption must come to a steady state; to be sustainable, this must happen before our natural capital is consumed beyond its ability to support our economy.

Without a doubt, the ecosystems and mineral resources of a given country—its natural capital—are a major element in the wealth of that country. In recent years, the World Bank's Environment Department has been working on ways of measuring the wealth of nations and has produced some insightful analyses, which we consider next.

Measuring the Wealth of Nations

What are the components of a nation's wealth? In a document titled *Where Is the Wealth of Nations? Measuring Capital for the 21st Century*, the World Bank indicates three major components of capital that determine a nation's wealth: produced capital, natural capital, and intangible capital (Figure 2–5).[2]

---

[1]Thomas Prugh, Robert Costanza, and Herman E. Daly, *The Local Politics of Global Sustainability* (Washington, DC: Island Press, 2000).

[2]World Bank, *Where Is the Wealth of Nations? Measuring Capital for the 21st Century*. (Washington, D.C.: World Bank, 2006).

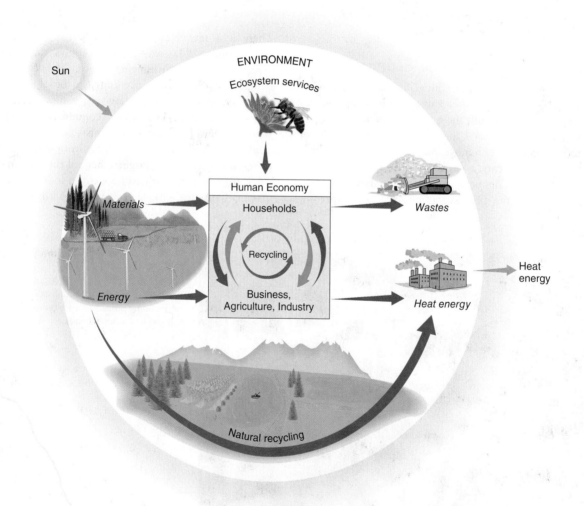

Figure 2–4 Ecological economists' view of economic activity. The human economy is embedded in the natural world. Energy flows through the system, arriving from the Sun and leaving in the form of heat. Other ecosystem goods and services are used by the human economy, and materials are recycled.

(a) Produced capital

(b) Natural capital

(c) Intangible capital

Figure 2–5 The wealth of nations. Three major components of a nation's capital determine its wealth: **(a)** produced capital, **(b)** natural capital, and **(c)** intangible capital, which largely consists of human capital such as knowledge and institutions, represented here by students on a university campus.

Produced capital. Produced capital is all of the human-made buildings and structures, machinery and equipment, vehicles and ships, highways and power lines, monetary savings and stocks, and so forth that are essential to the production of economic goods and services. Produced capital is the most easily measured of the three components of a nation's wealth. Produced capital assets have very clear income-earning potential, but they may also be subject to obsolescence and must be renewed continually. For example, a clothing factory produces a flow of goods destined for consumers, but the machinery wears out over time, and the building itself ages. Thus, we have *income*—a flow of goods—and *depletion*, which is referred to as *capital consumption*.

Natural capital. Natural capital refers to the goods and services supplied by natural ecosystems (renewable) and the mineral resources (nonrenewable) in the ground. Renewable natural capital (*ecosystem capital*) includes forests, fisheries, agricultural soil, water resources, and the like, which can be employed in the production of a flow of goods (e.g., lumber, fish, grain). In addition to producing goods, natural capital provides a flow of ecosystem services such as the production of oxygen, the breakdown of wastes, and the regulation of climate (Chapters 1, 5). Renewable natural capital is subject to depletion, but it has the capacity to yield a sustained income if it is managed wisely. Nonrenewable natural capital, such as oil and mineral deposits, is also subject to depletion, but it provides no services unless it is extracted and converted into some useful form.

Some components of natural capital are easier to measure than others. For instance, the lumber in a forest can be reckoned in board feet; it is more difficult to assess the impact of a forest on the local climate and the forest's capacity to sustain a high level of biodiversity. Unfortunately, the World Bank report did not assess the service components of natural capital because of difficulties in measuring them. Instead, the authors calculated mainly the actual resource inputs into production, such as mineral assets and forests as a source of timber.

Intangible capital. Intangible capital has three components. The first is **human capital**, which refers to the population and its physical, psychological, and cultural attributes, such as innate talents, competencies, and abilities. To this is added the value imparted by formal or informal education, which enables people to acquire skills that are useful in an economy. Investments in health and nutrition also contribute to increases in human capital.

The second component is **social capital**, the social and political environment that people create for themselves in a society. Social capital includes formal structures and relationships as defined by government, the rule of law, the court system, and civil liberties, as well as the relationships of people as they associate into religious or ethnic affiliations or join organizations in which they have a common interest.

The third component is **knowledge assets**, the codified or written fund of knowledge that can be readily transferred to others across space and time. These assets are useful only if people have access to them; hence, the vital importance of libraries, schools, universities, and the Internet.

Measuring True Economic Progress

The **gross national product**, or **GNP**, is the sum of all goods and services produced (really, consumed) by a country in a given time frame. The GNP was once the most commonly used indicator of the economic health and wealth of a country. Now, however, most countries use the **gross domestic product**, or **GDP**. The GDP is the GNP minus net income from abroad. As a per capita (per person) index, the GDP is often used to compare rich and poor countries and to assess economic progress in developing countries. An important aspect of calculating a *net* GDP is accounting for the assets that are used in the production processes. Buildings and equipment, for example, are essential to production, but they gradually wear out or depreciate. Their depreciation (called *capital depreciation*) is charged against the value of production; that is, an accounting of capital depreciation is subtracted from the production of goods and services in order to generate a net GDP.

Problems with Using the GDP. Unfortunately, the GDP measurement is flawed because it does not include goods created and services performed by a family for itself. Thus, subsistence farming in developing countries is not measured, and this omission surely produces a falsely lower GDP for these countries. The GDP also omits the natural services provided by ecosystems. For example, even though clean air is as important as medicine in treating asthma, only the medicine is included in the GDP.

In addition, the economists who invented the GNP never took into account the depreciation of natural capital. Such depreciation was regarded as *external* to a country's balance sheet. Following the classical view of economics, natural resources and their associated natural services were regarded as a "gift from nature." Therefore it is possible for a nation to cut down a million acres of forest and count the sale of the timber on the income side of the GDP ledger, while including only the depreciation of chain saws and trucks on the expense side. Completely hidden from this accounting is the loss of all the natural services once performed by the forest; incredibly, the disappearance of the million acres of forest is shown as an economic asset. As long as this discrepancy remains, it is possible for nations to underestimate the value of their natural resources. They are able to remove forests, deplete fisheries, lose soil by intensive farming, and degrade rangelands by overgrazing and still *account for these activities as economic productivity.* The GDP can increase due to production by industries that deplete and pollute the environment and then increase again with cleanups of that pollution—a double benefit!

Correcting this system of accounting would require calculating the current market value of natural assets and considering it to be part of the stock of a nation's wealth, as recommended by the World Bank. To this value would be added the value of the natural services performed by the ecosystems in which the resources are found (omitted by the World Bank analysis). The natural assets and the services they perform are the nation's natural capital. Other economists have produced reports that agree—when the nation's natural resources are drawn down, the depreciation represented by the loss of natural capital should be entered into the ledger as the GDP is calculated. Similar arguments can be made for human resources as a measure of a nation's wealth. A more accurate GDP would have to take all of these depreciations into account.

Other Possible Indicators. If the GDP is simply unable to measure true economic progress, some other index is necessary. There are several other ways to measure development. The **Index of Sustainable Economic Welfare (ISEW)** is one. This index takes into account a wide array of economic activities and their effects on human well-being, because some economic activity is negative. The **Genuine Progress Indicator (GPI)** is a measure of economic progress that extends the idea of the ISEW. Like the ISEW, the GPI is calculated by assuming that some kinds of economic activity are positive and sustainable and others are not, even if the latter are included in the GDP. Some of the positive factors in the GPI—such as labor that goes into housework, parenting, and volunteering—are not calculated in the GDP. Negatives in the GPI include factors such as the costs of crime, loss of leisure time, costs of pollution, depletion of nonrenewable resources, and loss of farmland. Figure 2–6 contrasts the ISEW and GPI of Finland with the GDP during the same period. Over the years, the per capita GPI has remained fairly constant, while the GDP has risen sharply, suggesting that the benefits of GDP growth are increasingly offset by the rising environmental and social costs of economic activity.

Other economists have recommended using the **Human Development Index (HDI)**, a somewhat simpler measure of human flourishing used by the UN. The HDI reflects health indicators, life expectancy, and literacy rates. However, one problem with both the GDP and HDI is that they do not require natural capital to be maintained, or that wealth be measured to include sustainability.

Figure 2–7 shows the Human Development Index and per capita resource use, or ecological footprint, of different countries. Environmental scientists often refer to the amount of resources required to support a person or group of people as their *footprint*, an idea that we will consider in greater detail later (Chapter 8). The dotted horizontal lines

Figure 2–6 Three measures of progress. A comparison of the GDP, GPI, and ISEW of Finland from 1945 to 2010; units are in thousands of euros per capita. The GDP (gross domestic product) is the conventional measure of economic progress used by economists. The GPI (Genuine Progress Indicator) is a measure of progress that includes negatives, similar to the ISEW (Index of Sustainable Economic Welfare). Note how the GPI and ISEW are falling behind the GDP.

(*Source:* Redrawn from J. Hoffren, "Future Trends of Genuine Progress in Finland," *Trends and Future of Sustainable Development,* Tampere 9-10, June 2011, http://www.docstoc.com/docs/84610133/Energy-and-material-flows.)

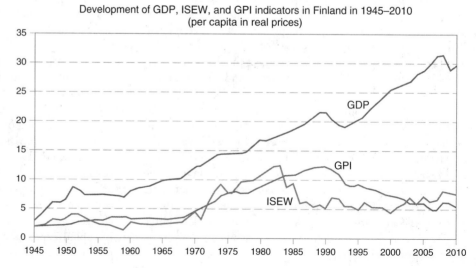

Development of GDP, ISEW, and GPI indicators in Finland in 1945–2010 (per capita in real prices)

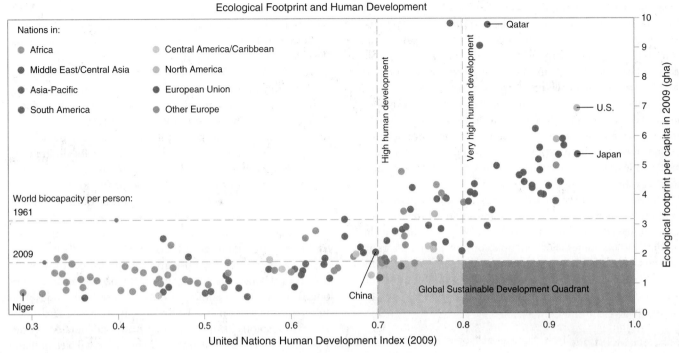

Figure 2–7 Finding a sustainable space. The horizontal axis is the Human Development Index (HDI); the vertical axis is the per capita ecological footprint. Both of these axes show components of human impact on the environment. Dots represent countries, color-coded by region. The horizontal dotted lines represent *biocapacity,* a measure of ecosystem capital per person. The blue area in the lower right is the only part of the graph that could be considered sustainable.

(*Source:* © 2015 Global Footprint Network, *Annual Report 2012*, Oakland, CA: GFN, p. 41. www.footprintnetwork.org)

UNDERSTANDING THE DATA

1. Why is the world biocapacity per person lower in 2009 than in 1961? What does that change do to the sustainability quadrant?
2. Why do three countries in the Middle East have the highest ecological footprints?

---

on the graph show *biocapacity,* a per capita measure of the availability of ecosystem goods and services. Note that biocapacity declined between 1961 and 2009, making the "safe space" for humanity smaller. The idea of a safe space for humanity is discussed elsewhere (Chapter 3).

Environmental Accounting. Environmental economists have concluded that valuation of ecosystem services is critical. In 2007, environment ministers from the leading global economies, coordinated by the UN Environment Program (UNEP), initiated a study intended to show how to apply economic thinking to the ecosystem services that are so beneficial to societies. That study, *The Economics of Ecosystems and Biodiversity* (TEEB), has produced seven reports, culminating in a 2010 final report that shows "how economic concepts and tools can help equip society with the means to incorporate the values of nature into decision making at all levels," to quote from the preface.[3] Such environmental accounting requires the accurate valuation of ecosystem services and the inclusion of their depreciation in economic models. For example, ARIES

(Artificial Intelligence for Ecosystem Services) is a Web-based technology that allows users to rapidly assess and evaluate ecosystem services in order to make better environmental decisions or track the value of ecosystem goods and services. ARIES is funded by the National Science Foundation through researchers at the University of Vermont, Conservation International, and Earth Economics. This technology has helped groups in locations from Puget Sound to Madagascar.

The use of economics to conserve resources we once took for granted will only increase in the future. A 2014 study estimated that ecosystems services contribute "more than twice as much to human well-being as global GDP." They estimated the annual global loss of ecosystem services due to changes in land use at $4.3–20.2 trillion.[4]

Essential Conditions for Development. There are many reasons why some countries have undergone rapid development and others have not; population growth is one that is cited in this text (Chapter 9). Perhaps the most important reasons relate to the major institutions that coordinate social and political life and that make up the social capital

---

[3]TEEB, *The Economics of Ecosystems and Biodiversity: Mainstreaming the Economics of Nature: A Synthesis of the Approach, Conclusions and Recommendations of TEEB,* 2010.

[4]R. Costanza et al., "Changes in the Global Value of Ecosystem Services," *Global Environmental Change* 26 (2014): 152–158.

of a society. For example, the definition of rights, the enforcement of those rights, and the facilitation of economic exchange are all considered essential to the successful development of human capital and produced capital. A well-developed body of law, an honest legal system, broad civic participation, and a free press are essential for maintaining the rights of all people within a society. A well-developed market economy and free entry into and exit from markets for people and businesses are conducive to development. Functioning communication and transportation networks and viable financial markets are also needed to sustain a market economy.

In societies where such resources are not in place, corruption, inefficiency, injustice, and intolerance often prevail. Continuing poverty and environmental degradation are the predictable outcomes. For example, one of the major components of intangible capital as measured by the World Bank is the "rule of law index." Denmark scores an 88 out of 100 on this index, while Pakistan, with its corruption and human rights abuses, scores 36.

CONCEPT CHECK How might an economist include the cost of the over-use of water in the price of a water-intensive product such as cotton? What are the downsides to this economic approach? ☑

2.3 Economy, Environment, and Ethics

To achieve long-term sustainability, we will need to live within boundaries in which people have sufficient human development, and populations and per capita resource use are low enough to maintain global ecosystem capital. There are a number of reasons why this is hard to accomplish. In addition to resource limitations, inequality in access to resources makes environmental sustainability difficult to achieve. None of the resources that are crucial to an economy are evenly distributed among nations or within nations. Great differences in wealth exist among the nations of the world, something we will see many times in other chapters. Any economy comes with ethical dilemmas, some due to these differences in wealth and opportunity, and some because what benefits one person economically may harm another, as we have seen already in several cases. In order to live sustainably, we need to address these ethical concerns.

Human Rights and Resources

Simply having natural capital and being able to transform it into produced capital does not guarantee human development for everyone in a society. For example, Guatemala has a great deal of natural capital, including highly productive agricultural land and the largest contiguous forest in the Americas, except for the Amazon. Yet Guatemala has the fourth highest level of chronic malnutrition in the world; the indigenous Mayan peoples have even higher rates of malnutrition. Causes of malnutrition include poverty and diets low in fresh foods with protein and vitamins. This chronic malnutrition hurts

the poor and costs the Guatemalan economy an estimated $8.4 million per day in lost wages, illness, loss of schooling, and other costs to society. Sometimes urban leaders who make decisions do not understand the lives of those in rural poverty and consequently don't make policies to address real problems. Part of the issue, however, is that Guatemala's economy is focused on export crops. Many people work for others, growing crops that they cannot afford to eat. In some cases, the best land is used to grow cash crops and only the poorest land is used to grow the maize that is eaten by local families.

While the free market has given many people a way to make a living, not everyone has the means to be a part of that market. The operation of the market does not reflect the needs of those who lack money. Often, it isn't simply that people cannot purchase what they need, but that their livelihoods are removed by the economic activities of others. When economic activities use up scarce resources or disperse pollution through the environment, the people most benefiting from these economic activities are rarely the ones who suffer the most from their effects. This is particularly true when indigenous peoples interact with the outside world. This inequity is a central ethical dilemma we will see in many later chapters. Thus an interest in human rights means an interest in the economic and political rights of the poor, including indigenous groups. These concerns overlap substantially with concern about environmental sustainability.

Intergenerational Equity

What we have just discussed is a matter of *intra-generational equity*—the "golden rule" of making possible for others what is possible for you. All forms of current assets need to be managed well to improve the well-being of people who already exist. Sustainable development is also about intergenerational equity—meeting the needs of the present without compromising the ability of future generations to meet their needs.

Economists point to some problems with this perspective, however. Economics deals with future resources by applying what is called the **discount rate**, the rate used for finding the present value of some future benefit or cost. The discount rate often approximates current interest rates. For example, a dollar is worth more to us today than it will be five years from now, even without inflation, because we can put the dollar to use now and can earn interest on that dollar if we invest it. At a 5% rate of interest, in five years our dollar would be worth $1.28. Turning this idea around, a dollar five years from now is worth only $0.78 today.

Following this reasoning, it can be concluded that some resources or benefits are worth more now than they will be in the future. This concept can be applied to a stand of trees. The owner has several options: (1) cut all the trees and sell the timber today; (2) hold onto the trees in the hope that they will bring a better price at some time in the future; or (3) harvest the trees on a sustainable basis, spreading out profits over time. If it turns out that the trees grow more slowly than current interest rates, the economically profitable decision would be to cut them now and invest the proceeds.

Thus using the discount rate can be a problem for natural capital, because in many cases it would appear to argue for maximizing short-term profit over sustainable, long-term use of the resource. For example, if we let future generations cope with global climate change, we can continue to enjoy the present benefits of the unrestrained use of fossil fuels. The same can be said about other resources, such as swordfish: We can harvest them now, and when they're gone, we can switch to some other species of fish. Unfortunately, the *self-interest* of present individuals and generations is often at odds with the longer-term *sustainable interests* of a community or a society. Perhaps it would be more accurate to say that the interests of the wealthy today are often considered to trump the interests of the poor, either now or in the future. To make things a bit more challenging, some have argued that conserving resources for the future could actually be thought of as putting the interests of future generations above those of the poor today. In answer, it can be said that both sets of interests are legitimate.

Waste

Wealth is often accompanied by waste. The way economies work for most goods and services, if a consumer can pay for a product, he or she is free to use it or even to waste it. There is no law against purchasing a shirt and throwing it away after wearing it only once, or taking a huge plate of food from a buffet and then throwing most of it away. However, when something is both necessary and scarce, societies may move to a different way of operating. Food rationing during wartime and bans on outdoor water use during a drought are examples.

According to some estimates, about 40% of the food produced in the United States is discarded as waste; this waste occurs in all parts of food production, from spoilage in agricultural fields to food that is discarded after preparation **(Figure 2–8)**. Laws designed to protect public health have unexpected consequences: expiration dates on food and laws requiring restaurants to throw out food once it has been served (even if it has not been touched) increase the level of food waste. The production of food requires a great deal of water and energy (Chapter 12). While it might be impossible to ship your half-eaten sandwich to someone who is hungry, wasting less food generally brings down food prices and reduces the amount of land, water, and energy used in food production.

The United Nations has identified food waste as a problem whose solution is critical to alleviating poverty. Many people are working on solutions. The City of London, for example, started an innovative program to help reduce food waste in restaurants. Participating restaurants offered special portion sizes, revised the way they served side dishes, and used left-over inventory to make soups. The pilot program helped 15 small businesses save £100,000 ($170,000) and eliminated 1.75 tons of food waste. This was accomplished without passing new laws, because reducing waste was both smart and financially sound. However, sometimes we need environmental public policy, which makes possible both the present and future well-being for human societies.

CONCEPT CHECK Guatemala has high levels of malnutrition, even though the country has an abundant supply of natural resources. What are two reasons for this situation? ☑

Figure 2–8 Food Waste. As much as 40% of the food in the United States is discarded as waste. The United Nations estimates that the global cost of food waste is $750 billion per year.

2.4 Environmental Public Policy

Environmental public policy includes all of a society's laws and agency-enforced regulations that deal with that society's interactions with the environment. Public policies are developed at all levels of government: local, state, federal, and international. The purpose of environmental public policy is *to promote the common good.* Just what the common good consists of may be a matter of debate, but at least two goals stand out: the improvement of human welfare and the protection of the natural world. Environmental public policy addresses two sets of environmental issues: (1) the prevention or reduction of air, water, and land pollution and (2) the use of natural resources such as forests, fisheries, oil, and land. All public policy is developed in a sociopolitical context that we will simply call *politics.*

Human societies and their economic activities have the potential for doing great damage to the environment, and that damage has a direct impact on present and future human welfare (see **Table 2–1** on the following page.). The effects of pollution and the misuse of resources are seen most clearly in those parts of the world where environmental public policy is not well established. As Table 2–1 shows, millions of deaths and widespread disease are directly traceable to degraded environments. The costs to human welfare are felt in the areas of health, economic productivity, and the ongoing ability of the natural environment to support human life. Therefore, laws to protect the environment are not luxuries; they are part of the foundation of any well-organized human society and essential for achieving sustainability.

Table 2–1	Principal Health and Productivity Consequences of Poor Environmental Management	
Environmental Problem	**Effect on Health**	**Effect on Productivity**
Water pollution and water scarcity	More than 3 million deaths and billions of illnesses a year are attributable to water pollution and poor household hygiene caused by water scarcity.	Fisheries are declining; rural household time (time spent fetching water) and municipal costs of providing safe water are increasing; water shortages constrain economic activity.
Air pollution	Many acute and chronic health impacts exist: excessive levels of urban particulate matter and smoky indoor air are responsible for 2 million premature deaths annually.	Acid rain and ozone have harmful impacts on forests, agricultural crops, bodies of water, and human artifacts.
Solid and hazardous wastes	Diseases are spread by rotting garbage and blocked drains. Risks from hazardous wastes are typically local, but often acute.	Groundwater resources are polluted and rendered unusable for irrigation or domestic use.
Soil degradation	Effects include reduced nutrition for poor farmers on depleted soils and greater susceptibility to drought.	Some 23% of land used for crops, grazing, and forestry has been degraded by erosion. Field productivity losses in the range of 0.5–1.5% of the gross national product are common on tropical soils.
Deforestation	Localized flooding leads to death and disease.	Effects include reduced potential for sustainable logging and for prevention of erosion, increased watershed instability, and diminished carbon storage capability for forests. Nontimber forest products are also lost.
Loss of biodiversity	Effects include the potential loss of new drugs.	Ecosystem adaptability is reduced, and goods and services are lost.
Atmospheric changes	Such changes result in possible shifts in vector-borne diseases, risks from climatic natural disasters, and skin cancers attributable to depletion of ozone shield.	Coastal investments are damaged by the rise in sea level; regional changes in agricultural productivity occur.

Sources: World Bank, *World Development Report, 1992.* New York: Oxford University Press, 1992; *Millennium Ecosystem Assessment*, 2005; UNEP Global Environment Outlook GEO₄, 2007.

Policy in the United States

The U.S. Constitution is widely regarded as responsible for establishing a very effective form of government. The conveners of the Constitution decided on three branches of government, all of which play a role in establishing and enforcing environmental public policy (Figure 2–9). The **legislative branch**, or the Congress, consists of the Senate and the House of Representatives. All of these legislators are elected by the people from the 50 states. The Congress has the power to establish laws, which must pass both houses of the legislature and then go to the **executive branch** of the government for approval by the president. A host of administrative agencies report to the president. We will encounter many of these agencies as we examine different environmental problems. The **judicial branch** of government is responsible for interpreting the law, from the Constitution and its amendments to the many pieces of legislation passed by the Congress.

Rules and Regulations. Environmental *public-policy law* is the responsibility of the Congress. Once a piece of legislation is passed by the House of Representatives and the Senate and

is signed into law by the president, it is implemented by the appropriate government agencies. The EPA and most other federal agencies are required to develop *rules* and *regulations* for issues only broadly covered in the laws; it is these rules that put the bite in the laws passed by Congress. In making rules, agencies are required to solicit public comments, evaluate the comments, and adjust the rules, depending on the validity of the comments. Finally, agencies publish their new rules in the *Federal Register* for all interested parties to see.

One final role of the Congress is to pass appropriations for the various programs that have already been authorized by law. The executive branch (the president) draws up a budget asking for appropriations for all of the government agencies and programs, and the Congress decides how much of what the president asks for will actually get funded. Thus, the two times the Congress can influence a given issue are when it is formulated into law and when it receives appropriations. Appropriations must be authorized every year, and new laws can always amend or abolish existing laws, so there are many opportunities for the political system to influence environmental public policy.

Capitol Building

White House

Supreme Court Building

Legislative
Congress

House of   Senate
Representatives

Executive
President
Vice President
Cabinet
Administrative Agencies

Judicial
Supreme Court
Federal Courts
Circuit and District Courts

Figure 2–9 The three branches of the United States government. The legislative, executive, and judicial branches provide a balance of power as laws are enacted, administered, and sometimes reviewed.

State and Local Levels. Policies to solve local and regional problems are developed at the state and community levels. States replicate much of the structure of the federal government, with legislatures, governors, a judiciary, and various agencies. Cities and towns employ conservation commissions, planning boards, health boards, water and sewer commissions, and the like. Cities and towns often establish zoning regulations to protect citizens from haphazard and incompatible land uses. Some municipalities and regions go further and establish broader policies, such as "smart growth" to address the problems of urban sprawl. Indeed, states and municipalities frequently initiate experiments in environmental public policy that lead the way to changes in national policy. California, for example, has passed a number of regulations that tackle environmental issues, such as banning incandescent bulbs and plastic single-use bags; it was also a pioneer in promoting auto emissions standards and efficiency standards for appliances (see *California's Green Economy,* p. 39).

Policy Options: Market or Regulatory?

The objective of environmental public policy is to change the behavior of polluters or resource-users so as to benefit public welfare and the environment. This is usually accomplished by using either a direct regulatory approach, also called a *command-and-control* strategy, or a *market-based* approach to set prices on pollution and the use of resources. Many policies include elements of each.

Regulatory Policies. Most of the laws that give the U.S. Environmental Protection Agency (EPA) its authority prescribe command-and-control solutions. These laws direct the EPA to regulate pollution problems, often specifying many of the details to be used. For example, legislators may choose to set standards on pollutants that protect the health of the most vulnerable members of the population. This is the case with the basic criteria covering air pollutants (see Chapter 19). To meet these standards, regulations are established, and polluters are required by law to comply with the regulations. One of the shortcomings of the regulatory approach is that if a polluter is told to use a particular technology or is given a cap on emissions, there is no incentive for the polluter to invest in technologies that would keep pollution at lower levels. A better approach might be to set a standard (for water quality, for example) and let the polluter decide how best to achieve the standard; another approach would be to provide incentives for investments that reduce pollution (Chapter 22).

Market-Based Policies. In contrast, market-based policies have the virtues of simplicity, efficiency, and, in theory, equity. All polluters are treated equally and may choose their responses on the basis of economic principles having to do with profitability. With a market-based approach, there are strong incentives to reduce the costs of using resources or of generating pollution. The cap-and-trade system for SO_2 emissions under the Clean Air Act of 1990 is a good example of the market approach to pollution control (see Chapter 19). User fees (pay as you throw) for the disposal of municipal solid waste are another (Chapter 21).

A new market-based initiative for conserving the services provided by ecosystems is **payments for ecosystem services (PES)**. Forest, farm, and wetlands owners hold properties that provide valuable services to societies, such as sequestering carbon, preserving biodiversity, and preventing soil erosion. Yet landowners are seldom compensated for these services. A PES program uses some form of environmental accounting, and those who benefit from the service pay the landowner to maintain the service and not convert the land to some environmentally undesirable use. Often the government does the paying on behalf of the society, as in the Conservation Reserve Program, where farmers are paid to protect the open space and wetlands they own (Chapter 11).

Whether regulation-based or market-based, public policy does not appear out of nowhere. We will now examine how public policy is developed.

CONCEPT CHECK Some parts of the Clean Air Act impose fines for emitting pollutants. Is this an example of a command-and-control or a market-based approach to regulation? ☑

Figure 2–10 The policy life cycle. Most environmental issues pass through a policy life cycle, and each issue is accorded a different degree of political "weight" as it moves through each stage of the cycle. In initial stages there is more dissention or disagreement about the reality of the problem or what society needs to do. The final result is a policy that has been incorporated into the society and a problem that is under control.

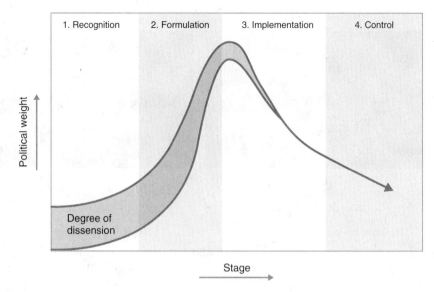

The Public Policy Life Cycle

When specific problems are addressed through public policy in democratic societies, the development of that policy often takes a predictable course, which we will call the *policy life cycle*. The typical policy life cycle has four stages: recognition, formulation, implementation, and control (Figure 2–10). Each of the stages carries a certain amount of political weight, which varies over time and is represented in the figure by the course of the life-cycle line. We will use the discovery of ozone depletion in the stratosphere as a case study to illustrate the policy life cycle.

Recognition Stage. In the early 1970s, chlorofluorocarbons (CFCs) were hailed as miracle chemicals: virtually inert, nontoxic, nonflammable, noncorrosive. They were ideal aerosol propellants, were extremely useful as refrigerants, and were effective as cleaning agents for electronics. Soon they were appearing in thousands of products and applications, including telecommunications, food processing, and cosmetics.

While working at the University of California at Irvine in 1972, physical chemist Sherwood Rowland and postdoctoral student Mario Molina (Figure 2–11) decided to investigate the question of what happens to CFCs in the atmosphere. The two scientists reasoned that CFCs would not break down unless they entered the stratosphere, where solar radiation is strong enough to break the tough molecules. They hypothesized that the breakdown of CFCs would release chlorine and that this chlorine would in turn react with ozone molecules in the stratosphere (Chapter 19). Rowland and Molina suddenly realized that the process they had uncovered had the potential to reduce the stratospheric ozone layer that blocked ultraviolet radiation from entering the lower atmosphere, where it could seriously damage biological tissues. They published their results in 1974 and began calling public attention to the dangers of CFCs. The chemical industry responded by ridiculing their work and verbally attacking the scientists wherever they spoke.[5]

The *recognition stage* of public policy begins with the early perceptions of an environmental problem, often coming as a result of scientific research. Scientists then publish their findings, and the media pick up the information and popularize it. During the recognition stage, dissention is high. Opposing views on the problem surface. Eventually, the problem gets attention from some level of the government, and the possibility of addressing it with public policy is considered.

Formulation Stage. In 1976 the National Academy of Sciences appointed a committee to look into the claims of Rowland and Molina. The committee issued a report that confirmed the dangers of CFCs and recommended banning the chemicals. The EPA considered the evidence and banned the use of CFCs in spray cans in 1978. CFCs continued to be used in increasing amounts, however, in other applications. Then,

Figure 2–11 Mario Molina and Sherwood Rowland. These two physical chemists discovered the threat posed by chlorofluorocarbons to the ozone layer in the stratosphere. Their work led to the Montreal Protocol, an international agreement to phase out the use of ozone-destroying chemicals around the world. They were awarded the 1995 Nobel Prize in chemistry for their work.

---

[5] Mario Molina and F. S. Rowland, "Stratospheric Sink for Chlorofluoromethanes: Chlorine Atom-Catalysed Destruction of Ozone," *Nature* 24, no. 9 (June 28, 1974): 810–812.

in 1985, a British team working in Antarctica reported a huge "hole" in the ozone layer over that continent. Further research showed that the ozone hole was due to the high release of chlorine from CFCs and to subsequent ozone destruction. The world scientific and environmental organizations began calling for global action. This led to a meeting convened by the United Nations in Montreal, Canada, in 1987. There, member nations agreed to scale back CFC production 50% by 2000.

Thus, the *formulation stage* is a stage of rapidly increasing political weight. The public is now aroused, and debate about policy options occurs. The political battles may become fierce, as questions dealing with regulation and who will pay for the proposed changes are addressed. Media coverage is high, and politicians begin to hear from their constituencies and lobbyists. During the formulation stage, policy makers consider the "Three Es" of environmental public policy: **effectiveness** (whether the policy accomplishes what it intends to do in improving the environment), **efficiency** (whether the policy accomplishes its objectives at the least possible cost), and **equity** (whether the policy parcels out the financial burdens fairly among the different parties involved).

Implementation Stage. The ozone hole continued to grow, however, so the signers of the Montreal Protocol agreed to speed up the phase-out of CFCs, moving the date for completion to 1996. The major chemical companies were already developing more benign substitutes for CFCs. Several other chemicals with ozone-destroying potential were added to CFCs as substances to be regulated.

At this point, the policy had reached the *implementation stage*. In this stage, policy has been determined, and the focal point moves to regulatory agencies. By now, the issue is not very interesting to the media, and the emphasis shifts to the development of specific regulations and their enforcement. Industry learns how to comply with the new regulations.

Control Stage. The final stage in the policy life cycle is the *control stage*. Today CFCs are no longer being manufactured and released into the atmosphere. Substitutes for them are readily available. The size of the ozone hole has stopped increasing, and the amount of CFCs in the stratosphere is starting to decline. The Montreal Protocol stands as a remarkable achievement of international diplomacy and concord. Sherwood Rowland and Mario Molina were awarded the Nobel Prize in chemistry in 1995 for their discovery. By this point, years have passed since the early days of the recognition stage. Problems are rarely completely resolved, but the environment is improving, with things moving in the right direction. Policies (and their derived regulations) are broadly supported and often become embedded in the society.

At any one time, different problems will be in different stages of the life cycle. Figure 2–12 shows the life-cycle stages of a number of environmental problems in the industrialized countries. Because new public policies can be costly, countries in different stages of economic development will be in different stages of the policy life cycle for a given problem.

Economic Effects of Environmental Public Policy

Most environmental policies impose significant economic costs that must be paid by some segment of society. For instance, we establish policies that control pollution because their benefits are judged to outweigh their costs—that is, the public welfare is improved, and the environment is protected. Who pays the costs? The equity principle implies that those benefiting from the policies—the customers of businesses whose activities are regulated and the people whose health is protected by laws against polluting—should pay, through taxes or higher charges; such costs are passed along to consumers, who are purchasing the product or service at its true cost (or closer to true).

On the other hand, some policies have relatively little or no direct *monetary cost*—they do not require major investments of administration or resources. For example, removing subsidies to special interests and denying special access to national resources saves money. It can result in a more efficient and equitable operation of the economy and can protect the environment.

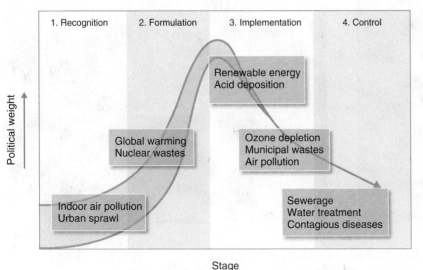

| 1. Recognition | 2. Formulation | 3. Implementation | 4. Control |

Renewable energy
Acid deposition

Global warming
Nuclear wastes

Ozone depletion
Municipal wastes
Air pollution

Indoor air pollution
Urban sprawl

Sewerage
Water treatment
Contagious diseases

Political weight

Stage

Figure 2–12 Environmental problems in the policy life cycle. Different environmental problems are in different stages of the policy life cycle in industrial societies.

In the United States, the fees charged for using federal lands are often significantly less than the fees charged by private landowners. This amounts to a subsidy for grazing cattle or harvesting timber on federal lands (Figure 2–13a). The real costs—and environmental consequences—of the products are not paid by the producers or the people buying the products. Instead, they are paid by all taxpayers and by future generations. One example of a person benefiting from a shared resource without paying the cost is rancher Cliven Bundy (Figure 2–13b), who has had a 20-year dispute with the federal Bureau of Land Management (BLM) in southeastern Nevada because he refuses to pay fees for grazing his cattle on federal lands. Bundy was in the news in 2014 over a standoff with federal agents attempting to remove his cattle.

Impact on the Economy. People who oppose environmental regulation have argued that environmental policies are excessive and bad for the economy—that our concern for environmental protection costs hundreds of thousands of jobs, reduces our competitiveness in the marketplace, or drives up the price of products and services. In recent years, these concerns have been investigated by many economists, and the general findings of their studies indicate that the "environment versus economy" trade-off is a myth. Even without considering the environmental benefits, economists found that environmental protection has no significant adverse effect on the economy; in fact, it has a pronounced positive effect. Here are some examples of what these studies have revealed.

- Estimated pollution-control costs are only 1.72% of value added. (Value added is the increase in value from raw materials to final production.) Industries are not going to shut down, move overseas, or lay off thousands of workers for costs of this size.

- Employers attribute only 0.1% of job layoffs to causes related to the environment.

- In general, states with the strictest environmental regulations also have the highest rates of job growth and economic performance. California established new energy efficiency policies in 1978; since then, its energy sector has added more than a million jobs, far outpacing other economic sectors. Indeed, California has set a remarkable pace in creating a green economy (see Sustainability, *California's Green Economy* p. 39).

- Healthier workers are more productive and spend less on medical care.

Environmental Protection and Jobs. Environmental protection is a huge industry. We need people to monitor soil and water quality, remove leaky underground oil tanks, protect wildlife habitat, and make sure cars produce fewer emissions. We need technologies to help us perform these tasks, and those technologies need to be invented and produced. The United States is the largest single producer of environmental technologies, accounting for 40% of the world's total market (about $782 billion); this industry provides 1.7 million workers with jobs in 117,000 U.S. enterprises. Worldwide, environmental technology is even larger.

On the whole, environmental public policy does not *diminish* the wealth of a nation; rather, it *transfers* wealth from polluters to pollution controllers and to less polluting companies. The environmental protection industry is a major job-creating, profit-making, sales-generating industry. We can only conclude that the argument that environmental protection is bad for the economy is simply unsound. Not only is environmental protection good for the economy, but environmental public policy is responsible for a safer, healthier, and more enjoyable environment.

CONCEPT CHECK   How does the degree of dissension about an issue change during the public policy life cycle? ☑

(a)

(b)

Figure 2–13 **Subsidized access to pubic resources.** (a) Timber harvesting in the Olympic National Forest, Washington. (b) Cliven Bundy, who was in a standoff with BLM federal agents in 2014 over his grazing of cattle on federal land. Both foresters and ranchers pay only a fraction of the market value of the resource they use.

SUSTAINABILITY
California's Green Economy

The global transition to a sustainable green economy from the current brown economy is under way. Globally, even nationally, it can be a hit-and-miss proposition, but one location in which this transition has been well documented is the state of California. Next 10, a San Francisco nonprofit, recently published a report on the green economy in that state.[6] Relative to 1995, jobs in the green economy have grown by 56%, while the total economy grew by 18%. The state employment department reported that people in more than 263,000 jobs spend half or more of their time on the production of green products or services, with an additional 170,000 people performing jobs that involve part-time activities in the green economy.

Just what are green jobs? Next 10 distinguished between a *core green economy* and an *adaptive green economy*. Core green economy businesses provide the products and services that deal in alternatives to carbon-based energy sources (e.g., wind, solar, biomass), conserve energy (e.g., building efficiency products and services) and other natural resources, and reduce pollution and recycle waste (e.g., emissions monitoring, recycling machinery). Adaptive green economic activities involve the businesses that are adjusting their processes to improve resource efficiency and change the consumption habits of households. The report shows that green economic activities employ people across every region of the state.

An example of a California green business is Sol Focus, located in Palo Alto. This company develops and manufactures innovative photovoltaic systems that are operating in several states and are being deployed in Spain, Greece, Italy, Portugal, and many other countries. The company, which holds more than 40 patents, specializes in the use of advanced optical systems that focus and convert large amounts of solar energy to electrical energy. One of its unique products is a system called concentrator photovoltaic (CPV) technology. This technology uses systems of cells employing reflective mirrors that concentrate solar energy up to 500 times, greatly reducing the number of solar cells needed to capture solar energy and also greatly reducing the cost of the collection panels. With 270 employees, Sol Focus is a great example of a green company that is making a difference in California and throughout the world.

[6]Next 10, *Many Shades of Green: Regional Distribution and Trends in California's Green Economy* (San Francisco: Next 10, 2010).

2.5 Cost-Benefit Analysis of Environmental Public Policy

Economics plays a strong role in the development of public policy. Those developing policies need to assess the potential benefits of laws and regulations and weigh them against the costs of their implementation. One of the key tools employed in this process is the cost-benefit analysis. In this section, we'll look at the details of cost-benefit analysis. In doing so, we will address the question of how to measure the cost-effectiveness of regulating pollution.

The president of the United States has the power to issue executive orders, which are usually directives to government agencies or departments designed to accomplish a presidential initiative. In 1981, President Ronald Reagan, concerned that the U.S. economy and business interests might be over-regulated, issued Executive Order 12291. This order required that all significant regulations (those that impose annual costs of at least $100 million) be supported by a cost-benefit analysis. Subsequent presidents, including President Obama, have issued similar executive orders that basically continue most of the policies of Reagan's order.

A **cost-benefit analysis** (or *benefit-cost analysis*) begins by examining the need for the proposed regulation and then describes a range of alternative approaches. Afterward, it compares the estimated costs of the proposed action and the main alternatives to the benefits that will be achieved. All costs and benefits are given monetary values (where possible) and compared by means of a *cost-benefit* (or *benefit-cost*) *ratio*. A favorable ratio for an action means that the benefits outweigh the costs, and the action is said to be cost effective. There is an economic justification for proceeding with it. If costs are projected to outweigh benefits, the project may be revised, dropped, or shelved for later consideration.

Cost-benefit analysis of environmental regulations is intended to build efficiency into policy so that society does not have to pay more than is necessary for a given level of environmental quality. If the analysis is done properly, it will consider *all* of the costs and benefits associated with a regulatory option. In so doing, it must address the problem of *external costs*.

External Costs

In economics, an **external cost** (or **externality**) is an effect of a business process that is not included in the usual calculations of profit and loss. For example, when a business pollutes the air or water, the pollution imposes a cost on society in the form of poor health or the need to treat water before using it. This is an *external bad*. When workers improve their job performance as a result of experience and learning, the improvement is not credited in the company's ledgers, although it is considered an *external good*.

Amenities such as clean air, uncontaminated groundwater, and a healthy ozone layer are not privately owned. In the absence of regulatory controls, there are no direct costs to a business for degrading these amenities. In other words, they are external costs. Therefore, there are no economic incentives to refrain from polluting the air or water. By including *all* of the costs and benefits of a project or a regulation, cost-benefit analysis effectively brings the externalities into the economic accounting.

Sometimes it is difficult to figure out where to put the cost of an externality. For example, international trade is important in improving the economy throughout the world. Yet the process of shipping goods is the biggest cause of the spread of invasive species, a problem that costs upwards of $300 billion a year in just six countries: the United States, the United Kingdom, South Africa, Australia, India, and Brazil. The cost of items that are shipped long distances ought to account for the likelihood that they will bring pest species to distant areas.

Environmental Regulations Impose Real Costs

The costs of pollution control include the price of purchasing, installing, operating, and maintaining pollution-control equipment and the price of implementing a control strategy. Even the banning of an offensive product has costs because jobs may be lost, new products must be developed, and machinery may have to be scrapped. In some instances, a pollution-control measure may result in the discovery of a less expensive way of doing something. In most cases, however, pollution control costs money. Thus, the effect of most regulations is to prevent an external bad by imposing economic costs that are ultimately shared by government, industry, and consumers.

Pollution-control costs tend to increase exponentially with the level of control to be achieved (Figure 2–14). That is, a partial reduction in pollution may be achieved by a few relatively inexpensive measures, but further reductions generally require increasingly expensive measures, and 100% control is likely to be impossible at any cost. Because of this exponential relationship, regulatory control often has to focus on an optimum cost-effectiveness, as Figure 2–14 indicates.

In most cases, pollution-control technologies and strategies are understood and available. Thus, equipment, labor, and maintenance costs can be estimated fairly accurately. Unanticipated problems that increase costs may occur, but as technology advances and becomes more reliable, experience is gained, and lower-cost alternatives frequently emerge. Figure 2–15 compares the costs and benefits of

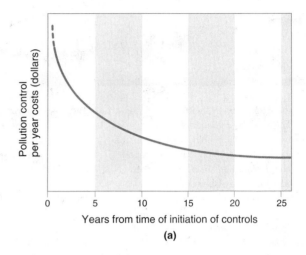

(a)

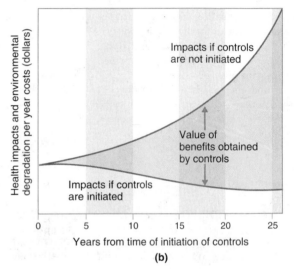

(b)

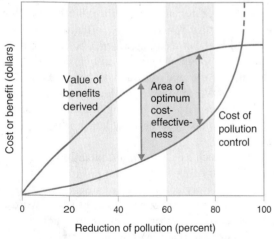

Figure 2–14 The cost-benefit ratio for reducing pollution. The cost of pollution control increases exponentially with the degree of control to be achieved. However, benefits to be derived from pollution control tend to level off and become negligible as pollutants are reduced to near or below threshold levels. The optimum cost-effectiveness is achieved at less than 100% control.

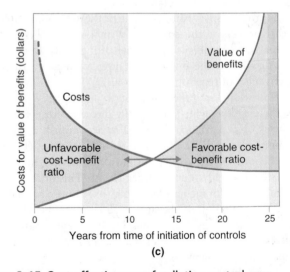

(c)

Figure 2–15 Cost-effectiveness of pollution control over time. (a) Pollution-control strategies generally demand high initial costs. The costs then generally decline as those strategies are absorbed into the overall economy. **(b)** Benefits may be negligible in the short term, but they increase as environmental and human health recovers from the impacts of pollution or is spared increasing degradation. **(c)** When the two curves are compared, we see that what may appear to be cost-ineffective expenditures in the short term (5–10 years) may in fact be highly cost-effective expenditures in the long term.

environmental regulations over time. The costs of pollution control are likely to be highest at the time they are initiated and then decrease as time passes.

Policy costs are often overestimated. A look at some historical examples reveals that industry reports predict higher costs of regulations than government studies do, and surprisingly, even government analysts often overestimate the costs of regulations. For example, industry sources claimed that the costs of eliminating CFCs from automobile air conditioners would increase the price of new cars by anywhere from $650 to $1,200. In reality, the price increase turned out to be much lower—from $40 to $400. One study of cost-benefit analyses revealed that overestimates of costs occurred in 12 of 14 regulations. The primary reason given for the overestimates is a failure to anticipate the impact of technological innovation. The costs of pollution control constitute a powerful incentive to make substitutions, to recycle materials, or to redesign industrial processes.

The Benefits of Environmental Regulation

The costs of improving the quality of the environment represent a major economic outlay. What benefits have we received in return?

The Regulatory Right-to-Know Act directs the Office of Management and Budget (OMB) to submit an annual report to the Congress containing, among other things, an estimate of the costs and benefits of federal rules. The 2014 study of costs and benefits during the decade from October 1, 2003, to September 30, 2012, showed that the benefits (those that could be measured in monetary units) far exceeded the costs. The benefits were estimated at between $147 billion and $800 billion, while industries, states, and municipalities spent an estimated $56 billion to $84 billion to comply with the rules. The OMB report demonstrated that the benefits of major regulations issued between 2000 and 2012 far exceeded the costs (Figure 2–16). The lion's share of both the costs and the benefits for 2012 was attributed to the EPA's regulation of sulfur dioxide in the air, as mandated by the Clean Air Act. Reductions in hospital visits, lost workdays, and premature deaths were benefits associated with reductions in this air pollutant.

Calculating Benefits. The benefits of regulatory policies are seldom as easy to calculate as the costs. Estimating benefits

is often a matter of estimating the costs of poor health and damages that *would have* occurred if the regulations had not been imposed (Figure 2–15b). For example, the projected environmental damage that would have resulted from a given level of sulfur dioxide emissions from a coal-fired power plant (an external bad) becomes a benefit (an external good) when a regulatory action prevents half of those emissions. Benefits include such things as improved human health, reduced corrosion and deterioration of materials, reduced damage to natural resources, preservation of aesthetic and spiritual qualities of the environment, increased opportunities for outdoor recreation, and continued opportunities to use the environment in the future. Examples of benefits are listed in Table 2–2, on the following page.

The values of some benefits can be estimated fairly accurately. For example, air-pollution episodes increase the number of people seeking medical attention. Because the medical attention provided has a known dollar value, eliminating a certain number of air-pollution episodes provides a health benefit of that value.

Shadow Pricing. Many benefits, however, are difficult to estimate. Putting a dollar value on maintaining the integrity of a coastal wetland or the enjoyment of breathing cleaner air is hard. One way people try to do so is to find out how much people are willing to pay to maintain these benefits. Economists use the concept of *shadow pricing*, which involves assessing what people *might* pay for a particular benefit if it were up to them to decide. For example, to evaluate the costs of noise nuisance from aircraft (an external bad) in the vicinity of Schiphol Airport in Amsterdam, researchers evaluated the housing prices in the vicinity of many airports around the world. They found that the closer the airport, and therefore the greater the noise, the lower the house value when compared with similar houses farther away. In this case, the *benefit* is seen as being free from the nuisance of low-flying aircraft, and people's willingness to pay is seen in the price of houses.

Shadow pricing becomes difficult when the analysis has to place a value on human life. Most of the pollutants to which people are exposed are hazardous, exacting a toll on health and life expectancy. To estimate the benefits of regulating such pollutants, it is necessary to calculate how

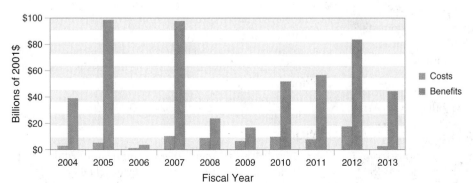

Figure 2–16 **Costs and benefits of major regulations, 2000–2013.** Clearly, the benefits of regulations far outweigh the costs of implementing the rules.

(*Source:* Office of Management and Budget, *2014 Report to Congress on the Benefits and Costs of Federal Regulations and Unfunded Mandates on State, Local and Tribal Entities, June 2014.*)

Table 2–2	Benefits That May Be Gained by the Reduction or Prevention of Pollution
Improved human health	Reduction and prevention of pollution-related illnesses and premature deaths Reduction of worker stress caused by pollution Increased worker productivity
Improved agricultural and forest production	Reduction of pollution-related damage More vigorous growth by removal of stress due to pollution Higher farm profits, benefiting all agriculture-related industries
Enhanced commercial or sport fishing	Increased value of fish and shellfish harvests Increased sales of boats, motors, tackle, and bait Enhancement of businesses serving fishers
Enhanced recreational opportunities	Direct uses, such as swimming and boating Indirect uses, such as observing wildlife Enhancement of businesses serving vacationers
Extended lifetime of materials and less necessity for cleaning	Reduction of corrosive effects of pollution, extending the lifetime of metals, textiles, rubber, paint, and other coatings Reduction of cleaning costs
Enhanced real estate values	Increased property values as the environment improves

many lives will be saved or how many people will enjoy better health. The lower the value that is put on a human life, the less likely it is that a regulation will pass the cost-benefit test. Finding a value for these benefits is fraught with ethical difficulties. Different approaches have resulted in a range of values for human life, from tens of thousands of dollars to $10 million per life saved. For example, the EPA calculated that new clean-air standards for ozone and particulates would prevent 18,700 premature deaths annually, and the monetary benefit of this prevention was valued at $110 billion (close to $6 million per life saved).

The values of nonhuman components of natural environments—a population of rare wildflowers, for example, or a wilderness site—are also difficult to value. The values of these things depend entirely on how willing people are to pay for their preservation, which may involve their own aesthetics, the difficulty of their financial situation, or the amount of information they have. When monetary values are assigned to their existence, the outcome will be strongly influenced by highly subjective elements in the analysis.

The Cost of Doing Nothing. Today many economists are trying to estimate the costs and benefits of various laws related to climate change adaptation and mitigation (covered in Chapter 18). However, doing nothing has costs as well. For example, economists have found that the government of Bangladesh is already spending $1 billion per year (6–7% of its annual budget) on climate change adaptations, such as resettling people whose homes are destroyed by rising seas. (This total does not include contributions from donor nations.) Thus the high costs of changing the way we use energy today have to be compared with the costs of not doing so.

Cost-Effectiveness Analysis

Cost-effectiveness analysis is another method of evaluating the costs of regulations. Here the merits of the goal are accepted, and the question is: How can that goal be achieved at the least cost? To find out, the costs of alternative strategies for reaching the goal are analyzed, and the least costly method is adopted.

Cost-effectiveness can be applied to the desired level of pollution control. As Figure 2–14 shows, a significant benefit may be achieved by modest degrees of cleanup. But note how differently the cost and benefit curves behave with increasing reduction of pollution. At some point in the cleanup effort, the lines cross and the costs exceed the benefits. Few additional benefits are realized when cleanup begins to approach 100%, yet costs increase exponentially. This behavior follows from the fact that living organisms—including humans—can often tolerate a threshold level of pollution without ill effect. Therefore, reducing the level of a pollutant below its threshold level will not yield an observable improvement. Optimum cost-effectiveness that meets the efficiency criterion for public policy is achieved at the point where the benefit curve is the greatest distance above the cost curve.

However, the perspective of time should be considered in calculating cost-effectiveness (Figure 2–15c). A situation that appears to be cost ineffective in the short term may prove extremely cost effective in the long term. This is particularly true for problems such as acid deposition and groundwater contamination from toxic wastes.

Progress

What has been the result of environmental regulation to date? Pollution of air and surface water reached critical levels in many areas of the United States in the late 1960s. Since that time, huge sums of money have been spent on pollution abatement. Cost-benefit analysis shows that, overall, these expenditures have paid for themselves many times over in decreased health care costs and enhanced environmental quality. As an example, consider the phase-out of leaded gasoline. The project cost about $3.6 billion, according to an EPA benefit-cost report. Benefits were valued at more than $50 billion, $42 billion of which were for medical costs that were avoided!

The following are some of the accomplishments of environmental public policy between 1980 and 2013, as reported by the EPA:

- Total emissions of six principal air pollutants decreased by 69% from 1980 to 2012.

- Average blood levels of lead for children have declined by more than 90% since 1980.

- Acid deposition decreased by 43–51% between 2000 and 2010.

- Releases of toxic chemicals decreased by 62% between 1986 and 2010.

- More than 436,406 leaking underground gasoline storage tanks were cleaned up between 1984 and 2013.

- Recycling of municipal solid waste increased from 7% in 1970 to 34.5% in 2012, diverting 86.6 million tons away from disposal in 2012.

- One 2012 estimate put the health care savings of the Clean Air Act at $22 trillion.[7]

Many of these environmental improvements occurred even as the GDP grew by 145%, the population rose by 40%, and the annual number of motor vehicle miles driven increased by 96%.

A cost-benefit analysis accompanies every new regulatory rule from the EPA and other federal agencies. Even when a rule is not classified as a major regulation, the accompanying documentation must demonstrate, by cost-benefit analysis, that the rule does not qualify as significant. Like it or not, cost-benefit analysis is now part of public policy in the United States.

CONCEPT CHECK Define cost-benefit analysis as it relates to environmental regulation. How does this method of analysis address external costs? How does it differ from cost-effectiveness analysis? ☑

2.6 Getting Society to Agree on Policy

Environmental concerns are often a battle of special interest groups. The issues are up for grabs, and protecting the environment often depends largely on the strength of nongovernmental organizations (NGOs). James Skillen, former executive director of the Center for Public Justice, says that this is a seriously flawed situation. Skillen believes that the protection of land, air, water, and the associated biota should

not have to depend on special interest groups for its advocacy. Rather, it is a matter of the *public good* and, as such, should be seen as a basic responsibility of government. To quote Skillen, "Recognition in basic law of the necessity of ecological health as the precondition of all public and market relations should become as fundamental as the recognition that certain human rights exist as the precondition of all public rules and market regulations."[8]

Moving environmental protection out of the arena of special interest groups and into law requires the efforts of citizens working together to improve the laws in their own countries. It also requires international regulations.

Citizen Involvement

As citizens, we are directly involved in public policy because we experience the outcomes of the policies. We pay for the benefits we receive through higher taxes and higher costs of the products we consume. We should care deeply about public policies. And the evidence of polls indicates that the American public does care. For example, a 2014 poll showed that American voters support the EPA in efforts to stop pollution from coal burning power plants even if it raises energy costs.[9]

What can you do to have an impact on environmental public policy? Here are some suggestions:

- Take note of the viewpoint of any political candidate on matters of environmental policy. Public concern about environmental policy can play an important role in the election or defeat of political candidates.

- Contact your legislators to inform them of your support for a particular environmental public policy under consideration by the Congress. You will very likely receive an acknowledgment of your concern. If enough people do the same, your legislator is likely to reflect your views on the issue.

- Become involved in local environmental problems. Local jurisdictions often have the power to make key decisions affecting the environment, and groups of citizens working together can often make a difference in how the decisions go.

- Become a member of an environmental NGO that informs its constituencies of environmental concerns around the country and lobbies for those concerns. There is a list of these organizations in Appendix A in the back of this book.

- Stay informed on environmental affairs; only when the public is informed is it likely that grassroots support will be maintained and environmental policy will really reflect public opinion. Incidentally, taking a course in environmental science is an excellent way to start getting informed!

---

[7]EPA, *The Benefits and Costs of the Clean Air Act from 1990 to 2020.* (Washington, D.C: U.S. EPA Office of Air and Radiation, 2011)

[8]James Skillen, How Can We Do Justice to Both Public and Private Trusts? Paper presented at the Global Stewardship Initiative National Conference, Gloucester, MA, October 1996. Presentation.

[9]Hart Research Associates and Public Opinion Strategies, NBC News/Wall Street Journal Poll, June 11–15, 2014, http://www.pollingreport.com/enviro.htm#AutoNumber91, accessed November 12, 2014.

Change the Economy to Green

Choices made by consumers can change the economy. When consumers decide to purchase products that have fewer negative effects on the environment, the outcome of their decisions can change the marketplace as effectively as laws and regulations. For example, deciding to purchase fish and forest products that are certified as being environmentally sustainable (Chapter 7) can lead to an increase in the number of environmentally sustainable products in the marketplace. Another way of supporting a more sustainable economy is to purchase products that are produced locally; such purchases help prevent the outsourcing of jobs and the need for long shipments that allow the influx of invasive species.

Green businesses and green investments move money that is already in the economy into products and services that are less environmentally damaging. For example, the Climate Works Foundation and the World Bank recently released a study promoting *climate-smart development*. The study showed that development projects that begin with the goals of increasing the economy and lowering the emission of pollutants can be financially sound and improve environmental quality and human health. Improving two problems with one solution promotes **co-benefits**, a better way to get the most out of problem-solving efforts.

Another way of promoting a green economy is through **social entrepreneurship**, the process of pursuing innovative solutions to social problems. Business owners typically measure their success by calculating the ratio of the money they make to the capital they invest. Social entrepreneurs also consider the value of their business to the common good. Social entrepreneurship often uses cutting edge social media such as the WIKI *Appropedia*, a sharing site for collaborative solutions to problems, including sustainability. Many social entrepreneurs support *open-source-appropriate technology*. These technologies are available free to the public and can be modified without the restrictions often imposed by patents, making them appropriate for use by people with little money.

Nonprofit organizations can also be environmentally conscious and entrepreneurial. One such organization is Adventurers and Scientists for Conservation, the dream of one hiker, Gregg Treinsh. This citizen-scientist group connects hikers with scientists; so far it has allowed more than 1,200 hikers to collect data on wilderness areas for 120 scientists. Similar ideas can be found in innovation networks such as Social and Environmental Entrepreneurs and Inhabit: Design for a Better World.

International Regulation

As difficult as it is to get nations to pass environmentally sound regulations, it is even harder to get countries to agree on sound environmental treaties. Table 2–3 lists some of the international laws or treaties that protect the environment;

Table 2–3 Selected International Laws and Treaties

Pollutants	Stockholm Convention on Persistent Organic Pollutants, Stockholm, 2001. Vienna Convention for the Protection of the Ozone Layer, Vienna, 1985, including the Montreal Protocol on Substances That Deplete the Ozone Layer, Montreal 1987.
Climate Change	Framework Convention on Climate Change (UNFCCC), New York, 1992, including the Kyoto Protocol, 1997.
Biodiversity and Endangered Species	International Convention for the Regulation of Whaling (ICRW), Washington, 1946. Convention on the International Trade in Endangered Species of Wild Flora and Fauna, (CITES), Washington, DC, 1973. Convention on the Conservation of Migratory Species of Wild Animals (CMS), Bonn, 1979. Convention on Biological Diversity (CBD), Nairobi, 1992.
Marine Environment	International Convention for the Prevention of Pollution of the Sea by Oil, London 1954, 1962, and 1969. Convention on the Prevention of Marine Pollution by Dumping of Wastes and Other Matter (London Convention), London, 1972. United Nations Convention on the Law of the Sea, LOS Convention, Montego Bay, 1982.
Nuclear	Comprehensive Test Ban Treaty, 1996.
Antarctica	Antarctic Treaty, Washington, DC, 1959.
Indigenous rights	Food and Agriculture Organization (FAO) International Undertaking on Plant Genetic Resources, Rome, 1983. International Tropical Timber Agreement, (ITTA), Geneva, 1994. Aarhus Convention, Convention on Access to Information, Public Participation in Decision-making and Access to Justice in Environmental Matters, Aarhus, 1998

we will refer to many of them in later chapters. As the world becomes more crowded and human consumption uses an even greater proportion of Earth's ecosystem services, international cooperation about global environmental issues will become even more critical.

Our global environment is intimately intertwined with human economies and systems of making laws and regulations. To move toward a sustainable future, we have to use the tools of accurate accounting in economics and wise decision making in public policy. We will also need innovative business and development solutions. As former EPA administrator Christine

Todd Whitman said in the introduction to the agency's *Draft Report on the Environment 2003*: "We are all stewards of this shared planet, responsible for protecting and preserving a precious heritage for our children and grandchildren. As long as we work together and stay firmly focused on our goals, I am confident we will make our air cleaner, our water purer, and our land better protected for future generations."

CONCEPT CHECK What are some of the ways ordinary consumers can change the brown economy to a green economy? List three ways. ☑

REVISITING THE THEMES

SOUND SCIENCE

Sound scientific research is critical to the development of effective environmental policy. The work of Sherwood Rowland and Mario Molina demonstrates the vital role science can play in the development of public policy—their scientific research led to the protection of Earth's ozone shield. Scientists can be pressured or misrepresented in the interest of politics, so it is essential for scientific work to avoid conflicts of interest. One component of the wealth of nations is *intangible capital*. Two elements of that capital are essential to maintaining science as a key to coping with environmental issues: education and the accumulation of scientific information. A truly wealthy nation will maintain a high level of scientific competence in its populace.

SUSTAINABILITY

Recent developments in China demonstrate how economic growth can come at the cost of environmental health. Ecological economists have made sustainability central to their economic theories. They rightly point out that the natural environment encompasses the economy. Therefore, the environment sets limits to global economic activities. A nation must manage its assets, but it can't do that unless it knows what they are and how to measure them. Depleting assets of any kind is a clear signal of unsustainable practice, and characteristic of the current business-as-usual

brown economy. A future sustainable *green economy* must focus on human and ecosystem well-being. It will value and preserve ecosystem services and goods. It will develop a host of "green" products that will revolutionize our technologies. It will discard the GDP as a measure of economic progress. Instead, it may even embrace the GPI or ISEW as better measures of economic well-being. Laws to protect the environment are essential for achieving sustainability. This is well illustrated by the accomplishments of environmental public policy as documented by the EPA and by the contrast to environmental conditions in places where the rule of law is not as strong.

STEWARDSHIP

Table 2–1 demonstrates the consequences of poor environmental stewardship; laws to protect the environment are not a luxury, but an essential foundation for a society. Over the last 40 years, U.S. public policy has accomplished much in achieving a safer and healthier environment. This represents our society's clear commitment to environmental stewardship. International treaties are also necessary for environmental protection. Internationally and nationally, stewardship demands that there be more equity between rich and poor. We need to be concerned about the rights of indigenous peoples and account for the value of noncash crops in our economic systems. Good stewardship of financial and ecosystem resources requires us to care about future generations and to properly account for changes to natural capital.

REVIEW QUESTIONS

1. What is one negative environmental effect of the rapid growth of the Chinese economy?

2. Three patterns of environmental indicators are associated with differences in the level of development of a nation. What problems decline, what problems increase and then decline, and what problems increase with the level of development?

3. Name three basic kinds of economic systems and explain how they differ.

4. What is the role of the World Trade Organization, and what have been some of the challenges to this role?

5. What are some characteristics of a sustainable economy?

6. Describe the three components of a nation's wealth. What insights emerge from the World Bank's analysis of the wealth of different groups of nations?

7. Why is the gross national product an inaccurate indicator of a nation's economic status? How can it be corrected?

8. Explain the importance of considering both intragenerational equity and intergenerational equity in addressing resource allocation issues.

9. What is the overall objective of environmental public policy, and what are the objective's two most central concerns?

10. Describe the various bodies responsible for U.S. environmental public policy at the federal, state, and local levels.

11. What are the advantages and disadvantages of the regulatory approach versus a market approach to policy development?

12. List the four stages of the policy life cycle, and show how the discovery of ozone layer destruction and the subsequent responses to it illustrate the cycle.

13. What is the conclusion of careful studies regarding the relationship between environmental policies, on the one hand, and jobs and the economy, on the other?

14. Give two examples of pollution control. For each control, identify at least one cost and one benefit.

THINKING ENVIRONMENTALLY

1. Consider the ultimate impact that environmental law has on the economy. Do short-term drawbacks justify long-term goals? If there is a conflict, which should come first—the economy or the environment?

2. When CFCs were found to cause the ozone hole, industry leaders said they could not be replaced in industrial applications. Later, they were replaced at a much lower cost than industry leaders had estimated. Think of an example of something today that seems impossible to do differently from the way we currently do it. What would it take to change the situation?

3. Examine the three types of capital that make up the wealth of nations. Which do you think is the most valuable? Justify your answer.

4. What different roles do you think the federal and state governments should have in environmental policy?

5. Suppose it was discovered that the bleach that is commonly used for laundry is carcinogenic. Referring to the policy life cycle, describe a predictable course of events until the problem is brought under control.

MAKING A DIFFERENCE

1. Our government makes the decisions that have the biggest long-term effects on the environment. Find a cause that you feel strongly about, and find a way to support it. Demonstrating is easier than it sounds—you don't necessarily have to drive a hundred miles to join in some kind of march. It's easy to join an online movement (for example, visit www .StopGlobalWarming.org) or write a letter to a senator.

2. Help call attention to instances of pollution in your community. You can talk to regulators in the county health department, the EPA, or whoever is in charge of monitoring pollution in your region. It's their job to listen.

3. Instead of taking a long-distance vacation and relaxing during your break, join a volunteer group and do something positive for your local environment. There are many opportunities to help out.

4. Find out the costs and benefits of your own lifestyle on a purely economic level. Decide what changes you can make to save you money and save the environment at the same time.

ECOLOGY: THE SCIENCE OF ORGANISMS AND THEIR ENVIRONMENT

Even if you've never stepped foot in one, you probably know that tropical rain forests are humid, warm, dense, and full of unusual plants and animals. You might not know, however, that they are home to a greater diversity of living things than anywhere else on Earth; that they are the result of millions of years of adaptive evolution; and that they store more carbon than is in the entire atmosphere. Nor can you see with your eyes how energy flows through these forests or how nitrogen and phosphorus are cycled and recycled. This knowledge comes from the work of many scientists who have been asking basic questions about how such natural systems function, how they came to be, and how they relate to the rest of the world.

How important is the knowledge that comes from these scientists? In this second part of the text, we'll see that information about how the natural world works is absolutely crucial to understanding how to manage and protect it. In Part Two, we will look at the science underlying how living things survive in the world and how matter and energy move throughout systems. We will use a hierarchical approach: we will start with the small and move to the large—from the molecules necessary for life and the solar energy that sustains it to how plants return to a habitat after a fire and why large regions of the world have different forms of vegetation, such as forests, desert, and grassland. Understanding environmental science begins with an understanding of how living organisms are structured, survive, relate to their environment (and each other), and change over time.

Chapter 3

Basic Needs of Living Things

The movie *March of the Penguins* brought the lives of emperor penguins (*Aptenodytes forsteri*) to the big screen and the public imagination. The largest of the penguins, emperor penguins are known for their lives in the harsh Antarctic. They live on a diet of fish, with some krill and the occasional squid or octopus, diving into the cold ocean waters from islands of sea ice. Those fish and squid feed on tiny animals called zooplankton, which in turn eat algae. Penguins also provide food for their own predators, the powerful leopard seals.

Much of the emperor penguins' year is spent at inland breeding grounds and in epic treks between them and the ocean. After mating, a female lays a single egg, which the male carefully protects under a flap of skin over his feet. The male spends the hard winter incubating the egg, surviving howling gales by huddling in a flock of similarly protective males. Meanwhile the female treks back to the ocean, where she swims and feeds. When she has packed on enough fat, the female treks back to the breeding ground. The parents then trade places, and the female takes over the chick rearing. Relieved of duty, the thin and worn male leaves for his own trip to feed in the ocean.

Antarctic Adaptations. Penguins have a number of specialized adaptations that allow them to survive in this harsh environment. Their heavy bones and dense, waterproof feathers allow them to swim through ice-cold water. In addition, their hemoglobin functions in the low-oxygen conditions, and they can shut down their metabolism while diving. On land, they breed in the coldest environment of any bird species. Their blubber, high feather density, and ability to maintain their body temperature through a wide range of air temperatures allow them to succeed under conditions that would kill most other animals. Yet in order to survive, grow, and reproduce, emperor penguins require certain conditions from their environment, such as particular sources of food and locations that are suitable for reproduction.

▲ Emperor penguins have basic needs for survival, as do their prey and predators.

A Changing World. Penguins face the loss of ocean fish from overfishing and changes in sea ice from climate change. In the future, it may be that their environmental requirements will not be met, but new evidence suggests that penguins may already be adapting their behavior to adjust to environmental change. In this chapter, we will follow the penguin and the other organisms in its food chain as examples of organisms living in their environment. All organisms get nutrients and energy from the environment and use them to build the cells and tissues of their bodies and to carry out activities such as moving, getting food, protecting themselves, reproducing, and getting rid of wastes.

Both the penguins and their prey have adaptations for dealing with the conditions under which they live. In this chapter, we will see how the basic functions of organisms fit into the scientific discipline of ecology. We will explore basic physical rules about matter and energy and how they control the lives of organisms in their environments. To do this, we will need to understand some basic chemistry and physics. At the chapter's end, we will look at the global movements of important nutrients and how those huge cycles both affect individual organisms and are affected by human activities.

3.1 Organisms in Their Environment

Emperor penguins are just one species in a family of about 20 species of flightless aquatic birds. Some, like the emperor penguin, live in extreme cold. Other species of penguins live in temperate waters, and one even lives near the equator. The areas in which each of these species lives is a function of several factors: the penguin's habitat requirements, its evolutionary history, and its interactions with other organisms. Penguins need certain resources to survive, such as food and a place to nest. Their food organisms, largely fish and squid, live on a diet of smaller animals, such as zooplankton, which are themselves supported by algae. Each of these parts of the food chain requires particular conditions, such as certain water temperatures, to survive. The distribution and abundance of the emperor penguin is a function of its requirements for growth and the history of the relationship between penguins and other species, including humans. To understand whether emperor penguins will thrive in the future, we need to turn to the discipline of ecology.

The Hierarchy of Ecology

Ecology (from *oikos*, meaning "house or home") is the study of all processes influencing the distribution and abundance of organisms and the interactions between living things and their environment. A very basic understanding of ecological terms and concepts will help later as we look at how changes in the environment affect living things.

Ecologists studying the conservation of emperor penguins ask an array of different questions. They may investigate what conditions the algae at the base of the food chain need, or the requirements of the penguins' fish and squid prey. They might ask what conditions allow penguins to grow most rapidly, or look for the factors that influence successful breeding. Any of these inquiries constitutes an ecological question, and several different branches of ecology exist to address them. Ecology can be described as a group of different, but related, disciplines: a hierarchy of studies. Scientists in different branches of the discipline might operate at different scales and ask different questions about the same system. The concept of ecological hierarchy is depicted in **Figure 3–1** on the next page.

As we investigate the science of ecology, we will look at these different levels in the hierarchy. We start by examining individual organisms at the species level. We will see how species are grouped into larger categories and how these groups interact. Species exist in specific habitats because those habitats provide the nutrients and other factors that a species needs to survive; we will look at how those needs are met. Lastly, we will take a broader view of matter and energy, and how energy and nutrients cycle through the various systems on Earth. All of these different processes are interlinked. Ecologists study the connections as well as the whole.

Species. A **species** is a group of individuals that share certain characteristics distinct from other such groups (robins versus redwing blackbirds, for example). Species are classified into closely related groups called *genera* (singular, *genus*), and they in turn are grouped into *families*. These are grouped into larger groups: *orders, classes, phyla* (singular, *phylum*), *kingdoms,* and *domains.* The official scientific name of any species is always Latin and has two parts: the genus name and the species epithet, or descriptive term. For example, the emperor penguin's name, *Aptenodytes forsteri*, tells us first the genus (*Aptenodytes*, meaning "without-wings-diver") and second a descriptive word (the species epithet, *forsteri*, named after explorer Johann Forster). All penguins are in the same family: Spheniscidae. Only the emperor penguin and one other species, the king penguin, are in the genus *Aptenodytes*, because they are considered to be more closely related to each other than to other penguins.

Sometimes it is difficult to determine exactly what constitutes a species. The biological definition of a species is *all members that can interbreed and produce fertile offspring.* Members of different species generally do not interbreed. This definition is good for many organisms, but does not work for those that do not need to mate to produce offspring. In those cases, scientists use a variety of other classification methods. New species can arise over periods of time through evolution, and species classifications are occasionally altered to reflect this.

Populations. Each species in a biotic community is represented by a certain **population**—that is, by a certain number of individuals that make up the interbreeding, reproducing

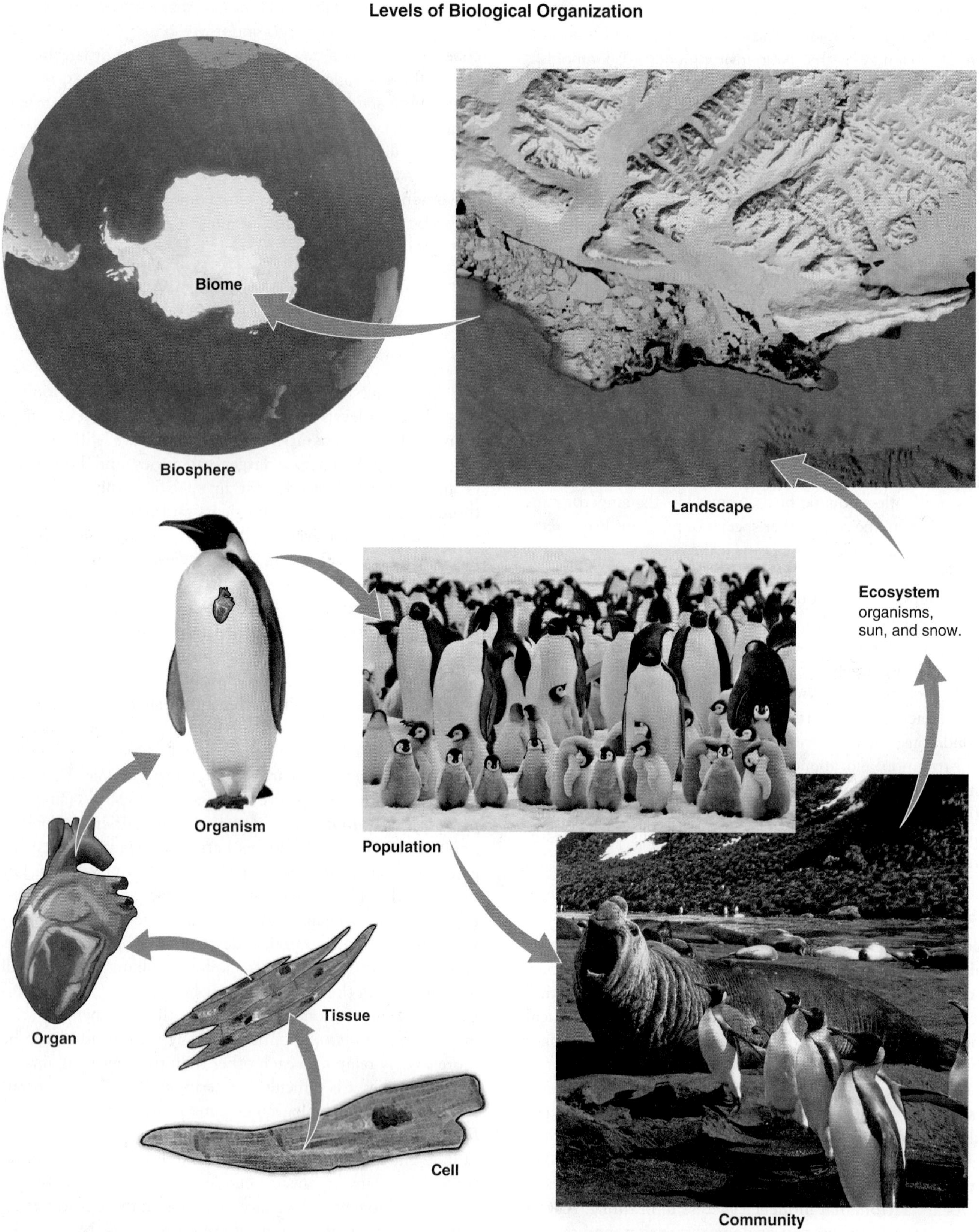

Levels of Biological Organization

Biome

Biosphere

Landscape

Ecosystem
organisms,
sun, and snow.

Organism

Population

Organ

Tissue

Cell

Community

Figure 3–1 **The hierarchy of life.** Organisms can be seen as a hierarchy of complex systems from subcellular components not shown here, to cells to tissues to organs to systems to whole individuals. Likewise, individuals function in a set of increasingly complex systems that make up the disciplines of ecology. These include populations, communities, ecosystems, and landscapes. The very largest regions are called biomes. Together, all living things make up the biosphere.

group. The distinction between population and species is that *population* refers only to those individuals of a certain species that live within a given area—for example, all of the gray wolves in Yellowstone National Park. *Species* is all inclusive, such as all of the gray wolves in the world.

Biotic Communities. The grouping of populations we observe when studying a forest, grassland, or pond is referred to as the area's **biota** (*bio* means "living") or **biotic community.** The biota includes all vegetation, from large trees through microscopic algae, and all animals, from large mammals, birds, reptiles, and amphibians through earthworms, tiny insects, and mites. It also includes microscopic creatures such as bacteria, fungi, and protozoans.

The particular kind of biotic community found in a given area is, in large part, determined by **abiotic** (nonliving, chemical, and physical) factors. Abiotic factors include the amount of water or moisture present, climate, salinity (saltiness), and soil type. Sometimes, we quickly refer to a community by its most obvious members, such as the dominant plants, usually indicators of the environmental conditions of a site. For example, we might describe a "hemlock/beech forest" or an "oak/hickory forest" (Figure 3–2).

The species within a community depend on one another. For example, certain animals will not be present unless certain plants that provide food and shelter are also present. In this way, the plant community supports (or limits by its absence) the animal community. In addition, every plant and animal species is adapted to cope with the abiotic factors of its region. For example, every species that lives in a temperate region is adapted in one way or another to survive the winter season, which includes a period of freezing temperatures (Figure 3–3). These interactions among organisms and their environments will be discussed later. For now, keep in mind that the populations of different species within a biotic community are constantly interacting with each other and with the abiotic environment.

Ecosystems. An **ecosystem** is an interactive complex of communities and the abiotic environment affecting them within a particular area. Forests, grasslands, wetlands, sand dunes, and coral reefs each contain certain species in particular spatial areas, and are all distinct ecosystems. Humans are parts of ecosystems, too, and some branches of ecology, such as urban ecology, focus on human-dominated systems. Other ecologists study the interaction between humans and nonhuman ecosystems, such as salt marshes or the open ocean, which are deeply affected by humans.

While it is convenient to divide the living world into different ecosystems, you will find that there are seldom distinct boundaries between them and that they are not totally isolated from one another. Many species occupy (and are a part of) two or more ecosystems or, as in the case of migrating birds, may move from one ecosystem to another at different times.

Transitional regions between ecosystems are known as **ecotones.** An ecotone shares many of the species and characteristics of both ecosystems it passes through. Ecotones may lack unique (and contain fewer overall) species when compared

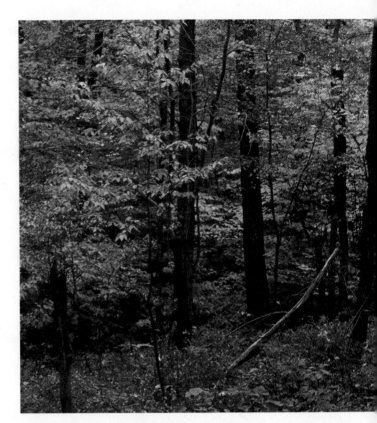

Figure 3–2 Predictable vegetation. An oak/hickory forest is dominated by oak and hickory trees, which define the type of forest habitat other species experience.

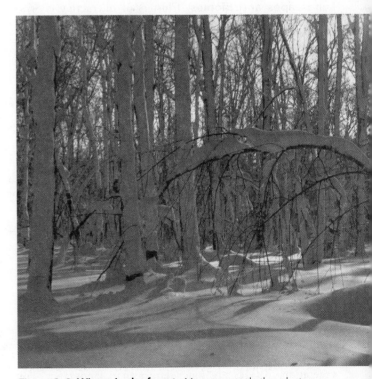

Figure 3–3 Winter in the forest. Many trees and other plants of temperate forests are so adapted to the winter season that they actually require a period of freezing temperature in order to grow again in the spring.

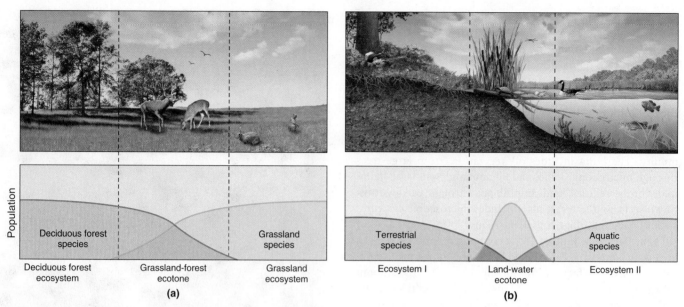

Figure 3–4 Ecotones. (a) Ecosystems are not isolated from one another. One ecosystem blends into the next through a transitional region—an ecotone—that contains many species common to both systems, though in lower numbers than in those systems. **(b)** An ecotone may create a unique habitat that harbors specialized species not found in either of the ecosystems bordering it. For example, cattails, reeds, and lily pads typically grow in the ecotone between land and freshwater ponds.

to bordering ecosystems, as in the ecotone between grassland and forest shown in Figure 3–4a. Alternatively, ecotones may house and support more different species than the ecosystems they pass through, as in the marshy ecotone that often occurs between the open water of a lake and dry land (Figure 3–4b).

Landscapes and Biomes. Clusters of interacting ecosystems like forests, open meadows, and rivers together constitute **landscapes.** Landscape ecologists study issues such as how much suitable habitat will be left for fish to lay eggs in a riverbed if the surrounding forest is clear-cut or how far squirrels will be able to travel if the forest is broken into small segments. Landscapes are usually on a scale humans can envision; for example, if you look out at the horizon and see a forest merge into a field or meadow, you are seeing a landscape. However, landscapes can also be observed on a large scale: one could also pull back from Earth and look at large areas of its surface that have similar characteristics (such as dominant vegetation). Many scientists use digital technology tools such as **geographic information systems** to study landscape features. They also use **remote sensing,** including images from satellites (Figure 3–5). A new area of ecology, **macrosystems ecology,** uses large amounts of data across many types of ecosystems to answer questions that are extremely wide in scope, such as questions about climate change or the effects of invasive species.

Tropical rain forests, grasslands, and deserts are **biomes** (discussed in Chapter 5). A biome is a large area of Earth's surface that shares climate and has similar vegetation. Biomes are much larger than landscapes. For example, grasslands (such as prairies) are geographical regions whose locations can be predicted by their rainfall and temperatures. On different continents, they may have

different species, but they are still grasslands, described by the dominant type of vegetation. As with ecosystems, there are generally no sharp boundaries between biomes. Instead, one grades into the next through transitional regions.

Major categories of aquatic ecosystems, such as estuaries, rivers, and the open ocean, are sometimes referred to as biomes. The characteristics of aquatic ecosystems are determined primarily by the depth, salinity, and permanence of the water in them (Chapter 5).

Biosphere. Regardless of how we choose to categorize and name different ecosystems, they are interconnected and interdependent. Terrestrial ecosystems are connected by the flow of rivers and by migrating animals. Sediments and nutrients

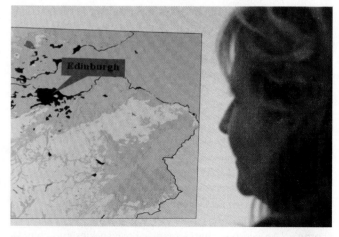

Figure 3–5 Landscape Ecology. Here a researcher demonstrates a mapping tool used in landscape ecology. This is a prototype of a geographical information system called Soilfit.

washing from the land may nourish or pollute the ocean. Seabirds and marine mammals connect the oceans with the land, and all biomes share a common atmosphere and water cycle. All the species on Earth live in ecosystems, and ecosystems are interconnected so that all living things form one huge system called the **biosphere**. This concept is analogous to the idea that the cells of our bodies form tissues, tissues form organs, organs are connected into organ systems, and all are interconnected to form the whole body.

CONCEPT CHECK What are three different ecological scales? Give an example of a question that a researcher could ask about emperor penguins at each scale. ☑

3.2 Environmental Factors

Let's look back at the emperor penguins in the Antarctic. Why are the penguins living there and not in another part of the world? The answer is both the environment and history. The penguins and their prey have evolved to live in the Southern Ocean that surrounds Antarctica, with the physical, chemical, and biological factors it contains. As already discussed, we describe these factors as either biotic (living) or abiotic (nonliving). The factors can be further characterized as either conditions or resources. **Conditions** are any factors that vary in space and time, but that are not used up or made unavailable to other species. These include temperature (extremes of heat and cold as well as average temperature), wind, pH (acidity), salinity (saltiness), and fire. **Resources** are any factors—biotic or abiotic—that are consumed by organisms. These include water, chemical nutrients (like nitrogen and phosphorus), light (for plants and algae), oxygen, food, and space for resting or nesting.

Some factors can be conditions in one situation and resources in another. For example, plants in the desert compete for water as a *resource*; however, in the ocean, water is a *condition* for which algae, fish, and penguins need special adaptations and for which they are not competing.

The degree to which each factor is present or absent profoundly affects the ability of organisms to survive. However, each species may be affected differently by each factor, and its response determines whether or not it will occupy a given region.

Optimums, Zones of Stress, and Limits of Tolerance

Different species thrive with different combinations of environmental factors. This principle applies to all living things, both plants and animals. For example, bamboo plants thrive in full sun and warm temperatures, with plenty of water and well-drained, nitrogen-rich soil. In contrast, desert cacti are generally adapted for low amounts of water, poor soil fertility, and extremes of heat and cold. Aquatic systems also vary in their environmental factors and residents—for example, the amounts of freshwater and saltwater determine the varieties of fish and other organisms present.

Organismal ecologists study the adaptations of organisms to specific environmental factors. Organisms can be grown under controlled conditions in which one factor is varied while other factors are held constant. Such experiments demonstrate that, for every factor, there is a certain level at which the organisms grow or survive most; an **optimum** (plural, *optima*). At higher or lower levels than the optimum, the organisms do less well, and at further extremes, they may not be able to survive at all. Figure 3-6 on the following page shows the optimum temperature for a species of bamboo.

It is common to speak of an *optimal range*, the range over which the best response occurs. The entire span that allows any growth at all is called the **range of tolerance**. The points at the high and low ends of the range of tolerance are called the **limits of tolerance**. Between the optimal range and the high or low limit of tolerance are **zones of stress**. Outside of the optimal range, the organisms experience increasing stress, until, at either limit of tolerance, they cannot survive.

A Fundamental Biological Principle. Based on the consistency of observations and experiments, the following is considered to be a fundamental biological principle: *Every species has an optimal range, zones of stress, and limits of tolerance with respect to every abiotic factor, and these characteristics vary between species.* For instance, some plants cannot tolerate any freezing temperatures; others can tolerate slight freezing; and some actually require several weeks of freezing temperatures to complete their life cycles. Some species have a very broad range of tolerance, whereas others have a much narrower range.

The range of tolerance for any factor affects more than just the growth of individuals. Because the health and vigor of individuals affect their ability to reproduce, the survival of the next generation is also affected, which in turn influences the population as a whole. Consequently, the population density (individuals per unit area) of a species is greatest where all conditions are optimal, and it decreases as any one or more conditions depart from the optimum. You expect to see the most bamboo in the places where all conditions are best for bamboo and less where conditions are not so favorable.

Law of Limiting Factors. In 1840, Justus von Liebig studied the effects of chemical nutrients on plant growth. He observed that restricting any one of the many different nutrients at any given time had the same effect: it limited growth. A factor that limits growth is called (not surprisingly) a **limiting factor**. If any one factor is outside the optimal range, it will cause stress and limit the growth, reproduction, or even survival of a population. Ecologists now refer to this observation as the **law of limiting factors** or Liebig's law of the minimum.

A limiting factor may be a problem of *too much* instead of a problem of *too little*. For example, plants may be stressed or killed not only by underwatering or underfertilizing, but also by overwatering or overfertilizing, which are common pitfalls for beginning gardeners. Note also that the limiting factor may change from one time to another. For instance, temperature may be limiting in the early spring, nutrients may be limiting in the winter, and water may be limiting if a drought occurs.

Figure 3–6 Survival curve. For every factor influencing growth, reproduction, and survival, there is an optimum level. Above and below the optimum, stress increases until survival becomes impossible at the limits of tolerance. The total range between the high and low limits is the range of tolerance.

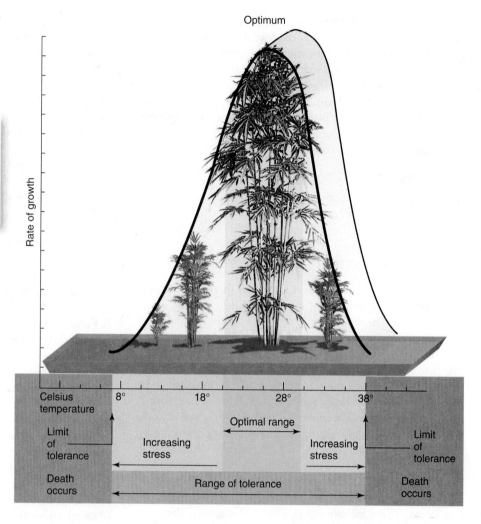

UNDERSTANDING THE DATA

1. If you were growing bamboo in a greenhouse, how would you expect its rate of growth to change if the temperature fell to 18°C?

2. What if the temperature rose to 40°C during a particularly hot day?

While one factor may be limiting at a given time, several factors outside the optimum may combine to cause additional stress or even death. In particular, pollutants may act in a way that causes organisms to become more vulnerable to disease or drought. Such cases are examples of **synergistic effects**, or **synergisms**, which are defined as two or more factors interacting in a way that causes an effect much greater than one would anticipate from each of the two acting separately.

Habitat and Niche. **Habitat** refers to the kind of place—defined by the plant community and the physical environment—where a species is biologically adapted to live. For example, deciduous forests, swamps, and grassy fields are types of habitats. Different types of forests (for instance, coniferous versus tropical) provide markedly different habitats and support different species of wildlife. Because some organisms operate on very small scales, we use the term *microhabitat* for things like puddles, spaces sheltered by rocks, and holes in tree trunks that might house their own small community (Figure 3–7).

Even when different species occupy the same habitat, competition may be slight or nonexistent because each species has its own niche. An **ecological niche** is the sum of all of the conditions and resources under which a species can live: what and where it feeds, what it feeds on, where it finds shelter and nests, and how it responds to abiotic factors.

Similar species can coexist in the same habitat but have separate niches. Competition is minimized because potential competitors are using different resources. For example, bats and swallows both feed on flying insects, but they do not compete because bats feed on night-flying insects and swallows feed during the day.

In the next section, we will see how organisms take these materials from the environment and use them to grow and maintain their bodies and do activities necessary for survival.

CONCEPT CHECK You fertilize your lawn with nitrogen. At first the grass grows rapidly, but then its growth slows down. How might this change relate to the Law of Limiting Factors? ☑

3.3 Matter in Living and Nonliving Systems

Living organisms need to be able to take in matter and energy from their environment to grow and function. For example, the algae in the ocean can survive only because they are able to take the Sun's energy and combine it with materials from the water around them to make the basic building blocks of their own tissues and to power their cellular activities. In turn, the algae are eaten by zooplankton, which are eaten by

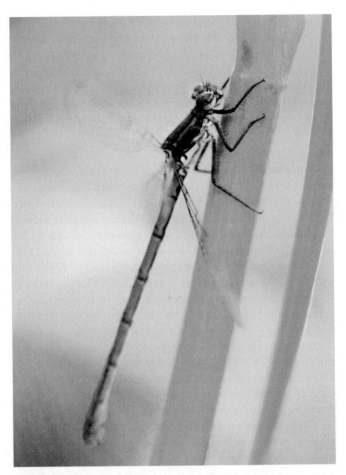

Figure 3–7 Habitat and niche. For part of its life cycle, this species of damselfly lives in a pool of water trapped by a type of tropical plant; the tiny pool is a microhabitat. The specific conditions required by the damselfly, such as the pool and the types of food it eats there or where it lays its eggs as an adult, are its niche.

fish, which in turn are eaten by penguins. Both the penguins and their prey get from their food the energy and matter necessary to build their own tissues and to do the work they need to do in each of their cells. To understand this simple ecological relationship, we need to know some basic chemistry and physics. Later, we will explore what happens to energy and matter as they move through the biosphere.

Basic Units of Matter

Matter is defined as anything that occupies space and has mass. This definition covers all solids, liquids, and gases as well as all living and nonliving things. Matter is composed of **atoms**—very small pieces—that are combined to form molecules, which in turn can be combined into more complex structures.

Atoms. The basic building blocks of all matter are atoms. Only 94 different kinds of atoms occur in nature, and these are known as the naturally occurring **elements**. Atoms are made up of protons, neutrons, and electrons, which in turn are made up of still smaller particles. For example, lead (Pb) is an element. An atom of lead is the smallest amount of lead

existing that has the characteristics of lead. A lead atom has a characteristic number of protons (positive particles), neutrons (neutral particles), and electrons (negative particles).

How can these relatively few building blocks make up the countless materials of our world, including the tissues of living things? Picture each kind of atom as a different-sized Lego® block. Like Legos, atoms can build a great variety of things. Also like Legos, natural materials can be taken apart into their separate constituent atoms, and the atoms can then be reassembled into different materials. All chemical reactions, whether occurring in a test tube, in the environment, or inside living things (and whether they occur very slowly or very quickly), involve rearrangements of atoms to form different kinds of matter.

Atoms do not change during the disassembly and reassembly of different materials. A carbon atom will always remain a carbon atom. In chemical reactions, atoms are neither created nor destroyed. The same number and kind of different atoms exist before and after any reaction. This constancy of atoms is regarded as a fundamental natural law, the **Law of Conservation of Matter**. Nuclear reactions differ from the chemical reactions we are discussing and can result in the splitting of an atom of one element into multiple atoms of another element. However, this is a very specific, rare instance and is not a chemical reaction. (Nuclear reactions will be discussed in Chapter 15.)

Now we turn our attention to the ways atoms are put together and how atoms are incorporated into organisms.

Molecules and Compounds. A **molecule** consists of two or more atoms (either the same kind or different kinds) bonded in a specific way. The properties of a material depend on the exact way in which atoms are bonded to form molecules as well as on the atoms themselves. A **compound** consists of two or more *different* kinds of atoms that are bonded. For example, the fundamental units of oxygen gas, which consist of two bonded oxygen atoms, are molecules (O_2) but not compounds. By contrast, when an oxygen atom binds with hydrogen atoms to create water, it is both a molecule and a compound (H_2O).

On the chemical level, the cycle of growth, reproduction, death, and decay of organisms is a continuous process of using various molecules and compounds from the environment (food), assembling them into living organisms (growth), disassembling them (decay), and repeating the process.

Four Spheres

During growth and decay, atoms move from the environment into living things and then return to the environment. To picture this process, think of the environment as three open, nonliving systems, or "spheres," interacting with the biosphere. The **atmosphere** is the thin layer of gases (including water vapor) separating Earth from outer space. The **hydrosphere** is water in all of its liquid and solid compartments: oceans, rivers, ice, and groundwater. The **lithosphere** is Earth's crust, made up of rocks and minerals. Matter is constantly being exchanged within and between these four spheres, as shown in Figure 3–8 on the following page.

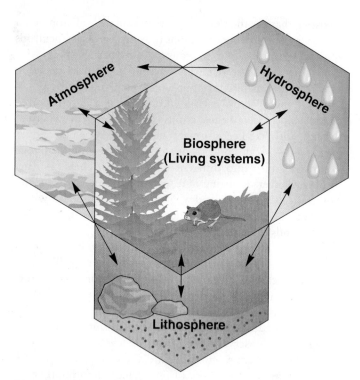

Figure 3–8 The four spheres of Earth's environment. The biosphere is all of life on Earth. It depends on, and interacts with, the atmosphere (air), the hydrosphere (water), and the lithosphere (soil and rocks).

(*Source:* Adapted from *Geosystems*, 5th ed., by Robert W. Christopherson, © 2005 Pearson Education, Inc.)

Atmosphere. The lower atmosphere is a mixture of molecules of three important gases—oxygen (O_2), nitrogen (N_2), and carbon dioxide (CO_2)—along with water vapor and trace amounts of several other gases. The gases in the atmosphere

are normally stable, but under some circumstances, they react chemically to form new compounds. For example, ozone is produced from oxygen in the upper atmosphere (Chapter 18). Plants take carbon dioxide in from the atmosphere, usually through their leaves. Animals usually take oxygen in through some type of specialized organ such as a lung, but some, like earthworms, can simply absorb oxygen through their skin.

Hydrosphere. While the atmosphere is a major source of carbon and oxygen for organisms, the hydrosphere is the source of hydrogen. Each molecule of water consists of two hydrogen atoms bonded to an oxygen atom, so the chemical formula for water is H_2O. A weak attraction known as *hydrogen bonding* exists between water molecules.

Water is an important molecule for living things and usually needs to be available in liquid form. Water occurs in three different states. At temperatures below freezing, hydrogen bonding holds the molecules in position with respect to one another, and the result is a solid **crystal** structure (ice or snow). At temperatures above freezing but below vaporization, hydrogen bonding still holds the molecules close but allows them to move past one another, producing the liquid state. Vaporization (evaporation) occurs as hydrogen bonds break and water molecules move into the air independently. As temperatures are lowered again, all of these changes of state go in the reverse direction. Generally, water undergoes melting and evaporation, but sometimes water molecules leave snow or ice and go directly into the air. This process is *sublimation* (Figure 3–9). Moving from one state to another either releases or requires a great deal of energy. This is one reason why many animals sweat to cool off: changing from

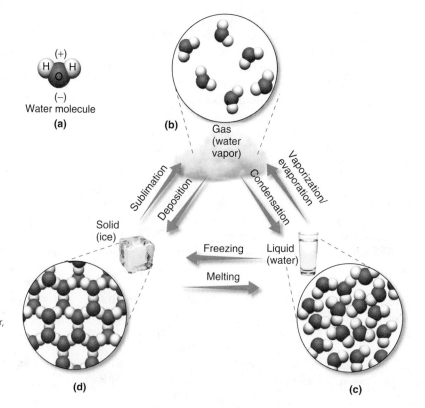

Figure 3–9 Water and its three states. (a) Water consists of molecules, each of which is formed when two hydrogen atoms bond to an oxygen atom (H_2O). **(b)** In water vapor, the molecules are separate and independent. **(c)** In liquid water, the weak attraction between water molecules known as hydrogen bonding gives the water its liquid property. **(d)** At freezing temperatures, hydrogen bonding holds the molecules firmly, giving the solid state—ice.

Figure 3–10 **Minerals.** Minerals (hard, crystalline compounds) are composed of dense clusters of atoms of two or more elements. The atoms of most elements gain or lose one or more electrons, becoming negative (−) or positive (+) ions, and form a predictable pattern held closely together. This photograph is of a mineral, gypsum, in a mine in Texas.

the liquid to gas state requires heat and removes it from the organism. Despite the changes of state, the water molecules themselves retain their basic chemical structure of two hydrogen atoms bonded to an oxygen atom; only the *relationship* between the molecules changes.

Lithosphere. All the other elements that are required by living organisms, as well as the approximately 72 elements that they do not require, are found in the lithosphere, in the form of rock and soil minerals. A **mineral** is a naturally occurring solid, made by geologic processes; it is a hard, crystalline structure of a given chemical composition (**Figure 3–10**). Most rocks are made up of relatively small crystals of two or more minerals, and soil generally consists of particles of many different minerals. Each mineral is made up of dense clusters of two or more kinds of atoms bonded by an attraction between positive and negative charges on the atoms (Appendix C).

Interactions. Air, water, and minerals interact with each other in a simple, but significant, manner. Gases from the air and ions (charged atoms) from minerals may dissolve in water. Therefore, natural water is inevitably a **solution** containing variable amounts of dissolved gases and minerals. This solution is constantly subject to change because various processes may remove any dissolved substances in it or additional materials may dissolve in it. Molecules of water enter the air by evaporation and leave it again via condensation and precipitation (see the hydrologic cycle, Chapter 10). Thus, the amount of moisture in the air fluctuates constantly. Wind may carry dust or mineral particles, but the amount changes constantly because the particles gradually settle out from the air. These interactions are summarized in **Figure 3–11**. The materials in these three spheres interact with the biosphere as living organisms take materials from the spheres and use them to create complex molecules in their bodies. We will discuss the process of building these molecules next.

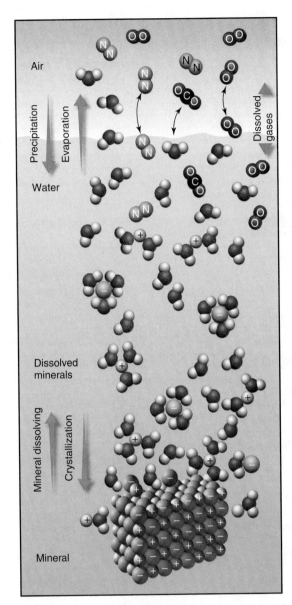

Figure 3–11 **Interrelationship among air, water, and minerals.** Minerals and gases dissolve in water, forming solutions. Water evaporates into air, causing humidity. These processes are all reversible: minerals in solution recrystallize, and water vapor in the air condenses to form liquid water.

Organic Compounds

Your body, like that of the emperor penguin, is composed of relatively large chemical compounds in a number of broad categories, such as proteins, carbohydrates (sugars and starches), lipids (fatty substances), and nucleic acids (DNA and RNA). These compounds usually contain six key elements: carbon (C), hydrogen (H), oxygen (O), nitrogen (N), phosphorus (P), and sulfur (S). Living things need other elements as well, sometimes in smaller, but no less important, amounts. **Table 3–1** on the following page lists the elements in living organisms.

Unlike the relatively simple molecules that occur in the environment (such as CO_2, H_2O, and N_2), the key chemical elements in living organisms bond to form very large, complex molecules. The chemical compounds making up the

Table 3–1 Elements Found in Living Organisms and the Locations of Those Elements in the Environment

Element (Kind of Atom)	Biologically Important Molecule or Ion in Which the Element Occurs[a]			Location in the Environment[b]		
	Symbol	Name	Formula	Atmosphere	Hydrosphere	Lithosphere
Carbon	C	Carbon dioxide	CO_2	X	X	X (CO_3)
Hydrogen	H	Water	H_2O	X	(Water itself)	
Atomic oxygen (required in respiration)	O	Oxygen gas	O_2	X	X	
Molecular oxygen (required in photosynthesis)	O_2	Water	H_2O		(Water itself)	
Nitrogen	N	Nitrogen gas Ammonium ion Nitrate ion	N_2 NH_4^+ NH_3	X	X X X	Via fixation X X
Sulfur	S	Sulfate ion	SO_4^{2-}		X	X
Phosphorus	P	Phosphate ion	PO_4^{3-}		X	X
Potassium	K	Potassium ion	K^+		X	X
Calcium	Ca	Calcium ion	Ca^{2+}		X	X
Magnesium	Mg	Magnesium ion	Mg^{2+}		X	X
Trace Elements[c]						
Iron	Fe	Iron ion	Fe^{2+}, Fe^{3+}		X	X
Manganese	Mn	Manganese ion	Mn^{2+}		X	X
Boron	B	Boron ion	B^{3+}		X	X
Zinc	Zn	Zinc ion	Zn^{2+}		X	X
Copper	Cu	Copper ion	Cu^{2+}		X	X
Molybdenum	Mo	Molybdenum ion	Mo^{2+}		X	X
Chlorine	Cl	Chloride ion	Cl^-		X	X

Note: These elements are found in *all* living organisms—plants, animals, and microbes. Some organisms require certain elements in addition to the ones listed. For example, humans require sodium and iodine.

[a]A molecule is a chemical unit of two or more atoms that are bonded. An ion is a single atom or group of bonded atoms that has acquired a positive or negative charge as indicated.

[b]X means that element exists in the indicated sphere.

[c]Only small or trace amounts of these elements are required.

tissues of living organisms are referred to as **organic**. Some of these molecules may contain millions of atoms. These molecules are constructed mainly from carbon atoms bonded into chains, with hydrogen and oxygen atoms attached. Nitrogen, phosphorus, or sulfur may be present also, but the key common denominator is carbon–carbon and carbon–hydrogen or carbon–oxygen bonds. **Inorganic**, then, refers to molecules or compounds with neither carbon–carbon nor carbon–hydrogen bonds. While this is the general rule, some exceptions occur; by convention, a few compounds that contain carbon bonds, such as carbon dioxide, are still considered inorganic.

All plastics and countless other human-made compounds are also based on carbon bonding and are, chemically speaking, organic compounds. To resolve any confusion this may cause, the compounds making up living organisms are referred to as **natural organic compounds** and the human-made ones as **synthetic organic compounds**. The term *organic* can have a completely different meaning, such as in *organic farming* (Chapter 12).

To summarize: the elements essential to life (C, H, O, and so on) are present in the atmosphere, hydrosphere, or lithosphere in relatively simple molecules. In the living organisms of the biosphere, these elements are organized into highly complex organic compounds. For example, algae in the Southern Ocean use the energy of sunlight to combine simple molecules from the water and air into the complex organic molecules that make up their cells. When algae are eaten, their organic molecules are digested and recombined to form the tissues of zooplankton. This process occurs again when zooplankton are eaten by fish, and once again when fish are eaten by emperor penguins. When the algae, fish or penguins die, the reverse process occurs through decomposition and decay. Each of these processes is discussed in more detail later in the chapter. Next, we discuss energy, the force that helps change chemical matter into substances that support life.

CONCEPT CHECK How is the process of dissolving salt (NaCl) in water (H_2O) different from the formation of water from oxygen (O_2) and hydrogen (H^+)? ☑

3.4 Matter and Energy

The universe is made up of matter and energy. Recall that matter is *anything that occupies space and has mass*. By contrast, light, heat, movement, and electricity do not have mass, nor do they occupy space. These are the common forms of energy that you are familiar with. What do the various forms of energy have in common? They affect matter, causing changes in its *position* or *state*. For example, the release of energy in an explosion causes things to go flying—a change in position. Heating water causes it to boil and change to steam, a change of state. On a molecular level, changes of state are actually also movements of atoms or molecules. For instance, the degree of heat energy contained in a substance is a measure of the relative vibrational motion of the atoms and molecules of the substance. Therefore, we can define **energy** as *the ability to move matter*.

Energy Basics

Energy can be categorized as either kinetic or potential (Figure 3–12). **Kinetic energy** is *energy in action or motion*. Light, heat, physical motion, and electrical current are all forms of kinetic energy. **Potential energy** is *energy in storage*. A substance or system with potential energy has the capacity, or potential, to release one or more forms of kinetic energy. A stretched rubber band has potential energy; it can send a paper clip flying. Numerous chemicals, such as gasoline and other fuels, release kinetic energy—heat, light, and movement—when ignited. The potential energy contained in such chemicals and fuels is called **chemical energy**.

Energy may be changed from one form to another in innumerable ways. Besides understanding that potential energy can be converted to kinetic energy, it is especially important

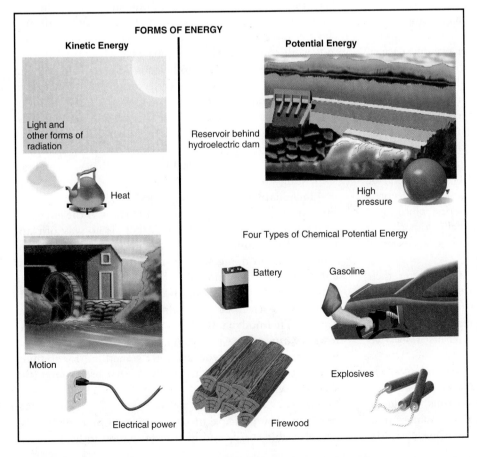

FORMS OF ENERGY

Kinetic Energy

Light and other forms of radiation

Heat

Motion

Electrical power

Potential Energy

Reservoir behind hydroelectric dam

High pressure

Four Types of Chemical Potential Energy

Battery

Gasoline

Firewood

Explosives

Figure 3–12 Forms of energy. Energy is distinct from matter in that it neither has mass nor occupies space. It has the ability to act on matter, though, changing the position or the state of the matter. Kinetic energy is energy in one of its active forms. Potential energy is the potential that systems or materials have to release kinetic energy.

Figure 3–13 **Energy can be transformed from one form to another.** Each time energy is transformed, some energy is transformed into heat (thermal) energy and is lost to the system's pool of usable energy. Heat dissipates from hotter to cooler areas, eventually radiating out into space. Thus energy transformations are never 100% efficient, except when the objective is to transform energy to heat.

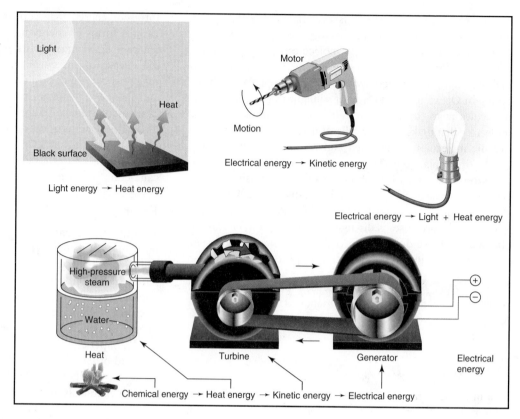

to recognize that kinetic energy can be converted to potential energy, illustrated in Figure 3–13. (Consider, for example, charging a battery or pumping water into a high-elevation reservoir.) We shall see later in this section that photosynthesis does just that.

Because energy does not have mass or occupy space, it cannot be measured in units of weight or volume. Instead, energy is measured in other kinds of units. One of the most common energy units is the **calorie**, which is defined as the amount of heat required to raise the temperature of 1 gram (1 milliliter) of water 1 degree Celsius. This is a very small unit, so it is frequently more convenient to use the kilocalorie (1 kilocalorie = 1,000 calories), the amount of thermal (heat) energy required to raise 1 liter (1,000 milliliters) of water 1 degree Celsius. Kilocalories are sometimes denoted as "Calories" with a capital "C." Food Calories, which measure the energy in given foods, are actually kilocalories. *Temperature* measures the molecular motion in a substance caused by the kinetic energy present in it.

If energy is defined as the ability to move matter, then no matter can be moved *without* the absorption or release of energy. Indeed, no change in matter—from a few atoms coming together or apart in a chemical reaction to a major volcanic eruption—can be separated from its respective change in energy.

Laws of Thermodynamics. Because energy can be converted from one form to another, numerous would-be inventors over the years have tried to build machines that produce more energy than they consume. One common

idea was to use the output from a generator to drive a motor that in turn would be expected to drive a generator to keep the cycle going and yield additional power in the bargain (a perpetual-motion machine). Unfortunately, all such devices have one feature in common: They don't work. When all the inputs and outputs of energy are carefully measured, they are found to be equal. There is no net gain or loss in total energy. This observation is now accepted as a fundamental natural law, the **Law of Conservation of Energy**. It is also called the **First Law of Thermodynamics**, and it can be expressed as follows: *Energy is neither created nor destroyed, but may be converted from one form to another.* What this law really means is that you can't get something for nothing.

Imaginative "energy generators" fail for two reasons: First, in every energy conversion, a portion of the energy is converted to heat energy. Second, there is no way of trapping and recycling heat energy without expending even more energy in doing so. Consequently, in the absence of energy inputs, any and every system will sooner or later come to a stop as its energy is converted to heat and lost. This is now accepted as another natural law, the **Second Law of Thermodynamics**, and it can be expressed as follows: *In any energy conversion, some of the usable energy is always lost.* Thus, you can't get something for nothing (the first law), and in fact, you can't even break even (the second law)!

Entropy. The principle of increasing entropy underlies the loss of usable energy and its transformation to heat energy.

Entropy is *a measure of the degree of disorder in a system,* so increasing entropy means increasing disorder. Without energy inputs, everything goes in one direction only—toward increasing entropy.

The conversion of energy and the loss of usable energy to heat are both aspects of increasing entropy. Heat energy is the result of the random vibrational motion of atoms and molecules. It is the lowest (most disordered) form of energy, and its spontaneous flow to cooler surroundings is a way for that disorder to spread. Therefore, the Second Law of Thermodynamics may be more generally stated as follows: *Systems will go spontaneously in one direction only— toward increasing entropy* (Figure 3–14). The second law also states that systems will go spontaneously only toward lower potential energy, a direction that releases heat from the systems.

It is possible to pump water uphill, charge a battery, stretch a rubber band, and compress air. All of these require the input of energy in order to increase the potential energy of a system. The verbs *pump, charge, stretch,* and *compress* reflect the fact that energy is being put into the system. In contrast, flow in the opposite direction (which releases energy) occurs spontaneously (Figure 3–15).

Whenever something gains potential energy, therefore, keep in mind that the energy is being obtained from somewhere else (the first law). Moreover, the amount of usable

Figure 3–14 Entropy. Systems go spontaneously only in the direction of increasing entropy. This spray can illustrates the idea. It took energy to concentrate a liquid in the can. Once it is dispersed, the energy is dissipated and entropy is increased.

energy lost from that "somewhere else" is greater than the amount gained (the second law). We now relate these concepts of matter and energy to organic molecules, organisms, ecosystems, and the biosphere.

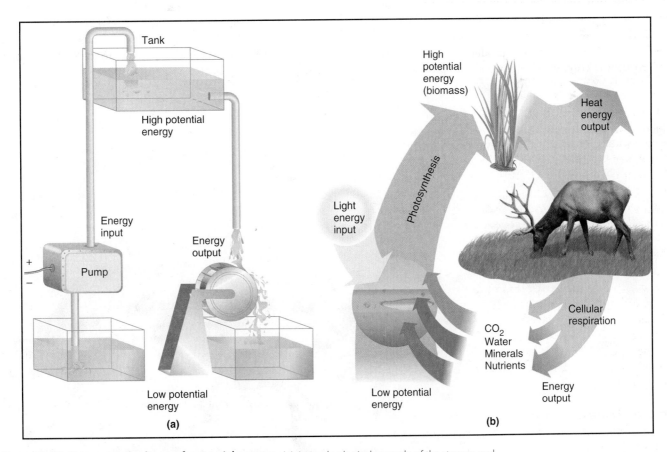

Figure 3–15 Storage and release of potential energy. (a) A simple physical example of the storage and release of potential energy. **(b)** The same principle applied to ecosystems.

Energy Changes in Organisms

Picture someone building a fire in a fireplace. The wood is largely composed of complex organic molecules. These molecules required energy to form, and therefore, their chemical bonds contain high potential energy. Lighting a fire causes a series of reactions that break these molecules and release energy. When released, the energy contained in those bonds can be used to do work. This chemical reaction involves **oxidation**, a loss of electrons, usually accomplished by the addition of oxygen (which causes *burning*). The heat and light of the flame are the potential energy being released as kinetic energy. In contrast, most inorganic compounds, such as carbon dioxide, water, and rock-based minerals, are nonflammable because they have very low potential energy. Thus, the production of organic material from inorganic material represents a gain in potential energy. Conversely, the breakdown of organic matter *releases* energy. Inside the cells of the emperor penguin eating squid, the reaction that takes place is similar to that of burning wood in a fireplace, except that it occurs in a set of small steps and is much more controlled.

Producers and Photosynthesis. The relationship between the formation and breakdown of organic matter, where energy is gained and released from chemical bonds, forms the basis of the energy dynamics of ecosystems. **Producers** (primarily plants and algae) make *high-potential-energy organic molecules* for their needs from *low-potential-energy raw materials* in the environment—namely, carbon dioxide, water, and a few dissolved compounds of nitrogen, phosphorus, and other elements (Figure 3–16). This "uphill" conversion is possible because producers use chlorophyll to absorb light energy, which "powers" the production of the complex, energy-rich organic molecules. On the other hand, **consumers** (all organisms that live on the production of others) obtain energy for movement and growth from *feeding on and breaking down organic matter made by producers* (Figure 3–17).

Plants and algae use the process of **photosynthesis** to make sugar (glucose, which contains stored chemical energy) from carbon dioxide, water, and light energy (Figure 3–16). This process, which also releases oxygen gas as a by-product, is described by the following chemical equation:

Photosynthesis
(An energy-demanding process)

$$6\,CO_2 \quad + \quad 12\,H_2O \quad \rightarrow \quad C_6H_{12}O_6 \ + \ 6\,O_2 \ + \ 6\,H_2O$$

carbon dioxide   water   light energy   glucose   oxygen   water
(gas)   (gas)
(low potential energy)   (high potential energy)

Chlorophyll in the plant cells absorbs the kinetic energy of light and uses it to remove the hydrogen atoms from water (H_2O) molecules. The hydrogen atoms combine with carbon atoms from carbon dioxide to form a growing chain of carbons that eventually becomes a glucose molecule. After the hydrogen atoms are removed from water, the oxygen atoms that remain combine with each other to form oxygen gas, which is released into the air. Water appears on both sides of the equation because 12 molecules are consumed and 6 molecules are newly formed during photosynthesis.

The key energy steps in photosynthesis take the hydrogen from water molecules and join carbon atoms together to form the carbon–carbon and carbon–hydrogen bonds of glucose. These steps convert the low-potential-energy bonds in water and carbon dioxide molecules to the high-potential-energy bonds of glucose.

Figure 3–16 Producers as chemical factories. Using light energy, producers make glucose from carbon dioxide and water, releasing oxygen as a by-product. Breaking down some of the glucose to provide additional chemical energy, they combine the remaining glucose with certain nutrients from the soil to form other complex organic molecules that the plant then uses for growth.

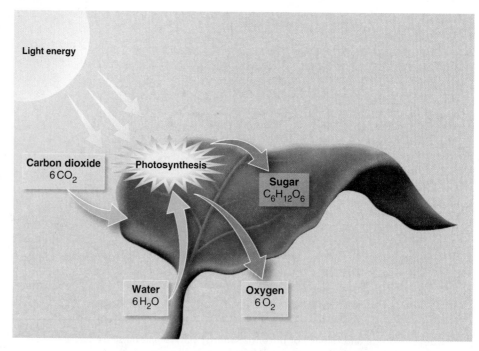

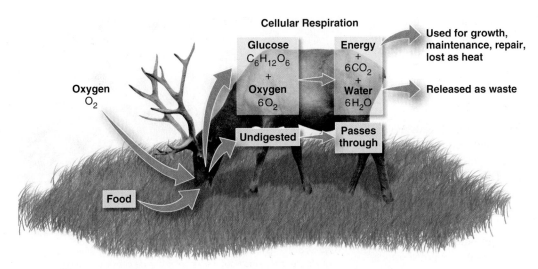

Figure 3–17 Consumers.
Only a small portion of the food ingested by a consumer is assimilated into body growth and repair. Some food provides energy in chemical bonds that are broken by the process of cellular respiration: this is the reverse process of photosynthesis, and waste products are carbon dioxide, water, and various mineral nutrients. A third portion is not digested and becomes fecal waste.

Within the Plant. The glucose produced in photosynthesis serves three purposes in the plant: (1) Glucose, with the addition of other molecules, can be the backbone used for making all the other organic molecules (proteins, carbohydrates, and so on) that make up the stem, roots, leaves, flowers, and fruit of the plant. (2) It takes energy to run the cellular activities of plants, such as growth. This energy is obtained when the plant breaks down a portion of the glucose to release its stored energy in a process called *cellular respiration*. (3) A portion of the glucose produced may be stored for future use. For storage, the glucose is generally converted to starch, as in potatoes or grains, or to oils, as in many seeds. This stored energy is what is available to be eaten by consumers (Figure 3–17).

None of these reactions, from the initial capture of light by chlorophyll to the synthesis of plant structures, takes place automatically. Each step is catalyzed by specific **enzymes**, proteins that promote the synthesis or breaking of chemical bonds. The same is true of cellular respiration.

Cellular Respiration. Inside each cell, organic molecules may be broken down through a process called **cellular respiration** to release the energy required for the work done by that cell. Most commonly, cellular respiration involves the breakdown of glucose, and the overall chemical equation is basically the reverse of that for photosynthesis:

Cellular Respiration
(An energy-releasing process)

$$C_6H_{12}O_6 + 6\,O_2 \rightarrow 6\,CO_2 + 6\,H_2O + \text{energy}$$

glucose   oxygen   carbon dioxide water
(high potential energy)   (low potential energy)

The purpose of cellular respiration is to release the potential energy contained in organic molecules to perform the activities of the organism. Note that oxygen is *released* in photosynthesis but *consumed* in cellular respiration to break down glucose to carbon dioxide and water. All organisms need to perform cellular respiration, even producers like plants and algae. In order to have the oxygen they need, animals such as deer grazing on grass (Figure 3–17) must obtain oxygen through some type of respiratory organ. This means that their ability to use food depends on the availability not only of food, but also of oxygen. Many animals, such as deer, use lungs to obtain oxygen and release carbon dioxide. Oxygen is more limited in water, and fish use gills to obtain it, but many aquatic places are severely oxygen limited.

In keeping with the Second Law of Thermodynamics, converting the potential energy of glucose to the energy required to do the body's work is not 100% efficient. The energy released from glucose is stored in molecules that work like compressed springs—that is, they store the energy so that the body can use it in cells later. But the number of "energy units" that come from respiration is not nearly as high as the amount of solar energy required for the photosynthesis of glucose. Cellular respiration is only 40–60% efficient. This is a great example of the Second Law of Thermodynamics at work. The rest of the energy is released as waste heat, and this is the source of body heat. This heat output can be measured in animals (cold-blooded or warm-blooded) and in plants.

Gaining Weight. The basis of weight gain or loss becomes apparent in this context. If you consume more calories from food than your body needs, the excess may be converted to fat and stored, and you gain weight as a result. In contrast, the principle of dieting is to eat less and exercise more, to create an energy demand that exceeds the amount of energy contained in your food.

The overall reaction for cellular respiration is the same as that for simply burning glucose. Thus, it is not uncommon to speak of "burning" our food for energy. There are other ways of releasing stored energy from food that do not require oxygen. Such anaerobic respiration includes fermentation, as in the process used to make wine. Yeast cells use anaerobic respiration to convert grape sugars into usable energy, releasing alcohol as a by-product. However, these methods are less efficient than oxidation, and organisms that use them do not thrive unless oxygen is severely limited.

Photosynthesis and cellular respiration are necessary for life on Earth to survive. Next we will see how the energy from the Sun that fuels photosynthesis moves through Earth's systems.

One-Way Flow of Energy

What happens to all the solar energy entering ecosystems? Most of it is absorbed by the atmosphere, oceans, and land, heating them in the process. The small fraction (2–5%) captured by living plants is either passed along to the consumers and **detritivores** (organisms that live on dead or decaying organisms) that eat them or degraded into the lowest and most disordered form of energy—heat—as the plants decompose. Eventually, all of the energy entering ecosystems escapes as heat. According to the laws of thermodynamics, no energy will actually be lost. So many energy conversions are taking place in ecosystem activities, however, that entropy is increased, and all the energy is degraded to a form unavailable to do further work. That heat energy may stay in the atmosphere for some period of time, but will eventually be re-radiated out into space. The ultimate result is that *energy flows in a one-way direction through ecosystems* and eventually leaves Earth; it is not lost to the universe, but it is no longer available to ecosystems, so it must be continually resupplied by sunlight.

Energy flow is one of the two fundamental processes that make ecosystems work. In contrast with the *flow* of energy, we talk about the *cycling* of nutrients and other elements, which are continually reused from those already available on Earth.

CONCEPT CHECK Explain how the growth of a tree can be a part of carbon storage although trees are eaten by herbivores. ☑

3.5 The Cycling of Matter in Ecosystems

Earlier we saw that the molecules that make up cells, which in turn make up tissues, contain certain key elements. For example, all organic molecules contain carbon; photosynthesis requires water (hydrogen and oxygen) and carbon dioxide (carbon and oxygen); potassium is part of the energy-holding mechanism in each cell; nitrogen is contained in every protein; and sulfur bonds help proteins stay in the right shape. There are many other necessary nutrients. Some are only necessary in very small amounts. For example, to make the protein hemoglobin in the blood of animals like the penguin, small amounts of iron are needed.

According to the Law of Conservation of Matter, atoms cannot be created or destroyed, so recycling is the only possible way to maintain a dynamic system. To see how recycling takes place in the biosphere, we now focus on the pathways of four key elements heavily affected by human activities: carbon, phosphorus, nitrogen, and sulfur. Because these pathways are circular and involve biological, geological, and

Figure 3–18 The lithosphere as a nutrient sink. When plants and animals die, elements from their bodies can be converted to rocks and stored underground. The carbon and the fossilized leaves in this piece of coal came from plants that grew in lush forests during the Coal Age, about 300 million years ago.

chemical processes, they are called **biogeochemical cycles.** Some processes in biogeochemical cycles, such as photosynthesis and respiration, occur rapidly. Others, such as the formation of coal from the remains of plants, take hundreds of millions of years (**Figure 3–18**). (The water, or hydrologic, cycle is covered in Chapter 10).

The Carbon Cycle

The global carbon cycle is illustrated in **Figure 3–19**. Boxes represent major *pools* of carbon, and arrows represent the movement, or flux, of carbon from one compartment to another. For descriptive purposes, it is convenient to start the carbon cycle with the *reservoir* of carbon dioxide (CO_2 molecules present in the air). Through photosynthesis and further metabolism, carbon atoms from CO_2 become the carbon atoms of the organic molecules making up a plant's body. The carbon atoms can then be eaten by an animal, such as the deer, and become part of the tissues of all the other organisms in the ecosystem. About half of the carbon-containing molecules are respired by plants and animals, and half are deposited to the soil (a large reservoir) in the form of **detritus** (dead plant and animal matter). Respiration by organisms in the soil that eat dead matter returns more carbon to the atmosphere (as CO_2).

The cycle is different in the oceans: Photosynthesis by phytoplankton and macroalgae removes CO_2 from the huge pool of inorganic carbonates in seawater, and feeding moves the organic carbon through marine food webs. Respiration by the biota returns the CO_2 to the inorganic carbon compounds in solution.

The carbon cycle includes several other important processes. The figure indicates two in particular: (1) diffusion exchange between the atmosphere and the oceans; and (2) the combustion of fossil fuels, which releases CO_2 into the atmosphere. Some geological processes of the carbon cycle are not shown in Figure 3–19. For example, the fossilization of dead plants and animals produced coal deposits in many

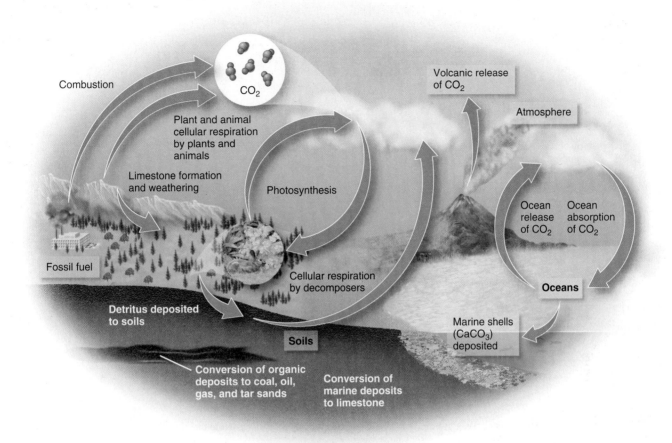

Figure 3–19 The global carbon cycle. Boxes in the figure refer to pools of carbon, and arrows refer to the movement, or flux, of carbon from one pool to another.

areas of Earth. This process removed vast amounts of carbon dioxide from the atmosphere and trapped it underground. Burning the coal and oil created by this process releases the CO_2 to the atmosphere. For another example, limestone (such as that formed by ancient corals) also keeps carbon out of circulation; however, the weathering of exposed limestone releases carbon into the aquatic system.

Because the total amount of carbon in the atmosphere is about 854 Gt (gigatons, or a billion metric tons) and photosynthesis in terrestrial ecosystems removes about 150–175 Gt/year, a carbon atom cycles from the atmosphere through one or more living things and back to the atmosphere about every five or six years.

Human Impacts. Human intrusion into the carbon cycle is significant. As we will see shortly, we are diverting (or removing) 40% of the photosynthetic productivity of land plants to support human enterprises. By burning fossil fuels, we have increased atmospheric carbon dioxide by 35% over preindustrial levels (Chapter 18). In addition, until the mid-20th century, deforestation and soil degradation released significant amounts of CO_2 into the atmosphere. However, more recent reforestation and changed agricultural practices have improved this somewhat.

The Phosphorus Cycle

The phosphorus cycle is similar to the cycles of other mineral nutrients—those elements that have their origin in the rock and soil minerals of the lithosphere, such as iron, calcium, and potassium. We focus on phosphorus because its shortage tends to be a limiting factor in a number of ecosystems and its excess can stimulate unwanted algal growth in freshwater systems.

The phosphorus cycle is illustrated in **Figure 3–20**, on the next page. Like the carbon cycle, it is depicted as a set of pools of phosphorus and fluxes to indicate key processes. Phosphorus exists in various rock and soil minerals as the inorganic ion phosphate (PO_4^{3-}). As rock gradually breaks down, phosphate and other ions are released in a slow process. Plants absorb PO_4^{3-} from the soil or from a water solution. Once the phosphate is incorporated into organic compounds by the plant, it is called **organic phosphate**. Moving through food chains, organic phosphate is transferred from producers to the rest of the ecosystem. As with carbon, at each step it is highly likely that the organic compounds containing phosphate will be broken down in cellular respiration or by decomposers, releasing PO_4^{3-} in urine or other waste material. Phosphate is then reabsorbed by plants to start the cycle again.

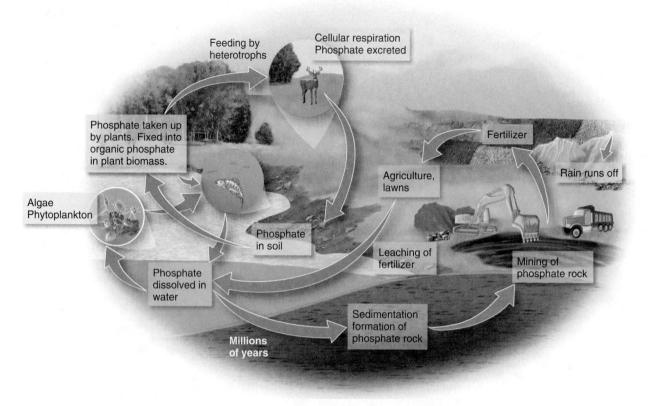

Figure 3–20 The global phosphorus cycle. Phosphorus is a limiting factor in many ecosystems. Note that the cycle is not connected to the atmosphere, which limits biosphere recycling.

Phosphorus enters into complex chemical reactions with other substances that are not shown in this simplified version of the cycle. For example, PO_4^{3-} forms solid, insoluble compounds with a number of cations (positively charged ions), such as iron (Fe^{3+}), aluminum (Al^{3+}), and calcium (Ca^{3+}). Phosphorus can bind with these ions to form chemical precipitates (solid, insoluble compounds) that are largely unavailable to plants. The precipitated phosphorus can slowly release PO_4^{3-} as plants withdraw naturally occurring PO_4^{3-} from soil, water, or sediments.

Human Impacts. The most serious human intrusion into the phosphorus cycle comes from the use of phosphorus-containing fertilizers. Phosphorus is mined in several locations around the world and is then made into fertilizers, animal feeds, detergents, and other products. Phosphorus is a common limiting factor in soils and, when added to croplands, can greatly stimulate production.

Unfortunately, human applications have tripled the amount of phosphorus reaching the oceans. This increase is roughly equal to the global use of phosphorus fertilizer in agriculture. Humans are accelerating the natural phosphorus cycle as it is mined from the earth and as it subsequently moves from the soil into aquatic ecosystems, creating problems as it makes its way to the oceans. There is essentially no way to return this waterborne phosphorus to the soil, so the bodies of water end up overfertilized. This leads in turn

to a severe water pollution problem known as eutrophication (Chapter 17). Eutrophication can cause the overgrowth of algae and bacteria and the death of fish.

The Nitrogen Cycle

The nitrogen cycle (Figure 3–21) has similarities to both the carbon cycle and the phosphorus cycle. Like carbon, nitrogen has a gas phase; like phosphorus, nitrogen acts as a limiting factor in plant growth. Like phosphorus, nitrogen is in high demand by both aquatic and terrestrial plants.

The nitrogen cycle is otherwise unique. Most notably, unlike in the other cycles, bacteria in soils, water, and sediments perform many of the steps of the nitrogen cycle.

The main reservoir of nitrogen is the air, which is about 78% nitrogen gas (N_2). This form of nitrogen is called **nonreactive nitrogen**; most organisms are not able to use it in chemical reactions. The remaining forms of nitrogen are called **reactive nitrogen** (Nr) because they are used by most organisms and can make chemical reactions more easily.

Plants in terrestrial ecosystems ("non-N-fixing producers" in Figure 3–20) take up Nr as ammonium ions (NO_4^+) or nitrate ions (NO_3^-). The plants incorporate the nitrogen into essential organic compounds such as proteins and nucleic acids. The nitrogen then moves from producers through consumers and, finally, to decomposers. At various points, nitrogen wastes are released, primarily as ammonium compounds.

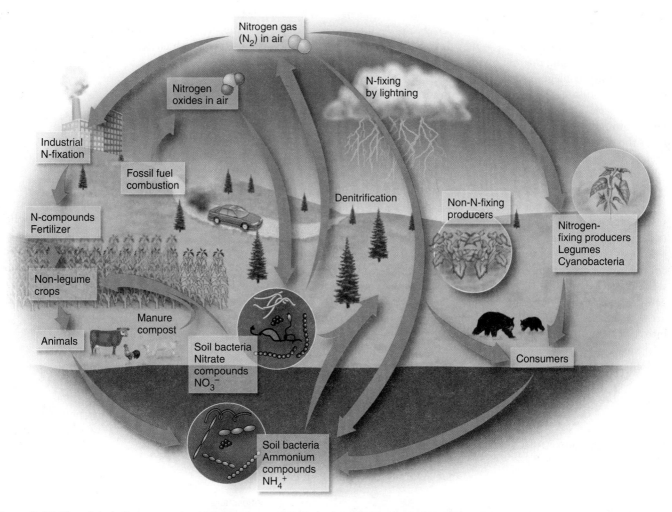

Figure 3–21 The global nitrogen cycle. Like phosphorus, nitrogen is often a limiting factor. Its cycle heavily involves different groups of bacteria.

A group of soil bacteria, the *nitrifying bacteria*, oxidizes the ammonium to nitrate in a process that yields energy for the bacteria. At this point, the nitrate is again available for uptake by green plants—a local ecosystem cycle within the global cycle. In most ecosystems, the supply of Nr is quite limited, yet there is an abundance of nonreactive N—if it can be accessed. The nitrogen cycle in aquatic ecosystems is similar.

Nitrogen Fixation. A number of bacteria and cyanobacteria (chlorophyll-containing bacteria, formerly referred to as blue-green algae) can use nonreactive N through a process called biological **nitrogen fixation**. In terrestrial ecosystems, the most important among these nitrogen-fixing microbes live in nodules on the roots of legumes, the plant family that includes peas and beans. The legume provides the bacteria with a place to live and with food (carbohydrates and proteins) and gains a source of nitrogen in return. From legumes, nitrogen enters the food web. The legume family includes a huge diversity of plants. Without them, plant production would be sharply impaired due to a lack of available nitrogen.

Three other important processes also "fix" nitrogen. One is the conversion of nitrogen gas to the ammonium form by discharges of lightning in a process known as *atmospheric nitrogen fixation*: the ammonium then comes down with rainfall. The second is the *industrial fixation* of nitrogen in the manufacture of fertilizer: the Haber-Bosch process, which converts nitrogen gas and hydrogen to ammonia, is the main source of agricultural fertilizer. The third is a consequence of the *combustion of fossil fuels*, during which nitrogen from coal and oil is oxidized; some nitrogen gas is also oxidized during high-temperature combustion. These last two processes lead to nitrogen oxides (NO_x) in the atmosphere, which are soon converted to nitric acid and then brought down to Earth as acid precipitation (Chapter 19).

Denitrification. **Denitrification** is a microbial process that occurs in soils and sediments depleted of oxygen. A number of microbes can take nitrate (which is highly oxidized) and use it as a substitute for oxygen. In so doing, the nitrogen is reduced (it gains electrons) to nitrogen gas and released back into the atmosphere. In sewage treatment systems, denitrification is promoted to remove nitrogen from wastewater before it is released into the environment (Chapter 20).

Human Impacts. Human involvement in the nitrogen cycle is substantial. Many agricultural crops are legumes (peas, beans, soybeans, alfalfa), so they draw nitrogen from the air, thus increasing the rate of nitrogen fixation on land. Crops that are nonleguminous (corn, wheat, potatoes, cotton, and so on) are heavily fertilized with nitrogen derived from industrial fixation. Both processes benefit human welfare profoundly. Also, fossil fuel combustion fixes nitrogen from the air (Figure 3–19). However, human actions more than double the rate at which nitrogen is moved from the atmosphere to the land.

The consequences of this global nitrogen fertilization are serious. Nitric acid (and sulfuric acid, produced when sulfur is also released by burning fossil fuels) has destroyed thousands of lakes and ponds and caused extensive damage to forests (Chapter 19). Nitrogen oxides in the atmosphere contribute to ozone pollution, global climate change, and stratospheric ozone depletion (Chapter 18). Surplus nitrogen has led to the "nitrogen saturation" of many natural areas, whereby nitrogen can no longer be incorporated into living matter and is released into the soil. Washed into surface waters, nitrogen makes its way to estuaries and coastal areas of oceans, where, just like phosphorus, it triggers a series of events leading to eutrophication, resulting in dead seafood, detriments to human health, and areas of oceans that are unfit for fish (Chapter 20).

This complex series of ecological effects is known as the **nitrogen cascade**, in recognition of the sequential impacts of Nr as it moves through environmental systems, creating problems as it goes.

The Sulfur Cycle

Sulfur is the last element we will investigate. Sulfur is important to living things because it is a component of many proteins, hormones, and vitamins. Sulfur is often linked in nature with oxygen atoms, as in sulfate (SO_4) (Figure 3–22). Most of Earth's sulfur is found in rocks and minerals, including deep ocean sediments. The weathering of rocks and volcanic activity sends sulfur into the atmosphere or soil. Sulfur also gets into the air when fossil fuels are burned and when mined metals are processed. In soil, sulfate can be taken up by plants and microorganisms. In air, sulfur dioxide (SO_2) contributes to acid rain when it combines with water vapor and forms sulfuric acid. Sulfates are added to water bodies as sulfur compounds fall from the atmosphere or are weathered from rocks. There they can be serious pollutants.

Human Impacts. The largest human impact of the sulfur cycle is the addition of sulfur oxides to the atmosphere and the addition of sulfates to water. Unlike phosphorus and

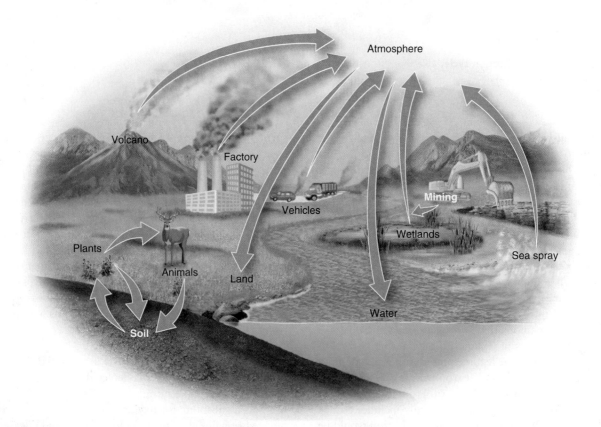

Figure 3–22 The global sulfur cycle. Sulfur spends little time in the atmosphere. Part of the cycle does rely on bacteria, but less than nitrogen's cycle. Like all three other cycles, parts of the sulfur cycle are sped up by human activity.

SUSTAINABILITY
Planetary Boundaries

Ancient peoples thought that Earth had an edge—if you took a ship too far, perhaps you and your crew would fall right off. Today, some scientists suggest that Earth has a metaphorical edge: a boundary where use of a particular resource becomes so great that it alters natural cycles and changes ecosystems. Reaching a point where dramatic changes are likely is called the tipping point, and the levels of change at which this might occur are called planetary boundaries, a term coined by Johann Rockström and his colleagues at Stockholm University, who published a paper entitled "Planetary Boundaries: Exploring the Safe Operating Space for Humanity." In that paper, they estimate that a safe boundary for nitrogen would be to limit industrial and agricultural fixation of nitrogen to 35 teragrams of nitrogen per year.[1] (A teragram is one billion kilograms.)

The innovation of the Haber-Bosch process for fixing nitrogen allowed humans to produce more food on the same amount of land than was possible previously. One estimate is that without nitrogen fertilizer, we would have needed four times the agricultural land in 2000 that we did in 1900 and half of all continental land would be needed for agriculture. The downside of so much fertilizer is that it washes into oceans, where it and other pollutants produce "dead zones" where oxygen is limited and most life cannot survive. Likewise, the use of fossil fuels has brought advantages to humans, but the acid rain created by the burning of these fuels kills lake fish.

Humanity might be reaching planetary boundaries or tipping points—where we suddenly experience ecological change that is difficult to reverse and that causes great changes to our ability even to feed ourselves. Just as a ship might have a lookout, scientists have an important role in helping us understand the world so that we can recognize when we are approaching or have passed planetary boundaries. If we have a way to estimate the consequences of what we are trying to do, such as this paper suggests, then both scientists and others will be involved in measuring those consequences and finding solutions for problems that arise. One place to start is to calculate your own nitrogen footprint. Just like a carbon footprint, a nitrogen footprint is calculated from your activities and tells you something about your own effect on the globe. You can find resource footprint calculators by searching on the Web. Scientists play a role in helping to make more efficient use of fertilizer and to make technologies to remove nitrogen from smoke in the burning of fossil fuels. These advancements, coupled with the will of individuals to consume less and be wise consumers, are parts of the solution to avoid taking our global ship across a planetary boundary.

---

[1]Rockström, J. et al., 2009. "Planetary Boundaries: Exploring the Safe Operating Space for Humanity." *Ecology and Society* 14(2): 32.

---

nitrogen, sulfur is not usually directly added to soils in order to improve their fertility. However, soil and water sulfate levels are increased by human action. Acid rain and water pollution are the major effects. For example, in heavily developed areas of the Everglades, the percent of sulfates in water is 60 times what it would normally be expected to be. Sulfate compound aerosols (small particles or drops) also play a role in the climate. They act to temporarily cool the atmosphere, although the compounds cause other problems when they fall to earth.

Comparing the Cycles

The four cycles we have looked at in depth differ in some important ways (Table 3–2). Carbon is found in large amounts in the atmosphere in a form that can be directly taken in by plants, so carbon is rarely the limiting factor in the growth of vegetation. Both nitrogen and phosphorus are often limiting factors in ecosystems. Phosphorus has no gaseous atmospheric component (though it can be found in airborne dust particles) and thus, unless added to an ecosystem artificially, enters very slowly. Nitrogen is unique because of the importance of bacteria in driving the cycle forward. Sulfur can get into the atmosphere, but only for a short time. It is also different from the others because of its concentration in mining runoff.

All four cycles have been sped up considerably by human actions. Nitrogen compounds in the atmosphere contribute to acid rain, and carbon dioxide is being moved

Table 3–2	Major Characteristics of the Carbon, Phosphorus, and Nitrogen Cycles		
Nutrient	**Major Source**	**Interesting Feature**	**Human Impact**
Carbon	Air	Taken in directly by plant leaves	Burning fuel moves it to air from underground
Phosphorus	Rock	No atmospheric component	Fertilizer use adds it to waterways
Nitrogen	Soil/Air	Bacteria drive the cycle	Fertilizer moves it to soil, burning moves it to air
Sulfur	Rock	Spends only a short time in the atmosphere	Burning moves it to air, rain and mining move it to soil and water

from underground storage in carbon molecules to the atmosphere, where it acts as a greenhouse gas. Both nitrogen and phosphorus are put on soil as fertilizer or get into water from sewage and runoff. Both act in water to promote overgrowth of algae (Chapter 20). Sulfur adds to acid rain and to water pollution.

Although we have focused on carbon, phosphorus, nitrogen, and sulfur, cycles exist for oxygen, hydrogen, iron, and all the other elements that play a role in living things. While the routes taken by distinct elements may differ, all of the cycles are going on simultaneously, and all come together in the tissues of living things. As elements cycle through ecosystems, energy flows in from the Sun and through the living members of the ecosystems. The links between these two fundamental processes of ecosystem function are shown in Figure 3–23.

In this chapter, we introduced the various fields of ecology and showed that the science of ecology encompasses living things and their relationships with each other and the environment. We looked at two ecological levels

in greater detail in this chapter—what is happening with individuals (organismal ecology) and what is happening to the large scale movement of matter and energy through the biosphere (ecosystem ecology). To understand the flow of matter and energy, we also included some background on the basics of atoms, molecules, and the laws of energy. Later, we will investigate the other subdisciplines of ecology introduced here, especially population ecology and community ecology.

CONCEPT CHECK Scientists believe that human activities have put the nitrogen cycle at a tipping point. List three ways that humans change the nitrogen cycle and three effects of those changes. ☑

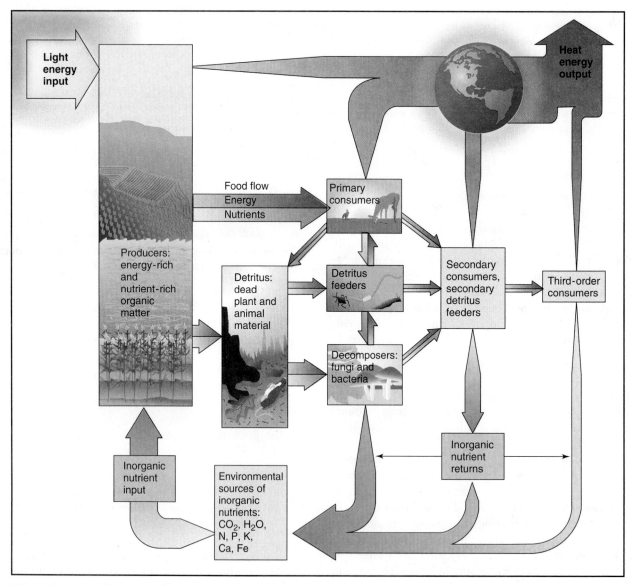

Figure 3–23 **Nutrient cycles and energy flow.** The movement of nutrients (blue arrows) and energy (red arrows) through an ecosystem. Nutrients follow a cycle, being used over and over. Light energy absorbed by producers is released and lost as heat energy as it is "spent."

REVISITING THE THEMES

SOUND SCIENCE

Ecology is the science of living things and the abiotic and biotic components interacting with them. Ecologists can study on a variety of levels. In this chapter, we looked at organismal ecology and ecosystem ecology—that is, how materials work in individual animals and how matter and energy flow throughout the world.

Both matter and energy are limited. Matter cycles in the form of molecules (sets of atoms) through living things and the abiotic environment. Energy flows in one direction—from the Sun, through activities on Earth, and back out into space. Using the emperor penguin and its food (fish and squid) as our examples, we demonstrated basic principles of organization within their bodies and the flow of energy and materials through their systems. The emperor penguin has a specific niche, or set of parameters, within which it needs to live. These resources and conditions include the amount of nesting areas and food it needs. Many of the penguin's prey fish eat algae, which contain chlorophyll, which is necessary for photosynthesis. In this chemical process, inorganic building blocks are turned into more-complex organic molecules whose bonds contain more chemical energy than their original components do. This energy can be released when zooplankton eat the algae and are in turn eaten by the fish and ultimately the penguin. The bodies of each of these organisms burn those molecules through a controlled process of cellular respiration. Energy is lost at each step of this energy path. In order to build the bodies of the algae, zooplankton, fish, and penguins, a number of nutrients need to be available. The cycles for four of these (carbon, phosphorus, nitrogen, and sulfur) were described. Humans affect these biogeochemical cycles by actions such as the burning of fossil fuels and the fertilization of agricultural land.

SUSTAINABILITY

The science in this chapter that leads to specific management actions is primarily that surrounding the needs of individual species in terms of habitat and niche, the consequences of limits on nutrients and energy, and the impacts of human effects on biogeochemical cycles. These ideas will be more thoroughly explored later. For instance, the science surrounding habitat and niche underlies our understanding of why some types of organisms are particularly vulnerable to extinction (Chapter 4). The science underlying the nitrogen and phosphorus cycles is important to management decisions about water pollution (Chapter 20). Finally, the science about carbon and nitrogen cycles aids in our understanding of air pollution and global climate change (Chapters 18 and 19). The sustainability essay provides a good example of how science is used to identify boundaries, and this can be used to manage our global environment—in this case, the nitrogen cycle.

The emperor penguin helps us look at basic processes within organisms and the flow of matter and energy. Later, we will look at ways individuals relate to members of their own species (population ecology) and ways species relate to other species (community ecology).

STEWARDSHIP

The emperor penguin was a focus throughout this chapter. It has a particular habitat, which is changing because of human activities. As stewards, researchers are trying to predict what will happen to penguins as sea ice and fish populations change. Human activity also speeds up biogeochemical cycles, and this can have global effects. All of us have to be stewards not only of the nutrients in our world, but also of the natural processes that move chemicals from one part of a cycle to another. Stewardship involves protecting not only resources but ecosystem functions. We will look at some of the ethical and justice implications of these scientific realities in other chapters. We will ask, "Why protect other species?" (Chapter 6). We will also look at the implications of energy loss at every transfer and apply those principles to human population growth and the need for food (Chapter 8).

REVIEW QUESTIONS

1. Distinguish between the biotic community and the abiotic environmental factors of an ecosystem.

2. Define and compare the terms *species*, *population*, and *ecosystem*.

3. Compared with an ecosystem, what are an ecotone, a landscape, a biome, and a biosphere?

4. How do the terms *organic* and *inorganic* relate to the biotic and abiotic components of an ecosystem?

5. What are the six key elements in living organisms?

6. What features distinguish between organic and inorganic molecules?

7. In one sentence, define *matter* and *energy*, and demonstrate how they are related.

8. Give four examples of potential energy. In each case, how can the potential energy be converted into kinetic energy?

9. State two energy laws. How do they relate to entropy?

10. What is the chemical equation for photosynthesis? Examine the origin and destination of each molecule referred to in the equation. Do the same for cellular respiration.

11. Describe the biogeochemical cycle of carbon as it moves into and through organisms and back to the environment. Do the same for phosphorus, nitrogen, and sulfur.

12. What are the major human impacts on the carbon, phosphorus, and nitrogen cycles?

THINKING ENVIRONMENTALLY

1. From local, national, and international news, compile a list of ways humans are altering abiotic and biotic factors on a local, regional, and global scale. Analyze ways in which local changes may affect ecosystems on larger scales and ways in which global changes may affect ecosystems locally.

2. Use the laws of conservation of matter and energy to describe the consumption of fuel by a car. That is, what are the inputs and outputs of matter and energy? (*Note:* Gasoline is a mixture of organic compounds containing carbon–hydrogen bonds.)

3. Using your knowledge of photosynthesis and cellular respiration, draw a picture of the hydrogen cycle and the oxygen cycle. (*Hint:* Consult the four cycles in the book for guidance.)

4. Look up everything you can find about iron seeding of the oceans, and decide whether you think the advantages outweigh the disadvantages.

MAKING A DIFFERENCE

1. Find the closest farm that does not use artificial fertilizer. Many are listed on Web sites under the terms "community supported agriculture" or "CSA." See if you can arrange a visit. See what they do to lower their impact on the nitrogen and phosphorus cycles and still maintain fertile soil.

2. Visit the EPA's Web site to find out how you can help protect your local watershed. Its project database allows you to get involved in monitoring, cleanups, and restorations in your area.

3. Take one of the biogeochemical cycles and try to track your impact on it. This is most easily done with carbon. Search on the Web under "carbon footprint."

Savanna
elephant
range

Kenya

Populations and Communities

African savanna elephants (*Loxodonta africana africana*) saunter through the glaring sun of the East African savanna. Trunks swaying side to side, they seek out shrubs and grasses. Nearby, the indigenous people graze their herds of rangy cattle on the sparse, tough grasses. These pastoralists view the elephants as competitors that reduce the amount of food available to their cattle. The herders resent the elephants' use of the land, while the villagers fear that the elephants will trample their crops.

Current Threats. Like many large, slow-growing animals, these elephants are increasingly rare. Numbering from 3 to 5 million at the start of the 20th century, today there are only 450,000 to 700,000. Loss of habitat is one reason for their decline. In addition, poachers kill them for their ivory, even though the ivory trade is illegal. As climate change increases drought, elephants have to migrate farther and face more conflicts with humans than before. In 2014, a team of researchers from Elephants Without Borders began using light aircraft to document every group of elephants in the 13 sub-Saharan countries where 90% of savanna elephants live. Their initial findings documented the heavy poaching of elephants in numerous places.

While the number of elephants is declining, another native species, the aggressive Sodom apple (*Solanum campylacanthum*), is increasing. Unfortunately for the herders, this weedy shrub is toxic to grazing animals such as cattle, sheep, and zebras. In Kenya, the eradication of the noxious Sodom apple is costing the government millions.

A Win-Win Solution? In spite of the problems caused by the ongoing changes in East Africa, there is a small amount of good news. Unlike grazing animals, browsers such as elephants and impalas can eat the Sodom apple. In fact, elephants have a voracious appetite for the shrub, which may provide a

▲ African savanna elephants can eat plants that are unpalatable to sheep and cattle.

win-win situation for pastoralists and conservationists. New research shows that the presence of elephants and impalas increases the availability of grass in some places where domestic and wild animals coexist.

The African elephant is a good example of the problems a declining species can face. Under changing environmental conditions, a species may adapt or find a new niche to exploit.

In this case, eating an invasive species opens up new resources for the elephant. But environmental changes may also put species in danger of extinction. In this chapter, we will look at how populations grow and what types of limitations populations can face. We will also see how populations of different species interact with each other, how species change over time, and how human activities can affect populations and groups of species.

4.1 Dynamics of Natural Populations

Like the African elephant in the savanna, each species in an ecosystem exists as a population; that is, each exists as a re-producing group. A **population** is a group of members of the same species living in an area, and the populations of different species living together in one area constitute a **community** (Chapter 3). Ecologists often want to know the pattern of change in the numbers of individuals in a population. In any population, this is easy to figure out if you have enough information. Births and immigration cause the population to grow, and deaths and emigration cause the population to shrink. This can be described by a simple equation:

(Births + Immigration) − (Deaths + Emigration)
= Change in population

This equation shows the change in the size of the population, or **population growth**. If births plus immigration are more or less equal to deaths plus emigration, the population is said to be in **equilibrium**. There is no change in the size of the population, so the population growth is zero. But often, populations are either increasing or decreasing—that is, their population growth is not zero. If you take the amount a population has changed and divide it by the time it had to change, you get the **population growth rate**. Imagine there is a very small population just coming into a new ecosystem. A scientist might want to know how many individuals there are right now, how fast the population will grow, or even what the future population size will be, using the current population and the growth rate. In order to calculate these parameters, scientists use population growth curves—graphs of how populations grow. We will look at several examples that represent broad types of growth patterns.

Population Growth Curves

In a natural setting, the growth of a population can follow several patterns. A constant rate of population growth is the simplest type of growth to model. Although constant growth rarely occurs in nature, it serves as a good comparison to other growth patterns because it is so simple. Constant population growth is calculated by this equation:

Starting population + (A constant × Time)
= Ending population

That is, the population at the end of a period of time is equal to the population at the beginning plus a constant multiplied by the number of time units. The blue line in **Figure 4–1** illustrates constant population growth.

Exponential Increase. Every species has the capacity to increase its population when conditions are favorable. For example, a beetle population growing in a bag of cereal could double every week (the first week 2, then 4, then 8, then 16 beetles). Such a series is called an **exponential increase**. In contrast to our first example of constant growth, exponential growth does not add a constant number of individuals to the population for each time period. Instead, the doubling time of the population remains constant. That is, if it takes two days to go from 8 beetles to 16 beetles, it will take two days to go from 1,000 beetles to 2,000 beetles. Such growth results in what is called a **population explosion**.

If we plot numbers over time during an exponential increase, the pattern produced is commonly called a **J-curve** because of its shape (see the green line in Figure 4–1). To calculate the growth rate of a population undergoing exponential growth, you need to know the current size of the population, the period of time over which the population will grow, and the number of offspring individuals can

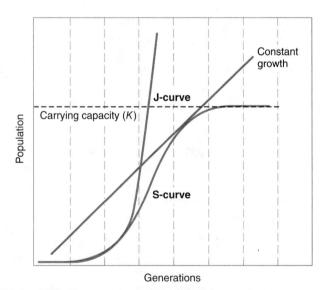

Figure 4–1 Three simple models of population growth.
Constant growth is represented by the blue line, exponential growth by the green line, and logistic growth by the orange line. The black dotted line represents the carrying capacity.

produce in a given amount of time if resources are unlimited. This last value is often designated r_{max}, or just r. The equation for exponential growth can be described as follows:

Starting population × A constant (e) multiplied by
    itself a certain number of times = Ending population

The number of times you multiply e by itself is equal to the value of r (the potential to produce offspring) multiplied by time. This means that under unlimited conditions, populations of organisms with a high r value will grow more rapidly than those with a low r value. Exponential growth can continue indefinitely only if there are no limits to the size the population can reach, but in nature this is extremely rare.

There are a few times when exponential growth occurs in nature. A beetle population getting into a big sack of cereal is a good example of a population explosion. However, exponential growth cannot continue indefinitely. There is an upper limit to the population of any particular species that an ecosystem can support. This limit is called the **carrying capacity** (K) (Figure 4–1). More precisely, the carrying capacity is the maximum population of a species that a given habitat can support without being degraded over the long term—in other words, a sustainable system.

Logistic Growth. As a population increases, one of two outcomes will eventually occur. One outcome is that the population will keep growing until it exhausts its essential resources, often food. Starvation or other problems will then cause the population to crash, or die off precipitously. Such a pattern of growth can be plotted as a J-curve followed by a steep drop, as shown by the green line in Figure 4–2.

Alternatively, some processes may slow population growth so that the size of the population levels off. The simplest version of this pattern is known as **logistic growth**. Plotting this type of growth on a graph results in an S-shaped curve, or **S-curve**, as shown by the orange line in Figure 4–1. The point at which the population levels off is the **carrying capacity** (K). The equation for logistic growth can be explained this way:

Starting population + (Reproductive capacity (r)
    × Population × A number that represents how far
    the population is from carrying capacity)
                    = Ending population

There are two things to notice about the logistic growth curve. First, as the population nears its carrying capacity, the growth of the population slows until the population size remains steady; at this time the population growth equals zero. Second, the maximum rate of population growth occurs halfway to carrying capacity. If, for example, you were to monitor the number of beetles in a sack of grain, you would find that the maximum production of new beetles occurs when the beetle population is at half its carrying capacity, not at a larger or smaller number.

Real-Life Growth. The population growth figures calculated using these simple models rarely match the exact growth figures found in real life, although many populations show J-shaped explosions followed by crashes (Figure 4–2). J-curves occur when there are unusual disturbances, such as the introduction of a foreign species, the elimination of a predator, or the sudden alteration of a habitat, all of which can allow the rapid growth of a population. This growth is, however, followed by some type of crash. In some cases, huge overshoots of carrying capacity (K) by one population result in a collapse of the habitat for many species.

Other populations, often those controlled by complex relationships among species, show some type of S-curve, followed by cycles of lower and higher numbers roughly around carrying capacity. Many shoot somewhat above carrying capacity and then a little below, and eventually they cycle around the carrying capacity. This is illustrated by the orange line in Figure 4–2.

Biotic Potential Versus Environmental Resistance

The rate at which members of a species reproduce under unlimited conditions (which we saw in the equations above as r) is a measure of its **biotic potential**—the number of offspring (live births, eggs laid, or seeds or spores set in plants) that members of a species can produce under ideal conditions. The biotic potentials of different species vary tremendously, averaging from less than one birth per year in certain mammals and birds to millions per year for many plants and some invertebrates. To have any effect on the size of subsequent generations, however, the young must survive and reproduce in turn. Survival through the early growth stages to become part of the breeding population is called **recruitment**.

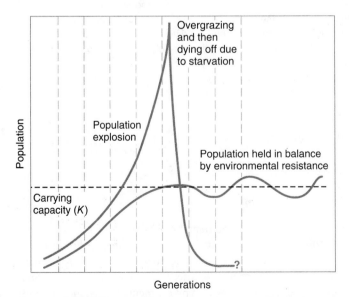

Figure 4–2 Real-life growth. In real populations, exponential growth is often followed by a crash (green line), and logistic growth is often modified so that the population fluctuates around K (orange line). Real-life populations often do not follow idealized equations, but such simple models help us to understand possible outcomes.

Sometimes changes in the environment offer a natural experiment from which scientists can learn valuable lessons. The elephants in Kruger National Park, South Africa, offer a good example.

For 60 years, the elephants in this park were protected from hunting. The ban on hunting allowed the elephant population to increase rapidly. Because elephants grow slowly and have only a few offspring that require a great deal of parental care, they are not usually prone to rapid population growth. Yet during this period, the elephant population grew exponentially, its graph resembling a J-curve. In part, this growth occurred because the elephants were not able to break into smaller groups and spread out across the landscape. Habitat loss limited them from the vast ranges over which they would have traveled, leaving them crowded together in the park. Eventually the elephants depleted their food resources, resulting in starvation and damage to the park ecosystem. The concept of the "balance of nature"—that somehow natural forces would cause the elephant population to even out—was not upheld because most of the elephants' natural predators had been reduced by hunting.

As scientists realized that the "experiment" had not worked as expected and the elephants and their ecosystem were in danger, managers had to come up with another plan for controlling the elephant population. This situation provided an opportunity for another type of experiment: In order to compare the effectiveness of several management strategies, the park was divided into regions with different management plans. From 1967 to 1994, more than 16,000 elephants were culled from the population. A much smaller number were moved to new locations outside of the park.

Unfortunately, hunting some of the adults left orphaned elephants and disrupted elephant

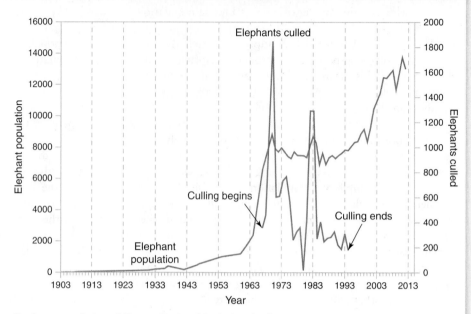

Elephant population of Kruger National Park, South Africa. The elephant population exploded after hunting was stopped and the migration of elephants was restricted. Between 1967 and 1994, culling (selected killing) was used to stabilize populations, which rapidly rose once culling was stopped.

UNDERSTANDING THE DATA

1. Which equation would you use to model the change in elephant population from 1923 to 1967?
2. How many elephants lived in the park in 1973? How many were culled? (*Hint:* Use the two scales.)

groups. In 1994, the practice of culling the herds was abandoned. After that, elephant numbers rose again. The impacts on biodiversity in the park, especially tree species, have been significant. Today wildlife managers are still attempting to control elephant populations and protect ecosystems in one of the only areas of the world where elephants aren't declining, but the problem of high elephant density is not yet solved. Strategies such as contraception and translocation have been less successful and more expensive than hoped.

The elephants in Kruger National Park illustrate how even large, slow-growing species can undergo exponential growth. They also illustrate the difference between local and global populations—elephants can be overabundant locally while they are quite rare globally. The story of the Kruger elephants also shows the difficulties of managing wildlife.

Source: H. Biggs, ed., *The Kruger Experience; Ecology and Management of Savanna Heterogeneity* (Washington, DC: Island Press, 2003) and www.elephantdatabase.org.

Environmental Resistance. Unlimited population growth is seldom seen in natural ecosystems because biotic and abiotic factors cause *mortality* (death) in populations. Among the biotic factors that cause mortality are predation, parasites, competition, and lack of food. Among the abiotic factors are unusual temperatures, moisture, light, salinity, and pH; lack of nutrients; and fire. The combination of all the biotic and abiotic factors that limit a population's increase is referred to as **environmental resistance** (Figure 4–3).

Sometimes environmental resistance lowers the rate of reproduction as well as causing mortality directly. For example, the loss of suitable habitat often prevents animals from breeding; certain pollutants also affect reproduction adversely. Such situations are defined as environmental resistance because they either block a population's growth or cause its decline. Additionally, environmental resistance can cause migration patterns to change, as when animals leave a drought-stricken area. Migration can lower or increase local populations.

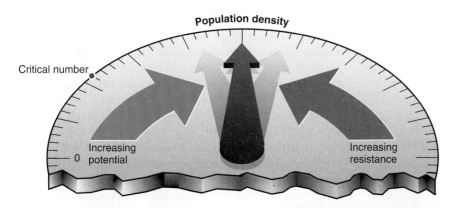

Population density

Critical number

0   Increasing
potential

Increasing
resistance

Biotic Potential

- Reproductive rate

- Ability to migrate (animals)
  or disperse (seeds)

- Ability to invade new habitats

- Defense mechanisms

- Ability to cope with adverse
  conditions

Environmental Resistance

- Lack of food or nutrients

- Lack of water

- Lack of suitable habitat

- Adverse weather conditions

- Predators

- Disease

- Parasites

- Competitors

Figure 4–3 Biotic potential and environmental resistance. A stable population in nature is the result of the interaction between factors tending to increase the population and factors tending to decrease the population. Human actions tend to decrease the amount of stable habitat and increase the amount of disturbed habitat, so some species are harmed while other species are benefited.

Reproductive Strategies. The interplay of environmental resistance and biotic potential drives the success of two common **reproductive strategies** in the natural world. The first is to produce massive numbers of young, but then leave survival to the whims of nature. This strategy often results in very low recruitment. Organisms with this strategy have life histories with rapid reproduction, rapid movement, and often a short life span. This strategy is highly successful if a species is adapted to an environment that can suddenly change and become very favorable, such as a rain-fed temporary pond. Organisms with this strategy are usually small and tend to have huge boom-and-bust populations. Because these organisms usually have a high *r*, they are called **r-strategists** or r-*selected species*. They are also called *weedy* or *opportunistic species*. One familiar *r*-strategist is the housefly, which multiplies very quickly but also has a high mortality rate.

In contrast, the second strategy is to have a much lower reproductive rate (that is, a lower biotic potential) but to care for the young until they can compete for resources with adult members of the population. The savanna elephant is a good example of this strategy. This strategy works best where the environment is stable and already well populated by the species. Organisms with such a strategy are larger, longer lived, and well adapted to normal environmental fluctuations. Because their populations are more likely to fluctuate around the carrying capacity, such species are sometimes called **K-strategists** or K-*selected species*. They are also called *equilibrial species* (Table 4–1).

Characteristics such as the age at first reproduction and the length of life are a part of an organism's **life history**—the progression of changes an organism undergoes in its life. Species such as the housefly and the elephant have very different life histories that reflect two extreme sets of adaptations. These life histories can be visualized in a graph called a **survivorship curve**, in which the number remaining from a group all born at the same time is shown decreasing over time until the maximum life span for the species

Table 4–1   General Characteristics of *r*-strategists and *K*-strategists

	r-strategists	*K*-strategists
Environment	Advantage if less stable	Advantage if more stable
Size	Smaller	Larger
Life span	Shorter	Longer
Age at first reproduction	Younger	Older
Offspring	More	Fewer
Parental care	Little or none	Long and involved
Population stability	Wild fluctuations	Mostly stable

is reached. Some species, like humans, have relatively low mortality in early life, and most individuals live almost the full natural life span for that species. This survivorship pattern is depicted as Type I in Figure 4–4. Other species, such as dandelions and oysters, have many offspring; most of their offspring die young, so only a few live to the end of a life span. Such a life history pattern is called Type III survivorship. Generally, *r*-strategists show the Type III pattern and *K*-strategists show the Type I pattern. Some species are not easily defined as either *r*-strategists or *K*-strategists. They may also have an intermediate survivorship pattern, depicted as Type II. Squirrels and corals are likely to fall into this intermediate category. Species actually form a continuum of these survivorship curves, much as they do in terms of the characteristics of *r*- and *K*-strategists.

We have examined species in decline, such as the savanna elephant. We will also study species experiencing rapid increase, such as pest species overtaking habitats (see Chapter 13). The reason we look at population growth equations is to show that there is a predictable connection between species characteristics, such as biotic potential (*r*), and patterns of growth lived out in different life histories. As we will see in Section 4.4, there are foreseeable ways that environmental pressures work on species with different biotic potentials.

There is also a predictable pattern to the way species are affected by large-scale human activities. While there are many exceptions, it is often the case that *r*-strategists become pests when humans alter the environment and that *K*-strategists become more rare, endangered, or extinct. Houseflies, dandelions, and cockroaches increase with human activity, while elephants, eagles, and oaks decline. Some exceptions to this rule include rare opportunistic species (*r*-strategists) that are so separated from new habitat that in spite of a high biotic potential, they cannot succeed.

In sum, whether a population grows, remains stable, or decreases is the result of interplay between its biotic potential and environmental resistance. In general, a population's biotic potential remains constant; it is changes in environmental resistance that allow populations to increase or cause them to decrease. Population balance is a *dynamic balance*, which means that additions (births and immigration) and subtractions (deaths and emigration) are occurring continually, and the population may fluctuate around a median. Some populations fluctuate very little, whereas others fluctuate widely. As long as decreased populations restore their numbers and the ecosystem's capacity is not exceeded, the population is considered to be at equilibrium.

CONCEPT CHECK Which kind of population growth curve would you expect to see in a population of squirrels in a forest? Which would you expect to see in ragweed growing on a construction site? Why? ☑

4.2 Limits on Populations

The carrying capacity of a habitat reflects how large a population can be sustained there, but it may not explain what is actually limiting the population. Sometimes limiting factors, such as crowding or disease, have a greater effect when there are more individuals in the population. The higher the **population density** (the number of individuals per unit area), the greater the role of that limiting factor. In other cases, the effect of a limiting factor is independent of the number of individuals in a population. For example, a tornado may reduce the number of trees in a forest, but the strength of the tornado has nothing to do with the density of the trees. We will see both of these types of limits in the next section.

Figure 4–4 Survivorship curves. Some organisms, like humans (Type I), experience low mortality throughout their lives, and most live to old age. Others (Type III), represented by an oyster, have many offspring, most of which die early. Still others are intermediate (Type II), here represented by a squirrel. Type I and Type III roughly correspond to the *K*-strategists and *r*-strategists.

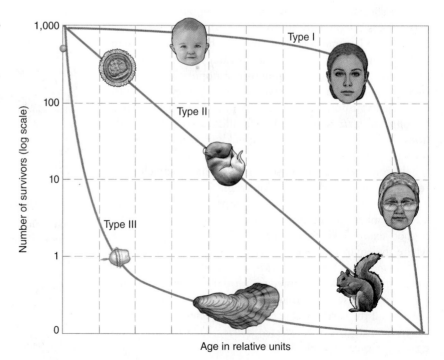

Number of survivors (log scale)

1,000

100

10

1

0

Type I

Type II

Type III

Age in relative units

Density Dependence and Independence

Rabbits thrive in a field with lush grass. As rabbits become more plentiful, foxes and coyotes find it easier to catch and eat more of them and less of something else. This slows the population growth of the rabbits. In this example, predation acts as a **density-dependent** limit on a population. A density-dependent limit is one that increases as population density increases—such as a disease or a food shortage. The environmental resistance caused by this limit increases mortality. Conversely, as population density decreases, the environmental resistance lessens, allowing the population to recover.

Other factors in the environment that cause mortality are **density-independent** limits—those whose effects are independent of the size of the population. For example, a fire that sweeps through a forest may kill all small mammals in its wake, regardless of their numbers. Such a fire is a density-independent cause of mortality. Although density-independent factors can be important sources of mortality, they are not involved in maintaining population equilibrium.

In the natural world, a population is subject to the sum of all the biotic and abiotic environmental factors around it. Many of these factors may cause mortality, but only those that are density dependent are capable of regulating the population, or keeping it in equilibrium. Because logistic growth occurs when populations become more crowded (and as they approach their carrying capacity), logistic growth is sometimes referred to as density-dependent population growth.

Environmental scientists distinguish between top-down and bottom-up regulation. *Top-down regulation* is the control of a population by predation, such as the control of rabbits by their predators, coyotes, or the control of deer by mountain lions (Figure 4–5a). In *bottom-up regulation,* the most important control of a population is the availability of some resource, often food. For example, the bighorn sheep in Figure 4–5b are limited by the availability of grass, a form of bottom-up regulation. In natural ecosystems, the same species may experience top-down and bottom-up control at different times.

Critical Number

We've discussed natural mechanisms that keep populations from growing indefinitely, but in nature, populations also decline. There is no guarantee that a population will recover from low numbers. Extinctions can and do occur. The survival and recovery of a population depend on a certain minimum population base, called the population's **critical number**. You can see a critical number at work in a pack of wolves, a flock of birds, or a school of fish. Often, the group is necessary to provide protection and support for its members. In some cases, the critical number is larger than a single pack or flock, because interactions between groups may be necessary as well. If a population is depleted below the critical number needed to provide such supporting interactions, the surviving members become more vulnerable, breeding fails, and extinction is almost inevitable.

The loss of biodiversity is one of the most disturbing global environmental trends (Chapter 1). Human activities are clearly responsible for the decline of many plants and animals. This is happening because many human impacts are not density dependent; they may even intensify as populations decline. Examples are alteration of habitats, introduction of alien species, pollution, hunting, and other forms of exploitation. In the United States, concern about the loss of species eventually led to the Endangered Species Act, which calls for the recovery of two categories of species. Species whose populations are declining rapidly are classified as **threatened**. If a population is near what scientists believe to be its critical number, the species may be classified as **endangered**. These definitions, when officially assigned by the U.S. Fish and Wildlife Service, set in motion a number of actions aimed at the recovery of the species in question (see Chapter 6).

CONCEPT CHECK Herons limit the number of frogs in a pond. Is this a bottom-up or a top-down control? Is it a density-dependent or a density-independent control? ☑

(a)   **(b)**

Figure 4–5 **Top-down and bottom-up control. (a)** Predators such as mountain lions control some deer populations, a form of top-down control. **(b)** The availability of grass controls the population of these bighorn sheep, a form of bottom-up control.

4.3 Community Interactions

Now that we have described population growth as a dynamic interplay between biotic potential and environmental resistance, we can look at the ways populations of different species living together in a biotic community interact with one another. The relationships between two species may be helpful, or positive (+); harmful, or negative (−); or neutral (0) for each species. These relationships are depicted in Table 4–2. Some of these positive and negative effects stem from a kind of **symbiosis**—a relationship in which the lives of two species are closely connected to each other. As we will see in several examples, symbionts have adaptations that make their relationship function.

The name of each type of interaction is given in the table. For example, a relationship in which one member benefits and the other is harmed (+ −) is **predation**. Parasitism is a subset of predation, as is herbivory (animals eating plants). A (− −) relationship is one in which both species are harmed. This would include competitive relationships where both species use a scarce resource. **Competition** occurs when two species or individuals rely on the same resource. In contrast, **mutualism** is a relationship between species in which both benefit (+ +). **Commensalism** occurs when one species is benefited and the other is not affected (+ 0).

The most important interactions are predation, competition, mutualism, and commensalism, but there are others. In **amensalism**, one species is unaffected and the other harmed (0 −) through an accidental interaction, such as an elephant stepping on a flower or a plant making a poison as a part of its normal metabolism that harms another organism. Similarly, it is theoretically possible to have a (0 0) interaction, called **neutralism**, in which neither species is harmed or benefited. Presumably the interaction would be a weak one.

Table 4–2 The Major Types of Interactions Between Two Species. Interactions can be defined by whether the impact on each species is positive, negative, or neutral.

Interaction Name	Effect on Species A	Effect on Species B	Example
Predation	+	−	Wolves eat moose
Competition	−	−	Two rabbit species eat the same grass
Mutualism	+	+	Lichens (fungus/alga)
Commensalism	+	0	Water buffalo/egret
Amensalism	0	−	An elephant tramples a plant
Neutralism	0	0	A fish in a stream and a sunflower in a valley

Predation

Now let's look in depth at predation, the most important interaction between species. In any relationship in which one organism feeds on another, the organism that does the feeding is called the **predator** and the organism that is fed on is called the **prey**. Predation is a conspicuous process in all ecosystems. Predation ranges from the predator-prey interactions between **carnivores** (meat eaters) and **herbivores** (plant eaters) to herbivores feeding on plants and **parasites** feeding on their hosts.

Parasites: A Special Kind of Predator. Parasites are organisms—either plants or animals—that become intimately associated with their "prey" and feed on it over an extended period of time, usually without killing it. However, sometimes they weaken it so that it becomes more prone to being killed by predators or adverse conditions. The plant or animal that is fed on is called the **host**. Parasites include a range of species, such as tapeworms, microscopic disease-causing bacteria, viruses, protozoans, and fungi (**Figure 4–6**). Parasites feed on their hosts in different ways. Tapeworms take energy from the intestine of a host animal, whereas parasitic plants tap into the sap of other plants.

Parasitic organisms affect the populations of their host organisms in much the same way that predators affect their prey—in a density-dependent manner. As the population density of the host increases, parasites and their vectors (agents that carry the parasites from one host to another, such as disease-carrying insects) have little trouble finding new hosts. Therefore, infection rates increase, causing higher mortality. Conversely, when the population density of the host is low, the transfer of infection is less efficient. This reduces the levels of infection and allows the host population to recover.

A tremendous variety of organisms may be parasitic. Various worms are well-known examples, but certain protozoans, insects, and even mammals (vampire bats) and plants (dodder) are also parasites. Parasitic fungi cause many plant diseases and some animal diseases (such as athlete's foot, Figure 4–6d). Indeed, virtually every major group of organisms has at least some members that are parasitic. Parasites may live inside or outside their hosts, as the examples shown in Figure 4–6 illustrate.

In medicine, a distinction is generally made between bacteria and viruses that cause disease (known as **pathogens**) and parasites, which are usually larger organisms. Ecologically, however, there is no real distinction. Bacteria are foreign organisms, and viruses are organism-like entities feeding on, and multiplying in, their hosts over a period of time. Therefore, disease-causing bacteria and viruses can be considered highly specialized parasites.

Regulation of Prey. Studies have shown that predators often regulate their herbivore prey; this is a type of top-down control. A well-documented example is the interaction between wolves and moose on Isle Royale, a 45-mile-long

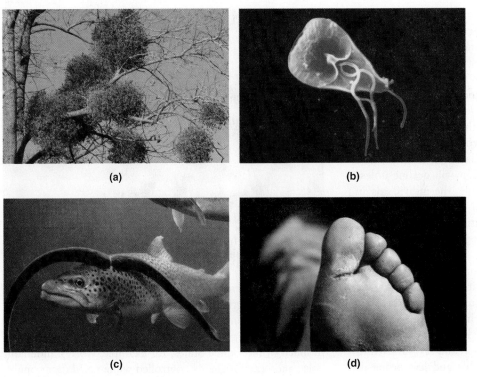

Figure 4–6 **Several types of parasites.**
Parasites may be animals, plants, fungi, or other organisms. Shown here are **(a)** mistletoe, a plant parasite on a tree host; **(b)** an intestinal parasite found in drinking water (*Giardia*) that infects human hosts; **(c)** lampreys, organisms that parasitize many types of fish, shown on a fish host; and **(d)** athlete's foot fungus on its host, a human foot.

island in Lake Superior that is now a national park. During a hard winter around 1900, a small group of moose crossed lake ice to the island and stayed. In the absence of predators, their population grew considerably. Then, in 1949, a small pack of wolves also managed to reach the island. The isolation of the island provided an ideal opportunity to study a simple predator-prey system, and in 1958, wildlife biologists began carefully tracking the populations of the two species (Figure 4–7). As the graph shows, a rise in the moose population is usually followed by a rise in the wolf population, followed by a decline in the moose population and then a decline in the wolf population. The data can be interpreted as follows: Fewer wolves represent low environmental resistance for the moose, so the moose population increases. Then the

abundance of moose represents optimal conditions (low environmental resistance) for the wolves, and the wolf population increases. More wolves mean higher predation on the moose (high environmental resistance); again the moose population falls. The decline in the moose population is followed by a decline in the wolf population, because there are fewer prey (high environmental resistance) for the wolves.

The dramatic fall in the moose population in 1996 cannot be attributed entirely to predation by the small number of wolves. That year deep snow (which makes it more difficult to get food) and an infestation of ticks caused substantial mortality. The sharp decline in the number of moose is thought to have been responsible for keeping the wolf population low, as there were few calves for them to catch.

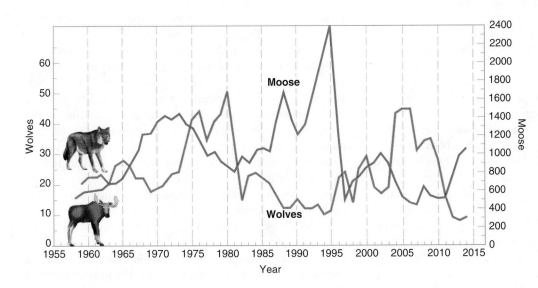

Figure 4–7 **Predator-prey relationships.** The fluctuations in the wolf and moose populations on Isle Royale are shown from 1955 to 2014.

Wolves are generally incapable of bringing down an adult moose in good physical condition. The animals they kill are the young and those weakened by another factor, such as sickness or old age.

The observation that wolves are usually incapable of killing moose that are mature and in good physical condition is extremely significant. This is what often prevents predators from eliminating their prey. As the prey population is culled down to healthy individuals who can escape attack, their predators are limited by the availability of their crucial food resource. The predator population will decline unless it can switch to other prey. Meanwhile, the individuals in the prey population are healthy and can readily produce the next generation. Thus, the regulation of populations in a predator-prey relationship is both top-down (on the prey) and bottom-up (on the predator).

A parasite may work in conjunction with a larger predator to control a given herbivore population. When a parasitic infection breaks out in a dense population of herbivores, predators can more easily remove individuals weakened by infection, leaving a smaller but healthier population. Relationships between a prey population and several natural enemies are generally much more stable and less prone to wide fluctuations than are those that involve only a single predator or parasite, because different predators and parasites come into play at different population densities.

Plant-Herbivore Dynamics. Herbivores prey on plants. Just as too many wolves can bring the moose population dangerously low, too many herbivores can do the same to their plant food. In turn, when herbivore populations are low, their food plants encounter low environmental resistance and can overshoot their carrying capacity. These imbalances can lead to problems in the surrounding ecosystem.

Overgrazing. If herbivores eat plants faster than they can grow, the plants will eventually be depleted, and animals that eat the plants will suffer. A classic example is the case of reindeer on St. Matthew Island, a 128-square-mile island in the Bering Sea. In 1944, a herd of 29 reindeer was brought to the island, where they had no predators. By 1963, this lack of predation had allowed the herd to increase to 6,000 individuals, although their food had nearly disappeared and most of the reindeer were malnourished. During the winter of 1963–1964, nearly the entire herd died of starvation; there were only 42 surviving animals in 1966.

Overgrazing can often occur in agricultural systems (Figure 4–8). The Food and Agriculture Organization of the United Nations (FAO) estimated that the world's cattle population increased from 720 million to 1.5 billion between 1950 and 2013; domestic sheep have also increased. These increases have caused overgrazing of more than 20% of rangeland.

Overgrazing demonstrates that no population can escape ultimate limitation by environmental resistance, although the form of environmental resistance and the consequences may differ. If an herbivore population is not held in check, it may explode, overgraze, and then crash as a result of starvation (exhibiting a J-curve and subsequent drop).

Figure 4–8 Plant-herbivore interaction: Overgrazing. Domestic animals such as these goats can overgraze land.

Predator Removal. Eliminating predators or other natural enemies of herbivores upsets plant-herbivore relationships, just as introducing an animal without natural enemies disrupts normal population controls on the introduced species. In much of the United States, for example, wolves, mountain lions, and bears originally controlled white-tailed deer populations. Many of these predators were killed because they were believed to be a threat to livestock and humans. At present, deer populations in many areas would increase to the point of overgrazing if humans didn't hunt them in place of their natural predators. Indeed, population explosions do occur where hunting is prevented.

Keystone Species. Sometimes the removal of a single species affects far more than just the other species with which it primarily interacts. For example, in the West Coast kelp forests, the sea otter (*Enhydra lutris*) eats sea urchins, which in turn eat kelp (Figure 4–9a). Because sea otters control the sea urchin population, hundreds of other species, including the young of many fish, are able to survive in the kelp forests. By the early 20th century, overhunting of sea otters had broken this food chain, leaving the sea urchin population unchecked. Sea urchins ate kelp down to the sea floor, creating areas called "urchin barrens" (Figure 4–9b). Subsequent protection of the otters has allowed them to return, and they are again protecting the entire habitat.

Scientists refer to the sea otter as a **keystone species**, in recognition of its crucial role in maintaining ecosystem biotic structure (in architecture, the keystone is a fundamental part of the support and structure of a building). The presence of the keystone species moderates other species that would otherwise take over the area (the sea urchin) or eat the most important species in the habitat (the kelp). Keystone species allow less-competitive species to flourish, thereby increasing overall diversity. Keystone species are such an integral part of their ecosystems that their removal can cause the collapse of the whole ecosystem.

The sea otters in the West Coast kelp forest are predators, but this is not always the case with keystone species. For example, prairie dogs are herbivores, but they serve as a keystone species by digging burrows that provide essential

(a)

(b)

Figure 4–9 **Keystone species.** **(a)** The sea otter eats sea urchins, which eat kelp in the Pacific Ocean along the Northwest Coast. **(b)** Without sea otters, the number of sea urchins increases, allowing them to overgraze the kelp. The resulting bare areas are called "urchin barrens," as seen in the foreground.

habitats for many other prairie species. Beavers, which are also herbivores, were once a dominant ecological force in eastern North America, creating openings in forests and wetlands that provided habitats for many other species. The trapping of beavers for the fur industry removed beavers from much of the North American landscape, to the detriment of species that rely on small wetland meadows for their survival.

Competition

Recall from Table 4–2 that there are species interactions in which both species are harmed. This result usually means that the species are competing for some scarce resource. We have discussed the concept of each species having a *niche*, the conditions and resources under which a species can live (Chapter 3). Species are said to have overlapping ecological niches when they compete with each other. Competition can also occur between members of the same species. Whether between species or between members of the same species, competition harms both participants. It lowers their fitness and their ability to produce offspring. In both types of competition, between and within species, you might expect that, over time, there would be a pressure to lower the overlap and have less negative interaction. Indeed, this is the case.

Intraspecific Competition. Competition among members of the same species is called **intraspecific competition**. This type of competition can occur over different types of resources.

Impact on the Species. In the short term, intraspecific competition can lead to the density-dependent regulation of a species' population through factors such as territoriality and self-thinning, a phenomenon whereby crowded organisms such as trees become less numerous as they grow bigger.

Intraspecific competition can also lead to long-term changes as the species adapts to its environment. The observation that individuals compete for scarce resources led Charles Darwin to identify the "survival of the fittest" as one of the forces in nature leading to evolutionary change in species. In any group of competing organisms, the individuals that are able to survive and reproduce, while others are not, demonstrate superior fitness to the environment. Indeed, every factor of environmental resistance is a selective pressure resulting in the survival and reproduction of those individuals with a genetic endowment that enables them to better cope with their surroundings. Those better able to compete are the ones who survive and reproduce, and their superior traits are passed on to successive generations. This is the essence of natural selection, which is discussed in Section 4.4.

Territoriality. In many species, some members defend a **territory**, or limited space. Because territoriality is competition for space with resources, a lack of suitable territories is a density-dependent limitation on a population—the larger the population grows, the more intense the competition for space becomes. The males of many species of songbirds claim a territory and defend it vigorously at the time of nesting. Their song warns other males to keep away **(Figure 4–10)**.

Figure 4–10 **Territoriality.** This male red-winged blackbird sings to defend his territory.

The males of many carnivorous mammals, including dogs, stake out a territory by marking it with urine; the smell warns others to stay away. If other males do encroach, there may be a fight, but in most species, a large part of the battle is intimidation—an actual fight rarely results in serious physical harm. While territoriality is usually associated with animals, some scientists consider plants, fungi, and other organisms to have territories under certain conditions.

Territoriality may seem negative, but in many cases, it ensures that some members of the population will get enough resources to survive. What if the population had no territories? If so, when resources were scarce, every encounter with another member of the species could end in a potentially lethal fight. Instead of some members getting enough resources to raise the next generation, it might be possible for all members to get only a portion of what they need, and a large percentage—or all—of the population would die. Even so, most species do not show territoriality. They may have other adaptations to lessen negative interactions with others of the same species, such as increased rates of dispersal.

Interspecific Competition. Now we turn to the effects of **interspecific competition,** the competition between members of different species. Consider a natural ecosystem that contains hundreds of species of plants, all competing for water, nutrients, and light. What prevents one plant species from driving out others? How do species survive competition? The most obvious answer is that sometimes they do not. If two species compete directly for the same resources, as sometimes occurs when a species is introduced from another continent, one of the two species generally perishes in the competition. This is the competitive exclusion principle. For example, the introduction of the European rabbit to Australia has led to the decline and disappearance of several small marsupial animal species due to direct competition for food and burrows.

In fact, competitive exclusion would be expected to be the norm in circumstances where a habitat is very simple and species require the same resources. However, in nature, we often see similar species occupying the same habitats without becoming extinct. This unexpectedly high biodiversity has intrigued scientists. G. E. Hutchinson, a famous limnologist (a scientist who studies freshwater systems), once wrote a paper asking about this phenomenon: "Homage to Santa Rosalia or Why Are There So Many Kinds of Animals?"[1] Hutchinson suggested, and further work has supported, that the answer lies in variability over space and time. Differences in topography, type of soil, and other factors mean that the landscape is far from uniform, even within a single ecosystem. Instead, it is composed of numerous microclimates or microhabitats. The specific abiotic conditions such as moisture, temperature, and light differ from location to location and through the course of time. This is true even in lakes, where tiny pockets of water may have different temperatures, chemistry, or light conditions. Often just when one competitor begins to get the upper hand, the seasons change and a different species benefits. Thus, the adaptation of a species to particular conditions enables it to thrive and overcome its competitors in one location or time, but not in another.

---

[1]*The American Naturalist*, vol. XCIII, no. 870, May–June 1959, 145–159.

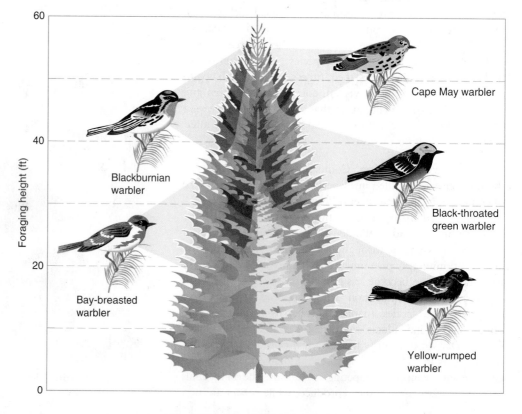

Figure 4–11 Resource partitioning. Five species of North American warblers experience less competition than it might first appear because they feed at different heights and in different parts of trees.

These examples simply mean that ecosystems are heterogeneous (variable) enough to support species with many different niches. Similarly, changes over time can allow multiple species to live in the same place. For example, many wildflowers of temperate deciduous forests (lady slippers, hepatica, trillium) sprout from perennial roots or bulbs in the early part of spring. While they would not be able to outcompete trees for light during the summer, these plants take advantage of the light that reaches the forest floor before the trees grow leaves.

Species that seem to be in competition can coexist in the same habitat, but have separate niches. Competition is minimized because potential competitors are using different resources. For example, woodpeckers, which feed on insects in deadwood, do not compete with birds that feed on seeds. Bats and swallows both feed on flying insects, but they do not compete because bats feed on night-flying insects and swallows feed during the day. Sometimes the "resource" can be the space used by different species as they forage for food. This is illustrated in Figure 4–11, which shows

five species of warblers that coexist in the spruce forests of Maine. The birds feed at different levels in the trees and on different parts of the trees. This division is called *resource partitioning*—the division of a resource and specialization in different parts of it. Subsequent study of these warblers has shown that when competition is more intense, they separate themselves even more.

All of these examples show that species can live in the same habitat because their adaptations, including behavior, limit competition. By limiting competitive interactions, each species can put more energy into reproduction. However, the reason that many species can share a habitat is not simply because their niches prevent them from competing. Instead, when species live in the same place, their characteristics (physical as well as behavioral) may eventually change so that they use only a portion of their potential niche (which they could have used all of when living alone). For more discussion on interspecific competition and character displacement, see Sound Science, *Studying Finches: The Life of a Scientist* and Figure 4–12.

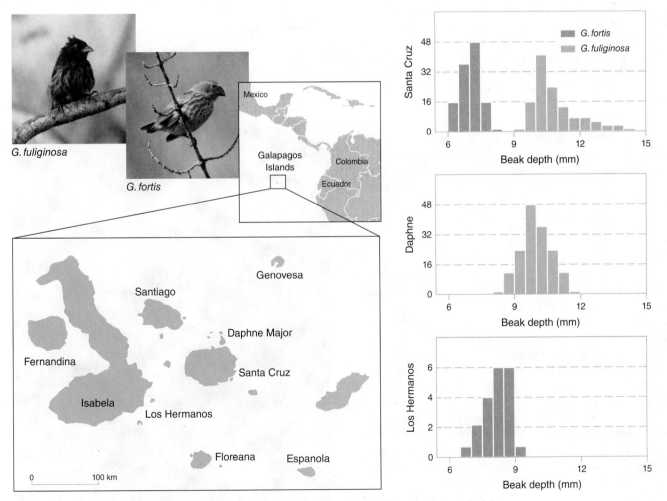

Figure 4–12 Character displacement/release. Finches of two species have different beak sizes when they occur together but do not when the species occur separately.

(*Source:* "Evolution of Character Displacement in Darwin's Finches" by Peter R. Grant and B. Rosemary Grant, from *Science*, July 2006, Volume 313(5784). Copyright © 2006 by AAAS. Reprinted with permission.)

(a)

(b)

Figure 4–13 **Mutualism.** **(a)** Pollinators and their flowering plants are mutualists. **(b)** These lichens are made up of fungi and algae living in mutualistic relationships.

Mutualism

In contrast to competition, mutualism is a relationship between two species from which both benefit. There are many mutualistic relationships. For example, the relationship between pollinators (such as bees, which receive nutrition from flowers) and the plants they pollinate (which receive pollination) is an example of mutualism (Figure 4–13a). Some scientists estimate that 70% of plant species have mutualistic relationships with fungi connected to their roots. The fungi benefit from the food produced by the plants; in exchange, the fungi make it easier for the plants to take in nitrogen and other nutrients from the soil. Lichens are a mutualistic relationship between a fungus and an alga (Figure 4–13b). One example of a mutualism between two animals is the relationship of the anemone fish and the sea anemone. The fish protects the anemone from predation by the butterfly fish, and in turn, the anemone's stinging tentacles (to which the anemone fish is immune) protect the fish from other predators.

Commensalism

Commensalism is more rare. It occurs when two species interact and one is benefited while the other is unaffected. The relationship between water buffalo and cattle egrets is an example: the grazer stirs up insect prey, and the birds follow it and eat the insects (Figure 4–14). Commensalism also includes organisms that hitch rides on other species or grow on them for support without harming them. Many orchids grow on the branches and trunks of large trees, but do not feed on them as parasitic plants do.

In contrast to commensalism is **amensalism**, where one species is harmed while the other is unaffected. Usually, this is accomplished by the natural chemical compounds produced by an organism, which are harmless to itself but harmful to other organisms around it. An example is the black walnut tree, which secretes a chemical compound from its roots that can harm or kill other plants in the area.

In summary, the relationships between two species or between two members of one species can be described by

the positive, negative, or neutral effects on the individuals. Ecologists study these potential outcomes. Relationships such as predation, competition, mutualism, and commensalism affect individuals, but are also part of the constant pressure that causes species to adapt to environmental change.

CONCEPT CHECK How might the removal of a predator harm a community of a species? How might the removal of a competitor affect the community? ☑

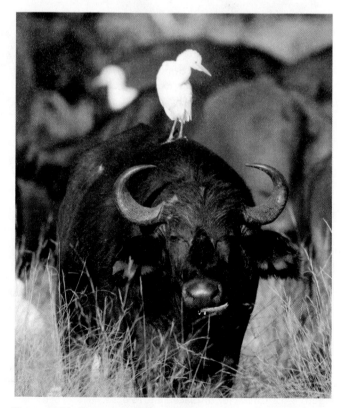

Figure 4–14 **Commensalism.** A cattle egret eats insects stirred up by a grazing water buffalo, which is neither harmed nor benefited by the egret's presence.

SOUND SCIENCE
Studying Finches: The Life of a Scientist

How do scientists know what happened to species in the past? For the most part, they look at what happens today. They look for physical patterns and mechanisms and at DNA and other parts of cells to see patterns that suggest relationships.

Darwin himself first observed evidence of past speciation. In 1835, during his famous voyage on the HMS *Beagle*, Darwin collected a number of finches from different islands of the Galápagos archipelago. Later, in consultation with ornithologist John Gould, Darwin speculated that differences in size and beak structure of the finches could have occurred after subpopulations of an original parental species were isolated from one another on separate islands.

Although Darwin himself did not work out the details of how the different Galápagos finches evolved, modern scientists have found what looks to be the most likely

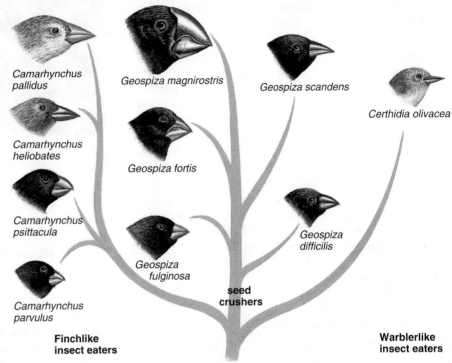

Some of Darwin's finches. The similarities among these birds attest to their common ancestor. Selective pressures to feed on different foods have caused modification and speciation in subpopulations. (*Source:* From *Biology: Life on Earth*, 4th ed., by Teresa Audesirk and Gerald Audesirk, copyright © 1994 by Prentice Hall, Inc., reprinted by permission of Pearson Education, Inc., Upper Saddle River, NJ 07458.)

Peter and Rosemary Grant

explanation. It is probable that at some time in the past, a few finches from the South American mainland were blown westward by a strong storm and became the first terrestrial birds to inhabit the relatively new (10,000-year-old) volcanic islands of the Galápagos. Later, some birds dispersed to nearby islands, where they were separated from the main population. These subpopulations encountered different selective pressures and became specialized for feeding on different resources, such as cactus fruit, insects, and seeds. In time, when the changed populations dispersed back to their original islands, they were different enough from the parent species that interbreeding among them did not occur, and the changed populations were distinguishable as new species.

Peter and Rosemary Grant have studied the Galápagos finches for decades. They

have investigated the fact that when the finch *Geospiza magnirostris* arrived on the same island as the similar species *Geospiza fortis*, the two species competed heavily for food seeds. Over 25 years of study, the scientists showed that the beak sizes of both species changed and competition was reduced as the finches began to eat slightly different-sized seeds. In populations where each species was alone, the beaks remained in a middle size. This is called character displacement, a physical change that lessens competition when two species co-occur (see Figure 4–12).

The Grants have spent six months of every year in the Galápagos since 1973 and have studied the fates of hundreds of birds, taking blood samples and measurements. This type of careful study is useful in understanding how evolution occurs.

4.4 Evolution as a Force for Change

Predation and competition are two mechanisms that keep natural populations under control. They also favor the survival of individuals with qualities that lower the impacts of negative interactions. Predators and prey become well adapted to each other's presence. Predators are rarely able to eliminate their prey species, largely because the prey has various defenses against its predators. Intraspecific competition is also a powerful force that can lead to improved adaptations of a species to its environment. Interspecific competition, by contrast, promotes adaptations that allow competing species to specialize in exploiting a resource. This specialization can lead to resource partitioning, which allows many potential competitors to share a basic resource—such as water in the soil. For example, different species of prairie plants have various kinds of roots, such as long tap roots or short fibrous roots, that allow them to use water in different areas of the soil, as depicted in **Figure 4–15**. In this section, we discuss how these various adaptations have come about.

Selective Pressure

Most young plants and animals in nature do not survive, instead falling victim to various environmental resistance factors. These factors—such as predators, parasites, drought, lack of food, and temperature extremes—are known as **selective pressures**. Selective pressures affect which individuals survive and reproduce and which are eliminated. For example, animals

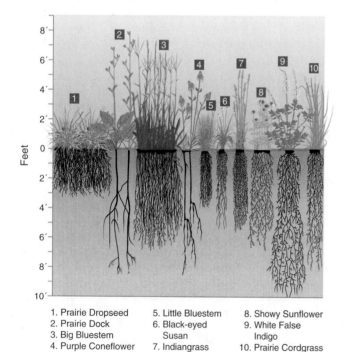

Figure 4–15 Resource partitioning in plants. Some species of prairie plants have short fibrous roots and others have long taproots. These differences allow the different species to make use of water in different areas of the soil. Researchers found greater plant diversity with greater soil depth because new niches open up for plants with longer taproots.

1. Prairie Dropseed
2. Prairie Dock
3. Big Bluestem
4. Purple Coneflower
5. Little Bluestem
6. Black-eyed Susan
7. Indiangrass
8. Showy Sunflower
9. White False Indigo
10. Prairie Cordgrass

Figure 4–16 Cryptic coloration. The color and shape of this Indonesian green leaf insect help protect it from predation.

with traits that allow them to escape from their predators (such as speed or coloration that blends in with the background) tend to survive and reproduce **(Figure 4–16)**. Those without such traits tend to become the predator's dinner. Any individual with a genetic trait that slows it down or makes it conspicuous will tend to be eaten. Thus, predators function as a selective pressure, favoring the survival of traits that enhance the prey's ability to escape or protect itself and causing the elimination of any traits handicapping the organism's escape. The need for food can also be seen as a selective pressure acting on the predator. Characteristics that enhance survival, such as keen eyesight and swift speed, benefit survival and reproduction. Sometimes predators use the ability to blend in as an advantage, too. The process of specific traits favoring the survival of certain individuals is known as **natural selection**.

Discovered independently by Charles Darwin and Alfred Russell Wallace, these concepts were first presented in detail by Darwin in his book *On the Origin of Species by Means of Natural Selection* (1859). The modification of the gene pool of a species by natural selection over many generations is the substance of **biological evolution**. Darwin and Wallace deserve tremendous credit for constructing their theory purely from their own observations, without any knowledge of genetics. The role of genetic information wasn't discovered until several decades later. Our modern understanding of DNA, mutations, and genetics supports the theory of evolution by natural selection.

Adaptations to the Environment

The selective pressures exerted by environmental resistance continually test the gene pool of a population. Indeed, virtually all traits of an organism can be seen as features that adapt the organism for survival and reproduction, or, in Darwinian terms, **fitness**. Such characteristics of organisms include the adaptations needed for coping with climatic and other abiotic factors; obtaining nutrients, energy, and water; defending against predation and disease-causing or parasitic organisms; finding or attracting mates (or pollinating and setting seed); and migrating and dispersing. These categories of adaptations are summarized in **Figure 4–17**.

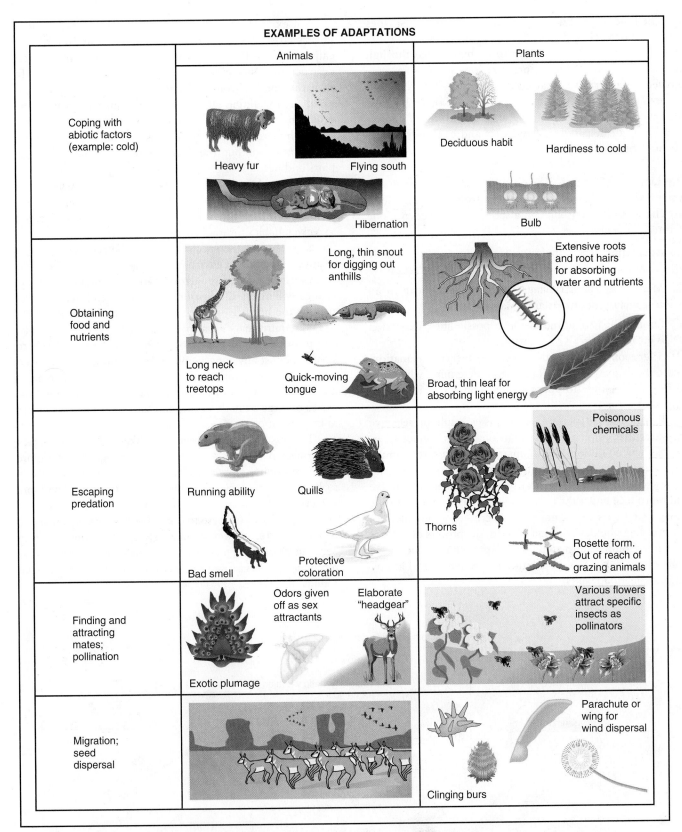

EXAMPLES OF ADAPTATIONS

	Animals	Plants
Coping with abiotic factors (example: cold)	Heavy fur / Flying south / Hibernation	Deciduous habit / Hardiness to cold / Bulb
Obtaining food and nutrients	Long neck to reach treetops / Long, thin snout for digging out anthills / Quick-moving tongue	Extensive roots and root hairs for absorbing water and nutrients / Broad, thin leaf for absorbing light energy
Escaping predation	Running ability / Quills / Bad smell / Protective coloration	Thorns / Poisonous chemicals / Rosette form. Out of reach of grazing animals
Finding and attracting mates; pollination	Exotic plumage / Odors given off as sex attractants / Elaborate "headgear"	Various flowers attract specific insects as pollinators
Migration; seed dispersal		Clinging burs / Parachute or wing for wind dispersal

Figure 4–17 Adaptation for survival and reproduction. The five general features listed in the left column are essential for the continuation of every species. Each feature is found in each species as particular adaptations that enable the species to survive or reproduce. Across the various species, a multitude of adaptations will accomplish the same function. Thus, a tremendous diversity of species exists, with each species adapted in its own special way.

The fundamental question about any trait is this: Does it increase survival and reproduction of the organism? If the answer is yes, the trait is likely to be maintained through natural selection. Consequently, various organisms have evolved different traits that accomplish the same function. For example, the ability to run fast, to fly, or to burrow, and protective features such as quills, thorns, or an obnoxious taste, all help reduce predation and can be found in various organisms.

The Limits of Change. The savanna elephant is facing tremendous loss of habitat and pressure from poaching. With any change in its environment, a species may experience one of three outcomes:

1. **Adaptation.** The population of survivors may gradually adapt to the new condition through natural selection.

2. **Migration.** The surviving population may move to an area where conditions are suitable.

3. **Extinction.** Failing the first two possibilities, extinction is inevitable.

Migration and extinction need no further explanation. The critical question is: What factors determine whether a species will be able to adapt to new conditions instead of becoming extinct?

Recall that adaptation occurs as a result of selective pressures that eliminate those individuals that cannot tolerate the new condition. For adaptation to occur, there must be some individuals with traits (alleles—variations of genes or new combinations of genes) that enable them to survive and reproduce under the new conditions. There must also be enough survivors to maintain a viable breeding population. If a breeding population is maintained, the process of natural selection should lead to increased adaptation over successive generations. If a viable population is *not* maintained at any stage, extinction results.

For example, by the early 1980s the California condor (a very large scavenger) had declined to about 20 birds because of deaths from poaching, lead poisoning, and habitat destruction. Only nine condors were left in the wild. To save the species from extinction, wildlife biologists captured the remaining wild condors and initiated a breeding program. As of August 2014, there were 434 condors, 202 of which were in captivity. This is still not a viable breeding population, but the wild condors are now forming pairs and nesting.

Keys to Survival. Four key factors affect whether or not a viable population of individuals is likely to survive new conditions: (1) geographical distribution, (2) specialization to a given habitat or food supply, (3) genetic variation within the gene pool of the species, and (4) the reproductive rate relative to the rate of environmental change.

Consider how these factors work. It is unlikely that any change will affect all locations equally. Therefore, a species such as the housefly—which is present over most of Earth, has no requirements for a specialized habitat or food supply, and possesses a high degree of genetic variation—is likely to survive almost any conceivable alteration in Earth's environment. The California condor, on the other hand, requires an extensive habitat of mountains and ravines, and the carcasses of large animals to eat, but has little genetic variation in its population.

The less vulnerable species are likely to be *r*-strategists (described in Section 4.1). In fact, *r*-strategists move more easily, reproduce more quickly, and are generally more likely to become pest species (Chapter 13) or invasive. The more vulnerable species are likely to be *K*-strategists. The factors affecting survival of these two species types are summarized in Figure 4–18. However, many species fall somewhere in between these extremes. In some cases, even very rapidly reproducing species can be vulnerable if there is enough environmental change.

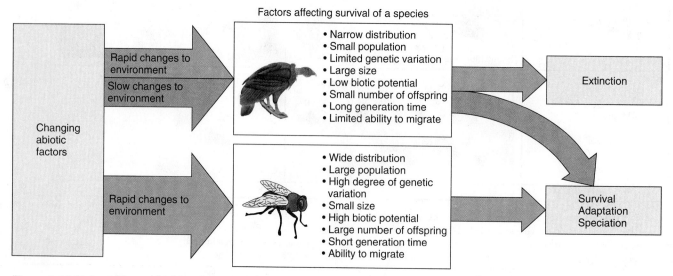

Figure 4–18 Vulnerable and highly adaptive species. Species differ broadly in their vulnerability to environmental changes. When confronted with rapid change in the environment, a housefly population is likely to explode and a condor population is likely to decline.

Genetic Change. Assuming the survival of a viable population, how fast can that population evolve further adaptations to enable it to better cope with new or changing conditions? Over the lifetime of the individual, there is no genetic change—and hence no genetic adaptation. In the population, genetic variation is brought about over generations by a number of mechanisms, including hybridization, mutation, and crossover (a process by which sections of chromosomes are swapped). Bacteria, for example, readily exchange small bits of genetic material with other bacteria, and recent studies have shown that microbes sometimes insert their genetic material into other organisms, including mammals, bringing new DNA variations into the genome. Genetic variation is the norm in living things.

The Evolution of Species. The newest estimates are that there are about 8.7 (+/− 1.3) million species of plants, animals, and microbes currently in existence, all living and functioning in ecosystems and all contributing to an amazing biodiversity. This estimate changes with new information because, of course, we haven't discovered most of the organisms that exist on Earth. However, biodiversity is diminishing because present species are becoming extinct faster than new species are appearing. In order to understand why, we need to understand how new species appear.

The infusion of new variations from mutations and the pressures of natural selection serve to adapt a species to the biotic community and the environment in which it exists. Over time, in this process of adaptation, the final "product"—giraffe, anteater, or redwood tree—may be so different from the population that started the process that it is considered a different species. This is one aspect of the process of speciation.

Two Species from One. The same process also may result in two or more species developing from one original species. There are only two prerequisites. The first is that the original population must separate into smaller populations that do not interbreed with one another. This **reproductive isolation** is crucial because if the subpopulations continue to interbreed, all the genes will continue to mix through the entire population, keeping it as one species. The second prerequisite is that separated subpopulations must be exposed to different selective pressures. As the separated populations adapt to these pressures, they may gradually become so different that they no longer belong to the same species. Thus, they are unable to interbreed with one another, even if they come together again later.

Consider the example of the arctic and gray foxes (Figure 4–19). Long ago, they may have been part of a single ancestral population that was eventually separated by glaciation. In the Arctic, selective pressures favor individuals that have heavier fur (which protects them from the cold); shorter tails, legs, ears, and noses (the shortness helps conserve body heat); and white fur (which helps the animals hide in the snow). In the southern regions, selective pressures favor individuals with thinner fur, which dissipates excessive body heat, and coloring that blends in with darker surroundings. Over many generations, the geographic isolation of the two subpopulations made it possible for one ancestral fox population to develop into two separate species.

New Species from Hybridization. As you might expect, we also see new species arise in real time today, and we have examples of species forming rapidly. The saltmarsh cordgrass, *Spartina angelica,* is one such species. It developed as a hybrid

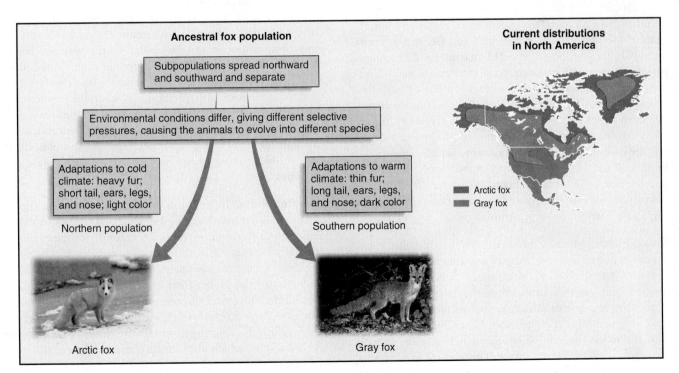

Figure 4–19 Evolution of new species. A population spread over a large area may face various selective pressures. If the population splits so that interbreeding does not occur, the different pressures may result in the subpopulations evolving into new species. The arctic fox and the more southerly gray fox illustrate this point.

of two other related species: *Spartina alterniflora*, a native of North America, and *Spartina maritima*, a native of England. The two do not usually interbreed. *Spartina alterniflora*, accidentally introduced to England's mud flats, did not thrive there, but when it bred with *Spartina maritima*, the hybrid daughter species (*S. angelica*) spread rapidly. It has a different number of chromosomes from either parent and does not interbreed successfully with either parent species. Now, where there were two similar species, there are three, and the daughter species has a different set of characteristics and can live in a slightly different niche than either parent species.

Descent with Modification. In sum, new species are not formed from scratch; they are formed only by the modification of existing species. Different selective pressures act on the constant variation of genetic material that naturally occurs. Combined with isolation and other processes (such as the unusual hybridization that occurred in the genus *Spartina*), new adaptations can arise. This concept is consistent with the observation that groups of closely related species are generally found in nature, instead of distinct species without close relatives. Patterns among larger groups suggest the same thing—that organisms can be grouped into large, closely related groups differentiated by older or newer characteristics.

Although it can be difficult to imagine, overwhelming evidence from today and patterns from the past inform us that the present array of plants, animals, and microbes has developed through evolution over long periods of time and in every geographic area on Earth. This process is the source of our current biodiversity.

Drifting Continents

The geographic isolation of populations is the foundation of the process of speciation. The slow movement of Earth's continents has allowed the isolation of broad groups of organisms. Using evidence from geology, German scientist Alfred Wegener proposed in 1915 that about 225 million years ago the continents were connected into a giant landmass called Pangaea. (A modern map of Pangaea is shown in Figure 4–20a.) This is part of a theory that the continents have been in motion since Earth formed. Wegener's theory was hotly debated for decades until, by the mid-20th century, the evidence for drifting continents became irrefutable. His theory has blossomed into a grand theory known as **plate tectonics**. This theory helps us to understand earthquakes and volcanic activity and is key to understanding the geographic distribution of present-day biota.

Moving Plates. Much of the interior of Earth is molten rock, kept hot by the radioactive decay of unstable isotopes remaining from the time when the solar system was formed (about 5 billion years ago). Earth's crust, the lithosphere—which includes the bottoms of oceans as well as the continents—is a relatively thin layer (ranging from 10 to 250 kilometers, or 16 to 160 miles) that can be visualized as huge slabs of rock floating on an elastic layer beneath. This is much like crackers floating next to each other in a bowl of soup. These slabs of rock are called **tectonic plates**. Some 14 major plates and a few minor ones make up the lithosphere (Figure 4–20b).

Within Earth's molten interior, hot material rises toward the surface and spreads out at some locations, while cooler material sinks toward the interior at other locations. Riding atop these convection currents, the plates move slowly, but inexorably, with respect to one another. The spreading process of the past 225 million years has broken up the original mass of Pangaea and brought the continents to their present positions. It also accounts for the other interactions between tectonic plates, such as volcanoes and earthquakes. The average rate of a plate's movement is about 6 centimeters (2.4 inches) per year, but over 100 million years, this adds up to almost 8,000 kilometers (5,000 miles) in the fastest-moving segments. The plate boundaries are locked by friction and hence are regions of major disturbances.

Adjacent tectonic plates move with respect to each other by separating (as in mid-ocean ridges), sliding past each other (at fault lines like the San Andreas Fault in California), or colliding. These plate movements have large environmental effects. A catastrophic earthquake off the Pacific coast of Tōhoku, Japan, on March 11, 2011, resulted from the Pacific plate sliding under the North American plate, forming a rift under the sea floor 300 kilometers (186 miles) long, pulling eastern Japan closer to North America by 4 meters (13 feet) and even slightly changing Earth's axis. The sudden uplift triggered a powerful **tsunami** (tidal wave) that displaced an enormous quantity of seawater and sent waves up to 10 kilometers (6 miles) inland in the Sendai area of Japan. The tsunami triggered a number of nuclear accidents at the Fukushima II Nuclear Power Plant (see Chapter 15). The earthquake and tsunami resulted in more than 15,845 deaths, with thousands more missing or injured.

Collisions and Geologic Features. The movements of tectonic plates shape our world. Over geologic time, plate collisions produce volcanic mountain chains and uplift regions into mountain ranges similar to the way the hoods of colliding cars crumple. An example of mountains caused by this type of movement is the Himalayas. The fact that volcanic eruptions and earthquakes continue to occur is evidence that tectonic plates are continuing to move today, as they have over millions of years. In fact, volcanoes and earthquakes mark the boundaries between the plates and have provided some of the best evidence for the plate tectonic theory.

Climate Shifts. In addition to the periodic catastrophic destruction that may be caused in localized regions by earthquakes and volcanic eruptions, tectonic movement may gradually lead to major shifts in climate. This can occur in three ways. First, as continents gradually move to different positions on the globe, their climates change accordingly. Second, the movement of continents alters the direction and flow of ocean currents, which in turn have an effect on climate. Third, the uplifting of mountains alters the movement of air currents, which also affects climate. (For example, see the rain-shadow effect in Figure 10–7.)

Increases in Diversity. Large-scale changes in Earth's crust have been part of evolutionary pressures that have increased

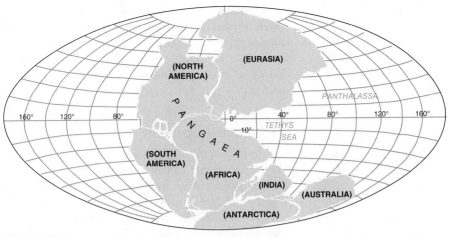

(a) 225 million years ago

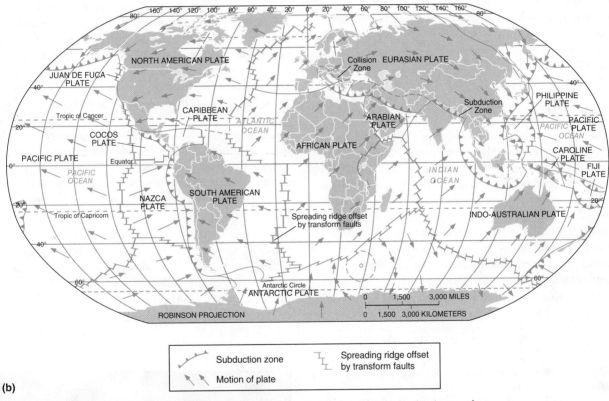

(b)

Figure 4–20 Drifting continents and plate tectonics. (a) Similarities in the types of rock, the distribution of fossil species, and other lines of evidence indicate that 225 million years ago all the present continents were formed into one huge landmass that we now call Pangaea. **(b)** The 14 major tectonic plates making up Earth's crust and their directions of movement.

(*Source:* From *Geosystems: An Introduction to Physical Geography*, 5th ed., by Robert W. Christopherson, copyright © 2005 by Prentice Hall, Inc., reprinted by permission of Pearson Education, Inc., Upper Saddle River, NJ 07458.)

diversity. As continents moved to their present position, they isolated organisms into broad regions. Sometimes scientists divide the world into six zoogeographic regions—areas of Earth's surface with distinctive animal groups. Australia, for example, is part of the Australasian region and is home to almost all of the world's marsupials—animals with pouches. Australia is located on the Indo-Australian tectonic plate. Its separation from other continents appears to be very ancient, which explains why so many species in the Australasian

region are unique: They have had millions of years to evolve separately from the species on other continental plates.

Sometimes the place where several tectonic plates meet will have an unusual mix of species. For example, the country of Indonesia, which consists of about 17,000 islands, is located on the junction of three tectonic plates—the Pacific, Eurasian, and Indo-Australian plates. The island of New Guinea (half of which belongs to Indonesia and half to the country of Papua New Guinea) is part of the

Indo-Australian plate and has plants and animals similar to those in Australia. Other islands in the region have biota more like those in parts of Asia. The island of Sulawesi lies right on a boundary between plates and has been formed by islands being pushed together. It is home to an odd combination of animals and plants with relatives in far distant parts of the world.

Because they have been isolated on islands, many Indonesian species have changed over time, developing adaptations not seen in their Asian or Australian relatives. Both Australia, long separated from other continents, and Indonesia have a high number of species that occur nowhere else in the world. Australia and Indonesia serve as examples of how long-term, large-scale changes, such as the movement of continents and the rise of islands, can increase the diversity of organisms.

CONCEPT CHECK Would you expect evolutionary change to be more rapid on an island chain or on a large, flat continent? Why? ☑

4.5 Implications for Management by Humans

Ecologists study populations and communities for a number of reasons. One is simply to further our understanding of the world. Another is to better manage natural resources by protecting declining species such as the African elephant and the California condor and by controlling pest species such as the Sodom apple and spotted knapweed. It is also important for us to study our own impacts on populations, as many of the causes of explosive growth or decline in populations stem from human actions, such as the removal of an important species or the introduction of non-native species (pests will be covered in Chapter 13 and biodiversity in Chapter 6).

Throughout this book, we will see many examples of species in decline. In later chapters we will discuss ways of conserving them, especially through captive breeding programs, habitat protection, and the prevention of the problems that cause them to decline. One such problem is introduced species.

Introduced Species

The Sodom apple described earlier is an "encroacher," a native species that can increase rapidly and take over ecosystems when some environmental factor changes. In the savanna, overgrazing by domestic animals and the loss of browsers such as elephants have opened up areas where the toxic plant thrives. This demonstrates how populations of certain species that have been limited by interactions in their niche expand rapidly when those limits are removed.

Humans have also altered population equilibria by introducing species from foreign ecosystems. These introductions alter community and population ecology relationships such as competition and predation. Over the past 500 years, and especially now that there is vast global commerce, thousands of species of plants, animals, and microbes have been accidentally or deliberately introduced onto new continents and islands. Most of the important insect pests we try to control—Japanese beetles, fire ants, apple snails, and gypsy moths, for example—are species introduced from other continents. A recent attempt to quantify the economic losses due to introduced species in the United States estimated the total cost to be in excess of $138 billion per year.

Invading Organisms. In 2008, researchers on Gough Island in the South Atlantic were shocked to discover one of the biggest reasons for the decline of nesting sea birds: predation by mice (Figure 4–21). Mice and rats have been introduced to thousands of oceanic islands by ships. Often, the descendants

Figure 4–21 Invasive animals. (a) Rodents such as this house mouse often overrun oceanic islands when they are introduced by ships. Invasive rodents are one of the chief causes of decline for sea birds. **(b)** This albatross chick is very vulnerable to rodent predators.

(a)

(b)

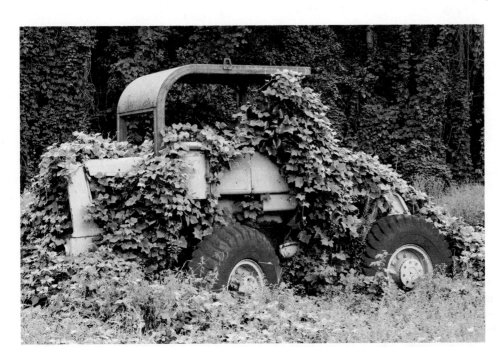

Figure 4–22 Invasive plant species. Kudzu overgrowing a tractor and trees.

of those first immigrants grow to be much larger than ordinary members of the species. Researchers knew there were many rodents on sea bird colonies; however, they did not understand the impact of the rodents until they used infrared cameras to capture night images of nesting birds. There they saw the mice directly preying on much larger baby birds. Restoration of sea bird populations requires removal of these mice; such removal has been found to be effective in trial cases.

While most people do not consider domestic cats to be pests, they are effective predators that have decimated small animal populations in places where they have been allowed to roam free. Indeed, cats are responsible for greatly diminished songbird populations in urban and suburban areas, including parks (Chapter 6).

Introduced plants can also cause ecological damage. One famous example is kudzu, a vigorous vine introduced to the United States from Japan in 1876. Initially used for animal fodder and erosion control, the quick-growing plant soon invaded forests. Today it occupies more than 3 million hectares (7 million acres) of the Deep South (Figure 4–22).

Invasive fungi and other microbes can wreak havoc in new environments. Prior to 1900, the dominant tree in the eastern deciduous forests of the United States was the American chestnut, which was highly valued for its high-quality wood and its prolific production of chestnuts, eaten by wildlife and people alike. In 1904, however, a fungal disease called the chestnut blight was accidentally introduced into New York. The fungus spread through the forests, killing nearly every American chestnut tree by 1950.

Exporting Problems. Invasive organisms may be exported as well as imported. In 1982, several species of jellyfish-like animals known as ctenophores were transported from the East Coast of the United States to the Black Sea and the Sea of Azov in Eastern Europe. In the Sea of Azov, the ctenophores shut down fisheries by killing larval fish and depriving larger fish of food (Figure 4–23). The ctenophores have also moved into the Mediterranean and Caspian Seas, along the Atlantic Coast of Europe, and have just begun to move into the North Sea.

Most invasive species are not problems in their native lands. In a new setting, however, their impact is quite different. Over long periods of time, the species within each ecosystem have developed adaptations to the other organisms within their own ecosystem, but these adaptations do not

Figure 4–23 Introduced species. Ctenophores, originating along the Atlantic Coast in the southern United States and introduced into Europe, have destroyed fishing in the Black Sea and the Sea of Azov.

prepare them to interact with species that have developed in other ecosystems.

Remedies. Invasive species are such a concern for the health of ecosystems that there are state and federal agencies working to slow their entry and spread. The National Invasive Species Information Center, for example, acts as a clearinghouse of information on invasive species in the United States. The United States has a number of regulations about invasive species, including the Nonindigenous Aquatic Nuisance Prevention and Control Act of 1990 and the National Invasive Species Act of 1996.

Another solution to the takeover by an invasive species is to introduce a natural enemy. Indeed, this approach has been used in a number of cases, including in that of spotted knapweed (see Sound Science, *The Biological Detective: The Case of Spotted Knapweed;* other approaches are discussed in Chapter 13). Unfortunately, the natural enemy that is introduced may become an invasive pest itself. In short, to prevent doing more harm than good, a great deal of research needs to be done before a natural enemy is introduced.

Ecological Lessons

The regulation of populations is a matter of complex interactions among the members of the biotic community, and these relationships are specific to the organisms in each particular ecosystem. Therefore, when a species is transported over a physical barrier from one ecosystem to another, it is unlikely to fit into the framework of relationships in the new biotic community. In most cases, it finds the environmental resistance of the new system too severe and dies out. No harm is then done. In a small percentage of cases, the transported species becomes invasive. (Invasive species will be discussed again in Chapters 6, 7, and 13.) Scientists use the lessons of population and community ecology to conserve species headed for extinction; they also use them to manage the species we harvest from wild populations.

CONCEPT CHECK Describe a human action that increases the size of a species' population and an action that decreases the size of a wild population. Give an example of each. ☑

SOUND SCIENCE

The Biological Detective: The Case of Spotted Knapweed

Millions of acres in the western United States are infested with a plant that is barely edible to grazers such as cattle, whose growth is harmed by the plant's toxins. Spotted knapweed (*Centaurea biebersteinii*), a member of the Aster family, was accidentally introduced into the United States as seeds from Europe in the late 1800s. It thrives in the West because it produces more seeds than other plants and its seeds can remain in the ground for five years, sprouting whenever there is an opening (high biotic potential). In addition, the plant has few natural predators (low environmental resistance).

Many herbivores eat specific plants and have adaptations that allow them to get around the plant's defenses. But when plants are moved into a new ecosystem, these herbivores do not usually travel with them. Without the limit of environmental resistance, the plant outcompetes other plants for light and water.

For years, scientists searched for a way to control knapweed. In the early 1970s, they thought they had the ideal solution. They introduced a gallfly that specializes in eating spotted knapweed, hoping to avoid the problems that occur when generalist herbivores that are moved from one part of the globe to another eat numerous local species. This solution

should have worked. The gallfly lays its eggs in the knapweed flowers. Larvae hatch and trigger the plant tissues to grow a gall, which protects the insect. The larvae overwinter in the galls and emerge in the spring. Because spotted knapweed plants put a great deal of energy into making the woody tissue in the galls, they make fewer seeds. In fact, the introduction of the gallflies did make the population of knapweed decline, but the decline was not as significant as expected. Why? Recently, scientists have discovered part of the answer to the mystery.

In a study reported in 2008 in the journal *Ecological Applications*, researcher Dean Pearson and his colleagues at the U.S. Forest Service's Rocky Mountain Research Station found a surprising relationship between the gallfly and deer mice. Gallfly larvae are eaten by local deer mice, which also eat seeds and other insects. Because there are so many spotted knapweed plants, there are plenty of gallflies. Because there are plenty of gallflies, there are plenty of deer mice. But the deer mice also eat the seeds of native plants. Because the native plants make many fewer seeds than the spotted knapweed, the remaining knapweed seeds can still outgrow the native plants. This

unexpectedly complex set of relationships means that even a biological control agent chosen because of its specificity to one host may have unintended consequences.

Source: D. E. Pearson and R. M. Callaway, "Weed Biocontrol Insects Reduce Native Plant Recruitment Through Second-Order Apparent Competition," *Ecological Applications 18* (2008): 1489–1500.

REVISITING THE THEMES

SOUND SCIENCE

The African savanna elephant is in decline. The elephant's population is controlled by human impacts such as poaching, habitat loss, and land use changes. The population of the native Sodom apple is increasing because of changes in the ecological controls with which it evolved; elephants may play a role in controlling this noxious weed. The kudzu plant and mice on islands are introduced exotic species, now living with species that have not evolved adaptations to them. These examples illustrate some of the ideas of the chapter, such as the limits to population growth, the relationships between plants and herbivores, the problem of introduced species, and the role humans can play in managing ecosystems.

The science of population and community ecology allows people to better understand why species live where they do and what factors affect their abundance. Populations can be controlled by abiotic factors such as precipitation or by a complexity of biological relationships, such as competition, predation, mutualism, and commensalism.

Species typically have one of two reproductive strategies in their life history. Opportunistic or *r*-selected species reproduce early, produce many small offspring, and do not care for their young, most of which die. In contrast, equilibrial or *K*-selected species have fewer offspring at older ages and care for their young. These differences correlate to differences in the ways in which populations grow: *r*-selected species often overshoot their carrying capacity and crash, while *K*-selected species are more likely to increase slowly and then fluctuate around their carrying capacity. Either way, all species are limited. When a few individuals start a new population, the growth may initially be exponential, but the population cannot maintain continual growth.

Finally, sound science shows us that biological systems change because they are constantly varying genetically and because changes in the environment over space and time exert different selection pressures. Sound science helps us to understand complex community relationships, as we saw in the case of the spotted knapweed.

SUSTAINABILITY

In this chapter, we saw that sustainable management means paying attention to the relationships in a community and to constraints such as carrying capacity. Sometimes it is difficult to find effective management practices, as was shown with numerous examples of human actions resulting in unexpected consequences, such as the attempts to control spotted knapweed. The movement of species from ecosystems where they have coevolved with other species to places lacking those evolutionary relationships often results in either the explosion of an unwanted pest or the collapse of another species. We have had to develop agencies and policies to minimize problems from the transport of species.

STEWARDSHIP

When questions arise about what's best for populations and communities, the answers indicate that what seems to benefit one species (importing a plant or animal, for example) may not benefit another. Human activities may change the environment so quickly or on such a large scale that some species cannot adjust to the changes, causing those species to decline. This is definitely the case with the African savanna elephant. Because disruptive changes can result from natural processes, such as forest fires or volcanoes, one might ask, "If change is natural, why are we concerned about species loss?" (This will be discussed further in Chapter 6).

Stewardship of populations requires that we understand them well, but it also may require actions such as setting aside habitat, reintroducing keystone species, removing invasive species, and making sure one species does not completely outcompete another. Stewardship of populations certainly involves trying to avoid moving weedy species around the globe and the over-exploitation of species we use as resources.

REVIEW QUESTIONS

1. What are three basic population growth curves? When might you see them in nature?

2. Define *biotic potential* and *environmental resistance*, and give factors of each. Which generally remains constant, and which controls a population's size?

3. Differentiate between the terms *critical number* and *carrying capacity*.

4. What is density dependence?

5. Explain the difference between *r*- and *K*-strategists. Where do these terms come from, and what are the characteristics of each broad category?

6. Describe the predator-prey relationship between the moose and wolves of Isle Royale. What other factors influence these two populations?

7. Distinguish between intraspecific and interspecific competition. How do they affect species as a form of environmental resistance?

8. What is meant by territoriality, and how does it limit the effects of competition in nature?

9. What problems arise when a species is introduced from a foreign ecosystem? Why do these problems occur?

10. What are selective pressures, and how do they relate to natural selection?

11. Describe several types of adaptations a species might have that would allow it to survive in a dry environment, escape predation, or lower competition with another species.

12. What factors determine whether a species will adapt to a change or whether the change will render it extinct?

13. How may evolution lead to the development of new species (speciation)?

14. What is plate tectonics, and how does this theory explain past movements of the continents? How have past tectonic movements affected the present-day distribution of plant and animal species on Earth's surface?

THINKING ENVIRONMENTALLY

1. Describe, in terms of biotic potential and environmental resistance, how the human population is affecting natural ecosystems.

2. Explain the key differences between *r*- and *K*-strategists, their life histories, and their reactions to environmental changes. Where do the *r* and *K* come from in population equations? What do they stand for?

3. Consider the various kinds of relationships humans have with other species, both natural and domestic. Give examples of relationships that (a) benefit humans but harm other species, (b) benefit both humans and other species, and (c) benefit other species but harm humans. Give examples in which the relationship may be changing—for instance, from exploitation to protection. Discuss the ethical issues involved in changing relationships.

4. Consider the research of the Grants over decades of work on the Galápagos. Why is such research important? How does this basic science benefit us as we try to protect ecosystem functions?

5. Describe how a human action such as removing a top predator or adding a species can have an impact on many species in an ecosystem.

MAKING A DIFFERENCE

1. If you have a garden, grow native plants instead of alien ones. Otherwise, seeds from introduced plants could escape into the surrounding ecosystems. Native grasses, flowers, shrubs, and trees are more likely to attract native birds, butterflies, and other insects.

2. Visit a nearby national park or nature reserve. Talk to the rangers to find out about any threatened species and how they are being protected. Get involved by volunteering at your local nature center or wildlife refuge. Go wildlife or bird watching in nearby parks. Wildlife-based recreation creates millions of jobs and supports local businesses.

3. If you are a hunter or fisher, use nonleaded shot and fishing gear so birds eating spent shot or lost sinkers will not suffer lead poisoning. Follow the regulations regarding seasons and catch limits, and report poachers to your local wildlife agency.

4. When choosing outdoor recreational activities, consider cross-country skiing and canoeing instead of snowmobiling or motor boating. These activities are quieter, and they increase your chances of seeing wildlife. Loud noises, especially in winter, disturb animals when they need to rest and conserve energy.

MasteringEnvironmentalScience®

Students Go to MasteringEnvironmentalScience for assignments, the eText, and the Study Area with animations, practice tests, and activities.

Professors Go to MasteringEnvironmentalScience for automatically graded tutorials and questions that you can assign to your students, plus Instructor Resources.

Ecosystems: Energy, Patterns, and Disturbance

Clouds of smoke and ash accompanied rumblings of the Earth as Iceland's volcano at Eyjafjallajökull ("EY-ya-fyat-lah-YOH-kuht") belched awake on April 14, 2010. The trembling that had affected the area over the previous month increased, and a billowing cloud of ash erupted into a plume that rose 9 kilometers (30,000 feet) into the air. The ash spread across Western Europe, clouding the sky and closing down all air travel for days; the largest air traffic disruption since World War II. Thousands of travelers were stranded as the volcano spewed particles and gases into the atmosphere. Ash fell on the ground for miles, making the grasses unfit for consumption by cattle and horses and making driving impossible. Even so, Eyjafjallajökull was a moderate volcano, without the massive lava flows and mudslides that accompany many volcanic eruptions. People were relatively safe and, except for the cessation of air travel, life returned to normal quickly.

Preparation. The eruption of Mount Pinatubo on June 15, 1991, in the Philippines was larger and had more-serious global effects. The eruption of this volcano, with no recorded history of eruptions, blew the top of the densely forested mountain into the air. This was the beginning of a process that spilled between 6 and 16 cubic kilometers of ash into the atmosphere, along with 20 million tons of sulfur dioxide and 10 billion metric tons of magma. Small particles known as aerosols were thrown into the air and traveled around the globe, blocking sunlight and lowering the global temperature by 0.5°C (0.9°F). More aerosols were produced by this volcanic eruption than any since Krakatoa in 1883. The sulfur found in the aerosols combined with water to form sulfuric acid, a component of acid rain. Fortunately, even though the volcano had not erupted in years prior to the event, scientists were able to predict that it would, and were able to prompt the largely successful evacuation of 60,000 people.

Chapter 5

Learning Objectives

5.1 Characteristics of Ecosystems: Describe how matter and energy flow through ecosystems by moving from one trophic level to another.

5.2 The Flow of Energy Through the Food Web: Explain three main ideas relating trophic pyramids with the relative numbers and biomasses of different levels in a food chain.

5.3 From Ecosystems to Global Biomes: Define and recognize characteristics of major broad regions called biomes, major aquatic regions, and factors that determine their placement on the globe.

5.4 Ecosystem Responses to Disturbance: Explain the effects of ecological disturbances, such as a fire or volcanic eruption, that are normal in ecosystems and can even be beneficial.

5.5 Human Values and Ecosystem Sustainability: Describe ways that humans alter ecosystem services, and explain why we need to manage ecosystems to protect their components from overuse.

▲ Iceland's Eyjafjallajökull volcano in April 2010.

Unfortunately, people did die in the aftermath, when typhoon rains and wet ash caused roofs to collapse, and crowded emergency housing allowed diseases to spread. Today, years after the eruption, parts of Mount Pinatubo still lie denuded as periodic landslides move material down the hillside.

Rebuilding. Volcanoes, triggered by the movements of tectonic plates, relieve pressure building under Earth's crust and spread minerals into the soil. The materials released by volcanoes are important parts of nutrient cycles. Volcanoes may build new lands altogether by causing islands, such as the Hawaiian Islands, to rise out of the ocean. Often, volcanic eruptions kill many of the organisms living in their immediate vicinity, but such organisms return and recolonize in a process of succession. In this chapter, we will look at ecosystems and how they function, including the processes that occur after a disturbance such as a wild fire, large flood, volcano, or retreat of a glacier. These processes offer us a chance to better understand the workings of an ecosystem.

Earlier, the focus was on the components of ecosystems (Chapters 3 and 4). We considered the environmental conditions and resources required by living things, the basic movement of energy and nutrients that sustain living things, and how populations work and relate to populations of other species. In this chapter, we will look at how basic processes such as photosynthesis and respiration in individuals play out in complex systems and how energy moves from one part of the community to another. We will see how concepts such as ranges of tolerance and limiting factors, which we applied to individuals, apply to large groups and control their distributions over large portions of Earth. We will also look at how ecosystems change over time as a result of disturbances such as volcanoes.

It is a popular belief that nature will be in balance if left alone. You will learn in this chapter, however, that an ecosystem is a dynamic system in which changes are constantly occurring. The notion of a balanced ecosystem, then, must be carefully defined. We will ask questions about balance, or equilibrium, in ecosystems: What does it mean for an ecosystem to be "in balance"? Are populations normally in a state of equilibrium? What happens when a major disturbance interrupts the "balance" in an ecosystem?

Our objective in this chapter is to examine more of the basic mechanisms that underlie the sustainability of all ecosystems, including those heavily influenced by human use. We will look at the values, goods, and services of ecosystems and their importance to humans. The chapter concludes with some perspectives on the challenge of managing ecosystems and a summary of ecosystem sustainability issues.

5.1 Characteristics of Ecosystems

Mt. Pinatubo is located on the island of Luzon in the Philippines, an area with one of the largest forests in the country. Luzon is also home to some of the few remaining individuals of the highly endangered Philippine eagle. We will look at this area to better understand the characteristics of ecosystems. The forest around Mt. Pinatubo, like all ecosystems, contains communities of interacting species and their abiotic factors. However, the species living in this ecosystem, or any other ecosystem, may function on very different scales, which makes it difficult to delineate a fixed boundary for ecosystems. For example, the lowland rain-forest ecosystem contains migratory birds that live there only seasonally. Some species inhabiting the ecosystem may also have little to no interaction with each other, such as the orchids high on branches of trees and organisms that live in leaf litter on the forest floor.

The rain forest around Mt. Pinatubo contrasts sharply with the treeless spaces of Iceland, where Eyjafjallajökull sent ash into the air. As we will see later in the chapter, what lives in an ecosystem is a function not only of characteristics of the regional climate, but of the landscape's history—including the actions of humans. Even so, there are predictable patterns to the distribution of ecosystems around the globe. Ecosystems that have a similar type of vegetation and similar climactic conditions are grouped into broader areas called *biomes*, such as the tropical rain-forest biome. (We will look at biomes in more depth in Section 5.4.)

The study of ecosystems allows us to understand concepts that cannot be predicted by understanding community interactions alone. For example, community relationships such as predation and parasitism (discussed in Chapter 4) provide the mechanisms for the flow of energy—as well as the flow of carbon, nitrogen, and phosphorus—through the natural world (discussed in Chapter 3). But these community relationships may not tell us about ecosystem properties, such as how much photosynthesis occurs in all plants or how much water can be absorbed by all of the plants at once. In this chapter, we will pull together the concepts presented in the previous two chapters, abiotic forces and biotic interactions, to look at characteristics of ecosystems—specifically, trophic levels, productivity, and consumption.

Trophic Levels, Food Chains, and Food Webs

Imagine a grassy meadow containing a variety of plants. As discussed in previous chapters, plants use energy from the Sun through the process of photosynthesis to produce complex chemical chains from more-simple molecules such as carbon dioxide and water. These plants can then be eaten by their predators, such as rabbits, mice, or even cattle (remember that herbivory is a subset of predation). In turn, the mouse could be eaten by one of its predators, a hawk or a fox; the squirrel eaten by a snake; and the cattle eaten by a human. Energy in the form of chemical bonds and nutrients in the form of molecules in the bodies of the prey are then used by the next **trophic level** (feeding level)—the predators—for energy and other physiological needs. This breakdown of chemical bonds to release energy is called cellular respiration. Recall that plants need to respire, too.

You can imagine two things from this. First, organisms can be linked into feeding chains—the grass to the mouse, to the snake, to the hawk, for example. Such **food chains** describe where the energy and nutrients go as they move from one organism to another. We usually refer to this as energy moving "up" the food chain. Second, you can imagine that the energy and nutrients are not all passed up the chain from one level to the next. Because of the energy laws we discussed, there is inefficiency every time one organism eats another (Chapter 3).

In natural ecosystems, food chains seldom exist as isolated entities. Mice feed on several kinds of plants, are preyed on by several kinds of animals, and so on. Some organisms, including humans, eat at more than one level in the food chain. Consequently, virtually all food chains are interconnected and form a complex web of feeding relationships—the **food web** (Figure 5–1).

Rather than simply asking what predators are eating snakes and how they affect the snake population or what mice are eating and how food availability affects mice, we ask different types of questions in ecosystem ecology—those that focus on trophic levels in general, in order to study bigger trends in the movement of energy and materials. Examples of these questions include the following: How much of the Sun's energy do the plants in the meadow trap? How much do they use, and how much is available to plant eaters? How many predators can the ecosystem support? In order to answer these questions, ecologists divide members in the food web into categories.

Trophic Categories

All organisms can be categorized as either autotrophs or heterotrophs, depending on whether they produce the organic compounds they need to survive and grow. Green plants, photosynthetic single-celled organisms, and chemosynthetic bacteria are **autotrophs** (*auto* = self; *troph* = feeding) because they produce their own organic material from inorganic constituents in their environment through the use of an external energy source. They are also referred to as producers. Organisms that must consume organic material to obtain energy are **heterotrophs** (*hetero* = other; *troph* = feeding). Heterotrophs may be divided into numerous categories, the two major ones being **consumers**, which eat living prey, and **decomposers**, which break down dead organic material. Together, producers, consumers, and decomposers produce food, pass it along food chains, and return the starting materials to the abiotic environment. Figure 5–2 (on the following page) defines the major trophic categories.

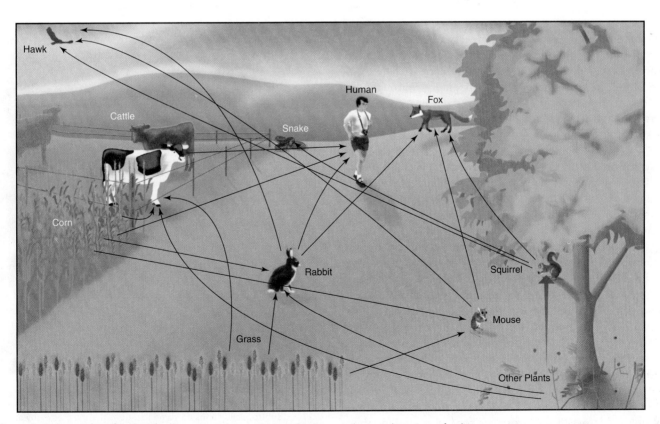

Figure 5–1 A meadow food web. The arrows show herbivores feeding on plants and carnivores feeding on herbivores. Specific pathways, such as that from nuts to squirrels to foxes, are referred to as *food chains*. A food web is the collection of all food chains, which are invariably interconnected.

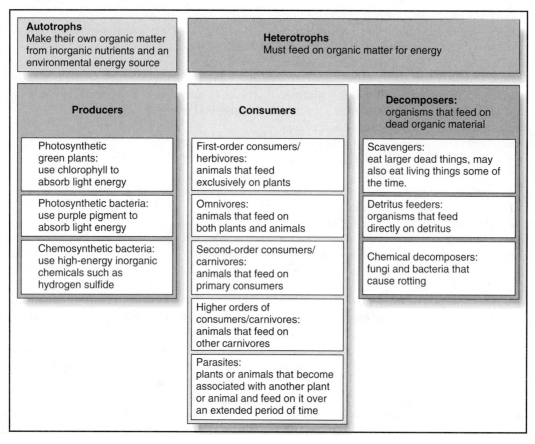

Autotrophs Make their own organic matter from inorganic nutrients and an environmental energy source	Heterotrophs Must feed on organic matter for energy	
Producers	**Consumers**	**Decomposers:** organisms that feed on dead organic material
Photosynthetic green plants: use chlorophyll to absorb light energy	First-order consumers/ herbivores: animals that feed exclusively on plants	Scavengers: eat larger dead things, may also eat living things some of the time.
Photosynthetic bacteria: use purple pigment to absorb light energy	Omnivores: animals that feed on both plants and animals	Detritus feeders: organisms that feed directly on detritus
Chemosynthetic bacteria: use high-energy inorganic chemicals such as hydrogen sulfide	Second-order consumers/ carnivores: animals that feed on primary consumers	Chemical decomposers: fungi and bacteria that cause rotting
	Higher orders of consumers/carnivores: animals that feed on other carnivores	
	Parasites: plants or animals that become associated with another plant or animal and feed on it over an extended period of time	

Figure 5–2 **Trophic categories.** A summary of how living organisms are ecologically categorized according to how they get their energy.

Producers. As we have seen, producers are organisms that capture energy from the Sun or from chemical reactions to convert carbon dioxide (CO_2) to organic matter (Chapter 3). On land, most producers are green plants. Through the process of photosynthesis, plants use light energy to convert CO_2 and water to organic compounds such as glucose (a sugar) and then release oxygen as a by-product. Plants use a variety of molecules to capture light energy in photosynthesis. The most predominant of these is **chlorophyll,** the green pigment that gives photosynthetic plants their color. In some producers, additional red or brown photosynthetic pigments may mask the green, as in red and brown algae.

Producers range in size from microscopic organisms that are not plants such as photosynthetic bacteria and single-celled algae, through medium-sized plants such as grass, daisies, and cacti, to gigantic trees. Every major ecosystem, both aquatic and terrestrial, has its particular producers that are actively engaged in photosynthesis. While plants are the best known producers, researchers are finding that more photosynthesis is done by microbes than had been realized. In the ocean, there are tiny producers called microplankton, nanoplankton, and the even smaller picophytoplankton (Figure 5–3). These autotrophs are eaten by tiny predators in a microscopic food chain. While they are small, their vast numbers make such tiny producers an important part of the ocean food web (Figure 5–4).

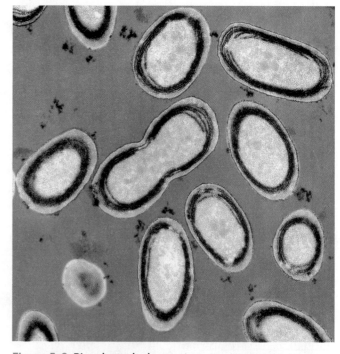

Figure 5–3 **Picophytoplankton.** These *Prochlorococcus* cyanobacteria are extremely small, photosynthetic microbes that live in tropical and subtropical oceans. In some locations there are as many as 100,000 cells per milliliter of sea water, making them possibly the most abundant photosynthetic organisms on Earth. Magnification: ×23,000 at 6 × 7cm size, colored TEM.

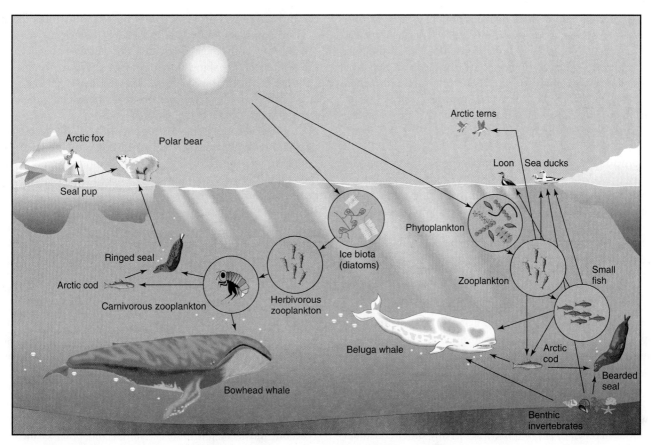

Figure 5–4 A marine food web. This simplified food web shows some of the relationships among organisms in the Beaufort Sea (north of Alaska and Canada) in spring. Zooplankton and phytoplankton are groups that include many species.

Not all producers use the Sun's energy as the basis of their activities. Certain bacteria, including some that live in deep-sea communities, are able to use the energy in certain inorganic chemicals to form organic matter from CO_2 and water. This process is called **chemosynthesis,** and the organisms using it are producers, too.

Producers are absolutely essential to every ecosystem. The photosynthesis and growth of producers such as green plants constitute the **primary production** of organic matter, which sustains all other organisms in the ecosystem.

Consumers. All consumers obtain their energy by feeding on organic matter. Consumers encompass a wide variety of organisms, ranging in size from plankton to blue whales. Among consumers are such diverse groups as protozoans, worms, fish, shellfish, insects, reptiles, amphibians, birds, and mammals (including humans).

Consumers can be divided into various subgroups according to their food source. Animals—as large as elephants or as small as mites—that feed directly on producers are called **first-order consumers** or **herbivores** (*herb* = grass). Animals that feed on first-order consumers are called **second-order consumers.** Thus, mice, which feed on vegetation, are first-order consumers, whereas snakes that feed on mice are second-order consumers (**Figure 5–5**). There may also be third-order, fourth-order, or even higher levels of consumers. Many

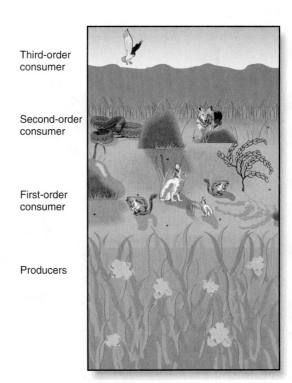

Figure 5–5 Trophic levels in a grassland. A grassland community showing the producers (plants) and the first-, second-, and third-order consumers (mouse, snake, and hawk, respectively).

animals occupy more than one position on the consumer scale. For instance, humans are first-order consumers when they eat vegetables, second-order consumers when they eat beef, and third-order consumers when they eat fish that feed on smaller fish that feed on algae. Second-order and higher-order consumers are also known as **carnivores** (*carni* = meat). Consumers that feed on both plants and animals are called **omnivores** (*omni* = all).

Decomposers. A very large proportion of the primary producer trophic level is not consumed in the grazing food web. As this material dies (leaves drop, grasses wither, and so on), it is joined by the fecal wastes and dead bodies from higher trophic levels and is referred to as *detritus*. This represents the starting point for a separate food web. In many cases, most of the energy in an ecosystem flows through the detritus food web (Figure 5–6).

Detritus is composed largely of cellulose because it consists mostly of dead leaves, the woody parts of plants, and animal fecal wastes. Nevertheless, it is still organic and high in potential energy for those organisms that can digest it—namely, decomposers. Scavengers, such as vultures, help break down large pieces of organic matter; detritus feeders, such as earthworms, eat partially decomposing organic matter; and chemical decomposers (various species of fungi and bacteria as well as a few other microbes) break down dead material on the molecular scale. Decomposers act as any other consumer does, using the organic matter as a source of both energy and nutrients. Some, such as termites, can digest woody material because they maintain decomposer microorganisms in their guts in a mutualistic, symbiotic relationship. The termite

(a detritus feeder) provides a cozy home for microbes (chemical decomposers) and takes in cellulose, which the microbes digest for both their own and the termites' benefit. The decomposers, breaking down detritus, serve as the primary consumers in this food chain. The detritus is similar to the "producer" trophic level, and the decomposers have their own predators that form higher trophic levels.

Most decomposers use oxygen for cellular respiration, which breaks detritus down into carbon dioxide, water, and mineral nutrients. However, some decomposers (certain bacteria and yeasts) meet their energy needs through the partial breakdown of glucose, which can occur in the absence of oxygen. This modified form of cellular respiration, called **fermentation,** results in end products such as ethyl alcohol, methane gas, and acetic acid. In nature, **anaerobic,** or oxygen-free, environments commonly exist in the sediments of lakes, marshes, and swamps and in the guts of animals, where oxygen does not penetrate readily. Methane gas is commonly produced in these locations. A number of large grazing animals, including cattle, maintain fermenting bacteria in their digestive systems in a mutualistic, symbiotic relationship similar to that just described for termites. As a result, both cattle and termites release methane.

As decomposers digest detritus, they break the bonds in large organic molecules, releasing chemical energy, which they use to live and grow. They also break the bonds holding carbon, nitrogen, phosphorus, sulfur, and other elements in the organic molecules, thereby releasing nutrients to their surroundings. The release of nutrients by decomposers is vitally important to primary consumers because it is the major source of nutrients in most ecosystems.

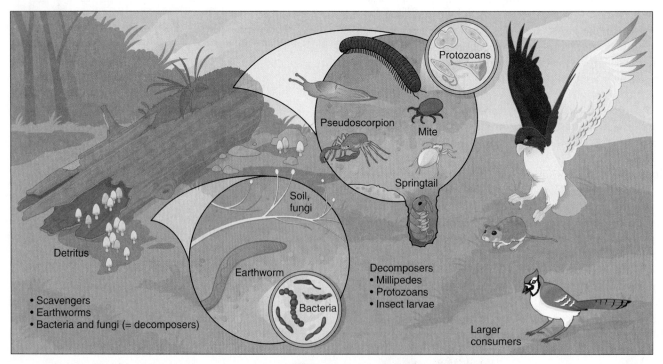

Figure 5–6 Detritus food web. Several types of decomposers work to break up large, small, and tiny bits of dead organic matter. These organisms in turn are eaten by other, higher-level consumers. In the detritus food web, the role of the producer is played by dead plant and animal matter.

Decomposers occur in the ocean as well as on land. Most of the decomposition in the ocean is carried out by *picoheterotrophs*—tiny, single-celled eukaryotes, bacteria, or members of Archaea, a primitive group of bacteria-like organisms. While it is difficult to estimate their abundance, picoheterotrophs are common in the deep ocean from the tropics to the poles. Because they occur throughout such a large volume of water, their impact on nutrient recycling, particularly the carbon cycle, is substantial.[1]

Limits on Trophic Levels

How many trophic levels are there? Usually, there are no more than three or four in terrestrial ecosystems and sometimes five in marine systems (Figure 5–4). This answer comes from straightforward observations. If you were to capture all of the organisms on each trophic level over a particular area or volume, dry them, and weigh them, you would find

[1] E. Sclaechter, "Pico Who? Small Things Considered" (2008): http://schaechter.asmblog.org/schaechter/2008/07/pico-who.html.

a pattern: The **biomass,** or total combined (net dry) weight, would typically be about 90% less at each higher trophic level for a terrestrial ecosystem. For example, if you dried and weighed all of the grass and other producers in an acre of grassland, you might have 907 kilograms (1 ton or 2,000 lb.) per acre. If you could capture, dry, and weigh all of the herbivores (everything from grasshoppers to rabbits) in the same acre, you would find the biomass of herbivores would be about 90.7 kilograms (200 lb.) per acre. Following that pattern, the same acre would yield about 9.7 kilograms (20 lb.) of second-order consumers (animals living on grasshoppers and rabbits) if you captured, dried, and weighed them. At this rate, you can't go through very many trophic levels before the biomass approaches zero. If you graphed the different levels of producer and consumer mass, you would have a **biomass pyramid** similar to the one shown in Figure 5–7. We'll see the reason for this pattern as we look at where energy and nutrients go.

CONCEPT CHECK What is the trophic level of a shrew eating a grasshopper? Can you figure out the trophic level of a shrew eating an earthworm? What might make that difficult? ☑

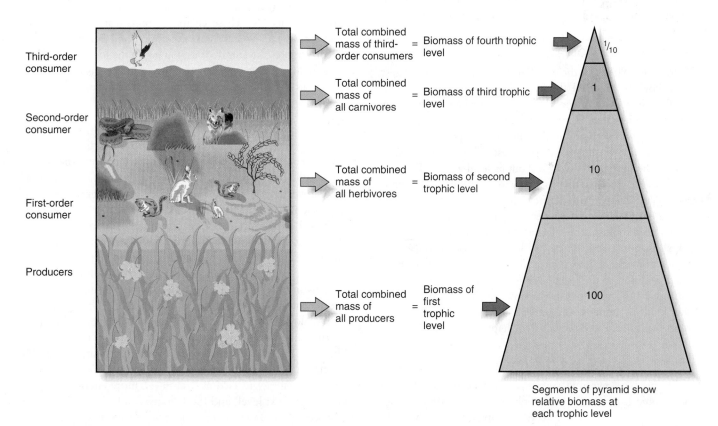

Figure 5–7 A biomass pyramid. A graphical representation of the biomass (the total combined mass of organisms) at successive trophic levels has the form of a pyramid, with about 10% of the energy contained in one trophic level being incorporated into the bodies of the next trophic level, giving each level a smaller biomass.

UNDERSTANDING THE DATA

1. The combined biomass of the producers in a meadow ecosystem is 1200 kilograms. What would you expect the biomass of the third-order consumers to be?

2. Why would it be very difficult for a predator to feed exclusively on eagles and hawks? Use the biomass pyramid to support your answer.

5.2 The Flow of Energy Through the Food Web

In most ecosystems, sunlight (or solar energy) is the initial source of energy absorbed by producers through the process of photosynthesis. The only exceptions are rare ecosystems, such as those near the ocean floor, where the producers are chemosynthetic bacteria. Primary production—the production of organic molecules from the Sun's energy through photosynthesis—captures only about 2%, at most, of incoming solar energy. Even though this seems like a small fraction, the resulting terrestrial net production—estimated at some 120 gigatons (billion metric tons, 1.1 billion U.S. tons) of organic matter per year—is enough to fuel all of the life in a biome. In a given ecosystem, the actual biomass of primary producers at any given time is referred to as the standing-crop biomass. Both biomass and primary production vary greatly in different ecosystems. For example, a forested ecosystem maintains a very large biomass compared with tropical grassland because so much carbon is captured in the wood of trees. However, the rate of primary production could be higher in the grassland; grazing animals might simply eat the produced organic matter. Consequently, standing biomass is not always a good measure of productivity.

The Fate of Food

Whereas 60–90% of the food that consumers eat, digest, and absorb is oxidized (burned) for energy, the remaining 10–40%, which is converted to the body tissues of the consumer, is no less important. This is the fraction that enables a body to grow, maintain, and repair itself. A portion of what is ingested by consumers is not digested, but simply passes through the digestive system and out as fecal wastes. For consumers that eat plants, this waste is largely **cellulose,** the material of plant cell walls. It is often referred to as fiber, bulk, or roughage, and some of it is a necessary part of the diet. The intestines need to push some fiber through them so that they can stay clean and open. Waste products can also include compounds of nitrogen, phosphorus, and any other elements present, in addition to the usual carbon dioxide and water. These by-products are excreted in the urine (or as similar waste in other kinds of animals) and returned to the environment. This process was illustrated previously (Chapter 3, Figure 3–17).

Energy Flow and Efficiency

As the biomass pyramid suggests, when energy flows from one trophic level to the next, only a small fraction is actually passed on. This is due to three things: (1) Much of the preceding trophic level is biomass that is not consumed by herbivores, (2) much of what is consumed is used as energy to fuel the heterotroph's cells and tissues, and (3) some of what is consumed is undigested and passes through the organism as waste (e.g., feces). Thus, there is a tremendous inefficiency at each trophic level, and energy is lost to the food web. In an ecosystem, therefore, only the portion of food that becomes the body tissue of the consumer can become food for the next

organism in the food chain. Incorporating matter and energy from a lower trophic level into the body of a consumer is often referred to as secondary production, and like primary production, it can be expressed as a rate (the amount of growth of the consumer, or consumer trophic level) over time.

The inefficiency of trophic levels means two things. First, individuals at higher levels in the biomass pyramid represent a greater amount of the Sun's energy for the same amount of body tissue. For example, it requires a great deal more photosynthesis to create a pound of hamburger than to create a pound of soybeans. This may be counterintuitive, because we just saw that there is less energy at each trophic level, but it requires a great deal more energy to produce a top-order consumer than a mid-level consumer and, likewise, more energy to produce a mid-level consumer than a producer of the same body weight. It usually also takes a longer time to produce a top-order consumer than a mid-level consumer or a producer. The top-order consumer also requires more water and other resources for a unit of body weight. In later chapters, as we look at food production in a crowded world, we will see why this matters.

Second, some materials are difficult to get rid of (such as chemicals that dissolve in fat) and therefore **bioaccumulate,** or build up in an organism's tissues, remaining in its body throughout its life. These chemicals remain in the bodies of predators at higher rates than in their prey and thus biomagnify (become more concentrated) as you go up the food chain. This is the reason, as we will see later, that some toxins build up in the tissues of higher-order consumers such as eagles and tuna.

Aquatic Systems

For simplicity, the focus of this chapter has been on terrestrial ecosystems. Keep in mind, though, that the same processes occur in aquatic ecosystems. As aquatic plants and algae absorb dissolved carbon dioxide and mineral nutrients from the water, they use photosynthesis to produce food and dissolved oxygen that sustain consumers and other heterotrophs. Likewise, aquatic heterotrophs return carbon dioxide and mineral nutrients to the aquatic environment.

There are two differences between aquatic and terrestrial systems in this regard, however. First, the transfer of energy is often more efficient in an aquatic system. There are also more cold-blooded animals here, and less energy is required to run the bodies of cold-blooded creatures than warm-blooded creatures. It also takes less energy to support body weight in water than on land or in the air. Consequently, when less energy is used at each level, more energy is available to the next level, and food chains can be longer.

Second, aquatic systems do not always have the same kind of biomass pyramid as terrestrial systems. If you could capture all of the organisms in part of an ocean ecosystem and dry and weigh them, you might sometimes find an inverse biomass pyramid, especially if you lumped some levels together (Figure 5–8). The most widely supported explanation for this generally inverted pyramid is the relatively short life cycles of the organisms at the bottom of the food chain. Unlike the trees and other perennial plants that dominate most terrestrial

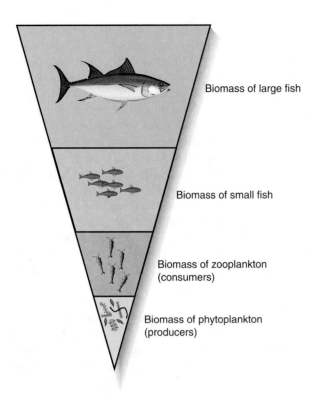

Figure 5–8 Reverse pyramid in aquatic systems. At certain times, some aquatic ecosystems have a small amount of algae and a much greater biomass of large fish. This is because the turnover is very high and life spans are so short at the lower trophic levels.

ecosystems, the producers in aquatic ecosystems are usually short-lived algae, while the animals at the top of the food chain, such as sharks and tuna, live far longer. The rapid growth and turnover of the algae allows them to produce a great deal of biomass. However, at any one moment, the biomass of the producers may sometimes be less than that of upper trophic levels.

CONCEPT CHECK Which is more energy efficient—feeding people with hamburgers or soyburgers? Why? ☑

5.3 From Ecosystems to Biomes

We have just seen how photosynthesis and basic energy laws translate into broad patterns in food chain lengths. Similarly, broad patterns in ecosystems translate into a pattern we see all over the globe: There is a predictable set of organisms that live under particular conditions, such as deserts occurring in dry areas and rain forests occurring in very wet areas. We can now use the concepts of optimums and limiting factors and the concepts of differences in light and productivity to gain a better understanding of why different regions may have distinct biotic communities, creating an amazing variety of ecosystems, landscapes, and biomes.

A **biome** can be defined as a large geographical biotic community, controlled by climate. Biomes are usually named after the dominant type of vegetation, such as deciduous forest or grassland. The area around Yellowstone National Park is typically called the Greater Yellowstone Ecosystem (Chapter 7). This ecosystem is contained within a vast geographical region,

or biome, called the northern temperate forest. Like ecosystems, biomes have fuzzy edges. Aquatic areas such as the tidal zone and the open ocean can also be characterized by temperature and light and divided into categories. These are not always called biomes, but function similarly.

The Role of Climate

The **climate** of a given region is a description of the average temperature and precipitation—the weather—that may be expected on each day throughout the entire year. Climates in different parts of the world vary widely. Equatorial regions are continuously warm, with little seasonal change in temperature. Above and below the equator, temperatures become increasingly seasonal (characterized by warm or hot summers and cool or cold winters); the farther we go toward the poles, the longer and colder the winters become, until at the poles it is perpetually cold. Likewise, colder temperatures are found at higher elevations, so that there are even snow-capped mountains on or near the equator (Figure 5–9).

Annual precipitation in an area also may vary greatly, from virtually zero to well over 250 centimeters (100 in.) per year. Precipitation may be evenly distributed throughout the year or concentrated in certain months, dividing the year into wet and dry seasons. A given climate will support only those species that find the temperature and precipitation levels within their ranges of tolerance. Population densities will be greatest where conditions are optimal and will decrease as any condition departs from the optimum (Chapter 3, Figure 3–7). A species will be excluded from a region (or local areas) where any condition is beyond its limit of tolerance. How will this variation affect the biotic community?

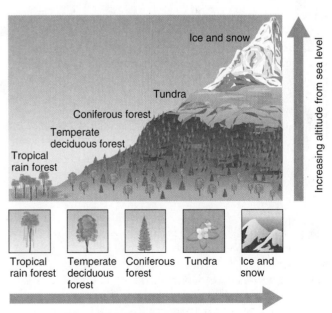

Figure 5–9 Effects of latitude and altitude. Decreasing temperatures, which result in the biome shifts depicted, occur with both increasing latitude (distance from the equator) and increasing altitude.

(*Source:* Redrawn from *Geosystems*, 5th ed., by Robert W. Christopherson. Copyright © 2005 by Pearson Prentice Hall, Inc., Upper Saddle River, NJ 07458.)

Terrestrial Biomes. Individual ranges of tolerance, particularly those for temperature and precipitation, determine where dominant species can live. In turn, the distribution of these species describes the placement of the biomes. To illustrate, let us consider some of the major types of biomes and their global distribution. Table 5–1 describes these terrestrial biomes and their major characteristics, and Figure 5–10 shows the distribution of the biomes as they occur globally.

Scientists estimate that a lack of water limits plant growth on 40% of Earth's land, temperature limits plant growth on 33% of the land, and a lack of sunlight limits growth on the remaining 27%.[2] Within the temperate zone (between 30° and 50° of latitude), the amount of rainfall is the key limiting factor. The **temperate deciduous forest** biome is found where annual precipitation is 75–200 centimeters (30–80 in.); *temperate rain forests* occur in a few areas where the rainfall is even higher. Where rainfall tapers off or is highly seasonal (25–150 cm or 10–60 in. per year), the **grassland** and **prairie** biomes are found; *savannas* are tropical grasslands with a few trees or shrubs. Regions receiving an average of less than 25 centimeters (10 in.) per year are occupied by a **desert biome.**

The effect of temperature, the other dominant parameter of climate, is largely superimposed on that of rainfall. That is, 75 centimeters (30 in.) or more of rainfall per year will usually support a forest, and generally temperature will determine the type of forest that grows there. For example, broad-leaved evergreen species, which are extremely vigorous and fast growing but cannot tolerate freezing temperatures, predominate in the **tropical rain forest.** By dropping their leaves and becoming dormant each autumn, deciduous trees are well adapted to freezing temperatures. Therefore, wherever rainfall is sufficient, deciduous forests predominate in temperate latitudes. Most deciduous trees, however, cannot tolerate the extremely harsh winters and short summers that occur at higher latitudes and higher elevations. These northern regions and high elevations are occupied by the **coniferous forest biome,** because conifers are better adapted to those conditions.

Temperature by itself limits forests only when it becomes low enough to cause **permafrost** (permanently frozen subsoil). Permafrost prevents the growth of trees, because roots cannot penetrate deeply enough in these conditions to provide adequate support. However, a number of grasses, clovers, and other small flowering plants can grow in the topsoil above permafrost. Consequently, where permafrost sets in, the coniferous forest biome gives way to the **tundra biome** (Table 5–1). At still colder temperatures, the tundra gives way to permanent snow and ice cover.

---

[2]R. Lindsay, "Global Garden Gets Greener," *Earth Observatory* (2003): http://earthobservatory.nasa.gov/Features/GlobalGarden/8/20/2014

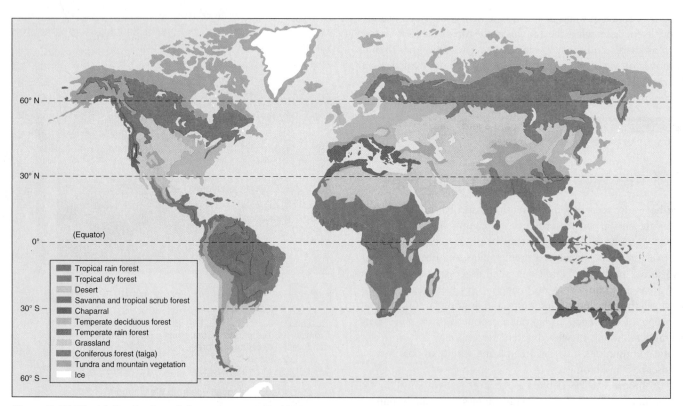

Figure 5–10 World distribution of the major terrestrial biomes. The biome of a region is largely determined by climate, but there are no clear boundaries between biomes. On this map, savanna and tropical scrub forest are grouped together, but other biologists consider scrub forests to be an extension of the tropical dry-forest biome.

(*Source: Biology: Life on Earth with Physiology,* by G. Audesirk, T. Audesirk, and B. Byers. Copyright © 2010 by Pearson Education. Reprinted with permission.)

Table 5-1 Major Terrestrial Biomes

Biome	Climate and Soils	Dominant Vegetation	Dominant Animal Life	Geographic Distribution
Deserts	Very dry; hot days, cold nights; rainfall less than 10 in. (25 cm)/yr.; soils thin and porous	Widely scattered thorny bushes and shrubs, cacti	Rodents, lizards, snakes, numerous insects, owls, hawks, small birds	North and southwest Africa, parts of the Middle East and Asia, Southwestern North America
Grasslands and Savannas	Seasonal rainfall, 10 to 60 in. (25–152 cm)/yr.; fires frequent; soils rich and often deep	Grass species, both tall and short; in savannas, shrubs and scattered trees are interspersed with grasses	Large grazing mammals (bison, wild horses; kangaroos); prairie dogs, coyotes, lions; termites important	Central North America, central Asia, subequatorial Africa and South America, southern India, northern Australia
Chaparral	Seasonal; mild, wet winters; hot, dry summers; rainfall 15–40 in. (38–100cm)/yr.; fires frequent; soil dry	Dense thickets of fire-adapted species; shrubs with small evergreen leaves, scrub oaks,	Jackrabbits, chipmunks, coyotes, mule deer, snakes, lizards, quail, other birds, insects	Regions around the Mediterranean Sea, Southern California, Baja Peninsula, Chile, South Africa, southwestern Australia
Tropical Rain Forests	Nonseasonal; annual average temperature 28°C; rainfall frequent, heavy; average over 95 in. (240 cm)/yr.; soils thin, poor in nutrients	High diversity of broad-leafed evergreen trees, dense canopy, vines	Enormous biodiversity; exotic, colorful insects, amphibians, birds, snakes; monkeys, tigers, jaguars	Northern South America, Central America, western central Africa, islands in Indian and Pacific Oceans, Southeast Asia
Tropical Dry (Deciduous) Forest	Hot with wet and dry seasons; rainfall 10–80 inches (25–200 cm)/yr; transition zone between rain forest and savanna	Broad-leafed deciduous trees (teak, mahogany, acacia, figs, oaks) and palms	Monkeys, deer, bats, rodents, large cats, parrots and many other birds	Mexico and western Central America, Caribbean, southeastern Brazil, eastern Africa, Madagascar, India, Southeast Asia, northern Australia
Temperate Forests	Seasonal; below freezing in winter; summers warm; rainfall from 30–80 in. (76–200 cm)/yr. in deciduous forests; higher in temperate rain forests	Broad-leafed deciduous trees, some conifers; shrubby undergrowth, ferns, mosses	Squirrels, raccoons, opossums, skunks, deer, foxes, black bears, snakes, amphibians, many soil organisms, birds	Western and central Europe, eastern Asia, eastern North America; temperate rain forests along west coasts of North and South America and Tasmania
Coniferous Forests	Seasonal; winters long, cold; precipitation light in winter, heavier in summer; soils acidic, with leaf litter	Cone-bearing trees (spruce, fir, pine, hemlock), some broad-leafed deciduous trees (birch, maple)	Large herbivores (mule deer, moose, elk); mice, hares, squirrels; lynx, bears, foxes; nesting area for migrating birds	Northern portions of North America, Europe, Asia, extending southward at high elevations
Tundra	Bitter cold, except for short growing season with long days and moderate temperatures; precipitation low; soils thin—and below, permafrost	Low-growing sedges, dwarf shrubs, lichens, mosses, and grasses	Year round: lemmings, arctic hares, arctic foxes, lynx, caribou; summers: abundant insects, many migrant shorebirds	North of the coniferous forest in Northern Hemisphere, extending southward at elevations above the coniferous forest

Note: See http://earthobservatory.nasa.gov/Experiments/Biome/index.php and click on the links for different biomes for photos and climatic data.

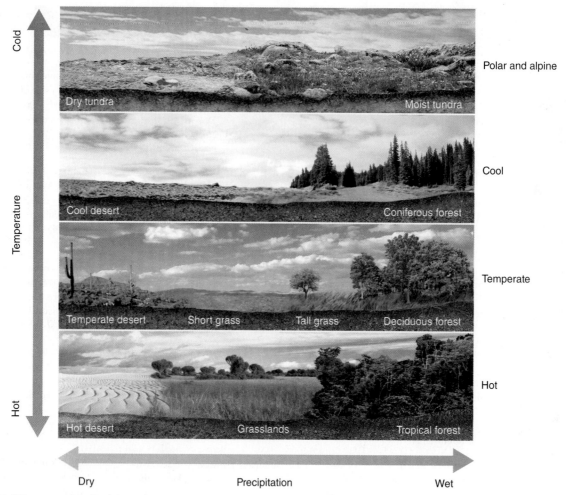

Figure 5–11 Climate and major biomes. Moisture is generally the overriding factor determining the type of biome that may be supported in a region. Given adequate moisture, an area will likely support a forest. Temperature, however, determines the kind of forest. The situation is similar for grasslands and deserts. At cooler temperatures, there is a shift toward less precipitation because lower temperatures reduce evaporative water loss. Temperature becomes the overriding factor only when it is low enough to sustain permafrost.

(*Source:* Redrawn from *Geosystems,* 5th ed., by Robert W. Christopherson. Copyright © 2005 by Pearson Prentice Hall, Inc., Upper Saddle River, NJ 07458.)

Any region receiving less than 25 centimeters (10 in.) of rain per year will be a desert, but the unique plant and animal species found in hot deserts are different from those found in cold deserts. A summary of the relationship between biomes and temperature and rainfall conditions is given in **Figure 5–11**. The average temperature for a region varies with both **latitude** (distance from equator) and **altitude** (distance from sea level), as shown in Figure 5–9.

Examples of Aquatic Systems. There are major categories of aquatic and wetland ecosystems that are similar to terrestrial biomes. Aquatic systems are determined primarily by the depth, salinity, and permanence of water in them. Among these ecosystems are lakes, marshes, streams, rivers, estuaries, bays, and ocean systems. As units of study, these aquatic systems may be viewed as ecosystems, as parts of landscapes, or as major biome-like features such as seas or oceans. **Table 5–2** lists several major aquatic systems and their primary characteristics.

Microclimate and Other Abiotic Factors

A specific site may have temperature and moisture conditions that are significantly different from the overall climate of the region in which it is located. For example, a south-facing slope, which receives more direct sunlight in the Northern Hemisphere, will be relatively warmer and hence drier than a north-facing slope. Similarly, the temperature range in a sheltered ravine will be narrower than that in a more exposed location, and so on. The conditions found in a specific localized area are referred to as the **microclimate** of that location.

Conditions also change with elevation. If you hike up a mountain located in a grassland, you may pass through forests, and, above the tree line, you may find areas of permafrost. This pattern of changing habitat with elevation means that the tops of mountains function ecologically as if they were more polar than their actual latitude (Figure 5–9).

Table 5–2 Major Aquatic Systems

Aquatic Systems	Major Environmental Parameters	Dominant Producers	Dominant Animal Life	Distribution
Lakes and Ponds (freshwater)	Bodies of standing water; low concentration of dissolved solids; seasonal vertical stratification of water	Rooted and floating plants, phytoplankton	Zooplankton, fish, insect larvae, turtles, ducks, geese, swans, wading birds	Physical depressions in the landscape where precipitation and groundwater accumulate
Streams and Rivers (freshwater)	Flowing water; low level of dissolved solids; high level of dissolved oxygen; often turbid from runoff	Attached algae, rooted plants	Insect larvae, fish, amphibians, otters, raccoons, wading birds, ducks, geese	Landscapes where precipitation and groundwater flow by gravity toward oceans or lakes
Inland Wetlands (freshwater)	Standing water, at times seasonally dry; thick organic sediments; high nutrients	Marshes (grasses, reeds); cattail swamps (water-tolerant trees); bogs (sphagnum moss, shrubs)	Amphibians, snakes, numerous invertebrates, wading birds, alligators, turtles	Shallow depressions, poorly drained, often occupy sites of lakes and ponds that have filled in
Estuaries (mixed)	Variable salinity; tides create two-way currents; often rich in nutrients, turbid	Phytoplankton, rooted grasses (salt-marsh grass), mangrove swamps in tropics with salt-tolerant trees, shrubs	Zooplankton, rich shellfish, worms, crustaceans, fish, wading birds, sandpipers, ducks, geese	Coastal regions where rivers meet the ocean; may form bays behind sandy barrier islands
Coastal Ocean (saltwater)	Tidal currents promote mixing; nutrients high	Phytoplankton, large benthic algae, turtle grass, symbiotic algae in coral	Zooplankton; rich bottom fauna of worms, shellfish, crustaceans; coral colonies, jellyfish, fish, turtles, gulls, terns, sea lions, seals, dolphins, penguins, whales	From coastline outward over continental shelf; includes kelp "forests" and eelgrass beds
Coral Reefs (saltwater)	Clear, shallow water in the tropics; high light penetration supports a great diversity of life	Red and green algae that live symbiotically in the bodies of coral animals; many other algae	Coral animals, zooplankton, sea worms, sea anemones, sea stars, wide diversity of fish	Along coasts and islands in the tropics, especially in the West Pacific, Caribbean and around Australia
Open Ocean (saltwater)	Great depths (to 11,000 m, 36,100 ft.); all but upper 200 m (656 ft.) dark and cold; poor in nutrients except in upwelling regions	Phytoplankton, sargassum weed	Zooplankton and fish adapted to different depths; seabirds, whales, tuna, sharks, squid, flying fish	Covering 70% of Earth, from edge of continental shelf outward
Hydrothermal Vents (saltwater)	Great depths; dark and hot; water exposed to volcanic rock may reach temperatures of 750°F (400°C); mineral-rich water often acidic	Chemosynthetic organisms; some are symbionts in other organisms	Extremophiles (organisms that live in extreme conditions), bacteria, giant tube worms, fish, crabs	Rare; along deep ocean trenches or rifts between tectonic plates

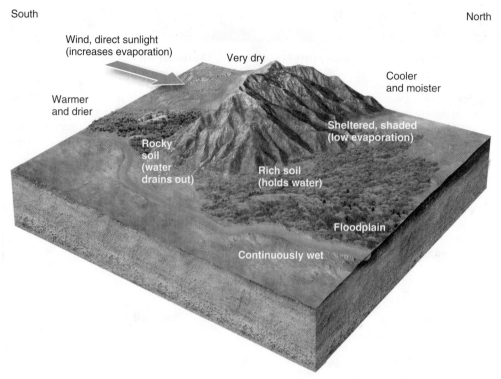

Figure 5–12 Microclimates. Abiotic factors such as terrain, wind, and type of soil create different microclimates by influencing temperature and moisture in localized areas.

Soil type and topography contribute to the diversity found in a biome, because these two factors affect the availability of moisture. In the eastern United States, for example, oaks and hickories generally predominate on rocky, sandy soils and on hilltops, which retain little moisture, whereas beeches and maples grow on richer soils, which hold more moisture, and red maples and cedars inhabit low, swampy areas. Figure 5–12 illustrates several factors that influence microclimate. In the transitional region between desert and grassland, with 25–50 centimeters (10–20 in.) of rainfall per year, a soil capable of holding water will support grass, but a sandy soil with little ability to hold water will support only desert species.

Biome Productivity

All biomes have varying levels of primary productivity. Those that have high productivity may help support organisms from other biomes. For example, many seabirds migrate through salt marshes and live off the vast productivity there before traveling to other places. Highly productive biomes, like rain forests, also remove carbon dioxide from the atmosphere and trap it (Chapter 18). Biomes with high productivity aren't better than others, though. Low-productivity biomes can have other important roles, such as habitat for rare species.

Why are some biomes or types of ecosystems more productive than others? Figure 5–13 presents (a) the average annual net primary productivity, (b) the percentage of different ecosystems over Earth's surface, and (c) the

percentage of global net primary productivity attributed to 19 of the most important ecosystems. Some key relationships between abiotic factors and specific ecosystems can be seen in the data. Tropical rain forests both are highly productive and contribute considerably to global productivity; they cover a large area of the land and are characterized by ideal climatic conditions for photosynthesis—warm temperatures and abundant rainfall. The open oceans cover 65% of Earth's surface, so they account for a large portion of global productivity, yet their actual rate of production is low enough that they are veritable biological deserts. Primary production in the oceans is limited by the scarcity of nutrients—a good lesson in the significance of limiting factors. The seasonal effects of differences in latitude can also be seen by comparing productivity in tropical, temperate, and boreal (coniferous) forests.

The biosphere consists of a great variety of environments, both aquatic and terrestrial. In each environment, plant, animal, and microbial species are adapted to all the abiotic factors. In addition, they are adapted to each other, in various feeding and nonfeeding relationships. Every ecosystem is tied to others through exchanges of air, water, and minerals common to the whole planet.

CONCEPT CHECK   A single species of prickly pear cactus grows in Arizona, in the plains of Montana, and along the coast of Rhode Island. Are these three places in the same biome? If not, why do you think this cactus is able to grow in all of them? ☑

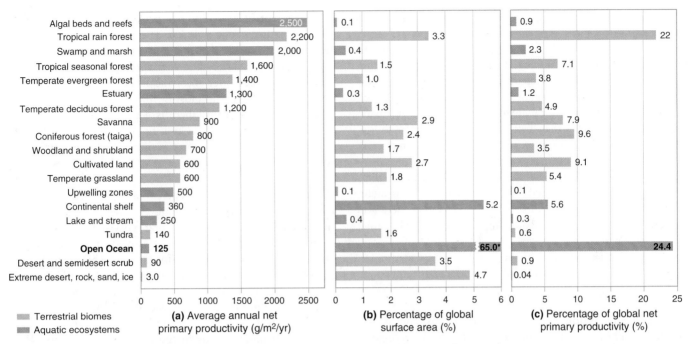

Figure 5–13 Productivity of different ecosystems. (a) The average annual net primary productivity of different ecosystems, **(b)** the percentage of different ecosystems over Earth's surface area, and **(c)** the percentage of global net primary productivity. The row for open ocean in **(b)** shows that the open ocean is such a high percentage of Earth's surface that it is off the scale relative to all of the other ecosystems.

(Source: Biology: Concepts and Connections, 4th Edition, by Neil A. Campbell, Jane B. Reece, Lawrence G. Mitchell and Martha R. Taylor. Copyright © 2003 by Pearson Education. Reprinted with permission.)

UNDERSTANDING THE DATA

1. According to the graphs, algal beds and reefs are the ecosystems with the highest rate of primary productivity, yet they contribute little to the total global primary productivity. Can you explain why?

2. What would happen to Earth's primary productivity if large areas of cultivated land reverted back to grassland?

5.4 Ecosystem Responses to Disturbance

People often picture ecosystems at equilibrium, as stable environments in which species interact constantly in well-balanced predator-prey and competitive relationships. This idea has led to the popular idea of "the balance of nature": natural systems maintain a delicate balance over time that lends them great stability. A more recent understanding of ecosystems, however, shows that they operate in a dynamic, changing way. Species' abundances and ecosystem characteristics are not unchanging. Rather, the landscape is composed of a shifting mosaic of patches, much like patches of lava and unburned areas after a volcano. Volcanoes such as those in Iceland and the Philippines give us a great example of how ecosystems respond to a **disturbance**—a significant change that kills or displaces many community members.

Ecological Succession

After a disturbance like a fire or volcanic eruption, new plants grow from the soil, new trees sprout, and eventually the original forest stands are replaced by new and different stands. First, however, these communities go through a series of intermediate stages. This transition from one biotic community to another is called **ecological succession**. Ecological succession occurs because the physical environment is gradually modified by the growth of a biotic community itself, such that the area becomes more favorable to another group of species and less favorable to the current occupants. In the past, succession was viewed as a very predictable and orderly process through a series of stages to a stable end point. That view is useful for a general understanding of how ecosystems change. However, new work, particularly through large-scale studies at Long Term Ecological Research sites, suggests that succession is more complicated, something we will discuss shortly.

In the most basic version of succession, pioneer species—the first to colonize a newly opened area in the aftermath of a fire, flood, volcanic lava flow, or other large disturbance—start the process. As these grow, they create conditions that are favorable to other kinds of colonizers, and succession is driven forward by the changing conditions that pave the way for other species. Driving succession forward by improving conditions for subsequent species is known as **facilitation**.

The succession of species does not go on indefinitely. A stage of development is eventually reached in which there appears to be a dynamic balance between all of the species and the physical environment. In the past, this final state was called a **climax ecosystem** because the assemblage of species continues on in space and time. The term *climax ecosystem* is still used, but many ecologists are moving away from its use because they believe it implies more stability than actually occurs in nature. In the real world, even climax communities experience change. Also, within the area of a climax ecosystem there may be patches of disturbance that open spaces for new growth. Some parts of the ecosystem may be in an earlier stage of succession; sometimes there are several "final" stages, with adjoining ecosystems in the same environment at different stages.

Three classic examples of succession are presented next: primary and secondary succession on land and succession in aquatic ecosystems.

Primary Succession. If an area lacks plants and soil—such as might occur after volcanic eruption or the recession of a glacier—the process of initial invasion and progression from one biotic community to the next is called **primary succession.** An example is the gradual invasion of a bare rock or gravel surface by what eventually becomes a climax forest ecosystem. Bare rock is an inhospitable environment. It has few places for seeds to lodge and germinate or for seedlings to survive. However, certain species, such as some mosses, are able to exploit this environment. Their tiny spores (specialized cells that function reproductively) can lodge and germinate in minute cracks, and moss can withstand severe drying simply by becoming dormant. With each bit of moisture, moss grows and gradually forms a mat that acts as a sieve, catching and holding soil particles as they are broken from the rock or as they blow or wash by. Thus, a layer of soil gradually accumulates, held in place by the moss (Figure 5–14).

The mat of moss and soil provides a suitable place for seeds of larger plants to lodge. The larger plants in turn collect and build additional soil, and eventually there is enough soil to support shrubs and trees. In the process, the fallen leaves and other litter from the larger plants smother and eliminate the moss and most of the smaller plants that initiated the process. Thus, there is a gradual succession from moss through small plants and, finally, to trees or whatever the regional climate supports, yielding the biomes typical of different climatic regions.

Secondary Succession. When an area has been cleared by fire, human activity, or flooding and then left alone, plants and animals from the surrounding ecosystem may gradually reinvade the area—through a series of distinct stages called **secondary succession.** The major difference between primary and secondary succession is that secondary succession starts with preexisting soil. Thus, the early, prolonged stages of soil building are bypassed. Hence, the process of reinvasion begins with different species. The steps leading from abandoned agricultural fields in the eastern United States back to deciduous forests provide a classic example of secondary succession (Figure 5–15).

Figure 5–14 Primary succession on bare rock. Moss and lichens invade bare rock and act as collectors, accumulating a layer of soil sufficient for complex plants to become established.

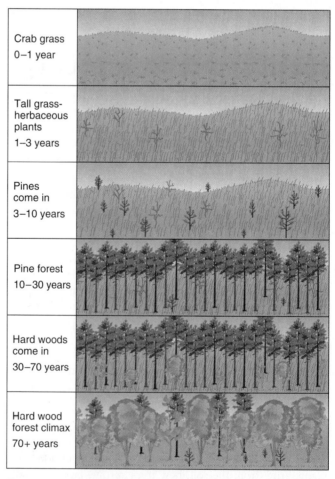

Crab grass
0–1 year

Tall grass-herbaceous plants
1–3 years

Pines come in
3–10 years

Pine forest
10–30 years

Hard woods come in
30–70 years

Hard wood forest climax
70+ years

Figure 5–15 Secondary succession. When an agricultural field in the eastern United States is abandoned, the communities of plants that invade the field typically follow these stages.

Figure 5–16 Aquatic succession. In this photograph, thick aquatic vegetation is filling in a shallow pond. Eventually, soil and dead plant matter may support terrestrial vegetation.

Aquatic Succession. Natural succession also takes place in lakes or ponds. Succession occurs because soil particles inevitably erode from the land and settle out in ponds or lakes, gradually filling them. Aquatic vegetation produces detritus that also contributes to the filling process. As the buildup occurs, terrestrial species from the surrounding ecosystem can advance in the ecotone and aquatic species must move farther out into the lake. In time, the shoreline advances toward the center of the lake until, finally, the lake disappears altogether (Figure 5–16). Here the climax ecosystem may be a bog or a forest. Disturbances such as droughts, floods, and erosion can send **aquatic succession** back to an earlier stage.

Disturbance and Resilience

In order for natural succession to occur, the spores and seeds of various invading plants and the breeding populations of various invading animals must already be present in the vicinity. Ecological succession is not a matter of new species developing or even old species adapting to new conditions; it is a matter of populations of existing species taking advantage of a new area as conditions become favorable. Where do these early-stage species come from if their usual fate is to be replaced by late-stage or climax species? They come from surrounding ecosystems in the early stages of succession. Similarly, the late-stage species are recruited from ecosystems in the later stages of succession. In any given landscape, therefore, all stages of succession are likely to be represented in the ecosystems. This is the case because disturbances constantly create gaps or patches in the landscape. When a variety of successional stages is present in a landscape, as opposed to one single climax stage, a greater diversity of

species can be expected; in other words, biodiversity is enhanced by disturbance.

Fire and Succession. Fire is an abiotic factor that has particular relevance to succession. It is a major form of disturbance common to terrestrial ecosystems. About 80 years ago, forest managers interpreted the potential destructiveness of fire to mean that all fire was bad, whereupon they embarked on fire prevention programs that eliminated fires from many areas. Unexpectedly, fire prevention did not preserve all ecosystems in their existing state. In pine forests of the southeastern United States, for instance, economically worthless scrub oaks and other broad-leaved species began to displace the more valuable pines. Grasslands were gradually taken over by scrubby, woody species that hindered grazing. Pine forests of the western United States that were once clear and open became cluttered with the trunks and branches of trees that had died in the normal aging process. This deadwood became the breeding ground for wood-boring insects that proceeded to attack live trees. In California, the regeneration of redwood seedlings began to be blocked by the proliferation of broad-leaved species.

Scientists now recognize that fire, which is often started by lightning, is a natural and important abiotic factor. As with all abiotic factors, different species have different degrees of tolerance to fire. In particular, the growing buds of grasses and pines are located deep among the leaves or needles, where they are protected from most fires. By contrast, the buds of broad-leaved species, such as oaks, are exposed, so they are sensitive to damage from fire. Consequently, in regions where these species coexist and compete, periodic fires tip the balance in favor of pines, grasses, or redwood trees. In relatively dry ecosystems, where natural decomposition is slow, fire may also help release nutrients from dead organic matter. Some plant species even depend on fire. The cones of lodgepole pine, for example, will not release their seeds until they have been scorched by fire (Figure 5–17).

Figure 5–17 Ground fire. Periodic ground fires are necessary to preserve the balance of pine forests. Such fires remove excessive fuel and kill competing species.

Ecosystems that depend on the recurrence of fire to maintain their existence are referred to as **fire climax ecosystems.** The category includes various grasslands and pine forests. Fire is being increasingly used as a tool in the management of such ecosystems. In pine forests, if ground fires occur every few years, relatively little deadwood accumulates. With only small amounts of fuel, fires usually just burn along the ground, harming neither pines nor wildlife significantly.

In the summer of 1988, fierce fires swept across Yellowstone National Park, burning thousands of acres and millions of trees. Drought and deadwood, which had accumulated during decades of fire suppression, combined to trigger a conflagration that was unprecedented in the history of the Park Service. Thousands of firefighters were involved in trying to stem the fires that swept across the landscape. These fires included fires on the ground and crown fires, which burn to the tops of canopy trees.

Crown fires sometimes occur naturally because not every area is burned on a regular basis. In fact, exceedingly dry conditions can make a forest vulnerable to crown fires even when no large amounts of deadwood are present. Even crown fires, however, serve to clear deadwood and sickly trees, release nutrients, and provide for a fresh ecological start. Burned areas soon become productive meadows as secondary succession starts anew. Thus, periodic crown fires create a patchwork of meadows and forests at different stages of succession that lead to a more-varied, healthier habitat that supports a greater diversity of wildlife than does a uniform, aging conifer forest.

Resilience. The ability of an ecosystem to return to normal functioning after a disturbance, such as a fire or flood, is called its **resilience.** Fire can appear to be a highly destructive disturbance to a forested landscape. Nevertheless, fire releases nutrients that nourish a new crop of plants, and in a short time, the burned area is repopulated with trees and is indistinguishable from the surrounding area. The processes of replenishment of nutrients, dispersion by surrounding plants and animals, rapid regrowth of plant cover, and succession to a forest can all be thought of as **resilience mechanisms** (Figure 5–18). Resilience helps maintain the sustainability of ecosystems. A good example is the rapid recovery of the forests around Mount St. Helens after its 1980 eruption (Figure 5–19).

Tipping Points. Ecosystem resilience has its limits, however. If certain species have been eliminated, natural succession may be blocked or modified. For example, Iceland was colonized by the Norse in the 11th century, and, as a result, trees were cut. By 1850, not a tree was left standing. Iceland had become a barren, tundra-like habitat (Figure 5–20). Succession was unable to proceed naturally, and Iceland has remained treeless. As in Iceland, some disturbances are so profound that they overcome normal resilience mechanisms and create an entirely new and far less productive ecosystem. This new ecosystem

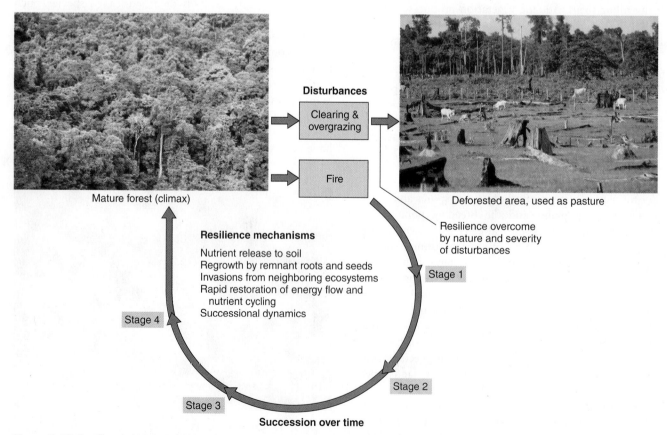

Mature forest (climax)

Disturbances

Clearing & overgrazing

Fire

Deforested area, used as pasture

Resilience overcome by nature and severity of disturbances

Resilience mechanisms

Nutrient release to soil
Regrowth by remnant roots and seeds
Invasions from neighboring ecosystems
Rapid restoration of energy flow and
    nutrient cycling
Successional dynamics

Stage 1

Stage 2

Stage 3

Stage 4

Succession over time

Figure 5–18 **Resilience in ecosystems.** Disturbances in ecosystems are usually ameliorated by a number of resilience mechanisms that restore normal ecosystem function in a short time.

Figure 5–19 Resilience after a volcanic eruption. The 1980 eruption of Mount St. Helens in Washington State killed most of the forest trees around it. By 1986, noble firs and other trees were returning to a slope on the mountain.

Figure 5–20 Iceland. Forests that originally covered much of this island nation were totally stripped for fuel in the 18th and 19th centuries. With natural succession impossible, Iceland has remained barren and tundra-like, as seen here.

can also have its own resilience mechanisms, which may resist any restoration to the original state. We can view this as a "tipping point"—a situation in an ecosystem where a small action catalyzes a major change in the state of the system.

Tipping points act as levers that can change a system to be better able—or less able—to function as an ecosystem. They stimulate the action of **feedback loops,** where system dynamics influence a system in ways that amplify (positive loop) or dampen (negative loop) the behavior of the system. For example, predator-prey cycles function as negative feedback loops that keep both populations from exploding in terms of numbers and then crashing. Removal of predators (to favor the prey species) would be a tipping point that could allow the positive feedback of prey reproduction to bring devastation to their plant food supply.

You might wonder why it matters if an ecosystem such as the forests of Iceland moves into an alternative state, such as an Iceland with no trees, and stays there. After all, don't treeless, cold places exist? As we will see below, ecosystems provide a number of services valuable to humans and other species. These include large-scale effects such as lowering flood surges, maintaining soil, maintaining even ground temperatures, and taking up carbon dioxide.

If just a few parts of the planet moved from one type of ecosystem to another, it really wouldn't matter. The problem is one of frequency. Large-scale human actions tend to lower ecosystem services such as flood control, wind-erosion control, and carbon dioxide storage. They tend to produce ecosystems either with very low productivity or with high primary productivity but low food chain length (such as a lake with too many nutrients). Such ecosystems tend to have less physical complexity and lower biological diversity. We will see this more when we discuss the problem of desertification (Chapter 11).

Long-Term Ecosystem Studies. A great deal of our understanding of ecosystem changes comes from studies of changes in fields after fires, on mountain slopes after volcanoes, and near retreating glaciers. However, as human-caused changes to

the environment have become global in scope, our studies have had to become longer and cover wider areas. Two ongoing research efforts of note help us to better understand ecosystems. The first is a system of ecosystems designated as Long-Term Ecological Research sites, funded by the National Science Foundation. These 26 sites are part of the largest ecological research network in the United States. They represent different types of ecosystems and different levels of human disturbance. Research projects at these sites focus on five core areas:

1. Pattern and control of primary production;

2. Spatial and temporal distribution of populations selected to represent trophic structure;

3. Pattern and control of organic matter accumulation in surface layers and sediments;

4. Patterns of inorganic inputs and movements of nutrients through soils, groundwater, and surface waters;

5. Patterns and frequency of site disturbances.

Some of these sites, such as the Hubbard Brook Experimental Station, which was critical to the discovery of the problem of acid rain, have been studied for decades.

The second ecosystem study initiative, funded by the National Science Foundation, is the National Ecological Observatory Network (NEON), a network operating in the United States and Puerto Rico. It is a system for collecting data about ecosystems and climate, pulling them together, and making them available for scientists to study. The data will be available to the public as well. NEON enables researchers to understand the impacts of climate change, land-use change and invasive species the scale of a continent, by providing the infrastructure for large, long term studies. Planned for years, construction on NEON infrastructure was begun in 2011; eventually 106 sites will be part of the NEON network.

CONCEPT CHECK What would you expect a landscape to look like if deforestation had reached a tipping point? Give an example where this has occurred. ☑

5.5 Human Values and Ecosystem Sustainability

Ecosystems have existed for millions of years, maintaining natural populations of biota and the processes that these carry out (which in turn sustain the ecosystems). One of the reasons for studying natural ecosystems is that they are models of sustainability. In this chapter, we started with trophic levels and then examined the flow of energy and nutrients in an ecosystem. These discussions have focused on how natural ecosystems work in theory. In reality, however, we depend on Earth's ecosystems for goods and services, also known as **ecosystem capital**. Is our use of natural and managed ecosystems a serious threat to their long-term sustainability? Many ecosystem scientists believe that it is.

Appropriation of Energy Flow

As we have seen, it is the Sun that energizes the processes of energy flow and nutrient cycling. Humans make heavy use of the energy that starts with sunlight and flows through natural and agricultural ecosystems. Agriculture, for example, provides most of our food. To produce food, we have converted almost 11% of Earth's land area from forest and grassland biomes to agricultural ecosystems. Grasslands sustain our domestic animals for labor, meat, wool, leather, and milk. Forest biomes provide us with 3.4 billion cubic meters (120 billion cubic feet) of wood annually for fuel, building material, and paper. Finally, some 12% of the world's energy consumption is derived directly from plant material, such as firewood and peat.

Calculations of the total annual global net primary production of land ecosystems range around 120 petagrams

SOUND SCIENCE
What Are the Effects of Reintroducing a Predator?

It is dusk in Yellowstone National Park. An elk stands at the edge of a stream. Suddenly the elk senses danger—a gray wolf is approaching. In a twist of fate, the elk charges its predator, chasing the wolf away. The elk is safe from being eaten, at least for the time being. Until recently this scene would have been impossible, because there were no wolves in Yellowstone. Hunters had removed them from the park. Removal of large predators such as wolves and mountain lions was once the norm in the country. Throughout the United States, wolves were so overhunted that they were in danger of extinction.

By the early 1990s, scientists and the public had begun to understand the importance of top predators, and there was an effort to bring the wolves back to the western states. One proposal was to reintroduce them to Yellowstone. This proposal was very unpopular with local ranchers, who worried the wolves would attack their herds of cattle and other animals. In spite of local concerns, a small reintroduction was begun in 1995–1996, when a group of 31 wolves were captured in Canada and released into Yellowstone.

During the time the wolves were gone, herbivores, especially elk, had increased dramatically, overeating vegetation in some areas. Researchers wondered if reintroducing wolves would lessen large herbivore grazing enough to allow natural vegetation to return. They anticipated wolves would have some effect on other herbivores, such as bison. They also expected a decline in coyotes, a smaller competitor, and possibly an increase in foxes. In fact, the reintroduction of wolves also changed the Yellowstone ecosystem in ways that surprised wildlife biologists.

After several years, scientists discovered that the aspen and willow trees were beginning to recover. The numbers of bison and beaver had increased, and there was less erosion into rivers and streams. The changes in aspen and willow distribution led some researchers to hypothesize that the plants were recovering because the wolves controlled elk behavior through a "landscape of fear." According to this hypothesis, the elk were so stressed by the wolves that they spent less time in the open by riverbanks and more time hidden in the forest. Some studies supported this interpretation.

More recent research into the behavior of wolves and elk, however, suggests that the effect of wolves on elk behavior is not clear. Elk do not appear to be very stressed. They do not change their behavior until wolves are close by (within a kilometer), which occurs infrequently—only every nine days or so. In addition, the fat content and pregnancy rates of female elk in areas where wolves are present

are similar to those of populations of elk where there are no wolves. As a result, the question of how wolves impact elk is still a matter under discussion. Even so, this is a good example of how scientists figure out whether one hypothesis is better than another.

In spite of the unresolved issues about elk behavior, it is clear that the reintroduction of wolves had far-ranging effects on the ecosystems of Yellowstone National Park. These changes affect the relationships among community members as well as ecosystem functions, such as water flow dynamics.

Source: Kauffman, Douglas, E. McWhirter, Michael D. Jimenez, Rachel C. Cook, John G. Cook, Shannon E. Albeke, Hall Sawyer, and P. J. White. "Linking Antipredator Behaviour to Prey Demography Reveals Limited Risk Effects of an Actively Hunting Large Carnivore." *Ecology Letters* 1023–1030, August 2013.

(1 petagram = 1 billion metric tons) of dry matter, including agricultural as well as the more natural ecosystems (Figure 5–21). In 1986, researchers calculated that humans appropriate around 30% of this total production for agriculture, grazing, forestry, and human-occupied lands.[3] More-recent calculations have fine-tuned this figure.[4] These estimates indicate that humans are using a large fraction of the whole. Further, because humans convert many natural and agricultural lands to urban and suburban housing, highways, dumps, factories, and the like, we cancel out an additional 8% of potential primary production. Thus, we appropriate about 40% of the land's primary production to support human needs. In using this much, we have become the dominant biological force on Earth, despite representing about 5% of the biomass.

Human use of land has become more efficient. An index called the **human appropriation of net primary productivity (HANPP)** is used to track human effects on ecosystems. Between 1910 and 2005, the human population quadrupled and economic activity increased 17-fold, but HANPP only doubled.[5] However, these measurements are uncertain, in part because some human activities, such as irrigating crops, can increase the productivity of land, even if the activity is unsustainable. Even so, the HANPP provides a useful way of picturing the global impact of human activities.

Involvement in Nutrient Cycling

Nutrients are replenished in ecosystems through the breakdown of organic compounds and the release of the chemicals that make them up—processes we have described as biogeochemical cycles. This maintains the sustainability of ecosystems indefinitely. However, human intrusion into these natural cycles is substantial, and it is bound to increase. As we have seen, our alterations of the nutrient cycles have long-term consequences (Chapter 3). Climate change is one consequence of these alterations. In China, for example, dramatic changes have caused the nitrogen cycle to reach a tipping point. In ground-breaking research, scientists found that the use of nitrogen fertilizer was no longer increasing crop yields, but was instead increasing the release of a greenhouse gas that drives climate change (Chapter 18).

Currently, global climate change is one of the most serious environmental issues facing society. The burning of fossil fuels has increased the level of carbon dioxide in the atmosphere to the point where, by the mid–21st century, it will have doubled since the beginning of the Industrial Revolution. Methane is also released from the drying and decomposition of areas that used to be permafrost and

[3] Peter Vitousek, Paul R. Ehrlich, Anne H. Ehrlich, and Pamela Matson, "Human Appropriation of the Products of Photosynthesis," *BioScience* 36, No. 6. (June 1986): 368–373.
[4] N. Ramankutty, A. T. Evan, C. Monfreda, and J. A. Foley, "Farming the Planet: 1. Geographic Distribution of Global Agricultural Lands in the Year 2000," *Global Biogeochemical Cycles* 22 (2008): GB1003.
[5] F. Klausmann et al. "Global human appropriation of net primary production doubled in the 20th century," *Proceedings of the National Academy of Sciences* 10, no. 25(2013): 10324–10329.

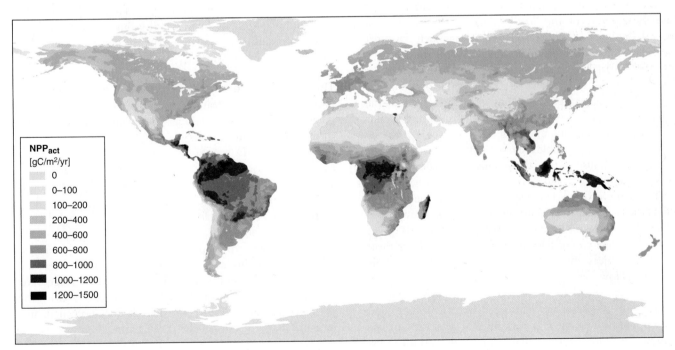

NPP$_{act}$
[gC/m^2/yr]

- 0
- 0–100
- 100–200
- 200–400
- 400–600
- 600–800
- 800–1000
- 1000–1200
- 1200–1500

Figure 5–21 Net primary productivity on land. Primary productivity varies greatly from biome to biome. NPP act represents the the amount of production left after the plants in a region use the energy they need to live, or the *net primary production* of the *actual* vegetation. The total global NPP is 120 billion metric tons per year (1 billion metric tons = 1 petagram).

(*Source:* H. Haberl et al. "Quantifying and mapping the global human appropriation of net primary production in Earth's terrestrial ecosystem," *PNAS.* 104: (2007) 12942–12947.)

from cattle. Because both carbon dioxide and methane are powerful greenhouse gases, a general warming of the atmosphere is occurring, leading to many other effects on Earth's climate (Chapter 20). Numerous ecosystems are being affected by the changes, and it is questionable whether they can function sustainably in the face of such rapid changes. The threat of disturbance on such an enormous scale is a major reason why many nations of the world have begun to take steps to reduce emissions of greenhouse gases.

Value of Ecosystem Capital

How much are natural ecosystems worth to us? To the poor in many developing countries, the products of natural ecosystems keep life going and may enable them to make economic gains that can lift them from deep poverty. Theirs is a direct and absolute dependence. For example, 1.6 billion people, many of them the world's poorest, depend on forests for much of their sustenance (in the form of fuelwood, construction wood, wild fruits, herbs, animal fodder, and bush meat). Their access to these ecosystem goods is often uncertain. Those of us in the developed world, however, are susceptible to a very different perspective: We often assign little value to natural ecosystems and are unconcerned about (and insulated from) human impacts on natural ecosystems. This perspective can have dangerous consequences.

Recall that we defined the goods and services we derive from natural systems as ecosystem capital (Chapter 1). In a first-ever attempt of its kind, a team of 13 natural scientists and economists collaborated in 1997 to produce a report titled "The Value of the World's Ecosystem Services and Natural Capital."[6] Their reason for making such an effort was that the goods and services provided by natural ecosystems are not easily seen in the market (meaning the market economy that normally allows us to place value on things) or may not be in the market at all. Thus, clean air to breathe, the formation of soil, the breakdown of pollutants, and similar items never pass through and thus are not given a value by the market economy. People are often not even aware of their importance. Because of this, these things are undervalued or not valued at all.

The major ecosystem services identified by the team, as well as the ecosystem functions that provide vital support to humans, are shown in Figure 5–22. The team made the point that it is useless to consider human welfare without these ecosystem services; thus in one sense, their value as a whole is infinite. However, the **incremental value** of each type of service can be calculated. That is, changes in the quantity or quality of a service may influence human welfare, and an economic value can be placed on that relationship. For example, removing a forest will affect the ability of the forest

to provide lumber in the future, as well as to perform other services, such as soil formation and promotion of the hydrologic cycle. The total economic value of these changes is the incremental cost of removing the forest.

Later attempts to determine the value of the world's ecosystem capital have been based on the same concepts. In 2014 the same authors recalculated ecosystem goods and services to be worth $125 trillion per year (in 2011 dollars) and estimated that changes in land use had caused global economy losses of $4.3 to $20.2 trillion per year.[7]

Using Ecosystem Value in Decision Making. The real power of the team's analysis lies in its use for making local decisions. For instance, the value of wetlands cannot be represented solely by the amount of soybeans that could be grown on the land if it were drained. Instead, wetlands provide other vital ecosystem services, and these should be balanced against the value of the soybeans in calculating the costs and benefits of a proposed change in land use. The bottom line of their analysis is, in their words, "that ecosystem services provide an important portion of the total contribution to human welfare on this planet." For this reason, the ecosystem capital stock (the ecosystems and the populations in them, including the lakes and wetlands) must be given adequate weight in public-policy decisions involving changes to that stock.

Despite the great value of ecosystem services, because they are outside the market and uncertain, they are too often ignored or undervalued, and the net result is human changes to natural systems whose social costs far outweigh their benefits. For example, coastal mangrove forests in Thailand are often converted to shrimp farms, yet an analysis showed that the economic value of the forests (for timber, charcoal, storm protection, and fisheries support) exceeded the value of the shrimp farms by 70%.

The Cost of Altering Ecosystems. In 2002, a new team looked at the 1997 team's calculation of value and examined the benefit-cost consequences of converting ecosystems to more-direct human uses (e.g., converting a wetland to a soybean field).[8] In every case, the net balance of value was a loss—that is, services lost outweighed services gained. The team examined five different biomes and found consistent losses in ecosystem capital, running at −1.2% per year, representing an annual loss of $250 billion through habitat conversion alone.

If the losses from conversion are so great, the authors asked, why is this conversion still happening? They suggested that the benefits of conversion were often exaggerated through various government subsidies, seriously distorting the market analysis. Thus, the market does not adequately measure many of the benefits from natural ecosystems, nor

[6] Robert Costanza et al., "The Value of the World's Ecosystem Services and Natural Capital," *Nature* 387 (1997): 253–260.

[7] Robert Costanza et al., "Changes in the Global Value of Ecosystem Services," *Global Environmental Change* 26 (2014): 152–158.
[8] Andrew Balmford et al., "Economic Reasons for Conserving Wild Nature," *Science* 297 (August 9, 2002): 950–953.

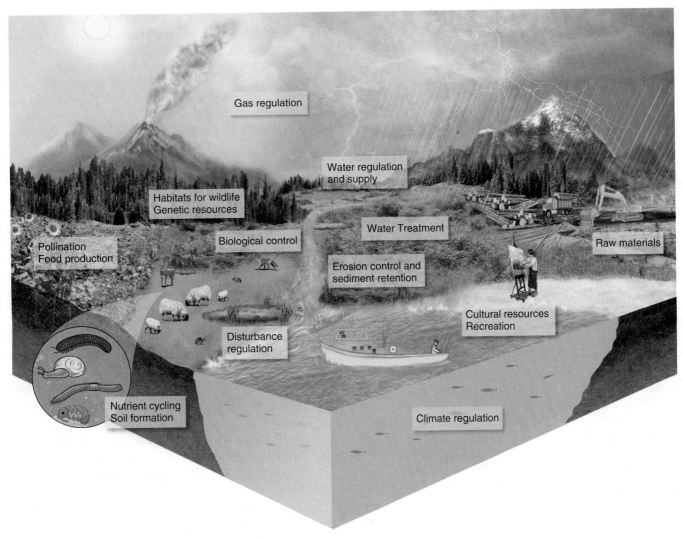

Figure 5–22 Ecosystem goods and services. The Millennium Ecosystem Assessment described ecosystem goods and services that we enjoy largely for free. The loss of these ecosystem features is an effect of environmental degradation and costs us billions of dollars.

does it deal well with the fact that many major benefits are realized on regional or global scales, while the benefits of conversion are local. A broad-based understanding of our tremendous dependence on natural systems should lead, logically, to proper management of those systems, a topic discussed later in this chapter.

Can Ecosystems Be Restored?

The human capacity for destroying ecosystems is well established. To some degree, however, we also have the capacity to restore them. In many cases, restoration simply involves stopping the abuse. For example, after pollution is curtailed, water quality often improves, and fish and shellfish gradually return to previously polluted lakes, rivers, and bays. Similarly, forests may gradually be restored to areas that have been cleared. Humans can speed up the process by seeding and planting seedling trees and by reintroducing populations of fish and animals that have been eliminated.

In some cases, ecosystems have been destroyed or disturbed to such an extent that they require the efforts of a new breed of scientist: the restoration ecologist. Unfortunately, sometimes the ecosystem is too greatly disturbed for scientists to repair. In a prairie, for example, lack of grazing by native herbivores, overgrazing by cattle or sheep, and suppression of regular fires can lead to much woody vegetation. Exotic species may abound, and there may be no remnants of the original prairie grasses and herbs on the site. Such sites may take intensive efforts to bring back any native species or may continue in an alternative ecosystem.

Why should we restore ecosystems? Restoration ecologists Steven Apfelbaum and Kim Chapman cite several important reasons.[9] First, we should do so for aesthetic reasons. Natural ecosystems are often beautiful, and the

---

[9] Steven I. Apfelbaum and Kim Alan Chapman, "Ecological Restoration: A Practical Approach," in *Ecosystem Management: Applications for Sustainable Forest and Wildlife Resources,* ed. Mark S. Boyce and Alan Haney (New Haven, CT: Yale University Press, 1997).

restoration of something beautiful and pleasing to the eye is a worthy project that can be uplifting to many people. Second, we should do so for the benefit of human use. The ecosystem services of a restored wetland, for example, can be enjoyed by present and future generations. The degradation of ecosystems robs them of their ecosystem capital, and restoring them returns their value. Finally, we should do so for the benefit of the species and ecosystems themselves. It can be argued that nature has value separate from human use and that people should act to preserve and restore ecosystems and species in order to preserve that value.

Managing Ecosystems

Virtually no ecosystem can escape human impact. We depend on ecosystems for vital goods and services—and by using ecosystems, we are in fact managing them. Forests are a good example. Management can be a matter of fire suppression (or, more recently, controlled burns), clear-cutting, or selective harvesting. The objective of management can be to maximize profit from logging or to maintain the forest as a sustainable and diverse ecosystem that yields multiple products and services. Good ecosystem management is based on understanding how ecosystems function, how they respond to disturbances, and what goods and services they can best provide to the human societies living in or around them. Since 1992, the U.S. Forest Service, the U.S. Bureau of Land Management, the U.S. Fish and Wildlife Service, the U.S. National Park Service, and the U.S. Environmental Protection Agency have all officially adopted ecosystem management as their management paradigm. What, then, is ecosystem management?

Ecosystem management is composed of several main principles. First, it looks at ecosystems on both small and large scales. For example, the impact of logging on a single river and the impact of that logging on the entire watershed are both considered in managing the ecosystem. Good management preserves the range of possible landscapes in an ecosystem instead of allowing one type to take over the area. In addition, management is not static—managers are given the opportunity to learn from experiments and to employ new knowledge from forestry and landscape science in their practice. Such "managers" can be volunteer land stewards or neighborhood groups, not just government agencies. As part of adapting to changing circumstances, management must also look at the human element of these ecosystems. The valuable goods and services of ecosystems are recognized, and through active involvement in monitoring and management, all **stakeholders** (all people with a stake in the ecosystem's health) are included as important elements in stewardship of the resources (see Stewardship, *Local Control Helps Restore Woodlands*). Above all, the paradigm incorporates the objective of ecological sustainability.

The U.S. Forest Service is responsible for managing some 192 million acres (77.7 million hectares) of national forests and is a key player in overseeing public natural resources. A decade ago, the Forest Service convened an

interdisciplinary committee of scientists for research-based recommendations on managing national lands. According to the committee's report, "the first priority for management is to retain and restore the ecological sustainability of watersheds, forests, and rangelands for present and future generations. The Committee believes that the policy of sustainability should be the guiding star for stewardship of the national forests and grasslands to assure the continuation of this array of benefits" (referring to the goods and services provided by the national lands).[10]

Internationally, the governing board of the Millennium Ecosystem Assessment process examined the findings of the more than 1,350 experts who produced the many reports of the assessment and issued its own statement of the key messages to be derived from the process. The title of this work is significant: *Living Beyond Our Means: Natural Assets and Human Well-Being.*[11] In it, the authors emphasized that all of us depend on nature and ecosystem services for our very survival. As stated in the first few lines of this document: "At the heart of this assessment is a stark warning. Human activity is putting such strain on the natural functions of Earth that the ability of the planet's ecosystems to sustain future generations can no longer be taken for granted." The following are some of the messages from the governing board:

- The loss of services derived from ecosystems is a significant barrier to the achievement of the Millennium Development Goals (now replaced by the Sustainable Development Goals) to reduce poverty, hunger, and disease.

- Measures to conserve natural resources are more likely to succeed if local communities are given ownership of them, share the benefits, and are involved in decisions.

- Even today's technology and knowledge can reduce considerably the human impact on ecosystems. They are unlikely to be deployed fully, however, until ecosystem services cease to be perceived as free and limitless and their full value is taken into account.

The Future

As the Millennium Ecosystem Assessment described, the future growth of the human population and its rising consumption levels will severely challenge ecosystem sustainability. By 2050, the world's population is expected to increase by about 2 billion people (see Chapter 8). The needs and demands of this expanded human population will create unprecedented pressures on the ability of Earth's systems to provide goods and services. Demands on agriculture will likely require at least 15% more agricultural land; irrigated land area is expected to

---

[10]Committee of Scientists, Department of Agriculture, *Sustaining the People's Lands: Recommendations for Stewardship of the National Forests and Grasslands into the Next Century* (1999). Available at www.fs.fed.us/news/science.

[11]Millennium Ecosystem Assessment, *Living Beyond Our Means: Natural Assets and Human Well-Being* (Statement of the Millennium Assessment Board, 2005). Available at www.millenniumassessment.org/en/Products.BoardStatement.aspx.

STEWARDSHIP
Local Control Helps Restore Woodlands

One way to promote good stewardship of resources is to involve the local stakeholders in decisions about the management of resources. Stakeholders are the people who may be affected by a given approach to environmental management. Primary stakeholders are those most dependent on a resource; often they depend upon the local ecosystem for their livelihood. Secondary stakeholders, such as government officials, conservationists, or scientists studying the resource, may live near the resource, but are not greatly dependent on it.

One place where poor, rural communities have been involved in natural resource management is the Shinyanga region of Tanzania. This dry region has a growing population, many of whom, like the man shown, are involved in pastoral agriculture. Extensive woodlands in this region once supported the Sukuma people with wood and forage for livestock. Traditionally, the Sukuma protected the woodlands in a system of vegetation enclosures called *ngitili*. This indigenous system prevented overgrazing and protected some forage for cattle for the end of the dry season.

During the mid-20th century, woodlands were cut and enclosures removed to control the tsetse fly. After independence in 1961, land was increasingly cleared for agriculture because of a growing population. Government policies exacerbated these trends. The effect was devastating to the environment: Land and soil were overused, wood was difficult to find, and wild fruit and plants became scarce, until the region was nicknamed "the desert of Tanzania."

In 1986, the Tanzanian government reversed its approach and promoted the revival of ngitili. But the government soon discovered the importance of involving the local community in decision making. The Sukuma villagers wanted some say in how forage enclosures were designed and run. They also wanted to be able to plant native species of plants that were grown in local nurseries. As a result of this dialog, the Tanzanian government allowed the Sukuma to exert ownership over the natural resources. Conservationists worked with villagers to develop effective management of the woodlands. This process involved both individuals and entire villages, where group ngitili were available for the local community.

Gradually, the dry, eroded landscape changed. Today, vegetation and wildlife have been restored on more than 350,000 hectares (860,000 acres). This restoration has greatly improved the livelihood of the Sukuma, with dividends in firewood, timber for construction, fodder for animals, and traditional foods. In this situation, the involvement of the primary stakeholders in decisions affecting local ecosystems was the key to restoration.

Sources: World Resources Institute. *World Resources 2005: The Wealth of the Poor: Managing Ecosystems to Fight Poverty* (Washington, D.C.: World Resources Institute, 2005).

Barrow, Ed. "300,000 Hectares Restored in Shinyanga, Tanzania—But What Did It Really Take to Achieve This Restoration?" *Sapiens* 7, no. 2. (2014), accessed December 26, 2014, http://sapiens.revues.org/1542.

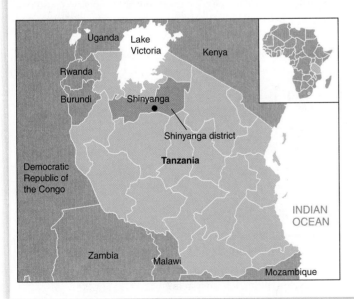

increase to about 60% of all land that has irrigation potential, greatly stressing an already strained hydrologic cycle. Such growing demands cannot be met without facing serious trade-offs between different goods and services. More agricultural land means more food, but less forest—and, therefore, less of the important services forests perform. More water for irrigation means more rivers will be diverted, so less water will be available for domestic use and for sustaining the riverine ecosystems.

If we clearly understand the consequences of simply continuing present practices, we may choose other alternatives. We need to understand how essential natural and managed ecosystems are for human well-being and to bring to bear the wisdom and political will necessary to promote their sustainability.

CONCEPT CHECK   What are two things managers can do to better care for ecosystems? ☑

REVISITING THE THEMES

SOUND SCIENCE

The natural processes in ecosystems drive their functions. Ecosystems are not static entities—change is one of their fundamental properties. Ecosystems can be grouped into large global areas called biomes, which are determined largely by precipitation and temperature. Ecosystems often include sub-ecosystems, caused by microclimates or other variations in the environment. Individual limits on the distribution of species determine which species live in each ecosystem.

Ecosystems can be characterized by their primary productivity and how energy and nutrients move through them. Food chains and webs contain several trophic levels. The first trophic level is composed of producers, which usually obtain their energy from the Sun; energy moves up through organisms that gain their energy from their prey. Ecosystems demonstrate several concepts we saw in earlier chapters. The concepts of energy and matter (and the inefficiency of their transfer) play out in forming the biomass pyramid and determining the length of food chains. The relationships between species determine the trophic levels in ecosystems.

A functioning ecosystem is resilient to disturbance—indeed, some ecosystems depend on disturbances to reset succession and maintain diversity. However, extremely large or unusual disturbances, such as the deforestation of Iceland or a volcanic eruption, may tip an ecosystem into an altered state from which it cannot recover. In some instances, this leads to much lower productivity or diversity; in other cases, sound environmental management can tip a degraded ecosystem back into a restored state. Scientists and others study the processes of restoring degraded ecosystems to health.

SUSTAINABILITY

Human impacts on the natural world have been so unprecedented and extensive that we crossed the line into unsustainable consumption some time ago; now we are depleting our stock of ecosystem capital instead of living off its sustainable goods and services. Converting ecosystems into farms and other human uses has frequently depleted their ability to provide us with goods and services. We are just beginning to calculate the lost economic value of these altered landscapes, not to mention their lost aesthetic and natural value. Any attempt at sustainability will need to include monitoring the world's ecosystems, setting international policies to protect them, and involving all stakeholders in their management. Basic science will inform much of the policy making and other solutions for sustainability discussed in later chapters.

STEWARDSHIP

Ecosystem management has shown us that we need to monitor the impact of our activities on ecosystems; although ecosystems are resilient, large disturbances can do irreparable damage to them. To be effective stewards we must first learn how ecosystems function. Then we must restrain our activities to allow ecosystems to continue to provide us with essential goods and services. Human agency has become the dominant force on Earth, affecting living systems all over the planet. In the process, we have the ability to drastically reduce biodiversity or to act as responsible stewards of our environment, as the example of locally controlled woodland enclosures in Tanzania showed. Ecosystem management is, above all, a stewardship approach to the natural world. As the U.S. Forest Service's Committee of Scientists put it, "the policy of sustainability should be the guiding star" for our actions, fittingly linking sustainability and stewardship.

REVIEW QUESTIONS

1. How did the volcanoes in Iceland and the Philippines change the environment to lesser or greater extents?

2. Name and describe the attributes of the two categories into which all organisms can be divided based on how they obtain nutrition.

3. Name and describe the roles of the three main trophic categories that make up the biotic structure of every ecosystem. Give examples of organisms from each category.

4. Give four categories of consumers in an ecosystem and the role that each plays.

5. Describe different members of the decomposition food web.

6. Differentiate among the concepts of *food chain*, *food web*, and *trophic levels*.

7. Relate the concept of the biomass pyramid to the fact that all heterotrophs depend on autotrophic production.

8. Describe how differences in climate cause Earth to be partitioned into major biomes.

9. What are three situations that might cause microclimates to develop within an ecosystem?

10. Identify and describe the biotic and the abiotic components of the biome of the region in which you live.

11. Define the terms *ecological succession* and *climax ecosystem*. How do disturbances allow for ecological succession?

12. What role may fire play in ecological succession, and how may fire be used in the management of certain ecosystems?

13. What is meant by ecosystem *resilience*? What can cause it to fail? How does this relate to environmental tipping points?

14. What is meant by the term *stakeholder*? How does ecosystem management involve stakeholders?

15. Succinctly describe ecosystem management. How is it related to sustainability?

16. Can ecosystems be restored? What has to happen for that to work?

17. How much of Earth's primary productivity is used or pre-empted by humans?

18. What are two ways that the reintroduction of wolves altered the ecosystems of Yellowstone National Park?

THINKING ENVIRONMENTALLY

1. Write a scenario describing what would happen to an ecosystem or to the human system in the event of one of the following: (a) all producers are killed through a loss of fertility of the soil or through toxic contamination, or (b) decomposers and detritus feeders are eliminated. Support all of your statements with reasons drawn from your understanding of the way ecosystems function.

2. Look at the following description of a broad global region, and describe what the biome name and main biota would be and how you know that: 40 centimeters (16 in.) of precipitation a year, seasons, soil frozen much of the year.

3. Consider the plants, animals, and other organisms present in a natural area near you, and then do the following: (a) Imagine how the area may have undergone ecological succession, and (b) analyze the population-balancing mechanisms that are operating among the various organisms. Choose one species, and predict what will happen to it if two or three other native species are removed from the area. Then predict what will happen to your chosen species if two or three foreign species are introduced into the area.

4. Consult the Web site www.ecotippingpoints.org, and read several of the success stories from the site. Compare the tipping points as presented, and explain the common properties of the tipping points.

5. Explore how the human system can be modified into a sustainable ecosystem in balance with (i.e., preserving) other natural ecosystems without losing the benefits of modern civilization.

MAKING A DIFFERENCE

1. Find the location of the national park or wilderness area that is nearest to the place where you live. If possible, make a plan to visit it. Find out what management issues the park has. Do they have advisories about fire risk, invasive species, or other ecosystem management issues?

2. Using an online map, find out what river drains your watershed and where that water goes. Is there a watershed stewardship group for the river? Who might be looking out for the protection of the ecosystem? If you are interested, organize a group volunteer activity.

3. In your area, is there a citizens' group for restoring any degraded ecosystems? There are many such small groups of citizens in the United States and elsewhere. See if you can link to a citizens' group. Volunteer if you can. One such activity is the annual CoastSweep, usually held in September, in which people all over the world clean up beaches. If you live near a coast, see if you can help.

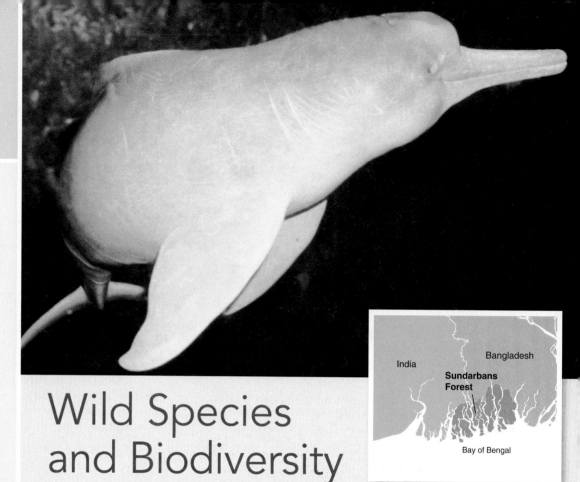

India

Bangladesh

Sundarbans Forest

Bay of Bengal

Wild Species and Biodiversity

A brown river winds slowly past mangroves filled with flocking birds, carrying with it plankton, crustaceans, and young fish that will support a vast food chain. A fisherman in a boat casts a net to the side. The place is the Sundarbans, the largest mangrove forest in the world. The long nose of a Ganges River dolphin breaks the water surface in the early morning mist. This endangered cetacean swims past mangrove roots and edges past a fishing net, its body narrowly avoiding entanglement.

Threatened. This area of the Bay of Bengal, near the crowded coast of Bangladesh, is home to populations of two endangered cetaceans: the Ganges River dolphin (*Platanista gangetica gangetica*) and a related subspecies, the Indus River dolphin (*Platanista gangetica minor*). These rare beasts are nearly blind and swim sideways to get hints of sunlight through the water as they find their prey by echolocation. Freshwater dolphins, like their saltwater cousins, live on fish. Because they live in rivers instead of oceans, they are in some of the most populous areas of the world. These rivers may be highly polluted, heavily divided by dams, and crowded with fishers.

River dolphins are among the most threatened large vertebrates in the world. In fact, another species of river dolphin, the Baiji, in China, was recently declared extinct, to the great sorrow of conservationists worldwide. The Baiji, nicknamed "the goddess of the Yangtze," declined because of loss of habitat, pollution, crowding, and injury from contact with fishing vessels, as well as overharvesting. The species, once spread throughout several large rivers in China, became rarer and rarer. In 2006, an extensive 45-day search by experts using sonar and other equipment failed to find any in the last areas the Baiji was known to inhabit, and it was declared extinct.

▲ The highly endangered Indus river dolphin lives in the mangrove forests of the Sundarbans, Bangladesh.

Areas of Importance. Recently, to protect the Ganges and Indus river dolphins from the same fate, Bangladesh has identified dolphin hot spots, or areas of importance, within the sprawling Sundarbans mangrove forest. Three of these areas are being set aside as breeding and feeding areas for dolphins. Intensive multicountry protections are needed to keep these unique animals from further decline. In an area already crowded with people, protecting species may seem impossible, but it is important. The effort of the Bangladeshi government is a hopeful move in the right direction.

The focus in this chapter is wild species. The world is experiencing a dramatic decline in wild organisms. This is such a concern that the United Nations has designated 2011–2020 the United Nations Decade of Biodiversity. While scientists can study individual species such as the Ganges River dolphin, they can also work on species decline in the context of biodiversity—the variety of life on Earth. In this chapter, we consider the importance of biodiversity as a whole, as well as that of individual species, and look at the public policies that seek to protect wild species and biodiversity.

6.1 The Value of Wild Species and Biodiversity

Recall that ecosystem capital is the sum of all the goods and services provided to human enterprises by natural systems (Chapter 1). This capital is enormously valuable to humankind. Robert Constanza and his coauthors estimated that the mean ecosystem capital in 1997 was $46.7 trillion (adjusted for 2009 dollars). In 2014, the same authors recalculated ecosystem goods and services at $125 trillion per year (in 2011 dollars).[1] (See Chapter 5.) The basis of much of our natural capital is ecosystems, and the basis of ecosystems is the plants, animals, and microbes—the wild species—that make them work. To maintain the sustainability of these ecosystems, their integrity must be preserved; this means maintaining their resilience, natural processes, and biodiversity, among other things. The protection of wild species and the ecosystems in which they live is therefore crucial.

Even those who agree on the need to protect wild species, however, may disagree on the kind of protection that should be afforded to them. Some want wildlife protected so as to provide recreational hunting. Others feel strongly that hunting for sport should be banned. Many have a deeper concern for the ecological survival of wild species, and they view the ongoing loss of biodiversity as an impending tragedy. In the developing world, however, many collect or kill wild species in order to eat or make money, so their personal survival is at stake. All of these attitudes stem from the values we place on wild species. Can the differing values be reconciled in a way that leads to sustainable management of these important natural resources?

Biological Wealth

The Millennium Ecosystem Assessment states that about 2 million species of plants, animals, and microbes have been examined, named, and classified, but many more species have not yet been found or cataloged. Estimates of the total number of species on Earth range widely, from 5 to 30 million. One recent estimate is around 8.7 million (+/− 1.3 million),

although there is still little agreement about the total. About 15,000 new species are described each year. Between 2010 and 2013, 441 new species were discovered in the Amazon, an area rich in new species. In 2013, a carnivore called the olinguito (*Bassaricyon neblina*) was described in the remote cloud forests of Ecuador; the olinguito was identified after scientists realized it was different from similar, closely related species (**Figure 6–1**).

These natural species of living things, collectively referred to as **biota,** are responsible for the structure and maintenance of all ecosystems. They, and the ecosystems that they form, represent the **biological wealth** that makes up most of the ecosystem capital that sustains human life and economic activity with goods and services. From this perspective, the biota found in each country can represent a major component of the country's total wealth. More broadly, this richness of living species constitutes Earth's biodiversity.

Humans have always depended on Earth's wild species for food and materials. Approximately 12,000 years ago,

Figure 6–1 Discovering new species. First identified in 2013, the olinguito (*Bassaricyon neblina*) from Ecuador was the first new carnivore to be described in the Western Hemisphere in 35 years.

---

[1] Robert Costanza et al., "Changes in the Global Value of Ecosystem Services," *Global Environmental Change* 26(2014): 152–158.

however, our ancestors began to select certain plant and animal species from the natural biota and propagate them, and the natural world has never been the same. Forests, savannas, and plains were converted to fields and pastures as the human population grew and human culture developed. In time, many living species were exploited to extinction, and others disappeared as their habitats were destroyed. The Center for Biological Diversity has reported that 631 species and subspecies went extinct in North America between 1642 and 2001. A recent report said that global biodiversity loss costs as much as $4.5 trillion a year.[2] We have been drawing down our biological wealth, with unknown consequences. One root source of this problem is the way we regard and value wild nature.

Two Kinds of Value

In the 19th century, hunters on horseback could ride out to the vast herds of bison roaming the North American prairies and shoot them by the thousands, often taking only the tongues for markets back in the east. Plume hunters decimated the populations of egrets and other shorebirds to satisfy the demands of fashion in the late 1800s. Appalled at this wanton destruction, 19th-century naturalists called for an end to the slaughter, and the U.S. public became sensitized to the losses. People began to see natural species as worthy of preservation, and naturalists began to look for ways to justify their calls to conserve nature. Thus, there was an emerging sense that species should not be hunted to extinction. But why not? Their problem then, and ours now, is to establish the thesis that wild species have some value that makes it essential to preserve them. If we can identify that value, then we will be able to assess what our moral duties are in relation to wild species.

Instrumental Value. Philosophers who have addressed this problem identify two kinds of value that should be considered. The first is **instrumental value**. A species or individual organism has instrumental value if its existence or use benefits some other entity (by providing food, shelter, or a source of income). This kind of value is usually anthropocentric; that is, the beneficiaries are human beings. Many species of plants and animals have instrumental value to humans and will tend to be preserved (conserved, that is) so that we can continue to enjoy the benefits derived from them.

Intrinsic Value. The second kind of value that must be considered is **intrinsic value**. Something has intrinsic value when it has value for its own sake; that is, it does not have to be useful to us to possess value. How do we know that something has intrinsic value? This is a philosophical question and comes down to a matter of moral reasoning. People often disagree about intrinsic value, as illustrated by the controversies about the rights of animals. Some people argue that animals have certain rights and should be protected

from being used for food, fur and hides, or experimentation. On the other side, some people hold that animals are simply property and have no rights at all. Others, probably the majority, hold an intermediate position: They agree that cruelty to animals is wrong, but they do not believe that animals should be given rights approaching those afforded to humans (Figure 6–2).

Many people believe that no species on Earth except *Homo sapiens* has any intrinsic value. Accordingly, they find it difficult to justify preserving species that are apparently insignificant. Still, in spite of the problems of establishing intrinsic value for species, there is growing support for preserving species that may appear useless to humans, even those that may never be seen by anyone except a few naturalists or taxonomists (biologists who classify organisms).

The value of natural species can be categorized as follows:

- Value as sources for food and raw materials,
- Value as sources for medicines and pharmaceuticals,
- Recreational, aesthetic, and scientific value, and
- Value for their own sake (intrinsic value).

Sources for Food and Raw Materials

Most of our food comes from agriculture, rather than being collected directly from wild populations, so it is tempting to falsely believe that food is independent of natural biota. In nature, both plants and animals are continuously subjected to the rigors of natural selection. Only the fittest survive. Consequently, wild populations have numerous traits for competitiveness, resistance to parasites, tolerance to adverse conditions, and other aspects of vigor.

In contrast, populations grown for many generations under the "pampered" conditions of agriculture tend to lose these traits because they are selected for production, not resilience. For example, a high-producing plant that lacks resistance to drought is irrigated, and the resistance to

Figure 6–2 Intrinsic value. Do species like this Blandings turtle have value separate from their value to humans? The question of intrinsic value is one of debate amongst philosophers.

---

[2] Pavan Sukhdev, *The Economics of Ecosystems and Biodiversity* (Interim Report funded by the German Ministry for Environment and the European Commission on the Economics of Ecosystems and Biodiversity). Presented at the UN's Convention on Biological Diversity, November 2008.

drought is ignored. Also, in the process of breeding plants for maximum production, virtually all genetic variation is eliminated. Indeed, the cultivated population is commonly called a cultivar (for *culti*vated *var*iety), indicating that it is a highly selected strain of the original species, with a minimum of genetic variation. When provided with optimal water and fertilizer, cultivars produce outstanding yields under the specific climatic conditions to which they are adapted. With their minimum genetic variation, however, they have virtually no capacity to adapt to any other conditions. For example, winter wheat varieties with high yields may be less winter hardy or more prone to pests and disease.

Wild Genes. To maintain vigor in cultivars and to adapt them to different climatic conditions, plant breeders comb wild populations of related species for the desired traits. When found, these traits are introduced into the cultivar through crossbreeding or, more recently, biotechnology (Chapter 12). For example, in the 1970s, the U.S. corn crop was saved from blight by genes from a related wild population of maize—that is, from natural biota. There are also tremendous opportunities for controlling pests by increasing genetic resistance with genes from wild biota (Chapter 13). If wild populations are lost, the options for continued improvements in food plants will be greatly reduced.

New Food Plants. The potential for developing new agricultural cultivars would also be lost if wild populations were destroyed. From the hundreds of thousands of plant species existing in nature, humans have used perhaps 7,000 in all, and modern agriculture has tended to focus on only about 30. Of these, three species—wheat, maize (corn), and rice—fulfill about 50% of global food demands. This limited diversity in agriculture makes modern plants ill suited to production under many different environmental conditions. For example, arid regions are generally considered to be unproductive without irrigation, but there are wild plants, including many wild species of the bean family, that produce abundantly under dry conditions.

Scientists estimate that 30,000 plant species with edible parts could be brought into cultivation. Many of these could increase production in environments that are less than ideal. For example, every part of the winged bean, a native of New Guinea, is edible: pods, flowers, stems, roots, and leaves. Recently introduced into many developing countries, this legume has already contributed significantly to improving nutrition. Loss of biological diversity undercuts similar future opportunities.

Woods and Other Raw Materials. Because species are selected from nature for animal husbandry, forestry, and aquaculture, essentially all the same arguments apply to those important enterprises, too. For example, nearly 3 billion people rely on wood for fuel for heating and cooking. Wood products rank behind only oil and natural gas as commodities on the world market. Demand for wood is increasing so rapidly that scientists are predicting a "timber famine" or "fuelwood crisis" in the next decade. Products

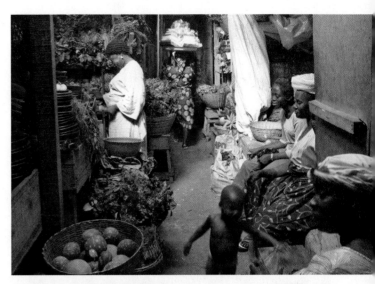

Figure 6–3 Forest products. These are all nonwood forest products developed to increase income in African forested areas. Such products are part of the instrumental value of tropical forests.
(*Source:* FAO.)

such as rubber, oils, nuts, fruits, spices, and gums also come from forests (Figure 6–3). All of these products are part of the value of wild species for humans, and forests are only one example of ecosystems that provide wild and natural living resources.

Banking Genes. Living organisms can be thought of as a bank in which the gene pools of all the species involved are deposited. Thus, biota is frequently referred to as a **genetic bank**. Because of concerns about agriculture, many scientists suggest that wild relatives of cultivated crops be conserved. In Kew, England, the Royal Botanic Gardens house the Millennium Seed Bank Partnership. Its goal is to house the seeds of 25% of the world's plants, some 75,000 species, by 2020. There is also the massive Svalbard Global Seed Vault, built into the permafrost deep in a mountain on a remote arctic island in Norway. This ultimate seed bank holds seeds that will act as a backup for the more than 1,000 seed banks around the world.

In a similar way, zoos act as a genetic bank for animals, and many zoos are active in conservation and captive breeding. In the United Kingdom, the Frozen Ark Project is collecting cells and DNA from species likely to go extinct in the near future. This collaborative effort with organizations around the world is designed to preserve genetic diversity while we try to stabilize species and slow currently high extinction rates.

Sources for Medicine

Earth's genetic bank is also important for the field of medicine. For thousands of years, the indigenous people of the island of Madagascar used an obscure plant, the rosy periwinkle, in their folk medicine. If this plant, which grows only on Madagascar, had become extinct before 1960, hardly anyone outside Madagascar would have cared. In the 1960s, however, scientists extracted two chemicals with

Figure 6–4 Medicinal plants. The rosy periwinkle, native to Madagascar, is a source of two anticancer agents that are highly successful in treating childhood leukemia and Hodgkin's disease.

medicinal properties—vincristine and vinblastine—from the plant (**Figure 6–4**). These chemicals have revolutionized the treatment of childhood leukemia and Hodgkin's disease. Before their discovery, leukemia was almost always fatal in children; today, with vincristine treatment, there is a 99% chance of remission.

The story of the rosy periwinkle is just one of hundreds. Chinese star anise, a tree in the magnolia family, yields star-shaped fruits that are harvested and processed to produce shikimic acid, the raw material for making Tamiflu®, an effective treatment against influenza viruses. Paclitaxel (trade name Taxol®), an extract from the bark of the Pacific yew,

has proven valuable for treating ovarian, breast, and small-cell cancers. For a time, this use threatened to decimate the Pacific yew because six trees were required to treat one patient for a year. Now, the substance can be extracted from the leaves of the English yew tree, or synthesized in a form that is easier for some patients to use.

Stories like these have created a new appreciation for the field of *ethnobotany*, the study of the relationships between plants and people. To date, more than 3,000 plants have been identified as having anticancer properties. Drug companies have been financing field studies of the medicinal use of plants by indigenous peoples (called *bioprospecting*) for years. The search for beneficial drugs in the tropics has led to the creation of parks and reserves to promote the preservation of natural ecosystems that are home to both the indigenous people and the plants they use in traditional healing. In Surinam, for example, a consortium made up of drug companies, the U.S. government, and several conservationist and educational groups has created the Central Surinam Reserve, a 4-million-acre (1.64-million-hectare) protected area. **Table 6–1** lists a number of well-established drugs that were discovered as a result of analyzing the chemical properties of plants used by traditional healers. Who knows what other medicinal cures are out there?

Recreational, Aesthetic, and Scientific Value

The species in natural ecosystems also provide the foundation for numerous recreational and aesthetic interests, ranging from sportfishing and hunting to hiking, camping, bird-watching, photography, and more (**Figure 6–5**). These activities involve a great number of people and represent a huge economic enterprise. One nonprofit, the Outdoor Foundation, produces an annual report on

Table 6–1 Modern Drugs from Traditional Medicines

Drug	Medical Use	Source	Common Name
Aspirin	Reduces pain and inflammation	*Filipendula ulmaria*	Queen of the meadow
Codeine	Eases pain; suppresses coughing	*Papaver somniferum*	Opium poppy
Ipecac	Induces vomiting	*Psychotria ipecacuanha*	Ipecac
Pilocarpine	Reduces pressure in the eye	*Pilocarpus jaborandi*	Jaborandi plant
Pseudoephedrine	Reduces nasal congestion	*Ephedra sinica*	*Ma-huang* shrub
Quinine	Combats malaria	*Cinchona pubescens*	Cinchona tree
Reserpine	Lowers blood pressure	*Rauwolfia serpentina*	Rauwolfia
Scopolamine	Eases motion sickness	*Datura stramonium*	Jimsonweed
Theophylline	Opens bronchial passages	*Camellia sinensis*	Tea
Tubocurarine	Relaxes muscles during surgery	*Chondrodendron tomentosum*	Curare vine
Vinblastine	Combats Hodgkin's disease	*Catharanthus roseus*	Rosy periwinkle

Figure 6–5 **Recreational, aesthetic, and scientific uses.** Natural biota provides numerous values, a few of which are depicted here.

outdoor recreation. In 2012, it reported that almost 150 million Americans (nearly half of the population) participated in outdoor recreation. This translated into nearly 12.4 billion outdoor excursions, with the most popular being jogging and trail running. Very likely, much of the public support for preserving wild species and habitats stems from the aesthetic and recreational enjoyment people derive from them.

Recreational and aesthetic values constitute a very important source of support for maintaining wild species, because activities inspired by these values also support commercial interests. **Ecotourism**—whereby tourists visit a place in order to observe wild species or unique ecological sites—represents the largest foreign-exchange-generating enterprise in many developing countries. Because some percentage of recreational dollars will be spent on activities related to the natural environment, any degradation of that environment affects commercial interests.

Scientific value comes from our need to have living things available so we can learn basic laws of nature, the way ecosystems function, and the way the world works generally. Societies have long held that such "pure science" is valuable. Many scientists are engaged in this endeavor by sheer curiosity about the world, including about the functioning of individual types of organisms. Realistically, though, most of the science done with these species aims for more pragmatic goals—new medicines, protection from gene loss in agriculture and forestry, and other outcomes already discussed.

A Cautionary Note. The use of wild species and biodiversity for instrumental value can raise some problems of justice as we navigate the needs of people with competing claims. For example, while the rosy periwinkle was a success story because of its effectiveness in combating cancer, little of the money returned to Madagascar, a very poor country. In other cases, such as the case of turmeric (a spice commonly used in India), ancient herbal remedies have been patented by large companies. While the patent on turmeric was eventually overturned, the patent on the rosy periwinkle was not, and money from the rosy periwinkle came to Madagascar only when the patent ran out. The indigenous people in the areas where biodiversity is highest may not be the ones to benefit from their resources. The Nagoya Protocol on Access and Benefit Sharing is a recent international treaty designed to protect the interests of indigenous peoples as products are discovered in wild sources (Section 6–4).

Similarly, ecotourism may bring money to poor areas of the world, but such tourism can also increase pollution, directly harm wildlife, or alter local societies in negative ways. For example, whale-watching boats on the St. Lawrence River disrupt the feeding of whales. Tourist boats on Mexican lakes frighten flocks of flamingoes and reduce their feeding time. Popular bird song apps for smart phones can distract birds so much that chicks may not get enough food from their parents. Ecotourism can marginalize indigenous people as part of the scenery, or they can remain poor while tourism operators gain wealth. Nonetheless, there is sustainable tourism, and people intending to travel in the hopes of promoting conservation can find tourism that is done well.

The Loss of Instrumental Value. Overall, a loss of biodiversity has a tremendous negative effect on the world in real, practical ways. The Economics of Ecosystems and Biodiversity (TEEB) report, released in May 2008, was the first international report to detail the economic and life-quality effects of biodiversity loss. Researchers have shown that the costs of the loss of biodiversity are higher for the world's poorest people. In fact, continued loss of biological diversity could halve the income of the poorest billion and a half people. One example of this loss of income occurred after a massive coral bleaching event in the Indian Ocean in 1998. It is estimated that the loss of food, jobs, tourism, and coastal protection will cost a total of $8 billion by 2018. Because the people who benefit most from the rapid use of resources are not the people who are most harmed by the loss of those resources, many ethicists believe such an outcome to be morally wrong. Thus, even the instrumental values of Earth's resources have an ethical component.

Value for Their Own Sake

The usefulness (instrumental value) of many wild species is apparent. But instrumental value alone may not be a compelling reason to protect many wild species. What about the species that have no obvious value to anyone—including many plant and animal species? Another strategy for preserving all wild species is to emphasize the intrinsic value of species rather than their unknown or uncertain ecological and economic values. From this viewpoint, the extinction of any species is an irretrievable loss of something valuable.

Environmental ethicists argue that the long-established existence of species means they have a right to continued existence. Living things have ends and interests of their own, and the freedom to pursue them is often frustrated by human actions. Do humans have the right to terminate a species that has existed for thousands or millions of years? Environmental philosopher Holmes Rolston III puts it this way: "Destroying species is like tearing pages out of an unread book, written in a language humans hardly know how to read, about the place where they live."[3] Some support this view by arguing that there is value in every living thing and that one kind of living thing (e.g., humans) has no greater value than any other. This argument, however, can lead to some difficulties, such as having to defend the rights of pathogens and parasites. A more common viewpoint held by ethicists is that, because humans have the ability to make moral judgments, they also have a special responsibility toward the natural world, and that this responsibility includes concern for other species.

Religious Support. Many ethicists find their basis for intrinsic value in religious thought. For example, in the Jewish and Christian scriptures, the book of Genesis expresses God's concern for wild species when he created them. Judeo-Christian scholars maintain that by declaring his creation good and giving it his blessing, God was saying that all wild things have intrinsic value and, therefore, deserve moral consideration and stewardship. In the story of Noah, described in the Hebrew Bible, God makes a covenant not only with humans, but also with all creatures, "every living thing." Similarly, in the Islamic tradition, the Quran (Koran) proclaims that the environment is the creation of Allah and should be protected because its protection praises the Creator. Many Native American religions have a strong environmental ethic (a system of moral values or beliefs); for example, the Lakota hold that humans and other forms of life should interact like members of a large, healthy family. Drawing on Hindu philosophy, the Chipko movement of northern India represents a strong grassroots environmentalism. Ethical concern for wild species underlies several religious traditions and represents a potentially powerful force for preserving biodiversity. Even if species have no demonstrable use to humans, it can still be argued that they have a value and a right to exist and that humans thus have a responsibility to protect them from harm that results from human actions.

The Land Ethic. Further, there is an ethic about the preservation of the ecosystems in which species live. This view of nature was most famously described by conservationist Aldo Leopold in an essay published in 1949 entitled "The Land Ethic." Leopold is known for being one of the first scientists to understand the importance of both fire and predators in maintaining ecosystem health. He advocated for protection of wild places—*wilderness*—as well as for better care of agricultural and other human-dominated land. "The Land Ethic," which was published in the collection *A Sand County Almanac*, was instrumental in generating interest in the protection of wilderness. In it, Leopold describes his viewpoint: "A thing is right when it tends to preserve the integrity, stability, and beauty of the biotic community. It is wrong when it tends otherwise."[4]

CONCEPT CHECK If you want to protect panda bears because they are cute and the world would be a lesser place without them (not bad reasons!), what types of values are you expressing? ☑

6.2 Biodiversity and Its Decline

In the introduction to this chapter, biodiversity was described as the variety of life on Earth. We can be more specific than that. The dimensions of biodiversity include the genetic diversity within a species and the variety of species as well as the range of communities and ecosystems. When scientists use the term *biodiversity*, however, sometimes they mean a characteristic that they can calculate and compare between habitats. In this definition, scientists use two measures to calculate biodiversity—the number of species and a measure of how "even" the species are. A habitat has low biodiversity if it is dominated by only one species, with few members of other species, because those few might easily leave or die out. Given the same number of species, diversity is considered to be higher in habitats where the dominance by any single species is low. In this chapter, our main focus is the diversity of species. Recall that over long periods of time, natural selection leads to speciation (the creation of new species) as well as extinction (the disappearance of species) (Chapter 4). Now, we are interested in how humans interact with wild species and impact biodiversity over a relatively short time.

How Many Species?

Almost 2 million species have been discovered and described, and certainly many more than that exist. Most people are completely unaware of the great diversity of species that occurs within any given taxonomic category. Groups especially rich in species are the flowering plants (about 224,000 species) and the insects (about 950,000 species), but even less-diverse groups, such as birds or ferns, are rich with species that are unknown to most people (Table 6–2). Groups that

[3]Holmes Rolston III, *Environmental Ethics: Duties to and Values in the Natural World* (Philadelphia: Temple University Press, 1988), 129.

[4]Aldo Leopold, "The Land Ethic," in *A Sand County Almanac and Sketches Here and There* (Oxford, England: Oxford University Press, 1949). Available at http://www.luminary.us/leopold/land_ethic.html.

Table 6–2 Number of Species on Earth. The number of known species (Catalogued), estimated species (Predicted), and standard error (SE) from surveys of terrestrial and ocean systems.

Species	Earth			Ocean		
	Catalogued	Predicted	±SE	Catalogued	Predicted	±SE
Eukaryotes						
Animalia	953,434	7,770,000	958,000	171,082	2,150,000	145,000
Chromista	13,033	27,500	30,500	4,859	7,400	9,640
Fungi	43,271	611,000	297,000	1,097	5,320	11,100
Plantae	215,644	298,000	8,200	8,600	16,600	9,130
Protozoa	8,118	36,400	6,690	8,118	36,400	6,690
Total	1,233,500	8,740,000	1,300,000	193,756	2,210,000	182,000
Prokaryotes						
Archaea	502	455	160	1	1	0
Bacteria	10,358	9,680	3,470	652	1,320	436
Total	10,860	10,100	3,630	653	1,320	436
Grand Total	1,244,360	8,750,000	1,300,000	194,409	2,210,000	182,000

Source: C. Mora, D. P. Tittensor, S. Adl, A. Simpson, and B. Worm, "How Many Species Are There on Earth and in the Ocean?" *PLoS Biol* 9, no. 8 (2014): e1001127. doi:10.1371/journal.pbio.1001127.

are conspicuous or commercially important—such as birds, mammals, fish, and trees—are much more fully explored and described than insects and very small organisms, such as soil nematodes, fungi, and bacteria. Taxonomists are aware that their work in finding and describing new species is incomplete. Estimates of unknown species keep rising as taxonomists explore more ecosystems. Whatever the number of species, the planet's biodiversity represents an amazing and diverse storehouse of biological wealth.

The Decline of Biodiversity

Biodiversity is declining in the United States as well as around the world. Because U.S. ecosystems are well studied, the trends that show themselves here (including the decline of many plant and animal species) may represent larger global trends. **Endemic** species (those found in only one habitat and nowhere else) are especially at risk. Some areas of the country and of the world are particularly vulnerable to species loss. These areas are often the focus of special conservation efforts.

North America. The biota of the United States is as well known as any. Even so, not much is known about most of the 200,000-plus species of plants, animals, and microbes estimated to live in the nation. At least 500 species native to the United States, including at least 100 vertebrates, are known to have gone extinct (or have "gone missing" and are

believed to have gone extinct) since the early days of colonization. An inventory of 20,897 wild plant and animal species in the 50 states, carried out in the late 1990s by a team of biologists from the Nature Conservancy and the National Heritage Network concluded that one-third are vulnerable (threatened), imperiled (endangered), or already extinct. In the United States, mussels, crayfish, fishes, and amphibians—all species that depend on freshwater habitats—are at greatest risk. In part, this potential for loss stems from the high number of endemic aquatic species in the American Southeast. For example, there are close to 200 species of freshwater mussels and clams in North America, giving these bivalves the greatest species diversity of any freshwater bivalve group in the world. Flowering plants are also of great concern, with one-third of their numbers in decline in North America.

Species populations are a more important element of biodiversity than even the species' existence. Populations of species that occupy different habitats and ranges contribute to biological wealth, as they provide goods and services important in ecosystems. Across North America, even populations of well-studied species are in decline. Commercial landings of many species of fish are down, puzzling reductions in amphibian populations are occurring all over the world, and many North American songbird species (such as the cerulean warbler, wood thrush, and scarlet tanager) are declining and have disappeared entirely from some locations. These birds are neotropical migrants; that is, they winter in the neotropics of Central and South America and breed in temperate North America. Other migratory groups have also suffered declines. By 2014, the precipitous 90% decline of monarch butterflies had made them world news, as they struggled to survive loss of habitat and exposure to chemicals.

Global Outlook. Worldwide, the loss of biodiversity is even more disturbing. One thorough analysis of biodiversity is the *Global Biodiversity Outlook,* published by the United Nations. Figure 6–6, from their 2010 assessment, shows groups from a comprehensive survey of more than 47,000 species, categorized by level of risk of extinction. The 2014 *Global Biodiversity Assessment 4* did not include the same analyses, but the trends were the same. Estimates of extinction rates in the past show alarming increases in loss of species. For mammals, the background (past) extinction rate was less than one extinction every thousand years, although rates were higher during five great extinction events. (Scientists estimate the background extinction rate from the fossil record of marine animals and from rates of change in DNA.) This rate can be compared with known extinctions of the past several hundred years (for mammals and birds,

20 to 25 species per hundred years and, for all groups, 850 species over about 500 years), indicating that the current rate of extinction is 100 to 1,000 times higher than past rates.

The International Union for Conservation of Nature (IUCN) keeps a running list of threatened species. In 2014, approximately 32% (2,030) of amphibian species, 22% (1,199) of mammal species, and 13% (1,373) of bird species were globally threatened.[5] The status of other groups of organisms is much less well known. The majority of threatened species are concentrated in the tropics, where biodiversity is so rich as to be almost unimaginable. Biologist Edward O. Wilson identified 43 species of ants on a single tree in a Peruvian rain forest, a level of diversity equal to the entire ant fauna of the British Isles.[6] Other scientists found 300 species of trees in a 1-hectare (2.5-acre) plot and almost 1,000 species of beetles on 19 specimens of a species of tree

---

[5] Red List Category summary for all animal classes and orders, IUCN Red List version 2011.2: Table 4a, accessed March 3, 2012, http://www.iucnredlist.org/documents/summarystatistics/2011_2_RL_Stats_Table4a.pdf.
[6] Edward O. Wilson, "The Arboreal Ant Fauna of Peruvian Amazon Forests: A First Assessment," *Biotropica* 19 (1987): 245–251.

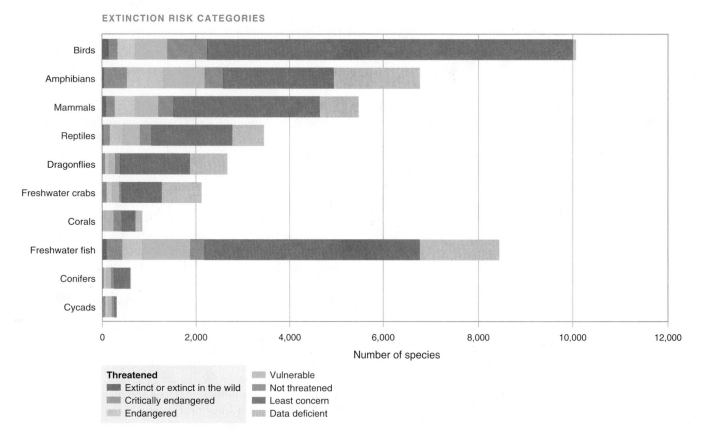

EXTINCTION RISK CATEGORIES

Threatened
■ Extinct or extinct in the wild
■ Critically endangered
▨ Endangered
▨ Vulnerable
■ Not threatened
■ Least concern
■ Data deficient

Figure 6–6 Who is going extinct? This bar graph shows groups from a comprehensive survey of more than 47,000 species, categorized by level of risk of extinction.

(Source: Global Biodiversity Outlook 3.)

UNDERSTANDING THE DATA

1. Which group has the largest number of extinct species?
2. If insects were included on the graph, which color on their bar would be largest? Why do you think so?

in Panama.[7] Unfortunately, the same tropical forests that hold such high biodiversity are also experiencing the highest rate of deforestation. Because the inventory of species in these ecosystems is so incomplete, it is virtually impossible to assess extinction rates in such forests. This uncertainty may contain a kernel of good news. Some scientists believe that we are overestimating the total number of species, and as a result, overestimating extinction rates in the areas where most habitat is being lost.[8] No one knows exactly how many species are becoming extinct, but the loss is real, and species loss represents a continuing depletion of the biodiversity of our planet.

Reasons for the Decline

The most famous extinction in American history was that of the passenger pigeon (*Ectopistes migratorius*), once the most common bird in North America (**Figure 6–7**). In the early and mid-1800s, the population of this bird was estimated at 3-5 million, but market hunting caused a rapid decline. By the end of that century, the wild population was gone; the last know wild bird was shot in 1900. The last passenger pigeon in captivity died in the Cincinnati Zoo in 1914. How could such a widespread creature go extinct so quickly?

In the distant past, extinctions were largely the result of processes such as climate change, plate tectonics, and even asteroid impacts. Current threats to biodiversity are sometimes described with the acronym **HIPPO**, which refers to five factors: habitat destruction, invasive species, pollution, population, and overexploitation. Many declining species experience combinations of several or all of these factors. Indeed, global climate change in particular increases the effects of some of these factors. Future losses will be greatest in the developing world, where biodiversity is greatest

and human population growth is highest. Africa and Asia have lost almost two-thirds of their original natural habitat. People's desire for a better way of life, the desperate poverty of rural populations, and the global market for timber and other natural resources are powerful forces that will continue to draw down biological wealth on those continents.

One key to slowing the loss of biodiversity lies in bringing human population growth down (Chapter 8). If the human population increases to 10 billion, as some demographers believe that it will, the consequences for the natural world may be frightening. Another key is to pull individual consumption down to sustainable levels in parts of the world with high consumption of energy and materials. These two concepts will be vital to solving several environmental issues and will be further discussed throughout the rest of the book.

We will now consider the five current threats to biodiversity that make up the acronym HIPPO.

Habitat Destruction. By far the greatest threat to biodiversity is the physical alteration of habitats through the processes of conversion, fragmentation, simplification, and intrusion. Habitat destruction has already been responsible for 36% of the known extinctions and is the key factor in the currently observed population declines. Natural species are adapted to specific habitats, so if the habitat changes or is eliminated, the species go with it.

Conversion. Natural areas are commonly converted to farms, housing subdivisions, shopping malls, marinas, and industrial centers. When a forest is cleared, for example, destruction of the trees is not the only result; every other plant or animal that occupied the destroyed ecosystem, either permanently or temporarily (e.g., migrating birds), also suffers. Global forest cover has been reduced by 40% already, and the decline continues. For example, the decline in North American songbird populations has been traced to the loss of winter forest habitat in Central and South America and the increasing fragmentation of summer forest habitat in North America. Agriculture already occupies 38% of land on Earth; croplands that replace natural habitats are quite inhospitable to all but a few species that tend to be well

---

[7]T. L. Erwin, "The Tropical Forest Canopy: The Heart of Biotic Diversity," in *Biodiversity*, ed. E. O. Wilson (National Academy Press, Washington, D.C., 1988), 123–129.

[8]M. J. Costello, R. M. May, and N. E. Stork, "Can We Name Earth's Species Before They Go Extinct?" *Science* 339, no. 6118 (2013): 413.

(a)

(b)

Figure 6–7 **The decline of the passenger pigeon. (a)** In the early 1800s, some flocks of passenger pigeons were so large that they darkened the sky as they passed overhead. Market hunting lead to their rapid decline. **(b)** The last passenger pigeon, a bird named Martha, lived at the Cincinnati Zoo. When she died in 1914, the species was extinct.

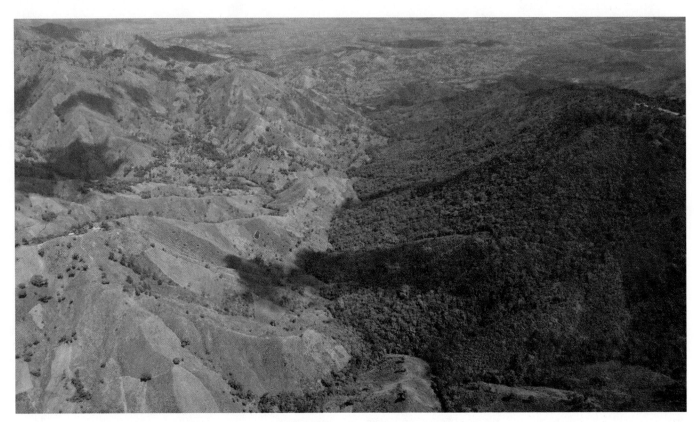

Figure 6–8 The border of Haiti and the Dominican Republic. Severe deforestation in Haiti, on the left, leads to habitat loss for many species. The forested area on the right is the Dominican Republic.

adapted to the new managed landscapes. Haiti provides one extreme example of habitat loss. Figure 6–8 shows the border between Haiti and the Dominican Republic, with Haiti on the left. Land use in Haiti has resulted in a tremendous loss of forests, something important not only in the context of loss of species, but also as a driver of soil erosion and water quality problems (as we will see in later chapters). In the Dominican Republic, social trends such as greater stability and democratization have helped the country avoid political unrest and environmental destruction.

Fragmentation. Natural landscapes generally have large patches of habitat that are well connected to other similar patches. Human-dominated landscapes, however, consist of a mosaic of different land uses, resulting in small, often geometrically configured patches that frequently contrast highly with neighboring patches (Figure 6–9). Small fragments of habitat can support only small numbers and populations of species, making them vulnerable to extermination. Species that require large areas are the first to go; those that grow slowly or have naturally unstable populations are also vulnerable.

Reducing the size of a habitat creates a greater proportion of **edges**—breaks between habitats that expose species to predators. Increased edge favors some species but may be detrimental to others. For example, the Kirtland's warbler is an endangered species that is highly dependent on large patches of second-growth jack pines in Michigan. The species is endangered because its habitat has been greatly fragmented, creating edges that favor the brown-headed cowbird, a nest parasite that can invade the forest and lay its

eggs in the nest of the rare warbler. Edges also favor species that prey on nests, such as crows, magpies, and jays. These predators are thought to be partly responsible for declines in many neotropical bird populations.

Roadways offer particular dangers to wildlife. The number of animals killed on roadways now far exceeds the number killed by hunters. As rural areas are developed, the increasing numbers of animals found on roadways are a serious hazard to motorists. With 4.1 million miles (6.6 million km) of paved roadways in the United States, more than a million animals a day become roadkill. Overpasses and tunnels are increasingly being built to provide wildlife with safe corridors. A 2008 study at Purdue University identified amphibians as especially affected by road mortality.[9] More than 95% of the roadkill that researchers identified over 11 months along a stretch of road in the Midwest were amphibians.

Simplification. Human use of habitats often simplifies them. Removing fallen logs and dead trees from woodlands for firewood, for example, diminishes an important microhabitat on which species ranging from woodpeckers to germinating forest trees depend. When a forest is managed for the production of one (or a few) species of tree, tree diversity declines, and as a result, so does the diversity of the plant and animal species that depend on the less favored trees. Streams are sometimes *channelized*—their beds are cleared of fallen trees and stony riffle areas (places with more rapidly

---

[9] D. J. Glista, T. L. DeVault, and J. A. DeWoody, "Vertebrate Road Mortality and Its Impact on Amphibians," *Herpetological Conservation and Biology* 3 (2008): 77–87.

Figure 6–9 **Fragmentation.** Development usually leads to the breaking up of natural areas, resulting in a mosaic of habitats that may not support local populations of species.

moving water), and sometimes the stream is straightened out by dredging the bottom. Such alterations inevitably reduce the diversity of fish and invertebrates that live in the stream.

Intrusion. Birds use the air as a highway, especially as they migrate north and south in the spring and fall. The clear surfaces of windows are one of the greatest dangers for birds, especially when the windows are lit from behind and birds are flying at night. Ornithologist Daniel Klem estimated that between 100 million and 1 billion birds die every year from crashing into glass windows.

Recently, telecommunications towers have presented a new hazard to birds. Although television towers have been around for decades, new towers are sprouting up on hilltops and in countrysides in profusion. Cell phone signal towers are also common. The lights often placed on these towers can attract birds, which usually migrate at night, and the birds simply collide with the towers and supporting wires. In 2000, a study estimated that the towers kill somewhere between 5 and 50 million birds a year.[10] Three organizations petitioned the Federal Communications Commission (FCC) to incorporate bird mortality in its decisions about new towers in 2002, but nothing was done. In 2008, a federal appeals court finally ruled that the FCC did in fact have to come up with a plan to protect birds. A 2011 federal policy provided hope that there would be an agreement to make towers more bird friendly.

Birds are just one example of organisms harmed by the intrusion of structures into habitats. Any solution to wild species' decline must include creative ways to lower the impact of human structures on other organisms.

Invasive Species. An exotic, or alien, species is one that is introduced into an area from somewhere else, often a different continent (see Chapter 4, where they were covered extensively). Because the species is not native to the new area, it is frequently unsuccessful in establishing a viable population and quietly disappears. Such is the fate of many pet birds, reptiles, and fish that escape or are deliberately released. Some exotic species establish populations that remain at low levels without becoming pests. Occasionally, however, an alien species finds the new environment to be very favorable, and it becomes an **invasive species** thriving, spreading out, and perhaps eliminating native species by predation or competition for space or food. Most of the insect pests and plant parasites that plague agricultural production were accidentally introduced (Chapter 13). Invasive species are major agents in driving native species to extinction and are responsible for an estimated 39% of all animal extinctions since 1600.

Accidental Introductions. As humans traveled around the globe in the days of sailing ships and exploration, they brought with them more than curiosity and dreams of exotic lands. They brought rats and mice to ports all over the world. As a result, rats are the most invasive mammal worldwide. The two most common invasive rat species are the

---

[10] G. Shire, *New Report Documents Hundreds of Thousands of Birds Killed at Communications Towers* (2000). Available from the American Bird Conservancy at www.abcbirds.org/newsandreports/releases/000629.html.

black rat (*Rattus rattus*, called the ship rat) and the brown rat (*Rattus norvegicus*, called the Norway rat). In the Middle Ages, fleas infesting black rats carried the bubonic plague bacterium to Europe, eventually killing a third of Europe's human population. Rats also eat crops, destroy property, and cause other harm to humans, but their effect on other species is profound. In terms of species loss, the arrival of rats is especially deadly to birds. According to one report, rats have been found preying on nearly one-quarter of seabird species. Thirty percent of seabirds are threatened with extinction, and rats are the biggest single cause.

Rats are only one example of accidental introduction of non-native species. **Figure 6–10** shows another: the red imported fire ant (*Solenopsis invicta*), which arrived in Mobile, Alabama, from South America, possibly in shipments of wood. The fire ant is a major pest across the southern United States, inflicting damage to crops and domestic animals. It also contributes to the decline of wild species. In Texas, it is estimated that fire ants kill a fifth of songbird babies before they leave the nest.

May I Introduce . . . Deliberate introductions sanctioned by the U.S. Natural Resources Conservation Service (in the interest of "reclaiming" eroded or degraded lands) have brought us kudzu and other runaway plants, such as the autumn olive and multiflora rose (Chapter 4). Salt cedar was introduced in the American Southwest for erosion control and has taken over riverbanks there. Many other exotic plants are introduced as horticultural desirables. *Fallopia japonica*, Japanese knotweed, is a good example. Tall, with attractive flowers, it originated in Asia and is now found in North America and Europe. It forms dense patches and outcompetes native plants. Unfortunately, its strong underground stem (rhizome) and root system cause the plant to spread even when it does not produce seeds.

There are specific instances of both deliberate and accidental introductions in **aquaculture**—the farming of shellfish, seaweed, and fish—which produces one-third of all seafood consumed worldwide. Parasites, seaweeds, invertebrates, and pathogens have been introduced together with the desired aquaculture species, often escaping from the pens or ponds and entering the sea or nearby river system. Sometimes the aquaculture species itself is a problem. Atlantic salmon, for example, are now raised along the Pacific Coast, and many have escaped from the pens and have been breeding in some West Coast rivers, where they compete with native species and spread diseases.

Over Time. The transplantation of species by humans has occurred throughout history, to the point where most people are unable to distinguish between the native and exotic species living in their lands. European colonists brought hundreds of weeds and plants to the Americas, and now most of the common field, lawn, and roadside plant species in eastern North America are exotics. For example, almost one-third of the plants in Massachusetts are alien or introduced—about 725 species, of which 66 have been specifically identified as invasive or likely to become so. In Europe, there are about 11,000 alien species, 15% of which harm biodiversity.

In North America, two invasive animals of particular concern are the Burmese python and the wild boar. Burmese pythons have been released into the wild by pet owners who no longer like them once they become large. There are now breeding populations of these snakes in the Florida Everglades, where they eat the native wildlife and are dangerous to humans (**Figure 6–11**). Wild boar were introduced by people who wanted to hunt and eat them. Today there are an estimated 4 million feral boar in the United States (**Figure 6–12**). They are dangerous and tear up soil, rooting up plants. One of the most destructive exotics is the house cat: Studies in the 1990s showed that the millions of domestic and feral cats in the United States kill an estimated 1 billion small mammals (chipmunks, deer mice, and ground squirrels, as well as rats) and hundreds of millions of birds

Figure 6–10 The red imported fire ant. This ant, *Solenopsis invicta*, has invaded the southern United States, where it kills other animals and sometimes starves them by eating their prey.

Figure 6–11 Burmese pythons invade the Florida Everglades. These predators were introduced by the pet trade and former pet owners. They are changing the ecology of the Everglades.

Figure 6–12 An invasive exotic species. Wild boar such as these in North Carolina have spread across the American South after escaping from captivity. Related to the domestic pig, they are larger and more aggressive.

annually. One report calculates the annual cost of invasive species in the United States to be $137 billion.[11]

Species originating in the Americas can cause havoc in other places, too. For example, the North American gray squirrel is invasive in Europe, where it outcompetes the native red squirrels and may be infecting them with a fatal virus. The gray squirrel is one of the 100 most invasive species worldwide.

Invasive Species and Trophic Levels. Plants provide food for herbivores that in turn provide food for carnivores. Non-native plants may be difficult for herbivores to eat and thus may keep energy and materials from passing up the food chain, even if a species has been in a new ecosystem for a long time. For example, the Norway maple was introduced to North America in 1756, but in spite of being in the United States for more than 250 years, these maples provide much less food for herbivores (like caterpillars) and their predators (like songbirds) than native trees do.

Pollution. Another factor that decreases biodiversity is pollution, which can directly kill many kinds of plants and animals, seriously reducing their populations. For example, nutrients (such as phosphorus and nitrogen) that travel down the Mississippi River from the agricultural heartland of the United States have created a fluctuating "dead zone" in the Gulf of Mexico, an area of more than 5,052 square miles (in 2014) where oxygen completely disappears from depths below 20 meters every summer[12] (see Chapter 20). Shrimp, fish, crabs, and other commercially valuable sea life are either killed or forced to migrate away from this huge area along the Mississippi and Louisiana coastline.

Pollution destroys or alters habitats, with consequences just as severe as those caused by deliberate conversions. Every oil spill kills seabirds and often sea mammals, sometimes by the thousands (water pollution is discussed further in Chapter 20). Some pollutants, such as the pesticide DDT,

can travel up food chains and become more concentrated in higher consumers (Chapter 5). Acid deposition and air pollution kill forest trees; sediments and large amounts of nutrients kill species in lakes, rivers, and bays; and global climate change, brought on by greenhouse gas emissions, is already affecting many species. Climate change is likely to accelerate in the next 50 years (see Chapter 18).

Pollution can spread disease. Human wastes can spread pathogenic microorganisms to wild species, a threat called pathogen pollution. For example, the manatee (an endangered aquatic mammal) has been infected by human papillomavirus, cryptosporidium, and microsporidium, with fatal outcomes. Pollution can also disrupt hormones and other body functions. Fish exposed to polluted river water may develop tumors or deformed organs. Sometimes they develop both male and female tissues, a condition called "intersex" (Figure 6–13). The recent and rapid rise in the incidence of these deformities has been traced to habitats that have been altered by human use, specifically run-off, industry, and sewage in populated areas.

Figure 6–13 Intersex fish. Up to 80% of male fish in some rivers are found to have a condition where they also have female tissues, such as eggs developing in their testes. This appears to result from pollution.

[11] D. Pimental, L. Lach, R. Zuniga, and D. Morrison, "Environmental and Economic Costs of Nonindigenous Species in the United States," *BioScience* 50 (2000): 53–65.

[12] D. Conners, "2011 Gulf of Mexico Dead Zone Smaller Than Scientists Predicted. EarthSky (2011): http://earthsky.org/earth/2011-gulf-of-mexico-dead-zone-smaller-than-scientists-predicted.

Population. Human population growth puts pressure on wild species through several mechanisms—through direct use of wild species, through conversion of habitat into agriculture or other use, through pollution, and through having to compete for resources with growing human populations (human population trends will be described in depth in Chapters 8 and 9). Overconsumption is tied to overpopulation and to the concept of overexploitation—the last letter in the HIPPO acronym—which will be discussed next. Even if each person using a resource used a modest amount, large numbers of people might still use too much, leading to a total overconsumption. Thus, overpopulation is a problem even if individuals use few resources. However, a small group of people can also overuse resources. Indirectly, overconsumption can drive pollution, conversion of habitat, and other effects that harm wildlife. In a world in which some individuals own literally billions of times the money and resources that others own, people with the most consumptive lifestyles have a disproportionate effect on the environment. Differences in level of consumption and in sheer numbers of people drive some of the tensions among the high-, middle-, and low-income countries (see Chapter 8).

Overexploitation. The overharvest of a particular species is known as overexploitation. Removing whales, fish, or trees faster than they can reproduce will lead to their ultimate extinction. The extinction of the passenger pigeon is one compelling example. In addition, overuse of individual species harms ecosystems. Overexploitation is estimated to have caused 23% of recent extinctions (past 500 years) and is often driven by a combination of greed, ignorance, and desperation.

Poor resource management often leads to a loss of biodiversity. Forests and woodlands are overcut for firewood, grasslands are overgrazed, game species are overhunted, fisheries are overexploited, and croplands are overcultivated. These practices set into motion a cycle of erosion and desertification, with effects far beyond the exploited area (this problem is explored in depth in Chapter 7).

Trade in Wild Products. One prominent form of overuse is the trafficking in wildlife and in products derived from wild species. Much of this "trade" is illegal. Worldwide, the U.N. Environmental Programme estimates that it generates $7 to $23 billion a year, making it among the most lucrative illicit sources of income, after narcotics, counterfeiting, and human trafficking. This illegal activity flourishes because some consumers are willing to pay exorbitant prices for furniture made from tropical hardwoods (like teak), exotic pets, furs from wild animals, traditional medicines from animal parts, and innumerable other "luxuries," including polar-bear rugs, ivory-handled knives, and reptile-skin handbags. For example, a panda-skin rug can sell for up to $25,000 in the United States.

The plight of various species of rhinoceros demonstrates the harm caused by the illegal trade in wildlife products. In 2011, the West African rhino was declared extinct. In October 2011, the Javan rhino was declared extinct in mainland Asia, surviving in only a small population in Indonesia. Rhino species are all very rare because of the high value placed on their horns; poachers kill rhinos to remove their horns, which are prized in Asian medicine and used for ornamental knife handles in the Middle East. Even preemptive removal of the rhinos' horns by wildlife managers does not protect the animals, because poachers kill them for the remaining bases of their horns. In addition, the dehorned animals do not survive well in the wild. Wildlife officials post guards around the rarest animals (**Figure 6–14**) or move them to safer locations. One South African Group, the Rhino Rescue Project, attempts to make poaching less appealing by de-valuing horns. Veterinarians drill small holes into the horns of sedated rhinos and put in a substance toxic to humans that will make the horn impossible to use in alternative medicine. Microchips are added and sometimes dye. Another solution is aggressive advertising against wild animal parts. An advertising campaign in Vietnam saw demand for rhino horn drop a third between 2013 and 2014. Protecting rhinos, like many other creatures, will take coordinated approaches.

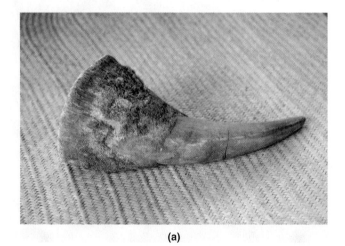

(a)

(b)

Figure 6–14 Black rhino horn and white rhino. (a) Rhino horn is prized in traditional Asian medicine and as ornamentation. **(b)** Only five northern white rhinos still exist in the world. This one in Kenya is guarded by armed rangers to protect it from poachers.

STEWARDSHIP
Lake Erie's Island Snake Lady

A three-foot-long grayish-tan snake basks on a rock at the shoreline of an island in Lake Erie, Pennsylvania. As the snake warms itself, the sunlight glints on its scales. Slowly it moves, one coil sliding over another, and quietly glides into the water. Head still, it watches a small round goby fish (*Neogobius melanostomus*). Suddenly, it strikes. The goby caught, the snake swallows slowly and slithers off to resume basking in the Sun. The snake is the Lake Erie water snake *Nerodia sipedon insularum*, an endemic to the area. It was once endangered and limited to only 1,000 individuals and was subject to conservation measures.

Out of the vegetation steps a woman in field gear. Kristen Stanford, known as the "Island Snake Lady," is the water snake recovery plan coordinator for this rare reptile, working out of the Ohio State University Stone Laboratory. Over the past several years, she and her colleagues caught, measured, tagged, and protected water snakes; as a result, the small snake recovered in numbers.

The hard work of scientists, policy makers, and enthusiasts paid off. The snake, returning to a population of 12,000 individuals, was taken off of the threatened species list in 2011. Species are placed on the threatened and endangered species lists when they are already so rare that, for many, recovery is a long,

drawn-out process. The rapid recovery of the snake is great news for conservationists.

What makes populations of one species increase rapidly and causes populations of another to decline? How do species relate to the other species around them? In the case of the water snake, intensive human killing and loss of habitats were the main causes of decline. Ironically, the presence of large numbers of invasive goby fish helped the snake come back from the brink of extinction. This unexpected benefit is not the norm for invasive species, however.

Many people find it hard to feel concerned about a declining snake. They may feel that snakes are threatening or are unattractive. Many people need a reason to care for a species in decline and find it much easier to relate to whales, a beautiful bird like the whooping crane, or perhaps the majestic Bengal tiger. These species are considered "charismatic" because of their wide appeal. But many of the world's creatures that are disappearing are the small and seemingly unimportant parts of the ecosystem, like a rare water snake or small beetles. Why care about them? Conservationists suggest several reasons for concern about the loss of species. For people like Kristen Stanford, protection of species, even the

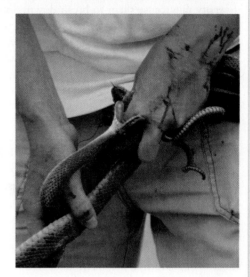

Lake Erie Water Snake

small ones, is worth the cold, the bitten hands, and the wet feet that come with being "the snake lady."

Sources: Water Snake, Lake Erie. (2001). Ohio Public Library Information Network. http://www.oplin.org/snake/fact%20pages/water_snake_lake_erie/water_snake_lake_erie.html. March 6, 2012.

Michael Scott, The Plain Dealer. "Lake Erie water snakes are making a comeback." Published Sunday, June 08, 2008, http://blog.cleveland.com/metro/2008/06/snakes.html.

Trade in Exotic Pets and Plants. The fad for exotic pets and houseplants is a growing problem. In many cases, these plants and animals are taken from the wild; only a small fraction of them survive the transition. When animals and plants are removed from their natural breeding populations, these individuals are no better than dead when it comes to maintaining the species. At least 40 of the world's 330 known species of parrots are threatened because of this fad. Trade in wild animals can be very profitable, with some Indonesian and South American parrots selling for up to $10,000 in the United States. Many pet stores carry imported species. Not purchasing wild-caught species is one way individuals can help preserve biodiversity.

In 1992, the United States, the largest importer of exotic birds, adopted the Wild Bird Conservation Act, a law designed to stop the wild capture of declining birds, uphold international treaties in wildlife trade, and support sustainable breeding programs. In 2004, more than 200 nongovernmental organizations signed the European Union Wild Bird Declaration, a call for wild bird protection in the European Union (EU). In a huge step forward in bird conservation, in 2007 the EU decided unanimously to prohibit the import of wild birds.

A Double Whammy—Combining Effects. The HIPPO acronym is generally helpful, but the decline of many species is the result of a combination of factors. Climate change, for example, can overlay many other problems, forcing species to leave their ranges or stressing them so they struggle with other issues. One good example of combined effects is colony collapse disorder, the sudden death of honeybee colonies, which has wiped out up to 90% of the hives owned by some beekeepers. The cause of this disorder is still a mystery, but is thought to be related to stress from pesticides overlaying a set of other problems, including mites, a parasitic fly, fungal and viral infections, and malnutrition from being exposed to monocultured plants. An EPA task force and scientists around the globe are trying to figure it out.

Consequences of Losing Biodiversity

Why is the loss of biodiversity a concern? The answer goes back to the values of wild species. Their intrinsic value is self-explanatory: Species are valuable regardless of human benefit. We humans also need biodiversity; this is instrumental value. We use wild species for food, fiber, fuel, and as a source of medicines; we value them aesthetically and want recreation in natural habitats. We need biodiversity to maintain healthy ecosystems, which bring a great deal of economic benefit.

The complexity of ecosystems means that the loss of diversity may cause them to break down. Sometimes the loss of a species affects ecosystems' functions that we take for granted. One example is the loss of vultures in India. In the 1980s, India's nine species of vultures were the most numerous raptors in the world, with a combined population of about 40 million. Vultures ate the carcasses of dead animals, including cattle. The use of the drug Diclofenac on cattle nearly wiped out the vultures. In just a couple of decades, the vulture populations declined precipitously, with some losing more than 99% of their members. Without vultures, the carcasses of dead animals accumulated, polluting the water, spreading disease, and attracting wild dogs. In 2006, India banned the use of Diclofenac in cattle. Although all the species of Indian vultures are now endangered, their populations may be slowly increasing.

Can ecosystems lose some species and still retain their functional integrity? Research indicates that the extinction of some species will not lead to ecosystem decline. This has risks, however, because it is possible to lose keystone species—species whose role is absolutely vital to the survival of many other species in an ecosystem (see Chapter 4). For example, by felling trees and building dams, beavers make wetland habitats that undergo succession as beavers move throughout the landscape, increasing ecosystem diversity in a region.

Many declining species are *K*-strategists—large, long-lived organisms that are relatively old at first reproduction and provide a high degree of parental care for their offspring (Chapter 4). Thus, they are vulnerable to rapid changes in the environment. Often they have the greatest demands for unspoiled habitat. Thus, wolves, elephants, tigers, moose, and other large animals are considered "umbrella" species for whole ecosystems.

Those species least likely to be harmed by human actions include widely distributed species with small size, rapid reproduction, short generation time, low parental care of offspring, and an ability to migrate (characteristics of *r*-strategists). Such species (roaches, houseflies, and others) are most likely to become pest species. Thus, the overarching effect of human disturbance of the environment is not only to decrease species that we may like, but also to increase a number that will become major pests.

Moving Forward

We do not really know what we are losing when species become extinct. One thing is certain: The natural world is less beautiful and less interesting as a result. We might be inclined to feel overwhelmed by the significant decline in species and in the abundance of organisms. But there are glimmers of hope. Sometimes species we thought were extinct are found not to be. Two of these are the Beck's petrel, not seen for 80 years before being found by an ornithologist in 2008, and the New Guinea big-eared bat, not seen for 120 years and thought extinct but found in Papua New Guinea in 2012. Sometimes new populations of very rare species are discovered, as happened in the case of the highly threatened grey-breasted parakeet. Only about 300 of these birds are known to live in the wild, all in northeastern Brazil. Like these parakeets, most rediscovered species remain on the brink of extinction, but every piece of good news is welcomed.

As we saw in Stewardship, *Lake Erie's Island Snake Lady,* sometimes the hard work of scientists and conservation volunteers pays off in the recovery of a rare and declining species. Once in a while, conservationists catch a break, as they did in 2008, when a rare Vietnamese turtle, one of the last four of its species in the world, was swept from a lake by a flood and caught by a fisherman who was going to sell the reptile to a restaurant for soup. Conservationists were able to purchase the 68-kilogram (150-lb.) turtle and return it to the lake.

While success in protecting species begins with the courage and determination of individuals, conservation requires coordinated efforts over regional, national, and international scales, involving the work of both scientists and policy makers. Sometimes new protections emerge from a change in policy. Ironically, the conservation of birds was enhanced in the EU when concern over avian flu made it possible to pass regulations limiting the import of wild birds. Sometimes scientific accomplishments make conservation doable. For example, 2008 saw the first live rhinoceros birth from frozen semen, and black-footed ferrets, an extremely rare weasel-like animal, have been born from sperm stored from fathers who had died eight years before (Figure 6–15). These breakthroughs in captive breeding may save species. In the next section, we will see more of the efforts scientists and others are putting into saving wild species.

Figure 6–15 Captive breeding with new techniques. Captive breeders are now able to use frozen semen to produce black-footed ferrets, an extremely rare animal.

CONCEPT CHECK What are the five main causes of the loss of biodiversity? Which of these caused the extinction of the passenger pigeon? ☑

6.3 Saving Wild Species

Current efforts to save wild species and preserve biodiversity involve basic science, individuals, corporations, public policy, and decisions at the local, national, and international levels (Table 6–3). Scientists are on the front lines of biodiversity protection, as they often have a better knowledge of what is out there in the world and whether or not it is declining. However, scientists cannot stop biodiversity loss by themselves. That requires laws, as well as the enforcement of those laws. Protection of biodiversity also requires stepping back to look at the big picture, seeing how people relate to the rest of the natural world, and examining how we can do so in more sustainable ways.

The Science of Conservation

The scientific projects that have been used to save the Lake Erie water snake, rhinos, and black-footed ferrets are just a few of many research activities in conservation biology, the branch of science that is most focused on the protection of populations and species. Conservation biologists use techniques such as captive breeding, telemetry (transmitting data through remote communication), and tracking devices in order to learn more about the natural history of particular species and to protect those that are vanishing. Getting accurate population counts can be extremely difficult. Scientists study the relationships among species and between biodiversity as a whole and ecosystem functions.

There is another critical element in the science of protecting species—**taxonomy**. Taxonomy is the cataloging of species and the naming of new ones. It allows us to understand what species are out there and to identify those in the most trouble. The number of taxonomists in developed countries has been declining, a trend that makes understanding biodiversity more difficult. At the same time, more taxonomists are being trained in developing countries, a positive trend that should help future research.

Nonprofit Efforts

One way individuals contribute to conservation of biodiversity is through the funding of nonprofit organizations such as the World Wildlife Fund, Audubon Society, and many others. With some fundraising campaigns, people are able to "sponsor" individual whales, tigers, polar bears, and other creatures considered "charismatic"—that is, widely appealing. The efforts to protect these species, which often require large areas of habitat, protect other species as well. Land conservation is used to protect multiple species at the same time (discussed in more detail in Chapter 7). The oldest regional land conservancy in the United States is the Trustees of Reservations in Massachusetts.

Zoos have a special role in biodiversity conservation. Not everyone believes it is a good thing for animals to be kept in zoos, away from their natural habitat. However, in many cases, zoos have a two-fold role in protecting species. First, they actively educate the public so people care about species protection. Second, many are involved in research and in captive breeding programs. For some species, the breeding done at zoos and other captive breeding facilities is the difference between survival and extinction. Botanical

Table 6–3 Actions Taken to Protect Species by Individuals and Different Groups

Group	Type of action	Example
Individual	Support policy	Vote to support biodiversity
	Smart consumerism	Purchase sustainable materials, use fewer resources
	Citizen science	Great Sunflower Project
Business	Sustainable practices	Coffee plantations with plants for birds
	Smart investment	Go into green technologies and businesses
Nonprofit	Land conservation	Protection of marine reserves/wildlife
	Captive breeding	Zoo programs
	Education	Environmental education
	Sustainable development	Working with local groups for conservation goals
Science	Basic research in natural history, taxonomy, captive breeding techniques	Rhino breeding
	Develop wildlife forensics	Scottish Wildlife Forensics labs
	Monitor species	IUCN Red List
	Control invasive species, and better understand other causes of species decline	Study effects of house cats or cell phone towers on birds
Government	Make sound policies for local and national needs	Lacey Act, Marine Mammal Act, Endangered Species Act,
	Develop international treaties	CITES, Convention on Biodiversity

gardens can have the same role for plants. Of course, breeding many species solely in captivity is not a sustainable long-term solution for species loss, but it may bridge the gap while humans figure out how to live more sustainably.

Individuals and Corporations

Individuals can directly support conservation by being careful in their consumption of resources, use of chemicals, and driving habits, by not purchasing exotic species or products made from endangered species, and through other personal choices. However, many of the actions individuals take are through support of nonprofits, pushing for policy actions, and pushing corporations to work in ways that are more biodiversity-friendly. For example, consumers are insisting on more organic foods and certification of wild-caught fish or forest products as sustainable (some of these approaches are discussed in Chapter 7).

Consumers also encourage businesses to care about species. Many businesses are beginning to realize that biodiversity loss could harm their profits. Biodiversity-friendly investing is on the rise as well. A survey in 2009 of global CEOs found that 27% expressed concern about the impacts of biodiversity loss on their business prospects.

Individuals can act as "citizen scientists" and take part in massive information-gathering efforts such as the Great Sunflower project, which monitors pollinators (Figure 6–16), the Firefly Project (which tracks firefly abundance), the annual Audubon Society Christmas Bird count (in which volunteers count birds), Project Budburst (a project that tracks when plants change seasonally), and others.

Governments: Local, State, and National Policies

Public policies, as well as the agencies that make and support them, are necessary for the protection of species.

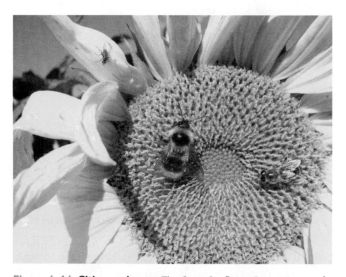

Figure 6–16 Citizen science. The Great Sunflower Project is one of many that use network of amateur scientists to collect data on wild species. In this case, they study pollinators, which are in decline.

Governments from local to international are involved in monitoring, conserving, and protecting biodiversity. If we want to save wild species, we first need to ask who owns wildlife. In the United States, property owners do not own the wildlife living on their lands; wildlife resources are public resources, protected under the Public Trust Doctrine. The government holds these resources in trust for all its people and is obliged to provide protection for these resources. This protective role is usually exercised by state fisheries and wildlife agencies. Sometimes the law mandates federal jurisdiction, as in the case of endangered species. Game animals were some of the first animals people recognized as having the potential to be overharvested; consequently, they are a group that receives special oversight, even if they are not currently rare.

Game Animals in the United States. Game animals are those animals traditionally hunted for sport, meat, or pelts. In the early days of the United States, there were no restrictions on hunting, and a number of species were hunted to extinction (the great auk, heath hen, and passenger pigeon) or near extinction (the bison and wild turkey). As game animals became scarce in the face of unrestrained hunting, regulations were enacted. State governments, backed up by the federal government, enacted laws establishing hunting seasons and bag limits and hired wardens to enforce the laws. Some species were given complete protection in order to allow their populations to build up to numbers that would once again allow hunting. The wild turkey, for example, was hunted to the brink of extinction. It made a slow comeback by the 1930s as a result of hunting restrictions and returned strongly after reintroduction into areas it had once inhabited. The turkey is now found in 49 states. Unfortunately, the turkey was also introduced to western states, where it was not native, and it is now a pest in those areas. Ideally, in the future it would be protected in states where it is native and removed from states where it is invasive.

Hunting and Conservation. Using hunting and trapping fees as a source of revenue, state wildlife managers enhance the habitats that support important game species. In spite of what seems on the surface to be destruction of wildlife, hunting has many positive aspects. Besides the fees hunters pay, many hunters belong to organizations dedicated to the game they are interested in hunting. Organizations such as Ducks Unlimited raise funds that are used for the restoration and maintenance of natural ecosystems vital to the game they are interested in hunting.

Defenders of hunting and trapping argue that their prey are often animals that lack natural predators and would increase to the point of destroying their own habitat, something that frequently happens in the United States with larger animals such as deer and elk. For example, there are so many deer in Pennsylvania that, in 2004, they were estimated to be at three times the natural carrying capacity; the large number of deer damage the ecosystems they inhabit as well as the other organisms living there.

Figure 6–17 **Wild cougar.** As humans encroach on wild habitat, animals come into human areas. This cougar was on a suburban roof in California. It was shot with a tranquilizer. Returning big predators to ecosystems is not likely to happen in human-inhabited areas.

Many people feel that encouraging hunting because of a lack of other predators may not be getting at the fundamental problems with species loss. They argue for the return of predators such as grizzly bears, mountain lions, and wolves in order to restore more natural checks and balances to ecosystems. Because humans do not always get along well with top predators, such reintroductions are unlikely to occur on a large scale (Figure 6–17).

Interactions Between Humans and Wildlife. Because wild animals are in the public trust, their numbers are managed by government agencies. Now that many populations are increasing, the government sometimes has to limit them. Many nuisance animals are thriving in highly urbanized areas, creating various health hazards. Opossums, skunks, bears, and deer are attracted to suburban and urban areas by opportunities for food; unsecured garbage cans and pet food left outside will ensure visits from raccoons. In 2012, the Centers for Disease Control and Prevention reported 6,162 cases of rabid animals in the United States, 59% of which were raccoons and bats.[13]

One highly controversial effort to control unwanted animals is carried out by the Wildlife Services agency of the U.S. Department of Agriculture. At the request of livestock owners, farmers, homeowners, and others, Wildlife Services routinely uses poisons, traps, and other devices to remove nuisance wildlife. This usually means killing them; 4 million animals were killed in 2014. Of these, most are native species such as raccoons or skunks, but others, such as wild hogs and starlings, are exotic invasive species. Wildlife

Services plays the important role of keeping both invasive species and native species from overtaking human areas and limiting negative human–wildlife interactions.

Protecting Endangered Species

In colonial days, huge flocks of snowy egrets inhabited the coastal wetlands and marshes of the southeastern United States. In the 1800s, when fashion dictated fancy hats adorned with feathers, egrets and other birds were hunted for their plumage. By the late 1800s, egrets were almost extinct. Concerned about the loss of birds, some people began to form state Audubon societies. Conservationists such as Harriet Hemenway and Minna Hall, two prominent Boston socialites, began a campaign to shame "feather wearers" and end the practice. The campaign caught on, and gradually attitudes changed; new laws followed. Government policies that protect animals from overharvesting are essential to keep species from the brink of extinction. Since the 1800s, several important laws have been passed to protect a wide variety of species.

Lacey Act. Florida and Texas were the first states to pass laws protecting plumed birds. Then, in 1900, Congress passed the **Lacey Act,** forbidding interstate commerce in illegally killed wildlife, making it more difficult for hunters to sell their kill. Since then, numerous wildlife refuges have been established to protect the egrets' breeding and migratory habitats. With millions of people visiting these refuges and seeing the birds in their natural locales, attitudes have changed significantly. Today, the thought of hunting these birds would be abhorrent to most of us, even if official protection were removed. Thus protected, egret populations were able to recover substantially. In the meantime, the Lacey Act and its amendments have become the most

---

[13]J. D. Blanton et al., *Rabies Surveillance in the United States During 2012.* Available at http://avmajournals.avma.org/doi/pdf/10.2460/javma.243.6.805.

Table 6–4	Selected Federal and International Conservation Regulations	
Regulation	**Implemented**	**Key Provisions**
U.S. Regulations		
Lacey Act	1990	Forbids interstate commerce in illegally killed wildlife, restricts injurious wildlife; amended to include plants
Endangered Species Act	1973	Protects species at risk of extinction; amended to protect critical habitat
Marine Mammal Protection Act	1972	Prohibits killing of marine mammals in U.S. waters or by U.S. citizens on high seas
International Regulations		
Convention on International Trade in Endangered Species of Wild Fauna and Flora (CITES)	1975	Prohibits international trade in endangered species
Convention on Biodiversity	1993	Promotes conservation of species, sustainable use of species, and equitability in development of biological resources

important piece of legislation protecting any wildlife from illegal killing or smuggling. Under the act, the U.S. Fish and Wildlife Service (USFWS) can bring federal charges against anyone violating a number of wildlife laws. There have been a number of amendments to the Lacey Act, including those that have broadened it to include plants. The act is also used to prohibit the movement of "injurious" species, including invasive species. In 2012, the U.S. Fish and Wildlife Service banned the importation and interstate transportation of four species of snakes, including the Burmese python, because of their negative impact on the Florida Everglades.[14] Table 6–4 summarizes the Lacey Act and some of the other federal and international regulations that are now used to protect wildlife.

For examples of the work of the U.S. Fish and Wildlife Service: In 2011, special agents worked on more than 13,565 violations of the Lacey Act and other wildlife-related laws. Their efforts resulted in fines of more than $8 million and a total of 87 years of jail time for those convicted. In one case, commercial fishermen in Virginia were convicted of illegally harvesting and selling 65,000 pounds of striped bass taken from the Potomac River. As wildlife smuggling increases and budgets tighten, USFWS agents work harder to catch people harming the rarest species.

Endangered Species Act. Congress took another major step when it passed a series of acts to protect endangered species. The most comprehensive and recent of these acts is the **Endangered Species Act** of 1973 (reauthorized in 1988). Recall that an **endangered species** is a species that has been reduced to the point where it is in imminent danger of becoming extinct if protection is not provided (Figure 6–18). Genetically distinct populations (subspecies) such as the Florida panther and the peninsular bighorn sheep may also be protected. Further, the act provides for the protection of **threatened species,** which are judged to be in jeopardy but not on the brink of extinction. When a species is officially recognized as being either endangered or threatened, the law specifies substantial fines for killing, trapping, uprooting (in the case of plants), modifying significant habitat of, or engaging in commerce in the species or its parts. The legislation forbidding commerce includes wildlife threatened with extinction anywhere in the world.

The Endangered Species Act is administered by the U.S. Fish and Wildlife Service for terrestrial and freshwater species (and a few marine mammals) and by the National Oceanic and Atmospheric Administration (NOAA) Fisheries Service for marine and anadromous species (those that migrate into freshwater). There are three crucial elements in the process of designating a species as endangered or threatened:

1. **Listing.** Species may be listed by the appropriate agency or by petition from individuals, groups, or state agencies. Listing must be based on the best available information and must not take into consideration any economic impact the listing might have.

2. **Critical Habitat.** When a species is listed, the agency must also designate as *critical habitat* the areas where the species is currently found or where it could likely spread as it recovers. A 1995 Supreme Court decision made it clear that federal authority to conserve critical habitats extended to privately held lands.

3. **Recovery Plans.** The agency is required to develop recovery plans that are designed to allow listed species to survive and thrive.

As of August 2014, 1,552 U.S. species were listed for protection under the act; recovery plans are in place for 1,146 of them. In addition, 142 candidate species are waiting to be listed.

---

[14]U.S. Fish & Wildlife Service Office of Law Enforcement, *FY 2012 Annual Report* (2014). Available at http://www.fws.gov/le/aboutle/annual.htm.

(a) Pine Barrens Tree Frog (*Hyla andersonii*)

(b) Haleakala Silversword (*Argyroxiphium sandwicense*)

(c) Red-cockaded Woodpecker (*Leuconotopicus borealis*)

(d) Karner Blue Butterfly (*Lycaeides melissa samuelis*)

(e) Swamp Pink (*Helonias bullata*)

(f) Spix's Macaw (*Cyanopsitta spixii*)

(g) Devil's Hole Pupfish (*Cyprinodon diabolis*)

(h) Arabian (White) Oryx (*Oryx leucoryx*)

(i) Sunda Pangolin (*Manis javanica*)

Figure 6–18 Endangered species. Shown are some examples of endangered species—species whose populations in nature have dropped so low that they are in imminent danger of becoming extinct unless protection is provided. The Arabian oryx **(h)** was recently delisted as an endangered species, but wild populations are still only 1,000 individuals.

Conflicting Values. Since its enactment, the Endangered Species Act has been a political lightning rod. It was scheduled for reauthorization in 1992, but because of political battles, it has been operating on year-to-year budget extensions. Opposition to the act comes from development, timber, recreational, and mineral interests, as well as other groups. Organizations such as the American Land Rights Association

and the American Forest Resource Council continually lobby for weakening or abolishing the act because they believe it limits the rights of landowners to log, construct buildings, and carry out other activities.

Proponents and opponents of the act view its success rate very differently. Some critics claim that the Endangered Species Act has been a failure because only a few of the

1,500 listed species (Table 6–5) have recovered and been taken off the list (33 species had been recovered by 2014). In response, proponents state that the two main causes of extinction—habitat loss and invasive species—are on the increase. Furthermore, species are listed only when they have already reached a dangerously low population, which means that they almost certainly would have gone extinct without protection; the populations of some 41% of the species on the list have stabilized or increased since being listed. We could also look at the number that have not gone extinct, when all of the evidence suggested they would. Then the success rate of the act would be 99%.

Other critics of the Endangered Species Act believe that it does not go far enough. A major shortcoming is that protection is not provided until a species is officially listed as endangered or threatened and a recovery plan is established.

For half of the protected vertebrate species, fewer than 1,000 individuals remained in each population at the time of listing; for half of the listed plant species, fewer than 120 individuals remained. Currently, the list's backlog includes some 142 candidate species, all of which are acknowledged to be in need of listing. In the first 20 years following the passage of the Endangered Species Act, 114 U.S. species went extinct or missing, which means that they haven't been seen despite at least 10 years of survey. The majority of these never made it to the list and became extinct during a delay in listing procedures.

Some opponents believe that the critical habitat designation puts unwanted burdens on property owners and has little merit in conserving species. One proposed alternative law, for example, would have completely eliminated critical habitat protections, replacing them with a statement that

Table 6–5   Federal Listings of Threatened and Endangered U.S. Plant and Animal Species, 2014

Summary of Listed Species, Listed Populations,[a] and Recovery Plans[b] as of July 10, 2014

	United States			Foreign			Total Listings (U.S. and Foreign)	U.S. Listings with Active Recovery Plans[c]
	Endangered	Threatened	Total Listings	Endangered	Threatened	Total Listings		
Mammals	71	20	91	256	20	276	367	72
Birds	81	18	99	214	17	231	330	86
Reptiles	14	23	37	70	20	90	127	36
Amphibians	21	14	35	8	1	9	44	17
Fishes	83	71	154	11	1	12	166	103
Clams	75	13	88	2	0	2	90	71
Snails	34	12	46	1	0	1	47	29
Insects	60	10	70	4	0	4	74	40
Arachnids	12	0	12	0	0	0	12	12
Crustaceans	22	3	25	0	0	0	25	18
Corals	0	2	2	0	0	0	2	0
Animal Subtotal	473	186	659	566	59	625	1,284	474
Flowering Plants	688	153	841	1	0	1	842	641
Conifers and Cycads	2	1	3	0	2	2	5	3
Ferns and Allies	28	2	30	0	0	0	30	26
Lichens	2	0	2	0	0	0	2	2
Plant Subtotal	720	156	876	1	2	3	879	672
Grand Total	1,193	342	1,535	567	61	628	2,163	1,146

Source: Department of the Interior, U.S. Fish and Wildlife Service, Division of Endangered Species, July 10, 2014.

[a]A listing has an E or a T in the "status" column of the tables in 50 C.F.R. §17.11(h) or 50 C.F.R. §17.12(h) (the "List of Endangered and Threatened Wildlife and Plants").
[b]Sixteen animal species (11 in the United States and 5 in foreign countries) are counted more than once in the above table, primarily because these animals have distinct population segments (each with its own individual listing status).
[c]Recovery plans pertain only to U.S. species.

recovery plans must identify areas of "special value" to the protection of the species; nothing in this designation would be mandatory. However, science suggests that critical habitat matters. Species whose critical habitat has been protected have been twice as likely to be on the road to recovery as those whose habitat has not been protected. In fact, the current approach is to protect the critical habitat of multiple species at once, an ecosystem approach.

Powerful Protection. In the final analysis, the Endangered Species Act is a formal recognition of the importance of preserving wild species, regardless of whether they are economically important. Species listed as endangered or threatened have legal rights to protection. The act is something of a last resort for wild species, but it embodies an encouraging attitude toward nature that has now become public policy. Many consider it to be the most powerful law in the United States for the protection of species.

Seeing Success

Efforts to protect species—such as the Lacey Act, the Marine Mammals Protection Act, the Endangered Species Act, and many other laws designed to keep our biotic heritage from going extinct—have resulted in some successes. The Lake Erie water snake is one such example. Another example of a species restored to parts of its former range because of these protection and recovery efforts is the gray wolf. Other species have also benefited significantly from these laws.

Birds of Prey. Both bald eagles and peregrine falcons have been removed from the Endangered Species List because their numbers have recovered significantly (Figure 6–19). Both are large raptors—predatory birds at the top of the food chain—that were driven to extremely low numbers because of the use of the pesticide DDT from the 1940s through the 1960s. Carried up to these predators through the food chain, DDT caused a serious thinning of the birds'

eggshells that led to nesting failures (see Chapter 1). In the early 1970s, DDT use was banned in both the United States and Canada, and the stage was set for the birds to recover. Although both raptors are now delisted, the Migratory Bird Treaty Act still protects them, and the Bald and Golden Eagle Protection Act protects the eagle.

Fly Away Home. A few species have gained exceptional public attention, and heroic efforts have been mounted to save them. Efforts to save the whooping crane, for example, include virtually full-time monitoring and protection of the migratory wild flock; from a low of 14 cranes in 1939, the population had increased to 304 in the spring of 2014. Scientists have used creative approaches to helping these wild birds survive, in some cases teaching them to migrate (Figure 6–20). Non-migratory flocks have been reintroduced in central Florida and Louisiana. Although each of these flocks is extremely vulnerable to environmental changes, conservationists remain hopeful.

Another success is the first delisted fish, the Oregon chub (*Oregonichthys crameri*), a small minnow that lives exclusively in Oregon's Willamette Valley. As a result of non-native predators and wetland drainage, its population had dwindled to fewer than 1,000 fish in eight locations. The fish was listed as endangered in 1993. Comprehensive efforts to protect the fish, involving private landowners, state and federal agencies, university researchers, and conservation groups have been critical to its recovery.

Bald eagles, peregrine falcons, whooping cranes, and the Oregon chub are examples of successful conservation. Although these species still need help, their numbers are clearly increasing. These examples show that current laws and the hard work of many can make a difference in slowing the rate of extinction.

CONCEPT CHECK Compare the roles of the Lacey Act and the Endangered Species Act in the protection of species. ☑

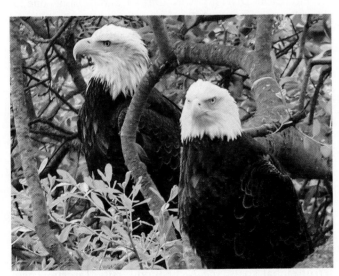

Figure 6–19 **The American bald eagle.** The Endangered Species Act worked well in bringing this magnificent national symbol to the point where it was dropped from the list.

Figure 6–20 **Whooping cranes and pilot.** A pilot in an ultra-light aircraft leads a flock of whooping cranes in their migration from Wisconsin to Florida.

6.4 Protecting Biodiversity Internationally

Serious efforts are being made to preserve biodiversity around the world, especially in the tropics, where so much of the world's biodiversity exists. These protective efforts require an immense amount of coordination among local, state, and federal authorities. The National Biological Information Infrastructure is a collaborative program that makes protecting species easier for agencies in the United States, but the United States also has to coordinate with the rest of the world. These partnerships not only create treaties, but also monitor species, share scientific advances, and find solutions to the needs of people who are clashing with wild species.

International Developments

There are numerous international developments for the protection of species. Some, like the International Union for Conservation of Nature (IUCN, formerly called the World Conservation Union), are organizations that monitor the successes and failures of our conservation efforts. Others are organizations that help coordinate scientists or policy makers around the globe. For example, the Invasive Species Specialist Group (ISSG) is a part of the Species Survival Commission of IUCN. The ISSG is made up of invited world experts who act as advisors and maintain a global invasive species database. The policy and treaty makers have to hammer out documents like the Convention on Biological Diversity. Finally, there has to be funding for species protection, some of which is provided by the Critical Ecosystem Partnership Fund and other foundations.

The Red List. The IUCN maintains a "Red List" of threatened species. Similar to the U.S. Endangered Species List, the Red List uses a set of criteria to evaluate the risk of extinction for thousands of species throughout the world. The list is updated frequently, and there were 22,176 species of plants and animals on it in 2014. Once published in book form, the Red List is now available in searchable electronic format on the Internet. Each species on the list is given its proper classification, and its distribution, documentation, habitat, ecology, conservation measures, and data sources are described.

The IUCN is not actively engaged in preserving endangered species in the field, but its findings are often the basis of conservation activities throughout the world, and it provides crucial leadership to the world community on issues involving biodiversity. One of its efforts was to help establish the Trade Records Analysis for Flora and Fauna in International Commerce (TRAFFIC) network, which monitors wildlife trade and helps to implement the decisions and provisions of the Convention on International Trade in Endangered Species of Wild Fauna and Flora (CITES).

CITES. Endangered species outside the United States are only tangentially addressed by the Endangered Species

Act. However, under the leadership of the United States, CITES was established in the early 1970s. CITES is not specifically a device to protect rare species—instead, it is an international agreement (signed by 169 nation-states) that focuses on trade in wildlife and wildlife parts. The treaty recognizes three levels of vulnerability of species—the highest being species threatened with extinction—and covers 5,000 animals and 28,000 plants, fungi, and other species. Restrictive trade permits and agreements between exporting and importing countries are applied to those species, sometimes resulting in a complete ban on trade if the nations agree. Every two or three years, the signatory countries meet at a Conference of the Parties. The latest such conference was held in 2013 in Bangkok, Thailand.

Perhaps the best-known act of CITES was to ban the international trade in ivory in 1989 in order to stop the rapid decline of the African elephant (from 2.5 million animals in 1950 to about 609,000 in 1989 and around 470,000 in 2008). To combat the illegal trade in ivory, officials are using DNA testing to determine the origin of ivory (see Sound Science, *Using DNA to Catch Wildlife Criminals*). A report released in 2014 declared that African elephants had reached a tipping point. Each year, 35,000 elephants are killed. Fewer are born than die, a formula for disaster. Any plan to protect African elephants needs to enable people in Africa to manage wildlife without overexploitation, persuade others to stop using ivory, and use science to stop ivory crime. These changes require a world outcry against ivory collection.

Convention on Biological Diversity. Although CITES provides some protection for species that might be involved in international trade, it is inadequate to address broader issues pertaining to the loss of biodiversity. With the support of the United Nations Environment Programme (UNEP), an ad hoc working group proposed an international treaty that would move to conserve biological diversity worldwide. After several years of negotiation, the Convention on Biological Diversity was drafted and became one of the pillars of the 1992 Earth Summit in Rio de Janeiro. The Biodiversity Treaty, as the convention is called, was ratified in December 1993 and is now in force. One hundred ninety-two states and the European Union are treaty members. As originally organized, the treaty addresses three complementary objectives: (1) the conservation of biodiversity, (2) the sustainable use of biodiversity services, and (3) the equitable sharing of the use of genetic resources found in a country. The treaty establishes a secretariat and designates the Conference of the Parties (the treaty members) as the governing body. The Conference of the Parties has established programs to guide international work in major biomes, including agricultural, dry-land, coastal, marine, and island ecosystems.

At the 2010 meeting of the convention held in Nagoya, Aichi Prefecture, Japan, world governments developed a set of global goals, the **Aichi Biodiversity Targets**. These five strategic goals and 20 targets form the basis of a plan to help nations better protect biodiversity by 2020. The

SOUND SCIENCE
Using DNA to Catch Wildlife Criminals

Scotland: A group of men stand around, holding shovels, nets, and the leashes of dogs wearing radio collars. A local police officer stops them, wondering what they are up to. They claim to be hunting rabbits, but the police think otherwise. In other years, there might not be anything for the police to do, but now, it is easier to gather evidence. The police swab the mouths of the dogs, and their suspicions are confirmed when the gathered specimens are found to have traces of fox DNA.

Sound like an odd episode of *CSI: Wildlife?* You are right! But in this real-life drama, Scotland's own Wildlife DNA Forensic Unit came to the rescue to give the police the data they needed to apprehend criminals. The illegal fox hunters pleaded guilty, and the work of the first dedicated wildlife DNA forensic testing facility in Europe was showcased.

New technologies can often help in crimes against wildlife, much as it is widely used to solve crimes against humans. In Italy, scientists were able to show that the blood on a knife owned by a suspected poacher had come from a specific poached wild boar. Scientists at the Center for Conservation Biology at the University of Washington have worked to find the origin of contraband ivory shipped from Africa to Singapore. Were the elephants slaughtered for their tusks all from one place or scattered throughout Africa? They compared DNA from the ivory with DNA patterns found in elephant dung collected all over the African continent. By doing so, they were better able to identify where the poaching had occurred. In this instance, they found that the origin of the tusks was a small area centered in Zambia. The group's efforts could be increased to

stop poaching. They were able to do the same with a second shipment of contraband ivory, detected in a shipping container in Malaysia. In this case, the elephants came from another region. Knowing how wildlife traffickers are attacking rare species is important to their protection.

Wildlife forensics is increasing as wildlife trafficking increases. Some people believe that the solution is not to ban traffic in rare species outright, but to "farm" them so that limited legal amounts could be harvested. This is a controversial approach, however. Regardless of how we try to protect them, crimes against wild species populations occur, and when they do, it is helpful to have cutting-edge science and technology to unravel the secrets.

Source: http://conservationbiology.net/research-programs/tracking-poached-ivory-2/

first of the strategic goals is "to address the underlying causes of biodiversity loss by mainstreaming biodiversity across government and society."[15] After the joint international meeting, individual countries agreed to form national biodiversity strategies and action plans within two years. National reports on those strategies and plans were first submitted in 2014.

Some examples of the Aichi Biodiversity Targets are:

- At least halve and, where feasible, bring close to zero the rate of loss of natural habitats, including forests
- Establish a conservation target of 17% of terrestrial and inland water areas and 10% of marine and coastal areas
- Restore at least 15% of degraded areas through conservation and restoration activities

The secretariat of the Convention on Biological Diversity and the UNEP jointly published the report *Global Biodiversity Outlook 1, 2, 3,* and 4 in 2001, 2006, 2010, and 2014, respectively. *Global Biodiversity Outlook 4* summarizes the state of the Aichi Biodiversity Targets midway to the 2020 end point. Some positive findings were that scientific knowledge is increasing and certification programs for aquaculture are improving. One target (11) is likely to be met: At least 17% of terrestrial and inland water areas have been conserved. However, coral reef

degradation, nutrient pollution, and habitat fragmentation are getting worse.

The U.S. and the Convention on Biological Diversity. President Clinton signed the Convention on Biological Diversity in 1993, but it was not ratified by the U.S Senate, leaving U.S. involvement in a state of limbo. However, many of the provisions of the treaty are being carried out by different U.S. agencies. The National Biological Service, for example, is working on several of the treaty's provisions. The United States continues to send delegations to the Conference of the Parties and participates in other related meetings, such as those dealing with genetically modified organisms.

The Nagoya Protocol. During the 2010 meeting of the Convention on Biological Diversity in Nagoya, Japan, the international community agreed to a new treaty, the Nagoya Protocol on Access to Genetic Resources and the Fair and Equitable Sharing of Benefits Arising from their Utilization, usually called the Nagoya Protocol. The purpose of this treaty was to support one of the major goals of the Convention on Biological Diversity—to protect biodiversity in a fair way. The treaty states that companies from wealthy nations cannot mine the genetic resources of other countries and then patent products that indigenous people have used for millennia. The protocol recognizes the value of traditional knowledge and includes incentives for the protection of biodiversity.

Critical Ecosystem Partnership Fund. The Critical Ecosystem Partnership Fund emerged in August 2000. Jointly sponsored by the World Bank, Conservation International,

---

[15] http://www.cbd.int/sp/targets/

the government of Japan, the MacArthur Foundation, and the Global Environment Facility, the fund provides grants to NGOs and community-based groups for conservation activities in biodiversity hot spots—34 regions, making up just 2.3% of Earth's land surface (Figure 6–21), in which 75% of the most threatened mammals, birds, and amphibians are located. While many scientists agree that focusing protection on the most biodiverse areas is a good way to protect species, there is concern that areas that have fewer species but are highly productive (such as salt marshes) or are otherwise very valuable to conservation may be ignored. Some argue for protection of these "cold spots" as well.[16]

Stewardship Concerns

What we do about wild species and biodiversity reflects on our wisdom and our values. Wisdom dictates that we take steps to protect the biological wealth that sustains so many of our needs and economic activities. Our values come from our view of species and their instrumental and intrinsic worth. Our values also reflect what we think about the importance of justice—whether we believe, for example, that the people who most benefit economically from the use of

---

[16]P. Karieva and M. Marvier, "Conserving Biodiversity Coldspots," *American Scientist* 91, no. 4 (2003): 344.

an ecosystem should be the same people who bear the risks of its decline, or whether we value maintaining choices and options for future generations.

Wisdom. The scientific team that put together the UN Global Biodiversity Assessment focused on four themes in its recommendations:

1. **Reforming policies that often lead to declines in biodiversity.** Many governments (including that of the United States) subsidize the exploitation of natural resources and agricultural activities that consume natural habitats.

2. **Addressing the needs of people who live adjacent to or in high-biodiversity areas or whose livelihood is derived from exploiting wild species.** Although protecting biodiversity benefits entire societies, the people who are closest to the protected areas are key to the success of efforts to maintain sustainability of natural resources. These people must be given ownership rights or at least communal-use rights to natural resources, and they must be involved in the protection and management of wildlife resources.

3. **Practicing conservation at the landscape level.** Many wild species require a variety of habitats or larger contiguous habitats than small local or regional conserved areas afford. For example, to address the needs of

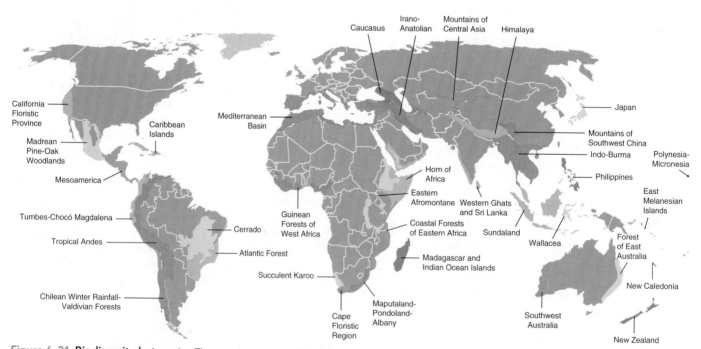

Figure 6–21 Biodiversity hot spots. These regions are considered the "emergency rooms" of planetary biodiversity. Twenty-one of these spots are being funded by the Critical Ecosystem Partnership Fund.
(*Source:* Redrawn from Conservation International.)

UNDERSTANDING THE DATA

1. What characteristics do the locations of the hot spots have in common?

2. Why are there so few hot spots in North America and Northeast Asia?

migratory birds, summer and winter ranges and flyways between them must be considered.

4. **Promoting research on biodiversity.** Far too little is known about how to manage diverse ecosystems and landscapes, and much remains to be learned about the species within the boundaries of different countries. Toward that end, a new field called biodiversity informatics has been established. Its goal is to put taxonomic data and other information on species online.[17]

Values and Economics. Our values also are demonstrated by our approach to wild species. To protect their value, we need to make the conservation of species a part of the economy. We have seen the costs involved with invasive species and the loss of commercially important species. Yet it also costs money to conserve species. Usually the expenditures to protect species must take place now, while the money we will save by protecting them becomes available later. In addition, the present costs are obvious, while the later savings are less obvious. In 2012, scientists estimated that it would cost $76 billion annually to reach the Aichi targets and protect species. Although that is a great deal of money, it is much less than the estimated costs of losing those species and the ecosystems they maintain. To put this in perspective, the cost of this investment in natural capital would be about one fifth of what the world spends on soft drinks annually.

At the start of the chapter, we learned about breeding reserves for river dolphins in Bangladesh, developed through the efforts of conservation biologists. The success of other conservation efforts shows us that we can indeed stem the tide of species declines with hard work. Thomas Jefferson once said, "For if one link in nature's chain might be lost, another might be lost, until the whole of things will vanish by piecemeal."[18] As this quote suggests, loss of wild species and their associated genetic and ecosystem diversity could lead to a breakdown in the natural functioning of ecosystems. For this reason, strategies to address the loss of biodiversity must focus on protecting whole ecosystems, as we will see in a later chapter (Chapter 7).

CONCEPT CHECK Compare the main objectives of the CITES Treaty and the Convention on Biological Diversity. ☑

---

[17] F. A. Bisby, "The Quiet Revolution: Biodiversity Informatics and the Internet," *Science* 89, 5488 (2000): 2309–2312.

[18] Thomas Jefferson, "Memoir on the Discovery of Certain Bones of a Quadruped of the Clawed King in the Western Parts of Virginia," *Transactions of the American Philosophical Society* vol. A (1799): 246–260.

REVISITING THE THEMES

SOUND SCIENCE

Sometimes we do not know enough science to manage species well. Fortunately, many universities provide excellent programs in wildlife management. The Millennium Ecosystem Assessment and the Global Biodiversity Outlook reports, as well as annual reports by the IUCN, pull together the solid work of thousands of scientists. Conservation biology is a science that is focused on the protection of living things. The work of conservation biologists helps us to understand why some species are going extinct, when they are recuperating, and sometimes when wildlife-related crimes have occurred. Sometimes scientists work together with "citizen scientists," networks of observers, to get more information. Other times they use cutting-edge technology such as DNA analysis to catch criminals. Furthermore, the science of taxonomy is foundational to the protection of species, and making sure we have scientists ready to study has to be part of the solution to biodiversity decline.

SUSTAINABILITY

Because ecosystem capital is of such high value to human enterprises, the pathway to a sustainable future depends on protecting the wild species that make ecosystems work. Such protection is afforded by local and federal policies and international treaties. Sustainability is enhanced when we take steps to encourage healthy populations of organisms; if they thrive, so do their natural ecosystems. Sometimes, however, it is necessary to remove animals (such as deer) from populations when their numbers increase too much. Long-term plans for sustainability have to address the root issues that cause biodiversity declines, including those caused by growing human populations and overconsumption.

STEWARDSHIP

The reestablishment of the Lake Erie water snake is good stewardship at work. Stewardship places a high value on wild species. Although instrumental value is sufficient reason for preserving many species, perceiving intrinsic value might be the nearest to a pure stewardship ethic because it works out primarily for the sake of the species. Although some people dislike snakes, their protection is important for ecosystems. Other species, such as river dolphins, are easy to like, and people rally for their protection.

The protection of biodiversity is tied to the ways people value wild species. These values can include instrumental or utilitarian values as well as intrinsic values. Wild species and their genes represent biological wealth, providing the basis for the goods and services coming from natural ecosystems and sustaining human life and economies. Species can also be thought of as a genetic bank, capable of being drawn on when new traits are needed. Losing biodiversity eventually leads to a biologically impoverished future.

REVIEW QUESTIONS

1. Define *biological wealth* and apply the concept to the human use of that wealth.

2. Compare instrumental value and intrinsic value as they relate to determining the worth of natural species. Where does Leopold's idea of the land ethic fit into these two categories?

3. What are the four categories into which the human value of natural species can be divided? Give examples of each one.

4. What means are used to protect game species, and what are some problems emerging from the adaptations of many game species to the humanized environment?

5. What is the Lacey Act, why was it needed, and how is it used to protect wild species?

6. How does the Endangered Species Act preserve threatened and endangered species in the United States?

7. What is one endangered species that has benefited from protection under the Endangered Species Act?

8. What is biodiversity? What do scientists need to know to calculate it for a habitat?

9. What do the letters in the acronym HIPPO stand for? What is an example of each?

10. Give several examples of ways habitat change can either harm species or aid them.

11. What are IUCN, CITES, the Nagoya Protocol, and the Convention on Biological Diversity? How do their roles differ?

12. What are the Aichi Biodiversity Targets?

THINKING ENVIRONMENTALLY

1. Some people argue that each individual animal has an intrinsic right to survival. Should this right extend to plants and microorganisms? What about the *Anopheles* mosquito, which transmits malaria? What about the bacteria that cause typhoid fever? Defend your position.

2. Choose an endangered or threatened animal or plant species, and use the Internet to research what is currently being done to preserve it. What dangers is your chosen species subject to?

3. Some critics of the Endangered Species Act think it is too strict, while other critics think it is not strict enough. Investigate one example of the application of the law—for example, to gray wolves—and decide what you think ought to be done.

4. Log onto Conservation International's biodiversity hot spots Web page and research one of the many hot spots. What is being done to protect biodiversity in this location?

MAKING A DIFFERENCE

1. Instead of devoting your whole yard to grass, convert some of your yard to native plants. The National Wildlife Federation has a great deal of information on turning your backyard into wildlife habitat and even a certification program.

2. Support groups that conserve and protect wildlife through the political process. Groups like the Center for Biological Diversity and others will alert members when there are political actions that they can support.

3. Consider volunteering or interning for a program that conserves wildlife. The Audubon Society has volunteer opportunities, as do many other wildlife conservation groups. These opportunities are found on the Internet.

4. Be careful about purchasing products that might come from a wild population. For example, it is better to buy pets bred in captivity rather than wild-caught pets.

MasteringEnvironmentalScience®

Students Go to MasteringEnvironmentalScience for assignments, the eText, and the Study Area with animations, practice tests, and activities.

Professors Go to MasteringEnvironmentalScience for automatically graded tutorials and questions that you can assign to your students, plus Instructor Resources.

The Value, Use, and Restoration of Ecosystems

Bright sunlight filters through shallow water, revealing the colors of a tropical reef. Fish flit by the branching arms of a coral. A parrotfish grazes a piece of seaweed, and a lobster hides at the base of a craggy rock. As the Sun warms the clear water, divers swim by. An endangered Nassau grouper floats in an eddy, eyeing a smaller fish. A non-native lionfish snaps out, catching prey. It's another day in a tropical paradise. Suddenly, a SCUBA diver snatches the venomous lionfish and heads for the surface with the catch.

Today is the Lionfish Derby, held in the Florida Keys and run by the Reef Environmental Education Foundation (REEF). In the Derby, teams compete for cash and prizes to capture these invasive fish and remove them from this critical habitat. Researchers will use data from the captured fish to better understand their biology and ecology. Lionfish lack predators in the Caribbean and eat the young of other fish. The capture removes some of the invaders from a habitat that is important to many other species. In 2013, the Florida Keys Lionfish Derby teams netted 12,012 fish! While lionfish derbies will not remove this invasive predator entirely, scientists believe that continually removing as many as possible will keep populations low enough to avoid reef damage.

Citizen Scientists. On a similar day elsewhere in the Caribbean, SCUBA divers move methodically through the water, recording the fish they see. They note the condition of coral reefs and record the presence of fish species, both herbivorous and carnivorous. This is also a REEF initiative: the Volunteer Survey Project. The SCUBA divers, trained volunteers in a network of "citizen scientists," will submit the information they collect, to be used by scientists who track what is happening with coral reef fish. Using these data, scientists have already been able to figure

▲ A diver observes an invasive lionfish in a Caribbean coral reef.

out a few things: Fish are declining, which affects the health of the coral reef. As herbivorous fish decline, algae and water plants can grow over the corals. Some of these produce toxic chemicals and kill the coral. As carnivorous fish decline, coral-eating sea stars are free to take over. The invasive lionfish can oust other native reef fish. Researcher Christopher Stallings of the Florida State University Coastal and Marine Lab is one of the scientists working with REEF. His work with the large database on fish in the Caribbean showed that some large-bodied fish such as the Nassau grouper have completely disappeared in certain highly populated areas.[1]

Critical Ecosystems. Around the world, coral reefs are critical ecosystems. They act as nurseries for marine fish and as an early warning system for habitat degradation. Because they

need clean, clear water and grow relatively slowly, corals show us when marine systems are in decline. Physical breakage, pollution, heat, and invasive species combine to kill off these colonial animals and their algal symbionts. A coral reef is a good example of an ecosystem that provides humans with goods and services and that needs monitoring, protection, and sometimes restoration. The dive teams competing in the Lionfish Derby and the SCUBA divers recording species while they dive are ways that ordinary people can be involved in the care of an ecosystem. One key to saving wild species, and protecting the human well-being that depends on them, is to preserve their habitats—the ecosystems in which they are found (Chapter 6).

The focus of this chapter is on the natural ecosystems that directly sustain human life and economies. We will pay special attention to those ecosystems that are in the deepest trouble. These ecosystems, the species in them, and the goods and services they generate are identified as ecosystem capital, a major component of the natural capital wealth of nations (Chapters 1, 2).

---

[1]C. Stallings, "Fishery-Independent Data Reveal Negative Effect of Human Population Density on Caribbean Predatory Fish Communities," *PLoS ONE* 4(5) (2009): [e5333].

7.1 Ecosystem Capital and Services

Earth's ecosystems vary greatly in the makeup of their species, but exhibit common functions, such as energy flow and the cycling of matter. The major types of terrestrial and aquatic ecosystems—the biomes—reflect the response of plants, animals, and microbes to different climatic conditions. For the purposes of this chapter, Earth's land area can be divided into five broad categories: forests and woodlands, grasslands and savannas, croplands, wetlands, and deserts and tundra (Figure 7–1a). The oceanic ecosystems can be categorized as coastal ocean and bays, coral reefs, and open ocean (Figure 7–1b). These eight terrestrial and aquatic systems provide all of our food; much of our fuel; wood for lumber and paper; fibers—leather, furs,

and raw materials for fabrics; oils and alcohols; and much more. The world economy and human well-being directly depend on the exploitation of the **natural goods** that can be extracted from ecosystems; the Millennium Ecosystem Assessment team refers to this function of ecosystems as their *provisioning service*.

Ecosystems also perform a number of other valuable services as they process energy and circulate matter in the normal course of their functioning (see Figure 5–22). *Regulating services* are benefits derived from the regulation of ecosystem processes, such as flood control or climate moderation. *Cultural services* comprise nonmaterial benefits people gain from ecosystems, including spiritual enrichment, knowledge, and aesthetic enjoyment. Finally, ecosystems provide services that maintain themselves, such as photosynthesis. These are called *supporting services*.

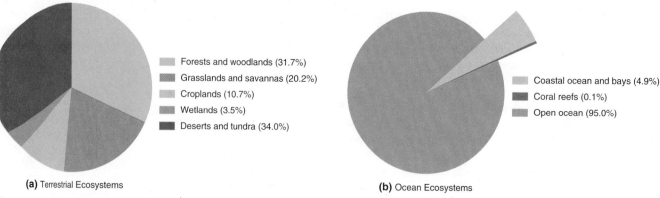

(a) Terrestrial Ecosystems

Forests and woodlands (31.7%)
Grasslands and savannas (20.2%)
Croplands (10.7%)
Wetlands (3.5%)
Deserts and tundra (34.0%)

(b) Ocean Ecosystems

Coastal ocean and bays (4.9%)
Coral reefs (0.1%)
Open ocean (95.0%)

Figure 7–1 Relative areas of terrestrial and ocean ecosystems. (a) Forests, which are the most productive ecosystems on land, occupy almost as much area as the least productive ecosystems, deserts and tundra. **(b)** In the oceans, the low-productivity open ocean dominates. Coral reefs represent only 0.1% of the whole, yet they are ecologically and economically important.

Table 7–1 shows combinations of services provided to societies by a spectrum of ecosystems. In their normal functioning, all natural and altered ecosystems perform some or all of these natural services, year after year. A team of natural scientists and economists calculated the total value of a year's global ecosystem goods and services to be $41 trillion (Chapter 5), but we receive them free of charge.

Ecosystems as Natural Resources

If the natural services performed by ecosystems are so valuable, why are we still draining wetlands and removing forests? The answer is clear: A natural area will receive protection only if the value a society assigns to the services it provides in its natural state is higher than the value the society assigns to converting it to a more direct human use. Natural ecosystems and the organisms in them are commonly referred to as *natural resources*. As resources, they are expected to produce something of value, and the most commonly understood value is economic value. By referring to natural systems as resources, we are placing them in an economic setting, so it becomes easy to lose sight of their ecological value. To avoid making this mistake, we use the concept of *ecosystem capital* instead of natural resources.

Table 7–1 Services from Various Types of Ecosystems

	Mountain and Polar	Forest and Woodlands	Inland Rivers and Other Wetlands	Drylands	Cultivated	Urban Parks and Gardens	Coastal	Island	Marine
Air quality regulation						√			
Biofuels					√				
Carbon sequestration		√							
Climate regulation (local and widespread)	√	√		√		√	√		√
Cultural heritage				√	√	√			
Disease regulation and medicines		√	√		√				
Dyes					√				
Education						√			
Erosion control									
Fiber	√			√	√		√		
Flood regulation		√	√				√		
Food	√	√	√	√	√		√	√	√
Fresh and drinking water	√	√	√		√			√	
Fuelwood and timber		√		√	√		√		
Nutrient cycling			√		√		√		√
Pest regulation					√				
Pollution regulation			√						
Recreation, ecotourism, and aesthetic values	√	√	√	√	√	√	√	√	√
Sediment retention and transport			√						
Waste processing							√		
Water regulation						√			

Source: Millennium Ecosystem Assessment, "Living Beyond Our Means: Natural Assets and Human Well-Being" (Statement of the Millennium Ecosystem Assessment Board, 2005).

Valuing Natural Capital. Markets provide the mechanism for assessing economic value, but do poorly at placing monetary values on ecosystem services. People and institutions may express their preferences for different ecosystem services via the market, an *instrumental value* (Chapter 6). Instrumental valuation works fairly well for the provisioning services provided by ecosystems (goods such as timber, fish, and crops). It is far more difficult to put prices on many of the regulating, cultural, and supporting services provided by natural ecosystems. These services are **public goods,** things that are not depleted as people use them and that cannot be effectively limited in order to charge money; for example, fresh air is a public good. Public goods are not usually provided by markets but contribute greatly to people's well-being. In fact, they are absolutely essential.

Regulating and cultural services provide benefits that are experienced over a large area; the loss of these services is also experienced regionally. For example, a large forested area modifies the climate in a region and absorbs and holds water. A small forest can be cut and the timber sold, providing income to the owners with no noticeable impact on the climate. Cutting a larger area of forest will provide an immediate economic benefit to the owners but will also cause a long-term loss of regulating and supporting services, affecting the well-being of people in the formerly forested area and well beyond. In this case, short-term profits are made at the long-term expense of many.

Some natural ecosystems are maintained in a natural or semi-natural state because that is how they provide the greatest economic value for their owners. For example, most of the state of Maine is owned by private corporations that exploit the forests for lumber and paper manufacturing; however, if Maine experienced a population explosion and corporations could sell their land to developers for more money than could be gained from harvesting timber, forested lands would quickly become house lots. **Figure 7–2** shows a common scene, a forest converted to a housing development. Similar conversions can be pictured: tropical forests to grasslands for grazing, agricultural fields to suburbs, and wetlands to vacation developments.

Some natural ecosystems are publicly owned (state and federal lands) or cannot be owned at all (ocean ecosystems). These ecosystems are still considered natural resources and may be subject to economically motivated exploitation. Sustainable exploitation of such systems will maintain the natural services they perform.

Future Pressures. As the human population continues to rise, pressures on ecosystems to provide resources will increase: more food, more wood, more water, and more fisheries products will be needed. The sustainable exploitation of ecosystem capital will become more difficult. The main objective of the thorough work of the Millennium Ecosystem Assessment was to evaluate the health of the services that people derive from the natural world. The assessment team found that 15 of the 24 services they evaluated were in decline worldwide, 5 were in a stable state, and 4 were increasing at the expense of other services (Chapter 1). Fortunately, ecosystems can sustain continued exploitation and also be restored from destructive uses that have degraded them.

CONCEPT CHECK List at least three ecosystems services that forests provide and three that urban parks provide. How are the services provided by these ecosystems similar? ☑

7.2 Types of Ecosystem Uses

An ecosystem's remarkable natural capacity to regenerate makes it (and its biota) a **renewable resource**. In other words, an ecosystem has the capacity to replenish itself despite certain quantities of organisms being taken from it. This renewal can go on indefinitely—it is sustainable. Recall that every species has the biotic potential to increase its numbers and that, in a balanced population, the excess numbers fall prey to parasites, predators, and other factors of environmental resistance (Chapter 4). It is difficult to find fault with activities that effectively put some of this excess population to human use. The trouble occurs when users such as hunters, fishers, or loggers take more than the excess and deplete the breeding population.

Figure 7–2 Conversion of forest to suburban development. Trees are cut to build housing developments such as this one in Loudoun County, Virginia. Similar conversions occur on grasslands and coastal wetlands.

Conservation Versus Preservation

Conservation of natural biota and ecosystems does not imply *no* use by humans whatsoever, although this may sometimes be temporarily expedient in a management program to allow a certain species to recover its numbers. Rather, the aim of conservation is to *manage or regulate use* so that it does not exceed the capacity of the species or system to renew itself. Conservation is capable of being carried out sustainably, and when sustainability is adopted as a principle, conservation has a well-defined goal.

Preservation is often confused with conservation. The objective of preservation is to ensure the continuity of species and ecosystems, regardless of their potential utility. Effective preservation often precludes making use of the species or ecosystems in question. For example, it is impossible to maintain old-growth forests and at the same time harvest the trees. Thus, a second-growth forest can be conserved (trees can be cut, but at a rate that allows the forest to recover), but an old-growth forest must be preserved (it must not be cut down at all).

Patterns of Human Use of Natural Ecosystems

Before examining the specific biomes and ecosystems that are under the highest pressure from human use, we need to consider a few general patterns of how people use natural resources.

Consumptive Use. People in rural areas make use of natural lands or aquatic systems that are in close proximity to them. When people harvest natural resources in order to provide themselves with food, shelter, tools, fuel, and clothing, they are engaged in **consumptive use**. Thus, people hunt for game (Figure 7–3) and fish and gather fruits and nuts in order to meet their food needs; they also collect natural products such as firewood, as well as wood or palm leaves to construct shelters. People may barter or sell such goods in exchange for other goods or services in local markets, but most consumptive use involves family members engaged in hunting and gathering to meet their own needs. This kind of exploitation usually does not appear in the calculations of the market economy of a country. Even so, this form of "wild income" is highly important to the poor; some 1.2–1.4 billion people, many of them poor, depend on forests for some of their income.

Dependence on consumptive use is most commonly associated with the developing world, but it also exists in the more rural areas of the developed world, where people depend on wood for fuel or wild game for food. Approximately 2.4 billion people rely on wood for heating fuel, some 40% of the population of less developed countries, and many use wood to boil water or cook food. Their well-being would be diminished if they were denied access to these "free" resources.

In many parts of Africa, wild game, or "bush meat," provides a large proportion of the protein needs of people. This can represent 80% of the fat and protein in the diet of rural Africans.[2] Without these resources, people's living standard would decline and their very existence could be threatened. Unfortunately, this is a largely unregulated practice and often involves poaching in wildlife parks. According to the Convention on International Trade in Endangered Species (CITES), poaching has contributed to the decline of 30 endangered species. Making matters worse, commercial hunters have recently turned what was once a matter of local people using wildlife for food into an unsustainable industry that supplies cities with tons of "fashionable" meat. For example, primate meat (e.g., chimpanzee) has been found in markets in New York, London, Paris, and other cities.

Productive Use. In contrast to the consumptive use of resources, **productive use** is the exploitation of ecosystem resources for economic gain, as when products like timber

---

[2]R. Nasi et al., *Conservation and Use of Wildlife-Based Resources: The Bushmeat Crisis* [PDF] (Montreal, Canada: Secretariat of the Convention on Biological Diversity; Bogor, Indonesia: Center for International Forestry Research (CIFOR), 2007), Technical Series no. 33.

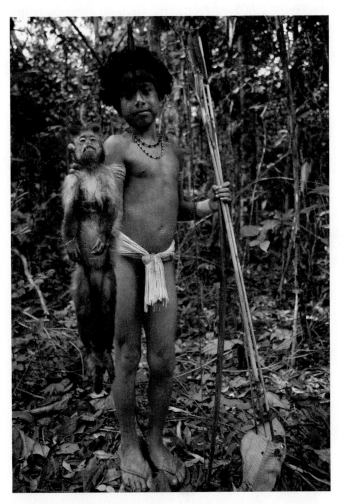

Figure 7–3 Consumptive use. Wild game is an important food source for millions. This young hunter of the Xingu tribe of the Amazon Basin in Brazil has captured a monkey, which he and his family will eat.

Figure 7–4 Productive use. These lobsters trapped by these fishers will be sold in a fish market, a productive use of ecosystem goods.

and fish are harvested and sold for national or international markets (Figure 7–4). Productive use is an enormously important source of revenue and employment for people in every country. The global fisheries catch in 2012, which reached 93 million metric tons (102 million tons), was valued at $130 billion. More than 1.2 billion hectares of forests are managed for wood production. Nearly half of the wood removed, however, is used for fuel, so a great deal of the value of forests occurs on the consumptive, noneconomic side. Planted forests are on the increase, shifting timber harvest from natural to planted forests and making conservation of some natural forests easier.

Another productive use is the collection of wild species of plants and animals for cultivation or domestication. Wild species may provide the initial breeding stock for commercial plantations or ranches; they may also be used as sources of genes to be introduced into existing crop plants or animals to add desirable genetic traits, such as resistance to disease. Wild species also continue to provide sources of new medicines (Chapter 6).

Property Rights. Consumptive and productive uses of natural ecosystem resources are the consequence of the rights of **tenure,** or property rights, over the land and water holding those resources. There are four kinds of tenure: (1) *private ownership,* which restricts access to natural resources; (2) *communal ownership,* which permits use of natural resources by members of the community; (3) *state ownership* of natural resources, which implies regulated use; and (4) *open access,* in which natural resources can be used by anyone. In all of these forms of tenure there is the potential for either sustainable or unsustainable use of the resource.

The concepts of private and communal land ownership often come into conflict in developing countries

where indigenous people, who may have lived in areas for millennia, cannot document their ownership of the land. Who owns the ecosystem products and services that outsiders want to develop? In 2014, the World Bank proposed changes to its funding process that make it easier for governments to ignore the rights of indigenous people. For example, governments can override the rights of hunter-gatherer tribes to allow others to log on their ancestral lands or can force evictions of indigenous people for the conservation of watersheds. As a result of these tensions, the perspective of combining stewardship with protecting the rights of others is vital. Are public policies that encourage stewardship, sustainability, and respect for human rights in place? Is the local or national government presence weak or corrupt? Are sound scientific principles employed in management strategies? Keep these questions in mind as we next examine two patterns that are commonly seen when people exploit natural resources.

Maximum Sustainable Yield. The central question in managing a renewable natural resource is this: How much use can be sustained without undercutting the capacity of the species or system to renew itself? A scientific model employed to describe this amount of use is the **maximum sustainable yield** (**MSY**): the highest possible rate of use that the system can match with its own rate of replacement or maintenance. MSY is often used to describe the harvesting of natural resources such as fish and forest products. The concept of MSY can also apply to the maintenance of parks, air quality, water quality and quantity, and soils—indeed, the entire biosphere. *Use* can refer to the cutting of timber, hunting, fishing, park visitations, the discharge of pollutants into air or water, and so on. Natural systems can withstand a certain amount of use (or abuse, in terms of pollution) and still remain viable. However, a point exists at which increasing use begins to destroy the system's regenerative capacity. Just short of that point is the MSY.

Optimal Population. Imagine trying to harvest the fish in a pond. To calculate the MSY of the fish population, you need to know the **carrying capacity** of the ecosystem—the maximum population the ecosystem can support on a sustainable basis. If the number of fish in your pond is well below carrying capacity, you won't be able to remove very many, because the population is low and there aren't many there to reproduce (Figure 7–5a). Allowing the population to grow will increase the number of reproductive individuals and thus the yield that can be harvested. However, as the population approaches the carrying capacity of the ecosystem, new individuals must compete with older individuals for food and living space. As a result, recruitment (survival of new individuals in the population) will fall drastically (Figure 7–5b). You will not be able to harvest many fish if you have many small individuals who are competing with each other for food. When a population is at or near the ecosystem's carrying capacity, production—and hence sustainable yield—can be increased by thinning the population so that competition is reduced and optimal growth and reproductive rates are achieved. Theoretically, the **optimal population** for harvesting the MSY is just halfway to the carrying capacity.

The practical use of the logistic model of growth is complicated by the fact that the carrying capacity may be unknown or may vary over long periods of time. Replacement of harvested individuals may also vary from year to year, because some years are particularly favorable to reproduction and recruitment, while others are not. Human impacts, such as pollution and other alterations of habitats, adversely affect reproductive rates, recruitment, carrying capacity, and, consequently, sustainable yields. For these reasons, managing natural populations to achieve the MSY is fraught with difficulties. Furthermore, accurate estimates of the size of the population and the recruitment rate must be made continually, and these data are hard to come by for many species.

Precautionary Principle. The conventional approach to managing harvests has been to use the estimated MSY to set a fixed quota. In the management of fisheries, this quota is called the **total allowable catch** (**TAC**). But if the data on population and recruitment are inaccurate, it is easy to overestimate the TAC, especially when there are pressures from fishers (and the politicians who represent them) to maintain a high harvest quota. In the face of such uncertainties, and also in response to repeated cases of overuse and depletion of resources, resource managers have been turning to the **precautionary principle**. That is, where there is uncertainty, resource managers must favor the protection of the living resource. Thus, exploitation limits must be set well enough below the MSY to allow for uncertainties. Those using the resource ordinarily want to push the limits higher, however, so the result is frequent conflicts between users and managers. These conflicts are considered in greater detail in the discussion of fisheries later in this section.

Using the Commons. A resource that is owned by a group of people in common, or by no one (open access), is known as a common-pool resource, or **commons**. Strictly speaking, a commons is an open-access resource that can be used by someone without subtracting from its use by others; for example, knowledge is a commons. Examples of natural common-pool resources include U.S. federal grasslands, on which private ranchers graze their livestock; coastal and open-ocean fisheries used by commercial fishers; groundwater drawn for farms; and the atmosphere, which everyone breathes and which is polluted by everyone who drives motor vehicles. Sustainability implies that common-pool resources will be maintained so as to continue to yield benefits not just for the present, but also for future users.

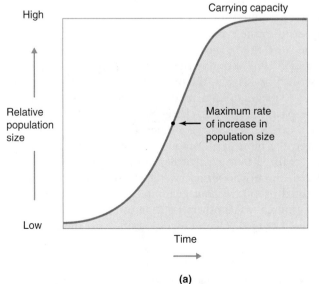

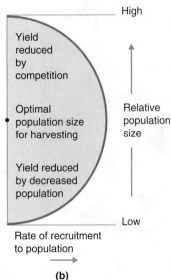

Figure 7–5 Maximum sustainable yield. The maximum sustainable yield occurs when the population is at the optimal level, meaning the rate of population increase is at a maximum. **(a)** The logistic curve of population size in relation to carrying capacity. **(b)** Recruitment plotted against population size, showing the effects of competition and decreased population levels.

UNDERSTANDING THE DATA

Historically, the population of a particular fish species hovered at 100,000 individuals. The population is now 10,000. How much would the population have to change to be at its most productive level?

SUSTAINABILITY
How Much for That Irrigation Water?

The dry bed of the Santa Cruz River in Arizona lies 20 feet below the bank sides, filled with water only in large floods. It wasn't always so. In the late 1880s, a local farmer dug deep ditches and tapped into the underground water moving below the shallow riverbed. Fierce storms sent raging torrents through the channels, eroding the riverbed and stranding farm fields above the precious water. The dry deep river of today and the increased cost of irrigation both result from human actions of the past.

For years, humans have used the natural world as a place to obtain goods and services, often without paying for them. How much is flood control worth? How much is the irrigation from a river worth? To estimate the value of goods and services—things like photosynthesis and water cycling, we need to know how ecosystems work and how our activities affect the rest of the system. That is part of what scientists do. We also have to determine how much the price of our activities and things we produce reflects the effect of our actions on ecosystems. This comparison is done primarily by economists. Policy makers help figure out how we can act together to change the way things are done (Chapter 2), in order to incorporate the real-world cost of ecosystem capital into our human decisions.

How do natural and social scientists determine the value of ecosystems? Some valuation is dependent on what stakeholders think. Researchers doing a study on the Santa Cruz

The Santa Cruz River faces pressures from drying, erosion, and invasive species.

River asked participants, "Why do people think rivers are important? What do they value about rivers?" Respondents valued water quality, the presence of water (in a river where stretches sometimes dried up altogether), lush vegetation, and habitat for birds and fish. They also valued human connections to the river, use by others downstream, and the recharge of underground water and other ecosystem services. The results show that the people around the Santa Cruz River want the river to have a role in their daily lives and desire its health. Of course, putting economic value on only the ecosystem services people recognize, value, and are willing to

pay for is a faulty approach, as we often do not know all the important features of ecosystems. However, it is a beginning and often provides useful information to policy makers.

Sources: Fisher, B., et al. "Ecosystem Services and Economic Theory: Integration for Policy-Relevant Research." *Ecological Applications* 18(8) (2008): 2050–2067.

Ringold, P. L., J. Boyd, D. Landers, and M. Weber. Report from the Workshop on Indicators of Final Ecosystem Services for Streams. 56p. EPA/600/R-09/137. July 2009.

Weber, M., et al. Why Are Rivers Important? Ecosystem Service Definition via Qualitative Research: A Southwestern Pilot Study. Oral Presentation. December 9, 2010. http://conference.ifas.ufl.edu/aces10/Presentations/Thursday/C/AM/Yes/0845%20MattWeber.pdf.

Figure 7–6 Common pasture. These cattle are grazing in Dartmoor National Park, in Devon, England. If the use of common land is unregulated, it will tend to be overgrazed.

Tragedy of the Commons. The exploitation of common-pool resources can present serious problems and may lead to the eventual ruin of the resource—a phenomenon called the "tragedy of the commons," taken from the title of a 1968 essay by the biologist Garrett Hardin.[3] As described by Hardin, the original commons were pastures around villages in England (Figure 7–6). The king provided these pastures without charge to anyone who wished to graze cattle. Herders were quick to realize that whoever grazed the most cattle stood to benefit the most. Even if they realized that a particular commons was being overgrazed, those who withdrew their cattle simply sacrificed personal profits without guaranteeing the conservation of the resource, while others went on using that pasture until it was damaged. One herder's sacrifice became another's

[3]Garrett Hardin, "The Tragedy of the Commons," *Science* 162 (1968): 1243–1247.

gain, so when individuals made a cost-benefit analysis they chose to use the pasture. Consequently, herders would add to their herds until the commons was totally destroyed. The lesson of this story is that where there is no management of the commons, the outcome is inevitable—the pursuit of individuals' own interests will lead to a tragedy for all.

Hardin's essay applies to a limited, but significant set of problems that arise when there is open access to a common-pool resource but no regulating authority (or the regulating authority is ineffective) and no functioning community. Exploitation of the resource then becomes a free-for-all in which profit is the only motive. One example of the tragedy of the commons was the extinction of the passenger pigeon, caused by the unregulated hunting of wild flocks of birds (Chapter 6). Coastal and offshore fisheries have consistently demonstrated the reality of the tragedy of the commons, with stocks of desirable fish declining all over the world. The tragedy can be avoided only by limiting freedom of access.

Limiting Access. One arrangement that can mitigate the tragedy is private ownership. When a renewable natural resource is privately owned, access to it is restricted and, in theory, it will be exploited in a manner that guarantees a continuing harvest for its owner(s). This theory does not hold, however, when an owner maximizes immediate profit and then moves on. Modern corporations lend themselves to this strategy: For example, a timber company may take over a natural resource and clear-cut the forest in order to pay off debts incurred in the process of the takeover. Other owners, though, will be in for the long run and manage the resources more responsibly because it is in their best interests to do so.

Where private ownership is unworkable or lacking, the alternative is to regulate access to the commons. Regulation should allow for: (1) protection, so that the benefits derived from the commons can be sustained; (2) fairness in access rights; and (3) mutual consent of the regulated. Such regulation can be the responsibility of the state, but it does not have to be. In fact, the most sustainable approach to maintaining the commons may be local community control, wherein the power to manage the commons resides with those who directly benefit from its use.

Because unregulated commons tend to be degraded, some have argued for the privatization of most resources. Others have argued for the concept of "the common heritage of human kind."[4] In this view, some resources belong to all of us and to future generations and have to be regulated accordingly. There is a much controversy about which approach will lead to the most sustainable use of resources.

Public Policies. To achieve the objectives of conservation when harvesting living resources, it is important to consider the concept and limitations of the MSY, as well as the social and economic factors causing the overuse and degradation that diminish the sustainable yield. After reviewing these factors, public policies that effectively protect natural resources can be established and enforced. Some principles that should be embodied in these policies are presented in Table 7–2. When policies based on principles like these are in place, it is possible for people to continue to put natural resources to sustainable consumptive and productive uses.

CONCEPT CHECK When factories, businesses, vehicles, and homes in a city contribute to smog, what is the common resource? In what way is the situation a "tragedy of the commons"? ☑

---

[4]UNESCO, *Declaration on the Responsibilities of the Present Generations Toward Future Generations,* accessed January 20, 2015, http://portal.unesco.org/en/ev.php-URL_ID=13178&URL_DO=DO_TOPIC&URL_SECTION=201.html.

Table 7–2   Principles That Should Be Incorporated into Public Policies to Protect Natural Resources
1. Natural resources cannot be treated as open-access commons. Habitats and species should be held by private, communal, or state entities that are responsible for their exploitation and sustainability.
2. Scientific research should be employed to assess the health of the resource and to set sustainable limits on its use.
3. To accommodate uncertainty, the precautionary principle should be used in setting limits for exploitation.
4. Regulations should be enforced; the resource must be protected from poaching.
5. Economic incentives that encourage the violation of regulations should be eliminated.
6. Subsidies that encourage exploitation of the resource should be removed.
7. All stakeholders should be involved in resource-related decisions; those living close to the resource should share in profits from its use.
8. Security of tenure is crucial for poor people who depend on the resource for their well-being.

7.3 Terrestrial Ecosystems Under Pressure

Today human domination of the landscape is pervasive, and ecosystems have been converted or domesticated in vast areas of the planet. Even where systems appear natural, it is likely that humans have killed larger animals, suppressed wildfires, controlled rivers, introduced exotic species, and harvested wood. Elsewhere we have converted grasslands and forests into grazing lands and crop cultivation, so that almost half the land surface is now used in the production of crops and animal products. To be sure, domesticated ecosystems still provide provisioning and regulating and cultural services, and may do so sustainably, but they do so at our consent and largely for our benefit.

Although human activities affect virtually all biomes and ecosystems, some are under more pressure than others. This section looks at forests and grasslands, the most important biomes in terms of economic significance and the potential for active human management.

Forest Ecosystems

Forests are lands where trees are the dominant life form and produce a canopy with more than 10% cover. They are the normal ecosystems in regions with year-round rainfall that is adequate to sustain tree growth. **Figure 7–7** shows the locations of the world's four major forest biomes.

To evaluate trends in forested areas, the UN's Food and Agricultural Organization (FAO) conducts periodic assessments of forest resources and compares the data with previous information, a giant undertaking. In 2005, the world's forest cover was 3.8 billion hectares (9.4 billion acres), or 30% of total land area.[5] An additional 1.4 billion hectares (3.5 billion acres) were classified as "other wooded land," or "woodlands," where the tree cover is less than 10% or the trees and shrubs grow no taller than 5 meters.

Forest Goods and Ecosystem Services. Forests are the most productive ecosystems on land, accounting for 50% of terrestrial plant productivity. They also provide a number of valuable goods, including lumber, wood for making paper, fodder for domestic animals, fibers, gums, latex, fruit, berries, nuts, and fuel for cooking and heating. In fact, forests provide fuel, food, medicine, and income for more than 1.6 billion people, mostly in the developing countries. In many rural areas, wood is the primary source of energy; 2.4 billion people use wood as the fuel for cooking.

About 30% of the world's forests are managed for the production of forest products. The most important forest product is wood for industrial use. Approximately 3.4 billion cubic meters (4.4 billion square yards) of wood are harvested annually from the world's forests for fuel, wood, and paper (**Figure 7–8**). The formal forestry sector employs 13.2 million people directly, but other forest-related jobs are estimated to support more than 41 million people around the globe. Because forests are already so heavily used, an increased need for forest products cannot safely be met solely by increased cutting of forests. Instead we need greater efficiency, less waste, and more recycling of forest products.

---

[5]E. J. Lindquist et al., "Global Forest Land-Use Change, 1990–2005," *FAO Forestry Paper No. 169* (Rome, Italy: Food and Agriculture Organization of the United Nations and European Commission Joint Research Centre, 2012).

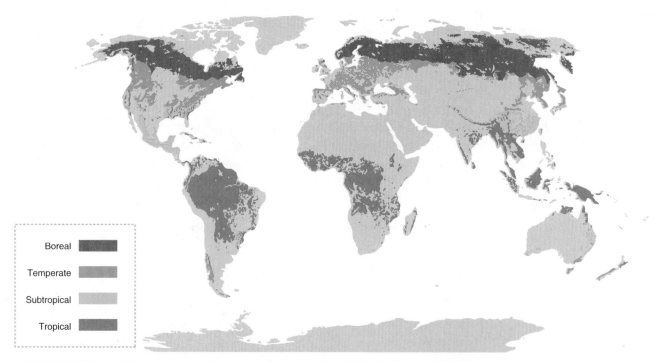

Boreal

Temperate

Subtropical

Tropical

Figure 7–7 World forest biomes. The FAO *Global Forest Resources Assessment 2000* produced this map showing the locations of the world's major forests.

Figure 7–8 **The scale of forestry.** It's hard to picture how much wood we use. This massive pile of logs on a wharf in Wellington, New Zealand, gives an indication of the scale of some logging operations.

Forests and woodlands perform a number of vital natural services. Among other things, they conserve biodiversity, prevent erosion, moderate flooding, store nutrients, moderate regional climates, and provide recreational opportunities. They also store carbon, which slows climate change (Chapter 18). In fact, the world's forests are estimated to absorb about 2 pentagrams of carbon per year, 30% of the carbon emissions caused by humans annually; this is approximately as much as the oceans absorb! In spite of their enormous value, forest ecosystems are under threat.

Threats to Forest Ecosystems. Even though forests are highly productive ecosystems, it has always been difficult for humans to exploit them for food. Consequently, forestlands are often converted into pastures or replaced with cultivated plants that are used directly for food. Indeed, when European colonists came to the Western Hemisphere, their first task was to clear the forests so they could raise crops on agricultural land. Once the forests were cleared, continual grazing or plowing effectively prevented the trees from re-growing. In addition to the removal of trees, forests can be threatened by pollution, climate change, and invasive species.

Deforestation. The largest threat to forests is the process of **deforestation,** the removal of forest and its replacement by another land use. Forests are being cut down at an unsustainable rate. A 2014 report on deforestation in the largest tracts of forest landscape found that over 100 million hectares (247 million acres, an area the size of Germany) had been degraded since 2000. Such forest loss is uneven, with most deforestation occurring in developing countries. Forest cover in the developed countries has remained stable recently, in part because developed countries import forest products. One study showed that a third of the deforestation in eight rapidly deforesting countries was due to the production of four commodities—beef, soybeans, palm oil, and wood products—and largely driven

by export markets.[6] The cultivation of palm oil is the largest driver of Indonesian deforestation; this deforestation includes vast forest fires that produce smog and greenhouse gases. The annual net changes in forest area in major world regions are shown in **Figure 7–9.**

---

[6]M. Persson, S. Henders, and T. Kastner, *Trading Forests: Quantifying the Contribution of Global Commodity Markets to Emissions from Tropical Deforestation - Working Paper 384* (Center for Global Development, 2014), http://www.cgdev.org/publication/trading-forests-quantifying-contribution-global-commodity-markets-emissions-tropical.

CHANGE IN FOREST AREA BY REGION, 1990–2010

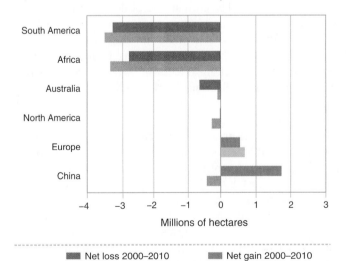

Net loss 2000–2010   Net gain 2000–2010
Net loss 1990–2000   Net gain 1990–2000

Figure 7–9 **Annual net changes in forest area from 2000 to 2010.** In some regions, the amount of forested land increased, but other regions lost forests. Virtually all of the losses occurred in the tropical forests of South America and Africa.

(*Source:* Food and Agriculture Organization [FAO] of the United Nations, *Global Forest Resources Assessment 2010,* xvi.)

Clearing a forest has huge consequences for the land and its people. In the short term, the land may yield crops or support livestock, more people may be fed, and more profit may accrue to the owners, but these effects do not last. When deforested, the overall productivity of the area is reduced. Local people lose their livelihood because the land no longer yields forest products. The loss of trees and leaf litter reduces the standing stock of nutrients and biomass. The soil begins to dry and erode. The hydrologic cycle is changed as water drains off the land instead of being released by transpiration through the leaves of trees or percolating into groundwater. These changes can alter the regional climate, making it drier and hotter. In dry regions, the result of deforestation may be *desertification,* the spread of deserts (Chapter 11). On a global scale, removed forests no longer act as a carbon sink for the removal of CO_2 from the air. Finally, biodiversity is greatly diminished.

Loss of forest health. Cutting forests is not the only way that humans adversely affect forests. Air pollution and climate change stress trees, while invasive species decrease biodiversity. Insect pests, disease, and fires cause significant damage. For example, in the United States and Canada, the mountain pine beetle has killed more than 11 million acres of forest since the 1990s. While temperate forests such as those in North America show signs of degradation, the most troubling changes have been seen in the tropics.

Case Study: Tropical Forests. Tropical forests are a storehouse of biodiversity, providing habitat for millions of plant and animal species. They are also crucial in maintaining Earth's climate, serving as a major sink for carbon and restraining the buildup of global carbon dioxide. Nevertheless, the removal of tropical forests continues at a rapid rate. Between 1960 and 1990, 20% of the tropical forests (over 445 million hectares, or 1.1 billion acres) were converted to other uses; this is equivalent to two-fifths of the land area of the United States. The current rate of deforestation is lower than in the 1990s, but it is still a major cause for concern.

Several factors contribute to the ongoing loss of tropical forests. The clearing of forests for subsistence agriculture is especially intense in Africa, where population growth is unrelenting. Cattle production is the largest cause of deforestation in the Amazon region. In Brazil, 200 million cattle live on 200 million hectares (about 494 million acres), an area about a quarter the size of the continental United States. Many governments in the developing world promote deforestation by encouraging the colonization of forested lands. Indonesia, which has the highest rate of deforestation in the world, has embarked on a program of resettlement and intends to convert 14 million hectares of forest into agricultural land between 2010 and 2020. International companies are logging millions of acres of tropical forest in smaller countries such as Belize, Suriname, and Papua New Guinea, where regulation is weak and corruption is widespread.

Tropical forests offer a cautionary tale of how *not* to care for an important ecosystem. Yet it is unreasonable to expect the countries of the world—especially the developing countries—to forego making use of their forests.

Solutions to Forest Problems. Despite the numerous threats to forests, there are also many positive trends. In the United States, where we have cut all but 5% of the forests that were here when the colonists first arrived, second-growth forests have regenerated wherever forestlands have been protected from conversion to croplands or house lots. There are more trees in the United States today than there were in 1920. In the East, some second-growth forests have aged to the point where they look almost as they did before their first cutting, except for the absence of the American chestnut and elm, two once-dominant trees that have largely died out because of invasive diseases (Chapter 4). Now scientists are trying to breed disease-resistant American chestnuts and elms in order to restore these forests further.

Improving forest management is part of the solution to pressures on forests. In addition to the careful use of forest resources, forest management may include protecting some forests from use. If forests are not cleared and converted to other uses, and if they are protected from stressors such as pollution and disease, they can be managed to yield a harvest of wood for fuel, paper, and building materials indefinitely. Even so, forest-management practices vary greatly in their impact on the forest itself and on surrounding ecosystems.

Silviculture. The practice of forest management with the objective of producing a specific crop (hardwood, softwood, pulp, wood chips) is called **silviculture.** Several practices are employed, with quite different impacts on the total forest ecosystem involved. Because trees take from 25 to 100 years to mature to harvestable size, the normal process of management is a rotation, which is a cycle of decisions about a particular stand of trees from its early growth to the point of harvest. Two options are open to the forester: even-aged management and uneven-aged management.

In even-aged management, trees of a fairly uniform age are managed until large enough to harvest, at which point they are cut down, and the area is replanted, with the objective of continuing the cycle in a dependable sequence. Fast-growing but economically valuable trees such as Douglas fir are favored for this kind of management. Harvesting is often accomplished by **clear-cutting**—removing an entire stand at one time. Even-aged management creates a fragmented landscape and has serious effects on biodiversity and adjacent ecosystems.

Uneven-aged management of a stand of trees results in a more diverse forest and lends itself to different harvesting strategies. One harvesting strategy is **selective cutting,** in which mature trees are removed in small groups or singly, leaving behind trees that can reseed themselves. Selective cutting allows the forest to maintain its biodiversity and normal ecosystem functions. Another strategy is **shelter-wood cutting,** which involves cutting groups of mature trees over longer periods (10 or 20 years), so at any time there are enough trees to provide seeds and to give shelter to growing seedlings. Like selective cutting, this method requires active management and skill, but unlike clear-cutting, it leaves a functional ecosystem standing. In effect, it should lead to a sustainable forest.

Sustainable Forestry. A common objective of forestry management is **sustained yield,** which means that the production of wood is the primary goal and the forest is managed to harvest wood continuously without being destroyed. The concept of sustained yield is similar to MSY in that its objective is to maximize harvest rates, but it ignores the ecosystem properties of forests. **Sustainable forest management,** by contrast, considers the functions of the entire ecosystem. The United Nations Conference on Environment and Development (UNCED) of 1992 described sustainable forest management this way: "Forests are to be managed as ecosystems, with the objectives of maintaining the biodiversity and integrity of the ecosystem, but also to meet the social, economic, cultural and spiritual needs of present and future generations."

Certification. The certification of forest products makes it possible for consumers to choose wood products that have been harvested sustainably. One certification program is run by the Forest Stewardship Council (FSC), an independent nonprofit organization that grew out of an alliance of industry representatives, forest scientists, and environmental organizations such as the National Wildlife Federation and the World Wide Fund for Nature. The mission of the FSC is to promote sustainable—or, as they call it, "responsible"—forestry. By 2014, more than 183 million hectares (452 million acres) of forest in 80 countries had been certified as using sustainable practices.

Managing Public Forests. About 80% of the world's forests are publicly owned; some are managed locally and others by national governments. Overall, governments spend more on forestry than they collect in revenue.[7] On average, governments spend about $7.50 per hectare to manage timber and build and maintain logging roads, and collect about $4.50 per hectare. This amounts to a massive subsidy of logging. However, some forests are managed to conserve biodiversity or to protect other ecosystem services (Figure 7–10). Worldwide, 13.5% of forests are in legally established protected areas. This proportion represents an increase, an acknowledgment by many countries of the important non-timber services of forests. China in particular has increased its protection of forests in order to slow the spread of deserts and soil loss (Chapter 11).

Many nations are preserving forests and putting them to use as tourist attractions. This practice can often generate much more income than logging. Putting forests under the control of indigenous villagers is an important part of forest protection. Where this practice has been implemented, forests have fared better than where they have been placed under state control. Sometimes such indigenous use occurs in areas called *extractive reserves* that yield non-timber goods. Some 12.3 million hectares (30.4 million acres) of the Brazilian Amazon have been set aside as extractive reserves.

Forests in Developing Nations. There are encouraging trends in forest management in the developing world—trends that indicate that those in power are paying more attention to the importance of forest goods and services (other than just wood) and the needs of indigenous people. These actions include practicing sustainable forest management, designating forest areas for conservation, and establishing more-sustainable plantations of trees for wood or other products, such as cacao and rubber (Figure 7–11).

The widespread practice of converting tropical forests into palm oil plantations has led to the destruction of rain forests, including the critical habitat of the Sumatran orang-utan. Protests of the companies that invest in or use palm oil has forced the largest palm oil producer to promise to stop its destructive practices, although it has not yet halted them.

In some cases, the value of ecosystem services such as climate regulation justifies forest protection. For example, the United Nations has a program to protect forests in order to absorb carbon emissions. This program creates a financial value for the carbon stored in forests, providing an incentive to reduce emissions from forested lands; in this way, developing nations do not have to bear the burden of protecting forests on their own.

---

[7]FAO, *Global Forest Resources Assessments, Key Findings 2010,* 9, accessed January 20, 2015, http://www.pefc.org/images/stories/documents/external/KeyFindings-en.pdf.

DESIGNATED FUNCTIONS OF FORESTS, 2010

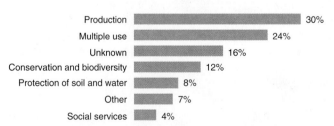

Figure 7–10 Designated uses of forests. Forests are managed for different ecosystem services.

(*Source:* Food and Agriculture Organization (FAO) of the United Nations, *Global Forest Resources Assessment 2010,* xxvi.)

Figure 7–11 A rubber tree plantation in Vietnam. The understory of plantation forests is less complex than that of natural forests and supports less biodiversity.

Grassland Ecosystems

Although forests are heavily used, another group of of ecosystems is even more heavily affected by human activity—the grasslands. Indeed, grassland ecosystems may be the most over-exploited ecosystems on the planet.

Grasslands are found in regions with moderate rainfall and are usually characterized by well-developed soils. Grasslands that occur in temperate regions are called *prairie, veldt, steppe,* or *pampas.* In the tropics, grasslands with scattered trees are called *savannas* (Chapter 5). The moderate rainfall and deep soils of grasslands provide excellent conditions for agriculture, particularly for the growth of grains such as wheat and corn. Unlike forests, grasslands have short food chains in which plants support large herbivores. Because they can support herds of grazing animals, grasslands are often used as pasture for cattle and other domestic animals that provide meat and other products.

In the United States, 80% of the native grasslands have been converted to agriculture and other purposes. This conversion has had negative effects on grassland natives such as antelope, elk, bison, and grassland birds. It has also increased erosion, lowered carbon storage, and lowered ground water recharge. Plowing, overgrazing, and over-irrigation cause erosion and salinization (Chapter 11). Grasslands around the world have experienced similar degradation. Because the rainfall in grasslands is usually less than that in forests, grasslands are particularly susceptible to the process of desertification. (These issues are described in Chapters 10, 11, and 12.)

> **CONCEPT CHECK** The primary degradation of forests is caused by overlogging. What is another cause of forest degradation? ☑

7.4 Ocean Ecosystems Under Pressure

Like forests and grasslands, ocean ecosystems are essential to human thriving and the health of the biosphere. Oceans cover 75% of Earth's surface and contain 1.3 billion cubic kilometers of water in a giant set of connected systems. Ocean and coastal ecosystems provide priceless goods and services that enable commerce and enhance human wellbeing. From ancient times, people living in coastal regions have relied on oceans and estuaries as a source of fish and other foods, as a means of transportation and shipping, and for a great variety of ecosystem services. Coasts are centers of recreation and tourism. In general, population densities are higher along the coast than inland; it is estimated that 23% of the world's population lives close to a coast (within 100 kilometers) and also close to sea level (at elevations less than 100 meters). Today, 22 of the largest 32 cities are located on estuaries.

As vast as the oceans are, human activities have impacted them, often severely (Figure 7–12). Several recent reports have highlighted the plight of oceans and the current worldwide effort to protect ocean ecosystems.[8] Let's look at the extent and value of ocean goods and services, threats to ocean ecosystems, and possible solutions to those threats.

---

[8]FAO, *State of World Fisheries and Aquaculture: Opportunities and challenges, 2014* (Rome: FAO-UN, 2015), http://www.fao.org/3/a-i3720e.pdf Jan.20; and IPSO, *State of the Ocean Report, 2013,* accessed January 20, 2015, http://www.stateoftheocean.org/research.cfm.

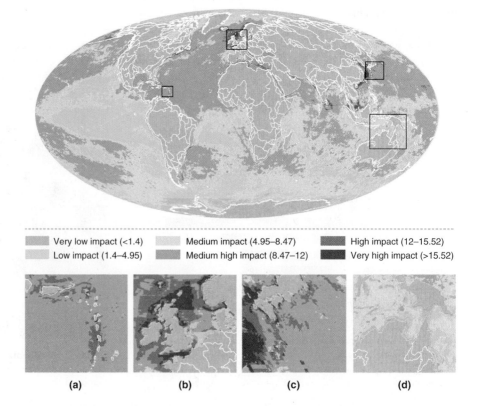

Figure 7–12 Human impact on ocean ecosystems. This map indicates the highest-impact areas in the eastern Caribbean **(a)** the North Sea **(b)** and Japanese waters **(c)**, and a least-impact area in the vicinity of northern Australia **(d)**.

(*Source:* Halpern, Benjamin S., et al. "A Global Map of Human Impact on Marine Ecosystems. *Science* vol. 319, no. 5865 (2008): 948–952.)

Very low impact (<1.4)
Low impact (1.4–4.95)
Medium impact (4.95–8.47)
Medium high impact (8.47–12)
High impact (12–15.52)
Very high impact (>15.52)

(a)   (b)   (c)   (d)

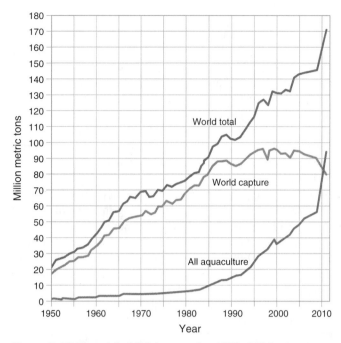

Figure 7–13 The global fish harvest for 1950–2012. The global fish catch plus aquaculture equals the world total. Note that aquaculture has provided all of the increase since 1995.

(*Source:* Data from FAO. *The State of the World's Fisheries and Aquaculture, 2010.*)

Ocean Goods and Ecosystem Services

Oceans provide a huge range of products that humans use, fish in particular. The world's fisheries provide employment for at least 200 million people and account for more than 15% of the total protein consumed by humans. The term *fishery* can refer to either a particular aquatic area or to a group of fish or shellfish being exploited. The total recorded harvest from marine fisheries and fish farming is reported annually by the FAO. The harvest has increased remarkably since 1950, when it was just 20 million metric tons **(Figure 7–13)**.

Marine ecosystems provide us with numerous vital ecosystem services as well. The ocean is important in the regulation of energy movement and nutrient cycling. An estimated 70–80% of the oxygen we breathe comes from the photosynthetic efforts of tiny marine algae. These organisms are an important part of the carbon cycle. Carbon dioxide in the atmosphere is also absorbed directly into the water. The proximity of huge water bodies moderates the temperature along coasts, and brings water to land in places where ocean breezes pass. The vast conveyor belts of the deep ocean pull cold waters toward the equator, while surface currents carry warm water northward, moderating the climate as they pass continents. The warmth of the jet stream, for example, makes the beaches of New Jersey such good swimming areas. Coastal habitats such as mangrove forests, sea grass beds, and salt marshes protect coasts from large storms and surges.

Threats to Ocean Ecosystems

The oceans are in trouble. Some of the major threats to them are the overharvest of fish, the loss of coastal habitats and coral reefs, and localized pollution such as trash, oil, and sediment. Global changes such as climate change, ocean acidification, and deoxygenation (from nutrient pollution) make it hard for ocean creatures to survive. Finally, lack of governance and regulation has allowed criminal activity. Let's briefly look at several of these ecosystem stressors.

Overfishing. Recent reports suggest that approximately half of the world's fish stocks are being fished at their maximum sustainable level. About one third are being fished beyond that level, some to commercial extinction. The basic problem of the fisheries is that there are too many boats chasing too few fish. The high technology of modern fishing fleets gives fish little chance to escape. Equipped with sophisticated electronic equipment that can find fish at great depths and trawlers with 200-foot-wide openings that scoop up everything on and close to the bottom, the fishing industry has become too efficient **(Figure 7–14)**. Such trawling can destroy coral reefs and other critical ocean habitats.

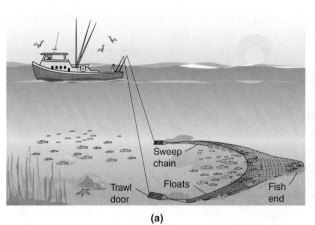
(a)

(b)

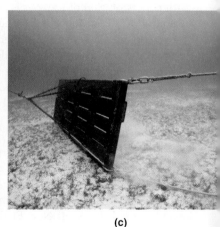
(c)

Figure 7–14 Trawling destroys bottom habitats. Bottom trawling is often used to harvest ground fish. Because it degrades the ocean bottom, this method of harvest has been compared to clear-cutting forests. **(a)** A ship drags a trawling net across the ocean floor. The steel trawl doors weigh up to 5 tons, and the distance between the doors is as much as 200 feet. **(b)** A Malaysian coral reef that has not been exposed to trawling. **(c)** After trawling, a once vibrant coral reef has become a pile of rubble that supports little biodiversity.

For years, parts of the oceans that were more than 12 nautical miles from a nation's coast were considered an international commons. By the end of the 1960s, however, international fishing fleets equipped with modern fish-finding technology were overfishing and seriously depleting fisheries in numerous regions of the sea. In the mid-1970s, as a result of agreements forged at a series of UN Conferences on the **Law of the Sea,** nations extended their limits of jurisdiction to 200 miles offshore. Since many prime fishing grounds are located between 12 and 200 miles from shore, this action effectively removed most fisheries from the international commons and placed them under the authority of particular nations.

To manage fisheries within the 200-mile limit, the U.S. Congress passed the **Magnuson-Stevens Fishery Management and Conservation Act** of 1976. This legislation established eight regional *management councils* made up of government officials and industry representatives. The councils are responsible for creating management plans for their regions, while the National Marine Fisheries Service (NMFS) provides advice and scientifically based information on assessing stocks; NMFS can reject elements of the plans submitted to it by the councils. The Magnuson-Stevens Act was amended in 1996 and again in 2006 to promote more sustainable fishing, including calling for better protection of habitat and promoting catch and release fishing. Despite these measures, the situation of overfishing in U.S. waters remains dire.

Globally, the world fish catch may appear stable, but many species and areas are overfished. The FAO has concluded that 52% of fish stocks are at their maximum sustainable yield, 16% are overexploited and will likely decline further, and 8% are depleted and are much less productive than they once were. Much of the catch (one-third) is of such low commercial value that it is used only for fishmeal and oil production. In addition to direct fishing pressure, many ocean species are in decline because people fishing for another species accidentally catch them. This **bycatch** can sometimes be larger than the intended catch, and wastes the lives of billions of organisms annually.

The FAO has also concluded that the capture fisheries are at their upper limit, as indicated by the long-term catch data. Restoration of fisheries is critical both to target species and to the ocean food web. Fisheries scientists reason that if overexploited and depleted fisheries were restored, the catch could increase by at least 10 million metric tons. That could happen, however, only if fishing for many species were temporarily eliminated or greatly reduced and the fisheries were managed on a more sustainable basis.

Case Study: Atlantic Cod.

For hundreds of years, cod, haddock, and flounder (called ground fish because they feed on or near the bottom and are caught by bottom trawling) were the mainstay of the fishing industry in the northeast Atlantic, particularly on the Grand Banks off Newfoundland and Georges Bank off New England. Recent events in both fisheries illustrate the tragedy of the commons and the impacts of modern fishing woes.

The Grand Banks were once the richest fishing grounds in the world. Unfortunately, however, the cod there were overexploited until the number of cod dropped so low that the fishery had to be closed. The biggest reason was an increase in fishing pressure, particularly after World War II, as technology allowed larger vessels to capture more fish and allowed fish to be found more easily. Industrial-sized factory ships from Russia, Japan, and other countries captured cod right outside of Canadian waters. At the same time, more and more fish were caught by Canadian fishing vessels. In addition, cod fishing contributed to the accidental catch of other fish, including capelin, an important cod food source. Changes in the food chain contributed to the precipitous decline of a once abundant resource. By 1991, the cod population had dropped to one one-hundredth of its former size. In 1992, the fishery was closed, costing the jobs of 35,000 fishers. So far, the cod population has not rebounded.

Events on Georges Bank, New England's richest fishing ground, parallel those of their neighbor to the north (Figure 7–15). In the early 1960s, ground fish accounted for two-thirds of the fish population on the bank. Since that time, ground fish in this fishery have plummeted to historically low levels. How did this occur? After the passage of the Magnuson-Stevens Act in 1976, fishing on Georges Bank increased in intensity, in part because the law encouraged the increased use of the fish resource. Increased fishing pressure resulted in a decline in the desirable species. The New England Fishery Management Council (NEFMC) set Total Allowable Catch (TAC) quotas, but fishers claimed that the TAC quotas were set too low and successfully argued for an indirect approach that employed mesh with openings of a size that allowed smaller fish to escape. In less than 10 years, the number of boats fishing Georges Bank doubled; the results were disastrous, as indicated by the cod landing data. In response to this rapid decline, NEFMC discarded the mesh-size approach and initiated a number of regulations designed to take some pressure off the Georges Bank cod. These regulations restricted the number of days that vessels could spend at sea, closed areas of the bank to fishing, and set TAC quotas at levels designed to reduce the harvest. In addition, the NEFMC bought back fishing permits and paid fishers to scrap their boats. In spite of these measures, cod landing consistently exceeded the TAC quotas by 150–200%. In 2004–2005, when the stock had plummeted to its lowest level ever, restrictions finally led to a much-reduced harvest. Despite these restrictions, the cod failed to rebound. In 2013 the cod harvest was reduced even further because fish stocks were not recovering. The outlook for North Atlantic cod as an economic resource is not good.

Ecosystem Changes. One of the effects of overfishing is the removal of large fish from ocean food webs. Because large and slow-growing fish take longer to replace, ocean fish are changing to smaller individuals and those that reproduce more rapidly. As fishers capture too many of one species, they move on to the next. First, they overfish the top predators—such as tuna, marlin, and swordfish—then they take fish that are smaller and lower on the food chain. This process of moving from species to species as each type is depleted is known as "fishing down the food chain." It means that commercial fishing has made long-lasting changes to basic ecological processes in marine ecosystems.

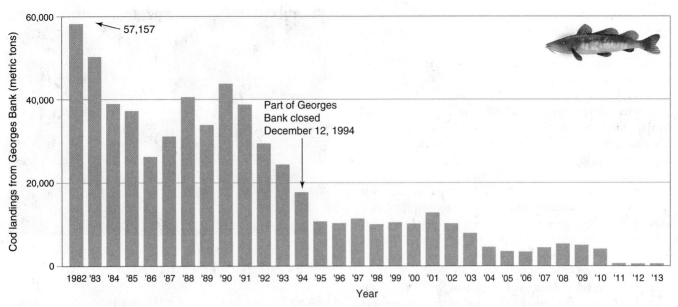

Figure 7–15 **Fishery in distress.** Data showing cod landings from Georges Bank, 1982–2013. (*Source:* NOAA, 2014.)

Loss of Coastal Habitats. Coastal zones protect the land from the impacts of storm surges and tides. They also support some of the most productive ecosystems on the planet and produce a great deal of food harvested by humans. Over the years, human development of the coastal zone has resulted in the degradation and loss of large portions of these valuable ecosystems. Salt marshes and mangrove swamps are particularly vulnerable. About half of the world's salt marshes have been lost, although they are highly productive fish habitat, important for waterfowl, and protect coasts from erosion.

Since 1983, half of the world's 18 million hectares (45 million acres) of mangroves have been cut down. Mangrove trees grow in warm coastal regions around the world. Mangroves have the unique ability to take root and grow in shallow marine sediments. There they protect the coasts from erosion and storm damage and provide a refuge and nursery for many marine fish, habitat for wildlife, fuel wood for people, and food for small livestock. Despite these benefits, mangroves are under assault from coastal development, logging, and shrimp aquaculture.

In Southeast Asia, large swaths of mangroves have been replaced with shrimp farms (Figure 7–16). The benefit is local, short term, and specific to the shrimp farmer, who is likely to be from outside the community, while the loss of services is more regional, long term, and diffuse to all those living near the coast. Recently the conversion of mangrove forests to shrimp farms has largely stopped because construction costs are high and the farms do not do well in the long term; most shrimp farms operate for only three to five years before they are abandoned because of diseases in the water. But mangrove loss from conversion to rice agriculture, fuel, wood pulp, and other uses has continued. The vast mangrove forests of the Sundarbans (Chapter 6) are declining rapidly.

Many areas of Southeast Asia hit by the 2004 tsunami had been deprived of their protective mangrove forests;

areas without mangroves were more heavily damaged than those with intact mangroves. In one study on mangroves in Tanzania, scientists found that the Saadani National Park, which offered the protection of a mangrove environment along the coast, improved the local economy. Although people using mangroves for wood suffered a short-term loss, in the long term the presence of the mangroves increased the quantity of fish and native shrimp available to the community.[9]

Local and international pressure to stop the destruction of mangroves is growing. Developing nations that once

---

[9]C. G. McNally, E. Uchida, and A. J. Gold, "Biodiversity Conservation and Poverty Traps Special Feature: The Effect of a Protected Area on the Tradeoffs Between Short-Run and Long-Run Benefits from Mangrove Ecosystems," *Proceedings of the National Academy of Sciences*, 2011 108 (34): 13945.

Figure 7–16 **Shrimp farms in Thailand.** Carved out of mangrove swamps, these shrimp farms provide only a fraction of the value of the services that were performed by the intact mangroves.

considered mangroves to be useless swampland are now beginning to recognize the value of the natural services they perform. Like terrestrial forests, mangroves absorb carbon dioxide. Because mangroves are so carbon rich, restoring them would be a good way to combat climate change.

Threats to Coral Reefs. In a band from 30° north to 30° south of the equator, coral reefs occupy shallow coastal areas and form atolls, or islands, in the ocean. Coral reefs (such as the Caribbean reefs discussed at the chapter's start) are among the most diverse and biologically productive ecosystems in the world. The reef-building coral animals live in a symbiotic relationship with photosynthetic algae called **zooxanthellae**; because these algae need adequate sunlight, corals are found only in water shallower than 75 meters (245 feet).

Coral reefs provide many benefits to local people. As ocean waves travel toward land, they first strike the coral reefs, which lessens the power of the waves and protects the land shoreward of the reefs. The great variety of fish and shellfish that live on reefs are important sources of food and trade for local people. Coral reefs also attract tourists who enjoy the warm shallow waters and the wildlife on the reefs, generating an estimated $30 billion per year in tourism and fishing revenues.

In recent years, coral reefs in areas close to human population centers have shown signs of deterioration. It is estimated that 20% of coral reefs have been destroyed by human activities. The threats are both global and local. Islanders and coastal people in the tropics often scour the reefs for fish, shellfish, and other edible sea life, sometimes using dynamite or cyanide to flush out fish, both very dangerous practices. Bottom trawling is also very destructive.

Rising levels of atmospheric carbon dioxide are causing ocean water to become acidified. **Ocean acidification** occurs as seawater near the surface of the ocean takes up the increasing amount of CO_2 pumped into the atmosphere by humans; as seawater becomes more acidic, the concentration of carbonate ions in the water is reduced. This, in turn, makes it more difficult for coral animals to build their calcium carbonate skeletons. Scientists point out that persistent high summer temperatures and the gradual acidification of the oceans could permanently wipe out coral reefs over vast areas of the shallow tropical oceans—an enormous loss of biodiversity and economic value. The widespread loss of corals due to bleaching (Figure 7–17) appears to come from a loss of the algal symbionts in corals as a result of a number of stressors, including heat.

Ocean Pollution. The ocean is the great repository of human effluent, plastic bags, oil, and sediment from farms, all of which make their way to the sea. Nutrient runoff from land is causing the loss of oxygen in vast areas of ocean. These dead zones cannot support most life (see Chapter 20). Plastics in the ocean are wreaking havoc on ocean food webs (discussed in both Chapters 20 and 21). The biggest changes in the ocean, though, relate to global changes in the water itself. Climate change is warming water, changing currents,

Figure 7–17 Bleached coral. This coral in the Red Sea shows the effects of excessively warm temperature.

melting ice, and causing sea levels to rise. It is also causing animals to change their ranges.

Lawlessness on the High Seas. The 1982 United Nations Convention on the Law of the Sea set the high seas (outside of the jurisdiction of individual countries) as a common resource for all nations. Since it was too difficult to fish there, the high seas acted as a large marine reserve. However, the lack of governance on the high seas has led to problems of lawlessness. Illegal, unreported, and unregulated fishing fleets abound. These fleets are often involved in criminal activities, such as black markets in regulated fisheries, trafficking of people, modern slavery, terrorism, and the smuggling of drugs and weapons. One study showed that 20–32% of wild-caught seafood imported into the United States comes from illegal, unregulated, or unreported fishing.[10]

Part of the problem is that laws have not kept up with technology. Today extensive fishing and even deep-sea mining occur in international waters. We need new agreements to protect the biodiversity of the high seas and manage the human activities there.

Aquaculture: A Mixed Bag

A worldwide increase in **aquaculture**, the farming of aquatic organisms, is helping fill in the gaps left by limited wild fisheries. Modern aquaculture includes species ranging from seaweed to bivalve mollusks, such as clams and oysters, to freshwater and marine fish. The extent of aquaculture has increased dramatically in the past 20 years. The growth of aquaculture in developing countries is helping to alleviate poverty and improve the well-being of rural people. In the case of species such as oysters, aquaculture is helping to restore ecosystem functions that were lost when wild populations were overharvested.

Despite its benefits, aquaculture also causes some threats to ocean ecosystems. For example, shrimp farming in many

[10]Pramod, Ganapathiraju, Katrina Nakamura, Tony J. Pitcher, and Leslie Delagran, "Estimates of Illegal and Unreported Fish in Seafood Imports to the U.S.A.," *Marine Policy* 48, (2014), 102–113.

tropical areas destroys mangrove habitats and accelerates the decline of fish species that depend on the mangroves for a part of their life cycle. In addition, carnivorous species such as shrimp and salmon are usually fed fish products from species such as whiting and herring. It is inefficient to feed one edible fish to another in order to feed humans. Indeed, salmon and shrimp farms consume several times as much fish protein as they produce! Other concerns about aquaculture include the spread of diseases from farmed to wild populations and the pollution caused by releasing nutrients into the ocean.

As more and more coastal areas are developed, near-shore waters are becoming less and less available for aquaculture. One solution to this problem could be open-ocean aquaculture, which is the farming of coastal species in ocean waters several kilometers from the shoreline. This is a new practice in North America, but even in Europe and other areas where it has been long practiced, total production levels are far lower than they could be.

Solutions to Ocean Problems

Ocean problems are significant, but international efforts to solve them have been launched. At the 2002 Johannesburg World Summit, delegates agreed to establish a network of protected marine areas by 2012 and to restore depleted global fisheries to achieve an MSY "on an urgent basis" by 2015. Scientists study the impact of marine reserves, organizations certify sustainable seafood, and the Global Ocean Commission has made a series of recommendations about protecting the resources of the high seas.

Case Study: Blue Whales. Today the great whales are among the most endangered species in the oceans, partly as a result of their slow growth, low reproductive rate, and need for an extended period of parental care. Whales were once harvested for their oil; more recently they have been hunted for their meat, which is considered a delicacy in Japan and a few other countries. Until the late 1980s, these giants of coastal waters and the open ocean were heavily depleted by overfishing.

In 1974, the International Whaling Commission (IWC), an organization of nations with whaling interests, decided to regulate whaling according to the principle of maximum sustainable yield. Whenever a whale species dropped below a minimum population size, the IWC banned the hunting of that species in order to allow its population to recover. At that time, three species—the right whale, bowhead whale, and blue whale—were at very low levels and were immediately protected. Because of difficulties in obtaining reliable data and enforcing catch limits, the IWC took more-drastic action and placed a moratorium on the harvesting of all whales beginning in 1986. The moratorium has never been lifted; however, some limited whaling by Japan, Iceland, and Norway continues, as does harvesting by indigenous people in Alaska, Russia, and Greenland. Despite these restrictions, the populations of some whales, such as the North Atlantic right whale, remain dangerously low.

Since the 1970s, blue whales in the eastern Pacific appear to have recovered to 97% of their historical population

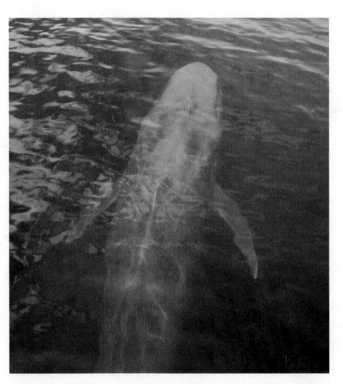

Figure 7–18 Blue whale recovery. Blue whales, such as the one shown here, were hunted until they were in danger of extinction. Since hunting was banned, their population in the eastern Pacific Ocean has been increasing.

(Figure 7–18). With only 2,200 individuals, that population is not large; because blue whales are so large (larger than any other animal) the carrying capacity of their environment is not high. Although these whales remain vulnerable, their recovery demonstrates how effective management can result in positive change.

Marine Reserves. The most direct path to the restoration of a fishery or a damaged habitat is likely to be the creation of a marine reserve, or marine protected area (MPA). Marine reserves are areas along coasts or in the open ocean that have been closed to all commercial fishing and mineral mining. These reserves often allow less-harmful activities, such as small-scale fishing, tourism, and recreational boating, although some prohibit all such activities. The benefits of MPAs include basic protection of these vital ecosystems, increased tourism, and the spillover of fish and larvae to surrounding fisheries.

The number of MPAs is growing, now numbering more than 5,000 around the world. More than 27% of the world's coral reefs are in MPAs, although not all of them are rigorously managed. In 2014, President Obama announced plans for the Pacific Remote Islands National Marine Monument, which will be the largest protected area on the planet.

Seafood Certification. Like forest product certification, seafood certification can help consumers select seafood products that have been harvested sustainably. Such certification makes the chains of ownership transparent, allowing buyers to evaluate the sources of foods. Consumers can make better choices and pressure can be brought to bear on companies. The Marine Stewardship Council is one of the better-known certifiers of seafood.

A Coherent Global Plan for Oceans. The Global Ocean Commission, based at Somerville College, Oxford, UK, is an independent commission formed as an initiative of The Pew Charitable Trusts and supported by Adessium Foundation, Oceans 5, and The Swire Group Charitable Trust. In June 2014, the commission released a report entitled *From Decline to Recovery: A Rescue Package for the Global Ocean,* which outlined eight proposals for the recovery of oceans.[11] These proposals include:

1. *Have a UN Sustainable Development Goal (SDG) for the oceans.* The 17 SDGs identified by the UN set the stage for world efforts to help the poorest move out of poverty while still protecting the world (Chapters 2, 9). Goal 14 is sustainable use of the oceans.

2. *Govern the high seas.* We need to develop and enforce treaties that protect biodiversity on the high seas.

3. *Stop overfishing.* We need to end subsidies for fishing and increase transparency about fishing, even registering all fishing boats worldwide.

4. *End illegal, unreported, and unregulated fishing (IUU).* Illegal fishing has the potential to cause the extinction of species. Setting limits for extremely large fishing vessels is essential to ending overfishing.

5. *Keep plastics out of the ocean.* We need to set plastic-reduction targets and provide incentives for changing what we do on land to keep plastics out of the ocean.

6. *Set safety and environmental standards for the offshore extraction of oil and gas.* Drilling and mining on the sea floor have the potential to cause industrial accidents. Setting and enforcing standards for these activities is crucial.

7. *Create a global ocean accountability board.* This board would monitor progress toward the goals determined by the Global Ocean Commission.

8. *Create a high seas regeneration zone.* Protecting large areas where fish are not harvested will protect those resources for future generations.

Ocean ecosystems are changing dramatically. Because many of these changes are negative, it might be easy to feel overwhelmed. Yet at the same time, many people are working for change, including the famous marine scientist and explorer Sylvia Earle. Earle founded the organization Mission Blue in order to establish marine protected areas, called "hope spots," saying, "I wish you would use all means at your disposal . . . to ignite public support for a global network of marine protected areas, hope spots large enough to save and restore the ocean, the blue heart of the planet."

CONCEPT CHECK What are two major ways oceans are degraded? ✓

---

[11]Global Ocean Commission, *Decline to Recovery: A Rescue Package for the Global Ocean,* 2014.

7.5 Protection and Restoration

The time has passed when the loss of ecosystems could be excused on the grounds that there are suitable substitute habitats just over the hill. With the rising human population, industrial expansion, and pressure to convert natural resources to economic gain, there will always be reasons to exploit natural ecosystems. The last resort for many species and ecosystems is protection by law in the form of national parks, wildlife refuges, and reserves. The World Database on Protected Areas (WDPA) for 2010 recognizes 161,000 "protected areas," sites whose management is aimed at achieving specific conservation objectives, in accordance with policies of the Convention on Biological Diversity. These sites, representing 13% of Earth's land area, cover a spectrum of protection from strict nature preserves to managed resource use. However, the establishment of a protected area is not always followed by effective management.

Public and Private Lands in the United States

The United States is unique among the countries of the world in having set aside a major proportion of its landmass for public ownership. Nearly 40% of the country's land is publicly owned; this land is managed by state or federal agencies for a variety of purposes, excluding development. The distribution of federal lands, shown in Figure 7–19, is greatly skewed toward the western states, a consequence of historical settlement and land distribution policies. Although most of the lands in the East and Midwest are in private hands, forests and other natural ecosystems still function all across these regions.

One of the first people to champion the preservation of natural areas in the United States was John Muir (1838–1914). A naturalist and writer, Muir spent a great deal of time in the mountains of California and founded the Sierra Club, now one of the country's largest conservation organizations. Muir wrote movingly about the aesthetic and spiritual values of nature, helping to persuade President Theodore Roosevelt and others to protect wild places. Muir's efforts led to the establishment of two national parks—Yosemite and Sequoia. Today Muir is often called "the father of the national parks."

At the beginning of the 20th century, President Theodore Roosevelt saw the destruction of prairies, bison, and birds in the western expanses. He created the national parks, wildlife refuges, and other federal land protection categories, eventually protecting approximately 230 million acres of public land. Other parks and public lands were added throughout the 20th century.

In March 2009, President Obama signed the Omnibus Public Lands Management Act into law. This sweeping legislation protected an additional 2.1 million acres of wilderness, increased the wild and scenic river system by 50%, and established new national parks and monuments. It also established a new federal land system, the National Landscape

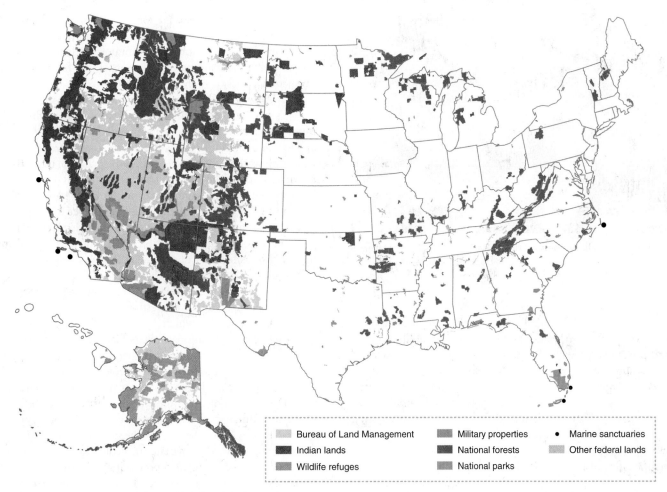

Figure 7–19 Distribution of federal lands in the United States. Because the East and Midwest were settled first, federally owned lands are concentrated in the West and Alaska.

(*Source:* Council on Environmental Quality, *Environmental Trends*, 1989.)

Conservation System, under the control of the Bureau of Land Management. This 27-million-acre system ties together 887 areas of high ecological and cultural value, some of the "crown jewels" of national lands. Some of these sites had already been designated as national monuments, historic trails, conservation areas, or wilderness.

Wilderness. Land given the greatest protection is designated as **wilderness.** Authorized by the Wilderness Act of 1964, this classification includes 110 million acres in more than 700 locations—more than 4% of the land area of the United States. The act provides for the permanent protection of these undeveloped and unexploited areas so that natural ecological processes can operate freely. Permanent structures, roads, motor vehicles, and other mechanized transport are prohibited. Timber harvesting is excluded. Some livestock grazing and mineral development are allowed where such use existed previously; hiking and similar activities are also allowed. Areas in any of the federally owned lands can be designated as wilderness; the Bureau of Land Management, National Park Service, Forest Service, and Fish and Wildlife Service all manage wilderness lands.

National Parks and National Wildlife Refuges. The **national parks,** administered by the National Park Service, and the **national wildlife refuges,** administered by the Fish and Wildlife Service, provide the next level of protection to 84 million and 150 million acres, respectively. Here the intent is to protect areas of great scenic or unique ecological significance, protect important wildlife species, and provide public access for recreation and other uses.

The dual goals of protecting and providing public access often conflict with each other. Parks and refuges draw so many visitors (more than 310 million visits a year) that the protection of natural sites can be threatened. At a number of the most popular parks, auto traffic has been replaced by shuttle vehicles because of the problems of pollution and habitat destruction accompanying the use of motor vehicles within the parks. All of the federal agencies involved are trying to cope with the rising tide of recreational vehicles such as snowmobiles and off-road vehicles used on most federal lands.

Increasingly, agencies, environmental groups, and private individuals are working together to manage natural sites as part of larger ecosystems. For example, the Greater Yellowstone Coalition has been formed to conserve the ecosystem that surrounds Yellowstone National Park. Yellowstone

Figure 7–20 Yellowstone National Park. Famous for its boiling pools, geysers, and wildlife such as the bison shown here, Yellowstone National Park is part of the Greater Yellowstone Ecosystem.

and Grand Teton National Parks form the core of the ecosystem, to which seven national forests, three wildlife refuges, and private lands are added to form the greater ecosystem of 18 million acres (Figure 7–20). The coalition acts to restrain forces threatening the ecosystem. Thus, it has challenged the Forest Service's logging activities, addressed residential sprawl and road building on private lands, and curtailed the shooting of bison that wander off the parklands.

This cooperative approach is important for the continued maintenance of biodiversity because so much of the nation's natural lands remain outside of protected areas. Cooperation may also help restrict development up to the borders of the parks and refuges, keeping them from becoming fairly small natural islands in a sea of developed landscape.

National Forests. Forests in the United States represent an enormously important natural resource, providing habitat for countless wild species as well as supplying natural services and products. These forests range over 740 million acres, of which about two-thirds are managed for commercial timber harvest. Almost three-fourths of the managed commercial forestland is in the East and privately owned; most of the remainder is in the West and is administered by a number of government agencies, primarily the Forest Service and the Bureau of Land Management. The Forest Service is responsible for managing

193 million acres of national forests. The Bureau of Land Management oversees 258 million acres that are a great mixture of prairies, deserts, forests, mountains, wetlands, and tundra, as well as 16 national monuments. Most federal timber production comes from the national forest lands.

Multiple Use. In the 1950s and 1960s, the Forest Service's management principle was *multiple use*, which allowed for a combination of extracting resources (grazing, logging, and mining), using the forest for recreation, and protecting watersheds and wildlife. This principle came from the first head of the Forest Service, Gifford Pinchot, who believed that the management of public lands should allow human use. The intent was to achieve a balance among these uses, but in practice multiple use emphasized extractive uses, allowing the ongoing exploitation of public lands by private interest groups such as ranchers, miners, and especially the timber industry.

To harvest timber, government foresters select tracts of forest they judge ready for harvesting and then lease the tracts to private companies, which log and sell the timber. The timber harvest from national forests has been a controversial issue. Historic data on timber sales (Figure 7–21) reflect major shifts in logging. Note the rapid rise following World War II to satisfy the housing boom, and later periods of rise and fall related to the level of environmental concern of the time.

A forest-management strategy called **new forestry** was adopted by the Forest Service in the 1990s. This practice is directed more toward protecting the ecological health and diversity of forests than toward producing a maximum harvest of logs. New forestry involves cutting trees less frequently (at 350-year intervals instead of every 60 to 80 years), leaving wider buffer zones along streams to reduce erosion and protect fish habitats, leaving dead logs and debris in the forests to replenish the soil, and protecting broad landscapes. According to a recent Forest Service analysis, growth of forests has exceeded harvests for several decades in both national and private forests.

The Roadless Controversy. Logging roads contribute to the destruction of forests and streams. The Roadless Area Conservation Rule of 2001 was designed to increase the protection of the wildest places by placing a moratorium on new logging roads in some areas. Many people objected to the moratorium. They believed that it would halt timber harvests and prevent access by off-road vehicles in

Figure 7–21 National forest timber sales. Timber sales over the years reflect changes in resource needs and political philosophy.

(*Source:* U.S. Forest Service. FY 1905–2013 National Summary Cut and Sold Data and Graph. Accessed October 22, 2014, available from http://www.fs.fed.us/forestmanagement/documents/sold-harvest/reports/2013/2013_Q1-Q4_CandS_SW.pdf.)

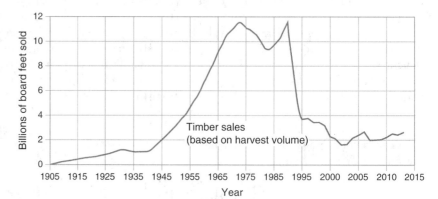

roadless areas (they were right!). Some 58 million acres of forestland were affected by the moratorium. In the years 2005–2011, there were several reversals and reinstatements of the rule, until a 2011 decision made the roadless rule the law of the land, although there have already been court challenges. Currently, this rule protects 58 million acres of forests.

Fires. The years 2006 and 2007 set new records for forest fires, with 9.9 million acres burning the former year and 9.3 million acres the latter. With a decade of drought in the western states (2014 was one of the worst on record for California) and serious drought in the South, it is no surprise that forest fires have become more extensive. A hot debate has focused on why so much forestland bears *fuel loads*—dead trees and dense understories of trees and other vegetation—and what to do about it. Most observers believe that a combination of fire suppression and logging practices are responsible for the high fuel loads. Ecologists remind us that wildfires can also be seen as natural disturbances, rather than ecological disasters—a confirmation of the dynamic nonequilibrium of ecosystems (Chapter 5).

Protecting Nonfederal Lands. From 1961 to 2014, voters in New Jersey protected 1.2 million acres of open space through public funding. A Massachusetts report showed that for every dollar the state invested in land conservation, four dollars were returned to the economy in natural goods and services.[12] These examples show that Americans are willing to spend money to protect nonfederal lands and that they gain a good deal from such conservation.

Often, landowners and townspeople want to protect natural areas from development but are wary of turning the land over to a government authority. One creative option is the **private land trust**, a nonprofit organization that will accept either outright gifts of land or **easements**—arrangements in which the landowner gives up development rights into the future but retains ownership of the parcel. The land trust may also purchase land to protect it from development. The land trust movement is growing considerably: There were 429 land trusts in the United States in 1980 and more than 1,723 in 2012, as reported by the Land Trust Alliance, an umbrella organization in Washington, DC, serving local and regional trusts. These trusts protect almost 47 million acres of land. The Nature Conservancy, a national and international land trust, conserves 15 million acres in the United States and also protects 102 million acres in other countries.

International Ecosystem Protection

Around the world, governments, private citizens, and nonprofit organizations are working to save ecosystems. There are also international treaties to protect ecosystems. The UNESCO World Heritage sites include coral reefs, forests, and other ecosystems that are considered to be of world-class importance. These sites often receive greater protection by the countries in which they are found than they would otherwise. Areas that are designated Ramsar Wetlands of International Importance are often centers for ecotourism.

Ecosystem Restoration

In recent years, the need for the restoration of damaged ecosystems has become clear, and a new subdiscipline with that objective—called **restoration ecology**—has been developed. The intent of ecosystem restoration is to repair damage to specific lands and waters so that normal ecosystem integrity, resilience, and productivity return. The ecological problems that can be mended by restoration include those resulting from soil erosion, surface strip mining, draining of wetlands, hurricane damage, agricultural use, deforestation, overgrazing, desertification, and the eutrophication of lakes. Because of the complexity of natural ecosystems, the practical task of restoration is often difficult. A thorough knowledge of ecosystem and species ecology is essential to the success of restoration efforts.

A great increase in restoration activity has occurred during the past 30 years, spurred on by federal and state programs. Most restoration projects are small and local, while the destruction of ecosystems occurs on a wider scale. However, there are some large-scale restoration. One of these is the largest wetland restoration in the world—the restoration of the Florida Everglades.

Freshwater Wetland: The Florida Everglades. The Florida Everglades, sometimes called a "river of grass," is America's largest subtropical wilderness (Figure 7–22). This network of wetland landscapes once occupied the southern half of Florida. Although development and the draining of wetlands have reduced the Everglades to half its original size, it is still an amazing and unique ecosystem, harboring

Figure 7–22 The Florida Everglades. This aerial view shows the meandering waterways that are necessary to functioning of the Everglades ecosystem. This ecosystem was degraded by the channelization of rivers and streams, the excessive withdrawal of water, and pollution. The ongoing restoration of the Everglades is the largest wetland restoration project in the world.

---

[12]The Trust for Public Land, *The Return on Investment in Parks and Open Space in Massachusetts*, 2013, 50.

a great diversity of wildlife. It is home to 16 wildlife refuges and four national parks, the crown jewel being Everglades National Park.

Water flows south from Lake Okeechobee by way of a 40-mile-wide, 100-mile-long shallow wetland leading to Florida Bay. During the 1900s, Floridians controlled water flow with a system of 1,000 miles of canals, 720 miles of levees, and hundreds of water control structures (locks, dams, and spillways), bounded on the east by a string of cities and their suburbs. In its center, 550,000 acres of croplands were created, forming a lucrative sugarcane industry. Now, water shortages in the winter leave too little for the natural ecosystems; in the summer rainy season, too much water is diverted to the Everglades, causing it to fill with pollutants. Sea level rise erodes the southern marshes.

In 2000, the U.S. Congress authorized the Comprehensive Everglades Restoration Plan (CERP), a plan to restore the South Florida Everglades ecosystem. CERP is the largest hydrologic restoration project ever undertaken in the United States and will require $10.5 billion and more than 35 years to complete. CERP is being managed by the U.S. Army Corps of Engineers, the federal agency that was originally responsible for creating much of the system of water control that made the restoration necessary. To restore the Everglades, 240 miles of levees and canals will be removed and a system of reservoirs and underground wells will be created to capture water for release during the dry season. The new flowage is designed to restore the river of grass, thereby restoring the 2.4 million acres of Everglades—not to its original state, but at least to a healthy system.

Estuary: The Chesapeake Bay. The Chesapeake Bay, surrounded by Maryland and Virginia, is a large estuary in a heavily populated area. Once producing a bounty of shrimp,

SOUND SCIENCE

Restoration Science: Learning How to Restore

Conservationist Aldo Leopold sat at a table in a converted chicken coop, writing about the value of nature. His essays on the damage inflicted by the unthinking use of ecosystem resources and his scientific papers on the impacts of fire and predator suppression were to change the course of history.

Around him lay acres of degraded farmland, its sandy soil depleted by years of unrelenting use, its sparse grasses standing as tufts across the eroded fields. 1n 1935, Leopold had bought this land near the Wisconsin River and begun its restoration. For years, he brought friends and students to the farm, encouraging them to sow seeds of native plants and promote the use of fire to restore the prairie. Gradually, the weedy plants were outcompeted by native grasses, shrubs, and wildflowers. Wild lupine vied for space with purple coneflower and wild indigo. Native butterflies returned, along with prairie voles and 13-lined ground squirrels. The restoration of his farm paralleled the restoration of native prairie at the University of Wisconsin Madison Arnold Arboretum, begun by Leopold and other scientists. It was the beginning of the field of restoration ecology.

The potential for the restoration of any ecosystem rests on three conditions:

1. Abiotic factors must have remained unaltered; or if not, it must be possible to return them to their original state.

2. Viable populations of the species formerly inhabiting the ecosystem must still exist.

3. The ecosystem must not have been upset by the introduction of foreign species that cannot be eliminated and may preclude the survival of reintroduced native species. If these conditions are met, revival efforts have the potential to restore the ecosystem to some semblance of its former state.

Aldo Leopold (1887–1948)

Today restoration ecologists use a set of actions described in a primer on ecological restoration.* These actions involve taking an *inventory*, developing a *model* of the structure and function of the desired ecosystem, setting goals for the restoration efforts, designing an *implementation plan,* and carrying out the plan. For example, to restore a prairie, it may be necessary to remove all herbaceous vegetation and then plow and plant the land with native grasses. To maintain the prairie, it may be necessary to burn it regularly or to introduce bison. Clear *performance standards* should be stated, along with the means of evaluating progress. Finally, the results of the implementation should be monitored and strategies developed for the long-term protection and maintenance of the restored ecosystem.

Leopold practiced these strategies on his own land in Wisconsin, bringing an abandoned and eroded farm back to health. In his essays in *A Sand County Almanac* as well as in other essays, Leopold wrote of this goal of understanding the natural world and caring for it. His essays also developed the ethical principles that underlie current efforts to protect species and their habitats (Chapter 6). Leopold died trying to control a brush fire, but his work lives on in the field of restoration ecology.

* Society for Ecological Restoration, International Science and Policy Working Group, *The SER International Primer on Ecological Restoration (2004)*, available at www.ser.org./.

Figure 7–23 Restoration of the Chesapeake Bay. The drainage basin of the Chesapeake Bay covers parts of six states. Pollution in runoff has killed fish and shellfish in the bay, but restoration efforts are beginning to yield results. These students from Ohio University are working on a restoration project at Poplar Island, Maryland.

Figure 7–24 Restoration of the Rio Grande. Overconsumption of water from the Rio Grande causes stretches of the river to periodically run dry. Recently the coordinated efforts of nonprofits and government agencies in Colorado, New Mexico, and Texas have been helping to restore water flows and riverbank vegetation.

oyster reefs, fish, crab, and water birds, the bay has been badly degraded by coastal development and runoff pollution from upstream watershed areas. The vast oyster reefs that once filtered and cleaned the bay water are largely gone. Fish and crab have declined. In 2014 a hot summer and late rains that washed pollutants into the bay caused a dead zone (Chapter 20) of low oxygen a cubic mile in volume. Nonetheless, the Chesapeake Bay Foundation has worked on a massive restoration effort involving pollution prevention programs, replanting of sea grasses, restoring wetlands, planting forest buffers, recreating passage for fishes, restoring oyster reefs, and managing crab and fish industries (Figure 7–23).

River: The Rio Grande. The Rio Grande in the southern United States is widely recognized as an endangered river (Figure 7–24). The nearly 2,000-mile long river has lost the large native sturgeon, eel, and red horse fish that used to inhabit it, along with at least four small native fish. Whooping cranes, which used to frequent the river, have not been present since 1998. In some dry spells, stretches of the river go dry. These problems result from dams and the excessive removal of river water as it wends through arid lands from the Rocky Mountains to the Gulf of Mexico.

Fortunately, a variety of restoration efforts are helping the Rio Grande to function as a healthier river system. One part of the restoration involves altering water management at dams to allow a more natural water flow beneath them. In the headwaters, locals are working to prevent stream bank erosion, remove ineffective water diversion dams, and restore wildlife habitat. Further down the river in New Mexico, the Army Corps of Engineers is conducting the Middle Rio Grande Ecosystem Restoration project to restore a more natural flow to a section of river. It is removing exotic vegetation, jetties, and other impediments, and reestablishing natural water forms such as oxbow lakes.

Tropical Forest: Brazil's Atlantic Coastal Forest. The Atlantic Coastal Forest of South America stretches from Brazil to northern Argentina, Paraguay, and Uruguay (Figure 7–25). This region has some of the highest animal and plant diversity in the world, with 2,200 species of birds as well as an abundance of mammals, reptiles, and amphibians, and 20,000 species of plants—8% of all the plants on Earth. The region also has 200 endemic birds and 21 endemic primates. Because of logging, invasive species, and conversion for development and agriculture, only 7–10% of the forest remains, and what remains is highly fragmented. The nonprofit Instituto Terra has been working since 1999 to restore 630 hectares (1,556 acres) of forest, establishing local forest plantings, a native plant nursery, and educational facilities, since ecological restoration in Brazil is in its infancy.

Figure 7–25 Restoration of Brazil's Atlantic Coastal Forest. Originally an area of extreme biodiversity, less than 7% of this ecosystem remains. Much of the forest was cut and replaced with sugar cane plantations. Restoration efforts such as this tree planting are helping to reestablish some of the native vegetation.

Figure 7–26 Restoration in the White Carpathians. The White Carpathians region in Central Europe has diverse grasslands and woodlands. This Protected Landscape Area is in the White Carpathian Mountains, Zitkova, Czech Republic.

Figure 7–27 Coral reef restoration. The restoration of this coral reef near Pulau Badi, Indonesia, is being funded by Mars, Incorporated.

Grasslands: White Carpathians, Czech Republic. Grasslands receive less attention than many other ecosystems, but they provide essential habitats for many plants, birds, and roving herbivores. The White Carpathians Protected Landscape Area (Figure 7–26) is a biosphere reserve in the Czech Republic that is especially known for its plant diversity, featuring forests interspersed with flowery meadows. It has a high diversity of native plants, including the greatest number of orchid species in Central Europe. Before restoration, much of grasslands in this region had been degraded. In one restoration, an area around 500 hectares (1,235 acres) was reseeded with a high diversity of native seeds and then mown regularly. The protection and restoration of degraded sections of this ecosystem is important to the biodiversity of Central Europe.

Coral Reefs: Mars Pulau Badi Project. The Coral Triangle encompasses a roughly triangular area of the oceans surrounding Indonesia, Malaysia, Philippines, Papua New Guinea, Solomon Islands, and Timor Leste. This region includes 5.7 million square kilometers of ocean and helps sustain 120 million people. The region is known for its high biodiversity, with more than 500 species of corals and many creatures that depend on them, including 3,000 fish species. However, 95% of the coral reefs in the Coral Triangle have been damaged or destroyed, largely by destructive fishing, coastal development, and climate change.

Since 2007, Mars, Incorporated, has sponsored the largest coral reef restoration in the world off the island of Pulau Badi in Indonesia. The ongoing project involves reseeding the reef with structures on which coral fragments grow, establishing a marine protected area, and creating jobs by starting seahorse and abalone farming. Figure 7–27 shows a diver placing corals on the reef. The multifaceted approach and attention to the needs of locals has made this restoration a success.

Final Thoughts

Ecosystems everywhere are being exploited for human needs and profit. It is certain that greater pressures will be put on natural ecosystems as the human population continues to rise. These pressures must be met with increasingly effective protective measures if we want to continue to enjoy the goods and services provided by ecosystems. The problems are most difficult in the developing world, where poverty forces people to take from nature in order to survive. If natural areas are to be preserved, the needs of people must be met in ways that do not involve destroying ecosystems. People must be provided with alternatives to overexploitation, a situation requiring both wise leadership and effective international aid.

In April 2000, when he was UN Secretary-General, Kofi Annan included in his millennium report to the General Assembly a stern warning that far too little concern was being given to the sustainability of our planet: "If I could sum it up in one sentence, I should say that we are plundering our children's heritage to pay for our present unsustainable practices. . . . We must preserve our forests, fisheries, and the diversity of living species, all of which are close to collapsing under the pressure of human consumption and destruction . . . in short, we need a new ethic of stewardship."

The Billion Tree Campaign demonstrates such a stewardship ethic. Begun under UN sponsorship in 2006, the campaign to plant one billion trees overran its goal in 18 months, then becoming the Seven Billion Tree Campaign. In 2011, the campaign was turned over to the Plant for the Planet Foundation, and today more than 12 billion trees have been planted. From individuals to heads of states, people in 155 countries are planting trees, acting as stewards for the climate and for peoples' needs.

CONCEPT CHECK What are two advantages of setting aside land and limiting use in order to protect ecosystems? What are two disadvantages? ☑

REVISITING THE THEMES

SOUND SCIENCE

Ensuring sustainable harvests in forests requires thorough assessments of tree growth and a balancing of the needs of ecosystems with those of the harvesters. Calculating a total allowable catch in fisheries requires solid scientific data on fish populations and an understanding of the MSY model. These limits are difficult, if not impossible, to accomplish for mobile creatures such as fish and whales. Fisheries scientists are constantly challenged to assess fish stocks accurately, and their recommendations can often be trumped by political and economic decisions, as is well illustrated in the plight of the New England fisheries.

The restoration of ecosystems requires a great deal of sound science in the form of species and ecosystem ecology. It also requires the ability to make midcourse corrections, which projects like the Everglades restoration will require in the future. Scientists promote the discipline of restoration ecology in order to achieve greater levels of success in restorations. To test hypotheses, scientists use data from experiments and data collected by citizen scientists, as the example of the REEF data collection showed.

SUSTAINABILITY

The protection of ecosystems requires intelligent policies and practices. Often this requires coordination, as an unregulated commons almost always tends to overuse. This is why we have the International Whaling Commission, for example. Unfortunately, there are many cases of unsustainable use: the taking of "bush meat" by commercial hunters, the cyanide poisoning of coral reefs, the exploitation of a commons where there is open access to all, and the deforestation in tropical rain forests, to name a few. The principle of sustainability is built into the model of the maximum sustainable yield, but in the practical management of fisheries, this is often difficult to achieve. For this reason, it is wise to use the precautionary principle when setting thresholds for resource use. Sustainable forestry is not so difficult, as long as it is focused on managing the forests as functioning ecosystems and biodiversity is respected. In much of the developing world, sustainable use of resources cannot be achieved without taking into account the needs of local people and involving them in management decisions. Incorporating ecosystem services into our calculations will make protection of ecosystems more economically viable. Such an approach requires knowledge of the value of ecosystem capital.

STEWARDSHIP

Stewardship care of ecosystems means consciously managing them so as to benefit both present and future generations. Sustainable consumptive and productive use of ecosystem resources involves tenurial rights over them; exercising these rights should be a matter of stewardship embodied in rules and policies. Table 7–2 presents a set of principles that promote such stewardship. When forests and fisheries are under the control of the people who most directly depend on their use, stewardship can flourish. Management of the fisheries exhibits both good and poor stewardship in different regions. Individuals can show stewardship by acting as citizen scientists or informed consumers or by helping nonprofits such as land trusts. Restoration ecology also involves stewardship as people work to restore areas that have been degraded.

Finally, it is heartening to see the former UN secretary-general endorsing a stewardship ethic as an important step in preserving the natural resources that are under tremendous pressure from present generations. The Billion Tree Campaign is a tangible demonstration of that ethic.

REVIEW QUESTIONS

1. How did individuals act to help scientists in the Caribbean Sea?

2. What are some goods and services provided by natural ecosystems?

3. Compare the concept of ecosystem capital with that of natural resources. What do the two reveal about values?

4. Compare and contrast the terms *conservation* and *preservation*.

5. Differentiate between consumptive use and productive use. Give examples of each.

6. What does maximum sustainable yield mean? What factors complicate its application?

7. What is the tragedy of the commons? Give an example of a common-pool resource, and describe ways of protecting such resources.

8. When are restoration efforts needed? Describe efforts under way to restore the Everglades.

9. Describe some of the findings of the most recent FAO Global Forest Resources Assessments. What are the key elements of sustainable forest management?

10. What is deforestation, and what factors are primarily responsible for deforestation of the tropics?

11. What is the global pattern of exploitation of fisheries? Compare the yield of the capture fisheries with that of aquaculture.

12. How did the 1996 and 2006 Magnuson-Stevens Fishery Conservation and Management Reauthorization Act change the original Magnuson-Stevens Act?

13. How do land trusts work, and what roles do they play in preserving natural lands?

14. How are coral reefs and mangroves being threatened, and how is this destruction linked to other environmental problems?

15. Compare the different levels of protection versus use for the different categories of federal lands in the United States.

16. Describe the progression of the management of our national forests during the past half-century. What are current issues, and how are they being resolved?

THINKING ENVIRONMENTALLY

1. Both consumptive use and productive use of natural ecosystems are necessary for high-level human development. To what degree should consumptive use hold priority over productive use? Think about more than one resource (lumber, bush meat, etc.).

2. Consider the problem presented by Hardin of open access to the commons without regulation. To what degree should the authorities limit the freedom of the use of these areas?

3. Consider the benefits and problems associated with coastal and open-ocean aquaculture. Is aquaculture a useful practice overall? Justify your answer.

4. Kofi Annan stated that we are in need of a "new ethic of stewardship." On what principles should this new ethic be built?

MAKING A DIFFERENCE

1. To determine the impact you have on the ecosystems around you, calculate your ecological footprint (several sites on the Internet can help with this). What can you do to lower your ecological footprint?

2. Consider Michael Pollan's work *The Omnivore's Dilemma*. The book deals with America's eating habits, government regulations regarding food production, farming techniques, and similar topics. Decide which eating habits are ecologically sustainable and which are not. Go to the Internet and search for sites that offer ecosystem-friendly recipes.

3. By transporting firewood from its native area, you may run the risk of spreading diseases and alien insect species that can damage trees and overrun whole areas of forest. You can avoid this by using firewood harvested locally or, if you need to carry firewood a distance, by finding wood marked with a U.S. Department of Agriculture (USDA) tag confirming that the logs are safe to move.

4. Organize a volunteer tree-planting group. You can purchase young trees at a nursery, college agricultural department, or city government. To raise money for a large project, try to find individual or corporate sponsors.

5. When you visit parks and other natural areas, stay on the designated trails. Avoid making new trails that might lead to erosion, and never disturb nesting birds or other wild animals raising their young.

THE HUMAN POPULATION AND ESSENTIAL RESOURCES

It's summer, and a weary farmer encourages a group of workers as they harvest garlic and lay it out to dry. Nearby at a brightly colored barn, cars of customers begin to arrive to pick up bags of fresh vegetables. The scene is a modern small farm in New England with a community supported agriculture (CSA) program. Some of the workers are volunteers; the customers have all purchased shares in the crop. The farm also hosts open houses and family days. Farming is adapting to changes in population patterns such as urbanization and smaller families. Across the world in Kenya, rapid human population growth has driven subdivision of land so extreme that farmers eke out a living on plots smaller than a hectare (2 acres). The population of the globe is now more than 7.2 billion people, and the movement of people from rural areas to urban centers continues unabated.

Human population is expected to pass 10 billion by 2100, bringing with it increased needs and changes in the ways we do things. The ever-growing number of humans, increasing urbanization, increasing consumption rates, and growth of older populations are trends in human populations that interact with the environment. Patterns in human population growth affect wild species and the ecosystems on which we depend, as well as the land, soil, and water we use to produce food.

In Part Three, we will consider how human populations are similar to and different from populations of other species (Chapter 8). We will see how changes in technology have affected our population growth and how that growth affects the rest of the world. We will also look at the demographic transition and what causes countries to move through it (Chapter 9). We will then look at the critical resources humans need, how we manage and use these resources, and their relationship to population growth (Chapters 10–13). It is necessary to look at population and consumption concerns in tandem with resource use and pollution because population growth and its twin, consumption, affect every environmental issue.

Chapter 8

Learning Objectives

8.1 Humans and Population Ecology: Explain how humans, like other organisms, are subject to natural laws and ecological processes. Describe some significant differences between humans and other creatures in their ability to change their own world.

8.2 Population and Consumption— Different Worlds: Explain the relationship between income and fertility in countries around the world.

8.3 Consequences of Population Growth and Affluence: Describe the likely outcome of unlimited population growth and the unlimited use of natural resources. Explain why both population growth and consumption patterns must be addressed for stewardship of resources to occur.

8.4 Projecting Future Populations: Explain how age structure, population momentum, and the demographic transition help social scientists understand populations and predict future population trends in developing and developed countries.

The Human Population

Sweden. The cold winter wind cuts through the streets of Stockholm, Sweden, as Anje, a 19-year-old university student, hurries home to the warmth of her apartment, where she will settle in for an evening at her computer and books, studying for a degree in banking. She will stop for a fast-food meal on the way. Like many in the European Union (EU), Anje gets a free university education. She plans to travel extensively when her studies are over and then work for an international firm. Anje works hard because jobs will be scarce and she wants to be established someday. Although she is dating, Anje has no plans to marry in the near future. She is likely to marry late, if she does, and have children later than many people in the world. When she is ready to have children, Anje is likely to have good benefits, with medical care and maternity leave. Because she lives in a society in which the number of elderly people is increasing, Anje is likely to care for her elderly parents, but good social and medical services will keep them active long into their 70s or even their 80s.

Indonesia. A world away, in the heat of bustling Jakarta, Indonesia, 19-year-old Atin rides a motorcycle home with her brother. They will stop at a market for chicken and noodles and come home to the small house they share with their parents and three younger siblings. Atin and her family have a few luxuries, such as a cell phone and a small television. A medical emergency could set them back, though, and Atin is worried about her father's coughing. Like millions of Indonesian men, he smokes local clove and tobacco cigarettes. Atin's family has struggled through financial insecurity. Her father lost a stable job and took a series of temporary jobs, and their mother began to work part time. Atin dropped out of high school to help her family. Now Atin and her brother are working to help support the family. Atin hopes to marry a young man her family knows, if they can get enough money together. Like many Indonesians, she will most likely have a smaller family than her parents did.

Burkina Faso. In the dry season, Burkina Faso (in West Africa) is a place of dust and Sun. Awa, a 19-year-old married mother of two, walks with her children along a dusty road toward her home. She is carrying a large container of water on her head. Awa's husband works on a farm in neighboring Côte d'Ivoire (Ivory Coast), where he makes a small amount of money and sends it home. Neither Awa nor her husband can read, and there is no nearby medical facility. She is worried because she is pregnant and has been feeling very ill. Awa and her husband had another child who died shortly after birth, and Awa is afraid that this coming baby will not survive either because she is sick and still nursing another child. Her feet ache and she is very tired, but there is little to be done about it. She sometimes cares for her mother and for the children of her sister, who died of AIDS. Her four surviving siblings have similar lives and live nearby. They all farm during the rainy season, growing cassava, yams, peanuts, and maize. A few members of her extended family work on a cotton farm or have other relatives working in the Ivory Coast.

Anje, Atin, and Awa are three of millions of young people living around the world. Their life circumstances give them different health concerns, access to education, and medical care. They are also likely to have different family sizes and to use natural resources differently. In the 20th century, the global human population experienced an unprecedented explosion, more than tripling its numbers. Now, the rate of growth is slowing down, but the increase in absolute numbers continues to be substantial—and in some parts of the world, there isn't much slowing.

Remarkable changes in technology and substantial improvements in human well-being have accompanied this growth. In just the past 55 years (1960–2015), average income per capita has more than tripled, global economic output has risen more than tenfold (from $10 trillion to over $100 trillion), life expectancy has risen from 53 to 69 years, and infant mortality has been cut by nearly two-thirds (from 115 to 42 deaths per thousand live births). Even so, Anje, Atin, and Awa do not live in a world where people experience average conditions. The remarkable improvements in living conditions are not experienced equally. Extreme poverty is still widespread. More than 1.2 billion people live on less than $1.25 per day, and the income gap between the richest and the poorest countries is enormous.

Humans are animals, and population ecologists can study human populations and ask many of the same questions they would if they were studying wolves or elk. But humans are also exceptional. They can make decisions about their own fertility, they have ethical systems that inform their decisions, and they can use technology to prolong life and increase food production. Even more unusual, humans have cultural constructs, such as money. Money and its economic institutions mean that survival is driven not only by evolutionary fitness, but also by the economic circumstances into which people are born.

In this chapter, we will put the stories of people like Anje, Atin, and Awa into the context of the dynamics of population growth and the related context of consumption and its social and environmental consequences. A continually growing population is unsustainable, as is a continually increasing consumption of natural resources, so the focus in this and the next chapter is population and consumption stability—and what is required to get there.

8.1 Humans and Population Ecology

Humans are part of the natural world, and human populations are subject to processes such as birth and death. Figure 8–1 shows human population growth over the past 2,000 years. The long, slow incline is followed by a rapid rise, giving what looks like a J-shaped curve (Chapter 4). On closer inspection, however, the past few decades have shown decreasing rates of population growth. Projections—possible future scenarios determined by assumptions about changes we will see—suggest that the global human population may level off, so the pattern in 2100 will be more like the S-shaped curve that characterizes logistic growth (Chapter 4).

The field of collecting, compiling, and presenting information about human populations like that presented in Figure 8–1 is called **demography;** the people engaged in this work are **demographers.** Demographers study population processes such as migration or changes in fertility and mortality. In environmental science, biological science quickly becomes intertwined with other fields, so social demographers often include economic, cultural, social, and biological factors in their analyses of populations. Consequently, we

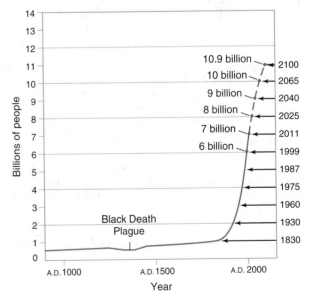

Figure 8–1 World population over the centuries. Throughout most of history the human population grew slowly, but in modern times it has expanded greatly.

(*Sources:* Basic plot from Joseph A. McFalls, Jr., "Population: A Lively Introduction," *Population Bulletin* 46, no. 2 (1991): 4; updated from UN Population Division, medium projection, 2012.)

cannot discuss human population ecology without quickly discussing differences in wealth or health care. Demographers use a set of specialized terms, described in Table 8–1, which we will use throughout the chapter.

r- or K-Strategists

We have contrasted organisms such as cockroaches, which breed rapidly and have boom-and-bust populations, with organisms such as elephants, which maintain smaller but more stable population sizes (Chapter 4). Cockroaches are described as "r-selected species," or "r-strategists," because they have a high r, or reproductive potential. They are contrasted with "K-selected species," or "K-strategists," which remain close to K (carrying capacity) and are characterized by long life spans, older age at first reproduction, fewer offspring, and high parental care of offspring. These characteristics form a continuum, with some organisms not easily placed in either group.

When we ask "Are humans r- or K-strategists?" it becomes clear that humans are like other organisms, but also significantly different. Just looking at the population growth curve of the past thousand years (Figure 8–1), we might be tempted to conclude that we are looking at a J-shaped curve representing the exponential growth typical of weedy species that move from place to place quickly but cannot outcompete other organisms over the long term (r-strategists). However, humans have characteristics such as high parental care and late reproduction that we would expect of the more stable K-strategists.

How can humans have the rapid population growth curve of a weedy, explosive species but the life history of a slower-growing, equilibrial species (Chapter 4)? The answer lies in the fact that in some ways humans are critically different from other organisms. Even though we humans are a part of ecosystems, some human characteristics make our role in ecosystems different from that of other organisms.

Remember that a *population* is an interbreeding group of a particular species in the same area. We do not typically talk about a widespread species as having a "global population," because local populations rise and decline according to local conditions. Drought in one area might be offset by good conditions in another. In contrast, we do talk about the "human population" and mean the whole global population. Prior to the modern era, human populations acted separately—disease, plenty, and scarcity were local effects, and populations rose and fell with surrounding conditions. Several social and cultural changes have altered the way human populations interact; these changes have also altered the ecology of the human population. In the modern age, humans act both as local populations, with population characteristics such as family size and life expectancy, and as part of a global whole. We humans are unique in our ability to regulate our reproduction, use fire, store food for long times, and adapt our environment with technology so we can live in more places. Consequently, population ecology can be applied to humans, but not in exactly the same way it is applied to other organisms.

Table 8–1	Demographic Terms
Term	**Definition**
Growth rate (annual rate of increase or decrease)	The rate of growth of a population, as a percentage. Multiplied by the existing population, this rate gives the net yearly increase (or decrease, if negative) for the population.
Total fertility rate	The average number of children each woman has over her lifetime, expressed as a rate based on fertility occurring during a particular year.
Replacement-level fertility	A fertility rate that will only replace a woman and her partner; theoretically 2.0, but actually slightly higher because of mortality and some individuals' failure to reproduce.
Infant mortality	The number of infant deaths per thousand live births.
Population profile (age structure)	A bar graph plotting numbers of males and females for successive ages in the population, starting with the youngest at the bottom.
Population momentum	The effect of current age structure on future population growth. Young populations will continue growing even after replacement-level fertility has been reached, due to reproduction by already-existing age groups.
Crude birthrate	The number of live births per thousand in a population in a given year.
Crude death rate	The number of deaths per thousand in a population in a given year.
Epidemiologic transition	The shift from high death rates to low death rates in a population as a result of modern medical and sanitary developments.
Fertility transition	The decline of birthrates from high levels to low levels in a population.
Demographic transition	The tendency of a population to shift from high birth and death rates to low birth and death rates as a result of the epidemiologic and fertility transitions. The result is a population that grows very slowly, if at all.

Revolutions

The changes that have allowed humans to become such a dominant part of the landscape are so extreme that they can be considered revolutionary. Here we will look at some of the revolutions that have brought humans from a number of isolated populations—ruled primarily by local conditions—to the global population of today. These include the development of agriculture, the Industrial Revolution, the development of modern medicine, the green agricultural revolution of the 1970s, and the current environmental revolution. In each case, we will see how new ways of doing things altered the effect of high populations.

Neolithic Revolution. Natural ecosystems have existed on Earth for hundreds of millions of years, while humans are relative newcomers to the scene. Anatomically modern humans most likely arose in Africa around 200,000 years ago. The earliest fully modern humans in Europe appeared some 40,000 years ago, during the end of a period known as the Upper *Paleolithic* (around 50,000 to 10,000 years ago). Archaeological evidence suggests that Paleolithic humans lived in small tribes of hunter-gatherers, catching wildlife and collecting seeds, nuts, berries, roots, and other plant foods (Figure 8–2). Settlements were small and groups moved on as resources were used up. Populations could not expand beyond the sizes that natural food sources supported, and deaths from predators, disease, and famine were common.

About 12,000 years ago, humans in the Middle East began to develop animal husbandry and agriculture. The development of agriculture provided a more abundant and reliable food supply, but it was a turning point in human history for other reasons as well. Because of its profound effect, it is referred to as the **Neolithic Revolution.** Agriculture allowed long-term or permanent settlements and the specialization of labor. There was more incentive and potential for technological development, such as better tools and better means of transporting water and other materials. Trade with other settlements began, and commerce was born. In addition, living in settlements allowed a greater storage of food and the development of preservation techniques, so food stored from one year was available to help people in years with less food production, evening out the risk from year to year. The greater stability of food supplies supported population growth, which in turn supported expanding agriculture. In short, modern civilization and population growth had their origins in the invention of agriculture about 12,000 years ago.

Industrial Revolution. For about another 11,000 years, the human population increased and spread out over the planet. Agriculture and natural ecosystems supported the growth of civilizations and cultures, which increased their knowledge and mastery over the natural world. With the birth of modern science and technology in the 17th and 18th centuries, the human population—by 1800 almost a billion strong—was on the threshold of another revolution: the **Industrial Revolution** (Figure 8–3). The Industrial Revolution and its technological marvels were energized by fossil fuels—first coal, then oil and gas. These fuels allowed people to use energy from the Sun that was trapped by plants a long time ago. This energy allowed humans to do work they could not have done if they had relied on human power—or even on the power of domesticated animals.

The harnessing of this extra energy allowed people to produce more food and numerous other products, but this

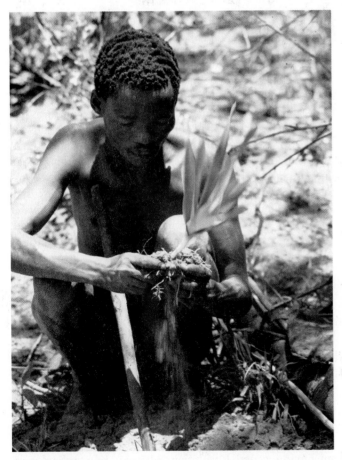

Figure 8–2 Foraging for food. Before the advent of agriculture, all human societies had to forage for their food, as this resident of the Sahel region in Africa is doing; agriculture allowed people to settle in cities and to divide labor.

Figure 8–3 Industrial Revolution. During the Industrial Revolution, coal and other fossil fuels were used to power machinery; economic growth and pollution were the consequences. This art of the 1890s shows a farmer using a steam-powered plow to prepare a field in the Dakotas.

increased production came with some unexpected costs. The burning of fossil fuels is similar to the process of cellular respiration: It breaks chemical bonds to release stored energy and releases water and carbon dioxide as waste. Fossil fuels contain other chemicals as well—such as sulfur and nitrogen compounds, mercury, lead, and arsenic—all of which can contribute to the pollution of land, air, and water.

Pollution and exploitation took on new dimensions as the industrial powers turned to the extraction of raw materials from all over the world. In time, every part of Earth was affected by this revolution. As a result, we now live in a time of population growth and economic expansion accompanied by environmental problems (Chapter 1).

Medical Revolution. One of the main reasons for the slow and fluctuating population growth prior to the early 1800s was the prevalence of infectious diseases that were often fatal, such as smallpox, diphtheria, measles, and scarlet fever. These diseases hit infants and children particularly hard, so it was not uncommon for a woman who had seven or eight live births to have only two or three children reach adulthood. In addition, epidemics of diseases such as typhus, cholera, and the black plague of the 14th century eliminated large numbers of adults. Prior to the 1800s, therefore, the human population was essentially in a dynamic balance with natural enemies—mainly diseases—and other aspects of environmental resistance. High reproductive rates were largely balanced by high mortality, especially among infants and children, producing a low rate of population growth.

In the late 1800s, scientists such as Louis Pasteur and others demonstrated that many diseases were caused by infectious agents (now identified as various bacteria, viruses, and parasites) and that these pathogens were often transmitted via water, food, insects, and rodents. Soon, vaccinations against different diseases were developed, and whole populations were immunized against such scourges as smallpox, diphtheria, and typhoid fever (Figure 8–4). At the same time, cities and towns began treating their sewage and drinking water. Later, in the 1930s, the discovery of penicillin (the first in a long line of antibiotics) resulted in cures for often-fatal diseases such as pneumonia and blood poisoning. Improvements in nutrition began to be significant as well. In short, better sanitation, medicine, and nutrition brought about spectacular reductions in mortality—especially among infants and children—while birthrates remained high. From a biological point of view, the human population began growing almost exponentially, as does any natural population once it is freed from natural enemies and other environmental restraints.

The Green Revolution. After the revolutions in industry and medicine, human populations increased dramatically. Concerns about whether we would be able to feed the rapidly burgeoning global population led to intense efforts to find new ways to increase agricultural efficiency. The development of chemical pesticides in World War II, along with an increase in irrigation and fertilizer use, dramatically increased crop yields. These trends were enhanced by the development of high-yield crops, which yielded more per acre than other

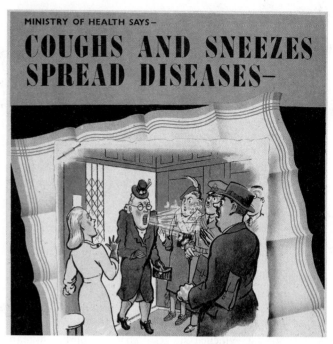

Figure 8–4 Preventing the spread of infectious disease. This cartoon is from a British Ministry of Health public health campaign emphasizes the importance of preventing the spread of germs. Improved hygiene and other aspects of modern medicine prolonged lives and caused populations to increase.
(*Source:* The National Archives/SSPL/Getty Images.)

crops. These advances in agriculture increased crop yields dramatically and allowed many countries to provide food for their rapidly growing populations (Figure 8–5).

Unfortunately, the industrialized agriculture that allowed for such high yields came at significant costs, much as the Industrial Revolution had. Industrialized agriculture allowed people to use resources, particularly soil and underground water, more rapidly than they were replaced, leading to an increase in soil erosion, soil and water pollution, and the loss of native plant varieties. (These problems will be discussed in some detail when we talk about soil in Chapter 11, water in Chapter 10, and food production in Chapter 12.) In addition, the use of pesticides initially saved many human lives and protected crops. In the long term, however, the overuse of pesticides resulted in **pesticide resistance,** the ineffectiveness of a pesticide when the target organisms are no longer affected by it. A current worldwide interest in sustainable agriculture is an attempt to rectify some of these problems. However, as we will see in Section 8.3, population growth is pressuring us to increase, rather than decrease, high-yield agriculture.

The Newest Revolution. In the past, humans experienced revolutions with new technological breakthroughs. Now, we have the Internet, computers, nanotechnology, robotics, and solar and other technologies. Many of these can be put to use in what some hope will be a real "green" revolution—an **Environmental Revolution.** More-efficient technologies, better urban and regional planning, policy and industrial changes, and the personal decisions of billions of people will be required to drive this revolution.

Figure 8–5 Green Revolution. Farmers in Punjab, India, harvest wheat. Fertilizer, irrigation, pesticides, and new varieties of crops are all part of industrialized agriculture, which feeds growing populations but also leads to the pollution of soil and water.

Humans are part of natural ecosystems, but they are different enough from other organisms to make their population ecology global and to allow them to change their world extensively. They have characteristics of both r- and K-strategists. We see the J-shaped curve for human population growth because, at different times in history, we have increased the number of people the world can support by changing the way we do things. Our past growth might seem to suggest that natural laws do not apply to humans. In the next section, we will consider whether the concepts of carrying capacity and limits apply to human populations.

Do Humans Have a Carrying Capacity?

It is challenging to determine a carrying capacity for humans, a fact that makes it difficult to predict the impact of the growing human population on the environment and human well-being. If people were close to starving many years ago and yet today there are many more people, people live longer, and proportionately fewer are extremely poor, how can we talk about humans reaching a carrying capacity? In fact, some people claim that the whole idea of a carrying capacity does not apply to humans. Such people point out that every time in the past that it was thought humans had hit a carrying capacity, a problem has been solved to overcome it. The economist Julian Simon, for example, has argued that humans will overcome Earth's limits by colonizing space.[1]

Most ecologists and demographers, however, would disagree. In their view, an escape into space is unlikely to solve the world's problems for the vast majority of humans, and humans are subject to limits and natural laws of population growth. How then do we reconcile natural laws with the pattern of human population growth in the past 2,000 years? Throughout history, humans have improved their survival rate, increased their populations, and increased their life span by doing some things other organisms cannot do (or cannot do on the scale we can). These improvements have essentially increased the carrying capacity for our species. For example, we were once limited by what we could capture or hunt in one season, but the development of agriculture and trade removed that limit. We were limited by our energy requirements, but the use of fossil fuels gave us a new resource. We were limited by child mortality, but medical breakthroughs lowered child mortality and populations exploded.

This pattern of population growth is exactly what you would expect to see under the law of limiting factors—each time a limit is removed, the population increases until there is another limiting factor. Simon represents a group that believes that humans will always be able to overcome limiting factors. However, in the past, increases have come at the expense of using up nonrenewable resource, such as fossil fuels. Increases in population have also come at the expense of creating another set of limits—this time, pollutants. The limiting factors that humans face in the 21st century include pollution from our own activities, a limit on agricultural land, a limit on ocean fish, and trade-offs between using land and water for human activities or maintaining ecosystems that provide us with vital services such as photosynthesis and flood control. In addition, Simon's argument assumes that the natural world has no value except as resources for humans. That is, the rest of nature has no *intrinsic* value (Chapter 6).

Because humans seem able to increase their carrying capacity in the short term—while potentially making trade-offs that will be paid for in the long term—determining a carrying capacity for the global human population is difficult. Demographer Joel Cohen explained the reason in his book *How Many People Can the Earth Support?*[2] Cohen said that there isn't a clear-cut answer to the question of human carrying capacity because part of the answer depends on what living standard we are willing to accept. For example, if we never ate

[1]Julian Simon, *The Ultimate Resource* (Princeton, NJ: Princeton University Press, 1983).

[2]Joel Cohen, *How Many People Can the Earth Support?* (New York: Norton, 1995).

meat, did not rely on wild organisms, and were willing to never travel, heat our homes, or use paper, Earth could support more people. On the other hand, if everyone on Earth wanted a 21st-century American way of life, we would need multiple Earths to support the global population we already have. Because so much of the answer depends on standard of living expectations, Cohen argued, it is very difficult to determine an actual carrying capacity. We can't figure out a carrying capacity for humans in the same way we do for fish, ducks, or flies, because we have (and want) more choices.

While it is difficult to reach agreement on a concrete number for the human carrying capacity, the world's scientific societies believe that there is some limit to human population growth. In 1993, representatives of the National Academies of Sciences from 58 countries met in New Delhi, India, to agree on and sign a statement about world population. They said, among other things, that "the earth is finite and that natural systems are being pushed ever closer to their limits."[3]

Scientists also believe that it is important to try to estimate what the limits of carrying capacity might be from an ecosystem perspective. In 2004, two professors in the Department of Spatial Economics at the Free University in Amsterdam, the Netherlands, reported on their meta-analysis (an analysis of multiple other analyses) of 69 papers on world population and carrying capacity, looking to see if there were any central trends. They identified a number of factors that could limit human populations, including "water availability, energy, carbon, forest products, nonrenewable resources, heat removal, photosynthetic capacity, and the availability of land for food production."[4] They concluded that the best estimates for global human carrying capacity centered on 7.7 billion people. Such analyses are difficult to do. While regional carrying capacities have been calculated, a similar global calculation has not been made recently. However, other types of analyses have reached similar conclusions.

Planetary Boundaries. In 2009, a team of 28 leading academics headed by Johan Rockström, executive director of the Stockholm Environment Institute, proposed a way of thinking about what parameters would need to be in place for humans to survive what they called "planetary boundaries" for a "safe operating space for humanity." The scientists claimed that there are a series of limits that act as tipping points; passing these limits would keep humans from surviving (Chapter 5). Within them is a "safe space."[5] Both population growth and overconsumption could push us past our boundaries. Estimates suggest that of nine boundaries they outlined, three (climate change,

biodiversity loss, and flow of nutrient cycles) may have already been crossed. To survive, we will have to find ways to meet human needs without pushing those boundaries, including the remaining six—ocean acidification, stratospheric ozone depletion, global freshwater use, land-system change, aerosol loading, and chemical pollution. (For more discussion, see Stewardship, *Lessening Your Ecological Footprint.*)

Picking up the Population Pace. From the dawn of history until the beginning of the 1800s, the human population increased slowly and variably, with periodic setbacks. It was roughly 1830 before world population reached the 1 billion mark. By 1930, however, the population had doubled to 2 billion. It reached 3 billion in 1960, and by 1975 it had climbed to 4 billion. Thus, the population doubled in just 45 years, from 1930 to 1975. Then in 1987, just 12 years later, it crossed the 5 billion mark. In 1999, the world population passed 6 billion, and it hit 7 billion in late 2011. On the basis of current trends (which assume a continued decline in fertility rates), the UN Population Division (UNPD) medium projection predicts that the world population will pass 8 billion in 2025, 9 billion in 2040, 9.55 billion around 2050, and 10 billion around 2065 (Figure 8–1). Future growth and projections will be described in Section 8.4.

One important result of the 2004 meta-analysis was that the best predictions of how large our human populations will get by 2050 are greater than the best estimates of our carrying capacity. Stabilizing population growth is part of establishing a sustainable society.

CONCEPT CHECK Explain how humans seem like both *r*-strategists and *K*-strategists. ✓

8.2 Population and Consumption—Different Worlds

To understand what is likely to happen in the next decades, we need to understand the lives of people like Anje, Atin, and Awa. Although globalization is part of the modern world, people in wealthy, middle-income, and poor countries live in radically different economic and demographic conditions. In fact, even within a single country there can be tremendous disparities in wealth and in quality of life.

Rich Nations, Middle-Income Nations, Poor Nations

The World Bank, an arm of the United Nations, divides the countries of the world into three main economic categories, according to average per capita gross national income (Figure 8–6):

1. **High-income, highly developed, industrialized countries.** This group (population 1.05 billion in 2014) includes the United States, Canada, Australia, New Zealand, the

[3]Population Summit of the World's Scientific Academies, 1993. National Academy of Sciences, National Academy of Engineering, Institute of Medicine (SEM), 7. Available at http://books.nap.edu/openbook.php?record_id=9148&page=7.

[4] J. Van Den Bergh and P. Rietveld, "Reconsidering the Limits to World Population: Meta-Analysis and Meta-Prediction," *BioScience* 54, no. 3 (2004): 195.

[5]J. Rockström et al., "Planetary Boundaries: Exploring the Safe Operating Space for Humanity," *Ecology and Society* 14, no. 2 (2009): 32, http://www.ecologyandsociety.org/vol14/iss2/art32/.

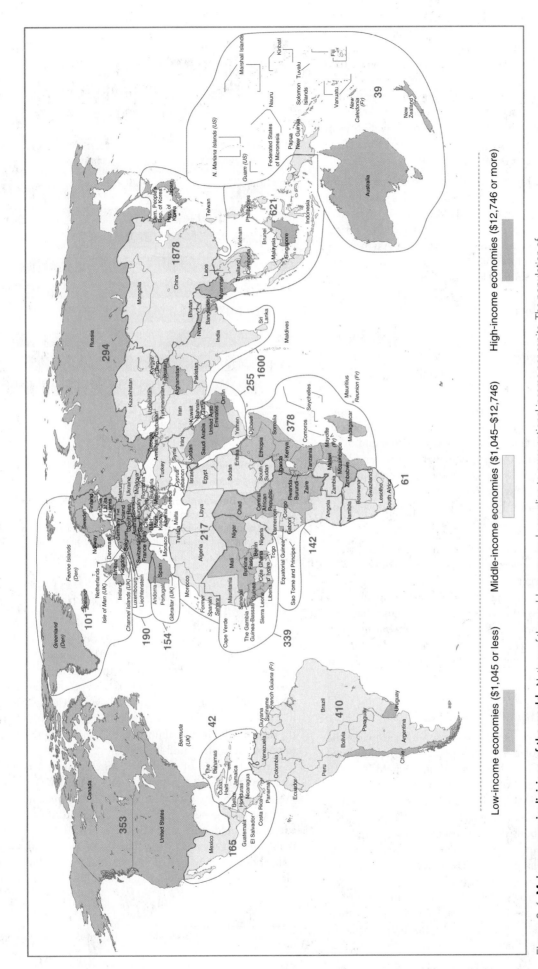

Figure 8–6 Major economic divisions of the world. Nations of the world are grouped according to gross national income per capita. The population of various regions (in millions) is also shown by magenta lines and numbers.

Low-income economies ($1,045 or less)

Middle-income economies ($1,045–$12,746)

High-income economies ($12,746 or more)

(*Sources:* World Bank Country and Lending Groups, accessed February 4, 2015, http://data.worldbank.org/about/country-and-lending-groups#; *World Population Data Sheet* (Washington, DC: Population Reference Bureau, 2014).)

countries of western Europe, Japan, Republic of Korea, Singapore, Taiwan, Israel, and several Arab states (2014 gross national income per capita $12,746 and above; average of $43,864). Sometimes the high-income nations are further divided into those belonging to the Organisation for Economic Co-operation and Development (OECD), an economic collaboration between primarily European countries, and non-OECD countries. Sweden, where Anje lives, is a high-income country in the OECD.

2. **Middle-income, moderately developed countries.** This group (population 4.97 billion) includes mainly the countries of Latin America (Mexico, Central America, and South America), northern and southern Africa, China, Indonesia and other southeastern Asian countries, many Arab states, eastern Europe, and countries of the former U.S.S.R. The average per capita income for this group in 2014 was $4,721. This category is further divided by the World Bank into lower middle-income and upper middle-income countries (2014 gross national income per capita: $1,045–$4,125 for the former category, $4,125–$12,746 for the latter). Indonesia, home to Atin and her family, is in the lower middle-income group.

3. **Low-income, developing countries.** This group (population 0.85 billion in 2014) is composed of countries in eastern, western, and central Africa, India and other countries of southern Asia, and a few former Soviet republics (2014 gross national income per capita: less than $1,045, average of $664). Burkina Faso, where Awa and her family live, is a low-income developing country.

The high-income nations are commonly referred to as **developed countries,** whereas the middle- and low-income countries (a much larger number) are often grouped together and referred to as **developing countries.** Of course, all countries are in the process of developing in some sense, but here the term refers to economic changes. The terms *more developed countries, less developed countries,* and *third-world countries* are being phased out, although you may still hear them used.

Moving Up: Good News

Some of the news about population is good. Worldwide, the low-income percent of the world's population has declined dramatically since 1998, falling from about 60% to less than 12%. This is a reduction of about 80% in only a few years. Until recently, living in a low-income country was the norm for humans, but now it is not. This change is part of the reason people talk about the "environmentalist's paradox": Some measures of human development are increasing at the same time that serious problems are arising in the environment (Chapter 1).

One of the improved measures is the movement of many countries from the lowest income status even as their populations are rapidly rising. The World Bank explains, "Developing countries have increased their share of the

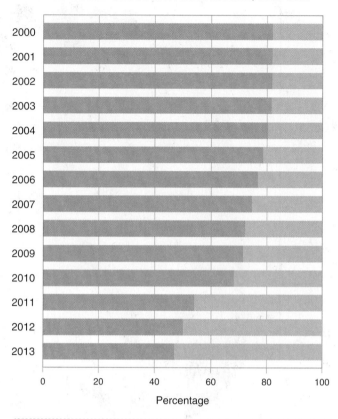

SHARE OF WORLD GROSS NATIONAL INCOME, 2000–2013

Figure 8–7 Between-country inequality. Low- and middle-income countries are gaining some in the share of world gross national income, but their populations are increasing more rapidly than those of high-income countries. Inequality between countries remains stark.

(*Source:* World Bank, 2014, http://data.worldbank.org/news/2010-GNI-income-classifications.)

- ■ High-income countries
- ■ Low- & middle-income countries

global economy by growing faster than rich countries, on average 6.8% per year compared to only 1.8% for high-income economies over the 2000 to 2010 period."[6] The movement of millions of people from the extremely low-income to the middle-income category is very good news, and an increasing percentage of the world's wealth is held by those in low- and middle-income countries (Figure 8–7). Even so, the disparity in the distribution of wealth among the countries of the world is still mind-boggling.

The disparity of wealth is difficult to understand just by looking at general income figures. The UN Development Program (UNDP) uses the *Human Development Index* as a measure of general well-being, based on life expectancy, education, and per capita income. The UNDP also uses the *Human Poverty Index,* based on additional information about literacy and living standards, to make a more direct measurement of poverty in both low- and high-income

[6]"Changes in Country Classifications," 2011, http://data.worldbank.org/news/2010-GNI-income-classifications.

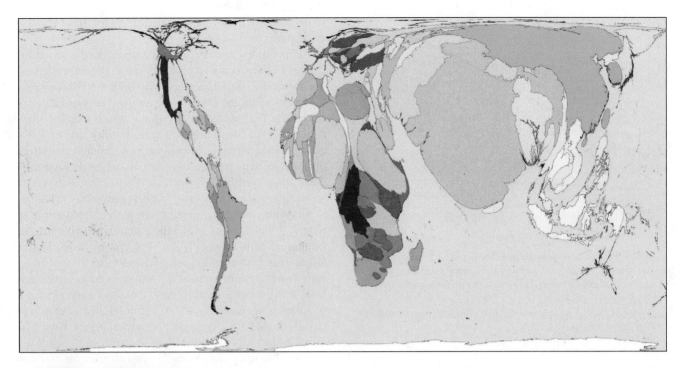

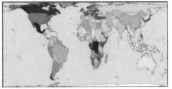

Land area

Technical notes
• Data are from the United Nations Development Programme's 2004 Human Development Report.
• The United Nations Development Programme uses one Human Poverty Index for poorer territories and another for richer territories. The scores of the latter can be divided by 10 so the indices are comparable.

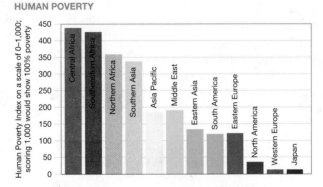

HUMAN POVERTY

Figure 8–8 Another way to picture poverty. The territory size shows the number of poor people in the country. On the human poverty index, Burkina Faso ranks highest and Sweden ranks lowest.

(*Source:* SASI Group (University of Sheffield) and Mark Newman (University of Michigan).)

countries (Figure 8–8). About 12% of the people in developed countries are unable to afford adequate food, shelter, or clothing, although these rates vary a good deal, between 6% and 21% in one study.[7] In contrast, about 50% of those in developing countries are unable to afford these necessities.[8] Anje, in Sweden, lives in a country with one of the lowest human poverty indexes, while Awa, in Burkina Faso, lives in a country with one of the highest. These data, available yearly in the UNDP Human Development Report, have been used to focus attention on the most deprived people in a country, to help countries develop appropriate policies, and to mark progress toward sustainable development.

Population Growth in Rich and Poor Nations

One of the reasons we spend so much time looking at wealth distributions in a study of environmental science is because population growth is related to poverty, environmental degradation, and other development measures. If you compare population growth in developed and developing countries, you find a discrepancy that parallels the great difference in wealth between these groups of countries. According to the UNPD, high-income countries such as Anje's Sweden, with a combined population of 1.05 billion in 2014, are growing at a rate of 0.1% annually. These countries will add less than 1 million to the world's population in a year. The middle-income countries, such as Atin's Indonesia, and the low-income countries, such as Awa's Burkina Faso, together had a 2014 total population of 5.89 billion. The populations of the middle-income countries are increasing at a rate of 1.4% annually, while

[7]OECD, "Crisis Squeezes Income and Puts Pressure on Inequality and Poverty, 2013, http://www.oecd.org/els/soc/OECD2013-Inequality-and-Poverty-8p.pdf.

[8]Pew Research, "Chapter 2: Personal Economic Conditions," *Pew Research Global Attitudes Project,* May 13, 2013, accessed October 28, 2014, http://www.pewglobal.org/2013/05/23/chapter-2-personal-economic-conditions/.

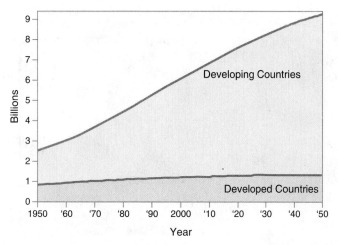

Figure 8–9 Population growth in different world regions. Because of larger populations and higher birthrates, developing countries represent an ever-growing share of the world's population. This trend is expected to continue.

(*Source*: UN Population Division, *World Population Prospects: The 2012 Revision* (New York: United Nations, 2010).)

those of the low-income countries are increasing at a rate of 2.4% annually, adding 81 million people per year. Consequently, more than 98% of world population growth is occurring in the developing (middle- and low-income) countries (Figure 8–9). In fact, population growth is highest in the least developed countries. As a result, the increase in wealth in low- and middle-income countries does not lead

to an increase in per capita wealth.[9] What lies behind this discrepancy?

Population growth occurs when births outnumber deaths. In the absence of high mortality, the major determining factor for population growth is births, conventionally measured using the **total fertility rate**—the average number of children each woman in a population has over her lifetime (Figure 8–10). Theoretically, a total fertility rate of 2.0 will give a stable population, because two children per woman, on average, will replace two parents when they eventually die. A total fertility rate greater than 2.0 will result in a growing population, because each generation is replaced by a larger one. Barring immigration, a total fertility rate less than 2.0 will lead to a declining population, because each generation will eventually be replaced by a smaller one. Given that infant and childhood mortality is not, in fact, zero and that some women do not reproduce, **replacement-level fertility**—the fertility rate that will just replace the population of parents—is 2.1 for developed countries and slightly higher for developing countries, which have higher infant and childhood mortality.

Total fertility rates have dropped all over the world, but most dramatically in the high-income countries. Total fertility rates in developed countries have declined over the past two centuries to the point where they now average 1.7, with

[9]Data Finder, 2014 World Population Data Sheet, http://www.prb.org/DataFinder/Topic/Rankings.aspx?ind=16&loc=243,245.

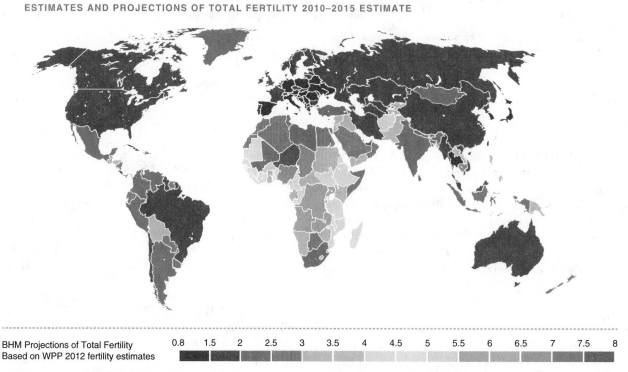

ESTIMATES AND PROJECTIONS OF TOTAL FERTILITY 2010–2015 ESTIMATE

BHM Projections of Total Fertility
Based on WPP 2012 fertility estimates

| 0.8 | 1.5 | 2 | 2.5 | 3 | 3.5 | 4 | 4.5 | 5 | 5.5 | 6 | 6.5 | 7 | 7.5 | 8 |

Figure 8–10 Total fertility rates around the world. Fertility rates around the world vary widely. The highest fertility rates occur in sub-Saharan Africa.

(*Source*: UN Population Division, *World Population Prospects: The 2012 Revision* (New York: United Nations, 2008).)

some as low as 1.1 (Table 8–2, Figure 8–10). The total fertility rate of the United States fell during the recent economic recession, but is higher than most developed countries (1.9 in 2014), and the U.S. population continues to increase because of immigration. When countries industrialize and have modern medicine, they typically undergo these changes, described as the **demographic transition**. This will be described in Section 8.4 when we discuss how demographers project future population trends.

In developing countries, fertility rates have come down considerably in recent years, to an average of 2.5, although some developing countries have rates as high as 7 or more (Table 8–2). Thus, the populations of developing countries will continue growing, while the populations of developed countries will stabilize or decline, although immigration can change that. As a consequence, the percentage of the world's population living in developing countries—already 85%—is expected to climb to more than 90% by 2075 (Figure 8–9). Nevertheless, it is not just the developing countries that have problems.

Different Populations, Different Problems

As we saw in the discussion of carrying capacity, both the number of people and their standards of living play a role in whether we achieve sustainability or experience environmental degradation. Some time ago, ecologist Paul Ehrlich and physicist John Holdren proposed a formula to account for the human factors that contribute to environmental deterioration and the depletion of resources. They reasoned that human pressure on the environment was the outcome of three factors: *population, affluence,* and *technology.* They offered the following formula:

$$I = P \times A \times T$$

According to this equation, called the **IPAT formula**, environmental impact (I) is equal to the population (P), multiplied by affluence and consumption patterns (A), multiplied by the level of technology of the society (T). The IPAT formula fits well with the concept of carrying capacity. Given the high level of technology in the industrialized countries and the affluent lifestyle that accompanies it, a fairly small population can have a very large impact on the environment.

Recently there have been attempts to improve the IPAT formula. The technology part of the equation has always been difficult to quantify. Some technologies improve environmental conditions and others harm it. For example, gasoline-powered automobiles pollute the atmosphere, while

Table 8–2 Population Data for Selected Countries

Country	Total Fertility Rate 2013	Population 2014	Population 2050-projected
World	2.5	7,238,183,566	9,683,476,862
Least Developed	4.3	915,817,323	1,854,951,000
Burkina Faso	5.9	17,886,000	46,574,000
Afghanistan	5.1	31,281,000	56,511,000
Iraq	4.1	35,111,000	80,547,000
Egypt	3.5	87,900,000	145,968,000
Haiti	3.4	10,753,000	16,822,000
Indonesia	2.6	251,452,000	365,323,000
India	2.4	1,296,245,000	1,656,919,000
China	1.6	1,364,072,000	1,311,782,000
More Developed	1.6	1,248,958,069	1,308,512,404
United States	1.9	317,731,000	395,284,000
Sweden	1.9	9,689,000	11,444,000
Canada	1.6	35,549,000	48,384,000
Japan	1.4	127,060,000	97,076,000
Italy	1.4	61,339,000	63,465,000
Taiwan	1.1	23.5 million	21.0 million

Source: Data from *2014 World Population Data Sheet* (Washington, DC: Population Reference Bureau, 2014).

catalytic converters reduce the amount of air pollution. Many technologies do both. For example, manufacturing solar panels produces pollution, but the renewable energy they capture replaces the use of fossil fuels. Some economists have rewritten the IPAT formula to separate the effects of technology (T) into two components: consumption per unit of GDP (How much do you use as your economy increases?) and impact per unit of consumption (How inefficient or how negative is what you are doing?). They call this formula ImPACT.[10]

Effect of Wealth. Sometimes people think that wealth will solve environmental problems—that is, the rich can afford the luxury of caring for the environment. Wealthy countries, the thinking goes, can more easily afford the technologies to lower air pollution, clean water, and take care of sewage, for example (Chapter 2). Give the poor quick ways to improve their economies, and eventually they will also take care of the environment.

It is true that the passage of many environmental laws has improved the environment in the United States and western Europe. It is also true that wealth is required to build technologies to deal with the needs of large groups of people with minimal environmental degradation. But the relationship between economic wealth and environmental health is not clear-cut for two reasons. First, although wealth can help solve problems such as infectious diseases in water, other problems, such as municipal waste, increase with wealth. Second, wealth allows people to care for the environment in their area, but doing so often pushes environmental problems to other, poorer places. The extraction of resources and the manufacture of goods in other countries for use in high-income countries have impacts in many other parts of the world.

Ecological Footprint. Some people argue that because the places with the highest-density populations—such as the Netherlands, Singapore, Hong Kong, and Japan—have healthy, long-lived, wealthy human populations, overpopulation is not real. If they can be so crowded, why can't the whole world be like that? But densely populated, wealthy areas such as Singapore use resources and affect ecosystems over a much wider area. People living in densely populated areas have impacts in the far reaches of the land and seas. These effects can be illustrated with an ecological or environmental footprint. An **environmental footprint** is an estimate of the amount of land and ocean an individual or population requires to provide the resources it uses and to absorb the wastes it produces. The environmental footprint of Hong Kong, for example, is many times greater than the size of the metropolitan region (see Stewardship, *Lessening Your Ecological Footprint.*)

Inequalities Within Countries. The developing countries, especially the low-income countries, do have a population problem, and it is making their progress toward sustainable development that much more difficult. Their needs are great: economic growth, more employment, wise leaders, effective public policies, fair treatment by other nations, and, especially, technological and financial help from the wealthy nations. Meeting these needs is a matter of justice, one of the key components of stewardship.

Even this description is too simplistic because within each nation there is a tremendous difference in wealth and poverty levels and individual experiences. India, for example, has hundreds of millions of extremely wealthy people, while hundreds of millions more live in extreme poverty. Some middle-income countries have a rising middle class with consumption patterns similar to those in the United States, Japan, and Europe. This group is called "the consumer class." China and India have more than 20% of the world's total population with this increased standard of living, even though there are many extremely poor people in each of these two countries. In China alone, there are more than 300 million people in the rising middle- and upper-class groups. Africa, where several of the poorest countries are located, has a rising middle class of 123 million, which is expected to grow rapidly.

The World Bank measures **inequality** in each country with the **Gini index,** a way of quantifying differences in wealth within countries. (Countries often calculate the Gini values during a census, so the values are not all from the same year, but they do not change rapidly, so comparison is still valuable.) Sweden has one of the lowest values (26 in 2008) and South Africa one of the highest (65 in 2011). Burkina Faso (39 in 2009) and Indonesia (36 in 2010) each have a Gini index in the middle; these countries have few ultra-rich citizens.

The United States (Gini index 42), an extremely wealthy country, also has large disparities between groups in terms of health care and access to services. Because of these disparities, it ranks lowest among the developed countries in life expectancy, in spite of spending more per person on health care than any other country. The United States has more children in poverty and more people in prison than any other high-income country. Income inequality can cause social unrest. Between 1968 and 2010, the percent of wealth owned by the top 1% increased from 28% to 37%.[11] This change contributed to the countrywide protests from 2011 onward called the "Occupy Wall Street" movement.

We have seen that simple statements about the life experiences of high-, middle-, and low-income countries may not apply to every person. The disparities between rich and poor within a country also matter. Differences between countries are not the only differences that justice requires us to address.

[10]Richard York, Eugene A. Rosa, and Thomas Dietz, "STIRPAT, IPAT, and ImPACT: Analytic Tools for Unpacking the Driving Forces of Environmental Impacts," *Ecological Economics* 46 (2003): 351–365.

[11]Philip Vermeulen, "Working Paper Series: How Fat Is the Top Tail of the Wealth Distribution?" (July 2014): 25, accessed August 10, 2014, https://www.ecb .europa.eu/pub/pdf/scpwps/ecbwp1692.pdf.

Enter Stewardship. Despite having fairly stable populations, the developed countries have their own problems with consumption, affluence, damaging technologies, and burgeoning wastes. These issues must be addressed to achieve sustainability, and this, too, requires wise leaders and effective public policies. It also requires willpower, both individually and as a society, and an ethical basis for caring for others and for the rest of the natural world.

Fortunately, the environmental impacts of affluent lifestyles may be moderated somewhat by practicing environmental stewardship. For example, suitable attention to wildlife conservation, pollution control, energy conservation and efficiency, and recycling can lower the negative impact of a consumer lifestyle. Such changes alter the technology part of the ImPACT equation, lessening the effect of our use of resources.

Some people argue that population growth is the main problem, others claim that our consumption-oriented lifestyle is chiefly to blame, and still others maintain that our inattention to stewardship is the prime shortcoming. In order to make the transition to a sustainable future, however, all three areas (and more) must be addressed. That is, the population must stabilize (a demographic transition), consumption must decrease (a resource and technology transition), and stewardship action must increase (a political and sociological transition).

CONCEPT CHECK What is the difference between the IPAT and ImPACT summaries of environmental effect? If you install a solar panel to offset some of your use of energy, how would these indexes change? (Assume you live in an affluent society.) ☑

8.3 Consequences of Population Growth and Affluence

Expanding populations and increasing affluence—it sounds like trouble for the environment, and it is. It also means trouble for people in high-, middle-, and low-income countries. In countries experiencing rapid population growth, people overuse land, and many flee to cities in the hopes that there will be new opportunities there. These hopes are not necessarily fulfilled. In high-income countries, increases in industrialization have resulted in huge amounts of wealth, but have also caused problems, including the transfer of environmental degradation to places where those most responsible for generating it do not have to experience its price.

Consequences of Rapid Growth

Prior to the Industrial Revolution, many humans survived through subsistence agriculture. Families lived on the land, raised livestock, and produced enough crops for their own consumption and perhaps some extra to barter for other essentials. Forests provided firewood and structural materials for housing. With small, isolated, and relatively stable populations, this system was generally sustainable. As the older generation passed away, the land and natural systems could still support the next generation. Indeed, many cultures sustained themselves in this way over thousands of years.

After World War II, modern medicines—chiefly vaccines and antibiotics—were introduced into developing nations, whereupon death rates plummeted and populations grew rapidly. Today 75% of the world's poor in developing

STEWARDSHIP
Lessening Your Ecological Footprint

Some estimates conclude that the average American places at least 20 times the demand on Earth's resources—including its ability to absorb pollutants—as does the average person in Bangladesh, a poor Asian country. Major world pollution problems—such as the depletion of the ozone layer, the impacts of global climate change, and the accumulation of toxic wastes in the environment—are largely the consequence of the high consumption associated with affluent lifestyles in the developed countries.

How might Atin, Awa, and Anje affect the environment? Anje in Sweden probably uses fewer resources and has a smaller footprint than her cousin in the United States, because the EU has more regulations on products and because Europe is densely populated and distances between cities are much smaller.

Public transportation, biking, smaller vehicles, and other lower-impact services are the norm. However, Europeans still have very high impacts on the world in comparison to citizens of low-income countries. Awa in Burkina Faso likely has a lower footprint from consumer impacts, but is likely to have more children and might cause more water pollution because sanitary facilities are not available. Overall, her use of resources will be much lower than that of Anje. Atin in Indonesia is likely to have consumption levels that are greater than those of Awa and lower than those of Anje.

If you search the Internet for "ecological footprint," you can calculate your own, one of the first steps in figuring out how to be more sustainable. A footprint calculator is likely to ask you to answer questions about your

housing, use of energy, purchasing habits, recycling, eating habits, and other behaviors. Then it will use this information to calculate how many Earths we would need if everyone lived at your standard of living. The calculator will suggest steps to reduce your footprint based on your habits. You can also calculate specific types of footprints such as nitrogen, carbon, and water. If Anje calculated her footprint, it would be less than a college student in the United States because she would be less likely to have a car and more likely to recycle and to take public transportation. Awa would require fewer resources because she uses much less energy, does not travel much, and grows her own food. Figuring out our contribution to the overuse of resources is the first step to change!

Figure 8–11 Small-scale agriculture. Seventy percent of the world's poor live in rural areas and are heavily dependent on small-scale agriculture. With just 2.5 acres, this Kenyan man farms the way his grandparents did, but is struggling to do so.

countries live in rural areas. Most are engaged in small-scale agriculture, but only 3.8% of official development assistance goes to agriculture. Rapid growth has negative consequences on a population that is largely engaged in farming a few acres or less of land (Figure 8–11). When farms become too small to support the next generation, only a few basic options are possible, all of which are being played out to various degrees by people in these societies. They can reform the system of land ownership, intensify the cultivation of existing land to increase the production per unit area, and open up new land to farm. They may also engage in illicit activities for income, move to other countries (either legally or illegally), or move to the cities and seek employment. Each of these options has consequences.

Changing Land Ownership. Two patterns of agricultural land ownership that have historically kept rural peoples in poverty are collectivization (the gathering of farmers into group farms) and ownership by the wealthy few. When China abandoned collective agriculture in 1978 and assigned most agricultural land to small-scale farmers, farm output grew more than 6% a year for the next 15 years. Ownership by the wealthy few was a common result of colonialism in the 19th and 20th centuries. At one point, 87% of South Africa's land was available only to whites or was owned by the government, leaving a much smaller amount of poor-quality land for use by people of color. Rising rural populations in developing countries has increased the need to reform the systems of land ownership. However, resolving these kinds of inequities can be difficult.

Intensifying Cultivation. The introduction of highly productive varieties of food grains in the Green Revolution had a dramatic beneficial effect in supporting the growing population, but has also caused some problems (Chapter 12). For example, intensifying cultivation means working the land harder. Plots have been put into continuous production with no time off. The results have been a deterioration of the soil, decreased productivity (ironically), and erosion. Given the countertrends

of rapidly increasing population and the deterioration of land from over-cultivation, food production per capita in Africa is currently decreasing and may decrease in China in the near future. Agriculture in Burkina Faso, Awa's country, is limited by drought and poor soil. Agriculture is often intensified in order to increase cash crops such as cotton and sugar cane—changes that may or may not help local farmers.

Opening Up New Lands. Opening up new lands for agriculture may sound like a good idea, but there is really no such thing as "new land." Most of the land that is useful for agriculture is already in production. Opening up new land means converting natural ecosystems to agricultural production, resulting in the loss of the goods and services those ecosystems contribute. Even then, converted land is often not well suited for agriculture. Figure 8–12 depicts an aerial view of small agricultural plots around the world. These landscape views enable us to see how intensive agriculture already is and how much land has been converted to its use. The use of steep hillsides, marginal lands, or cut forests for planting is not a long-term solution.

Illegal Activities. Anyone who doesn't have a way to grow sufficient food must gain enough income to buy it. Sometimes desperate people get food by poaching wildlife. Others get money through illegal activities, such as raising drug-related crops (Chapter 6). In Indonesia, Atin's home, there is a great deal of poaching of wildlife, including sea turtles, parrots, and many other species.

Migration. When food, water, or other resources are in short supply, people often migrate to new areas. Hundreds

Figure 8–12 Conversion of land for agriculture. This is a set of six satellite and computer generated images of fields in Minnesota (upper left), Kansas (upper middle)—where center pivot irrigation is responsible for the round field patterns—and northwest Germany (upper right). Near Santa Cruz, Bolivia (lower left), the fields radiate from small communities. Outside of Bangkok, Thailand (lower middle), rice paddies are skinny rectangular fields. In the Cerrado in southern Brazil (lower right), inexpensive flat land has resulted in large field sizes. Scale: covers an area of 10.5 × 12 km.

(*Source:* NASA.)

of millions of people in developing nations migrate to cities within their own country in search of employment and a better life. This **urbanization** is one of the biggest trends in population today. In 2008, the world passed a landmark—more than half of the population lived in cities. This trend is expected to continue, so that by 2050 urban dwellers will account for 70% of the global population. In 2014, there were 31 metro-regions that qualified as "megacities" of 10 million or more—the top 10 of which are shown in Figure 8–13—but most urbanites live in cities of less than half a million. Most of the net growth of the next 50 years in the developing countries will be absorbed in urban areas. About one-third of that urban growth is expected to occur in China and India.

Many people in poorer countries believe they can improve their well-being by migrating to a wealthier country. Each year, 3.5 million people move to more developed regions. But immigration has its problems, too. Immigrants are likely to live in poor housing and be subject to environmental degradation. Prejudice against foreigners is common, and there is difficulty accommodating rapid change in cultures, especially in countries with strong ethnic and cultural homogeneity.

Much migration is not from a poor country to a wealthy one, but from one low- or middle-income country to another, especially in areas with civil strife. Many such migrants qualify as *refugees*, people who leave their country to escape war, persecution, or natural disasters. (Refugees who stay within their own countries are called *internally displaced persons*.) Refugees are often placed in temporary refugee camps, where diseases and hunger often take a terrible toll on human life. Sometimes landless poor people simply move into areas of forest, cutting the trees to make small farms.

Current U.S. immigration laws officially admit around 675,000 permanent immigrants per year, with certain exceptions for close family members. This rate of immigration makes the United States the highest immigrant-receiving country in the world. More than a million people receive a green card, or permanent residence status, each year.[12] Refugee immigrants can be accepted in addition to the number of officially accepted immigrants; the president and Congress decide the number of such exceptional cases. The United States also has a high number of illegal immigrants, something that is extremely sensitive politically.

Challenges to Governments. In many developing countries, population growth and migration to urban areas are outpacing economic growth and the provision of basic services. Pakistan's population, for example, increased almost fivefold between 1960 and 2010; those decades of 3.2% annual growth have kept the country from alleviating poverty.[13] It is difficult for Pakistan to provide for its growing numbers.

---

[12]C. Nwosu, J. Batalova, and G. Auclair, "Frequently Requested Statistics on Immigrants and Immigration in the United States," Migration Policy Institute, accessed October 28, 2014, http://www.migrationpolicy.org/article/frequently-requested-statistics-immigrants-and-immigration-united-states.

[13]Mary Kent and Carl Haub, "The President of Pakistan on the Need to Slow Population Growth in the Muslim World," *Population and Development Review* 31, no. 2 (2005): 399.

(a)

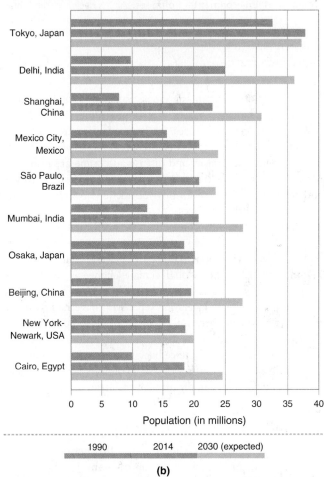

| | 1990 | 2014 | 2030 (expected) |

(b)

Figure 8–13 Growing cities. Since 1990, cities in the developing world have grown phenomenally, and a number of them are now among the world's largest. **(a)** Slums on the outskirts of São Paulo, Brazil. Thirty-two percent of the city's population lives in these blighted areas. **(b)** The top 10 world metropolitan areas in 2014.

(*Source:* Data for part (a) from UN Population Division, *World Urbanization Prospects: The 2012 Revision* (New York: United Nations, 2013).)

UNDERSTANDING THE DATA

1. According to the graph, which cities have the potential to be the largest in another 10 years? Which are the most stable?

2. Are any of the cities likely to shrink? Why or why not?

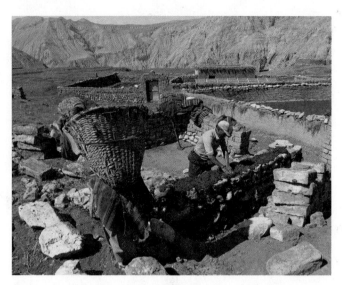

Figure 8–14 Environmental Refugees. After a decade of drought, these villagers from Dhe, Nepal, are moving their whole village to a place that is less dependent on snowmelt. Here they are using mud bricks to build pens for their animals.

The most rapidly expanding cities, especially in sub-Saharan Africa, have fallen so far behind in providing basic services that they are getting worse, not better. Streets are potholed, sanitation and drinking water are poor, electricity and telephone service are erratic, and disease and crime are rampant. Many are forced to live in sprawling, wretched shantytowns and slums that do not even provide adequate water and sewers, much less other services.

Although there are connections between rapid population growth in the developing world and environmental decline, the most pressing issue is the massive poverty that afflicts the low-income countries. Global climate change is making the people of these nations even more vulnerable. For example, Figure 8–14 shows villagers in Nepal who have had to move their entire village because of drought. The 2014 Human Development Report calls for collective action to increase *resilience,* the ability to bounce back from disasters, among the poor nations. A different set of issues connects population and environmental decline in the affluent countries.

Consequences of Affluence

The United States has the dubious distinction of leading the world in the consumption of many resources. We consume the largest per capita share of several major commodities including aluminum, coffee, copper, corn, lead, oil, oilseeds, natural gas, rubber, tin, and zinc.[14] We also have high per capita consumption of many other items, such as meat. The average American eats about twice the global average of meat, although per capita beef consumption in the United States has fallen by 36% since 1976 and

worldwide meat consumption is rising rapidly.[15] We nearly lead the world in paper consumption, using 504 pounds (229 kg) per person per year.[16] According to the Greendex, a consumer tracking index that National Geographic calculates for different countries, Americans are more likely than those in comparison countries to have energy-saving dishwashers, to have insulation in their homes, and to buy used items, but less likely to use public transportation, to buy local food, to eat home-cooked food, to eat fruits and vegetables, or to use their own reusable grocery bags when shopping.[17]

All of these factors and many more like them contribute to the unusually high environmental impact that those of us living in the United States make on the world. In addition, the United States continues to have a higher rate of population increase than most other developed countries; much of this growth is from immigration. Because of our high per capita resource use, population growth in the United States has a greater effect on global consumption and the environment than population growth in other places. During the 20th century, the United States tripled its population and increased its per capita use of raw materials 17 times.[18]

Despite the adverse effects of affluence, increasing the average wealth of a population can affect the environment positively. An affluent country such as ours provides amenities like safe drinking water, sanitary sewage systems and sewage treatment, and the collection and disposal of refuse. Thus, many forms of pollution are held in check, and the environment improves with increasing affluence. If we can afford gas and electricity, we are not destroying our parks and woodlands to obtain firewood. In short, we can afford conservation and management, better agricultural practices, and pollution control, thereby improving our environment.

Still, because the United States consumes so many resources, we also lead the world in the production of many pollutants. For example, by using such large quantities of fossil fuel (coal, oil, and natural gas) to heat and cool our homes, drive our cars, and generate electricity, the United States is responsible for a large share of the carbon dioxide being released into the atmosphere. Oil spills are a by-product of our appetite for energy. Similarly, emissions of chlorofluorocarbons (CFCs) that have degraded the ozone layer, emissions of chemicals that cause acid rain, and emissions of hazardous chemicals are all largely the by-products of affluent societies.

High individual consumption places enormous demands on the environment. The world's wealthiest 20% of

[14]P. Harrison, F. Pearce, and P. Raven, "Atlas of Population and the Environment Part 2" (AAAS 2001), accessed October 28, 2014, http://atlas.aaas.org/?part=2.

[15]R. Ferdman, "How Much Your Meat Addiction is Hurting the Planet," Wonblog, *The Washington Post,* (June 30, 2014), http://www.washingtonpost.com/blogs/wonkblog/wp/2014/06/30/how-much-your-meat-addiction-is-hurting-the-planet/.
[16]FAO, *FAO Yearbook of Forest Products 2012,* Rome: FAO. Pg 186.
[17]National Geographic, Greendex, accessed October 28, 2014, http://environment.nationalgeographic.com/environment/greendex/.
[18]U.S. Census Bureau, "Statistical Abstract of the United States" (April 2004); L. A. Wagner, *Materials in the Economy: Material Flows, Scarcity and the Environment* (2002): U.S. Geological Survey Circular 1221, http://pubs.usgs.gov/circ/2002/c1221/c1221-508.pdf.

people (some located in low- and middle-income countries) are responsible for 76% of all private consumption, while the poorest 20% consume only 1.5%. As a consequence of their high rate of consumption, most of the world's major fisheries are either fully exploited or overexploited and old-growth tropical forests are being clear-cut and turned into chips to make paper. Metals are mined, timber harvested, commodities grown, and oil extracted, all far from the industrialized countries where a disproportionate amount of these goods are used. Every one of these activities has a significant environmental impact. As increasing numbers of people strive for and achieve greater affluence, it seems more than likely that such pressures will mount.

Sustainable Consumption

One way of generalizing the effect of affluence is to say that it enables the wealthy to clean up their immediate environment by transferring their wastes to more distant locations. It allows them to obtain resources from distant locations, so they neither see nor feel the impacts of getting those resources. However, affluence also provides people with opportunities to exercise lifestyle choices that are consistent with the concerns for stewardship and sustainability. Tracking consumption patterns and trying to achieve sustainable consumption levels need to be part of our cultural responses to environmental degradation. One way governments could curb consumption would be by eliminating subsidies for consumer products and activities, such as purchasing suburban homes or subsidizing mining, forestry, or grazing (Chapters 7, 11). Some estimates of these subsidies are as high as $1 trillion globally each year.[19]

With this picture of population growth and its twin consumption and their impacts in view, the next section discusses future population trends, how demographers identify them, and the biggest population trend—the demographic transition.

CONCEPT CHECK Compare the environmental outcomes of adding 10 people to the populations of the United States, China, and Burkina Faso. ☑

8.4 Projecting Future Populations

During the 1960s, the world population growth rate peaked at 2.1% per year, after having risen steadily for decades. Following this peak, it began a steady decline (Figure 8–15). Twenty years later, the number of persons added per year peaked at 87 million. Why did the number of persons added to the population each year continue to increase after the rate of growth had begun to decline? The answer is that even though the population growth rate was in decline, there were increasing numbers of people who could reproduce. The total number of persons added every year did not go down

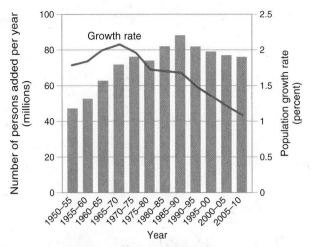

Figure 8–15 World population growth rate and absolute growth. Declining fertility rates in the past three decades have resulted in a decreasing rate of population growth, shown by the blue line. Even so, absolute numbers, represented by the orange bars, are still increasing by more than 70 million persons per year.

(*Source:* Data from Shiro Horiuchi, "World Population Growth Rate," *Population Today* (June 1993): 7, and (June/July 1996): 1; updated from UN Population Division, *World Population Prospects: The 2010 Revision* (New York: United Nations, 2010).)

until the growth rate fell low enough to offset the increase in reproducing people.

The declines in both the population growth rate and the number of persons added per year are primarily a consequence of the decline in the total fertility rate—the average number of babies born to a woman over her lifetime. In the 1960s, the total fertility rate was an average of 5.0 children per woman; it has since steadily declined to its present value of 2.6 children per woman. The age at which a woman first reproduces also affects how rapidly a population grows. A population in which women have children at a young age will grow more quickly than one in which women are slightly older when they reproduce, even if women have the same number of children. That is, there are more generations alive at the same time when women reproduce young. Thus, total fertility rate is the most important, but not the only, factor determining population growth.

Population Profiles

In studying population growth, demographers must consider more than just the increase in numbers, which is simply births minus deaths. They must also consider how the number of births ultimately affects the entire population over the **longevity,** or lifetimes, of the individuals. To determine patterns in population growth, demographers analyze the data collected through a census of the entire population, using a combination of household questionnaires and estimates for groups such as the homeless. In the United States and most other countries, a detailed census is taken every 10 years. Between censuses, the data may be adjusted using information regarding births, deaths, immigration, and the aging of the population.

---

[19]World Watch Institute, "State of Consumption Today," accessed October 28, 2014, http://www.worldwatch.org/node/810#1.

POPULATION BY AGE GROUPS AND SEX

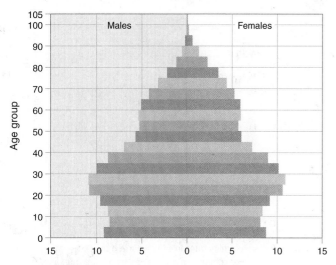

(a) United States of America: 1985

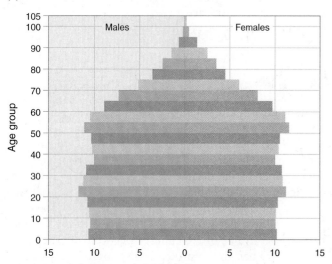

(b) United States of America: 2014

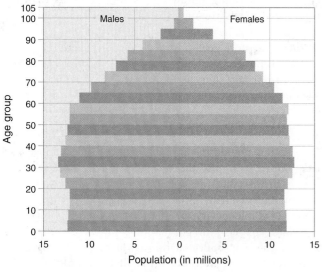

(c) United States of America: 2050

Figure 8–16 Population profiles of the United States. The age structure of the U.S. population **(a)** in 1985, **(b)** in 2014, and **(c)** projected to 2050.

(*Source:* U.S. Census Bureau, International Data Base.)

Demographers use the data for a population to construct a **population profile,** a bar graph depicting the number or proportion of people at each age; males and females are shown separately. **Figure 8–16** shows examples of such graphs. A population profile shows the **age structure** of the population—that is, the number of people in each age group at a given date. It is a snapshot of the population at a given time. Leaving out the complication of migration for the moment, each bar in a profile represents one *cohort* (group of the same age) of the population.

Figure 8–16 shows population profiles of the United States for 1985 and 2014 and a projected profile for 2050. As you look from one graph to the next, you can see a particular group as it ages. In high-income countries such as the United States, the proportion of people who die before age 60 is relatively small. Therefore, the population profile below age 60 is an "echo" of past events that affected birthrates. Figure 8–16a shows, for example, that smaller numbers of people were born between 1931 and 1935 (ages 50–68 in 1985). This drop was a reflection of lower birthrates during the Great Depression. The dramatic increase in people born in 1946 and for 14 years thereafter (ages 54–68 in 2014, Figure 8–16b) is a reflection of returning veterans and others starting families and choosing to have relatively large numbers of children following World War II—the "baby boom." The large baby-boom generation is now in middle age and retirement. The general drop in numbers of people born from 1961 to 1976 (ages 38–53 in 2014) resulted from declining fertility rates—the "baby bust." The rise in numbers of people born in more recent years (ages 24–33), termed the "baby-boom echo," is due to the large baby-boom generation producing children, even though the actual total fertility rate remained near 2.0. These changes in fertility are shown in **Figure 8–17.**

Predicting Populations

Current population growth in a country is calculated from three vital statistics: births, deaths, and migration. Demographers used to make population *forecasts.* Most often, they took current birth and death rates, factored in expected immigration and emigration, and then extrapolated the data into the future. They were nearly always wrong. For example, in 1964, the total fertility rate in the United States was around 3.0, although it was trending downward. The U.S. Census Bureau assumed that the total fertility rate would stay between 2.5 and 3.5 until 2000. On the basis of that assumption, the agency forecast a 2000 population of 362 million—81 million more than the actual number. The agency totally missed the baby bust of the late sixties and seventies.

Nowadays, demographers only make *projections,* and they cover their bases by making their assumptions about fertility, mortality, and migration very clear. Extrapolating the trend of lower fertility rates leads to the UN Population Division's projection that the global human population will reach 9.6 billion by 2050; it will not have leveled off at that time, but is likely to level off at some time in the 22nd century. This projection is the UN's *medium* scenario;

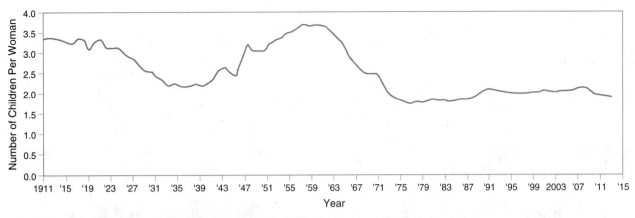

Figure 8–17 Total fertility rate in the United States, 1911–2012. The changes in fertility led to the baby boom and the baby bust in the United States. The rate now hovers slightly below the replacement level.

(*Source: 2012 World Population Data Sheet* (Washington, DC: Population Reference Bureau, 2012).)

other scenarios are based on different assumptions about fertility. **Figure 8–18** depicts the UN Population Division's projections until 2050 under four different sets of assumptions. The assumption of declining fertility rates is crucial; if current fertility rates remain unchanged (*constant* in the figure), the 2050 population will be 11.1 billion.[20] To achieve the medium scenario, people in the high-fertility areas of the world will need to continue to purposefully lower their fertility rate. The projected leveling off over 10 billion raises serious concerns about whether Earth can sustain such numbers.

Note that all of the UN's projections in Figure 8–18 assume fertility rates will fall. In fact, if the current rate of fertility is maintained, the human population will hit 28.6 billion by 2100. Instead, fertility rates in high-fertility countries are likely to fall, while those in extremely low-fertility

countries are likely to rise slightly. The medium scenario assumes that world fertility will drop from 2.55 children per woman to slightly more than 2 children per woman well before 2050. The high-fertility scenario assumes a total fertility rate half a child above that of the medium scenario, and the low-fertility scenario assumes a total fertility rate half a child below it. The constant fertility line in Figure 8–18 (orange) shows what will happen if the fertility rates around the world do not change. Note how each fertility assumption generates profoundly different world populations. Also note that none of these projections, even the one based on a drop in total fertility to 1.5 children per woman, will lead to a 2050 population that is smaller than today's population.

Population Projections for Developed Countries. In Sweden, a developed country in northern Europe, women have had a stable fertility rate for some time. This is reflected in the population profiles of Sweden in **Figure 8–19**, on the following page. Sweden has a boom group that will be getting older by 2050, but also has a stable production of new

---

[20]United Nations, Department of Economic and Social Affairs, Population Division, *World Population Prospects: The 2012 Revision, Volume I: Comprehensive Tables* (2011): ST/ESA/SER.A/313.

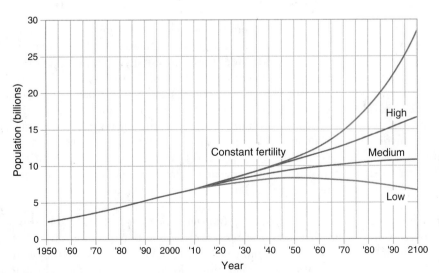

Figure 8–18 United Nations population projections. The most recent population projections show the outcomes of different fertility assumptions. The *Constant* projection assumes that fertility rates remain at their present level (2.6 children per woman). The *Medium* projection assumes a gradual decline in fertility in the developing countries, where 85% drop below replacement-level fertility by 2050 and the world total fertility rate is 2.02 children per woman. This projection is given the highest probability by the UN Population Division. The *High* projection assumes fertility rates at 1/2 child greater than the *Medium* projection, and the *Low* projection assumes fertility rates at 1/2 child lower than the *Medium* projection. All of these projections include the impact of the AIDS epidemic on increasing mortality.

(*Source:* UN Population Division, *World Population Prospects: The 2012 Revision* (New York: United Nations, 2012).)

POPULATION OF SWEDEN BY AGE GROUPS AND SEX

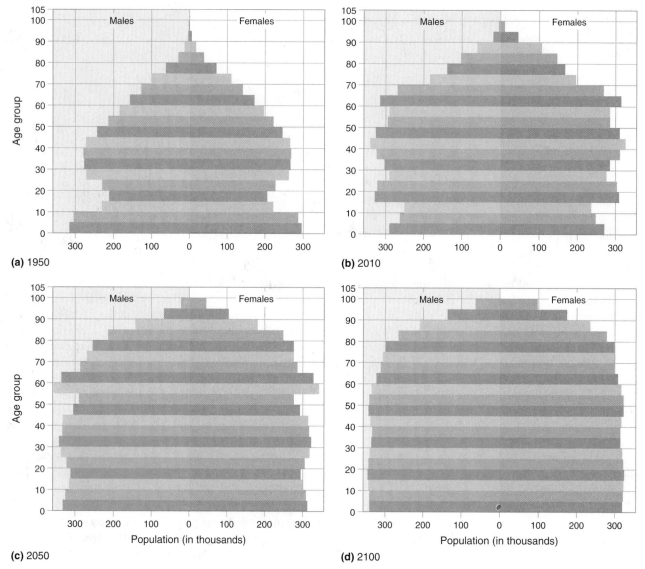

(a) 1950

(b) 2010

(c) 2050

(d) 2100

Population (in thousands)

Figure 8–19 Projecting future populations: A developed country. A population profile of Sweden, representative of a highly developed country, in four times. Note how stable the population becomes. (*Source:* U.S. Census Bureau, International Data Base.)

youth. In some countries, such as Italy (with a fertility rate of 1.3) or Japan, the number of workers for each dependent (child, elderly, disabled, or otherwise unable to work) will decline. In the case of these countries, it is because an aging population is not replaced by new babies.

Graying of the Population. For the next 20 years, Italy and Japan's populations will be **graying**, a term used to indicate that the proportion of elderly people is increasing. In 2011, about one fifth of Italians were age 65 or older; by 2050, about one third are expected to be 65 or older. Overall, a net population decrease is expected to occur. Unless a smaller population is the goal, it might be wise to encourage, and to provide incentives for, Italian couples to bear more children. In fact, many European governments have policies to encourage women to have more children, with mixed results so far. The very low fertility rate and the expected declining population seen in Italy and Japan are typical of an increasing number of highly developed nations. Worldwide, the population of people older than 65 years is expected to rise from 531 million to 1.5 billion in 2050. The surge of older people retiring may overwhelm government pension systems; life expectancies near 80 and retirement around 60 mean decades of retirement income. A number of countries in the Eurozone experienced financial difficulties in 2011 when it became clear that the burden of their social programs, including pensions, was unsupportable. Japan is also struggling to figure out how to care for its elderly.

The Effect of Immigration. Europe as a whole is on a trajectory of population decline if population change is considered without immigration. One obvious solution is to allow more immigration. However, allowing the declining numbers of Italian or Japanese people to be replaced by immigrant populations has implications for their culture and religion. Such changes are sometimes difficult for the host society. In fact, Europe is in the midst of a migration wave. More than 47.3 million immigrants were living and working in the countries of the European Union in 2010, many of whom did not have resident status.

Population Projections for the United States. Projections of future populations in the United States are a function of what we know about fertility rate and immigration. On the basis of the lower fertility rate expected from the 1980s, the U.S. population had been projected to stabilize at between 290 million and 300 million toward the middle of the next century. However, fertility rates and immigration were higher than expected, so new estimates made in 2008 projected that the U.S. population would be 439 million by 2050 (Figure 8–20). Adjustments in 2012 reflected a slightly lower fertility rate, projecting 400 million by 2050. These projections show how differences in the total fertility rate and immigration profoundly affect estimates of future population growth. In light of the concerns for sustainable development, the population policy of the United States (and scenarios that could lead to lower future growth) is under heavy debate.

Population Projections for Developing Countries.

Developing countries are in a situation that is vastly different from that of developed countries. Fertility rates in developing countries are generally declining, but they are still well above replacement level. The average total fertility rate in developing countries is currently 3.0, excluding China, where it is 1.6. Because of decades of high fertility rates, the population profiles of developing countries have a pyramidal shape.

While highly developed countries are facing the problems of a graying population, the high fertility rates in developing countries maintain an exceedingly young population.

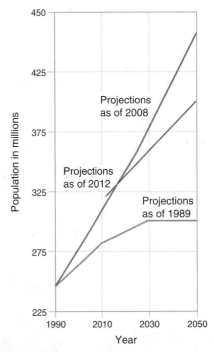

Figure 8–20 Population projections for the United States. Projections shift drastically with changes in fertility. Contrast the 1989 projection, based on a fertility rate of 1.8, the 2008 projection, based on an increased fertility rate of 2.1, and the 2012 projection, based on a current rate of 1.9.

(*Source:* U.S. Census Bureau, U.S. Population Projections.)

An "ideal" population structure, with equal numbers of persons in each age group and a life expectancy of 75 years, would have one fifth (20%) of the population in each 15-year age group. By comparison, in many developing countries, 40% to 50% of the population is below 15 years of age, whereas in most developed countries, less than 20% of the population is below the age of 15 (Table 8–3).

Table 8–3 Populations by Age Group

Region or Country	Percent of Population in Specific Age Groups			Dependency Ratio*
	<15	15 to 65	>65	
Sub-Saharan Africa	43	54	3	85
Latin America	27	66	7	51
Asia	25	68	7	47
Indonesia	29	66	5	51
Europe	16	67	17	49
Sweden	17	64	19	56
China	16	74	10	35
United States	19	67	14	49

Source: Data from *2014 World Population Data Sheet* (Washington, DC: Population Reference Bureau, 2014).

*Number of individuals below 15 and above 65, divided by the number between 15 and 65 and expressed as a percentage.

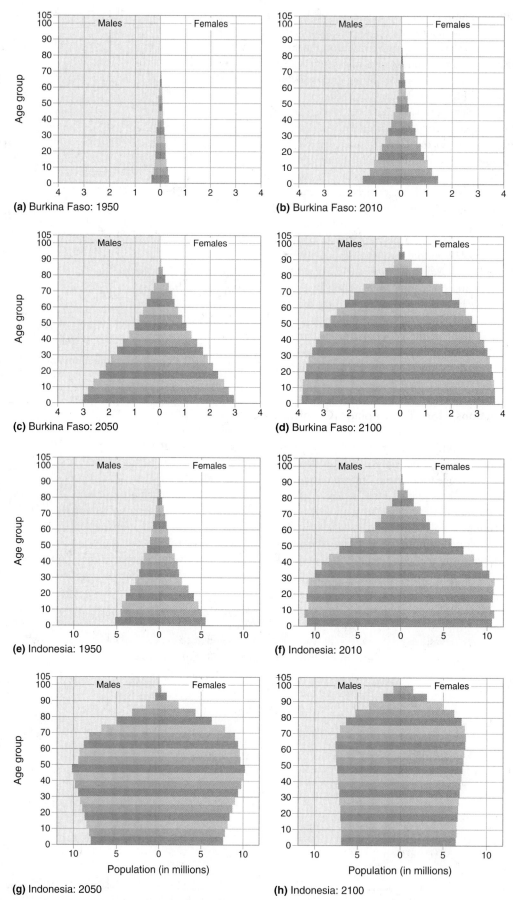

Figure 8–21 Projecting future populations: Indonesia and Burkina Faso. (a–d) Projections for Burkina Faso. **(e–h)** Projections for Indonesia.

(*Source*: U.S. Census Bureau, International Data Base.)

Burkina Faso and Indonesia. The demographics of Burkina Faso, Awa's home country, are typical of the lowest-income countries. Figure 8–21a–d shows a series of population profiles for Burkina Faso, which has a total fertility rate of 5.9. Even assuming that this high fertility rate will gradually decline by 2025, the population is likely to increase from 15.3 million (2010) to 23.4 million, yielding the profiles shown in Figure 8–21c and d. The country's high infant mortality rate (currently 91 per thousand) comes nowhere close to offsetting the high fertility rate. The impact of AIDS is also included in the UN projections. AIDS has increased child and young adult mortality and lowered average life spans in a number of countries. The pyramidal form of the profile is similar in 2010 and 2050, because, for many years, the rising generation of young adults produces an even larger generation of children. Thus, the pyramid gets wider and wider until the projected declining fertility rate begins to take effect.

Figure 8–21e–h depicts population profiles of Indonesia, Atin's home. While Indonesia is not well off, its population growth rate is lower than Burkina Faso's. Its large population may stabilize earlier.

Growth Impacts. What do these differing population structures mean in terms of the need for new schools, housing units, hospitals, roads, sewage facilities, telephones, and other services? One of the things they mean is that if a country such as Burkina Faso is simply to maintain its current standard of living, food production and the amount of housing and all other facilities must be almost doubled in as little as 25 years. Unfortunately, the population growth of a developing country can easily overwhelm its efforts to get ahead economically. Enormous growth is in store for the developing world, even assuming that fertility rates continue their current downward trend.

Even if it were possible to bring the fertility rates of developing countries down to 2.0 right now, population growth in those countries would not immediately stop because of *population momentum*. **Population momentum** refers to the effect of current age structures on future populations. In a young population, such as Burkina Faso's, momentum is *positive* because such a small portion of the population is in the upper age groups (where most deaths occur) and many young people are entering their reproductive years. Even if the rising generations have only two children per woman (replacement-level fertility), the number of births will far exceed the number of deaths. This imbalance will continue until the current children reach the Burkina Faso limits of life expectancy—50 to 60 years. In other words, only a population that has been at or below replacement-level fertility for decades will stop growing (achieve a stable population). Thus, the earlier the fertility rates are reduced, the greater the likelihood of ending population growth.

Europe's population, on the other hand, will soon begin to experience *negative* population momentum as a consequence of the low fertility of the past three decades. Of course, immigration would offset this effect.

The Demographic Transition

The concept of a stable, non-growing global population based on people freely choosing to have smaller families can become a reality, because it is already happening in developed countries. Early demographers observed that the modernization of a nation brings about more than just a lower death rate resulting from better health care. A decline in the fertility rate also occurs as people choose to limit the size of their families. Thus, as economic development occurs, human societies move from a primitive population stability, in which high birthrates are offset by high infant and childhood mortality, to a modern population stability, in which low infant and childhood mortality are balanced by low birthrates. A gradual shift in birth and death rates from the primitive to the modern condition in the industrialized societies is called the demographic transition (Figure 8–22, on the following page). The basic premise of the demographic transition is that there is a causal link between modernization and a decline in birth and death rates.

Birth and Death Rates. To understand the demographic transition, you need to understand the **crude birthrate (CBR)** and the **crude death rate (CDR)**. The CBR and the CDR are the number of births and deaths, respectively, per thousand of the population per year. By giving the data per thousand of the population, populations of different countries can be compared, regardless of their total size. The term *crude* is used because no consideration is given to what proportion of the population is old or young, male or female. Subtracting the CDR from the CBR gives the increase (or decrease) per thousand per year. Dividing this result by 10 then yields the *percent* increase (or decrease) of the population.

Number of births per 1,000 per year (CBR)		Number of deaths per 1,000 per year (CDR)		Natural increase (or decrease) in population per 1,000 per year			Percent increase (or decrease) in population per year
1,000	−	1,000	=	per 1,000	÷ 10 =		per year

A zero-growth population is achieved if, and only if, the CBR and CDR are equal and there is no net migration.

Epidemiologic Transition. Throughout most of human history, crude death rates were high—40 or more per thousand for most societies. By the middle of the 19th century, however, the epidemics and other social conditions responsible for high death rates began to recede and death rates in Europe and North America declined. The decline was gradual in the now-developed countries, lasting for many decades and finally stabilizing at a CDR of about 10 per thousand (Figure 8–21). At present, cancer, cardiovascular disease, and other degenerative diseases account for most mortality, and many people survive to old age. This pattern of change in mortality factors has been called the *epidemiologic* transition and represents one element of the demographic transition. (Epidemiology is the study of diseases in human societies.)

Fertility Transition. Another pattern of change over time can be seen in crude birthrates. In the now-developed countries, birthrates have declined from a high of 40 to 50 per thousand

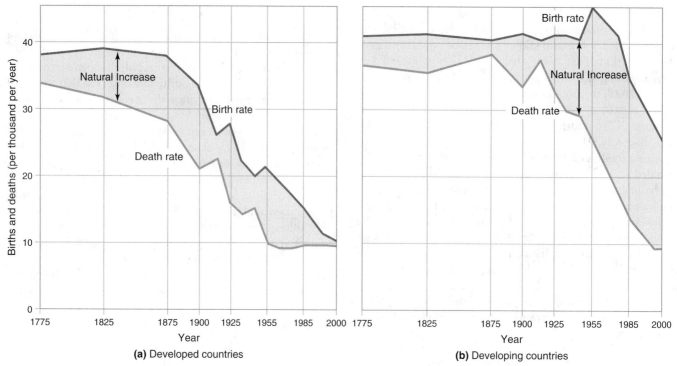

Figure 8–22 Demographic transition in developed and developing countries. (a) In developed countries, the decrease in birthrates proceeded soon after, and along with, the decrease in death rates, so very rapid population growth never occurred. **(b)** In developing countries, both birth and death rates remained high until the mid-1900s. Then the sudden introduction of modern medicine caused a precipitous decline in death rates. Birthrates remained high, however, resulting in very rapid population growth.

(*Source:* Redrawn with permission of Population Reference Bureau.)

to 9 to 12 per thousand—a *fertility* transition. As Figure 8–22 shows, this fertility transition did not happen at the same time as the epidemiologic transition; instead, it was delayed by decades or more. Because net growth is the difference between the CBR and the CDR, the time during which these two patterns are out of phase is a time of rapid population growth. The developed countries underwent such growth during the 19th and early 20th centuries and during the baby-boom years of the mid-20th century.

Phases of the Demographic Transition. The demographic transition is typically presented as occurring in the four phases shown in **Figure 8–23**. Phase I reflects a primitive stability resulting from a high CBR offset by an equally high CDR. Phase II is marked by a declining CDR—the epidemiologic transition. Because fertility and, hence, the CBR remain high, population growth accelerates during Phase II. The CBR declines during Phase III due to a declining fertility rate, but population growth is still significant. Finally, Phase IV is reached, in which modern stability is achieved by a continuing low CDR and an equally low CBR. Figure 8–23 also shows countries that typify each of the stages of the demographic transition.

Developed countries have generally completed the demographic transition, so they are in Phase IV. Developing countries, by contrast, are still in Phases II and III. In most

of these countries, death rates declined markedly as modern medicine and sanitation were introduced in the mid-20th century. Fertility and birthrates are declining but remain considerably above replacement levels. Therefore, populations in developing countries are still growing rapidly. In light of these circumstances, the key question is: What must happen in the developing countries so that they complete their demographic transition and reach Phase IV?

In this chapter, we looked at three young people—Anje in Sweden, Atin in Indonesia, and Awa in Burkina Faso. We saw that their life experiences were dramatically different from each other—in part because of broad differences between countries in wealth, health care, and other goods and services. Those differences in life experiences drive differences in consumption patterns and fertility—two parameters that need to be brought into line to have a sustainable world.

Next, we will investigate the factors that influence birth and death rates, as well as the contributions of economic development and family planning in bringing about the fertility transition (Chapter 9). In the process, we can begin to picture what sustainable development might look like for the developing world and what it will take to get there.

CONCEPT CHECK How does an increasing life span in a place like Indonesia affect the shape of the population pyramid, the dependency ratio, and population momentum? ☑

THE CLASSIC PHASES OF THE DEMOGRAPHIC TRANSITION

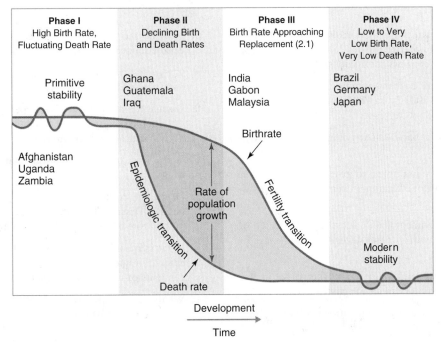

Figure 8–23 Phases of the demographic transition. The epidemiologic transition and the fertility transition combined to produce the demographic transition in the developed countries over many decades. (*Source:* Population Bulletin, Population Reference Bureau, http://www.prb.org/Publications/Datasheets/2011/world-population-data-sheet/popula.)

REVISITING THE THEMES

SOUND SCIENCE

Demographers use sound scientific methods to gather data about human populations and to identify and interpret trends in populations. Their work has clarified the nature of the demographic transition. Their population projections enable us to see what the future will look like if current trends continue.

SUSTAINABILITY

A sustainable world is a world whose population is generally stable—it cannot increase indefinitely, as the world's population currently is doing. World population may level off at 10 billion or even 9 billion, but we may be exceeding Earth's carrying capacity. Two or 3 billion more people will need to be fed, clothed, housed, and employed, all in an environment that continues to provide essential goods and services. Many agree that sustainability will be possible only if the additional people maintain a modest, low-consumption lifestyle, at a level that is lower than that currently experienced in the developed countries. People already in developed countries will need to make changes to lower their consumption so that others may have access to resources. These changes involve not only myriad personal decisions, but also coordinated action and the development of more efficient technologies.

In a sense, the United States is playing a small role in the process by accepting so many immigrants from developing countries—about 1% of the annual population growth in these countries. More significant is the impact of immigration on the United States itself, which is currently placing the country on a trajectory of indefinite population growth. Sustainability requires that we somehow stabilize population and consumption of resources both nationally and globally.

STEWARDSHIP

Resource stewardship requires us to promote justice and may require difficult decisions about what is right. Calling on the developing countries to keep down their future demands on the environment could be considered unjust, as so many people live in abject poverty while others consume resources at high rates. Still, there is no way for these countries to meet their future needs by mimicking what the developed world is currently doing with resources and the environment. Justice demands that both the rich and the poor countries reduce their environmental impact.

Stewardship requires both individual efforts and corporate action through policy. If enough people in the high- and middle-income countries practice environmental stewardship, the pathway to such a future will be easier. Stewardship can do much to alleviate the negative impacts of affluence and technology, prominent items in the IPAT formula. Population growth has to slow as well. Achieving that end without trampling on the rights of others is part of the job ahead of us. Immigration also presents ethical dilemmas as immigrants move to places that may or may not be prepared to receive them.

REVIEW QUESTIONS

1. In what ways is human population ecology similar to and different from that of other organisms? Why is it difficult to determine a carrying capacity for humans?

2. How has the global human population changed from prehistoric times to 1800? From 1800 to the present? What is projected over the next 50 years?

3. How does the World Bank classify countries in terms of economic categories?

4. What three factors are multiplied to give total environmental impact? Are developed nations exempt from environmental impact? Why or why not?

5. What are the environmental and social consequences of rapid population growth in rural developing countries? In urban areas?

6. Describe the negative and positive impacts of affluence (high individual consumption) on the environment.

7. What information is given by a population profile?

8. How do the population profiles and fertility rates of developed countries differ from those of developing countries?

9. Compare future population projections, and their possible consequences, for developed and developing countries.

10. Discuss the immigration issues pertaining to developed and developing countries. Which countries are sending the most immigrants and which are receiving the most?

11. What is meant by population momentum, and what is its cause?

12. Define the crude birthrate (CBR) and the crude death rate (CDR). Describe how these rates are used to calculate the percent rate of growth and the doubling time of a population.

13. What is meant by the demographic transition? Relate the epidemiologic transition and the fertility transition (two elements of the demographic transition) to its four phases.

14. What are examples of countries in different stages of the demographic transition?

THINKING ENVIRONMENTALLY

1. Why do some people see humans as not subject to the law of limiting factors? How would you respond to them?

2. List the parameters you think you would need to keep track of to make accurate predictions about future human population growth. For example, what conditions might drive mortality and fertility trends?

3. Consider the population situation in Italy or Japan: very low birthrates, an aging population, and eventual decline in overall numbers. What policies would you recommend to these countries, assuming their desire to achieve a sustainable society?

4. Consider the populations of Sweden, Indonesia, and Burkina Faso. How could people in each of these countries help those in the other countries to steward their resources better or to promote a sustainable future?

MAKING A DIFFERENCE

1. Calculate your ecological (environmental) footprint using a site such as http://www.myfootprint.org.

2. Come up with two things you can do to lower your resource consumption. They might include buying less of something you don't need or buying a product with less packaging. Agree to do it for one month, and decide whether you could continue longer.

3. Educate yourself on immigration policy. Decide whether you support immigration reform and what policies you would like to see in place. Then write to an elected official about it.

MasteringEnvironmentalScience®

Students Go to MasteringEnvironmentalScience for assignments, the eText, and the Study Area with animations, practice tests, and activities.

Professors Go to MasteringEnvironmentalScience for automatically graded tutorials and questions that you can assign to your students, plus Instructor Resources.

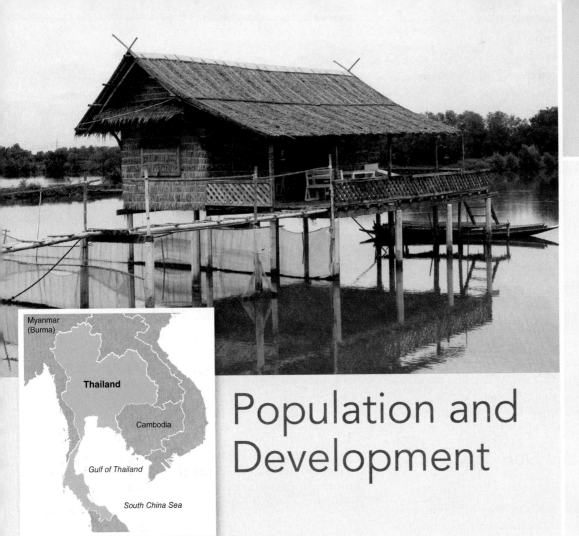

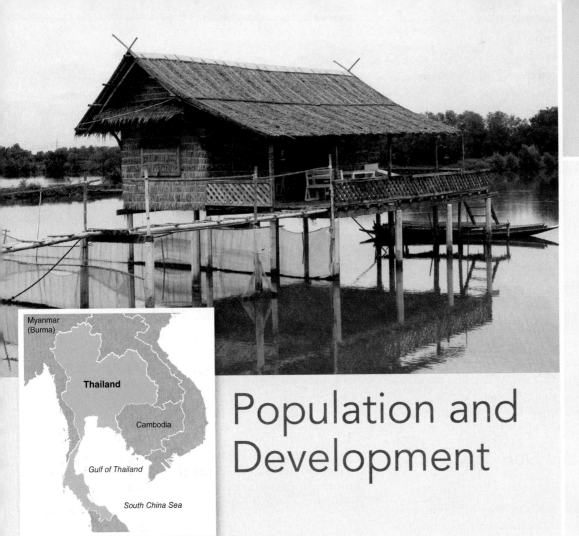

Population and Development

Learning Objectives

9.1 Predicting the Demographic Transition: Describe the main schools of thought about what causes the demographic transition.

9.2 Promoting Development: Explain how development, fertility, and environmental health relate to each other. Understand the Millennium Development Goals and the move to the Sustainable Development Goals. Describe how development can reduce fertility but also increase environmental stressors.

9.3 A New Direction: Social Modernization: Identify the five components of social modernization that countries focus on to achieve lower fertility rates.

The sun shines on the surface of the Chao Phraya River in Thailand. The water glistens as it runs past the rich fields in the central Thai region. In the villages by the river, extended families live in wooden structures on stilts, with water buffaloes and chickens safely housed beneath. Anong and her husband, Somchai, are part of such a family. They live with their two children, a boy and a girl, and Anong's elderly parents. While Somchai is the head of the family in most matters, Anong is in charge of finances and the household. Both work outside the home and earn a modest income. Both are literate. Although they live in only one room, the family has access to clean drinking water and sanitation. Both parents were raised in large families: Anong had seven siblings and Somchai five. Today, their family with only two children is more typical of Thai families.

Fertility Drop. Anong and Somchai's lives are typical of many in Thailand. Like the rest of their generation, they have experienced rapid change in their home country. Probably the biggest change is in family size. In a single generation, the country underwent one of the largest drops in fertility in world history. Between 1960 and 2014, the total fertility rate dropped from 6.4 children per woman to 1.6. At the same time, GNP per capita and life expectancy rose dramatically (from $1,059 to $7,274 and from 53 years to 69.9 years). This remarkable drop in population growth rate astonished the world. While the percentage of children in the population has declined, the percentage of elderly has increased. As a result, Anong and Somchai are juggling many demands, caring for both their children and Anong's parents.

▲ This Thai house on stilts has pens for aquaculture beneath it.

In 1960, when the Texas-sized country held 28 million people, demographers predicted that Thailand would fall into a poverty trap: Any increase in the economy would be eaten up by the needs of the growing population. Today, with population growth slowed and a population of about 70 million, the country has 25 million fewer people than demographers predicted in the 1970s. In fact, demographers predict that within a decade the population may begin to contract, a change that would bring other social changes.

Social Changes. How did Thailand stop its runaway population growth? Much was done through direct efforts to lower fertility: Voluntary contraception was increased, public health and awareness campaigns were undertaken, and health care clinics were opened, even in rural areas. For example, the public health campaign of one nonprofit group made contraception a familiar and humorous topic of everyday life.

But there were other factors as well. Modern Thai women now have higher freedom, equality, and literacy rates than Thai women of the past. In Thai culture, sons are not favored over daughters, so women are not under pressure to abort female fetuses or to have more children than they can support in order to produce sons. A nutrition program implemented by the government dramatically slashed child malnutrition. Contrary to some prevailing thought, the fertility transition Thailand experienced occurred even though the country was relatively poor, and even though most people remained rural.

While the drop in population growth prevented disaster, some of the economic growth and the support for millions more Thai citizens came at the expense of ecosystems. Thailand now has to contend with water and air pollution and the loss of biodiversity brought on by its activities. On the plus side, the country's ability to solve ecological problems is more likely with lower population growth. In the future, Thais will have to figure out how to navigate increased urbanization and an aging population.

This chapter follows up on the demographic information presented earlier (Chapter 8). We begin with a reassessment of the demographic transition.

9.1 Predicting the Demographic Transition

The *demographic transition* describes the shift from high birth and death rates to low birth and death rates that occurs in industrialized countries (Chapter 8). **Figure 9–1** plots the current birth and death rates of the major regions of the world to show where they fall in the demographic transition. Regions to the right of the vertical dividing line on this graph are on a fast track to completing the demographic transition or have already done so. Regions to the left of the line are in the middle of the demographic transition. These regions (half of the world) are mostly in their fifth decade of rapid population increase. It is as if they are trapped there, with serious consequences. One of the key questions scientists who study populations ask is, "What causes some countries to move through the demographic transition quickly and others to remain in the process indefinitely?"

Different Ways Forward

All countries trapped in long-term rapid growth would like to experience the economic development that has brought Thailand, South Korea, Indonesia, Malaysia, Brazil, and other low-income nations into the middle- and even high-income nation groups. If they did so, it is likely that they would then move through the demographic transition.

How did the now-developed nations come through the demographic transition without getting caught in the

Figure 9–1 World regions in the process of demographic transition. Crude birthrates (CBRs, orange) and crude death rates (CDRs, green) are shown for major regions of the world. A dividing line separates countries at or well along in the demographic transition from those still in the middle of the transition.

(*Source:* Data from *2014 World Population Data Sheet* (Washington, DC: Population Reference Bureau, 2014).)

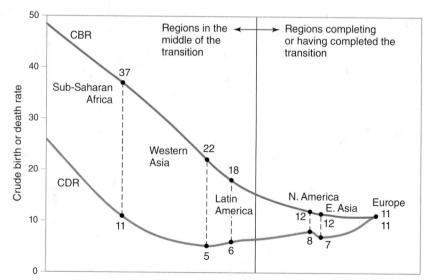

Progress in demographic transition

poverty-population trap? In these countries, the improvements in disease control that lowered death rates (the epidemiological transition) occurred gradually through the 1800s and early 1900s. Industrialization, which introduced factors that lowered fertility, occurred over the same period. Therefore, there was never a huge discrepancy between birth and death rates (see Figure 8–22). In contrast, modern medicine was introduced into the developing world relatively suddenly, bringing about a precipitous decline in death rates, while the fertility-lowering effects of development have been slow in coming. Figure 9–2 shows how extreme the differences in speed of the demographic transition can be.

Demographers and economists try to determine which factors are most likely to predict changes in populations and why. What causes educated women to have fewer children? If the GNP of a whole nation goes up, but inequality is high, will population growth rates still fall? Should governments simply try to increase the family economy for the poorest and assume fertility rates will fall, or do people need specific programs for family planning? The answers to these questions may vary by country.

Competing Ideas

What do the developing countries need to do to fast-forward their way through the demographic transition? From early on, two basic conflicting theories offered answers to this question:

1. Speed up economic development in the high-growth countries and population growth will slow down "automatically," as it did in the developed countries.

2. Concentrate on population policies and family-planning technologies to bring down birthrates.

Conferences on Population. These two schools of thought were reflected at two UN population conferences—one in Bucharest, Romania, in 1974 and a second in Mexico City in 1984. A conference in 1994 in Cairo, Egypt, and another in Stockholm, Sweden, in 2014, reflected some shifts in thinking. There were several other conferences in between.

Bucharest. At the Bucharest conference, the United States was a strong advocate of population control through family planning, while the developing nations argued that "development was the best contraceptive." Their resistance to family planning was bolstered by feelings that the developed world's promotion of population control was another form of economic imperialism or even genocide.

Mexico City. At the conference in Mexico City, the sides were somewhat reversed. Developing countries facing problems of excessive population growth were asking for more assistance with family planning. The United States, under pressure from anti-abortion advocates, took the position that development was the answer and terminated all contributions to international family-planning efforts, a policy that remained in effect until 1993. The other developed countries, however, remained convinced that family planning was essential and continued to support international efforts to help the developing countries implement policies designed to reduce fertility rates.

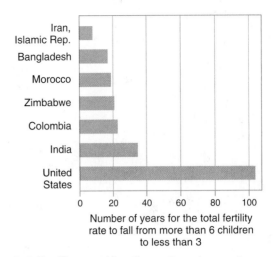

Figure 9–2 **Fertility transition times.** Countries experience very different rates of change in fertility.

(*Source:* www.gapminder.org.)

Cairo. At the International Conference on Population and Development (ICPD) in Cairo in 1994, poverty, population growth, and development were clearly linked, and resources and environmental degradation were an important focus. The ideological split between developed and developing countries was a thing of the past—all agreed that it was necessary to deal with population growth in order to make progress in reducing poverty and promoting economic development. The responsibility for bringing fertility rates down was placed firmly on the developing countries themselves, although the developed countries pledged to help with technology and other forms of aid. The participants also reached broad agreement on several main ideas. One of these was that development must be linked to a reduction in poverty. Another was that women's rights to health care, education, and employment were foundational to achieving slower population growth. Five-year anniversary reviews affirmed the universal commitment to the agreements forged at this conference.

Stockholm. The Sixth International Parliamentarians' Conference on the Implementation of the ICPD Programme of Action (IPCI/ICPD) took place in Stockholm, Sweden, in April 2014, 20 years after the plan of action developed at Cairo. At that meeting, representatives of different countries set a new series of goals for a post-2015 world. These goals focused on improving the lives of women and girls and increasing access to reproductive services, as well as improving health and education for all. ICPD Programme members unanimously adopted the Stockholm Statement of Commitment, a document with specific goals for nations about population, development, and human rights.

Large and Small Families

The fertility transition is the key element in the demographic transition. Many studies have shown that high fertility and poverty are linked, including a correlation

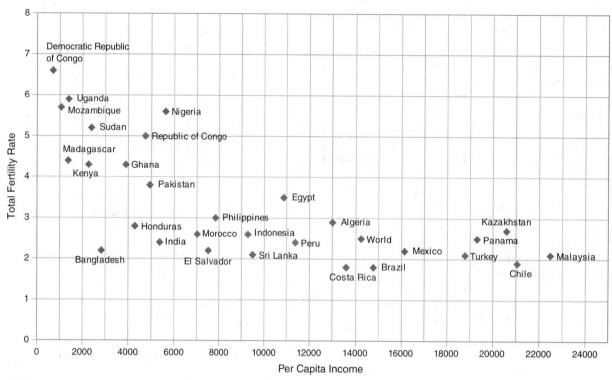

Figure 9–3 **Fertility rate and income in selected developing countries (2011).** There is a correlation between income (gross national income in purchasing power parity per capita) and lower total fertility. Factors that affect fertility more directly are health care, education for women, and the availability of contraception. (*Source: World Data Sheet 2014* and the World Economic Outlook Database, September 2014.)

between fertility rate and gross national income per capita, as shown in **Figure 9–3.** Countries on the far left side of this plot are all in the early or middle stages of the demographic transition; most are low-income economies. Those in the middle and to the right are middle-income economies that are moving through the transition. Thailand is somewhat of an exception to this pattern, because its fertility rate dropped while the country still had a relatively high level of poverty; in recent decades its economy has improved. Differences in fertility rates between countries with similar economies relate to differences in their societies and social supports.

To the observer in a developed country, it may seem strange that so many poor families in developing countries have large numbers of children. From our perspective, it is obvious that more children spread a family's income more thinly and handicap efforts to get ahead economically. In the United States, large families are rare. In other countries, large families are more common, though in countries with poor sanitation and little access to health care, children may not live to adulthood. The decision-making process that leads parents to produce large families is different in low-income countries than in the United States because the sociocultural conditions are very different.

Let's look at some examples. In western Uganda, most families are large and polygamous and make their living in agriculture. Because girls bring a bride price to their parents, both boy and girl children are valued. More than half of the women marry before the age of 18, and the average woman

has about six children.[1] Children are an asset in the workforce. Many people see the immediate need for help in the fields as more important than their children's long-term options, so most children do not receive education, a common trade-off in many poor rural areas **(Figure 9–4).**

Afghanistan is another good example. In 1950, the population was 8 million; in 2013, it was 31 million; and estimates for 2030 range between 43 million and 50 million. About 42% of Afghans are under the age of 15, and another 22% are between the ages of 15 and 24.[2] A number of factors combine to make life difficult for Afghan families: war, poverty, lack of human rights, uncertainty about the future, and a lack of education. A typical Afghan woman **(Figure 9–5)** might marry a man a dozen years older than herself and bear children as soon as she is able to get pregnant. She will have few rights to her own decision making, little schooling, and heavy social pressure to bear sons. She may not know about contraception or may live in a culture that doesn't accept it. Because of harsh conditions, both mothers and infants die in the birth process at higher rates than those in most of the world. In spite of all of these difficulties, fertility in Afghanistan is currently declining, although it is still an average of 5.4 children per woman. Between 2003 and 2012, the use of contraceptives rose from 5% to 22% in some rural areas.

---

[1]*The State of Uganda Population Report 2013.* Kampala: Republic of Uganda Population Secretariat, 138.

[2]CIA World Factbook, Afghanistan, 2014, https://www.cia.gov/library/publications/the-world-factbook/geos/af.html.

Figure 9–4 Children as an economic asset. In many developing countries, children perform adult work. This Peruvian child and adult are digging in a farm field.

Figure 9–5 Young parenthood. Young parenthood contributes to high rates of population growth. Marriage by young girls is common in Afghanistan, where this mother and child live.

Numerous studies reveal the following as primary reasons that the poor in developing countries have large families.

1. **Old age security.** A primary reason given by poor women in developing nations for having many children is to ensure that they will be cared for in old age.

2. **Infant and childhood mortality.** The common experience of children dying leads people to try to make sure that some of their children will survive to adulthood.

3. **Helping hands.** In subsistence-agriculture societies like rural Uganda, women do most of the work relating to the direct care and support of the family. Very young children help with chores and fieldwork. In short, children are seen as a productive asset.

4. **Status of women.** The traditional social structure in many developing countries discourages women from obtaining higher education, owning land or businesses, and pursuing many careers. Respect for a woman increases as she bears more children.

5. **Availability of contraceptives.** Poor women often lack access to reproductive health information, services, and facilities. Providing contraceptives to women is a major facet of family planning. Studies show a strong correlation between lower fertility rates and the percentage of couples using contraception (Figure 9–6). Worldwide about 220 million women would like to delay or prevent pregnancy but are unable to do so. Thus there is an unmet need for contraception.

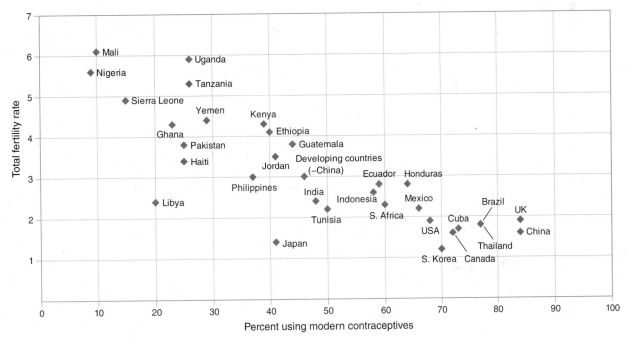

Figure 9–6 Fertility rate and prevalence of contraception in selected countries (2011). More than any other single factor, lower fertility rates are correlated with the percentage of the population using contraceptives.

(*Source:* Data from *2014 World Population Data Sheets* (Washington, DC: Population Reference Bureau, 2014).)

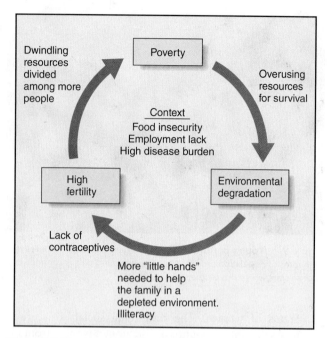

Figure 9–7 The poverty cycle. Poverty, environmental degradation, and higher fertility rates may become locked in a self-perpetuating, vicious cycle.

Fertility rates in developing countries remain high not because people in those countries are behaving irrationally or irresponsibly, but because the sociocultural climate in which they live has favored high fertility for years and because contraceptives are often unavailable. Sometimes the rise in population triggers a vicious cycle: poverty, environmental degradation, and high fertility drive one another forward (Figure 9–7). Increasing population density leads to a greater depletion of rural resources such as firewood and water, which in turn encourages couples to have more children to help gather resources. They are in a poverty trap. Rapid population growth can increase inequality in a society because the abundance of labor lowers the price of labor. This feeds the cycle, as inequality also promotes poverty.

The five factors supporting large families are common to pre-industrialized, agrarian societies. In contrast, industrialization, urbanization, and development bring conditions conducive to having small families. These conditions include the relatively high cost of raising children, the existence of pensions and a social-security system, opportunities for women to make money, access to affordable contraceptives, adequate health care, educational opportunities, and an older age at marriage.

Thus, it is not just economic development that leads to declining fertility rates. Rather, fertility rates decline insofar as development provides (1) security in one's old age apart from the help of children, (2) lower infant and childhood mortality, (3) universal education for children, (4) opportunities for education and careers for women, and (5) access to contraceptives and reproductive health services. These are all outcomes of public policies, and countries (such as Thailand) that have made them a high priority have moved into and through the demographic transition.

Demographic Dividend

As a country goes through the demographic transition, birthrates decline. As a result, the working-age population increases relative to the younger and older members of the population. This relationship is the **dependency ratio**, which is defined as the ratio of the nonworking population (under 15 and over 65) to the working-age population (Table 8–3). For a time, the society can spend less on new schools and more on factors that will alleviate poverty and generate economic growth—a **demographic dividend**. This demographic window is open only once, however, as the numbers of younger children decrease, and stays open for only a few decades, until the numbers of older people increase. Graying populations then increase the dependency ratio (see Sustainability, *Dealing with Graying Populations*). If countries have succeeded in increasing their economy, they will then be in a better position to care for dependents (Chapter 8). For countries well below replacement level, such as Japan, the fertility level may rebound slightly or immigration may increase to level the population out in the long term.

Figure 9–8 shows how the dependency ratios in several world regions have changed over time and are likely to change

Figure 9–8 The demographic window. As countries experience gradually decreasing fertility, dependency ratios decline and opportunities for development increase. The graph compares several developing regions with some developed countries.

(*Source: World Data Sheet 2011* and the World Economic Outlook Database, September 2011.)

UNDERSTANDING THE DATA

1. Where do you think the dependency ratio of South Asia will be in 2070?

2. How does the pattern of change in South Asia compare to that in East Asia and the Pacific?

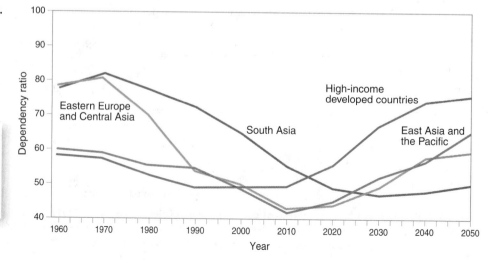

SUSTAINABILITY
Dealing with Graying Populations

One outcome of the demographic transition is a higher percentage of older people in a population. This population "graying" can be seen in Western Europe, Japan, and certain other countries. Graying populations may cause problems in countries with fertility rates that are below replacement level. As the population ages, the dependency ratio rises, placing a growing burden on people of working age, who must support both the young and the elderly, nonworking members of the population.

For some countries whose fertility rates dropped quickly, the solution is to encourage slightly larger families. France, for example, supports parents with programs such as family leave, preschool funding, and tax breaks for families with young children. Other countries accept increased immigration. Around the world, though, countries need to develop social structures to care for the elderly.

Japan is an excellent example of a country facing the challenge of a graying population. Although Japan has the world's third largest economy, its

population is declining and 25% of its population is elderly. By 2060, if current trends continue, 40% of Japan's population will be elderly. For many demographers, this seems a recipe for disaster. Yet

Graying Population in Japan. With the oldest population in the world, Japan is a leader in the production of innovations for elder care. Here a robot helps an elderly woman with shopping.

Japan's per capita income has been rising and Japan has the healthiest people in the world. While Japan's ratio of workers to dependents is similar to that of countries with high numbers of young people, Japan does not have to spend much on education. In the future, Japan may need to increase immigration in order to have enough workers.

Recently Japan's graying population has inspired a wave of innovation.[3] In fact, Japan has been a forerunner in the development of technologies to help elderly people. These include robots for grocery shopping and lifting objects, robotic pets, and technologies for tracking the health or movements of elders. These innovations will be useful as other countries around the world face demographic changes.

---

[3] F. Pierce, "Japan's Ageing Population Could Actually Be Good News," *The New Scientist* (2014): http://www.newscientist.com/article/dn24822-japans-ageing-population-could-actually-be-good-news.html.VCLMhEsvUrc.

---

in the future. A number of East Asian and Latin American countries, such as South Korea, Brazil, and Mexico, have taken advantage of this window. They have put in place social and economic policies that are consistent with poverty reduction and economic expansion, and as a result, they have experienced rapid economic development. The countries of South Asia (for example, Pakistan, India, and Bangladesh) are currently approaching this window of opportunity. Will they make the necessary investments in health, education, and economic opportunities to take full advantage of the demographic window? To do so, they must directly help the poorest people in their societies.

Case Studies: China and India

When a country has uncontrolled population growth, it isn't always obvious how to respond. Increasing modernization may help, but isn't guaranteed to work: a country may be headed into a poverty trap. If ignoring the problem isn't reasonable, and simply letting people die of starvation and disease is unethical (as most people would argue), then governments have to decide how strongly they will act to lower population growth. The most acceptable option—the one that is the theme of this chapter—is to create an economic and social climate in which people, of their own volition,

will desire to have fewer children. Along with such a climate, society will provide the means to enable people to exercise that choice.

Recent changes in China and India—the two most populous countries in the world—provide an instructive contrast. Both countries have experienced dramatic population growth and have rapidly expanding economies. Both countries also struggle to feed their populations, have more men than women, and suffer environmental degradation from activities designed to increase the economy. However, the social policies of these two countries are significantly different. China has a centralized government and a strong policy that allows only small families. Its fertility rate is low, its literacy rate is high, and its population is aging. India has a young population, higher fertility rates, and wide variation in family size. How have their policies caused their populations to change?

China. Several decades ago, China was in an intolerable position. Under the Communist leader Mao Tse-tung, the government encouraged people to have large families; in just a few decades China's population increased from 550 million to 900 million people. During the '40s and '50s, the government pursued a series of unfortunate plans for

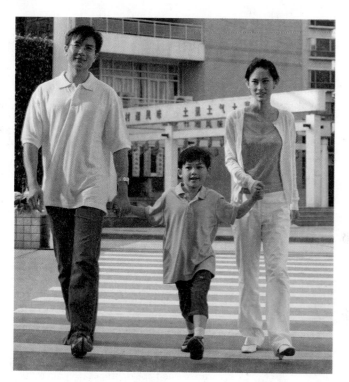

Figure 9–9 China's one-child law. China's one-child policy caused the country's population growth to slow dramatically. Here you see parents with their only child.

industry and agriculture that, along with natural disasters, led to famine. Between 1958 and 1960, crop production dropped by 70%, causing the greatest level of starvation in human history; between 20 million and 43 million people died. By the early '70s, China's leaders began a voluntary family-planning program of clinics and education, with some success.

In 1979, China introduced a policy of allowing families to have only one child. About 40% of married couples were obligated to this policy; others were allowed two or sometimes more children. The one-child policy affected mostly urban families (Figure 9–9). Exceptions were made for rural families, those who had daughters first, families in which both parents were only children, and ethnic minorities. To achieve its population goals, the government instituted an elaborate array of incentives and deterrents. Incentives included a single-child bonus (a monthly subsidy to one-child families), additional food rations, housing preferences for single-child families, better medical care, and job priority for only children. Disincentives for having an excessive number of children included taxes and repayment of the single-child bonus.

As a result of these policies, China saw a precipitous drop in its total fertility rate, from about 4.5 in the mid-1970s to 1.6 in 2014. Yet despite such controls, the population of China is still growing due to momentum and will continue to do so for some years. (China's current population is 1.4 billion, one-fifth of the world's people.)

One disturbing consequence of China's policy is a skewed ratio of males to females. Males are preferred in Chinese families, so some parents try to make sure that their one child is a male, through such extreme measures as selective abortion and even infanticide of female children. Now China anticipates 30 million excess males coming into adulthood by 2020, leaving many young men with no prospect of forming a family. This skewed sex ratio and the country's aging population have caused China to reconsider aspects of its one-child policy. In 2014, China relaxed the policy, allowing more parents to have a second child.

The one-child policy has been extremely controversial. Local governments in China, tasked with keeping population growth down, have carried out egregious human rights violations, such as razing the homes of people with multiple children and forced sterilizations and abortions. These actions have caused outrage in China and in the world community.

China's government pursued policies with which many people are uncomfortable. China's population policy has been a case study for environmental ethicists and a topic of passion for human rights workers since its inception, but it is the exception. Most countries in the world believe people should be able to make individual choices about family size.

India. In the nearly 70 years since India gained its independence from the United Kingdom, its population has more than tripled. With 1.25 billion people as of 2014 and a total fertility rate around 2.5, India is growing at a rate of 18 million per year and is expected to surpass China's population by 2025. In spite of many decades of family-planning programs, India's population has continued to grow, canceling out much of the country's gains in food production, health care, and literacy. Yet across India, especially in the south, fertility rates are dropping and poverty rates have fallen from 52% to almost 22% (Figure 9–10).

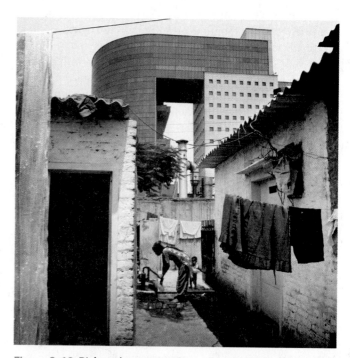

Figure 9–10 Rich and poor in India. A modern high-rise building towers above a poorer area of Gurgaon, on the southwest fringe of Delhi, highlighting India's huge contrast in wealth.

At this point, almost half of India's children under the age of five are malnourished, and two-thirds of its population live on less than $2 a day. India has the largest illiterate population in the world at 287 million.[4] With a per capita gross national income of $5,350 per year, India has a very long way to go to achieve the kind of development that characterizes the industrialized countries.

Population growth is quite different in the southern Indian state of Kerala. With a population of 33.8 million and an area about the size of Kentucky, Kerala is one of the most densely populated regions in the world. Like the rest of India, Kerala is crowded, per capita income is small, and food intake is low. However, people of Kerala have a life expectancy of 74 years, compared with 67 for all of India. Infant mortality and fertility rates are low and literacy is high. Access to education and health care is high and women are as well educated as men. Even though Kerala is a poor region, its people are well on the way to achieving a stable population, which is expected in 2021. Kerala's success in slowing population growth was a result of targeted development, not simply general economic growth. In Section 9.3, we will see some of the specific conditions under which a fertility transition is most likely for a low-income country.

CONCEPT CHECK What were the major causes of population change in Thailand, China, and India? What other factors were involved? ☑

9.2 Promoting Development

The good news is that many developing countries have made remarkable *economic* progress. The gross national products of some countries have increased as much as five-fold, bringing them from low-income to medium-income status, and some medium-income nations have achieved high-income status. There is still a great gap between rich and poor countries in income, as well as technology. But in the two decades after 1990, the percentage of people living in *extreme poverty*—those living on less than $1.25 per day—dropped from 36% to 18% of the world's population. By 2015, the percentage had dropped even further.

Tremendous *social* progress has been made in many developing countries, too. Since 1990, maternal mortality has dropped by 45%. Ninety percent of primary school-age children now attend school. Literacy rates, access to clean drinking water and sanitary sewers, and other indicators of development have generally improved. Furthermore, the fertility rates of most developing countries have declined, although they are still far above the replacement level (Table 9–1).

Nonetheless, there is still a long way to go. About one seventh of the world's population (1 billion people) live on less than $1.25 a day, nearly 1 billion lack access to toilets and must defecate outdoors, 1 billion lack access to clean water, more than 860 million live in urban slums under deplorable conditions, and more than 800 million lack sufficient food. There are also great disparities in economic progress among developing countries. Income growth and technological progress are lowest in three regions with continued high population growth and poverty: sub-Saharan Africa, the Middle East, and Latin America.[5]

Is the solution to these problems simply to facilitate economic growth and expect that the alleviation of poverty will follow? Not necessarily. As we have already seen, the wealth of a country may increase, but many people will remain in poverty if inequality is high. The United Nations Fund for Population Activities (UNFPA) puts it the following way in *State of World Population 2002*: "Development has often bypassed the poorest people, and has even increased their disadvantages. The poor need direct action to bring them into the development process and create the conditions for them to escape from poverty."[6] The alleviation of poverty was the explicit objective of the Millennium Development Goals.

---

[4]"India tops adult illiteracy: U.N. report" The Hindu, Jan 29, 2014, http://www.thehindu.com/features/education/issues/india-tops-in-adult-illiteracy-un-report/article5629981.ece

[5]Kavita Watsa, *Technology & Development: Findings from a World Bank Report: Global Economic Prospects 2008: Technology Diffusion in the Developing World* (Washington, DC: World Bank, 2008).

[6]UNFPA, *State of World Population 2002: People, Poverty, and Possibilities: Making Development Work for the Poor,* ed. Thoraya Ahmed Obaid (New York: UN Population Fund, 2002).

Table 9–1 Decline in Total Fertility Rates

	1985	1990	1995	2000	2005	2008	2011	2014
Africa	6.3	6.2	5.8	5.3	5.1	4.9	4.7	4.7
Latin America and the Caribbean	4.2	3.5	3.1	2.8	2.6	2.5	2.2	2.2
Asia (excluding China)	4.6	4.1	3.5	3.3	3	2.8	2.6	2.5
China	2.1	2.3	1.9	1.8	1.6	1.6	1.5	1.6
Developed countries	2	2	1.6	1.5	1.5	1.6	1.7	1.6

Source: Data from various *World Population Data Sheets* (Washington, DC: Population Reference Bureau).

The Millennium Development Goals

In 1997, representatives from the United Nations, the World Bank, and the Organization for Economic Co-Operation and Development (OECD) met to formulate a set of goals for international development that would reduce the extreme poverty in many countries and its various impacts on human well-being by 2015. The goals were sharpened and then adopted as a global compact by all UN members at the UN Millennium Summit in 2000. Table 9–2 lists the eight **Millennium Development Goals (MDGs)** and 17 targets; each goal also had indicators for measuring progress. The MDGs provided targets for developing countries working in partnership with developed countries and development agencies. An unprecedented amount of global attention was focused on the MDGs, including virtually every UN agency, the government agencies of many countries, and scores of NGOs.

Despite their wide support, the MDGs also had critics. For example, the MDGs were criticized because they focused on what donor countries ought to do. Critics

---

Table 9–2   Millennium Development Goals and Targets
(Conditions in 1990 are compared with those in 2015, the year for reaching the targets)

Goal 1. Eradicate extreme poverty and hunger.

- **Target 1:** Reduce by half the proportion of people whose income is less than $1 (now $1.25) a day. 2014: *Achieved.*
- **Target 2:** Reduce by half the proportion of people who suffer from hunger. *Not achieved, but declined from 24% to 14%.*

Goal 2. Achieve universal primary education.

- **Target 3:** Ensure that all children, boys and girls alike, are enrolled in primary school. *Not achieved, but reached 90%.*

Goal 3. Promote gender equality and empower women.

- **Target 4:** Eliminate gender disparities in primary and secondary education by 2005 and in all levels of education by 2015. *Gender parity achieved or close to being achieved in primary education.*

Goal 4. Reduce child mortality.

- **Target 5:** Reduce mortality rates by two-thirds for all infants and children under five. *Not achieved, but reduced by 50%.*

Goal 5. Improve maternal health.

- **Target 6:** Reduce maternal mortality rates by three-quarters. *Not achieved, but reduced by 45%.*
- **Target 7:** Achieve, by 2015, universal access to reproductive health. *Not yet achieved.*

Goal 6. Combat HIV/AIDS, malaria, and other diseases.

- **Target 8:** Have halted, by 2015, and begun to reverse the spread of HIV/AIDS. *Not yet achieved.*
- **Target 9:** Achieve, by 2010, universal access to treatment for HIV/AIDS for all those who need it. *Not yet achieved, but 6.6 million lives saved with treatment.*
- **Target 10:** Have halted, by 2015, and begun to reverse the incidence of malaria and other major diseases. *Close to being achieved.*

Goal 7. Ensure environmental sustainability.

- **Target 11:** Integrate the principles of sustainable development into country policies and programs, and reverse the loss of environmental resources. *Not achieved for several measures.*
- **Target 12:** Reduce by half the proportion of people without sustainable access to safe drinking water and basic sanitation. *Water safety achieved 5 years early; sanitation still not available for many.*
- **Target 13:** Achieve a significant improvement in the lives of at least 100 million slum dwellers by 2020. *Achieved, although slums continue to grow.*

Goal 8. Forge a global partnership for development.

- **Target 14:** Develop a trading system that is open, rule based, and nondiscriminatory—one that is committed to sustainable development, good governance, and a reduction in poverty. *Mixed achievement; trade liberalization occurred, but has slowed.*
- **Target 15:** With Official Development Assistance (ODA), address the special needs of the least developed countries (including the elimination of tariffs and quotas for their exports and an increase in programs for debt relief). *Partially achieved; the least developed countries are improving more rapidly than others, although still very poor.*
- **Target 16:** Deal with the debt problems of all developing countries by taking measures that will make debt sustainable in the long run. *Partially achieved; external service for debt in developing nations fell from 12% of export revenues to 3%.*
- **Target 17:** In cooperation with the private sector, make available the benefits of new technologies, especially information and communications technologies. *Dramatic increase: 40% of the world has Internet access and there are 7 billion cell phone subscriptions worldwide.*

Sources: United Nations, *Millennium Development Goals Report 2014* (New York, United Nations, 2014).

believed that the MDGs had been imposed on poorer communities, did not include the voices of the poorest people they were supposed to serve, and did not adequately address inequality.

A recent summary of global and regional progress toward the MDGs is found in *The Millennium Development Goals Report 2015*, produced by statisticians from a host of UN agencies. Table 9-2 provides a short assessment of the world's progress toward achieving the goals and targets. Progress toward some goals was largely successful. Between 1990 and 2015, at least 1.6 billion people gained access to safe drinking water, making Target 7c of the MDGs one of the first to be attained. Great strides were also made toward lowering the number of people in extreme poverty and increasing access to primary education, especially among girls. In all but two regions, primary school enrollment is now 90% or more; sub-Saharan Africa, which consistently ranks among the lowest regions on many quality-of-life measures, increased primary school enrollment from 58% to 76% in only a decade. The number of deaths from measles fell dramatically; about 80% of children in developing countries are now immunized (Figure 9–11). All regions of the world except for Oceania, sub-Saharan Africa, and Southern Asia saw drops in child mortality of 50% or more (Figure 9–12). In addition, many more women now have skilled attendants during childbirth and 81% of pregnant women have some prenatal care. In fact, the number of maternal deaths from childbirth was halved in less than 20 years.

Progress toward other goals and targets was not as successful. Nourishment of children, lowering maternal mortality, and access to sanitation were among them. The goal with the greatest number of targets where outcomes became worse rather than improving was Goal 7: *Ensure environmental sustainability*. Freshwater resources, forests, and fish stocks all declined, while species extinctions and greenhouse gases increased. On the positive side, however, the use of ozone-depleting CFCs declined, and many more people had access to safe drinking water.

Not all of the MDGs were achieved during the period from 2000 to 2015. Yet it is impressive that many critical areas of life for the poorest improved at the same time that the world population rose from 6 to 7.2 billion. Unfortunately, the environment was often harmed in doing so.

The Sustainable Development Goals

In 2012, just 20 years after the 1992 Rio Earth Summit at which a global agenda for sustainable development was first described, the UN Conference on Sustainable Development was held in Rio de Janeiro, Brazil. At this **Rio+20** meeting, world leaders, NGOs, and representatives of many stakeholder groups hashed out the goals described in a document called *The Future We Want*.[7] The participants agreed that whatever goals followed the end of the Millennium Development Goals had to focus on both development

---

[7]UN, *The Future We Want*, Outcomes Document, United Nations Conference on Sustainable Development, Rio De Janiero, Brazil, June 20–22, 2012, https://rio20.un.org/sites/rio20.un.org/files/a-conf.216l-1_english.pdf.pdf.

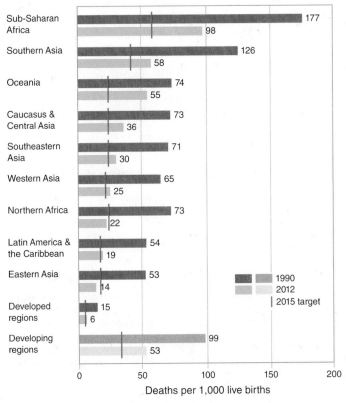

UNDER-FIVE MORTALITY RATE, 1990 AND 2012

Region	1990	2012
Sub-Saharan Africa	177	98
Southern Asia	126	58
Oceania	74	55
Caucasus & Central Asia	73	36
Southeastern Asia	71	30
Western Asia	65	25
Northern Africa	73	22
Latin America & the Caribbean	54	19
Eastern Asia	53	14
Developed regions	15	6
Developing regions	99	53

1990
2012
2015 target

Deaths per 1,000 live births

Figure 9–12 Millennium Development Goal 4, Target 5: *Reduce child mortality*. As a result of efforts toward this goal, deaths among children under five finally dropped below 10 million. About half of these deaths occur in sub-Saharan Africa.

(*Source: Millennium Development Goals Report 2014* (New York: United Nations, 2014).)

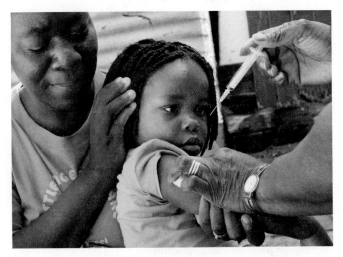

Figure 9–11 Measles vaccination. One success of the Millennium Development Goals has been the dramatic drop in measles. This child in a primary school in Uganda is being vaccinated for measles; today more than 80% of children in the world are vaccinated.

and sustainability. How could they do that while honoring peoples' rights of self-determination? Part of the answer came from a multiyear process during which groups and individuals around the world had an opportunity to express their opinions. As a result of this input, the core drafting team developed the **Sustainable Development Goals (SDGs)** for 2030. A preliminary version of these goals is shown in Table 9–3. The SDGs then went through an approval process by the UN member countries.

How do the SDGs differ from the MDGs? The SDGs are more numerous and complex than the MDGs; there were 8 MDGs, but there are 17 SDGs, each with numerous targets. The SDGs also have more parts focused on environmental sustainability. Goal 2 includes sustainable agriculture; Goal 6 includes clean water and sanitation; Goal 11 includes sustainable cities; Goal 12 addresses consumption; Goal 13 addresses climate change; Goal 14 is about protecting the oceans; and Goal 15 is about protecting terrestrial ecosystems. The inclusion of measures of environmental sustainability throughout goals on development demonstrates a recognition of the connection between environmental degradation and human poverty.

Like the MDGs, the SDGs may be very difficult to achieve. While the MDGs aimed to cut poverty in half by 2015, the SDGs aim to eradicate extreme poverty by 2030. One of the biggest lessons learned in planning the MDGs and SDGs is that countries cannot address global problems individually, especially when the land, atmosphere, and water are a heritage for all.

World Agencies at Work

Several international agencies coordinate development and help countries interact in an increasingly global economy. These include the World Bank and various UN agencies.

The World Bank. In 1944, delegates from 44 nations met in Bretton Woods, New Hampshire, to conceive a vision of development for the countries ravaged by World War II. They established the International Bank for Reconstruction and Development, which is now called the **World Bank**. The World Bank functions as a special agency under the UN umbrella, owned by the countries that provide its funds. As the more developed countries of Europe and Asia got back on

Table 9–3   Sustainable Development Goals. These goals will be accepted by UN member states and be active for the years 2015–2030.
Goal 1. End poverty in all its forms everywhere.
Goal 2. End hunger, achieve food security and improved nutrition, and promote sustainable agriculture.
Goal 3. Ensure healthy lives and promote well-being for all at all ages.
Goal 4. Ensure inclusive and equitable quality education and promote life-long learning opportunities for all.
Goal 5. Achieve gender equality and empower all women and girls.
Goal 6. Ensure availability and sustainable management of water and sanitation for all.
Goal 7. Ensure access to affordable, reliable, sustainable, and modern energy for all.
Goal 8. Promote sustained, inclusive and sustainable economic growth, full and productive employment and decent work for all.
Goal 9. Build resilient infrastructure, promote inclusive and sustainable industrialization and foster innovation.
Goal 10. Reduce inequality within and among countries.
Goal 11. Make cities and human settlements inclusive, safe, resilient, and sustainable.
Goal 12. Ensure sustainable consumption and production patterns.
Goal 13. Take urgent action to combat climate change and its impacts.
Goal 14. Conserve and sustainably use the oceans, seas, and marine resources for sustainable development.
Goal 15. Protect, restore, and promote sustainable use of terrestrial ecosystems, sustainably manage forests, combat desertification, and halt and reverse land degradation and halt biodiversity loss.
Goal 16. Promote peaceful and inclusive societies for sustainable development, provide access to justice for all, and build effective, accountable, and inclusive institutions at all levels.
Goal 17. Strengthen the means of implementation and revitalize the global partnership for sustainable development.

(*Source:* UN Report of Open Working Group of the General Assembly on Sustainable Development Goals, 2014.)

their feet, the World Bank directed its attention toward the developing countries.

With deposits from governments and commercial banks in the developed world, the World Bank now lends money to governments of developing nations at interest rates somewhat below the going market rates. In effect, the World Bank helps governments borrow large sums of money for projects they otherwise could not afford. It provides loans for a wide range of development activities, including activities that support the Millennium Development Goals and the Sustainable Development Goals. Annual loans from the World Bank have climbed steadily, from $1.3 billion in 1949 to $52.6 billion in 2013. Through its power to approve or disapprove loans and the amount of money it lends, the World Bank has become the major single agency providing aid to developing countries.

Over the years, a number of projects funded by the World Bank have been destructive to the environment and to the poorer segments of society. Critics cite many examples wherein the bank's projects exacerbated the cycle of poverty and environmental decline. For example, from 1963 to 1985 the World Bank funneled $1.5 billion into Latin America to clear millions of acres of tropical forests. Most of the cleared land was handed over to large cattle operations that produced beef for export. At that time, the World Bank procedures considered environmental concerns as an add-on rather than as an integral component of its policies.

In the 1990s, the World Bank pressured Bolivia to open its water sources to multinational purchasers. The thinking was that the local governments were corrupt and lacked the expertise to provide water, and that privatization would bring infrastructure and skills. The World Bank threatened to not renew a large loan unless water rights were privatized. This led to civil unrest when prices rose and poor people had limited access to water (Figure 9–13). In 2010, the World Bank loaned $3.75 billion to South Africa to build the world's fourth largest coal-fired power plant. The country had been in an electricity crisis, and major investments in electricity generation had not been made in decades. The plant will increase electricity production but greatly increase harmful environmental effects from coal use (Chapter 14).

In recent years, there has been a sea change in how the World Bank does business with the developing world. The bank helped initiate the Millennium Development Goals. As evidence of a change in its policies, the World Bank adopted a new environmental strategy, *Making Sustainable Commitments: An Environmental Strategy for the World Bank,* in 2002. The strategy "recognizes that sustainable development, which balances economic development, social cohesion, and environmental protection, is fundamental to the World Bank's core objective of lasting poverty alleviation." An analysis of more than 11,000 projects funded by the World Bank to address poverty reduction or biodiversity found that these goals were often mutually reinforcing. The analysis showed that, depending on the objectives, some

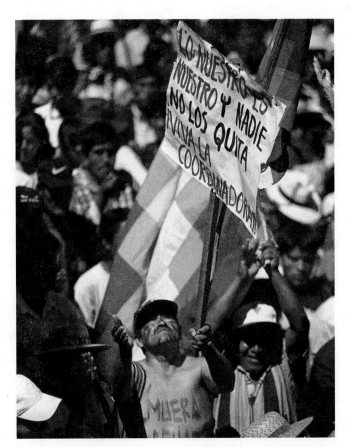

Figure 9–13 Bolivian water wars. This 2000 scene in Cochabamba, Bolivia, features people protesting the privatization of water. The sign says, "What's ours is ours and it cannot be taken away."

60% to 85% of all projects were rated as satisfactory or highly satisfactory.

UN Agencies Involved in Development. To quote its mission statement, the United Nations Development Programme (UNDP) is "the UN's global development network, advocating for change and connecting countries to knowledge, experience and resources to help people build a better life."[8] The agency has offices in 131 countries and works in 166. It helps developing countries attract and use development aid effectively. It also coordinates global and national efforts to reach the MDGs and the SDGs. The UN Food and Agricultural Organization (FAO) and the World Health Organization (WHO) are other important agencies whose work will be discussed in future chapters.

Although the World Bank and UN agencies such as the UNDP and the FAO can be criticized, they are often on the front line of trying to improve the lives of billions who suffer from the effects of poverty. We turn now to the debt crisis. To some extent, this crisis is an outcome of World

---

[8]UNDP Washington Representation Office, July 5, 2012, http://www.us.undp.org/WashingtonOffice/Mission.shtml.

Bank policies, but it is also a consequence of development aid in general.

The Debt Crisis

The World Bank, many private lenders, and other nations lend money to developing countries. In fact, China alone lends as much as the World Bank. Theoretically, development projects should generate revenues that are sufficient for the recipients to pay back their development loans with interest. Unfortunately, a number of things can prevent this outcome, including corruption, mismanagement, and honest miscalculations. Interest obligations climb accordingly, and any failure to pay interest is added to the debt, increasing the interest owed—the credit-debt trap. This cycle has caused developing countries to become increasingly indebted. Their total debt reached $4.8 trillion in 2012.[9] By 2014, developing nations were increasing their debt by $18.5 billion annually. Many countries have paid back in interest many times what they originally borrowed, yet the debt remains and they may never be able to repay their debts.

Short-Term Measures. The debt situation is an economic, social, and ecological disaster for many developing countries. In order to keep up even partial interest payments, poor countries often rely on short-term "solutions" that cause long-term problems. Three such "solutions" are described below.

Adopt austerity measures. Government expenditures are often reduced so that income can be used to pay interest on the debt. But what is cut? Usually, it is funds for schools, health clinics, police protection in poor areas, building and maintenance of roads in rural areas, and other goods and services that benefit not only the poor, but also the country as a whole. As a result of such actions, unemployment may rise (Figure 9–14).

Focus agriculture on growing cash crops for exports. Governments may focus agriculture on growing cash crops for export rather than for feeding the local population. Indigenous people often bear the brunt of such policies; sometimes their land is taken and used to grow crops for export.

Invite the rapid exploitation of natural resources for quick cash. Such exploitation may include logging forests or extracting minerals. With the emphasis on quick cash, few, if any, environmental restrictions are imposed. Thus, the debt crisis has meant disaster for the environment.

In essence, these measures are all examples of liquidating ecosystem capital to raise cash for short-term needs. They do not represent sustainability.

Debt Cancellation. The World Bank is addressing the problems of debt and poverty directly through two initiatives, one to provide small amounts of money to the poorest people and the other to relieve debt in the poorest counties. The Consultative Group to Assist the Poorest (CGAP) is designed to increase access to financial services for very poor households through **microlending**—the lending of small amounts of money to people with very little. The Heavily Indebted Poor Country (HIPC) initiative and its related program, Multilateral Debt Relief Initiative (MDRI), address the debt problem of the low-income developing countries, mostly from sub-Saharan Africa. To qualify, countries have to demonstrate a track record of carrying out economic and social reforms that lead to greater stability and alleviate poverty. By 2014, 36 countries had gone through the application process and, proving themselves stable, had received more than $75 billion in debt relief.[10] In most countries that received this debt relief, spending on education and health has risen substantially.

Development Aid

College students and their parents know all about the differences among scholarships, grants, and loans as they apply for financial aid. Paying back college loans is no picnic for the thousands of students who accept them in order to afford their education. It is all called "aid," but the best kind of aid is the grant or scholarship, which does not have to be repaid. Developing countries are well aware of the difference, too, and have been receiving grant-type aid from private and public donors that is essential to their continued development. How much aid have they been given?

Official Development Assistance. The best records are for official development assistance (ODA) from donor countries. In 2013, the total amount of ODA was $138.4 billion; this total had been increasing for several years, following declining aid in the 1990s and early 2000s. Aid rises and falls with the needs of low-income countries and also with

[9]The World Bank, **2014** International Debt Statistics, http://datatopics.worldbank.org/debt/ids/region/LMY.

Figure 9–14 Unemployment. Austerity measures can worsen unemployment, keeping national debt down but making the lives of people hard. Romania has unemployment up to 80% in some areas, especially among ethnic minorities such as these Sinti.

[10]IMF, Debt Relief Under the Heavily Indebted Poor Countries (HIPC) Initiative Factsheet, September 30, 2014, http://www.imf.org/external/np/exr/facts/hipc.htm.

PUBLIC AID AND PRIVATE REMITTANCES TO HAITI (MILLION $, INFLATION-ADJUSTED)

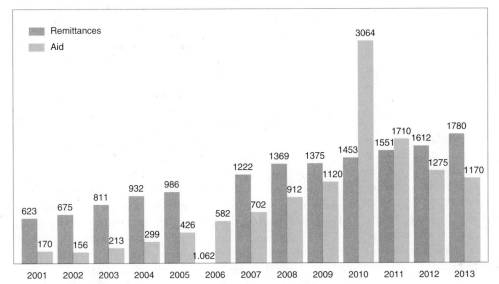

Figure 9–15 Haitian aid and remittances. Haiti has many citizens who live abroad and send money home. Aid and remittances are compared here. The spike in aid in 2010 reflects aid given after a large earthquake.

(*Source: World Bank DAC2, Center for Global Development 2014.*)

world events, conflicts, and donor interests. At the Rio Earth Summit of 1992, the aid target for donor countries was 0.7% of GNI. However, by 1997 actual aid was only 0.22%. It then began to rise, but slowly. In 2005, representatives from donor countries pledged to raise their ODA to 0.35% by 2010, well short of the 0.7% target, but a significant increase from what they were doing. By 2014, only six countries (Denmark, Luxembourg, the Netherlands, Norway, Sweden, and the United Kingdom) had exceeded this target. The average is only 0.3%.

Some of the shortfall in aid is caused by **donor fatigue**, which occurs when high-income countries have tired of handing out money. They see themselves asked for multimillion-dollar rescue packages, only to have a large percentage go to fill the pockets of corrupt leaders or to cover administrative costs. Steps have recently been taken to minimize this misdirection of funds. Allocation of aid is increasingly being tied to policy reform in the recipient countries.

Migration and Remittances. Some 232 million people live outside the country of their birth, and many more second-generation immigrants maintain ties to their native countries. Most of these migrants left their home country looking for employment; wage levels in developed countries are five times those of the countries of origin of most migrants. Migrants typically send much of their money home as **remittances**. This is a surprisingly large flow of funds—estimated at $436 billion in 2014 (just to developing countries) and more than double the amount of international aid flowing to those countries. There is no doubt that remittances alleviate the severity of poverty in the developing countries and can be thought of as a "do it yourself" kind of development aid. Figure 9–15 shows how important remittances are to the economy of Haiti.

Aid from the United States. How much aid does the United States give? In 2013, our ODA donations were $31.5

billion, the largest of any donor nation, but only 0.19% of our GNI. Norway, by contrast, gives the equivalent of 1.07% of its GNI in development aid. However, many private institutions and individuals in the United States also provide support to developing countries. As an example, Table 9–4 details the

Table 9–4	U.S. Economic Contributions to Developing Countries, 2010–2011*
Type of Funding	**Billions of $**
Official Development Assistance (ODA)*	30.9
Private philanthropy	39.0
Foundations	4.6
Corporations	7.6
Private and voluntary organizations	14.0
Volunteerism	3.7
Universities and colleges	1.9
Religious organizations	7.2
Remittances	100.2
Private capital investments	108.4
U.S. total economic engagement	278.5

*Report released in December 2014.

**Some ways to calculate U.S. government foreign aid report much larger numbers. For the most part, this is due to the inclusion of military aid. In 2011, the United States donated more than $15 billion in military assistance. Israel, Iraq, Afghanistan, and Pakistan are the largest recipients.

Source: The Index of Global Philanthropy 2013, Executive Summary, The Center for Global Prosperity (Hudson Institute, 2014), www.globalprosperity.org.

total U.S. contributions in fiscal year 2010–2011, including contributions from foundations, corporations, universities, religious groups, private organizations, and individuals in the United States. That year nongovernmental sources provided $39.0 billion in aid, outstripping our ODA. When remittances and private capital investments are added to these, the total economic contributions from the United States to developing countries amounted to $278.5 billion that year.

Global Financial Crisis. Developing world economies experienced record economic growth during 2002–2007, a result of increased export earnings and substantial foreign investment. The collective GNI of developing countries grew more than 5% per year, peaking at 8% in 2006 and well outpacing the developed countries. However, in late 2007, the U.S. housing bubble burst, leading to a collapse of financial markets that spread to other developed countries. The United States fell into a severe recession, and the recession spread globally.

The global crisis perturbed economies in developing countries in a number of ways, including results of less foreign investment in their markets and a reduction in their exports. These changes caused the growth of GNI in the developing countries to decline. Reduced aid from the developed countries, a decrease in remittances from foreign-born workers in wealthier countries, and an increase in worldwide food prices also hurt poor people around the globe.

Fragile States. Some of the world's poorest countries are designated **fragile states**. These conflict-prone countries lack a stable government that can protect its citizens and provide then with necessary services. For example, a country's government may be unable to protect its citizens from rogue groups during a war. In other situations, the government is corrupt or a military dictatorship.

Fragile states are home to 43% of the people living on less than $1.25 a day, those in the most extreme poverty. Much of the aid such countries receive is humanitarian and disaster relief, but what they most need is to get to a point where they can build infrastructure and social capital. Somalia, which has been at war and lacked a central government for 20 years, is a good example of a fragile state. In the Western Hemisphere, Haiti is a fragile state.

It is extremely difficult for aid agencies to work with fragile states because of the danger to aid workers. Such volatile situations are devastating for women and children. There are often floods of refugees, and rape and violence may be rife. Under such conditions, it may be nearly impossible to implement family planning or other long-term assistance programs. The greatest need is for basic, stable governance and security. When that is achieved other poverty issues may be tractable.

CONCEPT CHECK What are two differences between the Millennium Development Goals and the Sustainable Development Goals? ✓

9.3 A New Direction: Social Modernization

Even though foreign assistance is vital, the developing countries themselves must address the root causes of poverty. This usually means changes in public policy at more local levels. To illustrate, we will consider the relative roles played by development and family planning.

As the case of Thailand demonstrates, the shift from high to low fertility rates in the poorer developing countries does not require the economic trappings of a developed country. Instead, what is needed are efforts within the country, made on behalf of the poor, with particular emphasis on improving education, especially literacy and the education of girls and women; lowering infant mortality and improving life expectancy; making family planning available and affordable; enhancing employment opportunities; and managing resources to maintain ecosystem sustainability. As may be obvious, most of these activities are well aligned with the Sustainable Development Goals. That is, development targeted at alleviating the problems of the poorest people will also bring down population growth rates (Section 9.2). Similarly, the solutions to problems that contribute to population growth are the solutions to problems that cause people to have large families.

A great deal of focus is on women, because they not only bear children, but are also the primary providers of nutrition, childcare, hygiene, and early education. When women have more education and some income of their own, they are more likely to spend money on education and health care for children and less likely to have large numbers of children. Next we will briefly consider each part of social modernization.

Improving Education

Investing in the education of children (and adults who lack schooling) is a key element of the public-policy options of a developing country, one that returns great dividends (Figure 9–16). Providing education in developing countries

Figure 9–16 Education in developing countries. These girls are Afghan refugees in Pakistan, attending an NGO-funded school.

can be quite cost effective because any open space can suffice as a classroom and few materials are required. Evidence shows that providing secondary schooling for at least 50% of children pulls large segments of a poor country's population out of poverty and is an important factor in decreasing fertility and improving the health of children. In 1960, for example, Pakistan and South Korea had similar incomes and population growth rates (2.6%), but very different school enrollments—94% in South Korea versus 30% in Pakistan. Within 25 years, South Korea's economic growth was three times that of Pakistan's. Today the rate of population growth in South Korea has declined to 0.5% per year, while Pakistan's is still 2.2% per year. The lowest net enrollment of children in primary schools is in sub-Saharan Africa, where 33 million children are out of school, 56% of them girls.

Improving Health

The health care most needed by poor communities in the developing world is good nutrition and basic hygiene—steps such as boiling water to avoid the spread of disease and properly treating infections and common ailments such as diarrhea. (In developing countries, diarrhea is a major killer of young children. It is easily treated by giving suitable liquids, a technique called *oral rehydration therapy*.) The discrepancy in infant mortality rates (infant deaths per 1,000 live births) between developed countries (6/1,000) and developing countries (46/1,000) speaks for itself.

Life Expectancy. One universal indicator of health is human life expectancy. In 1955, average life expectancy globally was 48 years. Today, it is around 68 years for boys and 73 years for girls and rising.[11] This progress is the result of

---

[11]World Health Organization, World Health Statistics 2014, http://www.who.int/mediacentre/news/releases/2014/world-health-statistics-2014/en/.

social, medical, and economic advances in the latter half of the 20th century that have substantially increased the well-being of large segments of the human population. Recall that the epidemiologic transition is the trend of decreasing death rates that is seen in countries as they modernize (Chapter 8). Recall also that as this transition occurred in the now-developed countries, it involved a shift from high mortality rates (due to infectious diseases) to the present low mortality rates (due primarily to diseases of aging, such as cancer and cardiovascular diseases).

Despite the progress indicated by the general rise in life expectancy, there is a large difference in life expectancy between developed and developing countries. Women typically have longer life expectancies than men. Women in Spain have a life expectancy of 85 years, and women in Japan a life expectancy of 87 years, the longest life expectancy. Chad and Afghanistan have female life expectancies as low as 52 years; male life expectancies are even lower. Sierra Leone has the lowest life expectancies: 38 years for males and 46 years for females. The discrepancy between the most developed and least developed countries can be seen quite clearly in the distribution of deaths by main causes, shown in Figure 9–17. Infectious diseases were responsible for 47% of mortality in the least developed countries, compared with only 6% in the developed countries.

The solution is very evident: Primary health services need to be centered on people's needs, they must provide health security in the context of people's communities, and there must be universal access to primary health care. Disease prevention and the promotion of health are critical components of this primary health care. Health care in the developing world must emphasize maternal prenatal and postnatal care for mothers and children. Many government, charitable, and religious organizations are involved in providing basic health care; when this is extended to clinics in rural areas, it is one of the most effective ways of delivering primary health care to families.

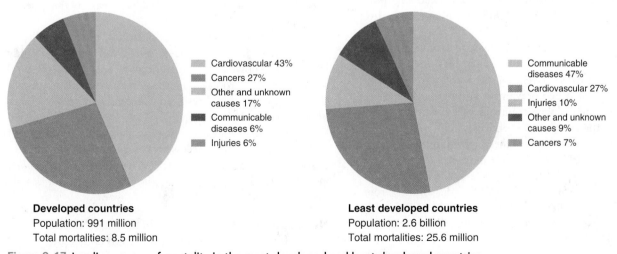

Developed countries
Population: 991 million
Total mortalities: 8.5 million

- Cardiovascular 43%
- Cancers 27%
- Other and unknown causes 17%
- Communicable diseases 6%
- Injuries 6%

Least developed countries
Population: 2.6 billion
Total mortalities: 25.6 million

- Communicable diseases 47%
- Cardiovascular 27%
- Injuries 10%
- Other and unknown causes 9%
- Cancers 7%

Figure 9–17 Leading causes of mortality in the most developed and least developed countries.
Infectious diseases are predominant in the least developed countries, while cancer and diseases of the cardiovascular system predominate in the highly developed countries.

(*Source:* Data from World Health Organization, *Projections of Mortality and Burden of Disease, 2002–2030* (2008).)

Reproductive Health. When women are better off, whole populations are improved. This is the focus of **reproductive health,** a fairly new concept in population matters. Reproductive health focuses on women and infants and, in some cases, on reproductive education and health for men. Key components include prenatal care, safe childbirth, postnatal care, information about birth-control options, prevention and treatment of sexually transmitted diseases (STDs), and prevention and treatment of infertility. Promotion of reproductive health also involves protecting women from violence and coercion. For example, reproductive health requires the elimination of violence against women (coercive sex and rape), sexual trafficking, and female genital mutilation (a traditional practice in some African societies).

Reproductive health care was strongly emphasized at the 1994 International Conference on Population Development held in Cairo, Egypt. It also underpins a number of the MDGs; Goal 5 (*Improve maternal health*) was expanded in 2005 to include reproductive health services. In the SDGs, Target 3.b of Goal 3 is similar (*Ensure universal access to sexual and reproductive health for all*).

HIV/AIDS. One of the greatest challenges to health care in developing countries is **acquired immunodeficiency syndrome (AIDS),** a sexually transmitted disease caused by the human immunodeficiency virus (HIV). The global epidemic of AIDS is most severe in many of the poorest developing countries, which are least able to cope with the consequences. This disease devastates the economies of communities by killing people in the prime of their working lives, increasing the dependency ratio as children are left in the care of other children or the elderly.

Most approaches to prevention involve getting people to change their sexual behavior—to practice safe sex, limit their sexual partners, and avoid prostitution. That these things can be accomplished is proven by the case of Uganda, where a 70% decline in HIV prevalence has been registered in recent years, linked to a 60% reduction in casual sex. The Ugandan government has mounted a national AIDS control program that includes a major education campaign.

Another effective preventive measure is male circumcision, which causes a 60% reduction in HIV risk in males involved in heterosexual relationships and protects more women than most other preventive strategies. The WHO strongly recommends that countries offer male circumcision as one of their anti-AIDS services.

Thailand once had the highest HIV infection rate in South Asia, but public education campaigns in the 1990s pulled the rate down, although it rose again in the 2000s. After a change in government in 2006, the country reinvigorated its public health efforts to combat HIV infections. An increase in funding from the developed countries has helped Thailand and other countries to combat this disease. Major sources of funding are the World Bank; the Global Fund to Fight AIDS, Tuberculosis and Malaria; the U.S. President's Emergency Plan for AIDS Relief; and local country budgets.

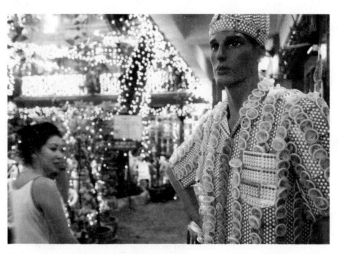

Figure 9–18 Family planning. Thailand reduced its high fertility in part through a humorous public education campaign. Cabbages and Condoms, shown here, is a restaurant chain whose profits fund the Population and Community Development Association, a Thai nonprofit.

Family Planning

Information on contraceptives, supplies of contraceptives, and treatments are the most direct ways to bring down high population growth. In poor nations, family planning is supported by agencies that in turn are supported by governments, international groups, and NGOs. The stated policy of family-planning agencies is to *enable people to plan their own family size—that is, to have children only if and when they want them.* In addition to helping people avoid unwanted pregnancies, family planning often involves diagnosing and overcoming fertility problems for those couples who are having reproductive difficulties.

Family planning is a critical component of reproductive health care (Figure 9–18). Those countries that have implemented effective family-planning programs have experienced the most rapid decline in fertility. In Thailand, for example, a vigorous family-planning program was initiated as part of a national population policy in 1971. Population growth subsequently declined from 3.1% per year to the present 0.5% per year.

In addition to stabilizing populations, the availability of contraceptives allows people to avoid suffering. One study in Tanzania showed an increase in contraceptive use in rural households after crop loss. Householders were able to prevent pregnancies that would result in children likely to die of malnourishment.[12] This strategy prevents a great deal of hardship but only works if contraception is available and affordable.

Women who lack access to contraception but want to postpone or prevent childbearing are said to have an *unmet need.* In developing countries, the unmet need for family planning ranges from 10% to 50% of women.

Negotiating Disagreements. The major document resulting from the 1994 Cairo population conference (ICPD

---

[12]S. Alam, "Economic shock of crop loss increases contraceptive use among women of rural Tanzania," (2013) Population Resource Bureau, http://www.prb.org/Publications/Articles/2013/tanzania-crop-loss.aspx.

Program of Action) states that family planning should focus on education and the provision of services directed at avoiding unwanted or high-risk pregnancies. It explicitly states that abortions should not be used as a means of family planning. This statement was neither aimed at eliminating abortion nor at supporting a right to abortion for women worldwide. Instead, it focused on maintaining support for family planning among stakeholders who disagreed about the controversial issue of abortion. Research shows that the rate of abortions is higher when family planning services such as contraception and education are unavailable than when they are available. It is hard to make policy on an international level because of social and individual disagreement about important issues. The ICPD Program of Action helped forge a middle ground between people with very disparate views and emphasized the importance of family planning services to women worldwide.

Planned Parenthood, which operates clinics throughout the world, is probably the best-known family planning agency. Another significant player is the United Nations Population Fund (UNFPA), which provides financial and technical assistance to developing countries at their request. The emphasis of UNFPA is on combining family planning services with maternal and child health care and expanding the delivery of such services in rural and marginal urban areas. Support for UNFPA and other family planning agencies is a contentious subject in the United States because of the strong opinions surrounding abortion.

U.S. Contributions to International Family Planning.
In 1984 and again in 2001, a presidential order prohibited the U.S. government from giving aid to foreign family-planning agencies if they provide abortions, counsel women about abortion if they are dealing with an unwanted pregnancy, or advocate for abortion law reforms in their own country. Because the United States had been the largest international donor to family-planning programs, these prohibitions were a blow to efforts to increase the availability of contraceptive services in developing countries. At one point, the United States withdrew its support for the entire 1994 ICPD Program of Action because it uses terms such as *reproductive services* and *reproductive health care*. Critics of this move pointed out that the Cairo agreement was critical to the fight against HIV/AIDS and prevented the very unwanted pregnancies that could lead to abortions. U.S. funding for family planning varies with political changes.

Employment and Income

Any economic system is based on the exchange of goods and services. In a simple barter economy, people agree on direct exchanges of certain goods or services. Barter economies are still widespread in the developing world. The introduction of a cash economy facilitates the exchange of a wider variety of goods and service, and everyone may prosper, as they have a wider market for what they can provide and a wider choice of what they can get in return.

In a poor community, people may have the potential to provide certain goods or services and may want other things in return, but there is no money to get the system off and running. In a growing economy, people who wish to start a new business generally begin by obtaining a bank loan to set up shop. In poor communities, however, it is very difficult to obtain bank loans. There are three main reasons why: Most banks consider the poor to be high credit risks; the poor often want smaller loans than commercial banks are willing to deal with; and many of the poor are women who are denied credit because of gender discrimination. Fortunately, this situation is changing.

In 1976, Muhammad Yunus (Figure 9–19), an economics professor in Bangladesh, conceived of and created a new kind of bank (now known as the **Grameen Bank**) that would engage in microlending to the poor. As their name implies, *microloans* are small—the cumulative average was $458 in 2015—and short-term, usually just four to six months. Nevertheless, they provide such basic things as seed and fertilizer for a peasant farmer to start growing tomatoes, pans for a baker to start baking bread, a supply of yarn for a weaver, tools for an auto mechanic, or an inventory of goods to start a grocery store. By 2015, nearly 6.2 million small loans had been made.

Yunus secured his loans by having the recipients form **credit associations**—groups of several people who agreed

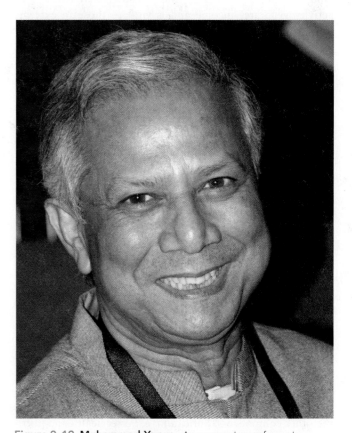

Figure 9–19 Muhammad Yunus. An economics professor in Bangladesh, Yunus created the Grameen Bank, which initiated the microlending movement. The Grameen Bank has loaned more than $6.6 billion over the past 35 years, and the microlending model has been duplicated in more than 40 countries. Yunus received the 2006 Nobel Peace Prize in recognition of the global impact of his work.

to be responsible for each other's loans. With this arrangement, the Grameen Bank has experienced an exceptional rate of payback—greater than 97%. Small-scale agriculture loans from the Grameen Bank have had outstanding results. In a rural area of Bangladesh, small loans, along with horticultural advice, are now enabling peasant farmers to raise tomatoes and other vegetables for sale to the cities. These people doubled their incomes in three years. Microlending has been found to have the greatest social benefits when it is focused on women.

The unqualified success of microlending in stimulating economic activity and enhancing the incomes of people within poor communities has been so remarkable that the Grameen concept has been adopted in more than 40 countries around the world (including the United States), with various modifications. In 2014, there were 10,000 microfinance institutions around the world. These efforts may do more than anything else to meet the MDGs and SDGs, as they empower the poor to improve their own well-being.

Resource Management

The world's poor depend on local ecosystem capital resources—particularly water, soil for growing food, and forests for firewood. Many lack access to enough land to provide an income and often depend on foraging in woodlands, forests, grasslands, and coastal ecosystems. This activity generates income and is vital to those in extreme poverty; 90% of the world's poorest people depend on forests for some of their sustenance (by extracting fuel wood, construction wood, wild fruits and herbs, fodder, or "bush meat" for both subsistence and cash). Forests, fisheries, reefs, grasslands, and waterways can all be common-pool resources for the poor (see Chapter 7). These can be a safety net, as well as an employment source. But if common-pool resources are not managed effectively, they are liable to be overused, especially when populations are increasing.

Protecting forest resources and tackling land degradation are two sustainable steps that can be addressed by country policies. One approach that has worked in Nepal is empowering the poor to manage community- or state-owned lands (Figure 9–20). There groups that use forests are given the right to own trees, but not the land. The groups develop forest-management plans, set timber sale prices, and manage the surplus income from the operations. There are now more than 14,000 user-groups managing 1.3 million hectares (3 million acres) of forest.

Putting It All Together

Each of the five components of social modernization—education, health, family planning, employment and income, and resource management—depends on and supports the other components. For example, better health and nutrition support better economic productivity, better economic productivity supports obtaining a better education, and a better education leads to a delay in marriage and the desire to have fewer children. The availability of family-planning services is essential to realizing the desire of parents to have

Figure 9–20 Improving resource management. These Nepalese women are planting saplings in a community reforestation project that will increase the amount of wood available for harvest.

fewer children. All the components work together to improve lives. Conversely, the lack of any component undercuts the ability to achieve all the other components.

Sustainable Development. Social modernization is entirely compatible with the MDGs and SDGs (Tables 9–2 and 9–3). As we move forward, we also need to protect valuable ecosystem services. Any form of modernization needs to be compatible with **sustainable development,** development that meets the needs of the present without compromising the ability of future generations to meet their own needs.[13]

Globalization. In their progression toward economic development and social modernization, the developing countries must confront the realities of globalization. The accelerating interconnectedness of human activities, ideas, and cultures is profoundly changing many human enterprises, with significant economic impacts. A small number of developing countries (China, Malaysia, and Indonesia, for example) have successfully entered the global market. Their exports have paved the way for economic progress, and these countries are becoming more connected to the economic- and information-rich high-tech world.

People in the poorest countries need to be included in a market economy. This requires jobs, and that often requires investment from outside agencies—businesses and NGOs with the knowledge and capital needed. In addition, there must be some hope of marketing the products, which means entering the global market on a level playing field. But at present the playing field is far from level because developed countries protect their exports with subsidies, quotas, and tariffs; these trade barriers work against the developing countries (Chapter 2). Goal 17 of the SDGs (*Strengthen the means of implementation and revitalize the global partnership for sustainable development*) calls for eliminating such barriers.

---

[13]World Commission on Environment and Development, *Our Common Future* (Oxford: Oxford University Press, 1987), 27.

Globalization can be harmful to the poor if it opens markets for agricultural products from the developed countries, which can ruin subsistence farming. Globalization can also widen the gap between the rich and poor within a society, as the wealthier members capitalize on the information flow and quick transfer of goods and services, leaving the poor farther behind. Aid can have a similar effect, undermining local economies. This is one reason that the 2005 Paris Declaration (an agreement on aid) emphasized that donor countries should buy local products where possible, rather than donating from abroad.

Critics of globalization claim that it does not seem capable of promoting financial stability and social justice. It also has the potential to dilute or even destroy cultural norms and religious ideals when it is uncritically welcomed into developing countries. Whether helpful or harmful, globalization is taking place, making it essential that the developing countries confront it head on and adapt its connections to their needs and cultures.

CONCEPT CHECK How does globalization make solutions to poverty more difficult? How does it make them easier?

REVISITING THE THEMES

SOUND SCIENCE

The sciences of demography, epidemiology, and sociology provide insights into the causes of the demographic transition and the demographic dividend. Demographers test hypotheses about what factors most predict, and in some cases cause, a drop in fertility. Medical advances and health care have helped bring many countries through the epidemiologic transition, yet there is still great need for basic health care in the poorer countries. Sound science has brought us improved AIDS treatment through anti-HIV medicines, but has not yet provided an effective vaccine. In some parts of the world, aging populations require a new set of medical and social supports.

SUSTAINABILITY

The demographic transition is essential for a sustainable future. This chapter describes what has to be done economically and socially to bring about sustainable development. Economic progress is being made in many developing countries, and it is accompanied by most of the elements of social modernization. However, economic progress is still lacking in many others. For these countries, social modernization is the most feasible approach, as it will bring fertility rates down and enable countries to open the demographic window that may make economic progress possible.

An element of development activities that is unsustainable is the debt built up over the years by many poorer developing countries. The repayment of these debts forces many developing countries to liquidate ecosystem capital in order to make their payments. In some cases, this problem has been alleviated by debt cancellation.

Sustainable development requires the integration of the principles of environmental sustainability into country policies and programs. Improving the management of environmental resources is an important component of social modernization. The 2012 Rio+20 UN conference and the Sustainable Development Goals reaffirmed the need to protect the environment even as we improve people's lives.

Social sustainability is another aspect of sustainability. Large-scale economic efforts by the World Bank and other organizations have sometimes harmed the poor and increased inequality. In order to be socially sustainable, any solution to economic or environmental issues needs to be just. This is one of the reasons the Grameen Bank, owned by the poor, has been successful.

STEWARDSHIP

Justice for the developing world was one of the key ethical issues identified in the discussion of stewardship in Chapter 1. *Distributive justice* is an ethical ideal that considers the gross inequality of the human condition as unjust—the inequality represented by the billion-plus people who live on less than $1.25 a day versus the income typical of people in the wealthy countries. Although equal distribution of the world's goods and services is neither attainable nor necessary, the existing poverty around the globe is an affront to human dignity. In order to address this injustice, the alleviation of poverty must be made a priority of public policy.

The case study of Thailand contrasts with those of China and India. All three countries have seen drops in high fertility, but Thailand achieved this change primarily through education, clinics, free contraception, and other parts of social modernization. China's drop in fertility involved some of these elements, but was also the result of involuntary policies associated with some serious human rights abuses. Because of the authoritarian approach taken by China, many people worldwide see Chinese population control as symbolic of the problems of government control. The fertility drop in India has not been as large as those in China and Thailand, but some Indian states, such as Kerala, have begun to stabilize their population, largely through social modernization.

REVIEW QUESTIONS

1. What are the two basic schools of thought regarding the demographic transition? How were these reflected in the three most recent global population conferences?

2. Discuss five specific factors that influence the number of children a poor couple may desire.

3. What are the MDGs? Where did they come from? Cite an example of one goal and a target used to measure progress in attaining it.

4. What are the SDGs? How do they compare to the MDGs?

5. What are two major agencies that promote development in poor nations? How do they carry out their work?

6. What is meant by the debt crisis of the developing world? What is being done to help resolve this crisis?

7. What is development aid, and how does it measure up against the need for such aid?

8. What are the five interdependent components that must be addressed to bring about social modernization?

9. Define *family planning*, and explain why it is critically important to all other aspects of development. What is meant by an unmet need for family planning?

10. What is microlending? How does it work?

11. What is the significance of globalization for economic and social development in the developing countries?

THINKING ENVIRONMENTALLY

1. Is the current world population below, at, or above the optimum? Defend your answer by pointing out things that may improve and things that may worsen as the population increases.

2. Suppose you are the head of an island nation with a poor, growing population and natural resources that are being degraded. What kinds of policies would you initiate to provide a better, sustainable future for your nation's people? What kind of help would you ask for?

3. List and discuss the benefits and harms of writing off debts owed by developing nations.

4. Look up the document *The Future We Want*. Describe what future you would want on a personal, regional, national, and global level for 30 years from now. How much depends on environmental health? How does the future you want match what other people around the globe hope for?

5. Do you believe contraceptives should be made available free of charge by governments, as this would have the potential to curb population growth and/or abortion rates? Defend your answer.

MAKING A DIFFERENCE

1. Think carefully about your own expectations for the number of children you will have. What concerns will you and your mate weigh as you plan your family?

2. Consider writing to an elected representative about an issue related to sustainable development, such as international aid, microloans, or support for human rights.

3. If you are not convinced of the profound impacts of poverty in the developing world, log on to the Millennium Project Web site and read about the Millennium Village movement (www.unmillenniumproject.org/mv/index.htm).

4. Encourage sustainability by buying products that originate from the developing world.

MasteringEnvironmentalScience®

Students Go to MasteringEnvironmentalScience for assignments, the eText, and the Study Area with animations, practice tests, and activities.

Professors Go to MasteringEnvironmentalScience for automatically graded tutorials and questions that you can assign to your students, plus Instructor Resources.

U.S. Drought Monitor September 2, 2014

Intensity:

Abnormally Dry
Moderate Drought
Severe Drought
Extreme Drought
Exceptional Drought

Water: Hydrologic Cycle and Human Use

Drought in the American Southwest. It's a tough time in America's almond-growing region—a section of California's Central Valley that produces more than 80% of the world's almonds. For several years this region has languished in deep drought, threatening the survival of the almond orchards. Almonds are a water-needy crop—it takes about 1.1 gallons of water to produce each nut. Indeed, almond trees require a great investment of water during all the years they are growing, including the first five years when they produce no nuts. The millions of gallons of water used to grow almonds cannot simply be turned off in a drought year. If farmers do not irrigate the almond trees, they will die, and the farmer will lose an investment of years.

California contains one sixth of the irrigated land in the country. Much of the water used for this irrigation has been diverted from far-away rivers. Those diversions endanger the salmon in the rivers, which require more water than is left for them. Some of the water used for irrigation is pumped from new wells that tap into ancient reserves of water that percolated underground millions of years ago. That groundwater is being withdrawn faster than any rain could replace it, depleting an underground resource owned by all.

Mega-Drought. California is not the only state in drought—the drought encompasses much of the Southwest. The drought, which has been going on since 2000, is now being called a "mega-drought." It is estimated to be the fifth most severe drought in that region since AD 1000. Stressed by the lack of water, trees in the forests are dying. The dry forests sit like tinder, waiting for a spark to set off a wildfire. The risk of wildfire is increased by the heavy load of unburned brush lying on the forest floor, the product of years of fire suppression. Large,

▲ These California almond trees are being removed because there is not enough water to irrigate them.

severe wildfires are now seven times more frequent than they were in the 1970s. Attempts to save water, dam rivers, pump water, and desalinate ocean water are all a part of the story of water in this region.

Around the World. The American Southwest is a microcosm of the world. All over, people are trying to balance the water supply to meet the needs for agriculture, drinking water, and biodiversity. In 2010 and 2011, southwestern China had its worst drought in 60 years. During that drought, the Chinese people who lived in the dry, wheat-growing areas relied on drinking water brought to them on trucks. Parts of Australia are also suffering severe drought, an effect scientists believe is made worse by climate change. Yet in other places, severe flooding pollutes waterways and sweeps away homes.

Today freshwater is a new form of gold, a precious commodity to be measured and traded. As we welcome new people into our crowded global community, finding clean, safe water for all will be one of our top priorities.

Chapter 10, the first of four chapters on essential resources, focuses on water. The concepts in this chapter are strongly intertwined with those we have already seen in chapters on population and development and with later chapters on soil, food production, and climate change.

10.1 Water: A Vital Resource

Water is absolutely fundamental to life as we know it. Happily, Earth is virtually flooded with water. A total volume of some 1.4 billion cubic kilometers (325 million mi^3) of water covers 75% of Earth's surface. About 97.5% of this volume is the saltwater of the oceans and seas. The remaining 2.5% is **freshwater**—water with a salt content of less than 0.1% (1,000 ppm). This is the water on which most terrestrial biota, ecosystems, and humans depend. Of the 2.5%, though, two-thirds is bound up in the polar ice caps and glaciers. Thus, only 0.77% of all water is found in lakes, wetlands, rivers, groundwater, biota, soil, and the atmosphere (Figure 10–1). Nevertheless, evaporation from the oceans combines with precipitation to resupply that small percentage continually through the solar-powered hydrologic cycle, described in detail in Section 10.2. Thus, freshwater is a continually renewable resource.

Streams, rivers, ponds, lakes, swamps, estuaries, groundwater, bays, oceans, and the atmosphere all contain water, and they all represent ecosystem capital—goods and services vital to human interests (see Table 5–2 for the array of aquatic ecosystems and their characteristics). They provide drinking water, water for industries, and water to irrigate crops. Bodies of water furnish energy through hydroelectric power and control flooding by absorbing excess water. They provide transportation, recreation, waste processing, and habitats for aquatic plants and animals. Freshwater is a vital resource for all land ecosystems, modulating the climate through evaporation and essential global warming (when water is in the atmosphere as water vapor). During the past two centuries, many of these uses have led us to construct a huge infrastructure designed to bring water under control. We have built dams, canals, reservoirs, aqueducts, sewer systems, treatment plants, water towers, elaborate pipelines, irrigation systems, and desalination plants. As a result, waterborne diseases have been brought under control in the developed countries, vast cities thrive in deserts, irrigation makes it possible to grow 40% of the world's food, and one-fifth of all electricity is generated through hydropower. These great benefits are especially available to people in the developed countries (Figure 10–2a).

In the developing world (Figure 10–2b), by contrast, 0.8 billion people still lack access to safe drinking water, 2.1 billion do not have access to adequate sanitation services, and more than 1.4 million deaths each year are traced to waterborne diseases (mostly in children under five). All too often in these countries, water is costly or inaccessible to the poorest in society, while the wealthy have it piped into their homes. In addition, because of the infrastructure that is used to control water, whole seas are being lost, rivers are running dry, millions of people have been displaced to make room for reservoirs, groundwater aquifers are being pumped down, and disputes over water have raised tensions from local to international levels. Today, freshwater is a limiting resource in many parts of the world and is certain to become even more so as the 21st century unfolds.

Feeding a growing world population will challenge our water management infrastructure, especially as cities and industries compete with agriculture for scarce water resources. The growing impact of climate change on the hydrologic cycle is in turn worsening the problems brought on by droughts and floods. Rainfall variability can reduce the economic growth of nations as floods devastate some croplands and droughts dry them up in other parts of the

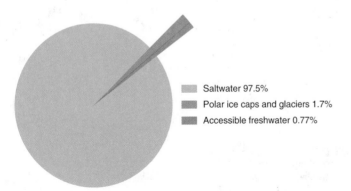

- Saltwater 97.5%
- Polar ice caps and glaciers 1.7%
- Accessible freshwater 0.77%

Figure 10–1 Earth's water. The Earth has an abundance of water, but terrestrial ecosystems, humans, and agriculture depend on accessible freshwater, which constitutes only 0.77% of the total.

(a)   (b)

Figure 10–2 Differences in availability. Human conditions range from **(a)** being able to luxuriate in water to **(b)** having to walk long distances and pull from deep pumps to obtain enough simply to survive. In many cases, however, the abundance of water is illusory because water supplies are being overdrawn.

world. Climate change also plays a role as warming increases evaporation and worsens the effects of droughts.

There are two ways to consider water issues. The focus in this chapter is on *quantity*—that is, on the global water cycle and how it works, on the technologies we use to control water and manage its use, and on public policies we have put in place to govern our different uses of water. Some of the terms we will be using to describe the water cycle are listed and defined in **Table 10–1**. Later we will focus on water *quality*—that is, on water pollution and its consequences, on sewage-treatment technologies, and on public policies for dealing with water pollution issues (Chapter 20).

CONCEPT CHECK As the polar ice caps melt, how do the relative percentages of Earth's saltwater and freshwater change? ☑

Table 10–1 Terms Commonly Used to Describe Water

Term	Definition
Water quantity	The amount of water available to meet demands.
Water quality	The degree to which water is pure enough to fulfill the requirements of various uses.
Freshwater	Water having a salt concentration below 0.1%. Rain, snow, rivers, lakes, and groundwater are freshwater.
Saltwater	Water that contains at least 3% salt (30 parts salt per 1,000 parts water), typical of oceans and seas.
Brackish water	A mixture of freshwater and saltwater, typically found where rivers enter the ocean.
Hard water	Water that contains minerals, especially calcium or magnesium, that cause soap to precipitate, producing a scale or scum.
Soft water	Water that is relatively free of minerals.
Polluted water	Water that contains one or more impurities, making it unsuitable for a desired use.
Purified water	Water that has had pollutants removed or is rendered harmless.
Storm water	Water from precipitation that runs off of land surfaces in surges.
Green water	Water in vapor form originating from water bodies, the soil, and organisms.
Blue water	Water in liquid form, including precipitation, renewable surface water runoff, and groundwater recharge.

10.2 Hydrologic Cycle and Human Impacts

Water moves through the land, oceans, and atmosphere in a continuous process called the hydrologic cycle. This cycle is represented in Figure 10–3. The basic cycle consists of water rising to the atmosphere through **evaporation** (the transformation of water molecules from a liquid to a gaseous state) and **transpiration** (the loss of water vapor as it moves from the soil through green plants and exits through leaf pores) and returning to the land and oceans through **condensation** (the formation of liquid water from a gaseous state) and **precipitation** (the release of water from clouds in the form of rain, sleet, snow, or hail).

Modern references to the hydrologic cycle distinguish between **green water** and **blue water**. Green water is water in vapor form, while blue water is liquid water wherever it occurs. Green water and blue water are vitally linked in the hydrologic cycle.

Evaporation, Condensation, and Purification

Recall that a weak attraction known as hydrogen bonding tends to hold water molecules (H_2O) together (Chapter 3). Below 32°F (0°C), the kinetic energy of the molecules is so low that the hydrogen bonding is strong enough to hold the molecules in place with respect to one another, and the result is *ice*. At temperatures above freezing but below boiling (212°F or 100°C), the kinetic energy of the molecules is such that hydrogen bonds keep breaking and re-forming with different molecules. The result is *liquid water*. As the water molecules absorb energy from sunlight or an artificial source, the kinetic energy they gain may be enough to allow them to break away from other water molecules entirely and enter the atmosphere. This is the process of evaporation, resulting in **water vapor**—water molecules in the gaseous state.

You will learn later that water vapor is a powerful greenhouse gas (Chapter 18). It contributes about two-thirds

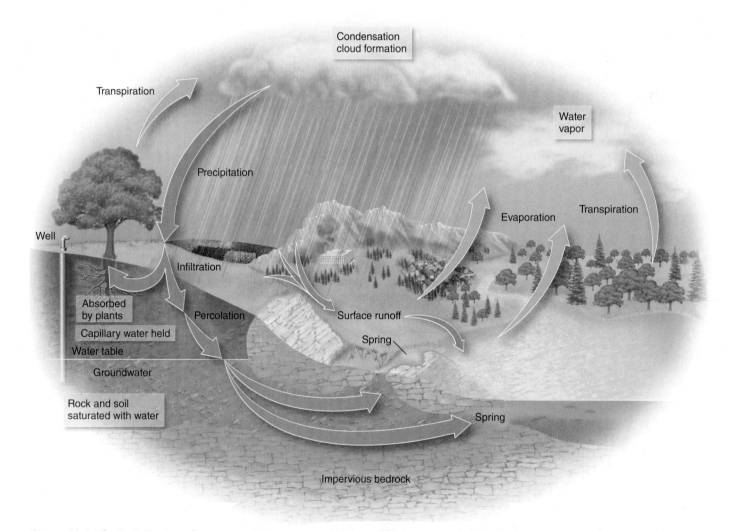

Figure 10–3 The hydrologic cycle. The Earth's freshwater is replenished as water vapor enters the atmosphere by evaporation and transpiration from vegetation, leaving salts and other impurities behind. As precipitation hits the ground, three pathways are possible: surface runoff, infiltration, and reabsorption by plants. Green arrows depict green water, present as vapor in the atmosphere; blue arrows show liquid water.

of the total warming from all greenhouse gases. The amount of water vapor in the air is the **humidity**. Humidity is generally measured as **relative humidity,** the amount of water vapor as a percentage of what the air can hold at a particular temperature. For example, a relative humidity of 60% means that the air contains 60% of the maximum amount of water vapor it could hold at that particular temperature. The amount of water vapor air can hold changes with the temperature. When warm, moist air is cooled, its relative humidity rises until it reaches 100%. Further cooling causes the excess water vapor to condense back to liquid water because the air can no longer hold as much water vapor (Figure 10–4).

Condensation is the opposite of evaporation. It occurs when water molecules rejoin by hydrogen bonding to form liquid water. If the droplets form in the atmosphere, the result is fog and clouds (fog is just a very low cloud). If the droplets form on the cool surfaces of vegetation, the result is dew. Condensation is greatly facilitated by the presence of **aerosols** in the atmosphere. Aerosols are microscopic liquid or solid particles (like hair spray) originating from land and water surfaces. They provide sites that attract water vapor and promote the formation of droplets of moisture. Aerosols may originate naturally, from sources such as volcanoes, wind-stirred dust and soil, and sea salts. Anthropogenic sources—sulfates, carbon, and dust—contribute almost as much as natural sources (more in some locations) and can have a significant impact on regional and global climates.

Purification occurs when water is separated from the solutes and particles it contains. The processes of evaporation and condensation purify water naturally. When water in an ocean or a lake evaporates, only the water molecules leave the surface; the dissolved salts and other solids remain behind in solution. When the water vapor condenses again, it is thus purified water—except for the pollutants and other aerosols it may pick up from the air. The water in the atmosphere turns over every 10 days, so water is constantly being purified. (The most chemically pure water for use in laboratories is obtained by distillation, a process of boiling water and condensing the vapor.)

Thus, evaporation and condensation are the source of all natural freshwater on Earth. Freshwater that falls on the land as precipitation gradually makes its way through aquifers, streams, rivers, and lakes to the oceans (blue water flow). In the process, it carries along salts from the land, which eventually accumulate in the oceans. Salts also accumulate in inland seas or lakes, such as the Great Salt Lake in Utah. The salinization of irrigated croplands is a noteworthy human-made example of this process (Chapter 11). Most of the freshwater from precipitation, however, is returned directly to the atmosphere by evaporation and transpiration.

Precipitation

Warm air rises from Earth's surface because it is less dense than the cooler air above. As the warm air encounters the lower atmospheric pressure at increasing altitudes, it gradually expands and cools—a process known as **adiabatic cooling**. When the relative humidity reaches 100% and cooling continues, condensation occurs and clouds form. As condensation intensifies, water droplets become large enough to fall as precipitation. **Adiabatic warming,** by contrast, occurs as cold air descends and is compressed by the higher air pressure in the lower atmosphere.

Precipitation on Earth ranges from near zero in some areas to more than 2.5 meters (100 in.) per year in others. The distribution depends on patterns of rising and falling air currents. As air rises, it cools, condensation occurs, and precipitation results. As air descends, it tends to become warmer, causing evaporation to increase and dryness to result. A rain-causing event that you see in almost every television weather report is the movement of a cold front. As the cold front moves into an area, the warm, moist air already there is forced upward because the cold air of the advancing front is denser. The rising warm air cools, causing condensation and precipitation along the leading edge of the cold front.

Two factors—global convection currents and rain shadows—may cause more or less continuously rising or falling air currents over particular regions, with major effects on precipitation.

Convection Currents. At the equator, the rays of the Sun are nearly perpendicular to the Earth's surface; at the poles, light rays strike the land at an angle. That difference causes the surface of the Earth to be hotter in some places than others and drives great air movements, called *global convection currents*. As the air at the equator is heated, it expands, rises, and cools; condensation and precipitation occur. The constant intense heat in these equatorial areas ensures that this process is repeated often, thus causing high rainfall, which, along with continuous warmth, supports tropical rain forests.

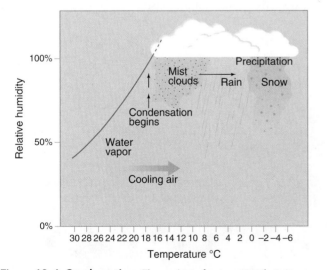

Figure 10–4 Condensation. The amount of water vapor that air can hold changes with the temperature. The red line follows an air mass that starts with 40% relative humidity (RH) at 30°C and is cooled. When the mass of air cools to 17°C, it has reached 100% RH, and condensation begins, forming clouds. If we started with a higher RH, clouds would form sooner when the air is cooled. Further cooling and condensation result in precipitation.

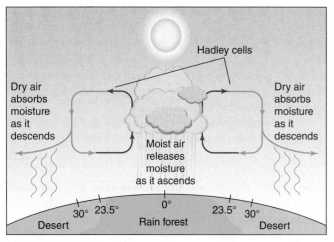

(a) Hadley cells at the equator

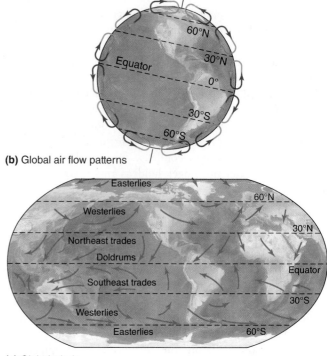

(b) Global air flow patterns

(c) Global winds

Figure 10–5 Global air circulation. (a) The two Hadley cells at the equator. **(b)** The six Hadley cells on either side of the equator, indicating general vertical airflow patterns. **(c)** Global wind patterns, formed as a result of Earth's rotation.

Rising air over the equator is just half of the convection current, however. The air, now dry, must come down again. Pushed from beneath by more rising air, it "spills over" to the north and south of the equator and descends over subtropical regions (25° to 35° north and south of the equator), resulting in subtropical deserts. The Sahara of northern Africa and the Kalahari of southern Africa are prime examples. The two halves of the system composed of the rising and falling air make up a **Hadley cell (Figure 10–5a** and **b).** Because of Earth's rotation, winds are deflected from the strictly vertical and horizontal paths indicated by a Hadley cell and tend to flow in easterly and westerly directions—the

easterly **trade winds** of the tropics, the mid-latitude westerlies, and the polar easterlies **(Figure 10–5c).**

Rain Shadows. When moisture-laden winds encounter mountain ranges, the air is deflected upward, causing cooling and high precipitation on the windward side of the range. As the air crosses the range and descends on the other side, it becomes warmer and increases its capacity to pick up moisture. Hence, deserts occur on the leeward (downwind) sides of mountain ranges. The dry region downwind of a mountain range is referred to as a **rain shadow (Figure 10–6).** For example, westerly winds, full of moisture from the Pacific Ocean,

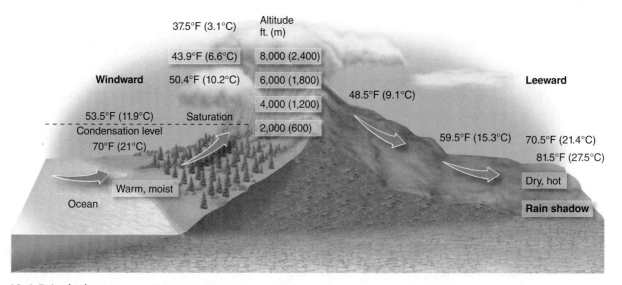

Figure 10–6 Rain shadow. Moisture-laden air cools as it rises over a mountain range, resulting in high precipitation on the windward slopes. Desert conditions result on the leeward side as the descending air warms and tends to evaporate water from the soil.

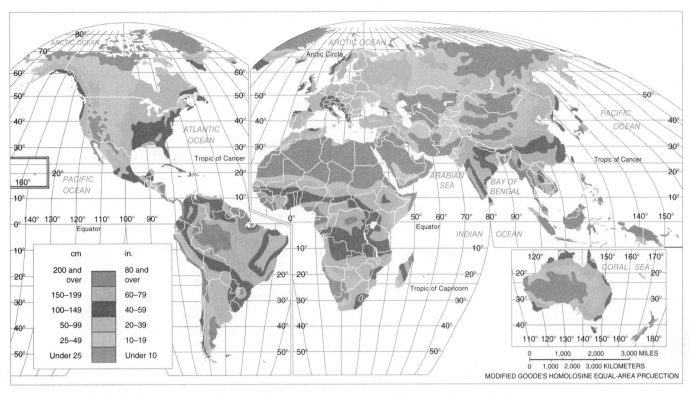

Figure 10–7 Global precipitation. Note the high rainfall in equatorial regions and the regions of low rainfall to the north and south.

(*Source:* Robert W. Christopherson, *Elemental Geosystems,* 7th ed., Pearson Education, Upper Saddle River, NJ.)

strike the Sierra Nevada Mountains in California. As the winds rise over the mountains, water precipitates out, supporting the lush forests on the western slopes. Immediately east of the southern Sierra Nevada, however, lies Death Valley, one of the driest regions of North America.

The general patterns of atmospheric circulation and the effects of mountain ranges combine to give great variation in the precipitation reaching the land (**Figure 10–7**). The distribution of precipitation in turn largely determines the biomes and ecosystems found in a given region (Chapter 5).

Groundwater

As precipitation hits the ground, it may either soak into the ground (**infiltration**) or run off the surface. The amount that soaks in compared with the amount that runs off is called the **infiltration-runoff ratio.**

Runoff flows over the surface of the ground into streams and rivers, which make their way to the ocean or to inland seas. All the land area that contributes water to a particular stream or river is referred to as the **watershed** for that stream or river. All ponds, lakes, streams, rivers, and other waters on the surface of Earth are called **surface waters.**

Water that infiltrates has two alternatives (Figure 10–3). The water may be held in the soil in an amount that depends on the water-holding capacity of the soil. This water, called **capillary water,** returns to the atmosphere either by evaporation from the soil or by transpiration through plants (a green water flow). The combination of evaporation and transpiration is referred to as **evapotranspiration.**

The second alternative is **percolation** (a blue water flow). Infiltrating water that is not held in the soil is called **gravitational water** because it trickles, or percolates, down through pores or cracks under the pull of gravity. Sooner or later, however, gravitational water encounters an impervious layer of rock or dense clay. It accumulates there, completely filling all the spaces above the impervious layer. This accumulated water is called **groundwater,** and its upper surface is the **water table** (Figure 10–3). Gravitational water becomes groundwater when it reaches the water table in the same way that rainwater is called lake water once it hits the surface of a lake. Wells must be dug to depths below the water table. Groundwater, which is free to move, then seeps into the well and fills it to the level of the water table.

Recharge. Groundwater will seep laterally as it seeks its lowest level. Where a highway has been cut through rock layers, you can frequently observe groundwater seeping out. Layers of porous material through which groundwater moves are called **aquifers.** It is often difficult to determine the location of an aquifer. Many times these layers of porous rock are found between layers of impervious material, and the entire formation may be folded or fractured in various ways. Thus, groundwater in aquifers may be found at various depths between layers of impervious rock. Also, the **recharge area**—the area where water enters an aquifer—may be many miles away from where the water leaves the aquifer. Underground aquifers hold some 99% of all *liquid* freshwater; the rest is in lakes, wetlands, and rivers.

Underground Purification. As water percolates through the soil, debris and bacteria from the surface are generally filtered out. However, water may dissolve and leach out certain minerals. Underground caverns, for example, are the result of the gradual leaching away of limestone (calcium carbonate). In most natural situations, the minerals that leach into groundwater are harmless. Thus, groundwater is generally high-quality freshwater that is safe for drinking. A few exceptions occur when the groundwater leaches minerals containing sulfide, arsenic, or other poisonous elements that make the water unsafe to drink.

Drawn by gravity, groundwater may move through aquifers until it finds some opening to the surface. These natural exits may be **seeps** or **springs**. In a seep, water flows out over a relatively wide area; in a spring, water exits the ground as a significant flow from a relatively small opening. As seeps and springs feed streams, lakes, and rivers, groundwater joins and becomes part of surface water. A spring will flow, however, only if it is *lower* than the water table. Whenever the water table drops below the level of the spring, the spring will dry up. Groundwater movement constitutes one of the loops in the hydrologic cycle.

Loops, Pools, and Fluxes in the Cycle

The hydrologic cycle consists of four physical processes: evaporation, condensation, precipitation, and gravitational flow. There are three principal loops in the cycle: (1) In the evapotranspiration loop (consisting of green water), the water evaporates and is returned by precipitation. On land, this water—the main source for natural ecosystems and rain-fed agriculture—is held as capillary water and then returns to the atmosphere by way of evapotranspiration. (2) In the surface runoff loop (containing blue water), the water runs across the ground surface and becomes part of the surface water system. (3) In the groundwater loop (also containing blue water), the water infiltrates, percolates down to join the groundwater, and then moves through aquifers, finally exiting through seeps, springs, or wells, where it rejoins the surface water. The surface runoff and groundwater loops are the usual focus for human water resource management.

In the hydrologic cycle, there are substantial exchanges of water among the land, the atmosphere, and the oceans; such exchanges are the cycle's fluxes. **Figure 10–8** shows

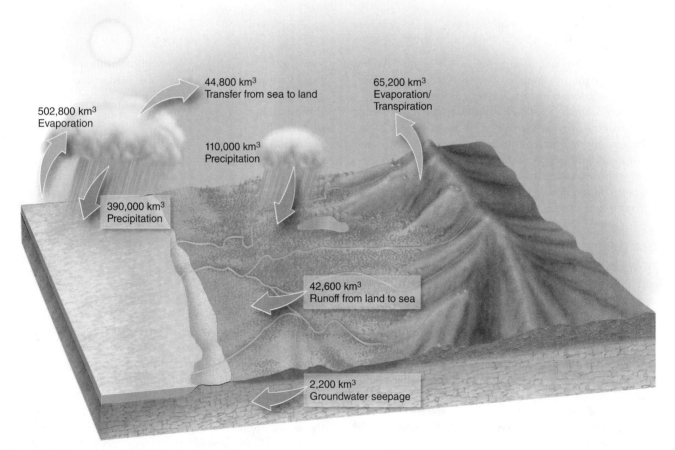

Figure 10–8 Yearly water balance. The data show the contribution of water from the oceans to the land via evaporation and precipitation, the movement of water from the land to the oceans via runoff and groundwater seepage, and the net balance of water movement between the land and oceans. Values are in cubic kilometers per year.

(*Source:* Figure 4.2 from *Earth Science*, 8th ed., by Edward Tarbuck and Frederick K. Lutgens, copyright © 1997 by the authors, reprinted by permission of Pearson Education, Inc., Upper Saddle River, NJ; revised with data from *GEO Yearbook 2003*, United Nations Environment Programme.)

estimates of these fluxes and of the sinks that hold water globally. About 97% of the Earth's water is in the oceans. Each year a volume of water evaporates off of the ocean surface that would cover it to 100 cm deep, about 425,000 cubic kilometers. Additional water evaporates from land and ice. All together the hydrologic cycle annually moves the equivalent of 25 times the water in the Great Lakes. Absent from Figure 10–8, however, are the human impacts on the cycle, which are large and growing.

Human Impacts on the Hydrologic Cycle

A large share of the environmental problems we face today stem from direct or indirect impacts on the water cycle. These impacts can be classified into four categories: (1) changes to Earth's surface, (2) changes to Earth's climate, (3) atmospheric pollution, and (4) withdrawals (or diversions) for human use.

Changes to the Surface of Earth. Imagine a temperate forest, its trees rooted in thick soil, drawing water from below, sending water up and out leaves, each tree stopping raindrops in a storm, letting them run down the trunk. The soil and leaf litter slow the water, letting it soak in. Then imagine cutting down the trees. What would happen? Put up a parking lot, a strip mall, a highway. Soon, the very processes that make up the water cycle will be altered. Alterations to the land surface stem primarily from removing vegetation, putting down impervious surfaces, and degrading wetlands.

Loss of Vegetation. In most natural ecosystems, precipitation is intercepted by vegetation and infiltrates into porous topsoil. From there, water provides the lifeblood of natural and human-created ecosystems. As forests are cleared or land is overgrazed and plants are not there to intercept rainfall, the pathway of the water cycle is shifted from infiltration and groundwater recharge to runoff, and the water runs into streams or rivers almost immediately. This causes floods, surface erosion, sediment, and other pollutants. Increased runoff means less infiltration and, therefore, less evapotranspiration and groundwater recharge. Lowered evapotranspiration means less moisture for local rainfall. The amount of groundwater may be insufficient to keep springs flowing during dry periods. Dry, barren, and lifeless streambeds are typical of deforested regions—a tragedy for both the ecosystems and the humans who are dependent on the flow.

Taming Rivers and Wetlands. Wetlands function to store and release water in a manner similar to the way the groundwater reservoir does, only on a shorter time scale. Therefore, the destruction of wetlands has the same impact as deforestation: Flooding is increased, droughts are also increased, and waterways are polluted. In the Midwest, many fields have been provided with underground drainage pipes (tiles) to move water more quickly so the fields can be planted. This action lowers the water table, and the result is predictable: Rainwater is moved more quickly from the land to the rivers to the ocean. Channelizing rivers (by straightening them) has much the same effect (Chapter 7).

Figure 10–9 Impervious surface. Impervious surfaces, such as those of roofs, roads, and parking lots, decrease the amount of water that can go into the soil and recharge groundwater and increase the amount of water and the speed with which water runs off surfaces.

Building Impervious Surface. Roads, buildings, parking lots, and other structures cover the soil, so even a slow steady rain cannot infiltrate.[1] In some places, impervious surfaces cover 90% of the land. The lower 48 states have more than 47,000 square miles of impervious surface, bigger than the size of Ohio (Figure 10–9).

Dams. One of the chief ways people try to lower flood risk is by building dams. Dams store water and slow the movement of water to the ocean. However, because they are not wetlands, they often do not result in increased groundwater recharge. In dry areas, the pooling of water in reservoirs can leave water exposed to evaporation. Ironically, evaporation from reservoirs can cause regional water loss even when the purpose of building the reservoir was to provide water during drought. Dams also cause some serious ecological problems. The Three Gorges Dam on the Yangtze River is a good example.

Flooding on the Yangtze River took more than 300,000 lives in the 20th century. This fact, combined with a need for power, prompted the building of one of the most spectacular dams in the world, the Three Gorges Dam. Completed in 2006, it is the largest hydroelectric project in the world (in terms of its maximum capacity), holding back 1.4 trillion cubic feet of water and generating as much as 84.7 billion kilowatt-hours of electricity annually, as much as the energy produced by burning 50 million tons of coal. The power generated raised the proportion of China's electricity generated from renewable sources from 7% to 15%. It will, for a time, bring much of the major flood potential of the Yangtze under control.

However, the dam has come with a high price. More than 1.3 million people have been displaced and relocated, and another 3 million to 4 million people will be relocated in coming years to make way for the 600-kilometer-long (410-mile-long) reservoir. The reservoir was 30% full in 2014 and is expected to

---

[1]Thomas R. Schueler, "The Importance of Imperviousness," reprinted in *The Practice of Watershed Protection* (Ellicott City, MD: Center for Watershed Protection, 2000).

be filled in 2017. As it fills, lowering the Yangtze River flow by 50%, saltwater has crept upriver at the mouth of the river, ships have been stranded, pollutants have accumulated, and many fish populations have been reduced due to interference with breeding cycles. Large reservoirs can increase earthquakes, a phenomenon called *reservoir-induced seismicity.* Between 2003 and 2009, the area around the Three Gorges Dam experienced 3,429 earthquakes, a 30-fold increase over the pre-dam period. In spite of the many drawbacks, Chinese officials continue with the project.

Moving Water to the Ocean. Whereas dams slow the movement of water, nearly all other human activities that change the surface of the land cause rainwater to flow to the ocean more quickly and lessen the amount going into groundwater. Deforestation, cultivation, the filling of wetlands, and the development of roads and other impervious surfaces increase the rate of runoff and erosion and reduce the rate of infiltration. These actions contribute to both floods and droughts and increase erosion (Chapter 11). Floods have always been common. In many parts of the world, however, the frequency and severity of flooding are increasing—not because precipitation is greater, but because land use has changed. One study of floods in 56 developing countries has confirmed the "sponge" effect of forests, concluding that continued loss of forests will lead to greater loss of lives and economies for the developing countries where floods are frequent. For example, extreme flooding in Bangladesh (a very flat country only a few feet above sea level) is now more common because the Himalayan foothills in India and Nepal have been deforested. Due to sediment deposited from upriver erosion, the Ganges River basin has risen 5–7 meters (15–22 ft.) in recent years. Floods are also increasing as a result of climate change. **Figure 10–10** shows the impact of floods that occurred in southeastern and central Europe in 2014, when thousands of people were displaced.

Mitigating Changes to Water Flow. Solutions to water problems caused by land use changes include changes in

Figure 10–10 Flooding in the Balkans. People were evacuated from their flooded houses in the town of Obrenovac in northern Serbia in 2014 when the region experienced its heaviest rainfall in a century.

requirements for new construction. In many places, developers are now required to build flood retention ponds near new construction. Although retention ponds lack the biodiversity of natural wetlands and often contain pollutants, they do stop floods and allow infiltration. The protection of wetlands and river vegetation has been important to slowing run-off. Another solution is to grow more vegetation in cities. Tree-lined streets, green parks, urban gardens, and rooftop plantings can increase soil infiltration and also lower heat. There is also a move to go to new types of surfaces, especially for driveways, that are more environmentally friendly. These solutions will be discussed under sustainable cities as well (Chapter 23).

Climate Change. There is now unmistakable evidence that Earth's climate is warming because of the rise in greenhouse gases, and as this occurs, the water cycle is being altered (Chapter 18). A warmer climate means more evaporation from land surfaces, plants, and water bodies because evaporation increases with temperature. A wetter atmosphere means more and, frequently, heavier precipitation and more flood events. In effect, global warming is believed to be speeding up the hydrologic cycle. Already, the United States and Canada have experienced a 5–10% increase in precipitation over the past century. Regional and local changes are difficult to project; different computer models predict different outcomes for a given region. However, global climate models project a radically changing hydrological world as the 21st century unfolds (Chapter 18). The climatologists calculate that a warmer climate will likely generate more hurricanes and more droughts; extended droughts in the western United States are likely to occur more frequently. They also calculate that many of the currently water-stressed areas like East Africa and the Sahel region will get less rainfall, which will bring large agricultural losses (Chapter 12).

Climate change has two other key impacts on water bodies. First, warm water has a greater volume. Warming the oceans contributes to sea level rise. Second is the movement of water from ice (the cryosphere) to liquid water. Glaciers and polar ice melt, and that water makes its way to the ocean. Both thermal expansion and melting ice contribute to the sea level rise that is already apparent. Melting ice adds freshwater, which is lighter than saltwater and does not mix well. The influx of great volumes of freshwater may affect ocean currents (see Sound Science, *What's New in the Water Cycle?*). Models suggest that oceans will rise about 2 to 3 feet (0.8 to 1 m) by 2100. While that does not sound like much, more than 630 million people live in low elevation areas, less than 33 feet (10 m) above sea level. You can imagine that many will have to move and others will experience flooding.

Atmospheric Particles. Tiny atmospheric particles (aerosols) form nuclei that enable water to condense into droplets. The more such particles there are, the greater is the tendency for clouds to form. Anthropogenic aerosols are on the increase, primarily in the form of sulfates (from sulfur dioxide in coal), carbon (in soot), and dust. They form a brownish

SOUND SCIENCE
What's New in the Water Cycle?

The water cycle is pretty straightforward on a first approximation. Every middle schooler has studied at least the basics. But there is still more to learn about this complex system. Scientists are particularly interested in determining how the global hydrologic system will be affected by human actions.

Peter Wadhams, a professor of oceanographic physics at the University of Cambridge, England, wanted to know more about the ocean and climate change. He wondered how ocean currents will be affected by changes in the water cycle as ocean temperatures rise. There are large ocean currents (thermohaline currents) that move vast quantities of water in a slow-moving "ocean conveyor." Wind and tides provide the energy for these currents. The speed, volume, and direction of currents can be affected by climate change because of two principles: cold water is denser than warm, and freshwater is less dense than saltwater. When deep water rises, it brings nutrients toward the surface; as surface water sinks, it brings oxygen down. Waters warmed in the tropics redistribute heat to the cold poles.

These currents help warm Europe and the east coast of North America. But there may be changes in the future. As glacial melt water and rain run-off from Siberia increase in the Arctic, cold freshwater forms a layer on top of the saltwater there. Enough freshwater might change the ocean conveyor, some scientists have hypothesized. In 2005, Wadhams and his colleagues investigated the cold dense waters of the Greenland Sea where columns of very dense water form.

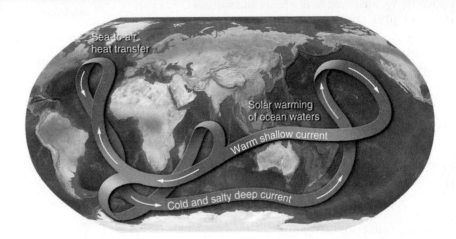

These "chimneys" (which can be 100 km in diameter and which last for months) carry water down more than 2,400 m towards the sea floor, contributing to the movement of masses of cold water in the deep ocean. But instead of the dozen gigantic rivers of water flowing downward that they had found other times, they found only two. Does this mean the deep ocean currents will slow? Is the change long term? Right now, Wadhams and others aren't sure what will happen. Some scientists say any shift in the thermohaline circulation is likely to occur slowly; others think a slowing of the conveyor could occur more abruptly.

In 2012, scientists from the University of Washington, including graduate student Sarah Purkey, explored the deepest, coldest waters on Earth—off of Antarctica. They found that those waters, dubbed the

Antarctic Bottom Water, have declined in volume by 60% (losing an equivalent of 50 times the flow of the Mississippi River) over the past several decades. The deep ocean is becoming less cold and dense, and scientists expect this to change the way the oceans work. Even though the basic hydrologic cycle is well known, the details of how it works and what changes we might expect in a changing environment are still part of today's science. Wadhams, Purkey, and all of their colleagues are on the cutting edge of sound science.

Sources: Wunsch, C. "What Is the Thermohaline Circulation?" *Science* 298(5596) (2002): 1179–1181. doi:10.1126/science.1079329. PMID 12424356.

Wadhams, P. "Convective Chimneys in the Greenland Sea: A Review of Recent Observations." *Oceanography and Marine Biology* 42, 1–27

Purkey, S. G., and G. C. Johnson "Global Contraction of Antarctic Bottom Water Between the 1980s and 2000s." *Journal of Climate*, 2012.

haze that is associated with industrial areas, tropical burning, and dust storms. Their impact on cloud formation is substantial, and where these aerosols occur, solar radiation to Earth's surface is reduced (so they have a cooling effect).

Their most significant impact, however, is on the hydrologic cycle. Because aerosols promote the formation of smaller-than-normal droplets in the clouds, aerosols actually *suppress* rainfall where they occur in abundance, even though they encourage cloud formation. As they do so, the atmospheric cleansing that would normally clear the aerosols is suppressed, and they remain in the atmosphere longer than usual. With suppressed rainfall comes drier conditions, so more dust and smoke (and more aerosols) are the result. The impact of aerosol pollution is just opposite to the impact of the greenhouse gases on climate and the hydrologic cycle.

Again, there are significant differences; the aerosol impact is more local, whereas the impact of greenhouse gases on climate and the hydrologic cycle is more global. In addition, the greenhouse gases are gradually accumulating in the atmosphere, while the aerosols have a lifetime measured in days.

Withdrawal for Human Use. Finally, there are many problems related to the fourth impact, diversion and withdrawals of water for human use. We will discuss these impacts next, as we look at how humans try to have enough clean freshwater—and not too much.

CONCEPT CHECK Where would you expect to see more green water (water vapor) forming—at the equator or 30 degrees north of the equator? ✓

10.3 Water: Getting Enough, Controlling Excess

Meeting water needs is a primary goal in our rapidly crowding world. We are seeing both stresses and successes. Between 1990 and 2012, 2 billion people were given access to clean drinking water, an achievement that met the Millennium Development Goal for water. However, water needs are very serious in many countries.

According to the Millennium Ecosystem Assessment, in 2005, humans were appropriating about 3,600 cubic kilometers of water per year, which represents 27% of the accessible freshwater runoff on Earth. Annual global withdrawal is expected to rise by about 10% every decade. While the human population tripled in the 20th century, use of water increased sixfold, much of that increase to enable water intensive higher-yielding agriculture.

The U.S. Geological Service periodically prepares estimates of major uses of freshwater in the United States (Table 10–2). It is significant that Americans use less water now than they did in 1980, even with population increases (Figure 10–11). Most of the water used in homes and industries is for washing and flushing away unwanted materials,

Table 10–2 U.S. Demands on Water*	
Use	Millions of Gallons (Liters) per Day
Consumptive	
Irrigation and other agricultural use	407 (1,540)
Nonconsumptive	
Electric power production	519 (1,964)
Industrial use	68 (257)
Public supply and self-supply	147 (556)

*About 15% of water withdrawals are saline, primarily used for thermoelectric power.

Source: M. A. Maupin, J. F. Kenny, S. S. Hutson, J. K. Lovelace, N. L. Barber, and K. S. Linsey, *Estimated Use of Water in the United States in 2010* (Circular 1405) (Reston, Va.: U.S. Geological Survey, 2014).

and the water used in electric power production is for taking away waste heat. These are **nonconsumptive water uses** because the water, though now contaminated with wastes, remains available to humans for the same or other uses if

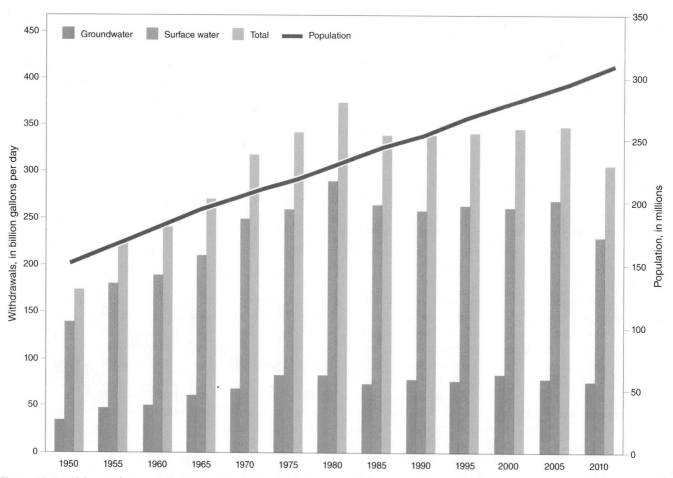

Figure 10–11 U.S. trends in population and freshwater withdrawals. Since 1980, Americans have been using less freshwater (86% of total water use), even though the population has increased substantially.

(*Source:* Maupin, M. A., J. F. Kenny, S. S. Hutson, J. K. Lovelace, N. L. Barber, and K. S. Linsey. *Estimated Use of Water in the United States in 2010* (Circular 1405) (Reston, VA: U.S. Geological Survey, 2014).)

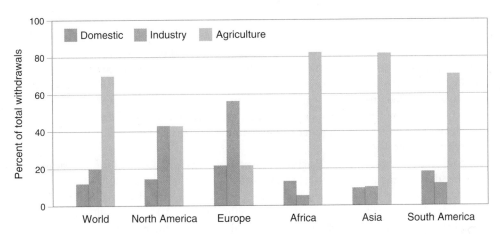

Figure 10–12 Regional usage of water. The percentage used in each category varies with the climate and relative development of the region. A less developed region with a dry climate (e.g., Africa) uses most of its water for irrigation, whereas an industrialized region (e.g., Europe) requires the largest percentage for industry.

(*Source:* FAO. AQUASTAT database, 2015. Accessed February 19, 2015.)

its quality is adequate or if it can be treated to remove undesirable materials. In contrast, irrigation is a **consumptive water use** because the applied water does not return to the local water resource. It can only percolate into the ground or return to the atmosphere through evapotranspiration. In either case, the water does reenter the overall water cycle but is gone from human control.

Worldwide, the largest use of water is for irrigation (70%), second is for industry (20%), and third is for direct human use (10%) **(Figure 10–12)**. These percentages vary greatly from one region to another, depending on precipitation and the degree to which the region is developed. Most increases in withdrawal are due to increases in irrigated lands. In the United States, irrigation accounts for nearly 80% of freshwater consumption.

Water is also essential for the process of *hydraulic fracturing*, or *fracking*, a water-intensive process for removing natural gas from shale (Chapter 14). Each gas well may require 4–6 million gallons of water; this water becomes contaminated with chemicals and remains underground.[2] According to a recent report, 40% of countries rich in shale deposits face water scarcity in their shale-deposit regions. These countries are having to choose between using water for fracking or for human consumption.

Sources

In the United States, about 25% of domestic water comes from groundwater sources and 75% from surface waters (rivers, lakes, reservoirs). Before municipal water supplies were developed, people drew their water from whatever source they could: wells, rivers, lakes, or rainwater. This approach is still used in rural areas in the developing countries. Women in many of these countries walk long distances each day to fetch water. Because surface waters and shallow wells often receive runoff, they frequently are polluted with various wastes, including animal excrement and human sewage likely to contain pathogens (disease-causing organisms). In

fact, an estimated 90% of wastewater in developing countries is released to surface waters without any treatment. Yet, unsafe as it is, this polluted water is the only water available for an estimated 1.1 billion people in less developed countries **(Figure 10–13)**. According to the UN World Health Organization, contaminated water is responsible for the deaths of more than 1.6 million people in developing countries each year, 90% of whom are children. Increasing water sanitation is part of the Sustainable Development Goals.

Technologies. In the industrialized countries, the collection, treatment, and distribution of water are highly developed. Larger municipalities rely primarily on surface-water sources, while smaller water systems tend to use groundwater. In the former, dams are built across rivers to create reservoirs, which hold water in times of excess flow and can be drawn down at times of lower flow. In addition, dams and reservoirs may offer power generation, recreation, and flood control.

Figure 10–13 Water in the developing world. In many villages and cities of the developing world, people withdraw water from rivers, streams, and ponds. These sources are often contaminated with pathogens and other pollutants. This Kenyan man is collecting water from a river used by people and animals.

Reservoirs created by dams on rivers are also major sources of water for irrigation. In this case, no treatment is required. However, water for municipal use is piped from the reservoir to a treatment plant, where it is treated to kill pathogens and remove undesirable materials, as shown in Figure 10–14. After treatment, the water is distributed through the water system to homes, schools, and industries. Wastewater, collected by the sewage system, is carried to a sewage-treatment plant, where it is treated before being discharged into a natural waterway (Chapter 20). Often, wastewater is discharged into the same river from which it was withdrawn, but farther downstream.

Whenever possible, both water and sewage systems are laid out so that gravity maintains the flow through the system. This arrangement minimizes pumping costs and increases reliability. On major rivers, such as the Mississippi, water is reused many times. Each city along the river takes water, treats it, uses it, treats it again, and then returns it to the river. In developing nations, the wastewater frequently is discharged with minimal or no treatment. Thus, as the water moves downstream, each city has a higher load of pollutants to contend with than the previous city had, and ecosystems at the end of the line may be severely affected by the pollution. Pollutants include industrial wastes as well as pollutants from households, because industries and residences generally utilize the same water and sewer systems.

Smaller public drinking-water systems frequently rely on groundwater, which often may be used with minimal treatment because it has already been purified by percolation through soil. More than 1.5 billion people around the world rely on groundwater for their domestic use. Both surface water and groundwater are replenished through the water cycle (Figure 10–3). Therefore, in theory at least, these two sources of water represent a sustainable or renewable (self-replenishing) resource, but they are not inexhaustible. Over the past 50 years, groundwater use has outpaced the rate of aquifer recharge in many areas of the world.

Is Bottled Water the Answer? One-half of Americans buy bottled water, which is sometimes seen as a solution to limited water availability, particularly after disasters. Sales of bottled water were $12 billion (about 10 billion gallons) in 2013 in the United States alone. Unfortunately, use of bottled water has a number of negative environmental effects. First, the bottling can occur on a large scale in areas where locals also want to use the water. Does a company have the right to pump out millions of gallons of water from an aquifer? This has been disputed. For example, after many years in the courts, in 2009, Michigan Citizens for Water Conservation won a lawsuit to keep Nestlé Waters North America, Inc., from pumping 400 gallons of water a minute from a groundwater supply in Michigan. The proposed pumping was drying a local stream. Nestlé was allowed to pump, but at a reduced rate. Bottled water also requires a great deal of energy, for the making of plastic bottles and the transportation of the product. Overall, bottled water may be good for emergencies but is not a good solution for a lack of drinking water generally.

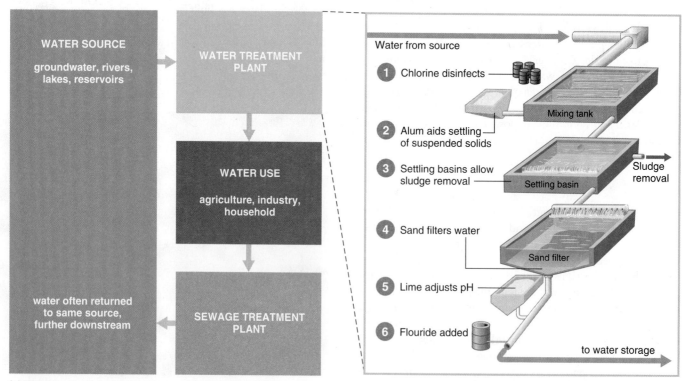

(a) Water is often taken from a river or reservoir, piped to the treatment plant, treated, used, then returned to the source.

(b) At the treatment plant, water is treated to kill pathogens and remove undesirable material.

Figure 10–14 **Municipal water use and treatment.** The stages of water treatment are outlined in the figure. Water is taken from a river or reservoir, piped to a treatment plant, treated with chemicals, filtered, adjusted for pH, and stored before being used.

Surface Waters

Human activities affect the flow of surface water in many ways. Surface waters are diverted for various uses, slowed and held in dams, and moved along swiftly by levees, impervious surfaces, channelized rivers, and the removal of vegetation. Of these activities, the use of dams is the most prominent for keeping and using surface water.

Dams and the Environment. To trap and control flowing rivers, more than 48,000 large dams (more than 50 feet high) have been built around the world. Dams together have a storage capacity around 16,000 km^3, serving 30–40% of the world's irrigated land and providing 19% of the world's total electricity supply. Of the 42,600 cubic kilometers of annual runoff in the hydrologic cycle, only 31% is accessible for withdrawal because many rivers are in remote locations and much of the remaining water is required for navigation, flood control, and the generation of hydroelectric power. Sixty-five percent of ocean bound freshwater is stopped by dams at some point.

Large dams also have an enormous direct social impact, leading to the displacement of an estimated 40 million to 80 million people worldwide (primarily in China and India) and preventing access by local people to the goods and services of the now-buried ecosystems.

Dam Impacts. The United States is home to some 87,000 dams. Of these, 44,000 are taller than 25 feet (7.6 m), and 6,400 are taller than 50 feet (15 m). **(Figure 10–15)**. These have been built to run mills, control floods, generate electric power, and provide water for municipal and agricultural use. These dams have enormous ecological impacts. When a river is dammed, valuable freshwater habitats, such as waterfalls, rapids, and prime fish runs, are lost. When the river's flow is diverted to cities or croplands, the waterway below the diversion is deprived of that much water.

The impact on fish and other aquatic organisms is obvious, but the ecological ramifications go far beyond the river. Wildlife that depends on the water or on food chains involving aquatic organisms is also adversely affected. Wetlands dry up, resulting in frequent die-offs of waterfowl and other wildlife that depended on those habitats. Fish such as salmon, which swim from the ocean far upriver to spawn, are seriously affected by the reduced water level and have problems getting around dams, even those equipped with fish ladders (a stepwise series of pools on the side of the dam, where the water flows in small falls that fish can negotiate). If the fish do get upriver, the hatchlings have similar problems getting back to the ocean. On the Columbia and Snake Rivers, juvenile salmon suffer 95% mortality in their journey to the sea as a result of negotiating the dams and reservoirs that block their way.

If a dam floods an area with dense vegetation, the water may become seriously polluted. Decaying vegetation alters water chemistry and causes pollutants such as mercury to leave the soil. This has been a tremendous problem with dams in tropical rain forests.

The Glen Canyon Dam on the Colorado River is an example of a large dam with significant environmental effects. The dam provides hydropower and created the second largest artificial lake in America: Lake Powell. An environmental impact study in the late 1980s and early 1990s concluded that the operation of the Glen Canyon Dam had seriously damaged the downstream ecology of the Colorado River and its recreational resources (hiking, river rafting). In response, Congress passed the Grand Canyon Protection Act in 1992, which required a process of "adaptive management" that would adjust water-release rates in a manner designed to enhance the ecological and recreational values of the river. This rule has helped balance the needs of different stakeholders.

Dams Gone. Increasingly, people are recognizing the environmental costs of dams and finding them to be unacceptable.

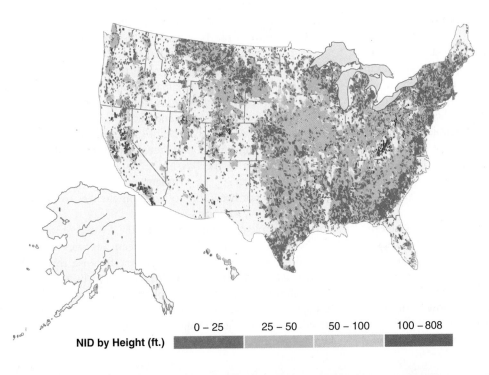

Figure 10–15 Number of dams in the United States. Each point represents a dam; the colors tell the height.
(*Source:* National Inventory of Dams Map.)

NID by Height (ft.) 0 – 25   25 – 50   50 – 100   100 – 808

(a)

(b)

Figure 10–16 Removal of a dam. (a) The Glines Canyon Dam on the Elwah River in Washington State. **(b)** Its 2014 removal was the largest dam removal yet in the United States. Gray sediment from behind the dam rushed downstream and reconfigured the river. Dam removal allowed salmon to spawn upstream.

Removing a dam is not easy, however. Legal complexities abound where long-existing uses of a dam (to control floods or to provide irrigation water or water for recreation) conflict with the expected advantages of removal (to reestablish historic fisheries such as salmon and steelhead runs or to restore the river for recreational and aesthetic use). Practical problems also exist. Yet in the last 20 years, 850 dams have been removed from streams and rivers in the United States. Figure 10–16 shows the successful removal of a dam on the Elwha River in Washington State.

Diversion of Surface Water. Humans often divert water, or move it to other places. Small-scale diversions include moving water though ditches or canals to irrigate fields. Although easy to use, irrigation ditches increase evaporation and often allow water to leak out of the ditch. In some places, disease vectors such as mosquitoes breed in irrigation ditches.

Other diversions move water over long distances. For example, the Chinese government is in the process of building the South-to-North Water Diversion Project, which will carry water from the Yangtze River and from rivers south of Beijing to the arid north. Initial flows began in 2014; eventually the diversion will carry 28 cubic kilometers (6.7 mi.^{3}) of water per year. China has discovered already, though, that large water diversion projects can bring unexpected environmental costs, including water pollution or the effects of water loss in the original rivers and lakes.

Impacts on Estuaries. When river waters are diverted on their way to the sea, serious impacts are seen in **estuaries**—bays and rivers in which freshwater mixes with seawater. Estuaries are among the most productive ecosystems on Earth; they are rich breeding grounds for many species of fish, shellfish, and waterfowl. As a river's flow is diverted (say, to irrigate fields), less freshwater enters and flushes the estuary. Consequently, the salt concentration increases,

profoundly affecting the estuary's ecology. This is exactly what is happening in San Francisco Bay, where the San Joachin and Sacramento Rivers meet and flow to the ocean. More than 60% of the freshwater that once flowed from rivers into the bay has been diverted for irrigation in the Central Valley (home to one sixth of all U.S. irrigated land) and for municipal use in southern California (25 million people served). In dry years, the farms and municipalities remove a third of the water that would normally flow through the delta into San Francisco Bay. Without the freshwater flows, saltwater from the Pacific has intruded into the bay, with devastating consequences. Chinook salmon runs have almost disappeared. Sturgeon, Dungeness crab, and striped bass populations are greatly reduced. Exotic species of plants and invertebrates have replaced many native species, and the tidal wetlands have been reduced to only 8% of their former extent.

Upriver from the bay, more than a thousand miles of levees protect farmlands that have sunk 20 feet or more below sea level due to compacting. If the levees were to collapse during an earthquake, massive amounts of saltwater would be sucked upriver, shutting down the huge pumps that divert freshwater to the cities and farms. In response to these problems, the CALFED Bay-Delta Program was established, to develop a long-term comprehensive plan that will restore ecological health in the Bay-Delta System. However, major stakeholders (farmers, government agencies, environmental groups) have not been able to agree on changes to the delta recommended by scientists.

The problem is not limited to the United States. The southeastern end of the Mediterranean Sea was formerly flushed by water from the Nile River. Because this water is now held back and diverted for irrigation by the Aswan High Dam in Egypt, that part of the Mediterranean is suffering severe ecological consequences (such as severe coastal erosion) as sediments normally flowing out to the coast are kept behind the dam.

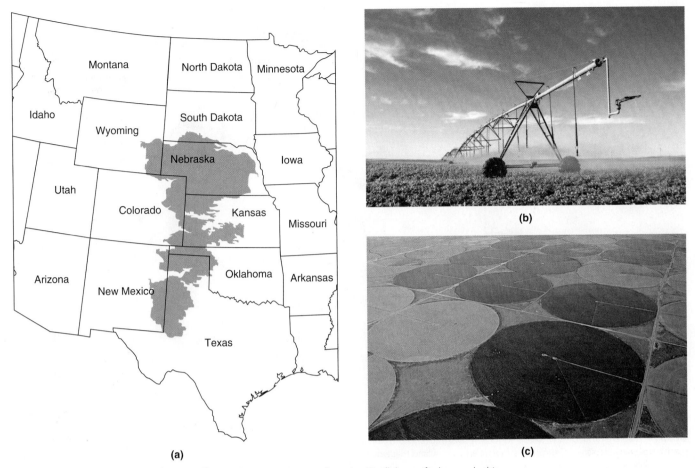

(a)

(b)

(c)

Figure 10–17 Exploitation of an aquifer. (a) Pumping up water from the Ogallala aquifer has made this arid region of the United States into some of the most productive farmland in the country. **(b)** Water is applied by means of center-pivot irrigation, in which the water is pumped from a central well to a self-powered boom that rotates around the well, spraying water as it goes. **(c)** An aerial photograph shows the extent of center-pivot irrigation throughout the region. Groundwater depletion will bring an end to this kind of farming.

Groundwater

Groundwater is the large unseen storage of water that accumulated slowly in porous rock over millennia. As mentioned before, some 99% of all liquid freshwater is in underground aquifers. Of this groundwater, more than three-fourths has a recharge rate of centuries or more. Such groundwater is considered "nonrenewable." Because groundwater supplies are vital for domestic use and especially for agriculture, they are a resource that requires management.

Groundwater Levels. The agriculture of the Midwestern United States makes the country the breadbasket for the world. Water to support that agriculture is mined from deep underground. Within the seven-state High Plains region of the United States, the Ogallala aquifer supplies irrigation water to 4.2 million hectares (10.4 million acres), one-fifth of the irrigated land in the nation (Figure 10–17). This aquifer is probably the largest in the world, but it is mostly "fossil water," recharged during the last ice age. The recent withdrawal rate has been about 10.2 billion cubic meters (8.3 million acre-ft.) per year, only one-tenth of the recharge rate. Water tables have dropped 30–60

meters (100–200 ft.) and continue to drop. Total depletion of the aquifer is greater than 340 cubic kilometers, a volume larger than the amount of water needed to cover New Mexico to a depth of 1 kilometer (about 2/3 mile). Irrigated farming has already come to a halt in some sections, and it is predicted that over the next 20 years another 1.2 million hectares (3 million acres) in this region will be abandoned or converted to dryland farming (ranching and the production of forage crops) because of water depletion. Figure 10–18, on the next page, depicts the overuse of water from several of the world's major aquifers.

Renewable groundwater is replenished by the percolation of precipitation water, so it is vulnerable to variations in precipitation. In tapping groundwater, we are tapping large, but not unlimited, natural reservoirs, contained within specific aquifer systems. Its sustainability ultimately depends on balancing withdrawals with rates of recharge. For most of the groundwater in arid regions, there is essentially no recharge. The resource must be considered nonrenewable. Like oil, it can be removed, but any water removed today will be unavailable for the future. Groundwater depletion has several undesirable consequences.

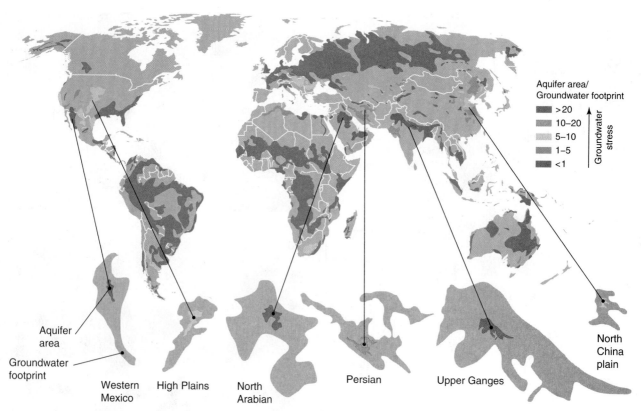

Figure 10–18 **Use of groundwater.** The map shows the size and shape of the major aquifers used in agriculture. Aquifers in blue have a withdrawal rate of less than 1, which means their water is being used at a sustainable rate. Below the map are groundwater footprints for six major aquifers from which water is being withdrawn more rapidly than it is being replenished. Footprints show the amount of land that would be needed to replace the water that is being withdrawn. For the large aquifers shown, the groundwater footprint was 3.5 times the size of the aquifers.

(*Source*: T. Gleeson et al. "Water balance of global aquifers revealed by groundwater footprint." *Nature* 488, 197–200.)

UNDERSTANDING THE DATA Which of the highlighted aquifers appears to have the most overdrawn water supply? (You may select one or two.) How can you tell?

Falling Water Tables. The simplest indication that groundwater withdrawals are exceeding recharge is a falling water table, a situation that is common throughout the world. Because irrigation consumes far and away the largest amount of freshwater, depleting water resources will ultimately have its most significant impact on crop production. Although running out of water is the obvious eventual conclusion of overdrawing groundwater, falling water tables can bring on other problems.

Diminishing Surface Water. Surface waters are affected by falling water tables. In various wetlands, for instance, the water table is essentially at or slightly above the ground surface. When water tables drop, these wetlands dry up, with the ecological results described earlier. Further, as water tables drop, springs and seeps dry up as well, diminishing even streams and rivers to the point of dryness. Thus, excessive groundwater removal creates the same results as the diversion of surface water.

Land Subsidence. Over the ages, groundwater has leached cavities in the ground. Where these spaces are filled with water, the water helps support the overlying rock and soil. As the water table drops, however, this support is lost.

Then there may be a gradual settling of the land, a phenomenon known as **land subsidence.** The rate of sinking may be 10–15 centimeters (6–12 in.) per year. In some areas of the San Joaquin Valley in California, land has settled as much as 9 meters (29 ft.) because of groundwater removal. Land subsidence causes building foundations, roadways, and water and sewer lines to crack. When it happens in cities, underground pipes break, causing leaks and wasting domestic water. Sewage pipes can also fracture, contaminating the groundwater. In coastal areas, subsidence causes flooding unless levees are built for protection. For example, where a 10,000-square-kilometer (4,000-mi.^{2}) area in the Houston–Galveston Bay region of Texas is gradually sinking because of groundwater removal, coastal properties are being abandoned as they are gradually being inundated by the sea. More than 80% of land subsidence in the United States results from groundwater withdrawal. Land subsidence is also a serious problem in many other places throughout the world.

Another kind of land subsidence, a **sinkhole,** may develop suddenly and dramatically (Figure 10–19). A sinkhole results when an underground cavern, drained of its supporting groundwater, suddenly collapses. Sinkholes may be

300 feet (91 m) across and as deep as 150 feet. The problem of sinkholes is particularly severe in the southeastern United States, where groundwater has leached numerous passageways and caverns through ancient beds of underlying limestone. An estimated 4,000 sinkholes have formed in Alabama alone, some of which have "consumed" buildings, livestock, and sections of highways.

Saltwater Intrusion. Another problem resulting from dropping water tables is **saltwater intrusion**. In coastal regions, springs of outflowing groundwater may lie under the ocean. As long as a high water table maintains a sufficient head of pressure in the aquifer, freshwater will flow into the ocean. Thus, wells near the ocean yield freshwater (Figure 10–20a). However, lowering the water table or removing groundwater at a rapid rate may reduce the pressure in the aquifer, permitting saltwater to flow back into the aquifer and hence into wells (Figure 10–20b). Saltwater intrusion is a serious problem in many European countries along the Mediterranean coast.

CONCEPT CHECK What are the positive effects of using dams to control the flow of rivers? What are some of the negative effects? ☑

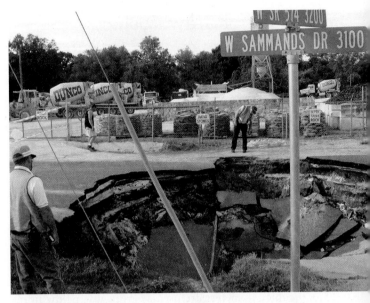

Figure 10–19 Sinkhole. The removal of groundwater may drain an underground cavern until the roof, no longer supported by water pressure, collapses. The result is the sudden development of a sinkhole, such as this one, which appeared in Plant City, Florida, on May 29, 2000.

Figure 10–20 Saltwater intrusion. In coastal areas, the removal of groundwater for human use can result in saltwater intrusion, which may make the water in the aquifer unfit for drinking. **(a)** Where aquifers open into the ocean, freshwater is maintained in the aquifer by the pressure of freshwater inland. **(b)** Excessive removal of water may reduce the pressure so that saltwater moves into the aquifer.

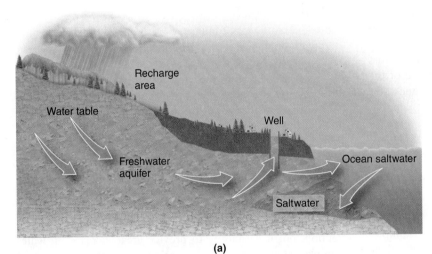

(a)

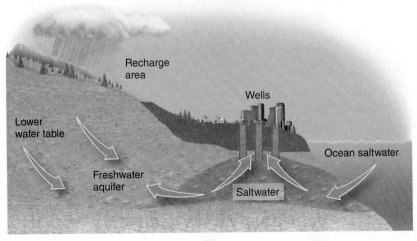

(b)

10.4 Water Stewardship, Economics, and Policy

The hydrologic cycle is entirely adequate to meet human needs for freshwater because it processes several times as much water as we require today. However, the water is often not distributed where it is most needed, and the result is a persistent scarcity of water in many parts of the world. In the developing countries, there is still a deficit of infrastructure, such as wells, water-treatment systems, and (sometimes) large dams, for capturing and distributing safe drinking water. Despite the growing negative impacts of overdrawing water resources, expanding populations create an ever-increasing demand for additional water for agriculture, industry, and municipal use.

Water for food dominates the current water demand, estimated at 6,800 cubic kilometers per year. Of this, some 5,000 cubic kilometers are used as green water flow in the world's rain-fed agriculture and 1,800 cubic kilometers as blue water flow in the form of irrigation water withdrawn from rivers, lakes, and groundwater. It is estimated that 4,500 cubic kilometers (of the total 6,800 km^3) are attributed to the developing countries to fulfill current food needs. If we assume that virtually all of the future population increase will be in the developing countries, then so also will be the increased need for food.

What are the possibilities for meeting existing needs and growing demands in a sustainable way? Planners call for a **Blue Revolution,** a radical change in the way we use water. There are only about five options: (1) capture more of the runoff water, (2) gain better access to existing groundwater aquifers, (3) desalt seawater, (4) conserve present supplies by using less water, and (5) make food production more efficient. We will now consider the advantages and disadvantages of these options.

Capturing Runoff

In some situations, people increase their supply of water by capturing rainwater before it reaches the land. For example, families may use rain barrels to capture runoff from their roofs. This water is often used for gardens. In regions with large seasonal fluctuations in rainfall and on islands with little groundwater, towns and villages may capture and store rainwater in large receptacles called *cisterns* (Figure 10–21).

Dams are the most widely used method of capturing surface water. China is the best example of both the pros and cons of dam building, which we have already discussed. In an effort to provide hydropower while cutting greenhouse gas emissions, China has committed to an aggressive program of dam building. New dam projects threaten China's remaining biodiversity.

In the United States, protection has been accorded to some rivers with the passage of the Wild and Scenic Rivers Act of 1968, which keeps rivers designated as "wild and scenic" from being dammed or affected by other harmful operations. The national system of wild and scenic rivers protects

Figure 10–21 A water catchment system. This house in Caye Caulker, Belize, diverts rainwater that falls on its roof into a black cistern, where it is stored for later use.

some 12,600 miles of our nation's rivers, less than 0.2% of the total length. The numerous large and small dams across the country have affected some 600,000 miles of our rivers, by comparison. Those rivers designated as wild and scenic are in many ways equivalent to national parks.[3]

Tapping More Groundwater

Already, more than 2 billion people depend on groundwater supplies. In many areas, groundwater use exceeds aquifer recharge, leading to shortages as the water table drops below pump levels. Groundwater depletion is considered the single greatest threat to irrigated agriculture, and it is happening in many parts of the world. India, China, West Asia, countries of the former Soviet Union, the western United States, and the Arabian Peninsula are all experiencing declining water tables. Calculations by the International Water Management Institute suggest that an expansion of irrigation on the order of 20% might be possible, but this will not meet the food needs in the developing countries by 2030.[4]

Exploiting renewable groundwater will continue to be an option, but it is unlikely to provide great increases in water supply because these same sustainable supplies, recharged by annual precipitation, are being increasingly polluted. Agricultural chemicals such as fertilizers and pesticides, animal wastes, and industrial chemicals readily enter groundwater, making groundwater pollution as much of a threat to domestic water use as depletion. For example, some 60,000 square kilometers of aquifers in western and central Europe will likely be contaminated with pesticides and fertilizers within the next 50 years. Also, India and Bangladesh are plagued with arsenic-contaminated groundwater, which has extracted a heavy toll in sickness and death for millions in those countries.

[3]"Wild and Scenic Rivers: Leading the Charge to Protect the Nation's Last, Best Rivers," American Rivers (September 2012), http://www.americanrivers.org/our-work/protecting-rivers/wild-and-scenic/.
[4]M. Falkenmark and J. Rockström, "The New Blue and Green Water Paradigm: Breaking Ground for Water Resources Planning and Management," *Journal of Water Resources Planning and Management* 132 (May/June 2006): 131.

Removing Salt from Seawater

The world's oceans are an inexhaustible source of water, not only because they are vast, but also because any water removed from them will ultimately flow back in. Not many plants can be grown in full seawater, however, and although researchers are developing salt-tolerant plants via genetic modification, we are not there yet. Thus, with increasing water shortages and most of the world's population living near coasts, there is a growing trend toward **desalination** (desalting) of seawater for domestic use. Between 15,000 and 20,000 desalination plants have been installed in some 150 countries, especially in the Middle East. From 2008 to 2013, the worldwide desalination capacity increased by 57%, largely as a result of responses to water shortages.

Two technologies—microfiltration, or *reverse osmosis*, and distillation—are commonly used for desalination. Smaller desalination plants generally employ microfiltration, in which great pressure forces seawater through a membrane filter that is fine enough to remove the salt. Larger facilities, particularly those that can draw from a source of waste heat (for example, from electrical power plants), generally use distillation (evaporation and condensation of vapor). Efficiency is gained by using the heat given off by condensing water to heat the incoming water. Even where waste heat is used, however, the costs of building and maintaining the plant, which is subject to corrosion from seawater, are considerable. The cost of desalinized water can be two to four times what most city dwellers in the United States currently pay, but it is still not a high price to pay for drinking water in areas of the world that are prone to drought; it is also much cheaper than bottled water.

In January 2008, the Tampa Bay seawater desalination plant was completed, signaling a new day for desalination (Figure 10–22). The plant is the country's first ever built to serve as a primary water source. The plant uses salty cooling water from a power plant and reverse osmosis to produce 25 million gallons of drinking water per day, providing Pasco County, Pinellas County, St. Petersburg, and other cities with some 10% of their drinking water. The Tampa Bay region has experienced chronic water shortages and is exploiting an opportunity that has attracted worldwide attention.

The largest desalination plant in the United States is the Carlsbad plant in southern California. Built between 2014 and 2016, it was designed to produce 50 million gallons of water per day for the San Diego area. Long-term drought in California and the increased efficiency of desalination technology made this reasonable. However, concerns about the plant's environmental impacts—especially its high use of energy and its effects on fish—remain.

Although the higher cost might cause some people to cut back on watering lawns and to implement other conservation measures, most people in the United States could afford desalinized water without unduly altering their lifestyles. For irrigating croplands, however, the higher cost of desalinized seawater would probably be prohibitive, except possibly for

Figure 10–22 Desalination plant. The Israel Ashkelon Seawater desalination plant is the largest reverse osmosis desalination plant in the world, producing 330,000 cubic meters of drinking water per day.

high-value cash crops such as greenhouse flowers or vegetables. Water conservation, which we will discuss next, is much more cost effective than desalination.

Using Less Water

A developing-nation family living where water must be carried several miles from a well finds that one gallon per person per day is sufficient to provide for all of its essential needs, including cooking and washing. Yet, a typical household in the United States consumes an average of 380 liters (100 gal.) per person per day. If all indirect uses are added (especially irrigation), this figure increases to 4,200 liters (1,100 gal.) per person per day. Similarly, a peasant farmer may irrigate by carefully ladling water onto each plant with a dipper, while typical modern irrigation floods the whole field. The way water resource planners think is beginning to change, however: Instead of asking how much water we "need" and where can we get it, people are now asking how much water is available and how can we best use it. In the United States, the rate of water use per capita has actually begun to drop, the result of some well-needed water conservation strategies already put in place. Some specific measures that are being implemented to reduce water withdrawals are described next. (Also see Stewardship, *The Energy-Water-Food Trade: Waste One, Waste Them All.*)

Agriculture. Agriculture is far and away the largest consumer of freshwater—some 40% of the world's food is grown in irrigated soils, which produce 100–400% higher yields than traditional farming. Most present-day irrigation wastes huge amounts of water, and in fact, half of it never yields any food. Where irrigation water is applied by traditional flood or center-pivot systems, 30–50% is lost to evaporation, percolation, or runoff. Several strategies have been employed recently to cut down on this waste. One is the surge flow method, in which computers control the periodic release of water, in contrast to the continuous-flood method. Surge flow can cut water use in half.

Drip, Drip. Another water-saving method is the drip irrigation system, a network of plastic pipes with pinholes that literally drip water at the base of each plant (Figure 10–23). Although such systems are costly, they waste much less water. Also, they have the added benefit of retarding salinization (see Chapter 11). Studies have shown that drip irrigation can reduce water use 30–70%, while actually increasing crop yields, compared with traditional flooding methods. Although drip irrigation is spreading worldwide, especially in arid lands such as Israel and Australia, 97% of the irrigation in the United States, and 99% throughout the world, is still done by traditional flood or center-pivot methods.

The reason that so few U.S. farms have changed over to drip irrigation systems is that it costs about $1,000 per acre to install them. In comparison, water for irrigation is heavily subsidized by the government, so farmers pay next to nothing for it. One recent U.S. program that was designed to save water backfired. Between 1996 and 2013, the federal government spent $1 billion to help farmers install more efficient irrigation equipment, expecting that the new equipment would save water. Instead, farmers expanded the amount of land under irrigation or planted crops that required more water. As a result, water supplies were not conserved.[5]

Treadle Pumps. In the developing countries, irrigation often bypasses the rural poor, who are at the greatest risk for hunger and malnutrition. Affordable irrigation technologies such as treadle pumps can be remarkably successful (Figure 10–24). In Bangladesh, low-cost treadle pumps enable farmers to irrigate their rice paddies and vegetable fields at a cost of less than $35 a system. The pump is worked with

the feet like a step exercise machine, is locally manufactured, and has been adopted by millions. The irrigation it affords makes it possible for farmers to irrigate small plots during dry seasons from groundwater lying just a few feet below the surface. More than 1.5 million treadle pumps have been sold in Bangladesh alone, increasing the productivity of more than 600,000 acres of farmland there. The pumps are now also in use in India, Nepal, Myanmar, Cambodia, Zambia, Kenya, and South Africa.

Enhance Soil Water Retention. We have already addressed the need and options for improvement in irrigation technology. However, most of the agriculture in developing countries is rain-fed, depending on green water flow from soil and resulting in evapotranspiration. Further, most of the soils are dryland soils, receiving only 10 to 30 inches of rainfall per year (Chapter 11). Crops raised on these soils are characteristically low in yield. Storing the seasonal rainfall with *check dams* (small rock or raised soil dams across seasonal water courses) increases yields. These dams recharge aquifers, restrain soil erosion, and provide for some irrigation water. Another strategy is to enhance rainfall retention, maintain soil fertility, and restrain soil erosion through conservation tillage, where crop residues are left in place and new crops are planted with little or no tilling.

---

[5]R. Nixon, "Farm subsidies lead to more water use," New York Times (June 7, 2013) http://www.nytimes.com/2013/06/07/us/irrigation-subsidies-leading-to-more-water-use.html?_r=1

Figure 10–23 Drip irrigation. Irrigation consumes the most water. Drip irrigation offers a conservative method of applying water.

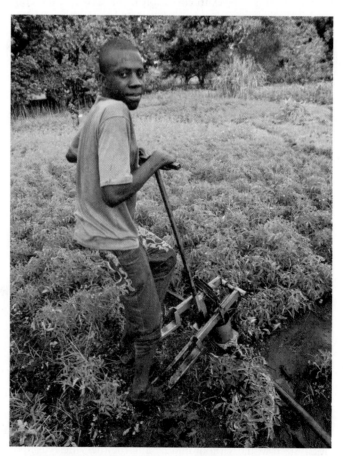

Figure 10–24 Treadle pump. This treadle pump in Zambia is operated like a step exercise machine. Millions of these pumps now allow rural farmers to raise crops during dry seasons.

Municipal Systems. The 100 gallons of water each person consumes per day in modern homes are used mostly for washing and removing wastes—that is, for flushing toilets (3–5 gallons per flush), taking showers (2–3 gallons per minute), doing laundry (20–30 gallons per wash), and so on. Watering lawns, filling swimming pools, and other indirect consumption only adds to this use.

Water conservation has long been promoted as a "save the environment" measure, though without much effect.

Now, numerous cities are facing the stark reality that it will be extremely expensive—and in many cases, impossible—to increase supplies by the traditional means of building more reservoirs or drilling more wells. The only practical alternative, they are discovering, is to take real steps toward reducing water consumption and waste.

Aging Infrastructure. In the United States, much of the water-delivery infrastructure built after World War II is coming to the end of its life and needs to be replaced. In many

STEWARDSHIP
The Energy-Water-Food Trade: Waste One, Waste Them All

Ever stand in a public restroom and see someone run the sink water and not turn it off? What about seeing someone leave their car idling? Or throwing food in the trash? Drive you crazy? It should! Because the way the world is intertwined, if you waste one (energy, water, or food), you waste the others or at least waste the opportunity to obtain them. It works this way:

It takes energy to get clean freshwater, especially if you live in a dry area or a place with water pollution. The water might have to be cleaned or moved long distances. Bottled water is even more energy intensive. For example, in California, the California State Water Project is the largest single user of energy in the state. One of its activities is moving water over the Tehachapi Mountains, which requires as much energy as a third of the household energy use in the region. Wasting water means you waste the energy it took to get it, clean it, and heat it, too. Because we need to be efficient with energy so we can avoid greater harm to the environment, water waste isn't just about the water—it hurts the whole system.

It takes water to get energy. This is true when water is used in cooling (nuclear plants), in coal mining, and in coal power plants.

The newest way of getting natural gas, called hydraulic fracturing, uses enormous amounts of water. There are more than 35,000 wells fractured in the United States each year, using an estimated 70 billion to 140 billion gallons of water. When the water is used, it is mixed with toxic chemicals and injected deep into the Earth, rather than returned to the regional water cycle. The use of water for fracking, as this process is called, directly competes with other uses of water such as agriculture and has been the source of lawsuits in dry states such as Wyoming and Montana.

Another example is biofuels. Crops used for biofuels such as ethanol are usually raised using land and water that could have been used

for food crops. Their use to fuel vehicles drove up the cost of basic food staples in 2007, contributing to the world food crisis (Chapter 12). Wasting energy might mean we have to obtain more energy and use more water, making it harder to produce adequate food.

We use water and energy to grow, store, and move food. Wasting food means that water and energy cannot be used for something else. It takes 1,300 liters of water to produce a kilogram of wheat, 3,000 liters for a kilogram of eggs, and 15,500 liters for a kilogram of beef. It also takes energy to store food and a great deal to transport it. The average American foodstuff travels 1,500 miles. As much as 40% of the energy used in food production is in the manufacture of pesticides and fertilizers (Chapters 12 and 13). But much of the recent increase in food energy cost is due to increased processing and packaging, such as for single-serving foods.

Food is often wasted. A Food and Agriculture Organization (FAO)–commissioned study found that 1.3 billion tons, roughly a third, of the food we produce for human consumption is wasted each year. This is the equivalent of half of the world cereals crop. Some food is lost during harvest or in storage or processing. Some is wasted in stores and at consumers' tables. Any way you work it, if we can save food, we can feed more people and use less water and energy.

Sources: NRDC. "Energy Down the Drain: The Hidden Costs of California's Water Supply." (2004): http://www.nrdc.org/water/conservation/edrain/edrain.pdf; EPA. Draft Plan to Study the Potential Impacts of Hydraulic Fracturing on Drinking Water Resources. Report Number: EPA/600/D-11/001, 2011.

Chapagain, A. K., and A. Y. Hoekstra. "Water Footprints of Nations." *Value of Water Research Report Series* (UNESCO-IHE) 6 (2004); Gustavsson, Jenny, Christel Cederberg, Ulf Sonesson, Robert van Otterdijk, and Alexandre Meybeck. *Global Food Losses and Food Waste*, FAO, 2011.

cities, water leaks out of aging pipes and land subsidence breaks underground water mains. Aging infrastructure is one of the biggest challenges in national water provision because of the high cost of replacing buried pipes.

Lawns. Grass is the most heavily irrigated crop. Cities in Arizona are paying homeowners to replace their lawns with **xeriscaping**—landscaping with desert species that require no additional watering—and the business is thriving. Many cities and towns ban certain uses of water when droughts reduce the available supply.

A Flush World. In 1997, the last phase of a regulation authorized by the 1992 National Energy Act took effect, and it became illegal to sell 6-gallon commodes. In their place is the wonder of the flush world: the 1.6-gallon toilet. Newer versions use even less water than the first, less effective low-flow toilets; the newest work perfectly well and save 10 gallons or more a day per person. In fact, the newest models are dual flush, with options for a 1.28-gallon (solids) or 0.8-gallon (liquids) flush. According to a recent estimate, U.S. regulations requiring the use of low-flow toilets have saved 18 trillion gallons of water over 20 years.[6] Los Angeles is offering the low-flow toilets free as part of its efforts to restore Mono Lake, which had lost a great deal of water. In 2011, the City of San Francisco found the downside of low-flow toilets: While the city had saved 20 million gallons of water a year (just from toilets!), the low-flow toilets weren't pushing the sewage through the pipes as quickly as before. To solve the problem, city engineers are planning to kill bacteria with chemicals, though other solutions have been identified.

Gray Water. It seems wasteful to raise all domestic water to drinking-water standards and then use most of it to water lawns and flush toilets. Increasingly, gray water recycling systems are being adopted in some water-short areas. **Gray water,** the slightly dirtied water from sinks, showers, bathtubs, and laundry tubs, is collected in a holding tank and used for such things as flushing toilets, watering lawns, and washing cars. Going further, a number of cities are using treated wastewater (sewage water) for irrigating golf courses and landscapes, both to conserve water and to abate the pollution of receiving waters (see Chapter 20). In Israel, 70% of treated wastewater is reused for irrigation of nonfood crops. If the idea of reusing sewage water turns you off, recall that *all* water is recycled by nature. There is hardly a molecule of water you drink that has not moved through organisms—including humans—numerous times. A number of communities are already treating their wastewater so thoroughly that its quality exceeds what many cities take in from lakes and rivers.

Virtual Water. Another option for reducing water loss in water-stressed regions is to import food or goods that require a great deal of water, to ship *virtual water,* as it is called. Many arid countries avoid using their own water supplies by importing farm products from regions that have excess freshwater. Indeed, an estimated 800 cubic kilometers of virtual water are traded yearly in the form of foodstuffs. However, many of the arid countries are also economically poor and have a limited ability to purchase great quantities of food. Furthermore, shipping takes energy, and as we have seen, wasting energy is eventually bad for water, too.

Consumers can help conserve water by purchasing products that have a low *water footprint,* a way of calculating the amount of water used to produce goods and services. The production of meat, for example, almost always has a larger water footprint than the production of grains or locally grown vegetables. Products that are heavily packaged or transported across the world require large water inputs. You can figure out your water footprint by using one of the calculators on the Web. Water footprints are sometimes divided into three parts: green water (rainwater), blue water (diverted surface water or pumped groundwater), and gray water (water required to dilute a pollutant to an acceptable level).

Public-Policy Challenges

The hydrologic cycle provides a finite flow of water through each region of Earth. When humans come on the scene, this flow is inevitably divided between the needs of the existing natural ecosystems and the agricultural, industrial, and domestic needs of humans, with the latter usually being met first. Researchers have estimated that humans now use 26% of total terrestrial evapotranspiration and 54% of accessible precipitation runoff. We are major players in the water cycle, but many facets of our water use are unsustainable. It takes public policy to strike a sustainable balance between competing water needs, and this is often sadly lacking.

Conflict. Maintaining a supply of safe drinking water for people is a high-priority issue. In addition, water for irrigation is vital to food production across much of the world. Often, these two demands come into conflict, generating what have been referred to as "water wars." One example was in Cochabamba, Bolivia, when the World Bank refused to renew loans unless the country opened water resources to foreign privatization. Investors were supposed to put money into providing clean drinking water, but some locals found they lost access to water and rates skyrocketed. This led to protests and the government eventually had to change plans. Figure 10–25 depicts protesters in California after a federal judge ordered that less water could be pumped from the Sacramento Delta. This ruling pitted environmentalists against farmers and urban residents in a state with continual water stress.

National Water Policy. Many countries have addressed their water resources and water needs with a national policy; not so the United States. The Clean Water Act and subsequent amendments authorize the U.S. Environmental Protection Agency (EPA) to develop programs and rules to carry out its mandate for oversight of the nation's water quality (considered in Chapter 20). The EPA, however, does not deal with water quantity. There is no federal agency to provide similar oversight for the water quantity issues discussed in this chapter.

---

[6]"18 Trillion Gallons of Water Saved During 20 Years of Low-Flow Toilet Regulations," available at http://www.americanstandard-us.com/pressroom/18-trillion-gallons-of-water-saved-during-20-years-of-low-flow-toilet-regulations/.

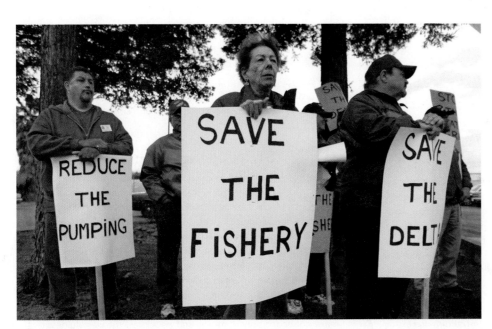

Figure 10–25 Water wars. Environmentalists in Sacramento, California, protest the overuse of water by industry that leaves less water for wild fish and river ecosystems.

Peter Gleick, a leading expert on global freshwater resources, addresses this public-policy need, pointing out that most water policy decisions in the United States are made at local and regional levels.[7] National policies, however, are needed to guide these lower-level decisions and also to deal with interstate issues and federally managed resources. The last time a water policy report was issued was in 1950, when President Harry S. Truman established a water resources policy commission and the commission gave its report. This commission needs to be revived or a new one needs to be established, and in either case, it must be given a mandate to collect data on water resources and problems and issue recommendations on how the federal government could facilitate water stewardship in the 21st century.

Key Issues. Any water resources policy commission needs to address several key ideas. Two of these are increasing efficiency and stopping subsidies. In the United States, scarcity and rising costs have led to significant gains in efficiency. As a result, water withdrawals peaked in the early 1980s and have leveled off since then (Figure 10–11). This change has been attributed to steps taken in all three sectors (domestic, industry, and irrigation) to reduce water losses. The United States still has a high per capita water use compared with the rest of the world. Also, water subsidies need to be reduced or eliminated. Policy makers have persistently subsidized water resources, especially for agricultural use, and the result has been enormous waste and inefficiencies.

Various policy approaches to water protection are possible, such as permit fees, "green" taxes, direct charges, and market-based trading (see Chapter 20). Public policy needs to focus on the watershed level as well. Watershed management must be integrated into the pricing of water. The City of New York, for example, found out that every dollar it invested in watershed management paid off handsomely, saving as much as $200 for new treatment facilities. One part of managing watersheds is to regulate dam operations so that river flow is maintained in a way that simulates natural flow regimes. When this happens, rivers below dams can sustain recreational fisheries and other uses, and the natural biota flourishes.

Finally, the United States must respond to the global water crisis with adequate levels of international development aid. By achieving the Millennium Development Goal of reducing by half the 1.1 billion people without access to safe drinking water, the world is saving millions of lives and billions of dollars annually. The next global effort regarding water will be to achieve Sustainable Development Goal 6: *Ensure availability and sustainable management of water and sanitation for all by 2030.*

International Action. The decade from 2005 to 2015 was designated the *UN International Decade for Action: Water for Life.* An agency called UN Water coordinated the decade's activities, addressing issues such as coping with water scarcity and pollution, empowering women and girls in water and sanitation, and managing water across boundaries. Every few years, there is an international meeting called the World Water Forum; the most recent was held in the Republic of Korea in 2015. With a large proportion of the developing world already experiencing shortages of clean water or water for agriculture, the population increases and pressure to develop these countries' resources that are certain to come in the next few decades will undoubtedly subject hundreds of millions to increased water stress. The problem is not that the Earth contains too little freshwater; rather, it is that we have not yet learned to manage the water that our planet provides.

CONCEPT CHECK What can individual citizens do to conserve water? What aspects of water conservation require government coordination? ☑

---

[7]Peter H. Gleick, "Global Water: Threats and Challenges Facing the United States," *Environment* 43 (March 2001): 18–26.

REVISITING THE THEMES

SOUND SCIENCE

What is known about the hydrologic cycle is the result of sound scientific research—research that is needed to make progress in managing water resources. Work performed under the Millennium Ecosystem Assessment has given us a better understanding of such matters as groundwater recharge, essential river flows, and the interactions between different land uses and water resources. The techniques of water treatment have been fine-tuned by scientific work as well as the development of reverse osmosis for desalination. Sound science is being used to figure out current questions, such as the how human activities are likely to affect the hydrologic cycle in the future. One we do not yet have an answer to is what is likely to happen with oceanic thermohaline circulation.

SUSTAINABILITY

The opening story showed how the current use of limited water supplies in California and the American Southwest is unsustainable. Still, the hydrologic cycle is a remarkable global system that renews water and does so sustainably. Groundwater—especially the fraction that is nonrenewable because it was formed centuries ago—is being used in an unsustainable manner. The consequences are immediate: Water tables disappear into the depths of the Earth, farms dry up because there is no water, and land subsidence ruins properties and groundwater storage itself. Conserving water through xeriscaping, low-flush toilets, gray water use, and more efficient food production and consumption is a sustainable path.

STEWARDSHIP

Exchanges among land, water, and atmosphere represent a cyclical system that has worked well for eons and that can continue to supply all our needs into the future—if we manage water wisely. However, we are changing the cycle in ways that reflect a lack of wisdom and a lack of resolve to meet people's needs in a just way. One way to remedy this is to look at water, energy, and food as interrelated. Increased efficiency and lower waste are necessary in all three areas if we are to provide water and food for more people. Stewardship of water resources means that we consider the needs of the natural ecosystems—and especially endangered species—when we make policy decisions. It means that we address the wastefulness that characterizes so much of our water use. It means in particular that we show concern and help the millions who lack adequate clean water and are suffering because of it. Some examples of good stewardship are the Glen Canyon Dam operation, the removal of outdated dams to restore river flow, the use of treadle pumps in many developing countries, and xeriscaping. In the United States, developing a national water policy would be a huge step in stewardship management.

REVIEW QUESTIONS

1. What are the factors contributing to the shortage of water in California?

2. Give examples of the infrastructure that has been fashioned to manage water resources. What are the challenges related to developing countries?

3. What are the two processes that result in natural water purification? State the difference between them. Distinguish between green water and blue water.

4. Describe how a Hadley cell works, and explain how Earth's rotation creates the trade winds.

5. Why do different regions receive different amounts of precipitation?

6. Define *precipitation, infiltration, runoff, capillary water, transpiration, evapotranspiration, percolation, gravitational water, groundwater, water table, aquifer, recharge area, seep,* and *spring.*

7. Use the terms defined in Question 6 to give a full description of the hydrologic cycle, including each of its three loops—namely, the evapotranspiration, surface runoff, and groundwater loops. What is the water quality (purity) at different points in the cycle? Explain the reasons for the differences.

8. How does changing Earth's surface (for example, by deforestation) change the pathway of water? How does it affect streams and rivers? Humans? Natural ecology?

9. Explain how climate change and atmospheric pollution can affect the hydrologic cycle.

10. What are the three major uses of water? What are the major sources of water to match these uses?

11. How do dams facilitate the control of surface waters? What kinds of impacts do they have?

12. Distinguish between renewable and nonrenewable groundwater resources. What are the consequences of overdrawing these two kinds of groundwater?

13. What are the five options for meeting existing water scarcity needs and growing demands?

14. Describe how water demands might be reduced in agriculture, industry, and households.

15. What is the status of water policy in the United States? Cite some key issues that should be addressed by any new initiatives to establish a national water policy.

THINKING ENVIRONMENTALLY

1. Imagine that you are a water molecule, and describe your travels through the many places you have been and might go in the future as you make your way around the hydrologic cycle time after time. Include travels through organisms.

2. Suppose some commercial interests want to create a large new development on what are presently wetlands. Describe a debate between those representing the commercial interests and those representing environmentalists who are concerned about the environmental and economic costs of development. Work toward negotiating a consensus.

3. Describe how many of your everyday activities, including your demands for food and other materials, add pollution to the water cycle or alter it in other ways. How can you be more stewardly with your water consumption?

4. An increasing number of people are moving to the arid southwestern United States, even though water supplies are already being overdrawn. If you were the governor of one of the states in this region, what policies would you advocate to address this situation?

5. Suppose some commercial interests want to develop a golf course on what is presently forested land next to a reservoir used for city water. Describe the impacts this development might have on water quality in the reservoir.

MAKING A DIFFERENCE

1. Since 1992, showerheads and faucets have been required to use no more than 2.5 gallons of water per minute. If you have older plumbing, it's possible that you may not meet these standards. Equipment such as aerating showerheads can help you reduce your water use (and your water bill) without decreasing effectiveness too much.

2. Consider taking a "navy" shower (so-called because ships at sea must conserve water): Wet down, then turn the shower off, then soap down, and then rinse down. You will save both water and energy this way.

3. Americans drink billions of single-use water bottles per year. In addition to the carbon miles it takes to ship these products, most Americans simply throw the recyclable containers away. A much more sustainable method of drinking is simply to fill reusable bottles with filtered tap water.

4. Old water heaters are most likely running at less than 50% efficiency. If this is the case for you, you can replace your water heater with a newer model or simply use a tankless heater, which heats water only when you need it.

5. A dishwasher can save more than 5,000 gallons of water per year as well as hundreds of hours of your time. Scraping your dishes first, running the washer only for full loads, and using the energy-saver settings can reduce water and energy usage even more.

Dust Storm Damage, 1930–1940

Area with most severe dust storm damage

Other areas damaged by dust storms

Soil: The Foundation for Land Ecosystems

"The nation that destroys its soil, destroys itself."

—Franklin D. Roosevelt[1]

A churning wall of soil, two miles high, bore across the plains. People caught outside fled for buildings, cloths held to their faces. Some faced with the blinding swirl lost their way and died, suffocated by dust. Dust was everywhere, driven by the howling winds: blinding, choking, filling cracks and windows. People wept, prayed, cried, despaired; the apocalypse had come. The day was Black Sunday, April 14, 1935. The dust storm was the worst of the "black blizzards" in the tragic period we call the American Dust Bowl. The almost-decade-long period of drought and dry storms was arguably the longest and worst environmental disaster in U.S. history. The storms carried tons of soil across seven states and left devastation in their wake. Some people fled the dust states or died. Others stayed and survived, only to find themselves coughing and sickly when storm after storm filled the air with dust, victims of a disorder called "dust pneumonia."

The Dust Bowl Begins. A drought beginning in 1931 precipitated the disaster. By 1934, the drought was profound: More than 75% of the country was in drought, and 27 states were severely affected. Although drought was the trigger, human policies and practices made the Dust Bowl worse. Rising wheat and cotton prices encouraged people to plant more and more land, even in areas more suited to ranging animals. During wet years, crops were bountiful, but plowing broke through the tough roots of prairie plants that held soil together, broke open the ground,

---

[1]Letter, February 26, 1937, to state governors. Available at http://www.presidency.ucsb.edu/ws/?pid=15373.

▲ A 2014 dust storm in Yuma, Arizona showed just a tiny picture of what the 1930s Dust Bowl was like.

and laid the dirt open to be carried away by the winds. Lack of trees or other windbreaks allowed soil to move freely. By 1935, estimates were that 850 million tons of topsoil had been blown off of the southern plains in that year alone. By the end of the Dust Bowl, many areas had lost 75% of their topsoil, once some of the richest soil in the world.

The Homestead Act of 1862, which enabled settlers to claim land and settle it, played a role in the disaster. By encouraging hundreds of thousands of small farms (160 to 320 acres) to be established between 1880 and 1920, the act ensured that too much land was cultivated in some areas of the country. Small farm owners were unlikely to let land lie fallow, diversify into pasture, plant crops with strips of vegetation in between, or farm less intensively. These homesteads were more vulnerable to financial ruin during drought. Homesteaders fled the southern plains in the largest migration in American history, described by John Steinbeck in his novel *The Grapes of Wrath*.

A dust storm that hit Phoenix, Arizona, on July 3, 2014, gave only the tiniest glimpse of such storms to modern Americans. With wind speeds of 50 miles per hour, clouds towered and airlines canceled flights. As was the case in the '30s, an ongoing regional drought contributed to the severity of the storm. Even so, the Phoenix storm did not rival the great dust storms of the Depression. Storms as severe as those in the Depression are rare now, in part because of better soil protection practices and the movement away from more intensive cultivation of marginal lands. Around the world, however, loss of soil continues to plague farmland and contribute to poor crops, water pollution, and the spread of deserts.

Past Neglect. Ninety percent of the world's food comes from land-based agricultural systems, and the percentage is growing as the ocean's fish and natural ecosystems are increasingly being depleted. Protecting and nurturing agricultural soils, which are the cornerstone of food production, must be a central feature of sustainability. Yet it is a feature that has been overlooked repeatedly in the past. The Dust Bowl is not the only time croplands have been severely degraded. The fall of the ancient Greek, Roman, and Mayan empires was more the result of a decline in agricultural productivity due to soil erosion than of outside forces.

As the world's population grows past 7.2 billion, croplands and grazing lands are being pressed to yield more crops. The result is *soil degradation,* in which soil becomes less capable of supporting plant growth. Throughout the world, soils are degraded by erosion, the buildup of salts, pollution from mining, and deforestation. These impacts contribute to water and soil pollution as well as the loss of ecosystem services from the ecosystems that are degraded. The landmark 1991 Global Assessment of Soil Degradation reported that 15% of Earth's land was degraded, mostly due to soil erosion. In 2013, a study by the FAO, called the Land Degradation Assessment in Drylands found that more than one-fourth of Earth's cropland is degraded.[2]

Our objective in this chapter is to develop an understanding of the attributes of soil required to support good plant growth, how these attributes may deteriorate under various practices, and what is necessary to maintain a productive soil. We will then look at what is being done nationally and internationally to rescue this essential resource and to establish practices that are sustainable.

[2]R. Biancalini, *Land Degradation Assessment in Drylands,* 2013 Food and Agriculture Organization of the United Nations: Rome.

11.1 Soil and Plants

Soil can be defined as solid material of geological and biological origin that is changed by chemical, biological, and physical processes, giving it the ability to support plant growth. A rich soil is much more than the dirt you might get out of any hole in the ground. Indeed, agriculturists cringe when anyone refers to soil as "dirt." The quality of soil can make the difference between harvesting an abundant crop and abandoning a field to weeds.

In addition to supporting the growth of crops, soil serves as a filter for water and plays key roles in Earth's nutrient cycles and in the exchange of atmospheric gases. Soil scientists still have a great deal to learn about soil (Figure 11–1). Researchers today are investigating questions such as, What factors can makes soil more fertile? How does the condition of soil affect public health? How can contaminated soil be improved? and How does soil affect climate change?

Soils are a fundamental part of terrestrial ecosystems and are home to one-quarter of the world's species. Recall that various detritus feeders and decomposers feed on the

Figure 11–1 Soil research. Soil research aims to protect and enhance soil productivity. This soil scientist is measuring soil radiation.

Figure 11–2 **Topsoil formation.**
Soil production involves a dynamic
interaction among mineral particles,
detritus, and members of the detritus food web.

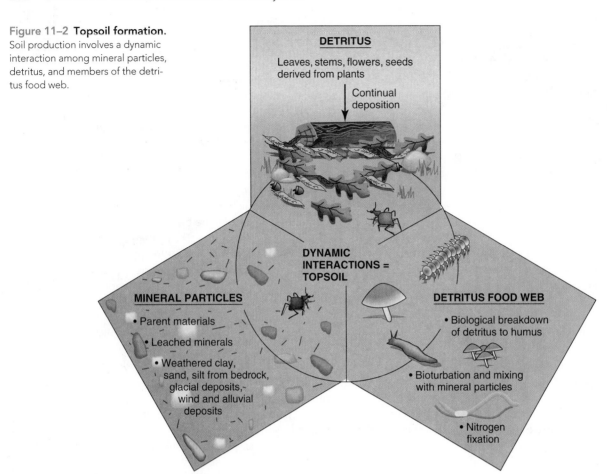

organic debris in an ecosystem (Chapter 4). Nutrients from the detritus are released into the soil and absorbed by producers, thus recycling the nutrients. In short, productive soil involves dynamic interactions among organisms, detritus, and mineral particles of soil (Figure 11–2).

Soil Characteristics

Most soils are hundreds or thousands of years old and change very slowly. Soil science is crossdisciplinary and is at the heart of agricultural and forestry practice. Many years of the study of soils have resulted in a system of classification of soil profiles and soil structure and a taxonomy of soil types from all over the world—as well as a large body of scientific literature investigating relevant topics. This section presents just enough of this information to enable you to open the door to soil science; you may even be motivated to follow up on your own in this fascinating field.

Soil Texture. The mineral material of soil, or **parent material,** has its origin in the geological history of an area. Parent material can be rock or sediments deposited by wind, water, or ice. Sooner or later parent material is broken down by natural **weathering** (gradual physical and chemical breakdown) to the point where it may be impossible to tell just what kind of rock the soil came from.

As rock weathers, it breaks down into smaller and smaller fragments—stones and smaller particles. These are

classified as sand, silt, and clay. **Sand** is made up of particles from 2.0 to 0.063 millimeters in diameter, **silt** particles range from 0.063 to 0.004 millimeters, and **clay** is anything finer than 0.004 millimeters (Figure 11–3). You can see the individual rock particles in sand, and you may be familiar with the finer particles called silt, but you would need a good microscope to see clay particles. Particles larger than sand are called gravel, cobbles, or boulders, depending on their size.

The sand, silt, and clay particles constitute the mineral portion of soil. **Soil texture** refers to the relative proportions of each type of particle in a given soil. If one predominates, soil is said to be sandy, silty, or clayey. You can determine the texture of a given soil by shaking a small amount of it with water in a large test tube then allowing the particles to settle. Because particles settle according to their weight, sand particles settle first, silt second, and clay last. (More precise measurements require a laboratory analysis.)

Soil scientists classify soil texture with the aid of a triangle that shows the relative proportions of sand, silt, and clay in a given soil (Figure 11–4). As the soil triangle shows, a soil that consists of roughly 40% sand, 40% silt, and 20% clay is called a **loam**. Loam containing organic matter is the ideal soil for many types of farming because it retains nutrients and some water but allows excess water to drain.

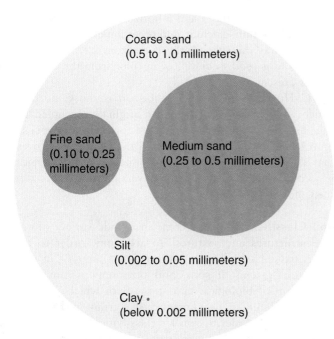

Figure 11–3 **The relative sizes of soil particles.** Clay particles are too small to see with the naked eye, while silt particles are slightly larger. Sand particles vary from small to large. Gravel is still larger.

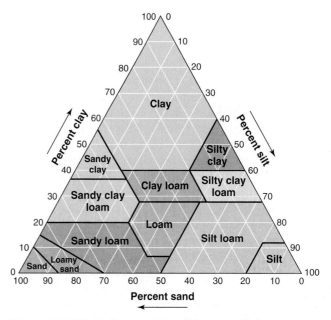

Figure 11–4 **The soil texture triangle.** Relative proportions of sand, clay, and silt are represented on each axis. For clay, read across horizontally; for silt, read diagonally downward toward the left from the right axis; for sand, read diagonally upward to the left from the bottom axis. If the triangle is read properly, the content of any soil should total 100%.

UNDERSTANDING THE DATA

1. What is the highest percentage of silt that a soil could contain and still be classified as a sandy clay?
2. A sample of soil is composed of 70% clay, 20% silt, and 10% sand. What type of soil is it?
3. Another sample of soil is composed of 60% silt, 10% sand, and 30% clay. What type of soil is it? ☑

Soil Properties. Soil scientists often ask questions about how the components of soil determine its properties. Three basic properties of matter determine how several important properties of soil are influenced by texture:

1. Larger particles have larger spaces separating them than smaller particles have. (Visualize the difference between packing softballs and packing golf balls in containers of the same size.)
2. Smaller particles have more surface area relative to their volume than larger particles have. (Visualize cutting a block in half again and again. Each time you cut it, you create two new surfaces, but the total volume of the block remains the same.)
3. Nutrient ions and water molecules tend to cling to surfaces. (This is why a nongreasy surface remains wet after you drain the water from it.)

These properties of matter profoundly affect such soil properties as water infiltration, the capacity to hold water and nutrients, and aeration. Table 11–1 shows the relationship between soil texture and these properties.

Soil texture also affects **workability**—the ease with which a soil can be cultivated. Workability has a tremendous impact on agriculture. Clayey soils are very difficult to work with because even modest changes in moisture content can turn them from too sticky and muddy to too hard and even bricklike. Sandy soils are very easy to work because they become neither muddy when wet nor hard and bricklike when dry.

Table 11–1 Relationship Between Soil Texture and Soil Properties

Soil Texture	Water Infiltration	Water-Holding Capacity	Nutrient-Holding Capacity	Aeration	Workability
Sand	Good	Poor	Poor	Good	Good
Silt	Medium	Medium	Medium	Medium	Medium
Clay	Poor	Good	Good	Poor	Poor
Loam	Medium	Medium	Medium	Medium	Medium

Soil Profiles. The processes of soil formation create a vertical gradient of layers that are often quite distinct. These horizontal layers are known as **soil horizons,** and a vertical slice through the different horizons is called a **soil profile** (Figure 11–5). A soil profile reveals a great deal about the factors that interact in the formation of soil.

O Horizon. The topmost layer, the O horizon, consists of dead organic matter (detritus) deposited by plants: leaves, stems, fruits, and seeds. Thus, the O horizon is high in organic content and is the primary source of energy for the soil community. Toward the bottom of the O horizon, the processes of decomposition are well advanced, and the original materials may be unrecognizable. At this point, the material is dark and is called **humus.**

A Horizon. The next layer in the profile is the A horizon, a mixture of mineral soil from below and humus from above. The A horizon is also called **topsoil.** Fine roots from the overlying vegetation permeate this layer. The A horizon is usually dark because of the humus that is present and may be shallow or thick, depending on the overlying ecosystem. Vital to plant growth, topsoil grows at the rate of an inch or two every hundred years.

E Horizon. In many soils, the next layer is the E horizon, where *E* stands for **eluviation**—the process of **leaching** (dissolving away) many minerals due to the downward movement of water. This layer is usually paler in color than the two layers above it.

B Horizon. Below the E horizon is the B horizon. This layer is characterized by the deposition of minerals that have leached from the A and E horizons, so it is often high in iron, aluminum, calcium, and other minerals. Frequently referred to as the **subsoil,** the B horizon is typically high in clay and is reddish or yellow in color from the oxidized metals present.

C Horizon. Below the B horizon is the C horizon, which is the parent material that originally occupied the site. This layer represents weathered rock, glacial deposits, or volcanic ash and usually reveals the geological process that created the landscape. The biological and chemical processes occurring in the upper layers have little effect on the C horizon.

Soil Classification. Soils come in a wide variety of vertical structures and textures. To give some order to this diversity, soil scientists have created a taxonomy, or classification system, of soils. Soil taxonomy is similar to the system of biological taxonomy, with which you may already be familiar. The most inclusive group in the taxonomy is the **soil order.** If you are classifying a soil, you find the order first and then work your way downward through the taxonomic categories (*suborders, groups, subgroups, families*) until you come to the soil *class* that best corresponds to the soil in question. There are literally hundreds of soil classes.

Figure 11–6 shows the worldwide distribution of the 12 major soil orders. The four major soil orders that are most important for agriculture, animal husbandry, and forestry are described below. A fifth type of soil that is particularly important to environmental health is also described.

Mollisols. Mollisols are fertile, dark soils found in temperate grasslands. They are the world's best agricultural soils and are encountered in the Midwestern United States; across temperate Ukraine, Russia, and Mongolia; and in the pampas (a broad region of grassy plains) in Argentina. They have a deep A horizon and are rich in humus and minerals; precipitation is insufficient to leach the minerals downward into the E horizon.

Alfisols. Alfisols are widespread, moderately weathered forest soils. Although not deep, they have well-developed O, A, E, and B horizons. Alfisols are typical of the moist, temperate forest biome and are suitable for agriculture if they are supplemented with organic matter or mineral fertilizers to maintain soil fertility.

Oxisols. The soils of tropical and subtropical rain forests are called oxisols. Such soils have a layer of iron and aluminum oxides in the B horizon and have little O horizon, due to the rapid decomposition of plant matter. Most of the minerals are in the living plant matter, so oxisols provide limited fertility for agriculture. If the forests are cut (as in "slash-and-burn" agriculture), a few years of crop growth can be obtained, but in time, the intense rainfall leaches the minerals downward, forming an impervious hardpan that resists further cultivation.

Aridisols. Aridisols are found in **drylands,** which include deserts as well as semiarid and seasonally dry areas. The paucity of vegetation and precipitation leaves aridisols relatively unstructured vertically. They are thin and light

O Horizon: Humus (surface litter, decomposing plant matter)

A Horizon: Topsoil (mixed humus and leached mineral soil)

E Horizon: Zone of leaching (less humus, minerals resistant to leaching)

B Horizon: Subsoil (accumulation of leached minerals like iron and aluminum oxides)

C Horizon: Weathered parent material (partly broken-down minerals)

Figure 11–5 **Soil profile.** This idealized soil profile shows the major horizons from the surface to the parent material.

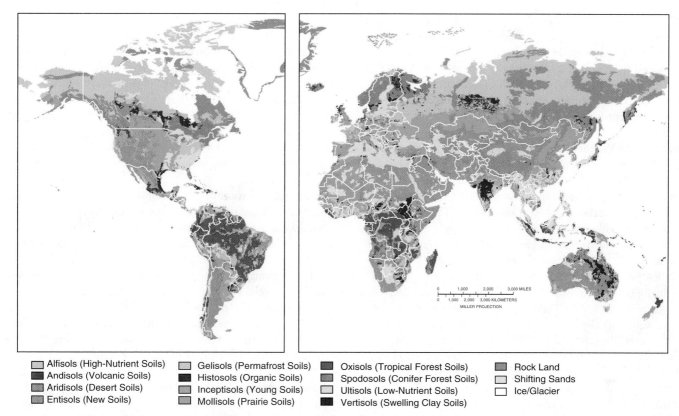

Figure 11–6 **Global map of soil orders.** The map shows the worldwide distribution of the 12 soil orders. (*Source:* U.S. Department of Agriculture's Natural Resources Conservation Services, World Soil Resources staff.)

Legend:
- Alfisols (High-Nutrient Soils)
- Andisols (Volcanic Soils)
- Aridisols (Desert Soils)
- Entisols (New Soils)
- Gelisols (Permafrost Soils)
- Histosols (Organic Soils)
- Inceptisols (Young Soils)
- Mollisols (Prairie Soils)
- Oxisols (Tropical Forest Soils)
- Spodosols (Conifer Forest Soils)
- Ultisols (Low-Nutrient Soils)
- Vertisols (Swelling Clay Soils)
- Rock Land
- Shifting Sands
- Ice/Glacier

colored and, in some regions, may support enough vegetation for rangeland animal husbandry. Irrigation used on these soils usually leads to salinization, as high evaporation rates draw salts to surface horizons, where they accumulate to toxic levels.

Hydric Soils. Another soil type—**hydric soil**—is important to ecosystems, although not necessarily good for agriculture. Hydric soils indicate a wetland or aquatic site. Hydric soils are not a soil class on their own, but are members of a number of other classes. Maps of hydric soils are used in delineating the boundaries of wetlands, in order to get permits for construction, and to protect wetlands and water bodies from the negative impacts of development.

The soil order *histosols* includes several types of hydric soils. They contain a buildup of undecayed vegetation called **peat** and hence are called peatlands. Peat contains a great deal of energy and can be dried and burned. Vast areas of northern zones have hydric soils, including soils in the order *gelisols*, that are permanently frozen at some depth, a condition called *permafrost.* One of the concerns about climate change is that warming will thaw permafrost and increase rates of decay. This would act as positive feedback, causing the decaying peat to release more greenhouse gases into the atmosphere (Chapter 18).

Soil Constraints. Soil scientists often investigate the relationship between soil and the plants it supports. Such scientists speak of *soil constraints,* meaning the various characteristics of soils that limit primary productivity, especially in agriculture and forestry. These constraints include poor soil drainage, low ability to retain nutrients, high levels of aluminum, low levels of phosphorous, soil cracking (from certain types of clays), the presence of soluble salts, shallowness, and a high likelihood of erosion. Let's look at the characteristics that enable soils to support plant growth, especially agricultural crops.

Soil and Plant Growth

For their best growth, plants need a root environment that supplies optimal amounts of mineral nutrients, water, and air (oxygen). The pH (relative acidity) and salinity (salt concentration) of soil are also critically important. **Soil fertility**—a soil's ability to support plant growth—often refers to the presence of proper amounts of nutrients, but it also includes a soil's ability to meet all the other needs of plants. Farmers speak of a given soil's ability to support plant growth as its **tilth.**

Mineral Nutrients and Nutrient-Holding Capacity. Rocks contain various mineral nutrients—phosphate (PO^{3-}_4), potassium (K^+), calcium (Ca^{2+}), and other ions—along with nonnutrient elements. Nutrients initially become available to roots through the weathering of rock. Weathering, however, is much too slow to support normal plant growth. The nutrients that support plant growth in natural ecosystems are supplied mostly through the breakdown and release, or recycling, of nutrients from detritus.

Nutrients may be washed from soil as water moves through it, a process called leaching. Leaching lessens soil fertility; it also contributes to pollution when materials removed from soil enter waterways. Consequently, soil's capacity to bind and hold nutrient ions until they are absorbed by roots is important. This property is referred to as the soil's **nutrient-holding capacity**.

Water and Water-Holding Capacity. The supply of water in soil is crucial to plant growth. Water is constantly being absorbed by the roots of plants, passing up through the plant and exiting as water vapor through microscopic pores in the leaves—a process called **transpiration**. The pores, called *stomata* (singular, *stoma*), permit the entry of carbon dioxide and the exit of oxygen in photosynthesis; however, the loss of water via transpiration through the stomata is dramatic. A field of corn, for example, transpires an equivalent of a field-sized layer of water 43 centimeters (17 in.) deep in a single growing season. Plants with inadequate water may wilt and eventually die.

Water is resupplied to soil by rainfall or irrigation. Three attributes of soil are significant in this respect (Figure 11–7). First is soil's ability to allow water to infiltrate or soak in. If water runs off the ground surface, it won't be useful to the plants. Worse, it may cause erosion.

The second attribute is the soil's ability to hold water after it infiltrates, a property called **water-holding capacity**. In soils with poor water-holding capacity, most of the infiltrating water percolates down below the reach of roots, thus becoming unavailable to plants. In contrast, soils with

good water-holding capacity act like sponges, holding a large amount of water and providing a reservoir from which plants can draw between rains. Sandy soils are notorious for their poor water-holding capacity, whereas clayey and silty soils are good at holding water.

The third critical attribute of soil is its ability to minimize **evaporative water loss**. This kind of evaporation depletes soil's water reservoir without serving the needs of plants. The O horizon functions well to reduce evaporative water loss by covering soil.

Aeration. The roots of plants require a constant supply of oxygen to carry out their metabolism. Land plants depend on soil being loose and porous enough to allow the diffusion of oxygen into and carbon dioxide out of the spaces in soil, a property called soil **aeration**. Overwatering fills the air spaces in soil, preventing adequate aeration. Plants that live in very wet soil have special adaptations that allow them to get oxygen.

Compaction, or packing of soil, which occurs with excessive foot or vehicular traffic, lowers soil air, reduces infiltration, and increases runoff. Again, soil texture strongly influences this property, as indicated in Table 11–1.

Relative Acidity (pH). The term **pH** refers to the acidity or alkalinity of any solution. A solution that is neither acidic nor alkaline is said to be neutral and has a pH of 7. The pH scale, which runs from 1 to 14, is discussed more fully later (Chapter 19). For now, it is important to know that different plants are adapted to different pH ranges. Most plants do best with a pH near neutral, but many plant species are

Figure 11–7 Plant-soil-water relationships. The Sun's energy drives transpiration by causing water to evaporate from the plant's leaves. When water evaporates, a vacuum is created that pulls more water up through the plant's tissues. The roots draw water from soil to replenish the evaporated water. Water lost from the plant by transpiration must be replaced from a reservoir of water held in soil. In addition to the amount and frequency of precipitation, the size of this reservoir depends on soil's ability to allow water to infiltrate, to hold water, and to minimize direct evaporation.

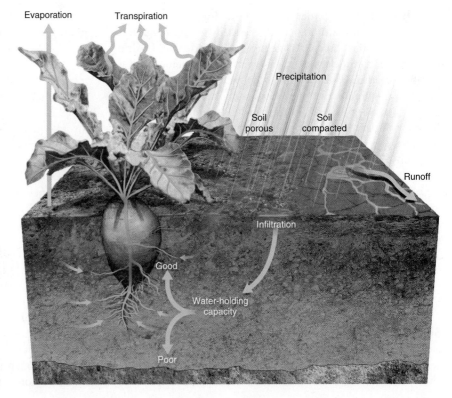

specialized for distinctly acidic or alkaline soils. Blueberries, for example, are acid lovers.

Salt and Water Uptake. A buildup of salts in soil makes it impossible for the roots of a plant to take in water. Indeed, if salt levels in soil get high enough, water can be drawn *out* of the plant (by osmosis), resulting in dehydration and death. Only plants with special adaptations can survive saline soils, and to date, none of those are crop plants. In Section 11.2 we will discuss how irrigation may lead to the accumulation of salts in soil (salinization).

Soil and Carbon Storage. Soils hold carbon from leaf litter and dead organisms. In fact, scientists estimate that soils hold as much as three times the amount of carbon held in the atmosphere and living plants.

Scientists are keenly interested in knowing what drives the rate at which the organic matter in soil decays. Scientists once thought that temperature was the main driving factor, but now know that factors such as soil moisture, physical separation of particles, and location of roots play a role. In some places, earthworms drive the rapid breakdown of forest litter. Microbial communities also play a role. A better understanding of these processes will help scientists make more accurate predictions about the carbon cycle and climate change. Scientists are now undertaking a long-term, large-scale study to determine how changes in leaf litter will combine with climate change and the effects of soil biota to change soils and, ultimately, the carbon cycle.[3]

---

[3]P. García-Palacios et al., "Climate and Litter Quality Differently Modulate the Effects of Soil Fauna on Litter Decomposition Across Biomes," *Ecology Letters* (2013), doi:10.1111/ele.12137.

The Soil Community

As we've seen, to support plant growth, soil must (1) have a good supply of nutrients and good nutrient-holding capacity; (2) allow infiltration, have good water-holding capacity, and resist evaporative water loss; (3) have a porous structure that permits good aeration; (4) have a pH near neutral; and (5) have low salt content. Moreover, these attributes must be sustained. How does a soil provide and sustain such attributes? Which is the best soil?

Recall the principle of limiting factors: The poorest attribute is the limiting factor (Chapter 3). The poor water-holding capacity of sandy soil, for example, may preclude agriculture altogether because soil dries out so quickly. The inability of clay soils to allow water infiltration or aeration prevents them from being good soils. As Table 11–1 indicates, the best soil textures are silts and loams because limiting factors are moderated in these two types of soil. Even though silts and loams are rated only "medium" for use in agriculture, their soil texture limitations are greatly improved by the organic parts of the soil ecosystem—detritus and soil organisms.

Organisms and Organic Matter in Soil. The dead leaves, roots, and other detritus accumulated on and in soil support a complex food web, including numerous species of bacteria, fungi, protozoans, mites, insects, millipedes, spiders, centipedes, earthworms, snails, slugs, moles, and other burrowing animals **(Figure 11–8)**. The most numerous and important organisms are the smallest—the bacteria; literally millions of bacteria can be seen and counted in a gram of soil.

Humus. As all these organisms feed, the bulk of the detritus is consumed through cellular respiration. Carbon

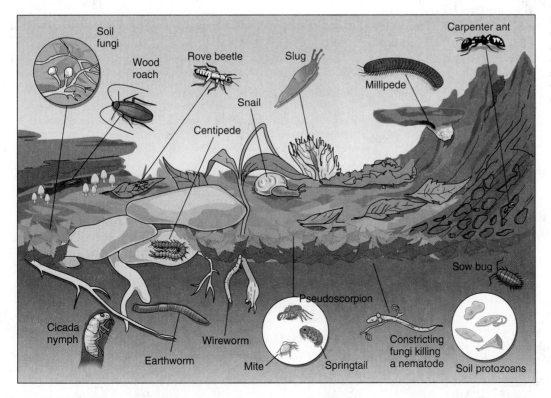

Figure 11–8 Soil as a detritus-based ecosystem. A host of organisms feed on detritus and burrow through soil, forming a humus-rich topsoil with a loose, clumpy structure.

Figure 11–9 Humus and soil structure. On the left is a humus-poor sample of loam. Note that it is a relatively uniform, dense clod. On the right is a sample of the same loam, but rich in humus. Note that it has a very loose structure, composed of numerous aggregates of various sizes.

Figure 11–10 The results of removing topsoil. This "soil" is nothing but gravel with topsoil removed. After 50 years, only lichens and a few stunted trees have developed. (The white ballpoint pen shows the scale of the photograph.)

dioxide, water, and mineral nutrients are released as by-products (see Chapter 4). However, each organism leaves a certain portion undigested; that is, a portion resists break-down by the organism's digestive enzymes. This residue of partly decomposed organic matter is humus, found in high concentrations at the bottom of the O horizon. A familiar example is the black or dark brown spongy material remaining in a dead log after the center has rotted out. Humus is good for plant growth because it has an extraordinary capacity for holding water and nutrients—as much as 100 times greater than the capacity of clay, on the basis of weight. **Composting** is the process of fostering the decay of organic wastes under more or less controlled conditions, and the resulting compost is essentially humus.

Soil Structure and Topsoil. As animals feed on detritus on or in soil, they often ingest mineral soil particles as well. For example, it is estimated that as much as 37 tons per hectare (15 tons per acre) of soil pass through earthworms each year in the course of their feeding. As the mineral particles go through the worm's gut, they become "glued" together by the indigestible humus compounds. Thus, earthworm excrements—or castings, as they are called—are relatively stable clumps of inorganic particles plus humus. The burrowing activity of organisms keeps the clumps loose. These loose clumps are a desirable component of **soil structure** (Figure 11–9). Whereas soil texture describes the size of soil particles, soil structure refers to their arrangement. A loose soil structure is ideal for infiltration, aeration, and workability. The clumpy, loose, humus-rich soil that is best for supporting plant growth is topsoil, represented in a soil profile as the A horizon (see Figure 11–4).

If plants are grown on adjacent plots, one of which has had all its topsoil removed, the results are striking: The yield from plants grown on subsoil is only 10% to 15% of that from plants grown on topsoil. In other words, a loss of all the topsoil would result in an 85% to 90% decline in productivity. The development of topsoil from subsoil or parent material is a process that takes hundreds of years or more.

Figure 11–10 shows a site where the topsoil was removed 50 years ago to provide fill for a highway. Although a lichen cover and some stunted trees have grown on the exposed gravel, there is still no topsoil there.

Interactions. There are some important interactions between plants and soil biota. A highly significant one is the symbiotic relationship between the roots of some plants and certain fungi called **mycorrhizae**. Drawing some nourishment from the roots, mycorrhizae penetrate the detritus, absorb nutrients, and transfer them directly to the plant (Figure 11–11). Thus, there is no loss of nutrients to leaching. Another important relationship is that of certain bacteria that add nitrogen to soil (Chapter 3). Not all soil organisms are beneficial to plants, however. For example, nematodes,

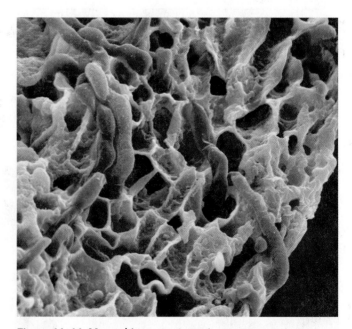

Figure 11–11 Mycorrhizae. Symbiotic fungi connect to the roots of plants; they help the plant by taking in nutrients and increasing root absorption, while getting protection and food. This scanning electron micrograph shows small fungal hyphae entering a plant root hair.

small worms that feed on living roots, are highly destructive to some agricultural crops. In a flourishing soil ecosystem, however, other soil organisms may control nematode populations.

There is still a great deal about soil communities that scientists do not understand. Scientists initiated the Global Soil Biodiversity Assessment, a collaborative effort to better understand the ecology of soils across the globe. Over time, this coordination should result in knowledge that we can use in protecting soils.

Soil Enrichment or Mineralization. You can now see how the aboveground and soil portions of an ecosystem support each other. The bulk of the detritus, which supports soil organisms, is from green-plant producers, so plants support soil organisms. In turn, as soil organisms feed on detritus, they create the chemical and physical soil environment that is beneficial to plant growth. Plants protect soil in two other important ways: The cover of living plants and detritus protects soil from erosion and reduces evaporative water loss. (These functions explain why it is desirable to keep organic mulch around vegetables growing in a garden.)

Unfortunately, it is all too easy to break the mutually supportive relationship between plants and soil. The maintenance of topsoil depends on the addition of detritus in sufficient quantity to balance losses. Without continual additions of detritus, soil organisms starve, and their benefit in keeping soil loose and nutrient rich is lost. Additional consequences can occur as well. Although resistant to digestion by soil organisms, humus does decompose at the rate of about 2% to 5% of its volume per year. As soil's humus content declines, the clumpy aggregate structure created by soil particles glued together with the humus breaks down. Water- and nutrient-holding capacities, infiltration, and aeration decline as well. This loss of humus and the consequent collapse of topsoil are referred to as the **mineralization** of soil because what is left is just the gritty mineral particles—sand, silt, and clay.

Thus, topsoil is the result of a dynamic balance between the addition of detritus and humus-forming processes, on the one hand, and the breakdown and loss of detritus and humus on the other (Figure 11–12). Soil deteriorates or is revitalized through loss or additions of organic matter.

Fertilizer. In agricultural systems, nutrients in topsoil are replenished with applications of **fertilizer**—material that contains one or more necessary nutrients, such as nitrogen or phosphorus. Fertilizer may be organic or inorganic. **Organic fertilizer** includes plant or animal wastes; manure and compost (rotted organic material) are two examples. Organic fertilizer also includes leguminous fallow crops (alfalfa, clover) or food crops (lentils, peas), which fix atmospheric nitrogen.

Inorganic fertilizers are chemical formulations of nutrients without organic components. Inorganic fertilizers provide necessary nutrients, but lack organic matter to support soil organisms and build the important soil structure we've just described. Loss of soil nutrients will be discussed shortly.

CONCEPT CHECK A shaded soil contains equal parts of sand, clay, and silt, as well as large amounts of leaf litter and fungi. What would you expect the three water-holding characteristics of this soil to be? ☑

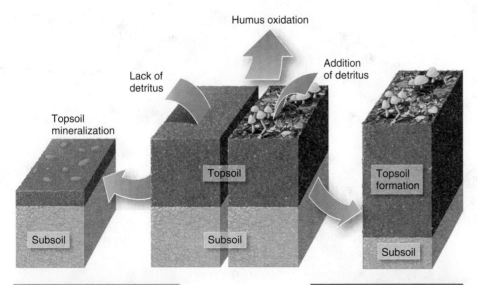

Loss of humus results in topsoil mineralization and consequent loss of:
• Water-holding capacity
• Nutrient-holding capacity
• Water infiltration
• Aeration

Gain of humus results in topsoil formation and consequent gain in:
• Water-holding capacity
• Nutrient-holding capacity
• Water infiltration
• Aeration

Figure 11–12 **The importance of humus to topsoil.** Topsoil is the result of a balance between detritus additions and humus-forming processes, as well as their breakdown and loss. If additions of detritus are insufficient, soil will gradually deteriorate.

11.2 Soil Degradation

In natural ecosystems, there is always a turnover of plant material, so new detritus is continuously being added to topsoil. However, when humans cut forests, graze livestock, or grow crops, the soil is at the mercy of their management or mismanagement. **Soil degradation** is a reduction in the capacity of soil to support plant life and to perform ecosystem functions. As the map in **Figure 11–13** shows, soil degradation is a worldwide problem (see Sound Science, *How Do We Know the Condition of Earth's Soils?*). One of the most critical forms of soil degradation is erosion.

Forms of Erosion

How is topsoil lost? The most pervasive and damaging force is **erosion,** which occurs when water and wind pick up particles of soil and carry them away. Erosion occurs any time soil is exposed to the elements. The removal may be slow and subtle, as when soil is gradually blown away by wind, or it may be dramatic, as when gullies are washed out in a storm.

In most natural terrestrial ecosystems, a vegetative cover protects against erosion. The vegetation intercepts falling raindrops, and the water infiltrates gently into the loose topsoil without disturbing its structure. With good infiltration, runoff is minimal. Runoff that does occur is slowed as the water moves through the vegetative or litter mat. Similarly, vegetation slows the velocity of wind and holds soil particles.

Splash, Sheet, and Gully Erosion. When soil is left bare and unprotected, it is easily eroded. **Splash erosion** occurs in storms as the impact of falling raindrops breaks up the clumpy structure of the topsoil. The dislodged particles wash into spaces between other aggregates, clogging pores and thereby decreasing infiltration and aeration. The decreased infiltration results in more water running off, carrying away the fine particles from the surface, a phenomenon called **sheet erosion.** As further runoff occurs, the water converges into rivulets and streams, which have greater volume, velocity, and energy and hence greater capacity to pick up and remove soil. The result is erosion into gullies, or **gully erosion** (Figure 11–14). This process of erosion can become a cycle that perpetuates itself.

Desert Pavement and Cryptogamic Crusts. Both wind and water erosion involve the differential removal of soil particles. The lighter particles of humus and clay are the first to be carried away, while rocks, stones, and coarse sand remain behind. Consequently, as erosion removes the finer materials, the remaining soil becomes progressively coarser—sandy, stony, and finally, rocky. Such coarse soils frequently reflect past or

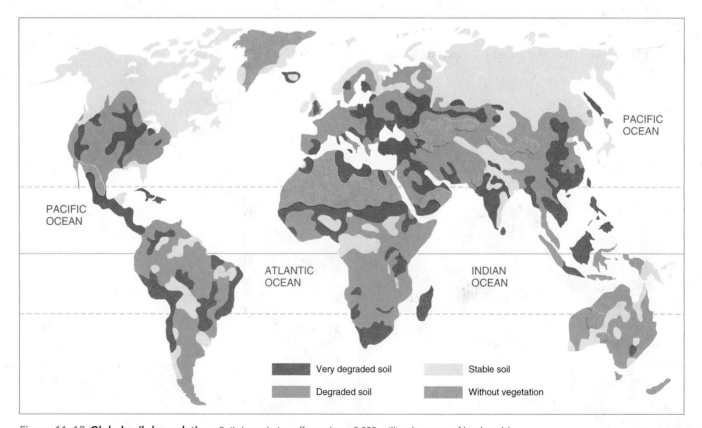

Figure 11–13 Global soil degradation. Soil degradation affects about 2,000 million hectares of land worldwide, including about 38% of the world's cropland. The colors on the map indicate the severity of degradation.

(*Source:* Philippe Rekacewicz, IAASTD—International Assessment of Agricultural Science and Technology for Development, UNEP/GRID-Arendal.)

ongoing erosion. Did you ever wonder why deserts are full of sand? The sand is what remains after the finer, lighter clay and silt particles have blown away. In some deserts, however, the removal of fine material by wind has left a thin surface layer of stones and gravel called a **desert pavement,** which protects the underlying soil against further erosion (Figure 11–15). Vehicular and pedestrian traffic damage this surface layer, allowing another episode of erosion to commence.

Soil with a crust hardened from drying may be colonized by certain kinds of primitive plants (algae, lichens, and mosses) called cryptogams. Their growth and colonization create a crust on the soil, called a **cryptogamic crust.** Such crusts have a positive ecological impact; they stabilize soil, slow erosion, and add nutrients through nitrogen fixation. However, they can also inhibit water infiltration and seed generation. Cryptogramic crust is readily broken up by livestock trampling or human intrusion. If the soil below is loosened, it can again be subject to wind and water erosion.

The Other End of the Erosion Problem. Water that is unable to infiltrate flows over the surface into streams and rivers, overfilling them and causing flooding. Eroding soil, called **sediment,** is carried into streams and rivers, clogging channels and intensifying flooding, filling reservoirs, killing fish, and generally damaging the ecosystems of streams, rivers, bays, and estuaries. In many regions of the world, excess sediments and nutrients resulting from erosion are the greatest source of surface-water pollution. Meanwhile, groundwater resources are also depleted because the rainfall runs off rather than recharging groundwater. This is an example of the tight connection between the health of soils and the availability and quality of water.

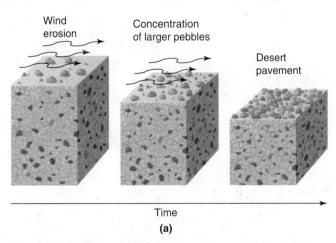

Figure 11–15 **Formation of desert pavement. (a)** As wind erosion removes the finer particles, the larger grains and stones are concentrated on the surface. The result is desert pavement, which protects the underlying soil from further erosion. **(b)** The Sinai Desert has a mixture of desert pavement, rock, and tracks from vehicles, which break up the desert pavement and initiate further erosion.

SOUND SCIENCE

How Do We Know the Condition of Earth's Soils?

To study weather, meteorologists use data from thousands of weather stations located around the globe. How do scientists study the world's soils? Like meteorologists, soil scientists need to gather reliable data from many different regions. At one point, regional governments collected the data on land use, agriculture, and soils, but these data were in different formats and used different vocabulary. Standardizing these data and providing the international community with information about soil resources has been the job of World Soil Information, an independent science-based foundation.

Global Assessments

The first worldwide assessment of land degradation was the 1991 Global Assessment of Soil Degradation (GLASOD), which concluded that 15% of Earth's soils were degraded. However, that report was compiled from limited data. GLASOD was followed by the Land Degradation Assessment in Drylands (LADA), a project launched by the UN Food and Agricultural Organization. This project used satellite remote-sensing techniques to produce an overall view of soil quality, called the Global Assessment of Land Degradation and Improvement (GLADA).[4] GLADA employed data from a 23-year time-series compiled by National Oceanic and Atmospheric Administration (NOAA) satellites and precipitation data from more than 9,300 land stations. One analysis combined plant productivity with rain-use efficiency to produce a map of regions that are worsening, calling them "hot spots," and regions that are showing improving trends, calling them "bright spots."

GLADA was most recently updated in 2010. One finding of this updated study was that land degradation in China is changing differently than expected. Degradation is most conspicuous in China's humid south and east, where urban development is intense. The condition of soils in the dry north and west were better than expected, the result of aggressive land protection and reforestation efforts by the Chinese government.

New Monitoring Techniques

The most recent technologies for monitoring soil use satellite images to estimate plant biomass. In one technique, satellites measure variations in light, which are used to calculate greenness. Greenness is then used as an index of the amount of vegetation in various regions; this, in turn, is used to assess land degradation because degraded land supports less vegetation. But in some biomes, even undegraded lands do not support a lot of vegetation, so the monitoring program measures changes, asking, "Is the ability of the land to produce vegetation declining?" Of course during certain times of year and during droughts, one would expect vegetation to decline, so the program must account for seasons and the effect of droughts to measure whether the soil is becoming more or less degraded.

The ability of scientists to accurately model what is happening to soil is rapidly improving. Such knowledge is essential to our stewardship of this precious resource.

---

[4]Z. G. Bai, R. de Jong, and G. W. J. van Lynden, An Update of GLADA—Global Assessment of Land Degradation and Improvement, ISRIC report 2010/08 (2010), ISRIC—World Soil Information, Wageningen.

Practices That Cause Erosion

Erosion is one of the main elements of soil degradation. Practices that expose soil to erosion include overcultivation, overgrazing, and deforestation, all of which are a consequence of unsustainable management practices. Yet as we will see, many of the impacts of these practices can be reversed.

Overcultivation. Traditionally, the first step in growing crops has been plowing to control weeds. The drawback is that the soil is then exposed to erosion by wind and water. Further, the soil may remain bare for a considerable time after planting and again after harvest. Plowing is frequently considered necessary to loosen soil to improve aeration and infiltration through it, yet all too often the effect is just the reverse. Splash erosion destroys soil's aggregate structure and seals its surface, so that aeration and infiltration are decreased. The weight of the tractors used in plowing compacts the soil. Plowing also accelerates the loss of water through evaporation.

Despite the harmful impacts inherent in cultivation, systems of *crop rotation*—a cash crop such as corn every third year, with hay and clover (which fixes nitrogen as well as adding organic matter) in between—have proved sustainable. If farmers abandon crop rotation, degradation and erosion exceed regenerative processes, and the result is a decline in the quality of soil. This is the essence of overcultivation.

Overgrazing. Grasslands that are too steep or receive too little rainfall to support cultivated crops have traditionally been used for grazing livestock. In fact, about two-thirds of dryland areas are rangelands. Unfortunately, such lands are often overgrazed. As grass production fails to keep up with consumption, erosion follows and the land becomes barren.

Overgrazing is not a new problem. In the United States in the 1800s, the American bison were slaughtered to starve out Native Americans and allow for stocking the rangelands with cattle. Cattle overgrazing became rampant, leading to erosion and encroachment by hardy desert plants such as sagebrush, mesquite, and juniper, which are not palatable to cattle. Western rangelands now produce less than half the livestock forage they produced before the advent of commercial grazing.

The broader ecological impact of overgrazing is significant. Livestock compete with native animals for forage, substantially reducing the populations of native animals. Livestock also pollute waterways with sediment and their waste. A study of rangeland management concluded that desertification affects some 85% of North America's dry-lands and that the most widespread cause of rangeland desertification is livestock grazing.[5]

In many cases, overgrazing occurs because rangelands are public lands that are not owned by the people who graze the animals. Where this is the case, herders who choose to have fewer livestock on the range sacrifice income, while others continue to overgraze the range. This problem, known as the "tragedy of the commons," was discussed earlier (Chapter 7).

Deforestation. Forest ecosystems are extremely efficient at holding and recycling nutrients and at absorbing and hold-ing water because they maintain and protect very porous, humus-rich topsoil. Investigators at Hubbard Brook Forest in New Hampshire found that converting a hillside from for-est to grassland doubled the amount of runoff and increased the leaching of nutrients many-fold. Much worse is what oc-curs if the forest is simply cut and its soil is left exposed. The topsoil becomes saturated with water and the muddy mass slides off the slope and into waterways, leaving only barren subsoil, which continues to erode.

Forests continue to be cleared at an alarming rate (Chapter 7). The problem is particularly acute when tropical rain forests are cut. Because of leaching by rainwater, tropi-cal soils (oxisols) are notoriously lacking in nutrients. When the forests are cleared, the thin layer of humus with nutrients readily washes away. Only the nutrient-poor subsoil is left.

Nutrient Mining

In agricultural systems, there is an unavoidable removal of nutrients from soil with each crop because the nutrients absorbed by plants are contained in the harvested material. This removal of nutrients, called **nutrient mining,** degrades the fertility of the soil. Therefore agricultural systems require inputs of nutrients to replace those removed with the har-vest. Nutrients are replenished with applications of organic or inorganic fertilizer.

Under intensive cultivation, nutrient content may be kept high with inorganic fertilizers, but mineralization, and thus soil degradation, proceeds. As the soil loses its nutrient-holding capacity, inorganic fertilizers are prone to leach into waterways, causing pollution (Chapters 12, 20). Often, chemical fertilizers are combined with crop irrigation, a combination that can cause further problems. The overuse of nitrogen-containing fertilizers is a leading cause of disrup-tions to the nitrogen cycle (Chapter 3). In addition, the pro-duction of many inorganic fertilizers is fossil-fuel intensive, contributing to global emissions of greenhouse gases.

Inorganic fertilizers have a valuable place in agriculture, but growers need to understand the different roles played by organic material and inorganic nutrients and then use each as necessary.

Figure 11–16 Dryland cultivation. Farmers in Jinan, China, till the semiarid desert soil and plant a crop in anticipation of some rainfall.

Drylands and Desertification

Dryland ecosystems—areas that are arid, semiarid, or season-ally dry—cover 41% of Earth's ice-free land and are found on every continent except Antarctica. They are defined by pre-cipitation, not temperature. Beyond the relatively uninhabited deserts, much of the land receives only 25 to 75 centimeters (10 to 30 in.) of rainfall a year, a minimal amount to support rangeland or nonirrigated cropland (Figure 11–16). Droughts are common in dryland areas. When the rainfall returns, however, vegetation often returns.

Regions that have sparse rainfall or long dry seasons sup-port grasses, scrub trees, or crops only insofar as soils have good water-holding and nutrient-holding capacity. Recall that clay and humus are the most important components of soil for both nutrient- and water-holding capacity. When clay and hu-mus particles are removed from soil, nutrients that are bound to those particles are removed as well. Even more serious, however, is the loss of the soil's water-holding capacity.

As the erosion of topsoil diminishes a soil's capacity to retain water and nutrients, drylands become deserts, both ecologically and from the standpoint of agricultural production. The term **desertification** is used to denote this process. Note that desertification doesn't mean advanc-ing deserts; rather, the term refers to a permanent reduc-tion in the productivity of the land. It is a process of land degradation with a number of possible causes, including extended droughts, overgrazing, erosion, deforestation, and overcultivation.

According to the Millennium Ecosystem Assessment, 10–20% of drylands (about 6 billion hectares) suffer from some form of land degradation. These lands are home to more than 2.3 billion people. Since drylands are commonly inhabited by some of the poorest people on Earth, promot-ing sustainable land management is essential to improving standards of living.

---

[5]Government Accountability Office, Livestock Grazing: Federal Expenditures and Receipts Vary, Depending on the Agency and the Purpose of the Fee Charged (September 30, 2005).

Figure 11–17 Flood irrigation. The traditional method of irrigation is to flood furrows between rows with water from an irrigation canal. This method is extremely wasteful because most water evaporates or percolates beyond the root zone.

Figure 11–18 Salinization. Millions of acres of irrigated land are now worthless because of the accumulation of salts left behind as water evaporates.

Irrigation, Salinization, and Leaching

How we use water has a huge effect on soil health, especially in agricultural systems. Nowhere is this more clear than with irrigation—supplying water to croplands by artificial means. One-sixth of all cultivated U.S. cropland is irrigated. Worldwide, irrigated acreage has risen dramatically in the past few decades. The FAO estimates that total irrigated land now amounts to just more than 300 million hectares (more than 740 million acres). Irrigation has dramatically increased crop production in regions that typically receive low rainfall and is a major part of the Green Revolution (Chapters 8, 12). However, irrigation is also a major contributor to land degradation and has a dramatic impact on regional water use (Chapter 10).

Methods of Irrigation. Traditionally, water has been diverted from rivers through canals and flooded through furrows in fields, a technique known as **flood irrigation** (Figure 11–17). Flood irrigation may cause the water table to rise, waterlogging the soiling and preventing crops from growing. An irrigation method that has become increasingly popular is **center-pivot irrigation,** in which water from a central well is pumped through a gigantic sprinkler that slowly pivots around the well (see Figure 10–17 b and c). In more recent years, the much more efficient **drip irrigation** method has been used on some crops; in drip irrigation, small tubes carry water directly to the plants.

Salinization. New irrigation projects continue to be built, but the gains from this expansion in acreage are now being offset by a negative trend: salinization. **Salinization** is the accumulation of salts in and on soil to the point where plant growth is suppressed (Figure 11–18). Salinization occurs because even the freshest irrigation water contains at least 200 to 500 ppm (0.02% to 0.05%) dissolved salts. Adding water to dryland soils also dissolves the high concentrations of soluble minerals that are often present in those soils as salts when water evaporates. Because it happens in drylands and renders the land less productive, salinization is considered a form of desertification.

Between 1995 and 2013, an estimated 17 million hectares (42 million acres) of the world's agricultural land was lost to salinization and waterlogging, a rate that translates into 4,600 football fields a day. In some places salinization reduces the yield of crops by as much as 60%.

Salinization can be avoided, and even reversed, if sufficient water is applied to leach the salts down through soil. However, attention must be paid to where the salt-laden water goes. For example, the Kesterson National Wildlife Refuge in California was destroyed by polluted irrigation drainage from selenium-enriched soils of the San Joaquin Valley. After receiving the salty drainage water for only three years, birds, fish, insects, and plants in the refuge began to die off. The refuge was declared a toxic waste site and has been drained and capped with soil. At least 14 other locations in the western United States have been identified as having the potential to accumulate toxic irrigation water, a process now known as the Kesterson Effect.

Leaching. In places that are heavily irrigated or have heavy rainfall, soils can lose nutrients and other chemicals to leaching. This is particularly true in tropical soils. Sometimes this is a problem because nutrients are leached out of soil; other times it is a problem because toxins are released from being bound to soil particles.

Some soil chemistry problems result from a change in soil pH. Acid rain, a by-product of burning fossil fuels, can lower soil pH. As acid rain percolates through soil, it can change the amount of attraction between some chemicals and soil particles. This change in pH can have two important effects. First, nutrients such as calcium, magnesium, and potassium that are bound to clay and humus can be separated from soil particles more easily and then leached away by rain. Second, a lower pH can mobilize toxins such as mercury or aluminum. Aluminum in soil is usually in the form of aluminum hydroxide, an insoluble compound. Acid rain breaks the chemical bonds in aluminum hydroxide, releasing toxic aluminum ions, a process that stunts the roots of plants, preventing them from absorbing nutrients.

Loss of Soil Biota

Soil health is largely regulated by detritus and microorganisms. The organic material in detritus holds water, binds mineral particles together, and slowly releases nutrients. Soil microbes help plants germinate, obtain nutrients, and perform other activities. Erosion, salinization, acid rain, and waterlogging can kill soil biota.

One new threat to soils is that of **nanoparticles**. Nanoparticles are used for a variety of purposes. For example, extremely small beads of silver are added to many products as an antimicrobial agent; you may be familiar with antimicrobial socks or similar products. Unfortunately, such nanoparticles can have negative effects on water and soil. A study in the *Journal of Hazardous Materials* in 2011 showed that nanoparticles can kill beneficial soil bacteria and reduce the ability of plants to take in nitrogen. However, the study looked at high concentrations of nanoparticles, and it is unclear how lower concentrations of nanoparticles will affect soils. One of the problems with nanoparticles is that they became widely used before sufficient studies of their environmental effects were completed.

Impacts of Mining

Agriculture is the human activity that probably has the greatest impact on soils, but other human activities also affect soils. For example, soils are affected by changes in land use. Construction projects remove vegetation and topsoil and often lead to erosion. Cutting down forests drives a great deal of erosion. Some of the most significant nonagricultural impacts on soil are the result of mining.

The scene after a strip-mining operation is completed is often similar to a moonscape—great piles of rocks with topsoil removed. Laws in some parts of the United States require these areas to be restored, though the success of such restoration varies. In the removed mountaintops of Appalachia, the plateaus left after coal mining are green but relatively unproductive (Chapter 14). The topsoil is gone, the remaining subsoil is too poor to support crops or most trees, and the vegetation that is planted is usually the non-native *Lespedeza cuneata*.

Mining is big business around the world. In addition to coal, people mine a variety of rocks and minerals such as gypsum, phosphates, salts, and metals, including a group of metals called **rare earth elements**. Rare earth elements are used in electronic devices and so are critical to modern activities. Despite their name, rare earth elements are relatively common in the Earth's crust, but are difficult to mine because they are not concentrated in the places where they are found. China mines 85% of the rare earth elements we use.

Mining removes vegetation and has huge impacts on soil erosion, water quality, and human health. Because of these connections, mining is of importance in our discussions on fossil fuels, environmental health, water pollution, and hazardous waste (Chapters 14, 17, 20, and 22). The first International Conference on Mining Impacts on the Human and Natural Environments was held in Florida in 2008. Since then, conferences on the environmental impacts of mining and the need for effective land restoration have been held around the world. In 2014, a conference in Beijing, China, covered topics relating to the impact of mining on the environment and successful mining land reclamation approaches. Hopefully, this is a direction scientists will pursue in order to lessen the impacts of mining on soil and water.

Creating New Land

Throughout history, societies have "created" new land. In many parts of Asia, people have built *terraces*—series of step-like fields and walls—into the sides of mountains to produce flat cropland. The Dutch claimed land from the sea by building levees and pumping out the saltwater. Many coastal cities, such as New York City and Boston, are built on filled-in wetlands. In fact, for many years, the U.S. Bureau of Land Reclamation helped to drain wetlands, which were considered wasted land. Now we know that such changes came with environmental costs, including the loss of habitats and ecosystem functions such as flood control (Chapter 10).

The country of China has created more land than any other country,[6] primarily along its coastlines. From 1949 until the 1990s, China reclaimed about 3 million acres from the China Sea. Today, China is also carrying out an aggressive program of inland land creation: The tops of mountains are being cut away and the material put into valleys to produce flat areas for city expansion, industry, and agriculture (Figure 11–19). Such mountaintop removal is done in coal mining areas in the United States (Chapter 14) as a part of mining operations, but the Chinese projects are occurring on an unprecedented scale. According to some estimates,

---

[6]P. H. Qian Li and Jianhua Wu, "Environment: Accelerate Research on Land Creation," *Nature* (June 2014), accessed November 20, 2014, http://www.nature.com/news/environment-accelerate-research-on-land-creation-1.15327.

Figure 11–19 Land creation. These machines are cutting down the tops of mountains to create a larger area for the city of Lanzhou in central China. Such large-scale efforts have serious environmental consequences.

700 mountains in central China will be leveled to produce hundreds of square miles of land.

These Chinese projects come with uncertain environmental costs. Air pollution, landslides, water pollution, and loss of ecosystems are all concerns. The city of Yan'an, in central China, expected to double its area by this process, is especially problematic because the mountains are comprised of compacted loess, a silt that will take years to settle before it can be used as a building site. Such large-scale changes to ecosystems are likely to produce unanticipated consequences. Part of sustainability is trying to avoid the negative consequences of actions we take to provide for growing populations.

CONCEPT CHECK Parks often have signs telling tourists to stay on paths. If people ignore these signs, how will the soils in desert areas be affected? How will the soils in wet, mountain areas be affected? ✓

11.3 Conserving and Restoring Soil

There is broad consensus that soils are vital to human societies and that they should be preserved. This is encouraging, and as long as this consensus translates into action, we will be moving in the direction of sustainability. Because topsoils grow so slowly (an inch every hundred years at best), sustainability means doing all we can to reduce soil erosion and prevent degradation.

Soil conservation must be practiced at multiple levels. The most important level is the individual landholder. Soil conservation is also affected by public policy. National farm policies play a crucial role in determining how soils are conserved. In addition, national policies governing the grazing of public lands, forestry practices, and mining can lead to either good or poor stewardship of soils.

Individuals and public policies must work in harmony to bring about effective, sustainable stewardship of soil resources. Let us examine briefly what is being done.

Helping Landholders in the United States

Those working on the land are best situated to put into practice the conservation strategies that will enhance their soils, bring better harvests, and improve their way of life. A great deal of traditional knowledge exists with landholders, who routinely practice soil conservation techniques.

Stopping Erosion. A number of techniques are widely employed to reduce erosion. Among the most conspicuous are contour farming and shelterbelts (Figure 11–20). In **contour plowing,** fields are plowed and cultivated at right angles to the contour of the slope, thereby slowing the downhill flow of water. **Shelterbelts** are rows of trees and shrubs planted beside fields to protect crops and livestock from wind and blowing snow. Erosion can also be prevented by growing grasses along waterways and strips of perennial plants along the sides of fields to filter runoff. (Also see Sustainability, *New Ditches Save Soil.*)

No-Till and Low-Till Agriculture. A technique that permits continuous cropping while minimizing soil erosion is **no-till farming,** depicted in Figure 11–21. In this technique, the field is first sprayed with herbicide to kill weeds. Next, a planting apparatus is pulled behind a tractor to accomplish several operations at once: A steel disk cuts a furrow through the mulch of dead weeds, drops seed and fertilizer into the furrow, and then closes it. At harvest, the process is repeated, and the waste from the previous crop becomes the detritus and mulch cover for the next. Thus, the soil is never left exposed, erosion and evaporation are reduced, and there is enough detritus, including roots from the previous crop, to maintain the topsoil. However,

(a)

(b)

Figure 11–20 Contour farming and shelterbelts. (a) Cultivation up and down a slope encourages water to run down furrows and may lead to severe erosion. The problem is reduced by plowing and cultivating along the contours at a right angle to the slope. In this photograph, strip cropping is also being utilized: The light green bands are one type of crop, the dark bands another. **(b)** Shelterbelts—belts of trees around farm fields—break the wind and protect soil from erosion, as seen here in New Zealand.

Figure 11–21 Apparatus for no-till planting. This no-till planter is planting corn over stubble with minimal disturbance to the protective vegetation layer on the surface of the soil. The apparatus creates furrows with cutting discs, applies herbicide and fertilizer, drops in seeds, and closes the furrows.

no-till farming requires the use of herbicides, which have their own problems and may increase the effects of pests that remain in crop residues from one season to another (Chapter 13).

No-till agriculture is now routinely practiced in areas of the United States. A variation on this theme is **low-till farming,** which is catching on in Asia. For example, to plant wheat after a rice crop has been harvested, only one pass over a field is made—to plant seed and spread fertilizer. In the past, farmers would plow 6 to 12 times to break up the muddy soil and dry it out for the wheat.

Natural Resources Conservation Service. The U.S. **Natural Resources Conservation Service (NRCS),** formerly the Soil Conservation Service, was established in response to the Dust Bowl tragedy of the early 1930s. Through a nationwide network of regional offices, the NRCS provides information regarding soil and water conservation practices to farmers and others. For example, federal extension service offices test and analyze soil samples and make recommendations for improvements. The NRCS performs an inventory of erosion losses in the United States. In 2010, the Natural Resources Inventory indicated that soil erosion on croplands has lessened in recent years, from 3.36 billion tons in 1982 to 1.72 billion tons in 2010.[7] This decrease is likely a consequence of improved conservation practices, such as shelterbelts and grassed waterways.

Ending Overgrazing. The restoration of rangelands benefits both wildlife and cattle production. One solution to overgrazing lies in better management. An NRCS program called the Conservation Stewardship Program provides information and support to enable ranchers who own their

lands to burn unwanted woody plants, reseed the land with perennial grass varieties that hold water, and manage cattle so that herds are moved to a new location before overgrazing occurs. Public policy, discussed later, also plays a role in ending overgrazing.

Soil Restoration. In some cases, degraded soils can be restored. Restoration usually involves the addition of organic material and the use of rooted plants to trap soils, as well as the introduction of soil biota. Compost, manure, other plant products, and earthworms may be used to enhance degraded soils. Sometimes contaminants such as toxic chemicals need to be removed from degraded soil. When organisms are used to remove contaminants, the process is called **bioremediation.** In the case of petroleum contamination, for example, soil microbes are used to break down the contaminant. In other situations, plants are used to take up pollutants.

U.S. Policy and Soils

Despite the crucial importance of saving agricultural land and forests, human activities continue to cause erosion, desertification, and salinization. Such land degradation is on a collision course with sustainability. Who is responsible for making changes? At the first level, it is the individual landowner, but landowners are subject to public policies, which is the second level of decision making for soil conservation.

Agricultural Policies. Farm policy in the United States was originally focused on increasing production, demonstrated by continuing high yields and surpluses. Today the federal government supports U.S. agriculture through a number of programs that maintain farm income and support of farm commodities. These programs provide benefits, but also have a number of environmental costs.

Farm Subsidies. Farmers are guaranteed price levels for many products, including grains, cotton, sugar, peanuts, dairy products, and soybeans. These subsidies help keep food prices from fluctuating year to year. Unfortunately, they can be bad news for the economy and the environment.[8] Subsidies encourage excessive use of pesticides and fertilizers, reduce crop rotation by locking farmers into annual crop support, and promote the continued drawdown of groundwater aquifers through irrigation. They also can promote the degradation of soil.

The Federal Agricultural Improvement and Reform Act (FAIR) of 1996 attempted to bring needed reforms. Subsidies and controls over many farm commodities were reduced or eliminated, giving farmers greater flexibility in deciding what to plant, but also forcing them to rely more heavily on the market to guide their decisions. The 2002 Farm Security and Rural Investment Act succeeded FAIR. It continued to subsidize a host of farm products, maintaining price supports and farm income for American farmers.

[7]U.S. Department of Agriculture, *Summary Report: 2010 National Resources Inventory* (Washington, DC: Natural Resources Conservation Service; and Ames, IA: Center for Survey Statistics and Methodology, Iowa State University, 2013), available at http://www.nrcs.usda.gov/Internet/FSE_DOCUMENTS/stelprdb1167354.pdf.

[8]Norman Myers and Jennifer Kent, *Perverse Subsidies: How Tax Dollars Can Undercut the Environment and the Economy* (Washington, DC: Island Press, 2001).

SUSTAINABILITY
New Ditches Save Soil

A hard rain pounds on a dusty farm field. Water races down a slope, running in rivulets through the rows, flowing over weeds at the field's edge, swelling the ditch that runs between it and the next field. In minutes the ditch is full, its banks covered with the rushing water. Thunder and lightning fill the sky overhead. The ditch's walls begin to crack, slumping into the torrent. Eddies of water carry brown soil away. The soil leaves the field and moves downstream—to a river that does not need this sediment, to clog an estuary, to cover organisms, and to remove a valued resource from the farm that needs it.

Farm ditches are a valuable way to control the flow of water. They keep fields from lying sodden after long rains, but they can also be channels of erosion, especially in a thundering summer downpour. Researchers in Iowa are trying to design a new kind of ditch—one that is effective under normal conditions but can also limit erosion during severe storms. They call this approach "two-stage ditch design." At the bottom of the ditch is a typical narrow channel. Partway up the ditch are two broad, vegetated areas (like benches) that lie on either side of the channel, enclosed by banks that go up to the level of the field. The flat, vegetated areas are designed to slow water and stop sediment during storms.

The ditches do more than save farmers' soil. The sediments carried in field runoff can overload waterways and kill vulnerable aquatic organisms. By keeping sediments out of streams and rivers, the ditches protect freshwater mussels from the effects of silt. Far away, less sediment and fewer nutrients will arrive to pollute the distant Gulf of Mexico, an indication of just how connected ecosystems are.

The two-stage ditch design was created by the Nature Conservancy and collaborating partners. General Mills, Cargill, and the Natural Resource Conservation Service provided funding. Other partners helped with testing, monitoring, and working with landowners. It has been used extensively in Ohio and other states. The two-stage drainage ditch is a good example of a low-tech solution to a common problem—one that will help limit erosion of valuable farm soil.

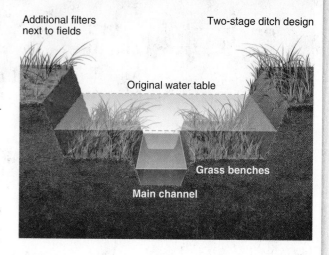

Additional filters next to fields

Two-stage ditch design

Original water table

Grass benches

Main channel

Farm Bills and Conservation. The U.S. Congress periodically passes legislation overseeing agriculture that includes incentives for soil conservation. In 1985, Congress passed the **Conservation Reserve Program,** under whose authority highly erodible cropland could be established as a "conservation reserve" of forest and grass. Farmers are paid each year for land placed in the reserve. Funding has been provided for two newer programs, the Wildlife Habitat Incentives Program and the Environmental Quality Incentives Program, both of which encourage conservation-minded landowners to set aside portions of their land or address pollution problems.

The 2002 Farm Bill initiated the Conservation Security Program (CSP), a voluntary program that encourages stewardship of private farmlands, forests, and watersheds by providing landowners incentives to conserve soil and water resources on their lands. This was renamed the Conservation Stewardship Program in the 2008 Farm Bill; more than $1 billion was provided for the new CSP. The 2014 Farm Bill updated the regulation of the Conservation Stewardship Program, reducing the funds provided for it and limiting it to 10 million new acres per year.

Policies for Grazing on Public Lands.
The U.S. Bureau of Land Management (BLM) leases grazing rights through 18,000 permits on 155 million acres of federal lands. The Forest Service leases grazing rights on an additional 95 million acres, with 6,165 permits issued. The income to the government is less than private landowners would be paid for grazing rights, though ranchers have to do more on federal lands, such as provide water and fences and get permits. Altogether these leases pay the government far less than the costs of administration: In 2013, for example, overstocking and the BLM subsidy of ranching on federal public lands cost $48.2 million, but brought in only $12.2 million in grazing fees to the federal government.

A worse problem is the ecological cost exacted by livestock to watersheds, streams, wildlife, and endangered species. In a variation of the tragedy of the commons, those who graze stock on public lands have little incentive to practice soil conservation.

In response to these issues, some ranchers and environmental groups propose that the government buy up some of the 26,000 existing grazing permits and "retire" those rangelands. The ranchers would be offered a generous payment for the permits they hold, and the land would be used for wildlife, recreation, and watershed protection. Another solution is to charge market value for grazing livestock on public lands.

Mining Regulations.
In many places, the land left after a mining operation lacks the topsoil necessary for plant

growth and contains toxic wastes. The U.S. Surface Mining Control and Reclamation Act of 1977 (SMCRA) was created to address the environmental impacts of strip mining. States are expected to regulate mines. Once mines are closed, the mined lands are supposed to be rehabilitated. Restoring such places can be very difficult, but is necessary for a healthy ecosystem.

Soil Conservation in the Developing World

In the developing world, as in the developed world, it is the individual landholders, farmers, and herders who hold the key to sustainable soil stewardship. They must be convinced that what they do will work, that it will be affordable, and that, in the long run, their own well-being will improve. Such persuasion may require help in the way of microlending, sound advice (often best coming from their peers), or simply encouragement to experiment (Chapter 9).

Empowering Farmers. One of the worldwide efforts to empower farmers to protect soil is the Sustainable Agriculture and Rural Development Initiative (SARD) facilitated by the Food and Agriculture Organization (FAO). SARD is an umbrella organization that coordinates efforts to help small farmers use sound agricultural practices and encourages sustainable development.

An example is found in the Bobo Diaulasso region of Burkina Faso, a rural community in Africa's drylands where intensive farming, driven by population growth, has led to a loss of soil fertility and declining crop yields. An FAO project is enabling a group of farmers there to adopt a range of sustainable practices, such as mixing plots of legumes and cereal grains to preserve soil fertility, cultivating soybeans to diversify income sources, and producing forage to feed livestock during the extended dry season. FAO hopes that the practices used there will serve as a model for implementation in other, similar situations. The increased production in Bobo Diaulasso is a prime example of stewardship in action.

The Keita Project. Preventing and reversing soil degradation is a key need of the more than 2 billion people who live in drylands, often among the poorest people on Earth and subject to climatic variations that can destroy their livelihood and well-being. Niger is one of the hottest countries of the world, with much of its land in the Sahara. The part of Niger that is south of the Sahara is in the Sahel, a climatic belt that runs across Africa from the Atlantic Ocean to the Red Sea. The Sahel is semiarid grassland, receiving 6–20 inches of rainfall a year, but less in drought years.

In 1982, the Italian government initiated a project in the Keita district of Niger, more than 4,000 square miles of plateaus, rocky slopes, and grasslands inhabited by 230,000 people. The project targeted the serious degradation brought on by overpopulation and the effects of soil and wind erosion. Some 41 dams were established to catch water from the summer rains, countless check dams were constructed to stop rainwater sheet erosion and provide soil and moisture

Figure 11–22 The Keita district of Niger. With the help of foreign donors, villagers in the drylands of Keita planted nearly 20 million trees to reverse desertification and improve food security.

for crops, and more than 18 million trees were planted (see Chapter 10). These anti-erosion projects took many years to complete. Local people, especially women, who were paid with food, carried out most of the work. The Keita district is now a flourishing environment for crops, livestock, and people (Figure 11–22). Desertification has been halted; it is also being actively monitored, in case it starts again. The people now have control over their plots of land, an important step toward long-term stewardship of local resources.

The Role of International Agencies

Internationally, there are few soil protection treaties, although a number of international efforts touch on soil protection. For example, in 2011 the Food and Agriculture Organization (FAO) and the European Commission initiated the Global Soil Partnership to promote joint efforts to protect soil resources and to restore the function of soils in ecosystems.[9] The United Nations declared 2015 to be the International Year of Soils and December 5 to be World Soil Day.

There have been international efforts to protect and restore soils by reversing deforestation and desertification. The United Nations established the Convention to Combat Desertification (UNCCD), which was signed and officially ratified by more than 100 nations in 1996. The goal of the UNCCD is to have a "land degradation neutral world," a term that came out of the Rio+20 meeting on development. Regular conferences and other UNCCD meetings have been concerned with such issues as the funding of projects to reverse land degradation, "bottom-up" programs that enable local communities to help themselves, and the gathering and dissemination of traditional knowledge on effective drylands agricultural practices. In 2014, India, recognizing the impact of desertification on its ability to provide food, began an effort to become land-degradation neutral by 2030.[10]

---

[9]Global Soil Partnership, accessed November 22, 2014, http://www.fao.org/globalsoilpartnership/en/.
[10]"India to Be 'Land Degradation Neutral' by 2030: Javadekar," *Zee News* (June 17, 2014), accessed November 21, 2014, http://zeenews.india.com/news/nation/india-to-be-land-degradation-neutral-by-2030-javadekar_940285.html.

One alliance, TerrAfrica, was formed in 2005 by several UN agencies and African countries to fight land degradation in sub-Saharan Africa. The alliance coordinates efforts toward arresting land degradation and promoting sustainable land management. For example, several TerrAfrica partners have been working with the government of Madagascar to reduce land degradation in its upland watersheds and southern arid lands.

Currently there are international collaborations among scientists, international agencies, and national governments to protect soil. The Economics of Land Degradation, for example, is a global initiative by public and private stakeholders aimed at making soil protection economically sustainable. One of the reasons for these collaborations is that many people have realized that our development goals are related to soil health. In the Sustainable Development Goals (SDGs), Goal 2 on food security and sustainable agriculture is directly related to soils. One of the targets for Goal 15 on terrestrial systems includes combatting desertification and ending land degradation. Ecosystem services of soil are important to achieving other SDGs, including those concerning climate change, forest conservation, and access to productive land.

At this point, most international soil protection efforts are focused on agriculture because soil health is so closely tied to food security. Attention to other issues, such as urban soils, soils degraded by mining or hazardous wastes, and the effect of climate change on soils, has lagged. In the future, soil health will probably get more international attention, because it is so closely connected to a number of the Sustainable Development Goals.

CONCEPT CHECK   How do no-till farming and contour plowing protect soil? ☑

REVISITING THE THEMES

SOUND SCIENCE

Soil science uses the principles of sound science to understand how soils are formed, how they are structured, and how they can be sustained. For example, scientific studies helped researchers design a more effective farm ditch. As with any science, however, it is possible to draw erroneous conclusions from data, so having more complete information increases our ability to make accurate predictions. The Land Degradation Assessment in Drylands (LADA) is a big step in that direction, as is the global effort to study soil degradation using satellite data. As we move forward, studies on the impacts of climate change and soils will be critical.

SUSTAINABILITY

Topsoils take hundreds of years to form and represent the most vital component of land ecosystems. Under most conditions, they are highly sustainable. Because of heavy human use for agriculture, grazing, and forestry and activities such as mining and urban development, however, lands are being degraded, causing the productivity of those ecosystems to decline. Drylands are particularly vulnerable. Some public policies, such as the fees charged for grazing livestock on public lands in the United States, promote the unsustainable use of public lands. The mountaintop flattening occurring in China is another unsustainable action that affects soils. On the other hand, many public policies promote conservation and the protection of land. For example, the UNCCD has been helping farmers adopt new, sustainable approaches to farming. Programs like TerrAfrica are spreading the word on these new approaches as well as on many sound traditional approaches that are not widely known.

STEWARDSHIP

Those who work the land or graze their livestock are stewards of the soil. So are miners and anyone else who uses land. They are temporary tenants whose lives will be far shorter than the time it took to form the soil. Their stewardship can lead to sustainable soil health or to further erosion and degradation. Soil is cared for at the local level. Soil stewardship takes place one field at a time, by the people who are most familiar with the land. In many countries, soil erosion is being reversed with practices such as no-till agriculture and contour-strip cropping. In the United States, the Conservation Stewardship Program encourages farmers to care for soils on their lands. Internationally, the Keita Project is a good example of land stewardship and restoration.

REVIEW QUESTIONS

1. Describe the events that led to the devastation of the Dust Bowl and the role played by human decision making.

2. Name and describe the properties of the three main components of soil texture.

3. What are the different layers, or horizons, in a soil profile? What makes each one distinct from the others?

4. Name and describe four major soil orders listed in this chapter. Where are they mainly found? How do hydric soils relate to soil orders?

5. What are the main things that plant roots must obtain from soil? Name and describe a process (natural or unnatural) that can keep plants from obtaining the amounts of each of these things they need for survival.

6. Describe the content and physical structure of the soils that best fulfill the needs of most crop plants.

7. Describe the roles that humus, detritus, and soil organisms play in creating a soil favorable to plants. Identify the relationships among these three factors.

8. What is meant by mineralization? What causes it? What are its consequences for soil?

9. How do water and wind affect bare soil? Define and describe the process of erosion.

10. What is meant by desertification? Describe how the process of erosion leads to a loss of water-holding capacity and, hence, to desertification.

11. What are drylands? Where are they found? How important are they for human habitation and agriculture?

12. How does mining affect soils?

13. What is meant by salinization, and what are its consequences? How does irrigation cause salinization?

14. Explain how soil biota can be killed by human actions and some of the effects.

15. What are at least two levels at which soil conservation must be practiced? Give an example of what is being done at each level.

THINKING ENVIRONMENTALLY

1. Why is soil considered to be a detritus-based ecosystem? Describe how the aboveground and belowground communities in an ecosystem are interrelated.

2. How would the pressure on urban soils differ from those in agricultural areas?

3. Critique the following argument: Erosion is always with us; mountains are formed and then erode, and rivers erode canyons. You cannot eliminate soil erosion, so it is foolish to ask people to do so.

4. The current system of management of grazing lands by the BLM is clearly unsustainable. What would you suggest in the way of public policy to make it sustainable?

MAKING A DIFFERENCE

1. To prevent soil erosion in your gardens, mulch twice a year. In the summer, mulch holds in moisture; in the winter it keeps plants warmer. It also keeps rain from washing away precious topsoil.

2. Don't use a leaf blower. It can erode topsoil more than even some of the most violent winds. It also burns fossil fuels.

3. Try mowing your fallen leaves to make shreds. Leave the shredded leaves in place and cover some with bark chips in garden beds to generate soil that is aerated, is rich with nutrients, and holds moisture better. Both of these practices combat erosion.

4. Composting is an excellent way to reduce waste and improve the tilth of soil. Two-thirds of solid waste is made up of organic material such as table scraps, trimmings from the yard, and paper products. Information on how to compost is available on the Internet. If you don't have enough space for traditional composting, you could try vermicomposting, which utilizes worms. Many municipalities have yard waste composting as well.

5. When you use fertilizers, use organic products and ones labeled "slow-release," because "quick-release" fertilizer can be easily washed out and pollutes groundwater.

Chapter 12

The Production and Distribution of Food

What does it take to get a MacArthur "Genius Grant" (a grant given by the John D. and Catherine T. MacArthur Foundation to people who show exceptional merit and promise for creative work)? For one famous athlete, it took a vision for a farm in the city. Will Allen, a former pro basketball player, had an idea for a way to provide fresh local food for the people of Milwaukee, Wisconsin, so he purchased a defunct plant nursery and began his foray into urban farming. The nonprofit he now directs, Growing Power, is a sustainable agriculture organization. It runs the last functioning farm within the city limits of Milwaukee, operates a 40-acre farm outside of the city, and has branched out as far as Chicago. The piece of the operation that most captures the imagination is the original farm on the north side of Milwaukee, where every piece of space is carefully used. On a 3-acre plot lies an intensive urban agricultural operation with greenhouses, fish, bees, chickens, and goats. The integrated system brings food to an estimated 10,000 people annually.

A Whole Life. Growing Power's Milwaukee farm meets needs other than simply providing fresh food. Each year the farm uses 400 tons of the city's food waste. It produces and sells worms, worm castings, compost, fruits and vegetables, goat milk and cheese, fish, and honey. It also provides job training for urban youth. And every year about 5,000 visitors come to learn about its activities.

This example of urban agriculture highlights the themes of this chapter: The future of agriculture will be different from agriculture of the past. It must produce enough food to support people and to be accessible, while lessening the negative effects on the environment caused by modern agriculture. There is a pressing need for change. Throughout the world, three decades of rapid population growth have left hundreds of millions of people dependent on imported food or food aid, while one in five in the developing world remains

▲ William Allen, at the urban farm Growing Power in Milwaukee, Wisconsin.

undernourished. As the world population continues its rise, no resource is more vital than food.

Food production connects to many other topics we address in this book, such as politics and economics, nitrogen and phosphorus cycles, population growth, poverty, water availability, soil, pests and their control, environmental health, climate change, water pollution, and sustainable cities. Food is central to our lives as humans and central to topics in environmental science. Wild foods, forest products, fisheries, and aquaculture are covered elsewhere. Here, we will look at three things necessary to provide nourishment for a burgeoning population and to protect the environment for the long term: the production of more food, the elimination of the harmful effects of food production on the environment, and the distribution of the right kinds of food to everyone on the planet.

12.1 Crops and Animals: Major Patterns of Food Production

By many measures, human societies have done very well at putting food on the table. More people are being fed, and with more nutritional food, than ever before. In the past 30 years, world food production has more than doubled, rising even more rapidly than the population. During this time, the daily amount of food available in the developing countries increased by 25%. A lively world trade in food-stuffs forms the bulk of economic production for many nations. How did we get to a place where we produce this much food? The answer lies in changes in agriculture. The Neolithic Revolution (Chapter 8), industrial agriculture, the development of high-yielding crops in the Green Revolution, and modern genetically modified crops have all played a role.

Some 12,000 years ago, the Neolithic Revolution saw the introduction of agriculture and animal husbandry, which probably did more than anything else to foster the development of human civilization. Virtually all of the world's major crop plants and domestic animals were established in the first thousand years of agriculture. In most of the developing world today, farmers continue to raise plants and animals by traditional methods that go back millennia. These subsistence farmers represent the great majority of rural populations.

Subsistence Agriculture in the Developing World

Subsistence farmers live on small parcels of land that provide them with enough food for their households and, it is hoped, a small cash crop. The food they produce and consume themselves is not counted in the global economy and is sometimes undervalued by others (Chapter 2). Subsistence farming is labor intensive and lacks practically all of the inputs of industrialized agriculture (Figure 12–1). In addition, it is often practiced on marginally productive land.

Typically, a family owns a small parcel of land for growing food and maintains a few goats, chickens, or cows. The system may be sustainable if crop residues are fed to livestock, the livestock manure is used as a fertilizer, and the family's nutrition is adequate. Such a family is making the best use of very limited resources. Unfortunately, subsistence agriculture is primarily practiced in regions experiencing the most rapid population growth, even though that kind of agriculture is best suited for areas with low population densities.

Worldwide, more than 500 million smallholder farms support more than 2 billion people. Such farms produce almost 80% of the food consumed in Asia and sub-Saharan Africa. Such agriculture has been called the "silent giant" that feeds most of the world's rural poor.

Successes of Subsistence Agriculture. The practice of subsistence agricultural varies with the local climate and

Figure 12–1 Subsistence farming. Subsistence farming feeds more than 2 billion people in the developing world. This farmer is plowing a rice paddy with water buffalo in southern Laos.

Figure 12–2 Slash-and-burn agriculture. A farmer watches a controlled fire in Laos, Southeast Asia. In some cases such agriculture can be a sustainable mix of forest and fields, but too much clearing is unsustainable.

with local knowledge. In some areas, subsistence agriculture involves shifting cultivation within tropical forests—often called slash-and-burn agriculture (Figure 12–2). Research has shown that this practice can be sustainable if it is done well. The cultivators create highly diverse ecosystems in which the cleared land supports a few years of crops and gradually shifts into agroforestry—a system of tree plantations with different ground crops employed as the trees grow. The reason this works, though, is because the local human population is low.

Difficulties with Subsistence Agriculture. Subsistence farmers live close to the edge, limited by the size of their farms, the availability of crops, animals, and tools, and other

factors such as rainfall and pests. They are often poor and lack the financial resources to survive periodic crop failures. Farmers reliant on small crops often lack access to markets and any type of insurance or credit. New technologies may increase crop yields for others, but not for those who are too poor to take advantage of them.

Subsistence farms are most common in Asia, Latin America, and especially in sub-Saharan Africa. There, agricultural production has lagged behind that of the rest of the developing world (Figure 12–3). Two-thirds of the people in sub-Saharan Africa are dependent on agriculture for their livelihood. Many countries in the region are dependent on imported food, and world food prices have been fluctuating.

The Development of Modern Industrialized Agriculture

At the start of the 19th century, the majority of people in the United States lived and worked on small farms. Human and animal labor turned former forests and grasslands into agricultural systems that produced enough food to supply a robust and growing nation (Figure 12–4a). Farmers used traditional approaches to combat pests and soil erosion: crops were rotated regularly, many different crops were grown, and animal wastes were returned to the soil. Farming, although difficult, was efficient enough to allow an increasing number of people to leave the farm and join the growing ranks of merchants and workers living in towns and cities.

In the mid-1800s, the Industrial Revolution came to the United States. This revolution transformed agriculture profoundly and greatly increased the efficiency of farming. Since the mid-1930s, the number of farms in the country has

Figure 12–3 The yield gap for corn. Subsistence agriculture often has lower crop yields. The difference in yield in terms of metric tons per hectare between sub-Saharan Africa and other regions continues to widen.

(*Source:* FAOSTAT.)

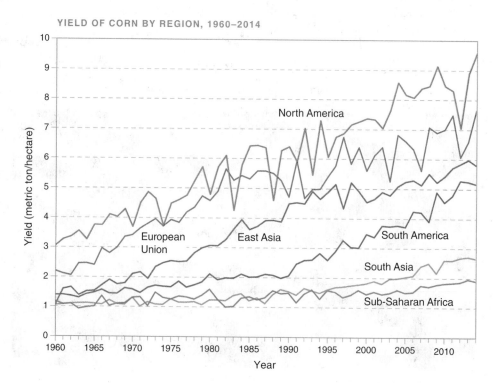

YIELD OF CORN BY REGION, 1960–2014

(a)

(b)

Figure 12–4 Traditional versus modern farming. **(a)** Traditional farming practiced in Arkansas. For hundreds of years, traditional practices on American farms supported the growing U.S. population. **(b)** Modern agricultural practice, illustrated by a combine loading corn into a tractor trailer in America's Midwest.

decreased by two-thirds (from 6.8 million to 2.1 million), while the size of farms has grown fourfold. (They now average 434 acres, or 176 hectares.) Today the remaining 2.1 million farms produce enough food for all the nation's needs, plus a substantial amount for trade on world markets (Figure 12–4b).

Virtually every industrialized nation has experienced this agricultural revolution. Crop production has been raised to new heights, doubling or tripling yields per acre (Figure 12–5). To understand how this tremendous increase in production was made possible, you need to understand the components of the agricultural revolution. These in turn reveal some of the problems brought on by the revolution.

Machinery and Infrastructure. In the modern farm, an incredible array of machinery handles virtually every imaginable aspect of working the soil: seeding, irrigating, weeding, and harvesting. This machinery has enabled farmers to cultivate far more land than ever, using less human labor.

The transformation of agriculture in the United States was greatly facilitated by the development of supporting infrastructure. These developments included new roads and rural electrification, irrigation facilities, agricultural programs in state universities, agricultural and soil extension services for farmers, the establishment of markets and efficient transportation of farm goods, banks and credit unions to provide farm loans, the formation of farm cooperatives, price support programs, and many other government subsidies of farming. Most of these developments were essential in achieving the agricultural revolution. However, many believe that the long-term maintenance of farm subsidies and reliance on fossil-fuel-driven machinery favors large corporate farms and works against the interests of small farms.

Land Under Cultivation. The United States has the largest area of arable land in the world. Cropland and pastureland comprise more than one billion acres (400 million ha). This is about 45% of the country's total land area. Essentially all of the good cropland in the United States is now under cultivation or held in short-term reserve.

Before 1960, much of the increased production in the United States came from bringing new land into production. Since then, increases in crop yields and consistent crop surpluses have lessened pressure to convert additional land to cropland, enabling farmers to be selective in the land they cultivate. The surpluses have also given farmers the opportunity to take erosion-prone land out of production. Under current farm policy, the **Conservation Reserve Program** reimburses farmers for "retiring" erosion-prone land and planting it with trees or grasses (Chapter 11). Recently some of this land has been brought back into production to meet a rising demand for corn used in ethanol production.

Globally, agriculture occupies about 38% of the land, a number that has remained relatively steady since 1994,

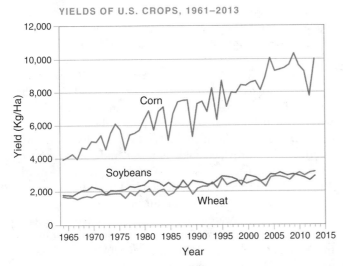

YIELDS OF U.S. CROPS, 1961–2013

Figure 12–5 U.S. crop yields. Corn, wheat, and soybean yields (measured as kilograms per hectare) have increased over time. (*Source:* FAOSTAT.)

although yields have increased. In the future, any expansion of cropland will come at the expense of forests and wetlands, which are both economically important and ecologically fragile.

Fertilizers and Pesticides. Farmers long ago learned that animal manure and other organic supplements to soil could increase their crop yields. When chemical fertilizers first became available for use, however, farmers discovered that they could achieve even greater yields. Moreover, the fertilizers were more convenient to use and more readily available than manure. When fertilizers were first employed, 15 to 20 additional tons of grain were gained from each ton of fertilizer used (Figure 12–6). Between 1950 and 1990, worldwide fertilizer use rose tenfold. Then, following a decline traced to hard times in the countries of the former Soviet Union after it broke up in the 1990s, the worldwide use of fertilizer resumed its rise and topped 176 million tons. It is now projected to reach 263 million metric tons by 2050. Most of the current increase in fertilizer use is in "underfertilized" countries such as China, India, and Brazil.

The manufacture of fertilizers uses energy from fossil fuels. As a result, fertilizer prices vary with the cost of fuel and other factors. For example, world fertilizer prices doubled in 2007, affecting the cost of food. They reached a high in 2008, and dropped afterward because of lower oil prices.

In traditional agriculture, pests such as insects and weeds can greatly reduce the yields of crops. The development of chemical pesticides provided significant control over pests, at least initially. However, due to natural selection, pests have become resistant to many pesticides. This leads to a cycle: More pesticides are used, and pests recover and increase. Currently, more than 113 weed species, 150 plant diseases, and 520 insects and mites are resistant to

pesticides. Between 10% and 16% of global crop production is now lost to pests, an amount that is expected to increase as pests become more resistant to pesticides and climate change allows pests to expand their ranges. The use of pesticides may also affect the health of humans and the environment. As with the use of farm machinery and fertilizer, the manufacture of pesticides depends on fossil fuel energy. Concerns about pesticides are discussed extensively elsewhere (Chapter 13).

Irrigation. Worldwide, irrigated acreage has doubled since 1950, while water withdrawals have tripled. Today irrigated acreage represents 18% of all cropland—some 683 million acres (277 million ha)—and produces 40% of the world's food. Irrigation is still expanding, but at a much slower pace because of limits on water resources; irrigation represents 70% of all water use. Recall that much current irrigation is unsustainable because groundwater resources are being depleted (Chapter 10). On as much as one-third of the world's irrigated land, agricultural production is being adversely affected by waterlogging and salinization—consequences of irrigating where there is poor drainage (Chapter 11). In addition, irrigation ditches are associated with the spread of as many as 30 diseases, because open water provides habitat for mosquito larvae, snails that carry parasites, and other vectors of disease.

The Green Revolution

The technologies that gave rise to the agricultural revolution in the industrialized countries were eventually introduced into the developing world. There, they gave birth to the remarkable increases in crop production called the **Green Revolution**.

High-Yielding Grains. Several decades ago, plant geneticists developed new varieties of wheat, corn, and rice that produced yields that were double to triple the yields of traditional varieties. In 1943, the Rockefeller Foundation sent agricultural expert Norman Borlaug and three other U.S. agricultural scientists to Mexico, with the objective of exporting U.S. agricultural technology to a developing nation that had serious food problems. Mexican wheat was well adapted to the subtropical climate, but it produced low yields and responded to fertilization by growing very tall stalks that were easily blown over. Using wheat from other areas of the world, Borlaug and his coworkers bred a dwarf hybrid with a large head and a thick stalk. The hybrid did well in warm weather when provided with fertilizer and sufficient water (Figure 12–7). The program was highly successful: By the 1960s, wheat production in Mexico had tripled. Mexico had closed the gap between food production and food needs and Mexican wheat appeared on the export market.

Research workers with the Consultative Group on International Agricultural Research (CGIAR) extended the work done in Mexico, introducing modern varieties of high-yielding wheat and rice to other developing countries. In one success story, Borlaug induced India to import hybrid

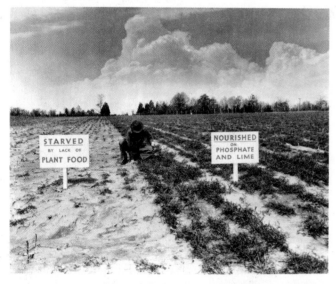

Figure 12–6 The effects of fertilizer. This 1942 photo shows the difference between fertilized and unfertilized crops and was part of a campaign to help farmers increase yields.

(*Source:* Franklin D. Roosevelt Presidential Library and Museum.)

Figure 12–7 Traditional versus high-yielding wheat. A comparison of an old variety of wheat with a new, high-yielding variety of dwarf wheat growing in Mexico.

Mexican wheat seed in the mid-1960s; in just six years, India's wheat production tripled. Many countries followed suit.

While the world population was increasing at its highest rate (2% per year), the worldwide production of rice and wheat increased 4% or more per year. In 35 years, global food production doubled. Borlaug was awarded the Nobel Peace Prize in 1970 in recognition of his contribution.

Worldwide Benefits. The Green Revolution revolutionized food production in the developing world between 1960 and 2000.[1] The early Green Revolution expanded food production in Asia and Latin America; the later years of the Green Revolution benefited Africa and the Middle East. High-yielding crop varieties are now cultivated throughout the world and have become the basis of food production in China, Latin America, the Middle East, southern Asia, and the western nations. Because the technology raises yields without requiring new agricultural lands, the Green Revolution has held back a significant amount of deforestation in the world.

Research on high-yielding crops has continued, and more varieties continue to be developed, released, and adopted by farmers. Recent research has focused on resistance to diseases, pests, and climatic stresses. Today, with the help of foreign investment, African crop scientists are working with conventional breeding techniques to improve traditional African food crops, an extension of Green Revolution efforts.

Costs of Modern Agriculture

Despite the great increases in production associated with industrialized agriculture and the Green Revolution, they are not a panacea for the world's food-population difficulties. Each of the modern agricultural technologies has externalities—costs not associated with the price of food (Chapter 2). These costs mean that ecosystems have been harmed and ecosystem capital has been lost. By 2004, the hidden costs of U.S. industrial agriculture were estimated at $5.7 billion to $16.9 billion per year.[2]

Special Conditions for High-Yield Crops. High-yielding varieties of crops can be hard to grow. In many locations, grains grow best on irrigated land, and water shortages have begun to occur as a result of this dependence (Chapter 10). In addition, modern varieties of grains require constant inputs of fertilizer, pesticides, and energy-intensive mechanized labor, all of which can be in short supply, or are used unsustainably. The increased use of fertilizers, pesticides, and irrigation have externalities that have already been described.

Monocultures. The Green Revolution and other aspects of industrialized agriculture increase **monocultures,** vast areas planted in just one crop. Monocultures have the disadvantage of making pests and diseases worse. Why? Large areas planted with a single crop allow insects and diseases to spread more rapidly than they could if the crop were interspersed with other plants that were not susceptible to the same diseases and pests.

Loss of Genetic Diversity. Over the long history of traditional agriculture, an enormous variety of domesticated plants and animals were developed; each breed of plant or animal had a distinctive genetic makeup that was well suited to a particular locale. The movement to industrialized agriculture has maximized yield at the cost of many of these "heritage breeds" or "heirloom varieties." In just the last 15 years, 190 breeds of farm animals have gone extinct. Thousands of wheat varieties have been lost.[3] In India, 90% of the traditional rice varieties are no longer grown on farms. In North America, a similar percentage of fruit and vegetable varieties have gone extinct. Sometimes even the flavor of food is lost as breeders achieve other gains, such as uniformity of shape or color.

The genetic diversity of heritage breeds is a valuable resource that can be used to develop crops and livestock that are better adapted to changing conditions. But once breeds are lost, this diversity is no longer available to meet our future needs. To prevent this loss of biodiversity, a number of seed banks around the world store the seeds of heirloom crops. On a smaller scale, one Indian farmer started a cooperative seed-sharing farm that raises and protects 930 different rice varieties on less than 3 acres.

[1]R. E. Evenson and D. Gollin, "Assessing the Impact of the Green Revolution, 1960 to 2000," *Science* 300 (May 2, 2003): 758–762.

[2]E. Tegtmeier and M. Duffy, "External Costs of Agricultural Production in the United States," *International Journal of Agricultural Sustainability* 2(1)(2012): 1–20.

[3]G. Varma, "Seed Savior," *Earth Island Journal* (2014), accessed December 8, 2014, http://www.earthisland.org/journal/index.php/eij/article/seed_savior/.

Animal Farming and Its Consequences

Raising livestock—sheep, goats, cattle, buffalo, and poultry—has many parallels to raising crops, and there are many connections between the two. One-third of the world's croplands are used to feed domestic animals; in the United States, 36% of cropland is used to feed animals. The care, feeding, and "harvesting" of some 4.7 billion four-footed livestock and almost 21.5 billion birds constitute one of the most important activities on the planet. The primary force driving this livestock economy is the large number of the world's people who enjoy eating meat and dairy products—most of the developed world and growing numbers of people in less developed nations. This growing appetite for meat, and especially for beef, has a significant impact on the environment (see Sound Science, *The Meat Footprint*).

There are multiple methods of raising livestock. Recent changes in the production of livestock include the development of concentrated livestock and poultry production and an increasing number of extremely large ranches, particularly in the tropics. In other places, such as rural societies in the developing world, livestock and poultry are raised on family farms or by pastoralists who are subsistence farmers. Overgrazing is a problem in many areas with high human populations and traditional herds.

Concentrated Livestock Production. In the developed world and increasingly in the developing world, large numbers of animals are raised in confinement in facilities called **concentrated animal feeding operations,** or CAFOs (Figure 12–8). More than half of the livestock in the United States are grown in CAFOs. Industrial-style animal farming such as CAFOs can damage the environment and human health in a host of ways.

One serious problem associated with CAFOs is the management of animal manure. In the United States, livestock produce as much as 1.2–1.4 billion tons of manure annually—between 3 and 20 times what humans produce—yet we do not have sewage treatment plants for livestock waste. Some of the waste from CAFOS leaks into surface waters, where it contributes to die-offs of fish, contamination with pathogens, and a proliferation of algae. The Environmental Protection Agency (EPA) asserts that animal-based agriculture is the most widespread source of pollution in the nation's rivers.

The crowded conditions in CAFOs allow diseases to incubate and spread among animals. Diseases may also spread from animals to humans, causing serious public-health concerns. For example, avian flu is a highly contagious disease that moves quickly through poultry that are grown at high densities. A 2014 outbreak of avian flu in the Fraser Valley of British Columbia, Canada, lead to the slaughter of 80,000 birds to prevent the disease from spreading; several countries banned the import of poultry products from that region. Further, many bacteria associated with animal farming are human pathogens, causing food-borne human diseases. The bacterium *salmonella* alone is estimated to cause losses of $2.65 billion per year in the United States.

Some people object to CAFOs because of the tight confinement of animals, which can leave them without room to turn around or, in some cases, without the ability to stretch their muscles; such facilities may require animals to lie in their own feces. Recent laws about CAFOs have focused on whether such treatment is humane.

Large Ranches. Large ranches occur around the world. The largest ranch in the world, Anna Creek Station in South Australia, covers 6 million acres, about the size of the nation of Israel. Many ranches are in the tropics. Ranches account for a great deal of the loss of rain forests. In the last 50 years, 17% of the Amazon River Basin forest has been lost, mostly due to conversion to cattle pasture, often after the land was briefly farmed for crops. Even though most of this land is best suited for growing rain forest trees, some of it supports rural populations of subsistence farmers who produce a diversity of crops. Much of the land, however, is held by relatively few ranchers who own huge spreads. The expansion of cattle production in the Amazon Basin is export driven; for example, Brazilian beef exports bring in more than $5.3 billion annually.

Figure 12–8 CAFOs. Concentrated animal feeding operations (CAFOs), like this cattle lot in Arizona, are common. These cattle are held in very tight quarters while they are fed grains to fatten them. This results in high concentrations of animal waste, sometimes leading to water pollution.

SOUND SCIENCE
The Meat Footprint

The juicy burger at your favorite diner, the steak at a top-end restaurant, the chicken fingers at a fast-food joint, the pork ribs at a barbeque—all contribute to an American diet with a high percentage of meat. The average American today eats about 57 pounds more meat per year than the average American in 1950. That increase in meat consumption occurred even as the number of vegetarians and vegans rose. Around the world, the consumption of meat has also risen, particularly in places like China where an emerging middle class has funds to enjoy richer foods. More land is used for livestock production than for any other type of food production. However, feeding people meat rather than plant products takes a great deal of energy and has environmental costs, externalities that we may not recognize.

Recently Gidon Eshel, an environmental scientist at Bard College, New York, and his colleagues estimated the cost of our increasing appetite for meat. Their paper, "Land, Irrigation Water, Greenhouse Gas, and Reactive Nitrogen Burdens of Meat, Eggs and Dairy Production in the United States," describes the effects of producing meat and dairy products on four environmental factors: the amount of land required, the amount of irrigation water used, the amount of greenhouse gases emitted, and the amount of nitrogen fertilizer used to grow the feed for livestock. The largest environmental effect of livestock production is the production of feed, including pastureland.

The authors concluded that the production of beef has a far greater environmental footprint than the production of other forms of meat. As their data show, beef production requires 28 times more land, 11 times more greenhouse gases, 5 times more irrigation water, and 6 times more nitrogen than the average of the other livestock categories. The authors acknowledge that in some places, cattle eat plants that humans cannot eat and that some livestock feed is made of plant parts such as leaves that are inedible to humans. Even taking that into account, however, eating beef has a disproportionately high environmental cost. Including the methane that cows release in their digestion of food would only have increased the disproportionately high effects of beef production.

Because of the high environmental costs of meat and dairy production, eating less meat, especially beef, is likely to be a part of successfully feeding the world of the future. In fact, a report by the UN Environmental Programme suggested that reducing meat consumption to 1940s levels (about a 60% drop for Americans) would be important to staying within sustainable cropland limits. As we see in this example, science plays an important role in helping us calculate the cost of human activities accurately, allowing us to make informed decisions about where to put our efforts.

Source: Eshel, G., A. Shepon, T. Makov, and R. Milo. "Land, Irrigation Water, Greenhouse Gas, and Reactive Nitrogen Burdens of Meat, Eggs, and Dairy Production in the United States." *Proceedings of the National Academy of Sciences* 111(33) (2014): 11996–12001.

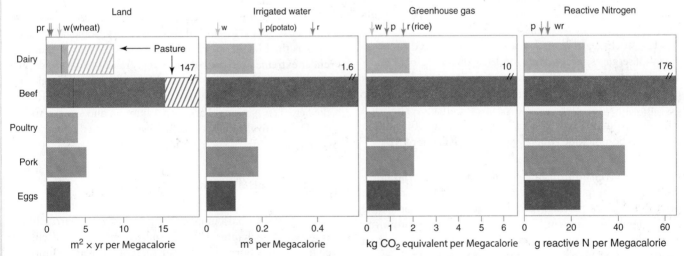

The Environmental Costs of Producing Meat. The graphs compare the environmental impact of the production of five types of animal-based foods. The resources needed to produce each food are listed per Megacalorie (1,000 kilocalories), about half of the daily energy requirement for an adult. (The nutritional unit 1 Calorie is equal to 1 kilocalorie). Arrows at the top show the environmental costs of growing wheat, rice, and potatoes. Pastureland is marked with hatch marks. For some factors, the value for beef is off the scale and the correct number is marked at the end of the bar.

UNDERSTANDING THE DATA

1. What is the ratio between the amount of nitrogen used to grow beef and that used to produce eggs?
2. Which is greater, the pastureland or the other land used for beef production?
3. Is there a difference in the amount of irrigation water needed for poultry and pork?
4. Which emits more greenhouse gases, producing a Megacalorie (1,000 kilocalories) of rice or of eggs?

Use of Antibiotics. Antibiotics are used heavily in animal husbandry. An estimated 80% of antibiotics used in the United States are used for livestock. These antibiotics are used to treat animal illnesses but are also added to animal feed, because sub-therapeutic doses encourage rapid tissue growth. The use of antibiotics is particularly high in CAFOs, because the animals are kept in such close quarters.

An unfortunate outcome of this widespread use is that many bacteria have developed resistance to antibiotics; such bacteria are far more difficult to treat when humans are infected.[4] Of most concern is the use of cephalosporins, a class of antibiotics used in the treatment of pneumonia, strep throat, and other human illnesses, especially in children. The USDA has placed restrictions on their use in the United States. In Canada, the practice of injecting chicken eggs with a type of cephalosporin was discontinued after a drug-resistant strain of *salmonella* infected birds and people alike.[5]

Climate Change. Because agro-ecosystems cover so much land, they are critical to changes in the climate. Indeed, agriculture releases more greenhouse gases than transportation does. Deforestation and burning in the tropics are estimated to release approximately 25% of the annual emissions of carbon dioxide released by humans, contributing a significant amount of carbon dioxide to the greenhouse effect. Yet carbon dioxide emissions represent only 27% of the greenhouse gases emitted by agriculture. Methane makes up about half of the greenhouse gases emitted by agriculture. In addition, the use of fertilizer releases the greenhouse gas nitrous oxide, which accounts for 29% of agricultural emissions.

Livestock rearing is also a significant source of other greenhouse gases, accounting for 14% of total global emissions.[6] Because their digestive process is anaerobic, cows emit methane through belching and flatulence. The anaerobic decomposition of manure also releases methane. The total methane produced by livestock is estimated at 100 million tons of carbon dioxide equivalent per year (Chapter 18).

Pastoralists. *Pastoralists* are subsistence farmers who move their cattle, sheep, or other grazing animals from place to place in search of pasture. Although their animals contribute to the methane problem, it is callous to fault pastoralists whose domestic animals enhance their diet and improve their quality of life. Many such animals live on plants humans cannot eat, and their manure is used for fertilizer and fuel. Livestock that are well managed can enhance the soil and enable rural farmers to maintain a balanced farm and grazing-land ecosystem, unless the herds overgraze local resources.

Figure 12–9 Heifer Project International. A Kenyan woman with a cow her family received from the Heifer Project. Heifer Project International provides animals for families in need. Participants agree to pass on the offspring of their livestock to others.

Farmers in developing countries now own two-thirds of the world's domestic animals, a trend that has made an important contribution to nutrition, especially for women and small children. Livestock can often be the savings for a family. In fact, one kind of sustainable development aid brought to rural families is the gift of a cow or a few goats or rabbits, as carried out by Heifer Project International (Figure 12–9).

In addition to livestock, pastoralists need secure access to land. Loss of lands results in food insecurity. Some pastoralists, such as the Maasai, an East African nomadic tribal group, are under extreme pressure because their grazing lands are desired by others. In 2014, the Tanzanian government reneged on a deal and threatened to turn Maasai ancestral lands bordering the Serengeti National Park into a hunting and safari park for the Dubai royal family's use. Indigenous rights activists were outraged, and the dispute is ongoing.

Biofuels and Food Production

Because of the accumulation of greenhouse gases in the atmosphere, global climate change is affecting virtually every country (Chapter 18). The most basic cause of climate change is the burning of fossil fuels. One way to mitigate climate change is to burn **biofuels,** or fuels derived from agricultural crops, instead of fossil fuels. The burning of biofuels releases no new carbon dioxide to the atmosphere. Biofuels are also a form of renewable energy.

The primary biofuel in production today is ethanol, which is a suitable substitute for gasoline. Ethanol is usually derived from the fermentation of cane sugar (in Brazil) or cornstarch (in the United States). A rising fraction (40% in 2013) of U.S. corn is devoted to meeting the growing demand from ethanol distilleries.

Corn is a staple food around the world. Many people were concerned that the diversion of corn to biofuel production lowered its availability for human consumption and

[4]2012 *Summary Report on Antimicrobials Sold or Distributed for Use in Food Producing Animals,* Food and Drug Administration, Department of Health and Human Services, 2014.

[5]M. Munro, "Canada's Chicken Farmers Ban Injections That Trigger Superbugs," Canada.com (April 17, 2014), accessed December 5, 2014, http://o.canada.com/news/national/canadas-chicken-farmers-ban-injections-that-trigger-superbugs.

[6]P. J. Gerber et al., *Tackling Climate Change Through Livestock—A Global Assessment of Emissions and Mitigation Opportunities* (Rome: Food and Agriculture Organization of the United Nations (FAO), 2013).

caused the rise in global food prices in the 2000s. Research by the International Food Policy Research Institute (IFPRI) suggested that perhaps 30% of the grain price increases between 2000 and 2007 was linked to biofuel production.

However, the relationship between using corn for biofuels and world food prices is not simple. Corn is already diverted to many other uses. The United States produces more than 32% of the world's corn, 45% of which is used for livestock, poultry, and fish feeds, while only 15% is used in human food and beverages. When corn is used for biofuel it is still useful for human and animal food; only the cornstarch is used for ethanol production, and the proteins, vitamins, and fiber are converted to high-energy animal feed, sweeteners, and corn oil. Many observers suggest that some limit be put on corn ethanol production, while making a greater investment in a new generation of biofuels based on grasses and timber waste (see Chapter 16) rather than on human or domestic-animal foods.

Part of the solution to feeding the modern world's growing population has been to use irrigation, fertilizers, pesticides, plant breeding, and industrial animal operations. Emerging technologies promise to change the face of agriculture even more.

CONCEPT CHECK   What were four components of the Green Revolution? ✓

12.2 From Green Revolution to Gene Revolution

Genetic engineering makes it possible to combine characteristics from genetically different organisms and to incorporate desired traits into crop lines and animals, producing *transgenic*, or **genetically modified** (GM), varieties. Genetic engineering is a radically different technology from the genetic research that produced the Green Revolution: Researchers are no longer limited to genes that already exist within a species or that could arise via mutation; it is now possible to exchange genes among bacteria, animals, and plants.

The most widely adopted GM products include those resistant to pests and those tolerant of herbicides. The first type is represented by crops with built-in resistance to insects that comes from genes taken from the bacterium *Bacillus thuringiensis*, abbreviated as Bt (Figure 12–10). The second type includes crops that are resistant to herbicides, including the chemicals glyphosate (usually the brand Roundup®) and glufosinate (several brands). Herbicide-resistant plants such as corn and soybeans allow farmers to spray herbicide without harming their crops and to employ no-till techniques.

In the years since bioengineered seeds became commercially available, farmers in the United States and many other countries have overwhelmingly turned to the transgenic breeds of soybeans, cotton, and corn. In the United States, about half of all land in production is planted with bioengineered crops. Worldwide, 433 million acres (175 million ha)

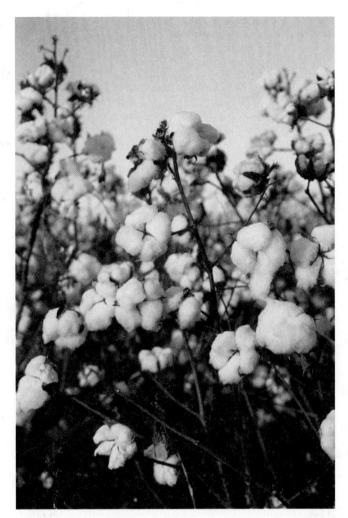

Figure 12–10 Bt cotton. Bioengineered cotton has a gene from a bacterium that helps it resist insect pests.

were planted with bioengineered crops in 2013. Figure 12–11, on the following page, shows the dramatic increase in the amount of farm land planted with GM crops. Most of this area is planted in just four crops—corn, soybeans, canola, and cotton. The use of bioengineered sugar beets is also increasing.

More recently, biotechnology has developed other genetically altered products, including strains of corn, potatoes, and cotton that are resistant to insects; rice resistant to bacterial blight disease; and sorghum (an important African crop) resistant to a parasitic plant called witchweed, which infests many crops in Africa. Genetic engineering has also produced trees and salmon that grow very quickly. One rapidly increasing trend is the development of "stacked" products, crops that contain two or more biotech genes, such as traits directed toward different insects or insect resistance plus herbicide tolerance.

Genetically modified organisms are under intense research and development, and the potential for transgenic crops and animals seems almost unlimited. Still, in spite of the obvious potential of these organisms, there are concerns about their development and use. Let's look at both sides of this controversy.

Figure 12–11 Global area of genetically modified crops. The plot shows the rapid increase in global areas planted with genetically modified crops (primarily cotton, soybeans, canola, and corn) for industrial (high-income) and developing countries as well as total global use.

(*Source:* Clive James. "Global Status of Commercialized Biotech/GM Crops." ISAAA, 2013.)

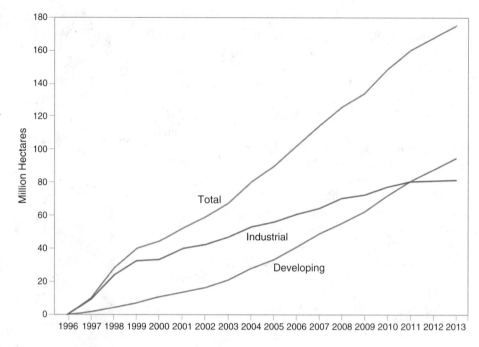

The Promise

The use of genetically modified crops and animals seems to be working, at least in the short term. A 2014 meta-analysis of 147 studies showed that overall, the use of GM soybean, corn, and cotton increased yields and farmer profits and decreased pesticide use.[7] Among the important possible environmental benefits of bioengineered crops are reductions in the use of pesticides because the crops are already resistant to pests, reductions in erosion because herbicide-resistant crops facilitate the use of no-till farming, and less environmental damage associated with bringing more land into production because existing agricultural lands can produce more food.

Crops for Developing Nations. The development of GM crops can help the developing world produce more food. Biotech crop research that can benefit developing countries is proceeding at a rapid pace. China, in particular, has emerged as a leader in plant biotechnology, having sequenced the rice genome. Among other aims, the objectives of this technology for developing nations are (1) to incorporate resistance to diseases and pests that attack important tropical plants; (2) to increase tolerance to environmental conditions that stress most plants, such as drought and high levels of salt; (3) to improve the nutritional value of commonly eaten crops; and (4) to produce pharmaceutical products in ordinary crop plants. Products under development include virus-resistant cassavas, sweet potatoes, melons, and papayas; insect-resistant eggplants (**Figure 12–12**); protein-enhanced corn and soybeans; fungus-resistant bananas; bananas and tomatoes that contain an antidiarrheal vaccine; and drought-tolerant sorghum and corn.

Pharma Crops. The biotech industry has recognized the advantages of producing pharmaceuticals by engineering genes for desirable products into common crop plants. Such bioengineered plants are called *pharma crops*. There have been field trials for crops genetically modified to produce hormones, enzymes, diagnostic drugs, and other substances, but none had yet come to market. One risk of such pharma crops is that they could contaminate food crops—some of the compounds being developed could be harmful if ingested by people or animals. In 2002, corn altered to produce a pig vaccine spread through fields in Iowa and Nebraska; intervention by the USDA kept the altered corn out of food. Because it seems unlikely that a foolproof system can be established, experts recommend producing pharmaceuticals with noncrop plants.

Figure 12–12 Genetically modified eggplant. The brinjal (Indian eggplant) on the left have been attacked by insects; those on the right have been genetically modified to resist insects attack. Such GM foods are controversial in India.

---

[7]W. Klumper and M. Qaim, "A Meta-Analysis of the Impacts of Genetically Modified Crops," *PLOS One*, November 3, 2014, doi:10.1371/journal.pone.0111629.

The Problems

Concerns about genetic engineering technology involve three main considerations: food safety, environmental problems, and justice issues.

Safety Issues. GM crops contain proteins from different organisms. Such proteins might pose an unexpected risk to the people who consume them. For example, tests showed that a Brazil-nut gene incorporated into soybeans was able to manufacture a protein in the soybeans that induced an allergic response in individuals who were allergic to Brazil nuts. Also, antibiotic-resistance genes are often incorporated into transgenic organisms. Health risks could arise if a separate process of gene transfer allowed antibiotic resistance to enter human systems or if a transgenic food product prevented an antibiotic from being used effectively. Another concern is that new combinations of genes could have unexpected health effects on consumers.

However, the World Health Organization, summarizing the finding of many researchers, has stated, "GM foods currently available on the international market have passed risk assessments and are not likely to present risks for human health."[8]

Environmental Concerns. Environmental concerns about genetically modified organisms (GMOs) have been on the radar of the scientific community for years. These concerns include the risks of developing resistant pests and superweeds and harming nontarget organisms, the loss of biodiversity, and the increase in the use of chemicals.[9]

Resistant Pests. One major environmental concern is the pest-resistant properties of GM crops. If pests have a broad exposure to a toxin incorporated into a plant, it is possible that they will develop resistance to the toxin and thus render it ineffective as an independent pesticide. The GM crop then loses its advantage.

Insect pests could also become resistant to GM plants that are supposed to be insect resistant. For example, Bt cotton is designed to resist the cotton boll weevil by producing a toxin once made by a bacterium. However, boll weevils that can survive on the engineered cotton now abound. When Chinese cotton farmers began turning to Bt cotton in the early 2000s, they found that they used much less pesticide, saving money, time, and their own health in the process. But these benefits did not last. In 2010, Chinese farmers suffered outbreaks of secondary pests (minor pests that became major in the absence of a competing pest), especially mirid bugs, which increased by twelvefold over their occurrence before the use of the Bt cotton (Figure 12–13).

Superweeds. Genes for herbicide resistance or for tolerance to drought and other environmental conditions can

Figure 12–13 Mirid bugs. Mirid bugs, such as this one, became a major pest on cotton in China only when Bt cotton resisted other competing insects and their natural predators were killed.

spread to ordinary crop plants or their wild relatives, possibly creating new "superweeds." Field studies indicate that this concern is valid. One study of genetically modified canola plants (used in animal feeds and vegetable oils) found that the vast majority of canola plants growing wild along roads in North Dakota had genes from GM canola crops. Glyphosate resistance has been found in a number of weeds that infest soybean, corn, and cotton fields. Because so much acreage is now planted with glyphosate-resistant crops, resistant weeds could spread widely and cause economic disaster for farmers.

Nontarget effects. Another concern is the ecological impact of GM crops on nontarget organisms. For example, the wind can carry pollen from Bt corn (resistant to the corn borer) to adjacent natural areas where the toxin may kill beneficial insects. This effect was shown to occur in laboratory tests with monarch butterflies. Later studies, however, indicated that Bt corn poses a low risk to the butterflies. The EPA cleared the corn for continued planting, although some scientists are still concerned.

Loss of Biodiversity. GM crops have increased the trend of earlier industrial agriculture: intensive monocultures and less genetic diversity, contributing to the loss of local plant varieties. Loss of animal diversity is also a concern.

The transgenic AquAdvantage® salmon, the first GM animal approved by the FDA for human consumption, presents another example of concerns about genetic diversity.[10] These farmed fish grow more quickly than native salmon and can aggressively outcompete them. Although the GM salmon are reared in tanks, scientific research has shown that if they were to escape into the wild, they could hybridize with brown trout and pass along their genes for rapid growth. While such an outcome has not yet occurred, the research suggests we should be cautious.

[8]World Health Organization, "20 Questions on Genetically Modified Foods/ Question 8," June 25, 2012, accessed December 5, 2014, http://www.who.int/ foodsafety/publications/biotech/20questions/en/.
[9]A. Snow et al., "Genetically Engineered Organisms and the Environment: Current Status and Recommendations," ESA Report, *Ecological Applications* 15(2): 377-404.

[10]K. Oke et al., "Hybridization Between Genetically Modified Atlantic Salmon and Wild Brown Trout Reveals Novel Ecological Interactions," *Proceedings of the Royal Society B* (2013): doi:10.1098/rspb.2013.1047.

Increase in Chemicals. Some pest-resistant GM crops clearly decrease the use of insecticides. But the jury is still out on whether GM crops that are glyphosate resistant (Round-Up Ready varieties) reduce the use of herbicides as they were originally expected to. Initially the spraying of glyphosate to kill weeds decreased the use of herbicides. But between 2001 and 2011, the use of herbicides increased by 26%.[11]

Access in the Developing World. In 2004, the UN Food and Agriculture Organization (FAO) gave its qualified approval to agricultural biotechnology in *The State of Food and Agriculture,* but concluded that the promise of the technology to alleviate global hunger and improve the well-being of farmers in developing countries was more theory than reality.[12] Several structural problems make this new technology less helpful than it could be. Issues regarding environmental justice are a significant part of the concern about genetically modified organisms.

For the first few years, nearly all GM crops were developed by large agricultural-industrial firms with a profit motive. Large seed companies purchased smaller seed companies, until just a few companies dominated the market. The GM crops were patented; farmers were forbidden by contract from propagating the seeds and had to purchase seeds annually. International trade agreements benefitted large companies at the cost of small farmers. This was a cause of dispute in Guatemala when indigenous groups fought for the right to plant and replant their own maize in spite of international trade regulations.

Farmers in developing countries are less able to afford the higher costs of the new seeds, which must be paid up front each year. Fortunately, some noncommercial and donor-funded laboratories are taking aim at this problem. Further, biotech research is booming in China, India, and the Philippines, where most of the fruits of the research are being made available to developing-world farmers. In addition, GM seeds are spreading rapidly through seed "piracy."

The experience of GM cotton growers in India illustrates a number of the concerns associated with GM crops. In 2002, India began planting Bt cotton and doubled its harvest. But 10 years later, farmers were frustrated. In six years, the crop had not increased, though the amount of GM crops planted had increased fourfold. In 2012, the harvest in most areas was half of what it was in 2011. The "white revolution" had not panned out the way farmers had hoped, at least in the regions without adequate irrigation or where land was marginal. Bt cotton required more nutrients, fertilizer, and water and was vulnerable to bacterial disease. In addition, many farmers could not afford the GM varieties, but seed for local varieties of cotton had disappeared.

Figure 12–14 GM crop protests. Protesters in France rip up genetically modified corn in protest.

GMO Controversy. All of these concerns, combined with the general fear of what unknown technologies can bring, have stimulated a rather heated controversy over the spreading use of genetically modified organisms (GMOs). Activists around the world have rallied to protest the development of such organisms and the "Frankenfoods" derived from them (Figure 12–14). The protests have been strongest in Europe, and some countries, such as Ireland, have banned GM crops.

The issue has been far less controversial in the United States. Because corn and soybeans are used in so many processed foods, an estimated 70% of all processed foods in the United States are thought to contain some genetically modified substances. Even so, there is controversy. Some U.S. consumer groups would like to see mandatory labeling of all foods derived from GMOs. Laws have been proposed, but such labeling is not yet required.

While the jury is still out on GM crops, countries everywhere are cautiously examining the technology, and many are moving forward. In 2013, 27 countries, including several in the European Union, were growing biotech crops. Most observers agree that biotechnological advances play a role in future food production. Consequently, the world community has been developing national and international policies about their use.

The Policies

In the United States, the EPA, the USDA, and the Food and Drug Administration (FDA) have regulatory oversight of different elements of the application of biotechnology to food crops. Concern over their oversight led to a number of reports from the National Academy of Sciences' National Research Council. One report, *Impact of Genetically Engineered Crops on Farm Sustainability in the United States* (2010), concluded that herbicide resistance in weeds was a real problem but

[11]Food and Water Watch, "Superweeds: How Biotech Crops Bolster the Pesticide Industry," (July 2013), accessed December 9, 2014, http://documents.foodandwaterwatch.org/doc/Superweeds.pdf#_ga=1.104040192.1025903332.1418136676.

[12]United Nations, Food and Agricultural Organization, *The State of Food and Agriculture 2003–2004: Agricultural Biotechnology: Meeting the Needs of the Poor?* (Rome: UN FAO, 2004).

that, overall, GM crops have been a benefit.[13] In sum, several reports endorsed the science, technology, and regulation of GM crops. At the same time, the reports called for more research on environmental and safety issues.

The Cartagena Protocol. In January 2000, the UN Convention on Biodiversity sponsored a conference to deal with international trade in GMOs. At issue was a fundamental difference in philosophy about how technologies, particularly new ones, should be regulated. Some felt that no one should be able to limit trade in a new product unless they had sufficient scientific evidence to know that it was harmful. Others believed that, because it takes a long time to do research, governments should be able to regulate trade when there was good reason to be concerned about a product before all of the research was completed.

After heated negotiations, the conference reached an agreement, called the **Cartagena Protocol on Biosafety.** Surprisingly, the agreement was welcomed by governments, the private sector, and environmental groups alike.

The Precautionary Principle. The Cartagena Protocol states that the "lack of scientific certainty due to insufficient relevant scientific information and knowledge . . . shall not prevent [a country] from making a decision" on the import of genetically modified organisms. The protocol gives countries the right to deny the entry of GMOs. However, countries cannot ban imports on a whim or simply to protect their own producers. The decision must be based on sound science (involving an assessment of the risks involved) and the broad sharing of information about the products. Thus, the agreement makes operational—for the first time in international environmental affairs—the **precautionary principle,** laid out in the 1992 Rio Declaration, which states that "Where there are threats of serious or irreversible damage, lack of full scientific certainty shall not be used as a reason for postponing cost-effective measures to prevent environmental degradation."[14] There were other stipulations in the protocol, including one about labeling shipments of GM products. The protocol was ratified by 147 countries (but not the United States) and became legally binding in 2003.

Table 12–1 lists a number of actions we take to increase food production as well as actions and conditions that eventually lower our ability to produce food. Solutions to the problems modern agriculture causes are going to require us to modify the way we produce food. We will consider this topic later in the chapter. First, let's look at food distribution. We already produce a lot of food, so we could ask, "Why doesn't everyone have healthy food?"

CONCEPT CHECK   How are genetically modified crops different from crops produced by earlier plant breeding techniques? ☑

---

[13] National Research Council, Committee on Genetically Modified Pest-Protected Plants, *Impact of Genetically Engineered Crops on Farm Sustainability in the United States* (Washington, DC: National Academy Press, 2000), 8.

[14] "Rio Declaration on Environment and Development 1992," Article 15, upheld in Cartagena Protocol, Articles 1, 10.6, and 11.8.

Table 12–1   Drivers of Food Availability. Some factors of modern agriculture increase food availability, and some result in effects that decrease food availability.

Increase food availability	Limit or decrease food availability
Green Revolution high-yield crops	need for water need for fertilizer
pesticide use	pest resistance secondary pest outbreaks
irrigation	salinization erosion overuse of ground water
fertilizers	disruption of nitrogen cycle nitrous oxide
conversion to farmland	loss of cropland to development use of poor-quality land erosion desertification
no-till farming with herbicides	herbicide resistance in weeds
GM crops with tolerance, high yield	resistance of pests to crops loss of genetic diversity
monocultures	vulnerability to pathogens and pests
use of fossil fuels, equipment	cost of fuel climate-change effects—plant stress from drought, insects damage from air pollution
shipping	energy and economic costs of shipping
refrigeration	energy and economic costs of refrigeration food loss in storage and transport
large farms	loss of small farmers
subsidies in wealthy countries	subsidies in some countries reduce food availability in other countries
emphasis on cash crops	loss of mixed crops loss of consumptive use poverty for some—lack of money to purchase food
concentrated animal feeding operations	pollution use of grains to feed animals use of foods as biofuels
fishing efficiency	overfishing

12.3 Food Distribution and Trade

In the past, the general rule for the availability of basic foods—grains, vegetables, meat, and dairy products—was *regional self-sufficiency*. Whenever climate, blight, or war interrupted the agricultural production of a region, the inevitable result was famine and death, sometimes on the scale of millions. Of course, refugees might flee a famine, and some cultures were naturally nomadic, so there was some moving around to escape famine. For the most part, however, natural disasters affected people locally but not people far away. This is one of the reasons the human population was relatively stable for thousands of years before the modern era.

Trade on ancient routes moved goods from place to place in a limited way. This was expanded when colonies were established in the New World. Timber, furs, tobacco, fish, sugar, coffee, cotton, and other raw materials began to flow back to the Old World to be exchanged for manufactured goods. With the Industrial Revolution, and especially with refrigeration and faster shipping, trade between nations intensified, and soon it became economically feasible to ship basic foodstuffs around the world. As the world trade in foodstuffs increased, the need for self-sufficiency in food diminished. Like other sectors of the economy, food has become globalized. In 2013, world trade in agricultural products was worth nearly $1.75 trillion.

Patterns in Food Trade

Given this global pattern, agricultural production systems supply more than a country's internal food needs. For many countries (especially those of the developing world), commodities such as coffee, fruit, sugar, spices, palm oil, cocoa, and nuts provide the only significant export products. This trade clearly helps the exporter, and it allows importing nations to enjoy foods that they are unable to raise themselves or that are out of season. In a market economy, the exchange works well only as long as the importing nation can pay for the food. Cash is earned by exporting raw materials, fuel, manufactured goods, or special commodities.

Grain on the Move. The most important foodstuffs on the world market are grains (corn, wheat, rice, barley, rye, and sorghum) and oilseeds, such as canola and soybeans. Some grain is imported by high- and middle-income countries to satisfy the rising demand for animal protein. Table 12–2 shows how the pattern of global trade in grain has changed over the past seven decades. In 1935, only Western Europe was importing grain; Asia, Africa, and Latin America were self-sufficient. By 1950, new patterns were emerging, and today the trade in grains—as well as other basic foodstuffs—has changed again. North America has become the major source of exportable grains—the world's "breadbasket" or, in another sense, the world's "meat market." The trade in oilseeds is similarly lopsided. In this case, the United States, Argentina, and Brazil have a corner on the export trade, and Europe and China are major importers. Most oilseed is used as a protein supplement in animal feeds.

Market Manipulation. Beginning in 2011, the government of Thailand, in order to support rural farmers, bought rice at above market value.[15] The Thai government withheld the rice from world trade in an attempt to drive up world prices. However, international traders quickly moved to purchase rice from India, Vietnam, and elsewhere as these countries increased production in response to the rice deficit. Thailand was left with storehouses of rice that it could not sell without steep losses. The Thai government had lost an estimated $4.4 billion. The program ran out of money and Thai farmers

---

[15]D. Kedmey, "How Thailand's Botched Rice Scheme Blew a Big Hole in Its Economy," *Time* (July 12, 2014), World.Time.com, accessed March 6, 2015, http://world.time.com/2013/07/12/how-thailands-botched-rice-scheme-blew-a-big-hole-in-its-economy/.

Table 12–2   World Grain Trade Since 1935

| | Amount Exported or Imported (Million Metric Tons)* | | | | | | | | | | |
Region	1935	1950	1960	1970	1980	1990	1995	2000	2004	2007	2013
North America	5	23	39	56	131	123	108	98	101	120	87.1
Latin America	9	1	0	4	–10	–12	–4	–16	–8	–7	6.1
Western Europe	–19	–22	–25	–30	–62	–10	21	13	10	0.2	43.2
Africa	1	0	–2	–5	–15	–31	–31	–44	–42	–51	–69.3
Asia	2	–6	–17	–37	–63	–82	–78	–74	–79	–72	–94.5
Australia and New Zealand	3	3	6	12	19	15	13	21	25	9	25.8

Source: FAOSTAT, FAO Food Outlook, June 2008.

*A minus sign in front of a figure indicates a net import; no sign indicates a net export.

lost their source of stable income. In 2014, the stockpile of rice in Thailand was estimated at 18 million metric tons. An audit showed that 70% of the rice was beginning to degrade and one fifth was inedible. The government moved to sell the stockpile even at a loss. Although this attempt at market manipulation was a failure, there are times when stockpiling grain is appropriate.

Keeping a Reserve. At no time in recent history has the world supply of grain run out. Worldwide, governments try to maintain 70 days worth of grain stored in reserves. Such reserves are often much higher, above 120 days during some years of the 1980s. In 2007, world grain reserves fell to 64 days, but rose shortly thereafter. In 2014, they reached 91 days.

Despite these reserves, many people still lack food. This is primarily because they cannot afford to purchase it. Even if food production increases as rapidly as the population grows, some people will still be poor. This problem is the chief cause of hunger in the world today.

The Global Food Crisis

Between 2006 and 2008, the prices of food commodities on the world market rose precipitously—on average, prices doubled (Table 12–3). As a result, the price of foods increased greatly. These increases triggered food riots and emergency measures in many countries. This global food crisis demonstrated the interconnection of the global trade in food.

Causes of the Food Crisis. One consequence of high commodity prices is that farmers get much higher prices for their crops; this is especially true for farmers in the higher-income countries that dominate the food export business. In contrast, farmers in developing countries are not connected to the world markets and therefore benefit little from high commodity prices. The developing countries whose food deficits force them to import commodities are faced with prices they can't afford. For most people in these countries, food expenditures are more than half of their income, and any increase in food price can result in an increase in hunger and malnutrition.

Responses to the Food Crisis. The United Nations addressed the issue at a meeting before the World Food Summit in Rome in June 2009. This summit was the third World Food Summit; summits in 1996 and 2002 had led to verbal commitments to eradicate hunger and malnutrition. Although climate change and bioenergy were addressed, the conference shifted much of its focus to the existing world food crisis.

One basic function of world conferences is to draw the attention of national leaders to international issues. As details of food shortages, high food prices, and shortfalls in aid were presented, the conference did indeed get the attention of national leaders. One response was that the World Bank began a new effort to provide immediate grants for food and agriculture in the poorest countries.

RIO+20 More recently, the 2012 UN Conference on Sustainable Development in Rio de Janeiro (called Rio+20) brought world leaders together to reassess progress 20 years after the 1992 Rio Earth Summit. The document produced at Rio+20, *The Future We Want*, highlighted seven themes: jobs, energy, cities, food, water, oceans, and disasters. Food and agriculture were central topics of the meeting and the outcomes document. At Rio+20, countries were given a challenge—to lower hunger to zero, a plan discussed shortly.

CONCEPT CHECK Why did the government of Thailand purchase and store large quantities of rice? Why did their scheme backfire? ☑

Table 12–3 Rise in World Market Prices of Selected Commodities

Commodity	Cost Basis	Average 2002–2006 Price	2007–2008 Price	2013–2014 Price
Wheat	$/ton	$168	$319	$242
Coarse grains	$/ton	113	181	166
Rice	$/ton	262	361	324
Oilseeds	$/ton	293	486	434
Vegetable oils	$/ton	587	1,015	764
Butter	$/100 kg	162	294	313
Whole-milk powder	$/100 kg	192	417	372

Source: OECD/FAO Agricultural Outlook 2008–2017.

Food Security

Bread for the World's Hunger Institute defines **food security** as "assured access for every person to enough nutritious food to sustain an active and healthy life."[16] Food security is the responsibility of three levels of society: family, the nation, and the global community. At each level, the players are part of a market economy as well as a sociopolitical system.

In a market economy, food flows in the direction of economic demand. Need is not taken into consideration. A cash economy, following the rules of the market, provides the *opportunity* to purchase food, but not the food itself. Where the economic status of the player is very low (a jobless parent, a poor country), another group or individual may be willing and able to provide the needed purchasing power or the food (but maybe not).

Family Food Security. Family members purchase, raise, and gather food from natural ecosystems or have the food provided by someone, usually in the context of the family. When families cannot do so, they need a *safety net*—policies or programs at the national, state, or local level whose objective is to meet the food security needs of all individuals in the society. Local food banks and community groups may provide food. In the United States, more than 50,000 local food charities help those who are hungry **(Figure 12–15)**.

State and National Food Safety Nets. On a state and national level, safety-net policies and programs include a variety of welfare measures, such as the Supplemental Nutrition Assistance Program and the Supplemental Security Income program, and voluntary aid through hunger-relief

---

[16]Bread for the World Institute, "Agriculture in the Global Economy Hunger 2003," *13th Annual Report on the State of World Hunger* (Washington, DC: Bread for the World Institute, 2003), 20.

organizations. For example, Feeding America, a network of food banks throughout the country, is part of the voluntary safety net that provides emergency food help for more than 25 million Americans. Feeding America retrieves food from food processors and suppliers and distributes it to the organization's network of food banks. One federal law—the Charity, Aid, Recovery, and Empowerment Act of 2003 (CARE Act)—allows family farmers, ranchers, and restaurant owners to deduct the costs of food donated to agencies such as Feeding America as a charitable contribution.

National *self-sufficiency in food*—enough food to satisfy the nutritional needs of all of a country's people—can be achieved by producing all the food people need or by buying the food on the world market. This goal implies that countries have policies that allow the poorest members into the market, such as a just land-distribution policy and a functioning market economy. Many nations, though, are not self-sufficient in food and turn to the global community for food aid and other forms of technical development assistance.

International Food Safety Net. A substantial amount of food aid flows from rich countries to food-poor countries. However, much more food is imported and paid for by the developing countries. Unfortunately, protective policies in developed countries result in tariffs—taxes the developed countries set on imported agricultural products—and in subsidies given to the agricultural sector. U.S. farm bills have continued the subsidies of most agricultural products. The European Union is notorious for its high tariff barriers and large subsidies for its agricultural sector, even though it has reformed the rules for subsidies.

Haiti is a good example of the problem of uneven trade policies. In 1995, it cut its tariff on rice from 50% to 3%, opening the country's markets to subsidized rice from the United States. By 2000, rice production in Haiti was cut in half, leaving thousands of small rice farmers without a means of livelihood. Haiti began to import rice. Haiti's problems were exacerbated by the 2010 earthquake. After the earthquake, several NGOs and scientists collaborated to try a new rice-growing initiative, which produced yields that were 64% higher, with fewer seeds and less water and fertilizer. However, uneven trading policies still make it difficult for Haitian farmers to survive in the world market.

There are other initiatives that can help the poorer countries become self-sufficient in food. One of the most important is relieving the debt crisis in developing countries (Chapter 9). Another factor in meeting global nutrition needs is the *trade imbalance* between the industrial and the developing countries. The globalization of markets is now working to the advantage of the developing countries, which have a surplus of labor and low labor costs. As a result, many industrial corporations have outsourced their manufacturing to developing countries, which is helping workers in developing countries improve their well-being.

Figure 12–15 Food safety net. When families have difficulty obtaining food, the local community is the first level of defense. These women are choosing produce from a church food bank.

CONCEPT CHECK How do tariffs and subsidies affect food prices and distribution? ☑

SUSTAINABILITY
Preventing Food Crises

On a cold day in February 1936, photographer Dorothea Lange was returning from a month-long assignment where she was photographing and documenting the plight of the poor in the Great Depression. As she passed a camp of ragged tents in Nipomo, California, she was only interested in returning home. But miles down the road, something called her to return to an enclave of pea pickers in tents, starving after the collapse of a pea crop. There she met and photographed the woman who became the iconic image of the effects of the Great Depression. A widow of 32, Florence Owens Thompson and her small children were living on vegetables scrounged from the frozen ground and birds her children had caught. Having sold her tires to buy food, she had nowhere to go and no plan to escape their deprivation.

Lange took only five photos, but they became the images that spoke of the suffering of farmworkers across the country. One known as "Migrant Mother" is probably the most famous picture of the time. Those photos woke the nation and food was brought to those pickers isolated in a barren field. The images Lange took gripped the imagination and caring of others.

Sometimes images of children starving wring our hearts and cause us to pour out kindness and wealth. But sometimes, when the problems are not easily resolved, such images fill people with a sense of despair. Instead of doing something, no matter how small, they do nothing.

Relief and development agencies know that to solve food shortages and prevent famine, food needs to be given when a drought or disaster is looming, not when children are already standing, blank eyed, with protruding bellies. In recent years, there have been several years of poor crops in the Sahel region of Africa. In 2010, 2011, and 2012, people who track crop outcomes predicted severe food shortages months before they occurred. They reported these findings in the news, but international aid from governments and charities was slow to come. Millions of people faced starvation, when their plight could have be lessened, if not avoided, by timely aid. One worker for Save the Children said, "The humanitarian system is like an ambulance—it is focused on disaster response, not prevention."[17]

In a sustainable world, prevention has to play a larger role. Dorothea Lange could not prevent the plight of the pea pickers in California. But she could help get them aid. Today's aid organizations are trying to prevent large-scale famines before they happen, though this can be difficult to do.

"Migrant Mother." This iconic photo of a migrant pea picker by photographer Dorothea Lange captured the despair of the Great Depression.

(*Source:* The Library of Congress.)

[17] Andrew Wardell, "Extreme Hunger in East Africa and the Sahel: Forewarned but Not Forearmed," Poverty Matters Blog, *The Guardian*, May 9, 2012, http://www.guardian.co.uk/global-development/poverty-matters/2012/may/09/extreme-hunger-east-africa-sahel.

12.4 Hunger, Malnutrition, and Famine

At a UN World Food Conference in 1974, delegates from all nations subscribed to the objective "that *within a decade* no child will go to bed hungry, that no family will fear for its next day's bread, and that no human being's future and capacities will be stunted by malnutrition." That was an unattained dream, though great strides have been made. In the Millennium Development Goals, the target was *to cut world hunger in half by 2015* in the developing countries. This meant a reduction in the proportion of hungry people from 20% to 10%. That target was met, although the food crisis of 2007–2008 added 75 million to the 848 million people in the world still suffering from hunger and malnutrition. The Sustainable Development Goals are poised to continue the effort to cut hunger. Let's look at hunger and malnutrition in the world.

Hunger and Nutrition

Hunger is the general term referring to a lack of basic food required to provide energy and to meet nutritional needs such that the individual is unable to lead a normal, healthy life. **Undernourishment** is the lack of adequate food energy (usually measured in Calories). **Malnutrition** is the lack of essential nutrients, such as specific amino acids, vitamins, and minerals, and can occur in people who eat nutritionally poor diets even if they have other food available.

Another nutritional disorder is **overnourishment,** a condition of eating too much. Overnourishment is becoming epidemic, even in many developing countries. Worldwide, more than 1.5 billion adults are overweight. This is partly the result of an *obesogenic* culture, one that promotes the consumption of calorie-rich foods and a sedentary lifestyle. More than one-third of the U.S. population is clinically obese (more than 30 pounds overweight). As the World Health

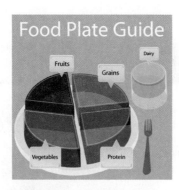

Food Plate Guide

GRAINS 7 ounces	VEGETABLES 3 cups	FRUITS 2 cups	MILK 3 cups	MEAT & BEANS 6 ounces
Make half your grains whole Aim for at least **3 1/2 ounces** of whole grains a day	**Vary your veggies** Aim for these amounts each week: **Dark green veggies** = 3 cups **Orange veggies** = 2 cups **Dry beans & peas** = 3 cups **Starchy veggies** = 6 cups **Other veggies** = 7 cups	**Focus on fruits** Eat a variety of fruit Go easy on fruit juices	**Get your calcium-rich foods** Go low-fat or fat-free when you choose milk, yogurt, or cheese	**Go lean with protein** Choose low-fat or lean meats and poultry Vary your protein routine—choose more fish, beans, peas, nuts, and seeds

Find your balance between food and physical activity

Be physically active for at least **30 minutes** most days of the week.

Know your limits on fats, sugars, and sodium

Your allowance for oils is **6 teaspoons a day.**

Limit extras—solid fats and sugars—to **290 calories a day.**

Your results are based on a 2200 calorie pattern.

This calorie level is only an estimate of your needs. Monitor your body weight to see if you need to adjust your calorie intake.

Figure 12–16 The Food Plate Guide. The new food guide produced by the U.S. Department of Agriculture to help people evaluate their food intake. To get a personalized recommendation, go to https://www.supertracker.usda.gov/default.aspx.

Organization describes the situation, "65% of the world's population lives in countries where overweight and obesity kill more people than underweight."[18]

To address nutritional problems in the United States, the Department of Agriculture (USDA) established a graphic of a food plate (Figure 12–16). The Food Plate Guide represents the kinds of foods that should be emphasized in a healthful diet—fruits and vegetables, whole grains, lean protein, and calcium-rich foods. Foods that are high in calories and saturated fats but have little nutritional value are to be avoided. Of course, healthy eating and exercise are both parts of a healthy lifestyle.

Extent and Consequences of Hunger

Accurate, reliable figures on the worldwide extent of hunger are unavailable, mainly because few governments document such figures. Hunger is often a seasonal phenomenon in rural areas that are supported by subsistence agriculture, as people are forced to ration their stored food in order to survive until the beginning of the next harvest. On the basis of household surveys, the FAO estimated in 2014 about 805 million people in the world are chronically undernourished, a drop of 120 million in less than five years.[19] This is great news, although the fact that hunger still occurs on a large scale should concern everyone.

Where? Two-thirds of the world's undernourished people live in Asia. Yet in recent decades, Asia has experienced a dramatic decline in undernourishment, with the exception of western Asia. The region with the highest percentage of undernourished people is sub-Saharan Africa, where one person in three is afflicted. When faced with such large numbers, it is easy to lose track of the fact that each person is a unique individual whose potential, hopes, and dreams are limited by this lack of food.

---

[18] World Health Organization, *2012 Obesity and Overweight Factsheet No. 311*, accessed March 6, 2012, http://www.who.int/mediacentre/factsheets/fs311/en/.

[19] World Food Programme, *Hunger Statistics 2014*, accessed December 12, 2014, http://www.wfp.org/hunger/stats.

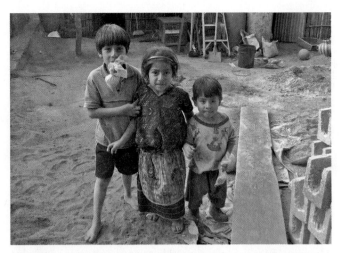

Figure 12–17 Malnourished children. These Guatemalan children are suffering from chronic malnourishment, a condition that can stunt their growth.

Consequences. The effects of undernourishment and malnutrition are greatest in preschool children and next greatest in women. Hunger can prevent normal growth in children, leaving them thin, stunted, and often mentally and physically impaired (Figure 12–17), a condition known as protein-energy malnutrition.

Undernourishment in early childhood can seriously limit growth and intellectual development throughout life. In Guatemala, for example, half of the children under five are chronically undernourished, with rates much higher among indigenous people. This state of malnourishment irreversibly stunts children and inhibits their physical and mental capabilities. As adults, their productivity may be reduced by as much as a third. UNICEF estimated that chronic malnutrition costs Guatemala $8.4 million a day because of sickness, repeated schooling, hospitalization, and lowered productivity.

Sickness and death are companions of hunger. Because poor nutrition lowers a person's resistance to disease, measles, malaria, and diarrheal diseases are common and are major causes of death in the malnourished and undernourished.

Root Cause of Hunger

As we have already seen, the root cause of hunger is poverty. Hungry and malnourished people lack either the money to buy food or enough land to raise their own. One FAO report on food insecurity states that "Most of the widespread hunger in a world of plenty results from grinding, deeply rooted poverty."[20]

Addressing hunger and the poverty that causes it will be essential to making progress toward all the other Sustainable Development Goals. One measure of progress toward reducing hunger is the proportion of children under five who are underweight because of chronic or acute hunger. Figure 12–18

compares data on the proportion of underweight children from 1990 and 2012, showing that several regions—eastern Asia, Latin America and the Caribbean, and northern Africa—were on target to reduce this measure of hunger 50% by 2015.

A number of countries significantly reduced the extent of poverty and hunger during the 1980s and 1990s. Indonesia has benefited from oil exports and Green Revolution technology, and it continues to put a major emphasis on rural development and the country's social infrastructure. Malawi has been able to double its agricultural productivity through a voucher program for fertilizer and seed, helping poor farmers feed their families and produce crops for the market. Thus it is possible for societies to make progress in reducing the extent of poverty and hunger. Yet in any given year, there are countries or even whole regions that face exceptional food emergencies. It is here that international responsibility comes most sharply into focus.

Famine

A **famine** is a severe shortage of food accompanied by malnutrition, epidemics, and a significant increase in the death rate. Famine is a clear signal that a society is either unable or

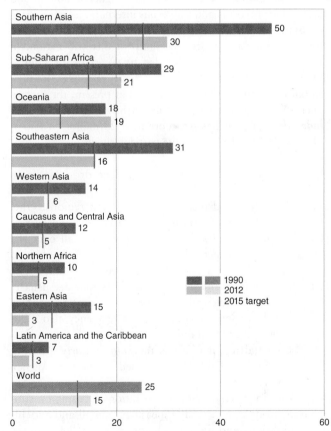

PROPORTION OF CHILDREN UNDER AGE FIVE MODERATELY OR SEVERELY UNDERWEIGHT, 1990 AND 2012 (PERCENTAGE)

Figure 12–18 Underweight children. The figure shows the percentages of children under age five who were underweight in 1990 and 2012. Millennium Development Goal 1 was to reduce the 1990 proportion 50% by 2015.

(*Source: Millennium Development Goals Report 2014* (New York: United Nations, 2014).)

[20]UN Food and Agriculture Organization, *State of Food Insecurity in the World 2002* (Rome: UNFAO, 2003).

unwilling to distribute food to all segments of its population. In recent years, two factors—drought and conflict—have been immediate causes of famines.

Drought. More than 70 million people who practice subsistence agriculture or tend cattle, sheep, and goats occupy the Sahel, the broad belt of drylands south of the Sahara in Africa. Although the region normally has enough rainfall to support dry grassland or savanna ecosystems, the rainfall is seasonal, undependable, and prone to failure. Drought and deforestation have brought on catastrophic desertification, which in turn has triggered massive dust storms that spread clouds of Sahara dust across the Atlantic Ocean and into the Northern Hemisphere.

Drought is blamed for famines that occurred in the Sahel in 1968–1974, 1984–1985, and again in 2010–2012. Beginning in 1965, the region experienced 20 years of subnormal rainfall, with tragic results. Crops withered, forage for livestock declined, watering places dried up, and livestock died. Both farmers and pastoralists began abandoning their land and migrating toward urban centers, where they ended up in refugee camps. Unsanitary conditions in the camps led to the spread of infectious diseases such as dysentery and cholera, and many thousands died before effective aid could be organized. The 1984–1985 famine is thought to have been responsible for almost a million deaths in Ethiopia alone. The number would have been higher if not for aid extended by Africans and numerous international agencies.

In the last few decades, rainfall in the Sahel has increased, although droughts are still the norm in this arid region. In 2003–2004, favorable weather led to desert locust infestations, which devastated crops in many regions. Although rainfall is normal in the western Sahel at present, the eastern Sahel—the Horn of Africa—remains in the grip of an extended drought that has put more than 23 million people at risk. The years 2010–2013 were particularly dry.

Warning Systems. In order to prevent droughts from leading to famines, two major efforts provide crucial warnings about food insecurity in different regions. One, the FAO's *Global Information and Early Warning System (GIEWS),* monitors food supply and demand all over the world, watching especially for those places where food supply difficulties are expected or are happening. The GIEWS issues frequent reports and alerts that are usually the result of its field missions or on-the-ground reports from other UN agencies and NGOs. The agency also relies on satellite imagery to monitor crops.

The second major effort is the *Famine Early Warning System Network (FEWS NET),* funded by the U.S. Agency for International Development. FEWS NET is a partnership of agencies involved in satellite operations focusing on sub-Saharan Africa. It continually measures rainfall and agricultural conditions and gives governments and relief agencies accurate and timely assessments of the status of food security in African countries.

Conflict. Devastating and prolonged civil warfare put millions of people at risk of famine. In the 1990s, war was the cause of the famines that threatened Ethiopia, Eritrea, Somalia, Rwanda, Sudan, Mozambique, Angola, and Congo. Continuing conflicts in Afghanistan, Sri Lanka, and Syria interfere with access to food, and a 30-year civil war in Guatemala helped to set the stage for the chronic malnutrition in that country, though it is not an actual famine.

Famines from drought and war are preventable. India, Brazil, Kenya, and southern Africa have coped with droughts in recent years by mobilizing effective relief in the form of food, clothing, and medical assistance. Unfortunately, international aid often comes after a food shortage has become a famine (see Sustainability, *Preventing Food Crises*).

Hunger Hot Spots

Much of Africa has experienced long-term and severe droughts as well as widespread civil conflict. Erratic weather patterns and a general warming trend, thought by some scientists to be the outcome of global climate change, have caused crop harvests to oscillate between average and poor. Climate change is expected to affect global food production worldwide, lowering food production in South and Southeastern Asia, for example, by as much as 50% over the next three decades. Other countries with recurrent hunger and food insecurity include North Korea, Haiti, and Bangladesh. In 2014, the FAO identified 36 countries, including 26 African countries in need, as being "in crisis and requiring external assistance."[21] The "external assistance" here is food aid; the reasons for the need vary from conflict to drought, floods, hurricanes, cyclones, and localized crop failures.

Food Aid

Widespread, severe hunger and famine are being averted today only because of food aid, though it does not always work as well as we would like. The World Food Program (WFP) of the United Nations coordinates global food aid, receiving donations from the United States, Japan, and the countries of the European Union. In addition, many nations conduct their own programs of food aid to needy countries. International food aid has fluctuated wildly. At about 6.2 million metric tons in 1996, it rose to 15 million metric tons in 1999 when the grain harvest in the Russian Federation collapsed and the United States increased aid. It dropped to less than 5.0 million metric tons in 2011 (Figure 12–19).

Food aid is distributed to countries all over the world, not just where famines are threatened. Food aid is negotiated on the world stage. In April 2012, a new Food Assistance Convention (FAC) was agreed on (replacing the Food Aid Convention of 1999). This international treaty requires members to provide a minimum amount of food assistance. The 2012 FAC also addressed concerns about the problems of food aid, including activities that foster dependence, and problems caused by aid that isn't useful to the recipients.

---

[21] UN Food and Agriculture Organization, *Crop Prospects and Food Situation, No. 1* (Rome: UNFAO, 2009).

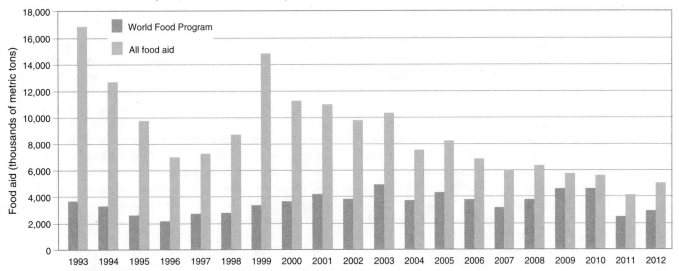

Figure 12–19 Global food aid. Global food aid from 1993 to 2012. Orange indicates food aid delivered by the World Food Program; green indicates food aid from all sources.

(*Source:* World Food Program annual reports, FAIS, UNFAO.)

When Aid Doesn't Help. Numerous humanitarian campaigns to end world hunger have been mounted in the past 50 years. The United States and Canada have been world leaders in giving away food (which is first purchased from farmers and, therefore, represents a subsidy). A number of serious famines have been moderated or averted by these efforts, and certainly the need for ongoing emergency food aid is undeniable. However, routinely supplying food aid in an attempt to alleviate chronic hunger in developing countries can be the worst thing to do.

Why is a routine supply of food aid a problem? Free or very cheap foreign food undercuts the local market. In effect, local farmers must compete economically with free or low-cost imported food. When they cannot earn a profit, local farmers stop producing and eventually enter the ranks of the poor. In the long run, the entire local economy deteriorates.

Another way that food aid fails results from the cost and time required to store and ship food. The U.S. food aid program, the largest humanitarian food assistance program in the world, gives much of its aid under regulations that require the purchase and shipment of U.S. food around the world.[22] Food produced in the United States moves more slowly, in some cases requiring 147 days rather than the 35 days it could have taken if locally sourced. Furthermore, a portion of the aid must be shipped on U.S. flagged vessels. Purchasing and shipping food from the United States costs a great deal more than purchasing food elsewhere. From 2003 to 2012, USAID spent $7.4 billion on food for aid and $9 billion on shipping and logistics. While most donor countries now use a system of cash or vouchers, the United States continues to use a system that wastes time and funds,

although the 2014 Farm Bill did include a few changes to make the system more efficient.

When It Does Help. The World Food Programme (WFP) is aware of the problem of local economy collapse and targets most of its food aid to emergency situations. Some aid goes to development projects, benefiting poor people in countries where food security is at risk and regions where malnutrition is severe. The intention of this aid is to free poor people of the need to provide food for their families and allow them to devote their efforts to other development activities. As an important component of this strategy, the WFP now makes 80% of its food purchases in the developing countries themselves. This is a win-win solution, where small-scale farmers receive income for their crops and needy people are fed. In fact, WFP has special programs designed to prevent hunger by better connecting poor farmers to the market.

The Goal Is Zero, the Way There Is Complex

Alleviating hunger is primarily a matter of addressing the absolute poverty that afflicts one of every five people on the planet. At the 2012 Rio+20 summit, UN Secretary-General Ban Ki-moon offered the Zero Hunger Challenge to the world. Ki-moon said that achieving zero hunger would lower conflict and increase productivity and prosperity. The challenge was focused on five objectives: 100% access to adequate food all year round; zero stunted children under two years old and no more malnutrition in pregnancy and early childhood; all food systems sustainable; 100% growth in smallholder productivity and income, particularly for women; and zero loss or waste of food, including responsible consumption. It is a tall order, but to get anywhere, we must begin by setting high goals.

[22]A. Sreevatsan and R. Andersson, "Aid Undermined," *Medill National Security Reporting Project 2014*, accessed December 9, 2014, http://foodaid.nationalsecurityzone.org/aid-undermined/.

Although food is our most vital resource, we do not treat it as a commons; it is not free to all who need it. Indeed, the production and distribution of food constitute one of the most important economic enterprises on Earth. What has not been done, however, is to bring the market economy under the discipline of sustainability. Short-term profit crowds out long-term sustainable management of natural resources.

We understand the situation well, but the solutions are not simple. The solutions lie in the realm of political and social action at all levels of responsibility. Given the current groundswell of concern about the environment and the global attention to the Millennium Development Goals and, we hope, the Sustainable Development Goals, there may never be a better time to turn things around and take more seriously our responsibilities as stewards of the planet and as our brothers' (and sisters') keepers.

CONCEPT CHECK What are three components of effective food aid? ✓

12.5 Feeding the World as We Approach 2030–2050

Looking ahead to 2020, we need to be able to feed an additional 700 million people and also make significant progress toward meeting the Sustainable Development Goal of eliminating existing hunger and malnutrition in the world—and do so without the unsustainable practices agriculture currently employs. Is this possible? What will we need to do as we look even further ahead, to 2030 and 2050?

Given the limits on suitable land for agriculture, there are several prospects for increasing the supply of food. We could continue to increase crop yields with new technologies. We could grow food crops on land that is now used for feedstock crops, biofuels, or cash crops. We could use new agricultural opportunities such as urban agriculture and mixed permacultures. We could also attempt to feed more people with the food we already produce by changing social structures so more people have a chance to enter the market—and by eating foods that require less fuel, water, land, and pesticides to produce.

Increasing Food Production

A dramatic rise in crop yields in the developing countries was the major accomplishment of the Green Revolution: Wheat yields tripled, rice and corn yields more than doubled, potato yields rose 78%, and cassava yields rose 36%, according to the UN Food and Agricultural Organization (FAO). Interestingly, grain yields have continued to rise in some of the *developed* countries, too. In France, for example, wheat yields have quadrupled since 1950, reaching more than 7.4 tons per hectare (2.93 tons per acre). In the same period, rice yields in Japan have risen 170%. Indeed, the genetic potential exists for wheat yields of 14 tons per hectare or higher, and rice yields can be as high as 13 tons per

hectare under the right conditions. Can we expect yields to continue to increase up to their genetic potential?

Environmental Limits. It turns out that the great differences in grain yields among regions have less to do with the genetic strains used and more to do with the weather. Egypt and Mexico, for example, irrigate their wheat and get higher yields, while U.S. wheat is rain-fed. Australian wheat yields are even lower, the result of the country's sparse rainfall. Once the agricultural land is planted with high-yielding strains and fertilized to the maximum, other factors—soil, rainfall, and available sunlight—limit productivity. These environmental limits are a reminder that agricultural sustainability is highly dependent on the conservation of soil and water (Chapters 10 and 11).

Doubly Green. Some people still see the Green Revolution as the way forward. Gordon Conway, president of the Rockefeller Foundation, has called for a "Doubly Green Revolution," by which he means, "a revolution that is even more productive than the first Green Revolution and even more 'green' in terms of conserving natural resources and the environment. During the next three decades, it must aim to repeat the successes of the Green Revolution on a global scale in many diverse localities and be equitable, sustainable, and environmentally friendly."[23]

New Developments in Breeding. One part of the next generation of agriculture is the development of *super green rice*. Developed using modern breeding techniques but not genetic engineering, super green rice is a group of about 250 varieties of rice bred to survive on marginal lands. These varieties may be tolerant of salt or drought or have other capabilities that allow them to survive in poor conditions. As a result, they require less fertilizer, water, and pesticides than GM and other high-yield crops commonly used in modern agriculture.

A new technique called **marker-assisted breeding** can speed up traditional plant breeding. In this technique, scientists use DNA sequencing to identify genes for desirable traits in crop plants or their wild ancestors. Then plants with the desired gene are crossbred with a modern crop breeding line. Screening the DNA of seedlings for the desired gene makes the process move more quickly. Crops being developed with this approach include high-calcium "super-carrots," potatoes with more starch, and pigeon peas with higher yields.

Expanding Our Diets. One way to increase the supply of food is to eat a broader range of foods. Insects, for example, are protein-rich and have always been a part of the diet in Asia, Latin America, and Africa. Worldwide, about 1,900 insect species are consumed by humans. Raising insects for food would provide jobs and lower food insecurity. Insects can also be fed to poultry and fish, thereby freeing up grain for human consumption.

---

[23]Gordon Conway, *The Doubly Green Revolution: Food for All in the 21st Century* (Ithaca, NY: Cornell University Press, 1999).

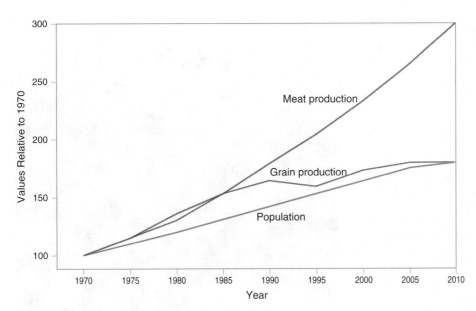

Figure 12–20 Global population and grain and meat production. Grain production has kept up with the population increase, and meat production has greatly outpaced it over the past 35 years. High meat consumption in the West makes it difficult to produce food sustainably.

(*Source:* FAOSTAT.)

Using Current Production More Effectively

One way to feed more people is to use what we are producing more efficiently. Several mechanisms for this are to divert less human food to animal feed and biofuels, to eat locally, and to reduce food waste.

Fewer Crops to Animal Food and Biofuels. We can switch from the production of feed grain and cash crops to food for people. Almost 70% of domestic grain in the United States is used to feed livestock. The percentages drop over other regions of the world in proportion to the economic level of the region. Sub-Saharan Africa and India, for example, use only 2% of their grain to feed livestock. The current trend, however, is in exactly the opposite direction (Figure 12–20). More people want to eat higher on the food chain. A more sustainable approach is to find a balance in which people have protein in their diets but are not heavily dependent on meat. In other words, people in developed countries need to eat lower on the food chain, allowing others to eat more.

Locavore, Anyone? Eating locally lowers the environmental cost of food because it reduces the need for transportation and refrigeration. It also reduces food waste, at least to some extent. **Locavores** are people who try to purchase or grow most of their food within their own region. Of course, many foods are not available in all seasons so foods have to be preserved in some way.

Community-supported agriculture programs (CSAs) are on the rise in American urban and suburban areas. People support CSAs by purchasing shares of the produce from local farms and often by volunteering (Figure 12–21). Because shares are purchased in advance, the risks of a poor crop are shared, so farmers need to borrow less money and are likely to weather difficult seasons more easily. CSA programs are popular because they provide fresh local produce and support for farms that might not be economically viable if they

had to compete with cheap food trucked long distances from industrial farms.

Waste Not, Want Not. Eliminating food waste would allow us to feed more people without using more resources. This important concept is covered elsewhere (Chapters 2, 10). Many of the causes of waste are systemic and not easily solved. As the example of rice stores in Thailand shows, storage of food has impacts on economic markets and includes the risk of spoilage. Sometimes there is a trade-off between public health concerns and reducing waste. For example, health regulations control what restaurants can do with leftover food and how long food can be on store shelves. Changes in these regulations could help us waste less of the food we raise.

Figure 12–21 Community-supported agriculture. This woman runs a CSA farm in West Falmouth, Maine. CSAs lower the financial risk for farmers and increase access to local produce.

Sustainable Agriculture

Modern agriculture has too many externalities to be sustainable. We cannot set up our most important systems (such as food production) to constantly damage the very ecosystems on which we depend. Feeding many people sustainably is going to take something different than more of the same. It will require sustainable agriculture, that is, agriculture that can be carried on indefinitely without environmental harm.

What would sustainable agriculture look like? It might include some of the large farms we are used to, though with fewer chemicals, fewer fossil fuels, and fewer of the most wasteful types of irrigation (as we have seen in other chapters). It would include various forms of organic farming. But it will also require wildly creative approaches very different from industrial agriculture, such as the intensive, mixed-product (polyculture) urban farm described in the story of Growing Power at the beginning of this chapter. Here are some trends in the future of food production.

Organic Farms. Organic farming has been around a long time and overlaps with many other sustainable practices. In the United States, the USDA can certify farms as organic. Certified organic farms have to meet strict criteria. They are limited to natural pesticide products and cannot use the artificial fertilizers typical of modern agriculture. Often, organic farm yields are somewhat lower and organic food prices higher. In one study of more than 322 research papers, scientists found that organic yields were 80% of conventional yields on average (with a great deal of variability). However, in intensive mixed cultures (such as the Growing Power farm), with practices where different types of crops are grown together (companion planting) and the natural enemies of pests are supported, yields can be high (Chapter 13).

In the United States, certified organic farms cannot use GM crops. This is a controversial issue. Many people who support organic farming also see GM crops (especially those modified for drought or salt tolerance or enhanced nutrition) as one of the tools we will need to use to feed the world. It is also difficult to keep organic farms GM free; farmers often find their crops contaminated with GM crop plants. As a result of some of these limitations, many farms that use less polluting methods are not certified as organic by the USDA.

Urban Agriculture. Urban farming occurs across the world and is part of the future of food production. Cuba is the poster country for urban farming and the only country in the world with extensive state-supported urban agriculture. Food shortages of the 1990s drove a grassroots effort to produce food in Cuban city neighborhoods. To meet the needs of small-scale urban farmers, the Cuban government created an Urban Agriculture Department. It provides expertise and places to buy seeds and equipment and supports infrastructure. In Havana alone, there are more than 30,000 people growing food in more than 8,000 farms and gardens (Figure 12–22). Cubans even grow rice in small plots.

Figure 12–22 Urban agriculture. A man thinning seedlings in an urban garden in Havana, Cuba.

Small-scale rice production now rivals the large-scale production of large farms in Cuba.

Urban Foraging. Another model of urban food production is foraging. **Urban foragers** find fruits, berries, and other foods available in the urban landscape. Sometimes they create urban food maps, which are available on the Web. There is a growing community of urban foragers.

In Seattle, community organizers designed a 7-acre mixed foraging and garden plot area called an **urban food forest**. Initially designed as a class project, this food forest (the Beacon Food Forest) expects to use grants to become the largest urban food forest on U.S. public land. The project will be realized over a period of years.

Algae to the Rescue. Imagine a world in which nutritious algae grow in fish tanks on nearly everyone's windowsills, farms in the ocean produce edible seaweeds, algae are grown in ponds and tanks for fish food, and vast bags of wastewater are treated with algae, which are then fed to fish (Figure 12–23). Algae in ditches at the edges of farm fields take up nutrients and provide a rich fertilizer for the fields. Algae are also used to produce biofuels.

That imagined future is now. Work is proceeding rapidly on all of these fronts to produce a world in which algae can be used in a wide array of ways. Food from algae is very

Figure 12–23 Algae for food. NASA has a project for algal treatment of wastewater. The algae are then fed to fish. Other researchers are investigating the use of algae as biofuels.

Figure 12–24 Permanent polyculture. Polycultures have a variety of perennial plants in an agricultural setting. On this Welsh farm, fruit trees, herbs, wildflowers, and vegetables grow together.

nutritious, high in omega-3 fatty acids, vitamins, and antioxidants. As sources of biofuel, algae are competitive because of their high oil content; as nutrient extractors, algae are efficient because of their rapid growth.

Dr. Aaron Baum is a good example of an algal entrepreneur. He is working with NASA, which has a large algae project (the OMEGA project) designed to use wastewater to raise algae for biofuel production. But Baum's major effort is with small food algae production. He designs kits for people to raise *Spirulina*, an extremely nutritious type of algae found in food supplements. Instead of taking a small-scale idea and trying to scale up to industrial levels, Baum is trying to make household-sized algae-growing apparatuses that can be used by any apartment dweller to grow healthy food.

Laboratory-Grown Meat. This high-tech world of food and fuel grown in vats or petri dishes may also soon include lab-grown meats. Eating meat is environmentally costly in part because we have to raise whole animals, a very inefficient effort. Scientists are working on raising meat products that can be grown in the lab, in the hopes of allowing us to eat meat without the downsides of current animal farming. This is a long way off, but may be important in the future.

Permanent Polycultures. In Salina, Kansas, Wes Jackson has spent a career working toward another model of sustainable agriculture. Instead of the conventional monoculture that is susceptible to insects and diseases and drought, Jackson and his colleagues are trying to create an agricultural system modeled on a native prairie. What they describe is **permanent polyculture**, a group of perennial plants of different species grown together that produce food or fuel. This is hard to achieve, in part because perennial plants put more energy into roots and leaves and less into the parts of the plant we eat. There are other models of permanent polyculture, such as the field shown in **Figure 12–24**. This type of research is important to a radically new approach to food production.

Policy Change

The future also needs to include policy and individual changes. U.S. farm policies currently encourage large farms, support fewer workers, discourage sustainable agriculture, and subsidize industrial agriculture to the detriment of the environment. Incorporating externalities such as the true cost of erosion, water pollution, or overuse of water resources into the price of food would go a long way to changing the way we produce food. Finding ways to support farmers in the United States without subsidizing large industrial agriculture and CAFOs at the expense of small or organic farmers and small farmers abroad would go a long way toward solving job and food insecurity. Providing means to get crop insurance for small farmers, including subsistence farmers, would help lessen their vulnerability to the environmental changes we anticipate with climate change. Furthermore, food aid needs to be structured in a way that does not undermine local economies or waste money on shipping.

On the international level, middle-income countries are home to the majority of hungry and malnourished people. Therefore efforts to increase economic development need to support individual access to food, not just the average wealth of nations. Efforts to address climate change are also essential to increasing food security (Chapter 18).

Individuals need to feel connected to the rest of the world and empowered to do something about the big issues of our time. One part of this is transparency from producers. Just as wood and fish are certified sustainable (Chapter 7) and food can be certified organic (Chapter 13), requiring transparency in food systems can help consumers make better decisions.

Food production is central to every topic in environmental science. In order to solve environmental problems, we will need to have new ways of approaching them.

CONCEPT CHECK What are the advantages of eating less meat and reducing food waste? How do the advantages of these measures compare with those of measures taken to produce more food? ✓

REVISITING THE THEMES

SOUND SCIENCE

The new high-yield grains that led to the Green Revolution were developed using science. The agricultural research that continues to seek ways to improve crop yields is also a form of science, as is the research that shows the environmental costs of conventional agriculture. Sound science is needed to provide an accurate evaluation of all genetically modified crops and foods before they are implemented. Most importantly, scientific methods and tools will be needed to find long-term solutions to food production. Scientific techniques are being used to grow algae for food and biofuels and to accomplish other sustainable agriculture goals.

SUSTAINABILITY

Global land and water used for agricultural crops and animal husbandry represent two of the most crucial and irreplaceable components of ecosystem capital. However, the land must be nourished; it is a renewable resource, but it can be eroded and worn down by unsustainable practices. Water can be overused. As populations increase, pressure will mount to turn more forests and wetlands into croplands, a transformation with trade-offs that are not usually beneficial due to the loss of the goods and services of the natural lands.

Many of the techniques of modern agriculture are not sustainable, including heavy use of fossil fuels, ancient groundwater, pesticides, artificial fertilizer, and monocultures. Many consider the diversion of food crops to biofuel production unsustainable. Likewise, some food aid is harmful to sustainability in developing countries, where self-sufficiency in food is essential to establishing a sustainable society. A number of technologies and approaches make up sustainable agriculture, the approach most likely to both feed populations and maintain ecosystem health.

STEWARDSHIP

The remarkable progress made in feeding the world's people is certainly good stewardship in action. This is especially true of the efforts made by national and international agencies and the many NGOs involved in food aid to identify those suffering from food insecurity and ensure that they have sufficient food for their needs. Thousands of organizations and people show that they care as this work goes on. In the United States, the Feeding America organization plays an important role in keeping people from going hungry in a land of plenty. The work of Borlaug and his team and of the CGIAR is particularly noteworthy because these groups have sought to improve the yields of important grains and make the new high-yielding varieties available to the developing countries.

Stewardship requires us to ensure that all people have access to crops and seeds, rather than allowing that power to be held by a few companies. We still need to be better stewards of the ecosystem services affected by unsustainable food production. In part, this will involve changing ways people relate to the market—a focus on access rather than production, on lowering waste and crop loss, and on diverting less food to livestock feed or biofuels.

REVIEW QUESTIONS

1. Describe the differences between subsistence and modern agriculture. What are the major components of the agricultural revolution? What are the environmental costs of each?

2. What is the Green Revolution? What have been its limitations and its gains?

3. How do sustainable animal farming and industrial-style animal farming affect the environment on different scales?

4. What are biofuels, and what are their impacts on food production?

5. What are the major advantages and problems associated with using biotechnology in food production?

6. Describe the patterns in grain trade between different world regions over the past 70 years. How do the food crisis of 2006–2008 and food shortages of 2010–2011 relate to these patterns?

7. Describe the levels of responsibility for food security. At each level, list several ways food security can be improved.

8. Define *hunger, malnutrition,* and *undernourishment.* What is the extent of these problems in the world? What is their root cause?

9. Discuss the causes of famine, severe hunger, and chronic malnutrition, and name the geographical areas most threatened by them.

10. How is relief from famine and severe hunger accomplished? Why does food aid sometimes aggravate poverty and hunger?

11. Define sustainability in an agricultural context. In what ways is modern industrial agriculture unsustainable?

12. What are at least four cutting-edge concepts in sustainable agriculture?

THINKING ENVIRONMENTALLY

1. Imagine that you have been sent as a Peace Corps volunteer to a poor African nation experiencing widespread hunger. Design a strategy for assessing the needs of the people and for contacting appropriate sources for help.

2. Calculate your "food footprint" by looking for one of several Web sites where this is calculated. Usually it gives a value of either the land or ocean area required to sustain you or, in some cases, your carbon dioxide production. How many Earths would it take to feed everyone if we all ate like you do?

3. Use the Web to evaluate the work of FEWS NET (www .fews.net) and the GIEWS (www.fao.org/GIEWS). How do these organizations function in bringing food aid to people with the greatest need?

MAKING A DIFFERENCE

1. Purchase local foods. If you have the time and inclination, try cooking more of your own food.

2. If you are cautious about genetically altered foods and you want to know whether a product is organic or not, check the barcode number on the package: If the first digit is 9, the food is organic. If that number is 4, the item was a result of conventional farming. If the number is 8, then it has been genetically engineered.

3. Plant your own garden; you control all of the chemical inputs (if any), and no carbon dioxide is emitted in getting the goods onto your plate.

4. Bring your lunch to work; you'll save your own money as well as the environment.

5. Demonstrate your concern for the hungry and homeless in your city by giving your time (or donating foodstuffs) to a local soup kitchen or food pantry.

Chapter 13

Learning Objectives

13.1 The Need for Pest Control: Define the major groups of pests and the different methods we use to control them.

13.2 Chemical Treatment: Promises and Problems: Explain the serious problems that accompany overuse of new and more effective chemical pesticides, such as DDT.

13.3 Alternative Pest Control Methods: Describe the major types of alternatives to using pesticides to control pests.

13.4 Making a Coherent Plan: Explain the main principles and give examples of the integrated pest management approach to reducing the use of pesticides.

13.5 Pests, Pesticides, and Policy: List and describe the federal and international policies for controlling pests and those that regulate the use of pesticides.

Pests and Pest Control

What do the city of Lubbock, Texas, and the New York City subway have in common? In 2014, they both dealt with outbreaks of bedbugs (*Cimex lectularius*). Bedbugs, once common in the United States, had been well controlled since World War II, but have resurged to epidemic proportions in recent years. Small blood-sucking wingless insects, bedbugs are a public health nuisance. Their bites cause itchy rashes, though they do not transmit other diseases. While bedbug-sniffing dogs can be used to pinpoint infestations, getting rid of the pests can be extremely difficult. Bedbugs lay a lot of eggs, can hide in small cracks, and can go a year without feeding. Once controlled by DDT, a pesticide no longer in use in the United States, bedbugs are finding the modern world to their liking. They are spread by an increase in domestic and international travel and can be found in hotels ranging from cheap to expensive. Many are resistant to the pesticides we do use. Some bedbugs in New York City are 264 times more resistant to a common pesticide than similar bedbugs in Florida. Once established, bedbugs are terribly difficult to eradicate.

One Particularly Bad Outbreak. In 2010, a particularly bad summer outbreak of bedbugs prompted five states to ask the Department of Defense for resources to get rid of the pests. The state of Ohio petitioned the Environmental Protection Agency (EPA) for permission to use a pesticide, Propoxur, for in-home use. The petition was denied because of concerns about the effect of the chemical on children, leaving Ohio in a quandry. Bedbug research found that the foggers, or pesticide bombs, commonly used in other types of domestic infestations are ineffective against the insects. In 2012, the Centers for Disease Control and Prevention (CDC) and the EPA gave a joint statement about bedbugs as a public health concern. Among the non-chemical means of controlling bedbugs, the CDC recommends vacuuming regularly, sealing cracks, treating with heat, removing clutter, and monitoring. A dry soil called diatomaceous earth, which is non-toxic but scratches the waxy coating found on bedbugs, can also be distributed.

▲ Bedbugs are a modern scourge, highly adapted to survive attempts to get rid of them.

A Great Illustration. Bedbugs are a great illustration of the issues surrounding controlling pests: the costs of pests, the difficulties with chemical controls, and the need for a combination of treatments used in concert. The heyday of chemical pesticides as a silver bullet is over due to their environmental effects and pesticide resistance. Now we have to use new and innovative approaches to pests, whether they are found in agricultural settings, in structures, in the flour on your pantry shelf, or in beds. In this chapter, we will connect the issues of pests in our daily lives to environmental science.

13.1 The Need for Pest Control

Since earliest times, humans have suffered frustration and losses brought on by destructive pests. To this day, farmers, herdsmen, and homeowners wage a constant battle against the insects, plant pathogens, and unwanted plants that compete with them for the biological use of crops, animals, and homes.

Pests are organisms that interfere with human endeavors. Pests include a wide variety of organisms—insects, weeds, molds, pathogens, and nuisance wild animals. Of course, this is a very human-centered approach; we call a species a pest if we do not like how it affects us. In reality, pests are not a scientific category. The emphasis in this chapter is on those pests that affect agriculture, households, stored products, wood products, and structures. Other chapters consider nuisance wild animals, non-native species in natural ecosystems, and pathogens (Chapters 6, 7, and 17).

Types of Pests

Pests are often defined by the type of activity they disturb or what they live on. *Agricultural pests* are organisms that feed on agricultural crops, ornamental plants, or animals. The most notorious of these organisms are insects, but certain fungi, viruses, worms, snails, rodents, and birds also fit into the category (**Figure 13–1a–c**). *Weeds* are plants that compete with agricultural crops, forests, and forage grasses for light and nutrients (**Figure 13–1d**). Some weeds poison cattle or have other serious effects; others simply detract from the appearance of lawns and gardens. Pests also destroy important crops. In the 1990s, Russia suffered a 70% yield loss from a potato fungus, *Phythophtora infestans*, considered to be global agriculture's worst crop disease. Another fungal pest, soybean rust, has spread throughout South America, causing $2 billion in losses in Brazil alone in 2004. A 2011 plague of mice in Australia caused more than $300 million in crop losses.

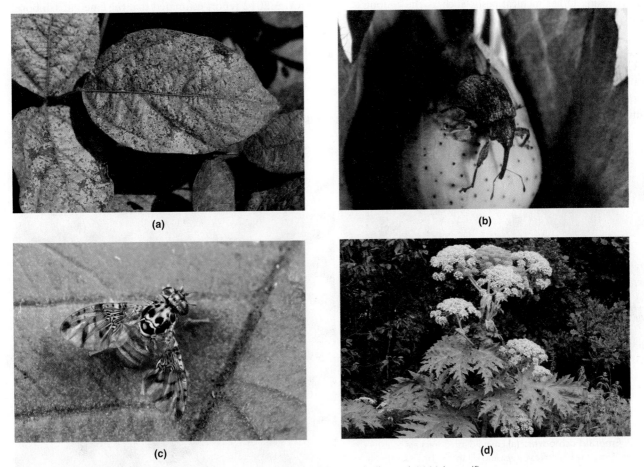

(a)

(b)

(c)

(d)

Figure 13–1 Agricultural pests and weeds. (a) Soybean rust. **(b)** Cotton boll weevil. **(c)** Male medfly. **(d)** Giant hogweed.

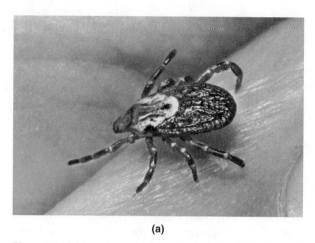

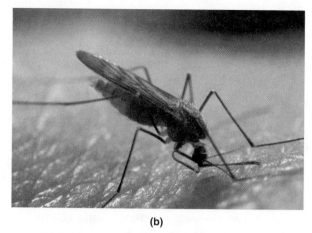

(a)

(b)

Figure 13–2 Veterinary and medical pests. (a) The dog tick is a veterinary pest. **(b)** The *Anopheles* mosquito can transmit malaria.

Veterinary pests affect domestic animals. These include the screwworm fly, for example, or fleas, ticks, and mites (Figure 13–2a). Often such pests are disease **vectors**. They may be similar to pests attacking humans, the *medical pests*. Houseflies are significant pests because they can carry diseases onto food, especially from manure sources. Darkling beetles are significant pests of poultry production worldwide; they carry human and animal diseases and destroy insulation. Mosquitoes (Figure 13–2b) carry dengue and yellow fevers and malaria; snails carry schistosomiasis, and chagas bugs carry a protozoan parasite. Other medical pests, such as the bedbug mentioned earlier, carry no diseases. The malaria-carrying mosquito, one of the most important medical pests, will be discussed at some length later (see Stewardship, *DDT for Malaria Control*) though many medical pests are discussed elsewhere (Chapter 17).

Forest pests destroy living trees and wood products. The mountain pine beetle is devastating to forest ecosystems

(Chapter 7). Other wood pests include beetles such as the emerald ash borer (Figure 13–3a). Originally from Russia, China, Japan, and Korea, the beautiful green beetle was first discovered in the United States in 2002. Since then it has spread to at least 24 states and parts of Canada. Its spread is monitored with purple traps visible along highways (Figure 13–3b). The beetle burrows into the layers directly under the bark of ash trees, destroying the layers and killing the trees. The borer threatens the nearly 7.5 billion trees (about 37.5 million in urban and suburban settings) of several ash species in North America. A 2012 report estimated that the emerald ash borer and another forest pest, the Asian longhorned beetle, cost $1.7 billion in local government expenditures annually and an additional $830 million in lost property values.

Stored-product pests live in items such as cereals and processed foods. Rice weevils and flour moths may be familiar pests in kitchen pantries. The Khapra beetle (*Trogoderma granarium*) is one you probably have not seen

(a)

(b)

Figure 13–3 Emerald ash borer. (a) Emerald ash borers are forest pests that tunnel into the bark of ash trees and kill them. **(b)** Traps for emerald ash borers are found along roads. They are important parts of monitoring the movement of the insects.

because of the efforts of federal agents to keep it out of the country. It is tiny; can eat many types of food; is resistant to many pesticides; and can even survive without food for long periods. Between 2001 and 2013, this beetle was intercepted about 200 times a year in various products at U.S. borders. Sometimes border agents use trained canines to detect such pests (**Figure 13–4**).

Biofouling organisms are organisms that settle on surfaces in aquatic environments. Some make shipping more expensive and less fuel efficient; others clog intake pipes and screens for industries. Anything in the water can be covered. Biofouling is described later (see Sound Science, *Marine Fouling Organisms: Keeping One Step Ahead of the Barnacles*). Other types of pests specialize on cloth, leather, feathers, paper, and various other products.

The Cost of Controlling Pests. Part of the credit for modern human prosperity can be attributed to pest control. Two of the most frequently used tools in this effort are **pesticides,** chemicals that kill animals and insects thought to be pests, and **herbicides,** chemicals that kill plants. The use of pesticides and herbicides is quite costly. Recently, the EPA estimated that efforts to control pests in the United States involved the use of 1.133 billion pounds (514,000 metric tons) of herbicides and pesticides annually (**Figure 13–5**), at a direct cost of $12.5 billion. According to the EPA, 5.21 billion pounds (2.4 million metric tons) of pesticides were used worldwide in 2007, at a cost of $39 billion.[1] Many of

Figure 13–4 Sniffing out pests. A U.S. Customs and Border Protection agriculture specialist works with an agricultural detector beagle. This team works to find invasive species or banned products, some containing insects known to be harmful to U.S. agriculture.

the changes in agricultural technology, such as monoculture and the widespread use of genetically identical crops, have boosted yields. Yet despite the use of pesticides, the proportion of crops lost to pests has increased. Depending on the crop and region, the losses to pests range from 20% to more than 50%.[2] During the past half-century, the use of herbicides and pesticides multiplied manyfold, leading to an unsustainable dependency on them.

[1]EPA Pesticide Sales and Usage 2006–2007 Market Estimates Report, 2011, available at http://www.epa.gov/opp00001/pestsales/07pestsales/table_of_contents 2007.htm.

[2]E. C. Oerke, "Crop Losses to Pests," *Journal of Agricultural Science 144* (2006): 31–33.

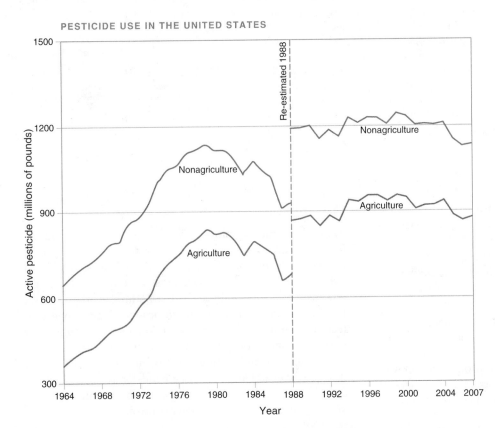

Figure 13–5 Pesticide use in the United States. Nonagricultural and agricultural uses of pesticides (active ingredients) are shown for the period 1964–2007. A change in the way totals were calculated means that values before and after 1988 are difficult to compare.

(*Source:* Data from the Environmental Protection Agency, Office of Pesticide Programs.)

SOUND SCIENCE

Marine Fouling Organisms: Keeping One Step Ahead of the Barnacles

Finding a home is a rite of passage for marine organisms that settle from a free swimming stage to a sessile (attached, sedentary) form. Barnacles are a good example; they have early life stages that move about in currents and eventually change to a form that must attach itself to a solid object in order to continue developing. This means that solid surface area is often the limiting factor for the growth of marine communities. Every ship that plies the oceans is not only a container for human use, but a moving habitat of usable real estate for small marine organisms.

These "biofouling" organisms settle on the underwater parts of ships, piers, and intake pipes for power plants, often ruining them. On the hulls of ships, biofouling increases drag, which may increase the use of fuel by as much as 40%, raising both the cost and environmental impacts of shipping. Biofouling is estimated to cost the shipping industry $60 billion a year. Governments and industry spend billions to prevent it. Solutions aren't obvious.

Antifouling measures (the prevention or removal of fouling organisms) often involve biocides: some type of toxin embedded on the surface such as in a paint. Unfortunately, tributylyin (TBT), considered the most toxic pollutant ever deliberately released into the ocean, is also the most commonly used in antifouling. You might expect that something that kills barnacles and algae would be harmful to oysters and mussels, and it is. TBT is used on more than 70% of the world's vessels because it is effective, but a ban on TBT coatings on new ships makes finding new solutions urgent. New research focuses on non-toxic coatings.

Some are polymers, which create a low-friction environment that is hard to cling to. Others make it energetically difficult for organisms to attach. They are made of chemicals that have alternating positive and negative charges that repel organisms trying to attach. Other research focuses on mimicking biology. Whales often have barnacles on them, but sharks don't. It looks like the tiny points that cover sharks may keep fouling organisms off. So a biomimetic (biology mimicking) film with tiny points similar to those in sharkskin is in the works. This is a type of nanotechnology that looks very promising.

Fouling organisms cost a great deal and waste fuel, increasing climate change. But it isn't obvious how to solve the problem without toxins; at least, it hasn't been until new research has provided alternatives. Finding these new solutions can be the fun part of the job for some scientists.

Sources: Vietti, P. "New Hull Coatings Cut Fuel Use, Protect Environment." Office of Naval Research (June 4, 2009). Accessed June 2012. http://www.eurekalert.org/pub_releases/2009-06/oonr-nhc060409.php.

Evans, S. M., T. Leksono, and P. D. McKinnell. "Tributyltin Pollution: A Diminishing Problem Following Legislation Limiting the Use of TBT-Based Anti-Fouling Paints." *Marine Pollution Bulletin* 30(1) (January 1995): 14–21.

Pests and Climate Change. It is hard to predict the exact effects climate change will have on pests, but scientists are trying to figure out trends as the world changes. Many insects, including several of medical or agricultural importance, are already increasing their ranges, and pathogens will respond to changes in precipitation, temperature, and weather events. In North America, ticks are increasing as a result of warmer temperatures. Researchers anticipate some increase in crop pests such as nematodes and weevils that attack the roots of the plantain, a crop in the tropics, as temperatures increase. In other places, drought may lower mosquito populations. Overall, scientists expect an increase in pests, but a great deal of research still needs to be done.

Different Philosophies of Pest Control

Medical practice employs several basic means of treating infectious diseases. One approach is to give the patient a massive dose of antibiotics or other medicines, hoping to either eliminate the pathogen or stop it before it can get established. Another approach is to stimulate the patient's immune system with a vaccine to produce long-lasting protection against any future invasion. Yet another is to prevent a patient from becoming ill in the first place by maintaining a healthy lifestyle and cutting down on activities that transmit disease (for example, by washing hands or avoiding sneezing on others). In practice, all of these means are often used to keep a particular pathogen under control.

The same basic philosophies govern the control of pests. Sometimes we use **chemical treatment**. Like the use of antibiotics and other medicines, chemical treatment seeks to eradicate or greatly lessen the numbers of the pest organism. Although it has had much success, this approach gives only short-term protection. Furthermore, the chemical often has side effects that are highly damaging to other organisms.

A different approach is **ecological pest control**. Like stimulating the body's immune system, this approach seeks to give long-lasting protection by developing control agents on the basis of knowledge of the pest's life cycle and its ecological relationships. Such agents, which may be other organisms or chemicals, work in one of two ways: Either they are highly specific for the pest species being fought, or they manipulate one or more aspects of the ecosystem. Ecological control emphasizes the protection of people and domestic plants and animals from damage from pests rather than eradication of the pest organism. Thus, the benefits of pest control can be obtained while maintaining the integrity of the ecosystem.

These two philosophies are combined in the approach called **integrated pest management (IPM)**. IPM is an approach to controlling pest populations by using all suitable methods—chemical and ecological—in a way that brings about long-term management of pest populations with minimal environmental impact. This approach is increasing in usage, especially where pesticides are seen as undesirable because of health risks and in developing countries where the cost of pesticides is often prohibitive.

CONCEPT CHECK Why does ecological pest control lead to fewer pests than chemical treatments do? ✓

13.2 Chemical Treatment: Promises and Problems

Pesticides are categorized according to the group of organisms they kill. There are insecticides for insects, rodenticides for mice and rats, fungicides for fungi, and herbicides for plants. None of these chemicals, however, is entirely specific to the organisms it was designed to control; they can all pose hazards to other organisms, including humans.

Development of Chemical Pesticides and Their Successes

Finding effective materials to combat pests is an ongoing endeavor. The early substances used to control agricultural pests, frequently called **first-generation pesticides**, included toxic heavy metals such as lead, arsenic, and mercury. Scientists now recognize that these substances may accumulate in soils, inhibit plant growth, and poison animals and humans. In addition, toxic heavy metals lose their effectiveness as pests become increasingly resistant to them. For example, in the early 1900s, citrus growers were able to kill 90% of injurious **scale insects** (minute insects that suck the juices from plant cells) by placing a tent over an infested tree and piping

in deadly cyanide gas for a short time. By 1930, this same technique killed as little as 3% of these pests.

The next step had its origins in the science of organic chemistry. During the 19th century, chemists synthesized thousands of organic compounds, but for the most part, these compounds sat on shelves because uses for them had not yet been found. By the 1930s, however, with agriculture expanding to meet the needs of a rapidly increasing population and with first-generation pesticides failing, farmers needed something new. In time, *second-generation pesticides*, as they came to be called, were developed as a result of synthetic organic chemistry. One of the most famous is the pesticide DDT.

The DDT Story. In the 1930s, a Swiss chemist named Paul Müller began systematically testing some organic chemicals for their effect on insects. In 1938, he hit on the chemical dichlorodiphenyltrichloroethane (DDT), a chlorinated hydrocarbon that had first been synthesized some 50 years earlier. Just traces of DDT killed flies in Müller's laboratory.

DDT appeared to be nothing less than the long-sought "magic bullet," a chemical that was extremely toxic to insects and yet seemed nontoxic to humans and other mammals. It was inexpensive to produce and effective against a broad spectrum of insect pests (Figure 13–6). In addition, it was persistent (it did not break down readily in the environment) and, hence, provided lasting protection. (DDT belongs to a group of chemicals now called *persistent organic pollutants*.) This persistence lowered cost by eliminating both the material and the labor expense of repeated treatments.

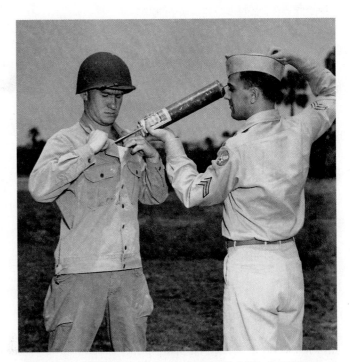

Figure 13–6 DDT in war. During and after World War II, DDT was sprayed to kill vectors of typhus, malaria, and other diseases with great success. This image from 1945 shows a U.S. soldier spraying DDT directly on a fellow soldier.

In War... During World War II, DDT quickly became indispensible. The military used it to control body lice that spread typhus fever. As a result, World War II was one of the first wars in which fewer people died of typhus than of battle wounds. On the island of Saipan, DDT helped in the fight against dengue fever. The World Health Organization (WHO) of the United Nations used DDT throughout the tropical world to control mosquitoes and greatly reduced the number of deaths caused by malaria. There is little question that DDT saved millions of lives. In fact, the virtues of DDT were so outstanding that Müller was awarded the Nobel Prize for medicine in 1948 for his discovery.

And in Peace. Postwar uses of DDT expanded dramatically. The chemical was sprayed on forests to control spruce budworm, on salt marshes to kill nuisance mosquitoes, on suburbs to control the beetles that spread Dutch elm disease. DDT even proved highly effective in controlling agricultural insect pests. Indeed, DDT was so effective that many crop yields increased dramatically. Growers could ignore other, more painstaking methods of pest control such as crop rotation and the destruction of old crop residues. They could grow more productive but less resistant varieties. They could grow certain crops in a broader range of conditions. In short, DDT gave growers more options for growing the most economically productive crop.

The success of DDT led to the development of a great variety of synthetic organic pesticides. Currently, the U.S. Environmental Protection Agency (EPA) Pesticide Program regulates more than 18,000 pesticide products, most of which are synthetic organic pesticides. Examples and some characteristics of these insecticides and herbicides are listed in Table 13–1.

Problems Stemming from Chemical Pesticide Use

Overuse of chemical pesticides has led to a wide range of environmental and human health problems. Human health effects are of particular concern, especially when pesticides get into water bodies and groundwater. People who work directly with pesticides, especially with inadequate safety gear, and children, whose developing bodies are heavily affected by toxins, are among the most vulnerable.

Adverse Human Health Effects. As with all toxic substances, pesticides can be responsible for both acute and chronic health effects.

Acute Effects. According to the American Association of Poison Control Centers, more than 88,690 persons called poison control centers after having been exposed to pesticides in the United States during 2012. Those with the most exposure were farm workers or employees of pesticide companies who came in direct contact with the chemicals. Different pesticides affect different human systems: glyphosate poisoning, for example, can lead to nausea, abdominal pain, shock, and respiratory failure; pyrethrin poisoning leads to allergic reactions, seizures, pneumonia, and coma.

There is no global documentation of pesticide poisoning; however, the WHO has recently proposed standards for diagnosis and identification of acute pesticide poisoning cases. The WHO has estimated that there are 1 million serious poisoning episodes per year leading to hospitalization, but numbers are far higher when detailed household surveys are taken. Most of these cases occur in developing countries, where regulations and training are often lax and information on pesticides may be poorly understood. The use of

Table 13–1 Characteristics of Some Pesticides

Insecticide Class	Examples	Toxicity to Mammals	Persistence
Organophosphates	Parathion, malathion, phorate, chloropyrifos	High	Moderate (weeks)
Carbamates	Carbaryl, methomyl, aldicarb, aminocarb, carbofuran	High to moderate	Low (days)
Chlorinated hydrocarbons	DDT, toxaphene, dieldrin, chlordane, lindane	Relatively low	High (years)
Pyrethroids	Permethrin, bifenthrin, esfenvalerate, decamethrin	Low	Low (days)
Herbicide Class	**Examples**	**Effects on Plants**	
Triazines	Atrazine, simazine, cyanizine	Interfere with photosynthesis, especially in broadleaf plants	
Phenoxy	2-, 4-D; 2-, 4-, 5-T; methylchlorophenoxybutyrate (MCPB)	Cause hormonelike effects in actively growing tissue	
Acidamine	Alachlor, propachlor	Inhibit germination and early seedling growth	
Dinitroaniline	Trifluralin, oryzalin	Inhibit cells in roots and shoots; prevent germination	
Thiocarbamate	Ethylpropylthiolcarbamate (EPTC), cycloate, butylate	Inhibit germination, especially in grasses	
Phosponate	Glyphosate	Inhibit amino acid synthesis; kill a wide variety of plants	

Source: Private Pesticide Applicator Training Manual, 1993, U.S. Environmental Protection Agency, Region VIII.

Figure 13–7 Pesticides used unsafely. This farmer in Thailand is spraying pesticide on soybeans, without the benefit of protective gear.

pesticides by untrained persons is considered to be the major cause of these poisonings (Figure 13–7), but in many cases, children and families come in contact with the pesticides through aerial spraying, the dumping of pesticide wastes, incorrect storage of pesticides in the home, or the use of pesticide containers to store drinking water.

Chronic Effects. Pesticides are applied to fields and orchards to reduce pest damage to crops. They are also used to protect harvested food so that it is brought undamaged to market. Because of the wide-ranging use of pesticides by most farmers, consumers are inevitably exposed to pesticide residues on their food and farmers are exposed to low levels even when they avoid acute exposure. Symptoms of pesticide poisoning (headaches, blurred vision, fainting, and nausea) are often similar to other illnesses, so exposure may not be diagnosed. The public-health concern is that pesticides might have chronic effects including the potential for causing cancer, as indicated by animal testing. In fact, evidence of carcinogenicity has been observed for many pesticides. Epidemiological evidence has implicated organochlorine pesticides in various cancers, including lymphoma and breast cancer.

Other chronic effects include dermatitis, neurological disorders including permanent brain and nerve damage, birth defects, and infertility. For example, male sterility has been clearly linked to dibromochloropropane (DBCP), once used to control nematodes but now banned. Two disorders recently associated with pesticide use are suppression of the immune system and disruption of the endocrine system. New epidemiological evidence suggests that long-term exposure to pesticides can trigger Parkinson's disease in some people. Workers in India and Russia exposed to pesticides

experienced abnormally low white blood cell counts. (White blood cells are crucial elements of the immune system.)

Endocrine Disruption. Laboratory tests have shown that a number of pesticides, including atrazine and alachlor (herbicides) and DDT, endosulfan, diazinon, and methoxychlor (insecticides), interfere with reproductive hormones. These **endocrine disruptors** are described at length elsewhere (Chapter 22). A rise in the incidence of breast cancer among humans and reports of abnormal sexual development in alligators, fish, and other animals have suggested the possibility that very low levels of a number of chemicals are able to mimic or disrupt the effects of estrogenic hormones (Chapter 10). Estrogneic hormones are sexual hormones that are highly potent at low concentrations. The Food Quality Protection Act (FQPA) of 1996 directed the EPA to develop procedures for testing chemicals for endocrine disruption activity, which it has done through the Endocrine Disruptor Screening Program (EDSP).

Adverse Environmental Effects. The story of DDT is a good illustration of the environmental impacts of pesticides as well.

There Go the Birds. In the 1950s and 1960s, ornithologists (people who study birds) observed drastic declines in populations of many species of birds that fed at the tops of food chains. Fish-eating birds such as the bald eagle and osprey (Figure 13–8) were close to extinction. Investigators at the U.S. Fish and Wildlife National Research Center near Baltimore, Maryland, showed that the problem was reproductive failure. The eggs were breaking in the nest before hatching. Furthermore, the eggs contained high concentrations of dichlorodiphenyldichloroethylene (DDE), a product of the partial breakdown of DDT by the animal's body. DDE interferes with calcium metabolism, causing birds to lay thin-shelled eggs.

Because of accumulation, small, seemingly harmless amounts received over a long period of time may reach

Figure 13–8 Osprey. Populations of fish-eating birds such as the osprey (shown here feeding its young), brown pelican, and bald eagle were decimated in the 1950s and 1960s by the effects of widespread spraying with DDT. With the banning of the pesticide, these populations have greatly recovered. The bald eagle was taken off the endangered species list in 1994.

toxic levels in one organism. This phenomenon—referred to as **bioaccumulation**—can be understood as follows: Many synthetic organic chemicals are highly soluble in lipids (fats or fatty compounds), but less soluble in water. In the body, they are stored in lipids rather than being excreted by the kidneys. Thus, synthetic organics such as pesticides and their breakdown products that are absorbed with food or water are trapped and held by the body's lipids, while the water and water-soluble wastes are passed in the urine. The body cannot fully metabolize them and has no mechanism to excrete them, so they gradually accumulate in the body and may produce toxic effects.

Up the Chain. The bioaccumulation that occurs in individual organisms may be compounded at higher trophic levels (Chapter 5). Each organism accumulates the contamination from its food, so the concentration of contaminants in its body is many times higher than that in its food. This effect concentrates contaminants as they go up the food chain. Eventually, this multiplying effect, **biomagnification**, causes extremely high levels of contaminants in the top levels of the biomass pyramid (Chapter 5). Figure 13–9 shows how DDT and its metabolic products worked their way up the food chain to the top predators.

Silent Spring. In the 1950s, Rachel Carson, a U.S. Fish and Wildlife Service biologist and an accomplished science writer, began reading the disturbing scientific accounts of the effects of DDT and other pesticides on wildlife. By 1962, Carson finished a book, *Silent Spring*, that documented the effects of the almost uncontrolled use of insecticides across the United States. Its basic message was that, if insecticide use continues as usual, there might someday come a spring with no birds—and with ominous consequences for humans as well.

Silent Spring became an instant best seller and triggered a debate that has not ended. Representatives of the agricultural and chemical industries claimed that the book was an unreasonable and unscientific account that, if taken seriously, would halt human progress. At the same time, the book was hailed as an unparalleled breakthrough in environmental understanding.

By the early 1970s, concerns about environmental and long-term health effects led to the banning of DDT in the United States and most other industrialized countries. Numerous other chlorinated hydrocarbon pesticides (for example, chlordane, dieldrin, endrin, and heptachlor) were also banned because of their propensity for

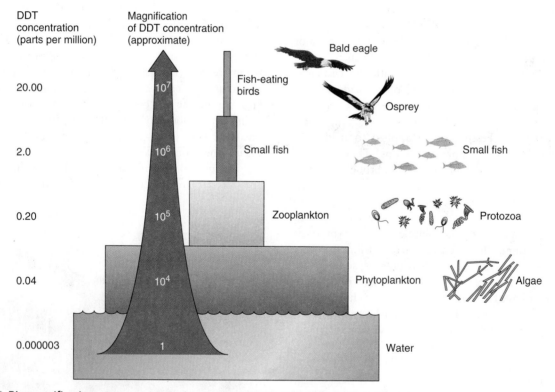

Figure 13–9 Biomagnification. Organisms on the first trophic level absorb the pesticide from the environment and accumulate it in their bodies (bioaccumulation). Each successive consumer in the food chain accumulates the contaminant to yet a higher level. Thus, the concentration of the pesticide is magnified manyfold as it moves up the food chain. Organisms at the top of the chain are likely to accumulate toxic levels.

UNDERSTANDING THE DATA

1. You have gone fishing in a lake where the concentration of a persistent pollutant in the water is 0.17 parts per billion. You catch a predatory fish that eats smaller fish; the predatory fish is at the same trophic level as fish-eating birds. Use the ratios in the diagram to calculate the likely concentration of the pollutant in the fish.

2. Does the concentration exceed 0.0025 parts per million (the upper limit for fish meat considered to be edible)?

bioaccumulation in the environment and suspected potential for causing cancer. In the years since the banning of DDT, the bird species that were adversely affected have recovered.

Rachel Carson is credited with stimulating the start of the modern environmental movement and the creation of the EPA. *Silent Spring* has become a classic, and the regulation of insecticides and other toxic chemicals is in a sense a monument to Rachel Carson, who died of cancer only two years after her book was published. (See Chapter 1 for more about Rachel Carson.)

Development of Resistance by Pests. One fundamental problem with chemical pesticides is that they gradually lose their effectiveness. Over the years, it becomes necessary to use larger and larger quantities, to try new and more potent pesticides, or to do both to obtain the same degree of control. Synthetic organic pesticides fare no better than first-generation pesticides in this respect. For example, in 1946, 2.2 pounds (1 kg) of pesticides provided enough protection to produce about 60,000 bushels of corn. By 1971, it took 141 pounds (64 kg) to produce the same amount, and losses due to pests actually *increased* during the intervening years. Resistance is not limited to agricultural pests. Stored-product pests are resistant to a fumigant used in shipments, and many medical and veterinary pests have developed pesticide resistance as well. In 2011, poison-resistant "supermice" were discovered in Spain and Germany. They appeared to have resulted from the breeding of mice from Algeria and continental Europe.

STEWARDSHIP
DDT for Malaria Control

The use of DDT for controlling malaria is one of those situations where stewardship decisions involve conflicting views of the common good. Malaria exacts a terrible toll in death and disease: 124 million to 283 million bouts of sickness and 584,000 deaths in the year 2013. The good news is that we met the Millennium Development Goal of halting the spread of malaria and reducing its incidence by 2015. In fact, malarial parasite infection rates dropped 26% between 2000 and 2013. However, the bad news is that we thought malaria would have been wiped out long ago. Fifty years ago, malaria seemed to be on the way out as new synthetic drugs killed the protozoan parasites that caused the disease and DDT wiped out mosquitoes everywhere. But resistance emerged in both parasites and mosquitoes, leaving us with fewer control options. DDT is one of a group of 12 persistent organic pollutants targeted for phasing out by the United Nations Environment Program, but it is still used against malarial mosquitoes in parts of the world because it is effective and inexpensive.

About 20 species of *Anopheles* mosquito carry malaria, which itself is caused by four different but related parasite species. Members of the *Anopheles* genus are night hunters, biting people as they rest or sleep. Many people in countries with malaria cannot afford bed nets to keep mosquitoes away. In about 11 countries, DDT is still the main line of defense against the mosquito, in part because it is inexpensive. The typical application involves spraying inside houses and eaves at a concentration of 2 g/m², once or twice a year. Even

where mosquitoes are resistant to DDT, it acts as a repellent or irritant that keeps the mosquitoes from staying around. In 2006, the WHO reversed a 30-year policy and officially put DDT on a par with bed nets and antimalarial drugs as major tools for controlling malaria.

One major downside of this approach is the difficulty of keeping DDT out of agricultural fields, where it can enter food chains and poison top predators. Another downside is the high exposure of some individuals when DDT is sprayed indoors. Alternative pesticides, such as synthetic pyrethroids, are already available, but resistance to these is appearing, too. An integrated public-health approach is needed

wherein mosquito breeding places are eliminated, bed nets are employed, and malaria cases are treated promptly to intercept the life cycle of the parasite. However, it is rare for all of these measures to be present in developing-country locales.

There is no magic bullet. The story of DDT and malaria illustrates the history of problems with chemical pesticides: development of resistance by pests, resurgences and secondary-pest outbreaks, and adverse environmental and human health effects. These problems can be extremely widespread and persistent and have led to an increased need for alternative methods of pest control.

Evolution at Work. Resistance builds up because pesticides destroy the sensitive individuals of a pest population, leaving behind only those few that already have some resistance to the pesticide. Resistance develops most rapidly in *r*-strategists with high reproductive capacity (Chapter 4). A single pair of houseflies, for example, can produce several hundred offspring that may mature and reproduce themselves only two weeks later. Consequently, repeated pesticide applications result in the unwitting selection and breeding of genetic lines that are highly resistant to the chemicals that were designed to eliminate them. The Colorado potato beetle, for example, developed resistance to 52 compounds across all chemical insecticide classes, including cyanide, in 50 years. In short, pesticide resistance simply confirms that natural selection is a powerful evolutionary force. Many major pest species are resistant to all of the principal pesticides. Plant pests can also become resistant to herbicides, as previously described (Chapter 12).

Resurgences and Secondary-Pest Outbreaks. The second problem with the use of synthetic organic pesticides is that, after a pest has been virtually eliminated with a pesticide, the pest population not only recovers, but also explodes to higher and more severe levels. This phenomenon is known as **resurgence**. To make matters worse, small populations of insects that were previously of no concern because of their low numbers may suddenly start to explode, creating a new problem called a **secondary-pest outbreak**. In China,

the extended use of Bt cotton led to a secondary outbreak of mirid bugs, insects which previously had not been major pests. (Chapter 12). Unfortunately, the species appearing in secondary-pest outbreaks quickly became resistant to pesticides, thus compounding the problem.

Some scientists use the term *pesticide treadmill* to describe attempts to eradicate pests with synthetic organic chemicals. Overuse of chemicals increases resistance and secondary-pest outbreaks, which lead to the use of new and larger quantities of chemicals, which in turn lead to more resistance and more secondary-pest outbreaks. The same can occur on weeds or stored-product pests. The use of glyphosate on GM crops and the resistance of weeds to this powerful herbicide is a good example (Chapter 12).

The chemical approach breaks down because it ignores basic ecological principles. It assumes that the ecosystem is a static entity in which one species, the pest, can simply be eliminated. In reality, the ecosystem is a dynamic system of interactions, and a chemical assault on one species perturbs the system and produces other, undesirable effects. Populations of plant-eating insects are frequently held in check by other insects or spiders that parasitize or prey on them (Figure 13–10). Pesticide treatments often have a greater impact on these natural enemies than on the plant-eating insects they are meant to control. Consequently, with their natural enemies suppressed, the populations of both the original target pest and other plant-eating insects explode.

Figure 13–10 Invertebrate food chains.
(a) Food chains exist among insect pests, their food and their predators, just as they do among higher animals. **(b)** Ant lions are the larvae of lace-winged insects in the family Myrmeleontidae. This ant lion is using its powerful jaws to capture an ant. **(c)** Spiders eat many pest species. This huntsman spider is eating a grasshopper.

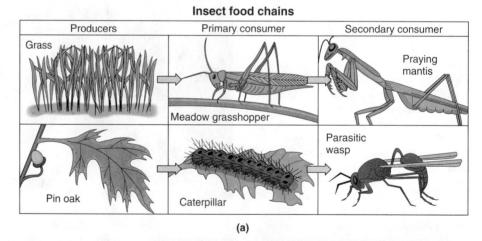

Insect food chains

Producers	Primary consumer	Secondary consumer
Grass	Meadow grasshopper	Praying mantis
Pin oak	Caterpillar	Parasitic wasp

(a)

(b)

(c)

Nonpersistent Pesticides: Are They the Answer? A key characteristic of chlorinated hydrocarbons is their persistence; that is, they take a long time to break down. Lingering in the environment for years, they can contaminate many organisms and become biomagnified. DDT, for example, has a half-life on the order of 20 years. Because the persistent pesticides have been banned, the agrochemical industry has substituted nonpersistent pesticides for the banned compounds. Synthetic organic phosphates such as malathion, parathion, and chlorpyrifos, as well as carbamates such as aldicarb and carbaryl, have been used extensively in place of chlorinated hydrocarbons (see Table 13–1). These compounds are potent inhibitors of the enzyme cholinesterase, which is essential for proper functioning of the nervous system in all animals. They break down into simple nontoxic products within a few weeks after their application. Thus, there is no danger of their migrating long distances through the environment and affecting wildlife or humans long after being applied.

Toxicity. For several reasons, however, nonpersistent pesticides are not as environmentally sound as they might appear. First of all, they are persistent enough to harm non-target organisms. The total environmental impact of a pesticide is a function not only of its persistence, but also of its toxicity, its dosage, and the location where it is applied. Many of the nonpersistent pesticides are far more toxic than DDT. They present a significant hazard to agricultural workers and others exposed to these pesticides. For instance, the organophosphates are responsible for an estimated 70% of all pesticide poisonings. The law requires the EPA to develop new health-based standards that address the risk of children's exposure to such pesticides, which has led to the banning of several chemicals from a range of home garden and agricultural uses.

Bird Kill. Nonpersistent pesticides may still have far-reaching environmental impacts. For example, carbofuran is a pesticide widely used on alfalfa, rice, and soybeans. It is spread on the ground and absorbed into plants through their roots. Unfortunately, it is highly toxic to birds and other vertebrates. In the '80s, the EPA estimated that 1 million to 2 million birds were killed annually by carbofuran. In one incident, 800 to 1,200 ducks and geese died after treatment of an alfalfa field with carbofuran. Because of its toxicity to humans and wildlife, the EPA announced in 2008 that it would no longer allow residues of carbofuran on any food, essentially revoking all uses. The EPA decision was based primarily on the grounds that the pesticide posed an unacceptable risk to toddlers.

Another hazard of nonpersistent pesticides is that desirable insects may be just as sensitive to them as pest insects are. Bees, for example, which play an essential role in pollination, are highly sensitive to them. Thus, the use of these compounds creates an economic problem for beekeepers and jeopardizes pollination. Also, regular spraying of neighborhoods with malathion to control mosquitoes leads inevitably to great declines in butterflies and fireflies. Many pesticides also harm predatory insects and spiders, which would otherwise keep pests down.

Finally, nonpersistent chemicals are just as likely to cause resurgences and secondary-pest outbreaks as persistent pesticides are, and pests become resistant to nonpersistent pesticides just as readily.

Synergistic Effects. As scientists try to figure out what is happening in ecosystems, the answer may be complicated by multiple factors that work together to create an unexpected outcome, something called a **synergistic effect.** A possible example is the mystery of what is killing pollinators. Pollinators are in decline worldwide. Some of the deaths are a phenomenon known as honeybee colony collapse disorder (CCD), which came into the news in 2006–2007 when beekeepers discovered their hives dying and their worker bees mysteriously gone (Figure 13–11). Meanwhile, other pollinators such as bumblebees and various types of solitary bees are also in decline. Scientists know CCD is not a simple case of acute pesticide poisoning because in the case of acute poisonings, worker bees are found dead at the hive. In CCD cases, the worker bees are nowhere to be found.

Scientists hypothesize that CCD is the result of the synergistic effects of a mixture of pests, pathogens, and chemicals that weaken or confuse bees. Several parasitic and pathogenic factors may be involved, including the invasive varroa mite, viruses, a gut parasite, poor nutrition, and stress. Pesticides may interact with these other factors. Several recent studies support the involvement of pesticides in the pollinators' decline. In one study on honeybees, researchers found that bees exposed to nectar containing a type of pesticide called a neonicotinoid had more difficulty finding their way back to the hive than those exposed to a control. In another study, bumblebees treated with neonicotinoids had lower growth rates and produced fewer queens than unexposed bees. In yet another study, honeybees were found to eat less food and perform fewer communication dances when exposed to pesticides. In solving the mystery of CCD and other problems with pollinators, scientists are finding that chemicals may contribute to many other problems, causing a perfect storm of variables.

CONCEPT CHECK How does resistance to pesticides evolve? What are two ways of slowing the development of pesticide resistance in an insect population? ☑

Figure 13–11 Pollinators in decline. Colony collapse disorder (CCD) killed the bees in this hive in Florida. CCD has caused a massive, worldwide die off of bees that has puzzled scientists.

13.3 Alternative Pest Control Methods

Numerous ecological and biological factors affect the relationship between a pest and its host. *Ecological control* seeks to manipulate one or more of these natural factors so that crops and people are protected without jeopardizing environmental and human health. This natural approach depends on an understanding of the pest and its relationship with its host and with its ecosystem. The more we know about the organisms involved, the greater are our opportunities for ecological control.

The life cycle typical of moths and butterflies is shown in Figure 13–12. Many groups of insects have a similarly complex life cycle. The development of each stage may be influenced by numerous abiotic factors, and at each stage, the insect may be vulnerable to attack by a parasite or predator. The proper completion of each stage depends on internal chemical signals provided by insect hormones. Locating mates, finding food, and other behaviors depend on external chemical signals. All these findings suggest ways in which pest populations may be controlled without resorting to synthetic chemical pesticides.

Next we will consider the four general categories of ecological pest control: **cultural control**, control by **natural enemies**, **genetic control**, and **natural chemical control**.

Cultural Control

A cultural control is a nonchemical alteration of one or more environmental factors in such a way that the pest finds the environment unsuitable or is unable to gain access to its target. This type of control can be practiced on all scales—by a homeowner, a farmer, or even a government.

Control of Pests Affecting Crops, Gardens, and Lawns. Weed problems in lawns are frequently a result of cutting the grass too short. If grass is left at least 3 inches (8 cm) high, it will usually maintain a dense enough cover to keep out most crabgrass and other noxious weeds. Thus, monitoring the height of grass is a form of cultural weed control. In gardens, a clean mulch of material such as grass clippings or hay will keep down the growth of weeds while also protecting the soil from drying and erosion. A gardener may also control pests by avoiding plants that act as attractants or by planting ones that act as repellants for certain pests. Roses attract several kinds of common pests, while marigolds and chrysanthemums are justly famous for repelling insects.

Some parasites require an alternative host, so they can be controlled by eliminating that host (Figure 13–13). Also, hedgerows, fencerows, and shelterbelts can provide refuges where natural enemies of pests (birds, amphibians, praying mantises, and so on) can be maintained.

Figure 13–12 Complex life cycle of insects. Like the moth shown here, most insects have a complex life cycle that includes a larval stage and an adult stage. Biological control methods recognize the different stages and attack the insect, using knowledge of its specific needs and behaviors at each stage.

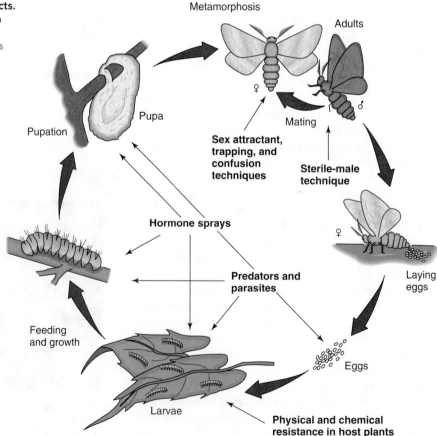

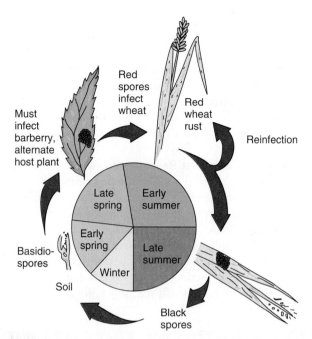

Figure 13–13 Cultural control. Part of the life cycle of wheat rust, a parasitic fungus that is a serious pest on wheat, requires that the rust infest barberry, an alternate host plant. The elimination of barberry in wheat-growing regions has been an important cultural control.

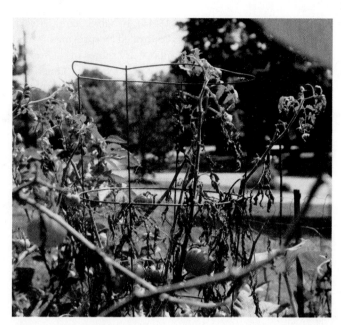

Figure 13–14 Tomato wilt. This tomato plant has been devastated by tomato wilt. Growing tomatoes in the same plot maintains wilt spores in the soil, guaranteeing that wilt will strike the plants every year. (*Source:* Author's garden.)

Crops. Insects and the spores of plant pathogens may overwinter or complete part of their life cycle in the dead leaves, stems, or other plant residues that remain in the fields after harvesting. Therefore it is important to manage any crop residues that are not harvested. Plowing under or burning the material may be quite effective in keeping pest populations to a minimum. Crop rotation—the practice of changing crops from one year to the next—may provide control because pests of the first crop cannot feed on the second crop (Figure 13–14) and vice versa. Similarly, breaking up monocultures with mixed plantings lowers the crop loss to any one pest. Trap cropping involves planting a pest's preferred host plant near a crop to lure pests away from the crop. Finally, refuge plants (untreated plants near treated crops) allow natural enemies to survive. Natural enemies of pests are maintained in the uncultivated strips. This approach is also used with insect-resistant GM crops to keep all of the insects from being exposed to the GM crop and becoming more resistant to the crop (Chapter 12).

Stopping Imports. Most of the pests that are hardest to control were unwittingly imported from other parts of the world. Therefore, it is important to keep would-be pests out of the country. This is a major function of U.S. Customs and Border Protection and of the agriculture departments of some states. Biological materials that may carry pest insects or pathogens are either prohibited from crossing the border or subjected to quarantines, fumigation, or other treatments to ensure that they are free of pests. A recent example is the Asian longhorned beetle, which arrived in packing crates from China. The beetle infests healthy maple, birch, poplar, and ash trees and eventually kills them. As a result of this pest, new regulations require that wooden crates and pallets from China be kiln-dried or chemically treated. The Animal and Plant

Health Inspections Service (APHIS) of the USDA has the job of tracking and stopping the flow of pests into the United States.

Cultural Control of Pests Affecting Humans. We routinely practice many forms of cultural control against diseases and parasitic organisms. For instance, disposing properly of sewage and avoiding drinking water from unsafe sources are cultural practices that protect against waterborne disease-causing organisms. Brushing hair, bathing, laundering clothes and linens, and cleaning rugs and furniture are all cultural practices that eliminate head and body lice, fleas, bedbugs, and other parasites. Bed nets, which are widely used in the tropics, offer a barrier that keeps mosquitoes away from sleepers. Special covers for mattresses can cut down on bedbugs and mites. Garbage disposal, housekeeping, and sealing cracks in buildings help keep down populations of roaches, mice, rats, flies, mosquitoes, and other pests. Sanitation in handling food as well as refrigeration, freezing, canning, and drying of foods are cultural controls that inhibit the growth of organisms that cause rotting, spoilage, and food poisoning.

If these practices are compromised, as they usually are in any major disaster, there is the very real danger of widespread mortality resulting from outbreaks of parasites and diseases (see Chapter 17).

Control by Natural Enemies

There are four well-defined types of natural enemies: predators, parasitoids, pathogens, and herbivores. This type of control is used all over the world. In the eastern United States, tiny predator beetles have been imported in an attempt to control the invasive, non-native wooly adelgid, which is killing off hemlock trees. In Africa, mealybugs are controlled by parasitic wasps; these types of wasps are also effective in controlling caterpillar populations (Figure 13–15). In sub-Saharan Africa, the swarming desert locust was controlled by spraying infested fields with "Green Muscle," a mix of dry spores of the fungus *Metarhizium anisopliae* with oil. Water hyacinth, which has blanketed many African lakes, is coming under control following the introduction of Brazilian weevils. More than 30 weed species worldwide are now limited by insects introduced into their habitats.

Protect the Natives. The challenge involved in using natural enemies lies in finding organisms that provide control of the target species without attacking other, desirable species. Entomologists estimate that of the 50,000 known species of plant-eating insects that have the potential for being serious pests, only about 1% are actually causing problems. The populations of the other 99% are held in check by one or more natural enemies. Therefore, the first step in using natural enemies for control should

be *conservation*—protecting the natural enemies that already exist. This means avoiding the use of broad-spectrum chemical pesticides, which may affect natural enemies of the target species even more than they do the target pests themselves. Restricting the use of broad-spectrum chemical pesticides will often allow natural enemies to reestablish themselves and control *secondary pests*—those that become serious problems only after the use of pesticides.

Import Aliens as a Last Resort. Effective natural enemies are not always readily available. In many cases, the absence of a pest's natural enemies is the result of accidentally importing the pest without also importing its natural enemies. Quite often, effective natural enemies have been found by systematically combing the home region of an introduced pest to find its various predators or parasites.

Once an enemy species has been identified, it must be carefully tested before intentionally releasing it, something that has

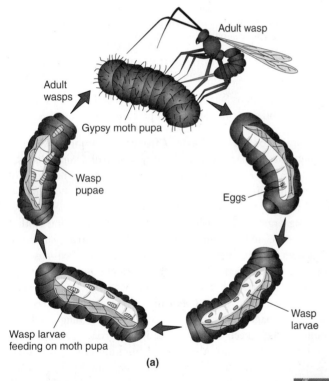

(a)

(b)

(c)

Figure 13–15 Parasitic wasps. (a) The life cycle of the parasitic wasp that uses the gypsy moth as its host. **(b)** A wasp depositing eggs in a gypsy moth larva. **(c)** Another insect parasite, a braconid wasp, lays its eggs on the pest known as the tomato hornworm, which is the larva of the sphinx moth. The wasp larvae feed on the caterpillar, and shortly before the caterpillar dies, they emerge and form the cocoons seen here.

Figure 13–16 **Cane toad.** Introduced into Australia to control beetles in sugar cane fields, the cane toad (*Bufo marinus*) quickly became a pest.

Figure 13–17 **Hessian fly.** This pest of wheat is controlled by maintaining wheat plants that produce a chemical that is toxic to the fly.

not always been done well. The cane toad (Figure 13–16), imported into Australia in 1935 to control beetles in sugar cane fields, overran Queensland, eating, outcompeting, and poisoning many native species without ever controlling the beetles that prompted its original release. On the other hand, the cassava mealybug, *Phaenococcus manihoti*, ran rampant across Africa, destroying as much as 50% of a crop that was a staple for 200 million people, until it was successfully controlled by the introduced parasitoid wasp *Epidinocarsis lopezi*. Modern releases are much more likely to succeed than earlier releases because of better science and more rigorous controls.

Plant Breeding

Most plant-eating insects and plant pathogens attack only one species or a few closely related species. This specificity implies a genetic incompatibility between the pest and any species that are not attacked. Most genetic-control strategies are designed to develop genetic traits in the host species that provide the same incompatibility—that is, resistance to attack by the pest. The technique has been used extensively in connection with plant diseases caused by fungal, viral, and bacterial parasites.

Unfortunately, the battle between agricultural scientists and plant diseases never ends. In the 1950s, plant breeders developed varieties of wheat that were resistant to wheat stem rust, a parasitic fungus that used to devastate fields of wheat. However, a new strain of the fungus appeared in Africa in the early 2000s and is spreading eastward into Asia, infecting many varieties of wheat in the developing world.

Other plants are bred to produce substances that are lethal or at least repulsive to the would-be pest. One example is the resistance of new varieties of wheat to the Hessian fly (Figure 13–17). The fly lays its eggs on wheat leaves, and the larvae move into the main stem, weakening it. Scientists at the University of Kansas developed a variety of wheat that produces a chemical that is toxic to the Hessian fly; this reduced losses to this pest to less than 1% where resistant varieties have been planted for several years.

Plants can be bred to have physical traits that provide physical barriers and structural traits that impede the attack of a pest. For example, hooked hairs on the leaf surfaces of some plants tend to trap and hold insects, and glandular hairs that exude a sticky substance fatally entrap others (Figure 13–18).

Biotechnology. As we have seen elsewhere, biotechnology techniques can be used to incorporate pest control into plants (Chapter 12). One promising strategy is to incorporate the gene for the protein coat of a plant virus into the plant itself. When the plant manufactures the virus protein coat, the plant becomes resistant to infection by the real virus. In this way, crop plants have been made resistant to more than a dozen plant viruses. Another strategy is to craft a chemical that interferes with an insect's normal cell function, silencing a gene that controls molting and sidetracking that process—thus, a gene-silencing pesticide.

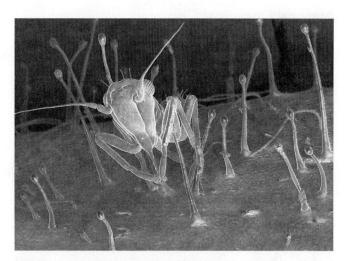

Figure 13–18 **Alfalfa glandular hairs.** This scanning electron micrograph shows an immature potato leafhopper trapped by sticky glandular hairs on the stem of a resistant alfalfa strain.

Figure 13–19 Bioengineered potatoes. A researcher holds two plants of a potato cultivar Fortuna. The left plant is not genetically modified, while the plant on the right is resistant to *Phytophora infestans*, a plant pest.

Bt. The bacterium *Bacillus thuringiensis* (Bt) produces a potent protein that kills the larvae of a number of plant-eating insects but is harmless to mammals, birds, and most other insects. Scientists have engineered the gene that produces this protein into a number of plants, including cotton, potatoes, and corn (Figure 13–19). Unfortunately, resistance to Bt by some pests and outbreaks of secondary pests are dampening the success of Bt cotton (Chapter 12). Anticipating the development of resistance because of the heavy exposure of pests to the Bt toxin, the EPA has required seed companies to verify that farmers plant 20% of their fields with non-Bt versions of the crop. The non-Bt fields provide a refuge for pests that are susceptible to the Bt toxin, allowing these insects to interbreed with potentially resistant colleagues from the Bt acreage and, hopefully, dilute resistance genes in the insect populations. Bt toxin is also used directly as a natural chemical control, described next.

Natural Chemical Control

Natural chemicals can be compounds that are used directly as biocides (such as Bt toxin), pesticides made from plants, chemicals used to mimic communication chemicals, or chemicals that boost immunity.

Some natural chemicals mimic those used to communicate internally or between members of a species. Some species communicate via **pheromones**—chemicals secreted by one individual that influence the behavior of another individual of the same species. Once identified and synthesized, pheromones may be used to trap or confuse pests. In the trapping technique, the pheromone is used to lure males or females into traps or into eating poisoned bait (Figure 13–20). In the confusion technique, the pheromone is dispersed over the field in such quantities that males become confused, cannot find females, and thus fail to mate. Research has identified a large number of pheromones, and these natural chemicals are now key tools employed in IPM on many insects, such

as the cotton boll weevil and the pink bollworm for cotton, the codling moth for pears and apples, the tomato pinworm for tomatoes, and several bark beetles that infest forest trees, among others.

Hormones are chemicals produced in organisms that provide "signals" that control developmental processes and metabolic functions. Hormones or hormone mimics can be used to disrupt the life cycle of a pest. Two advantages of such natural chemicals are that they are nontoxic and that they are highly specific to the pest in question. (They do not affect the natural enemies of the pest to any appreciable extent.) If the affected pest is eaten by another organism, it is simply digested. For example, an insecticide called Mimic® is a synthetic variation of the molting hormone of insects. Mimic® triggers the molting process to begin in insect larvae but not to continue, which eventually kills the larva. Mimic is specific to moths and butterflies and is viewed as a potent agent in the control of the spruce budworm, the gypsy moth, the beet armyworm, and the codling moth—all highly devastating pests.

Natural chemicals can also be used on crops to boost plant defenses. When plants (including tomato, maize, sweet pepper, and wheat) are grown from seeds dipped in jasmonic

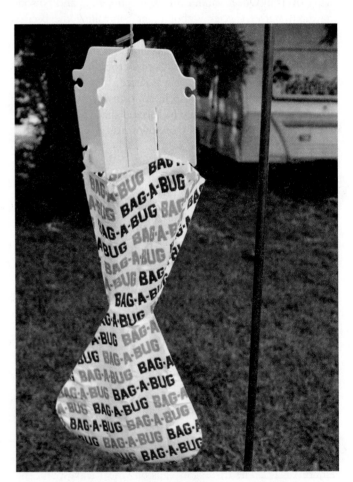

Figure 13–20 Trapping technique. Adult Japanese beetles are lured into a trap by scented bait they find attractive. Once inside the trap, baffles prevent the beetles from escaping. These traps are available in hardware stores and are quite effective. (Japanese beetles are notorious lawn grubs as larvae; as adults, they are voracious consumers of garden flowers and vegetables.)

acid, a naturally occurring plant compound, aphid attacks and caterpillar damage decline. These early results may lead to promising new directions. The spraying of Bt toxin into water or onto crops rather than incorporation into genetically modified plants is another example of natural chemical control.

Sterile-Insect Techniques

Another strategy for controlling insects involves flooding a natural population with sterile males that have been reared in laboratories. Combating the screwworm fly provides a prime illustration of this *sterile-insect technique*. This fly, which is closely related to the housefly, lays its eggs in open wounds of cattle and other animals, causing pain, infection, and often death. Early in the 20th century, the problem became so severe that cattle ranching from Texas to Florida and northward was becoming economically impossible.

Screwworm flies have two essential features that make control possible: their populations are never very large, and female flies mate only once. If a female mates with a sterile male, no offspring are produced. Today sterile males are routinely used to control this pest. After huge numbers of screwworm larvae are grown in laboratories, the resulting pupae are subjected to just enough high-energy radiation to render them sterile. These sterilized pupae are then air-dropped into the infested area. Ideally, 100 sterile males are dropped for every normal female in the natural population, giving a 99% probability that wild females will mate with one of the sterile males. This technique was successful in eliminating the screwworm fly from Florida, Mexico, and most of Central America. The sterile-insect technique has saved billions of dollars for the cattle industry.

The sterile-insect technique has been used successfully on a number of agriculturally important pests. Recently, the technique was used to eradicate the tsetse fly from the island of Zanzibar, thus eliminating the disease trypanosomiasis (sleeping sickness), which the tsetse fly carries.

CONCEPT CHECK   List three possible ways to use natural chemicals to deter insect pests. ☑

13.4 Making a Coherent Plan

The increasing availability of alternative pest control methods and the accumulating evidence of the failures of chemical treatment have led to a growing movement away from the use of pesticides, particularly on foods. This movement must contend with a strong tendency on the part of farmers to employ pesticides.

Pressures to Use Pesticides

A species becomes a pest only when its population multiplies to the point of causing significant damage. Natural controls are generally aimed at keeping pest populations below damaging levels, not at total eradication of the populations. By keeping pest populations down, natural controls avert significant damage while preserving the integrity of the ecosystem.

The question to be asked when facing any pest species is: Is the species causing significant damage? Damage should be deemed significant only when the economic losses due to the damage considerably outweigh the cost of applying a pesticide. This point is called the **economic threshold** (Figure 13–21). If significant damage is not occurring, natural controls are already operating, and the situation is probably best left as is. Spraying with synthetic chemicals at this stage is more than likely to upset the natural balance and make the situation worse through resurgences. It is also not cost effective.

If, however, significant damage *is* occurring, a pesticide treatment may be in order. A grower who believes that his or her plantings are at risk may resort to **insurance spraying** (the use of pesticides to prevent losses to pests) such as occurs on European apple crops, where the most important threats are mildew and scab—airborne diseases that, once established, cause significant economic damage. Farmers typically employ insurance spraying to control these diseases.

Consumers also put indirect pressure on growers to use pesticides because they select the best-looking fruits and vegetables, leaving the remainder to be sold at lower prices or to be discarded. Growers know that blemished produce means less profit, so they indulge in **cosmetic spraying**—the use of pesticides to control pests that harm only the item's outward appearance. Cosmetic spraying accounts for a significant fraction of pesticide use, does nothing to enhance crop yield or nutritional value, and results in an increase in pesticide residues remaining on the produce.

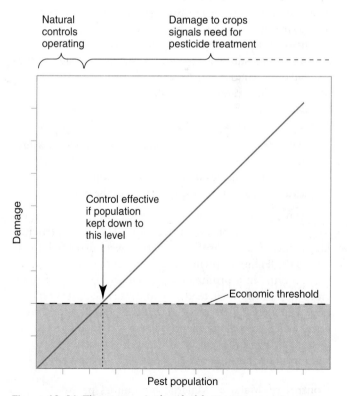

Figure 13–21 The economic threshold. The objective of pest control should not be to eradicate a pest totally. All that is needed is to keep population levels below the economic threshold—the point at which the damage caused by a pest considerably outweighs the cost of applying a pesticide.

Integrated Pest Management

IPM aims to minimize the use of synthetic organic pesticides without jeopardizing crops. This is made possible by addressing all the interacting sociological, economic, and ecological factors involved in protecting crops. With IPM, the crop and its pests are seen as part of a dynamic ecosystem. The goal is not eradicating the pests, but maintaining crop damage below the economic threshold. The EPA uses a four-tiered approach in describing IPM:

1. **Set action thresholds.** With the economic threshold in mind, IPM identifies a point at which pest populations or environmental conditions indicate that some control action is needed.

2. **Monitor and identify pests.** Pest populations are monitored, often by persons employed by local agricultural extension services or farm cooperatives or by persons acting as independent consultants. These **field scouts** are trained in identifying and monitoring pest populations (traps baited with pheromones are used for this purpose) and in determining whether the population exceeds the economic threshold.

3. **Prevent pests.** Cultural and biological control practices form the core of IPM techniques. Practices such as crop rotation, polyculture (instead of monoculture), the destruction of crop residues, the maintenance of predator populations, refuge plants, and carefully timed planting and fertilizing are basic to IPM. "Trap crops" are often used. That is, early strips of a crop such as cotton are planted to lure existing pests and then the pests are destroyed by hand or by the limited use of pesticides before they can reproduce. For aquatic disease vectors such as mosquitoes, getting rid of breeding sites such as old tires, containers, water-filled ruts, and other breeding grounds is important. For rodents, making sure food and garbage are not available is part of prevention.

4. **Control pests.** If the preceding steps indicate that pest control is needed in spite of the preventive methods, measures are taken to decrease the pest population. Here selected brands of pesticides may be used in quantities that do the least damage to the natural enemies of the pest.

Making the economic benefits of natural controls known to growers is an important aspect of IPM. Because pesticides increased yields and profits when they were first used, many farmers still cling to them, believing that they offer the only way to bring in a profitable crop. The rising costs of pesticides, along with their tendency to aggravate pest problems, have eliminated their economic advantage. Pest-loss insurance, which pays the farmer in the event of loss due to pests, has enabled insured growers in developed countries to refrain from unnecessary and costly insurance spraying.

Control of Malaria. Malarial mosquitoes provide a good example of the value of research in IPM. Control now consists of a combination of cultural controls such as pesticide-impregnated nets and removal of mosquito breeding sites along with chemical sprays, Bt, and encouragement of natural enemies such as birds and predatory insects. There is hope that current research will yield even more solutions. In the case of pesticide-resistant mosquitoes, scientists have found that a pathogenic fungus can kill mosquito larvae. Other research has shown that infection with a bacterium that is found in many insects doesn't kill malarial mosquitoes but keeps them from passing along the malarial parasite as effectively. Recently, scientists have been able to rear large numbers of mosquitoes in a lab environment cheaply in order to try the sterile-male technique. In other research, scientists have found a virus that kills older mosquitos. Only older mosquitoes can pass on malaria. Consequently, killing them as adults but before they are ready to transmit the disease, rather than constantly killing the youngest stages, will cut malarial transmission and at the same time slow the development of pesticide resistance.

IPM in Indonesia. In the developing world, governmental agricultural policy often determines the extent to which IPM is adopted. Governments and aid agencies usually subsidize the purchase of pesticides, thereby strongly encouraging growers to step onto the pesticide treadmill. By contrast, Indonesia has provided a viable IPM model for other rice-growing countries. Faced with declining success in controlling the brown plant hopper (Figure 13–22), Indonesian rice growers have switched from heavy pesticide spraying to a light spraying regime that preserves the natural enemies of the insect. The success of the program can be traced to close cooperation between the Indonesian government and the FAO (Figure 13–23). FAO workers conducted training sessions for farmers that lasted throughout the growing season. Eventually, more than 200,000 farmers were taught about the rice agro-ecosystem and IPM techniques, and these farmers formed a corps that could in turn teach others.

The economic and environmental benefits of the Indonesian IPM program have been remarkable. The government has saved millions of dollars annually by not purchasing pesticides, farmers have not had to invest in pesticides and spraying equipment, the environment has been spared the application of thousands of tons of pesticide, fish are once again thriving in the rice paddies, and the health benefits of reduced pesticide use have been spread from applicators to consumers and wildlife. The FAO is sponsoring similar IPM training programs in many other developing countries.

Figure 13–22 Brown plant hoppers. Immature brown plant hoppers (*Nilaparvata lugens*), shown on the stem of a rice plant.

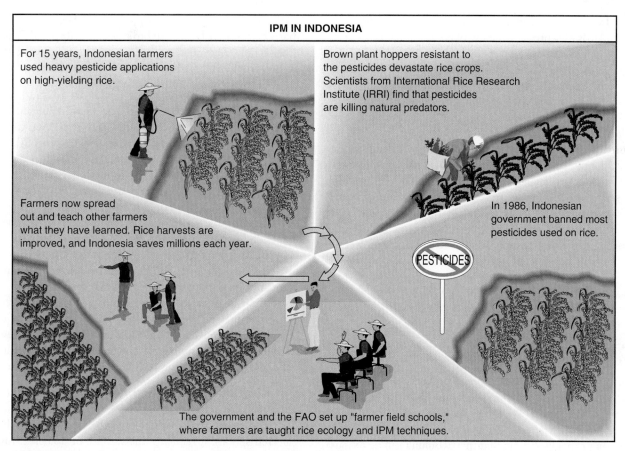

IPM IN INDONESIA

For 15 years, Indonesian farmers used heavy pesticide applications on high-yielding rice.

Brown plant hoppers resistant to the pesticides devastate rice crops. Scientists from International Rice Research Institute (IRRI) find that pesticides are killing natural predators.

In 1986, Indonesian government banned most pesticides used on rice.

PESTICIDES

Farmers now spread out and teach other farmers what they have learned. Rice harvests are improved, and Indonesia saves millions each year.

The government and the FAO set up "farmer field schools," where farmers are taught rice ecology and IPM techniques.

Figure 13–23 Integrated pest management. IPM has helped Indonesian rice farmers bring the brown plant hopper under control after years of frustration on the pesticide treadmill.

Organically Grown Food

Have you ever purchased meat, fruit, or vegetables with a label like the one shown in Figure 13–24? For consumers concerned about pesticides in their food, **organic food** is an attractive option. Many farmers are turning away from the use of pesticides, chemical fertilizers, antibiotics, and hormones as they raise grains, vegetables, and livestock for the organic food trade. Typically, organic farms are small, employ traditional farming methods with diverse crops, and are tied to local economies. For the farmers, the organically raised crops often represent lower crop yields, but also lower expenses. Sales of organics have increased rapidly over the past 15 years. In the United States, the sale of organically grown food has passed the $33 billion mark (up from $1 billion in 1990) and is more than $63 billion worldwide. Organic foods are more costly, but recent tests have shown that these foods are far less likely to have any pesticide residues, and those that do have residues contain much lower amounts than conventionally grown foods. Most residues in organic foods came from older, persistent pesticides, such as DDT, lingering in the soil.

USDA Organic. The certification of organic foods has been a contentious process. In 1990 Congress passed the **Organic Foods Protection Act**, which established the National Organic Standards Board (NOSB), under USDA auspices. After many years of work, proposed guidelines for certification from the NOSB were aired in early 1998. Possibly in response to lobbying from the food industry, the board allowed genetically engineered foods, irradiated foods (foods subjected to radiation to kill bacteria), and crops fertilized with sewage sludge to be included as organic. A firestorm of protest from the organic-food community ensued. The NOSB went back to work, and in late 2002, the USDA published its final standards for certifying organic foods. As a result, no product involving genetically engineered or irradiated foods or foods

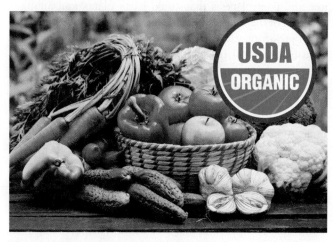

Figure 13–24 Organic foods. This symbol on food signals that the food or crop has been produced on a certified organic farm.

fertilized with sewage sludge can be certified as 100% organic. The standards also prohibit the use of conventional pesticides, antibiotics, growth hormones, and chemical fertilizers. To market certified organic foods, farmers now must have their operations scrutinized by USDA-approved inspectors. Only then may they use the USDA seal on their products. If the food is at least 95% organic, it may also display the seal, but may not claim to be 100% organic.

Because it is expensive and relatively difficult to be certified organic, many farmers who use significantly less pesticides and artificial fertilizers are not certified. In addition, some organic farms are extremely large and distant, so the environmental costs of transportation, storage, and packaging of foods remain. For this reason, buying local and supporting less toxic farming benefit the environment even if organic certification is not available.

CONCEPT CHECK   What are three problems organic farming might help solve? What is one problem it might increase? ☑

13.5 Pests, Pesticides, and Policy

When you go to the grocery store and purchase food, most likely you assume it will have minimal pesticide residues. You might even think, "Well, they wouldn't sell it if it wasn't safe." Likewise, when you go to a plant nursery and buy a plant, you probably assume that they wouldn't sell a noxious weed pest that is likely to invade your state and cost farmers millions of dollars. Both of these assumptions are at least partially true. While the fact that something is sold doesn't guarantee its safety, the United States has a wide range of guidelines followed by state and federal agencies to monitor, eradicate, and prevent pests and to protect food safety. There are also international treaties designed to lower the impact of pests and prevent the misuse of dangerous pesticides. Let's look at regulation of pests and then pesticides.

Regulation of Pests

In 1972, the United States Animal and Plant Health Inspection Service (APHIS), a part of the USDA, was established to monitor and control pests that could harm plants and animals. The jobs of the agency have changed over time. It has consolidated jobs formerly done by multiple agencies, including the Bureau of Animal Husbandry and various plant bureaus. It includes regulation of animals used in research and exhibition and by dealers, a function created after the Animal Welfare Act of 1966.

One of the jobs of APHIS is to monitor potential pests in order to act quickly and prevent infestations that would require more pesticide use later (Figure 13-25). One example has been its intensive monitoring of soybean rust, a fungus that has devastated soybean crops in Brazil. When the pathogen was blown into the United States by winds from Hurricane Irene and began to spread northward, APHIS had to act quickly. The agency set up a continual Web-based tracking system with records of all counties where the pest was discovered. APHIS organized governmental agencies, universities, and farmers to track the pest, using *sentinal plots,* small fields planted early that were rigorously surveyed to track fungal spread. This and other actions were effective, and the environment was different enough from its native Brazil that the fungus has never reached the epidemic proportions here that it has there.

Plant Protection Act. Plants have been regulated in the United States since the 1912 Plant Quarantine Act, the 1957 Federal Plant Pest Act, and the Federal Noxious Weed Act of 1974. These functions were consolidated into the Plant Protection Act of 2000, which, among other things, allowed APHIS to crack down on illegal smuggling of plants and agricultural products that could harbor pests.

One program APHIS runs is the noxious weed program. Plants designated as noxious weeds cannot be moved across state lines, imported, or exported. States often have their own noxious weed lists and may have additional regulations. Plants not on the noxious weed list may also be pests.

APHIS also tracks plant pests and regulates imports and exports to prevent movement. In 2012, APHIS was carefully regulating the movement of ash tree products and firewood to allow interstate commerce but halt the spread of the emerald ash borer, a juggling act that makes shipping difficult.

Diseases, Bioterrorism, and GM Organisms. Over the years, APHIS' role has expanded to include monitoring for pests that might attack wildlife and the monitoring of bioengineered species. It regulates the welfare of domestic animals and protects U.S. plant and animal agriculture from incoming disease. One APHIS program tracks the efforts to eradicate diseases such as cattle brucellosis, bovine tuberculosis, and pseudorabies in swine. APHIS also tracks plant diseases. In 2003, APHIS personnel detected a bacterial disease of geraniums in greenhouses in four Midwestern states. A devastating disease that could ruin potato and tomato crops, this bacterium had not been detected in the United States since 1999. APHIS agents determined that the disease had been imported in geranium cuttings from Kenya. Geranium imports from Kenya were halted until the outbreak could be contained.

Figure 13–25 Catching them early. On December 23, 2013, one adult coconut rhinoceros beetle, an invasive pest on palms, was caught in a trap in Honolulu, Hawaii. Others were soon discovered. APHIS is coordinating a response to eradicate the pest before it can spread.

APHIS oversees the Agricultural Bioterrorism Protection Act of 2002, which regulates laboratories and vaccine companies that use, possess, or transport toxins and agents that could be harmful to humans, plants, or animals. The act aims to prevent the release of devastating diseases from a research lab or medical facility. APHIS also has the job of assessing the safety of bioengineered organisms. In 2012, it completed assessments of the possible effects of glyphosphate-resistant sugar beets. The assessments are held open to public review and then APHIS makes a regulatory determination.

U.S. federal agencies collaborate with other countries to prevent the spread of pest organisms. The International Plant Protection Convention (IPPC) ratified by the United States in 1972 set up an international system to quarantine products to prevent pest spread. One provision of the 1995 United Nations Convention on the Law of the Sea was to prevent the introduction of marine species to new environments. Other international agreements such as the Agreement concerning Cooperation in the Quarantine of Plants and their Protection against Pests and Diseases of 1959 and the Agreement on the Application of Sanitary and Phytosanitary Measures of 1994 help prevent the spread of plant diseases and other pests.

Regulation of Pesticides

Regulation of pesticides focuses on three main areas: safety and testing of pesticides, protection of pesticide workers, and protection of consumers from products that may have pesticides on them. As is the case with regulation of potential pests, there are numerous historic federal regulations about pesticides, which have been pulled together into modern laws. There are also international treaties.

Federal Regulations. The Federal Insecticide, Fungicide, and Rodenticide Act (FIFRA), established in 1947 and amended in 1972, is administered by the EPA, which has total jurisdiction over the manufacture, sale, use, and testing of pesticides. FIFRA requires manufacturers to register pesticides with the EPA. As part of registration, it must test for toxicity to animals (and, by extrapolation, to humans) and set usage standards. For example, chlordane, which was used for 40 years as a pesticide on crops, lawns, and gardens, was tested and shown to be acutely toxic to animals in high doses. Chronic exposure led to damaging effects on most vital systems and caused liver cancer in animals. Acute and chronic exposure of humans to chlordane showed similar results, except for the development of cancer. On the basis of these results, the EPA canceled all uses of chlordane for food crops in 1978, but allowed continued use against termites until 1988, when all uses of the chemical in the United States were banned. The law also stipulates that the EPA must determine whether a product "generally causes unreasonable adverse effects on the environment," and if it so determines, the product may be restricted or banned. This was the basis for the banning of DDT.

The rapid increase in the number of chemicals used as pesticides makes their regulation difficult. The EPA's Pesticide Program places high priority on registering "reduced-risk" pesticides, especially those that promise to replace more toxic chemicals. Many are **biopesticides**—naturally occurring substances that control pests. These include **microbial** pesticides such as Bt, **plant-incorporated protectants** (in GM crops) such as corn bioengineered to manufacture Bt in its tissues, or biochemical substances that are naturally occurring, such as pheromones (e.g., Mimic). These pesticides can usually be registered more rapidly than other kinds of pesticides. The EPA has recently been under fire for giving provisional registration to hundreds of pesticides without requiring them to be fully tracked or tested.

FFDCA. Protecting consumers from exposure to pesticides on food was a contentious and confusing process for many years. Three agencies are involved: the EPA, the Food and Drug Administration (FDA), and the Food Safety and Inspection Service of the USDA. Basically, the EPA sets the allowable tolerances for pesticide residues, based on its assessment of toxicity, and the FDA monitors and enforces the tolerances on all foods except meat, poultry, and egg products, which are managed by the USDA.

The original enabling act was the Federal Food, Drug, and Cosmetic Act (FFDCA) of 1938. For many years, one clause of the FFDCA, the so-called **Delaney clause**, was highly controversial. The clause stated, "No [food] additive shall be deemed to be safe if it is found to induce cancer when ingested by man or animal." Because pesticides represent one of the largest categories of toxic chemicals to which people are exposed, this clause prohibited many pesticides from being used on foodstuffs when those pesticides had been found to cause cancer in laboratory tests with animals. In essence, the law stated that, if a given pesticide presents *any* risk of cancer, *no detectable residue* may remain on the food. Consequently, as analytical chemical techniques became more and more sensitive, more and more pesticides were banned for use on food.

FQPA. The National Research Council examined this dilemma and recommended in 1987 that the anticancer clause be replaced with a *negligible risk standard*, based on risk-analysis procedures that had been developed subsequent to the Delaney clause. After years of debate, Congress passed the **Food Quality Protection Act (FQPA)** in 1996 and did away with the Delaney clause and amended both FIFRA and FFDCA. The major requirements of the act include: setting "a reasonable certainty of no harm" as the safety standard for substances applied to foods; giving special consideration to the exposure of young children to pesticide residues; defining a risk of more than one case of cancer per million people as the allowable upper limit over long periods of exposure to chemicals; requiring evaluation of all possible sources of exposure to a given pesticide, not just exposure from food; reassessing older products according to the new requirements within 10 years; and applying the same standards to raw and processed foods (previously regulated separately).

Care for Kids. One of the emphases in the new legislation was protection of children. The focus is on children both because they typically consume more fruits and vegetables per unit of body weight than do adults and because studies have shown that children are frequently more susceptible to carcinogens and neurotoxins. Accordingly, the FQPA requires the EPA to add a tenfold safety factor in assessing children's risks from pesticides.

International Pesticide Regulation

U.S. laws and international treaties both regulate the way the United States relates to the rest of the world regarding pesticides. The United States currently exports thousands of tons of pesticides (active ingredients) each year, some of which are banned in the United States itself. A number of factors—low literacy, weak laws, unfamiliarity with equipment—increase the potential for serious exposure to toxic pesticides in the developing countries. FIFRA requires "informed permission" prior to the shipment of any pesticides banned in the United States. This permission comes from the purchaser. The EPA must then notify the government as to the identity of the importing company. Internationally, there are treaties that regulate information about pesticides.

Treaties and Guidance. Two international treaties and one guidance document provide a framework for pesticides in the international arena. The **Rotterdam Convention** was adopted in 1998 and went into effect in February 2004. It is a treaty ratified by 70 countries that promotes open exchange of information about hazardous chemicals. Central to the convention is the process of **prior informed consent (PIC)**, whereby exporting countries inform all potential importing countries of actions they have taken to ban or restrict the use of pesticides or other toxic chemicals and label them clearly. Governments in the importing countries, UN agencies, and exporting countries communicate so everyone can make decisions with the information they need.

The *International Code of Conduct on the Distribution and Use of Pesticides* is a worldwide guidance document that addresses conditions of safe pesticide use. The most recent version was approved by the FAO in November 2002. The code makes clear the responsibilities of both private companies and countries receiving pesticides in promoting their safe use. In 2004, the **Stockholm Convention**, a treaty designed to eliminate or restrict persistent organic pollutants, came into effect. By 2013, 178 countries and the European Union were parties to the convention; the United States remained a notable non-ratifier. Countries are expected to ban 9 of the 12 most dangerous chemicals (the "dirty dozen"), use DDT only for malaria control, and limit the accidental production of two pesticide breakdown products, dioxins and furans.

The Future. These international efforts are leveling the safety field for pesticide use globally and pulling the worst chemicals out of the global stream, at least slowly. Pesticide use illustrates two trends we see in many environmental science issues: the problems of cumulative impacts and of unintended consequences (Chapter 1). As with many environmental issues, significant progress in both pest and pesticide control has benefited from grassroots action and pressure from public-interest groups and NGOs, such as the Consumers Union and the Pesticide Action Network. These and other groups are pressing for continued movement in the direction of safe and sustainable pest prevention and management. Care for the welfare of our fellow human beings and all other creatures is necessary to solve pest problems in a fashion that is sustainable in the long term.

CONCEPT CHECK What is the role of APHIS in controlling pests? What is the role of the EPA? ☑

REVISITING THE THEMES

SOUND SCIENCE

It was sound science that uncovered the connections between DDT and the decline of many predatory birds—biomagnification was a great surprise to everyone. Some amazing applications of sound science are seen in the imaginative work on alternative pest controls, such as the sterile-male technique, the use of pheromones to control pests, studies that have uncovered natural enemies of pests, research into a wide range of ways to control malaria-carrying mosquitoes, and other parts of integrated pest management. Agricultural scientists are constantly challenged to keep up with new strains of plant pests by developing resistant varieties of the crops they attack.

SUSTAINABILITY

Ecological control is a sustainable approach to controlling pests, while the chemical treatment approach guarantees that pesticides will need to be used indefinitely. Pest resistance, resurgences, and secondary outbreaks have been the norm. Ecological control methods provide the means to achieve control of most pests in a sustainable way, especially when they are controlled by natural enemies. IPM presents a reasonable alternative for managing pests by limiting the use of pesticides. The organic food movement is proof that food can be grown without chemicals and, for consumers and the environment, is a great way to keep pesticides out of the natural world (and the refrigerator). Reducing or eliminating the use of pesticides and keeping food safe and healthy are two of the pillars of the sustainable-agriculture movement. (See Chapter 12.)

STEWARDSHIP

The widespread use of DDT during and just after World War II saved millions of lives. However, the continued, often indiscriminate, use of this persistent pesticide and others like it led to trouble in the environment. The careful work of Rachel Carson in documenting this trouble in the 1960s was a landmark in environmental stewardship. Yet the continued use of DDT to control malaria-bearing mosquitoes shows that discerning the common good is not always easy. Because stewardship also includes human welfare and health, the focus of the federal and international regulations on the protection of children and agricultural workers is an example of stewardship. It can be argued that the work of the EPA in complying with federal regulation is stewardship in action. The agency is an aggressive promoter of reduced use of the more toxic pesticides.

REVIEW QUESTIONS

1. Define *pests*. Why do we control them? Describe at least four types.

2. Discuss two basic philosophies of pest control. How effective are they?

3. Explain the pros and cons of using the synthetic organic pesticide DDT against malarial mosquitoes.

4. What adverse environmental and human health effects can occur as a result of pesticide use?

5. Define *bioaccumulation* and *biomagnification*.

6. Why are nonpersistent pesticides not as environmentally sound as first thought?

7. Describe the four categories of natural, or biological, pest control. Cite examples of each and discuss their effectiveness.

8. Define the term *economic threshold* as it relates to pest control. What are cosmetic spraying and insurance spraying?

9. What are the four steps in IPM? Explain how IPM worked in Indonesia.

10. What is organic food, and how is it now certified?

11. How does APHIS help control pests in the United States? What international regulations help control the movement of pests across borders?

12. How does FIFRA attempt to control pesticides? What new perspective does the FQPA bring to the policy scene?

13. Describe how the Stockholm Convention, Code of Conduct, and Rotterdam Convention help regulate international use of pesticides.

THINKING ENVIRONMENTALLY

1. Consider the ethics of the United States' exportation of pesticides. A large portion of this exported material is banned in the United States itself. If the choice were up to you to continue this or put it to an end, what would you do? Remember that such pesticides have at least a positive immediate benefit to these countries. Are they a necessary evil?

2. Almost one-third of the chemical pesticides bought in the United States are for use in houses and on gardens and lawns. What should manufacturers and users do to ensure the limited and prudent use of pesticides?

3. Should the government give farmers economic incentives to switch from pesticide use to IPM? How might that work?

4. Investigate how bugs and weeds are controlled on your campus. Are IPM techniques being used? Organic fertilizers? If not, consider lobbying for their use.

5. Imagine you're the mayor of a mid-sized suburban town where mosquitoes invade from neighboring marshes. What kind of restrictions would you place on local pesticide use? Defend your answer.

MAKING A DIFFERENCE

1. When you plant a garden, select hardy varieties of plants that are known to flourish in your particular region. Pests are less likely to destroy these types because they grow much faster than they can be eaten, even without pesticides.

2. Try planting some herbs in your garden because they often have many pest-oriented benefits. They are very likely to attract the predatory insects that will naturally take out the pests. Sage, basil, chives, yarrow, cilantro, and garlic are all great choices.

3. Another way to bring in the predatory insects is to buy them. A great number of companies now breed and sell species such as praying mantises, lacewings, and ladybugs, all of which help get rid of your pest problems.

4. To help keep a pesticide-free lawn looking green, let the grass grow a little longer than usual. This way it will need less water and food. Select a mower that mulches grass and leaves; lawn clippings provide nitrogen, and fallen leaves are great for conditioning soil.

HARNESSING ENERGY FOR HUMAN SOCIETIES

How essential are energy sources and services to human societies? Industrialized countries are so completely dependent on fossil fuels and nuclear energy that we cannot imagine an economy apart from these energy sources. Developing countries are following the same example, hoping to enhance the well-being of their people. They have a long way to go; more than 1.3 billion people lack access to electricity, and another billion have unreliable service. Our pursuit of energy sources and services exacts an underappreciated toll in human life and well-being. The processes of energy exploration, extraction, transportation, and power generation carry risks that often result in accidental deaths. They also cause air and water pollution, which pose risks to human health.

A Difficult Situation. Continued production of fossil fuels will require more difficult methods of extraction. These methods can damage fragile ecosystems and lead to the loss of biodiversity, while burning fossil fuels adds the environmental costs of climate change. Nuclear power remains controversial, its promise overcast by the significant risks from radioactive waste and leaks.

There is good news, too. Wind turbines are appearing all over the landscape. Solar panels adorn rooftops, and solar "farms" are harvesting the Sun's limitless energy. Biofuels like ethanol are powering millions of vehicles. Regulations are mandating major increases in vehicle fuel economy, and new standards are making energy use far more efficient. Renewable energy sources are on the move, already supplying 13% of electrical generation in the United States and 22% of the global electricity supply.

Can we chart a path to a secure energy future? One approach is "business as usual"—pump oil, drive gas-guzzling vehicles, gouge mountains for coal, and lace the countryside with natural-gas wells and pipelines. An alternative—the sustainable future—requires us to focus our energy policy on greater energy efficiency and renewable energy. The challenge is here, and the choice is ours to make. Part Four describes what is going on in this important area.

Energy from Fossil Fuels

Approximately 300 million years ago, the Gulf of Mexico was formed by a sinking of the sea floor. In subsequent geological eras, organic matter accumulated in sediments that were deposited in the Gulf, where it was buried and converted into oil and gas. Today, large deposits of oil and gas lie below the warm waters of the Gulf. Petroleum exploration began in the Gulf in earnest after World War II, and 1947 marked the first offshore well drilled out of the sight of land, under 20 feet of water. Since then, some 15,000 oil and gas wells have been drilled in the Gulf of Mexico, in depths of up to 10,000 feet below the surface and a further 13,000 feet below the sea floor.

Oil and gas production in the Gulf has been prolific, currently accounting for 30% of U.S. yield. However, much of the oil and gas that can be extracted from shallow wells has already been withdrawn from the Gulf, so the industry has been forced to drill in deeper waters and deeper strata to find oil. To get at the deep oil and gas, increasingly expensive and elaborate rigs are required, platforms that float on the surface and support all the necessary equipment for drilling in deep water, housing a crew, and providing a helicopter pad. One such rig was *Deepwater Horizon*; on April 20, 2010, that rig had a crew of 126.

The Blowout at Macondo. Drilling at the BP-owned Macondo Well was begun in October 2009 by another rig. The *Deepwater Horizon* moved to the site in February 2010 and successfully completed the drilling. By April 20, the crew of the rig had inserted a cement barrier over the well surface preliminary to abandoning the well for the time being. Such a barrier must be able to withstand high pressures because the oil and gas accessed by the boring are themselves under high pressure. Suddenly, the well began blowing seawater and drilling mud up the string of connected drill casings to the rig platform, forming a geyser 240 feet into the air. The drilling mud was followed by a slushy combination of mud, oil, and gas, which ignited and exploded. The explosion and subsequent fire on the rig killed 11 workers and injured many more before they could be evacuated. The rig continued to burn until, 36 hours later, it sank into the Gulf.

▲ Fire boats spray water on the burning remnants of the Deepwater Horizon oil rig.

Under pressure from below, oil began to gush out of the well on the Gulf floor and continued to do so for 87 days until the well was capped off. Close to 5 million barrels of oil flowed up into the Gulf of Mexico, making an awful mess—forming a huge slick, contaminating some 625 miles of Gulf coastline, and killing countless sea turtles, birds, fish, and smaller marine organisms. It was one of the largest oil spills in world history.

An intense cleanup began immediately; much oil on the surface was simply burned, sending tons of black soot into the atmosphere. Some oil was pumped up from the bottom in the vicinity of the well, some was siphoned by booms and boats on the surface, but the vast bulk of the spill either evaporated at the surface of the Gulf or was broken down by bacteria in the water column and along the shores. These natural forces were aided by the high volatility of the Gulf oil and the warm temperatures of Gulf water. Nevertheless, the Deepwater Horizon spill was an economic disaster for the Gulf Coast fishing and tourism industries during the spring and summer of 2010 and afterwards. Much of this loss is being covered by a $20 billion trust fund created by BP at the demand of President Barack Obama.

Costs of Fossil Fuels. Ultimately the environmental consequences of oil spills, drilling operations, and coal mining are some of the many costs of doing business in a fossil fuel–driven economy. Americans use almost 19 million barrels of oil every day. In future chapters, you will learn some of the other costs, such as air pollution, which is affecting human and environmental health, and rising greenhouse gases, which are leading to global climate change.

This chapter presents an overview of the sources of the fossil fuels that support transportation, homes, and industries. We discuss the consequences of our dependence on fossil fuels and then close the chapter with a look at public policy.

14.1 Energy Sources and Uses

Every day, you use energy—the part of the physical world that allows objects to move and to do work (see Chapter 3). Some energy comes from the center of our planet as it cools, but the ultimate source of most of the energy on Earth is the Sun, where it is released by nuclear reactions. This energy travels to Earth as radiation and then flows through Earth's systems. Recall that plants trap the Sun's energy to make complex molecules (chemical energy), which they use to carry out their life functions. Animals and other consumers get their energy by eating plants or other consumers.

We humans get our energy from the food we eat, which is used to build tissues and move our bodies. In addition to the energy in our food, we use a variety of energy sources to power machinery, heat our homes, communicate, grow crops, travel from place to place, and perform the many other activities of daily life. These energy sources include fossil fuels as well as renewable energy sources, such as solar and wind energy. The energy sources we choose to use have a profound impact on our lives, our society, and the environment.

Energy Transformations

As energy flows through a system, it can change from one form to another. **Primary energy sources** can be used in the form in which they are found in nature; the Sun's radiation, geothermal energy, coal, wood, and nuclear energy are primary energy sources. **Secondary energy sources**, such as electricity, need to be made from primary sources; thus secondary energy sources are really energy holding systems. Differences in efficiency and cost drive decisions about which energy sources we use.

Efficiency of Energy Transformations. In every energy transformation, some usable energy is lost to the system; this loss is a direct result of the Second Law of Thermodynamics (Chapter 3). Lost energy is released as heat; on Earth, that energy is eventually released into space. As we will see throughout the chapter, the concept of the inefficiency of energy transfer at every level of a system is critical to understanding the pros and cons of different energy sources. For example, burning coal releases a certain amount of energy. Transforming that energy into electricity, transporting the electricity over wires, and using it to produce light involves losses of energy at each step in the flow of energy. If a primary source of energy (such as coal) is turned into a secondary source of energy (such as electricity), there is more energy lost than if the primary energy source could be used directly.

In some situations, energy can be used more efficiently by combining several activities in one structure. For example *district heating*—a system in which a large, central plant heats several residential and commercial spaces—is usually more efficient than having many smaller heating systems in separate locations. The most efficient of these systems is district heating with **combined heat and power (CHP)**, also called *cogeneration*. CHP produces electricity and heat simultaneously.

Costs of Obtaining Energy. In this chapter we will discuss the costs for obtaining different types of fuel. These include direct financial and energy costs as well as environmental costs during extraction, refinement, transportation, and use. One way to measure the cost of obtaining a source of energy is to calculate the ratio of energy returned to energy invested, or EROI. (For a longer discussion, see Sound Science, *Energy Returned on Energy Invested*.) We will see that some fossil fuels are relatively easy to obtain, which is part of the reason why we are heavily dependent on them. We will also see how economic factors such as the cost of extraction and the supply of a fuel make prices rise or fall. Yet other costs, such as the pollution of land, water, and air are largely externalities in the economic system. Let's look at a little bit of history in order to understand how we got to a situation where fossil fuels dominate the global economy and environment.

A Brief History of Energy Use

Throughout human history, the advance of technological civilization has been tied to the development of energy sources. In early times (and even today in less developed regions), the major energy source was muscle power. Some people lived in relative luxury by exploiting the labor of others—slaves, indentured servants, and minimally paid workers. Human labor was supplemented by the use of domestic animals for agriculture and transportation, by water or wind power for milling grain, and by the Sun.

In the late 1700s, the breakthrough that launched the Industrial Revolution was the development of the steam engine. In a steam engine, water is boiled in a closed vessel to produce high-pressure steam, which pushes a piston back and forth in a cylinder. In turn, the piston turns a wheel to drive the machinery. The steam engine solved the problem of finding a steady, portable power source. Steam engines rapidly became the power source for steamships, steam shovels, steam tractors, steam locomotives, and stationary engines to run sawmills, textile mills, and virtually all other industrial plants.

At first, the major fuel for steam engines was firewood. Then, as demands for energy increased and firewood around industrial centers became scarce, coal was substituted. By the end of the 1800s, coal had become the dominant fuel, and it remained so into the 1940s. In addition to being used as fuel for steam engines, coal was widely used for heating, cooking, and industrial processes. In the 1920s, coal provided 80% of all energy used in the United States.

Although coal and steam engines powered the Industrial Revolution, which greatly improved life for most people, there were many drawbacks. The smoke and fumes from the numerous coal fires made air pollution in cities far worse than anything seen today. Coal is also notoriously hazardous to mine and dirty to handle, and burning coal results in large quantities of ash that must be removed. As for steam engines, because of the size and bulk of the boiler, the engines were heavy and awkward to operate (Figure 14–1). Frequently, the fire had to be started several hours before the engine was put into operation in order to heat the boiler sufficiently.

In the late 1800s, the simultaneous development of three technologies—the internal combustion engine, oil-well drilling, and the refinement of crude oil into gasoline and other liquid fuels—combined to provide an alternative to steam power. The replacement of coal-fired steam engines and furnaces with petroleum-fueled engines and oil furnaces was an immense step forward in convenience. Also, the air quality improved greatly as cities were gradually rid of the smoke and soot from burning coal. (It was only in the 1960s, with the tremendous proliferation of cars, that pollution from gasoline engines became a problem.) Further, the gasoline-driven internal combustion engine provided a valuable power-to-weight advantage that allowed rapid advances in technology. A 100-horsepower gasoline engine weighs only a tiny fraction of what a 100-horsepower steam engine and its boiler weigh, and jet engines have an even greater power-to-weight ratio. Automobiles and other forms of ground transportation would be cumbersome, to say the least, without this power-to-weight advantage, and airplanes would be impossible.

Figure 14–1 Steam power. Steam-driven tractors like this 1915 Case marked the beginning of the industrialization of agriculture. Today, tractors half the size do 10 times the work. Modern tractors are also much easier and safer to operate.

Economic development throughout the world since the 1940s has depended largely on technologies that consume fuels refined from crude oil. By 1951, crude oil became the dominant energy source for the United States. Since then, it has become the mainstay not only of the United States (making up 36% of total U.S. energy consumption), but also of most other countries, both developed and developing. Crude oil currently provides about 33% of the total annual global energy consumption (Figure 14–2). In the countries of eastern Europe and in China, coal is still the dominant fuel. Coal ranks second globally, at 30%, and provides 20% of U.S. demand.

Natural gas, the third primary fossil fuel, is often found in association with oil or during the search for oil. Natural gas is largely methane, which produces only carbon dioxide and

GLOBAL PRIMARY ENERGY SUPPLY

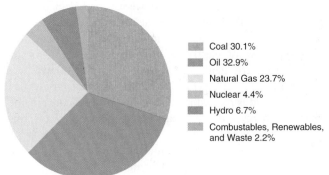

- Coal 30.1%
- Oil 32.9%
- Natural Gas 23.7%
- Nuclear 4.4%
- Hydro 6.7%
- Combustables, Renewables, and Waste 2.2%

Figure 14–2 Global primary energy supply. The world consumed some 12,730 million tons of oil equivalent in 2013. The sources of the total primary energy supply are shown here. Note that 87% comes from fossil fuels.

(*Source:* Data from BP Statistical Review of World Energy, 2014, 41.)

water as it combusts; thus, it burns more cleanly than coal or oil. In terms of pollution, it is the most desirable fuel of the three. Despite the obvious fuel potential of natural gas, at first there was no practical way to transport it from wells to consumers. Any gas released from oil fields was (and in many parts of the world still is) flared—that is, vented and burned in the atmosphere, a tremendous waste of valuable fuel. Gradually, the United States constructed a network of pipelines connecting wells with consumers. With the completion of these pipelines, the use of natural gas for heating, cooking, and industrial processes escalated rapidly because gas is clean, convenient (no storage bins or tanks are required on the premises), and relatively inexpensive. Currently, natural gas satisfies about 30% of U.S. energy demand and 24% of the global demand.

Thus, three fossil fuels—crude oil, coal, and natural gas—provide 86% of U.S. energy consumption and 87% of the world's consumption. The remaining percentages are furnished by nuclear power, hydropower, and renewable sources. The changing pattern of energy sources in the United States is shown in Figure 14–3. The picture is similar for other developed countries of the world, though the percentages differ somewhat depending on the energy resources of the country. With the help of fossil fuel energy, people and goods have become so mobile that the human community has in fact become a global community. Adding to this linkage is the explosion in information storage and flow, made possible by digital devices and the Internet—all powered by electricity.

Electrical Power Production

A considerable portion of the energy we gain from fossil fuels is used to generate **electrical power**, defined as the amount of work done by an electric current over a given time. The electricity itself is an **energy carrier**—it transfers energy from a primary energy source (coal or waterpower, for instance) to its point of use. Electricity defines our modern technological society, powering appliances and computers in our homes, lighting up the night, and enabling the global Internet and countless other vital processes in business and industry to operate. Approximately 32% of fossil fuel production is now used to generate electricity in the United States; in 1950, the figure was only 10%.

Generators. Electric generators were invented in the 19th century. In 1831, English scientist Michael Faraday discovered that passing a coil of wire through a magnetic field causes a flow of electrons—an electrical current—in the wire. An **electric generator** is basically a coil of wire that rotates in a magnetic field or that remains stationary while a magnetic field is rotated around it. The process converts mechanical energy into electrical energy. Remember, any time energy is converted from one form to another, some is lost through resistance and heat (in accordance with the Second Law of Thermodynamics; see Chapter 3). Also, some energy is lost in transmitting the electricity over wires from the generating plants to end users. In the end, it takes three units of primary energy to create one unit of electrical power that can be put to use. This apparently inefficient proposition is justified because electrical power is so useful and indispensable.

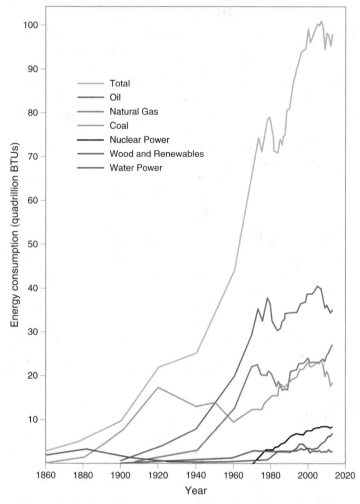

Figure 14–3 Energy consumption in the United States. Total consumption and major primary sources from 1860 to 2013 are shown here in quadrillion Btus (one Btu is the amount of heat required to raise one pound of water one degree Fahrenheit). Note how the mix of primary sources has changed over the years and how the total amount of energy consumed has continued to grow. Note also the skyrocketing increase in the use of oil after World War II (1945) as private cars and car-dependent commuting became common.

(*Source:* Data from Energy Information Administration, U.S. Department of Energy, *Monthly Energy Review*, February 2015.)

In the most widely used technique for generating electrical power, a primary energy source is employed to boil water, creating high-pressure steam that drives a **turbine**—a sophisticated paddle wheel—coupled to the generator (Figure 14–4a). The combined turbine and generator are called a **turbogenerator**. Any primary energy source can be harnessed for boiling the water. Coal, oil, and nuclear energy are most commonly used at present, but burning refuse, solar energy, and geothermal energy (heat from Earth's interior) may be more widely used in the future.

In addition to steam-driven turbines, gas- and water-driven turbines are used. In a gas turbogenerator, the high-pressure gases produced by the combustion of a fuel (usually natural gas) drive the turbine directly (Figure 14–4b). For hydroelectric power, water under high pressure—at the base of a dam or at the bottom of a pipe beginning at the

MAJOR METHODS OF GENERATING ELECTRICITY

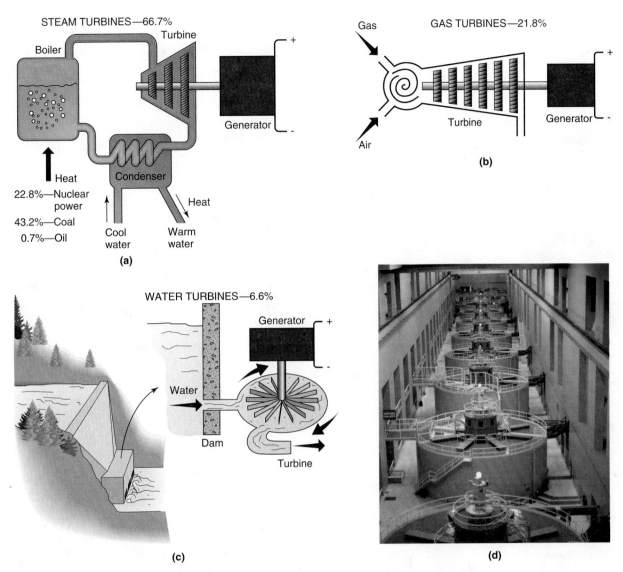

Figure 14–4 Electrical power generation. Most electricity is produced commercially by driving generators with **(a)** steam turbines, **(b)** gas turbines, and **(c)** water turbines. The percentage of the U.S. electricity supply derived from each source (as of 2013) is indicated. Smaller amounts of power (4.1%) are now also coming from wind energy, solar thermal energy, and photovoltaics (solar cells). **(d)** Each generator in this array in a hydroelectric plant is attached by a shaft to a turbine propelled by water.

(*Source:* Data from Energy Information Administration, U.S. Department of Energy, *Annual Energy Review* 2013, Lawrence Livermore National Laboratory, October 2014.)

top of a waterfall—is used to drive a hydroturbogenerator (Figure 14–4c–d). Wind turbines are also coming into use. A utility company supplying electricity to its customers will employ a mix of power sources, depending on the cost and availability of each and the demand cycle.

There are currently nearly 7,000 power plants in the United States producing 4.1 billion megawatt-hrs. of electricity. Figure 14–5, on the next page, shows the remarkable shifts that have occurred in U.S. sources of electricity over the last 70 years. Although few new coal plants have been built in recent years, coal-burning power plants stay on line for decades; coal still provides more electrical power than any other source. Note, however,

the exceptional rise in recent years of natural gas-powered plants and the appearance of renewable power sources.

Fluctuations in Demand. Where does your electricity come from? It depends on what's available. Most utility companies are linked together in what are called *pools*. In New England, for example, ISO New England ties together some 350 electric-power-generating plants through 8,000 miles of transmission lines, called the **power grid**. The utility is responsible for balancing electricity supply and demand, regardless of daily or seasonal fluctuations. The generating capacity of the ISO New England utilities is currently 31,000 megawatts (MW), produced by a variety of energy

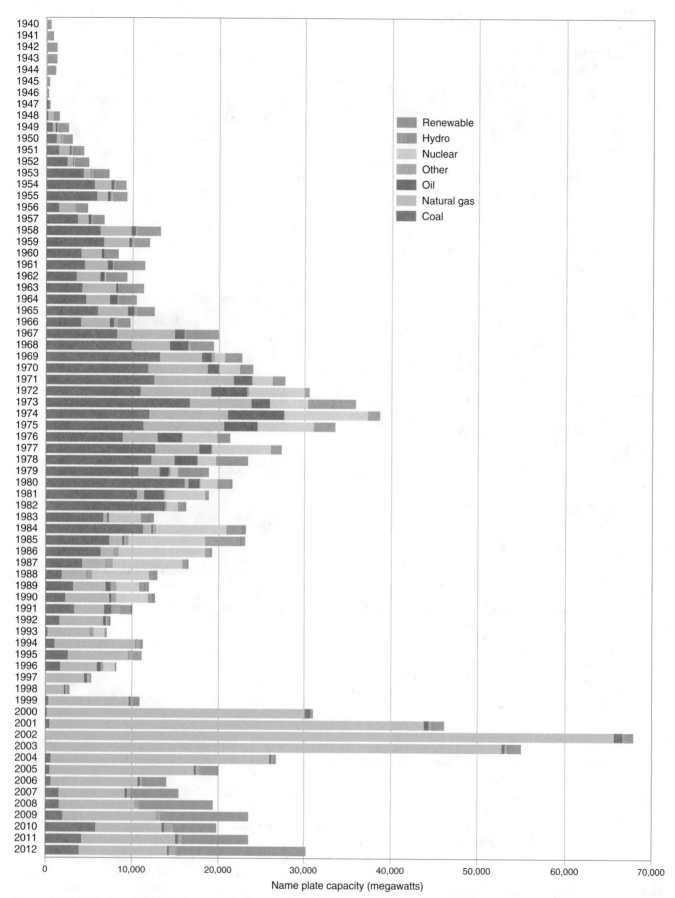

Figure 14–5 U.S. electric generating capacity by year in service, 1940–2012. Remarkable changes have occurred over the years in the construction of power plants, with many decades of coal plants, two decades of nuclear plants, and, most recently, a huge increase in natural gas and renewable sources.

(*Source:* U.S. Energy Information Administration: EIA-860 Annual Electric Generator Report, 2013.)

Table 14–1 Energy Sources for Generating Electrical Power (% of total)

Source	New England	United States	World
Nuclear power	33.2	22.8	11.0
Natural gas	37.0	21.8	23.0
Coal	5.6	43.2	40.0
Oil	8.9	0.7	4.0
Hydro	7.5	6.6	17.0
Wood, refuse, renewables	7.8	4.9	5.0

Sources: ISO New England, 2014; U.S. Energy Information Administration, 2013; Shift Project Data Portal, 2012.

sources, as shown in Table 14–1. One MW is enough electricity to power 800 homes, on average. The greatest demands on electrical power in New England come in the summer months, when air conditioning use is at its highest. The all-time peak load (highest demand) record of 28,130 MW was set on August 2, 2006, a very hot day in New England. A power pool must accommodate daily and weekly electricity demand as well as seasonal fluctuations.

Demand Cycle. The typical pattern of daily and weekly electrical demand in the United States is shown in Figure 14–6. The baseload represents the constant supply of power provided by the large coal-burning and nuclear plants with capacities as high as 1,000 MW. As demand rises during the day, the utility draws on additional plants that can be turned on and off—the intermediate and peak-load power sources that represent the utility's reserve capacity. These are

the utility's gas turbines and diesel plants, or the utility may use pumped-storage hydroelectric power (water is pumped to a reservoir using off-peak power and is allowed to flow through turbines when needed). Occasionally, a deficiency in available power will prompt a brownout (a reduction in voltage) or a blackout (a total loss of power), temporarily disrupting whole regions. Such events are most likely to occur during times of peak power demand.

Summer heat waves represent the greatest source of sudden pulses in the demand for electricity. As the climate continues to grow warmer in many parts of the United States, more and more regional utilities are being pushed to the edge of their ability to satisfy summer electricity demand, although they still have some reserve. An even more serious problem is the antiquated system that controls the power transmission grid connecting power sources to users.

Blackout. Power brownouts and blackouts are a threat to an economy that is now highly dependent on computers and other electronic devices (brownouts are a cutback in power, blackouts are total power failures). On August 14, 2003, the largest blackout in U.S. history hit eight eastern states and parts of Canada, imposing a $30 billion cost on the economies of the two countries. The blackout began when several power lines in Ohio brushed against trees and shut down. Transmission lines around the failure cut out, a local power plant shut down, more and more lines and plants dropped out, and within eight minutes, 50 million people across eight states and two Canadian provinces were without power. Power outages and interruptions cost the United States an estimated $150 billion a year.

What is needed to prevent blackouts and brownouts is a self-healing **smart grid**, one that is able to monitor problems in real time, react to trouble quickly, and isolate trouble areas so that a cascading failure like the one in 2003 would be impossible. The technology for this fix is available, but funding is a huge problem. However, the utilities are busy installing one component of a smart grid—*smart meters*—by the millions. The meters give instant feedback on electricity use to homeowners and the utility control center. Another component would be *smart sensors* at many points on the grid; the sensors would provide feedback on transformers and power lines, allowing instant shunting of power away from breaks and between transformers. A pilot installation of smart grid technology in Boulder, Colorado, in 2008 was highly successful. The government has invested $4.5 billion in smart grid projects, matched by $5.5 billion from private sources.

Clean Energy? There was a time when electrical power was promoted as being the ultimate clean, nonpolluting energy source. This is true at the point of its use: *Using* electricity creates essentially no pollution but *generating* electricity does. For a time, electricity was relatively inexpensive, and electric heating systems were installed in many new homes. This is no longer true. The public knows that electricity is (1) an expensive way to heat homes and (2) generated from fossil fuels and nuclear energy. Coal-burning power plants are the major source of electricity in the United States but

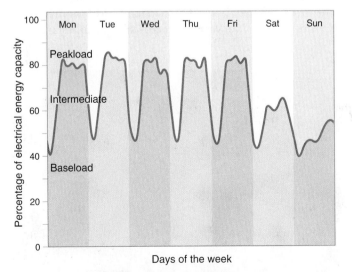

Figure 14–6 Weekly electrical demand cycle. Electrical demand fluctuates daily, weekly, and seasonally. Less energy is used at night and on weekends. A variety of generators must be employed to meet base, intermediate, and peak electricity needs.

are heavily implicated in acid deposition and global climate change (Chapters 18 and 19). Nuclear power is a much-mistrusted technology because of the potential for accidents and the problem of disposal of the nuclear waste products. (Nuclear power is discussed in Chapter 15.) Additional adverse effects of using electrical energy apply to other parts of the fuel cycle, such as mining coal and processing uranium ore. The pollution is simply transferred from one part of the environment to another. The only sources of energy that produce less pollution are the renewables—hydroelectricity, solar, wind, and biomass—and these have their own limitations (Chapter 16).

The problem of transferring pollution from one place to another becomes even more pronounced when you consider efficiency; for example, the production of electricity from burning fossil fuels has an efficiency of only 30–35%. The energy loss is accounted for partly by some heat energy from the firebox going up the chimney and partly by a large amount of heat energy remaining in the spent steam as it leaves the turbine. These unavoidable energy losses, called **conversion losses**, are a consequence of the need to maintain a high temperature difference between the incoming steam and the receiving turbine so as to maximize efficiency. Heat energy will go only toward a cooler place, so it cannot be recycled into the turbine. The most common practice is to dissipate it into the environment by means of a **condenser** (a device that turns turbine exhaust steam into water). In fact, the most conspicuous features of coal-burning and nuclear power plants are the huge cooling towers used for that purpose (Figure 14–7). Another significant energy loss occurs as the electricity is transmitted and distributed to end users by connecting wires.

An alternative to cooling towers is to pass water from a river, a lake, or an ocean over the condensing system (Figure 14–4a). With this approach, however, the waste heat energy is transferred to the body of water, with the result that all the small planktonic organisms drawn through the condensing system with the water are cooked and the warm water added back to the waterway may have deleterious effects on the aquatic ecosystem. Thus, waste heat energy discharged into natural waterways is referred to as **thermal pollution**.

Electricity is a form of energy that is indispensable to modern societies as they are currently structured. Worldwide, the electrical power industry produces more than 22,000 billion kilowatt-hours a year and, at the same time, generates a challenging array of environmental problems.

Matching Energy Sources to Uses

In addressing whether energy resources are sufficient and where additional energy can be obtained, it is necessary to consider more than just the source of energy because some forms of energy lend themselves well to one use, but not to others. For example, the vast majority of cars and trucks traveling on U.S. highways have internal combustion engines that require liquid fuels. Almost all other machines used for moving, such as tractors, airplanes, locomotives, and ships, also depend on liquid fuels. An enormous infrastructure has been set up to supply these vehicles with gasoline and other liquid fuels. Crude oil, together with much smaller amounts of biofuels, is needed to support this transportation system. Coal and nuclear power can do little to mitigate the demand for crude oil because they are suitable only for boiling water to drive turbogenerators, which will do essentially nothing for our transportation system. Although this situation might change with more widespread use of electric vehicles, their practicality for long distances still requires a technological breakthrough in the form of a low-cost, lightweight battery that can store large amounts of power. In addition, electric propulsion would not be practical for trucking and aviation.

Primary energy use is commonly divided into four categories: (1) transportation, (2) industrial processes, (3) commercial and residential use (heating, cooling, lighting, and appliances), and (4) the generation of electrical power, which goes into categories (2) and (3) as a secondary energy use. The major pathways from the primary energy sources to the various end uses are shown in Figure 14–8. Transportation, which represents 27% of energy use, depends on petroleum (and some ethanol), whereas nuclear power, coal, and water power are limited to the production of electricity. Natural gas and oil are more versatile energy sources, and significant

Figure 14–7 Cooling towers. A coal-fired power plant is equipped with cooling towers to dissipate the heat from the spent steam after it has passed through the turbines. Members of the public often mistake such cooling towers for nuclear reactor containment vessels.

shifts have occurred in the use of these sources in recent years. For example, today only 1% of U.S. electricity is generated by oil, whereas 35 years ago it was about 15%.

Figure 14–8 also indicates the high proportion of consumed energy that is lost as heat ("waste energy"), rather than being used for its intended purpose—some 59%. Some waste is inevitable, as dictated by the Second Law of Thermodynamics, but the current losses are much greater than necessary. The efficiency of energy use can be doubled for some uses, including cars that get twice as many miles per gallon and appliances that consume half as much power. In general, increasing the efficiency of energy use would have the same effect as increasing the energy supply, at far less expense. Such increases in efficiency represent a largely untapped "reserve" of energy that is saved, the **conservation reserve** discussed in more detail in Section 14.5.

CONCEPT CHECK Of the various energy sources used to generate electricity, those that power steam turbines are used most often. Why might that be? ☑

ESTIMATED U.S. ENERGY USE IN 2013

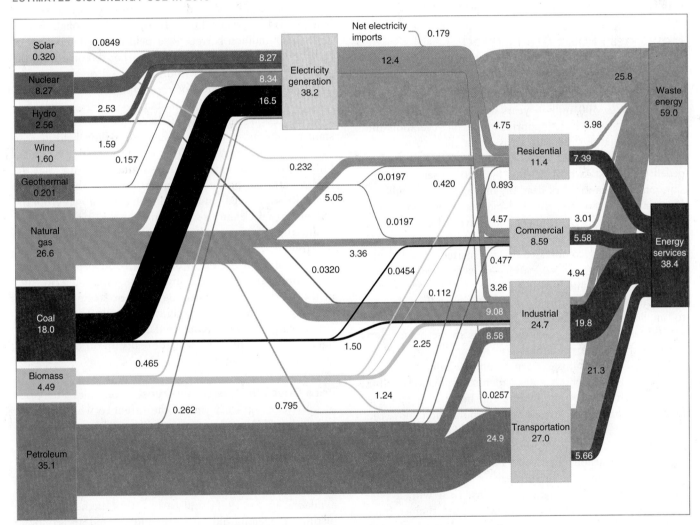

Figure 14–8 Pathways from primary energy sources to end uses in the United States. Only major pathways are shown. Note that end uses are connected to primary sources in specific ways. Note also the large percentage of energy that is wasted as it is converted to heat and lost. All energy is expressed in quads, or quadrillion Btus (10^{15} Btus). The total U.S. energy use in 2013 was an estimated 97.4 quads.

(*Source:* Lawrence Livermore National Laboratory. Data from Energy Information Administration, U.S. Department of Energy, *Annual Energy Review* 2013, October 2014.)

UNDERSTANDING THE DATA

1. What is the total amount of electricity generated from all the energy sources in the United States? What percentage of this electricity is never used?

2. In which sector is the use of energy most *inefficient*: residential, commercial, industrial, or transportation? How can you tell?

3. If electric vehicles become more common, which value will increase?

14.2 Exploiting Crude Oil

The United States has abundant reserves of coal, adequate reserves of natural gas (at least for the time being), and the potential for the expansion of nuclear power. Therefore, no shortages loom in these areas now. (Note that the environmental concerns regarding coal and nuclear power are being set aside for the moment.) Petroleum to fuel our transportation system, however, is another matter. The U.S. crude-oil supplies fall far short of meeting demand; we now depend on foreign sources for half of our crude oil. Foreign sources come at increasing costs, such as trade imbalances, military actions, economic disruptions, and oil spills. Let us examine the situation, starting with where crude oil comes from and how it is exploited.

How Fossil Fuels Are Formed

Crude oil, coal, and natural gas are called *fossil fuels* because they are derived from the remains of living organisms (Figure 14–9). In the Paleozoic and Mesozoic geological eras, 100 million to 500 million years ago, freshwater swamps and shallow seas, which supported abundant vegetation and phytoplankton, covered vast areas of the globe. Anaerobic conditions in the lowest layers of these bodies of water impeded the breakdown of detritus by decomposers. As a result, massive quantities of dead organic matter accumulated. Over millions of years, this organic matter was gradually buried under layers of sediment and converted by microbial action, pressure, and heat to coal, crude oil, and natural gas. Coal is highly compressed organic matter—mostly leafy material from swamp vegetation—that decomposed relatively little. Sediments buried 7,500 to 15,000 feet are in a temperature range that breaks down large organic molecules to smaller carbon molecules, producing crude oil. Below 15,000 feet, molecular breakdown leads to the one-carbon compound methane, or natural gas.

The crude oil that is extracted from the ground is a dark mixture of hydrocarbons and contaminants, such as sulfur and nitrogen. Before this mixture can be used, it must be refined to separate the various hydrocarbons and remove the contaminants. In the refining process, crude oil is distilled, or heated to high temperatures so it vaporizes. Because the various hydrocarbon chains are of different lengths, they vaporize and condense at different temperatures. As they condense, they form separate fractions, each with a particular weight and viscosity. The lighter and thinner fractions include kerosene and gasoline. Diesel fuel is heavier and thicker than gasoline; wax and asphalt are even heavier and thicker. Further chemical processes can be used to turn some of the heavier compounds into lighter ones, which are in greater demand.

Crude-Oil Reserves Versus Production

The total amount of crude oil remaining in the ground represents Earth's oil **resources**. Finding and exploiting these resources is the challenge. The science of geology provides information about the probable locations and extents of ancient shallow seas. On the basis of their knowledge and their field experience, geologists make educated guesses as to where oil or natural gas may be located and how much may be found. These estimates are termed **undiscovered resources**. The estimates may be far off the mark because there is no way to determine whether undiscovered resources actually exist except by exploratory drilling.

Proved Reserves. If exploratory drilling confirms the presence of oil, further drilling is conducted to determine the extent and depth of the **oil field**. From that information, a fairly accurate estimate can be made of how much oil can be economically obtained from the field. That amount then becomes **proved reserves**. Amounts are given in barrels of oil (1 barrel = 42 gallons), and the content of each reserve field is expressed in probabilities. Thus, a 50% probability, or P50, of 700 million barrels for a field means that a total of 700 million barrels is just as likely as not to come from the field. The final step, the withdrawal of oil or gas from the field, is called **production** or *recovery*. Both terms refer to the extraction of materials from Earth.

Production from a given field cannot proceed at a constant rate because crude oil is a viscous fluid held in pore spaces in sedimentary rock, just as water is held in a sponge. Initially, the field may be under so much pressure that the first penetration produces a gusher. Gushers, however, are generally short lived, and the oil then seeps slowly from the sedimentary rock into a well from which it is pumped out. Ordinarily, conventional pumping (**primary recovery**) can remove only 25% of the oil in an oil field. Further removal, of up to 50% of the oil in the field, is often possible, but it is more costly because it involves manipulating pressure in the oil reservoir by injecting brine or steam (**secondary recovery**), which forces the oil into the wells. In some fields, even greater recovery (**enhanced recovery**) is obtained by injecting carbon dioxide, which breaks up oil droplets and enables them to flow again.

Price Drives the Use of Reserves. Economics—the price of a barrel of oil—determines the extent to which reserves are exploited. No oil company will spend more money to extract oil than it expects to make selling the oil. At $13 per barrel of crude oil, the price in the late 1990s, it was economical to extract no more than 25–35% of the oil in a given field. Thus, an increase in price makes more reserves available. In the 1970s and early 1980s, higher oil prices justified reopening old oil fields and created an economic boom in Texas and Louisiana. Then an economic recession occurred in the late 1980s, and a drop in oil prices caused the fields to be shut down again. Following that, oil prices began to rise in 1998, reached a peak of $145 a barrel in mid-2008, and then began a sharp decline as global oil demand dropped—a consequence of a severe global economic crisis. As the global economic picture began to improve, oil prices rose again, until by mid-2011 oil was once again more than $100 a barrel. However, increased oil production and a switch to natural gas in some sectors caused the price of oil to continue to change, dropping precipitously in 2014.

FORMATION OF FOSSIL FUELS

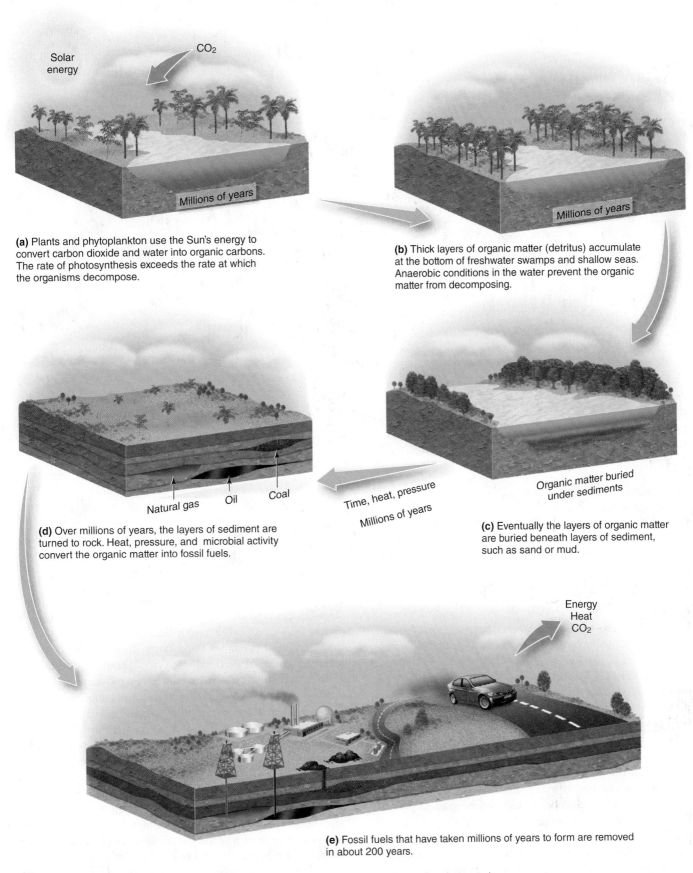

(a) Plants and phytoplankton use the Sun's energy to convert carbon dioxide and water into organic carbons. The rate of photosynthesis exceeds the rate at which the organisms decompose.

(b) Thick layers of organic matter (detritus) accumulate at the bottom of freshwater swamps and shallow seas. Anaerobic conditions in the water prevent the organic matter from decomposing.

(d) Over millions of years, the layers of sediment are turned to rock. Heat, pressure, and microbial activity convert the organic matter into fossil fuels.

(c) Eventually the layers of organic matter are buried beneath layers of sediment, such as sand or mud.

(e) Fossil fuels that have taken millions of years to form are removed in about 200 years.

Figure 14–9 Energy flow through fossil fuels. Coal, oil, and natural gas are derived from biomass that was produced many millions of years ago. Deposits are finite, and because formative processes require millions of years, they are nonrenewable.

Figure 14–10 **Oil production and consumption in the United States.** Four stages can be seen: (1) Up to 1970, the discovery of new reserves allowed production to parallel increasing consumption. (2) From 1970 to 1973, a lack of new oil discoveries caused production to decline while consumption continued to climb, resulting in a vast increase in oil imports. (3) In the late 1970s to early 1980s, high oil prices promoted both lowered consumption and increased production—which included bringing the Alaskan oil field into production. (4) From 1986, when oil prices declined sharply, until the present, consumption has again increased while production has resumed its decline, until very recently.

(*Source*: Data from Energy Information Administration, U.S. Department of Energy, *Annual Energy Review 2010*, October 2011.)

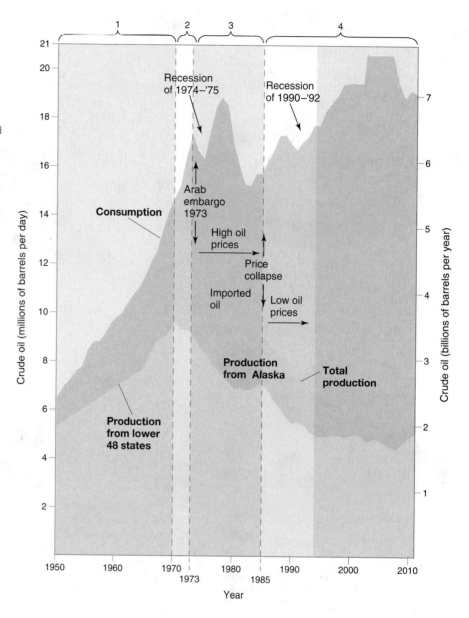

U.S. Reserves and Importation

Up to 1970, the United States was fairly oil independent, producing about two-thirds of oil consumed and keeping imported oil prices low (Figure 14–10). In 1970, however, new discoveries fell short, so production decreased, while consumption continued to grow rapidly. This downturn vindicated the predictions of the late geologist M. King Hubbert, who proposed that oil exploitation in a region would follow a bell-shaped curve. Hubbert observed the pattern of U.S. exploitation and predicted that U.S. production would peak (now called the *Hubbert peak*) between 1965 and 1970. At that point, about half of the available oil would have been withdrawn, and production would decline gradually as reserves were exploited. Figure 14–10 demonstrates that a Hubbert curve can be applied to U.S. oil production.

To fill the energy gap between rising consumption and falling production, the United States depended increasingly on imported oil, primarily from the Arab countries of the Middle East. European countries and Japan did likewise.

The Oil Crisis of the 1970s. It was obvious that the United States and most other highly industrialized nations were becoming dependent on oil imports. In a classic economic move, a group of predominantly Arab countries known as the Organization of Petroleum Exporting Countries (OPEC) formed a cartel and agreed to restrain production in order to get higher prices. Imported oil began to cost more in the early 1970s, and then, in conjunction with an Arab-Israeli war in 1973, OPEC initiated an embargo of oil sales to countries that gave military and economic support to Israel (such as the United States). Spot shortages quickly occurred, which escalated into widespread panic and long lines at gas stations. We were willing to pay almost anything to have oil shipments resumed—and pay we did: OPEC resumed shipments at a price of $12 a barrel ($66 in 2015 dollars), more than four times the previous price of $2.30 a barrel.

Then, by continuing to limit oil production all through the 1970s, OPEC was able to keep supplies tight enough to force prices even higher. In the early 1980s, a barrel of oil delivered to the United States cost $36 ($90 in 2015 dollars), and the United States was paying accordingly. The high energy costs were devastating to the economy; inflation and unemployment rose and the country experienced several recessions. The purchase of foreign oil became, and remains, the single largest factor in our **balance-of-trade deficit**—where we buy more goods from other countries than we sell as exports.

Adjusting to Higher Prices. In response to the higher prices, the United States and other industrialized countries made a number of adjustments during the 1970s. The following are the most significant changes made by the United States:

- We increased domestic production by stepping up oil exploration, building the Alaska pipeline, and reopening old oil fields.

- Congress took steps to decrease consumption by setting standards for automobile fuel efficiency at 27.5 miles per gallon or mpg (cars averaged 13 mpg in 1973), lowered speed limits to 55 miles per hour, promoted higher efficiencies for building insulation and appliances, and began to develop alternative energy sources.

- To protect against another OPEC boycott, we created a strategic oil reserve in underground caverns in Louisiana. The current stockpile is 696 million barrels of oil, equivalent to about 37 days of consumption at 19 million barrels per day.

- We encouraged oil exploration and production in a number of non-OPEC countries, especially those in the Western hemisphere.

The results of the new actions were not immediate, nor could quick returns have been expected. For example, it takes at least three to five years to design a new car model and start production. Then another six to eight years pass before enough new models are sold and old ones retired to affect the overall highway "fleet."

As we entered the 1980s, however, these efforts were definitely having an impact. The United States continued to depend substantially on foreign oil, but consumption was declining, and production, up after the Alaska pipeline was

completed, was holding its own (see Figure 14–10). Elsewhere in the world, significant discoveries in Mexico, Africa, and the North Sea made the world less dependent on OPEC oil. As a result of all these factors, plus OPEC's inability to restrain its own members' production, world oil production exceeded consumption in the mid-1980s, and there was an oil "glut."

Supply exceeded demand, with predictable results. In 1986, world oil prices crashed from the high $20s to $14 per barrel. From that time until late 1999, oil prices fluctuated in the range from $10 to $20 per barrel, almost as low as they were in the early 1970s. The lower oil prices apparently benefited consumers, industry, and oil-dependent developing nations, but they hardly solved the underlying problem.

Back to Old Ways. Whereas high oil prices stimulated constructive responses, the collapse in oil prices undercut those responses:

- Oil exploration was curtailed, and production from older fields was terminated.

- Conservation efforts and incentives were abandoned. Government standards for automobile fuel efficiency were frozen at 27.5 mpg. Speed limits were raised to 65 and even 75 mph.

- Congress terminated tax incentives and other subsidies for the development or installation of alternative energy sources.

Production of crude oil in the United States continued to drop as proved reserves were drawn down. At the same time, consumption of crude oil rose again because of increases in the number of cars on the highways and the average number of miles per year that each car is driven. Adding to that consumption, consumers began buying vehicles that were less fuel efficient. Minivans, four-wheel-drive sport-utility vehicles (SUVs), light trucks, and other less fuel-efficient models constituted more than half of the new-car market. Our dependence on foreign oil rose after the mid-1980s. Recently, however, intensive oil production in the United States has lessened our dependence. Today less than 50% of the crude oil we use comes from imports (Figure 14–11).

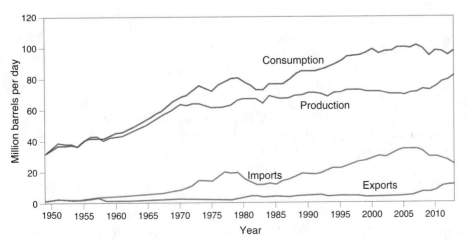

Figure 14–11 Consumption, domestic production, imports, and exports of petroleum products. After the oil crises of the 1970s, the United States became less reliant on foreign oil. Since the mid-1980s, consumption of foreign oil has been rising, although recently it has fallen.

(*Source:* Energy Information Administration, U.S. Department of Energy, *Annual Energy Review* 2014, December 2014.)

Figure 14–12 The cost of fossil fuel imports. (a) The cost in dollars per barrel (today's dollars) and net imports; **(b)** gasoline prices in today's dollars.

(*Source:* Annual Energy Overview 2013. U.S. Energy Information Administration.)

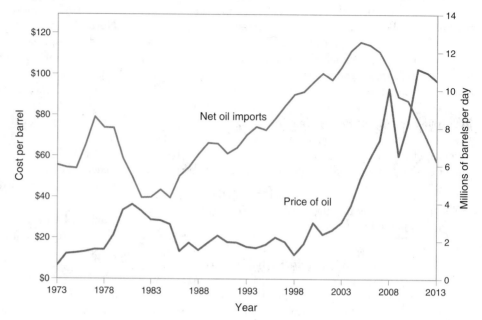

(a) NET OIL IMPORTS AND PRICE OF OIL

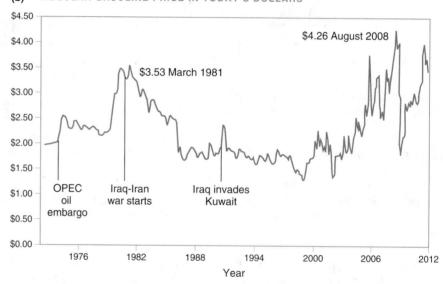

(b) REGULAR GASOLINE PRICE IN TODAY'S DOLLARS

Back to the Future. The collapse in oil prices set the stage for another developing crisis: a rise in oil prices that began in late 1998. OPEC was concerned about the continuing oil glut and its low prices, so it decided to cut production just as East Asia was coming out of a serious recession. The rapid rise in demand brought on a shortfall in the oil supply, and the price of crude oil rose precipitously (Figure 14–12a). With demand outpacing supply, oil prices kept rising, peaking in mid-2008. Homeowners were strapped with high fuel-oil prices, commuters faced gasoline prices in the $4 range (Figure 14–12b), people cut back on driving, economic growth began to slacken, and only the oil industry prospered as record profits poured in. Internationally, food prices rose to unprecedented levels, forcing many in the developing world into dependency and hunger. In the United States, shoppers purchased smaller vehicles, sales of hybrid vehicles blossomed, and the auto industry began closing factories

that produced SUVs, vans, and trucks. Once again, Congress responded, raising the fuel efficiency standard, mandating the increasing use of renewable fuels, and establishing new standards for energy efficiency.

In the fall of 2008, the U.S. housing market collapsed and the bottom dropped out of the stock market. Soon the countries of the world slid into a recession that was called the worst in 75 years. The impact on oil prices was profound: Oil demand dropped, supply exceeded demand, and the price dropped back below $50 a barrel by December of the same year. Gasoline prices dropped down into the $2 a gallon range, and homeowners facing the winter could anticipate much lower fuel-oil prices. Oil-exporting countries suddenly faced a sharp decline in their revenues. However, prices have risen and fallen since then. International factors, such as political unrest in several oil-producing countries and large oil purchases by China, caused prices to increase. More recently, the extraction of oil trapped in shale in the United States has led to lower prices. Because of the complexity of the international market, the future price of oil is hard to predict.

The Consequences of U.S. Dependency

What problems does our dependency on foreign oil present? Such a dependency presents problems at three levels: (1) the costs of purchasing oil, (2) the risk of supply disruptions due to political instability in the Middle East, and (3) ultimate resource limitations.

Costs of Purchase. The United States exports a small amount of oil and imports a great deal. The cost of this reliance on foreign oil is an increased trade deficit. In some years, our net oil trade deficit accounts for as much as two-thirds of our total deficit. The oil trade deficit reflects the volume of imports and the price of oil at the time: A high oil price increases the deficit, and a low price decreases it. After increases in the early 2000s, by 2014 our oil trade deficit had been decreasing for several years.

Payments for foreign oil have significant consequences for our economy. The price we pay at the pump is basically the same whether the oil is produced here or abroad. If the oil is produced at home, however, the money we pay for it circulates within our own economy, providing jobs and producing other goods and services.

Finally, there are the environmental costs of oil spills such as the *Deepwater Horizon* accident and of other forms of pollution from the drilling, refining, and consumption of fuels—although estimating total damage in dollars is sometimes difficult.

Persian Gulf Oil. Our dependence on foreign oil poses risks to our national security. The Middle East is a politically unstable region of the world. As noted earlier, it was the unexpected Arab boycott that plunged us into the first oil crisis in 1973. Maintaining access to Middle Eastern oil is also a major reason for our efforts toward negotiating peace agreements between various parties in that region. Recognizing this political instability, President Jimmy Carter in January 1980 stated that the United States would use military force if needed to ensure our access to Persian Gulf oil (the so-called **Carter Doctrine**). This doctrine was tested in the fall of 1990 when Saddam Hussein of Iraq invaded Kuwait, then producing 6 million barrels of oil per day. The U.S.-led Persian Gulf War of early 1991 threw Hussein's armies out of Kuwait and gave the United States an established military presence in the Persian Gulf region. It was this presence, however, that angered the radical Islamic terrorist Al Qaeda organization and contributed to the horrific September 11, 2001, attacks on the World Trade Center and the Pentagon.

Two More Wars. In response to 9/11, U.S. and British forces invaded Afghanistan, which had been harboring Osama bin Laden and elements of Al Qaeda. The purpose of the war was to capture bin Laden, destroy the Al Qaeda training camps, and overthrow the ruling Taliban regime. Although Afghanistan is not an oil-producing country, it

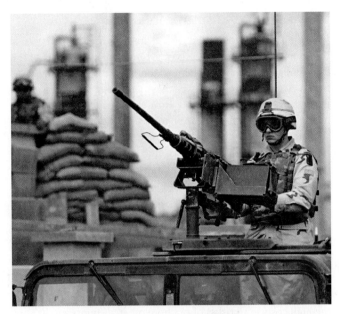

Figure 14–13 The U.S.-Iraqi war and oil. A soldier from the U.S. 173rd Airborne stands guard at a refinery near the Baba Gurgur oil field in Kirkuk, Iraq.

can be argued that the Afghan war was a consequence of our presence in the Persian Gulf region. The Afghan war was formally ended in December 2014, although U.S. troops were expected to remain there until 2016. It was America's longest war, but did not result in a clear end of the Taliban regime.

In 2003, a U.S.–British coalition invaded Iraq in order to overthrow Saddam Hussein's rule and eliminate suspected weapons of mass destruction (**Figure 14–13**). Some observers suggest that oil was an important part of the picture. Although it is clear that the United States and Britain did not simply commandeer the Iraqi oil, they have spent billions to restore Iraq's oil industry—and the United States and other oil-hungry Western countries are eager customers for the oil now that it is flowing again at a rate of 2.9 million barrels per day. Those countries rely on Persian Gulf oil for half of their oil imports. It is fair to say that military costs associated with maintaining access to Persian Gulf oil represent a substantial U.S. government subsidy of our oil consumption, one that has also been costly in human lives.

Other Sources. The United States has turned more and more to non–Middle Eastern oil suppliers. Canada is now the number-one exporter of oil to the United States, with about 40% of the market. Mexico and Colombia are also major sources of oil. Western Hemisphere sources now provide about 52% of our imports. The relative political stability and proximity of these sources are good reasons for the United States to turn to them and away from the Middle East. Russia, a number of former Soviet republics, and several African countries (especially Nigeria) are important new sources of oil. However, most of these other sources, like the United States, are on the downside of the Hubbert curve and are seeing production declining. The conclusion? We will continue to depend on OPEC and the Persian Gulf for a substantial share of our oil imports.

Oil Resource Limitations and Peak Oil

Crude-oil production in the United States is decreasing because of diminishing domestic reserves. Moreover, on the basis of the "Easter-egg hypothesis," there is dim hope for major new U.S. finds. According to this hypothesis, as your searching turns up fewer and fewer eggs, you draw the logical conclusion that nearly all the eggs have been found. In terms of oil, North America is already the most intensively explored of any continental landmass. The last major find was the Alaskan oil field in 1968. Discoveries since then have been small in comparison, with increasing numbers of dry holes in between. Indeed, much of the "new" oil in the United States is coming from innovative computer mapping of geological structures and horizontal drilling technology that enable oil to be identified and withdrawn from small isolated pockets that were previously missed within old fields. The *Deepwater Horizon* oil leak was a consequence of the need to develop wells in deeper water and deeper below the sea floor in order to keep production up. In spite of the high costs (and risks) of offshore oil, it accounts for 30% of domestic production at present. Oil resources like deep Gulf oil and potential resources (undiscovered) below the Arctic Ocean will be exploited only if prices are high enough to pay for exploration and production (see Sound Science, *Energy Returned on Energy Invested*).

How much oil is still available for future generations? More than 1,150 billion barrels (BBs) of oil have already been used, almost all in the 20th century. Currently, the nations of the world are using about 92 million barrels per day. Rising demand by the developing countries has more than counterbalanced a slight decline in the demand for oil in some developed countries resulting from weak economies and greater energy efficiency, so the global demand for oil continues to rise. How long can the existing oil reserves provide that oil? As with many issues, it depends upon whom you believe.

Hubbert's Peak. Some years ago, oil geologists Colin Campbell and Jean Laherrère provided an analysis that put the world's proved reserves at about 850 BBs.[1] This is much lower than the 1,688 BBs claimed by more recent oil industry reports (Table 14–2). On the basis of their estimates and the known amount already used, the two geologists calculated a Hubbert curve indicating that peak oil production has already occurred. Princeton professor and former oil geologist Kenneth Deffeyes used the Hubbert technique in his book *Beyond Oil: The View from Hubbert's Peak* to draw the same conclusions.[2] Even if these scientists are wrong and the oil industry estimates are right, it would delay the peak by only a decade at most. The Association for the Study of Peak Oil (ASPO), founded by Colin Campbell, believes that the peak for conventional oil is already behind us (Figure 14–14). If not, it is likely to be close, from all reports.

---

[1]Colin J. Campbell and Jean H. Laherrère, "The End of Cheap Oil," *Scientific American*, March 1998.
[2]Kenneth Deffeyes, *Beyond Oil: The View from Hubbert's Peak* (New York: Hill and Wang, 2005).

Table 14–2 Proved World Reserves and Annual Use of Fossil Fuel Reserves, 2014

| Region | Reserves | | |
	Petroleum (billion barrels)*	Natural Gas (trillion ft³)*	Coal (billion metric tons)*
North America	230	413	245
South and Central America	330	271	15
Europe and Eurasia	148	2,000	311
Middle East	795	2,835	1.0
Africa	130	502	32
Asia Pacific	42	537	288
Total	1,688	6,558	892

| | Use | | |
	Petroleum	Natural Gas	Coal
World use, 2013	33.2 billion barrels	118.2 trillion ft³	5.47 billion metric tons
World use, 2013 (quads)	189 quads	120.6 quads	96 quads

Source: BP Statistical Review of World Energy, 2014.

*The fuels may be compared by calculating their energy content in British thermal units (Btus) and then converting these figures to quadrillion Btus or quads, where 1 quad = 10^{15} Btus. One billion barrels of petroleum yield about 5.7×10^{15} Btus, or 5.7 quads. One trillion cubic feet of natural gas yield about 1.02×10^{15} Btus, or 1.02 quads. One billion metric tons of coal yield about 17.65×10^{15} Btus, or 17.65 quads.

GLOBAL OIL AND GAS PRODUCTION PROFILES

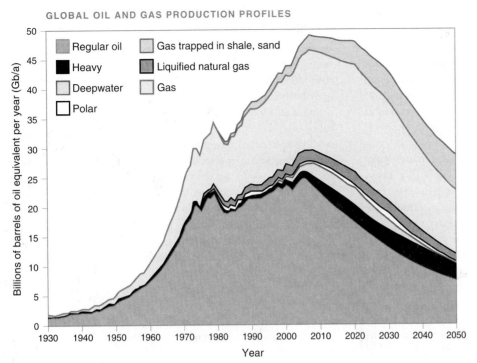

Legend:
- Regular oil
- Heavy
- Deepwater
- Polar
- Gas trapped in shale, sand
- Liquified natural gas
- Gas

Figure 14–14 Hubbert curve of oil and gas production. The Association for the Study of Peak Oil and Gas believes that combined oil and gas production has already peaked.

(*Source:* ASPO Newsletter No. 100, April 2009. http://aspoireland.files.wordpress.com/2009/12/newsletter100_200904.pdf. October 20, 2011.)

Downward Trajectory. At the current rate of use (which is expected to rise), the proved reserves can supply only 43 years of oil demand. New discoveries continue to add to proved reserves, but they are only a fraction of our annual use. The remaining oil will be more costly to extract, and at some point, producers will close wells rather than sustain losses on their oil production. Thus, the 21st century will be an era of declining oil use.

As oil becomes scarcer, and developing countries such as China and India demand more of it, the United States needs to be careful about its dependence on foreign oil. We can become more independent in three ways: (1) We can increase the fuel efficiency of our transportation system, which uses most of the oil; (2) we can develop alternatives to fossil fuels, many of which are ready now; and (3) we can use the other fossil fuel resources we have to make fuel for vehicles. This leads us to some *non-conventional sources of oil*—oil sand and oil shale—which are abundant in the United States and Canada.

Oil Sand and Oil Shale

Oil sand is a sedimentary material containing bitumen, an extremely viscous, tarlike hydrocarbon. When oil sand is heated, the bitumen can be "melted out" and refined in the same way as crude oil can. Northern Alberta, in Canada, has the world's largest oil-sand deposits, and Canada is commercially exploiting them because the cost has been competitive with oil prices. Canadian oil sands currently yield 1.6 million barrels of oil per day (584 million barrels per year).

The United States is eager to buy the oil, which represents more than 10% of our imported oil. Oil from the Canadian oil sands is piped to refineries in Oklahoma and Illinois via the Keystone pipelines **(Figure 14–15)**.

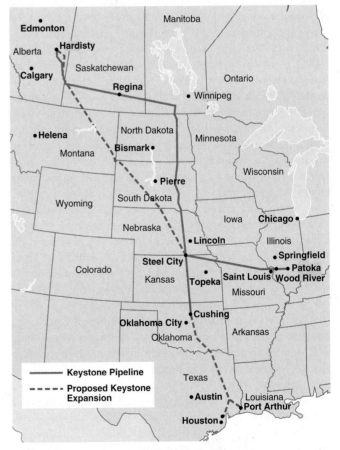

Figure 14–15 Keystone pipelines. The existing Keystone pipelines carry oil from Alberta oil sands to Oklahoma and Illinois. The proposed Keystone XL pipeline (Proposed Keystone Expansion on the map) is a larger line that can carry oil to Texas and the Gulf.

(*Source:* U.S. State Department.)

These pipelines have a limited capacity, however, and TransCanada Keystone has proposed a longer and larger pipeline, the Keystone XL, that would run 1,700 miles through seven U.S. states to Port Arthur, Texas, on the Gulf coast. This proposed extension is controversial. Proponents cite the large number of jobs the pipeline construction would provide, property taxes to the communities it would pass through, and the energy security that Canadian oil provides to the United States.

Objections to the Keystone XL project are many: leaks in the pipeline are bound to occur; the pipeline would cross ecologically sensitive lands; and it would also cross over major recharge areas of the Oglalla aquifer, an extremely important source of irrigation water to Midwestern states. (In 2010, corrosive oil from oil sands broke a pipeline owned by a Canadian energy company, polluting the Kalamazoo River in Michigan, even though the company had known about the corrosion of the pipes for five years.) In addition, Canada does not guarantee that the oil carried through this extension will be sold to the United States. Opponents are also concerned about the substantial environmental damage associated with the extraction of oil from the Canadian oil sands; 110,000 acres of boreal forest and wetlands have already been heavily disturbed. In addition, the extraction process generates enormous quantities of greenhouse gases.

Oil shale is a fine sedimentary rock containing a mixture of solid, waxlike hydrocarbons called *kerogen*. The United States has extensive deposits of oil shale in Colorado, Utah, Texas, and Wyoming. When shale is heated to about 1,100°F (600°C), the kerogen releases hydrocarbon vapors that can be recondensed to form a black, viscous substance similar to crude oil, which can then be refined into gasoline and other petroleum products. However, about a ton of oil shale yields little more than half a barrel of oil. The mining, transportation, and disposal of wastes necessitated by an operation producing, for instance, a million barrels a day (just 5% of U.S. demand) would be a Herculean task, to say nothing of its environmental impact.

Nevertheless, because the deposits contain an estimated 800 billion barrels of oil, the rising costs of importing oil make it inevitable that every domestic source will be considered. In November 2008, the Interior Department's Bureau of Land Management (BLM) finalized rules for how commercial oil-shale leases will be regulated, opening the way for oil companies to begin serious consideration of developing oil from the shale deposits. The process faces stiff opposition because of its potential to damage air quality and groundwater; any oil-shale development is subject to National Environmental Protection Act analysis (environmental impact statements) and must get local and state approval.

Worldwide, the combination of oil shale and oil sand has been estimated to hold several trillion barrels of oil. Large-scale commercial development of oil sands has begun, but oil shale will remain undeveloped until oil prices rise. However, shale has recently begun yielding natural gas.

CONCEPT CHECK How would a drop in the price of oil be expected to affect the excavation of oil sands? ☑

14.3 Drilling for Natural Gas

In the United States, natural gas is gaining popularity as a fossil fuel of choice. Natural gas has a number of benefits over other fossil fuels. Burning gas in power plants releases only half as much CO_2 as burning coal to produce the same amount of electricity. Natural gas is also cleaner than either oil or coal, releasing far fewer pollutants when it is burned.

Almost equal quantities of natural gas go to industrial, residential, and commercial use and electrical power generation (Figure 14–8). As with oil, the cost of natural gas is subject to market fluctuations in supply and demand, and prices often rise and fall with seasonal changes. At the 2014 rate of use (24.4 trillion cubic feet), proved reserves in the United States (308 trillion cubic feet) hold only a 12-year supply of gas. However, new drilling is opening up more deposits.

Shale Gas and Fracking. Until recently, most natural gas was mined from fields associated with oil and coal (Figure 14–16). As a fossil fuel, gas production in the United States peaked in 1970 and gradually declined. Proved reserves were not increasing as rapidly as domestic use. All this has changed, however. Using new techniques adapted from oil extraction, an abundant source of natural gas trapped in shale has dramatically changed the natural gas picture in the past few years. New shale gas resources provide an additional supply of gas, bringing the estimated reserves to 2,300 trillion cubic feet, or a total of about 100 years at today's usage rate. This is a conservative estimate. The United States is blessed with an abundance of shale gas basins, the largest being the huge Marcellus basin extending from New York to Tennessee.

Some natural gas has always been found around shale rock formations, tapped by conventional vertical wells, but enormous quantities of gas remained trapped in the shale itself. The new shale mining technique involves

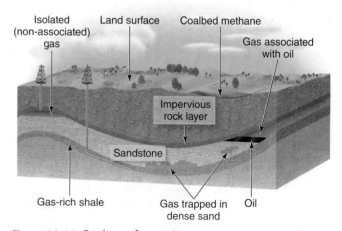

Figure 14–16 Geology of natural gas resources. Until recently, most natural gas was accessed via vertical drilling into reservoirs associated with coal and oil. Natural gas is also found in isolated beds. Natural gas trapped in shale and in very densely packed sand or silt has recently been developed using horizontal drilling and "fracking," or injecting hydraulic fluid under high pressure to open up the shale and release the gas.

(*Source:* U.S. Energy Information Administration.)

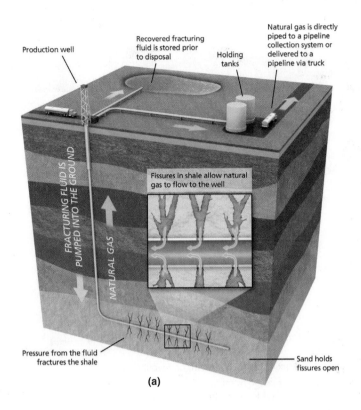

(b)

Figure 14–17 Hydraulic fracturing. Fracking can be used to extract either natural gas or oil from porous shale. **(a)** The fracking process starts with the injection of a fluid containing water, sand, and chemicals into shale. The pressure of the fluid fractures the rocks, allowing bubbles of trapped gas or oil to move to the surface. **(b)** These fracking wells and storage pools are located on the Roan Plateau in Colorado.

drilling a vertical well into the shale and then turning the drill horizontally for distances up to a mile (Figure 14–17). Then, at regular intervals along the horizontal pipe, hydraulic fluids containing sand are forced into the pipe by huge pumps, fracturing the shale and releasing gas into spaces propped open by sand. The fluid, and then the gas, flows back up the well. The technique is called hydraulic fracturing, or **fracking**; the amount of hydraulic fluids pumped into just one well can exceed a million gallons at a single stage, and a number of stages can be employed over time at the well.

The new shale sources have risen to 20% of U.S. gas production, and one consequence of this new abundance is lower prices. However, there are some serious environmental concerns with shale gas and fracking. Although the hydraulic fluids used are mostly water, other chemicals are added to the water, and much of the fluid comes back up to the surface in the process, requiring holding ponds that over time could leak into groundwater. The injected water could also contaminate underground drinking water sources at other stages in the process.

A 2014 report on the environmental effects of fracking by the New York State Department of Health described a wide array of public health consequences resulting from fracking, such as air and water pollution. One less well-known consequence of deep-well injection of wastewater was an increase in the number of earthquakes. New York State's ban on fracking was, in part, a response to this report.[3]

Because of the increasing fleet of natural gas power plants (Figure 14–5), gas is expected to surpass coal if supplies continue to grow as expected and prices stay low. And because of its extensive use in residential and commercial heating, natural gas has been replacing oil as well. But if we need to move to an economy that is not based on fossil fuels, this reliance on natural gas can be only a temporary stage.

Global Resources. Worldwide, the estimated natural-gas resource base is even more plentiful (Table 14–2) than in the United States. Almost four times as much equivalent gas (that is, equivalent in energy) is likely to be available as oil. Much of it, however, is relatively inaccessible. Gas must either be pumped through a pipeline (a process limited by the length of the line) or be subjected to high pressures so that it remains a liquid at room temperature (liquefied natural gas, or LNG). It can then be shipped in large tankers; currently, eight LNG terminals in the United States handle these ships. Thirteen more have been approved but are not yet built. Constructing new LNG terminals has proven to be difficult, as the public views the terminals and tankers as major security and safety hazards. With domestic gas supplies now more abundant, new LNG terminals may not be needed.

Natural Gas–Run Cars. Natural gas can also meet some of the fuel needs for transportation. With the installation of a tank for compressed gas in the trunk and some modifications of the engine fuel-intake system—costing several thousand dollars—a car will run perfectly well on natural gas. Natural gas is a clean-burning fuel, producing carbon dioxide and water but virtually no hydrocarbons or sulfur oxides. Natural gas is in use in the United States by many buses and a number of private and government car fleets,

---

[3]*NY State Department of Health, A Public Health Review of High Volume Hydraulic Fracturing for Shale Gas Development*, 2014, 176.

SOUND SCIENCE
Energy Returned on Energy Invested

Consider an oil field where all you have to do is drill a hole and oil comes gushing up. This enterprise requires little energy invested (equipment manufacture, workers, transportation of oil) compared with energy returned in the form of the barrels of oil captured. The ratio of energy returned to energy invested (**EROI**) in this case could be 100:1. That's excellent. Over time, however, the oil flow diminishes and soon has to be coaxed out with secondary or tertiary recovery methods. The EROI declines, and after a while, it drops to 1:1; there is no more point in extracting oil from that well. This is a simple illustration of an important consideration; namely, that some forms of energy are more economical to capture than others.

Oil exploitation in the United States began with very high EROI values, perhaps 100:1. Then, over time, it became increasingly expensive to coax up the oil or to import it (see graph). Now, the EROI ratio is closer to 5:1 in the United States and 10:1 in Saudi Arabia. Oil from oil sands does even worse, at 2.9:1. Other fossil fuels do better: coal at 80:1, natural gas at 15:1. We continue to use oil, however, because it is the only energy source that supports transportation. Natural gas may be favored over coal because of pipelines and of the much lower cost of gas turbines versus coal-fired power plants. So there are other considerations besides EROI that factor into the choice of energy.

As the graph shows, some renewable energy sources do well (hydroelectric,

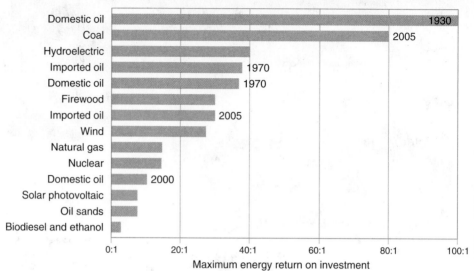

Fossil fuels and renewable energy sources are plotted as ratios of maximum energy return to investment. Because this ratio changes over time, values for some fuels represent more than one year.

(*Source:* Figure 10 in Charles A. S. Hall and John W. Day Jr. "Revisiting the Limits to Growth after Peak Oil." *American Scientist 97*, May–June 2009, 230–237.)

wind), and some do not (solar photovoltaic, biodiesel, and ethanol). If time is factored in (duration of the energy-capturing device in the case of wind and solar), these sources do better. And again, biodiesel and ethanol have a special appeal because they are liquid fuels for transportation. However, the EROI consideration helps to explain why the world isn't just rushing to renewables, as the fossil fuels are presently a more efficient energy source. However, as with the choices of fossil fuels, there are other considerations

besides EROI to factor in. For example, the fossil fuels all release the greenhouse gas CO_2 when burned to produce energy. A future where fossil fuels are consumed until they become economically impractical is an unthinkable future because of global climate change. We will need a renewable energy economy long before that happens. EROI does, however, provide some guidance for choosing renewable energy sources. Right now, wind energy looks like the best choice.

but use is restricted due to the limited number of natural-gas service stations. Detroit automakers have stopped selling natural gas–powered cars, but they are abundant in Latin America and Asia-Pacific. Worldwide, some 18 million natural gas vehicles were in operation in 2014, many of them buses; such vehicles are expected to number 25 million by 2019. It will take stronger public-policy support for this option to grow in the United States; until that happens, the natural-gas option will likely stay on the back burner.

Synthetic Oil. With the aid of the Fischer-Tropsch process, developed in the 1920s, natural gas can be chemically converted to a hydrocarbon that is liquid at room temperature and pressure—basically, a synthetic oil. Researchers are hotly pursuing ways to convert natural gas to synthetic

oil more efficiently. Presently, the fuel produced is only about 10% more expensive than oil. Several major oil companies are spending billions on gearing up refineries to convert gas to liquid diesel and home-heating fuels and researchers are seeking an appropriate catalyst that can convert natural gas (methane) to methanol (methyl alcohol), a liquid. These options promise to make the abundant, but inaccessible, "stranded" natural gas deposits suddenly very valuable.

Applying natural gas to the transportation sector—both directly as a fuel and indirectly as a synthetic oil—may extend the oil phase of our economy, but it will not be a sustainable solution because it is still a fossil fuel.

CONCEPT CHECK What are two benefits of using natural gas? What are two concerns? ☑

14.4 Mining Coal

Coal is the world's most abundant fossil fuel. The leading producer of coal is China (3.6 billion tons in 2013), where a booming economy has stimulated energy demand. Most of this is burned to produce electricity; coal provides nearly 80% of China's electricity. In the United States, 45% of electricity comes from coal-fired power plants. In cost per British thermal unit (Btu), coal is much less expensive than natural gas or oil. Current U.S. reserves are calculated to be 237 billion tons, and at the 2013 rate of production (0.98 billion tons/year), the supply could last hundreds of years. Unlike oil and natural gas, we produce more coal each year than we use, and we export about 6% of coal production annually.

Mining coal can be hazardous, and injuries and fatalities are not uncommon—in 2013, a methane explosion in a Turkish coal mine killed more than 300 miners and trapped 600 others underground. Although such accidents are rare, coal mining comes with many dangers. The most serious of these is the occupational disease called black lung, or coal workers' pneumoconiosis (CWP), acquired by inhaling coal dust. Fatalities and hospitalizations from this disease have risen in recent years, prompting the federal government to post new regulations designed to lower the risks posed by coal dust.

Coal can be obtained by surface mining (67%) or underground mining (33%). Both methods have substantial environmental impacts. Land subsidence and underground fires often occur in conjunction with underground mines. The town of Centralia, Pennsylvania, for example, has been abandoned and bought by the government because of a coal fire that started 40 years ago and could burn for another 100 years. Coal fires around the world produce almost as much carbon dioxide as do all the cars and trucks in the United States, contributing significantly to the world's greenhouse gas emissions. In underground mining, at least 50% of the coal must be left in place to support the roof of the mine, which is one reason why mining companies have turned to surface mining.

Surface Mining

In surface, or strip, mining, dynamite is used to break overlying layers, and then gigantic power shovels turn aside the rock above the coal seam and remove the coal. For example, **mountaintop removal mining** began in the southern Appalachians in the late 1970s as a more economical way to get at the valuable low-sulfur coal found in seams up to 1,000 feet down. The Whitesville mine (Figure 14–18) is only one of many that have devastated parts of West Virginia, Kentucky, Tennessee, and Virginia. Roughly 7%—more than 470 mountains—of the 12 million acres of mountainous land in these states has been or will soon be disturbed by mountaintop removal. Already, more than 700 miles of streams have been completely buried, and another 800 miles have been degraded by the waste products that leach from the mine waste. (This process is similar to the mountaintop removal process used in China to create new land; see Chapter 11.)

Lawsuits in the late 1990s over the impact of mine waste held the practice down, but a rule by the Bush administration changed all that. In 2002, mining debris was reclassified from "waste" to "fill," thus allowing companies free rein to dump the debris into stream valleys. A number of grassroots organizations, including the Ohio Valley Environmental Coalition, Mountain Justice, Appalachian Voices, and Christians for the Mountains, have been formed to combat mountaintop removal mining. They face a formidable array of supporters of the coal industry who argue that the procedure is legal, generates thousands of jobs, and provides the country with cheap energy. Indeed, the state of West Virginia supplies about 15% of U.S. coal production.

In recent years, mountaintop removal mining has come under greater regulatory control. Citing the Clean Water Act, the EPA issued "final guidance" on the mining practices; these guidelines strengthen and clarify the permitting process and give greater protection to waterways affected by mining operations. This action was prompted by a widespread

Figure 14–18 Aerial view of mountaintop coal removal. This coal mining operation in Whitesville, West Virginia, is managed by Massey Energy. The tops of mountains are stripped away and dumped in valleys to get at the coal below.

failure of the mining companies to provide water discharge monitoring reports and the use of a fast-track permitting process that led to violations of the Clean Water Act.

Coal Power

Virtually all of the coal the United States produces is used to generate electricity. Once it is mined, coal is transported to large power plants. A typical 1,000-MW plant burns 8,000 tons of coal a day—a mile-long train's worth. Combustion of this much coal releases 20,000 tons of carbon dioxide and 800 tons of sulfur dioxide (most of which is prevented from entering the atmosphere if the power plant is well provided with scrubbers; see Chapter 19). Then there is the waste: 800 tons of fly ash and 800 tons of boiler residue ash daily, requiring a special landfill.

A 2008 spill of coal ash slurry in a containment area of a coal plant run by the Tennessee Valley Authority in Kingston, Tennessee, sent 1.1 billion gallons (4.2 million cubic meters) across 300 acres of land and into adjacent streams. This spill caused the EPA to reevaluate coal ash storage and begin a long process of developing a regulation to protect the ground and water from such contamination. This regulation was finalized in 2014.

Coal-burning power plants are responsible for more than generating electricity; each year pollution from them takes an estimated 13,200 lives and costs more than $100 billion in health care expenditures. The main reason for these impacts is older power plants that lack pollution controls, exempted by grandfather clauses in older Clean Air Act legislation. The EPA invoked the Clean Air Act "good neighbor rule," which prohibits downwind pollution, to issue its Cross-State Air Pollution Rule. This rule required heavy reductions in sulfur dioxide and nitrogen oxides in older power plants, effective in 2014. However, much of the impact of coal-burning power plants is their effect on greenhouse gas emissions, discussed next.

Climate Issues

Coal combustion is the world's largest source of CO_2, which is the greenhouse gas most significantly forcing climate change (see Chapter 18). If progress is to be made in reducing CO_2 emissions, coal should be on the way out. In reality, new coal-burning power plants are coming on line as older ones are being phased out. The new plants are more efficient and are equipped with the latest pollution controls but lack any control of CO_2 emissions. Yet it turns out that there are ways to continue to use coal and, at the same time, greatly reduce CO_2 emissions, but the technology is still being developed.

The government-and-industry-sponsored Clean Coal Power Initiative (CCPI) program is the successor to the Clean Coal Technology (CCT) program. The CCT supported a spectrum of experimental technologies applicable to existing power plants and to a new generation of coal-fired power plants. The objectives of both programs have been to remove pollutants from coal before or after burning and to achieve higher efficiencies so that greenhouse gas emissions

are reduced. Another technology developed under the program is the **integrated gasification combined cycle (IGCC) plant**. In IGCC, coal is mixed with water and oxygen and heated under pressure to produce a synthetic gas, or **syngas**, which is burned in a gas turbine to produce electricity. Sulfur and other contaminants are removed before burning. The CO_2 stream from the process is concentrated; one syngas plant in North Dakota pipes the CO_2 to an oil field, where it is injected into oil wells to enhance recovery.

Clean coal technologies would be an improvement over the current ones. But they still involve substantial greenhouse gas emissions and all of the drawbacks of mining the coal.

In 2014, the EPA proposed a Clean Power Plan, a regulation to lower greenhouse gas emissions from power plants to 30% below 2005 levels by 2030. The EPA estimated that this rule would result in climate and health benefits worth $55–$93 billion in 2030; those health benefits would include the prevention of 2,700–6,600 premature deaths and 140,000–150,000 asthma attacks in children.[4]

CONCEPT CHECK What are two advantages and two disadvantages of using coal as a source of energy? ✓

14.5 Energy Policy

Because energy supply and demand are nationwide issues, only federal policies have the sweeping power to effect changes that will make a difference. Public policy can bring about major improvements in energy use. During the period from 1973 to 1995, the United States saw an 18% reduction in energy growth, saving the economy $150 billion a year in energy expenditures. This reduction was due to the steps taken in response to the oil crisis of the 1970s. But as that crisis faded into history, the country slipped back into old ways. The state of affairs clearly demanded a new energy policy.

Supply-Side and Demand-Side Policies

In order to have secure energy and be less reliant on imported oil, both *supply-side* and *demand-side* policy options are possible. Supply-side initiatives increase available fossil fuels, whereas demand-side policies lower the use of fossil fuels through efficiency or alternative energy sources.

Supply-side policies dealing with fossil-fuel energy are *business-as-usual* solutions to our energy security issues. They do little to reduce our vulnerability to oil price disruptions, terrorist actions, and global climate change. Demand-side policies, by contrast, have the advantage of reducing our energy needs and making it possible to move to a future in which renewable energy can take over. Such policies will decrease our vulnerability to terrorists and to disruptions of the oil market. They will also save money and reduce pollution from burning fossil fuels.

---

[4]EPA, "Fact Sheet: Clean Power Plan Overview: Cutting Carbon Pollution from Power Plants," December 23, 2014, http://www2.epa.gov/carbon-pollution-standards/fact-sheet-clean-power-plan-overview.

Overview of Major Energy Policies

In recent years, the federal government has used both supply-side and demand-side policies to influence our use of energy. Table 14–3 and Table 14–4 (on the following page) compare features of three recent federal energy policies.

After years of debate, Congress passed the **Energy Policy Act of 2005 (PA)**, which was intended to establish U.S. energy policies for years to come. The PA acted on recommendations from the Bush administration to increase the energy supply, but also included a mixture of other initiatives. Then a sea change in environmental policy led to the **Energy Independence and Security Act of 2007 (EI)**, which addressed mainly the demand side. After 2008, the new Obama administration used both supply and demand policies to influence energy policy. The **American Recovery and Reinvestment Act of 2009** directed $90 billion into clean energy. The Recovery Act and various Obama administration actions related to energy policy are referred to as **O&R**.

Current U.S. energy policy focuses on lowering dependence on foreign oil, making sure our energy supply is secure, and solving climate change. To achieve these goals, we use a combination of approaches, some of which are better for the environment and some of which are worse.

Supply-Side Policies. To increase supply, the federal government has pursued actions that allow for faster permitting, increased exploration, and greater use of public lands in exploring and developing domestic sources of oil and gas. Two very contentious issues in increasing domestic sources have been the approval of new oil pipelines and the possibility of opening the Arctic National Wildlife Refuge (ANWR) to drilling. The O&R approved an oil pipeline into Minnesota and Wisconsin to transport oil from Alberta's tar sands, approved drilling of exploratory wells in the Arctic Ocean, and advocated for approval of the Keystone XL pipeline for oil from Canada's tar sands. The increasing use of coal has been

Table 14–3 Supply-side strategies in three recent U.S. energy policies.

	Supply-Side Policies		
Feature	Energy Policy Act of 2005 (PA)	Energy Independence and Security Act of 2007 (EI)	Obama administration actions and Recovery Act of 2009 (O&R)
Exploring and developing domestic sources of oil and gas	Promoted taking an inventory of offshore oil and gas resources but backed away from opening the Arctic National Wildlife Refuge (ANWR) to drilling.		Approved an oil pipeline into Minnesota and Wisconsin to transport oil from Alberta's tar sands; approval of the controversial Keystone XL line from Canada to Texas is pending. Approved drilling of exploratory wells in the Arctic Ocean. Took steps to increase federal protection of an oil-rich region of ANWR.
Increasing the use of the vast coal reserves for energy	Promoted energy from coal via clean coal technologies, loan guarantees, and research incentives.		Opened up new federal lands in Wyoming's Powder River Basin to coal mining.
Continuing subsidies to oil and nuclear industries	Guaranteed $6.9 billion in tax incentives and billions more in loans for these industries.	Recommended cutting oil and gas subsidies in order to fund a proposed $467 billion job-creation package. This would bring in $41 billion over a decade.	Included a broad repeal of tax incentives for oil and gas in the original legislation, but only a limited repeal was enacted, enough to offset some estimated costs.
Removing environmental and legal obstacles to energy development	Provided tax credits for hydro-power plants and stream-lined permitting for drilling.		
Providing access to remote sources of natural gas	Required reporting of progress in developing a natural-gas pipeline to bring Alaskan natural gas to the lower 48 states, established federal control over the siting and permitting of new LNG terminals.		Recommended expediting the construction of a natural-gas pipeline to bring Alaskan natural gas to the lower 48 states.

another controversial element of supply-side policy. The O&R opened up new federal lands in Wyoming's Powder River Basin to coal mining, potentially yielding 2.35 billion tons of coal. The PA promoted energy from coal via "clean coal" technologies. Several of these provisions have been hotly contested by people concerned about their environmental effects.

Federal laws have continued subsidies to oil and nuclear industries, an approach not supported by environmentalists or advocates for small government. The PA guaranteed $6.9 billion in tax incentives and billions more in loans. In contrast, the EI included a broad repeal of tax incentives for oil and gas in the original legislation, designed to offset the cost of many of the new elements of the law. Opposition to the repeal forced the removal of this option. The O&R recommended cutting oil and gas subsidies in order to fund a proposed $467 billion job-creation package.

Demand-Side Policies. Lowering demand requires more efficient technology and changes in behavior. This is done through a number of incentives and disincentives—"carrots" and "sticks." Some changes have the effect of removing economic externalities by making environmentally damaging activities reflect their true cost. For example, in 2014 the EPA began to regulate emissions of greenhouse gases from coal-fired gas plants. This was a huge step in reflecting the true cost of coal on the environment.

Table 14–4 shows several ways the federal energy laws promote energy policies to lower demand. Increasing the mileage for motor vehicles, for example, results from raising the **corporate average fuel economy (CAFE)** standards, the fuel efficiency requirements for vehicles. The most recent energy policy mandated a CAFE target of 38 mpg for cars and 28 mpg for light trucks by 2016. New standards have also been issued for heavy trucks and buses, designed to increase fuel economy by 25%. This conservation reserve has already been tapped to some extent, because auto manufacturers had already doubled the average fuel efficiency of cars from 13 to 27.5 mpg. Federal law also grants consumer tax credits for fuel-efficient hybrid and other advanced-technology vehicles.

Table 14–4 Demand-side strategies in three recent U.S. energy policies.

Feature	Energy Policy Act of 2005 (PA)	Energy Independence and Security Act of 2007 (EI)	Obama administration actions and Recovery Act of 2009 (O&R)
Increasing the mileage standards for motor vehicles	Authorized continued study of the CAFE standards, granted consumer tax credits for fuel-efficient hybrid and other advanced-technology vehicles.	Raised the CAFE standards for the combined fleet of cars and light trucks gradually, so as to meet a target of 35 mpg by model year 2020.	This goal has been superseded by rules from the Obama administration that mandate higher CAFE targets.
Increasing the energy efficiency of lighting, appliances, and buildings	Energy Star program addresses energy efficient buildings; provides tax breaks for those making improvements to the energy efficiency of their homes (see Chapter 16).	Required a 30% efficiency increase for lightbulbs; did not ban the incandescent bulb entirely.	Extended the Energy Star program to many more facilities; provided tax credits for improving energy efficiency in buildings; developed more efficiency standards for appliances; promoted plans for a smart grid; invested in weatherizing low-income homes.
Encouraging industries to use combined heat and power (CHP) technologies	Took no significant action; amended a previous act that required utilities to buy power from CHP plants.	Defined CHP as a suitable form of energy savings, making it eligible for energy efficiency grants and research proposals.	
Promoting greater use of non-fossil fuel sources of energy (nuclear and renewable energy)	Included a wide range of incentives to stimulate nuclear power: tax credits for generated electricity, insurance against regulatory delays, and federal loan guarantees. Renewables: tax credits for wind, solar, and biomass energy and funding for research and development of renewable energy initiatives.	Established a renewable fuel standard, encouraging the production of biofuels.	Poured some $90 billion into clean energy, including subsidies to home renewable energy systems and businesses, electric car development, advanced biofuels, smart electric meters, factories to manufacture renewable energy components, and many others (Chapter 16).

Lighting, appliances, and buildings. Demand can be lowered by increasing the energy efficiency of lighting, appliances, and buildings. The PA provided tax breaks for manufacturers of energy efficient appliances; encouraged the EPA to continue its Energy Star program, which also addresses energy efficient buildings; and provided tax breaks for those making energy efficiency improvements to their homes (see Chapter 16). The EI required a 30% efficiency increase for lightbulbs. Compact fluorescent lightbulbs (CFLs) are between 20% and 25% efficient (Figure 14–19), whereas incandescent lightbulbs are only 5% efficient. This act also set new efficiency standards for appliances (such as refrigerators, freezers, washers, dishwashers), addressed energy efficiency in buildings and industry, and encouraged the development of energy efficient "green" commercial buildings. The Recovery Act extended the EPA's successful Energy Star program to many more facilities; provided tax credits for improving energy efficiency in buildings; developed more efficiency standards for appliances; promoted plans for a smart grid; and invested in weatherizing low-income homes.

Combined heat and power. U.S. energy policies encourage industries to use combined heat and power (CHP) technologies. In CHP technology, a factory or large building installs a small power plant that produces electricity and heats the building with the "waste" heat. Such systems can achieve an efficiency of 80%. A new energy-saving technology—the combined-cycle natural-gas unit—has been increasingly employed to generate electricity. Two turbines are employed in this unit: The first turbine burns natural gas in a conventional gas turbine (Figure 14–4b); the second turbine is a steam turbine (Figure 14–4a) that runs on the excess heat from the gas turbine. The use of this technology boosts the efficiency of the conversion of fuel energy to electrical energy from around 30% to as high as 50%. The cost of combined-cycle systems is half that of conventional coal-burning plants, and the pollution from such systems is much lower. A large proportion of new units being built or planned for the future consists of combined-cycle systems.

Renewable energy. Several federal laws promote the greater use of non-fossil fuel sources of energy (nuclear and renewable energy). The PA included a wide range of incentives to stimulate nuclear power: tax credits for generated electricity, insurance against regulatory delays, and federal loan guarantees. The act also authorized $4.5 billion in support of renewable energy and efficiency, with tax credits for wind, solar, and biomass energy and major funding for research and development of renewable energy initiatives.

The EI established a renewable fuel standard, encouraging the production of biofuels. The Recovery Act poured some $90 billion into clean energy, including subsidies to renewable energy systems for homes and businesses, electric car development, advanced biofuels, smart electric meters, and factories to manufacture renewable energy components. These and other renewable energy provisions will be covered later (Chapter 16).

Conservation reserve. In rethinking our energy strategies, we need to realize that energy has no real value by

Figure 14–19 Energy-efficient lightbulbs. Replacing standard incandescent lightbulbs with screw-in fluorescent bulbs (shown here) can cut the energy demand for lighting by 75–80%. The higher cost of the fluorescent bulbs is more than offset by the greater efficiency and much longer lifetime of the bulb.

itself. Its only worth is in the work it can do. Thus, what we really want is not energy, but comfortable homes, transportation, manufactured goods, and so forth. As a result, many energy analysts argue that we should stop thinking in terms of where we can get additional supplies of fossil fuels. Instead, we should be thinking of ways to satisfy these needs with a minimum expenditure of energy and the least environmental impact. Imagine discovering a new oil field having a production potential of at least 6 million barrels a day—three times the capacity of the Alaskan field, even at its height. Furthermore, imagine that this newfound field is inexhaustible and that exploiting it will not adversely affect the environment. Of course, such an oil field does not exist, but saving energy through greater efficiency is the equivalent of exploiting an untapped reserve—a *conservation reserve.*

Increasing the Cost of Carbon. A number of economists and environmental groups suggest that we should be paying more for gasoline and other fossil fuels. In most other developed countries, gasoline is so heavily taxed that it costs consumers upward of $6 per gallon—unlike U.S. prices, which were below $3 for 14 of the past 20 years. If the price of fossil fuels included the externalities of their production and consumption, consumers would have an incentive to use less. Thus, increasing the price of these fuels is a demand-side policy. Two strategies for achieving this goal are a carbon tax and a cap-and-trade approach. Both strategies also come up in any discussion of climate change (Chapter 18).

Carbon tax. A carbon tax is a tax levied on all fuels according to the amount of carbon dioxide that they produce when they are extracted, transported, and consumed. Such a tax would provide incentives to use renewable energy sources, which would not be taxed, and disincentives to consume fossil fuels. A number of European countries have already adopted such a tax, but in the United States such a tax is controversial.

Cap and trade. Other economists suggest using a process in which a national government or international body sets a cap on the amount of pollutant allowed into the environment. Different groups are allotted amounts they can produce. They may buy and sell (trade) those rights. A company that invests in very efficient technologies would be able to sell its permits; those that do not would be forced to buy permits. Periodically the total allotment drops, forcing permit holders to become more efficient. The European Union Emission Trading Scheme is an example of a cap-and-trade program. The United States used a similar approach to reduce the emissions of sulfur and nitrogen oxides that caused acid rain (Chapter 19).

State Policies. In addition to federal regulations, states may enact laws that have broad energy impacts. In 2011, for example, California set a higher standard for lightbulb efficiency and enacted a law requiring 33% of state energy supplies to be renewable by 2020; then governor Jerry Brown said this Renewable Energy Portfolio would enable California to lead the country in energy policy. In 2013, California set in place a policy to require utilities to develop energy-storage capacities to lower the risk of power interruptions. In 2014, New York State banned fracking for natural gas. States may also increase taxes on gasoline and other fossil-fuels.

International Policies. Countries cooperate internationally in energy policy. The most obvious way is through climate change treaties and nuclear treaties (Chapters 16 and 18).

Final Thoughts

While conservation of fossil fuels is extremely important, there are two major pathways for developing a low-carbon energy future: pursuing nuclear power and promoting renewable-energy applications. Both could be expanded to meet further energy needs if suitable technological solutions and public acceptance (in the case of nuclear power) or pressure and government support (in the case of renewable energy) are provided. (Chapter 15 focuses on the nuclear option, and Chapter 16 focuses on renewable options).

CONCEPT CHECK   Why are supply-side policies aimed at reducing U.S. dependence on foreign oil usually considered to be environmentally unsustainable? ✓

REVISITING THE THEMES

SOUND SCIENCE

Science and technology have flourished in the age of fossil fuels, not least because they have devoted much research toward developing new and often ingenious ways of extracting the energy from those fuels. The need today, however, is for scientists and engineers to point the way toward using fossil fuels with greater efficiency and leaving as much fossil fuel in the ground as is possible. Major scientific tasks include the development of technologies to increase the efficiency of energy use and the accurate assessment of the impacts of using fossil fuels.

SUSTAINABILITY

The 19th, 20th, and 21st centuries have benefited immensely from the fossil fuels buried in the Earth over past eons. These are, however, nonrenewable resources. They will not be available for very long, as human history goes. We are currently traveling on an unsustainable track, powered by fuels that are running out. Sustainable options are available, but we are not putting our best efforts there. The most serious consequence of our current track is that we continue to pump greenhouse gases, especially CO_2, into the atmosphere, where they are already affecting the global climate and promise to do so at an accelerating rate as we burn ever more fossil fuels. Eventually, human civilization will have to make a transition to a low-carbon economy. The sooner this becomes a reality, the better off the entire planet will be.

STEWARDSHIP

The fossil fuel business is costly in human health and welfare, but we are so deeply committed to using fossil fuels that we simply chalk these concerns up to the cost of doing business. Further, what concern, if any, should we have for future generations? During the 20th century, we used half of the available oil, and our use is accelerating. A thoughtful stewardship of available resources should motivate us to refrain from pumping all the oil and natural gas to satisfy present demands and to plan for the future generations that will be facing far greater difficulties than we are as they struggle to complete the transition to renewable energy.

REVIEW QUESTIONS

1. What happened in the Gulf of Mexico in April 2010, and what was the cause? The consequences?

2. How have the fuels to power homes, industry, and transportation changed from the beginning of the Industrial Revolution to the present?

3. What are the three primary fossil fuels, and what percentage does each contribute to the U.S. energy supply?

4. Electricity is a secondary energy source. How is it generated, and how efficient is its generation?

5. Name the four major categories of primary energy use in an industrial society, and match the energy sources that correspond to each.

6. What is the distinction between *undiscovered resources* and *proved reserves* of crude oil, and what factors cause the amounts of each to change?

7. What did the United States do in the early 1970s to resolve the disparity between oil production and consumption? What events caused the sudden oil shortages of the mid-1970s and then the return to abundant, but more expensive, supplies?

8. What is the Carter Doctrine, and how does it relate to wars in the Persian Gulf?

9. What is meant by Hubbert's Peak, and how does it apply to the United States and the world?

10. What is the impact of Canadian oil sands on oil reserves and supply to the United States?

11. What new technology has revolutionized the natural gas scene in the United States?

12. Describe surface mining as it is practiced in the southern Appalachians.

13. What phenomenon may restrict the consumption of fossil fuels, especially coal, before resources are depleted?

14. Compare the Energy Policy Act of 2005, the Energy Independence and Security Act of 2007, and Obama administration initiatives relative to supply- and demand-side strategies.

15. To what degree can energy conservation serve to mitigate energy dependency? What are some prime examples of energy conservation?

THINKING ENVIRONMENTALLY

1. Statistics show that many developed countries use far less energy per capita than we in the United States do. Explain why this is so.

2. Predict what gasoline prices will do in the next 10 years, and rationalize your prediction.

3. Suppose your region is facing power shortages (brownouts). How would you propose solving the problem? Defend your proposed solution on both economic and environmental grounds.

4. List all the environmental impacts that occur during the "life span" of coal—that is, from mining to the electricity in your home or dorm.

5. Explore the pros and cons of the Keystone XL pipeline, fracking for shale gas, and mountaintop removal coal mining.

MAKING A DIFFERENCE

1. An easy first step for reducing your energy use is replacing your conventional lightbulbs with compact fluorescents. You've probably heard that before, but have you done it? Fluorescent bulbs do contain a small amount of mercury, so if you accidentally break one, it would be wise to open your windows; and when you clean up the glass, don't use your bare hands. The mercury levels are low enough that the last traces will vaporize in a matter of hours.

2. When you are in the market for transportation, choose a vehicle that gets high mileage per gallon in anticipation of the higher fuel costs that will certainly occur before the vehicle wears out. If you can afford it, purchase a hybrid vehicle.

3. To further cut down on power, turn your thermostat down a few degrees in the winter and up a few degrees in the summer. Wear warm or light clothes to help make up for the temperature difference. One-sixth of power use in the United States is due to air-conditioning.

4. Buy a power strip. This makes it easy to turn things off without having to go to the trouble of unplugging them. Some strips can even detect idle current and shut them off without the need to flip any switches.

Learning Objectives

15.1 Nuclear Energy in Perspective: Explain how nuclear power was born in the United States, and outline the role of nuclear energy in the United States and the world.

15.2 How Nuclear Power Works: Review the basics of nuclear power, and compare the benefits and disadvantages of nuclear power with those of coal power.

15.3 The Hazards and Costs of Nuclear Power Facilities: Summarize the essentials of radioactivity and nuclear wastes; evaluate nuclear power in the light of high-level wastes, nuclear accidents, and economic considerations.

15.4 More Advanced Reactors: Describe how fast-neutron reactors are used to reprocess spent nuclear fuel, and assess the potential for fusion as a source of energy.

15.5 The Future of Nuclear Power: Examine the impact of global climate change on the nuclear energy option, and assess the future prospects for nuclear power.

Nuclear Power

Earthquakes are common in Japan, which sits on the Pacific Rim, an area well known for its seismic activity. On March 11, 2011, the Pacific tectonic plate slipped under the plate that lies under northern Japan, 42 miles offshore and 15 miles below the surface (see Chapter 4, Figure 4–21, for plate tectonic map). The resulting upthrust triggered an earthquake of magnitude 9.0, the most severe in Japan's recorded history, and a tsunami that hit the northern Japanese coast with a wall of water 10–50 feet high. The earthquake was damaging, but the Japanese infrastructure was built with earthquakes in mind and the damage was not devastating. The tsunami, however, hit the coast within 10–30 minutes of the earthquake, sweeping over tsunami seawalls meant to protect cities and towns and pushing buildings, vehicles, and debris inland at a frightening speed. People had little time to escape the incoming waves; the death toll was close to 20,000. More than 1 million buildings were damaged or destroyed, and the economic losses from the disaster were estimated to be more than $300 billion.

Japan's 55 nuclear power plants are located on the coastline where cooling water is available. When the earthquake struck, 11 of the plants shut down automatically. However, several of the plants were on the east coast, and those closest to the epicenter of the earthquake suffered damage first from the earthquake and then from the tsunami. The Fukushima Daiichi power station, with six nuclear reactors, took a direct hit. Seawater swept by the reactors, cut off all electrical power, and demolished the backup generators and diesel fuel tanks. All of the reactors automatically shut down when the quake hit, but even with the nuclear fission halted, radioactive decay in the fuel rods generated heat and needed to be kept cool. With no cooling water available, three of the reactors suffered meltdown of the nuclear fuel in the reactor vessels. Several units were rocked by explosions set off by hydrogen gas generated by extreme temperatures in the fuel rods, stripping the hydrogen from surrounding water. All of these events released radioactive

▲ A huge tsunami wave at Miyako City, Japan, on March 11, 2011.

fission products into the air. Authorities proclaimed a mandatory evacuation zone for a 12-mile (20-km) radius from the stricken plant; 75,000 people were moved to emergency housing.

Containment. The next day, plant workers began to inject seawater into the reactor vessels to cool them down. Another problem became evident when pools holding spent fuel rods also lost water and began releasing radiation. These too eventually received cooling from seawater. Within days, electrical power was restored to the facility, but conditions in the vicinity of the reactors and pools were so radioactive that workers could not get near enough to make repairs. Some 800 workers were involved in efforts to bring the reactors under control, working short periods in order to limit their exposure to the considerable radiation. The 12-mile exclusion zone received significant radioactive fallout from the stricken plant; it is likely that this will become a "dead zone." Workers erected a fabric cover over the stricken units to contain airborne radiation emanating from the reactors. New cooling systems for the reactors have been established, but temperature levels in the reactors and fuel pools were not brought down below the boiling point until six months after the disaster. Ultimately, the objective of this phase of containment was to achieve *cold shutdown*—where the temperature at the base of the reactor vessels is below 200°F and pressure is at 1 atmosphere.

By the end of 2011, heat from radioactive decay in the fuel and vessels was under control and no more radioactive vapors were escaping. However, the decommissioning of the plant is expected to take 30 to 40 years. By 2015, there were still 6,000 workers in the plant every day, wearing heavy gear to protect them from the dangerous conditions. On a positive note, no citizens or plant workers have died from radiation exposure, though a number of workers at the site received significant radiation doses. This is in huge contrast to the death toll from the tsunami.

Consequences. Fallout from the Fukushima Daiichi nuclear disaster was not all radioactive. Shortly after the disaster, Japan's prime minister, Naoto Kan, announced that his government would cancel plans to build 14 more nuclear power plants, which aimed to bring Japan from its current 30% of electricity generated by nuclear power to 50%. Other countries have responded in divergent ways: China is moving full speed ahead with its plans to increase its reliance on nuclear power but, like virtually every country with nuclear power plants, is assessing safety conditions at all of its existing plants and those under construction. Germany and Switzerland announced plans to phase out nuclear energy, and Italy has put its plans to build new reactors on hold. The United States, like China, has ordered a safety review of all existing nuclear power plants.

Japan's nuclear disaster is a reminder that nuclear power carries with it risks of events that have low probabilities of occurring but huge consequences if they do. Swedish Nobel physicist Hannes Alfvén said it well, in reference to nuclear power: "No acts of God can be permitted."[1]

The 2011 nuclear disaster in Japan is one of the most devastating incidents in the history of nuclear power. In this chapter, we will delve into that history through the lenses of science and public policy as the many concerns about nuclear power are raised and discussed.

---

[1]"Energy and Environment," *Bulletin of the Atomic Scientists* (May 1972), 6.

15.1 Nuclear Energy in Perspective

Limits to fossil fuel reserves, the environmental costs of their extraction, transport, and use, and the long-term threat of climate change all suggest that an energy future based on fossil fuels is not an option. To achieve a carbon-neutral future, one in which we are not putting more carbon dioxide into the atmosphere than ecosystems can absorb, it seems prudent to do everything possible to develop non-fossil-fuel sources of energy. Nuclear power is an alternative that contributes few emissions to global warming, and there is sufficient uranium to fuel nuclear reactors well into the 21st century, with the possibility of extending the nuclear fuel supply indefinitely through reprocessing technologies. Thirty countries now have nuclear power plants in operation. In some of these countries, nuclear power generates more electricity than any other source. Is nuclear energy the key to the future of global energy? Is nuclear energy the best route to a sustainable relationship with our environment?

The History of Nuclear Power

Geologists have known for years that fossil fuels would not last forever. Sooner or later, other energy sources would be needed. It was anticipated that nuclear power could produce electricity in such large amounts and so cheaply that we would phase into an economy in which electricity would take over virtually all functions, including the generation of other fuels, at nominal costs. Following World War II, determined to show the world that the power of the atom could benefit humankind, the U.S. government embarked on a course to lead the world into the Nuclear Age.

The Nuclear Age. And so the government moved into the research, development, and promotion of commercial nuclear power plants (along with the continuing development of nuclear weaponry). Using this research, companies such as General Electric and Westinghouse constructed nuclear power plants that were ordered and paid for by utility companies, with assurances from the federal government, via the Price-Anderson Act of 1957, that these corporations and utilities would be covered by insurance for any legal

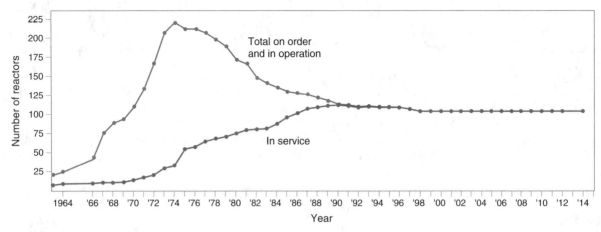

Figure 15–1 Nuclear power in the United States. After the early 1970s, when orders for plants reached a peak, few utilities called for new plants, and many canceled earlier orders. Nevertheless, the number of reactors in service increased steadily as plants under construction were completed. The number of operating reactors peaked at 112 and is holding between 100 and 104.

(*Source:* Data from U.S. Department of Energy.)

liabilities incurred. The Atomic Energy Commission, now the Nuclear Regulatory Commission (NRC), an agency in the U.S. Department of Energy (DOE), set and enforced safety standards for the operation and maintenance of the new plants, as it does today.

In the 1960s and early 1970s, utility companies moved ahead with plans for numerous nuclear power plants (Figure 15–1). By 1975, 53 plants were operating in the United States, producing about 9% of the nation's electricity, and another 170 plants were in various stages of planning or construction. Officials estimated that by 1990 several hundred plants would be online and by 2000 as many as a thousand would be operating. A number of other industrialized countries got in step with their own programs, and some less developed nations were going nuclear by purchasing plants from industrialized nations.

Curtailed. After 1975, however, the picture changed dramatically. Utilities stopped ordering nuclear plants, and numerous existing orders were canceled. In some cases, construction was terminated even after billions of dollars had already been invested. Most strikingly, the Shoreham Nuclear Power Plant on Long Island, New York, after being completed and licensed at a cost of $5.5 billion, was turned over to the state in the summer of 1989, to be dismantled after generating electricity for only 32 hours. The reason was that citizens and the state deemed that there was no way to evacuate people from the area should an accident occur. Similarly, just a few weeks earlier, California citizens voted to shut down the Rancho Seco Nuclear Power Plant, located near Sacramento, after a 15-year history of troubled operation. In the United States, 28 units (separate reactors) have been shut down permanently for a variety of reasons.

Currently . . . At the end of 2014, 100 nuclear reactors were operating in the United States. The Watts Bar Unit 1 reactor in Tennessee was the last unit to come online, beginning operations in February 1996; Unit 2—partially completed,

but halted in 1988—is again under construction and will be the first reactor to be completed in the 21st century in the United States (Figure 15–2). The 100 existing reactors (the most in any country) generate 20% of U.S. electrical power. However, in looking ahead, the U.S. Energy Information Administration projected an increase in nuclear generating capacity because of government incentives to stimulate the nuclear power industry. These increases have not yet been seen, in part because the retirement of aging reactors has somewhat offset the opening of new reactors.

Figure 15–2 The Watts Bar Nuclear Generating Station. Located in Spring City, Tennessee, Unit 1 of this power plant was the last reactor to come online in the United States (in 1996). Unit 2 construction was halted in 1988 but has since resumed, with completion expected in 2016.

Global Picture. After a catastrophic nuclear power plant accident at Chernobyl in the Soviet Union in April 1986, nuclear power was rethought in many countries (see Section 15.3). Yet the demand for electricity is more robust than ever, and the other means of generating electricity have their own problems. Coal-powered electricity produces more greenhouse gases and other pollutants than does any other form of energy. Oil and natural-gas supplies are more limited, and oil is vital to transportation and home heating. Hydroelectric power is already heavily developed. Wind and solar power, though increasing rapidly, are still only a small fraction of the total electricity generated.

Therefore, nuclear power plants continue to be built. Including those in the United States, the world has a total of 437 operating nuclear reactors, with an additional 69 under construction. Nuclear power now generates about 13.5% of the world's electricity, but dependence on nuclear energy varies greatly among the 30 countries operating nuclear power plants (Figure 15–3). France now produces 74% of its electricity with nuclear power and has plans to go to more than 80%. China and India, the world's population giants, also are investing heavily in nuclear energy, China with 26 and India with 6 new units under construction. If the safety and economic issues can be resolved, can nuclear power supply future energy needs? In order to respond intelligently to this issue, we must start with a clear understanding of what nuclear power is.

CONCEPT CHECK Although many countries are building nuclear reactors, the number of active nuclear power plants has been relatively stable for years. Why hasn't the number of active plants been increasing? ☑

15.2 How Nuclear Power Works

The objective of nuclear power technology is to control nuclear reactions so that energy is released gradually as heat. As with plants powered by fossil fuels, the heat energy produced by a nuclear plant is used to boil water and produce steam, which

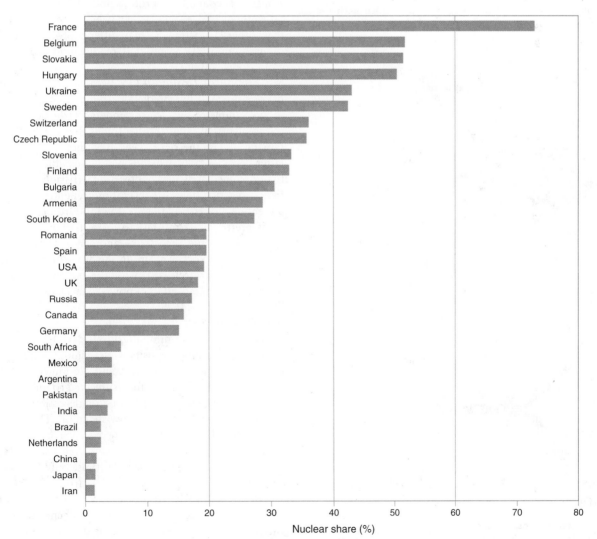

Figure 15–3 Nuclear share of electrical power generation. In 2013, those countries lacking fossil fuel reserves tended to be the most eager to use nuclear power.

(*Source:* Data from International Atomic Energy Agency.)

then drives conventional turbogenerators. Nuclear power plants are baseload plants (see Chapter 14); they are always operating unless they are being refueled, and they are large plants, generating up to 1,400 MW.

From Mass to Energy

The release of nuclear energy is completely different from the burning of fuels or any other chemical reactions discussed so far. To begin with, materials involved in chemical reactions remain unchanged at the atomic level, though the visible forms of these materials undergo great transformation as atoms are rearranged to form different compounds. For example, the carbon in gasoline is still carbon after it is burned and comes out of the tailpipe as carbon dioxide. Nuclear energy, however, involves changes at the atomic level through one of two basic processes: fission and fusion. In **fission**, a large atom of one element is split to produce two smaller atoms of different elements (Figure 15–4a). In **fusion**, two small atoms combine to form a larger atom of a different element (Figure 15–4b). In both fission and fusion, the mass of the product(s) is less than the mass of the starting material, and the lost mass is converted to energy in accordance with the law of mass-energy equivalence ($E = mc^2$), first described by

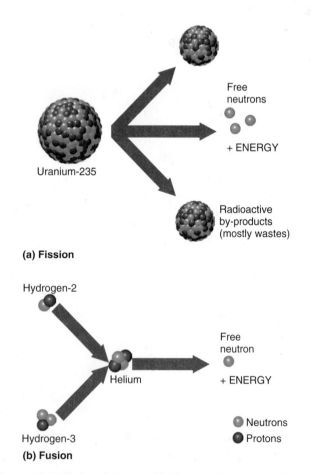

(a) Fission

(b) Fusion

Figure 15–4 Nuclear fission and fusion. Nuclear energy is released from either **(a)** fission, the splitting of certain large atoms into smaller atoms, or **(b)** fusion, the fusing together of small atoms to form a larger atom. In both cases, some of the mass of the starting atom(s) is converted to energy.

Albert Einstein. Because the small loss of mass is multiplied by the speed of light squared, the amount of energy released by this mass-to-energy conversion is tremendous: The sudden fission or fusion of just one kilogram of material, for example, releases the devastatingly explosive energy of a nuclear bomb.

The Fuel for Nuclear Power Plants. All current nuclear power plants employ the fission (splitting) of uranium-235. The element uranium, which occurs naturally in various minerals in Earth's crust, exists in two primary forms, or isotopes: uranium-238 (^{238}U) and uranium-235 (^{235}U). **Isotopes** of a given element contain different numbers of neutrons but the same number of protons and electrons. Many elements exist in more than one isotopic form.

The number 238 or 235 that accompanies the chemical name or symbol for uranium is called the **mass number** of the element. It is the sum of the number of neutrons and the number of protons in the nucleus of the atom. All atoms of any given element must contain the same number of protons, so variations in mass number represent variations in numbers of neutrons. ^{238}U contains 92 protons and 146 neutrons; ^{235}U contains 92 protons and 143 neutrons. Although all isotopes of a given element behave the same chemically, their other characteristics may differ profoundly. In the case of uranium, ^{235}U atoms will readily undergo fission, but ^{238}U atoms will not.

Fission of Uranium-235. Because ^{235}U is an unstable isotope, a small but predictable number of the ^{235}U atoms in any sample of the element undergo radioactive decay and release neutrons, among other things. If one released neutron moving at just the right speed hits another ^{235}U atom, the latter becomes ^{236}U, which is highly unstable and undergoes fission immediately into lighter atoms (**fission products**). The fission reaction gives off several more neutrons and releases a great deal of energy (Figure 15–4a). Ordinarily, these neutrons are traveling too fast to cause fission, but if they are slowed down and then strike another ^{235}U atom, they can cause fission to occur again. In this way, more neutrons and more energy are released, with the potential to repeat the process. A domino effect, known as a **chain reaction**, may thus occur (Figure 15–5a).

Nuclear Fuel. To make nuclear fuel, uranium ore is mined, purified into uranium dioxide (UO_2), and enriched. Significant uranium deposits in the United States are found in Wyoming and New Mexico and are mined underground or in open pits. The ore is then subjected to **milling**, where it is crushed, treated chemically, and rendered into **yellowcake**, so called because of its yellowish color. The yellowcake, about 80% UO_2, is then delivered to a conversion facility for purification and enrichment. Because 99.3% of all uranium found in nature is ^{238}U (thus, only 0.7% is ^{235}U), **enrichment** involves separating ^{235}U from ^{238}U to produce a material containing from 3% to 5% of ^{235}U (and the rest ^{238}U). Since these two isotopes are chemically identical, enrichment is based on their slight difference in mass. The technical difficulty of enrichment is the major hurdle that prevents less-developed countries from advancing their own nuclear capabilities.

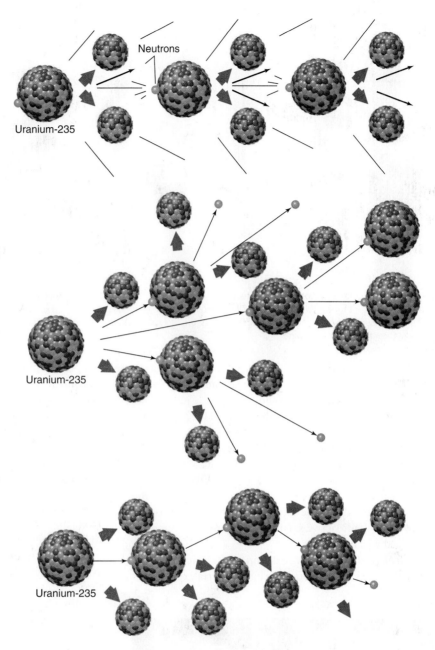

(a) Chain Reaction. A simple chain reaction. When a ^{235}U atom fissions, it releases two or three high-energy neutrons, in addition to energy and split "halves." If another ^{235}U atom is struck by a high-energy neutron, it fissions, and the process is repeated, causing a chain reaction.

Neutrons

Uranium-235

(b) Self-Amplifying Chain Reaction. A self-amplifying chain reaction leading to a nuclear explosion. Because two or three high-energy neutrons are produced by each fission, each may cause the fission of two or three additional atoms.

Uranium-235

(c) Sustaining Chain Reaction. In a sustaining chain reaction, the extra neutrons are absorbed in control rods, so that amplification does not occur.

Uranium-235

Figure 15–5 **Fission reactions.**

An Atomic Bomb. When ^{235}U is highly enriched, the spontaneous fission of an atom can trigger a self-amplifying chain reaction. In nuclear weapons, small masses of virtually pure ^{235}U or other fissionable material are forced together so that the two or three neutrons from a spontaneous fission cause two or three more atoms to undergo fission; each of these in turn triggers two or three more fissions, and so on. The whole mass undergoes fission in a fraction of a second, releasing all the energy in one huge explosion (Figure 15–5b).

The Nuclear Reactor. A nuclear reactor for a power plant is designed to sustain a continuous chain reaction (Figure 15–5c). Control is achieved by enriching the uranium to 3–5% ^{235}U. The modest 3–5% enrichment will not support the amplification of a chain reaction into a nuclear explosion. However, in the process of fission, some of the

faster neutrons are absorbed by ^{238}U atoms, converting them into ^{239}Pu (plutonium), which then also undergoes fission when hit by another neutron. So much plutonium is produced in this way that at least one-third of the energy of a nuclear reactor comes from plutonium fission.

Moderator. A chain reaction can be sustained in a nuclear reactor only if a sufficient mass of enriched uranium is arranged in a suitable geometric pattern and is surrounded by a material called a **moderator**. The moderator slows down the neutrons that produce fission so that they are traveling at the right speed to trigger another fission. In slowing down the neutrons, the moderator gains heat. In nuclear plants in the United States, the moderator is near-pure water, and the reactors are called light-water reactors (LWRs). The term *light water* denotes ordinary water (H_2O), as opposed to the substance known as *heavy water*, or deuterium oxide (D_2O); deuterium is an isotope of hydrogen.

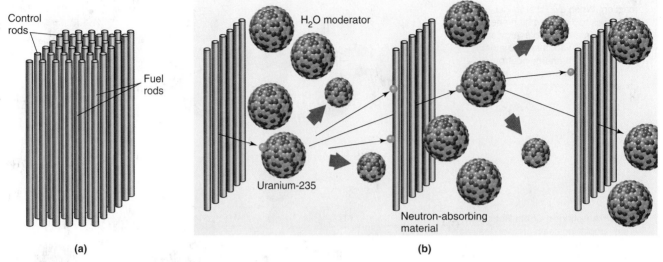

Figure 15–6 A nuclear reactor. (a) In the core of a nuclear reactor, a large mass of uranium is created by placing uranium in adjacent tubes, called fuel rods. The rate of the chain reaction is moderated by inserting or removing rods of neutron-absorbing material (control rods) between the fuel rods. **(b)** The fuel and control rods are surrounded by the moderator fluid, near-pure water.

Fuel Rods. To achieve the geometric pattern necessary for fission, the enriched uranium dioxide is made into pellets that are loaded into long metal tubes. The loaded tubes are called **fuel elements** or **fuel rods** (Figure 15–6a). Many fuel rods are placed close together to form a reactor core inside a strong reactor vessel that holds the water, which serves as both moderator and coolant (Figure 15–6b). Over time, fission products that also absorb neutrons accumulate in the fuel rods and slow down the rate of fission and heat production. The highly radioactive spent-fuel rods are then removed and replaced with new ones.

Control Rods. The chain reaction in the reactor core is controlled by rods of neutron-absorbing material, referred to

as **control rods**, inserted between the fuel rods. The chain reaction is started and controlled by withdrawing and inserting the control rods as necessary. Consequently, a nuclear reactor is simply an assembly of fuel rods, moderator-coolant, and movable control rods (Figure 15–7). As the control rods are removed and a chain reaction is initiated, the fuel rods and the moderator become intensely hot.

The Nuclear Power Plant. In a nuclear power plant, heat from the reactor is used to boil water to provide steam for driving conventional turbogenerators. One way to boil the water is to circulate it through the reactor, as in the **boiling-water reactor.** Two-thirds of the power plants in the United States, however, are **pressurized-water reactors** (Figure 15–8) employing a double loop. In these, the moderator-coolant water is heated to more than 600°F by circulating it through the reactor, but it does not boil because the system is under very high pressure (155 atmospheres). The superheated water is circulated through a heat exchanger, boiling other, unpressurized water that flows past the heat exchanger tubes. This action produces the steam used to drive the turbogenerator.

Loss-of-Coolant Accident (LOCA). The double-loop design of the pressurized-water reactor isolates hazardous materials in the reactor from the rest of the power plant. However, both reactor types have one serious drawback: If the reactor vessel should crack, the sudden loss of cooling water from around the reactor, called a **loss-of-coolant accident (LOCA),** could result in the core's overheating. The sudden loss of the moderator-coolant water would cause fission to cease, because the moderator would no longer be present. Nevertheless, the fuel core could still overheat, because 7% of the reactor's heat comes from radioactive decay in the newly formed fission products. In time, the uncontrolled decay would release enough heat energy to melt the materials in the core, a situation called a

Figure 15–7 A nuclear reactor. Technicians ready the core housing to receive uranium fuel rods in this nuclear reactor.

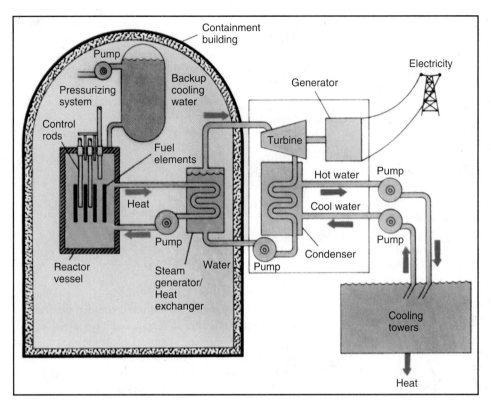

Figure 15–8 **Pressurized nuclear power plant.** The double-loop design isolates the pressurized water from the steam-generating loop that drives the turbogenerator. One loop is completely contained within the containment building (left). This loop, under high pressure so that the water does not boil, absorbs the heat from the nuclear reactions and transfers it to the steam generator/heat exchanger. There, water in the second loop is heated to boiling and circulated to the turbogenerator (right), where the steam turns a turbine to generate electricity. The steam is condensed by cool water, and the hot water is circulated to cooling towers.

meltdown. Then the molten material falling into the remaining water could cause a steam explosion. To guard against all this, backup cooling systems keep the reactor immersed in water should leaks occur, and the entire assembly is housed in a thick concrete containment building (Figure 15–8).

In the Fukushima Daiichi nuclear plants, cooling water was no longer being circulated because the pumping mechanism had no power. A meltdown occurred in three of the units, and explosions resulted as hydrogen was separated from the superheated water (a chemical reaction where the water is split into its constituent hydrogen and oxygen).

Comparing Nuclear Power with Coal Power

Nuclear plants are baseload plants that provide the foundation for meeting the daily and weekly electrical demand cycle. If nuclear power plants are phased out, one option is replacing such large plants with coal-burning plants. The United States has abundant coal reserves, but is burning coal the course of action we wish to pursue? Nuclear power has some decided environmental advantages over coal-fired power (**Figure 15–9**). Comparing a 1,000 MW nuclear plant with a coal-fired plant of the same capacity, each operating for one year, we find the following characteristics:

• **Fuel needed.** The coal plant consumes 2–3 million tons of coal. If this amount is obtained by strip mining, significant environmental destruction and acid leaching will result. If the coal comes from deep mines, there will be human costs in the form of accidental deaths and impaired health. The nuclear plant requires about 30 tons of enriched uranium, obtained from mining 75,000 tons of ore, with much

less harm to humans and the environment. The fission of about 1 pound (0.5 kg) of uranium fuel releases the energy equivalent to burning 50 tons of coal.

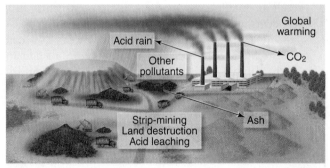

Coal

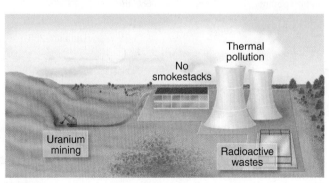

Nuclear

Figure 15–9 **Nuclear power versus coal-fired power.** Both methods of generating electricity have environmental advantages and disadvantages. The nuclear power option assumes perfect containment of radioactivity and the availability of some method for storing and disposing of nuclear waste.

- **Carbon dioxide emissions.** The coal plant emits more than 7 million tons of carbon dioxide into the atmosphere, contributing to global climate change. The nuclear plant emits none while it is producing energy, but fossil fuel energy is used in the mining and enriching of uranium, the construction of the nuclear plant, the decommissioning of the plant after it is shut down, and the transportation and storage of nuclear waste. Because a coal plant also needs to be constructed and the coal and waste ash must be transported, much the same can be said for the coal plant.

- **Sulfur dioxide and other emissions.** The coal plant emits more than 300,000 tons of sulfur dioxide, particulates, mercury, and other pollutants, leading to acid rain and health-threatening air pollution (if it is a new plant, most of these pollutants will be captured by scrubbers). The nuclear power plant produces no acid-forming pollutants or particulates.

- **Radioactivity.** A coal plant releases 100 times more radioactivity than does a nuclear power plant because of the natural presence of radioactive compounds (uranium, thorium) in the coal. The nuclear plant releases low levels of radioactive waste gases.

- **Solid wastes.** The coal plant produces about 600,000 tons of ash that require disposal. (Much of it is reused in cement production.) The nuclear plant produces about 250 tons of highly radioactive wastes that require storage and ultimate safe disposal. (Safe disposal is an unresolved problem.)

- **Accidents.** A worst-case accident in the coal plant could result in fatalities to workers and a destructive fire. Accidents in a nuclear plant can range from minor emissions of radioactivity to catastrophic releases that can lead to widespread radiation sickness, scores of human deaths, untold numbers of cancers, and widespread, long-lasting environmental contamination.

It is the disposal of radioactive wastes and the potential for accidents that have led to public concern over nuclear power. Can these problems be overcome, or are other energy options—natural-gas-powered boilers or renewable energy—less problematic and less costly than nuclear power?

CONCEPT CHECK Which source of power has a bigger effect on the atmosphere—coal power or nuclear power? ☑

15.3 The Hazards and Costs of Nuclear Power Facilities

Assessing the hazards of nuclear power requires an understanding of radioactive substances and their danger.

Radioactive Emissions

When uranium or any other element undergoes fission, the split "halves" are atoms of lighter elements, such as iodine, cesium, strontium, cobalt, or any of some 30 other elements. These newly formed atoms—the direct products of the fission—are generally unstable isotopes of their respective elements. Unstable isotopes (called **radioisotopes**) become stable by spontaneously ejecting subatomic particles (alpha particles, beta particles, and neutrons), high-energy radiation (gamma rays and X-rays), or both. Radioactivity is measured in **curies** (after Marie Curie, the scientist who first coined the term *radioactivity*); one gram of pure radium-226 gives off one curie per second, which is approximately 37 billion spontaneous disintegrations into particles and radiation. The particles and radiation are collectively referred to as **radioactive emissions**. Materials in the reactor may also be converted to unstable isotopes and become radioactive by absorbing neutrons from the fission process **(Figure 15–10)**. These indirect products of fission, along with the direct products, are the **radioactive wastes** of nuclear

Figure 15–10 Radioactive wastes and radioactive emissions. Nuclear fission results in the production of numerous unstable isotopes, designated radioactive waste. These isotopes give off potentially damaging radiation until they regain a stable structure.

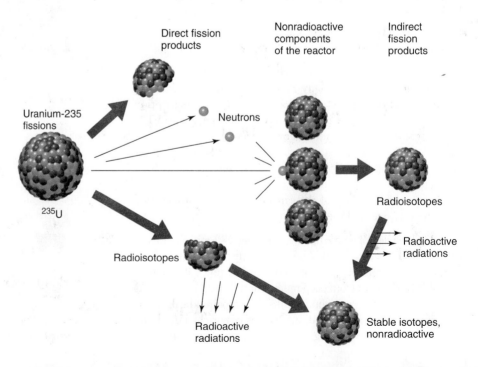

Direct fission products

Nonradioactive components of the reactor

Indirect fission products

Uranium-235 fissions

Neutrons

^{235}U

Radioisotopes

Radioisotopes

Radioactive radiations

Radioactive radiations

Stable isotopes, nonradioactive

power. The direct products of fission generate **high-level wastes**, so called because they are highly radioactive. The indirect products of fission, such as the reactor materials, are much less radioactive and are classified as **low-level wastes**, a category that also includes materials from hospitals and industry.

Biological Effects. A major concern of the public regarding nuclear power is that large numbers of people may be exposed to low levels of radiation, thus elevating their risk of cancer and other disorders. Is this a valid concern?

Radioactive emissions can penetrate biological tissue, resulting in radiation exposure. This exposure is measured as absorbed dose, using joules per kilogram (J/kg)—energy imparted (J) per mass of body tissue (kg). Doses are reported in Sieverts (Sv); the old unit *rem*, still in use by some agencies, is the equivalent of 0.01 Sv.

As radiation penetrates tissue, it displaces electrons from molecules, leaving behind charged particles, or ions, so the emissions are called ionizing radiation. The process of ionization may involve breaking chemical bonds or changing the structure of molecules in ways that impair their normal functions. Ordinarily, the radiation is not felt and leaves no visible mark unless the dose is high.

High Dose. In high doses, radiation may cause enough damage to prevent cell division. This is why, in medical applications, radiation can be focused on a cancerous tumor to destroy it. However, if the whole body is exposed to such levels of radiation (over 1 Sievert is considered a high dose), a generalized blockage of cell division occurs, preventing the normal replacement or repair of blood, skin, and other tissues. This result is called radiation sickness and may lead to death a few days or months after exposure. Very high levels of radiation may totally destroy cells, causing immediate death.

Low Dose. In lower doses, radiation may damage DNA, the genetic material inside cells. Cells with damaged (mutated) DNA may then begin dividing and growing out of control, forming malignant tumors or leukemia. If the damaged DNA is in an egg or a sperm cell, the result may be birth defects (mutations) in offspring. The effects of exposure to radiation may go unseen until many years after the event (10 to 40 years is typical). Other effects include the weakening of the immune system, mental retardation, and the development of cataracts.

Exposure. Health effects are directly related to the level of exposure. There is broad agreement that doses between 100 and 500 milliSieverts (1 milliSievert, or mSv = 1/1,000 of a Sievert) result in an increased risk of developing cancer. Evidence for this hypothesis comes from studies of patients with various illnesses who were exposed to high levels of X-rays in the 1930s, before the potential harm of such radiation was understood. People in these groups subsequently developed higher-than-normal rates of cancer and leukemia. It is noteworthy that the Japanese Health Ministry has raised the legal limit for workers at the Fukushima power plant from 100 to 250 milliSieverts, making it possible for them to spend more time working at the stricken plant, but also increasing their risk of developing cancer.

Is radiation dangerous even at doses below 100 milliSieverts? Some scientists point to the ability of living cells to repair small amounts of damage to DNA and believe that there is a threshold of radiation below which no biological effects occur. A study by the National Research Council (NRC)[2] found that there is no safe level of radiation, dismissing the idea of a threshold. Federal agencies employ the concept of no threshold and assume a direct relationship between the amount of radiation and the incidence of cancer. The NRC report predicted that 1 in 1,000 individuals would develop cancer from an exposure to 10 mSv; a tenfold increase in exposure would increase the risk to 1 in 100. Federal standards set 1.7 mSv/yr as the maximum exposure permitted for the general populace per year, except for medical X-rays.

Sources of Radiation. Nuclear power is by no means the only source of radiation, as Table 15–1 shows. There is also normal background radiation from radioactive materials, such as uranium and radon gas that occur naturally in the Earth's crust and cosmic rays from outer space. For most people, background radiation is the major source of radiation exposure. In addition, we deliberately expose ourselves

---

[2]National Research Council, *Health Risks from Exposure to Low Levels of Ionizing Radiation: Beir VII-Phase 2* (Washington, DC: National Academy of Sciences, 2005). Available at fermat.nap.edu/execsumm_pdf/11340.

Table 15–1 Relative Doses from Radiation Sources

Source	Dose
All sources (avg.)	3.6 mSv/year
Cosmic radiation at sea level*	0.26 mSv/year
Terrestrial radiation (elements in soil)	0.26 mSv/year
Radon in average house	2 mSv/year
X-rays and nuclear medicine (avg.)	0.50 mSv/year
Consumer products	0.11 mSv/year
Natural radioactivity in the body	0.39 mSv/year
Mammogram	0.30 mSv
Chest X-ray	0.10 mSv
CT scan of abdomen	10 mSv
Continued fallout from nuclear testing	0.01 mSv/year
Living near a nuclear power station (assuming perfect containment)	< 0.01 mSv/year
Federal standards for the general populace	1.7 mSv/year

Source: Data from U.S. Nuclear Regulatory Commission.

*Increases 0.005 mSv for every 100 feet of elevation.

UNDERSTANDING THE DATA According to Table 15–1, which of these pairs of sources releases more radiation: (a) ten chest X-rays or one CT scan; (b) the radon in an average house or three mammograms?

to radiation from medical and dental X-rays and, more and more frequently, from CT scans, by far the largest source of human-induced exposure. For the average person, these deliberate exposures equal one-fifth the exposure from background sources. The average person in the United States receives a dose of about 3.6 mSv per year. Thus, the argument becomes one of relative hazards. That is, does or will the radiation from nuclear power significantly raise radiation levels and elevate the risk of developing cancer?

During normal operation of a nuclear power plant, the direct fission products remain within the fuel elements and the indirect products are maintained within the containment building, which houses the reactor. No routine discharges of radioactive materials (other than permissible minor stack emissions) into the environment occur. Even very close to an operating nuclear power plant, radiation levels from the plant are lower than normal background levels. A radiation detector will pick up more radiation from the ground or the concrete of a basement floor than it will when held within 150 yards of a nuclear power plant. Careful measurements have shown that public exposure to radiation from normal operations of a power plant is less than 1% of natural background radiation.

The main concern about nuclear power, therefore, should not focus on a plant's normal operation. Instead, the real problems arise from the storage and disposal of radioactive wastes and the potential for accidents.

Radioactive Wastes and Their Disposal

To understand the problems surrounding nuclear waste disposal, you must understand **radioactive decay**, a process in which unstable isotopes, as they eject particles and radiation, become stable and cease to be radioactive. As long as radioactive materials are kept isolated from humans and other organisms, the decay proceeds harmlessly.

Half-Life. The rate of radioactive decay is such that half of the starting amount of a given isotope will decay within a certain period. In the next equal period, half of the remainder (half of a half, which equals one-fourth of the original) decays, and so on. The time for half of the amount of a radioactive isotope to decay is known as the isotope's **half-life**. The half-life of an isotope is always the same, regardless of the starting amount.

Each particular radioactive isotope has a characteristic half-life. The half-lives of various isotopes range from a fraction of a second to many thousands of years. Uranium fission results in a heterogeneous mixture of radioisotopes, the most common of which are listed in Table 15–2. Much of this material—in particular, the remaining ^{235}U and the plutonium (^{239}Pu) that has been created by the neutron bombardment of ^{238}U—can be recovered and recycled for use as nuclear fuel in an operation called **reprocessing**. Although Britain and France have a large reprocessing operation, U.S. policy prohibits the practice because of concerns over plutonium and atomic weapons (plutonium can be more easily fabricated into nuclear weapons than can uranium). A move to a lower carbon footprint may require rethinking that policy.

Nuclear wastes include spent fuel as well as less dangerous wastes. A single nuclear power plant produces about

Table 15–2	Common Radioactive Isotopes Resulting from Uranium Fission and Their Half-Lives
Short-Lived Fission Products	**Half-Life (days)**
Strontium-89	50.5
Yttrium-91	58.5
Zirconium-95	64.0
Niobium-95	35.0
Molybdenum-99	2.8
Ruthenium-103	39.4
Iodine-131	8.1
Xenon-133	5.3
Barium-140	12.8
Cerium-141	32.5
Praseodymium-143	13.6
Neodymium-147	11.1
Long-Lived Fission Products	**Half-Life (years)**
Krypton-85	10.7
Strontium-90	28.0
Ruthenium-106	1.0
Cesium-137	30.0
Cerium-144	0.8
Promethium-147	2.6
Additional Product of Neutron Bombardment	**Half-Life (years)**
Plutonium-239	24,000

20 metric tons of spent fuel annually, accumulating to a worldwide total of 71,780 metric tons of spent fuel over four decades. Low-level radioactive wastes include items that have come into contact with radioactive materials and are themselves now radioactive, such as water, concrete, tools, and used protective clothing. These items must be disposed of as nuclear waste, but the time until their radioactivity returns to safe levels may be as short as decades.

The development of nuclear power went ahead without ever fully addressing the issue of what would be done with the radioactive wastes. Proponents of nuclear power generally assumed that the long-lived high-level wastes could be solidified, placed in sealed containers, and buried in deep, stable rock formations (geologic burial) as the need for such containment became necessary. However, this has not yet happened, anywhere.

Thus, the problem of nuclear waste disposal is twofold:

- **Short-term containment.** Storage during this period allows the radioactive decay of short-lived isotopes (Table 15–2). In 10 years, fission wastes lose more than 97% of their radioactivity. Wastes can be handled much more easily and safely after this loss occurs.

- **Long-term containment.** The Environmental Protection Agency (EPA) had recommended a 10,000-year minimum to provide protection from the long-lived isotopes. However, some isotopes will remain radioactive for hundreds of thousands of years. Recognizing this, Congress extended the protection planning horizon to 1 million years.

Tanks and Casks. For short-term containment, spent fuel is first stored in deep swimming pool–like tanks on the sites of nuclear power plants. The water in these tanks dissipates waste heat (which is still generated to some degree) and acts as a shield against the escape of radiation. The storage pools can typically accommodate 10 to 20 years of spent fuel. However, the storage pools at U.S. nuclear plants had been filled to 50% of their capacity by 2004; by 2015 there was little remaining capacity. After a few years of decay, the spent fuel may be placed in air-cooled dry casks for interim storage until long-term storage becomes available (Figure 15–11). The casks are engineered to resist floods, tornadoes, and extremes of temperature. Currently, 70 facilities (mostly nuclear plants) in the United States are licensed to employ dry storage, with the prospect of many more being licensed as pool storage capacity is exhausted.

High-Level Nuclear Waste Disposal. The United States and most other countries that use nuclear power have decided that geologic burial is the safest option for disposing of the highly radioactive spent fuel, but no nation has developed plans to the point of actually carrying out the burial.

Most nuclear nations have not even been able to find a site that may be suitable for receiving wastes. Where sites have been selected, questions about their safety have su‌⸝‌ The basic problem is that no rock forma‌⸝ anteed to remain stable and dry for tens alone millions) of years. Everywhere scient‌⸝ evidence of volcanic or earthquake activity leaching within the past 10,000 years or so, that it may occur again in a similar period. If occur, the still-radioactive wastes could escape ronment and contaminate water, air, or soil, wi‌⸝ effects on humans and wildlife.

Yucca Mountain. In the United States, e‌⸝ cate a long-term containment facility have been by a severe "not in my backyard" (NIMBY) sy‌⸝ number of states, under pressure from citizens, ha legislation categorically prohibiting the disposal o‌⸝ wastes within their boundaries. In the meantime, to select and develop a long-term repository has increasingly critical, given the buildup of spent fuel i‌⸝ nuclear power plant. The **Nuclear Waste Policy Act o**‌⸝ committed the federal government to begin receiving nuclear waste from commercial power plants in 1998. At the end of 1987, Congress called a halt to the debate and selected a remote site, Yucca Mountain, in southwestern Nevada (Figure 15–12), to be the nation's civilian nuclear waste disposal site. Two other sites being considered were rejected.

Not surprisingly, Nevadans fought this selection, passing a law in 1989 that prohibits any storage of high-level radioactive waste in the state. The federal government has the power to override state prohibitions, though, and the Nevada site has undergone intensive study and exploratory construction over the past 25 years, at a cost of $10 billion. In July 2002, the federal government officially designated the Yucca Mountain facility as the nation's nuclear waste repository. The DOE submitted a license application for operation

Figure 15–11 Dry cask storage. These dry storage casks contain spent fuel assemblies generated at the James A. Fitzpatrick nuclear power plant in Scriba, New York. The assemblies had been stored for more than 30 years in the spent fuel pool on the plant site. Each dry cask has a 3½-inch steel liner surrounded by 21 inches of reinforced concrete and weighs 126 tons.

Figure 15–12 Yucca Mountain, Nevada. This remote ridgeline in Nevada was chosen as the nation's sole high-level nuclear waste depository. The site is on federally protected land, approximately 100 miles northwest of Las Vegas.

of the Yucca Mountain repository to the NRC in June 2008, and NRC staff recommended that the agency adopt DOE's environmental impact statement for the project.

Critics of the site and DOE's plans for disposal question if any degree of safety can be guaranteed, given the possibilities of earthquakes, volcanic activity, or other changes in geologic time. Indeed, studies began to suggest that water would leak through the mountain into the repository, corrode the fuel assemblies stored there, and then release radioisotopes into groundwater. When it was ruled that the site would have to be guaranteed safe for a million years, things began to look bad for the NRC's plans for long-term storage at Yucca Mountain.

Yucca Rejected. Eventually, the Obama administration decided to terminate the Yucca Mountain program, citing health and safety concerns as the reason. Instead, the administration appointed a Blue Ribbon Commission to recommend a new strategy for the nuclear power plant waste dilemma. The commission issued its report early in 2012, with the following key recommendations:

- There needs to be a new approach to selecting future nuclear waste management facilities, a transparent one based on consent of the states and local communities.

- A new federal corporation should plan and manage the waste disposal program, independent of the NRC.

- There must be a prompt effort to develop a new permanent geologic disposal facility. Even if Yucca Mountain is used, the inventory of spent nuclear fuel will soon exceed the amount that could be stored at that location.

- Prompt efforts are needed to develop one or more consolidated interim storage facilities; after cooling in containment pools for a decade or so, the spent fuel would be transferred to dry casks at interim sites in one or several central locations.[3]

In spite of the downsides, in 2014, the Nuclear Regulatory Commission released a report stating that Yucca Mountain did meet the criteria for long-term sustainability. Congress may decide to push the project forward.

One radioactive waste repository has already been constructed and is operating to receive defense-related wastes—the Waste Isolation Pilot Plant (WIPP) near Carlsbad, New Mexico. The site is above a huge salt deposit, and the wastes are conveyed to salt caves 2,150 feet underground. In this case, the city of Carlsbad was happy to have the depository, which is accompanied by significant economic benefits. The site was opened in 1999, and has already received more than 10,000 shipments of wastes totaling more than 105,000 cubic yards. These wastes are largely plutonium-contaminated by-products of nuclear weapons plants and laboratories around the country. In February of 2014, the site had two accidents. In one, a truck caught on fire. In another, an air monitor alarm went off and a small amount of radioactivity was released; the cause of the accident has not been determined. Although no

one was exposed to more radiation than is typical of a chest X-ray, such incidents highlight the need for extremely high safety measures in a large waste facility.

Radioactive wastes from weapons manufacture in the United States and Russia are discussed in Sound Science, *Radioactivity and the Military;* except for the **Megatons to Megawatts** program converting Russian weapons uranium to power-plant fuel, it is not a pretty picture.

Nuclear Power Accidents

Although the waste issues are a serious concern, it is the potential for accidents that generates the greatest concern over nuclear power.

Three Mile Island. On March 28, 1979, the Three Mile Island Nuclear Power Plant near Harrisburg, Pennsylvania, suffered a partial meltdown as a result of a series of human and equipment failures and a flawed design. The steam generator (Figure 15–8) shut down automatically because of a lack of power in its feedwater pumps, and eventually, a valve on top of the generator opened in response to the gradual buildup of pressure. Unfortunately, the valve remained stuck in the open position and drained coolant water from the reactor vessel. There were no sensors to indicate that this pressure-operated relief valve was open. Operators responded poorly to the emergency, shutting down the emergency cooling system at one point and shutting down the pumps in the reactor vessel. One instrument error compounded the problem: Gauges told operators that the reactor was full of water when, actually, it needed water badly. The core was uncovered for a time and suffered a partial meltdown. Ten million curies of radioactive gas were released into the atmosphere.

The drama held the whole nation—particularly the 300,000 residents of metropolitan Harrisburg, who were poised for evacuation—in suspense for several days. The situation was eventually brought under control, and no injuries or deaths occurred, but it would have been much worse if the meltdown had been complete. The reactor was so badly damaged and so much radioactive contamination occurred inside the containment building that the continuing cleanup proved to be as costly as building a new power plant. There are no plans to restart the reactor. GPU Nuclear, operator of the plant, has since paid out $30 million to settle claims from the accident, although the company has never admitted the existence of any radiation-caused illnesses.

Chernobyl. Prior to 1986, the scenario for a worst-case nuclear power plant disaster was a matter of speculation. Then, at 1:24 AM local time on April 26, 1986, events at a nuclear power plant in Ukraine (then a part of the Soviet Union) made such speculation irrelevant (Figure 15–13). Since that day, Chernobyl has served as a horrible example of nuclear energy gone awry.

Here's how it happened: While conducting a test of standby diesel generators, engineers disabled the power plant's safety systems, withdrew the control rods, shut off the flow of steam to the generators, and decreased the flow of coolant water in the reactor. However, they did not allow for

---

[3]Lee H. Hamilton and Brent Scowcroft, chairmen, *Draft Report of the Blue Ribbon Commission on America's Nuclear Future*, July 2011 (November 1, 2011), www.brc.gov.

Figure 15–13 Chernobyl. An aerial view of the Chernobyl reactor disaster. This disaster was a serious nuclear power accident, directly killing at least 31 people and putting countless others in the surrounding countryside at risk for future cancer deaths.

the radioactive heat energy generated by the fuel core; lacking coolant, the reactor began to heat up. The extra steam that was produced could not escape and had the effect of rapidly boosting the energy production of the reaction. In an attempt to quell the reactor, the engineers quickly inserted the carbon-tipped control rods. The carbon tips acted as moderators, slowing down the neutrons that were produced in the reaction. The neutrons, however, were still speedy enough to trigger more fission reactions, and the result was a split-second power surge to 100 times the maximum allowed level. Steam explosions then blew the 2,000-ton top off the reactor, the reactor melted down, and a fire was ignited in the graphite that was being used as a moderator, burning for days. At least 50 tons of dust and debris bearing 100–200 million curies of radioactivity in the form of fission products were released in a plume that rained radioactive particles over thousands of square miles, 400 times the radiation fallout from the atomic bombs dropped on Hiroshima and Nagasaki in 1945.

The Consequences. As the radioactive fallout settled, 135,000 people were evacuated and relocated. The reactor was eventually sealed in a sarcophagus of concrete and steel. A barbed-wire fence now surrounds a 1,000-square-mile exclusion zone around the reactor site. The soil remains contaminated with radioactive compounds, yet for years 2,000 Ukrainian workers were bused in daily to work at the remaining reactor in the Chernobyl complex. In December 2000, the last reactor was permanently shut down.

Only two engineers were directly killed by the explosion, but 28 of the personnel brought in to contain the reactor in the aftermath of the explosion died of radiation within a few months. Over a broad area downwind of the disaster, buildings and roadways were washed down to flush away radioactive dust. Even with these precautions, many people in or near the evacuation zone were exposed to high levels of radiation, especially the short-lived radioisotope iodine-131.

A report by a team of experts convened by United Nations agencies has provided the most reliable assessment of the impacts of the accident on the surrounding populations.[4] The main health impact has been an outbreak of thyroid cancer among children who drank milk contaminated with radioactive iodine. Thousands of cases have been diagnosed in this group, with numerous deaths. Several thousand additional deaths due to cancer are expected, but the increase will be hard to detect because of the hundreds of thousands of cancer deaths from other causes to be expected in the affected populations. However, 22 years after the accident, Russian television's *Health* program reported increases in thyroid cancer some 20 to 70 times above normal in many regions of the country and recommended that people between the ages of 20 and 40 from regions around the Ukraine seek ultrasound thyroid examinations.

The environmental effects of the Chernobyl disaster are complex. Immediately after the disaster, many plants and animals died, including an entire forest of pines. But later, trees that were more tolerant of radiation grew and animals returned. (Many animals remain healthy even when they are exposed to radiation and experience mutations.) Ironically, the evacuation of humans from the area allowed populations of elk, deer, bear, numerous birds, and the rare Przewalski's horse to thrive (Figure 15–14).

Could It Happen Here? Are we in the United States in danger of such an explosion occurring at a nuclear power plant? Nuclear scientists argue that the answer is no because U.S. power plants have a number of design features that should make a repeat of Chernobyl impossible. For one thing,

---

[4]The Chernobyl Forum: 2003–2005, *Chernobyl's Legacy: Health, Environmental and Socio-Economic Impacts, and Recommendations to the Governments of Belarus, the Russian Federation and Ukraine,* 2nd rev. version (Vienna, Austria: International Atomic Energy Agency, 2005). Available at www.iaea.org/Publications/Booklets/Chernobyl/chernobyl.pdf.

Figure 15–14 Ecosystem Resilience at Chernobyl. Although radiation initially killed off whole forests of pine, many other plants, and numerous animals, ecosystems around Chernobyl have rebounded in subsequent decades. These rare wild horses are thriving in part because the area is too dangerous for human habitation.

SOUND SCIENCE
Radioactivity and the Military

Some of the worst failures in handling radioactivity have occurred at military facilities in the United States and in the former Soviet Union. A recent calculation taken from government figures indicated that the U.S. nuclear program has, over the years, sickened 36,500 Americans and killed more than 4,000, most of these in connection with the manufacture and testing of nuclear weapons (most of the sickness and death is due to cancer). Further, liquid high-level wastes stored at many U.S. facilities have leaked into the environment and contaminated wildlife, sediments, groundwater, and soil. Activities at these sites have been shrouded in secrecy; only recently have documents revealing past accidents and radioactive releases been declassified and made available to the public. Deliberate releases of uranium dust, xenon-133, iodine-131, and tritium gas have been documented at Hanford, Washington; Fernald, Ohio; Oak Ridge, Tennessee; and Savannah River, South Carolina. Cleaning up these sites is now the responsibility of the DOE, which has already spent more than $50 billion on this problem. The Hanford, Washington, site is the largest of the contaminated sites and will cost an estimated $113 billion more by 2090—probably the largest environmental project the country will ever undertake.

From Russia, with Curies

USSR military weapons facilities were even more irresponsible. The worst was a giant complex called Chelyabinsk-65, located in the Ural Mountains. For at least 20 years, nuclear wastes were discharged into the Techa River and then into Lake Karachay. At least 1,000 cases of leukemia have been traced to radioactive contamination from the Chelyabinsk facility. Even today, standing for one hour (thus receiving 6 Sv of radiation) on the shore of Lake Karachay will cause radiation poisoning and death within a

An explosion at Soviet Russia's military weapons facility at Chelyabinsk so heavily contaminated the region that many local villages had to be abandoned and the residents resettled. This was the village of Metlino.

week. The lake dried up in the summer of 1967, and winds blew radioactive dust with 5 million curies across the countryside, contaminating hundreds of thousands of people. This lake is considered to be the most polluted spot on Earth. Russian authorities have filled the lake with hollow concrete blocks, rocks, and soil, and plants now spread their greenery over this permanently contaminated site.

Megatons to Megawatts

The end of the Cold War has brought the welcome dismantling of nuclear weapons by the United States and the nations of the former Soviet Union. Thousands of nuclear weapons are being dismantled, many as a result of earlier agreements between U.S. president Bill Clinton and Russian president Boris Yeltsin. The agreements include a joint closure of all remaining plutonium weapons production reactors, funds to aid Russia in destroying missile silos and dismantling nuclear submarines and bombs, and vast cuts in the nuclear arsenals of the two former enemies. The radioactive components of the weapons must be disposed of where they are safe from illegal access and will pose no danger to human health for centuries to come—a daunting task.

One element of the U.S.–Russian agreements, the Megatons to Megawatts Program, involves a government-industry partnership wherein a private U.S. company oversees the dilution of weapons-grade uranium to the much lower enrichment of power-plant uranium, sells it to U.S. nuclear power plants, and pays the Russian government a market price. For two decades, about 10% of U.S. electricity was produced from uranium purchased through this program. The last shipment, containing 500 tons of bomb-grade uranium, arrived in the United States in December 2013.

The closing down of much of the nuclear warfare enterprise is certainly a welcome development, but the problems of security and disposal of the wastes will remain as a long-term legacy of the Cold War.

---

the Chernobyl reactor used graphite as a moderator rather than the water used in LWRs. Further, LWRs are incapable of developing a power surge more than twice their normal power, well within the designed containment capacity of the reactor vessel. Finally, LWRs have more backup systems to prevent the core from overheating, and these reactors are housed in thick, concrete-walled containment buildings designed to withstand explosions such as the one that occurred in the Chernobyl reactor, which had no containment building.

LWRs are not immune to accidents, however, the most serious being a core meltdown as a result of a total loss of coolant.

This is what happened at three of the reactors at the Fukushima Daiichi power plant as a result of the earthquake and tsunami off the coast of Japan on March 11, 2011 (Figure 15–15).

Fukushima Daiichi. The opening story of this chapter describes what can go wrong with nuclear power in the face of unexpected natural disaster. There was no anticipation of such an intense earthquake accompanied by a strong tsunami and, as a result, no way to accurately protect the power plant. Power was lost from the electrical grid and backup diesel generators, leaving only eight hours of batteries to

Figure 15–15 Fukushima Daiichi power plant. The power plant has six reactors, four of which were damaged by nuclear meltdowns and/or hydrogen explosions. This photo shows the wrecked building housing the No. 4 reactor.

provide power for cooling the reactors. There was no way to provide cooling to spent fuel once the water drained out of their pools. There was no way to vent dangerous gases from the containment buildings. Again, could such an accident happen in the United States? The Union of Concerned Scientists (UCS), an NGO that has spent years evaluating nuclear power plant safety, believes that it could.[5] A task force organized by the NRC disagrees.[6]

The real cost of the accidents at Three Mile Island, Chernobyl, and Fukushima must be reckoned in terms of public trust. Public confidence in nuclear energy plummeted after the first two accidents, which pointed to human error as a highly significant factor in nuclear safety—and human error is something the public understands well. Fukushima demonstrated nuclear power's vulnerability to disasters, and the public also understands natural disasters. Although the Fukushima disaster was the largest nuclear accident ever to occur in Japan, the country had previously experienced several smaller accidents at nuclear power plants that had already eroded public confidence. After Fukushima, operations were suspended at many of Japan's nuclear reactors. Because of the danger of such disasters, the future safety of nuclear power must be a major concern.

Safety and Nuclear Power

As a result of Three Mile Island and other, lesser incidents in the United States, the NRC has upgraded safety standards not only in the technical design of nuclear power plants,

but also in maintenance procedures and in the training of operators. Thus, proponents contend, nuclear plants were designed to be safe in the beginning, and now they are safer than ever. Some proponents of nuclear power claim that we now have the technology to build inherently safe nuclear reactors, designed in such a way that any accident would result in an automatic quenching of the chain reaction and suppression of heat from nuclear decay. In reality, however, eliminating all risk of radioactive release is an impossible expectation.

Passive Safety. Nuclear scientists have designed a new generation of nuclear reactors with built-in passive safety features rather than the active safety features found in current reactors. **Active safety** relies on operator-controlled actions, external power, electrical signals, and so forth. As accidents have shown, operators may override such safety factors, and electricity, valves, and meters can fail or give false information. **Passive safety**, by contrast, involves engineering devices and structures that make it virtually impossible for the reactor to go beyond acceptable levels of power, temperature, and radioactive emissions; their operation depends only on standard physical phenomena, such as gravity and resistance to high temperatures.

New Generations of Reactors. Nuclear scientists distinguish four generations of nuclear reactors. *Generation I* reactors are the earliest, developed in the 1950s and 1960s; few are still operating. The majority of today's reactors are *Generation II* vintage, the large baseline power plants, of several designs. *Generation III* refers to newer designs with passive safety features and much simpler power plants—among them the so-called **advanced boiling-water reactors (ABWRs)**, depicted in Figure 15–16 on the following page. General Electric has submitted a design for an **economic simplified boiling water reactor** for NRC approval. This new design has many more passive safety features than General Electric's current ABWR. In the event of a loss-of-coolant accident, two separate passive systems would allow water to drain by gravity to the reactor vessel.

In East Asia, the nuclear reactor design of choice is now the ABWR, which produces 1,400 MW of electricity. The new plants are designed to last for 60 years, can be constructed within five years, and are simple enough that a single operator can control them under normal conditions. They generate electricity at a cost competitive with the costs of fossil fuel options.

Generation IV plants are now being designed and will likely be built within the next 20 years. One example of a Generation IV reactor is the pebble-bed modular reactor (PBMR), which will feed spherical carbon-coated uranium fuel pebbles gradually through the reactor vessel, like a gumball machine. The PBMR will be cooled with fluidized helium, an inert gas, which will also spin the turbines. These reactors will be small, will produce about 160 MW of power, and are expected to be cheap to build and inexpensive to operate. The modules can be built in a factory and shipped to the location of the power plant.

[5]Union of Concerned Scientists, "Can It Happen Here?" *Catalyst*, Summer 2011: 7–9.
[6]Nuclear Regulatory Commission, Recommendations for Enhancing Reactor Safety in the 21st Century (Near-Term Task Force, July 12, 2011). November 3, 2011, www.psr.org.

Figure 15–16 **Advanced Boiling-Water Reactor.** The core is surrounded by three concentric structures: a reactor pressure vessel, in which heat from the reactor boils water directly into steam; a concrete chamber (outlined with heavy black line) and a water pool, which together contain and quench steam vented from the reactor in an emergency; and a concrete building, which acts as a secondary containment vessel and shield. Any excessive pressure in the reactor will automatically open valves that release steam into a quenching pool, reducing the pressure. Water from the quenching pool can, if necessary, flow downward to cool the core.

(*Source:* "Advanced Light-Water Reactors," by M. W. Golay and N. E. Todreas. Copyright 1990 by *Scientific American*. All rights reserved.)

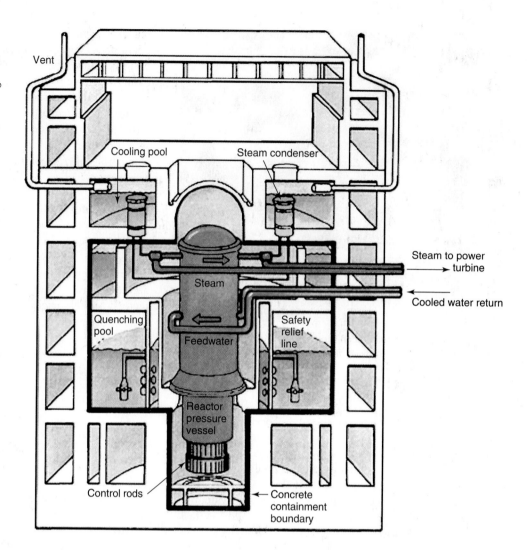

One motive behind the new designs is to restore the public's confidence in nuclear energy. Previously, nuclear proponents had emphasized the very low probabilities of accidents. As we have seen, though, improbable events can happen, and when they happen to nuclear power plants, the consequences can be dreadful.

Terrorism and Nuclear Power. What would have happened if one of the terrorist groups that took over four airliners on September 11, 2001, had targeted a nuclear power plant instead of the World Trade Center? This question has raised great concern in many venues, from the NRC, to Congress, to local hearings in municipalities that host nuclear plants. After some uncertainty following September 11, the general consensus is that a jetliner could not penetrate the very thick walls of the containment vessel protecting the reactor. After September 11, the NRC immediately beefed up the requirements for security around every power plant—adding more guards, more physical barriers, and vehicle checks at greater distances from the plant—as well as keeping the shoreline near the plants off limits.

A subsequent National Academy of Sciences (NAS) report concluded that the most vulnerable locations for terrorist attack

are the spent-fuel pools.[7] The NAS recommended policies that would protect the pools from a loss-of-coolant-pool-water event brought on by terrorist activity.

New Safety Issues. The Fukushima Daiichi nuclear disaster has generated new concerns about the safety of both current and possible future nuclear power reactors. In the United States, the NRC has deployed a defense-in-depth philosophy that covers everything from design and operation, to precautions against failures and mistakes, containment against the release of radioisotopes in case of accidents, and emergency plans to protect citizens. Nevertheless, the Near-Term Task Force felt that the NRC's approach could be strengthened by:

- Clarifying and solidifying the regulatory framework that governs power plants. This would correct the current patchwork of regulations and voluntary industry initiatives.

- Upgrading units that might be affected by earthquakes and flooding events, and considering responses to events

---

[7]National Research Council, *Safety and Security of Commercial Spent Nuclear Fuel Storage: Public Report* (Washington, DC: National Academies Press, 2006). Available at www.nap.edu/catalog/11263.html.

that are highly unlikely but would have catastrophic repercussions if they happened.

- Improving the venting and prevention against potential hydrogen gas release.
- Strengthening the on-site emergency capabilities against failures of electrical power to the plants.[8]

The UCS analysis agreed with these recommendations and added a further one:

- Require plant owners to move spent fuel after five years to dry cask storage; this lowers the risk of overheating in the pools in case of lost cooling water (as in Fukushima).[9]

The potential for accidents and terrorist acts certainly raises questions about a more widespread promotion of nuclear power. However, as Figure 15–1 shows, utilities in the United States were already turning away from nuclear power before Three Mile Island and Chernobyl. The reasons were mainly economic.

Economics of Nuclear Power

What was behind the cancellation of so many nuclear power plants (more than 100)? First, projections of future energy demands were overly ambitious, so a slower growth in the demand for electricity postponed orders for all types of power plants. Second, the costs for nuclear power were higher than expected, especially relative to other fuel sources. Nuclear power plants and their required safety features are expensive to build. Because many of the costs of fossil fuels are externalized, the costs of using fossil fuels are artificially low. Some energy companies had expected a carbon tax to raise the cost of fossil fuels and make nuclear power more competitive, but that has not occurred, at least in the United States (Chapter 18).

Third, public protests frequently delayed the construction or start-up of a new power plant. Such delays increased costs still more because the utility was paying interest on its investment of several billion dollars, even when the plant was not producing power. As these costs were passed on, consumers became yet more disillusioned with nuclear power.

Finally, safety systems may protect the public, but they do not prevent an accident from financially ruining the utility. Because radioactivity prevents straightforward cleanup and repair, an accident can convert a multibillion-dollar asset into a multibillion-dollar liability in a matter of minutes, as Three Mile Island demonstrated. Thus, nuclear power involves a financial risk that utility executives have been hesitant to take.

The Vermont Yankee plant, which closed in 2014, provides an example of the complex factors affecting nuclear power plants. One of the reasons for its closure was economic—it is more expensive to produce electricity with nuclear power than with fossil-fuel powered plants because there is no carbon tax on the emission of greenhouse gases. Another reason was anti-nuclear activism, resulting from fears about radioactivity.

Figure 15–17 Maine Yankee. The Maine Yankee Nuclear Power Plant in the process of being decommissioned, in 2003. The decommissioning is now complete, involving the removal of 200,000 tons of solid waste by rail, truck, and barge.

Decommissioning. At some point, nuclear power plants need to be closed down, or *decommissioned*, a complex process. Decommissioning can be extremely costly. Decommissioning Maine Yankee (Figure 15–17) cost $635 million. (By comparison, the plant cost just $231 million to build in the late 1960s.) The site now houses some 64 dry casks containing the spent-fuel assemblies, which await disposal at their final resting place, wherever that happens to be. The owners of the plant are allowed to bill former customers to recover some costs—a "stranded cost" nobody likes. Faced with these costs, some utilities are opting to repair older plants in spite of the high costs of such repairs. Other closed reactors are on sites currently occupied by active units. The old, closed-down reactors are simply defueled and allowed to sit until the day when all the reactors are closed and the entire site must be decommissioned. In one closed reactor (Humboldt Bay, California), only the nuclear components were removed, and the plant was converted to a conventional natural-gas plant.

Longevity? Originally, it was thought that nuclear plants would have a lifetime of about 40 years. It now appears that lifetimes will be considerably less for many of the older plants. For example, the government of Ontario decided in 2005 to shut down two 540-MW nuclear reactors more than a decade before their scheduled retirement. The reactors had been idle since 1997 because of corrosion in the pipes carrying heavy coolant water from the reactor core. The repair bill was estimated to be $1.6 billion, approximately the cost of a new reactor.

The lifetimes of nuclear power plants are often shorter than originally expected due to embrittlement and corrosion.

---

[8]Task Force, ix.
[9]UCS, 8.

Embrittlement occurs as neutrons from fission bombard the reactor vessel and other hardware. Gradually, this neutron bombardment causes the metals to become brittle enough that they may crack under thermal stress—for example, when emergency coolant waters are introduced in the event of a loss-of-coolant accident. When the reactor vessel becomes too brittle to be considered safe, the plant must either be shut down or be repaired at great cost.

Corrosion is a normal consequence of steam generation. Very hot, pressurized water flows from the core into the steam generator through thousands of 3/4-inch-diameter pipes immersed in water (Figure 15–8). The water inside and outside these pipes contains corrosive chemicals that, over time, cause cracks to develop in some of the pipes. Cracked pipes are "repaired" by plugging them, cutting the overall plant power. In March 1995, using a new high-tech probe, officials discovered that up to half of the steam generator pipes in the Maine Yankee plant in Wiscasset, Maine, had developed cracks, some penetrating 80% of the pipes' thickness. As a result, the plant was shut down after only 24 years of operation.

Sometimes the lifespan of nuclear plants can be extended by upgrading equipment and replacing parts that are wearing out. Such lifespan extensions can help reduce the cost of power.

New Life? Until the Fukushima disaster, nuclear power was being viewed more favorably by policy makers, as well as by many in the environmental community, in spite of the economic issues. Most of the operating nuclear power plants in the United States were built before global climate change was recognized as a serious problem, but concerns about the greenhouse gas emissions of conventional coal-fired power plants have brought many to reconsider the advantages of nuclear power. This has come into sharp focus as the time has been drawing near for the retirement of a number of older nuclear plants, most of which will reach the end of their operating licenses within the next 10 years. As plants reach the end of their operating licenses (after 30 years or so), owners are routinely applying for license extensions. After an NRC review, these are awarded, and many plants now are licensed to operate for 60 years. Most of the new reactors planned are to be built on the sites of existing nuclear power plants. In the United States at least, nuclear power has a good track record, and the global warming imperative plus our growing need for electricity suggest that nuclear power may continue its revival.

CONCEPT CHECK Compare the disasters at Chernobyl and Fukushima. What are two similarities and two differences? ☑

15.4 More Advanced Reactors

Uranium—especially ^{235}U—is not a highly abundant mineral on Earth. At the height of optimism about nuclear energy in the 1960s, when as many as 1,000 plants were envisioned by the turn of the century, it was forecast that shortages of ^{235}U would develop. Breeder reactors, which utilize chain reactions, were seen as the solution to this problem.

Breeder (Fast-Neutron) Reactors

Recall from Section 15.2 that when a ^{235}U atom fissions, two or three neutrons are ejected. Only one of these neutrons needs to hit another ^{235}U atom to sustain a chain reaction; the others are absorbed by something else. The breeder reactor is designed so that (nonfissionable) ^{238}U absorbs the extra neutrons, which are allowed to maintain their high speed. (Such reactors are now called **fast-neutron reactors**.) When this occurs, the ^{238}U is converted to plutonium (^{239}Pu), which can then be purified and used as a nuclear fuel, just like ^{235}U. Thus, the fast-neutron reactor converts nonfissionable ^{238}U into fissionable ^{239}Pu. Because the fission of ^{235}U generally produces two neutrons in addition to the one needed to sustain the chain reaction, the fast-neutron reactor may produce more fuel than it consumes. Furthermore, 99.3% of all uranium is ^{238}U, so converting that to ^{239}Pu through fast-neutron reactors effectively increases nuclear fuel reserves more than a hundredfold.

In the United States and elsewhere, small fast-neutron reactors are operated for military purposes. France, Russia, and Japan are currently the only nations with commercial fast-neutron reactors, and these have not been in continuous use. Fast-neutron reactors present most of the problems and hazards of standard fission reactors, plus a few more. The consequences of a meltdown in a fast-neutron reactor would be much more serious than those of a meltdown in an ordinary fission reactor, due to the large amounts of ^{239}Pu, with its half-life of 24,000 years. In addition, because plutonium can be purified and fabricated into nuclear weapons more easily than can ^{235}U, the potential for the diversion of fast-neutron reactor fuel to illicit weapons production is greater. Hence, the safety and security precautions needed for fast-neutron reactors are greater. Also, fast-neutron reactors are more expensive to build and operate. However, they have the ability to extract much more energy from recycled nuclear fuel, and they produce much less high-level waste than do conventional nuclear plants.

Reprocessing

The United States has enough uranium stockpiled to satisfy current industry needs. Although the global supply of uranium is not a problem right now, a number of countries are reprocessing spent fuel through chemical processes, recovering plutonium and mixing it with ^{238}U to produce mixed-oxide (MOX) fuel that is about 5% plutonium and suitable for further use in many nuclear power plants. In this way, some of the high-level waste is reused and does not add to the stockpile of waste.

The Bush administration DOE established a program called the Advance Fuel Cycle Initiative (AFCI), moving toward establishing reprocessing in the United States as a way of reducing our growing inventory of spent fuel. However, the Obama DOE canceled this initiative because of concerns about security, and redirected AFCI into a program that simply provides funding for academic research—on fuel separation and transmutation of plutonium to useful fuels, among other topics. One outcome of AFCI is the International Framework for Nuclear Energy Cooperation, a partnership of 25 countries aiming to improve the proliferation-resistance of the nuclear fuel cycle and work toward cooperative reprocessing of used fuel in fast reactors. It remains to be seen if this partnership develops into something more substantial than just consulting and research, however.

Fusion Reactors

The vast energy emitted by the Sun and other stars comes from fusion (Figure 15–4b). The Sun is composed mostly of hydrogen. Solar energy is the result of the fusion of hydrogen nuclei (protons) into larger atoms, such as helium. Scientists have duplicated the process in the hydrogen bomb, but hydrogen bombs hardly constitute a useful release of energy. The aim of fusion technology is to carry out fusion in a controlled manner in order to provide a practical source of heat for boiling water to power steam turbogenerators.

The d–t Reaction. Because hydrogen is an abundant element on Earth (there are two atoms of it in every molecule of water) and helium is an inert, nonpolluting, nonradioactive gas, hydrogen fusion is promoted as the ultimate solution to all our energy problems; that is, fusion affords pollution-free energy from a virtually inexhaustible resource: water. However, most current designs do not use regular hydrogen (^{1}H), but rather employ the isotopes deuterium (^{2}H) and tritium (^{3}H) in what is called the **d–t reaction.** Deuterium is a naturally occurring, nonradioactive isotope that can be extracted in almost any desired amount from the hydrogen in seawater. Tritium, by contrast, is an unstable, gaseous, radioactive isotope that must be produced artificially. Radioactive and difficult to contain, tritium is a hazardous substance. As a result, fusion reactors could easily become a source of radioactive tritium leaking into the environment unless effective (and costly) designs are used to prevent that.

In the present state of the art, fusion power is still an energy consumer rather than an energy producer. The problem is that it takes an extremely high temperature—some 100 million degrees Celsius—and high pressure to get hydrogen atoms to fuse. In the hydrogen bomb, the temperature and pressure are achieved by using a fission bomb as an igniter. Getting to the high temperature is only one part of the problem; extracting the heat from the reacting region is also a daunting task, given the large amount of energy released from the fusion reactions.

Two Ways of Controlling Fusion. Efforts are well under way to demonstrate that controlled fusion is possible. One method is to focus an array of powerful laser beams on a pellet containing hydrogen isotopes (d–t). The laser beams would crush the core of the pellet with such energy that the hydrogen atoms inside would fuse together (ignite) and release energy. This method is employed at the U.S. National Ignition Facility (NIF) in Livermore, California, a $3.5 billion facility completed in 2009 and getting close to achieving better than break-even results—releasing more energy than the energy of the laser beams. This is a demonstration project, however, and even if it works, it would generate only about

3.5 kilowatts of heat, enough to power one home. The scale-up needed to accomplish laser-driven fusion power is intimidating, but the Japanese and Europeans also have plans to work on this basic method.

A second process is the Tokamak design, in which ionized hydrogen is contained within a magnetic field while being heated to the necessary temperature. Some controlled fusion has been achieved in the Tokamak devices, and in late 1994, a Tokamak facility at the Princeton Plasma Physics Laboratory in Princeton, New Jersey, achieved 10.7 MW of fusion power in 0.27 seconds of reaction. As yet, however, the break-even point has not been reached: More energy is required to run the magnets than is obtained by fusion.

An international research team from seven countries is working on a design for an International Thermonuclear Experimental Reactor, a prototype fusion reactor of the Tokamak variety. The facility, under construction in Cadarache, France, is expected to cost more than $20 billion, and the reactor is expected to produce 10 times as much energy as it consumes. Recently, the United States rejoined the project, some 10 years after withdrawing because of concerns about costs and the science itself.

No matter what design is used, developing, building, and testing a fusion power plant will require many more years and many billions of dollars. This assumes that the engineering difficulties will be overcome, which is by no means a given. Thus, fusion is, at best, a very long-term and uncertain option.

CONCEPT CHECK   What are the main difficulties in developing a way to use nuclear fusion as a source of power? ☑

15.5 The Future of Nuclear Power

With fossil fuels contributing to global climate change and with alternatives such as solar and wind power in their early stages, the long-term outlook for energy is discouraging (Chapter 16). Using nuclear power seems to be an obvious choice; however, there continues to be organized opposition to nuclear energy, and it is still a costly technology, highly dependent on government patronage. After the Fukushima Daiichi power-plant disaster, most countries are examining their nuclear options, as we have seen. However, a total of 69 nuclear reactors are currently under construction (see **Figure 15–18**, on the following page) and 437 are still operating, generating 13.5% of the world's electricity. Russia and other eastern European countries are holding onto their existing nuclear power plants and building more. A new plant is being built in Finland, and the United Kingdom is planning a new generation of reactors to replace its aging fleet. Asian countries are welcoming nuclear power, with China, South Korea, and India seemingly intent on using it to fulfill a major share of their energy needs. The International Atomic Energy Agency predicts that the Fukushima disaster will slow growth in nuclear energy but not reverse it. The agency also predicts that nuclear power will double its production of energy by 2030.

Figure 15–18 Nuclear power reactors under construction, 2014. Worldwide, 69 reactors were under active construction in 2014, with an electrical capacity of 68 Gigawatts.

(*Source:* Data from International Atomic Energy Agency.)

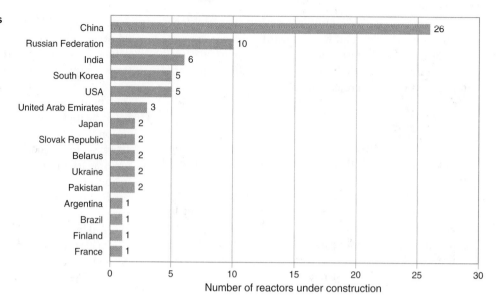

Figure 15–18 Nuclear power reactors under construction, 2014. Worldwide, 69 reactors were under active construction in 2014, with an electrical capacity of 68 Gigawatts.

(*Source:* Data from International Atomic Energy Agency.)

Opposition

Opposition to nuclear power is based on several premises:

- People have a general distrust of technology they do not understand, especially when that technology carries with it the potential for catastrophic accidents or the hidden, but real, capacity to induce cancer.

- Problems involving lax safety, operator failures, and cover-ups by nuclear plants and their regulatory agencies have occurred in the United States, Canada, and Japan.

- The problems of high construction costs and unexpectedly short operational lifetimes have already been mentioned. Thus, the economic argument can also be used to oppose nuclear power.

- The nuclear industry has repeatedly presented nuclear energy as extremely safe, arguing that the probabilities of accidents occurring are very low. However, when accidents do occur, probabilities become realities, and the arguments are moot.

- Nuclear power plants are viewed as prime targets for terrorist attacks. Critics argue that, therefore, it only makes sense to reduce the number of such targets, not add more of them.

- There remains the crucial problem of disposing of nuclear waste. Siting a long-term nuclear waste repository has been a difficult political as well as technological problem in country after country.

Opponents of nuclear power also cite the basic mismatch between nuclear power and the energy problem. The main energy problem for the United States is an eventual shortage of crude oil for transportation purposes, yet nuclear power produces electricity, which is not yet used much for transportation. Consequently, nuclear power simply competes with coal-fired power in meeting the demands for baseload electrical power. Given the high costs and additional financial risks of nuclear power plants, coal is cheaper, and the United States does have abundant coal reserves.

Nonetheless, there are still the environmental problems of mining and burning coal, especially global climate change. Burning coal emits more CO_2 per unit of energy produced than any other form of energy does (Chapter 14). If the long-term environmental costs were factored into the price of coal, that price would be considerably higher than it is now.

In the final analysis, nuclear power may be a unique energy option, but it is also controversial. Interestingly, public opinion about nuclear power has been shifting. For a time, polls indicated that Americans viewed nuclear power as increasingly important, in the light of surging prices for oil and gas. Some leaders in the environmental field are viewing nuclear energy more favorably, given the serious consequences for global climate change of continued use of fossil fuels. However, shortly after Fukushima, an *ABC News/ Washington Post* poll indicated that 64% of Americans oppose any new nuclear construction. Opposition continues because of the perceived risks to both human health and the environment and because of the mismatch to our most critical energy problem—an impending shortage of fuel for transportation. Opponents don't believe that the benefits are worth the risks.

Rebirth of Nuclear Power

If nuclear energy is to have a brighter future, it will be because we have found the continued use of fossil fuels to be so damaging to the atmosphere that we have placed limits on their use but have not been able to develop adequate alternative energy sources. Nuclear supporters point out that U.S. nuclear power plants prevent the annual release of 590 million metric tons of carbon dioxide, as much as that released by 113 million cars, 1 million metric tons of sulfur dioxide, and half a million metric tons of nitrogen oxides. Also, electric cars and plug-in hybrids are now available from major car manufacturers. If this catches on, nuclear-generated electricity can indeed be substituted for oil-based fuels.

Observers agree that if the rebirth of nuclear power is to come, a number of changes will have to be made:

- The public must be convinced that every possible measure has been taken to prevent terrorist attacks and sabotage.

- The industry's manufacturing philosophy will have to change to favor standard designs and factory production of the smaller reactors instead of the custom-built reactors presently in use in the United States. Recently, the DOE selected three designs for development, one of which—the Westinghouse advanced pressurized-water reactor—is a 600 MW unit similar to the one pictured in Figure 15–16. The unit, designed for modular construction, relies primarily on passive safety.

- The framework for licensing and monitoring reactors must be streamlined, but without sacrificing safety measures. This has largely been accomplished with the Energy Policy Act of 2005. The rules allow the NRC to approve a plant design prior to the selection of a specific site.

- The waste dilemma must be resolved. With the Yucca Mountain waste repository now in limbo, resolving the dilemma is becoming more difficult. The recommendations of the Blue Ribbon Commission should be adopted.

Nuclear Energy Policies. President George W. Bush made expanding nuclear energy a major component of his energy policy by signing into law the **Energy Policy Act of 2005,** which included a tax credit for the first 6,000 MW of new nuclear capacity for the first eight years of operation, insurance to protect companies building new reactors from the risk of regulatory delays, and federal loan guarantees for up to 80% of construction costs of new power plants. The act also authorized funds for the planning and construction of a nuclear plant that would produce hydrogen from water (the Next Generation Nuclear Plant), a key part of a coming fuel-cell technology for automobiles (see Chapter 16).

The newer **Energy Independence and Security Act of 2007** addressed nuclear power only tangentially; the legislation had a provision to promote American-built nuclear power plants in overseas markets.

In his 2011 State of the Union address, President Obama included nuclear energy as one of the components of a national goal of generating 80% of U.S. electricity from "clean energy sources" by 2035. Today the United States is the world's largest producer of nuclear power. However, because of the low cost of coal and natural gas in 2014, experts expect only four more nuclear reactors to come online by 2020. Although future trends are still uncertain, many experts believe that the use of nuclear power will increase.

CONCEPT CHECK What are two worldwide trends that are likely to increase the use of nuclear power? What two trends would be likely to decrease its use? ☑

REVISITING THE THEMES

SOUND SCIENCE

The science behind nuclear power was developed during World War II and was used to produce the two atomic bombs that ended the war with Japan in 1945. Nuclear physicists were eager to show that atomic energy could be put to beneficial uses, however, and with that in mind, the technology for nuclear power plants was developed. Sound science was, and continues to be, essential for this enterprise, especially in regard to improvements in safety and efficiency. Nevertheless, scientific uncertainties remain: For example, what is a safe level of exposure to radioactivity? How can nuclear wastes be stored safely for thousands of years to come? Can nuclear reactors be designed to be accident proof? Can fusion energy ever become a commercial reality? Answers to these questions will challenge scientists for years to come.

SUSTAINABILITY

Many experts believe that nuclear power has a role to play in getting us to a sustainable energy future. But because uranium must be mined and is not an abundant mineral, supplies will not last indefinitely. Reprocessing fuel can extend the life of nuclear power, but it carries risks that may be unacceptable. Still, nuclear fission may be a way of bridging the gap between fossil fuel energy and renewable energy. The most important aspect of nuclear fission is that it does not release greenhouse gases and would be capable of providing most of our electrical power, thus enabling us to avoid relying on coal.

The current dilemma is how to deal with the buildup of the highly radioactive spent-fuel wastes. Storing them in pools and casks on the grounds of the power plants is unsustainable. Consequently, countries using nuclear power must find ways to dispose of the wastes safely, which means making sure that they are contained for thousands of years.

STEWARDSHIP

It was a major political decision to develop nuclear power for generating electricity. The technology is so expensive and fraught with risk that only government-level agencies can manage the development and ongoing regulation of nuclear power. The roles of the NRC and the EPA are crucial in protecting the U.S. public from the hazards and risks of keeping nuclear power in operation and safely handling its wastes. New reports from the government agencies provide useful recommendations for handling nuclear wastes and protecting against accidents like the Fukushima Daiichi disaster in Japan.

Government subsidies and guarantees remain in place for this enterprise, and given its potential for massive consequences if things go wrong, that may continue to be necessary. International agreements, such as those between the United States and Russia to dismantle nuclear warheads, point to the importance of maintaining close political ties among all countries with nuclear weapons capabilities as well as those with nuclear power plants (because these can be a source of weapons-grade nuclear material).

REVIEW QUESTIONS

1. Describe the impacts of the earthquake and tsunami that rocked Japan on March 11, 2011, on the Fukushima Dai-ichi nuclear power plants, and on the Japanese people.

2. Compare the outlook regarding the use of nuclear power in the United States and globally in each of the following time periods: 1960s, 1980s, and early 21st century.

3. Describe how energy is produced in a nuclear reaction. Distinguish between fission and fusion.

4. How do nuclear reactors and nuclear power plants generate electrical power?

5. What are radioactive emissions, and how are most humans exposed to them?

6. How are radioactive wastes produced, and what are the associated hazards?

7. Describe the current practices for dealing with spent fuel rods.

8. What problems are associated with the long-term containment of nuclear waste? What is the current status of the disposal situation?

9. Describe what went wrong at Three Mile Island, Chernobyl, and Fukushima.

10. What features might make nuclear power plants safer?

11. What are the four generations of nuclear reactors? Describe the advantages of Generations III and IV.

12. Discuss economic reasons that have caused many utilities to opt for coal-burning plants rather than nuclear-powered plants.

13. How do fast-neutron and fusion reactors work? Does either one offer promise for alleviating our energy shortage?

14. Explain why nuclear power does little to address our largest energy shortfall.

15. Discuss changes in nuclear power that might brighten its future.

THINKING ENVIRONMENTALLY

1. Evaluate the risks of nuclear power. Are we overly concerned or not concerned enough about nuclear accidents? Could an accident like Chernobyl or Fukushima happen in the United States?

2. Would you rather live next door to a coal-burning plant or a nuclear power plant? Defend your choice.

3. Global climate change has been cited by nuclear power proponents as one of the most important justifications for further development of the nuclear power option. Give reasons for their belief, and then cite some reasons that would counter their argument.

4. Go to the Web site www.ap1000.westinghousenuclear.com to investigate the reactor described there. Where is it being designed, and where will it likely be used in the United States? What are its features?

MAKING A DIFFERENCE

1. Pay a visit to a nuclear power plant, and explore the ways the facility (a) is protected against terrorist attack, (b) continually trains the plant's operators, (c) keeps the public informed in the event of a nuclear accident, and (d) handles its high-level wastes.

2. Visit the radiology department of your local hospital, and ask a radiologist to explain the different ways radiation is used in the hospital's medical procedures and how (or whether) the hospital tries to limit the radiation exposure of patients. Ask the radiologist what the hospital does with its low-level radioactive waste.

Renewable Energy

Visit Germany and you'll see medieval towns and fast cars on the autobahn. You'll also hear talk of another German distinctive: *Energiewende*, or energy transformation, the country's policy of transition to renewable energy sources, energy efficiency, and sustainable development. To meet this goal, the country is investing in new energy technologies, with a goal of 80% of the nation's power to be supplied by renewables by 2050.

The city of Freiburg im Breisgau is on the forefront of this massive transformation. Established in the Middle Ages, nearly 900 years ago, the city thrived during the Renaissance. Today, Freiburg is a prototype of a new Renaissance— a shift away from fossil fuels and nuclear power to other sources of energy. Solar panels cover its houses, schools, and train station; energy-efficient trams run on the streets; and nearly everyone works to conserve energy. Between 1992 and 2011, the city cut its carbon dioxide emissions by 20%.

Freiburg exemplifies the changes occurring as Germany reaches for its lofty ideal. Today Germany leads the world in the use of renewable energy sources such as solar and wind power. Most of the time, 25–30% of the country's energy is produced from renewables, more than double the percentage in the United States. During several peak days in 2014, an amazing 75% of Germany's domestic electricity came from renewable energy sources.

A decentralized approach. Germany's Renewable Energy Act guarantees that electricity from renewables, including that from small suppliers, has priority access to the electrical grid. Even individuals with solar panels can sell electricity to the grid. This access is an incentive for local groups to build solar parks and wind farms. By 2011, one German village was producing 300% more energy than it needed and selling the excess to the national grid. Of course, there are other ways to support the energy transformation, especially by improving the efficiency of energy conversions. To this end, some local power plants produce heat and power at the same time, thereby increasing efficiency.

▲ Solar panels on the roofs of houses in Freiburg, the greenest city in Germany.

Dramatic changes don't come without difficulties.

Despite their benefits, all renewable energy sources have drawbacks. There are, for example, costs involved in switching systems, as well as environmental costs for their production and use. The intermittent supply of solar and wind power limits the use of these renewables: solar panels trap energy only on sunny days, while wind turbines generate electricity only when the wind blows. There are also difficulties in coordinating systems when there is an energy glut. Fortunately, a number of engineers and inventors are seeking solutions to these problems. One German inventor has developed a battery that is better at storing the energy produced by sunlight through the night.

Overall, Germany's investment in *Energiewende* has lowered its greenhouse gas emissions as well as its reliance on imported natural gas. The cost of renewable energy in Germany has fallen so that it is now at parity with that of fossil fuels. Although electric bills in Germany are high, an expense that has a dampening effect on industry, the transition has driven research and provided jobs. Today Germany still uses coal and natural gas in power plants, and its vehicles still run on gas and diesel. However, the country is making great strides in its decades-long transition.

In this chapter, we will look at different sources of renewable energy, their pluses and minuses, and the roles they may play in a sustainable future. Figuring out the best way to use these resources sustainably will be part of our global task.

16.1 Strategic Issues

The objective of this chapter is to give you a greater understanding of the potential for capturing energy from sunlight, wind, biomass, flowing water, and other sources. In 2013, renewable energy provided 10% of U.S. primary energy (Figure 16–1) and 14% of electric power generation; President Barack Obama has called for 80% of the nation's electricity to be generated from clean energy (non-fossil fuel) sources by 2035. Worldwide, renewable energy supplies 11% of total energy use and about 20% of electric power. (Table 16–1). The UN has targeted a reduction in fossil fuel energy of 80% by 2050, meaning that other sources would have to supply that much. These are ambitious goals, and they raise some significant questions.

Why?

Why should we want to replace fossil fuels with renewable sources of energy in just a matter of decades? The most salient reason is global climate change (see Chapter 18). The use of fossil fuels has loaded the atmosphere with the greenhouse gas carbon dioxide (CO_2), raising the concentration of that one gas by more than 40% to its present concentration of 400 parts per million (ppm). As a result, global temperatures and sea level have already risen, with serious environmental consequences. If we were to continue to increase our use of fossil fuels at the present rate, atmospheric CO_2 would be over 600 ppm by mid-century and well over 1,000 ppm by 2100, with global temperature and sea level increases that would be catastrophic. Most international agencies have agreed that to avoid catastrophic climate change we must achieve a stable CO_2 level of 450 ppm by mid-century. This is why there is a call to reduce our use of fossil fuels by 80% before 2050.

There are other reasons. Oil and gas reserves will only last a matter of decades, so we are going to have to find new energy sources anyway, and in this century (Chapter 14). By contrast, the supply of renewable sources is not being depleted, although there are limits to the rate at which they can

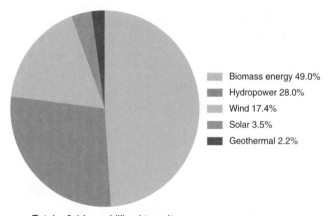

RENEWABLE-ENERGY USE IN THE UNITED STATES IN 2013

- Biomass energy 49.0%
- Hydropower 28.0%
- Wind 17.4%
- Solar 3.5%
- Geothermal 2.2%

Total = 9.14 quadrillion btu units
(= 10% of U.S. energy use in 2013)

Figure 16–1 Renewable-energy use in the United States. A mix of sources of renewable energy provided 10.0% of the nation's energy used in 2013.

(*Source:* Data from Department of Energy, Energy Information Agency, August 2014.)

Table 16–1	Renewable Electric Power at End of 2013, GW Power Capacity	
Technology	**World**	**United States**
Hydropower	1,000	78
Wind power	318	61.1
Biomass power	88	15.8
Solar PV	139	12.1
Geothermal power	12	3.4
Solar thermal power	3.4	0.9
Ocean (tidal) power	0.5	~0
Total Renewable Power capacity	1,560	172

Source: REN21. Global Status Report. 2014.

be used (Figure 16–2). Also, there are 1.3 billion people who currently do not have access to electricity and many more whose access is intermittent at best. Renewable energy sources are well suited to supplying power in areas not served by a central electrical grid. And there is the matter of other pollutants, such as sulfur dioxide, mercury, and nitrogen oxides, all of which are given off as we burn fossil fuels and all of which contribute significantly to poor health and fatalities.

Getting It Done

Is it possible to accomplish a shift to renewables in so short a time span? Here we encounter some significant problems. Fossil fuels are great sources of concentrated energy, and we access them, distribute them, and employ them in a massive infrastructure. Our modern industrial society "has been built around, by and for fossil fuels," as one commentator put it.[1] It took 70 to 100 years to build the infrastructure of our fossil fuel economy, and we have only a few decades to rebuild much of this infrastructure around clean energy sources. (Experts consider nuclear power to be a clean energy source because it does not generate CO_2, although it

does produce hazardous wastes and is not renewable.) This transition will be nothing short of a revolution, one that is possibly as large as the Industrial Revolution, with its shift from work carried out by humans and animals to work done by machines. Clearly, the manufacture of wind turbines, PV panels, transmission lines, concentrated solar power plants, electric vehicles, and recharging stations would have to take place in a relatively short period of time, and this would require massive government support.

Another barrier to accomplishing this transition is government subsidies for fossil fuels. Between 2003 and 2013, global subsidies to fossil fuels were $775 billion, while those for renewables were only $88 billion. In the United States, fossil fuel subsidies are $38 billion annually.[2] Even so, new government subsidies for renewables are helping to support the development of new technologies. An additional barrier is the fluctuating price of fossil fuels, which makes investment in renewables difficult. The oil production glut and price crash of 2014 made alternative energy sources less competitive.

The global shift from fossil fuels to renewable or clean energy sources is starting to happen, so the question is, will it

---

[1]Roberts, David. "Direct subsidies to fossil fuels are the tip of the (melting) iceberg." *Grist*, October 26, 2011.

[2]Ibid.

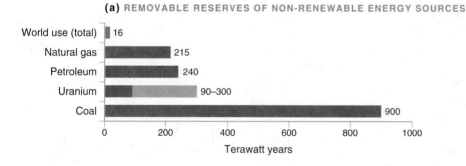

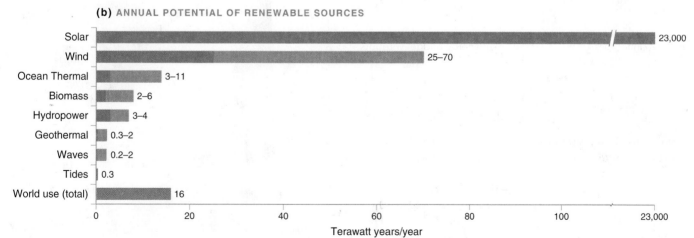

Figure 16–2 The supply of renewable and non-renewable sources of energy. Renewable energy sources have the potential to provide a great deal more energy for human use than they currently do. **(a)** The removable reserves of non-renewable energy sources in terawatt years are compared with **(b)** the annual potential of renewable sources in terawatt years/year.

(*Source:* Data from R. Perez and M. Perez. "A fundamental look at energy reserves for the planet," *Solar Update: IEASHC Newsletter* (2009) 50:2–3.)

happen fast enough? There are some encouraging developments that suggest it is possible, but much more has to be done:

- Venture capitalists are sinking a lot of money into green technologies; globally, $214 billion was invested in 2013 into wind farms, solar power, electric cars, and other renewable technologies.

- Thirty-eight states now have some form of a renewable portfolio standard (RPS) or alternative energy portfolio standard (AEPS), which requires electric utilities to generate a certain amount of electricity from sources other than fossil fuels; for example, California must generate 33% of its electricity from renewable sources by 2021.

- Natural gas is becoming more available and can fill a "bridge energy" role to substitute for coal (because coal generates twice as much CO_2 as natural gas per kW of electricity) and to work in tandem with renewable sources that are intermittent in their delivery, to smooth out gaps in electricity generation.

- It may be possible to continue to use coal *if*, and *only if*, the CO_2 generated in the coal power plant is captured from the waste stream and sequestered, a process known as carbon capture and sequestration (CCS). The technology is known, but the process adds a significant cost to the electricity generated.

- It is possible to hold greenhouse gas emissions to a limit of 450 ppm CO_2— if we stop building more fossil fuel infrastructure and allow the present infrastructure to live out normal lifespans. In other words, it is the future coal-powered plants and gasoline-powered automobile manufacturing plants that will commit the world to exceeding safe limits of greenhouse gas emissions.

We can now turn our attention to the different forms of renewable energy. As we saw in Figure 16–2, which compares the amount of energy that is potentially available from fossil fuels with that from various sources of renewable energy, the supply of solar energy far exceeds that of all other energy sources.

CONCEPT CHECK Does the discovery of new oil reserves make it unnecessary for us to move to an energy budget without fossil fuels? Why or why not? ☑

16.2 Putting Solar Energy to Work

Before turning our attention to the practical ways that solar energy is captured and used, let us consider some general concepts of solar energy.

Solar energy originates with thermonuclear fusion in the Sun. (Importantly, all the chemical and radioactive by-products of the reactions remain behind on the Sun.) The solar energy reaching Earth is radiant energy, entering at the top of the atmosphere at 1,366 watts per square meter, the **solar constant**. This energy ranges from ultraviolet light to visible light and infrared light (heat energy) **(Figure 16–3)**. About half of this energy actually makes it to Earth's surface, 30% is reflected, and 20% is absorbed by the atmosphere.

Thus, full sunlight can deliver about 700 watts per square meter to Earth's surface when the Sun is directly overhead. At that rate, the Sun can deliver 700 MW of power (the output of a large power plant) to an area of 390 square miles (1,000 km²). The total amount of solar energy reaching Earth is vast—almost beyond belief. Just 40 minutes of sunlight striking the land surface of the United States yields the equivalent energy of a year's expenditure of fossil fuel. The Sun delivers 10,000 times the energy used by humans.

Using some of this ample supply of solar energy will not change the basic energy balance of the biosphere. Solar energy absorbed by water or land surfaces is converted to heat and eventually lost to outer space. Even the fraction that is absorbed by vegetation and used in photosynthesis is ultimately given off again in the form of heat as various consumers break down food (Chapter 3). Similarly, if humans were to capture and obtain useful work from solar energy, it would still ultimately be converted to heat and lost in accordance with the Second Law of Thermodynamics (see Chapter 3). The overall energy balance would not change.

Although solar energy is an abundant source, it is also diffuse (widely scattered), varying with the season, latitude, and atmospheric conditions. The main challenge in using solar energy is one of taking a diffuse and intermittent source and concentrating it into a form, such as fuel or electricity, that can be used to provide heat and run vehicles, appliances, and other machinery. This requires us to solve problems involving the *collection*, *conversion*, and *storage* of solar energy. Also, in the final analysis, overcoming such hurdles must be cost effective. In the sections that follow, you will see how we can overcome or, even better, sidestep these three problems to meet our energy needs in a cost-effective manner.

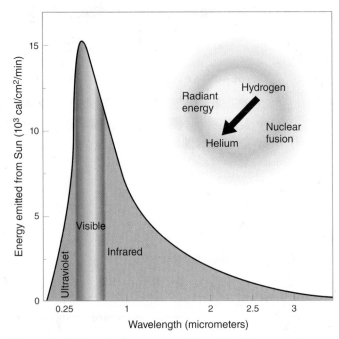

Figure 16–3 The solar-energy spectrum. Equal amounts of energy are found in the visible-light and the infrared regions of the spectrum. In the Sun, the nuclear fusion of hydrogen to form helium emits a spectrum of radiant energy that reaches Earth.

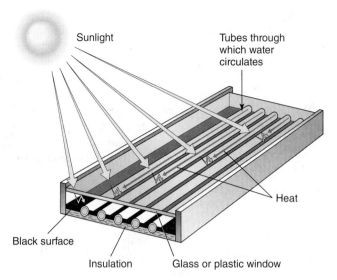

Figure 16–4 The principle of a flat-plate solar collector. As sunlight is absorbed by a black surface, it is converted to heat. A clear glass or plastic window over the surface allows the sunlight to enter the collector, which traps the heat. Air or water is heated as it passes over and through tubes embedded in the black surface.

Solar Heating of Water

Solar hot-water heating is already popular in warm, sunny climates. A solar collector for heating water consists of a thin, broad box with a glass or clear plastic top and a black bottom in which water tubes are embedded (Figure 16–4). Such collectors are called **flat-plate collectors**. Faced toward the Sun, the black bottom gets hot as it absorbs sunlight—similar to how black pavement heats up in sunlight—and the clear cover prevents the heat from escaping, as in a greenhouse. Water circulating through the tubes is heated and conveyed to a tank, where it is stored.

In an *active system*, the heated water is moved by means of a pump. In a *passive* solar water-heating system, the system is mounted so that the collector is lower than the tank. Thus, heated water from the collector rises by natural convection into the tank, while cooler water from the tank descends into the collector (Figure 16–5). The tank usually has a source of auxiliary heat (electric or gas) in order to get the temperature to a desired level or to provide heat when solar energy is insufficient.

In temperate climates, where water in the system might freeze, the system may be adapted to include a heat-exchange coil within the hot-water tank. Then antifreeze fluid is circulated between the collector and the tank. Worldwide, there are approximately 60 million solar hot-water systems, but this is still only a small fraction of the total number of hot-water heaters. The reason is the initial cost; in the United States, solar heaters cost 5 to 10 times as much as gas or electric heaters. However, over time, the solar system costs less to operate than an electric or a gas system. In China, more than 18 million households get their hot water from solar thermal systems; China leads the world in this application of solar energy.

Solar Space Heating

Sunlight may also be used to heat the air in buildings. This is called *solar space heating*, a type of passive solar heating. Inexpensive, even homemade, flat-plate collectors similar to those used to heat water can be used for space heating. All that is needed is to have air circulate through the collector box. Again, efficiency is gained if the collectors are mounted, to allow natural convection to circulate the heated air into the space to be heated (see Figure 16–6, on the next page).

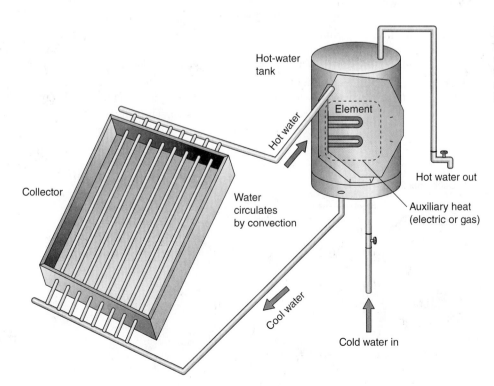

Figure 16–5 Solar water heaters. In nonfreezing climates, simple water-convection systems may suffice. In freezing climates, an antifreeze fluid is circulated. Solar heat is augmented by an auxiliary heat source in the hot-water tank.

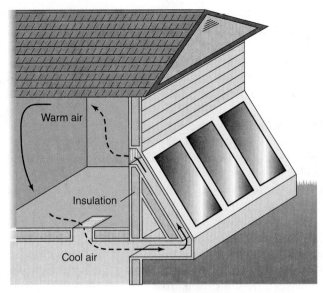

Figure 16–6 Passive hot-air solar heating. Many homeowners could save on fuel bills by adding homemade solar collectors like the one shown here. Air heated in the collector moves into the room by passive convection.

Building = Collector. The greatest efficiency in solar space heating is gained by designing a building to act as its own collector. The basic principle is to have windows facing the Sun. In the winter, because of the Sun's angle of incidence, sunlight can come in and heat the interior of the building (Figure 16–7a). At night, insulated curtains or shades can be pulled down to trap the heat inside. A well-insulated building, with appropriately made doors and windows, acts as its own best heat-storage unit. Other systems for storing heat, such as tanks of water or masses of rocks, have not proved to be cost effective. In the summer, excessive heat can be avoided by using an awning or overhang to shield the window from the high summer Sun (Figure 16–7b).

Appropriate landscaping can contribute to the heating and cooling efficiency of both solar and nonsolar space-heating designs, thereby conserving energy. In particular, deciduous trees or vines on the sunny side of a building can block much of the excessive summer heat while letting the desired winter sunlight pass through. An evergreen hedge on the shady side can provide protection from the wind.

Energy Stars. In almost any climate, a well-designed passive solar-energy building can reduce energy bills substantially, with an added construction cost of only 5–10%. Because about 25% of our primary energy is used for space and water heating, proper solar design, broadly adopted, would create enormous savings on oil, natural gas, and electrical power. In 2001, the Environmental Protection Agency (EPA) extended its **Energy Star Program** to buildings and began awarding the Energy Star label to public and corporate edifices. Buildings awarded the label use at least 40% less energy than others in their class and must meet a number of other criteria to qualify. By 2014, more than 25,500 buildings had earned the label—a major contribution to **energy conservation**.

Backup Systems. A common criticism of solar heating is that a backup heating system is still required for periods of inclement weather. Good insulation is a major part of the answer to this criticism. People with well-insulated solar homes find that they have minimal need for backup heating. When backup is needed, conventional forms of heating are used to augment the solar heating system. In any case, the criticism concerning the need for backup heating misses the point: The objective of solar heating is to reduce our dependency on conventional fuels. Even if solar heating and improved insulation reduced the demand for conventional fuels by a mere 20%, that would still represent sustainable savings of 20% of the traditional fuel and its economic and environmental costs.

Solar Production of Electricity

Solar energy can also be used to produce electrical power, providing an alternative to coal and nuclear power. There are two distinct approaches: photovoltaic systems and concentrated solar power.

Photovoltaic Cells. A solar cell—more properly called a **photovoltaic, or PV, cell**—looks like a simple wafer of material with one wire attached to the top and one to the bottom (Figure 16–8). As sunlight shines on the wafer, it puts out an amount of electrical current roughly equivalent to that emitted by a flashlight battery. Thus, PV cells collect light and convert it to electrical power in one step. The cells, measuring about four inches square, produce about one watt of power. Some

Figure 16–7 Solar building siting. In contrast to utilizing expensive and complex active solar systems, solar heating *can* be achieved by suitable architecture and orientation of the home at little or no additional cost. **(a)** Large, Sun-facing windows permit sunlight to enter during the winter months. Insulating curtains or shades are drawn to hold in the heat when the Sun is not shining. **(b)** Suitable overhangs, awnings, and deciduous plantings prevent excessive heating in the summer.

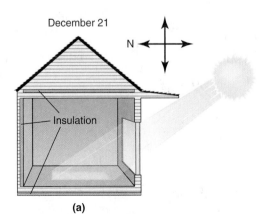

(a)

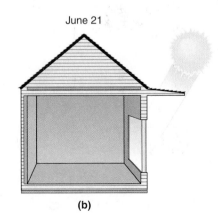

(b)

40 such cells can be linked to form a module that generates enough energy to light a lightbulb. Varying amounts of power can be produced by wiring modules together in panels.

How They Work. The simple appearance of PV cells belies a highly sophisticated level of science and technology. Each cell consists of two very thin layers of semiconductor material separated by a junction layer. The lower layer has atoms with single electrons in their outer orbital that are easily lost. The upper layer has atoms lacking electrons in their outer orbital; these atoms readily gain electrons. The kinetic energy of light photons striking the two-layer "sandwich" dislodges electrons from the lower layer, creating a current that can flow through a motor or some other electrical device and back to the upper side. Thus, with no moving parts, solar cells convert light energy directly to electrical power, with an efficiency of 15–20%.

Because they have no moving parts, solar cells do not wear out. However, deterioration due to exposure to the weather limits their life span to about 30 years. The major material used in PV cells is silicon, one of the most abundant elements on Earth, so there is little danger that the production of PV cells will ever suffer because of limited resources.

Uses. PV cells are already in common use in pocket calculators, watches, and numerous toys. Panels of PV cells provide power for rural homes, irrigation pumps, traffic signals, radio transmitters, lighthouses, offshore oil-drilling platforms, Earth-orbiting satellites, and other installations that are distant from power lines. It is not hard to imagine a future in which every home and building has its own source of pollution-free, sustainable electrical power from an array of PV panels on the roof. Current installations in many countries and 43 states in the United States employ *net metering*, where the rooftop electrical output is subtracted from the customer's use of power from the power grid.

Inverters. The most complicated (and yet necessary) component of a PV system is the **inverter**. The inverter (Figure 16–9) acts as an interface between the solar PV modules and the electric grid or batteries. It must change the incoming direct current (DC) from the PV modules to alternating current (AC) compatible with the electricity coming from the grid and the devices that will be powered by the PV system. It must also act as a control, able to detect and respond to fluctuations in voltage or current on either DC or AC, and it must be robust enough to withstand the high temperatures of an attic (where rooftop PV systems are involved). An inverter for a rooftop system costs several thousand dollars, but in most cases, these inverters are eligible for local and national subsidies. Fortunately, the cost of inverters has come down over time.

Cost. The cost of PV power (cents per kilowatt-hour) is the cost of the installed PV systems divided by the total

Figure 16–9 PV system inverter. Inverter (on right) and controllers for a PV system at Channel Islands National Park, California. The inverter is a device that connects the PV panels of a rooftop system to the electric grid or devices powered by PV electricity. It converts the direct current coming from PV panels to alternating current and also acts as a control for the system.

Figure 16–8 Photovoltaic cell. A photovoltaic cell powers this portable solar panel charging a radio on a camping trip in Sequoia National Park, California.

amount of power they may be expected to produce over their lifetime (currently as low as 12 cents per kilowatt-hour). This cost is quite comparable to that of other power alternatives (8–16 cents per kilowatt-hour for residential electricity in the United States). The first PV cells had a cost factor several hundred times that of electricity from conventional power stations transmitted through the power grid, so those cells were used mainly in areas far from the grid. PV power had its first significant application in the 1950s, in the solar panels of space satellites. This application started the development cycle rolling. As more efficient cells and less expensive production techniques evolved, costs came down dramatically. In turn, applications, sales, and potential markets expanded, creating the incentive for further development. Today the PV-power industry is the fastest-growing energy technology industry in the world. In 2013, the industry shipped solar panels with a total capacity of 37 gigawatts (GW). Existing solar PV panels now have the capacity to turn sunlight into more than 137 GW of electricity (Figure 16–10).

Utilities. The utility companies are moving toward large-scale PV installations (Figure 16-11). At the end of 2014, there were 40 plants worldwide with capacities greater than 20 MW, with 31 more planned or under construction. The largest one in the world is the 550 MW Topaz Solar Farm in California, which provides electricity for 160,000 homes. This record won't last long, as there are much larger ones under construction. In some cases, PV is getting the nod over concentrated solar power facilities (discussed later), because it can be added in smaller increments, allowing limited transmission lines to handle the voltages.

However, the most promising future for PV power is in the installation of PV panels on rooftops, where a huge amount of unused space is readily available. For residential use, utilities in many states have established programs that provide incentives to customers to install 2- to 4-kilowatt (kW) PV systems on their roofs. A 2 kW system provides about half of the annual energy needs of a residence; if the homeowner is generating more than needed, the utility buys the excess electricity at the retail rate. Even though the rooftop

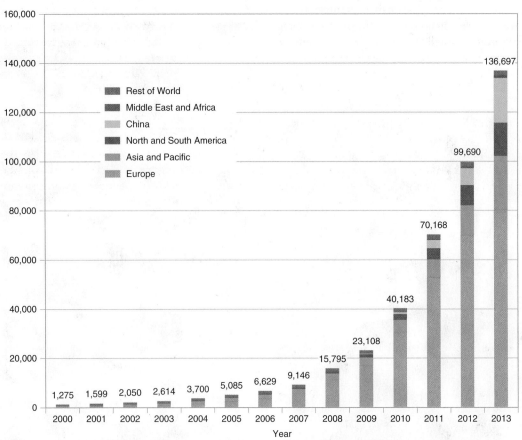

EVOLUTION OF GLOBAL CUMULATIVE INSTALLED CAPACITY 2000–2013 (MW)

Figure 16–10 The market for PV panels. Global sales of PV panels have increased dramatically, approaching 137,000 MW by 2013.

(*Source:* EPIA.)

> UNDERSTANDING THE DATA
>
> 1. Between 2010 and 2011, which region had the greatest percentage of increase in installed solar capacity?
> 2. Which region do you think had the greatest percentage increase between 2014 and 2015? How do you know?

Figure 16–11 PV power plant. The Copper Mountain Solar Facility in Nevada consists of 775,000 solar panels and has an installed capacity of 48 MW.

systems represent a loss in power purchases, the utilities benefit because they avoid the need to build expensive new power plants. A federal investment tax credit of 30% of the system cost was enacted in the Emergency Economic Stabilization Act of 2008, providing incentives to homeowners to go solar. Sunny California passed the 1,000 MW (1 GW) mark in late 2011. In 2013 alone, the worldwide capacity of solar rooftop power increased by 40 GW. The world leader in installed rooftop PV is Germany, with more than 37 GW installed capacity.

New PV Technologies. To compete with electricity from conventional sources, the cost of solar cells needs to be low. At least four new technologies may drive the cost even lower: (1) thin-film PV cells, in which cheap amorphous silicon or cadmium telluride is used instead of expensive silicon crystals, and the film can be applied as a coating on roofing tiles or glass; (2) a 3-D silicon solar cell that captures up to 25% of the incoming light energy; (3) glass coated with light-absorbing dyes that transmit energy to solar cells on the edge of the glass, while allowing much of the light to pass through to conventional solar panels; (4) flexible plastic polymer cells. All of these techniques are either in production or close to it.

Concentrated Solar Power (CSP). If you have ever focused sunlight through a magnifying glass to burn a hole in a leaf, you have used CSP. Several technologies convert solar energy into electricity by using reflectors (or concentrators) to focus sunlight onto a receiver that transfers the heat to a conventional turbogenerator. These devices work well only in regions with abundant sunlight and plenty of space; currently they are producing power in the U.S. Southwest and in Spain. A major drawback to this technology is the need for a lot of water to cool the steam that is generated. Nevertheless, CSP is growing rapidly, now at 1.8 GW generating capacity globally, with a number of projects under construction. The two methods currently in the lead for CSP are the solar (or parabolic) trough and the power tower.

Solar Trough. One method that is proving to be cost effective is the **solar-trough** collector system, so named because the collectors are long, trough-shaped reflectors tilted toward the Sun **(Figure 16–12)**. The curvature of the trough is such that all of the sunlight hitting the collector is reflected onto a pipe running down the center of the system. Oil or some other heat-absorbing fluid circulating through the pipe is thus heated to very high temperatures. The heated fluid is passed through a heat exchanger to boil water and produce steam for driving a turbogenerator. This technology has the capability of storing some thermal energy for release at night.

Solar-trough is the most developed of the CSP technologies, and most new CSP plants are solar-trough. The largest solar-trough plant in the world is Solana, a 250 MW facility located in the Arizona desert.

Power Tower. A **power tower** is an array of Sun-tracking mirrors that focuses the sunlight falling on several acres of land onto a receiver mounted on a tower in the center of the area **(Figure 16–13)**. The receiver transfers the heat collected to a molten-salt liquid, which then flows either to a heat exchanger to drive a conventional turbogenerator or to a tank at the bottom of the tower to store the heat for later use. At present, the Sierra SunTower is the only power tower operating in the United States, generating 5 MW of electricity.

Figure 16–12 Solar thermal power in southern California. A solar-trough power plant. Sunlight striking the parabolic-shaped mirrors is reflected onto the central pipe, where it heats a fluid that is used in turn to boil water and drive turbogenerators.

Figure 16–13 Sierra SunTower. Sun-tracking mirrors are used to focus a broad area of sunlight onto a molten-salt receiver mounted on the two towers. The hot salt is stored or pumped through a steam generator, which drives a conventional turbogenerator.

The Future of Solar Energy

Solar electricity is growing at a phenomenal rate—the solar power business is currently a $60 billion industry and is expected to reach $137 billion by 2020. The industry is hard-pressed to keep up with current demand. Solar energy does have certain disadvantages, however. First, the available technologies are still more expensive than conventional energy sources, though solar sources are getting close to parity with conventional ones. Second, solar energy works only during the day, so it requires a backup energy source, a storage battery, or thermal storage for nighttime power. Also, the climate in many parts of the world is not sunny enough to use solar energy in the winter. These drawbacks should be weighed against the hidden costs that are not included in the cost of power from traditional sources: air pollution, strip mining, greenhouse gas emissions, and nuclear waste disposal, among others.

Matching Demand. The Sun provides power only during the day, but 70% of electrical demand occurs during daytime hours, when industries, offices, and stores are in operation. Thus, considerable savings still can be achieved by using solar panels just for daytime needs and continuing to rely on conventional sources at night. In particular, the demand for air-conditioning, which, after refrigeration, is the second largest power consumer, is well matched to energy from PV cells. In the long run, the nighttime load might be carried by forms of indirect solar energy, such as wind power and hydropower (discussed later in the chapter).

About 66% of U.S. electrical power (51% worldwide) is currently generated by coal-burning and nuclear power plants (Chapter 14, Table 14–1). Therefore, the development of solar and wind power can be seen as gradually reducing the need for coal and nuclear power. Unlike new nuclear or coal power plants, which take years to plan and build, solar or wind facilities can become operational within months of the decision to build them. The utility does not have to guess what power demands will be in 10 or 15 years. With solar or wind power, a utility can add capacity as it is needed, make relatively small investments at any given time, and have those investments start paying back almost immediately. Thus, this approach involves much less financial risk for both the utility and its consumers.

One further advantage of solar and wind facilities is their relative invulnerability to terrorist attack compared with oil and gas pipelines and large power plants, especially nuclear facilities. Furthermore, solar and wind facilities produce less hazardous waste than traditional power plants.

Solar power has also proved suitable for meeting the needs of electrifying villages and towns in the developing world. Where centralized power is unavailable, rural electrification projects based on PV cells are already beginning to spread throughout the land (see Sustainability, *Transfer of Energy Technology to the Developing World*). The cost of the alternative—putting roughly 1.3 billion people who are still without electricity on the **power grid** (and thus requiring central power plants, high-voltage transmission lines, poles, wires, and transformers)—is unimaginable.

SUSTAINABILITY
Transfer of Energy Technology to the Developing World

The nations of the industrialized Northern Hemisphere have achieved their level of development by using energy technologies based largely on fossil fuels (and, to a lesser extent, nuclear energy). These nations' development took place during a time when fossil fuels were inexpensive. Only as those fuels became more expensive did the nations of the North begin to get serious about technologies that would make energy use more efficient. The specter of global climate change has brought a new perspective to the need to wean the industrialized North away from fossil fuel energy in the 21st century.

The same traditional fossil fuel–based technologies have been adopted by the developing nations of the Southern Hemisphere, except that these nations are lagging behind in their ability to implement the technology on a large scale. As discussed earlier in the chapter, however, the fossil fuel–based energy pattern of the 20th century will have to be replaced largely with renewable-energy systems in

the 21st century. If the United States and other developed nations are willing to play a significant role in helping the developing nations, it is questionable whether the same developmental path the North took should be promoted. Instead, there is the opportunity to engage in some "leapfrogging" technology transfer: The North can put its development dollars into such technologies as photovoltaics, wind turbines, and efficient public transport systems for the cities.

The solar route is especially attractive for many of the climates in the developing world, where an estimated 1.3 billion people lack electricity. For example, more than 300,000 PV systems installed in Kenya over a 35-year period now provide power to 1% of the rural Kenyan population. The systems, with costs ranging from $300 to $1,500, are marketed to people with incomes averaging less than $100 per month. Only the solar panels are imported; local companies provide the batteries and other system components. People use

the energy for lighting, television, and radio. Annual sales in Kenya exceed 30,000 systems. However, the population is growing, so the proportion of solar power used remains low. Financing has been accessible. Some of the financing originates with development agencies from the North, but the majority is Kenyan.

Unfortunately, technological transfer is not always easy. The difficulty of obtaining, installing, and maintaining high-quality PV products has lowered the benefit of this effort to Kenyans. Many of the installed PV systems have broken and have not been repaired. As a result, the Kenyan government is now requiring training and licensing of PV technicians. Japan and other countries are providing training, with the expectation that Kenyans will eventually take over the training. This type of education is a part of increasing a country's human capital, a key component of development. The relatively modest success of solar power in an area of the world with abundant solar potential suggests that there is no magic solution to technology transfer.

Drawbacks of Solar Power. The chief drawback of solar power is that sunlight is not continuously available, so solar power plants must be supplemented with other sources of energy. While passive solar power is low tech, active solar power, especially PV panels, requires manufacturing processes that are polluting and has a high cost of installation. The widespread use of solar panels will eventually require systems for their recycling, in part because they contain mercury, lead, and cadmium.

Another concern about solar towers is the incineration of birds that fly in the path of the concentrated radiation. The Ivanpah Solar Electric Generation System, a plant in the California desert southwest of Las Vegas that opened in 2014, was estimated to have killed close to 1,000 birds in just one year. The plant is working on deterrents to prevent birds from coming into the dangerous air space. Despite these drawbacks, the abundance of solar energy makes it likely that solar power will be a large part of our energy future.

CONCEPT CHECK Compare the scale of operation for PV solar panels and power towers. What causes the difference? Which technology is appropriate for small-scale installations? ☑

16.3 Indirect Solar Energy

Water, fire, and wind have provided energy for humans throughout history. Dams, firewood, windmills, and sails represent indirect solar energy because energy from the Sun is the driving force behind each. (Recall from Chapter 10 that solar energy powers the evaporation in the hydrologic cycle and also drives the convection currents that produce the wind.) What is the potential for expanding these options from the past into major sources of sustainable energy for the future?

Hydropower

Early in technological history, it was discovered that the force of falling water could be used to turn paddle wheels, which in turn would drive machinery to grind grain, saw logs into lumber, and run simple machines. The modern culmination of this use of waterpower, or hydropower, is huge hydroelectric dams, where water under high pressure flows through channels, driving turbogenerators (**Figure 16–14**). The amount of power generated is proportional to both the height of the water behind the dam—that is what provides the pressure—and the volume of water that flows through. (Chapter 10 describes the role played by dams in trapping and controlling rivers for flood control, irrigation, and hydropower.)

About 6.6% of the electrical power generated in the United States currently comes from hydroelectric dams, most of it from about 300 large dams concentrated in the Northwest and Southeast. Worldwide, hydroelectric dams have a generating capacity of 1,000 GW, with more very large projects in the construction or planning stage. Hydropower generates about 18% of electrical power throughout the world and is by far the most abundant form of renewable energy in use.

Trade-offs. Dams—especially large ones—are often controversial; they provide important benefits but also involve some serious consequences. The most obvious *advantages* are these:

- Hydropower eliminates the cost and environmental effects of fossil fuels and nuclear power. Hydroelectric plants have longer lives than fuel-fired ones; some plants have been in service for 100 years.

- Dams provide flood control and also provide irrigation water for agriculture. Some 40% of the world's food is generated from the irrigation facilitated by the reservoirs behind dams.

- Reservoirs behind dams often provide recreational and tourist opportunities.

- Dams can be employed in **pumped-storage power plants**; during the night, when electricity demand is low, the dam's hydrogenerator pumps water up to an elevated reservoir through a tunnel. At times of high demand, the water is returned to the lower reservoir through the tunnel, where it spins turbines and generates electricity.

There are also some serious *disadvantages*, however:

- The reservoir created behind the dam inevitably submerges farmland or wildlife habitats and perhaps towns or land of historical, archaeological, or cultural value.

- Dams and the large reservoirs created behind them often displace rural populations. In the past 50 years, as many as 40 million to 80 million people have been forced to move to accommodate the rising waters of reservoirs. In some places people are adequately compensated, but in others they are not.

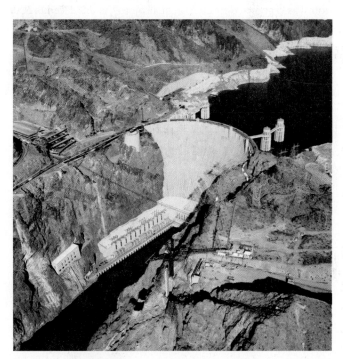

Figure 16–14 Hoover Dam. About 6.6% of the electrical power used in the United States comes from large hydroelectric dams such as this one on the Colorado River in Nevada.

- Dams impede or prevent the migration of fish, even when fish ladders are provided. Federal surveys show that fish habitats are suffering in the majority of the nation's rivers because of damming.

- Dam reservoirs increase the area of water exposed to the Sun, increasing the rate of evaporation. About 170 cubic kilometers of water, nearly 7% of the water consumed by humans, evaporates from dam reservoirs each year. This evaporation reduces the amount of water available for irrigation and downstream ecosystems and can increase the salinity of water supplies.

More Dams? In the United States, there are 87,000 dams that are six feet high or more, and an estimated 2 million smaller structures. As a result, only 2% of the nation's rivers remain free flowing, and many of these are now protected by the Wild and Scenic Rivers Act of 1968, a law that effectively gives certain scenic rivers the status of national parks. The United States and most developed countries have already brought their hydropower to capacity, and the trend is in the direction of removing many of the dams that impede the natural flow of rivers.

Nonetheless, hydropower is expanding rapidly and may double in capacity in the next decade, largely because of the construction of new dams in the developing world. Proposals for new dams outside of the United States are embroiled in controversy over whether the projected benefits justify the ecological and sociological trade-offs. For example, the Nam Theun 2 Dam, a 1,075 MW project on a tributary of the Mekong River in Laos, is expected to help the economic development of that country. The project received final approval in 2005 from the World Bank and Asian Development Bank for guarantees and loans, and construction was completed in 2010. The project has displaced 6,200 people and has inundated 175 square miles (450 km^2) of the Nakai Plateau. In addition, most of the power will be sold to Thailand (only 75 MW are to be channeled to Laos at first). International Rivers (IR) is monitoring the implementation of the project. Charges and countercharges have flown back and forth between the World Bank and IR concerning the impacts of the dam, especially on downstream communities. It is too soon to issue a verdict on the dam's overall effects, but the controversy highlights the concerns about dams and the sometimes negative impact of the World Bank.

Another huge dam, completed in 2006, is the Three Gorges Dam on the Yangtze River in China (see Chapter 10). The largest dam ever built (1.4 miles long), Three Gorges has displaced some 1.2 million people in order to generate 22,000 MW of electricity and control the flood-prone river. Offsetting the human and economic costs is the fact that it would take more than a dozen large coal-fired power plants to produce the same amount of electricity.

A special World Commission on Dams, convened to examine the impacts and controversies surrounding large dams, reported its findings in 2000. The commission began with the assertion that dams are only a means to an end, the end being "the sustainable improvement of human welfare." The report found dams to be a mixed blessing, concluding that large dams should be built only if no other options exist.[3] Following up the World Commission work, the United Nations Environment Program (UNEP) established the Dams and Development Forum, which meets annually to participate in debate and decision making over dam development in various countries. Many more dams will be built, largely in the developing countries, especially China, Brazil, Turkey, and India. Despite their significant environmental costs, these dams will certainly play a role in the energy future of those countries.

Wind Power

Once a standard farm fixture, windmills fell into disuse as transmission lines brought abundant lower-cost power from central generating plants. Not until the energy crisis and rising energy costs of the 1970s did wind begin to be seriously considered again as a potential source of sustainable energy. Growing rapidly, wind power now supplies around 2.6% of the world's electricity, but there are no technical, economic, or resource limitations that would prevent wind power from supplying at least 12% of the world's electricity by 2020. In Denmark, which is hard pressed by a lack of fossil fuel energy, wind now supplies nearly 40% of the electricity. China is the world leader in wind energy, with almost 100 GW of capacity.

A new feature of the landscape is the "wind farm," a collection of wind turbines that can generate power while the land beneath them is being used for traditional agricultural purposes. Some of these installations are huge: the Horse Hollow Wind Farm near Abilene, Texas, consists of 421 turbines with a capacity of 735 MW (Figure 16–15). The largest wind farms in the world are farm conglomerates in China. The Gansu Wind Farm in northwest China is expected to be producing 20 GW of electricity by 2020.

In the past five years, global wind power capacity has increased at a rate of 20% per year, to 318 GW at the start of 2014. Wind is the second-largest component of renewable electricity in the world (behind hydropower), because it has become economically competitive with conventional energy

---

[3]*Dams and Development: A New Framework for Decision-Making* (The Report of the World Commission on Dams, Earthscan Publications, 2000).

Figure 16–15 Horse Hollow Wind Farm. Located near Abilene, Texas, this wind farm consists of 421 turbines with a capacity of 735 MW.

sources, and countries all over the world are encouraging the installation of wind turbines by setting targets and providing subsidies (Figure 16–16).

Design. Many different designs of wind machines have been proposed and tested, but the one that has proved most practical is the age-old concept of wind-driven propeller blades. The propeller shaft is geared directly to a generator. (A wind-driven generator is more properly called a wind turbine than a windmill.) As the reliability and efficiency of wind turbines have improved, the cost of wind-generated electricity has decreased. Wind farms are now producing pollution-free, sustainable power for around five cents per kilowatt-hour, a rate that is competitive with the rates of traditional sources. Moreover, the amount of wind that can be tapped is immense. The American Wind Energy Association calculates that wind farms located throughout the Midwest could meet the electrical needs of the entire country, while the land beneath the turbines could still be used for farming. Hundreds of new turbines are sprouting from Midwestern farms, and the farmers are paid royalties ($3,000 to $5,000 per turbine per year) for leasing their land to wind developers. The U.S. Department of Energy (DOE) is promoting a scenario where wind energy will provide 20% of U.S. electricity needs by 2030—an ambitious, but by no means impossible, target.

Drawbacks. Despite its advantages, wind power does have some drawbacks. First, wind is an intermittent source of energy—its intensity varies greatly—so backup or storage must be considered. Turbines must be located where the wind is strongest and most even, but such areas may be far from users of electricity.

Second is aesthetic considerations: One or two wind turbines can be charming, but a landscape covered with them can be visually unappealing to some. More importantly, wind turbines are noisy. The large turbines make enough noise that they are not best situated near residences or in relatively pristine environments.

Third, wind turbines can kill birds and bats. These deaths are increasing with the rapid expansion of wind power. Taller turbines, more common in extremely large wind farms, kill more birds. The tens of thousands of birds and bats killed by wind turbines pales in comparison to the hundreds of millions of birds killed annually by cats, automotive traffic, and glass windows. However, in areas where turbines are concentrated, deaths from turbines can be high, especially for large soaring birds, such as hawks and condors. Locating wind farms on migratory routes or near critical habitats for endangered species can cause problems.

Offshore Wind Farms. Plans for offshore wind farms are putting a new spin on wind energy. Denmark already has many offshore turbines, as do the Netherlands, the United Kingdom, and Sweden. Belgium, France, and Germany are also in the development stage. The primary lure of offshore

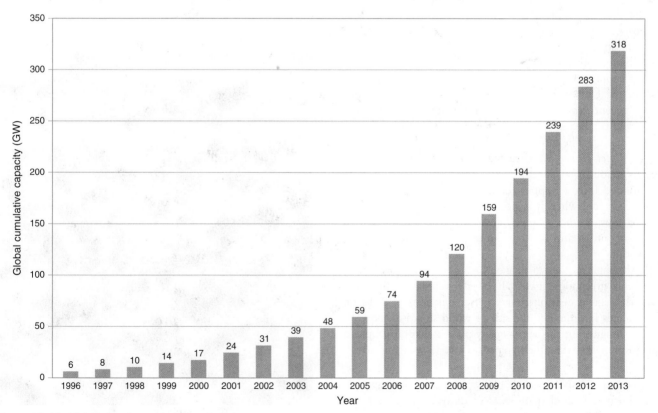

Figure 16–16 Global development of wind energy. As the cost of wind energy has dropped to the level of being very competitive with conventional energy sources, wind turbine capacity has risen dramatically.

(*Source:* Data from Global Wind Energy Council.)

wind is the dependability and strength of the maritime winds. Siting is perhaps easier, too, as the turbines are usually far enough offshore to have much less of a visual impact than onshore sites, and land does not have to be purchased.

Whether offshore or onshore, the global drive to install wind power is generating electricity that is emission free; in fact, the 43,000 MW of new wind turbines installed in a single year displace 125 million tons of carbon dioxide, which is equivalent to the emissions from 65 coal-fired power plants.

Biomass Energy

Firewood is the oldest source of energy; humans have been burning it for heat since the beginning of history. But there is nothing like a new name to put life into an old concept. Thus, "burning firewood" has become "utilizing biomass energy." **Biomass energy** means energy derived from present-day photosynthesis. Like wind power and hydropower, biomass is an indirect form of solar energy. As Figure 16–1 indicates, biomass energy leads hydropower in renewable-energy production in the United States. (Most uses of biomass energy are for heat.)

In addition to burning wood in a stove, major means of producing biomass energy include burning municipal wastepaper and other organic waste, generating methane from the anaerobic digestion of manure and sewage sludges, running power plants on wastes from timber operations, and producing alcohol from fermenting grains and other starchy materials.

Burning Firewood. Wherever forests are ample relative to the human population, firewood—or fuelwood, as it is called—can be a sustainable energy resource. Fuelwood is the primary source of energy for heating and cooking for some 2.6 billion people, amounting to about 9% of total energy use from all sources.

In the United States, woodstoves have become popular in recent years. About 2.5 million homes rely entirely on wood for heating, and another 9 million use wood for some heating. The most recent development in woodstoves is the *pellet stove*, a device that burns compressed wood pellets made from wood wastes **(Figure 16–17)**. The pellets are loaded into a hopper (controlled by computer chips) that feeds fuel when it is needed. The stove requires very little attention and burns efficiently, relying on one fan to draw air into it and another fan to distribute heated air to the room.

Although the design of wood stoves has improved, they still emit greenhouse gases. If poorly ventilated, they can produce a great deal of smoke. EPA regulations of 2014 on particulates have limited the sale of wood stoves to more efficient designs, but this leaves many poor and rural people in the position of heating their homes with non-compliant stoves.

Fuelwood Crisis? In developing counties, most people who burn fuelwood to meet their daily needs obtain the wood by foraging in local forests. Gathering wood, converting wood into charcoal, and selling the wood products in urban areas are important income-producing activities

for many in developing countries. All these activities have the potential to degrade local forests and woodlands. Yet as development progresses, people have more disposable income and shift up the "energy ladder" from wood to charcoal to fossil fuels.

The use of fuelwood peaked in the 1990s and is now declining. This shift may protect forest resources in the short term, but it is not sustainable because the replacement fuels are usually fossil fuels. Highly efficient stoves made for rural areas are a part of the solution to the problem of the overuse of wood. They also help reduce indoor and outdoor air pollution, topics that are discussed elsewhere (Chapter 17, Chapter 19). Another part of the solution is to burn waste materials rather than wood.

Burning Wastes. Facilities that generate electrical power from the burning of municipal wastes (waste-to-energy conversion) are discussed in another chapter (Chapter 21). Many sawmills and woodworking companies are now burning wood wastes, and a number of sugar refineries are burning cane wastes to supply all or most of their power. In Virginia, three 63 MW coal-fired power plants are being converted to biomass power, burning primarily wood waste from timber operations; the first was converted in 2013. Power from these sources may not meet more than a small percentage of a country's total electrical needs, but it represents a productive and relatively inexpensive way to dispose of biological wastes as well as to prevent more fossil fuel combustion.

Figure 16–17 Pellet stove. Pellet stoves use pelletized wood waste to achieve an efficient burn that requires much less care than burns from conventional woodstoves.

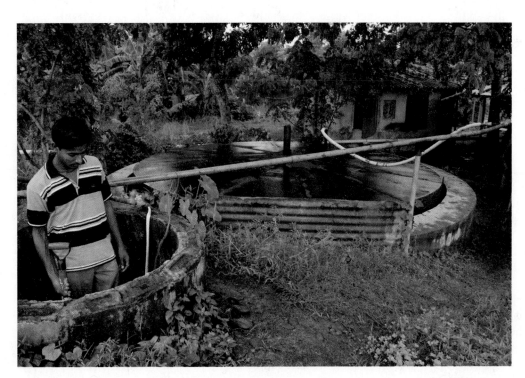

Figure 16–18 **Biogas power.** Animal wastes are introduced to this unit in rural India and mixed with water. The wastes then decompose under anaerobic conditions, producing biogas. The dome over the concrete structure holds the gas. Biogas digesters can range from small ones used for rural families to large ones used to collect energy from wastes at huge livestock or regional composting operations.

Producing Methane. The anaerobic digestion of sewage sludge yields *biogas,* which is two-thirds methane, plus a nutrient-rich treated sludge that is a good organic fertilizer (examined further in Chapter 20). Animal manure can be digested likewise. When the disposal of manure, the production of energy, and the creation of fertilizer can be combined in an efficient cycle, great economic benefit can be achieved. In China and India, millions of small farmers maintain a simple digester in the form of an underground brick masonry fermentation chamber with a fixed dome on top for storing gas (Figure 16–18). Agricultural wastes, such as pig manure and cattle dung, are put into the chamber and diluted with water. The anaerobic digestion of the wastes produces biogas, which is used for cooking, heating, and lighting. The residue slurry makes an excellent fertilizer. The concept has been expanded throughout the developing world to provide an alternative to fuelwood. Currently, some 34 million households employ these devices. Both India and China provide subsidies for these family-scale biogas plants.

Large, regional biogas plants are used to generate electricity, especially in Asia. In Camellia, Australia, a suburb of Sydney, 80,000 tons of food waste are collected every year from supermarkets, restaurants, and households; this waste is digested to produce enough electricity to power 3,600 homes. In the United States, methane released from landfills and commercial composting operations is being captured and burned to produce electricity. A proposed food composting operation in Tulare, California, will use discarded food as a source of biogas for transportation, discussed next.

CONCEPT CHECK What is one advantage and one disadvantage of wind power? Of biomass energy? ☑

16.4 Renewable Energy for Transportation

Declining oil reserves and the enormous impact of transportation's demands for oil on our economy and on the global climate suggest that the most critical need for a sustainable energy future is a new way to fuel our vehicles. Renewable energy is already answering a small but growing part of this need in the form of biofuels. In the long term, the answer may come in the form of hydrogen and fuel cells.

Biofuels

Complex organic matter from plants and animal wastes can be processed to make fuels for vehicles. Currently, two fuels are seeing rapid expansion globally: **ethyl alcohol** (ethanol) and **biodiesel**. High oil prices, strong farmer support, government subsidies, and environmental concerns have led to a major expansion of global ethanol and biodiesel production since 2000. Although these two biofuels currently provide only 2.5% of global transportation fuels, this is just the beginning, as new technologies and government support promise to make biofuels a major competitor for fossil fuels. The current sources of biofuels are agricultural and food commodities.

Ethanol. Ethanol is produced by the fermentation of starches or sugars. The usual starting material is corn, sugarcane, sugar beets, or other grains. Fermentation is the process used in the production of alcoholic beverages. The new part of the concept is that, instead of drinking the brew, we distill it and burn it in our cars. However, production costs make ethanol more expensive than gasoline unless oil sells for more than $55 a barrel. To stimulate the industry in the United States, federal tax credits have been extended to biofuels, which make them more competitive. **Gasohol**, which is 10% ethanol and 90% gasoline, is now marketed in most parts of the country.

Farm Products. In 2013, 14.3 billion gallons (340 million barrels) of ethanol for fuel, the equivalent of 194 million barrels of oil, were produced in the United States (ethanol contains about two-thirds the energy of regular gasoline, gallon for gallon). One-third of the nation's corn crop is dedicated to this use; however, an ethanol factory also produces corn oil and livestock feed. A variety of tax credits help ethanol compete with gasoline, and a **renewable fuel standard (RFS)** requires a minimum volume of renewable fuel in gasoline. The U.S. policy mandated a level of 9 billion gallons of renewable fuel in 2008 (all corn-based ethanol), to rise to a limit of 36 billion gallons in 2022, but the increase beyond 2015 must be met by using sources other than corn.

Brazil has also invested heavily in fuel ethanol; there the feedstock is sugarcane, which is much less costly to grow than corn. Brazil used to lead the world in ethanol production, but has been surpassed by the United States, using corn as the basis of ethanol production.

Can we expect ethanol production from corn to make serious inroads on the oil import and greenhouse gas emissions problems? Currently, biofuels represent about 8% of U.S. fuel consumption. This could rise to 20% if the country's entire corn crop were devoted to ethanol production—an impossible assumption, given the importance of corn for animal feed and export. With some 84 million acres devoted to corn harvest in the United States, there is little suitable farmland for significant expansion.

Another issue is the energy efficiency of corn-based ethanol. Fossil fuel energy is required to produce the corn (fertilizer, pesticides, farm machinery), to transport it, and to operate the ethanol plant. Using ethanol instead of gasoline reduces greenhouse gas emissions, mile for mile, by only 12–18%. If new land is cleared of its forests or grasslands to plant corn for biofuels, the carbon emissions from the land-use change should be added to the debit from fossil-fuel greenhouse gas emissions. When this is done, corn-based ethanol is a net loser in its impact on climate change.[4] Thus the use of corn-based ethanol is more damaging to the environment than people initially thought it would be. In addition, the use of corn as fuel rather than as food for livestock and humans has contributed to rising food prices. Because 60 countries have RFS targets, diversion of food to fuel is a significant concern (Chapter 12).

Second-Generation Biofuel. A more likely long-run technology for producing ethanol is the use of cellulosic feedstocks such as agricultural crop residues, grasses (such as switchgrass), logging residues, fast-growing trees, and fuelwoods from forests. Ethanol produced this way is called *second-generation biofuel* (current starch-based processes

Figure 16–19 Miscanthus x giganteus. This hardy perennial grass grows to a height of four meters and is a prime candidate for a second-generation cellulosic biofuel.

produce *first-generation biofuel*). One crop with exceptional potential is Miscanthus (Figure 16–19), a giant perennial grass that can be grown in poor soils and that outproduces corn and other grasses.

With the use of enzymes to break down the cellulose to sugars, demonstration facilities have shown that cellulosic feedstock technology can be a cheaper and more energy-efficient process than corn-based production. Although the technology is still in the initial stages, three large cellulosic biofuel plants opened in 2014 and others are planned. The RFS requires that all of the increase in the target biofuel after 2015 be derived from second-generation biofuel sources. The advancement of the industry depends on tax credits, which are necessary to research but can change with government budgeting priorities.

Biodiesel. Imagine a truck exhaust that smells like french fries. That is not only an improvement over diesel exhaust, but also an indication that the truck is burning biodiesel, made largely from soybean oil. Sales of biodiesel have reached 1.8 billion gallons, which is competitive with petroleum diesel fuel because of a hefty $1-per-gallon production tax credit. Still, this is only 1% of the diesel fuel used for transportation in the United States. Standard biodiesel is a 20% mix of soybean oil in normal diesel fuel, but practically any natural oil or fat can be mixed with an alcohol to make the fuel; one excellent source is recycled vegetable oil from frying. However, there are competing uses for the vegetable oils.

There are some offbeat technologies surfacing to produce biodiesel, such as a factory in Missouri that took in truckloads of turkey wastes from a turkey-processing plant and turned them into biofuels. Problems with odor made the plant owners switch to using other starting materials, such as used fryer oil, but the process can be done with almost any organic waste imaginable. In another technology, genetic engineering is being applied to algae to induce them to produce oils and complex alcohols. Significant money is being invested in these enterprises, and commercial-scale production is only a few years away.

---

[4]Sims R., et al., "Transport," in *Climate Change 2014: Mitigation of Climate Change. Contribution of Working Group III to the Fifth Assessment Report of the Intergovernmental Panel on Climate Change* (Cambridge: Cambridge University Press, 2014).

Hydrogen: Highway to the Future? Over the past decade, there has been increasing interest in powering cars with hydrogen. Conventional cars with internal combustion engines can be adapted to run on hydrogen gas in the same manner as they can be adapted to run on natural gas. (Hydrogen is not really a fuel, but an energy carrier, like electricity; it must be generated using some other energy source.) Neither carbon dioxide nor hydrocarbon pollutants are produced during the burning of hydrogen. Indeed, the only major by-product of H_2 combustion is water vapor:

$$2H_2 + O_2 \rightarrow 2H_2O + \text{energy}$$

(Some nitrogen oxides will be produced because the burning still uses air, which is nearly four-fifths nitrogen.)

A Good Idea, but . . . If hydrogen gas is so great, why aren't we using it? The answer is that there is virtually no hydrogen gas on Earth. Any hydrogen gas in the atmosphere has long since been ignited by lightning and burned to form water. And although many bacteria in the soil produce hydrogen in fermentation reactions, other bacteria are quick to use the hydrogen because it is an excellent source of energy. Thus, there are abundant amounts of the *element* hydrogen, but it is all combined with oxygen in the form of water (H_2O) or other compounds.

Although hydrogen can be extracted from water via **electrolysis** (the reverse of the preceding reaction), electrolysis requires an *input* of energy. Hydrogen can also be derived chemically from hydrocarbon fuels such as methane, petroleum oil, and methyl alcohol, but these processes are also energy losers; that is, more energy is put into the process than is contained in the hydrogen. Moreover, because these compounds contain carbon, it would be better to use them directly than to obtain hydrogen from them and use it. Therefore, the use of hydrogen as a truly clean energy carrier must await an energy source that is itself suitably cheap, abundant, and nonpolluting—like solar energy.

Plants Do It. The "holy grail" of research and eventual commercialization is to accomplish what plant cells do: produce fuel from water using solar energy. To date, two ways have been proposed for mimicking this process. The first is to employ a catalyst to do what the photosynthetic pathway does. Two elements are required: a collector that converts photons into electrical energy (electrons), and an **electrolyzer** that uses the electron energy to split water into hydrogen and oxygen. Recently, researchers from the Massachusetts Institute of Technology (MIT) described a process that employs a water-oxidizing electrolyzer catalyst made from cobalt and a heavy metal alloy, all common chemicals.[5] The process uses light energy captured by a silicon solar cell and generates hydrogen and

oxygen in either a wired or a wireless cell. It operates in neutral water and ambient temperature and pressure, and achieves a solar-to-fuels efficiency of 4.7% (wired) and 2.5% (wireless). This may be the key step needed to get to that holy grail.

Electrolysis. The second approach is simply to pass an electrical current through water and cause the water molecules to dissociate. Hydrogen bubbles come off at the negative electrode (the cathode), while oxygen bubbles come off at the positive electrode (the anode). (You can demonstrate this process for yourself with a battery, some wire, and a dish of water. A small amount of salt placed in the water will facilitate the conduction of electricity and the release of hydrogen.) The hydrogen gas is collected, compressed, and stored in cylinders, which can then be transported or incorporated into a vehicle's chassis. One important drawback to this technology is that it is difficult to store enough hydrogen in a vehicle to allow it to operate over long distances. Also, compressing the hydrogen requires energy, thus reducing the energy efficiency involved. As an alternative, the hydrogen can be combined with metal hydrides, which can absorb the gas and release it when needed, but finding a hydride that is not too heavy has been a challenge. Another pathway that holds promise is to convert the hydrogen into liquid formic acid (HCOOH) by combining it with carbon dioxide. The hydrogen is then released by use of a catalyst.

Solar Energy to Hydrogen. A rather elaborate plan has been suggested for taking solar energy–produced hydrogen for vehicles to its logical conclusion. The plan would start by building arrays of solar-trough or photovoltaic generating facilities in the deserts of the southwestern United States, where land is cheap and sunlight plentiful. The electrical power produced by these methods would then be used to produce hydrogen gas by electrolysis. The most efficient way to move the hydrogen is through underground pipelines; there are already hundreds of miles of such pipelines through which hydrogen is transported for use in chemical industries. Although this would be an expensive part of a hydrogen infrastructure, the technology for it is already well known.

Model U. The Ford Motor Company unveiled in the United States a "concept car" it has dubbed the Model U. The Model U sports a 2.3-liter internal combustion engine that runs on hydrogen and is equipped with a hybrid electric drive. The car uses four 10,000 psi tanks to store hydrogen, enough to range 300 miles before a fill-up. And there's the problem: Before such cars can become feasible, there must be a hydrogen-fueling infrastructure. Once that infrastructure is in place, cars like the Model U would likely be replaced by a completely new way of moving a vehicle: by means of fuel cells.

Fuel Cells. An alternative to burning hydrogen in conventional internal combustion engines uses hydrogen in **fuel cells** to produce electricity and power the vehicle with an electric motor. Fuel cells are devices in which hydrogen or some other fuel is chemically combined with oxygen in

---

[5]Steven Y. Reece et al., "Wireless Solar Water Splitting Using Silicon-Based Semiconductors and Earth-Abundant Catalysts," *Science* 334: 645–648. November 4, 2011.

a manner that produces an electrical potential rather than initiating burning (Figure 16–20a). Emissions from fuel cells consist solely of water and heat. Because fuel cells create much less waste heat than conventional engines do, energy is transferred more efficiently from the hydrogen to the vehicle—at a rate of 45–60% versus the 20% for current combustion-engine vehicles. To power a vehicle, hundreds of fuel cells are combined into a fuel-cell stack. The vehicle must also include a hydrogen storage device, a device to force oxygen into the fuel cells, and a cooling system (fuel cells generate heat). Fuel cells are now powering 150 buses in many cities in the United States and around the world. The fuel-cell stack develops more than 200 kW of power from hydrogen in tanks mounted on the roof of the bus.

"Concept" vehicles operating on fuel cells have been developed by all the major car manufacturers. Honda, however, is the first manufacturer to commit to a dedicated fuel-cell vehicle assembly line. The Honda FCX Clarity (Figure 16–20b) is rolling off the assembly lines, but

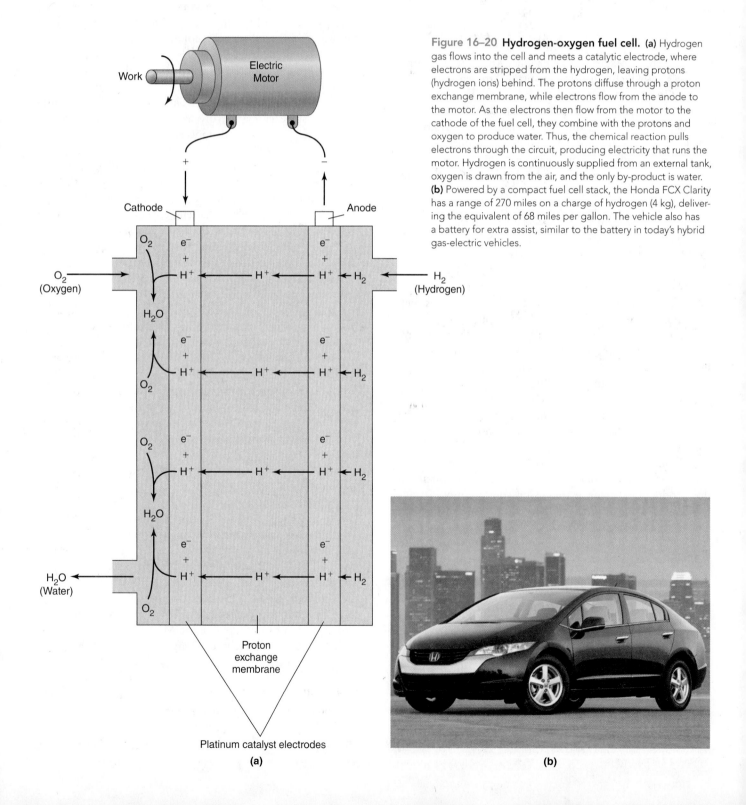

Figure 16–20 Hydrogen-oxygen fuel cell. (a) Hydrogen gas flows into the cell and meets a catalytic electrode, where electrons are stripped from the hydrogen, leaving protons (hydrogen ions) behind. The protons diffuse through a proton exchange membrane, while electrons flow from the anode to the motor. As the electrons then flow from the motor to the cathode of the fuel cell, they combine with the protons and oxygen to produce water. Thus, the chemical reaction pulls electrons through the circuit, producing electricity that runs the motor. Hydrogen is continuously supplied from an external tank, oxygen is drawn from the air, and the only by-product is water. **(b)** Powered by a compact fuel cell stack, the Honda FCX Clarity has a range of 270 miles on a charge of hydrogen (4 kg), delivering the equivalent of 68 miles per gallon. The vehicle also has a battery for extra assist, similar to the battery in today's hybrid gas-electric vehicles.

(a)

(b)

relatively few have been sold. Of course, we would need many fueling stations for hydrogen to be used widely. In the United States, California is leading the way by building hydrogen fueling stations.

If a hydrogen infrastructure is put in place and vehicles powered by fuel cells are available, we will have entered a **hydrogen economy**—one in which hydrogen becomes a major energy carrier. With solar- and wind-powered electricity also in place, we would no longer be tied to nonrenewable, polluting energy sources. Still, the obstacles in the way of the hydrogen economy are formidable, and many believe it will be decades before it even has a chance of happening. Global climate concerns convey an urgency to resolving this issue, and many believe that the best pathway is greater use of hybrid vehicles and other technologies that are already available.

CONCEPT CHECK What is one advantage and one disadvantage of fuel cell technology? ☑

16.5 Additional Renewable-Energy Options

In addition to water, wind, biomass, and hydrogen energy, several other renewable-energy options have been developed or are in the concept stage.

Geothermal Energy

In various locations in the world, such as the northwestern corner of Wyoming (Yellowstone National Park), there are springs that yield hot, almost boiling, water. Natural steam vents and other thermal features are also found in such areas. They occur where the hot, molten rock of Earth's interior is close enough to the surface to heat groundwater, particularly in volcanic regions. **Geothermal energy**

uses such naturally heated water or steam to heat buildings or drive turbogenerators. In 2014, geothermal energy had more than 12,000 MW of electrical power capacity (equivalent to the output of 11 large nuclear or coal-fired power stations), in countries as diverse as Nicaragua, the Philippines, Kenya, Iceland, and New Zealand. The largest single facility is in the United States at The Geysers, 70 miles (110 km) north of San Francisco (Figure 16–21). With more than 3,400 MW from geothermal energy, the United States is the world leader in the use of this energy source; new projects in the development stage will increase this to more than 4,500 MW. As impressive as this application of geothermal energy is, nearly double the amount is being used to directly heat homes and buildings, largely in Japan and China.

In Australia, Europe, and Japan, **enhanced geothermal systems (EGS)** are being used to generate electricity. EGS involves drilling holes several miles deep into granite that holds temperatures of 400 degrees or more. Pressurized water is then injected into one hole, where it absorbs heat from the rock. The heated water turns into steam and flows up another shaft to a power plant, where it generates electricity. This underground heat is widespread in the United States, unlike the present limited geothermal sources. An MIT team proposed a $1 billion investment to exploit the EGS potential in the United States, predicting that by 2050 some 100 GW of electrical power could be developed in this way.

Heat Pumps. A less spectacular, but far more abundant, energy source than the large geothermal power plants exists anywhere pipes can be drilled into the Earth. Because the ground temperature below six feet or so remains constant, the Earth can be used as part of a heat exchange system that extracts heat in the winter and uses the ground as a heat sink in the summer. Thus, the system can be used for heating and cooling and does away with the need for separate furnace and

Figure 16–21 Geothermal energy. The Geysers, located in northern California, is the largest complex of geothermal power plants in the world.

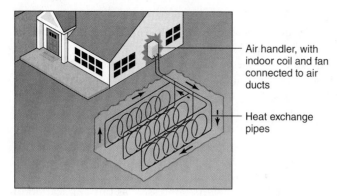

Figure 16–22 Geothermal heat pump system. The pipes buried underground facilitate heat exchange between the house and the Earth. This system can either cool or heat a house and can be installed almost anywhere, though at a higher initial cost than a conventional heating, ventilation, and air-conditioning system.

Air handler, with indoor coil and fan connected to air ducts

Heat exchange pipes

air-conditioning systems. As **Figure 16–22** shows, a **geothermal heat pump (GHP)** system involves loops of buried pipes filled with an antifreeze solution or a refrigerant, circulated by a pump connected to the air handler (a box containing a blower and a filter). The blower moves house air through the heat pump and distributes it throughout the house via ductwork. In most cases, the up-front costs of the heat pump system are repaid in six to eight years, and then the savings really begin.

Four elementary schools in Lincoln, Nebraska, installed GHP systems for their heating and cooling. Their energy cost savings were 57% compared with the cost of conventional heating and cooling systems installed in two similar schools. Taxpayers will save an estimated $3.8 million over the next 20 years with the GHP systems. According to the EPA, GHP systems are by far the most cost-effective, energy-saving systems available. Although they are significantly more expensive to install than conventional heating and air-conditioning systems, they are trouble-free and save money over the long run. New heat pump installations in the United States are growing rapidly, with more than 1 million in operation in 2014.

Energy from the Oceans

Over the years, experts have considered the oceans as a possible source of energy. Energy might come from the movement of water in tides and waves or from thermal gradients in the water column. Unfortunately, these energy sources are little used because they are hard to harness or are only available in limited places.

Tidal and Wave Power. A phenomenal amount of energy is inherent in the twice-daily rise and fall of the ocean tides, brought about by the gravitational pull of the Moon and the Sun. Many imaginative schemes have been proposed for capturing this pollution-free source of energy.

The most straightforward idea is the **tidal barrage,** in which a dam is built across the mouth of a bay and turbines are mounted in the structure. The incoming tide flowing through the turbines generates power. As the tide shifts, the blades are reversed so that the outflowing water continues to generate power.

In about 30 locations in the world, the shoreline topography generates tides high enough—20 feet (6 m) or more—for this kind of use. Large tidal power plants already exist in two of those places: France and the Bay of Fundy in Canada. The Annapolis Tidal Generating Station in the Bay of Fundy, has operated since 1984, at 20 MW capacity. Thus, this application of tidal power has potential only in certain localities, and even there, it would not be without adverse environmental impacts. The barrage would trap sediments, impede the migration of fish and other organisms, prevent navigation, alter the circulation and mixing of saltwater and freshwater in estuaries, and perhaps have other, unforeseen ecological effects.

Other techniques are being explored that harness the currents that flow with the tides. For example, in 2006, Verdant Power began a project to harness the tidal energy of the East River in New York City. The project has passed its first demonstration phase, connecting six turbines to local electric customers. Verdant plans to install 30 turbines that will connect to the grid and generate 1 MW of power. San Francisco is eyeing a similar scheme to tap the energy of the 400 billion gallons of water that flow under the Golden Gate Bridge with every tide.

Standing on the shore of any ocean, an observer would have to be impressed with the energy arriving with each wave, generated by offshore winds. It might be possible to harness some of this energy, but the technological challenge is daunting. There are a number of proposed mechanisms for capturing wave energy, but this technology is only in the early research-and-development stage.

Ocean Thermal-Energy Conversion. Over much of the world's oceans, there is a thermal gradient of about 20°C (36°F) between the surface water heated by the Sun and the colder deep water. **Ocean thermal-energy conversion (OTEC)** is an experimental technology that uses this temperature difference to produce power. Unfortunately, various studies indicate that OTEC power plants have little economic promise—unless, perhaps, they can be coupled with other, cost-effective operations. For example, in Hawaii, a shore-based OTEC plant uses the cold, nutrient-rich water pumped from the ocean bottom to cool buildings and supply nutrients for vegetables in an aquaculture operation, in addition to cooling the condensers in the power cycle. Interest in duplicating such operations is minimal at present.

CONCEPT CHECK Is it likely that tidal power and ocean thermal energy will ever provide us with a large amount of energy? Why or why not? ☑

16.6 Policies for a Sustainable-Energy Future

We have examined our planet's energy resources, requirements, and management options (Chapters 14–16). The global issues are clear: Fossil fuels, especially oil and natural gas, will not last long at the current (and increasing) rate of consumption. Even more important, every use of fossil fuels produces unhealthy pollutants and increases the atmospheric burden of carbon dioxide, accelerating the already-serious climate changes brought on by rising levels of greenhouse gases. A sustainable-energy future does not include significant fossil-fuel energy. Global targets, therefore, are (1) stable atmospheric levels of greenhouse gases, especially carbon dioxide, and (2) a sustainable-energy development-and-consumption pattern that is based on renewable energy.

Our main focus in this chapter has been on sustainable energy *development*, but there is another energy strategy that focuses on energy *consumption*—namely, energy conservation and efficiency. Energy saved is energy that does not have to be developed or paid for. Earlier we laid out two policies that have energy conservation as a goal (Chapter 14). These were increasing the mileage standards for motor vehicles and increasing the energy efficiency of lighting, appliances, and buildings. The Energy Star program, described earlier in this chapter, encourages municipalities and businesses to achieve energy savings. We turn now to public policies regarding energy, including those that encourage the use of renewable energy and energy conservation. Energy policies are enacted at the national and international levels.

U.S. Energy Policy

The United States is heavily dependent on imported oil for transportation, although U.S. production of shale oil has increased dramatically in recent years (Chapter 14). Importing oil hurts the economy and causes an *energy security* problem: Not only are we subject to the whims of oil producers, but we are also vulnerable to terrorist attacks on various facets of our energy infrastructure.

In recent years, the U.S. government has responded to these concerns by enacting various laws and regulations. Previously we examined policies pertaining to fossil fuels and nuclear power in the Energy Policy Act of 2005 and the Energy Independence and Security Act of 2007 (Chapters 14 and 15). These laws have become the major expression of federal energy policy, though many other laws promoting renewable energy and conservation have been enacted. The American Recovery and Reinvestment Act of 2009 gave a huge boost ($43 billion) to renewable energy and energy conservation, with the intent of doubling the energy from renewable sources by 2012. Although this aggressive goal was not achieved, the use of renewable energy has increased dramatically.

Supply-Side Provisions. A number of supply-side provisions have been enacted into law (Chapter 14). These include:

- the establishment of renewable fuel standards (RFS) for ethanol and biodiesel;
- funding for research and development of renewable energy;
- tax credits for electricity generated with wind, hydropower, geothermal, and biomass;
- tax credits for investment in geothermal heat pumps, solar water heating, biofuels, and PV systems.

The federal government has not established a renewable portfolio standard (RPS). Such a standard would require utilities to provide an increasing percentage of power from non-hydroelectric renewable sources—wind, solar, and biomass. However, 38 states and the District of Columbia have implemented RPS requirements.

Demand-Side Provisions. A variety of provisions that focus on energy conservation and efficiency have also been enacted into federal law. These include:

- raising the corporate average fuel economy (CAFE) standards for the combined fleet of cars and light trucks;
- requiring an increase in the efficiency of lightbulbs;
- encouraging Energy Star efficiency appliances;
- extending the Energy Star energy-efficiency program for buildings;
- tax credits for many types of improvements to homes and commercial buildings that enhance energy efficiency;
- promoting the development of hybrid and hydrogen fuel cell vehicles;
- encouraging waste-energy recovery (see Chapter 21).

International Energy Policy

Most energy policy is enacted by individual countries. At least 64 countries have a national target for renewable energy. As we have seen, Germany has a national energy policy with goals for several decades that are designed to move that country to a high use of renewables.

International cooperation can also increase the use of renewable energy. At this point, there is no overarching international energy treaty, but treaties about limiting greenhouse gas emissions have been negotiated with varying degrees of success. Although most parts of the world do not have comprehensive multi-national energy policies, the world is working on international agreements about climate change, discussed later (Chapter 18).

The most organized group of countries on the issue of energy policy is the European Union. A 2007 treaty included language that empowered the EU to form a coherent plan for promoting energy efficiency, developing renewable energy, lowering greenhouse gas emissions, and operating an energy market. Many European countries are concerned about their reliance on Russian natural gas. In 2014, EU member states and several other countries (together called the *Energy Union*) unveiled an energy plan to become more energy independent and diversified.

Latin American and Caribbean countries have attempted to produce a joint energy policy similar to that of the EU. At present they have not reached an agreement, largely because their approaches to energy policies differ; the countries are divided on whether to promote fossil fuels or renewables.

The Global Economy

Whether the world moves to more sustainable energy use depends in part on the economics of energy. For the world to switch to more sustainable practices, the cost of wasting energy and using fossil fuels needs to be greater than the cost of conservation and renewable energy sources. When oil prices fell dramatically in 2014, consumers used more gasoline than before; as a result, the financial viability of renewables was in danger of collapse.

Whether we move in a sustainable direction depends in part on how the rest of the world helps developing countries expand their energy sources. In some cases, the World Bank and International Monetary Fund have supported projects such as large coal-burning power plants and hydropower structures that are very damaging to the environment. In the long term, such projects are not sustainable. In contrast, Germany provides us with an example of a country that is making a coordinated transition to a sustainable-energy future. Although all alternative energy sources have costs and benefits, a diversified energy plan, increased efficiency, and efforts to lessen the environmental costs of energy production are all necessary parts of a more sustainable world.

CONCEPT CHECK What is a renewable fuel standard (RFS)? How does an RFS promote the use of renewable energy? ☑

REVISITING THE THEMES

SOUND SCIENCE

Some elegant technologies have been discussed in this chapter, developed by scientists in research laboratories and then applied to solving some very practical problems. PV devices, for example, are being intensely researched in order to bring down the costs and make solar energy competitive with conventional energy sources. Research at MIT has produced a real breakthrough in the discovery of a catalyst that can split water, imitating photosynthesis. Scientists and engineers are carrying out research on many renewable-energy options, including solar-trough and solar-dish systems, hydrogen technologies, improved PV devices, low-wind-speed turbines, and a number of biomass technologies. This is science at its best, used to develop new ideas that may revolutionize the energy scene.

SUSTAINABILITY

At the risk of repetition, global civilization is on an unsustainable energy course. Our intensive use of fossil fuels has created a remarkable chapter in human history, but we are still terribly dependent on maintaining and increasing our use of oil, gas, and coal. The results have already been seen in rising atmospheric greenhouse gases and the climate changes they are producing. Future use of fossil fuel will bring on more-intense climate changes, and long before we burn all the available oil, gas, and coal, we will have been forced to turn to other energy sources. Renewable-energy sources derived from the Sun can provide a sustainable supply of energy. Yet every source of energy has costs: Some renewable sources cause environmental damage, and lessening that damage has to be part of planning for the future. Germany, for example, has embarked on a transition to a renewable-energy economy, powered by renewable-energy technologies. The country is now working toward solutions to the problems associated with such a transition. Other countries are in different stages of attempting such a transition.

STEWARDSHIP

Stewardship of energy includes the careful and efficient use of energy resources. Government policies are rapidly moving forward in encouraging renewable energy by providing subsidies for homeowners and businesses to embrace renewable-energy options. Policies are also improving the energy efficiency and conservation picture. After all, energy saved is energy that does not have to be developed. Picture solar panels on every home, church, hospital, and school roof as well as wind turbines on college campuses, grazing lands, and offshore wind farms. Many nongovernmental organizations are involved with programs that distribute PV and biogas systems to people in rural developing-country communities. The impact is literally being felt around the globe.

REVIEW QUESTIONS

1. Why is it important for renewable-energy sources to replace fossil fuels? What are the prospects for getting it done?

2. How much solar energy is available, and what happens when it is used? What are some problems with harnessing solar energy?

3. How do active and passive solar hot-water heaters work?

4. How can a building best be designed to become a passive solar collector for heat?

5. How does a PV system work, and what are some present applications of such cells?

6. What has been happening in recent years in the PV market? How are utilities encouraging PV power?

7. Describe two concentrated solar power systems, how they work, and their potential for providing power.

8. What is the potential for developing more hydroelectric power in North America versus developing countries, and what are the impacts of such development?

9. Where is wind power being harvested, and what is the future potential for wind farms?

10. What are some ways of converting biomass to useful energy, and what is the potential environmental impact of each?

11. Which biofuels are used for transportation, and what is the potential for increasing biofuels in the United States?

12. How can hydrogen gas be produced via the use of solar energy? How might hydrogen be collected and stored to meet the need for fuel for transportation in the future?

13. How do fuel cells work, and what is being done to adapt them to power vehicles?

14. What is geothermal energy, and what are two ways it is being harnessed?

15. Describe current policy for renewable energy and energy efficiency as an outcome of recent government programs.

THINKING ENVIRONMENTALLY

1. What are some of the problems that need to be overcome before solar power can become a thoroughly mainstream form of energy production? What would make solar power more sustainable? Defend your answer.

2. What kinds of power systems could be used as backups for solar heaters in the event of bad weather?

3. How much promise does hydrogen power hold? Is its eventual sustainability just a myth or something to be pursued?

4. Imagine you are charged with building a facility run entirely by renewable resources (with the possible exception of backup power). How would you evaluate which sources have the most potential?

5. Audit all of the energy needs in your daily life. How much energy do you consume? Describe how each need might be satisfied by a renewable-energy option. (Check the Internet for tools for this exercise.)

MAKING A DIFFERENCE

1. Many everyday electrical products can now be powered by photovoltaic (PV) cells. You can find solar-powered chargers for MP3 players, cell phones, digital cameras, and even laptops. While this power may not be as dependable as that from a wall charger, it certainly provides a sustainable alternative while the Sun shines (especially useful on road and camping trips, where conventional power isn't available).

2. If you live close to work, consider not driving there at all and instead taking a bicycle. The benefits are endless; great exercise, low emissions, and no money toward gas are just the beginning.

3. Take advantage of the following services from your electric company: energy conservation, home-energy audits, and subsidized compact fluorescent lighting.

4. Consult the free *Greentips* newsletter of the Union of Concerned Scientists on its Web site (www.ucsusa.org), and follow the advice given in this publication for reducing your energy footprint.

5. Adopt ways to use less fuel and electricity both at home and where you work. (Turn thermostats up in the summer and down in the winter, turn off lights when you leave a room, wear sweaters, use energy-efficient lightbulbs, and so on.)

MasteringEnvironmentalScience®

Students Go to MasteringEnvironmentalScience for assignments, the eText, and the Study Area with animations, practice tests, and activities.

Professors Go to MasteringEnvironmentalScience for automatically graded tutorials and questions that you can assign to your students, plus Instructor Resources.

PART FIVE

POLLUTION AND PREVENTION

An industrial plant sends a steady stream of exhaust gases into the air, and the wind carries the pollutants away. If it didn't, life near the plant would be nasty. But where is "away"? Is it the next county, state, or country? The upper atmosphere? In reality, there is no "away" for the pollution of land, water, and air resulting from the activities of the 7.2 billion people inhabiting Earth. We are fouling our own nest, our only planet.

In his book *Red Sky at Morning*, James Gustave Speth describes four long-term trends in pollution that marked the 20th century:

First, *the volumes of pollutants released increased by huge amounts*—so much so that the developed countries finally began to legislate limits on many pollutants.

Second, *pollution evolved from gross excesses to microtoxicity*, where major releases of toxic gases and waterborne pathogens were replaced by smaller but much more toxic releases of organic chemicals and trace metals such as dioxin and mercury.

Third, *industrial pollution shifted from developed countries to developing* countries. China, for example, is reaping the benefits of rapid industrial growth and at the same time suffering from massive and health-damaging pollution.

Fourth, *the effects of pollution shifted from local to global*. Air pollution, resulting from the use of fossil fuels and industrial chemicals, has brought acid rain on a regional scale and climate change and ozone depletion on a global scale.[1]

In this part of the book, we begin with a look at the links between environmental hazards such as pollution and disease organisms and human health. The chapters that follow examine the major forms of pollution, from global and regional air pollutants, to water pollutants, to municipal and industrial wastes. Each chapter discusses the public policies crafted to deal with the problems caused by pollution. Without exaggeration, the problems of pollution pose some of the most difficult and important issues facing human society today.

---

[1]James Gustave Speth, *Red Sky at Morning* (New Haven: Yale University Press, 2004).

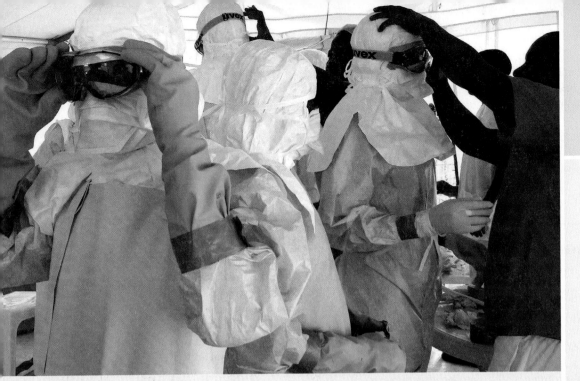

Environmental Hazards and Human Health

You may have followed the story in the news. In early 2014, an outbreak of the Ebola virus spread through several West African nations. Ebola is a type of hemorrhagic fever—a viral disease that causes bleeding from the mouth, nose, ears, and internal organs. It is a very serious, often fatal disease. Certain animals, particularly fruit bats, can carry the virus without becoming sick. Occasionally the virus is transmitted to humans, who then transmit it to one another through bodily fluids such as sweat and blood.

A Challenging Crisis. It is difficult to control the spread of the disease because the virus mutates rapidly, there is no vaccine, and there are no specific drugs for its cure. The death rate is so high that it is extremely difficult to study. Although Ebola does not kill the millions that malaria does, an outbreak is so serious that it requires the coordinated effort of the entire global health community. When the 2014 outbreak occurred, the World Health Organization and national governments tried to stop it as soon as possible. This was difficult to do in the countries of Sierra Leone, Liberia, and Guinea, where the outbreak was the greatest because of the limited health care infrastructure: the countries had few medical clinics and health care workers and little medicine to treat the victims of the disease.

An International Response. Over the next several months, more than 22,000 people contracted the disease, and more than 8,800 died. Doctors and nurses from around the world came to volunteer. Some died. Travel was restricted in the areas with the greatest infection rates, although some individuals who were infected, particularly volunteer health workers, were moved to Europe or the United States for treatment. The outbreak of Ebola also threatened the food security of millions of people living in the region. People bought food and hoarded it, driving up food prices. Labor shortages on farms caused harvests to decline, and people in quarantined areas lost income. Despite these disruptions, measures to

▲ Medical workers from the nonprofit Doctors Without Borders suit up to treat Ebola in Liberia.

control the infection were put in place. Because of the coordinated work of dedicated health workers, the outbreak was contained to only a few countries and did not spread across the globe.

The Ebola outbreak demonstrates the close connection between human health and the environment. In this chapter we will examine the nature of environmental hazards and the consequences of exposure to them. Then we will consider some of the pathways whereby humans encounter the hazards, interventions that can alleviate the risks to health, and how public policy and risk assessment can improve our management of environmental hazards. We begin with some basics about how public health officials measure health.

17.1 Human Health, Hazards, and the Environment

The study of the connections between hazards in the environment and human disease and death is often called **environmental health**. Note that it is *human* health that is the issue here, not the health of the natural environment. Environmental health is a subset of **public health,** an endeavor to prevent disease, promote health, and protect the public from hazards.

Measures of Public Health

The World Health Organization (WHO) defines **health** as "a state of complete physical, social, and mental well-being, and not merely the absence of disease or infirmity."[2] Health has many dimensions—physical, mental, emotional, and spiritual. Unfortunately, it is virtually impossible to measure all these dimensions of health for a society. Instead, our study of environmental health will focus on disease, and we will consider health to be simply the absence of disease. **Epidemiology** is the study of the presence, distribution, and prevention of disease in populations.

Morbidity and Mortality. Two measures are used in studies of disease in societies: *morbidity* and *mortality*. **Morbidity** is the incidence of disease in a population; it is commonly used to trace the presence of a particular type of illness, such as Ebola or influenza. **Mortality** is the incidence of death in a population. Records are usually kept of the cause of death, making it possible to analyze the relative roles played by infectious diseases and noninfectious diseases, such as cancer and heart disease.

Life Expectancy. One universal indicator of health is human life expectancy. In 1955, the average global life expectancy was 48 years. Today it is 70 years and continuing to rise; it is expected to reach 73 years by 2025. This progress is the result of social, medical, and economic advances in the past 60 years that have increased the well-being of larger and larger segments of the human population. Recall the *epidemiologic transition*: the trend of decreasing death rates that is seen in countries as they modernize (Chapter 8). As this transition occurred in the developed countries, it involved a shift from high mortality (due to infectious diseases) toward the present low mortality rates (due primarily to diseases of aging, such as cancer and cardiovascular diseases).

Figure 17.1 shows male and female life expectancies at birth for a range of countries. Some of the differential in life expectancies is due to the fact that the various countries have undergone the epidemiologic transition to different degrees, with very different consequences. Recently many countries have experienced an increase in life expectancy; part of this positive trend can be attributed to government actions taken to achieve the Millennium Development Goals and Sustainable Development Goals, including those regarding clean water and access to sanitation.

Despite the progress indicated by the general rise in life expectancy, one-sixth (almost 9 million) of the world's annual deaths are those of children under the age of five in low-income countries. A high proportion (36%) of the mortality in low-income countries is caused by common infectious diseases (as opposed to 6% in the high-income countries). On the other hand, economic growth resulting from intensive agriculture and industrialization is usually accompanied by environmental hazards (see Chapters 12, 19, and 22). Thus both poverty and affluence involve hazards that affect public health, a topic we will consider later in the chapter.

Hazards and Risk

In this chapter, we are concerned with the environment in the context of human life—the physical, chemical, and biological setting where people live. Thus, the home, neighborhood, and workplace, as well as the air, water, food, and even the climate constitute elements of the human environment. Within the human environment, there are *hazards* that can make us sick, shorten our lives, or contribute to human misfortune in other ways. In the study of environmental health, a **hazard** is anything that can cause (1) injury, disease, or death to humans; (2) damage to personal or public property; or (3) deterioration or destruction of environmental components.

The existence of a hazard does not mean that undesirable consequences inevitably follow. Instead, we speak of the connection between a hazard and something happening because of that hazard as a **risk,** defined as the probability of suffering injury, disease, death, or some other loss as a result of exposure to a hazard. The notion of **vulnerability** is also important; some people are more vulnerable to certain risks than are others. Therefore, an appropriate notation of risk should be

$$\text{Risk} = \text{Hazard} \times \text{Vulnerability}$$

---

[2]World Health Organization Glossary, available at www.who.int/health.systems performance/docs/glossary.htm.

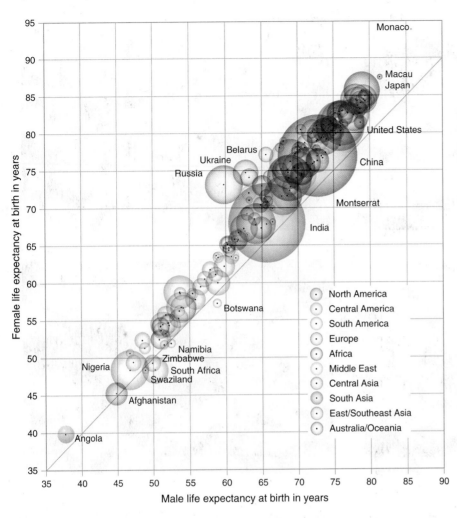

Figure 17–1 Life expectancies around the world. Average life expectancies vary from country to country. The size of the bubble indicates the relative size of a country's population. If males and females in a country have the same life expectancy, the country's circle is centered on the line; a position above the line indicates that the life expectancy of females is greater than that of males.

(*Source:* Data from 2013 CIA factsheet, figure from Creative Commons.)

UNDERSTANDING THE DATA

1. Which country has a greater overall life expectancy, China or the United States?

2. How do the life expectancies of Russia and India most differ?

3. What does the concentration of gray circles indicate?

For example, the presence of the Ebola virus in an area is a *hazard* that presents the *risk* of humans contracting the disease and dying. People working with the sick are much more *vulnerable* than are others. Risks are expressed as probabilities, so the analysis of the probability of suffering some harm from a hazard becomes a problem for scientists or other experts to address, and the assessment of risk becomes the basis of policy decisions.

This chapter will focus on hazards in the environment. Environmental health practitioners often divide environmental hazards into three groups: chemical, physical, and biological. A fourth group—cultural hazards—is not always considered to be environmental, but is useful for comparison. We could also regard the lack of access to necessary resources such as food and water to be a hazard to public health, but such hazards are covered elsewhere (Chapters 9 and 12). As we will see, some health problems result from a combination of factors.

Chemical Hazards

Many different kinds of chemicals are hazardous to health. Some hazardous chemicals occur naturally, but many others are produced by humans. Industrialization has brought with it a host of technologies that employ chemicals such as

cleaning agents, pesticides, fuels, paints, medicines, and those used directly in industrial processes such as tanning leather and refining oil. The manufacture, use, and disposal of these chemicals often bring humans into contact with them. People may be exposed to them by eating contaminated food or drinking contaminated water, by breathing contaminated air, by absorbing the chemicals through the skin, or by using the chemicals directly or accidentally.

Toxicity. Toxicity is the condition of being harmful, deadly, or poisonous. Even though certain substances in the air, soil, and water may be toxic, the link between their presence and the development of a health problem is harder to establish than with infectious diseases. Often the best that can be done is to establish a statistical correlation between exposure levels and the development of an adverse effect. The science of **toxicology** studies the impacts of toxic substances on human health and investigates the relationships between the presence of such substances in the environment and health problems. (Chapter 22 discusses toxic substances and toxicology in greater detail.)

The toxicity of a substance depends on the degree of exposure and on the *dose* of the toxic substance—the amount actually absorbed by a sensitive organ. Toxicologists distinguish between acute and chronic exposure to chemical

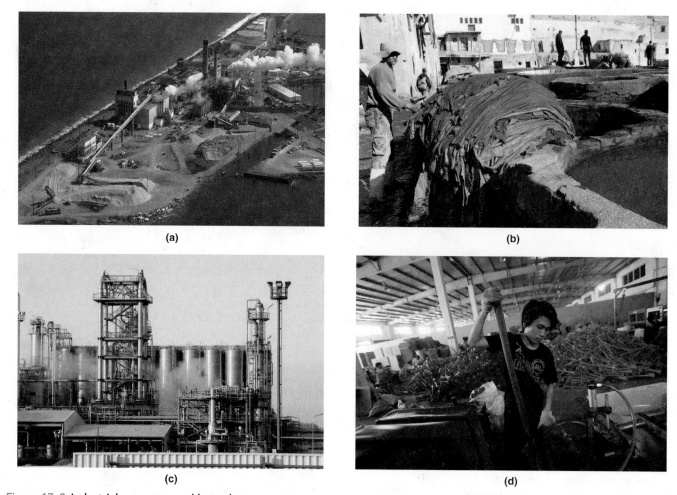

Figure 17–2 Industrial processes and hazards. Many industries produce chemicals that contribute to the impact of toxic substances on human health. Some examples of hazardous workshops are **(a)** a paper mill in Port Angeles, Washington; **(b)** a tannery in Fez, Morocco; **(c)** an oil refinery; and **(d)** a plastics factory in China.

hazards. In **acute** exposure, a substance brings on life-threatening reactions within a period of hours or days. **Chronic** exposure occurs when a substance causes the gradual deterioration of a variety of physiological functions over a period of years. Many chemicals, such as heavy metals, organic solvents, and pesticides, are hazardous to human health at very low concentrations. Episodes of acute poisoning are easy to understand and are clearly preventable, but the long-term exposure to low levels of many substances presents the greatest challenge to our understanding.

For most substances, there is a threshold below which no toxicity can be detected (see Figure 22–1). Different people have different thresholds of toxicity for given substances. For example, children are often at greater risk than adults because children are growing rapidly and are incorporating more of their food (and any contaminants that are in it) into new tissue. Embryos are even more sensitive. Substances transmitted across the mother's placenta can have grave impacts on the embryo's development, especially in the early stages.

Some substances cause mutations and others disrupt development. These are called *mutagens* and *teratogens* respectively, and are most dangerous to developing fetuses, young children, or to reproductive-age adults. They cause birth defects and increase the rates of miscarriage.

Carcinogens. Figure 17–2 shows a number of industrial processes that produce hazardous chemicals, including toxins and carcinogens. *Carcinogens* are cancer-causing agents. Because cancer often develops over a period of 10 to 40 years, it is frequently hard to connect the cause with the effect. The U.S. Department of Health and Human Services lists 47 chemicals that are now known to be human carcinogens and 183 more chemicals that are "reasonably anticipated to be human carcinogens" because of animal tests.[3] The agency also lists six biological agents and six types or sources of radiation that are known to be carcinogens.

A cancer is a cell line that has lost its normal control over growth. In a malignancy, the cells grow out of control and form tumors that may spread to other parts of the body. Recent research has revealed the presence of as many as three dozen genes that can cause cancerous cells to develop.

---

[3]U.S. Department of Health and Human Services, *Report on Carcinogens, Thirteenth Edition*, October 2, 2014, U.S. Department of Health and Human Services, Public Health Service, National Toxicology Program.

These genes are like ticking time bombs that may or may not go off. Even if cancer develops, it can be a process with several steps, often with long periods of time in between. *Environmental* carcinogens are either chemicals that are able to bind to DNA and prevent it from functioning properly or agents, such as radiation, that can disrupt the structure of DNA. When such an agent initiates a mutation in a cell, it can take upward of 40 years for the cell to accumulate the sequence of mutations that leads to a malignancy.

Cancer takes a very large toll in the developed countries, accounting for one-fourth of mortality (see Chapter 9). In 2014, an estimated 23% of the total deaths in the United States were traced to cancer. Even though it is largely a disease of older adults, cancer can strike people of all ages. Between 1991 and 2011, U.S. cancer deaths declined by 22%, largely because of a drop in smoking, early cancer detection, and better cancer treatments. The best strategy, however, is prevention, so great medical attention is now given to the environmental carcinogens and cultural habits, such as tobacco use, that are clearly related to cancer development.

Heavy Metals. The heavy metals lead and mercury deserve special mention as environmental hazards. Both are quite toxic.

Lead affects children more than adults, damaging their kidneys and brains, and impairing mental development. High lead levels can lead to coma and death. Unfortunately, prior to our modern knowledge of its danger, lead was widely added to paint and gasoline, and industries such as smelters spewed lead into the air. Today half a million children in the United States have elevated levels of lead in their blood, but this represents an improvement. Between 1980 and 1999, the EPA regulated the level of lead in gasoline so well that transportation-related emissions of lead dropped by 95%. During the same time, the level of lead in the air decreased by 94%. The Lead Contamination Control Act of 1988 allowed the U.S. Centers for Disease Control and Prevention (CDC) to develop the Childhood Lead Poisoning Prevention Program. Lead is still found in contaminated soils and in paint in old houses, but lead levels in children are dropping with lowered lead exposure (Figure 17–3).

Mercury poisoning can cause neurological dysfunction, developmental disabilities, and death; developing fetuses and young children are particularly vulnerable. Mercury occurs naturally in the soil and water; it is spread into the air by coal-burning power plants and industries, as well as by volcanoes. Mercury is also used in mining for gold and silver, especially in small mines. Worldwide, such mining is the leading cause of mercury pollution. One extremely toxic form of mercury, methyl mercury, can travel up the food chain; one of the primary ways people encounter it is by eating contaminated fish (see Chapter 22). The EPA regulates mercury pollution in order to lower U.S. emissions.

Biological Hazards

Human history can be told from the perspective of the battle with pathogens such as viruses and bacteria. It is a story of epidemics such as the black plague and typhus, which ravaged Europe in the Middle Ages, killing millions in every city, and of smallpox, which swept through the New World. The story takes a positive turn in the 19th century with the first vaccinations and the "golden age" of bacteriology, when scientists discovered most of the major bacterial diseases and brought bacteria into laboratory culture. The 20th century brought the advent of virology (the study of viruses and the treatment of the diseases they cause), the discovery of antibiotics, and immunizations that led to victories over polio and many other childhood diseases and the global eradication of smallpox. During the last half of the 20th century, molecular biology became a powerful tool in the battle against disease.

Diseases are inevitable components of our environment. Many are there regardless of our human presence, and others are uniquely human pathogens whose access to new, susceptible hosts is mediated by the environment. Today, approximately one-fifth of all human deaths are caused by

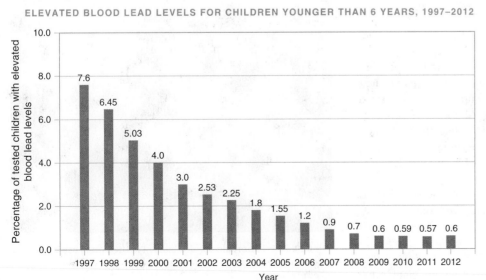

ELEVATED BLOOD LEAD LEVELS FOR CHILDREN YOUNGER THAN 6 YEARS, 1997–2012

Figure 17–3 Elevated blood lead levels in U.S. children. The percentage of children whose blood lead level is greater than or equal to 10 micrograms per deciliter (μ/dL) has dropped dramatically, the result of regulations that have lowered children's exposure to lead. Worldwide, lead remains one of the most significant causes of intellectual disability because of its effects on developing brains. (*Source:* CDC National Center for Environmental Health national surveillance data, http://www.childtrends.org/wp-content/uploads/2013/05/81_Blood_Lead_Levels.pdf.)

Table 17–1	Global Mortality from Major Infectious Diseases, 2013	
Cause of Death	**Estimated Yearly Deaths***
Respiratory infections	3,819,000
HIV/AIDS	2,661,360
Diarrheal diseases	1,445,800
Tuberculosis	1,250,000
Malaria	1,169,500
Selected Tropical Diseases	
Leishmaniasis	51,563
Dengue	14,660
Schistosomiasis	11,650
Trypanosomiasis (sleeping sickness)	9,110
Intestinal roundworms	2,675

*Total deaths from infectious diseases in 2013 are estimated at 16.0 million by the WHO.
Source: Data from Institute for Health Metrics and Evaluation (IHME). GBD Database. Seattle, WA: IHME, University of Washington, 2014. Available at http://www.healthdata.org/search-gbd-data?s=global%20mortality. Accessed January 30, 2015.

infectious and parasitic diseases. Table 17–1 shows the leading infectious diseases; the greatest mortality is caused by respiratory infections, HIV/AIDS, diarrheal diseases, tuberculosis, and malaria.

Respiratory Infections. Acute respiratory infections, including pneumonia, diphtheria, influenza, and streptococcal infections, are the leading causes of death from infectious disease. Pneumonia is by far the most deadly of these. In developed countries, bacterial and viral respiratory infections represent the most common reasons for visits to the pediatrician.

Influenza, or the "flu," is a viral illness that affects the respiratory system. In 1918, a deadly strain of influenza spread around the world, eventually killing 50 million people. The virus that caused this worldwide epidemic, or **pandemic**, likely originated in pigs and spread to humans through farmers who had contact with infected pigs. Although we have not had an epidemic as serious since then, regular flu outbreaks continue to occur. (The connection between diseases and animals will be described further in Section 17.2.)

Diarrheal Diseases. Each year, diarrheal diseases kill more than 760,000 children under age five. Some of these deaths are the consequence of ingesting food or water contaminated with bacterial pathogens from human wastes, such as *Salmonella, E. coli, Campylobacte,* and *Shigella.* One-third of these deaths are traced to a highly contagious

virus—**rotavirus**—that infects virtually every child between the ages of three months and five years. Rotavirus is rarely diagnosed, yet it causes most cases of severe diarrhea in young children. Rotavirus infections are more likely to be fatal in low-income countries than in wealthy countries because of differences in treatment: Severe diarrhea often leads to dehydration, which is quickly life threatening. Dehydration can be reversed with *rehydration therapy,* the replacement of lost fluids with balanced saline solutions. Such treatment is readily available in developed countries, but is often unavailable in developing countries. Fortunately, one rotavirus infection usually induces lifetime immunity, and vaccines are now able to prevent most severe rotavirus infections.

HIV/AIDS. Acquired immune deficiency syndrome (AIDS) is caused by the human immunodeficiency virus (HIV). The virus is spread in bodily fluids, including blood, semen, and breast milk. The virus leads to a compromised immune system, making its victims vulnerable to many other infections. This disease is taking a terrible toll in the developing world, where more than 90% of HIV-infected people live (Chapter 9). Even so, new treatments are enabling people with HIV to survive for years and lowering the rate of transmission from mothers to babies.

Tuberculosis. Although AIDS has overtaken tuberculosis as the disease that causes the most adult deaths worldwide, tuberculosis (TB) continues to be a major killer. Today close to one-third of the world's people are infected with the bacteria *Mycobacterium tuberculosis* (Figure 17–4); perhaps 10% of these will develop a life-threatening case of TB during their lifetime. Treatment with antibiotics reduced the incidence of TB during the mid-20th century, but in recent years it has reemerged. Some of the resurgence can be traced to the AIDS epidemic, which allows TB to flourish. Much of the disease's comeback, however, is the result of multi-drug-resistant strains of the microbe.

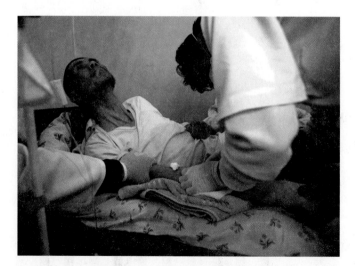

Figure 17–4 Tuberculosis. This patient in Uzbekistan has a form of tuberculosis that is multi-drug resistant, requiring him to take as many as 20 pills at a time.

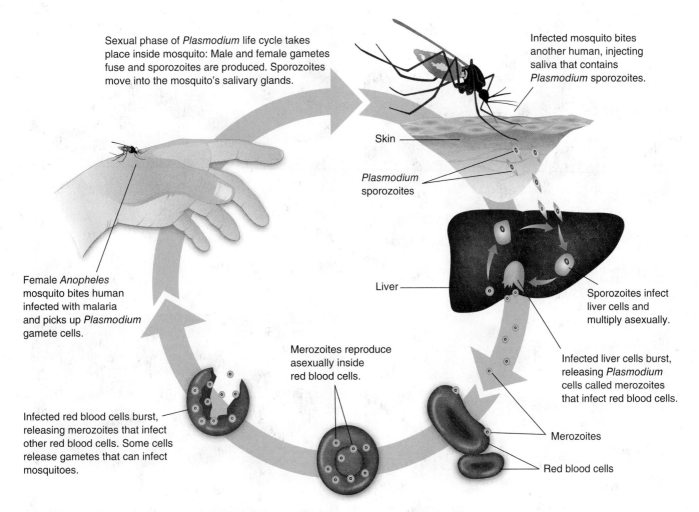

Sexual phase of *Plasmodium* life cycle takes place inside mosquito: Male and female gametes fuse and sporozoites are produced. Sporozoites move into the mosquito's salivary glands.

Infected mosquito bites another human, injecting saliva that contains *Plasmodium* sporozoites.

Skin

Plasmodium sporozoites

Liver

Sporozoites infect liver cells and multiply asexually.

Infected liver cells burst, releasing *Plasmodium* cells called merozoites that infect red blood cells.

Merozoites

Red blood cells

Merozoites reproduce asexually inside red blood cells.

Female *Anopheles* mosquito bites human infected with malaria and picks up *Plasmodium* gamete cells.

Infected red blood cells burst, releasing merozoites that infect other red blood cells. Some cells release gametes that can infect mosquitoes.

Figure 17–5 Life cycle of the malarial parasite. The complex life cycle of the protozoan parasite *Plasmodium*, which causes malaria in humans, involves the *Anopheles* mosquito and a human host. Humans who are infected with the parasite suffer from fever, chills, and anemia. Severe cases of malaria can be fatal.

Malaria. Of the infectious diseases present in the tropics, malaria is by far the most serious, accounting for an estimated 198 million cases each year and 584,000 deaths. Caused by protozoans of the genus *Plasmodium*, malaria is transmitted by mosquitoes of the genus *Anopheles* (Figure 17–5). The disease begins with an infected mosquito biting a person. From there, the protozoans invade the liver and then the red blood cells, where they multiply, destroying the cells. The protozoans then invade other red blood cells. This chain of events occurs in cycles, leading to periodic episodes of fever, chills, and malaise. Those infected with the protozoans also suffer from severe anemia, high fever, headaches, and vomiting.

Physical Hazards

Many physical hazards, including natural disasters, are the outcome of meteorological, hydrological, or geological forces. Human activities can also cause physical hazards, such as traffic accidents and exposure to radiation.

Natural Disasters. Natural disasters, including heat waves, severe storms, floods, landslides, forest fires, earthquakes,

and volcanic eruptions, take a toll on human life and property. Every year, such disasters kill hundreds of thousands and cost millions or even billions of dollars. Table 17–2 records the mortality from the major natural disasters in just one year, 2013–2014.

Table 17–2	Mortality from Major Natural Disasters, 2013–2014		
Country	**Type of Disaster**	**Date**	**Deaths**
Philippines	Tropical cyclone	8/11/2013	6,293
India	Flood	6/27/2013	6,054
Pakistan	Earthquake	9/24/2013	399
China (P. Republic)	Earthquake	3/8/2014	617
United States (Arizona)	Forest Fire	12/7/2013	19

Source: EM-DAT: The OFDA/CRED International Disaster Database, www.emdat.be, Université Catholique de Louvain, Brussels (Belgium).

Figure 17–6 Dangerous physical hazards. Natural disasters bring death and destruction in their wake. **(a)** On May 22, 2011, the city of Joplin, Missouri, was almost leveled by an EF-5 tornado that killed 160. **(b)** Heavy winter storms in the Pacific Northwest caused this tributary of the Columbia River to rampage. **(c)** A magnitude 7.0 earthquake destroyed the Haitian capital Port-au-Prince on January 12, 2010; more than 220,000 lives were lost. **(d)** A huge offshore earthquake triggered a devastating tsunami in Japan on March 1, 2011, killing more than 20,000.

Tornadoes. Tornadoes are among the most destructive forces known in nature. Spawned from the severe weather accompanying thunderstorms, they may develop winds reaching as high as 300 miles per hour. Because of the tendency for cold, dry air masses from the north to mix with warm, humid air from the Gulf of Mexico, the central United States generates more tornadoes than anywhere else on Earth. Each year, an average of 780 tornadoes strike the United States. The year 2011 was particularly bad: In April, a series of tornadoes swept across the southeastern United States, killing 322 people; in May, another tornado nearly leveled Joplin, Missouri, killing 160 (Figure 17–6a).

Floods. Floods are often the result of intense storms, but may also result from a long period of steady rain (Figure 17–6b). In 2010, huge floods covered one-fifth of Pakistan; 20 million people were affected and 2,000 died. In recent years, floods caused by Typhoon Haiyan (2012) and Typhoon Utor (2013) devastated the Philippines. Floods are often very costly. The year 2011 was a record year for insurance loss ($105 billion), when there were major floods in Thailand and Australia and a record flood in the Mississippi River basin. Floods are often accompanied by landslides, which can wipe out houses and roads.

Fires. Forest fires and fires in human-built settings may be caused by lightning or human actions. Fires can have a significant impact on human health, particularly in causing respiratory illness. Fires are a common occurrence in certain ecosystems, such as pine woodlands and grasslands; the organisms that live in such ecosystems are usually adapted to the presence of fire (Chapter 5).

Earthquakes and Tsunamis. Most earthquakes are triggered by the movement of Earth's tectonic plates. Although they may occur in the interior of continents, earthquakes are most common along the edges of the oceanic and continental plates. Some earthquakes are the result of human activities, such as hydraulic fracturing (Chapter 14) and the building of dams (Chapters 7, 16). The effects of large-scale earthquakes, such as one that hit Haiti in 2010, can last for years (Figure 17–6c).

When an earthquake occurs below the ocean, it can trigger a vast wave called a *tsunami* that hits coastal areas. A 2004 tsunami in the Indian Ocean was one of the deadliest natural disasters in recorded history. The 2011 earthquake in Japan triggered a tsunami that killed nearly 20,000 people (Figure 17–6d); it also caused part of a nuclear power plant to collapse, allowing radiation to leak into the environment (Chapter 15).

Volcanoes. Although volcanic eruptions are not common, the hot ash, lava, and noxious gases associated with some eruptions can be deadly and cause significant damage to property. Regions with active volcanoes, such as Iceland, Hawaii, and some Caribbean islands, need to have volcano disaster plans.

Radiation.

Radiation occurs in both high-energy (ionizing) and lower-energy (non-ionizing) forms. Ionizing radiation includes radiation from nuclear power, X-rays, and cancer treatments. In the last several decades, a decline in the ozone layer has allowed more of the Sun's ultraviolet radiation to reach Earth's surface; this increase has caused a dramatic increase in skin cancer. The effects of ionizing radiation from radon gas and the dangers of exposure to nuclear radiation are discussed in other chapters (Chapters 15 and 19). Non-ionizing radiation, which is used in lasers, microwaves, and radio waves, poses less risk to human health.

Cultural Hazards

Some 35% of all deaths in the United States can be traced to **cultural hazards** (Figure 17–7). People often engage in behaviors that expose them to hazards. They may, for example, smoke cigarettes, eat too much, drive too fast, sunbathe, consume alcoholic beverages, use harmful or addictive drugs, engage in risky sexual practices, get too little exercise, or choose hazardous occupations. Although the WHO and the U.S. Centers for Disease Control do not consider most of these to be environmental health factors, many cultural hazards are closely connected with environmental health factors (see Stewardship, *A Cultural Hazard Worsens Other Risks*).

Obesity is a cultural hazard that is linked to numerous health problems, including diabetes and heart disease. Currently, one-third of all U.S. adults are obese and another third are overweight. Obesity is a problem in poor nations as well as in the developed world. In 2014, more than 1.9 billion of the world's adults were overweight, 600 million of whom were obese. Worldwide, the conditions of obesity and overweight are linked to more deaths than the condition of underweight.

The connection between behavior and risk is especially lethal in the case of acquired immune deficiency syndrome (AIDS), caused by the human immunodeficiency virus

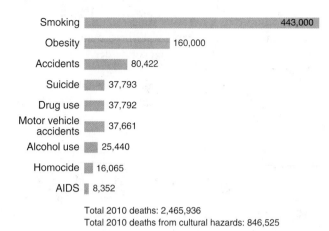

Total 2010 deaths: 2,465,936
Total 2010 deaths from cultural hazards: 846,525

Figure 17–7 Deaths from various cultural hazards in the United States, 2010. Many hazards in the environment are a matter of culture—lifestyle, personal choices, and society. Other hazards are due to accidents and the lethal behavior of other people.

(*Sources:* Data from the National Vital Statistics Reports, January 11, 2012; smoking deaths and obesity deaths from the Centers for Disease Control and *Journal of the American Medical Association.*)

(HIV). High-risk sexual behavior and the use of intravenous drugs are risk factors for HIV/AIDS. Having HIV is, in turn, a risk factor for drug-resistant TB infection. TB infections are also more likely when people are exposed to indoor and outdoor air pollution, a connection that highlights the complex interactions between cultural and other hazards.

Some cultural hazards, such as living in a high-crime area, are not really a function of individual choice. For other cultural hazards, individual choice is only a part of the picture. Being sedentary, for example, is linked with a number of negative health outcomes. Living in a place where it is unsafe to go outside, or having a job that requires long hours of sitting, can make healthy activity more difficult. Designing living places that support healthy behaviors is part of creating sustainable communities (Chapter 23).

Pollution

Many of the hazards we encounter, especially chemical and biological hazards, could be considered pollutants. The U.S. Environmental Protection Agency (EPA) defines **pollution** as "the presence of a substance in the environment that, because of its chemical composition or quantity, prevents the functioning of natural processes and produces undesirable environmental and health effects." Any material that causes pollution is called a **pollutant**.

Almost anything may be a pollutant, including mercury from mining operations, cholera in unsanitary water, or radiation from nuclear waste. Even light and noise can be pollutants. The only criterion is that the addition of a pollutant results in undesirable changes. The impact of an undesirable change may be largely aesthetic, such as the unsightliness of roadside litter. The impact may be on ecosystems as a whole—the die-off of fish or forests, for example. Or the impact may be on human health, such as human wastes contaminating water supplies. The effect of a pollutant may be as direct as a poison, or as indirect as a chemical destroying ozone and indirectly increasing cancer rates. The effects of pollution can range from very local (the contamination of an individual well) to global (adding ozone-depleting chemicals to the atmosphere).

We tend to think of pollution as the introduction of human-made materials into the environment, but undesirable changes may be caused by the introduction of too much of what are otherwise natural compounds, such as fertilizer nutrients introduced into waterways and carbon dioxide introduced into the atmosphere.

Pollutants are almost always the by-products of otherwise worthy and essential activities such as producing crops, creating comfortable homes, providing energy and transportation, and manufacturing products. An overview of the categories of pollutants that result from various activities is shown in **Figure 17–8**. Pollution problems have become more pressing because our growing population and expanding per capita consumption have increased the number and amounts of by-products released into the environment. In addition, many widely used materials, such as aluminum cans, plastic packaging, and synthetic chemicals, are **nonbiodegradable**. That is, they resist breakdown by decomposers and consequently accumulate in the environment. To live sustainably, we need to adapt the means of meeting our present needs so that by-products will not jeopardize present and future generations.

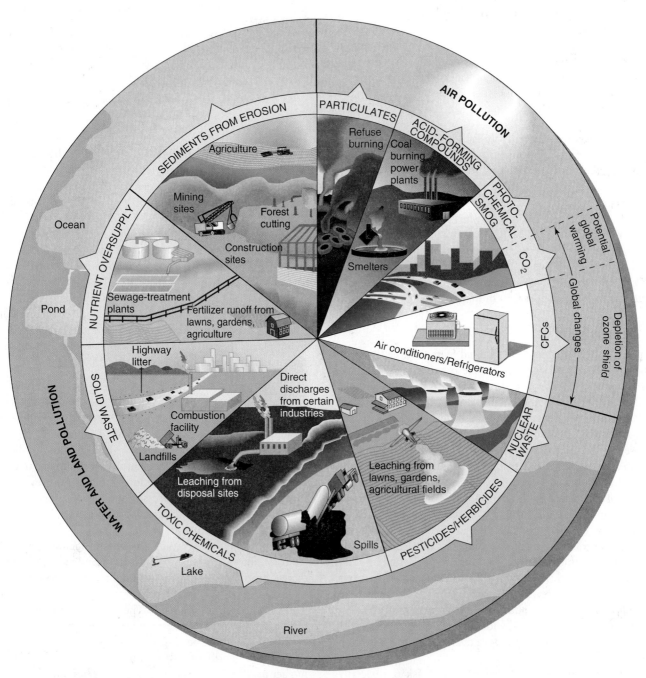

Figure 17–8 Categories of pollution. Pollution is an outcome of otherwise worthy human endeavors. Major categories of pollution and some activities that cause them are shown here.

Protecting Environmental Health

In the United States, the federal agency with the primary responsibility for protecting the health and safety of the public is the **Centers for Disease Control and Prevention (CDC)**, an agency within the Department of Health and Human Services. The CDC gathers data and provides information on health issues to state and local health care providers. Each state has a health department, and most municipalities have health agents, too. The primary responsibility for health risk management and prevention resides with the CDC and the state public-health agencies. These agencies are responsible for monitoring certain diseases and environmental hazards and for controlling epidemics; for example, they may require public-health measures such as immunizations and quarantines. In addition, the Food and Drug Administration (FDA) plays a role in protecting food supplies, and the Environmental Protection Agency (EPA) plays a role by regulating environmental pollutants.

Virtually every country has a ministry of health similar to the CDC that acts on behalf of its people to manage and minimize health risks. The health policies that are put into place, however, are subject to limitations of information and funding. Ideally, a health ministry should direct its limited resources toward strategies that will accomplish the greatest risk prevention.

Fortunately, all countries have access to the programs and information of the World Health Organization (WHO), a UN agency established in 1948. The WHO is centered in Geneva, Switzerland, and maintains six regional offices around the world. The agency is staffed by health professionals and other experts and is governed by the UN member states through the World Health Assembly. Much of what the WHO does is related to public health, such as tracking diseases and monitoring trends in health.

CONCEPT CHECK What is the difference between a hazard and a risk? ☑

STEWARDSHIP
A Cultural Hazard Worsens Other Risks

Lifestyle choices such as driving fast, imbibing alcohol, and climbing mountains involve a significant risk of accident and death. But one cultural hazard—tobacco use—is particularly problematic. In the United States, tobacco use is the leading preventable cause of death, linked to about 450,000 deaths annually. Every year U.S. adults who smoke cigarettes cost the United States $170 billion in health care and another $151 billion in lost productivity. In some developed countries, tobacco use is declining, but it remains high in the former socialist countries of Eastern Europe and is on the increase in most developing countries. The World Health Organization (WHO) puts the number of smokers worldwide at more than 1.1 billion, with 6 million deaths per year due to cigarette smoking.

Cigarette smoking has been clearly correlated with cancer and other lung diseases; it kills by impairing respiratory and cardiovascular functioning. Smoking can also exacerbate the effects of a number of other environmental health hazards. Smokers living in polluted air experience a much higher incidence of lung disease than do smokers living in clean air—a synergistic effect. Smoking increases the risk of black lung disease for coal miners and the risk of lung disease for workers who are exposed to asbestos. Smoking is linked to half of the tuberculosis deaths in India. In 1993, the EPA classified *environmental tobacco smoke*, meaning secondhand smoke, as a known human carcinogen; this move meant that exposure to smoking by others was clearly an environmental hazard.

In the United States, several measures have been taken to reduce the use of tobacco. Raising cigarette taxes has been shown to be one of the most effective ways to reduce smoking, especially in young people, so tobacco taxes are high. Warning labels are required on smoking materials, advertising has been curtailed, and smoking has been banned in most public places. The 2009 Tobacco Control Act gave the Food and Drug Administration the authority to regulate tobacco. Since these antismoking efforts began, the number of adult smokers in the United States has dropped from 42% to 19.8% of the population.

In 2003, the World Health Assembly adopted the WHO Framework Convention on Tobacco Control, a treaty that has since been signed by 173 countries. In 2008, the WHO initiated a public health campaign to combat the global smoking epidemic. That campaign focuses on monitoring tobacco use and prevention policies, protecting

Activists in Seoul, South Korea, march to celebrate World No Tobacco Day on March 31, 2011.

people from tobacco smoke, warning against the use of tobacco, banning tobacco advertising, making smoking more costly, and helping smokers to quit. However, governments collect some $133 billion in tobacco excise tax revenues while spending less than $1 billion on tobacco control, making the elimination of tobacco a conflict of interest. In order to be good stewards of public health funds and stewards of our own bodies, we need to continue to push back against tobacco use because of its impact on other health problems.

17.2 Pathways of Risk

The environmental hazards identified in Section 17.1 are responsible for untold human misery and death. But if many of the consequences of exposure to hazards are preventable, how is it that the hazards turn into mortality statistics? In other words, what are the pathways that lead to human deaths from risks of infection, exposure to chemicals, and vulnerability to physical hazards? If prevention is the goal—and it should be—this knowledge is crucial.

A report by the WHO identifies a small number of environmental risk factors that increase the probability of adverse health outcomes and need to be priorities for the future in public health.[4] Some major public-health risks are associated with the lack of an important resource such as food, clean water, or contraception. Several other public-health concerns, such as tobacco use, are associated with individual choices but interact with environmental variables. Figure 17–9 compares several of the top environmental risk factors to the cultural hazard of smoking, which by far outweighs the other risks. We will consider only a few examples of risk factors, because a thorough examination of this topic would take up volumes.

Exposure to Biological Hazards

Many biological hazards are infectious diseases. Depending on the pathogen causing them, infectious diseases may be transmitted through the air, by unclean water or contaminated food, by animals, by direct contact with other people, or by contact with infected bodily fluids, such as blood and semen. Such means of transmission can be described as pathways of risk.

Unsafe Water. The WHO describes unsafe water as one of the greatest environmental risks. Many microbial diseases are transmitted through contaminated water, and some of these diseases, such as cholera and typhoid fever, are deadly.

The contamination of water is usually the result of inadequate hygiene or inferior sewage treatment. These conditions, in turn, may be the result of a lack of education or a lack of the resources needed to build effective sanitation systems. According to the United Nations' 2014 report on the Millennium Development Goals, an estimated 2.6 billion people are still without adequate sanitation, and 884 million lack access to safe drinking water.

Industrial Food Production. Food- and water-borne illnesses are not unique to the developing world. Outbreaks of food-borne diseases are common in developed countries as well. The CDC estimates that every year 48 million Americans are sickened by contaminated food and 3,000 die. Species of *Salmonella*, bacteria that infect the intestinal tract and bring on diarrhea, fever, and cramps, are responsible for more cases of illness than any other food-borne pathogen (Figure 17–10). The factory-like conditions under which much of our food is processed provide an opportunity for *Salmonella* and other pathogens to contaminate food on a wide scale, resulting in diarrheal outbreaks and major recalls of food products. In 2010, an outbreak of *Salmonella* infections led to a nationwide recall of 500 million eggs. It was occurrences like this that brought about passage of the **FDA Food Safety Modernization Act** of 2010, expanding the FDA's authority over the food processing industry. Food safety continues to be a battle; a 2014 *Salmonella* outbreak connected to bean sprouts caused 115 cases of illness in 12 northeastern states.

Tropical Conditions. If you picture life in the tropics as a blissful paradise, you might be surprised to learn that the tropics are home to a wide variety of biological hazards not found

---

[4]World Health Organization, "Priority Risks and Future Trends," in *Environment and Health in Developing Countries,* World Health Organization, 2014, available at http://www.who.int/heli/risks/ehindevcoun/en.

Smoking — 6.0
Natural disasters — 0.02
Climate change impacts — 0.15
Lead exposure — 0.23
Unintentional poisonings — 0.35
Urban air pollution — 0.8
Road traffic accidents — 1.2
Malaria — 1.2
Indoor smoke — 1.6
Unsafe water and sanitation — 1.7

Deaths (millions per year)

Figure 17–9 Global risk factors. The World Health Organization identified a number of environmental risk factors that were priorities for the future. The graph compares the worldwide deaths caused by these factors to those caused by the cultural hazard of smoking.

(*Source:* World Health Organization. "Priority Risks and Future Trends." In *Environment and Health in Developing Countries.* World Health Organization, 2014. Available at http://www.who.int/heli/risks/ehindevcoun/en.)

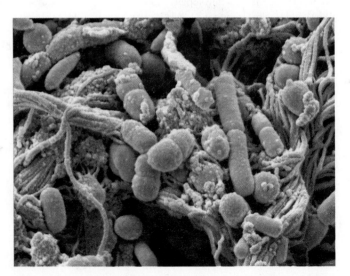

Figure 17–10 *Salmonella* bacteria. *Salmonella* is the most common food-borne disease in the United States. In this photomicrograph, the bacteria are proliferating on chicken skin.

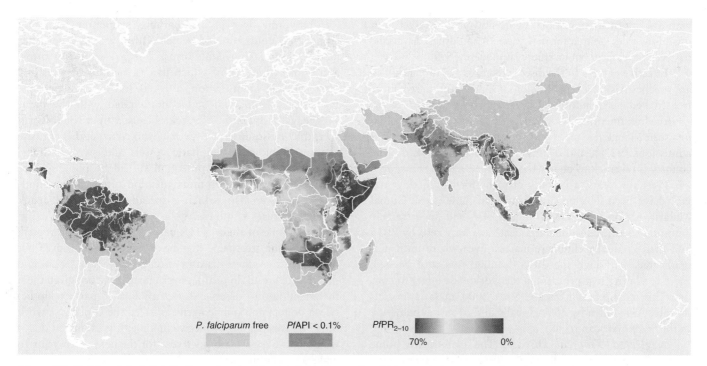

P. falciparum free   PfAPI < 0.1%   PfPR₂₋₁₀

70%   0%

Figure 17–11 The global distribution of malarial parasites. *Plasmodium falciparum* is the most deadly and widespread of the three major species of malaria parasites. This map shows the rate of infection in 2- to 10-year-olds in 2010. That year there were an estimated 216 million cases of malaria and 655,000 deaths.

(*Source:* Malaria Atlas Project. Used with permission. http://www.map.ox.ac.uk/about-map/open-access. Accessed January 16, 2012.)

in temperate regions. For example, wet tropical climates provide ideal conditions for the year-round spread of insect-borne diseases. Mosquitoes are **vectors** (organisms that transmit infectious agents) for several deadly diseases, including malaria, yellow fever, dengue fever, Japanese encephalitis, and West Nile virus. The WHO describes the complex relationship between tropical diseases and the environment, saying "Poorly designed irrigation and water systems, inadequate housing, poor waste disposal and water storage, deforestation and loss of biodiversity, all may be contributing factors to the most common vector-borne diseases."[5]

Malaria. The most serious and widespread tropical disease is malaria, caused by protozoans of the genus *Plasmodium.* Figure 17–11 shows the global distribution of the most deadly of these protozoans. Because the pathway to malarial infection is through mosquito bites, most attempts to control the disease have been aimed at **vector control,** such as attacking the *Anopheles* mosquito with the use of insecticides. Unfortunately, some mosquitoes have developed resistance to all of the pesticides employed, so eradication has remained an elusive goal. In some countries, DDT is still sprayed on walls of huts and houses, but its use in this manner is highly controversial because of the pesticide's harmful environmental effects (see Stewardship, *DDT for Malaria Control,* in Chapter 13). One preventive measure that has been highly successful is the placement of insecticide-treated netting over beds. Children provided with such bed nets

experience a substantial reduction in mortality from malaria and other diseases (Figure 17–12).

Efforts to combat malaria also include **treatment strategies,** or methods of curing people infected with the protozoans. Such treatment is difficult because *Plasmodium* protozoans have developed resistance to one treatment drug after another, so new ones have to be developed regularly. In 2002, molecular biologists successfully sequenced the genomes of the *Anopheles* mosquito and the most lethal malaria parasite, *P. falciparum.* Armed

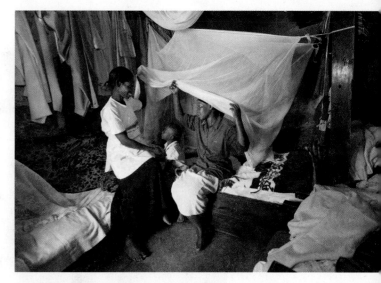

Figure 17–12 Insecticide-treated bed net. A home health worker installs a bed net in a home in Kampala, Uganda.

[5]Ibid.

with this information, researchers are now able to target both organisms in the search for new vaccines and drugs. A vaccine has been developed, but its efficacy is modest (55%).

Effective control of the disease requires a combination of approaches. Using a combination of long-lasting insecticide-treated bed nets, indoor DDT spraying, and quick access to drug treatment, the southern Africa nation of Namibia has seen a huge decline in cases of malaria (83%), hospital admissions (92%), and deaths (96%) in just 10 years. The country is on track to be malaria-free by 2020.

Over the years, a number of plans for conquering malaria have been developed; the latest of these is the Global Malaria Action Plan (GMAP). This plan had an extremely ambitious goal: reduce malaria deaths to near zero by 2015 and then gradually eliminate malaria everywhere until it is eradicated. Although the 2015 deadline was not reached, there has been great progress in the worldwide effort to conquer this most difficult disease. Since 2000, malaria mortality rates have fallen by 26% globally and by more than 54% in the WHO African region.

Neglected Tropical Diseases. The tropics are home to a number of less well-known diseases called **neglected tropical diseases (NTDs)**; they are called *neglected* because there has not been significant international funding for their control and treatment. Almost everyone in deep poverty in tropical regions is infected with at least one NTD. Even if people don't die from them, these diseases can cause tremendous suffering.

The transmission of the NTDs is closely linked to the living conditions in many low-income tropical countries. People contract them by walking without shoes in places with human or animal feces, from drinking polluted water, and from bites from insect vectors. For example, *schistosomiasis* is a disease caused by parasitic flatworms that live in the veins of the intestine and bladder, causing pain and internal bleeding. These flatworms also live in freshwater snails and their larvae swim in water; people contract the flatworms by drinking, bathing, swimming, or doing laundry in contaminated water. Other tropical parasites include *Ascariasis* roundworms, *Trichuriasis* whipworms, and hookworms, all of which live in the human intestinal tract, causing malnutrition, anemia, and other symptoms. People get them from soil contaminated with the parasites' eggs. *Lymphatic filariasis* is caused by a roundworm spread by mosquitoes; the people who contract this parasite experience blocked lymphatic vessels and swollen limbs. *Trachoma* is a bacterial infection spread by poor hygiene and houseflies; this common infection often causes blindness. *Onchocerciasis*, which is spread by the bite of black flies, also causes blindness.

Every one of these NTDs can be treated and prevented; the drugs for treatment are inexpensive and effective, but perhaps because these diseases infect people who are poor and do not often result in death, governments have not given them the attention they deserve. A recently formed partnership between donors and public agencies called the Global Network of NTDs is beginning to receive funding. It is estimated that it would cost \$2–\$3 billion per year for the next seven years to make a serious impact on these diseases.[6]

There is some very good news about one former scourge: Guinea worm disease. Just two decades ago, some 3.5 million people in 21 countries were infested with this roundworm, acquired by drinking water infested with larvae. Public-health efforts to keep Guinea worms out of drinking water have been instrumental in combatting the disease (Figure 17–13). The Carter Center has carried out a successful campaign to eradicate this disease, now down to some 1,000 cases in four African countries. The disease is so close to eradication that many in public health would like to see a major effort to get rid of it once and for all.

---

[6]Peter Jay Hotez, "A Plan to Defeat Neglected Tropical Diseases," *Scientific American* 302 (January 2010): 90–96.

Figure 17–13 Filtering water to eradicate Guinea worms. Maintaining clean water is critical to the eradication of Guinea worm disease. Here a Guinea worm eradication team teaches villagers in Ghana how to filter water.

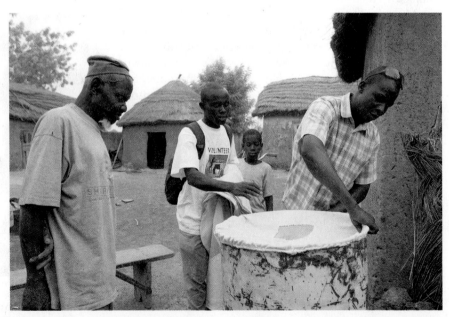

Interactions with Animals. Diseases that can be transmitted between animals and humans are said to be **zoonotic**. One study found that out of 1,415 pathogens able to infect humans, 61% were zoonotic. Zoonotic pathogens are one source of new, or *emerging*, diseases.

Humans that eat bush meat or live in close contact with wild animals may contract diseases from them. For example, the Nipah virus, first observed in Malaysia in 1999, is endemic in fruit bats (Figure 17–14). People and pigs become infected by contact with fruit contaminated with bat urine and feces. The virus has killed more than 200 people and has shown signs of human-to-human transmission in some cases. As discussed in the chapter opener, some African fruit bats harbor the virus that causes Ebola.

Modern intensive methods of rearing animals can also facilitate the transfer of diseases from animals to humans. The globalization of travel makes the movement of these diseases more rapid. In 2003, a strain of influenza called the avian flu first appeared in poultry. The disease eventually spread to 63 countries, killing or leading to the culling of 400 million domestic poultry and countless wild birds and causing $20 billion in losses. The disease also spread to humans through contact with infected birds, with 582 cases and 343 deaths. In 2009, the H1N1 strain of swine influenza initiated a pandemic in humans; although this strain of the flu was a relatively mild disease, it caused more than 18,000 deaths. Many other emerging diseases are connected to environmental changes that bring people into contact with animals.

Exposure to Chemical Hazards

Each year, unintentional poisonings kill 355,000 people globally, and many more people experience nonlethal poisoning effects. How are people exposed to chemical substances that can bring them harm? Let's look at several of the ways.

Industrial Exposure. The extraction of minerals and industrial manufacturing are major routes of exposure to toxic substances, as is the use of toxic chemicals such as pesticides. The dumping of wastes exposes many to harmful chemicals. In low-income countries, people are exposed to airborne pollutants in occupations such as informal recycling, where plastics and metals are burned or melted over open fires. While this topic is covered elsewhere (Chapters 12, 19), it would be hard to overemphasize how the spreading of chemicals throughout the environment has changed ecosystems and harmed human health. Lead is a good example of a widespread contaminant. The WHO estimates that lead poisoning kills 230,000 people globally each year and that one-third of all children experience cognitive effects from lead exposure.

Mining is a particularly dangerous industry that exposes workers to both chemical and physical hazards, such as the collapse of structures (also see Chapter 22). The mercury used in mining is one hazard, but contamination from toxic mine tailings is a more common concern in large mines. For example, the town of Picher, Oklahoma, was once in the national center of lead and zinc mining. Underground cave-ins and piles of toxic mine tailings made it a danger zone.

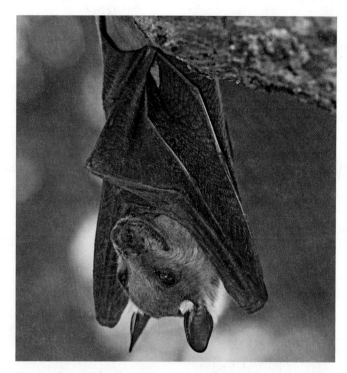

Figure 17–14 Animal vectors. Interaction with wild animals is one pathway for the transmission of human diseases. Fruit bats such as this one may carry the viruses for Ebola and Nipah without dying.

By 2009, many of its buildings were near collapse and 34% of its children were found to have lead poisoning. The site was evacuated and declared uninhabitable by the state of Oklahoma. The state purchased the town and the site was declared a Superfund site (Chapter 19).

One newer concern stems from the chemicals in the water that is pumped into the ground during hydraulic fracturing, or *fracking* (see Chapter 14). Several recent studies have suggested that fracking chemicals have negative health effects. Nevertheless, some companies involved in fracking have resisted reporting which chemicals they are pumping into the ground, claiming that this information is a trade secret. This means the EPA and local public-health officials do not know which chemicals may be contaminating the groundwater. This issue is sure to be the subject of legal actions in the future.

Outdoor Air Pollution. Airborne pollutants are a particularly challenging set of chemical hazards to control, as they are difficult to measure and difficult to avoid. Humans breathe an average of 30 pounds (14 kg) of air into their lungs each day. Living in cities exposes people to industrial pollution and pollution from transportation such as carbon monoxide, ozone, smog, and particulates. Air pollution is estimated to kill 800,000 people annually and is discussed at length elsewhere (Chapter 19).

Indoor Air Pollution. Much attention is given to air pollutants released into the atmosphere—and rightly so. However, *indoor* air pollution can pose an even greater risk to health. The air inside the home and workplace often contains much higher levels of hazardous pollutants than outdoor air does.

In the developed countries, the overall indoor air pollution problem is threefold. First, increasing numbers and types of products and equipment used in homes and offices give off hazardous fumes. Second, buildings have become increasingly well insulated and sealed; hence, pollutants are trapped inside. Third, people spend more time indoors than out. The average person spends 90% of his or her time indoors.

The sources of indoor air pollution are quite varied, as evidenced in **Figure 17–15.** Sometimes mold and other pollutants accumulate to such an extent that a building is uninhabitable, a condition called "sick building syndrome."

In the developing world, at least 3 billion people rely on biofuels such as wood and animal dung for cooking and heat. The WHO reports that the pollution from burning these fuels contributes to nearly 1.6 million deaths annually. Fireplaces, cooking fires, and stoves are often improperly ventilated (if at all), exposing the inhabitants to very high levels of pollutants, such as carbon monoxide, nitrogen and sulfur oxides, soot, and benzene. These exposures cause acute respiratory infections in children; chronic lung diseases, including asthma and bronchitis; lung cancer; and birth-related problems. One study in Mexico showed that women who were continually exposed to indoor smoke had 75 times the risk of developing a chronic lung disease as women who weren't exposed. Solutions to this problem include the use of ventilated stoves that burn more efficiently (see Chapter 1) and the conversion from biofuels to bottled gas or liquid fuels such as kerosene.

Asthma and Air Pollution. Asthma is currently at epidemic proportions in the United States, afflicting 25 million people each year. Asthma attacks the respiratory system so that air passages tighten and constrict, causing wheezing, tightness in the chest, coughing, and shortness of breath. In acute attacks, the onset is sudden and sometimes life threatening. This condition is worsened by exposure to both indoor and outdoor air pollution. Substances that can trigger asthma attacks include dust, animal dander, mold, second-hand smoke, swimming pool chlorine, and various gases and particles. About half of asthma is traced to allergies, but the

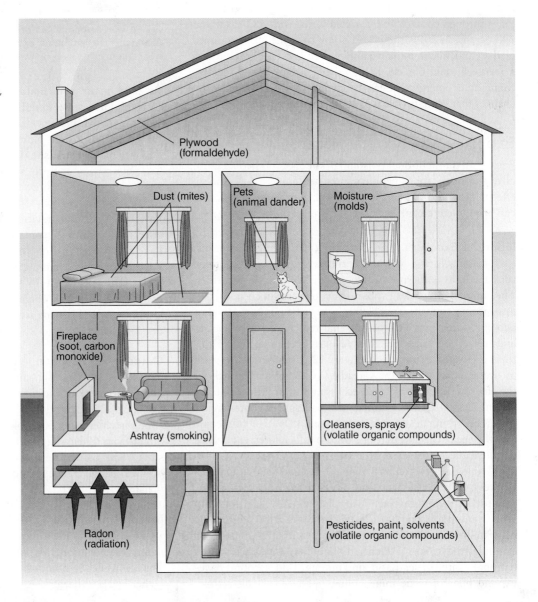

Figure 17–15 Indoor air pollution. Pollutants originate from many sources and can accumulate to unhealthy levels, leading to asthma and "sick-building syndrome" (characterized by eye, nose, and throat irritations; nausea; irritability; and fatigue) as well as other serious problems.

Plywood (formaldehyde)

Dust (mites)

Pets (animal dander)

Moisture (molds)

Fireplace (soot, carbon monoxide)

Ashtray (smoking)

Cleansers, sprays (volatile organic compounds)

Radon (radiation)

Pesticides, paint, solvents (volatile organic compounds)

remainder has other, unknown causes. Through its Indoor Environments Division Web site, the EPA provides helpful guidelines for controlling many of the substances that trigger asthma in schools, homes, and other indoor environments.

Oddly, asthma is much more common in developed countries, in spite of the greater exposure to indoor air pollution in developing countries. Recent research has suggested that frequent parasitic worm infections stimulate a well-regulated anti-inflammatory network in the immune systems of most people in the developing countries, protecting them against many allergic diseases such as asthma. In the developed countries, where such parasitic infections are uncommon, this protection is lacking, so allergic disorders are much more common. It is interesting that children who grow up on farms in developed countries also experience much lower frequency of asthma and allergies. These observations have led to what is called the **hygiene hypothesis:** Immune systems need to encounter microbes (and worms!) when they are young in order to keep inflammatory responses such as asthma under control. Apparently, too much cleanliness can make you sick![7]

Exposure to Physical Hazards

How are people exposed to physical hazards? They may encounter them anywhere, but are particularly at risk as they travel and in certain kinds of workplaces. Some of the greatest risks are associated with natural disasters. Fortunately, we can take action to minimize these risks.

Collisions with Motor Vehicles. Our modern means of transportation move us quickly, but also pose significant risks to our safety. Every year collisions with motor vehicles kill 1.2 million people and injure many more. Pedestrians and bicyclists are particularly vulnerable to injuries from interactions with vehicles; crowded or narrow roads that lack safety features such as lights, wide shoulders, guard rails, biking or walking lanes, and smooth surfaces contribute to the risk. Much of the world lacks such structures, making travel a dangerous activity.

Workplace Exposure. In many occupations, workers are exposed to physical hazards such as loud noises and radiation; they may also be at risk of accidental injury. In the United States, such workplace hazards are monitored by laws and subject to certain regulations. In countries that lack such laws, workers are exposed to more hazards with fewer protections. The dangers of the workplace environment are a subset of environmental hazards, which are discussed further in a later chapter (Chapter 22).

Harm from Natural Disasters. Each year, natural disasters affect more than 100 million people. Disasters inevitably trigger emergency relief efforts and subsequent humanitarian assistance and reconstruction. The effects of natural disasters provide a good example of the difference between being exposed to a hazard and experiencing actual harm.

Location. Much of the loss brought on by natural disasters is a consequence of poor environmental stewardship and lack of planning. Hillsides and mountains are deforested, leaving the soil unprotected; people build homes and towns on floodplains; villages are nestled up to volcanic mountains; cities are constructed on known geological fault lines; and coastal marshes and mangrove forests are replaced by house lots. Populations along the U.S. "hurricane coasts" (Gulf of Mexico and southern Atlantic) continue to rise in spite of the many hurricane strikes. Living in floodplains is also risky, as is building in fire-prone areas.

Risk-Reduction Strategies. Much of the human death and misery brought on by natural hazards could be prevented by rigorous *disaster risk-reduction strategies*. Such strategies include strong building codes and effective warning systems. Anyone who has lived in the American Midwest is probably familiar with the sirens and the radio, TV, and phone alerts that sound tornado warnings, and of the need to get to a place to shelter when the danger of tornadoes is high. These alerts are part of the region's strategy for reducing the harm caused by frequent tornadoes.

Tsunami warning systems, consisting of a network of sensors and communications infrastructure, can also be very effective **(Figure 17–16)**. Such systems have existed for the Pacific Ocean since 1965, but they were not installed in the Indian Ocean until 2006, after the 2004 tsunami that

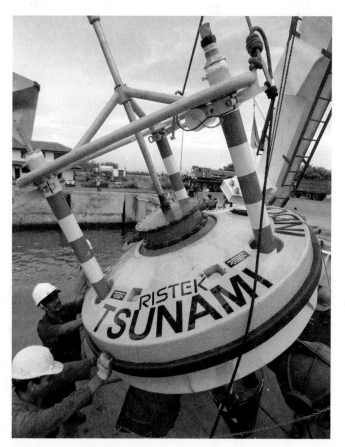

Figure 17–16 Reducing the risks of natural disasters. This sensing buoy is part of a tsunami warning system off of the coast of Indonesia. Data is beamed to satellites, and communicated to people in coastal zones.

---

[7]Mitch Leslie, "Gut Microbes Keep Rare Immune Cells in Line," *Science* 335 (March 23, 2012): 1428.

devastated that region. In 2011, a tsunami warning was broadcast widely on Japan's east coast immediately following the offshore earthquake and saved many thousands of lives.

The Indian state of Odiasha has recently employed a wide range of strategies to protect the public from cyclones, which are tropical storms with hurricane-force winds. Among other actions, Odiasha has built cyclone shelters and coastal embankments, developed evacuation plans, and installed warning systems. This preparation has paid off. In both 1999 and 2013, the state experienced category 5 cyclones. But while the 1999 disaster cost 10,000 lives and $4.5 billion in damages, the more recent disaster cost only 40 lives and $700 million. This change was because of the state's concerted effort to reduce the risks associated with storms.

The importance of risk-reduction strategies can be seen by comparing the outcomes of the Japanese earthquake of 2011 and the Haitian earthquake of 2010. Both were powerful earthquakes, but the damage they inflicted on the two countries was significantly different. Why? Earthquakes are common in Japan, a wealthy country that has been able to develop and enforce strong building codes. As a result, most Japanese buildings are designed and built to withstand the impact of severe earthquakes. When the magnitude 9.0 earthquake struck, most of the Japanese buildings swayed, but only a small portion of them collapsed. Most of the fatalities associated with the earthquake were due to the tsunami that accompanied it. In contrast, the city of Port-au-Prince, Haiti, lacked earthquake-resistant buildings; more than 250,000 buildings were significantly damaged and more than 100,000 people died.

A Framework for Action. In January 2005, 168 country delegations met in Kobe, Japan, for the UN World Conference on Disaster Reduction. At that conference, delegates adopted the *Hyogo Framework for Action* (HFA), a global blueprint for disaster risk reduction efforts for the coming decade. The framework identified five priorities for action:

1. **Make disaster risk reduction a priority:** Ensure that disaster risk reduction is a national and a local priority with a strong institutional basis for implementation.

2. **Know the risks and take action:** Identify, assess, and monitor disaster risks—and enhance early warning.

3. **Build understanding and awareness:** Use knowledge, innovation, and education to build a culture of safety and resilience at all levels.

4. **Reduce risk:** Reduce the underlying risk factors. For example, most of the fatalities in earthquakes are caused by damaged or collapsing buildings. Strict building codes in earthquake-prone areas can greatly reduce the death toll.

5. **Be prepared and ready to act:** Strengthen disaster preparedness for effective response at all levels.

A midterm review of the HFA indicated that significant progress has been made, although that progress is uneven: the more vulnerable segments of society still lack the resilience to hazards called for by the plan, in part because risk reduction is more difficult in poor areas.

The Risks of Being Poor

Poverty increases the risk of a number of environmental hazards. In general, poor people are exposed to more environmental hazards than wealthy people, and thus are at greater risk of being harmed. Poor people are more likely than wealthy people to be exposed to toxic chemicals and disease organisms; they are also more likely to lack access to disaster warning systems. The people who are dying from infectious diseases are those who lack access to adequate health care, clean water, nutritious food, healthy air, sanitation, and shelter. The relationship between poverty and environmental hazards also moves in the other direction—those who have been harmed by environmental hazards may be pushed into poverty (see the poverty cycle in Chapter 2).

Poverty and Longevity. Overall, the wealthier a country becomes, the healthier its population. People in high-income countries tend to live longer, and when they die, the chief causes of death are the degenerative diseases of old age, such as cancer, heart disease, and stroke. By contrast, only one-third of the deaths in the low-income countries are indicative of people who are living out their life spans.

The availability of medical treatment is one reason for this difference. In 2014, the WHO estimated that high-income countries had an average of 90 nurses and midwives for every 10,000 people, while low-income countries had only two such experts for the same number of people. Another reason for this discrepancy is that some parts of developing nations lack the resources to ensure safe drinking water and sanitation. In addition, many of the people in developing nations are exposed to the high levels of indoor air pollution associated with burning wood, dung, and coal in simple cooking stoves.

Poverty and Malnutrition. Being underweight is a risk factor for many hazards. The leading reason for children being underweight is malnutrition, and the primary cause of malnutrition is poverty (see Chapter 12). Not surprisingly, the poorest fifth of the world's population shows the highest prevalence of underweight children. (To learn about another cause of malnourishment, see Sound Science, *Water Pollution Drives Malnutrition in India.*)

The Role of Education. Education plays a role in the relationship between wealth and health. Advances in socioeconomic development and advances in education are strongly correlated. As people—especially women—become more educated, they use their knowledge to secure relief from hazards. They may, for example, improve their hygiene, immunize their children, or recognize dangerous symptoms such as dehydration and seek oral rehydration therapy.

Priorities and Inequalities. Nutrition, education, and the general level of wealth in a society do not tell the whole story, however. Even in developed countries, poor people may have less education, be more exposed to crime, have little access to healthy food, or have less access to health insurance. In addition, they may be more exposed to specific environmental hazards such as pollution (see Chapter 22).

SOUND SCIENCE
Water Pollution Drives Malnutrition in India

India's economy has been developing rapidly for decades. Even so, the country has more children suffering from the effects of long-term malnutrition than any other nation. Almost 48% of Indian children under five—61.7 million of them—have stunted growth caused by a lack of nutrients. These children face lifelong health difficulties and lowered cognitive function. Why is this problem so pervasive? To many, a lack of food has seemed the obvious answer. Several key initiatives, including the Zero Hunger Challenge (an initiative of the UN), a joint report by UNICEF, the WHO, and the World Bank, and a 2012 summit of world leaders from private and public sectors, have proposed solutions to malnutrition. Each has assumed that the stunting is due to an inadequate availability of food.

Recent research, however, suggests that stunting occurs in spite of the fact that some households have adequate food. In fact, India has a much higher rate of stunting than many much poorer countries, including Cambodia, Chad, and Haiti. And it isn't only poor Indians who show this effect. One-third of the children of the wealthiest Indians are stunted. Thus

Villagers in Gujarat, India, wash their vessels in a stagnant pond that is also used by farm animals.

economic growth for Indians does not necessarily reduce the rate of malnutrition.

One recent study investigated an alternative hypothesis: *Children in India are stunted because of chronic illness due to poor water sanitation; their bodies are weakened by disease, so they cannot use the food they eat.* India is densely populated and more than half of the population lack access to toilets. Open defecation is very common. No Indian city has a comprehensive sewage treatment system; waterways are full of human waste. The constant exposure of children to water-borne diseases causes many to suffer from diarrhea and vomiting. Even if they are fed nutritious food, the constancy of illness leaves them unable to absorb the nutrients they need.

Malnutrition resulting from poor sanitation is a good example of the connectivity of environmental issues. In this case, the growth of children is affected by chronic illnesses, which in turn are affected by household economics and views about the importance of sanitation. Poor sanitation has other effects, too. A recent World Bank report estimated that in one year alone, India lost $53.8 billion in economic activity because of poor sanitation, nearly 7% of the nation's GDP. In 2014, the Indian government launched the "Clean India Campaign," an effort to provide every Indian household with access to a toilet by 2019. The building of infrastructure needs to be accompanied by a large-scale educational campaign emphasizing the connection between sanitation and health.

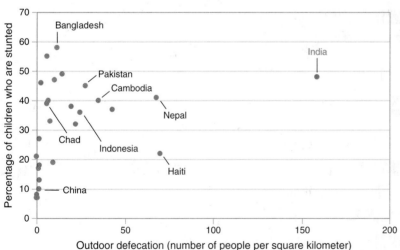

CHILDHOOD STUNTING AND POOR SANITATION

y-axis: Percentage of children who are stunted (0 to 70)
x-axis: Outdoor defecation (number of people per square kilometer) (0 to 200)

Labeled points: Bangladesh, Pakistan, Cambodia, Nepal, India, Chad, Indonesia, Haiti, China

Eighty-one percent of all childhood stunting occurs in 10 countries. While the main cause of stunting is a lack of food, water-borne diseases contribute to the problem, especially in countries where many people lack toilets and must defecate outdoors.

(*Source:* Open defecation data from Joint Monitoring Program WHO/UNICEF Water and Sanitation country reports; stunting data from UDAID *Demographic and Health Surveys* and UNICEF *Improving Child Nutrition, 2013*; density data from WorldAtlas.com.)

UNDERSTANDING THE DATA Compare the density of outdoor defecation in Bangladesh and Haiti with that in India. What is the relationship between these densities and the degree of stunting in these countries?

Source: G. Harris, "Poor sanitation in India may afflict well-fed children with malnutrition" *New York Times* July 13, 2014.

A study of American cities showed that income inequality, and not just average income, was a predictor of mortality rates. That is, cities with higher income inequality had worse health outcomes, even if they were wealthier on average.

On the other hand, a nation with a modest per capita income may choose to put its resources into improving the health of its population rather than into other activities. Countries such as Costa Rica, China, and Sri Lanka have a much longer life expectancy and lower infant mortality than their gross domestic product would predict. They have accomplished this enviable record by focusing public resources on activities such as immunization, the upgrading of sewer and water systems, and land reform.

The Effects of Climate Change

The WHO identifies climate change as a major risk to public health, not only because of the direct effects of warming on people, but also because of its indirect effects on other health risk factors. People in a warmer world will experience more of the hazards of heat waves, which are particularly severe in crowded cities. As the world becomes warmer, tropical insect vectors will expand their ranges. Agricultural production may fall in some areas, affecting food availability.

It is also likely that severe weather events will increase in frequency. Scientists are still determining to what extent global climate change is behind current natural disasters, especially intense hurricanes, tornadoes, floods, and droughts. All the evidence suggests that climate change will intensify future natural disasters as temperatures and sea levels rise (Chapter 18).

CONCEPT CHECK What are three pathways that increase the likelihood of a person being harmed by a chemical hazard? By a biological hazard? ☑

17.3 Risk Assessment

In the developed countries, our lives are freer of hazards than ever before. Even so, our society still faces many hazards and the risks that they present to our health. For our own self-interest, we should know about the risks to which we are subjected. Such knowledge enables us to make informed choices about these risks. When risks are thoroughly understood, strategies for reducing disease and death are far easier to identify and prioritize. If governments are to act as stewards of the health of their people, they desperately need the kind of information a thorough evaluation of health risks can bring them.

Risk assessment is the process of evaluating the risks associated with a particular hazard. Figure 17–17 shows a number of risks, expressed as years of lost life expectancy, for men and women in Denmark. Risk may also be expressed as

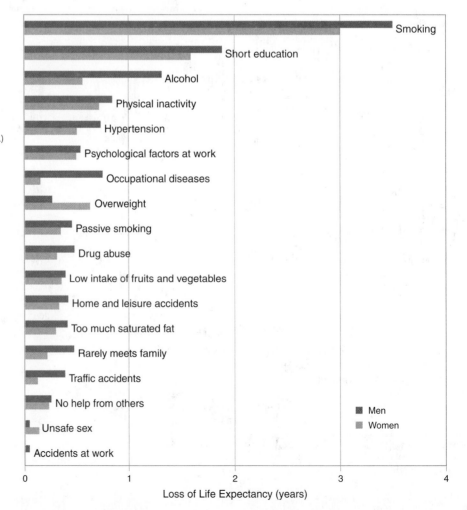

Figure 17–17 Loss of life expectancy from various risks. One way to illustrate the relative risks posed by different hazards is to rank them in terms of years of lost life associated with each risk. The data show the loss of life expectancy for men and women in Denmark.

(*Source:* Knud Juel, Jan Sorensen, and Henrik Bronnum-Hansen. Risk Factors and Public Health in Denmark, Summary Report. Copenhagen: National Institute of Public Health and University of Southern Denmark, 2007.)

the probability of dying from a given hazard. For example, the annual risk of dying from smoking may be expressed as 1.5 per 1,000 (Table 17–3). That is, in the course of a year, 1.5 people out of every 1,000 who smoke will die as a result of their habit. For men, this is equivalent to about 3.5 years of lost life expectancy. In other words, men who never smoke will live an average of 3.5 years longer than those who do.

Risk assessment is an important tool in the development of public policy. For the Environmental Protection Agency (EPA), risk assessment is a major way of applying science to the hard problems of environmental regulation. Many other U.S. agencies use risk assessment to address a wide variety of hazards. The WHO now recommends risk assessment as an ideal way to promote the health of people everywhere.

Risk Assessment by the EPA

Risk assessment began at the EPA in the mid-1970s as a way of addressing the cancer risks associated with pesticides and toxic chemicals. It soon developed into a science-based methodology directed toward understanding how pollutants affect human health and the environment. As currently performed at the EPA, human health risk assessment has four steps: hazard identification, dose-response assessment, exposure assessment, and risk characterization (Figure 17–18). The EPA also performs **ecological risk assessment (ERA)** in its responsibility for protecting natural organisms and ecosystems from the adverse effects of human activities. The process used for ERA is quite similar to that used for human health risk assessment.

Hazard Identification: Is This Really a Hazard? **Hazard identification** is the process of examining evidence linking a potential hazard to its harmful effects. In the case of accidents, the link between car crashes and deaths is obvious. In these cases, historical data, such as the annual highway death toll, are useful for calculating risks.

In other cases, the link is not so clear because there is a time delay between the first exposure and the final outcome. For example, it is often very difficult to establish a link between exposure to certain chemicals and the development of cancer some years later. In cases where the linkage is not obvious, researchers may need data from epidemiological studies or animal tests. An **epidemiological study** tracks how a sickness spreads through a community. To find a link between cancer and exposure to some chemical, an epidemiological study would examine all the people especially exposed to the chemical and determine whether that population has a statistically higher cancer rate than the general population.

Ask the Rats. We do not want to wait 20 years to find out that a new food additive causes cancer, so we turn to **animal testing** to find out now what *might* happen in the future. A test involving several hundred animals (usually mice or rats) takes about three years. If a significant number of the animals develop tumors after having been fed the substance being tested,

Hazard	Annual Risk*		
Cancer	1.8	per	1,000
Smoking	1.5	per	1,000
Motor vehicle accidents	1.2	per	10,000
Drug use	1.2	per	10,000
Suicide	1.1	per	10,000
Firearm injuries	1.0	per	10,000
Accidental poisoning	9.9	per	100,000
Alcohol use	8.3	per	100,000
Falls	8.1	per	100,000
HIV/AIDS	3.1	per	100,000
Work injuries	1.7	per	100,000
Drowning	1.2	per	100,000
Hang-gliding	8.6	per	1,000,000
Rock climbing	3.2	per	1,000,000
Choking by food	2.8	per	1,000,000
Air transport	2.3	per	1,000,000
Excessive heat	1.2	per	1,000,000

Table 17–3 Some Common Hazards Ranked According to Their Degree of Risk

*Probability of dying

Source: Data from National Vital Statistics Reports 59:4, March 16, 2011, and Health and Safety Executive, UK, September 4, 2012.

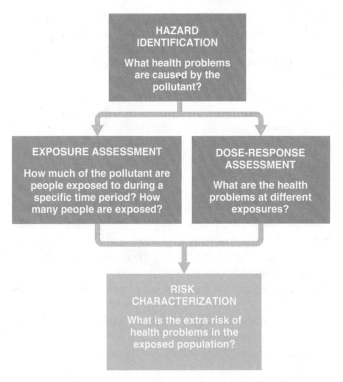

Figure 17–18 **Risk assessment as performed by the EPA.** The process of risk assessment involves four steps.

(*Source:* EPA, http://epa.gov/risk/risk-characterization.htm. January 19, 2011.)

then the substance is either a possible or a probable human carcinogen, depending on the strength of the results.

Three objections to animal testing have been raised: (1) rodents and humans may have very different responses to a given chemical; (2) the doses used on the animals are often much higher than those to which humans may be exposed; and (3) some people believe it is unethical to use animals for such purposes. Although there are obvious differences between rodents and humans, all chemicals shown by epidemiological studies to be human carcinogens are also carcinogenic to test animals, suggesting that animal tests have some predictive value for humans.

Just Test People? In the past, chemical manufacturers that were interested in assessing the safety of pesticides (in order to have their products approved for use) paid human subjects to test the products. Remarkably, such tests are far less expensive than the usual animal testing. A National Academy of Sciences panel addressed this topic, stating that the use of human volunteers "to facilitate the interests of industry or of agriculture" is unjustifiable. The EPA banned such tests on children and established new, more stringent ethical standards to protect adult subjects.[8]

Other Tests. Information about the physical and chemical properties of a substance can be used to predict its toxicity. For example, ToxCast™ is an automated technology that exposes living human cells to chemicals and then screens them for any toxic effects; this rapid process is now being used to test up to 10,000 chemicals. Initial tests using chemicals already subjected to animal testing and epidemiological studies have indicated that this approach has great promise in reducing the time it takes to evaluate the risk posed by a chemical hazard.

In the end, hazard identification takes a *weight of the evidence* approach to determining the potential ill effects of a chemical or process. Standard descriptors are used in its conclusions, such as "likely to be carcinogenic to humans." Thus, hazard identification tells us that we *may* have a problem.

Dose-Response and Exposure Assessment: How Much for How Long? When animal tests or human studies show a link between exposure to a chemical and an ill effect, the next step is to analyze the relationship between the concentrations of the chemical in the test (the dose) and both the incidence and the severity of the response. This process is called **dose-response assessment.** The projected effects of different concentrations of the chemical are shown on a *dose response curve* (Figure 17–19). It is at this point that vulnerability may be introduced; for example, it is known that children are more vulnerable to some cancer-causing chemicals because they are growing. From this information, projections are made about the number of cancers that may develop in a human population exposed to different doses of the chemical.

After completing a dose-response assessment, scientists carry out an **exposure assessment,** which considers the effect of the length of exposure. This procedure involves identifying

[8]EPA, "Expanded Protections for Subjects in Human Research Studies," February 16, 2011, http://www.epa.gov/oppfead1/guidance/human-test.htm, accessed January 19, 2012.

human groups already exposed to the chemical, learning how their exposure came about, and calculating the doses and duration of their exposure.

Risk Characterization: What Does the Science Say? The final step, **risk characterization,** is to pull together all the information gathered in the first three steps in order to determine the risk and its accompanying uncertainties. Here is where science is able to inform the risk manager. Commonly, risk is expressed as the probability of a fatal outcome due to the hazard, as in Table 17–3.

The EPA conducts an evaluation of risk information on hazards that chemicals pose to human health. The agency employs its **Integrated Risk Information System (IRIS)** on science-based information gathered from assays and epidemiological studies and publishes IRIS assessments after they have been thoroughly reviewed in draft form. Information on more than 550 chemical substances can be found in the IRIS database. Another EPA agency, the **National Center for Environmental Assessment (NCEA)** is the EPA's program for assessment of substances and processes that are widely released into the environment, such as common air pollutants. The NCEA has published 178 risk assessments on subjects from aerosols to watersheds.

Public-Health Risk Assessment

Until recently, most risk assessments were conducted as a function of environmental regulation. Then the WHO broke new ground in the area of risk assessment with its 2002 *World Health Report.*[9] A 2009 WHO report updated this

[9]World Health Organization, *The World Health Report 2002: Reducing Risks, Promoting Healthy Life* (Geneva: World Health Organization, 2002), www.who.int/whr/2002/en.

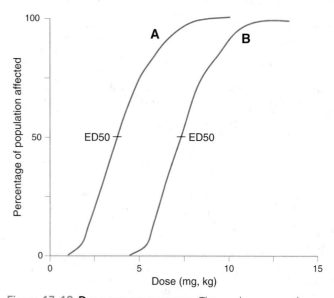

Figure 17–19 Dose response curves. The graph compares the toxicity of two chemicals. The line for each chemical shows the percentage of the population affected at various doses, or concentrations of the chemical. ED50 is the dose at which 50% of the population is affected. Compound A is more toxic than Compound B, because the dose required for a response is lower.

effort.[10] The WHO effort began by looking at all of the risk factors commonly known to be responsible for poor health and mortality and then asking this question: Of all the disease burden in this population, how much could be caused by this [particular] risk?

Criteria for Attention. The CDC tracks a number of risk factors in the United States, while the WHO tracks risk factors on a worldwide basis. Certain risk factors have been chosen for special attention, on the basis of the following criteria:

- The factor must have a high potential for a global impact on public health and mortality.

- The factor must have a high likelihood of causality—it must be possible to identify the cause-effect relationships between the factor and poor health.

- There must be reliable data on the prevalence of the risk factor and its relationship to disease and mortality.

- It should be possible to develop countrywide policies to substantially lessen the impact of the risk factor.

The DALY. In addition to mortality, public-health experts consider other measures of public health, including longevity and the prevalence of disabilities. The WHO reports use a measurement called the *disability-adjusted life year,* or *DALY.* One DALY represents the loss of one healthy year of a person's life. This measure assesses the burden of a disease in terms of life lost due to premature mortality and time lived with a disability. The WHO reports reveal some remarkable differences in risk factors in different groups of countries (Table 17–4).

Stewards of Health. Public-health ministries and agencies are the primary stewards of the health and welfare of a country's people. In the United States, the prevention of exposure to hazards includes the efforts of agencies such as the EPA. To quote the 2002 *World Health Report,* "Governments are the stewards of health resources. This stewardship has been defined as 'a function of a government responsible for the welfare of the population and concerned about the trust and legitimacy with which its activities are viewed by the citizenry.'"[11]

Governments can accomplish much in the arena of preventive health, but they must nurture their scarce resources carefully. Each risk factor calls for a different suite of interventions. As a country evaluates its highest health risks, it will often have to prioritize its risk-prevention strategies.

Risk Management

In human health and environmental risk assessment, the analysis of hazards and risks is a task that falls primarily to the scientific community. The task is incomplete, however, if no further consideration is given to the information gathered.

Risk management involves (1) a thorough review of the information pertaining to the hazard in question and a risk

Table 17–4 Ten Leading Selected Risk Factors as Percentage Causes of Disease Burden, Measured in DALYs*

Low-income countries	Percentage of total	DALYs (millions)
Underweight	9.9%	82
Unsafe water, sanitation, and hygiene	6.3%	53
Unsafe sex	6.2%	52
Suboptimal breastfeeding	4.1%	34
Indoor smoke from solid fuels	4.0%	33
Vitamin A deficiency	2.4%	20
High blood pressure	2.2%	18
Alcohol use	2.1%	18
High blood glucose	1.9%	16
Zinc deficiency	1.7%	14
Middle-income countries		
Alcohol use	7.6%	44
High blood pressure	5.4%	31
Tobacco use	5.4%	31
Overweight and obesity	3.6%	21
High blood glucose	3.4%	20
Unsafe sex	3.0%	17
Physical inactivity	2.7%	16
High cholesterol	2.5%	14
Occupational risks	2.3%	14
Unsafe water, sanitation, and hygiene	2.0%	11
High-income countries		
Tobacco use	10.7%	13
Alcohol use	6.7%	8
Overweight and obesity	6.5%	8
High blood pressure	6.1%	7
High blood glucose	4.9%	6
Physical inactivity	4.1%	5
High cholesterol	3.4%	4
Illicit drugs	2.1%	3
Occupational risks	1.5%	2
Low fruit and vegetable intake	1.3%	2

*Disability-adjusted life years: one DALY = loss of one healthy year of a person's life.
Source: World Health Report, 2009.

[10]World Health Organization, *Global Health Risks 2009: Mortality and Burden of Disease* (Geneva: World Health Organization, 2009).

[11]World Health Organization, *World Health Report 2002.* (Geneva: World Health Organization, 2002).

characterization of that hazard and (2) a decision as to whether the weight of the evidence justifies a regulatory action. Public opinion can play a powerful role in both processes. In general, however, a regulatory decision will hinge on one or more of the following considerations:

- Cost-benefit analysis. This type of analysis compares costs and benefits relative to a technology or proposed chemical product; if done fairly, the analysis can make a regulatory decision very clear cut. When the EPA issued new regulations for air-pollution particulate matter in 2006, the benefits in human health far outweighed the regulatory costs.

- Risk-benefit analysis. A decision can be made with regard to the risks versus the benefits of being subjected to a particular hazard, especially when those benefits cannot be easily expressed in monetary values. The use of medical CT scans is a good example. Because they represent a radiation exposure, the scans carry a calculable risk of cancer, but the benefit derived when you scan the brain for a suspected tumor is much greater than the cancer risk involved. However, there is concern about repeated CT scans because one abdominal scan is equivalent to 500 chest X-rays.

- Public preferences. As we will see, people have a much greater tolerance for risks that they feel are under their own control or are voluntarily accepted.

Risk management has been thoroughly incorporated into the EPA's policy-making process. Its most common use has been in the design of regulations. The process has enabled the agency to target appropriate hazards for regulation and to determine where to aim the regulations: at the source, at the point of use, or at the point of disposal of the hazardous agent. Policy makers often adopt a particular risk level as a standard against which they can measure new risks.

Risk Perception

The public's concern about environmental problems can be traced to the fear of hazards that pose a risk to human life and health. People may perceive that their lives are more hazardous than ever before, but that is not true. In fact, our society is freer from hazards than it ever has been, as is evidenced by increased longevity. Why, then, do people protest against nuclear power plants, waste sites, and pesticide residues in food when, according to experts, these hazards pose extremely small risks? The answer lies in people's **risk perceptions**—their intuitive judgments about risks. In short, people's perceptions are not consistent with the results coming from a scientific analysis of risk. There are some good reasons for this discrepancy—and some lessons to be learned from it.

Hazard Versus Outrage. The reason for the inconsistency between public perception and actual risk calculations, according to Peter Sandman of Rutgers University, is that the public perception of risks is often more a matter of *outrage* than one of *hazard*.[12] Sandman holds that, while the term *hazard* expresses primarily a concern for morbidity and mortality, the term *outrage* expresses a number of additional concerns:

- Lack of familiarity with a technology. Examples include how nuclear power is produced and how toxic chemicals are handled.

- Extent to which the risk is voluntary. Research has shown that people who have a choice in the matter will accept risks roughly 1,000 times as great as when they have no choice. People are much more accepting of a risk if they are in control of the elements of that risk, as in driving an automobile, swimming, and eating.

- Public impression of hazards. Accidents involving many deaths (as at Bhopal) or a failure of technology (for example, at Three Mile Island) are thoroughly imprinted into public awareness by media coverage and are not quickly forgotten (Chapters 15 and 22).

- Overselling of safety. The public becomes suspicious when scientists or public-relations people play up the benefits of a technology and play down its hazards.

- Morality. Some risks have the appearance of being morally wrong. If it is wrong to subject humans to pesticide testing, the notion that the chemical companies clearly benefit from the testing is unacceptable. A moral imperative demands action, even if it is costly.

- Fairness. The benefits and the risks should be connected. If the benefits go to someone else, why should you accept any risk?[13]

The Media's Role. The public perception of risk is strongly influenced by the media, which are far better at communicating the outrage elements of a risk than they are at communicating the hazard elements. Public concern over mad cow disease received extensive media coverage and had a high "outrage quotient," and beef from Canada was banned in 2003 when the disease was found in two animals. However, cigarette smoking, which causes 480,000 deaths a year in the United States alone, receives minimal media attention because it is not "news." Hence, there is no outrage factor. Indeed, it has been suggested that, if all the year's smoking fatalities occurred on one day, the media would go ballistic and smoking would be banned the next day!

Public Concern and Public Policy. Generally speaking, *public concern*, rather than cost-benefit analysis or risk-benefit analysis conducted by scientists, often drives public policy. The EPA's funding priorities are set largely by Congress, which is supposed to be responsive to its constituency.

---

[12]Peter M. Sandman, "Risky Business," *Rutgers Magazine* (April/June 1989): 36–37.

[13]V. Covello and P. Sandman, "Risk Communication, Evolution and Revolution," in *Solutions to an Environment in Peril*, ed. A. Wolbarst (Baltimore, MD: Johns Hopkins University Press, 2001), 164–178.

Unfortunately, if public concern is the primary impetus for public policy, some serious risks may get less attention than they deserve. In particular, environmental concerns are commonly perceived as much less important than they really are because of the public's preoccupation with other, more immediate issues, such as the economy and jobs. This difference underscores the importance of *risk communication,* a task that should not be left to the media. For example, the value of ecosystems and their connections to human health and welfare need far greater emphasis in the public consciousness, and this responsibility falls to the scientific and educational communities as well as to government agencies.

Studies have shown that the most effective risk communication occurs by starting with what people already know, discovering what they need to know, and then tailoring the message so that it provides people with the knowledge that helps them make a more informed evaluation of risks. To be successful, risk communication should also involve dialogue among all the parties involved—policy makers, the public, experts, and other interested groups.

CONCEPT CHECK What are three reasons a person might perceive a risk to be greater than it is? To be less than it is? ☑

REVISITING THE THEMES

SOUND SCIENCE

The continuing battle against environmental health problems such as cancer and poisoning requires the sustained attention of dedicated research laboratories around the world. This is modern science serving the interest of humankind. Better science allows us to warn against natural disasters and to understand how multiple factors play a role in malnutrition. In the case of an outbreak like Ebola, epidemiologists team with others to determine how to limit the disease and treat those who are ill. Risk assessment utilizes the scientific community in the final step of the process, at which risk characterization takes place. Once scientists have evaluated the risk, it is up to others to manage the risk. The WHO states that the basis for the major risk factors plaguing most societies has been scientifically established. There is more work to do, of course, but the failure to prove scientific causality cannot be an excuse for inaction by those responsible for people's health.

SUSTAINABILITY

A foundational strategy of sustainable development is the reduction of poverty, in part because the poor are susceptible to many hazards. Numerous risk factors are a consequence of the lack of access to clean water, healthy air, sanitation, nutritious food, and health care. Infectious diseases are a tremendous

impediment to economic progress in poor countries. It is hard to see how these countries can make much progress when malaria, tuberculosis, and HIV/AIDS take such a heavy toll on their children and young adults, although better treatment is causing some of these diseases to become less common. The success in the battles against Guinea worm and some other diseases is encouraging. Several of the Millennium Development Goals and Sustainable Development Goals target environmental health, which involves everything from indoor air pollution to warning systems for natural disasters. Another part of environmental health is for those with wealth to live sustainably.

STEWARDSHIP

Governments are the stewards of their country's health resources. Health ministries have a difficult task if they are not given the funding and personnel to carry out their mission. Fortunately, organizations such as the WHO can provide the vital information governments need to address their country's greatest health problems. The WHO exemplifies the stewardship concept of a community of nations acting together to bring care and information to those whose health is at risk. Those who work alongside the WHO—the health ministries, health-related nongovernmental organizations, and health care providers—are the hands and hearts of the world community.

REVIEW QUESTIONS

1. Define *public health, morbidity, mortality,* and *epidemiology.*

2. Differentiate between a hazard and a risk. Where does vulnerability fit in?

3. What are four categories of human environmental hazards? Give examples of each.

4. What are cultural hazards? What is an example of a cultural hazard with environmental components?

5. How does tobacco use increase vulnerability to other hazards?

6. Define *pollution, pollutant,* and *nonbiodegradable.*

7. Describe the public-health roles of the CDC and the WHO. What are the roles of the FDA and the EPA?

8. List five environmental risk factors that are responsible for global mortality and disease.

9. What is the significance of malaria worldwide, and what are some recent developments in the battle against this disease?

10. What are four ways that poverty affects mortality and morbidity?

11. Outline the four steps of risk assessment used by the EPA.

12. What methods are used to test chemicals for their potential to cause cancer?

13. Describe the process of risk assessment for public health recently carried out by the WHO. What is a DALY?

14. What are several differences between public risk perception and scientific risk assessment?

THINKING ENVIRONMENTALLY

1. Consider Table 17–4. For each group of countries, discuss the risk factors that are primarily a consequence of human choice. Pick one of these factors, and describe how you might proceed to bring it under control in an appropriate country.

2. What global accountability is there for natural disaster relief? Consider the example of devastating weather in developing countries that is triggered by climate change (which could be instigated by wealthier countries).

3. Consult the CDC Web site (www.cdc.gov), and investigate one of the emerging infectious diseases or another disease listed there. What are the risks you take, and what precautions should you take, if traveling to an area where the disease has been found?

MAKING A DIFFERENCE

1. Monitor the hazards in your own life, and take steps to reduce high-risk behavior.

2. Install a carbon monoxide detector if you don't already have one. Also check your home's radon levels.

3. Steer clear of phthalates, found in nearly three-quarters of today's beauty products. Phthalates are endocrine disruptors, substances that can obstruct hormone functions, damage organs, and cause infertility, miscarriage, birth defects, and breast cancer.

4. Six percent of an average American's day is spent on the road, but that time makes for 60% of their fine particle pollution intake. Driving less is not only good for the environment, but also good for you!

5. Instead of using toxic chemicals to clean house, consider homemade combinations of white vinegar, castile soap, baking soda, and water.

MasteringEnvironmentalScience®

Students Go to MasteringEnvironmentalScience for assignments, the eText, and the Study Area with animations, practice tests, and activities.

Professors Go to MasteringEnvironmentalScience for automatically graded tutorials and questions that you can assign to your students, plus Instructor Resources.

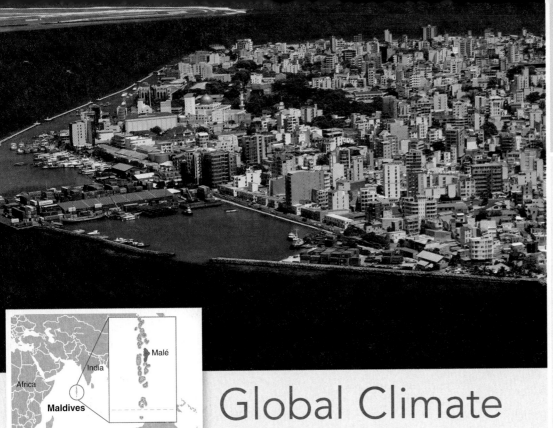

Global Climate Change

Learning Objectives

18.1 Atmosphere, Weather, and Climate: Describe the structure of the atmosphere; distinguish between weather and climate; explain how the greenhouse effect warms the Earth and identify the greenhouse gases.

18.2 The Science of Climate Change: Summarize the evidence that human activities are causing climate change; evaluate the various greenhouse gases and their effects on present and future climates.

18.3 The Effects of Climate Change: Explain how climatologists model Earth's climate and predict future outcomes; describe the impact of climate change on natural and human systems.

18.4 Controversy and Ethics: Explain why climate change is controversial; describe the ethical principles underlying the need for a response to climate change.

18.5 Responding to Climate Change: Define mitigation, adaptation, and geoengineering; describe measures that can be taken to mitigate and adapt to climate change and evaluate their effectiveness.

Mohamed Nasheed, then president of the Maldives, stood in front of the delegates to a UN meeting, speaking passionately about how climate change was already affecting his tiny nation. The Maldives is an archipelago of 1,190 islands in the Indian Ocean that some 320,000 people call home. Many of the islands are atolls (ring-shaped islands formed of coral), and others are little more than sandbars. The capital, Malé, is a square mile of densely packed buildings surrounded by a seawall. The average elevation in the islands is a mere 1 meter (3 feet) above sea level; recently, seafront erosion has claimed many beaches and some buildings.

The Maldives, Nasheed told the assembly, is like the proverbial canary in the coal mine. The impact of rising seas on these small, vulnerable islands should alert the world to an impending crisis. To prompt world action, Nasheed promised that his country would lead the way by becoming carbon neutral by 2020. Like many leaders, he had to help his people adapt to the warming climate and help them make changes to prevent further global effects. Nasheed was working on a sovereign-wealth fund that would enable the Maldivians to purchase land somewhere (India? Sri Lanka?); the fund was to be financed by the nation's tourism industry. Unfortunately, Nasheed's tenure was shortened by political unrest, so he was not able to attain all of his lofty goals.

Rising Sea Level. Over in the South Pacific, residents of numerous island nations are bracing for the rising sea level that is already submerging parts of their lands. Kiribati and Tuvalu are two island nations whose existence is threatened by environmental changes linked to global warming; as saltwater encroaches on the limited freshwater that supports plant life on the islands, it is likely that the entire populations of these two countries—some 110,000—will, like the Maldivians, be forced to emigrate.

▲ Malé, the capital of the Maldives, is under threat from rising seas caused by ocean warming.

Scientists point out that the populations of these isolated islands will be merely the first wave of climate change refugees unless the global community of nations takes rapid steps to curb the fossil fuel emissions that are largely responsible for global climate change. There is scientific consensus that Earth's systems are already changing in response to the ever-increasing levels of carbon dioxide in the atmosphere. The relentless rise of sea level, now at 3.2 millimeters per year, is accelerating as the oceans continue to warm and ice sheets and glaciers melt. Estimates of the total increase in sea level by the year 2100 range from a low of 26 centimeters to a high of 82 centimeters.[1] The warming climate is also increasing

the risk of extreme weather events, such as severe storms and droughts.

Difficult Choices. We are in the predicament of having to make far-reaching decisions about how to protect ourselves, including the most vulnerable among us, and future generations from dangerous environmental conditions. To make these decisions, we need to understand what is happening with our climate. How do scientists study our climate? Indeed, what controls the climate?

To answer these questions, we will first investigate the structure of the atmosphere and how it brings us our weather and climate. Then we will consider the scientific evidence for global climate change and examine projections of future conditions. We will conclude by looking at various actions that can be taken now and in coming decades to address global climate change.

---

[1] IPCC, "Summary for Policymakers," in *Climate Change 2013: The Physical Science Basis, Contribution of Working Group I to the Fifth Assessment Report of the Intergovernmental Panel on Climate Change*, ed. T. F. Stocker, et al. (Cambridge: Cambridge University Press, 2013).

18.1 Atmosphere, Weather, and Climate

In 2013, after sifting through thousands of studies about global warming, scientists from the **Intergovernmental Panel on Climate Change (IPCC)** published its Fifth Assessment Report (AR5) on climate change, an incredibly detailed and comprehensive study. In this report, the IPCC concluded that *the warming of Earth's climate system is unequivocal.* The warming of the globe has a ripple effect that results in an array of changes in the atmosphere, ocean, and land, including increased evaporation and precipitation, higher-energy storms, melting of ice, sea level rise, and other effects that we call **climate change**. Scientific evidence also indicates that a significant portion of these changes is **anthropogenic**, or the result of human activities. How can human activities have such a profound effect on global systems? To understand how this is happening, we will begin by examining the factors that control Earth's climate.

Layers of the Atmosphere

You can think of the atmosphere-ocean-land system as an enormous weather engine, driven by solar power and strongly affected by the Earth's rotation and its tilted axis. The interactions of these factors produce the day-to-day changes in the atmosphere called weather as well as the longer-term patterns of Earth's climates.

The atmosphere is a collection of gases that gravity holds in a thin envelope around Earth. Figure 18–1 shows the vertical structure of the atmosphere. Table 18–1 compares the characteristics of the lowest two layers—the troposphere and the stratosphere.

The lowest layer, the **troposphere,** ranges in thickness from 10 miles (16 km) in the tropics to 5 miles (8 km) in higher latitudes. The gases within the troposphere are responsible for moderating the flow of energy to Earth and are involved with the biogeochemical cycling of many elements and compounds—oxygen, nitrogen, carbon, sulfur, and water, to name the most crucial ones. This layer contains practically all

---

Table 18–1   Characteristics of the Troposphere and Stratosphere

Troposphere	Stratosphere
Extent: Ground level to 10 miles (16 km)	Extent: 10 miles to 31 miles (16 km to 50 km)
Temperature normally decreases with altitude, down to −70°F (−59°C)	Temperature increases with altitude, up to +32°F (0°C)
Much vertical mixing, turbulent	Little vertical mixing, slow exchange of gases with troposphere, via diffusion
Substances entering may be washed back to Earth	Substances entering remain unless attacked by sunlight or other chemicals
All weather and climate take place here	Isolated from the troposphere by the tropopause

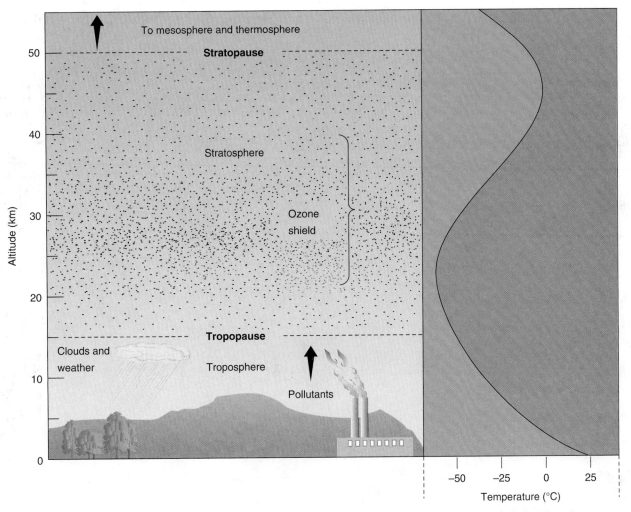

Figure 18–1 Structure and temperature profile of the atmosphere. The diagram on the left shows the layers of the atmosphere and the ozone shield, while the plot on the right shows the vertical temperature profile.

of the water vapor and clouds in the atmosphere; it is the site and source of our weather.

Except for localized temperature inversions, the troposphere gets colder with altitude. Air masses in this layer are well mixed vertically, so pollutants can reach the top within a few days. Substances entering the troposphere—including pollutants—may be changed chemically and washed back to Earth's surface by precipitation (Chapter 19). Capping the troposphere is the **tropopause**, a boundary region where air shifts from cooling with height and begins to warm.

Above the tropopause is the **stratosphere**, a layer within which temperature *increases* with altitude, up to about 40 miles above the surface of Earth. The temperature increases primarily because the stratosphere contains ozone (O_3), a form of oxygen that absorbs high-energy radiation emitted by the Sun. Because there is little vertical mixing of air masses in the stratosphere and no precipitation from it, substances that enter it can remain there for a long time. Beyond the stratosphere are two more layers, the **mesosphere** and the **thermosphere**, where the ozone concentration declines and only small amounts of oxygen and nitrogen are found.

Weather and Climate

The winter of 2015 brought record-breaking cold and snows to New England. Meanwhile the Pacific Northwest was experiencing unusual warmth. On a global scale, the National Climate Data Center reported that February 2015 was the second hottest February on record. That winter an unusually warm mass of Pacific Ocean water was controlling regional winds and altering weather patterns. The difference between local experience and long-term global patterns is the difference between weather and climate.

The day-to-day variations in temperature, air pressure, wind, humidity, and precipitation—all mediated by the atmosphere—constitute **weather**. Weather is variable, short-term, and local, and may not represent larger regional or global trends. In contrast, **climate** is the average temperature and precipitation expected throughout a typical year in a given region; it is the result of long-term weather patterns in a region, patterns that involve decades or longer (Chapter 5). Climate is controlled by a number of factors including the amount of radiation from the Sun, the tilt of the Earth, the chemistry of the atmosphere, and the movement and mixing of ocean water. The scientific study of the atmosphere—both of weather and of climate—is *meteorology*.

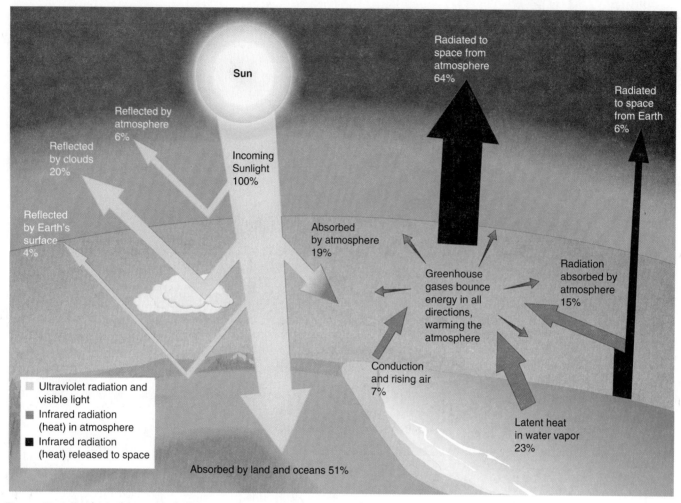

Figure 18–2 Earth's solar energy balance. Energy from the Sun travels to Earth in the form of short-wave (high-energy) radiation. Nearly one-third (30%) of this radiation is reflected back into space. The remainder is absorbed by the oceans, land, and atmosphere. Greenhouse gases absorb energy in the atmosphere and radiate it in all directions. Eventually, this absorbed energy is radiated back to space as long-wave (lower–energy) radiation, also called infrared radiation or heat.

Earth's Energy Balance

Energy from the Sun enters the atmosphere in the form of visible and ultraviolet light (shortwave radiation). It then takes a number of possible courses (Figure 18–2). Some radiation (30%) is reflected by the atmosphere, clouds, and Earth's surface, but most is absorbed by the atmosphere, oceans, and land, which are heated in the process, producing long-wave, infrared radiation (heat). In the atmosphere, energy can be absorbed and interact with molecules of gases that bounce it back and forth before it is reradiated into space.

If there were no atmosphere, Earth would be very cold because nothing would trap the energy from the Sun; heat would quickly be lost to space. Instead, molecules in the atmosphere trap energy by bouncing it in different directions; some of this energy bounces back and forth between the atmosphere and the surface of the land and water. This energy warms the Earth's surface and lower atmosphere. This natural process is called the greenhouse effect. A similar process occurs in a car on a sunny day. When the car's windows are closed, sunlight coming in through the windows is absorbed by the car's seats and other interior surfaces. This light energy is converted into heat energy, which is given off in the form of infrared radiation. The infrared radiation is blocked by glass, so it cannot leave the car. The trapped heat energy causes the interior air temperature to rise.

Forcing Factors. How warm the atmosphere becomes depends on the interactions of *internal factors* such as the oceans, the gaseous makeup of the atmosphere, snow cover, and sea ice, as well as *external factors* such as solar radiation, the Earth's rotation, and slow changes in our planet's orbit. **Radiative forcing** is the influence that any particular factor has on the energy balance of the atmosphere-ocean-land system. The effect of a *forcing factor* can be *positive,* leading to warming, or *negative,* leading to cooling. Forcing is expressed in units of watts per square meter (W/m^2); averaged over the entire planet, solar radiation entering the atmosphere is about 340 W/m^2. It is this radiation that is acted on by the various forcing factors.

Greenhouse Gases. Water vapor, CO_2, and certain other gases in the atmosphere serve as positive forcing factors. These gases play a role analogous to that of the windows in our car, or the glass roof in a greenhouse, hence the term **greenhouse gases.** Light energy that comes through the atmosphere is absorbed and converted to heat energy at the planet's surface. The infrared heat energy radiates back upward through the atmosphere and into space. The greenhouse gases in the troposphere absorb some of the infrared radiation and reradiate it back toward the surface; other gases (N_2 and O_2) in the troposphere do not.

This greenhouse effect was first recognized in 1827 by the French scientist Jean-Baptiste Fourier and is now firmly established. The greenhouse gases are like a heat blanket, insulating Earth and delaying the loss of infrared energy (heat) to space. Without this insulation, average surface temperatures on Earth would be about $-19°C$ instead of $+14°C$, and life as we know it would be impossible. Therefore, our global climate depends on Earth's concentrations of greenhouse gases. If these concentrations increase or decrease significantly, their influence as positive forcing agents will change, and our climate will change accordingly.

Cooling Processes. Earth's atmosphere is also subject to negative forcing factors. For example, on average, clouds cover 50% of Earth's surface and reflect some 20% of solar radiation into space before it ever reaches the ground. The reflectivity of a surface to sunlight is called **albedo.** The albedo of the planet as a whole prevents a certain amount of warming in the first place.

Clouds. The albedo effect of thick, low-lying clouds can be readily appreciated if you think of what happens when a large cloud passes between you and the Sun. The radiation that was heating you and your surroundings is suddenly intercepted higher up in the atmosphere, and you feel cooler. In contrast, high, wispy clouds have a positive forcing effect, absorbing some of the solar radiation and releasing some infrared radiation. Recent work indicates that the current net impact of clouds is a slightly positive forcing.

The Cryosphere. Snow, glaciers, ice sheets, and sea ice (the *cryosphere*) reflect sunlight, contributing to the planetary albedo and cooling the planet. Arctic warming has reduced this effect in recent decades, as open water and darker, unfrozen ground absorb more sunlight than ice and snow.

Aerosols. Aerosols are microscopic liquid or solid particles originating from land and water surfaces (Chapters 10, 19). Scientists have found that industrial aerosols (from ground-level pollution) play a significant role in canceling out some of the warming from greenhouse gases. Sulfates, nitrates, dust, and soot from industrial sources and forest fires enter the atmosphere and react with compounds there to form a high-level aerosol haze. This haze reflects and scatters some sunlight; the aerosols also contribute to the formation of clouds, which likewise increase the planetary albedo. *Sooty* aerosols (from fires) have an overall warming effect, but industrial *sulfate* aerosols create substantial cooling, mainly through their tendency to increase global cloud cover.

Volcanoes. Volcanic activity can also lead to planetary cooling. When Mount Pinatubo in the Philippines erupted in 1991, some 20 million tons of particles and aerosols entered the atmosphere, causing a significant drop in global temperature as radiation was reflected and scattered away. This global cooling effect lasted until the volcanic debris was finally cleansed from the atmosphere by chemical change and deposition, a process that took several years. Some sulfate aerosols from volcanic sources enter the stratosphere, where they have a longer-term cooling effect; these aerosols have increased since 2000, effectively reducing some of the global warming that otherwise might have occurred.[2]

Solar Variability. Because the Sun is the source of radiant energy that heats Earth, any variability in the Sun's radiation reaching Earth is likely to influence the climate. Direct monitoring of solar irradiance goes back only a few decades, but it has confirmed that there is an 11-year cycle involving solar radiation increases during times of high sunspot activity. Sunspots are known to block cosmic ray intensity, and because cosmic rays help seed clouds, the diminished cloud cover allows more solar radiation to reach Earth's surface. Variations in solar forcing are thought to be partly responsible for the warming trend in the early 20th century.

The Net Result . . . Global atmospheric temperatures are a balance between the positive and negative forcing from natural causes (natural greenhouse gases, clouds, volcanoes, solar irradiance) and anthropogenic causes (sulfate aerosols, soot, ozone, increases in greenhouse gases). The net result varies, depending on one's location. The various forcing agents bring us year-to-year fluctuations in climate and make it difficult to attribute any one event or extreme season to human causation.

Moving Air

Some of the energy that is released from Earth's surface is transferred to the atmosphere. Thus, air masses grow warmer at the surface of Earth and tend to expand, becoming lighter. The lighter air then rises, creating *vertical* air currents. On a large scale, this movement creates major *convection currents* (Chapter 10, Figure 10–5). Air must flow in to replace the rising warm air, and the inflow leads to *horizontal* airflows, or wind. The ultimate source of the horizontal flow is cooler, denser air that sinks, gains heat, and reduces relative humidity as it descends.

Global Convection Cells. The natural movement of air that derives from the uneven heating of Earth's surface at different latitudes produces the large-scale convection cells called the Hadley cells (Figure 10–5). These major flows of air create regions of high rainfall (equatorial) and deserts (25° to 35° north and south of the equator). Together with the rotation of the Earth, the Hadley cells produce the prevailing easterly and westerly winds in different areas of the globe.

Local Convection Currents. On a smaller scale, convection currents bring us the day-to-day changes in our weather.

---

[2]S. Solomon et al., "The Persistently Variable 'Background' Stratospheric Aerosol Layer and Global Climate Change," *Science* 333, no. 6044 (2011): 866–70, doi:10.1126/science.1206027.

Figure 18–3 A convection cell. Driven by solar energy, convection cells produce the main components of our weather as evaporation and condensation occur in rising air and precipitation results, followed by the sinking of dry air. This process moves masses of air poleward. Combined with the rotation of Earth, these movements produce global winds.

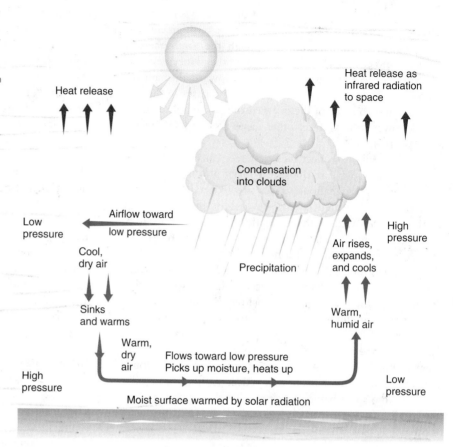

Weather reports inform us of regions of high and low pressure, but where do these come from? Rising air (due to solar heating) creates high pressure up in the atmosphere, leaving behind a region of lower pressure close to Earth. Conversely, once the moist, high-pressure air has cooled by reradiating heat to space and losing heat through condensation (thereby generating precipitation), it flows horizontally toward regions of sinking, cool, dry air (where the pressure is lower). There the air is warmed at the surface and creates a region of higher pressure (Figure 18–3). The differences in pressure lead to airflows, which are the winds we experience. Winds tend to flow from high-pressure regions toward low-pressure regions.

Jet Streams. The large-scale air movements of Hadley cells are influenced by Earth's rotation from west to east. This creates the trade winds and the general flow of weather from west to east in temperate regions. Higher in the troposphere, Earth's rotation and air-pressure gradients generate veritable rivers of air, called **jet streams,** that flow eastward at speeds of more than 300 mph and meander considerably. Jet streams are able to steer major air masses in the lower troposphere. One example is the polar jet stream, which steers cold air masses into North America when it dips downward in latitude.

Put Together . . . . Air masses of different temperatures and pressures meet at boundaries we call **fronts,** which are regions of rapid weather change. Other movements of air masses due to differences in pressure and temperature include hurricanes and typhoons and local, but very destructive, tornadoes.

There are also major seasonal airflows called **monsoons,** which are created by major differences in cooling and heating between oceans and continents. The summer monsoons of the Indian subcontinent are famous for the beneficial rains they bring and notorious for the devastating floods that can occur when the rains are heavy.

Putting these movements all together—the general atmospheric circulation patterns and the resulting precipitation, the wind and weather systems generating them, and finally the rotation of Earth and the tilt of the planet on its axis, which creates the seasons—yields the general patterns of climate that characterize different regions of the world, and ultimately, the biomes and ecosystems in each region.

Oceans, Atmosphere, and Climate

More than two-thirds of Earth is covered by oceans, making ours mostly a water planet. Because we live on land, it is hard for us to imagine the dominant role the oceans play in determining both our weather and climate. Nevertheless, the oceans are the major source of water for the hydrologic cycle and the main source of heat entering the atmosphere. The evaporation of ocean water supplies the atmosphere with water vapor, and when water vapor in the atmosphere condenses, it supplies the atmosphere with heat—the *latent heat of condensation* (Figure 10–4). The oceans also play a vital role in climate because of their innate *heat capacity*—their ability to absorb energy when water is heated. Indeed, the entire heat capacity of the atmosphere is equal to that of just the top 3 meters of ocean water. It is this property that makes the climate in coastal areas milder than that in areas inland. Finally, through the movement of currents, the oceans convey heat energy throughout the globe.

Thermohaline Circulation. Large-scale ocean currents are driven by a pattern of **thermohaline circulation,** where *thermohaline* refers to the effects that temperature (*thermo-*) and salinity (*-haline*) have on the density of seawater. This pattern, called the **Meridional Overturning Circulation (MOC),** is like a giant, complex conveyor belt, moving water masses from the surface to deep oceans and back again (**Figure 18–4**). A key part of this circulation is in the high-latitude North Atlantic, where salty water from the Gulf Stream moves northward on the surface and is cooled by Arctic air currents. This cooling increases the density of the water. As water evaporates, the salinity of the surface water increases, a process that also increases its density. Eventually the denser water sinks to depths of up to 4,000 meters, forming the **North Atlantic Deep Water (NADW).** This cold, deep-water current spreads southward through the Atlantic to the southern tip of Africa, where it is joined by cold Antarctic waters. Together, the two streams spread northward into the Indian and Pacific Oceans as deep currents. Gradually, the currents slow down and warm, becoming less dense and welling up to the surface, where they are further warmed and begin to move as surface waters back again toward the North Atlantic. This movement transfers enormous quantities of heat toward Europe,

providing a climate that is much warmer than the high latitudes there would suggest.

The circulation of ocean water is actually more complex than the simple conveyor-belt model. Eddies and oceanic winds inject and subtract water masses at many points in the overall circulation. Yet it is clear that the turnover does occur—sinking at the poles forms deep waters, and warm water does return northward to affect the climate of northern Europe. If large amounts of cold freshwater (from ice melt or precipitation) were to enter Arctic waters, the thermohaline circulation could slow or stop; fortunately, there is scant evidence that such an event is likely to occur.

Ocean-Atmosphere Oscillations. Regional weather is notoriously erratic on time scales from a few days to decades. Studies of a number of ocean-atmosphere processes—processes that produce erratic weather on a global scale—have provided new insights into natural climate variability. These processes are becoming better understood, and are beginning to provide explanations for some previously puzzling developments.

One such process is the **North Atlantic Oscillation (NAO),** in which atmospheric pressure centers switch back and forth across the Atlantic, altering the patterns of winds and storms. Researchers have found evidence that a mid-century cooling was the result of shifting ocean circulation linked to the NAO. There is also a warm-cool cycle, the **Interdecadal Pacific Oscillation (IPO),** that swings back and forth across the Pacific over periods of several decades.

In the *El Niño/La Niña* Southern Oscillation (ENSO), major shifts in atmospheric pressure over the central equatorial Pacific Ocean lead to a reversal of the trade winds. *El Niños* tend to encourage warming, while *La Niñas* do the opposite. Incredibly, these reversals dominate the weather over much of the planet, and their effects can last for a year or two. The 1997–2000 ENSO was the most intense of recent decades, spawning powerful global weather patterns that caused more than $36 billion in damages and more than 22,000 deaths. The year 2010 had a strong *El Niño*, but 2011 was a strong *La Niña* year, with corresponding warmer and cooler global temperatures. In the following years, these cycles declined.

CONCEPT CHECK   What two processes increase the density of ocean water in the North Atlantic? How does this higher density affect the movement of ocean water? ☑

18.2 The Science of Climate Change

Over Earth's long history, global and regional climates have changed many times. Today, however, the best evidence tells us that climates are being changed by human actions. Scientists have concluded that the atmosphere is warmer, the oceans are warmer, there is more precipitation, there are more extreme weather events such as hurricanes, floods, droughts, and heat waves, glaciers are melting, and the sea level is rising.

The IPCC report stated that it is *very likely* (a 90–100% probability) that most of the warming is caused by increased

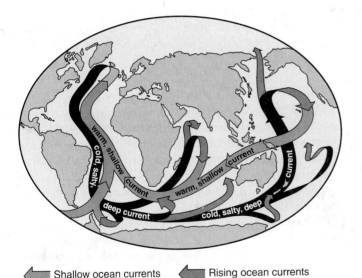

Shallow ocean currents   **Rising ocean currents**
Deep ocean currents   **Descending ocean currents**

Figure 18–4 The Meridional Overturning Circulation. Salty water flowing to the North Atlantic is cooled and sinks, forming the North Atlantic Deep Water system. This deep flow extends southward and is joined by Antarctic water, whereupon it extends into the Indian and Pacific Oceans. Surface currents then proceed in the opposite direction, returning the water to the North Atlantic.

(*Source:* Adapted from Figure 3.15 from *Our Changing Planet,* 2nd ed., by Fred T. MacKenzie. Copyright © 1998 by Prentice Hall, Inc. Reprinted by permission of Pearson Education, Inc., Upper Saddle River, NJ 07458.)

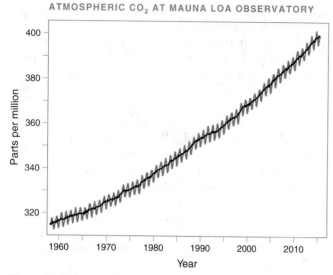

ATMOSPHERIC CO₂ AT MAUNA LOA OBSERVATORY

Figure 18–5 Atmospheric carbon dioxide concentrations. The concentration of CO_2 in the atmosphere fluctuates between winter and summer because of seasonal variation in photosynthesis. The average concentration is increasing due to human activities—in particular, burning fossil fuels and deforestation.

(*Source:* NOAA Earth System Research Laboratory, 2011.)

concentrations of anthropogenic (human-made) greenhouse gases in the atmosphere. The most significant anthropogenic greenhouse gas is carbon dioxide (CO_2), which has risen by 40%—from 280 parts per million (ppm) to over 400 ppm—since the Industrial Revolution began, much of it since 1960 (Figure 18–5). The source of this added CO_2 is no surprise: Most of it comes from burning fossil fuels, although some also comes from agricultural wastes, deforestation, and other sources. Let's examine how the IPCC reached these conclusions.

How to Get Expert Information

The scientific investigation of climate change is a worldwide endeavor involving thousands of scientists. *Climatologists,* the scientists who study climate, study the atmosphere, but they also study ocean currents, ice, rocks, and other natural features to find evidence of the many processes that determine climate. Because the field of climate change is relatively young and because new technologies are being invented that allow us to measure temperatures and other factors more precisely and to model complex climate systems more accurately, the field is exciting and ever changing. The large scale of this endeavor means that it requires funding from governments and tracking across national boundaries.

As the evidence of climate change began to grow, the scientific community realized that there was a critical need for climate research to be consolidated and synthesized so scientists around the world could access consistent and accurate data. In response to this need, the UN Environment Program (UNEP) and the World Meteorological Society established the Intergovernmental Panel on Climate Change (IPCC) in 1988. The mission of the IPCC has been to provide accurate information that will lead to an understanding of human-induced climate change.

The work of the IPCC has been guided by two basic questions:

Risk assessment—How much is the climate system changing, and how do the changes affect society and ecosystems?

Risk management—How can we manage the climate system through adaptation and mitigation?

The IPCC has three working groups: one to assess the scientific issues (Working Group I), another to evaluate the impact of global climate change and the prospects for adapting to it (Working Group II), and a third to investigate ways of mitigating the effects (Working Group III).[3] The latest IPCC assessment working groups consist of more than 850 authors and review editors representing 85 countries. These experts are unpaid and participate at the cost of their own research and other professional activities.

The IPCC reports contain user-friendly summaries that are useful places to start reading. They also contain much longer reviews of all the scientific literature and are a valuable tool for those wanting more detailed information. The IPCC released the Fifth Assessment Report (AR5) during 2013–2014. *Climate Change 2014: Synthesis Report* was the final volume, and contained key findings of the working groups.

There are other places to look for expert information. In 2014, the American Association for the Advancement of Science (AAAS), the largest professional scientific society in the United States, released a report on climate change entitled *What We Know.*[4] Written for the public, this short report addresses many of the concepts in the longer IPCC report. Many of the concepts are also covered in the *National Climate Assessment 3*, published by the U.S. Global Change Research Program.[5]

The following discussion of climate change makes use of the reports from the IPCC and other expert groups. We begin with a look into the past, which harbors "records" of climate change.

Climates in the Past

Searching the past for evidence of climate change is a major scientific enterprise, one that becomes more difficult the farther into the past we try to search. Systematic records of the factors making up weather, including temperature, precipitation, and storms, have been kept for little more than 130 years. Nevertheless, these records clearly inform us that our climate is far from constant.

The record of surface temperatures, from weather stations around the world and from literally millions of observations of temperatures at the surface of the sea, tells an interesting story (Figure 18–6). Since 1880, the global average temperature has shown periods of cooling and warming. During the 20th century, two warming trends occurred, one from 1910 to 1945

[3]See, for example, Susan Solomon et al., Eds., *Climate Change 2007: The Physical Science Basis,* Contribution of Working Group I to the Fourth Assessment Report of the Intergovernmental Panel on Climate Change (New York: Cambridge University Press, 2007).

[4]Mario Molina et al., *What We Know: The Reality, Risks and Response to Climate Change* (Washington, DC: AAAS, 2014).

[5]J. M. Melillo et al., eds., *Climate Change Impacts in the United States: The Third National Climate Assessment* (Washington, DC: U.S. Global Change Research Program, 2014).

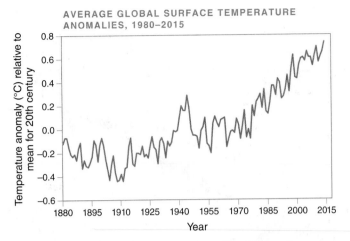

AVERAGE GLOBAL SURFACE TEMPERATURE ANOMALIES, 1980–2015

Figure 18–6 Surface temperature anomalies. The graph compares the observed, globally averaged annual surface temperatures of the land and ocean to the mean value for the 20th century (0.0).

(*Source:* National Climatic Data Center, 2015)

and a second from 1976 until the present. In each decade since 1976, the global average temperature has increased an average of 0.28°C (0.50°F) over land and 0.11°C (0.20°F) over the ocean. Every year since 1976 has had an average global temperature higher than the long-term global average.

Further Back. Observations on climatic changes can be extended much further back in time with the use of **proxies**—measurable records that can provide data on factors such as temperature, ice cover, and precipitation. For example, historical accounts suggest that the Northern Hemisphere enjoyed a

warming period from AD 1100 to 1300. This was followed by the "Little Ice Age," between AD 1400 and 1850. Additional proxies include tree rings, pollen deposits, changes in landscapes, marine sediments, corals, and ice cores.

Some recent research on ice cores has provided a startling view of a global climate that oscillates according to several cycles. This work has also afforded evidence that remarkable changes in the climate can occur within as little as a few decades. Ice cores in Greenland and the Antarctic have been analyzed for thickness and gas content, especially the content of carbon dioxide (CO_2) and methane (CH_4), two greenhouse gases. Ice cores have also been tested for the relative proportions of the **isotopes** of certain elements; *isotopes* are forms of an element in which the atoms have more or less than the usual number of neutrons. Isotopes of oxygen, as well as isotopes of hydrogen, behave differently at different temperatures when condensed in clouds and incorporated into ice.

The long-term record based on these analyses indicates that Earth's climate has oscillated between ice ages and warm periods **(Figure 18–7a)**. During major ice ages, huge amounts of water were tied up in glaciers and ice sheets, and the sea level was lower by as much as 400 feet (120 m). Ice cores from Antarctica show eight glacial periods and eight interglacial periods over the past 800,000 years. During the glacial periods, temperatures and greenhouse gases were low; during the warm interglacial periods, just the opposite was true. The greenhouse gas and temperature data match exactly. Carbon dioxide levels ranged between 150 and 280 ppm.

The most likely explanation for these major oscillations is the periodic variations in Earth's orbit, which cause the

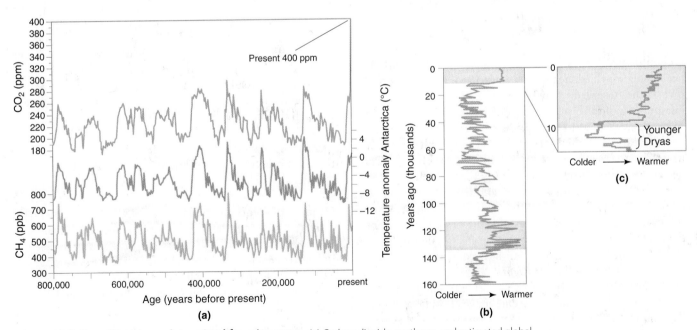

Figure 18–7 Past climates, as determined from ice cores. (a) Carbon dioxide, methane, and estimated global temperature from Antarctic ice cores for the past 800,000 years, compared with the past century. Note how closely the temperature and greenhouse gas concentrations coincide. **(b)** Temperature patterns of the past 160,000 years, demonstrating climatic oscillations. **(c)** Higher resolution of the past 12,000 years. The Younger Dryas cold spell occurred at the start of this record. Note how rapidly the cold spell dissipated, at around 10,700 years before the present.

(*Sources:* **(a)** The Bohr Institute; **(b)** & **(c)** Adapted from Robert W. Christopherson, *Geosystems: An Introduction to Physical Geography,* 5th ed. (Upper Saddle River, NJ: Pearson Prentice Hall, 2005).)

distribution of solar radiation over different continents and latitudes to vary substantially. These periodic oscillations, called **Milankovitch cycles** (after the Serbian scientist who first described them), take place at intervals of 100,000, 41,000, and 23,000 years.

Rapid Changes. Superimposed on the major oscillations is a record of rapid climatic fluctuations during periods of glaciation and warmer times (Figure 18–7b). One such rapid change, called the **Younger Dryas** event (*Dryas* is a genus of Arctic flower), occurred toward the end of the last ice age. Earth had been warming up for 6,000 years and then plunged again into 1,150 years of cold weather. At the end of this event, 11,700 years ago, Arctic temperatures rose 7°C in 50 years. The impact on living systems must have been enormous. One of the reasons scientists study ancient climates is to figure out whether such rapid changes could occur today. Scientists hypothesize the Younger Dryas event may have been triggered by a change in ocean currents, volcanoes, or an impact or giant burst of air from a near-Earth comet. It is still difficult to know whether such dramatic change is likely in our future.

Evidence of Climate Change Today

There is much natural variation in weather from year to year, and local temperatures do not necessarily follow globally averaged ones. Since the mid-1970s, the average global temperature has risen about 0.6°C (1.1°F), almost 0.2°C per decade (Figure 18–6). The warming is happening everywhere, but is greatest at high latitudes in the Northern Hemisphere. Because of the continued increases in anthropogenic greenhouse gases in the atmosphere, the observed warming is considered to be the consequence of an enhanced greenhouse effect.

Ocean Heat Content. The upper 3,000 meters of the ocean have warmed measurably since 1955, a warming that dwarfs the observed warming of the atmosphere, accounting for 90% of the heat increase of Earth systems in the past several decades. In a key analysis of Earth's energy imbalance, researchers found that at the end of the 20th century, heat was stored near the surface of the ocean.[6] By the first decade and a half of the 21st century, that heat had moved to lower layers in the Atlantic and southern oceans, making the warming at the surface of the ocean less dramatic. Over the past decade, the ocean has absorbed essentially all of the heat not yet seen in the atmosphere. A fascinating story about the measurement of ocean heat content is described in Sound Science, *Getting It Right*.

This ocean's absorption of heat energy has long-term and short-term consequences. One long-term consequence is the impact of this stored heat as it eventually comes into equilibrium with the atmosphere, raising temperatures over land and in the atmosphere even more. The immediate

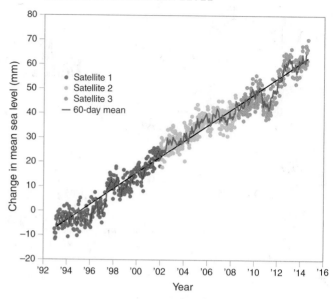

CHANGE IN GLOBAL SEA LEVEL

Figure 18–8 The rise in global mean sea level. Satellite radar altimeters have provided estimates of global sea level rise since 1993. The current rate is 3.2 mm/year. Note the 6 mm decline in 2010–2011, a consequence of the strong *La Niña* that produced massive rainfall events over land. Note, too, that the sea level started back up in mid-2011 as the *La Niña* began to subside.

(*Source:* Colorado University Sea Level Research Group, 2011. Used with permission.)

consequence is the rise in sea level that occurs because of thermal expansion. Sea level rose about 15 cm (6 inches) during the 20th century, a rate of 1.5 millimeters per year. It is now rising more rapidly, at a rate of 3.2 millimeters per year (Figure 18–8). About half of this rise is due to thermal expansion, and the remainder results from the melting of glaciers and ice caps.

The IPCC AR5 documents a general loss of land ice as glaciers and ice caps have thinned and melted. Sea ice is also melting. However, because it is already floating on the ocean, melting sea ice does not cause the sea level to rise, except through thermal expansion. Significantly, historical sea level data indicate that the sea level was relatively stable until 100 years ago; the modern rise in sea level is a new phenomenon that is consistent with the known pattern of anthropogenic climate warming.

Other Observed Changes. The IPCC AR5, the National Climate Assessment, and other reports document a number of recent changes, all of which are consistent with the picture of climate change caused by an increase in greenhouse gases. Many of these changes are listed here; note that most of them have potentially serious consequences for human and natural systems.

- Broad increases in warm temperature extremes and decreases in cold temperature extremes have been documented.

- Heat waves, like those that scorched Europe in 2010 and California and Australia in 2013 and 2014, are increasing in intensity and frequency.

---

[6]James Hansen et al., "Earth's Energy Imbalance: Confirmation and Implications," *Science* 308 (June 2005): 1431–35.

SOUND SCIENCE
Getting it Right

Josh Willis works at NASA's Jet Propulsion Laboratory and specializes in measurements of the ocean's response to global warming. He publishes his results in peer-reviewed journals such as the *Journal of Geophysical Research*. In a 2004 paper, he presented work that showed the ocean heat content increasing between 1993 and 2003, in agreement with what climatologists were observing in their measurements of atmospheric temperature. This is no trivial matter; the oceans absorb some 90% of the heat from global warming. However, Willis then did a follow-up study with colleague John Lyman, updating the time series from 2003 to 2005, showing that the oceans suddenly began to cool, and in a large way. This finding alarmed many climatologists. It also fed the climate change skeptics, who pounced on the data as proof that global warming wasn't happening.

One climatologist who was especially dubious about Willis's new data was Takmeng Wong, who was working at NASA's Langley Research Center. Wong had been measuring the exchange of energy between the Earth and space, and his data showed that the amounts of entering and outgoing energy at the top of the atmosphere were gradually climbing out of balance: More energy was entering than leaving. This was true for the 2003–2005 time period as well, and Wong politely asked Willis to reexamine his data because there was a clear discrepancy between the two studies.

When Willis measured an even deeper ocean cooling in 2006, he began to entertain serious doubts about his data. The source of the temperature data being employed by Willis was a flotilla of Argo floats—underwater robots that sink to 2,000 meters, drift with the ocean currents, and then pop up to the surface, taking water temperatures as they ascend. They then transmit the data to a satellite. Willis found that, when tested, many new Argo floats were giving bad data—in fact, indicating cooler temperatures than other devices being employed. Willis also consulted an independent set of data on satellite measurements of sea level. When water heats up, it expands, and the sea level data were showing a continual rise. One final factor in the data relates to a different data set on ocean temperatures from devices called XBTs, which are dropped into the ocean and relay temperatures as they sink. An independent paper had shown that the XBT data from previous years were too warm.

When Willis corrected his ocean warming time series with the too-warm XBT measurements and the too-cool Argo measurements, the indications of ocean cooling went away. The warm pulses of the 1970s and 1980s disappeared, and the time series showed a gradual warming. Taken together, ocean warming, sea level rise, and energy exchange with space are now telling a consistent story, one of a warming that can be attributed only to rising greenhouse gas emissions.

Two conclusions can be drawn from this story. First, science is always changing as we get more and more information, but it is also self-correcting, as scientists see problems and solve them. Second, an unexpected finding may be the result of the methods used to collect data, in this case in measurements.

Source: Rebecca Lindsey, Correcting Ocean Cooling, NASA Earth Observatory, November 5, 2008.

Josh Willis of the NASA Jet Propulsion Laboratory. Willis is receiving the Presidential Early Career Award for Scientists and Engineers from John Holdren (White House Office of Science and Technology Policy Director) and Lori Garver (NASA Associate Administrator).

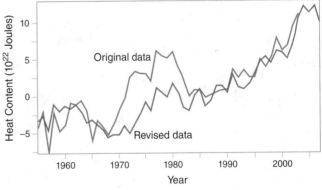

Heat capture by the oceans. The graph shows the changes in heat content of the upper 2,290 feet (700 m) of the world's oceans, as plotted by the Global Oceanographic Data Archaeology and Rescue Project (GODAR). Heat content is measured in joules (J). Two plots are seen: an original data plot and a revised data plot after corrections were made.

(*Source: Rebecca Lindsey, Correcting Ocean Cooling, Earth Observatory, November 5, 2008.*)

- Droughts are increasing in frequency and intensity; the fraction of land area experiencing drought has more than doubled over the past 30 years, leading to a global decline in net primary productivity and significant reductions in corn and wheat crops. The increased forest fire activity in the western United States is clearly related to droughts, warmer temperatures, and earlier springs.

- Patterns of precipitation are changing, with greater amounts from 30° N and 30° S poleward and with smaller amounts from 10° N to 30° N. Also, there have been widespread increases in heavy precipitation events, leading to more frequent flooding. In the United States, the greatest increase in storms with heavy precipitation has occurred in the Northeast and Midwest.

- There are also seasonal changes: In the Northern Hemisphere, spring comes earlier and fall comes later, shifting ecosystems and populations out of sync. More tree deaths and forest insect damage have also been traced to the warmer temperatures.

- While overall global temperature has increased a moderate 1.1°F, temperatures in Alaska, Siberia, and northwestern Canada have risen 5°F in summer and 10°F in winter. Spring comes two weeks earlier in the Arctic than it did at the turn of the 21st century. Permafrost has been melting over the polar regions—lifting buildings, uprooting telephone poles, and breaking up roads.

- Rising Arctic temperatures have brought about a major shrinkage of Arctic sea ice, with an accelerating decrease in the volume and extent of ice (Figure 18–9). This shrinkage is part of a positive feedback loop: As sea ice melts, it exposes dark water, which absorbs more energy than the white ice; this energy warms the water, causing more ice to melt. Similarly, thinner ice is more prone to breakage by the wind, and smaller chunks of ice melt more quickly.

- The Greenland Ice Sheet, a huge reservoir of water, is melting and calving icebergs at unprecedented rates, undoubtedly a response to the rise in temperatures in the northern latitudes. A net loss of 530 cubic miles of water occurred between 2003 and 2011. Record warm summers have accelerated Greenland ice melt; the largest loss was in 2012–2013. The two-mile-high ice sheet contains enough water to raise the sea level 23 feet (7 m).

- Temperatures in Antarctica, which holds enough ice to raise the sea level by 185 feet (57 m), have risen by more than 0.5°C over the past 55 years. The West Antarctic Ice Sheet is the continent's most vulnerable ice mass because it is grounded below sea level. Ocean water is constantly in contact with the edges of this ice sheet, and the ice sheet has been shrinking as the water eats away at the ice.

- Glaciers have resumed a retreat that has been documented for more than a century, but that slowed in the 1970s and 1980s. The World Glacier Monitoring Service has followed the melting and slow retreat of mountain glaciers all over the world. The data collected show an acceleration of glacial "wastage" since 1990 that is likely a response to recent rapid atmospheric warming (Figure 18–10). Loss of mountain glaciers is especially serious, as these glaciers are often critical sources of water for arid lands below. Glacier meltwater is often stopped by temporary dams; when these dams burst, the resulting floods and landslides may kill people living in their paths.

- As sea temperatures increase, populations of marine fishes are shifting northward, impacting commercial fisheries. Changing sea temperatures have also affected the distribution of phytoplankton production over major areas of the ocean.

(a) EXTENT OF SEA ICE, SEPTEMBER, 2014

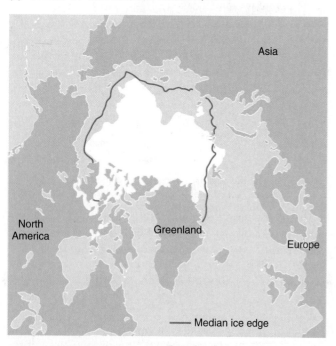

(b) AVERAGE MONTHLY ARCTIC SEA ICE EXTENT SEPTEMBER 1979–2014

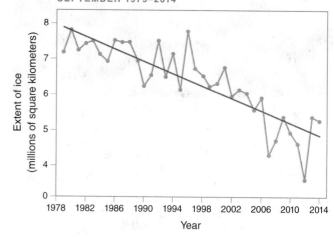

Figure 18–9 Decline of Arctic sea ice. (a) The extent of sea ice in September 2014 was 5.02 million square kilometers (1.94 million square miles). The orange line is the median ice extent for September from 1979 to 2014. **(b)** Tracking with satellites shows the 36-year downward decline in the Arctic sea ice at the end of the northern summer. The decline is now 13.3% per decade.

(*Source:* National Snow and Ice Data Center University of Colorado, Dec 2014.)

(a)

(b)

Figure 18–10 Shrinkage of the Muir glacier. Since 1941, the Muir glacier, a tidewater glacier in Alaska's Glacier Bay, has retreated more than 7 miles (12 km) and thinned by more than 2,600 feet (800 m). Once covered with ice and snow **(a)**, the land now holds trees and shrubs **(b)**.

(*Source*: NSIDC/WDC for Glaciology, compiled 2002, updated 2006, *Online Glacier Photograph Database*, Boulder, CO: National Snow and Ice Data Center/World Data Center for Glaciology, digital media.)

- The oceans have absorbed about 30% of the CO_2 emissions from human sources, slowing the rise of atmospheric CO_2 considerably. But in doing so, the oceans are experiencing a decrease in pH, a process known as **ocean acidification** (Chapter 20). Although the decrease is only about 0.1 pH unit to date, there is serious concern about the future state of seawater pH in light of continued anthropogenic CO_2 production. At some level, ocean acidification will make the warm oceans inhospitable to coral reefs; the coral animals depend on an equilibrium for carbon minerals that is affected by acidic conditions. The bottom line: Human use of fossil fuels has begun to change not only the temperature, but also the chemistry of the surface oceans.

Given the accumulating evidence that the climate is changing and that the change is already causing an array of changes in biological, physical, and chemical systems on Earth, it is time to consider the causes of these changes in more detail.

Rising Greenhouse Gases

According to the IPCC AR5 Synthesis Report, "It is *extremely likely* that more than half of the observed increase in global average surface temperature from 1951 to 2010 was caused by the anthropogenic increase in greenhouse gas concentrations and other anthropogenic forcings together."[7] Greenhouse gas concentrations have continued to increase.[8] The most important of these anthropogenic greenhouse gases is carbon dioxide (CO_2).

Carbon Dioxide. More than 100 years ago, the Swedish scientist Svante Arrhenius reasoned that differences in CO_2 levels in the atmosphere could greatly affect Earth's energy budget. Arrhenius suggested that, in time, the burning of fossil fuels might change the atmospheric CO_2 concentration, though he believed that it would take centuries before the associated warming would be noticeable. He was not concerned about the impacts of this warming, arguing that such an increase would be beneficial. (He lived in Sweden!)

Monitoring Atmospheric Carbon Dioxide. How do scientists know that the amount of CO_2 in the atmosphere is rising? In 1958, Charles Keeling began measuring CO_2 levels on Mauna Loa, in Hawaii. Measurements there have been recorded continuously, and they reveal a striking increase in atmospheric levels of the gas (Figure 18–5). The concentrations increased exponentially until the energy crisis in the mid-1970s, rose at a rate of 1.5 ppm/year for several decades, and in the 2000s began rising at a rate of 2.0 ppm/year and higher. The data also reveal an annual oscillation of 5–7 ppm, which reflects seasonal changes of photosynthesis and respiration in terrestrial ecosystems in the Northern Hemisphere. When respiration predominates (late fall through spring), CO_2 levels rise; when photosynthesis predominates (late spring through early fall), CO_2 levels fall. Most salient, however, is the relentless rise in CO_2. As of the end of 2014, atmospheric CO_2 levels were at 400 ppm, 40% higher than they were before the Industrial Revolution and higher than they have been for more than 800,000 years (Figure 18–7a). Thus, our insulating blanket is thickening, and it is reasonable to expect that this will have a warming effect.

Sources of Carbon Dioxide. As Arrhenius suggested, the obvious place to look for the source of increasing CO_2 levels is our use of fossil fuels. Every kilogram of fossil carbon burned produces 3.7 kilograms of CO_2 that enters the atmosphere (each carbon atom in the fuel picks up two oxygen atoms in the course of burning and becoming CO_2). In 2013, almost 10 billion metric tons (gigatons, or Gt) of fossil fuel carbon (GtC) was released, an annual increase

[7]R. K. Pachauri and A. Reisinger, eds., *Climate Change 2014: Synthesis Report* (contribution of Working Groups I, II, and III to the Fifth Assessment, Report of the Intergovernmental Panel on Climate Change, IPCC, Geneva, Switzerland, 2014).

[8]S. A. Montzka et al., "Non-CO_2 Greenhouse Gases and Climate Change," *Nature* 476 (August 4, 2011): 43–50.

ESTIMATES OF GLOBAL CARBON EMISSIONS FROM FOSSIL-FUEL COMBUSTION AND CEMENT MANUFACTURE

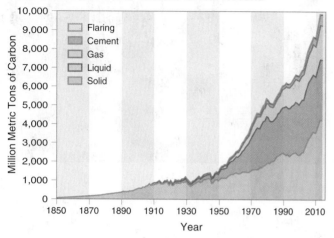

Figure 18–11 **Global carbon emissions from fossil fuel combustion.** Data from 1850 to 2013, from the Carbon Dioxide Information Analysis Center (CDIAC). The rise in last two decades is conspicuous.

(*Source:* CDIAC, Oak Ridge National Laboratory, 2014.)

SOURCES OF CARBON EMISSIONS

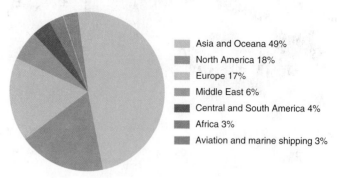

Figure 18–12 **Sources of carbon dioxide emissions from fossil fuel burning.** Total emissions of carbon in 2013 were approximately 9.9 billion metric tons (9.9 GtC). Total emissions of CO_2 (which weighs more than carbon alone) were 36 billion metric tons.

(*Source:* Data from U.S. Energy Information Administration, December 2014.)

of 2.5% (Figure 18–11). This carbon was added to the atmosphere as CO_2. About half of this CO_2 comes from industrialized countries; about 5% comes from cement production and is usually included in figures reported for fossil fuel emissions (Figure 18–12). Carbon dioxide is estimated to account for about 20% of greenhouse warming, at current levels.

Climate change skeptics sometimes claim that much of the new CO_2 comes from natural sources. This claim can be put to rest by considering the ^{14}C signature of atmospheric CO_2. ^{14}C is a radioactive isotope of carbon that is formed in the atmosphere by the bombardment of nitrogen atoms by cosmic rays; it decays over time to form nitrogen atoms. Fossil fuels contain no ^{14}C because of radioactive decay (the half-life of ^{14}C is 5,730 years). If significant amounts of CO_2 originating from fossil fuel combustion are added to the atmosphere, the ^{14}C in the atmosphere should be declining, and it clearly is (Figure 18–13).

It is estimated that land use changes such as deforestation add another 0.9 GtC annually to the carbon already coming from fossil fuel combustion. Over the past 50 years, the release of carbon from the use of fossil fuels and deforestation has more than tripled. Fortunately, *sinks* remove more than half of these annual carbon emissions.

Carbon Sinks. Careful calculations show that if all the CO_2 emitted from burning fossil fuels accumulated in the atmosphere, the concentration would rise by 3 ppm or more per year, not the actual 2 ppm charted in Figure 18–5. Thus, there must be carbon sinks that absorb CO_2 and keep it from accumulating at a more rapid rate in the atmosphere. There is broad agreement that the oceans serve as a sink for about 30% of the CO_2 emitted; some of this is due to the uptake of CO_2 by phytoplankton, and some is a consequence of the undersaturation of CO_2 in seawater. There are limitations to the ocean's ability to absorb CO_2, however, because only the top 300 meters of the ocean are

in direct contact with the atmosphere. (Deep ocean layers do mix with the upper layers, but the mixing time is more than 1,000 years.) And, as the oceans acidify (because they are absorbing CO_2), their capacity to continue this absorption will diminish.

The seasonal swings illustrated in Figure 18–5 show that biota can influence atmospheric CO_2 levels. Measurements indicate that terrestrial ecosystems also serve as carbon sinks. Indeed, these land ecosystems apparently have stored a net 15%, a figure that includes the losses from deforestation. As a result, terrestrial ecosystems, and especially forests, are increasingly valued for their ability to sequester carbon. A simple model of the major dynamic pools and fluxes of carbon is presented in Figure 18–14.

Other Greenhouse Gases. Water vapor, methane, nitrous oxide, ozone, and chlorofluorocarbons (CFCs) also absorb

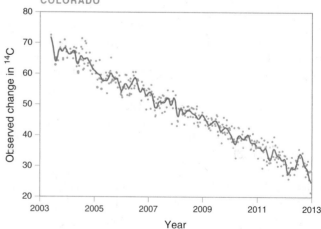

Figure 18–13 **^{14}C trend in the atmosphere.** The change in ^{14}C is the ratio of ^{14}C to normal ^{12}C. The smaller the ratio, the fewer ^{14}C atoms are in the air. The decreasing trend is due to ^{14}C depleted CO_2 emitted from burning fossil fuels.

(*Source:* NOAA Earth System Research Laboratory, 2014.)

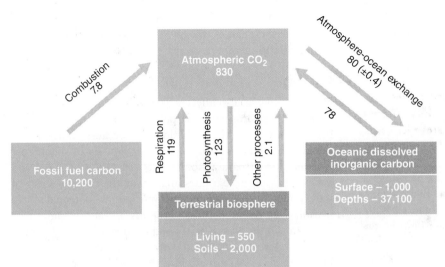

Figure 18–14 Global carbon cycle. Data are given in GtC (billions of metric tons of carbon). Pools are in the boxes, and fluxes (GtC per year) are indicated by the arrows.

(*Source:* U.S. DOE. 2008. *Carbon Cycling and Biosequestration: Report from the March 2008 Workshop,* DOE/SC-108, U.S. Department of Energy Office of Science, 2–3. http://genomicscience.energy.gov.)

infrared radiation and add to the insulating effect of CO_2 (Table 18–2). Most of these gases are increasing in concentration, raising the concern that future warming will extend well beyond the calculated effects of CO_2 alone.

Water Vapor and Clouds. Water as vapor (a gas) or in clouds (an aerosol) absorbs infrared energy and is the most abundant greenhouse gas. Water vapor accounts for about 50% of the greenhouse effect, and clouds an additional 20%. Although water vapor plays an important role in the greenhouse effect, its concentration in the troposphere is quite variable. Through evaporation and precipitation (the hydrologic cycle), water undergoes rapid turnover in the lower atmosphere (see Chapter 10). In the last several decades, however, meteorologists have found a general increase in the amount of water vapor in the atmosphere.

Methane. Methane (CH_4) is the third-most-important greenhouse gas. A molecule of methane is 25 times more effective at absorbing infrared radiation than is a molecule of CO_2 (Chapter 14). In natural ecosystems, methane is a byproduct of the microbial fermentation of plant material, largely in wetlands; it is also generated by the microbial digestion of plant material in the stomachs of ruminants.

Anthropogenic sources of methane include livestock (particularly cattle, which are ruminants), manure, rice cultivation, landfills, coal mines, and the production and transmission of natural gas. Although methane is gradually destroyed in reactions with other gases in the atmosphere, it is being added to the atmosphere faster than it can be broken down. Like CO_2, methane is more abundant than it has been for at least the past 800,000 years.

Nitrous Oxide. Nitrous oxide (N_2O) levels have increased some 19% during the past 200 years and are still rising at a rate of 0.7 ppb per year (Chapters 12, 14). Sources of the gas include agriculture, the oceans, and the burning of biomass; lesser quantities come from industry and the burning of fossil fuels. N_2O is produced in agriculture via anaerobic denitrification processes, which occur wherever high levels of nitrogen (a major component of fertilizers) are available in soils (see Chapter 3). The buildup of N_2O is particularly unwelcome because its long residence time (114 years) makes the gas a problem not only in the troposphere, where it contributes to warming, but also in the stratosphere, where it contributes to the destruction of ozone.

Table 18–2 Anthropogenic Greenhouse Gases in the Atmosphere

Gas	Average Preindustrial Concentration (ppb)*	Approximate Current Concentration (ppb)	Radiative Forcing (W/m²)	Average Years in Atmosphere
Carbon dioxide (CO_2)	280,000	399,320	1.88	100–300
Methane (CH_4)	714	1,800	0.50	12
Nitrous oxide (N_2O)	270	326	0.17	120
Chlorofluorocarbons and halocarbons	0	1.3	<0.2	12–3,200

Sources: IPCC AR5; recent data from CDIAC.

*Parts per billion; 1,000 ppb = 1 part per million (ppm).

Ozone. Depending on its location in the atmosphere, ozone (O_3) can have either a warming or cooling effect. In the troposphere, ozone acts as a greenhouse gas, but in the stratosphere it has a cooling effect. However, the warming effect predominates. The greatest source of tropospheric ozone is anthropogenic, through the action of sunlight on pollutants (Chapter 19). Major sources are automotive traffic and burning forests and agricultural wastes. Scientists estimate that the concentration of ozone in the atmosphere has increased 36% since 1750.

CFCs and Other Halocarbons. Emissions of halocarbons (Chapter 19) are entirely anthropogenic. Like nitrous oxide, halocarbons are long lived and contribute to both global warming in the troposphere and ozone destruction in the stratosphere. Used as refrigerants, solvents, and fire retardants, some halocarbons have 10,000 times the capacity to absorb infrared radiation that CO_2 does. The rate of production of halocarbons has declined due to adherence to the Montreal Protocol, and the concentration of halocarbons in the troposphere began declining in 1990 and is now only one-tenth of its former level.

The Atmosphere's Control Knob. A report from the Goddard Institute for Space Studies examined the role of the different greenhouse gases in maintaining temperatures on Earth.[9] Water vapor, which contributes the lion's share of the greenhouse effect, readily *condenses* and can be removed from the atmosphere. In contrast, CO_2 and the other *noncondensing* gases are the essential components of global warming because their "forcing" role sustains the current levels of water vapor and clouds. Without the stability of these noncondensing greenhouse gases, the water vapor and clouds could not maintain our current greenhouse

---

[9]Andrew A. Lacis et al., "Atmospheric CO_2: Principal Control Knob Governing Earth's Temperature," *Science* 330 (October 15, 2010): 356–59.

conditions. Thus, CO_2, because it is the most abundant noncondensing gas, is the key gas, the "control knob" that governs Earth's temperature.

Changes in Other Forcing Factors

Human activities also affect several other forcing factors. For example, the aerosols in air pollution have a cooling effect because they reflect the Sun's radiation. Because human activities are releasing more than double the amount of sulfate that is coming from natural sources, its effect is substantial and persistent. Climatologists estimate that the cooling effect of these pollutants has counteracted much of the potential global warming in recent decades. Progress in curbing industrial pollutants in the United States and Europe has caused the impact of anthropogenic aerosols to decline over those regions. However, industrial growth in China and India has created a major rise in aerosols over Asia, causing cooling there.

Black soot from anthropogenic sources, largely fires and exhaust from diesel engines, reduces the albedo of the cryosphere, including glaciers and ice sheets. By darkening snow and ice, soot promotes the absorption of radiant energy rather than its reflection. Changes in land use, such as deforestation, also affect albedo.

Solar radiation also varies over time, and increases are thought to contribute to global warming. **Figure 18–15** shows our understanding of the relative impacts of the various radiative forcing factors. Positive forcing factors include a variety of greenhouse gases and black carbon on snow, which decreases the albedo, or reflectivity of Earth's surface. Ozone in the troposphere is a positive forcing factor, but ozone in the stratosphere is a negative forcing factor. It is evident that the effect of anthropogenic forcing is far greater than the effect of natural changes in solar radiation.

CONCEPT CHECK What human activities are the major sources of CO_2 emissions? What other human activities contribute to global climate change? ☑

Figure 18–15 Radiative forcing factors. Positive and negative factors contribute to changes in Earth's energy balance. The graph presents the amount of radiative forcing due to human factors and natural changes in solar radiation. Forcing values for the year 2005 are compared to baseline values (0) for the pre-industrial era (1750). Values are in watts per square meter.

(*Source:* Leland McInnes, Creative Commons)

UNDERSTANDING THE DATA

1. Which of the forcing factors has the greatest warming effect? Cooling effect?
2. How does the effect of ozone in the stratosphere differ from its effect in the troposphere?

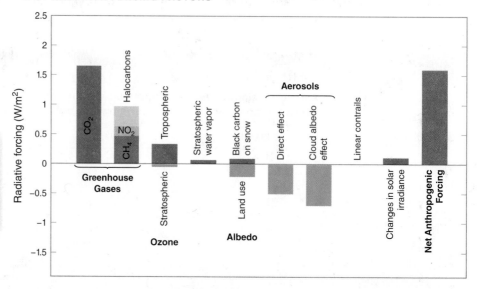

RADIATIVE FORCING FACTORS

18.3 The Effects of Climate Change

Higher temperatures, rising sea levels, heat waves, droughts, intense precipitation events, shifts in seasons, melting ice sheets, and Arctic thawing—these are happening right now, even as greenhouse gas emissions are continuing to rise. The International Energy Agency projects a growing demand for fossil fuels as populations and economies grow. If global fossil fuel use continues on its present trajectory, emissions of greenhouse gases will increase 35% by 2030. By 2050, emissions will increase 100%. Given the huge reserves of coal, ample supplies of natural gas, and remaining oil reserves, there is no reason to doubt this projection. However, it is clear that commitment to this future use of fossil fuels also will commit us to greater changes in climate. How much change, and when, would be a matter of guesswork if it were not for the use of climate models.

Modeling Global Climate

Computer models are essential tools for exploring Earth's weather and climate. Modern weather forecasting employs powerful computers that are capable of handling large amounts of atmospheric data and applying mathematical equations to model the processes taking place in the atmosphere, oceans, and land. Forecasts of conditions for 72 hours model Earth's weather. Climatologists employ similarly powerful computers to explore Earth's climate. Their computer models combine global atmospheric circulation patterns with ocean circulation, radiation feedback from clouds, and land surface processes to produce **atmosphere-ocean general circulation models (AOGCMs)** that are capable of simulating long-term climatic conditions. Climatologists use AOGCMs to explore the potential effects of rising concentrations of greenhouse gases.

How accurate are these AOGCMs? The best test of such models is the degree to which they can simulate the present-day climate. One study compared some 50 climate models developed over the past two decades, concluding that the most recent models have reached an unprecedented level of realism.[10] AOGCMs are now able to reproduce changes in surface temperatures over continents and over time scales of decades. The models used for the IPCC AR5 are able to model the effects of large volcanoes accurately as well. Yet while these models can present global patterns realistically, the details of regional changes from these models are very uncertain.

Figure 18–16 contrasts the data generated by computer models with actual data for the period from 1860 to 2010. Note that models using anthropogenic forcing factors are much closer to the actual measurements, showing the impact of human activities on the climate.

---

[10]Thomas Reichler and Junsu Kim, "How Well Do Coupled Models Simulate Today's Climate?" *Bulletin of the American Meteorological Society* 89, No. 3 (March 2008): 303–331.

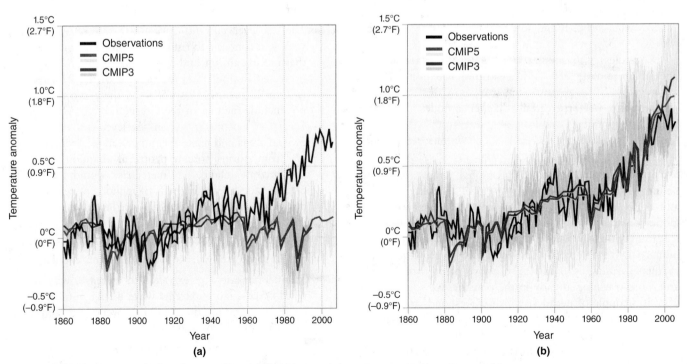

(a)   **(b)**

Figure 18–16 Comparing results of computer models with actual data. Scientists used computer models (CMIP3 and CMIP5) to predict global surface temperature anomalies, or differences from the mean surface temperature during the period from 1860 to 2010. They then compared these predictions to the actual observed data for the period. **(a)** Models that used only natural factors did not reflect the significant warming trend at the end of the century. **(b)** Models that used both natural and anthropogenic forcing factors were much closer to the actual observed values.

(*Source:* M. Mann and L. Kump. *Dire Predictions.* Pearson: London, 68–69)

Different Pathways to 2100

Climatologists use AOGCMs to project how changing concentrations of greenhouse gases will affect the climate in the 21st century and beyond. In IPCC AR5, researchers investigated scenarios representing four different sets of assumptions about the level of anthropogenic emissions. These scenarios, called **representative concentration pathways (RPCs)**, projected significantly different levels of radiative forcing by the year 2100. (Recall that radiative forcing is the amount of additional energy taken up by Earth's system as a result of changes to the greenhouse effect, expressed in watts per square meter, W/m^2.) The RCP representing the lowest increase in the emissions of greenhouse gases projected that radiative forcing in 2100 would be 2.6 W/m^2 more than preindustrial levels. The other three scenarios projected additional radiative forcing of 4.5, 6.0, and 8.5 W/m^2. All four scenarios projected a rise in CO_2 concentrations and a corresponding rise in mean global temperature.

Figure 18–17 compares the world of today with the expected outcomes of RCP 2.6 W/m^2 and 8.5 W/m^2 by the year 2100. RCP 2.6 assumes that we keep greenhouse gas emissions very low. To reach no higher than that level, we would need to stop increasing emissions immediately. RCP 8.5 assumes a high rate of population growth and little technological change. In each scenario, the uncertainty of the predictions is represented by a colored band around the line. After mid-century, the two scenarios diverge so that their uncertainties do not overlap, meaning that these are truly different trajectories. Note that in RCP 8.5, the world is much warmer by the end of the century. This shows how small differences in the next few decades result in much greater differences in the following years. Right now, our emissions are on track with RCP 8.5.

Regardless of the amount of warming, the pattern will be uneven across the globe. Future warming is expected to occur in geographical patterns similar to those of recent years: The warming will be greatest at higher latitudes and in the interior of continents and least over the North Atlantic and the southern oceans. However, it will be warmer everywhere.

Stopping at 450 ppm. Rising global temperatures are linked to two major impacts: *regional climatic changes* and *a rise in sea level*. Earth's history tells us that with 3°C global warming, equilibrium sea level could be at least 80 feet (25 m) higher than today. This is clearly dangerous to human civilization and should be avoided. However, with current warming at about 0.8°C above preindustrial levels and more warming already in the pipeline, we could get to 2°C in a couple of decades, which corresponds to a stable concentration of CO_2 (with the other greenhouse gases included) in the atmosphere of around 450 ppm. To achieve stabilization at 450 ppm will require major changes in our energy systems. The current carbon emissions of 36 $GtCO_2$ per year will have to fall to 7 $GtCO_2$ per year by 2100 and then drop even further. In order to stabilize the concentration of greenhouse gases in the atmosphere, emissions need to peak and decline.

Climate Commitment. Even if there were no increase at all in greenhouse gases, the current greenhouse gas concentration represents a *commitment* to further temperature increase and sea level rise due to the delay in the release into the atmosphere of energy stored in the oceans. That **climate commitment** is on the order of about 0.1°C per decade for the next several decades. The expected continued rise in sea level would cause disasters in most coastal areas, which are home to half the world's population and its business and commerce.

The Impact of Climate Change on Ecosystems

The changing climate will affect all of Earth's ecosystems. An understanding of these effects is vital because we depend on natural and agricultural ecosystems for a host of goods and services (see Chapter 7). The particular temperature and moisture regimes in different parts of the world have generated various types of ecosystems, reflecting the adaptations of plants, animals, and microbes to the prevailing weather patterns of a region. Changes in climate trigger changes in plant and animal behavior, such as the timing of flowering, migration, and dormancy. As the climate changes, can ecosystems change with it in such a way that the crucial support they provide us is not interrupted?

Some of the projected outcomes of climate change include:

- Because of the warming climate, snow cover and sea ice will continue to decrease, possibly fully opening up the Arctic Ocean by mid-century. Glaciers and ice caps will continue to shrink, resulting in a rise in sea level.

- The sea level, currently rising at a rate of 3.2 mm per year, will continue to rise as a consequence of continued thermal expansion and snow and ice melt. The range projected by the IPCC for the end of the 21st century is 7 to 23 inches (19 to 59 cm); these projections did not include melting of the Greenland and Antarctic Ice Sheets. A half-meter rise in sea level will flood many coastal areas and make them much more prone to damage from storms, forcing people to abandon properties and migrate inland. A 1-meter rise would flood 17% of Bangladesh, creating tens of millions of refugees and destroying half of the country's rice lands.

- As the upper layers of the North Atlantic Ocean warm and become fresher, the thermohaline circulation is expected to slow down, but no models show it collapsing.

- Storm intensities are expected to increase, with higher wind speeds, more extreme wave heights, and more intense precipitation; there will be fewer but more intense hurricanes in the Atlantic Basin.

- Because the oceans will be absorbing so much CO_2, they will continue to acidify, threatening coral reefs (which will already be stressed by warming tropical seas) and shellfish everywhere.

- Crop yields will increase in some places and decrease in others; overall yields are expected to decrease.

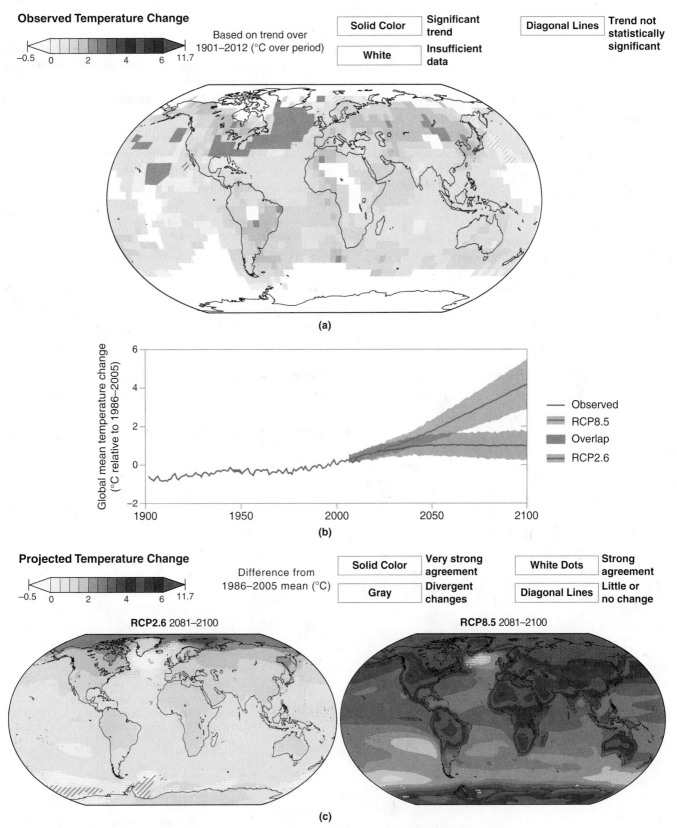

Figure 18–17 Observed and projected changes in annual average surface temperature. This figure informs understanding of climate-related risks in the WGII AR5. It illustrates temperature change observed to date and projected warming under continued high emissions and under ambitious mitigation. **(a)** Observed temperature change compared to historic record. **(b)** Two Representative Concentration Pathways, one high and one low (2.6 and 8.5 W/m²), lead to different global temperatures. **(c)** Projected global increases in temperature for Representative Concentration Pathways 2.6 and 8.5 W/m².

(*Source:* C. Field et al., eds. *Climate Change 2014: Impacts, Adaptation, and Vulnerability.* Contribution of Working Group II to the Fifth Assessment Report of the Intergovernmental Panel on Climate Change. Cambridge, England: Cambridge University Press, 2014, figure SPM. 4.)

- Outbreaks of cold air in the Northern Hemisphere will diminish by 50% or more. Growing seasons would be longer in middle to high latitudes, and temperatures would be more moderate in high-latitude countries such as Canada and Russia.

- Heat waves will become more frequent and longer lasting as the climate warms. Growing seasons will lengthen, and frost days will decrease in middle and high latitudes.

- Precipitation will decrease in already-dry regions and increase in already-wet regions; there will be an increase in extremes of daily precipitation. Areas of extreme drought will expand, affecting up to 30% of the world's land surface.

In general, global climate change will have a profound impact on Earth's ecosystems and individual species, the impacts accruing as the temperature warms over time (Figure 18–18). Outcomes will include damage to polar ecosystems, bleached and dying coral reefs, major loss of Amazon rain forests, widespread extinctions of species, and loss of ecosystem resilience.

Impact of Climate Change in the United States

The U.S. Global Change Research Program integrates the efforts of many federal agencies involved in the scientific effort to understand global climate change. One of the tasks of the Global Change Research Program has been to communicate the findings of the scientific community to policy makers and the broader public. The 2014 report, *Global Climate Change Impacts in the United States,* is the third National Climate Assessment and is available as an interactive document on the Internet.[11] It draws from 19 existing Global Change Research Program reports and other peer-reviewed scientific assessments, such as those from the IPCC. It provides a detailed look at the impacts of climate change on various regions of the United States and on various sectors, including transportation, agriculture, and human health. Highlights of the report include the following:

- Since record-keeping began, the average U.S. temperature has risen 0.7–1.1°C and will rise more; heat waves could become commonplace in the next 30 years.

- Precipitation increased an average of about 5% over the previous 50 years. Shifting patterns have generally made wet areas wetter, while dry areas have become drier. This is expected to continue.

- The Northeast will experience increased coastal flooding, extreme precipitation, and heat waves.

- The Northwest is experiencing changes in snow melt and stream flow.

- The Southeast will continue to struggle with heat waves and limited water as population growth also affects water use.

- The Southwest is experiencing the worst drought since 1900; projections suggest that droughts and wildfires could worsen across the region.

- In some parts of the Midwest and Great Plains, longer growing seasons and increased carbon dioxide are increasing crop growth. In other areas, increased evaporation and heat waves are stressing plants. As temperatures

---

[11]Thomas R. Karl, Jerry M. Melillo, and Thomas C. Peterson, eds., *Global Climate Change Impacts in the United States* (2014), http://nca2014.globalchange.gov/highlights/overview/climate-trends.

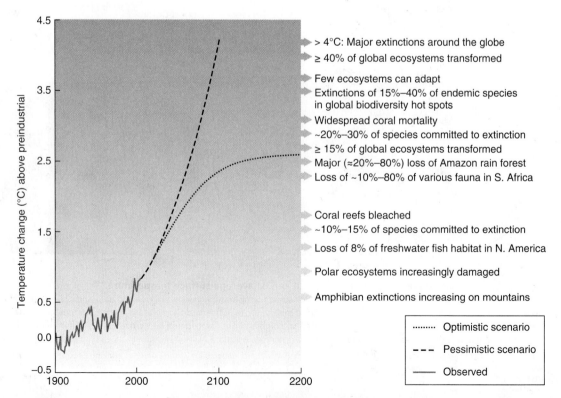

Figure 18–18 Impact of global warming on biological systems. As temperatures increase over time, the harm to biological systems will become greater. The optimistic scenario reflects emissions between the RCP 2.6 and 4.5 W/m² scenarios of the IPCC AR5. The pessimistic scenario is closest to the 8.5 W/m² scenario.

(*Source:* Kerr, Richard A. "How Urgent Is Climate Change?" *Science* 318, no. 5854 (November 2007). Copyright © 2007 by AAAS. Reprinted with permission.)

> 4°C: Major extinctions around the globe
≥ 40% of global ecosystems transformed

Few ecosystems can adapt
Extinctions of 15%–40% of endemic species in global biodiversity hot spots
Widespread coral mortality
~20%–30% of species committed to extinction
≥ 15% of global ecosystems transformed
Major (≈20%–80%) loss of Amazon rain forest
Loss of ~10%–80% of various fauna in S. Africa

Coral reefs bleached
~10%–15% of species committed to extinction
Loss of 8% of freshwater fish habitat in N. America
Polar ecosystems increasingly damaged
Amphibian extinctions increasing on mountains

Temperature change (°C) above preindustrial

.... Optimistic scenario
– – – Pessimistic scenario
——— Observed

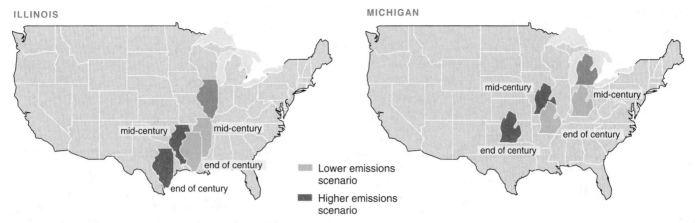

Figure 18–19 **Climate on the move in the Midwest.** Model projections of summer average temperature and precipitation changes in Illinois and Michigan are shown here for mid-century and end of century. Note that both states are projected to get considerably warmer and have less summer precipitation.

(*Source:* CSSP, *Global Climate Change Impacts in the United States: Unified Synthesis Product*, 2009.)

increase and precipitation decreases, the summer climate in Illinois and Michigan is predicted to shift southward and westward (Figure 18–19).

- During the past 50 years, sea levels along many U.S. coasts rose 5–13 centimeters (2–5 inches); coastal cities with some 40 million people are at risk because of the likely rise in sea level linked to 21st-century greenhouse gas emissions.

- Alaska will be affected by declining Arctic sea ice, thawing permafrost, and increased coastal erosion (Figure 18–20). Significantly, the impacts of climate change in Alaska are greater than those in any other region of the United States. Average temperatures are projected to rise 2.2–3.9°C by mid-century and could rise 4.4–7.2°C by the end of the century if global emissions continue on their present course.

Tipping Points?

Recall that risk is composed of both the likelihood that an event will occur and the severity of harm from that occurrence (Chapter 17). The AAAS report, *What We Know*,

Figure 18–20 **Regional impacts of climate change.** Alaska is already experiencing the effects of climate change, including the collapse of permafrost, rising sea levels, and increased damage from storms. Residents of Shishmaref, Alaska, are trying to move their village after storms have caused destruction, such as the collapse of this house.

described some rapid changes that would cause tremendous damage but have a low likelihood of occurring. The authors made the case that, just like a house fire, which you don't expect to occur but take actions to prevent and to insure yourself against, abrupt climate change deserves our attention. The main features of such abrupt change would be the collapse of the Greenland Ice Sheet, the slowing of the deep ocean conveyor belt, and the dramatic loss of Arctic sea ice.

Several of these events would trigger large-scale positive feedback loops. In the Arctic, up to 90% of the upper layer of permafrost would thaw, allowing vast areas of peat to decay, releasing methane and CO_2. Such warming could also release the methane that is now held in frozen hydrate crystals on the sea floor. The greenhouse gases released from both of these sources would cause the warming cycle to continue.

The idea that we could be close to a tipping point that would change the way large-scale cycles work is one of the reasons researchers have concluded that we have passed a planetary boundary on climate change (Chapter 2). This does not mean there is nothing we can do about it; it does mean that the solution has to involve burning less fossil fuel.

In conclusion, if nothing is done to reduce our emissions of greenhouse gases, computer modeling tells us that the 21st century will see climate changes that are dangerous, possibly catastrophic. Sea levels will keep rising, and there will be more frequent heat waves, changed weather patterns, and threats to natural ecosystems. In fact, these changes will go on for hundreds of years. Needless to say, this is not good news, and it has gotten the attention of the world. Efforts are under way to respond to these changes by reducing our greenhouse gas emissions and by finding ways to adapt to the effects of the coming climate changes. However, our civilization has no experience with a problem like this, and the way forward is still being debated.

CONCEPT CHECK

1. What is a climate commitment?

18.4 Controversy and Ethics

Today many people agree that we need a global response to climate change. But before we deal with these responses, we will consider two questions: Why is there still so much doubt about climate change? and What ethical concerns does climate change raise?

Why So Much Doubt?

The scientific consensus on global climate change is clear: Global warming is well under way, and the consequences of further climate change are potentially catastrophic. In the United States, however, public opinion polls reveal a high level of skepticism about global warming.[12] This attitude has puzzled scientists, who have seen the evidence for global warming growing stronger all the time.[13] Figure 18–21 illustrates the gap between the scientific consensus and public perception. Some of this discrepancy is caused by the fact that most people do not spend time examining the scientific evidence but instead form their opinions based on the views of organizations and media sources they trust. The high degree of public skepticism, then, can be traced to several factors:

- A campaign to undermine trust in climate science by organizations and corporations that stand to lose influence and profits if carbon emissions are curtailed.[14]

- A political divide in which some people connect the concept of climate change with political positions. This causes them to reject or accept climate science on the basis of unrelated politics.

- A general distrust of scientists, made worse by claims and counterclaims by those purporting to represent science

and by a desire for "balanced reporting" that gives voice to fringe speakers.

- Concern about the actions we may need to take in order to solve climate change problems. The fact that there is no single easy solution makes it harder for people to want to embrace this concern.

- The coincidence of an economic crisis in the United States that has focused the public's attention on economic issues and provided fuel for the campaigns determined to attack the public's trust in climate science and policy making.

If the stakes were not so high, one could pass off this skepticism as an aberration. However, given the high stakes, the denial of climate change lands us firmly in the domain of ethics.

Ethical Concerns

In a recent book, philosopher Stephen Gardiner refers to our flawed attempts to address climate change as an *ethical* failure—a "perfect moral storm."[15] Gardiner describes three converging factors as having deep moral significance: 1) Climate change is a *global* "storm," one where the wealthy nations are able to shape what is done about it, possibly to the detriment of the poorer nations. 2) Climate change is an *intergenerational* storm, where the present generation has the power to affect future generations, who have no voice in the matter. 3) Climate change is a *theoretical* storm, in that the threat of climate change is unprecedented and we lack adequate moral theories to give guidance for appropriate actions. The convergence of these three interacting "storms," says Gardiner, has led to our present lack of real progress in addressing the looming threat of global climate change.

Here are four ethical principles that speak to the issue:

The Precautionary Principle. As articulated by the 1992 Rio Declaration, the **precautionary principle** states, "In order to protect the environment, the precautionary approach shall be widely applied by States according to their capabilities.

---

[12]Pew Research Center, *More Say There Is Evidence of Global Warming* (2012), http://www.people-press.org/files/legacy-pdf/10-15-12%20Global%20Warming%20Release.pdf.

[13]John Cook et al., "Quantifying the Consensus on Anthropogenic Global Warming in the Scientific Literature," *Environmental Research Letters* 8, no. 2 (2014), doi:10.1088/1748-9326/8/2/024024.

[14]Naomi Oreskes and Erik M. Conway, *Merchants of Doubt* (New York: Bloomsbury Press, 2010).

[15]Stephen M. Gardiner, *A Perfect Moral Storm: The Ethical Tragedy of Climate Change* (New York: Oxford University Press, 2011).

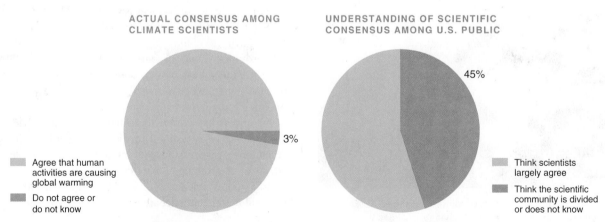

ACTUAL CONSENSUS AMONG CLIMATE SCIENTISTS

3%

- Agree that human activities are causing global warming
- Do not agree or do not know

UNDERSTANDING OF SCIENTIFIC CONSENSUS AMONG U.S. PUBLIC

45%

- Think scientists largely agree
- Think the scientific community is divided or does not know

Figure 18–21 A communication gap. Although 97% of climate scientists are convinced by evidence that Earth is warming and that this process is largely the result of human activities, in a 2012 poll, only about 45% of U.S. respondents thought this was the position of scientists.

Where there are threats of serious or irreversible damage, lack of full scientific certainty shall not be used as a reason for postponing cost-effective measures to prevent environmental degradation."[16] Employing this principle is like taking out insurance: We are taking measures to avoid a highly costly, but uncertain, situation (albeit one that is appearing quite likely). The risk of serious harmful consequences is real, the scientific consensus is strong, but the political response, as we will see, is still inadequate.

Polluter Pays Principle. According to the **polluter pays principle,** polluters should pay for the damage their pollution causes. This principle has long been part of environmental legislation. CO_2 is where it is because the developed countries (and now many developing countries) have burned so much fossil fuel since the beginning of the industrial era. It has become clear that CO_2 is a pollutant.

The Equity Principle. The people who produce the most greenhouse gas emissions are not the people who bear the bulk of the burden from climate change. This represents a fundamental inequity, generally divided by wealth and time. Currently, the richest 1 billion people produce 55% of CO_2 emissions, while the poorest 1 billion produce only 3%. In addition, our choices today limit the ability of future generations to make decisions about their standard of living. International equity and intergenerational equity are ethical principles that should compel the rich and privileged to care about those generations that follow—especially those in the poor countries who are even now experiencing the consequences of global climate change.

Care of the Most Vulnerable Principle. One ethical principle is provision for the most vulnerable, sometimes called a preferential option for the poor. If climate change is unchecked, every country will be affected, but the effects will be most disastrous for the world's poorest countries and people. Climate change will be much harder on the poor, because they are already struggling to meet their needs for shelter, food, employment, and health care. Thus climate change acts as a **risk multiplier** for other problems, making the effects of other risks greater. Here are some of the reasons:

- The poor, especially in the less developed countries, are especially vulnerable to disasters. They lack buffers to help them deal with crop failures, storms, and floods—the kinds of impacts that will increase in the changing climate.

- The poorer countries are less able to divert resources to adapt to the changing climate; they are already challenged to provide health care, education, and food.

- Attempts to mitigate climate change will have a greater impact on the poor and poor countries, especially if the cost of fuels rises.

- Rising sea levels and other effects of climate change will inevitably displace millions in the poorer countries; the flood of environmental refugees will challenge the ability of the international community to provide aid.

- The lack of resources, especially water and arable land, will lead to greater internal and international conflicts. The U.S. Department of Defense has identified climate change as a significant national security issue, since climate change may act as an accelerant of conflict.

Environmental Protection: A Human Right. The concept of treating environmental protection as a human right comes out of some of these ethical frameworks and is beginning to appear in international law. The 1998 *Aarhus Convention,* a treaty signed by the European Union and 46 other nations, gives rights to people to have access to information, participate in decision making, and have justice in issues of environmental quality. The treaty doesn't guarantee environmental quality, but it does say that if environmental laws are broken, people have a right to have redress. This is an important foundation for a just society.

CONCEPT CHECK What are two reasons that people have trouble believing climate change is real or needs attention? ☑

18.5 Responding to Climate Change

If we are convinced that climate change is a significant global problem that demands our attention, there are a number of actions that can be taken by individuals, nations, and the international community. We can take steps to reduce the emissions of greenhouse gases; this is the *mitigation* response. Mitigation efforts will reduce the magnitude of future climate change. However, since some climate change is already underway, we will also need to take steps to adjust to the coming climate change; this is the *adaptation* response.

Mitigation

The goal of mitigation is to reduce the rate at which greenhouse gases are added to the atmosphere and, eventually, to bring about a sustainable balance. Table 18–3, on the following page, presents a summary of the leading mitigation tools, many of which are already in place in the United States and globally. The majority of the options address the energy sector, where most of the anthropogenic greenhouse gases originate. What is needed, of course, is to "decarbonize" the energy economy, so that energy needs are met without fossil fuels. Every provision that promotes greater energy efficiency and the use of renewable energy sources is going to reduce greenhouse gas emissions. Recent legislation in the United States has made progress in energy efficiency and renewable energy (Chapters 14 and 16).

One way to reduce emissions is to stop burning fossil fuels at their source. Oil wells often have pockets of natural gas on top of them. This gas is routinely released and burned, a process called flaring.[17] Every year, routine gas

---

[16]UNEP. Rio Declaration on Environment and Development, Principle 15. 1992.

[17]World Bank, *Countries and Oil Companies Agree to End Routine Gas Flaring,* April 17, 2015, http://www.worldbank.org/en/news/press-release/2015/04/17/countries-and-oil-companies-agree-to-end-routine-gas-flaring.

Table 18–3   Options for Mitigation of Greenhouse Gas Emissions

Regulations

Cap-and-trade	Place a cap on greenhouse gas emissions; for CO_2, this can be accomplished by limiting the use of fossil fuels in industry and transportation. Rights to emit greenhouse gases are allocated to companies and are tradable in a market-based system. This has worked well in the EU, where countries are allocated permits and can trade with other countries. It is anticipated that the cap will be reduced gradually until a stabilized atmosphere is achieved. This option generates potentially large revenues, which can be used to alleviate some of the problems created by the program.
Subsidies	Remove fossil fuel subsidies such as depletion allowances, tax relief to consumers, and support for oil and gas exploration.
Mileage standards	Reduce the amount of fuels used in transportation by raising mileage standards, encouraging carpooling, and stimulating mass transit in urban areas.
Other efficiency standards	Levy a tax on all fuels according to the amount of CO_2 that they produce when consumed. Such a tax, proponents believe, would provide both incentives to use renewable sources, which would not be taxed, and disincentives to consume fossil fuels. This puts the polluter-pays principle into practice.
Carbon tax	Levy a tax on all fuels according to the amount of CO_2 that they produce when consumed. Such a tax could provide both incentives to use renewable sources, which would not be taxed, and disincentives to consume fossil fuels. This puts the polluter-pays principle into practice.

Alternative Energy

Renewable energy	Invest in and deploy an increasing percentage of energy in the form of renewable-energy technologies, including wind power, solar collectors, solar thermal energy, photovoltaics, biofuels, hydrogen-powered vehicles, and geothermal energy.
Nuclear power	Encourage the development of nuclear power, first resolving issues concerning cost-effectiveness, reliability, spent fuel, and high-level waste.

Carbon Capture

Sequestration	Sequester CO_2 emitted from burning fossil fuels at power plants by capturing it at the site of emission and pumping it into deep, porous underground formations.
Reforestation	Stop the loss of tropical forests and encourage the planting of trees and other vegetation over vast areas now suffering from deforestation. Trees and other vegetation provide a sink for carbon that removes CO_2 from the atmosphere.

Stabilizing Growth

Population growth	Reducing population growth would make a substantial contribution to reducing greenhouse gas emissions, especially in the economically robust developing countries.

flaring burns about 140 billion cubic meters of natural gas, emitting more than 300 million tons of CO_2 into the atmosphere, equivalent to the emissions of 77 million cars. The World Bank *Zero Routine Flaring by 2030* initiative is an attempt to eliminate the waste of natural gas.

Another way to mitigate emissions is to stop deforestation and the burning of tropical forests. An international program called Reducing Emissions from Deforestation and Forest Degradation (REDD) rewards developing countries for maintaining their forests.

Costs and Benefits. There are economic costs and benefits to mitigation efforts. The benefits come in the form of new industries and the jobs they create, savings from efficiency measures, and the ecosystem benefits from forests that are preserved and regrown. The costs include the costs of the infrastructure needed to increase efficiencies of energy use and to switch to lower-carbon technologies for power, heat, and transportation;

there are also costs involved in slowing or stopping deforestation. As calculated by the Stern Review, the net costs for stabilization at 550 ppm CO_2 will amount to around 2% of global GDP by 2050. However, the Stern Review also calculated the costs of inaction to be from 5% to 20% of GDP.

Contract and Converge. The phrase *contract and converge* describes a process by which the world will move from the current high emissions and high level of energy inequality to a future with lower emissions and greater energy equality. Our total use of fossil fuels needs to *contract*, so we leave much of our fossil fuels underground. We also need to *converge*— heavy emitters need to cut their fossil fuel use more and the poorest need to be allowed to increase theirs. This concept speaks to concerns about fairness: The developing countries' first priority should be to alleviate poverty, whereas the developed countries should be focusing on lowering environmental impact.

Adaptation

While mitigation strategies reduce the effects of climate change, adaptation strategies aim to reduce the risk associated with the various hazards likely to be caused by climate change. In adaptation, we anticipate harm to natural and human systems and plan responses to lessen the vulnerability of people, their property, and the biosphere to the coming changes. Thus, climate **adaptation** involves making adjustments in anticipation of changes that are likely to occur in the next decades, such as increasing drought and water shortages, increasingly severe storms, and rising sea level. For example, a city might recognize the risks associated with rising sea levels and rebuild its seaport to prepare for the changes.

Preparing for climate change will involve a wide range of actions, from planting mangroves in order to protect coastal zones to investing in infrastructure that can withstand rising temperatures. There are also social adaptations, such as increasing the availability of insurance for the severe storms and droughts expected to accompany climate change. Table 18–4 presents a number of adaptations appropriate for developed and developing countries.

International Efforts. One example of effective adaptation has already occurred in Ethiopia's Tigray province, a hot, drought-prone region once known for its frequent famines. Today, however, the desert is greener and food more abundant because of widespread community efforts to build *check dams,* which are small rock dams that terrace hillsides and allow water to seep into the ground, refilling aquifers. The numerous check dams reduce the region's vulnerability to the predicted impacts of climate change, such as rising temperatures and increasing drought (Figure 18–22).

Figure 18–22 Adapting to climate change in dry regions. These Ethiopian men are building small check dams, which will help to slow the flow of water, stop erosion, and increase the supply of groundwater.

Adaptations can increase the resilience of societies and decrease the negative outcomes of the changing climate. Even so, adaptation is a difficult process because it is aimed at a moving target—and one that is uncertain at best. The Green Climate Fund is one of a number of international funds that have been established to help the poorest countries address the need for climate adaptation.

Table 18–4 Adaptation Strategies

All Countries	
Agriculture	Farmers shift to climate-resistant crops or crops more appropriate to the new climate regime. Irrigation is expanded to areas once fed only by rain.
Structural concerns	Societies invest to reduce the costs and disruption from extreme weather events such as storms, floods, and droughts; investments might include seawalls, new reservoirs, and the re-vegetation of coastal wetlands.
Emergency preparedness	Societies invest in early-warning systems, high-quality climate information, intervention when needed; investments might include heat-wave warning systems and shelters for those displaced by severe weather events.
Reducing risks	Societies establish a financial safety net to help their poorest members, providing insurance that accurately reflects new risks to discourage poor choices.
Developing Countries	
Promoting development	Investments are made in health, education, economic opportunity, and food security to lower people's vulnerability to climate impacts and help them build adaptive capacity.
Controlling diseases	Malaria, tuberculosis, and emerging diseases are brought under control so that a changing climate does not increase their threat.
Enhancing economic progress	Societies support economic growth by promoting literacy, electronic communication, road infrastructure, and the diversification of economic activity.

Co-Benefits. Some climate mitigation and adaptation strategies solve other problems as well, a phenomenon called *co-benefits*. For example, reducing the use of inefficient small cook stoves, coal-fired power plants, and gas flaring at oil wells will also reduce the brown cloud pollution that harms crops and human health (Chapters 14, 17, 19). It is reasonable to start by pursuing efficiencies and solutions with co-benefits. But we may also have to take more dramatic steps, such as altering basic climate functions.

Geoengineering

A number of people have suggested that the problems associated with climate change could be addressed by altering Earth's processes. Such strategies are called **geoengineering** or, more narrowly, *climate intervention*. Most suggested geoengineering activities would either remove CO_2 from the atmosphere or modify Earth's albedo.

Sequestering Carbon Dioxide. Some geoengineering strategies involve removing CO_2 from the atmosphere and *sequestering* it, or storing it where it cannot easily return to the atmosphere. For example, scrubbers on power plants can sequester CO_2 released by burning fossil fuels before it is released into the atmosphere (Figure 18–23). One proposed method of sequestration would be to erect devices that would directly absorb CO_2 from the atmosphere so that it could be stored in deep reservoirs.

Changes in land management to enhance the value of forests and agricultural lands as carbon sinks would also sequester carbon. Charcoal or *biochar*—partially burned wood or other organic material—decays very slowly. After discovering that indigenous Amazonian peoples use biochar to enrich the soil in their fields, some researchers have suggested burying biochar as a way of sequestering carbon in soil. The newest research, however, suggests that biochar may be less effective than first hoped.

Because the natural weathering of rocks binds carbon dioxide, enhancing the rate at which rocks weather could be

another way to sequester carbon. Combining direct removal from air with biomass burning results in another strategy: bioenergy with carbon capture and sequestration. For example, people suggest growing algae in the ocean, burning it for energy, and capturing the carbon for underground storage before it escapes into the atmosphere. Another idea is to fertilize vast areas of the ocean with iron, thereby stimulating phytoplankton growth, which would remove CO_2 and sink to the depths.

Modifying Earth's Albedo. Other geoengineering strategies would reduce the amount of solar radiation reaching Earth's surface or cause the Earth's surface to reflect more light; both of these strategies would alter Earth's albedo, or reflectivity. Proposed approaches include injecting sulfate particles into the stratosphere, where they would act as reflective clouds; using satellites to release particles that would deflect the Sun's radiation; and increasing the reflectivity of marine clouds. Such approaches would be extremely expensive and would be likely to have unintended consequences, such as the alteration of weather patterns or the destruction of ecosystems. For example, injecting sulfate particles into the atmosphere would lower temperatures in the short term, but not in the long term. Sulfur oxides would also cause acid rain; in the long term, they might even *increase* global warming.

In 2015, the National Research Council released a report evaluating the promise of such climate interventions.[18] It concluded that some of the ideas are serious proposals, while others are clearly unworkable. According to that report, some strategies for sequestering carbon dioxide could be effective, but they would require international cooperation and are likely to be costly. In contrast, strategies for modifying Earth's albedo could be carried out by nations or even individual companies, but they would involve larger risks and thus present an array of ethical concerns. The report concluded, "climate intervention is no substitute for reductions in carbon dioxide emissions and adaptation efforts aimed at reducing the negative consequences of climate change."

International Efforts

Because climate change is global, any effective response to it must involve international cooperation. Mitigation in particular cannot be done by individual countries alone. In addition, funding for climate adaptation in the poorest countries will have to come from high-income countries. The international community is trying to reach agreement on how to proceed, an extremely difficult process. Consequently, we have a long history of international meetings that have led to a number of critical agreements. National representatives meet annually or more often to plan, negotiate, and agree to treaties about climate change. Some of these meetings are listed in Table 18–5. As the table shows, the process of international decision making does not always result in binding agreements, but that is the goal.

Figure 18–23 Carbon sequestration. These workers are using equipment that is capturing carbon dioxide from emissions at a power plant in Germany.

[18]NAS, *Climat Intervention: Carbon Dioxide Removal and Reliable Sequestration* (Washington, DC: National Academies Press, 2015); NAS, *Climate Intervention: Reflecting Sunlight to Cool Earth* (Washington, DC: National Academies Press, 2015).

Table 18–5   Selected International Meetings on Climate Change

Year	Location	Main Outcomes
1992	Rio de Janeiro	**The Framework Convention on Climate Change (FCCC)**, in which industrialized countries agreed to reduce emissions of greenhouse gases to 1990 levels by 2000. Although the goals were not achieved, the FCC serves as the foundation of international agreements on climate change.
1997	Kyoto, Japan	**Kyoto Protocol**, in which 38 nations agreed to reduce emissions of six greenhouse gases to 5.2% below 1990 levels by 2012. Developing countries not included.
2007	Bali, Indonesia	**Bali Roadmap**, for negotiation and planning. Agreed deep cuts in global emissions will be required.
2009	Copenhagen, Denmark	**Copenhagen Accord**, agreed to by 187 countries, pledges to act to limit global temperature increases to 2°C above pre-industrial levels. Commitment to set countrywide emissions targets for 2020. No binding agreement.
2010	Cancún, Mexico	Approved mitigation pledges. Established a **Green Climate Fund** of $30 billion to help the developing countries. No binding agreements on emissions.
2011	Durban, South Africa	**Durban Platform**, an agreement to begin negotiations that would lead to a legally binding agreement by 2015, to take effect by 2020.
2012	Doha, Qatar	Kyoto Protocol extended until 2020. Developing nations lobbied for the wealthier nations to set up a fund to pay for loss and damage caused by climate change.
2013	Warsaw, Poland	Meeting went poorly. The Green Climate Fund had not been funded as promised. Australia did not send any high-level representatives. Representatives of 132 poor countries walked out of the talks. Discussed the **Warsaw Mechanism**—a way of getting expertise and aid to nations suffering from sea level rise, desertification, droughts, and floods.
2014	Lima, Peru	Preparing for Paris, 2015.

Framework Convention on Climate Change. One of the five documents signed by heads of state at the UN Conference on Environment and Development's Earth Summit in Rio de Janeiro in 1992 was the **Framework Convention on Climate Change (FCCC)**. This framework serves as the foundation of international agreement about climate change. Its stated objective is to *stabilize the greenhouse gas content of the atmosphere at levels and on a time scale that would prevent dangerous anthropogenic interference with the climate system.*

In the FCCC, countries agreed to the goal of stabilizing greenhouse gas levels in the atmosphere, with all industrialized nations reducing their emissions to 1990 levels by the year 2000. Countries were to achieve the goal by voluntary means. Five years later, all the developed countries except those of the European Union had *increased* their greenhouse gas emissions by 7% to 9%, while the developing countries had increased theirs by 25%. It was obvious that the voluntary approach was failing.

Agreements to Cut Emissions. Prompted by a coalition of island nations threatened by rising sea levels, another international meeting was convened in Kyoto, Japan, in December 1997, to craft a binding agreement on reducing greenhouse gas emissions. At the meeting, 38 industrial and former Eastern Bloc nations agreed to reduce emissions of six greenhouse gases to 5.2% below 1990 levels, to be achieved by 2012; this agreement was called the **Kyoto Protocol**. However, the developing countries refused to agree to *any* reductions, arguing that the developed countries had created the problem and that it was only fair that the developing countries continue on their path to development as the developed countries had, energized by fossil fuels. This stance was in accord with the FCCC principle of "common but differentiated responsibilities." Under this principle, each nation must address climate change, but its priorities and efforts may differ according to national circumstances.

Even the targeted reductions of the Kyoto Protocol would have been insufficient to stabilize atmospheric concentrations of greenhouse gases. The IPCC calculates that it would take immediate reductions in emissions of at least 60% worldwide to stabilize greenhouse gas concentrations at today's levels. Since there is no chance that reductions of this size will occur, the concentration of greenhouse gases in the atmosphere will continue to rise. The major weakness of the Kyoto Protocol is that the world's largest greenhouse gas emitters—India, China, and the United States—are not participating. The United States withdrew from the treaty in 2001. Recently Canada, Japan, and Russia also withdrew, with the result that Kyoto now only covers 13% of global emissions. Nevertheless, Kyoto was an important first step toward binding agreements to stabilize the atmosphere.

In 2000 in Copenhagen, Denmark, 187 countries committed to setting emissions targets in the **Copenhagen Accord.** In 2011, at Durban, South Africa, delegates reached an agreement to begin negotiations that would lead to a legally binding agreement by 2015, to take effect by 2020. The agreement, called the Durban Platform, aims at ensuring that emissions reductions will meet the goal of keeping global temperature rise below 2°C. In 2012 in Doha, Qatar, the Kyoto Protocol was extended until 2020.

The Pledge Gap. At the Durban meeting, the United Nations Environment Program released a report that showed that the pledges of countries under the Copenhagen Accord were falling short of the target; thus, there is a *climate emissions gap* to be addressed, a gap of between 6 and 11 GtCO$_2$, depending on how well the pledges are kept (Figure 18–24).[19] The report states that to stay within the 2°C limit, emissions would have to peak before 2020, and that won't happen unless new emissions targets are established. It is possible to close the gap, but current pledges are insufficient.

Funding Low-Income Countries. As already described, low-income countries lack the resources to mitigate or adapt to climate change. Furthermore, they already suffer losses from the current effects of climate change. Consequently, developing countries have repeatedly asked for climate change–related aid. In 2010 in Cancún, Mexico, nations agreed to a Green Climate Fund of $30 billion to help the developing countries, and in 2012 in Doha, Qatar, developing nations lobbied for funds to help them pay for loss and damage due to climate change. However, by 2013, in a meeting in Bali, Indonesia, the developing nations were not being given the promised climate aid and 132 nations walked out of negotiations. Funding for climate change–related costs will continue to be a part of aid

for developing counties and an important part of meeting the Sustainable Development Goals.

Bilateral Meetings. Some discussions about climate change involve just a few nations. In 2014, the United States and China held separate climate talks, announcing a plan for both countries to curb emissions by 2030. Later in 2014, the United States and India had top-level discussions about climate change. Because India has a much higher percentage of people in poverty and lower per capita greenhouse gas emissions than the United States and even China, Indian leaders were unwilling to promise cuts in emissions.

National Legislation. Legislative action at the national level is critical to the reduction of emissions. A research team from the United Kingdom recently analyzed legislation in 66 countries that emit 90% of greenhouse gas emissions, including 155 laws related to climate change that were in effect in those countries.[20] The study was encouraging, finding that countries are using many approaches to reduce greenhouse gas emissions and prepare for the impacts of climate change. This brings us to the question: What has the United States been doing about global climate change?

U.S. Policy on Climate Change

Climate change policy in the United States comprises scientific research, national legislation, regulatory action, and state and local action. One major policy move was made in 2001 when President George W. Bush announced that the United States would withdraw from the Kyoto Protocol agreement. In June 2013, President Obama announced a Climate Action Plan to better prepare America for the effects of climate change.

Scientific Research. The *Global Change Research Act of 1990* established a policy of support for research on climate change; the act mandates an interagency approach, where all of the agencies and departments that might be involved in climate change research are to coordinate their activities and programs as part of the official Global Change Research Program (GCRP). Lead agencies in this effort have been the NOAA, NASA, the U.S. Geological Survey, and the EPA. One example of the GCRP's effort has been the 20 reports of the Climate Change Science Program. Hundreds of U.S. scientists have been in the lead in conducting research on climate change and publishing their work, especially as they have participated in the IPCC assessments.

National Legislation. While 66 countries have over 500 greenhouse gas emissions laws, the United States lags in climate-change legislation. In 2009, the House passed the *American Clean Energy and Security Act,* which would have established an emissions reduction and trading plan called cap-and-trade (see Table 18–4). Unfortunately, the bill died in the Senate. However, the *American Recovery and Reinvestment Act* has invested more than $43 billion in the support of renewable energy and energy efficiency programs,

---

[19]United Nations Environment Program, *Bridging the Emissions Gap: A UNEP Synthesis Report* (November 2011), accessed June 14, 2015, www.unep.org/publications/ebooks/bridgingemissionsgap/ .

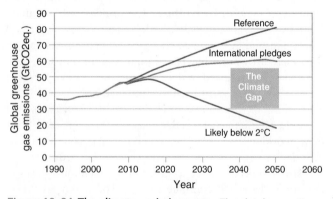

Figure 18–24 The climate emissions gap. The plot shows estimates of global greenhouse gas emissions in gigatons of CO$_2$ equivalents (other gases equated to CO$_2$). The blue reference line shows emissions growth in the absence of the targets and pledges, the orange line shows the pledges made by countries following the Copenhagen Accord, and the brown line shows the emissions pathway needed to keep the growth in global temperature below 2°C.

(*Source:* Climate Action Tracker, December 11, 2011. Copyright © 2009 by Ecofys and Climate Analytics.)

---

[20]Terry Townshend et al., "Legislating Climate Change on a National Level," *Environment* 53 (September/October 2011): 5–17.

which will certainly have an impact in lowering greenhouse gas emissions. Other legislative actions include laws that established subsidies and rebates for renewable energy sources. Disagreement within the U.S. Congress has made it difficult to accomplish more through legislation.

Regulatory Action. Under the Copenhagen Accord, the White House has committed the United States to a reduction of greenhouse gas emissions by 17% below 2005 levels, to be achieved by 2020. Major tools to accomplish this include all of the policies to promote renewable energy and energy conservation, especially the regulations that raise the corporate average fuel economy standards (Chapter 16). Most important, however, is the role of the EPA. In 2009, taking its cue from a 2007 Supreme Court ruling, the EPA determined that six key greenhouse gases pose a danger to public health and welfare, thus giving the agency a green light to regulate CO_2 and other greenhouse gases. This move has infuriated many industry groups and their political allies, but their efforts to block the EPA from this role have so far failed.

The EPA began its regulatory efforts with the *New Source Performance Standards*, which regulate the emissions of greenhouse gases from new fossil fuel–burning power plants (starting in 2013) and from existing power plants (starting in 2014). This permitting rule requires all new power plants to hold emissions to 1,000 pounds of CO_2 per megawatt hour, something that natural gas plants can accomplish but coal plants cannot easily do without new technology.

State and Local Action. Individual states are also taking action to address global climate change. Many states have set quantitative goals for reducing greenhouse gas emissions. Some states are adopting renewable portfolio standards and mandating the reporting of greenhouse gas emissions (see Chapter 16). Several regions have established emissions cap-and-trade programs similar to the Clean Air Act program that was successful in bringing down sulfur dioxide emissions. In addition, more than 1,000 mayors have pledged to move their cities to make greenhouse gas reductions under the Climate Protection Agreement.

In conclusion, global climate change is perhaps the greatest challenge facing human civilization in the 21st century. We all need to put our effort into finding solutions to the problems it presents.

CONCEPT CHECK Mitigation, adaptation, and bioengineering are three ways of addressing climate change. Give an example of each. ✓

REVISITING THE THEMES

SOUND SCIENCE

Our knowledge of climate change depends on the solid work of scientists from around the world. The IPCC, with the thousands of scientists involved, has performed a masterful assessment of the scientific basis of climate change. In the United States, the Global Change Research Program has produced a portfolio of excellent analyses of climate change and its many dimensions. Climate science highlights the changing nature of science, as we critique findings and solve problems in measurement and calculations. It also depends on modeling, a critical tool for projecting likely outcomes. Scientists are now looking at the technical possibilities of mitigation, adaptation, and geoengineering efforts. The best science suggests we cannot rely on geoengineering to replace climate change mitigation.

SUSTAINABILITY

Earth is in the midst of an unsustainable increase in atmospheric greenhouse gas levels, the result of our intense use of fossil fuels and deforestation. We have crossed a planetary boundary and need to reverse our course. Projections of business-as-usual fossil fuel use point to global consequences that are disastrous, but not inevitable. A sustainable pathway is still open to us, and it involves a combination of steps we can take to mitigate the emissions and bring the atmospheric concentration of greenhouse gases to a stable or even declining level; to achieve these lowered rates we may use geoengineering. We also need to adapt to changes that are likely to occur. Each time we have decisions to make, we can choose options that maintain resilience.

STEWARDSHIP

The stewardship of Earth's natural processes is not something we can ignore—we must face this "perfect moral storm." Several principles have been cited as arguments for effective action to prevent dangerous climate change: the precautionary principle, the polluter-pays principle, and the equity principle. If we care about our neighbors and our descendants, we will take action both to mitigate the production of greenhouse gases and to enable the poorer countries to adapt to unavoidable climate changes, even if taking such action is costly.

REVIEW QUESTIONS

1. What are the important characteristics of the troposphere? Of the stratosphere?

2. What are the main components of our weather, and how is weather related to climate?

3. How has global mean temperature changed since the mid-1800s?

4. How does the modern global climate compare with that of 400,000 and 800,000 years ago? How do we know?

5. What is the Meridional Overturning Circulation? How does it work? What are the impacts of the *El Niño/La Niña* Southern Oscillation (ENSO)?

6. What is radiative forcing? Describe some warming and cooling forcing agents.

7. What evidence of a warming Earth do global land and ocean temperatures provide?

8. Which of the greenhouse gases are the most significant contributors to global warming? How do they work?

9. Describe the data for atmospheric CO_2. What are significant sources and sinks for atmospheric CO_2? Why is CO_2 called "the atmosphere's control knob"?

10. How are computer models employed in climate change research? Describe several scenarios for 21st-century climate change.

11. What are the major IPCC projections for climate changes in the 21st century?

12. What does the IPCC 5th Assessment report say about sea level rise by 2100?

13. What is a representative concentration pathway? How does the +2.6 compare to the +8.5?

14. What are four effects of climate change that we are already seeing in the United States?

15. Give three ethical principles that can be brought to bear on "the perfect moral storm."

16. What are three mitigation steps that could be taken to stabilize the greenhouse gas content of the atmosphere?

17. What are three adaptation steps that could be taken to adjust to inevitable changes in climate?

18. What is geoengineering? Give two examples.

19. What is the Kyoto Protocol? What happened in Copenhagen? In Bali? What is the climate emissions gap?

20. What national efforts is the United States making to address climate change?

THINKING ENVIRONMENTALLY

1. What steps could you take to lower your climate impact? Check the Union of Concerned Scientists Web site for suggestions.

2. Look at the AAAS resource *What We Know* and the Third National Climate Assessment on the Web, and find out what climate impacts are expected for your region of the country.

3. What aspects of climate change are having the greatest international impacts now? What recent developments most increase the urgency of action in the international arena?

4. Trace the history of the international conferences and their outcomes. What are the major roadblocks to reaching an international agreement to control greenhouse gas emissions?

5. Determine which countries signed the last international climate agreement and compare their CO_2 emission amounts to the countries who did not sign it. What are the main areas of contention that are blocking any binding agreements?

MAKING A DIFFERENCE

Anything you can do to cut energy use will cut your CO_2 output. In general, cutting CO_2 will also cut your bills. Listed below are just a few examples.

1. Buy items with the smallest amounts of packaging. The more packaging products have, the more space they take up in the distribution trucks, and the more CO_2 emissions they create.

2. Washing your clothes in cold water instead of hot will keep 200 pounds of carbon dioxide from entering the atmosphere a year.

3. The worst form of transportation for the climate is driving alone in a car. Use public transportation when possible, bike, walk, and carpool, or buy a hybrid—there are many options! When deciding on a car, make fuel economy a high priority. Driving 25 fewer miles a week will earn you 1,500 fewer pounds of CO_2 per year.

4. Become active in asking your elected officials to address climate change.

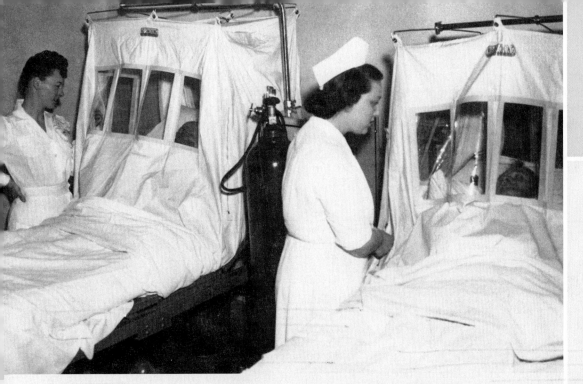

Atmospheric Pollution

On Tuesday morning, October 26, 1948, the people of Donora, Pennsylvania (population 13,000), awoke to a dense fog. Donora lies in an industrialized valley along the Monongahela River, south of Pittsburgh. On the outskirts of town, a sooty sign read, "DONORA: NEXT TO YOURS, THE BEST TOWN IN THE U.S.A." The town had a large steel mill that used high-sulfur coal and a zinc-reduction plant that roasted ores laden with sulfur. At the time, most homes in the area were heated with coal. At first, the fog did not seem unusual: Most of Donora's fogs lifted by noon, as the Sun warmed the upper atmosphere and then the land. This one, however, didn't lift for five days.

Sickness and Death. Through Wednesday and Thursday, the air began to smell of sulfur dioxide—an acrid, penetrating odor. By Friday morning, the town's physicians began to get calls from people in trouble. At first, the calls were from elderly citizens and those with asthmatic conditions. They were having difficulty breathing. The calls continued into Friday afternoon. People young and old were complaining of stomach pain, headaches, nausea, and choking. Work at the mills went on, however. The first deaths occurred on Saturday morning. By 10 AM, one mortician had nine bodies; two other morticians had one each. On Sunday morning, the mills were shut down. Even so, the owners were certain that their plants had nothing to do with the trouble. Mercifully, the rain came on Sunday and cleared the air—but not before more than 6,000 townspeople were stricken and 20 elderly people had died. During the next month, 50 more people died.

Memorial. Eventually it was determined that the deaths had been caused by a combination of polluting gases and particles, a thermal inversion in the lower atmosphere, and a stagnant weather system that, together, brought home the deadly potential of air pollution. Today, a historical marker in the town and a small storefront museum commemorate the Donora Smog of 1948. The lasting legacy of the smog and of those who suffered sickness and death are the state and national laws to control air pollution. Donora was a landmark event.

Chapter 19

Learning Objectives

19.1 Air-Pollution Essentials: Describe normal atmospheric cleansing and the formation of industrial and photochemical smog. Identify global trends in air pollution.

19.2 Major Air Pollutants and Their Sources: Identify the variety and sources of the major air pollutants, and classify them as primary or secondary pollutants.

19.3 Impacts of Air Pollutants: Explain how air pollutants impact human health and cause damage to agricultural crops and natural ecosystems.

19.4 Bringing Air Pollution Under Control: Compare trends in air pollution in the United States and the EU with those in developing nations. Explain how U.S. laws such as the Clean Air Act have helped reduce air pollution; describe the role of the Environmental Protection Agency in controlling air pollution.

19.5 Destruction of the Ozone Layer: Assess how chlorofluorocarbons and other gases have been implicated in the destruction of stratospheric ozone, and review the steps taken to bring back the ozone layer.

▲ Nurses treat victims of the lethal smog in Donora, Pennsylvania, in 1948.

During the last half century, air pollution in the United States and some other regions has decreased, while in other areas it has become considerably worse. A 2012 report by the World Health Organization attributed 4.3 million deaths per year to air pollution. Many of these deaths were the result of indoor air pollution, which is discussed elsewhere (Chapter 17). Earlier, we looked at the structure and function of the atmosphere and its links to weather and climate, and examined the effects of greenhouse gases on climate change (Chapter 18). In this chapter, we will consider the impacts of a host of other air pollutants on human health and the environment.

19.1 Air-Pollution Essentials

For a long time, we have known that the atmosphere contains numerous gases. The major constituents of the atmosphere are nitrogen (N_2), at a level of 78.08%; oxygen (O_2), at 20.95%; water vapor, ranging from 0% to 4%; argon (Ar), at 0.93%; and carbon dioxide (CO_2), at 0.04%. Smaller amounts of at least 40 *trace gases* are normally present as well, including ozone, helium, hydrogen, methane, nitrogen oxides, sulfur dioxide, and neon. In addition, the atmosphere contains *aerosols*: microscopic liquid and solid particles such as dust, carbon particles, pollen, sea salts, and microorganisms, originating primarily from land and water surfaces and carried up into the atmosphere.

Outdoor air is a valuable and essential resource, providing gases that sustain life and maintain warmth (Chapter 18). It also shields the Earth from harmful radiation. With the advent of the Industrial Revolution, the mixture of gases and particles in our atmosphere began to change, and we learned the hard way about air pollution.

Pollutants and Atmospheric Cleansing

Air pollutants are substances in the atmosphere—certain gases and aerosols—that have harmful effects. Three factors determine the level of air pollution:

- the amount of pollutants entering the air,
- the amount of space into which the pollutants are dispersed, and
- the mechanisms that remove pollutants from the air.

The lower atmosphere, called the *troposphere* (Figure 19–1), is the site and source of our weather and contains almost all of the water vapor and clouds of the atmosphere (Chapter 18). Pollutants that enter the troposphere are usually removed within hours or a few days unless they are carried up into the

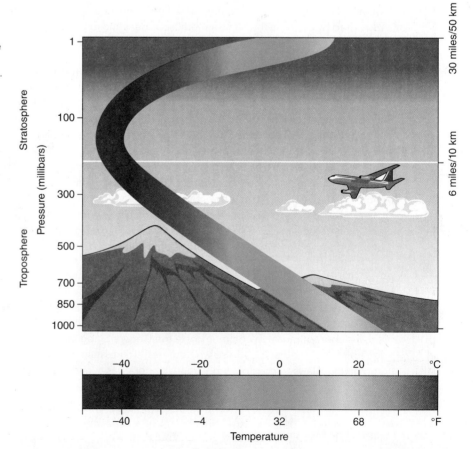

Figure 19–1 Layers of the atmosphere closest to Earth's surface. The troposphere extends from the surface to roughly 6 miles above the surface, and the stratosphere is above that. The colored band shows the average temperature of the atmosphere at different altitudes. In the troposphere, temperatures generally decrease with height, while in the stratosphere temperatures increase with height.

(*Source:* CCSP 1.1. *Temperature Trends in the Lower Atmosphere: Steps for Understanding and Reconciling Differences.* Thomas R. Karl, Susan J. Hassol, Christopher D. Miller, and William L. Murray, Eds., 2006. A Report by the Climate Change Science Program and the Subcommittee on Global Change Research, Washington, DC.)

upper troposphere by towering cumulus clouds. If that happens, they can be carried far and persist for many days before they, too, are removed by cleansing agents.

Some pollutants are resistant to most of the cleansing mechanisms and can be carried up into the next higher layer of the atmosphere, the *stratosphere*. The most notorious of these are the ozone-depleting chemicals (chlorine and bromine compounds), which we will consider in Section 19.5.

Atmospheric Cleansing. Environmental scientists distinguish between natural and anthropogenic (human-caused) air pollutants. For millions of years, volcanoes, fires, and dust storms have sent smoke, gases, and particles into the atmosphere. Trees and other plants also emit volatile organic compounds into the air around them. However, there are mechanisms in the biosphere that remove, assimilate, and recycle these natural pollutants. First, a naturally occurring compound, the **hydroxyl radical (OH)**, oxidizes many gaseous pollutants into products that are harmless or that can be brought down to the ground or water by precipitation (Figure 19–2). Sea salts—picked up from sea spray as air masses move over the oceans—are a second cleansing agent. These salts, now aerosols, act as excellent nuclei for the formation of raindrops. The rain then brings down many particulate pollutants (other aerosols) from the atmosphere to the ocean, cleansing pollutant-laden air. Third, sunlight breaks organic molecules apart. These three processes hold natural pollutants below toxic levels (except in the immediate area of a source, such as around an erupting volcano).

Many of the pollutants oxidized by the hydroxyl radical are of concern because human activities have raised their concentrations far above normal levels. Recent studies have shown that the hydroxyl radical plays a key role in the removal of anthropogenic pollutants from the atmosphere.

Highly reactive hydrocarbons are oxidized within an hour of their appearance in the atmosphere, and nitrogen oxides (NO_x) are converted to nitric acid (HNO_3) within a day. It takes months, however, for carbon monoxide (CO) to be oxidized by hydroxyl and years in the case of methane (CH_4). As Figure 19–2 shows, the photochemical breakdown of tropospheric ozone is the major source of the hydroxyl radical (not shown is a newly discovered additional source of the hydroxyl radical, from the photoexcitation of nitrogen dioxide). As we will see, higher levels of ozone result from higher concentrations of other polluting gases.

Recent research indicates that the atmospheric levels of hydroxyl are quite stable, suggesting that formation and breakdown are balanced.[1] Some hydroxyl is used up during reactions with CO and CH_4, but hydroxyl is recycled as it reacts with most other pollutants. This is good news for now, but researchers are concerned that future pollutants and climate change could affect the hydroxyl balance.

Smogs and Brown Clouds

Down through the centuries, the practice of venting the products of combustion and other fumes into the atmosphere remained the natural way to avoid their noxious effects within buildings. With the Industrial Revolution of the 1800s came crowded cities and the use of coal for heating and energy. It was then that air pollution began in earnest.

Industrial Smog. In *Hard Times*, Charles Dickens described a typical English scene: "Coketown lay shrouded in a haze of its own, which appeared impervious to the sun's rays. You

---

[1]S. A. Montzka et al., "Small Interannual Variability of Global Atmospheric Hydroxyl," *Science* 331 (January 7, 2011): 67–69.

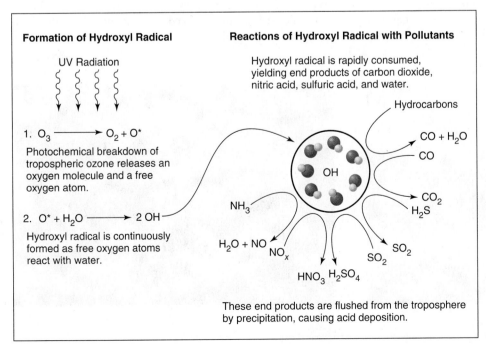

Formation of Hydroxyl Radical

UV Radiation

1. $O_3 \longrightarrow O_2 + O^*$
Photochemical breakdown of tropospheric ozone releases an oxygen molecule and a free oxygen atom.

2. $O^* + H_2O \longrightarrow 2\,OH$
Hydroxyl radical is continuously formed as free oxygen atoms react with water.

Reactions of Hydroxyl Radical with Pollutants

Hydroxyl radical is rapidly consumed, yielding end products of carbon dioxide, nitric acid, sulfuric acid, and water.

Hydrocarbons → $CO + H_2O$
→ CO
OH
→ CO_2
H_2S

NH_3
$H_2O + NO$
NO_x
SO_2
SO_2
HNO_3 H_2SO_4

These end products are flushed from the troposphere by precipitation, causing acid deposition.

Figure 19–2 **The hydroxyl radical.** This is a simplified model of atmospheric cleansing by the hydroxyl radical (OH). The first step is the photochemical destruction of ozone, which is the major process leading to ozone breakdown in the troposphere. The second step produces hydroxyl radicals, which react rapidly with many pollutants, converting them to substances that are less harmful (CO_2) or that can be returned to Earth via precipitation (HNO_3, H_2SO_4).

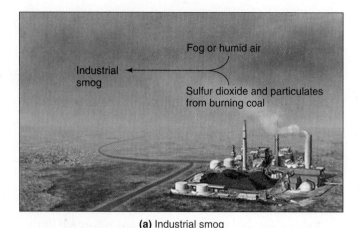

(a) Industrial smog

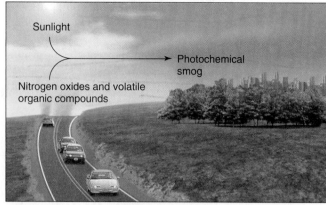

(b) Photochemical smog

Figure 19–3 **Industrial and photochemical smog.** (a) Industrial smog occurs when coal is burned and the atmosphere is humid. (b) Photochemical smog occurs when sunlight acts on vehicle pollutants.

only knew the town was there because there could be no such sulky blotch upon the prospect without a town." This shrouding haze became known as **industrial smog** (a combination of *sm*oke and f*og*), an irritating, grayish mixture of soot, sulfurous compounds, and water vapor (Figure 19–3a). This kind of smog continues to be found wherever industries are concentrated and where coal is the primary energy source; such smog is most likely to form in winter or in cool, foggy weather. The Donora incident is classic example. Currently, industrial smog can be found in the cities of China, India, Korea, and some eastern European countries.

Photochemical Smog. After World War II, the booming economy in the United States led to the creation of vast suburbs and a mushrooming use of cars for commuting to and from work. As other fossil fuels were used in place of coal for heating, industry, and transportation, industrial smog was largely replaced by another kind of smog. Increasingly, Los Angeles and other cities served by huge freeway systems began to be enshrouded daily in a brownish, irritating haze that was different from the more familiar industrial smog

(Figure 19–4). Weather conditions were usually warm and sunny rather than cool and foggy, and, typically, the new haze would arise during the morning commute and begin to dissipate only by the time of the evening commute. This **photochemical smog**, as it came to be called, is produced when several pollutants from automobile exhausts—nitrogen oxides and volatile organic carbon compounds—are acted on by sunlight (Figure 19–3b).

Inversions. Certain weather conditions intensify levels of both industrial and photochemical smogs. Under normal conditions, the daytime air temperature is highest near the ground, because sunlight strikes Earth and the absorbed heat radiates to the air near the surface. The warm air continues to rise, carrying pollutants upward and dispersing them at higher altitudes (Figure 19–5a). At times, however, a warm air layer occurs above a cooler layer. This condition of cooler air below and warmer air above is called a **temperature inversion** (Figure 19–5b). Inversions often occur at night, when the surface air is cooled by radiative heat loss. Such inversions are usually short lived, as the next morning's

(a)

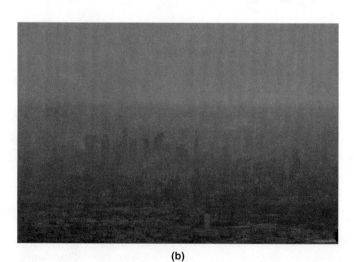

(b)

Figure 19–4 **Los Angeles air.** (a) Los Angeles on a good day. (b) LA under a blanket of photochemical smog.

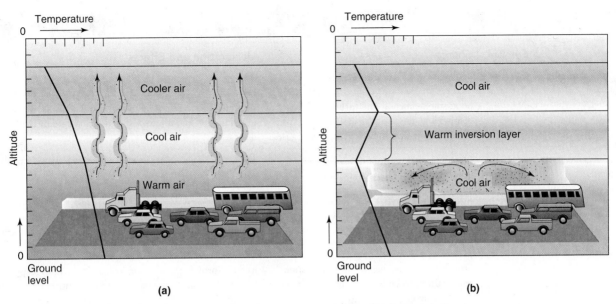

Figure 19–5 Temperature inversion. A temperature inversion may cause episodes of high concentrations of air pollutants. **(a)** Normally, air temperatures are highest at ground level and decrease at higher elevations. **(b)** In a temperature inversion, a layer of warmer air overlies cooler air at ground level.

sunlight begins the heating process anew and any pollutants that accumulated overnight are carried up and away. During cloudy weather, however, the Sun may not be strong enough to break up the inversion for hours or even days. A mass of high-pressure air may move in and sit above the cool surface air, trapping it. Topography can also intensify smog as cool ocean air flows into valleys and is trapped there by nearby mountain ranges. Los Angeles has both topographical features.

When such long-term temperature inversions occur, pollutants can build up to dangerous levels, prompting local health officials to urge people with breathing problems to stay indoors. For many people, smog causes headaches, nausea, and eye and throat irritation. It may aggravate preexisting respiratory conditions, such as asthma and emphysema. In some industrial cities (such as Donora), air pollution reached lethal levels under severe temperature inversions. These cases became known as *air-pollution disasters*. London experienced repeated episodes of inversion-related disasters in the mid-1900s. One episode in 1952 resulted in 4,000 pollution-related deaths.

Atmospheric Brown Clouds. A relative newcomer on the air-pollution scene is the **atmospheric brown cloud (ABC)**, a 1- to 3-kilometer-thick blanket of pollution. These vast, murky clouds are caused by pollution from forest fires, the burning of fossil fuels and farm wastes, and millions of small, inefficient household stoves. ABCs are found in regional hot spots in East Asia, South Asia, Southern Africa, and the Amazon Basin. The large quantities of black carbon and soot in ABCs reduce the amount of sunlight reaching the ground, a phenomenon called *global dimming* (see Sound Science, *Complex Clouds and Co-Benefits of Solutions*). Ozone and acid rain accompany ABCs.

The impacts of ABCs are serious: major dimming, reductions in rainfall, solar heating of the atmosphere and decreased reflection by snow and ice (both of which are contributing to a shrinking of the Hindu Kush–Himalaya–Tibetan glaciers). These polluted air masses also cause a weakening of the Indian summer monsoon; reduced crop yields due to the dimming and to ozone levels; and heightened acute and chronic health effects typical of exposure to industrial and urban air pollution. ABCs contribute to a global trend: increasing air pollution in many urban areas, especially in Asia. India in particular is suffering from the effects of air pollution. Thirteen of the 20 most polluted cities in the world are in India. Several of the others are in China (see Chapter 2).

Industrial smog from coal burning, photochemical smog from vehicular exhaust, and atmospheric brown clouds are all caused by the release of pollutants, and we now turn our attention to those pollutants.

CONCEPT CHECK What are the key characteristics of industrial smog, photochemical smog, and atmospheric brown smog? How does each form? ☑

19.2 Major Air Pollutants and Their Sources

In large measure, anthropogenic air pollutants are direct and indirect products of the combustion of coal, gasoline, other liquid fuels, and refuse such as waste paper, food, and plastics. These fuels and wastes are organic compounds. With complete combustion, the products of burning organic compounds are carbon dioxide and water vapor. Unfortunately, combustion is seldom complete, and many complex chemical

Table 19–1 Major Anthropogenic Air Pollutants and Their Effects

Name	Symbol	Source	Description and Effects
Primary Pollutants			
Suspended particulate matter	PM	Soot, smoke, metals, and carbon from combustion; dust, salts, metal, and dirt from wind erosion; atmospheric reactions of gases	Complex mixture of solid particles and aerosols; reduces lung function and affects respiration and cardiovascular disease
Volatile organic compounds	VOC	Incomplete combustion of fossil fuels; evaporation of solvents and gasoline; emission from plants	Mixture of compounds, some carcinogenic; major agent of ozone formation
Carbon monoxide	CO	Incomplete combustion of fuels	Invisible, odorless, tasteless gas; poisonous because of ability to bind to hemoglobin and block oxygen delivery to tissues; aggravates cardiovascular disease
Nitrogen oxides	NO_x	From nitrogen gas due to high combustion temperatures when burning fuels; also from wood burning	Reddish-brown gas and lung irritant; aggravates respiratory disease; major source of acid rain; contributes to ozone formation
Sulfur dioxide	SO_2	Combustion of sulfur-containing fuels, especially coal	Poisonous gas that impairs breathing; major source of acid rain; contributes to particle formation
Lead	Pb	Battery manufacture; lead smelters; combustion of leaded fuels and solid wastes	Toxic at low concentrations; accumulates in body and can lead to brain damage in children
Air toxics	Various	Fuel combustion in vehicles; industrial processes; building materials; solvents	Toxic chemicals, many of which are known human carcinogens (e.g., benzene, asbestos, vinyl chloride); Clean Air Act identifies 187 air toxics
Radon	Rn	Rocks and soil; natural breakdown of radium and uranium	Invisible and odorless radioactive gas can accumulate inside homes; second leading cause of lung cancer in United States
Secondary Pollutants			
Ozone	O_3	Photochemical reactions between VOCs and NO_x	Toxic to animals and plants and highly reactive in lungs; oxidizes surfaces and rubber tires
Peroxyacetyl nitrates	PAN	Photochemical reactions between VOCs and NO_x	Damages plants and forests; irritates mucous membranes of eyes and lungs
Sulfuric acid	H_2SO_4	Oxidation of SO_2 by OH radicals	Produces acid deposition; damages lakes, soils, artifacts
Nitric acid	HNO_3	Oxidation of NO_x by OH radicals	Produces acid deposition; damages lakes, soils, artifacts

substances are involved. The combustion of fuels and refuse creates a host of gaseous and particulate products that in turn create unwanted problems before they are cleansed from the air. Other processes producing air pollutants are the evaporation of volatile substances and strong winds that pick up dust and other particles.

Primary Pollutants

Table 19–1 presents the major sources, characteristics, and general effects of the major air pollutants. The first seven (**particulates, volatile organic compounds, carbon monoxide, nitrogen oxides, sulfur dioxide, lead,** and **air toxics**) are called **primary air pollutants** because they are the direct products of combustion and evaporation. Some primary pollutants may

undergo further reactions in the atmosphere and produce additional undesirable compounds, called **secondary air pollutants** (**ozone, peroxyacetyl nitrates, sulfuric acid,** and **nitric acid**). Sources of air pollution are summarized in Figure 19–6. Power plants are the major source of sulfur dioxide, particulates originate from all sorts of combustion, and transportation accounts for the lion's share of carbon monoxide and nitrogen oxides. This diversity suggests that strategies for controlling air pollutants will be different for the different sources.

When fuels and wastes are burned, particles consisting mainly of carbon are emitted into the air; these are the particulates we see as soot and smoke (Figure 19–7). In addition, various unburned fragments of fuel molecules are given off; these are the volatile organic compound (VOC) emissions. When carbon is not completely oxidized, it produces carbon

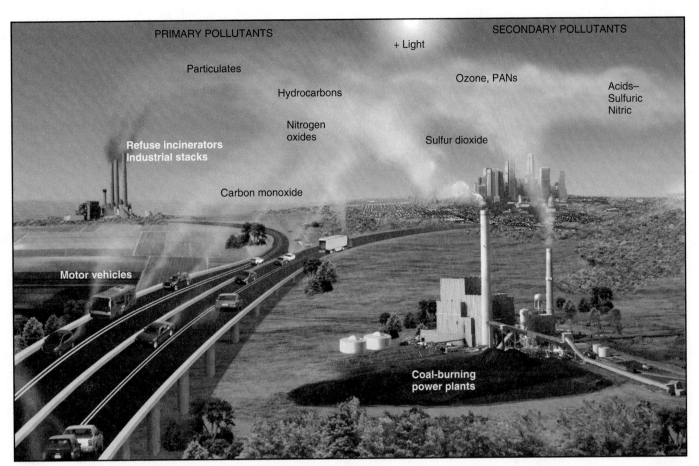

Figure 19–6 The prime sources of the major anthropogenic air pollutants. Industrial processes and transportation generate a toxic brew of primary and secondary air pollutants.

monoxide (CO) rather than carbon dioxide (CO_2), which explains why dangerous carbon monoxide fumes form when burning is inefficient.

Combustion takes place in the air, a mixture of 78% nitrogen and 21% oxygen. At high combustion temperatures, some of the atmospheric nitrogen gas is oxidized to form the

Figure 19–7 Industrial pollution. Because the stack of this wood-products mill in New Richmond, Quebec Province, Canada, lacks electrostatic precipitators, the plume contains substantial amounts of suspended particulate matter that can be seen from a great distance.

gas nitric oxide (NO). In the air, nitric oxide immediately reacts with additional oxygen to form nitrogen dioxide (NO_2) and nitrogen tetroxide (N_2O_4). These compounds are collectively referred to as the *nitrogen oxides*, or NO_x. Nitrogen dioxide absorbs light and is largely responsible for the brownish color of photochemical smog.

In addition to organic matter, fuels and refuse contain impurities and additives, and these substances are also emitted into the air during burning. Coal, for example, contains from 0.2% to 5.5% sulfur. In combustion, this sulfur is oxidized, giving rise to several gaseous oxides, the most common being sulfur dioxide (SO_2). Coal also contains heavy-metal impurities such as mercury, and refuse contains an endless array of impurities.

Tracking Pollutants. The Environmental Protection Agency (EPA) operates the Clearinghouse for Inventories and Emissions Factors, which tracks trends in national emissions of the primary pollutants from all sources. The EPA also follows air quality by measuring ambient concentrations (where *ambient* means "present in the air") of the pollutants at thousands of monitoring stations across the country. According to EPA data, 2013 emissions in the United States of the first five primary pollutants amounted to 116 million tons. By comparison, in 1970, when the first Clean Air Act became law, these same five pollutants totaled 301 million tons. The relative amounts emitted

Figure 19–8 U.S. emissions of five primary air pollutants, by source, for 2013. *Fuel combustion* refers to fuels burned for electrical power generation and for space heating. Note especially the different contributions by transportation and fuel combustion/electric utilities, the two major sources of air pollutants.

(*Source:* EPA National Emissions Inventory, 2014. http://www.epa.gov/ttn/chief/trends/index.html.)

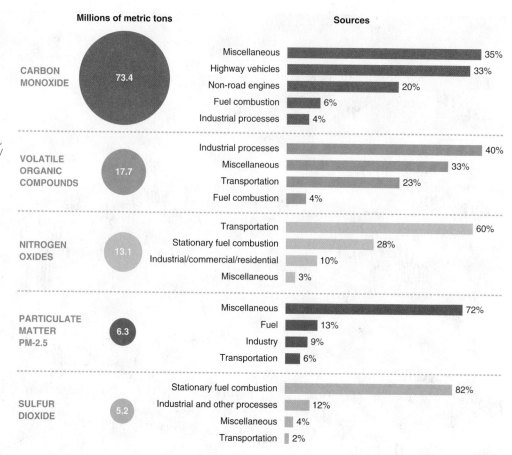

in the United States in 2013, as well as their major sources, are illustrated in **Figure 19–8**. In general, there has been a remarkable improvement in the rate of emissions over the past 40 years (**Figure 19–9**), reflecting the effectiveness of Clean Air Act regulations. This progressive improvement over time has occurred in spite of large increases in the gross domestic product, vehicular miles traveled, population, and energy consumption (**Figure 19–10**).

Getting the Lead Out. The sixth type of primary pollutant—lead and other heavy metals—is discussed separately because the quantities emitted are far less than the levels for the first five. Before the EPA-directed phaseout in the 1980s and 1990s, lead was added to gasoline as an inexpensive way to prevent engine knock. Emitted with the exhaust from gasoline-burning vehicles, lead remained airborne and traveled great distances before settling as far away as the glaciers of Greenland. Since the phaseout, concentrations of lead in the air of cities in the United States have declined remarkably. However, lead levels are elevated around airports where piston-engine aircraft operate, because aviation gasoline still uses leaded gasoline (jet fuel has no added lead). Between 1980 and 2013, lead levels in air fell 99%. At the same time, levels of lead in children's blood have dropped 92% (some children still get lead from dust and paint chips). Lead concentrations in Greenland ice have also decreased significantly. All these declines indicate that lead restrictions in the United States and most other Northern Hemisphere nations have had a global impact. The data indicate that we have reached

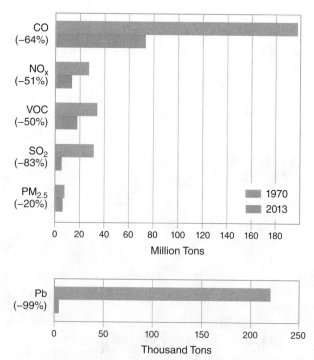

Figure 19–9 Clean Air Act impacts. Comparison of percent reductions in the emission of six criteria air pollutants between 1970 and 2013. PM$_{2.5}$ comparison values are from 1990, when data were first gathered, and 2013.

(*Source:* EPA National Emissions Inventory (NEI) Air Pollutant Emissions Trends Data.)

COMPARISON OF GROWTH AREAS AND EMISSIONS, 1980–2013

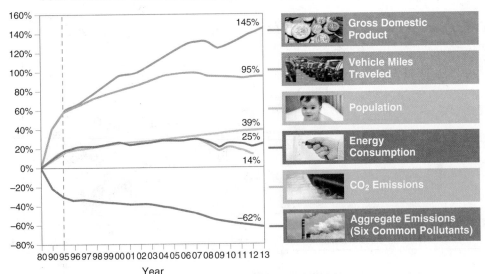

Figure 19–10 Comparison of growth versus emissions. Gross domestic product, vehicle miles traveled, energy consumption, CO_2 emissions, and population are compared with aggregate reductions in emissions of the six principal pollutants from 1980 to 2013.

(*Source:* EPA Air Trends, http://www.epa.gov/airtrends/aqtrends.html#emission.)

UNDERSTANDING THE DATA

1. How did CO_2 emissions in 2010 differ from those in 1980?
2. Since 1980, which has increased the most: CO_2 emissions, aggregate emissions, or energy consumption per person? Which has decreased the most?

a steady state in lead emissions and that any new reductions must target leaded aviation gasoline, which accounts for half of the current lead air emissions. Lead smelters and battery manufacturers also add to the burden of lead.

Toxics and Radon. As with lead, the concentrations of toxic chemicals and radon in the air are small compared with those of the other primary pollutants. Some of the air's toxic compounds—benzene, for example—originate with transportation fuels. Most, however, are traceable to industries and small businesses.

Radon, by contrast, is produced by the spontaneous decay of fissionable material in rocks and soils. Radon levels are highest in areas with certain types of bedrock, such as granite and shale. Radon escapes naturally to the surface and seeps into buildings through cracks in foundations and basement floors, sometimes collecting in the structures. Basements can be tested for the level of radon; the EPA standards limit radon exposure to 4 picocuries (a measure of radioactivity) per cubic liter. While radon can cause cancer, overexposure is avoidable with some simple changes, such as providing adequate ventilation in basements and sealing foundation cracks.

Secondary Pollutants

Ozone and numerous reactive organic compounds are formed as a result of chemical reactions between nitrogen oxides and volatile organic carbons. Because sunlight provides the energy necessary to propel the reactions, these products are also known as **photochemical oxidants**. Note that these reactions and the ozone they produce are entirely in the troposphere; ozone in the stratosphere is "good" ozone, protecting life from damaging ultraviolet (UV) radiation. In preindustrial times, ozone concentrations ranged from 10 to 15 parts per billion (ppb), well below harmful levels. Summer concentrations of ozone in unpolluted air in North America now range from 20 to 50 ppb. Polluted air may contain ozone concentrations of 150 ppb or more, a level considered quite unhealthy if encountered for extended

periods. EPA data indicate that ambient ozone levels, monitored across the United States at some 1,100 sites, have declined more than 30% since 1980. Los Angeles, the U.S. city with the highest level of ozone pollution, now has one third fewer days of unhealthy ozone levels than it had fifteen years ago. Even so, 108 million people in the U.S live in counties that still do not meet ozone standards.

Ozone Formation. A simplified view of the major reactions in the formation of ozone and other photochemical oxidants is shown in **Figure 19–11**, on the following page. Nitrogen dioxide absorbs light energy and splits to form nitric oxide and atomic oxygen; the oxygen atoms rapidly combine with oxygen gas to form ozone. If other factors are not involved, ozone and nitric oxide then react to form nitrogen dioxide and oxygen gas. A steady-state concentration of ozone results, and there is no appreciable accumulation of the gas (Figure 19–11a).

When VOCs are present, however, the nitric oxide reacts with them instead of with the ozone, causing several serious problems. First, the reaction between nitric oxide and the VOCs leads to highly reactive and damaging compounds known as peroxyacetyl nitrates, or PANs (Figure 19–11b). Second, numerous aldehyde and ketone compounds are produced as the VOCs are oxidized by atomic oxygen, and these compounds are also noxious. Finally, with the nitric oxide tied up in this way, the ozone tends to accumulate. Because of the complex air chemistry involved, ozone concentrations may peak 30 to 100 miles (50 to 160 km) downwind of urban centers where the primary pollutants were generated. As a result, ozone concentrations above the air quality standards are sometimes found in rural and wilderness areas.

Sulfuric acid and nitric acid are considered secondary pollutants because they are products of sulfur dioxide and nitrogen oxides, respectively, reacting with atmospheric moisture and oxidants such as hydroxyl. Sulfuric and nitric acids are the acids in acid precipitation. We now turn our attention to this important and widespread problem.

Major pollutants from vehicles

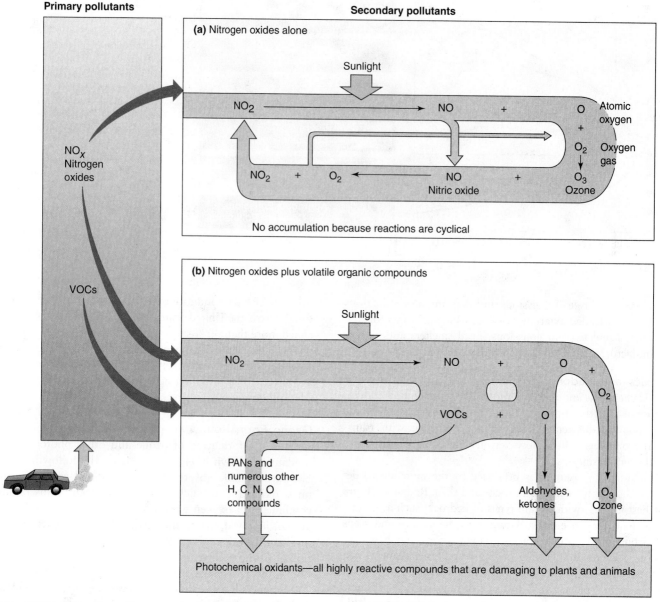

Figure 19–11 Formation of ozone and other photochemical oxidants. (a) Nitrogen oxides, by themselves, would not cause ozone and other oxidants to reach damaging levels, because reactions involving nitrogen oxides are cyclical. **(b)** When VOCs are also present, however, reactions occur that lead to the accumulation of numerous damaging compounds—most significantly, ozone, the most injurious.

Acid Precipitation and Deposition

Acid precipitation refers to any precipitation—rain, fog, mist, or snow—that is more acidic than usual. Because dry acidic particles are also found in the atmosphere, the combination of precipitation and dry-particle fallout is called **acid deposition**. In the late 1960s, Swedish scientist Svante Odén first documented the acidification of lakes in Scandinavia and traced it to air pollutants originating in other parts of Europe and Great Britain. Since then, careful monitoring has shown that broad areas of North America, as well as most of Europe and other industrialized regions of the world, have been experiencing precipitation that is between 10 and 1,000 times more acidic than usual. This

has affected ecosystems in diverse ways, as illustrated in Figure 19–12.

To understand the full extent of the problem, we must first understand some principles about acids and how we measure their concentration.

Acids, Bases, and pH. Acidic properties (for example, a sour taste and corrosiveness) are due to the presence of hydrogen ions (H^+, hydrogen atoms without their electron), which are highly reactive. Therefore, an **acid** is any chemical that releases hydrogen ions when dissolved in water. The chemical formulas of a few common acids are shown in Table 19–2. Note that all of them ionize—that is, their

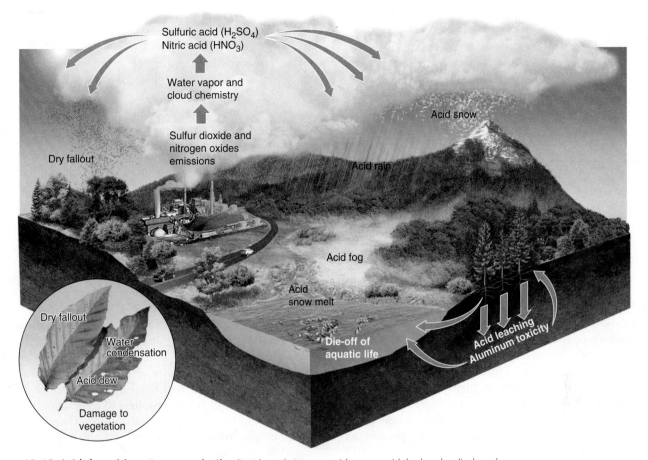

Figure 19–12 Acid deposition. Emissions of sulfur dioxide and nitrogen oxides react with hydroxyl radicals and water vapor in the atmosphere to form their respective acids, which return to the surface either as dry acid deposition or, mixed with water, as acid precipitation. Various effects of acid deposition are noted.

Table 19–2   Common Acids and Bases

Acid	Formula	Yields	H^1 Ion(s)	Plus	Negative Ion
Hydrochloric acid	HCl	$\rightarrow$	H^+	+	Cl^- Chloride
Sulfuric acid	H_2SO_4	$\rightarrow$	$2H^+$	+	SO_4^{2-} Sulfate
Nitric acid	HNO_3	$\rightarrow$	H^+	+	NO_3^- Nitrate
Phosphoric acid	H_3PO_4	$\rightarrow$	$3H^+$	+	PO_4^{3-} Phosphate
Acetic acid	CH_3COOH	$\rightarrow$	H^+	+	CH_3COO^- Acetate
Carbonic acid	H_2CO_3	$\rightarrow$	H^+	+	HCO_3^- Bicarbonate
Base	**Formula**	**Yields**	OH^2 **Ion(s)**	**Plus**	**Positive Ion**
Sodium hydroxide	NaOH	$\rightarrow$	OH^-	+	Na^+ Sodium
Potassium hydroxide	KOH	$\rightarrow$	OH^-	+	K^+ Potassium
Calcium hydroxide	$Ca(OH)_2$	$\rightarrow$	$2OH^-$	+	Ca^{2+} Calcium
Ammonium hydroxide	NH_4OH	$\rightarrow$	OH^-	+	NH_4^+ Ammonium

Figure 19–13 **The pH scale.** Each unit on the scale represents a tenfold difference in hydrogen ion concentration.

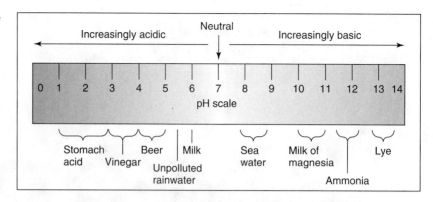

components separate—to create hydrogen ions plus a negative ion. The higher the concentration of hydrogen ions in a solution, the more acidic the solution.

A **base** is any chemical that releases hydroxide ions (OH^-, oxygen-hydrogen groups with an extra electron) when dissolved in water. (See Table 19–2.) The bitter taste and caustic properties of alkaline, or basic, solutions are due to the presence of hydroxide ions.

The concentration of hydrogen ions is expressed as **pH.** The pH scale goes from 0 (highly acidic) through 7 (neutral) to 14 (highly basic) **(Figure 19–13).** As the pH scale numbers go from 0 to 7, the concentration of hydrogen ions (H^+) decreases, and the solution becomes less acidic and more neutral. As the numbers go from 7 to 14, the concentration of hydrogen ions gets lower, but the concentration of hydroxide ions (OH^-) increases, and the solution becomes more basic. The numbers on the scale (0–14) represent the negative logarithm (power of 10) of the hydrogen ion concentration, expressed in grams per liter (g/L). For example, a solution with a pH of 1 has an H^+ concentration of 10^{-1} g/L, a pH of 2 has a concentration of 10^{-2} g/L, and so on.

Because numbers on the pH scale represent powers of 10, there is a *tenfold difference* between each unit and the next. For example, pH 5 is 10 times as acidic (has 10 times as many H^+ ions) as pH 6, and pH 4 is 10 times as acidic as pH 5.

Extent and Potency of Acid Precipitation. In the absence of any pollution, rainfall is normally somewhat acidic, with a pH of 5.6, because carbon dioxide in the air readily dissolves in, and combines with, water to produce carbonic acid. **Acid precipitation,** then, is any precipitation with a pH less than 5.5.

Since the middle of the 20th century, acid precipitation has been the norm over most of the Western world. The pH of rain and snowfall over a large portion of eastern North America used to be quite acidic, reflecting the west-to-east movement of polluted air from the Midwest and industrial Canada. Many areas in this region once regularly received precipitation having a pH of 4.0 and, occasionally, as low as 3.0 **(Figure 19–14a).** Acid precipitation has not disappeared in eastern North America, but pH values lower than 4.6 are now uncommon **(Figure 19–14b).** Acid precipitation was heavy in Europe in the latter half of the 20th century, from

the British Isles to central Russia. Like North America, acid precipitation has declined greatly in these regions. It is on the rise in China and India as these countries increasingly exploit their coal resources.

Sources of Acid Deposition. Chemical analysis of acid precipitation reveals the presence of two acids, sulfuric acid (H_2SO_4) and nitric acid (HNO_3). As we have seen, burning fuels produce sulfur dioxide and nitrogen oxides, so one source of the acid deposition problem is evident. These oxides enter the troposphere in large quantities from both anthropogenic and natural sources. Once in the troposphere, hydroxyl radicals (Figure 19–2) react with them to form sulfuric and nitric acids, which dissolve readily in water or adsorb to particles and are brought down to Earth in acid deposition. This usually occurs within a week of the oxides entering the atmosphere. Another important source of acidification is traced to the use of nitrogen fertilizers, made by the Haber-Bosch process, whereby atmospheric N_2 (a nonreactive form of nitrogen) is converted into other nitrogen compounds that can be used by plants (Chapter 3). As a result of this process, ammonia (NH_3) can escape into the atmosphere. Whether in the atmosphere, soil, or natural waters, ammonia is eventually oxidized to nitrate (NO_3^-) by bacterial action, a process that generates acidity.

Natural sources contribute substantial quantities of pollutants to the air, including 50 million to 70 million tons per year of sulfur dioxide (from volcanoes, sea spray, and microbial processes) and 30 million to 40 million tons per year of nitrogen oxides (from lightning, the burning of biomass, and microbial processes). Anthropogenic sources per year are estimated at 115 million tons of sulfur dioxide, 60 million to 70 million tons of nitrogen oxides, and 50 million tons of ammonia.[2] The vital difference between natural and anthropogenic sources is that anthropogenic sources are strongly concentrated in industrialized and agricultural regions, whereas the emissions from natural sources are spread out and are a part of the global environment. Levels of anthropogenic oxides have increased sixfold

[2]Karen C. Rice and Janet S. Herman, "Acidification of Earth: An Assessment Across Mechanisms and Scales," *Applied Geochemistry* 27 (2012): 1–14.

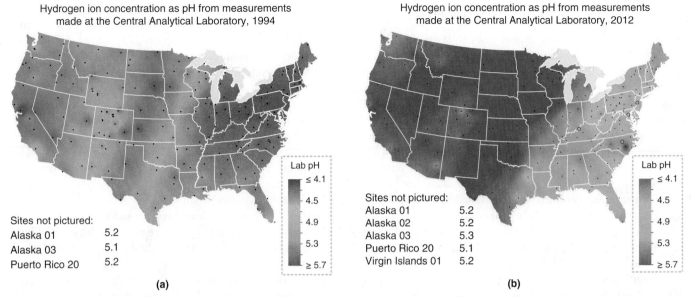

Hydrogen ion concentration as pH from measurements made at the Central Analytical Laboratory, 1994

Hydrogen ion concentration as pH from measurements made at the Central Analytical Laboratory, 2012

Lab pH
≤ 4.1
4.5
4.9
5.3
≥ 5.7

Sites not pictured:
Alaska 01   5.2
Alaska 03   5.1
Puerto Rico 20   5.2

Sites not pictured:
Alaska 01   5.2
Alaska 02   5.2
Alaska 03   5.3
Puerto Rico 20   5.1
Virgin Islands 01   5.2

(a)

(b)

Figure 19–14 Acid deposition in eastern North America improves. (a) 1994 rainfall pH distribution across the United States; dots show locations of measuring stations. **(b)** 2012 rainfall pH distribution.

(*Source:* National Atmospheric Deposition Program (NRSP-3). 2014. NADP Program Office, Illinois State Water Survey. 2204 Griffith Dr., Champaign, IL 61820.)

UNDERSTANDING THE DATA   In 2012, which states still had some regions where pH levels were lower than 4.5?

since 1900, while levels of natural emissions have remained fairly constant.[3]

As Figure 19–8 indicates, 5.2 million tons of sulfur dioxide were released into the air in 2013 in the United States; 92% was from fuel combustion (mostly from coal-burning power plants). Some 12.0 million tons of nitrogen oxides were released, 51% of which can be traced to transportation emissions and 41% to fuel combustion at fixed sites. In addition, 4.1 million tons of ammonia were released into the atmosphere from fertilizer use and livestock. Nitrogen oxides and ammonia are doubly damaging because the nitrogen contributes to eutrophication in aquatic ecosystems (Chapter 20). In the eastern United States, the source of much of the acid deposition was identified as the tall stacks of 50 huge, older coal-burning power plants located primarily in the Midwest. The tall stacks were built to alleviate local sulfur dioxide pollution at ground level. The good news is that many of these same plants are now reducing their emissions as a result of the Clean Air Act. Acid-generating emissions have decreased by 65% over the past 20 years, but there is obviously more that can be done. Of great significance, the pattern of sources has shifted from predominantly sulfur dioxide to nitrogen oxides and ammonia.

On a global scale, air quality is improving in some countries, or even some parts of countries, and worsening in others. Figure 19–15 illustrates changes in fine particulates,

one pollutant that represents a real danger to human health, a consideration we will discuss next as we look at the impacts of air pollution.

CONCEPT CHECK   Explain how secondary pollutants form. How do they differ from primary pollutants? ☑

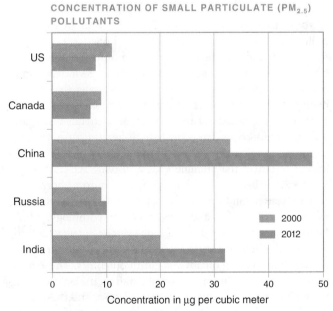

CONCENTRATION OF SMALL PARTICULATE (PM$_{2.5}$) POLLUTANTS

Concentration in μg per cubic meter

2000
2012

Figure 19–15 Changes in particulate pollution. Data from the WHO show how the concentration of small particulate air pollutants in five countries changed between 2000 and 2012.

(*Source:* Yale University Environmental Performance Index—Air Pollution Map. 2014.)

19.3 Impacts of Air Pollutants

Air pollution is an alphabet soup of gases and particles, mixed with the normal constituents of air. The amount of each pollutant present varies greatly, depending on its proximity to the source and various conditions of wind and weather. As a result, we are exposed to a mixture that varies in makeup and concentration from day to day—even from hour to hour—and from place to place. Consequently, the effects we feel or observe are rarely, if ever, the effects of a single pollutant. Instead, they are the combined impact of the whole mixture of pollutants acting over our life span up to that point.

For example, plants may be so stressed by pollution that they become more vulnerable to other environmental factors, such as drought or attack by insects. Humans may develop lung disease due to the combined effects of ozone and NO_x. Given the complexity of this situation, it is often difficult to determine the role of any particular pollutant in causing an observed result. Nevertheless, some significant progress has been made in linking cause and effect. Figure 19–15 compares the recent changes in air quality in several countries.

Human Health

The air-pollution disasters in Donora and London demonstrated that exposure to air pollution can be deadly. Every one of the primary and secondary air pollutants (Table 19–1) is a threat to human health, particularly the health of the respiratory system (Figure 19–16). Health impacts of air pollution include asthma, chronic bronchitis, and cardiovascular problems. *Acute* exposure to some pollutants can be life threatening, but many effects are *chronic,* acting over a period of years to cause a gradual deterioration of physiological functions and eventual premature mortality. Moreover, some pollutants are *carcinogenic,* adding significantly to the risk of lung cancer as they are breathed into the lungs.

Chronic Effects. Many people living in areas of urban air pollution suffer from chronic effects. Long-term exposure to sulfur dioxide can lead to bronchitis (inflammation of the bronchi). Chronic inhalation of ozone can cause inflammation and, ultimately, fibrosis of the lungs, a scarring that permanently impairs lung function. Carbon monoxide reduces the capacity of the blood to carry oxygen, and extended exposure to carbon monoxide can contribute to heart disease. Chronic exposure to nitrogen oxides impairs lung function and is known to affect the immune system, leaving the lungs open to attack by bacteria and viruses. Exposure to airborne particulate matter can bring on a broad range of health problems, including respiratory and cardiovascular pathology. Other factors—such as poor diet, lack of exercise, preexisting diseases, and individual genetic makeup—may add to the effects of pollution to bring on adverse health outcomes. Those most sensitive to air pollution are small children, asthmatics, people with chronic pulmonary or heart disease, and the elderly.

COPD. Exposure to one or a combination of these pollutants can lead to **chronic obstructive pulmonary disease** (COPD), a slowly progressive lung disease that makes it increasingly hard to breathe. Close to 24 million people in

the United States have impaired lung function, and COPD is the fourth leading cause of death (134,000 deaths/year). Worldwide, COPD afflicts up to 10% of adults ages 40 and older. The leading cause of this disease is smoking, but in many developing countries, the burning of wood and dung in indoor stoves for heating and cooking delivers much the same burden of pollutants as smoking to the women who tend the stoves and the children who are with them. COPD involves three separate disease processes: emphysema (destruction of the lung alveoli—see Figure 19–16), bronchitis (inflammation and obstruction of the airways), and asthma.

Asthma. Asthma is an immune-system disorder characterized by impaired breathing caused by constricted air passageways. Asthma episodes are triggered by contact with allergens (dust mites, molds, pet dander) and compounds in polluted air (ozone, particulate matter, SO_2). According to the National Center for Health Statistics, almost 439,000 hospitalizations for asthma occur yearly, together with 2.1 million visits to hospital emergency departments. The highest prevalence is among children from birth to 4 years of age. The incidence of asthma in the United States peaked in 2003 and has since been dropping; currently it affects 26 million people (including 7 million children).

Strong Evidence. Two key studies recently analyzed by the Health Effects Institute provide strong evidence of the harmful effects of fine particles and sulfur pollution. These studies followed thousands of adult subjects living in 154 U.S. cities for as long as 16 years. In the studies, higher concentrations of fine particles were correlated with increased mortality, especially from cardiopulmonary disease and lung cancer. The more polluted the city, the higher the mortality. The studies were used by the EPA to step up its regulatory attention to fine particles.

The United Nations Environmental Program has identified air pollution as the greatest worldwide environmental health hazard. Researchers at MIT estimated that health care expenses related to air pollution cost China $112 billion annually. Another report found that meeting EPA standards would save the United States $28 billion annually in reduced health care costs, missed work, school absences, and premature deaths.

Mercury. One of the heavy-metal pollutants, mercury, deserves special attention. It is a toxic element, one that binds to key proteins in the body and can lead to many neurological disorders. In fetuses and small children, mercury impairs brain development. Power plants are the major source of mercury reaching the environment. Once it is present in aquatic ecosystems, mercury is converted into a highly toxic form, methylmercury, which accumulates in fish as a consequence of bioaccumulation (Chapter 13). In tests, the EPA reported methylmercury concentrations that exceeded recommendations in the game fish of 49% of lakes and reservoirs. Every state has advisories warning consumers about the dangers of mercury in locally caught fish. As we will see, the EPA has recently stepped up the regulatory restrictions on mercury releases from power plants. Worldwide, mercury pollution is a significant problem (Chapters 17, 21).

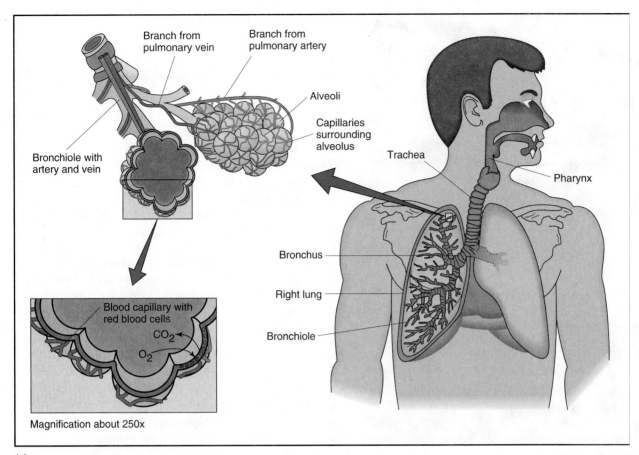

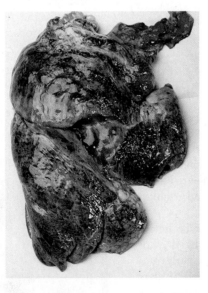

(b)

Figure 19–16 The respiratory system. (a) In the lungs, air passages branch and rebranch and finally end in millions of tiny sacs called alveoli. These sacs are surrounded by capillaries. As blood passes through the capillaries, oxygen from inhaled air diffuses into the blood from the alveoli. Carbon dioxide diffuses in the reverse direction and leaves the body in the exhaled air. **(b)** On the left is normal lung tissue, and on the right is lung tissue from a person who suffered from emphysema, a chronic lung disease in which some of the structure of the lungs has broken down. Cigarette smoking and heavy air pollution are associated with the development of emphysema and other chronic lung diseases.

Acute Effects. In severe cases, air pollution reaches levels that cause death, though such deaths usually occur among people already suffering from critical respiratory or heart disease. The gases present in air pollution are known to be lethal in high concentrations, but nowadays such concentrations occur only in cases of accidental poisoning. Nevertheless, intense air pollution puts an additional stress on the body, and if a person is already in a weakened condition (as are, for example, the elderly or asthmatics), this additional stress may be fatal. Most of the deaths in air-pollution disasters reflect the acute effects of air pollution. However, recent research shows that even moderate air pollution can cause changes in the cardiac rhythms of people with heart disease, triggering fatal heart attacks.

Carcinogenic Effects. The heavy-metal and organic constituents of air pollution include many chemicals known to be carcinogenic in high doses. According to the industrial reporting

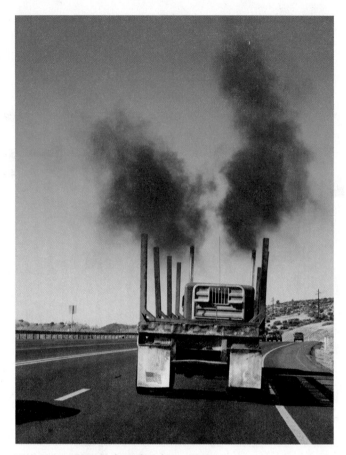

Figure 19–17 Diesel truck exhaust. Diesel truck exhaust is a major source of hazardous air pollutants as well as particulates, and the exhaust from diesel engines has been classified as a probable human carcinogen.

required by the EPA, 4.7 million tons of hazardous air pollutants (the air toxics—see Table 19–1) are released annually into the air in the United States. The presence of trace amounts of these chemicals in the air may be responsible for a significant portion of the cancer observed in humans. One major source of such carcinogens is diesel exhaust (Figure 19–17). The EPA has classified diesel exhaust as a likely human carcinogen and has established the National Clean Diesel Campaign to reduce emissions from the current fleet of diesel trucks and establish standards for new diesel-powered engines.

In some cases, exposure to a pollutant can be linked directly to cancer and other health problems by way of epidemiological evidence. One pollutant that clearly and indisputably is correlated with cancer and other disorders is **benzene**. This organic chemical is present in motor fuels and is also used as a solvent for fatty substances and in the manufacture of detergents, explosives, and pharmaceuticals. Benzene is released into the environment by the combustion of fossil fuels, particularly by the burning of coal and oil and in the emissions of motor vehicles. Benzene is also present in tobacco smoke, which accounts for half of the public's exposure to the chemical. The EPA has classified benzene as a known human carcinogen, linked to leukemia in persons encountering the chemical through occupational exposure. Chronic exposure to benzene can also lead to numerous blood disorders and damage to the immune system.

The Environment

Experiments show that plants are even more sensitive to air pollutants than are humans. Before emissions were controlled, it was common to see wide areas of totally barren land or severely damaged vegetation downwind from smelters and coal-burning power plants. The pollutant responsible was usually sulfur dioxide.

Crop Damage. Nowadays, damage to crops, orchards, and forests downwind of urban centers is caused mainly by exposure to ozone. The ozone gains access to plants through their stomata, the respiratory pores in all plant leaves. Symptoms of ozone damage include black flecking and yellowing of leaves (Figure 19–18). Crops vary in their susceptibility to ozone, but damage to many important crops (such as soybeans, corn, and wheat) is observed at common ambient levels of ozone (Figure 19–19). Unfortunately, ozone levels are highest in summer, when crops are growing, and rural areas often experience high ozone levels as ozone forms downwind of pollution sources. In the United States, ozone causes crop losses of \$1–3 billion annually. Worldwide, ozone exposure destroys 8–14% of the soybean crop. Much of the world's grain production occurs in regions that receive enough ozone pollution to reduce crop yields. India is experiencing severe ozone pollution. In 2005, ozone damaged an estimated 6.7 million tons of Indian rice, wheat, soybeans, and cotton; that wheat and rice alone would have been enough to feed 94 million people.[4]

Forest Damage. The negative impact of air pollution on wild plants and forest trees may be even greater than on agricultural crops.

---

[4]Sachin D. Ghude et al., "Reductions in India's Crop Yield Due to Ozone," *Geophysical Research Letters* (2014) 41: 5685–5691.

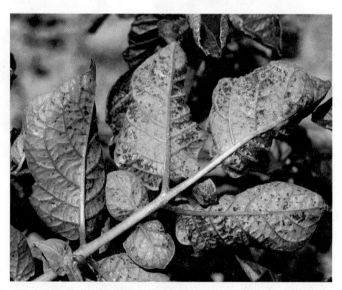

Figure 19–18 Ambient ozone injury. This potato leaf shows the black flecking symptomatic of ozone damage.

(*Source:* U.S. Department of Agriculture Agricultural Research Service. Effects of Ozone Air Pollution on Plants. 2009. www.ars.usda.gov/Main/docs.htm?docid=12462.)

PROPORTIONAL YIELD RESPONSE

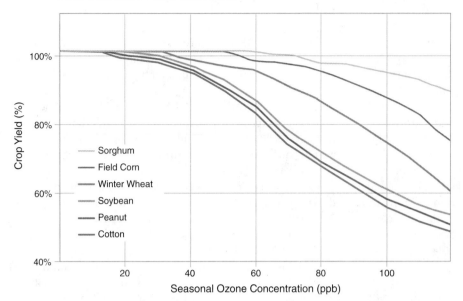

Figure 19–19 **Ozone impact on crop yields.** Some crop species react more strongly to ozone than others.

(*Source:* U.S. Department of Agriculture Agricultural Research Service: Effects of Ozone Air Pollution on Plants. 2009. www.ars.usda.gov/Main/docs.htm?docid=12462.)

Ozone Damage. Studies have shown that ozone damage to forest trees begins at about 40 ppb and gets more intense with higher levels. In California and southern Appalachian forests, ozone levels often reached 60 ppb and above, and seasonal losses in tree stem growth in the latter were measured at 30–50% for most species. Studies in European forests found that ozone stress affects photosynthetic carbon uptake, increases respiration of foliage, amplifies water loss, and generally disrupts carbon and nutrient metabolism. In sum, research has suggested that as temperatures climb and moisture is less predictable under conditions of expected global climate change, ozone levels will have an increasingly serious impact on forest growth.

Damage from Acid Deposition. From the Green Mountains of Vermont to the San Bernardino Mountains of California, the die-off of forest trees in the 1980s caused great concern. Red spruce forests were especially vulnerable. In New England, 1.3 million acres of high-elevation forests were devastated. Commonly, the damaged trees lost needles as acidic water drew calcium from them, rendering them more susceptible to winter freezing. Sugar maples, important forest trees in the Northeast, have shown extensive mortality, ranging from 20–80% of all trees in some forests.

Much of the damage to forests from acid precipitation is due to chemical interactions within the forest soils. Sustained acid precipitation at first adds nitrogen and sulfur to the soils, which stimulate tree growth. In time, though, these chemicals leach out large quantities of the buffering chemicals (usually calcium and magnesium salts). When these buffering salts no longer neutralize the acid rain, aluminum ions, which are toxic, are dissolved from minerals in the soil. The combination of aluminum and the increasing scarcity of calcium, which is essential to plant growth, leads to reduced tree growth. Research at the Hubbard Brook Experimental Forest in the White Mountains of New Hampshire has shown a marked reduction in calcium and magnesium in

the forest soils from the 1960s on, which is reflected in the amount of calcium in tree rings over the same period. The net result of these changes has been a decline in forest growth. Experimental liming of soils in forests showing maple tree dieback was able to restore the trees to health, and aluminum ion concentrations were seen to decrease greatly. Although acid deposition has declined since the 1980s, few studies have tracked the recovery of terrestrial ecosystems.

Impacts on Aquatic Ecosystems. Some 50 years ago, anglers started noticing sharp declines in fish populations in many lakes in Sweden, Ontario, and the Adirondack Mountains of New York State. As the ecological damage continued to spread, scientists looked for the cause. Since that time, studies have revealed many ways in which air pollution—especially acid deposition—alters and may destroy aquatic ecosystems.

The pH of an environment is extremely critical because it affects the function of virtually all enzymes, hormones, and other proteins in the bodies of all organisms living in that environment. Ordinarily, organisms are able to regulate their internal pH within the narrow limits necessary to function properly. A consistently low environmental pH, however, often overwhelms the regulatory mechanisms in many life forms, thus weakening or killing them. Most freshwater lakes, ponds, and streams have a natural pH in the range from 6 to 8, and organisms have adapted accordingly. The eggs, sperm, and developing young of these organisms are especially sensitive to changes in pH. Most are severely stressed, and many die, if the environmental pH shifts as little as one unit from the optimum.

As aquatic ecosystems become acidified, a shift occurs from acid-sensitive to acid-tolerant species. The acid-sensitive species die off, either because the acidified water kills them or because it keeps them from reproducing. Figure 19–12 shows that acid precipitation may leach aluminum and various heavy metals from the soil as the water percolates through it. Normally, the presence of these elements in the soil does not pose a problem because they are bound in insoluble mineral compounds and, therefore, are not absorbed by organisms. As the compounds are dissolved by low-pH water, however, the metals are freed. They may then be absorbed by and are highly toxic to both plants and animals. For example, mercury tends to accumulate in fish as lake waters become more acidic.

Acid deposition hit some regions especially hard in the latter part of the 20th century. In Ontario, Canada, approximately 1,200 lakes lost their fish life. In the Adirondacks, 346 lakes were without fish in 1990. In New England and the eastern Catskills, more than 1,000 lakes have suffered from recent acidification. The physical appearance of highly acidified lakes is deceiving. From the surface, they are clear and blue, the outward signs of a healthy condition. However, the only life found below the surface is acid-loving mosses growing on the bottom.

Neutralizing Capacity. As Figure 19–14a indicates, wide regions of eastern North America receive substantial amounts of acid precipitation, yet not all areas have acidified lakes. Apparently, many areas remain healthy, whereas others have become acidified to the point of becoming lifeless. How is this possible? The key lies in the system's **acid neutralizing capacity** (**ANC**). Despite the addition of acid, a system may be protected from changes in pH by a neutralizer—a buffer that, when present in a solution, has a large capacity to absorb hydrogen ions and thus maintain the pH at a relatively constant value.

Limestone (CaCO$_3$) is a natural acid neutralizer (Figure 19–20) that protects lakes from the effects of acid precipitation in many areas of the North American continent, particularly in the Midwest. Lakes and streams receiving their water from rain and melted snow that have percolated through soils derived from limestone will contain dissolved limestone. The regions that are sensitive to acid precipitation are those containing much granitic rock, which does not yield good ANC. The good news is that the sensitive regions

(New England, the Adirondacks, the Northern Appalachian Plateau) are recovering their ANC, and the percentage of acidic lakes has declined in the past 18 years.[5] The decline is consistent with the decrease in acid deposition in these regions, which has been brought about by the reductions of SO$_2$ and NO$_x$ emissions mandated by the Clean Air Act.

Effects on Materials and Aesthetics. Walls, windows, and other exposed surfaces turn gray and dingy as particulates settle on them. Paints and fabrics deteriorate more rapidly, and the sidewalls of tires and other rubber products become hard and checkered with cracks because of oxidation by ozone. Metal corrosion is increased dramatically by sulfur dioxide and acids derived from sulfur and nitrogen oxides, as are weathering and the deterioration of stonework.

Limestone and marble (which forms from limestone) are favored materials for the outsides of buildings and for monuments. The reaction between acid deposition and limestone is causing these structures to erode at a tremendously accelerated pace. Monuments that have stood for hundreds or even thousands of years with little change are now crumbling away (Figure 19–21). The corrosion of buildings, monuments, and outdoor equipment by acid precipitation costs billions of dollars for replacement and

[5]Executive Office of the President, National Acid Precipitation Assessment Program Report to Congress 2011: An Integrated Assessment, USGS, 2011, http://ny.water.usgs.gov.

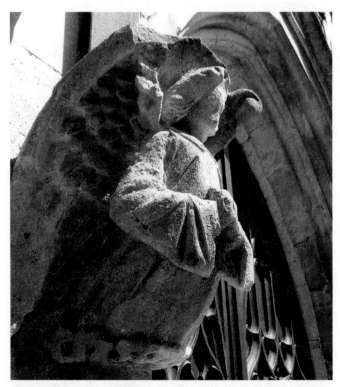

Figure 19–21 Effect of pollution on monuments. The corrosive effects of acids from air pollutants are dissolving away the features of many monuments and statues, as seen in the angel perched on the flying buttress of this church.

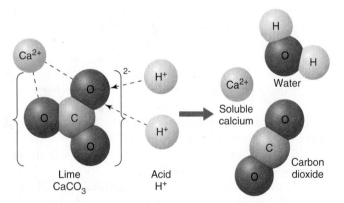

Figure 19–20 Acid neutralizing. Acids may be neutralized by certain nonbasic compounds. A chemical such as limestone (calcium carbonate) reacts with hydrogen ions as shown. Hence, the pH of a lake or river remains close to neutral, despite the additional acid, when it contains adequate acid neutralizing capacity.

repair each year in the United States. In an extreme case of corrosion, a 37-foot bronze Buddha in Kamakura, Japan, is slowly dissolving away as precipitation from Korea and China bathes the statue in highly acidic rain.

Visibility. A clear blue sky and good visibility can have a deep psychological impact on people. Can a value be put on these benefits? Many of us spend thousands of dollars and hundreds of hours commuting long distances to work so that we can live in a less polluted environment than the one in which we work. Ironically, the resulting traffic and congestion cause much of the very pollution we are trying to escape. In an encouraging move, the EPA established a Regional Haze Rule in 1999, aimed at improving the visibility at 156 national parks and wilderness areas. The regulations call on all 50 states to establish goals for improving visibility and to develop long-term strategies for reducing the emissions (especially particles) that cause the problem.

International Effects. Although atmospheric oxides in the United States have dropped significantly, pollutants still come back down as acid deposition, generally to the east of their origin because of the way weather systems flow. The deposits cross national boundaries, too. As a result, Canada receives half of its acid deposition from the United States, and Scandinavia gets it mostly from Great Britain and other western European nations. Japan receives the wind-borne pollution from widespread coal burning in Korea and China. For good reason, the problem of acid deposition has been addressed at national and international levels. Accordingly, we turn next to the efforts made to curb air pollution.

CONCEPT CHECK   Would you expect the effects of acid rain to be greater in an area where the bedrock is limestone (calcium carbonate) or where it is granite? Why? ☑

19.4 Bringing Air Pollution Under Control

By the 1960s, it was obvious that pollutants produced by humans were overloading natural cleansing processes in the atmosphere. The unrestricted discharge of pollutants into the atmosphere could no longer be tolerated. As a result, Congress produced several important regulations to control air pollution. Those regulations were highly effective. By 2014, scientists were finding that several critical air pollutants, including those that cause acid rain, were declining in the United States. The changes were so dramatic that rain was increasing over cities where smog had dampened it, and the waters of the Chesapeake Bay were getting cleaner. Although air pollution remains a problem, the improvement in air quality in much of the United States over the last few decades is one of the success stories of the modern environmental movement. Let's begin by looking at some of the public policies that brought about that improvement.

Clean Air Act

Federal legislation regarding air pollution was first introduced in 1955 as the **Air Pollution Control Act**. Several minor amendments followed until, under grassroots pressure from citizens, the U.S. Congress enacted a major bill, the **Clean Air Act Amendments of 1970 (CAA)**. Together with amendments passed in 1977 and 1990, this law, administered by the EPA, represents the foundation of U.S. air-pollution control efforts. The act called for identifying the most widespread pollutants, setting **ambient standards** (levels that need to be achieved to protect environmental and human health), and establishing control methods and timetables to meet standards.

Air Quality Standards. The CAA mandated the setting of standards for four of the primary pollutants—particulates, sulfur dioxide, carbon monoxide, and nitrogen oxides—and for the secondary pollutant ozone. At the time, these five pollutants were recognized as the most widespread and objectionable ones. Today, with the addition of lead, they are known as the **criteria pollutants** and are covered by the **National Ambient Air Quality Standards (NAAQS)** (Table 19–3). The primary standard for each pollutant is based on the presumed highest level that can be tolerated by humans without noticeable ill effects, minus a 10–50% margin of safety. For some of the pollutants, long-term and short-term levels are set. The short-term levels are designed to protect against acute effects, while the long-term

Table 19–3   National Ambient Air Quality Standards for Criteria Pollutants (NAAQS)

Pollutant	Averaging Time*	Primary Standard
PM$_{10}$ particulates	24 hours	150 μg/m^3
PM$_{2.5}$ particulates**	1 year	13 μg/m^3
	24 hours	35 μg/m^3
Sulfur dioxide	1 hour	0.075 ppm
	3 hours	0.5 ppm
Carbon monoxide	8 hours	9 ppm
	1 hour	35 ppm
Nitrogen dioxide	Annual mean	0.053 ppm
	1 hour	0.100 ppm
Ozone	8 hours	0.075 ppm***
Lead	3 months	0.15 μg/m^3

*The averaging time is the period over which concentrations are measured and averaged.
**PM$_{2.5}$ is the particulate fraction having a diameter smaller than or equal to 2.5 micrometers.
***New proposed standard is 0.065–0.070 ppm.
Source: U.S. EPA, Office of Air Quality Planning and Standards, *National Ambient Air Quality Standards.* October 2011. www.epa.gov/air/criteria.html.

standards are designed to protect against chronic effects and damage to crops, animals, vegetation, and buildings. The EPA is required to review the pollutants every five years and make adjustments if the science justifies changes.

Emission Standards. In addition, the CAA required the EPA to establish **National Emission Standards for Hazardous Air Pollutants (NESHAPs)** for a number of toxic substances (the air toxics). The Clean Air Act Amendments of 1990 greatly extended this section of the EPA's regulatory work by specifically naming 187 toxic air pollutants for the agency to track and regulate. The EPA tracks both the criteria pollutants and the hazardous air pollutants in a database, the National Emissions Inventory. This database serves as an essential source of information for the EPA in carrying out its regulatory responsibilities.

Control Strategy. The basic strategy of the 1970 Clean Air Act was to regulate the emissions of air pollutants so that the *ambient* criteria pollutants would remain below the primary standard levels. This approach is called **command-and-control** because regulations were enacted requiring industry to achieve a set limit on each pollutant through the use of specific control equipment. The assumption was that human and environmental health could be significantly improved by a reduction in the output of pollutants. If a particular region was in violation for a given pollutant, a local government agency would track down the source(s) and order reductions in emissions until the region came into compliance.

Unfortunately, this strategy proved difficult to implement. Most of the regulatory responsibility fell on the states and cities, which were often unable or unwilling to enforce control. In spite of these difficulties, however, total air pollutants were reduced by some 25% during a time when both population and economic activity increased substantially, and even in the noncompliant regions, the severity of air-pollution episodes lessened. Accordingly, some significant changes were made in the law and in the regulations that followed.

1990 Amendments. The Clean Air Act Amendments of 1990 (CAAA) targeted specific pollutants more directly and enforced compliance more aggressively, through such means as the imposition of sanctions. As with the earlier law, the states do much of the work in carrying out the mandates of the 1990 act. Each state must develop a State Implementation Plan (SIP) that is designed to reduce the emissions of every NAAQS pollutant whose control standard (Table 19–3) has not been attained. One major change was the addition of a permit application process: Polluters must apply for a permit that identifies the kinds of pollutants they release, the quantities of those pollutants, and the steps they are taking to reduce pollution. Permit fees provide funds the states can use to support their air-pollution control activities.

Under the CAAA, regions of the United States that have failed to attain the required levels must submit *attainment plans* based on **reasonably available control technology (RACT)** measures. Offending regions must convince the EPA that the standards will be reached within a certain time frame. The state SIPs and attainment performances are enforced by EPA sanctions that impose costly withholding of federal funds in case of failures.

Dealing with the Pollutants

The Clean Air Act and its amendments give the EPA broad authority to write regulations for reducing air pollutants, using a process known as rulemaking. The process involves a proposed rule stage and a final rule stage. The EPA will give notice first of a proposed rule, which allows a period of one to three months for public review and comments. Next, the agency is required to consider the comments received. When the final rule is published, responses to the comments are usually discussed, and any changes made on the basis of the comments are noted. Then the final rule is published in the Federal Register (available on the Internet), the official daily publication for a host of federal notices, regulations, executive orders, and other presidential documents. Once published, a final rule has the force of law until it is amended or repealed. We will consider the air pollution regulations by the different categories of pollutants.

Reducing Particulates. Prior to the 1970s, the major sources of particulates were industrial stacks and the open burning of refuse. The CAA mandated the phaseout of open burning of refuse and required that particulates from industrial stacks be reduced to "no visible emissions." To reduce stack emissions, industries were required to install filters, electrostatic precipitators, and other devices (though many facilities in place before the 1970 CAA were exempted). These measures markedly reduced the levels of particulates since the 1970s, but particulates continued to be released from steel mills, power plants, cement plants, smelters, construction sites, and diesel engines. Wood-burning stoves and wood and grass fires also contribute to the particulate load, making regulation even more difficult. Before 1987, the NAAQS particulate standard was "total suspended particulates." In 1987, this was replaced by the PM_{10} standard (all particles 10 micrometers or less in diameter), based on the understanding at the time of particles that could reach the interior regions of the respiratory tract.

The EPA added a new ambient air quality standard for particulates—$PM_{2.5}$—in 1997, on the basis of information indicating that smaller particulate matter (less than 2.5 micrometers in diameter) has the greatest effect on health. The coarser material (PM_{10}) is still regulated but is not considered to play as significant a role in health problems as do the finer particles. There is overwhelming evidence that the finer particles (2.5 microns or less) go right into the lungs when breathed. In response to this evidence, the EPA revised the particulate standard in 2006, reducing the amount of $PM_{2.5}$ to which a person could be exposed in 24 hours to 35 $\mu g/m^3$. In 2012, the EPA lowered the annual exposure standard to 12–13 $\mu g/m^3$.

SOUND SCIENCE
Complex Clouds and Co-Benefits of Solutions

One of the striking features of air pollution is its complexity. Atmospheric brown clouds (ABCs) provide an excellent example of this complexity. These huge clouds contain both sulfate aerosols and black carbon particles, such as those released by diesel fumes. The black carbon particles absorb energy, causing a warming effect. At the same time, the sulfate aerosols reflect light, which causes cooling. The sulfate aerosols also affect water droplets, causing the clouds to produce smaller droplets, so rain is less likely to form. These aerosols and particles are linked together in tiny droplets, so the processes of cooling and warming occur simultaneously throughout the clouds. The haze of the ABCs reduces the amount of light reaching the surface of the land and ocean (global dimming), thus decreasing temperatures and evaporation near the surface, while in the lower atmosphere ABCs tend to increase temperature.

These complex interactions mean that across the globe, scientists estimate that ABCs have a net effect of cooling, which means that they mask some of the warming effects of greenhouse gases. But on a local scale, particularly over China and India, these clouds have a net heating effect. The effects of ABCs on rainfall are also varied: They decrease rain and snowfall in critical areas such as the north of India, the Himalayas, and northern China while increasing precipitation in southeastern China. As you might imagine, predicting the impact of aerosols in clouds is one of the biggest problems climate scientists face!

It might be tempting to think we should leave the ABC clouds as they are, in case their cooling effect is offsetting some of the impacts of climate change. Some people have even suggested releasing sulfur aerosols into the stratosphere in order to reflect light and cool the planet! Unfortunately, that approach doesn't solve climate change and is likely to cause other problems.

Instead, we should be looking for ways to solve more than one problem at the same time; such added benefits are called *co-benefits* (Chapter 2). In one recent study, researchers found that the benefits of the Clean Air Act have included not only better health

from cleaner air, but also an improvement in the quality of water, as seen in the Chesapeake Bay.

On the global scale, efforts to solve the problem of climate change include reducing the burning of wood and fossil fuels. Such changes will improve smog as well.

In many developing nations, forests are burned to clear land for agriculture. In Indonesia, palm oil companies have been accused of setting illegal fires to clear vast tracts of forest for plantations; the drying peat that was beneath the forests continues to burn long after the original fire. In 2014, an Indonesian court fined a palm oil company for illegal habitat destruction. By 2014, at least 20 international food companies had pledged to produce and supply only "deforestation–free" palm oil. In 2015, two giant palm oil traders suspended purchases from a plantation developer who had repeatedly been in trouble for deforestation. Holding large corporations accountable is part of both clearing up ABCs and slowing climate change.

Investing in public transportation, cleaner power plants, alternative energy sources, and efficient cook stoves will lower the emission of greenhouse gases while also reducing the respiratory illnesses and crop losses that result from polluted air. The reverse is also true. Addressing crop loss by preventing air pollution will help solve the problems of acid rain and climate change. Preventing respiratory illness by controlling the production of smog will save crops and keep the planet from warming.

One recent scientific study showed that the plants growing in cities reduce air pollution more than scientists had realized. Planting

The sun shines dimly through the brown smog over the Taj Mahal in India.

grasses, ivy, and other plants removed NO_x and small particulates from the air; the plants also lowered ground-level ozone. In another study, scientists found that trees in London were able to remove particulates from the air and that planting a mix of trees around heavily polluted areas could improve air quality.

To address large-scale environmental issues such as ABCs and climate change, we need to seek solutions that solve multiple problems at once. Such solutions will help us achieve a sustainable future.

Sources: R. Agrawal et al. *Atmospheric Brown Clouds: Regional Assessment Report With Focus on Asia.* Nairobi, Kenya: UNEP, 2008.

University of Maryland Center for Environmental Science. "Clean Air Act Has Led to Improved Water Quality in the Chesapeake Bay Watershed." *ScienceDaily* (November 6, 2013).

American Chemical Society. "Green Plants Reduce City Street Pollution up to Eight Times More Than Previously Believed." *ScienceDaily* (July 18, 2012).

University of Southampton. "How Trees Clean the Air in London." *ScienceDaily* (October 5, 2011).

Figure 19–22 Air toxics monitoring station. A radiation monitoring device has been added to this air toxics monitoring station located on the grounds of a school in Anaheim, California.

Controlling Air Toxics. By EPA estimates, the amount of air toxics emitted annually into the air in the United States is around 4.7 million tons. These come from stationary sources (factories, refineries, power plants), mobile sources (cars, trucks, buses), and indoor sources (solvents, building materials). Cancer and other adverse health effects, environmental contamination, and catastrophic chemical accidents are the major concerns associated with this category of pollutants. Under the CAAA, Congress identified 187 toxic pollutants. It then directed the EPA to identify major sources of these pollutants and to develop **maximum achievable control technology (MACT)** standards, the NESHAPs. Besides affecting control technologies, the standards include options for substituting nontoxic chemicals, giving industry some flexibility in meeting MACT goals. State and local air-pollution authorities are responsible for seeing that industrial plants achieve the goals. In response to the CAAA, the EPA has mobilized more than 300 monitoring sites to track emissions of the affected substances (Figure 19–22). To reduce the contribution from vehicles, the agency requires cleaner burning fuels (reformulated gasoline) in urban areas. Industrial sources are being addressed by setting emission standards for some 82 categories of stationary sources, such as paper mills and oil refineries.

After years of piecemeal and delayed regulations, the EPA published final rules in December 2011 that for the first time forced the older coal- and oil-fired power plants to control their pollutants. The major focus of these new standards is mercury, a contaminant in coal and oil that is released into the air when these fuels are burned. The standards—the **Mercury and Air Toxics Standards**—require all coal- and oil-fired power plants to limit their emissions of mercury, acid gases, and other metallic toxic pollutants, preventing 90% of the mercury and 88% of acid gases from being emitted. Because of the controls needed to accomplish these reductions, emissions of SO_2 and fine particles are also greatly reduced. These new controls are estimated to save between $37–$90 billion and prevent up to 11,000 premature deaths per year; the estimated cost to industry to effect these controls is $9.6 billion per year. There have been a series of appeals to these rules, but as of 2014 the rules had been upheld.

Limiting Pollutants from Motor Vehicles. Cars, trucks, and buses release nearly half of the pollutants that foul our air. Vehicle exhaust sends out VOCs, carbon monoxide, and nitrogen oxides that lead to ground-level ozone and peroxyacetyl nitrates (PANs). Additional VOCs come from the evaporation of gasoline and oil vapors from fuel tanks and engine systems. The 1970 CAA mandated a 90% reduction in these emissions by 1975. This timing proved to be unrealistic, but enough improvements have been made over the years that today's new cars have actually achieved a 90% emissions reduction. This improvement is fortunate because driving in the United States has been increasing much more rapidly than the population has. Between 1970 and 2011, the number of vehicle miles increased from 1 trillion to 3 trillion miles per year, and between 1980 and 2010, the number of registered vehicles increased more than 56%. It is hard to imagine what the air would be like without the improvements mandated by the CAA.

Emissions. The reductions in automobile emissions have been achieved with a general reduction in the size of passenger vehicles, along with a considerable array of pollution-control devices, among which is one that affords the computerized control of fuel mixture and ignition timing, allowing more complete combustion of fuel and decreasing VOC emissions. To this day, however, the most significant control device on cars is the **catalytic converter** (Figure 19–23). As exhaust passes through this device, two different catalysts are at work: a reduction catalyst that converts NO_x emissions to harmless nitrogen gas and an oxidation catalyst that oxidizes unburned hydrocarbons and carbon monoxide to carbon dioxide.

All vehicles used for personal transportation, up to larger SUVs and passenger vans, are subject to specific emissions standards. These are the emissions that are tested by state-certified inspection stations, now required in most states.

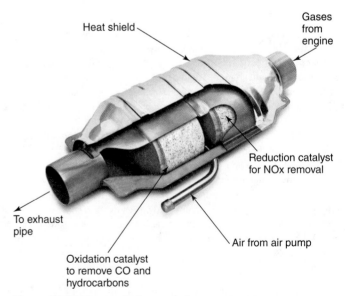

Heat shield

Gases from engine

Reduction catalyst for NOx removal

To exhaust pipe

Air from air pump

Oxidation catalyst to remove CO and hydrocarbons

Figure 19–23 Cutaway of a catalytic converter. Engine gas is routed through the converter, where catalysts promote chemical reactions that change the harmful hydrocarbons, carbon monoxide, and nitrogen oxides into less harmful gases, such as carbon dioxide, water, and nitrogen.

ADJUSTED FUEL ECONOMY BY MODEL YEAR

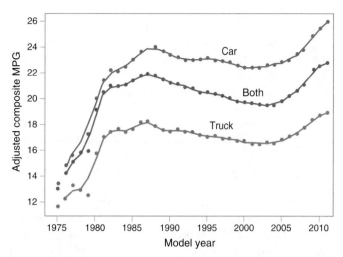

Figure 19–24 Fuel economy and light-duty vehicles. Changes in the fuel economy in model years from 1975 to 2011. "Truck" includes pickups, family vans, and SUVs.

(*Source:* EPA, Office of Transportation and Air Quality. Light-Duty Automotive Technology, Carbon Dioxide Emissions, and Fuel Economy Trends: 1975 Through 2011. EPA-420-R-12-001a, March 2012.)

Efficiency. Although the Clean Air Act does not address fuel efficiency, it is obvious that less fuel burned means fewer pollutants emitted. Two negative factors have affected fuel efficiency and consumption rates. First, the elimination of federal speed limits in 1996 reduced fuel efficiencies because of the higher speeds. Second, the sale of light trucks, including sport-utility vehicles (SUVs), minivans, and pickup trucks, surged in the 1990s and early 2000s. The EPA adjusts its estimate of the fuel economy of each model year based on real-world values for different types of vehicles, such as automobiles and light trucks. These fuel efficiencies have been increasing, in part because of government standards **(Figure 19–24)**.

CAFE Standards. Under the authority of the Energy Policy and Conservation Act of 1975 and its amendments, the National Highway Traffic Safety Administration was given the authority to set *corporate average fuel economy* (CAFE) standards for motor vehicles. The intention of the law was to conserve oil and promote energy security. The standard set in 1984 for passenger cars required a fleet average of 27.5 mpg. The standard for light trucks (pickups, SUVs, minivans) was set at 20.7 mpg and eventually raised to 22.2 mpg. The **Energy Independence and Security Act of 2007** mandated a new CAFE fuel economy standard of 35 mpg for the combined fleet of cars and light trucks, to be achieved by 2020. In 2009, the EPA set more stringent standards with an expectation that fuel economy standards would increase yearly. Greenhouse gas emission standards were also set and a new set of CAFE standards was rolled out: the combined fleet would have to achieve a 54.5 mpg and 163 g/mile CO_2 emissions target by 2025.

Getting Around. Even though many Americans continue to drive SUVs and larger automobiles, there are definite signs of change. One option for drivers who want better mileage is the **hybrid electric vehicle,** which combines a conventional gasoline engine and a battery-powered electric motor to achieve substantial improvements in fuel economy. Such vehicles have been marketed by car manufacturers, including Honda, Toyota, Ford, and GM, for more than a decade and are the most fuel-efficient cars in the United States. The Toyota Prius averages 46 mpg, while the Honda Insight gets 43 mpg, according to EPA figures.

As more and more of the new hybrids are sold, prices are expected to drop substantially. For a time, a federal tax credit was available for hybrid buyers that offset much of the cost differential. Many states still offer tax incentives for hybrid buyers.

Another trend is the development of plug-in hybrid cars and electric cars. The plug-in hybrid has a larger battery pack than the conventional hybrid and can be driven up to 40 miles on a charge; after that, a down-sized gasoline engine either recharges the battery or serves as the source of propulsion. Electric vehicles are just that—all electric, with rechargeable batteries that give the vehicle a range up to 100 miles, depending on the speed of driving. Both in-home and commercial charging stations are becoming available to enable speedy recharging of the batteries.

Managing Ozone. Because ozone is a secondary pollutant, the only way to control ozone levels is to address the compounds that lead to ozone formation. For a long time, it was assumed that the best way to reduce these levels was simply to reduce emissions of VOCs. The steps pertaining to motor vehicles in the CAAA addressed the sources of about half of the VOC emissions. Point sources (industries) accounted for another 45% of such emissions, and area sources (numerous small emitters, such as dry cleaners, print shops, and users of household products) represented the remaining 5%. These efforts have been effective. Since passage of the CAAA, VOC emissions have declined by 50%.

However, recent understanding of the complex chemical reactions involving NO_x, VOCs, and oxygen has thrown some uncertainty into the strategy of simply emphasizing a reduction in VOCs. The problem is that both NO_x and VOC concentrations are crucial to the generation of ozone (Figure 19–11). Either one can become the rate-limiting factor in the reaction that forms ozone. Thus, as the ratio of VOCs to NO_x changes, the concentration of NO_x can become the controlling chemical factor, pointing to the need to reduce NO_x emissions as well.

New Ozone Standards? A revised ozone standard was proposed in 1997, but a court battle delayed its implementation until 2004. Cost-benefit estimates indicated that the anticipated health benefits would far outweigh the costs of compliance. In 2008 the EPA lowered the ozone exposure standard, and in 2015 there was a proposal to lower it still further. As we find out more about the impact of ground-level ozone, our understanding of its dangers increases.

Down with NO_x. In a response to petitions from states in the Northeast that were having trouble achieving the ambient standards for ozone and particulates because of out-of-state emissions, the EPA has implemented regulations to reduce NO_x emissions from mobile sources, power plants, and large industrial boilers and turbines. These regulations require SUVs, minivans, and pickup trucks to be held to the same emissions criteria as passenger cars. In all, these changes represent an 83% reduction over the 2003 standards.

Protecting Downwind States. States in the eastern half of the country continue to bear the burden of pollutants that are carried long distances by wind and weather, making it more difficult for them to attain the NAAQS levels for ozone and particulate matter. To address this, the EPA announced a new rule in 2005 to accomplish further reductions of SO_2 and NO_x emissions. The **Clean Air Interstate Rule (CAIR)** set lower caps on SO_2 and NO_x in 28 states in the Midwest and East, many of which are upwind of states that have not attained the emissions standards for ozone and particulate matter. CAIR was expected to reduce SO_2 emissions by 50% beginning in 2010 and by 65% beginning in 2015. NO_x emissions were expected to be reduced by similar percentages. By 2012, SO_2 emissions had been reduced by 68% and NO_x emissions by 53%.

CAIR received a legal setback when a federal appeals court overturned the CAIR regulations. In response, the EPA, numerous states, and the Environmental Defense Fund successfully petitioned the court to reconsider its decision. In 2008, the court reversed its ruling and kept CAIR in force.

The Next Generation Legislation. In 2011, the EPA announced the **Cross-State Air Pollution Rule (CSAPR)** to replace CAIR. This rule addresses the same pollutants as CAIR and several additional states in the Midwest **(Figure 19–25)**. The rule required reductions in NO_x and SO_2 to begin in 2012 and tighten further in 2014, with the final reductions in power plant SO_2 emissions of 73% from 2005 levels and NO_x reductions of 54%. The rule involves some significant improvements in the manner in which upwind states are required to protect downwind states. The ruling

was challenged in 2012 and briefly suspended but ultimately upheld by the Supreme Court. In 2015, the EPA rolled out a new schedule for power plant compliance.

Cap-and-Trade. To accomplish the emissions reductions, both CAIR and CSAPR involve a system that was first implemented with the Acid Rain Program (ARP), called **cap-and-trade.** A target cap, or budget, is set by EPA for SO_2 or NO_x, in units of millions of tons of emissions. At the start, power plants are granted emission allowances based on formulas in the regulations. The plants are free to choose how they will achieve their allowances, whether by installing emissions control systems or purchasing allowances from utilities that are under their budgeted allowance.

The impacts of these regulations for human health are significant. The EPA projected that the regulations would prevent 13,000 to 34,000 premature deaths, 15,000 nonfatal heart attacks, and 400,000 cases of aggravated asthma in 2014. The monetary benefits were estimated at $120 billion to $280 billion per year, while the costs of implementation were projected to be approximately $2.4 billion per year.[6]

Coping with Acid Deposition

Scientists working on acid-rain issues in the 1980s calculated that a 50% reduction in acid-causing emissions in the United States would effectively prevent further acidification of the environment. This reduction was not expected to correct the already bad situations, but together with natural buffering processes, it was estimated to be capable of preventing further environmental deterioration. Because about 50% of acid-producing emissions came from coal-burning power plants, control strategies focused on these sources.

Political Developments. Although evidence of the link between power-plant emissions and acid deposition was well established by the early 1980s, no legislative action was taken until 1990. The problem was one of different regional interests. Throughout the 1980s, a coalition of politicians from the Midwestern states, representatives of high-sulfur coal producers, and representatives from the electric power industry effectively blocked all attempts at passing legislation that would take action on acid deposition.

On the other side were New York and the New England states, as well as most of the environmental and scientific communities, which argued that it was necessary to address acid deposition and that the best way to do so was to control emissions from power plants. Also, since 70% of Canada's acid-deposition problem came from the United States, diplomatic pressure toward a resolution was applied.

With the passage of the 1990 amendments to the Clean Air Act, the controversy became history. However, two decades of action were lost because of political delays. Eventually, Canada and the United States signed a treaty stipulating that Canada would cut its power plant SO_2 emissions by half and cap them at 3.5 million tons by the year 2000. Canada has done well; its

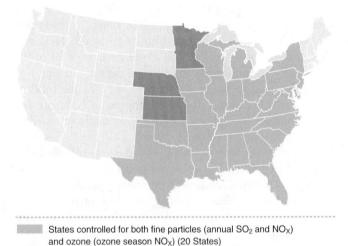

States controlled for both fine particles (annual SO_2 and NO_x) and ozone (ozone season NO_x) (20 States)

States controlled for fine particles only (annual SO_2 and NO_x) (3 States)

States controlled for ozone only (ozone season NO_x) (5 States)

States not covered by the Cross-State Air Pollution Rule

Figure 19–25 Map of states covered by the Cross-State Air Pollution Rule (CSAPR). The CSAPR is the 2011 EPA rule that replaces the Clean Air Interstate Rule (CAIR), designed to prevent downwind states from suffering from NO_x and SO_2 pollutants.

(*Source:* EPA CSAPR Fact Sheet. www.epa.gov/airtransport/statesmap.html. May 2, 2011.)

---

[6]EPA, Cross-State Air Pollution Rule (CSAPR), www.epa.gov/airtransport/. February 27, 2015.

emissions are now about 1.8 million tons and are expected to stay down. So has the United States.

Title IV of the Clean Air Act Amendments of 1990. One of the 1990 amendments to the Clean Air Act, Title IV, addressed the acid-deposition problem by mandating reductions in both sulfur dioxide and nitrogen oxide levels. The major provisions of the Title IV Acid Rain Program (ARP) included the following:

- Emissions of SO_2 from power plants were reduced to 10 million tons below 1980 levels. This involved setting a permanent cap of 8.9 million tons on such emissions.

- In a major departure from the command-and-control approach, Title IV promoted a free-market approach to regulation of SO_2. This was the original cap-and-trade program that the CAIR and CSAPR have been grafted on.

- New utilities do not receive allowances. Instead, they have to buy into the system by purchasing existing allowances. Thus, there are a finite number of allowances. In fact, some citizen's groups purchase allowances only to retire them, forcing companies to become more efficient.

- Nitrogen oxide emissions from power plants were regulated, primarily by regulating the boilers used by utilities and by mandating the continuous monitoring of emissions.

Accomplishments of Title IV. The utilities industry has responded to the new law with three actions:

1. Many utilities have switched to low-sulfur coal, available in Appalachia and the western United States.

2. Many older power plants have added scrubbers—"liquid filters" that put exhaust fumes through a spray of water containing lime. SO_2 reacts with the lime and is precipitated as calcium sulfate ($CaSO_4$). Because the technology for high-efficiency scrubbing is well established, more and more power plants are installing scrubbers, which have been required for all coal-burning power plants built after 1977.

3. Many utilities are trading their emission allowances. The purchase of allowances often represents a less costly way to achieve compliance than by purchasing low-sulfur coal or adding a scrubber. A typical transaction might involve a trade in the rights to emit 10,000 tons of sulfur dioxide at a cost of $19 a ton. The combination of approaches has created so many efficiencies that it has cut compliance costs to 10% of what was expected.

To carry out Title IV of the Clean Air Act Amendments, the EPA initiated a two-phase approach. Phase I aimed at a reduction of 3.5 million tons of SO_2 by 2000. Remarkably, the goal was met and even exceeded—at a cost far below the gloomy industry predictions (Figure 19–26). Phase II began in 2000 and targeted the remaining sources of SO_2 in order to reach the 8.9-million-ton cap by 2010. This goal was met for the first time in 2007, when power plant SO_2 emissions totaled 8.9 million tons of SO_2. The emissions continue to drop, falling to 5.17 million tons in 2010, as CAIR reductions have been implemented and as more and more power plants have switched from coal to natural gas. This is a decrease of 67% from 1990 levels (Figure 19–26).

The 2001 goal for a 2-million-ton reduction in NO_x emissions by fixed sources was met in 2000 and every year since; in fact, reductions in emissions of NO_x are more than double the Title IV reduction objective. The Acid Rain Reduction Program was responsible for a large portion of these reductions, but the CAIR has also played a significant role in recent years, as we have seen.

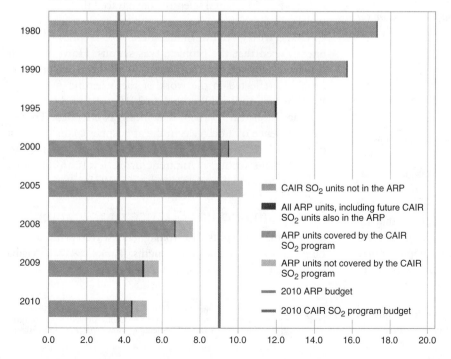

SO$_2$ EMISSIONS FROM CAIR SO$_2$ ANNUAL PROGRAM AND ACID RAIN PROGRAM SOURCES, 1980–2010

CAIR SO$_2$ units not in the ARP

All ARP units, including future CAIR SO$_2$ units also in the ARP

ARP units covered by the CAIR SO$_2$ program

ARP units not covered by the CAIR SO$_2$ program

2010 ARP budget

2010 CAIR SO$_2$ program budget

Figure 19–26 Power plant SO$_2$ emissions covered under the Acid Rain and CAIR programs. Emissions from all power plants were under ARP allowances for the first time in 2006 and even more so in 2007.

(*Source:* Executive Office of the President. National Acid Precipitation Assessment Program Report to Congress 2011: An Integrated Assessment. USGS, 2011. http://ny.water.usgs.gov.)

Field News. The news from the field reflects the trends. Under the Acid Rain Program, air quality has improved, with significant benefits to human health, reductions in acid deposition, some recovery of freshwater lakes and streams, and improved forest conditions. Concentrations of sulfate in rain and deposition on the land have shown a significant decline (41–69%) over a large part of the eastern United States in the past 18 years. The acidified streams and lakes of the Adirondacks and New England have seen some recovery, in the form of pH increases and increases in acid neutralizing capacity.

One reason for the slow recovery is the continued impact of nitrogen deposition. Apparently, nitrogen plays a much larger role in acid deposition than was once believed. In passing the CAAA, Congress did not set curbs on nitrogen emissions, but simply opted to reduce the emissions from fixed sources. The other reason for the slow recovery is the long-term buildup of sulfur deposits in soils. It takes a long time for residual sulfur to be flushed from natural ecosystems. Accordingly, most scientists believe that the recovery of the affected ecosystems will require further reductions in both sulfur and nitrogen emissions. Acid rain still falls on eastern forests, as Figure 19–14b indicates.

International Regulation of Air Pollution

Overall, the Clean Air Act has been a tremendous success in the United States. The Acid Rain Program alone reduced SO_2 by 40% over 20 years, as well as reducing NO_x, $PM_{2.5}$ particulates, and ozone. This occurred while output from coal-fired power plants increased by 25%. Yet as we have seen, global air pollution has skyrocketed. Air pollution problems cross boundaries and clearly affect people worldwide. Many countries lack the strong environmental regulations we have (although some have stronger regulations).

International treaties are one way of coordinating efforts to care for the global commons of the atmosphere. There are international treaties about climate change (Chapter 18) and the loss of the ozone layer (discussed shortly). The most important other air pollution treaty is *The Convention on Long-Range Transboundary Pollution* (LRTAP), which was rolled out in 1979. Forty-four countries, including the United States, have signed this treaty, which includes some recently added provisions limiting specific pollutants. In 1981, an existing international agreement about aviation (from 1944) was amended to limit aircraft engine emissions.

The European Union. Since the 1970s, the European Union has made air pollution a priority. Research there found that poor air quality is the leading environmental cause of premature death in the EU. While European countries are sovereign nations, they act in concert on many issues. For years they have had joint emissions standards and expectations that individual countries will monitor their own air and make plans to meet air quality standards. In 2013, the EU adopted a new *Clean Air Programme for Europe*, which includes plans to lower air pollution to specific targets by 2030.

Regional Air Pollution Networks. The United Nations Environmental Program (UNEP) has helped countries in several regions form networks to protect air quality. These include the Lusaka Agreement on air quality in southern Africa and similar agreements for eastern Africa and the combined region of central and western Africa. In Asia, the UNEP helped bring about the Malé Declaration on Air Pollution and a treaty on the control of haze pollution caused by clearing forests. In central Asia, the UNEP worked with countries to produce an environmental treaty considering air pollution as a major issue. In Latin America, it worked with the environmental ministers of countries to produce a joint statement on preventing air pollution. These international collaborations are moving air pollution out of the realm of separate nations and into the arena of coordination and cooperation among states.

Sustainable Development Goals and Air Quality. Air pollution comes up in larger international planning efforts as well. Chapter 9 of *Agenda 21,* the 1992 document from the Earth Summit in Rio de Janeiro, was titled "Protection of the Atmosphere." Although the Sustainable Development Goals do not have a specific goal for the protection of the atmosphere, Goal 3 (*Ensure healthy lives and promote well-being for all at all ages*) has this target: "By 2030, substantially reduce the number of deaths and illnesses from hazardous chemicals and air, water, and soil pollution and contamination." Achieving the targets for at least six of the other Sustainable Development Goals will require improved air quality.

CONCEPT CHECK What evidence do we have that the Clean Air Act was successful? ☑

19.5 Destruction of the Ozone Layer

The stratospheric ozone layer protects Earth from harmful ultraviolet radiation. The depletion of this layer is another major atmospheric challenge that is the result of human technology. Environmental scientists have traced the problem to a widely used group of chemicals: the **chlorofluorocarbons,** or **CFCs.** As with global warming, there were skeptics who were unconvinced that the problem was serious. However, scientists have achieved a strong consensus on the ozone depletion problem, reflected in the work of the Scientific Assessment Panel of the Montreal Protocol. In 2010, this panel concluded that the Montreal Protocol is working as intended: There is clear evidence of a decrease in the atmospheric concentration of ozone-depleting substances and some early signs of stratospheric ozone recovery. Antarctic ozone levels are projected to return to normal levels around 2060–2075.[7]

Radiation and Importance of the Shield

Solar radiation emits electromagnetic waves with a wide range of energies and wavelengths (Figure 19–27). Visible light is that part of the electromagnetic spectrum that can be

---

[7]World Meteorological Organization, Global Ozone Research and Monitoring Project—Report No. 52, Scientific Assessment of Ozone Depletion: 2010, http://ozone.unep.org/Assessment_Panels/SAP/Scientific_Assessment_2010/index.shtml.

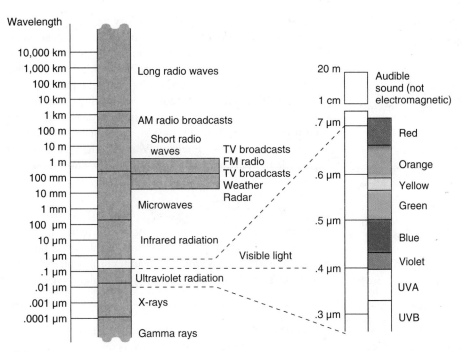

Figure 19–27 The electromagnetic spectrum. Ultraviolet radiation, visible light, infrared radiation, and many other forms of radiation are at different wavelengths on the electromagnetic spectrum.

UNDERSTANDING THE DATA

1. Which has a greater wavelength—infrared radiation or X-rays?
2. Roughly how many times greater is the wavelength of long radio waves than the wavelength of the longest gamma rays?

detected by the eye's photoreceptors. Ultraviolet (UV) wavelengths are slightly shorter than the wavelengths of violet light, which are the shortest wavelengths visible to the human eye. UVB radiation consists of wavelengths that range from 280 to 315 nanometers (0.28 μm to 0.32 μm), whereas UVA radiation is from 315 to 400 nanometers (0.32 μm to 0.40 mm). Because energy is inversely related to wavelength, UVB is more energetic and, therefore, more dangerous, but UVA can also cause damage.

Upon penetrating the atmosphere and being absorbed by biological tissues, UV radiation damages protein and DNA molecules at the surfaces of all living things. (This damage is what occurs when you get a sunburn.) If the full amount of UV radiation falling on the stratosphere reached Earth's surface, it is doubtful that any life could survive. We are spared the more damaging effects from UV rays because most UV radiation (more than 99%) is absorbed by ozone in the stratosphere. For that reason, stratospheric ozone is commonly referred to as the **ozone shield** (Figure 19–28).

UV Radiation and Human Health

Even the small amount (less than 1%) of UVB radiation that does reach Earth can cause serious health problems. The normal aging of skin—wrinkling, yellowing, and the development of irregular patches of heavily pigmented skin—is now known to be largely the result of damage caused by UV radiation from the Sun. So are eye damage, cataracts, and blindness. The most serious impact of chronic exposure to the Sun is skin cancer; more than 1 million new cases occur each year in the United States, according to the National Cancer Institute. Melanoma is the most deadly of the three forms of skin cancer attributable to UV exposure, as it metastasizes (spreads) easily. Melanoma is often traced to occasional sunburns during childhood or adolescence or to sunburns in people who normally stay out of the sun.

The effects of chronic exposure to the Sun are heightened by the loss of the ozone layer. The EPA publishes daily forecasts of UV exposure, issued by the Weather Service for regions across the country. The goal of this service is to remind people of the dangers of UV radiation and to prompt them to take appropriate actions, such as wearing sunglasses, protective clothing, and sunscreen, to avoid acute UV exposure.

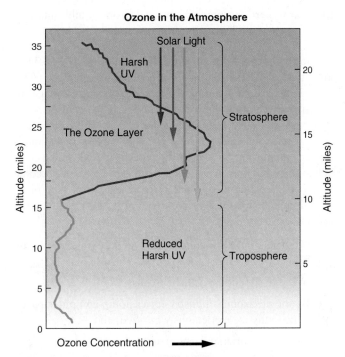

Figure 19–28 Ozone in the atmosphere. The vertical distribution of ozone in the troposphere (green) and stratosphere (red), showing the location of the stratospheric ozone layer that absorbs 99% of UV radiation.

(*Source:* Synthesis and Assessment Product 2.4. Report by the U.S. Climate Change Science Program and the Subcommittee on Global Change Research. November 2008.)

Formation and Breakdown of the Shield

Ozone is formed in the stratosphere when UV radiation acts on oxygen (O_2) molecules. The high-energy UV radiation first causes some molecular oxygen (O_2) to split apart into free oxygen (O) atoms, and these atoms then combine with molecular oxygen to form ozone via the following reactions:

$$O_2 + UVB \rightarrow O + O \qquad (1)$$

$$O + O_2 \rightarrow O_3 \qquad (2)$$

Not all of the molecular oxygen is converted to ozone, however, because free oxygen atoms may also combine with ozone molecules to form two oxygen molecules in the following reaction:

$$O + O_3 \rightarrow O_2 + O_2 \qquad (3)$$

Finally, when ozone absorbs UVB, it is converted back to free oxygen and molecular oxygen:

$$O_3 + UVB \rightarrow O + O_2 \qquad (4)$$

Thus, the amount of ozone in the stratosphere is dynamic. There is an equilibrium due to the continual cycle of reactions of formation (Eqs. 1 and 2) and reactions of destruction (Eqs. 3 and 4). Because of seasonal changes in solar radiation, ozone concentration in the Northern Hemisphere is highest in summer and lowest in winter. Also, in general, ozone concentrations are highest at the equator and diminish as latitude increases—again, a function of higher overall amounts of solar radiation. However, the presence of other chemicals in the stratosphere can upset the normal ozone equilibrium and promote undesirable reactions there.

Halogens in the Atmosphere. Chlorofluorocarbons (CFCs) are a type of halogenated hydrocarbon (see Chapter 22). CFCs are nonreactive, nonflammable, nontoxic organic molecules in which both chlorine and fluorine atoms have replaced some hydrogen atoms. At room temperature, CFCs are gases under normal (atmospheric) pressure, but they liquefy under modest pressure, giving off heat in the process and becoming cold. When they revaporize, they reabsorb the heat and become hot. These attributes led to the widespread use of CFCs (more than 1 million tons per year in the 1980s) for the following applications:

- In refrigerators, air conditioners, and heat pumps as the heat-transfer fluid;
- In the production of plastic foams;
- By the electronics industry for cleaning computer parts, which must be meticulously purified; and
- As the pressurizing agent in aerosol cans.

Rowland and Molina. All of the preceding uses led to the release of CFCs into the atmosphere, where they mixed with the normal atmospheric gases and eventually reached the stratosphere. In 1974, chemists Sherwood Rowland and Mario Molina published a classic paper (for which they were awarded the Nobel Prize in Chemistry in 1995) concluding that CFCs could damage the stratospheric ozone layer through the release of chlorine atoms and, as a result, UV radiation would increase and cause more skin cancer (their story is described in Chapter 2 to illustrate the public-policy life cycle).[8] Rowland and Molina reasoned that, although CFCs would be stable in the troposphere (where they have been found to last 70 to 110 years), in the stratosphere they would be subjected to intense UV radiation, which would break them apart, releasing free chlorine atoms via the following reaction:

$$CFCl_3 + UV \rightarrow Cl + CFCl_2 \qquad (5)$$

Ultimately, all of the chlorine of a CFC molecule would be released as a result of further photochemical breakdown. The free chlorine atoms would then attack stratospheric ozone to form chlorine monoxide (ClO) and molecular oxygen:

$$Cl + O_3 \rightarrow ClO + O_2 \qquad (6)$$

Furthermore, two molecules of chlorine monoxide may react to release more chlorine and an oxygen molecule:

$$ClO + ClO \rightarrow 2Cl + O_2 \qquad (7)$$

Reactions 6 and 7 are called the **chlorine catalytic cycle** because chlorine is continuously regenerated as it reacts with ozone. Thus, chlorine acts as a **catalyst,** a chemical that promotes a chemical reaction without itself being used up in the reaction. Because every chlorine atom in the stratosphere can last from 40 to 100 years, it has the potential to break down 100,000 molecules of ozone. Thus, CFCs are judged to be damaging because they act as transport agents that continuously move chlorine atoms into the stratosphere. The damage persists because the chlorine atoms are removed from the stratosphere only very slowly. Figure 19–29 shows the basic processes of ozone formation and destruction, including recent refinements to our knowledge of those processes that will be explained shortly.

EPA Action. After studying the evidence, the EPA became convinced that CFCs were a threat and, in 1978, banned their use in aerosol cans in the United States. Manufacturers quickly switched to nondamaging substitutes, such as butane, and things were quiet for several years. CFCs continued to be used in applications other than aerosols, however, and skeptics demanded more convincing evidence of their harmfulness.

Atmospheric scientists reasoned that any substance carrying reactive halogens to the stratosphere had the potential to deplete ozone. These substances include halons, methyl chloroform, carbon tetrafluoride, and methyl bromide. Chemically similar to chlorine, bromine also attacks ozone and forms a monoxide (BrO) in a catalytic cycle. Because of its extensive use as a soil fumigant and pesticide, bromine plays a major role in ozone depletion. (Bromine is 16 times as potent as chlorine in ozone destruction.)

[8]M. J. Molina and F. S. Rowland, "Stratospheric Sink for Chlorofluoro-Methanes: Chlorine-Atom Catalyzed Distribution of Ozone," *Nature* 249 (1974): 810–812.

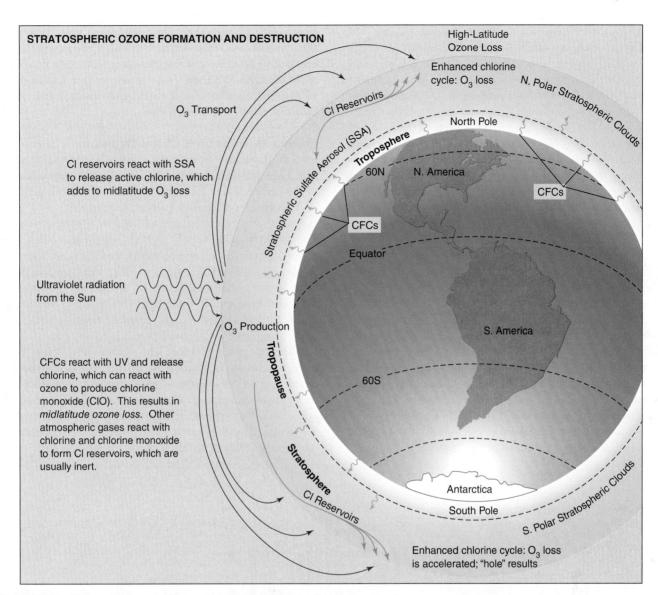

Figure 19–29 **Stratospheric ozone formation and destruction.** UV radiation stimulates ozone production at the lower latitudes, and ozone-rich air migrates to high latitudes. At the same time, CFCs and other compounds carry halogens into the stratosphere, where they are broken down by UV radiation, releasing chlorine and bromine. Ozone is subject to high-latitude loss during winter, as the chlorine cycle is enhanced by the polar stratospheric clouds. Midlatitude losses occur as chlorine reservoirs are stimulated to release chlorine by reacting with stratospheric sulfate aerosol.

The Ozone "Hole." In the fall of 1985, British atmospheric scientists working in Antarctica reported a gaping "hole" (actually, a serious thinning) in the stratospheric ozone layer over the South Pole. There, in an area the size of the United States, ozone levels were 50% lower than normal. The hole would have been discovered earlier by NASA satellites monitoring ozone levels except that computers were programmed to reject data showing a drop as large as 30% as being due to instrument anomalies. Scientists had assumed that the loss of ozone, if it occurred, would be slow, gradual, and uniform over the whole planet. The ozone hole came as a surprise, and if it had occurred anywhere but over the South Pole, the UV damage would have been extensive. As it is, the limited time and area of ozone depletion there have not apparently brought on any catastrophic ecological events so far.

News of the ozone hole stimulated an enormous scientific research effort. A unique set of conditions was found to be responsible for the hole. In the summer, gases such as nitrogen dioxide and methane react with chlorine monoxide and chlorine to trap the chlorine, forming so-called *chlorine reservoirs* (Figure 19–29) and preventing much ozone depletion.

When the Antarctic winter arrives in June, it creates a tightly swirling wind current in the stratosphere, with cold temperatures that promote clouds. The cloud particles provide surfaces on which chemical reactions release molecular chlorine (Cl_2) from the chlorine reservoirs.

When sunlight returns to the Antarctic in the spring, the Sun's warmth breaks up the clouds. UV light then attacks the molecular chlorine, releasing free chlorine and initiating the chlorine cycle, which rapidly destroys ozone. Eventually the wind current disperses, and ozone-rich air returns to the area.

Patches of ozone-depleted air move around, causing UV radiation to increase to 20% above normal in Australia. Television stations there now report daily UV readings and warnings for Australians to stay out of the sun. Scientists estimate that in Queensland, where the ozone shield is thinnest, two out of three Australians will develop skin cancer by the age of 70. People living in other parts of the Southern Hemisphere, such as New Zealand and Chile, also experience the effects of the thinner ozone shield. The ozone hole intensified during the 1990s and has leveled off between 22 million and 27 million square kilometers, an area as large as North America (Figure 19–30). Depletion of stratospheric ozone can occur in the Arctic as well, especially during cold winters.

Further Ozone Depletion. Ozone losses have not been confined to Earth's polar regions, though they are most spectacular there. A worldwide network of ozone-measuring stations sends data to the World Ozone Data Center in Toronto, Canada. Reports from the center reveal ozone depletion levels of 3.5% and 6% over the period 2006–2009 in the midlatitudes of the Northern and Southern Hemispheres, respectively. Polar ozone depletion and the subsequent mixing of air masses are responsible for much of this loss. There is virtually no ozone depletion over tropical regions. The good news from the network is that ozone levels are improving.

Is the ozone loss significant to our future? The EPA has calculated that the ozone losses of the 1980s will eventually have caused 12 million people in the United States to develop skin cancers over their lifetime and that 93,000 of these cancers will be fatal. Americans are estimated to have developed more than 900,000 new cases of skin cancer a year in the 1990s. The ozone losses of that decade allowed more UVB radiation than ever to reach Earth.

Coming to Grips with Ozone Depletion. The dramatic growth of the hole in the ozone layer (Figure 19–30) galvanized a response around the world. Scientists and politicians in the United States and other countries worked together to develop treaties designed to avert a UV disaster.

The Montreal Protocol. In 1987, under the auspices of its environmental program, the United Nations convened a meeting in Montreal, Canada, to address ozone depletion. Member nations reached an agreement, known as the **Montreal Protocol,** to scale back CFC production 50% by 2000. All 196 United Nations member countries have ratified the agreement.

The Montreal Protocol was written even before CFCs were so clearly implicated in driving the destruction of ozone and before the threat to Arctic and temperate-zone ozone was recognized. Because ozone losses during the late 1980s were greater than expected, an amendment to the protocol was adopted in June 1990. The amendment required participating nations to phase out the major

Figure 19–30 Ozone loss and extent of ozone hole. (a) The decline in ozone from 1979 to 2014 is shown in Dobson units for September 21 to October 16 each year (the average amount of ozone in the stratosphere is 300 Dobson units). **(b)** The size of the ozone hole is plotted for the same time frame, showing the origin of the hole and its leveling off in the last few years. No data were acquired in 1995, shown as a blue column in both plots.

(*Source:* NASA Ozone Hole Watch. http://ozonewatch.gsfc .nasa.gov/. February 28, 2015.)

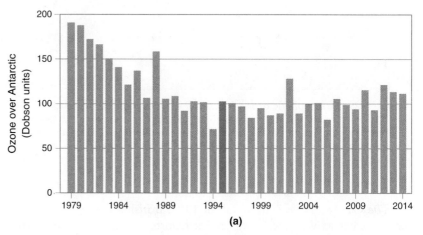

(a)

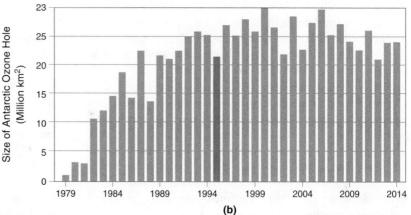

(b)

chemicals destroying the ozone layer by 2000 in developed countries and by 2010 in developing countries. In the face of evidence that ozone depletion was accelerating even more, another amendment to the protocol was adopted in November 1992, moving the target date for the complete phaseout of CFCs to January 1, 1996. Timetables for phasing out all of the suspected ozone-depleting halogens were shortened at the 1992 meeting.

Quantities of CFCs were still being manufactured after 1992 to satisfy legitimate demand in the developing countries. However, according to the Montreal Protocol, all manufacturing of CFCs was to have been halted by December 31, 2005. The results of the protocol can be seen clearly in Figure 19–31, which shows the time course of production of ozone-depleting substances (ODS) and the abundance of those chemicals in the stratosphere. Ozone depletion continues to occur because the ODS are long-lived in the stratosphere.

Action in the United States. The United States was the leader in the production and use of CFCs and other ODS, with DuPont Chemical Company being the major producer. Following years of resistance, DuPont pledged in 1988 to phase out CFC production by 2000. In late 1991, a company spokesperson announced that, in response to new data on ozone loss, DuPont would accelerate its phaseout. Many of the large corporate users of CFCs (AT&T, IBM, and Northern Telecom, for example) phased out their CFC use by 1994.

The Clean Air Act Amendments of 1990 also address this problem—in Title VI, Protecting Stratospheric Ozone. Title VI is a comprehensive program that restricts the production, use, emissions, and disposal of an entire family of chemicals identified as ozone depleting. For example, the program calls for a phaseout schedule for the hydrochlorofluorocarbons (HCFCs), a family of chemicals being used as less damaging substitutes for CFCs until nonchlorine substitutes are available. Halons—used in chemical fire extinguishers—were banned in 1994. Title VI also regulates the servicing of refrigeration and air-conditioning units.

The Montreal Protocol had its 25th anniversary in 2012. As a result of this and subsequent agreements, CFCs are no longer being produced or used. Substitutes for CFCs are readily available and, in some cases, are even less expensive than CFCs. The most commonly used substitutes are HCFCs, which still contain some chlorine and are scheduled for a gradual phaseout (by 2030). Although they contain chlorine, HCFCs are largely destroyed in the troposphere and are much less capable of destroying stratospheric ozone than CFCs. The most promising substitutes are hydrofluorocarbons (HFCs), which contain no chlorine and are judged to have no ozone-depleting potential. They do, however, have significant global warming potential. Because CFCs are potent greenhouse gases, phasing out CFCs and other ozone-depleting substances is helping the efforts to slow down global climate change (Chapter 18). Ironically, HFCs are likely to have a marked influence on global climate change if their use increases in the future.

Additional Ozone-Depleting Substances. Other ozone-depleting substances are still a concern. Methyl bromide continues to be produced and released into the atmosphere; the compound is employed as a soil fumigant to control agricultural pests. Under the Montreal Protocol, it was scheduled to be completely phased out by 2005 in the developed countries and 10 years later in the developing countries. In the meantime, bromine is likely to play a greater role in ozone depletion relative to chlorine. Responding to requests from the agricultural industry, the United States has repeatedly received "critical use" exemptions to the ban for use of this ozone-depleting chemical. However, the EPA is phasing out these uses; the 2012 level of use was just 4.6% of 1991.

Another ODS is nitrous oxide (N_2O). Figure 19–11a shows how nitric oxides are able to destroy ground-level ozone. This can also happen in the stratosphere, where N_2O is broken down into NO_x and the nitrogen oxides deplete ozone there, along with the chlorine and bromine

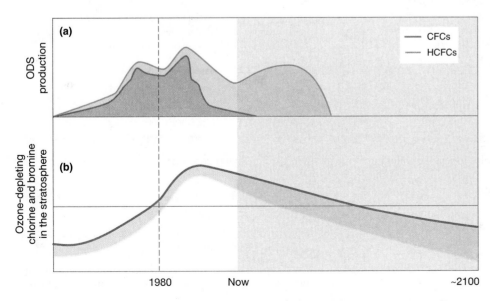

Figure 19–31 Production and presence of ozone-depleting substances in the atmosphere: Past, present, and future. (a) Production of ozone-depleting substances (ODS); CFCs are shown in green, and HCFCs, being used to replace CFCs, are shown in blue. **(b)** Relative abundances of ozone-depleting chlorine and bromine in the stratosphere. The impact of the Montreal Protocol and its amendments is clear.

(*Source:* UN Environmental Program, Scientific Assessment of Ozone Depletion: 2006.)

compounds. Even though the current ozone depletion in the stratosphere is largely due to accumulated chlorine and bromine compounds from past releases, nitrous oxide is the predominant anthropogenic ODS currently being released and is expected to remain the largest throughout the 21st century. Oddly, this substance is not yet addressed by the Montreal Protocol and its amendments.[9]

Final Thoughts. The ozone story is a remarkable episode in human history. From the first warnings in 1974 that something might be amiss in the stratosphere because of a practically

inert and highly useful industrial chemical, through the development of the Montreal Protocol and the final steps of CFC phaseout, the world has shown that it can respond collectively and effectively to a clearly perceived threat. The ozone layer is expected to recover completely by mid-century in the midlatitudes and a couple of decades later in the polar regions. The scientific community has played a crucial role in this episode, first alerting the world and then plunging into intense research programs to ascertain the validity of the threat. This is a most encouraging development as we look forward to the actions that must be taken during the rest of 21st century to prevent catastrophic global climate change.

CONCEPT CHECK How do the effects of ozone in the troposphere differ from the effects of ozone in the stratosphere? ✓

[9] A. R. Ravishankara et al., "Nitrous Oxide (N$_2$O): The Dominant Ozone-Depleting Substance Emitted in the 21st Century," *Science* 326 (October 2, 2009): 123–125.

REVISITING THE THEMES

SOUND SCIENCE

Atmospheric science has come a long way in the past half-century. Some of its triumphs include understanding causes and effects that are operative in photochemical smog, discovering the cleansing role of the hydroxyl radical, tracing the sources of lead and mercury in humans, working out causes and effects in the acid-rain story, and alerting the world to ozone shield destruction. Sound science is increasingly important in measuring ambient pollutant levels and in devising technologies to prevent more pollution. Many scientists have put their reputations and careers on the line as they have fought for policy changes to correct the human and environmental damage they have uncovered.

SUSTAINABILITY

Air pollution was on an unsustainable track, with industrial smog producing health disasters, photochemical smog making city living intolerable for many, crops and natural ecosystems deteriorating seriously, and the ozone shield being destroyed. Less obvious, but just as important, chronic exposure to air pollutants was responsible for poor health and increased mortality for millions. Are we now back on a more sustainable track? The important indicators would be long-term declines in the following: lead in the environment, annual amounts of emissions of pollutants, acid and

sulfate deposition in the Northeast, ambient levels of pollutants, asthma hospitalizations, and locations that are noncompliant. Most of these parameters are moving in the right direction, but much remains to be done. The ozone story represents a sustainable pathway. Levels of ozone-destroying chemicals are now declining, and it is very likely that the ozone layer will be "healthy" by midcentury. International trends in air pollution are mixed. While the regulations of the Montreal Protocol have lowered CFCs, many cities, especially in China and India, have excessive air pollution and its attendant effects on human health.

STEWARDSHIP

Stewardship means doing the right thing when faced with difficult problems and choices. It is fair to say that our society has responded well in caring for children threatened by high lead levels in their blood and, in general, in taking the costly actions we have documented in this chapter. Further, in addressing acid deposition, we have even shown care for the natural world. The need for stewardship is still very strong, however, and thankfully, there are many in nongovernmental and other organizations, and indeed many in the EPA, that are determined to stay the course and make sure that effective action is taken. Often, though, they have to do battle with industry lobbies and special interests that are concerned with the economic bottom line and that seem unable to embrace the common good represented by effective action to curb pollution.

REVIEW QUESTIONS

1. What naturally occurring cleanser helps to remove pollutants from the atmosphere? Where does it come from? What other mechanisms also act to cleanse the atmosphere of pollutants?

2. Describe the origins of industrial smog, photochemical smog, and atmospheric brown clouds. What are the differences in the cause and appearance of each?

3. What are the seven major primary air pollutants and their sources?

4. What are the major secondary pollutants, and how are they formed?

5. What is the difference between an acid and a base? What is the pH scale?

6. What two major acids are involved in acid deposition? Where does each come from?

7. What impacts does air pollution have on human health? Give the three categories of impact, and distinguish among them.

8. Describe the negative effects of pollutants on crops, forests, and other materials. Which pollutants are mainly responsible for these effects?

9. How can a shift in environmental pH affect aquatic ecosystems? What protects these ecosystems in some regions?

10. What are the criteria pollutants and the National Ambient Air Quality Standards, and how are the standards used?

11. Discuss ways in which the Clean Air Act Amendments of 1990 address the failures of previous legislation.

12. Describe the new Mercury and Air Toxics Standards and the impacts they will have on industry and on human health.

13. What are the major steps being taken to reduce emissions from passenger vehicles? What are the CAFE standards, and how are they changing?

14. How is a cap-and-trade system used to lower air pollution?

15. How do the global trends in air pollution differ from those in the United States?

16. Identify and describe several international efforts to limit air pollution.

17. How is the protective ozone shield formed? What causes its natural breakdown?

18. How do CFCs affect the concentration of ozone in the stratosphere and contribute to the formation of the ozone hole?

19. Describe the international efforts that are currently in place to protect the ozone shield. What evidence is there that such efforts have been effective?

THINKING ENVIRONMENTALLY

1. Motor vehicles release close to half the pollutants that dirty our air. What alternatives might be introduced to encourage a decrease in our use of automobiles and other vehicles with internal combustion engines?

2. How would our pattern of life be different in the absence of the Clean Air Act and its amendments? Write a short essay describing the possibilities.

3. The wind blows across Country A and into Country B. Country A has fossil fuel–burning power plants, but a low level of air pollution. Country B has power plants run by moving water (hydroelectric plants), but a higher level of pollution. What's happening? What could the two countries do to correct the situation? Explain how this pertains to international air-pollution laws.

4. A large amount of air pollution is attributed to manufacturing. Some would argue that it is not technologically possible, or economically efficient, to eliminate this pollution. How would you respond to this argument?

MAKING A DIFFERENCE

1. Avoid burning trash in your yard. It's illegal in many places and can lead to cancers and lung diseases.

2. Take advantage of public transportation. It helps ease urban congestion as well as atmospheric conditions in the area.

3. When you buy a new car, consider purchasing a hybrid-electric or an all-electric vehicle.

4. Discuss air-pollution issues with your employer. Many employers are realizing the benefits of considering air quality issues. Green travel plans, alternative-powered vehicles, and the provision of cycling facilities are just three ways employers can help to reduce their impacts on air quality, realize a financial savings, and maintain a healthy workforce.

MasteringEnvironmentalScience®

Students Go to MasteringEnvironmentalScience for assignments, the eText, and the Study Area with animations, practice tests, and activities.

Professors Go to MasteringEnvironmentalScience for automatically graded tutorials and questions that you can assign to your students, plus Instructor Resources.

Chapter 20

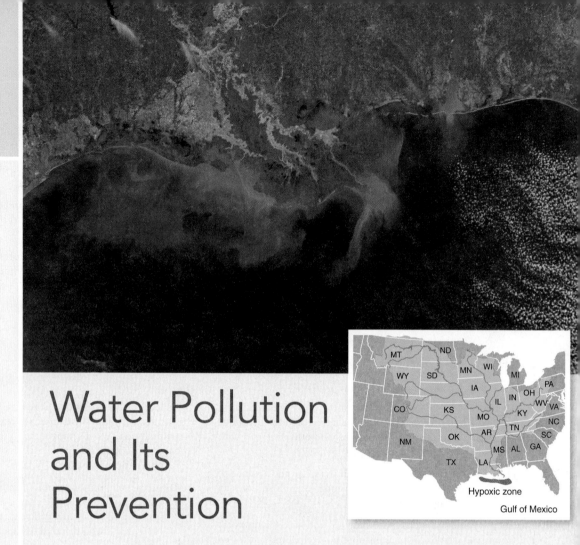

Hypoxic zone

Gulf of Mexico

Water Pollution and Its Prevention

The Mississippi River watershed encompasses 40% of the land area of the United States; the river collects water from this huge area and eventually delivers it to the Gulf of Mexico. Because much of the watershed is in America's agricultural heartland, the Mississippi River is a reflection of what happens on the nation's farms. During the latter half of the 20th century, more and more fertilizers were spread on agricultural fields, and more animals were raised on feedlots. At the same time, wetlands bordering river tributaries were drained and no longer intercepted agricultural runoff. The result was a tripling of soluble nitrogen delivered to the river and, eventually, to the Gulf of Mexico.

Dead Zone. The impact of this major flow of nitrogen to the Gulf of Mexico was first detected in 1974 by marine scientists who found areas in the gulf where oxygen had disappeared from bottom sediments and much of the water column above them. This absence of oxygen was deadly to the bottom-dwelling animals and any oxygen-breathing creatures in the water column that were not able to escape. At first, the hypoxic (lacking oxygen) area was thought to be a minor disturbance that would disappear seasonally. But then the hypoxic area suddenly doubled in size following the 1993 floods in the Midwest, and it continues to grow. The shrimpers and fishers who work these waters have learned to avoid the hypoxic area.

The Culprit. To the scientists working on the problem, the connection between nitrogen and hypoxia was clear. Nitrogen is a common limiting factor in coastal marine waters. Where abundant, it promotes the dense growth of phytoplankton—photosynthetic microorganisms that float in the water. Zooplankton—microscopic

▲ Sediment and nutrients from the Mississippi River watershed pour into the Gulf of Mexico.

animals in the water—rapidly consume the phytoplankton and multiply. As the abundant phytoplankton and zooplankton (and their fecal pellets) die or sink toward the bottom, they are decomposed by bacteria, a process that consumes dissolved oxygen. Eventually, the oxygen is used up, creating a dead zone that can extend up from the bottom of the sea to within a few meters of the surface.

Because the gulf's fishery is a $2.8 billion dollar enterprise, the problem eventually got national attention. In 1998, Congress passed the Harmful Algal Bloom and Hypoxia Research and Control Act, establishing an interagency task force of scientists charged with assessing the dead zone in the Gulf of Mexico. The Mississippi River/Gulf of Mexico Watershed Nutrient Task Force released its first report in May 2000, confirming the connection between nitrogen and the dead zone and laying out options for reducing the nitrogen load coming downriver. Two key strategies were the reduction of fertilizer use to greatly reduce farm runoff, and the restoration and promotion of nitrogen retention and denitrification processes in the basin.

In 2001, the task force proposed an action plan with the goal of reducing the size of the hypoxic area to 1900 square miles (5000 km^2) by 2015. Unfortunately, this goal was not met. In 2014, the dead zone occupied about 5,700 square miles (14,800 km^2).

The Gulf of Mexico dead zone shows that it is unreasonable to allow vital water resources to receive pollutants and then expect those resources to continue to provide us with their usual bounty of goods and services. Since the 1960s, the number of coastal dead zones has doubled every decade; worldwide, more than 550 are now known.

In this chapter, we will focus on water *quality*. (Chapter 10 dealt with water *quantity*—the global water cycle and water resources.) We will discuss eutrophication and other forms of water pollution and take a close look at wastewater treatment.

20.1 Perspectives on Water Pollution

In the early days of the Industrial Revolution, little attention was given to the pollution of lakes, rivers, and the coastal ocean. Industries emptied their wastes into the closest waterway, and the general means of disposing of human excrement was the outdoor privy (or behind the nearest bush). Seepage from the privy frequently contaminated drinking water and caused disease, especially in places where privies and wells were located near one another.

In the late 1800s, Louis Pasteur and other scientists showed that sewage-borne bacteria were responsible for many infectious diseases. This important discovery led to intensive efforts to rid cities of human excrement as expediently as possible. Cities already had drain systems for storm water, but using these systems for human wastes had been prohibited. With the urgency of the situation, however, minds quickly changed. The flush toilet was introduced, and sewers were tapped into storm drains. Thus, Western civilization adopted the one-way flow of flushing domestic and industrial wastewater into natural waterways. Introducing sewage into a waterway was symptomatic of an approach that believed "the solution to pollution is dilution."

As a result of this practice, receiving waters that had limited capacity for dilution became open cesspools of foul odors and filth as the overload of organic matter depleted dissolved oxygen and most aquatic life suffocated. For increasing distances around or downstream from the sewage outfall, the water became unfit for any recreational use. Any use of the water for human consumption required extensive treatment to remove pathogens.

Industrial wastes only added insult to injury. From the dawn of the Industrial Age to the relatively recent past, it was common practice to flush all waste liquids and contaminated water into sewer systems or directly into natural waterways. Many human health problems followed, but they either were not recognized as being caused by the pollution or were accepted as the "price of progress."

In the United States, the Federal Water Pollution Control Act of 1948 was the first federal water pollution legislation, providing technical assistance to state and local governments, but otherwise leaving things up to the states and local municipalities. In the 1950s, as industrial production expanded and synthetic organics came into widespread use, many streams and rivers essentially became open chemical sewers as well as sewers for human waste. These waters not only were devoid of life, but also were themselves hazardous. Finally, in 1969, the Cuyahoga River, which flows through Cleveland, Ohio, was carrying so much flammable material that it actually caught fire, which destroyed seven bridges before the fire burned itself out (Figure 20–1).

Figure 20–1 Cuyahoga River on fire. In 1969, the Cuyahoga River in Cleveland, Ohio, caught fire, burning seven bridges. This incident captured national attention and contributed to the public outrage over water pollution, leading to the Clean Water Act of 1972.

Figure 20–2 Point and nonpoint sources. Point sources are far easier to identify and correct than the diffuse nonpoint sources.

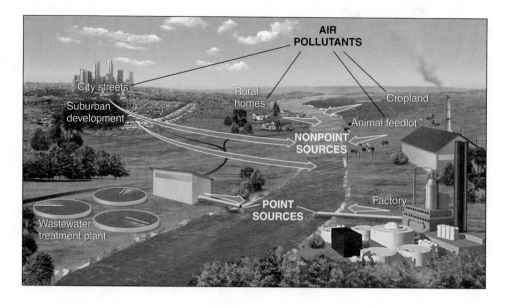

Worsening pollution (from both chemicals and sewage) and the increasing recognition of its adverse health effects finally created a degree of public outrage that pushed Congress to pass the **Clean Water Act of 1972 (CWA)**. This legislation gave the newly created Environmental Protection Agency (EPA) the responsibility to carry out a very ambitious goal—to restore and maintain the chemical, physical, and biological integrity of the nation's waters. The CWA was to achieve zero discharge of pollutants by 1985, a totally naïve (and impossible) task, given the nature of the pollutants involved. However, imperfect as it is, the Clean Water Act (including its amendments) is one of the most far-reaching and effective environmental laws ever enacted.

Water Pollution: Sources and Types

Water pollutants originate from a host of human activities and reach surface water or groundwater through an equally diverse host of pathways. Regulators distinguish between **point sources** and **nonpoint sources** of pollutants (Figure 20–2). Point sources involve the discharge of substances from factories, sewage systems, power plants, underground coal mines, and oil wells. These sources are relatively easy to identify and, therefore, easier to monitor and regulate than nonpoint sources, which are poorly defined and scattered over broad areas. Some of the most prominent nonpoint sources of pollution are agricultural runoff (from farm animals and croplands), storm-water drainage (from streets, parking lots, and lawns), and atmospheric deposition (from air pollutants washed to or deposited as dry particles on Earth). Pollution occurs as rainfall and snowmelt flow over and through the ground, picking up pollutants as they go.

Two basic strategies are employed in attempting to bring water pollution under control: (1) reduce or remove the sources and (2) treat the water before it is released so as to remove pollutants or convert them to harmless forms. Water treatment is the best option for point sources. Source reduction is the best option for nonpoint sources, although it can also be used for point sources.

Pathogens. The most serious and widespread water pollutants are infectious agents that cause sickness and death (see Chapter 17). The excrement from humans and other animals infected with certain **pathogens** (disease-causing bacteria, viruses, and other parasitic organisms) contains large numbers of these organisms or their eggs. Table 20–1 lists the most common waterborne pathogens. Even after symptoms of disease disappear, an infected person or animal may still harbor low populations of the pathogen, thus continuing to act as a carrier of disease. If wastes from carriers contaminate drinking water, food, or water used for swimming or bathing, the pathogens can infect other individuals (Figure 20–3).

Public Health. Before the connection between disease and sewage-carried pathogens was recognized in the mid-1800s, disastrous epidemics were common in cities.

Table 20–1	Pathogens Carried by Sewage
Disease	**Infectious Agent**
Typhoid fever	*Salmonella typhi* (bacterium)
Cholera	*Vibrio cholerae* (bacterium)
Salmonellosis	*Salmonella* species (bacteria)
Diarrhea	*Escherichia coli* (bacterium)
	Campylobacter species (bacteria)
	Cryptosporidium parvum (protozoan)
Infectious hepatitis	Hepatitis A virus
Poliomyelitis	Poliovirus
Dysentery	*Shigella* species (bacteria)
	Entamoeba histolytica (protozoan)
Giardiasis	*Giardia intestinalis* (protozoan)
Numerous parasitic diseases	Roundworms, flatworms

Figure 20–3 The Ganges River in India. In many places, the river is used simultaneously for drinking, washing, and the disposal of sewage, resulting in a high incidence of disease.

Epidemics of typhoid fever and cholera often killed thousands of people. Today public-health measures that prevent this disease cycle have been adopted throughout much of the world. The following measures are essential to the control of waterborne diseases:

- Purification and disinfection of public water supplies with chlorine or other agents
- Sanitary collection and treatment of human and animal wastes

- Maintenance of sanitary standards in all facilities in which food is processed or prepared for public consumption
- Instruction in personal and domestic hygiene practices

Standards regarding these measures are set and enforced by government public-health departments. A variety of other measures are enforced as well. For example, if bathing areas are contaminated with raw sewage, health departments close them to swimming.

Sanitation = Good Medicine. Many people attribute good health in a population to modern medicine, but good health is primarily a result of the prevention of disease through public-health measures. According to the 2014 report Progress on Drinking Water and Sanitation,[1] since 1990, 2 billion people have gained access to improved sanitation; 77 countries met the Millennium Development Goal of halving the number of people without access to safe drinking water and basic sanitation (Figure 20–4). Yet many regions still need to make significant progress. Every year, more than 2 million deaths (mostly among children under five) are traced to waterborne diseases. The Sustainable Development Goals about agriculture, cities, and management of resources all include targets to prevent water pollution. Moving forward toward 2030 will require providing

[1]Progress on Drinking Water and Sanitation: 2014 Update. UNICEF and World Health Organization. 2014. http://www.unicef.org/media/files/JMPreport2014.pdf

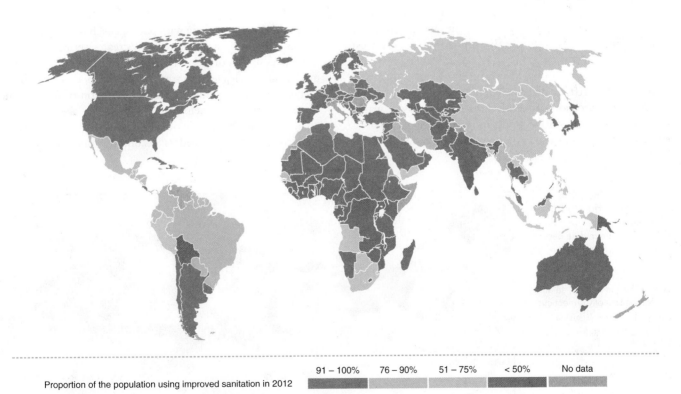

Proportion of the population using improved sanitation in 2012 — 91 – 100% | 76 – 90% | 51 – 75% | < 50% | No data

Figure 20–4 Worldwide distribution of improved sanitation. The map shows the proportion of the population in each country that had access to improved sanitation in 2012. The Millennium Development Goal of reducing by half the proportion of people without improved sanitation was not met, but efforts continue.

(*Source:* Progress on Drinking Water and Sanitation: 2014 Update. UNICEF and World Health Organization, 2014. www.unicef.org/gambia/Progress_on_drinking_water_and_sanitation_2014_update.pdf.)

Figure 20–5 Testing water for sewage contamination by the Millipore technique.
(a) A Millipore filter disk is placed in the filter apparatus. **(b)** A sample of the water being tested is drawn through the filter, trapping bacteria on the disk. **(c)** The disk is placed in a petri dish on a medium that supports the growth of bacteria and imparts a green color to fecal *E. coli* bacteria. Next, the dish is incubated for 24 hours, during which time the bacteria on the disk multiply to form a colony visible to the naked eye. **(d)** *Escherichia coli* bacteria are identifiable as the colonies with a metallic green sheen. The presence of *E. coli* indicates sewage contamination.

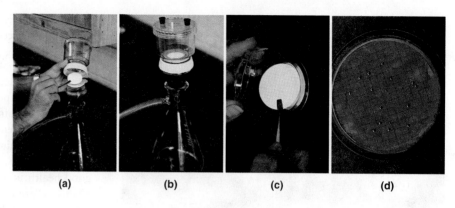

(a)   (b)   (c)   (d)

sanitation for the nearly one quarter of the world's people who currently lack toilets and must relieve themselves outdoors. (See Chapter 17, Sound Science, *Water Contamination Drives Malnutrition in India*.)

Largely because of poor sanitation regarding water and sewage, a significant portion of the world's population is chronically infected with various pathogens. Moreover, populations in areas where there is little or no sewage treatment are extremely vulnerable to deadly epidemics of diseases spread by sewage, especially cholera. The World Health Organization (WHO) reports numerous cholera outbreaks annually, and cases frequently number in the thousands, with hundreds of deaths. The 2010 earthquake in Haiti created huge refugee camps, where an outbreak of cholera brought 180,000 cases and 5,000 deaths. The source of the cholera there was traced to UN peacekeepers from Nepal, where the disease is endemic. Wastes from the peacekeepers' camps were simply emptied into the nearest watercourse.

Monitoring Pathogens. An important aspect of public health is the detection of sewage-borne pathogens in water. It is impossible to test for every pathogen that might be present; instead, scientists use an indirect method called the *fecal coliform test*. This test is based on the fact that large numbers of a bacterium called *Escherichia coli* (*E. coli*) are excreted with fecal material. In most situations, *E. coli* is not a pathogen, but it does serve as an indicator organism: Its presence indicates that water is contaminated with fecal wastes and that sewage-borne pathogens also may be present.

The fecal coliform test detects and counts the number of *E. coli* in a sample of water (Figure 20–5). To be safe for drinking, a 100 milliliter (0.4 cup) sample of water may not contain any *E. coli*. The same sample with up to 200 *E. coli* is considered safe for swimming. Beyond that level, a river or beach may be posted as polluted, and swimming and other direct contact should be avoided. By contrast, raw sewage (99.9% water, 0.1% waste) has *E. coli* counts in the millions.

Organic Wastes. Along with pathogens, human and animal wastes contain organic matter that creates serious problems if it enters bodies of water untreated. Other kinds of organic matter, such as leaves and trash, can enter bodies of water as a consequence of runoff. With the exception of plastics and some human-made chemicals, these wastes are biodegradable. As in the Gulf of Mexico dead zone (Figure 20–6), when bacteria and detritus feeders decompose organic matter in

Figure 20–6 The 2014 dead zone in the Gulf of Mexico. Each year oxygen disappears from middle depths to the bottom over thousands of square miles of the Gulf of Mexico. In 2014 the dead zone was 5,708 square miles, an area almost the size of Connecticut.

(*Source:* NOAA News, Aug. 4, 2014. With permission from LUMCON.)

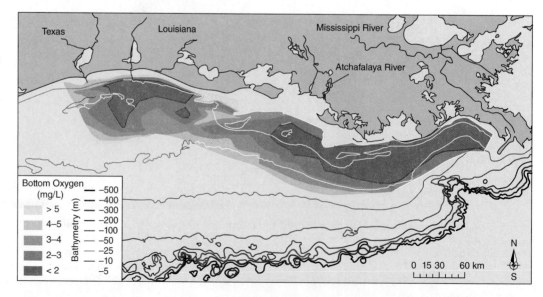

water, they consume oxygen gas dissolved in the water. The amount of oxygen that water can hold in solution is severely limited. In cold water, dissolved oxygen (DO) can reach concentrations up to 10 parts per million (ppm); much less can be held in warm water. Compare this ratio with that of oxygen in air, which is 200,000 ppm (20%), and you can understand why even a moderate amount of organic matter decomposing in water can deplete the water of its DO. Bacteria keep the oxygen supply depleted as long as there is dead organic matter to support their growth and oxygen replenishment is inadequate.

BOD. Biochemical oxygen demand (BOD) is a measure of the amount of organic material in water, stated in terms of how much oxygen will be required to break it down biologically, chemically, or both. The higher the BOD measure,

the greater the likelihood that dissolved oxygen will be depleted in the course of breaking down the organic material. A high BOD causes so much oxygen depletion that animal life is severely limited or precluded, as in the bottom waters of the Gulf of Mexico. Fish and shellfish are killed when the DO drops below 2 or 3 ppm, though some are unable to tolerate even higher DO levels. If all of the oxygen in the system is depleted, only bacteria can survive, using their abilities to switch to fermentation or anaerobic respiration (metabolic pathways that do not require oxygen). A typical BOD value for raw sewage would be around 250 ppm. Even a moderate amount of sewage added to natural waters containing at most 10 ppm DO can deplete the water of its oxygen and produce highly undesirable consequences well beyond those caused by the introduction of pathogens (Figure 20–7).

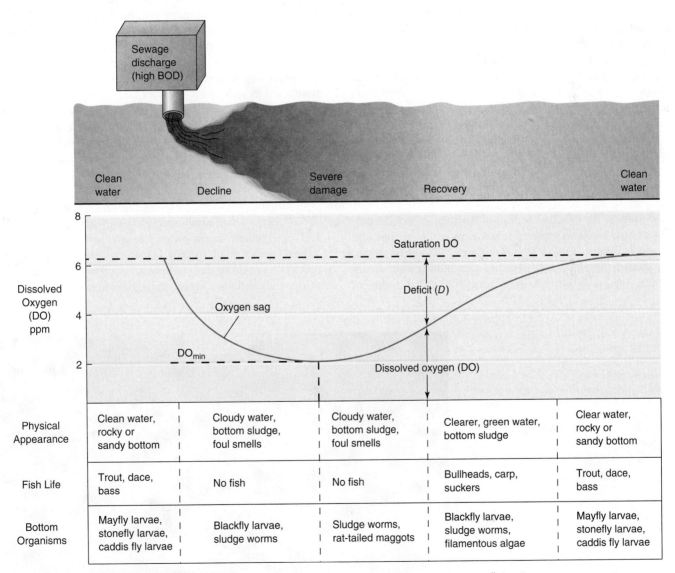

Figure 20–7 The oxygen sag curve. When sewage with a high BOD is discharged into a stream or small river, it creates an oxygen deficit and severely affects the stream's biology.

UNDERSTANDING THE DATA   When you sample the water in a river, your dissolved-oxygen meter reads 5 ppm. What aspects of the environment tell you whether the river is in the decline or recovery stage of the oxygen sag curve?

Chemical Pollutants. Because water is such an excellent solvent, it is able to hold many chemical substances in solution. Water-soluble **inorganic chemicals** constitute an important class of pollutants that include heavy metals (such as lead, mercury, arsenic, and nickel), acids from mine drainage (sulfuric acid) and acid precipitation (sulfuric and nitric acids), and road salts employed to melt snow and ice in the colder climates (sodium and calcium chlorides). The **organic chemicals** are another group of substances found in polluted waters. Petroleum products pollute many bodies of water, from the major oil spills in the ocean to the small streams receiving runoff from parking lots. Other organic substances with serious impacts are the pesticides that drift down from aerial spraying or that run off from agricultural fields and lawns, as well as the various industrial chemicals, such as polychlorinated biphenyls (PCBs), cleaning solvents, and detergents (Chapter 13).

Many of these pollutants are toxic even at low concentrations (Chapter 22). Some may become concentrated as they are passed up the food chain in a process called biomagnification (Chapter 13). Even at very low concentrations, they can render water unpalatable to humans and dangerous to aquatic life. At higher concentrations, they can change the properties of bodies of water so as to prevent them from serving any useful purpose except perhaps navigation. Acid mine drainage, for example, pollutes thousands of miles of streams in regions of the United States where coal and minerals are mined (Figure 20–8). Stream bottoms are coated with orange deposits of iron, and the water is so acidic that only a few hardy bacteria and algae can tolerate it.

Sediments. As natural landforms weather, especially during storms, a certain amount of sediment enters streams and rivers. However, erosion from farmlands, deforested slopes, overgrazed rangelands, construction sites, mining sites, stream banks, and roads can greatly increase the load of sediment entering waterways. Frequently, storm drains simply lead to the nearest depression or some natural streambed. Sediments (sand, silt, and clay) have direct and extreme physical impacts on streams and rivers. When erosion is slight, streams and rivers of the watershed run clear and support algae and other aquatic plants that attach to rocks or take root in the bottom. These producers, plus miscellaneous detritus from fallen leaves, support a complex food web of bacteria, protozoa, worms, insect larvae, snails, fish, crayfish, and other organisms, which keep themselves from being carried downstream by attaching to rocks or seeking shelter behind or under rocks. Even fish that maintain their position by active swimming occasionally need such shelter to rest (Figure 20–9a).

Suspended Load. Sediment entering waterways in large amounts has an array of impacts. Sand, silt, clay, and humus are quickly separated by the agitation of flowing water and are carried at different rates. Clay and humus are carried in suspension, making the water turbid and reducing the amount of light penetrating the water and, hence, reducing photosynthesis as well. As the material settles, it coats everything and continues to block photosynthesis. It also kills the animals by clogging their gills and feeding structures. The eggs of fish and other aquatic organisms are particularly vulnerable being smothered by sediment.

Bed Load. Especially destructive is the **bed load** of sand and silt, which is not readily carried in suspension, but is gradually washed along the bottom. As particles roll and tumble along, they scour organisms from the rocks. They also bury and smother the bottom life and fill in the hiding and resting places of fish and crayfish. Aquatic plants and other organisms are unable to reestablish themselves because the bottom is a constantly shifting bed of sand (Figure 20–9b and c). Modern storm-water management is designed to reduce the bed load, usually via storm drains that are periodically emptied of their sediment. Many newer housing developments include a storm-water retention reservoir, which

Figure 20–8 Acid mine waste. In August 2015, EPA workers who were attempting to clean up a closed gold mine in Colorado accidentally broke through a barrier, releasing three million gallons of acidic water into the Animas River. Here kayakers paddle across the polluted river. Acid mine drainage contains high concentrations of cadmium, lead, arsenic, sulfides, and iron.

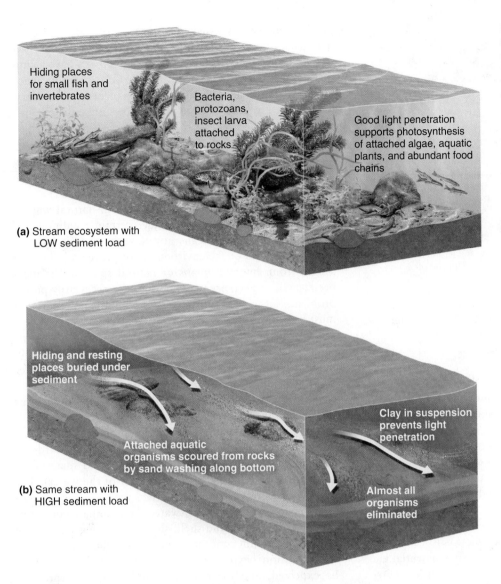

(a) Stream ecosystem with LOW sediment load

Hiding places for small fish and invertebrates

Bacteria, protozoans, insect larva attached to rocks

Good light penetration supports photosynthesis of attached algae, aquatic plants, and abundant food chains

Hiding and resting places buried under sediment

Attached aquatic organisms scoured from rocks by sand washing along bottom

Clay in suspension prevents light penetration

Almost all organisms eliminated

(b) Same stream with HIGH sediment load

Figure 20–9 Impact of sediment on streams and rivers. (a) The ecosystem of a stream that is not subjected to a large sediment bed load. **(b)** The changes that occur with large sediment inputs. **(c)** The Platte River at Lexington, Nebraska. The sandbars seen here constitute the bed load; they shift and move with high water, preventing the reestablishment of aquatic vegetation.

(c)

Figure 20–10 Storm-water management. Rather than letting excessive runoff from developed areas cause flooding and other environmental damage, runoff can be funneled into a retention pond, as shown here. Then it can drain away slowly, maintaining natural stream flow, or it can recharge groundwater. The retention pond may be designed to retain a certain amount of water and thus create a pocket of wildlife habitat in an otherwise urban or suburban setting.

is simply a pond that receives and holds runoff from the area during storms. Water may infiltrate the soil or may create a pocket of natural wetland habitat supporting wildlife (Figure 20–10).

Sediments do not receive the attention the news media give to hazardous wastes and certain other pollution problems, but erosion is so widespread throughout the world that few streams and rivers escape the harsh impact of excessive sediment loads.

Nutrients. Some of the inorganic chemicals carried in solution in all bodies of water are nutrients—essential elements required by plants. The two most important nutrient elements for aquatic plant growth are **phosphorus** and **nitrogen,** and they are often in such low concentrations in water that they are the limiting factors for phytoplankton or other aquatic plants. More nutrients mean more plant growth, so nutrients become water pollutants when they are added from point or nonpoint sources and stimulate undesirable plant growth in bodies of water. The most obvious point sources of excessive nutrients are sewage outfalls. As we will see in Section 20.2, it is costly to remove nutrients from sewage during the treatment process, so that is not always done. If sewage is not treated at all, higher levels of nutrients will enter the water, which is what's happening in much of the developing world.

Agricultural runoff is the most notorious nonpoint source of nutrients. The nutrients are applied to agricultural crops as chemical fertilizers and manure, are dissolved in irrigation water, and are present in crop residues. When nutrients are applied in excess of the needs of plants or when runoff from agricultural fields and feedlots is heavy, the nutrients are picked up by water that eventually enters streams, rivers, lakes, and the ocean. Other nonpoint sources of nutrients include acid rain and snow, lawns and gardens, golf courses, and storm drains.

Earlier we discussed the concept of safe planetary boundaries, the zone in which humans need to live in order to have a sustainable future (Chapter 1). Experts have identified the processes of the nitrogen cycle as one area in which human activities have already crossed a planetary boundary; we will need to back-track in order to be sustainable.

Water Quality Standards. When is water considered polluted? Many water pollutants—such as pesticides, cleaning solvents, and detergents—are substances that are found in water only because of human activities. Others, such as nutrients and sediments, are always found in natural waters, and they are a problem only under certain conditions. In both cases, *pollution* means any quantity that is harmful to human health or the environment or prevents full use of the environment by humans or natural species. The mere presence of a substance in water does not necessarily pose a problem. Rather, it is the *concentration* of the pollutant that must be of primary concern.

Criteria Pollutants. But what concentration is worrisome? To provide standards for assessing water pollution, the EPA has established the National Recommended Water Quality Criteria. On the basis of the latest scientific knowledge, the EPA has listed 167 chemicals and substances as **criteria pollutants.** The majority of these are toxic chemicals, but many are also natural chemicals or conditions that describe the state of water, such as nutrients, hardness (a general measure of dissolved calcium and magnesium salts), and pH (a measure of the acidity of the water). The list identifies the pollutant and then recommends concentrations for freshwater, saltwater, and human consumption (usually of fish and shellfish). Values are given for the **criteria maximum concentration (CMC),** the highest single concentration beyond which environmental impacts may be expected, and the **criterion continuous concentration (CCC),** the highest sustained concentration beyond which undesirable impacts may be expected. The criteria are *recommendations* meant to be used by the states, which are given the primary responsibility for upholding water-pollution laws. States and Native American tribes may revise these criteria, but their revisions are subject to EPA approval.

Drinking water standards are much stricter. For these, the EPA has established the Drinking Water Standards and Health Advisories, a set of tables that are updated periodically. These standards, covering some 94 contaminants, are enforceable under the authority of the **Safe Drinking Water Act (SDWA) of 1974.** They are presented as **maximum contaminant levels (MCLs).** To see how these two sets of standards work, consider the heavy metal arsenic.

Arsenic. Arsenic is listed as a known human carcinogen by the Department of Health and Human Services; it occurs naturally in groundwater, reaching high concentrations in some parts of the world. The CMC and CCC values for arsenic are 340 µg/L and 150 µg/L (1 µg/L = 1 part per billion) for freshwater bodies and 69 µg/L and 36 µg/L for saltwater bodies. The drinking-water MCL concentration, however, is 10 µg/L, and therein lies a story.

For years, the MCL concentration for arsenic was 50 µg/L, despite warnings by the U.S. Public Health Service and the WHO that this level was much too high because of the risk of cancer. (Both bodies recommended 10 µg/L.) From standard risk assessment processes, the cancer mortality risk was assessed at 1 in 100 for people drinking water regularly at the 50 µg/L level. This ratio is more than 100 times higher than the permitted risk for any other contaminant. There was a good deal of controversy about changing the MCL because of the cost of attaining lower concentrations, but in October 2001, EPA administrator Christine Whitman set the arsenic MCL at 10 µg/L.

Other Applications. Two important applications of water quality criteria are the **National Pollution Discharge Elimination System (NPDES)** and **Total Maximum Daily Load (TMDL)** programs. The NPDES program addresses point-source pollution and issues *permits* that regulate discharges from wastewater-treatment plants and industrial sources. The TMDL program, by contrast, evaluates all sources of pollutants entering a body of water, especially nonpoint sources, according to the water body's ability to assimilate the pollutant. For both of these programs, water quality criteria provide an essential means of assessing the condition of the receiving body of water as well as evaluating the levels of the various pollutant discharges. We will examine these two programs in more detail when we consider the problem of eutrophication.

The good news is that, according to the EPA, 92% of the people in the United States have access to drinking water that meets the *drinking water standards*. However, on the basis of data supplied to the EPA by the states, more than 42,000 rivers, lakes, and estuaries are not meeting the recommended *water quality standards,* and this excludes more than 60% of U.S. waters that have not been assessed at all. The major problems (in order) are pathogens, mercury, nutrients, other heavy metals, sediments, and oxygen depletion. As we proceed, we will address the specifics of public-policy responses to this serious national problem. We turn now to ways of dealing with the biological wastes that are a large part of our pollution problem.

CONCEPT CHECK What are three problems caused by the spillage of sewage into waterways? ☑

20.2 Wastewater Treatment and Management

The first wastewater-treatment plants in the United States were built around 1900. To alleviate the problem of sewage-polluted waterways, facilities were designed and constructed to treat the outflow before it entered the receiving waterway. However, the combined volumes of sewage and storm water soon proved impossible to handle. During heavy rains, wastewater would overflow the treatment plant and carry raw sewage into the receiving waterway. Gradually, regulations were passed requiring municipalities to install separate systems—**storm drains** to collect and drain runoff from precipitation and **sanitary sewers** to receive all the wastewater from sinks, tubs, and toilets in homes and other buildings. (Note the distinction between the two terms; it is incorrect to speak of storm drains as sewers.)

The ideal modern system, then, is one in which all wastewater is collected separately from storm water and is fully treated to remove pollutants before the water finally is released. But progress toward this goal has been extremely uneven. Up through the 1970s, even in the United States and other developed countries, thousands of communities still discharged untreated sewage directly into waterways, and for many more, the degree of treatment was minimal. Indeed, the increasing sewage pollution of waterways and beaches was the major impetus behind the passage of the Clean Water Act.

Before addressing methods of treating wastewater, let us clarify which contaminants and pollutants we are talking about.

The Pollutants in Raw Wastewater

A sewer system brings the outflow from all tub, sink, and toilet drains in homes and buildings together into larger and larger sewer pipes, just as the branches of a tree eventually come together into the trunk. The mixture that flows from the end of the "trunk" of the collection system is called **raw sewage** or **raw wastewater**. Because we use such large amounts of water to flush away small amounts of dirt, especially as we stand under the shower or just run the water, most of what goes down sewer drains is water. In the United States, raw wastewater is about 1,000 parts water for every 1 part waste—99.9% water to 0.1% waste.

Given our voluminous use of water, an average community of 10,000 persons will produce 1.5 million to 2.0 million gallons (6–8 million L) of wastewater each day. With the addition of storm water, raw wastewater is diluted still more. Nevertheless, the pollutants are sufficient to make the water brown and foul smelling.

The pollutants in wastewater generally are divided into the following four categories, which correspond to the techniques used to remove them:

1. **Debris and grit:** rags, plastic bags, coarse sand, gravel, and other objects flushed down toilets or washed through storm drains.

2. **Particulate organic material:** fecal matter, food wastes from garbage-disposal units, toilet paper, and other matter that tends to settle in still water.

3. **Colloidal and dissolved organic material:** very fine particles of particulate organic material, bacteria, urine, soaps, and detergents and other cleaning agents.

4. **Dissolved inorganic material:** nitrogen, phosphorus, and other nutrients from excretory wastes and detergents.

In addition to these four categories of pollutants, raw wastewater may contain variable amounts of pesticides, heavy metals, and other toxic compounds.

Removing the Pollutants from Wastewater

The challenge of treating wastewater is more than installing a technology that will do the job; it is also finding one that will do the job at a reasonable cost. A diagram of wastewater treatment procedures is shown in **Figure 20–11**. Note that the diagram shows both primary and secondary treatment.

Primary Treatment. Because debris and grit will damage or clog pumps and further treatment processes, removing this material is a necessary first step. Usually, removal involves two steps: the screening out of debris and the settling of grit. Debris is removed by letting raw sewage flow through a **bar screen**, a row of bars mounted about 1 inch apart. Debris is mechanically raked from the screen and taken to an incinerator. After passing through the screen, the water flows through a **grit chamber**—a swimming pool–like tank—in which velocity is slowed just enough to permit the grit to settle. The settled grit is mechanically removed from these tanks and taken to landfills.

After debris and grit are removed, the water then flows very slowly through large tanks called **primary clarifiers**. Because its flow is slow, the water is nearly motionless for several hours. The particulate organic material, about 30–50% of the total organic material, settles to the bottom, where it can be removed. At the same time, fatty or

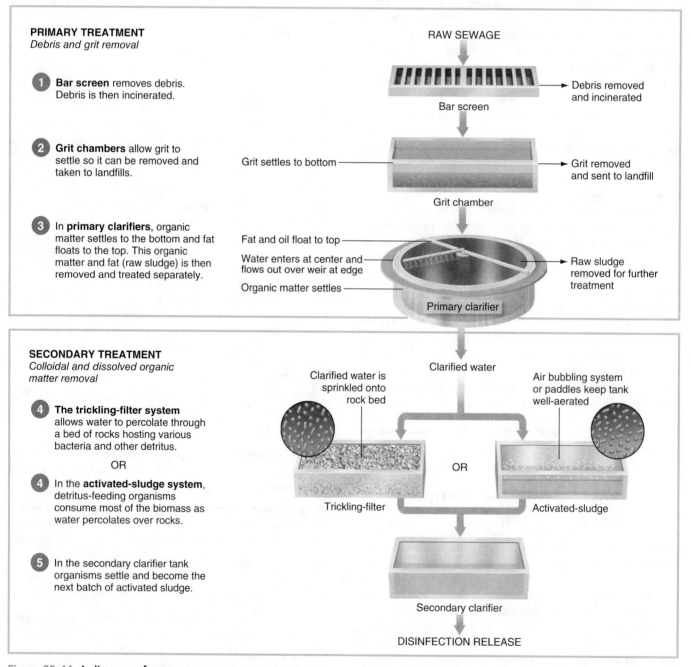

Figure 20–11 A diagram of wastewater treatment. Raw sewage moves from the grit chamber to undergo primary treatment, in which sludge is removed. The clarified water then undergoes secondary treatment (shown here as activated-sludge or trickling-filter treatment).

oily material floats to the top, where it is skimmed from the surface. All the material that is removed, both particulate organic material and fatty material, is combined into what is referred to as **raw sludge,** which must be treated separately.

Primary treatment involves nothing more complicated than putting polluted water into a tank, letting material settle, and pouring off the water. Nevertheless, such treatment removes much of the particulate organic matter at minimal cost.

Secondary Treatment. **Secondary treatment** is also called biological treatment because it uses organisms—natural decomposers and detritus feeders. Basically, an environment is created that enables these organisms to feed on the colloidal and dissolved organic material and break it down to carbon dioxide, mineral nutrients, and water via their cell respiration. The wastewater from primary treatment is a food- and water-rich medium for the decomposers and detritus feeders. The only thing that needs to be added to the water is oxygen to enhance the organisms' respiration and growth. Either of two systems may be used to add oxygen to the water: a *trickling-filter system* or an *activated-sludge system.*

In the **trickling-filter system,** the water exiting from primary treatment is sprinkled onto, and allowed to percolate through, a bed of fist-sized rocks 6–8 feet (2–3 m) deep (Figure 20–12). The spaces between the rocks provide good aeration. The organic material in the water, including pathogenic organisms, is absorbed and digested by decomposers and detritus feeders as it trickles by. This type of secondary treatment works best for small and medium-sized communities with large tracts of land.

The **activated-sludge system** (Figure 20–11, bottom) is a more efficient, and the most common, secondary-treatment system. Water from primary treatment enters a large tank that is equipped with an air-bubbling system or a rapidly churning system of paddles. A mixture of detritus-feeding organisms, referred to as **activated sludge,** is added to the water as it enters the tank, and the water is vigorously aerated as it moves through the tank. Organisms in this well-aerated environment reduce the biomass of organic material, including pathogens, as they feed. As the organisms feed they tend to form into clumps, called *floc,* that settle readily when the water is stilled. Then, from the aeration tank, the water is passed into a **secondary clarifier** tank, where the organisms settle out, and the water—now with more than 90% of the organic material removed—flows on. The settled organisms are pumped back into the aeration tank. (They are the activated sludge that is added at the beginning of the process.) Surplus amounts of activated sludge, which occur as populations of organisms grow, are removed and added to the raw sludge. This process removes the organisms and organic material, but leaves nutrients in the water.

Biological Nutrient Removal. The nutrient-rich water from activated-sludge systems has the potential to cause eutrophication if it is released into natural bodies of water. To avoid the problems of eutrophication, treatment plants can be upgraded to use a process of removing nutrients and oxidizing detritus called **biological nutrient removal (BNR).** BNR removes both nitrogen and phosphorus.

(a)

(b)

Figure 20–12 Trickling filters for secondary treatment. (a) The water from primary clarifiers is sprinkled onto, and trickles through, a bed of rocks 6–8 feet deep. **(b)** Various bacteria and other detritus feeders adhering to the rocks consume and digest organic matter as it trickles by in the water, which is collected at the bottom of the filters and goes on for final clarification and disinfection.

Nitrogen. Recall that, in the natural nitrogen cycle, various bacteria convert nutrient forms of nitrogen (ammonia and nitrate) back to non-nutritive nitrogen gas in the atmosphere through denitrification (Chapter 3, Figure 3–21). For the biological removal of nitrogen, then, the activated-sludge system is partitioned into zones, and the environment in each zone is controlled in a manner that promotes the denitrifying process, as shown in Figure 20–13, on the following page. The resulting inert nitrogen gas is vented into the air.

Phosphorus. In an environment that is rich in oxygen but relatively lacking in food (the environment of zone 3 in Figure 20–13), bacteria take up phosphate from the solution and store it in their bodies. Thus, phosphate is removed as the excess organisms are removed from the system. These organisms, together with the phosphate they contain, are

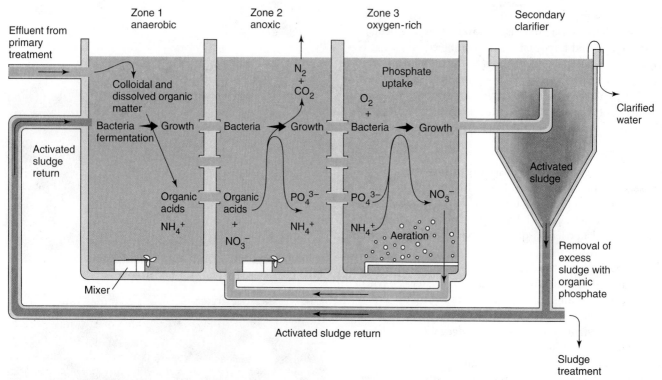

Figure 20–13 Biological nutrient removal (BNR). Secondary treatment—the activated-sludge process—may be modified to remove nitrogen and phosphate, while at the same time breaking down organic matter. For this BNR process, the aeration tank is partitioned into three zones; only zone 3 is aerated. As seen in the diagram, ammonium (NH_4^+) is converted to nitrate (NO_3^-) in zone 3. Recycled to zone 2, which is without oxygen gas (anoxic), the nitrate supplies the oxygen for cell respiration. In the process, the nitrate is converted to nitrogen gas, which is released into the atmosphere. Phosphate is taken up by bacteria in zone 3 and removed with the excess sludge.

added to, and treated with, the raw sludge, ultimately producing a more nutrient-rich, treated sludge product.

Various chemical treatments are often used as an alternative to BNR. One such process is simply to pass the effluent from standard secondary treatment through a filter of lime, which causes the phosphate to precipitate out as insoluble calcium phosphate. Another is to treat the effluent with ferric chloride, which produces insoluble ferric phosphate, or with an organic polymer, which gives rise to a floc. With these methods, most of the phosphorus is removed along with the sludge.

Final Cleansing and Disinfection. With or without BNR, the wastewater is subjected to a final clarification and disinfection. Although few pathogens survive the combined stages of treatment, public-health rigors still demand that the water be disinfected before being discharged into natural waterways. The most widely used disinfecting agent is chlorine gas because it is both effective and relatively inexpensive. However, chlorine gas is dangerous to work with. Sodium hypochlorite (Clorox®) provides a safer way of adding chlorine. Because chlorine reacts with organic chemicals, all chlorine treatments produce a variety of toxic disinfection by-products, such as chlorinated hydrocarbons. Some municipalities disinfect with chloramines, which produce fewer toxic by-products. Residual chlorine introduced to waterways is itself toxic and can harm natural biota. These impacts are minimized if the effluent is well diluted by natural water.

Other disinfecting techniques are coming into use. One alternative is ozone gas, which kills microorganisms and breaks down to oxygen gas, improving water quality in the process. Ozone is unstable and, hence, explosive, so it must be generated at the point of use, a step that demands considerable capital investment and energy. Another disinfection technique is to pass the effluent through an array of ultraviolet lights mounted in the water. The ultraviolet radiation kills microorganisms but does not otherwise affect the water.

Final Effluent. After all these steps, the wastewater from a modern treatment plant has a lower organic and nutrient content than many bodies of water into which it is discharged. BOD values are commonly 10 to 20 ppm, as opposed to 200 ppm in the incoming sewage. In other words, discharging the wastewater may actually help to improve the water quality in the receiving body. In water-short areas, there is every reason to believe that the treated water itself, with minimal additional treatment, could be recycled into the municipal water-supply system.

Bear in mind that the techniques we have described here represent the state of the art. Many cities, even in the developed world, are still operating with antiquated systems that provide lower-quality treatment. A few coastal cities still persist in discharging sewage that has received only primary treatment directly into the ocean, although in the United States this practice has almost disappeared.

Treatment of Sludge

Recall that the particulate organic matter that settles out or floats to the surface of sewage water in primary treatment forms the bulk of *raw sludge*, though the sewage also contains excesses from activated-sludge and BNR systems. Raw sludge is a gray, foul-smelling, syrupy liquid with a water content of 97–98%. Pathogens are certain to be present in raw sludge because it includes material directly from toilets. Indeed, raw sludge is considered to be a biologically hazardous material. However, as nutrient-rich organic material, it has the potential to be used as organic fertilizer if it is suitably treated to kill pathogens and if it does not contain other toxic contaminants.

Three commonly used methods for treating sludge and converting it into organic fertilizer are **anaerobic digestion, pasteurization,** and **composting.** Because this is a developing industry, it is unclear which method will prove most cost effective and environmentally acceptable over the long run. Also, none of these methods is capable of removing toxic substances such as heavy metals and nonbiodegradable synthetic organic compounds. The presence of such toxins can preclude the use of sludge as fertilizer.

Anaerobic Digestion. Anaerobic digestion is a process that allows bacteria to feed on the sludge in the absence of oxygen. The raw sludge is put into large airtight tanks called **sludge digesters** (Figure 20–14). In the absence of oxygen, a consortium of anaerobic bacteria breaks down the organic matter. The end products of this decomposition are carbon dioxide, methane, and water. Thus, a major by-product of anaerobic processes is **biogas,** a gaseous mixture that is about two-thirds methane. The other third is made up of carbon dioxide and various foul-smelling organic compounds that give sewage its characteristic odor. Because of its methane content, biogas is flammable and can be used for fuel. In fact, it is commonly collected and burned to heat the

Figure 20–14 Anaerobic sludge digesters. In these tanks at the Deer Island Sewage Treatment Plant, Boston, the bacterial digestion of raw sludge in the absence of oxygen leads to the production of methane gas, which is tapped from the tops of the tanks, and humus-like organic matter, which can be used as a soil conditioner. The egg-like shape of the tanks facilitates mixing and digestion.

sludge digesters because the bacteria working on the sludge do best when maintained at about 104°F (38°C).

After four to six weeks, anaerobic digestion is more or less complete, and what remains is called **treated sludge** or **biosolids,** consisting of the remaining organic matter, which is now a relatively stable, nutrient-rich, humus-like material suspended in water. Pathogens have been largely, if not entirely, eliminated.

Such treated sludge makes an excellent organic fertilizer that can be applied directly to lawns and agricultural fields in the liquid state in which it comes from the digesters, providing the benefit of both the humus and the nutrient-rich water. Alternatively, the sludge may be dewatered by means of belt presses, whereby the sludge is passed between rollers that squeeze out most of the water (Figure 20–15a). This leaves the organic material as a semisolid **sludge cake** (Figure 20–15b), which, after disinfection, is easy to store and to spread on fields with traditional manure spreaders.

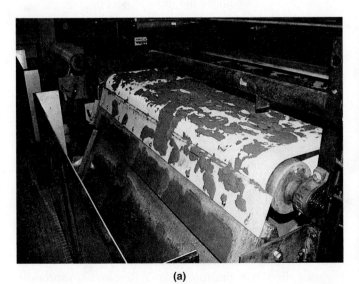

(a)

(b)

Figure 20–15 Dewatering treated sludge. (a) In a belt press, liquid sludge is run between canvas belts going over and between rollers, so that much of the water is pressed out. **(b)** The resulting "sludge cake" is a semisolid humus-like material that can be used as an organic fertilizer.

Pasteurization. After the raw sludge is dewatered, the resulting sludge cake may be put through ovens that operate like oversized laundry dryers. In the dryers, the sludge is pasteurized—that is, heated sufficiently to kill any pathogens (exactly the same process that makes milk safe to drink). The product is dry, odorless organic pellets. Milwaukee, Wisconsin, which has a particularly rich sludge resulting from the brewing industry, has been using this process for more than 60 years. The city bags and sells the pellets throughout the country as an organic fertilizer under the trade name Milorganite®.

Composting. Another process sometimes used to treat sewage sludge is composting. Raw sludge is mixed with wood chips or some other water-absorbing material to reduce the water content. It is then placed in **windrows**—long, narrow piles that allow air to circulate easily through the material and that can be turned with machinery. Bacteria and other decomposers break down the organic material to rich humus-like material that makes an excellent treatment for poor soil.

Alternative Treatment Systems

Despite the expansion of sewage collection systems, many homes in rural and suburban areas lie outside the reach of a municipal system. For these homes, *on-site* treatment systems are required. Currently, 25% of the U.S. population is served by such systems. The traditional and still most common on-site system is the **septic system** (Figure 20–16). Wastewater flows into the tank, where particulate organic material settles to the bottom. The tank acts like a primary clarifier in a municipal system. Water containing colloidal and dissolved organic material, as well as dissolved nutrients, flows into the gravel or crushed stone lining the leaching field and gradually percolates into the

soil. Organic material that settles in the tank is digested by bacteria, but accumulations still must be pumped out regularly. Soil bacteria decompose the colloidal and dissolved organic material that comes through the leaching field. Prerequisites for this traditional system are suitable land area for the drain field and subsoil that allows sufficient percolation of water.

Unfortunately, on-site systems frequently fail, resulting in unpleasant sewage backup into homes and pollution of groundwater and surface waters. Homeowners are often unaware of how the systems work, and most local regulatory programs do not require homeowners to be accountable for them. Substantial nutrient and pathogen pollution of groundwater and surface waters is traced to failed on-site systems. For this reason, the EPA has made several resources available to local and state agencies responsible for maintaining water quality: voluntary national guidelines, a handbook for management, and a manual for on-site wastewater systems.

Septic System Primer. The following are some simple suggestions for successful septic system maintenance:

- Use caution in disposing of materials down the drain. Do not dump coffee grounds, cat litter, diapers, grease, feminine hygiene products, pesticides, paint, gasoline, household chemicals (like chlorine-containing compounds), or other products that could either fill up or clog the septic tank or kill the bacteria that make the system work.

- Maintain your system by having it inspected regularly and pumped out at least every five years—more frequently if you have a larger family. Keep records of maintenance, and have a map of the location of your septic tank and the leaching field.

- If you have a garbage disposal unit, get rid of it; the added materials will shorten the life of the system and force it to be pumped out more frequently.

- Keep heavy equipment and vehicles off your system and leaching field.

Composting Toilet Systems. A viable and relatively inexpensive alternative to septic systems, especially for more remote and occasional use, is the **composting toilet** system (Figure 20–17). The objective of the system is to provide a sanitary means of treating human wastes that will destroy human pathogens and produce a stable humus-like end product that can be safely disposed of.

These systems require active management, and the end product must legally be buried or removed by a licensed septage hauler. They are capable of reducing toilet wastes to 10–30% of their original volume. Their most common use is in vacation cottages, roadside rest stops, and parks, but where building codes permit, they can also be employed in urban settings. The Institute of Asian Research in Vancouver, British Columbia, is a 30,000-square-foot office complex that is not connected to the city sewer system and uses only composting toilets to handle human wastes.

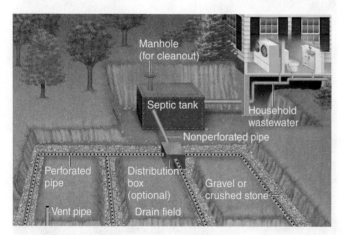

Figure 20–16 Septic system. Sewage treatment for a private home, using a septic tank and leaching field. Normally, the pipes and the tank are buried underground. They are shown uncovered here only for illustration.

(*Source:* http://www.catawbariverkeeper.org/issues/library-of-documents-on-sewage-issues-and-treatment/septictankdiagram.jpg/view?searchterm=septic+system.)

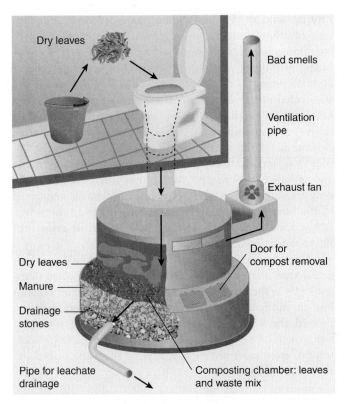

Dry leaves

Bad smells

Ventilation pipe

Exhaust fan

Door for compost removal

Dry leaves

Manure

Drainage stones

Pipe for leachate drainage

Composting chamber: leaves and waste mix

Figure 20–17 Composting toilet system. Composting toilets use little or no water. Leaves or hay are added to the solid and liquid wastes from the toilet, and the mixture decomposes to form a clean, safe compost that is later removed.

Using Effluents for Irrigation. The nutrient-rich water coming from the standard secondary-treatment process is beneficial for growing plants. The problem is that we don't want to put that water into waterways, where it will stimulate the growth of undesirable algae. But why not use it for irrigating plants we do want to grow? This is a way of completing the nutrient cycle. Indeed, the concept has been put into practice in a considerable number of locations as an alternative to upgrading the treatment to remove nutrients.

The nutrient-rich effluent from standard secondary treatment from St. Petersburg, Florida, was causing cultural eutrophication in Tampa Bay. Now, St. Petersburg uses the effluent to irrigate 4,000 acres (1,600 hectares) of urban open space, from parks and residential lawns to a golf course. Revenues from the water sales help offset operating costs. Similarly, Bakersfield, California, receives a $30,000 annual income from a 5,000-acre (2,000-hectare) farm irrigated with its treated effluent. And Clayton County, Georgia, is irrigating 2,500 acres (1,000 hectares) of woodland with partially treated sewage. Hundreds of similar projects are under way around the country.

A number of developing countries irrigate croplands with raw (untreated) sewage effluents. The crops respond well, but parasites and disease organisms can easily be transferred to farmworkers and consumers. Therefore, it is important to emphasize that only *treated* effluents should

be used for irrigation. In Matimangwe, Mozambique, sewage is treated in dry pits and turned into a rich fertilizer that dramatically increases the food harvest, using a latrine system called EcoSan. The wastes are mixed with soil and ash and left for eight months, while the lack of moisture and long incubation kill off pathogens and turn the wastes into a rich compost.

Reconstructed Wetland Systems. In treating wastewater, it is also possible to make use of the nutrient-absorbing capacity of wetlands in suitable areas and under suitable climatic conditions. The project may be part of a wetlands recovery program, or artificial wetlands may be constructed.

In the 1960s and 1970s, for example, much of the land around Orlando, Florida, which was originally wetlands, was drained and converted to cattle pasture. At the time, Orlando was discharging 20 million gallons (75 million L) per day of nutrient-rich effluent into the St. Johns River following secondary treatment. Through the Orlando Easterly Wetlands Reclamation Project, 1,200 acres (480 hectares) of pastureland were converted back to wetlands. The project involved scooping soil from pastures and building berms (mounds of earth) around them to create a chain of shallow lakes and ponds. In addition, 2 million wetland plants ranging from bulrushes and cattails to various trees were planted. The effluent entering the upper end now percolates through the wetlands for about 30 days before entering the St. Johns River virtually pure. The project has recreated a wildlife habitat that supports some threatened species, such as the snail kite.

Wetland systems can be designed for small as well as large areas and are becoming an increasingly popular alternative for small communities. The key to success for such systems is to ensure that they are kept in balance and not provided more input than they are able to handle.

Many aquatic systems are simply not able to act like wetlands and absorb extra nutrients without major changes in ecosystem function. This response, called eutrophication, is the most widespread water-pollution problem in the United States.

CONCEPT CHECK Why is the removal of pathogens and the breakdown of organic molecules insufficient in sewage treatment? ✓

20.3 Eutrophication

Recall that the term *trophic* refers to feeding (Chapter 5). Literally, *eutrophic* means "well nourished." Eutrophication is part of the process of natural succession (discussed in Chapter 5). Over periods of hundreds or thousands of years, ponds and lakes are gradually enriched with nutrients, leading to changes in the composition of the ecosystem. This *natural eutrophication* is a normal process. In contrast, the accelerated eutrophication caused by human activities is called **cultural eutrophication**.

Different Kinds of Aquatic Plants

To understand eutrophication, you need to be able to distinguish between *benthic* plants and *phytoplankton*.

Benthic Plants. **Benthic plants** (from the Greek *benthos*, meaning "deep") are aquatic plants that grow attached to, or are rooted in, the bottom of a body of water. As shown in Figure 20–18a, benthic plants may be categorized as **submerged aquatic vegetation (SAV)**, which generally grows totally under water, or **emergent vegetation**, which grows in water but with the upper parts emerging from the water.

To thrive, SAV requires water that is clear enough to allow sufficient light to penetrate to allow photosynthesis. As water becomes more *turbid* (cloudy), light is diminished. In extreme situations, it may penetrate to just a few centimeters beneath the water's surface. Thus, increasing turbidity decreases the depth at which SAV can survive.

Another important feature of SAV is that it absorbs its required mineral nutrients from the bottom sediments through the roots, just as land plants do. SAV is not limited by water that is low in nutrients. Indeed, enrichment of the water with nutrients is counterproductive for SAV because it stimulates the growth of phytoplankton.

Phytoplankton. **Phytoplankton** consist of numerous species of photosynthetic algae, protists, and chlorophyll-containing bacteria (cyanobacteria, formerly referred to as blue-green algae) that grow as microscopic single cells or in small groups, or "threads," of cells. Phytoplankton live suspended in the water and are found wherever light and nutrients are available (Figure 20–18b). In extreme situations, water may become pea-soup green (or tea colored, depending on the species involved), and a scum of phytoplankton may float on the surface and absorb essentially all the light. However, phytoplankton reach such densities only in nutrient-rich water, because, not being connected to the bottom, they must absorb nutrients from the water. A low level of nutrients in the water limits the growth of phytoplankton accordingly.

Considering the different requirements of SAV and phytoplankton, the balance between them is altered when nutrient levels in the water are changed. As long as water remains low in nutrients, populations of phytoplankton are suppressed, the water is clear, and light can penetrate to support the growth of SAV. As nutrient levels increase, phytoplankton can grow more prolifically, making the water turbid and thus shading out the SAV.

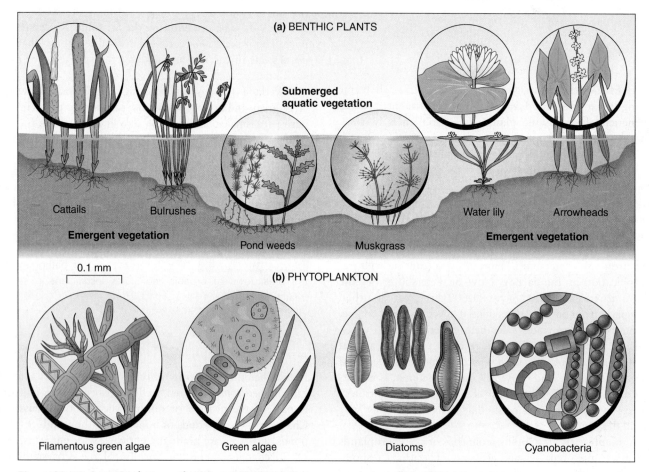

Figure 20–18 **Aquatic photosynthesizers. (a)** Benthic, or bottom-rooted, plants. These are subdivided into submerged aquatic vegetation (SAV) and emergent vegetation. **(b)** Phytoplankton, various photosynthetic organisms that are either single cells or small groups or filaments of cells, float freely in the water.

The Impacts of Nutrient Enrichment

A lake in which light penetrates deeply—one in which the bottom is visible beyond the immediate shoreline—is **oligotrophic** (low in nutrients). Such a lake is fed by a watershed that holds its nutrients well. A forested watershed, for example, holds nitrogen and phosphorus tightly and allows very little of those elements to enter the water draining into the streams and rivers. If the streams feed into a lake, it will reflect their low nutrient content.

In an oligotrophic lake, the low nutrient levels limit the growth of phytoplankton and allow enough light to penetrate to support the growth of SAV, which draws its nutrients from the bottom sediments. In turn, the benthic plants support the rest of a diverse aquatic ecosystem by providing food, habitats, and dissolved oxygen. Oligotrophic lakes are prized for their aesthetic and recreational qualities as well as for their production of game fish.

The Process of Eutrophication. As the water of an oligotrophic body becomes enriched with nutrients, numerous changes are set in motion. First, the nutrient enrichment allows the rapid growth and multiplication of phytoplankton, increasing the turbidity of the water. The increasing turbidity shades out the SAV that lives in the water. With the die-off of SAV, there is a loss of food, habitats, and dissolved oxygen from their photosynthesis.

Phytoplankton have remarkably high growth and reproduction rates. Under optimal conditions, phytoplankton biomass may double every 24 hours, a capacity far beyond that of benthic plants. Thus, phytoplankton soon reach a maximum population density, and continuing growth and reproduction are balanced by die-off. Dead phytoplankton settle out, resulting in heavy deposits of detritus on the lake or river bottom. In turn, the abundance of detritus supports an abundance of decomposers, mainly bacteria. The explosive growth of bacteria, consuming oxygen via respiration, creates an additional demand for dissolved oxygen. The result is the depletion of dissolved oxygen, creating hypoxic conditions, with the consequent suffocation of fish and shellfish. A dead zone has been created. A dead zone can be as limited as a small lake basin or as extensive as the dead zone in the Gulf of Mexico, but the cause is always nutrient enrichment. (The addition of nutrients may also have deleterious effects on wetlands such as salt marshes, as described in Sound Science, *Can Salt Marshes Absorb Our Nutrients?*)

SOUND SCIENCE
Can Salt Marshes Absorb Our Nutrients?

Plum Island, Massachusetts, is one of many barrier islands along the East Coast. From the air, the island looks like a long sandy finger pointing south. To the east is the Atlantic Ocean and to the west is the Great Marsh, the largest contiguous salt marsh in New England. Salt marshes are highly productive environments that serve as nurseries for young fish and other organisms. They also store floodwaters and protect coastal areas from storms. Healthy marshes also absorb carbon, an important part of the environmental sequestration of carbon emissions. Yet as sea levels rise, many salt marshes are declining.

Plum Island has been the home of a decade-long project on water pollution, a part of the Long Term Ecological Research network of U.S. sites. There, scientists ask a variety of questions, including: What happens when coastal communities allow nutrients to run off into salt marshes? Since nitrogen and phosphorus often make terrestrial plants grow more rapidly, you might expect that the nutrients would cause the marsh grasses to grow more quickly. A long-term experiment showed, however, that the reality of eutrophication is more complicated.

To find the effect of additional nutrients, researchers used microcomputers to add controlled amounts of nutrients to tidal creeks in the Plum Island estuary, enriching about 30,000 square meters of marsh regularly during the growing season. After nine years, the results were surprising. Initially, the salt marsh grasses along creek banks grew taller, as scientists had thought they would. However, the grass plants had smaller roots and rhizomes (underground stems). These short-rooted grasses often toppled over, pulling mud with them. The additional nutrients also caused the microbes in the marsh soil to speed up the process of decomposition, making the soil less stable. The result was the breakdown of creek banks, making the marsh mud and soil more vulnerable to erosion. Such changes would eventually make a salt marsh less capable of retaining floodwater or protecting the land from storms.

The Plum Island experiment was unusual because it was long term, large scale, and begun in a relatively unpolluted marsh area. While the enrichment did damage the marsh,

Researcher Linda Deegan shows how the tidal creek banks in the Plum Island marsh are slumping, a result of too many nutrients being added to the marsh.

it gave scientists an opportunity to learn something new. In contrast to short-term experiments involving the application of fertilizer to the surface of a marsh, the project mimicked real-world pollution—regular run-off into tidal creeks. Now we know that tidal marshes cannot simply absorb the nutrients that run into them, but must be protected from nutrient pollution.

Source: L. Deegan et al. "Coastal Eutrophication as a Driver of Salt Marsh Loss." *Nature* 490 (2012): 388–392.

In sum, *eutrophication* refers to this whole sequence of events, starting with nutrient enrichment and proceeding to the growth and die-off of phytoplankton, the accumulation of detritus, the growth of bacteria, and, finally, the depletion of dissolved oxygen and the suffocation of higher organisms (Figure 20–19). For humans, a eutrophic body of water is unappealing for swimming, boating, and sportfishing. If the lake is a source of drinking water, its value may be greatly impaired because phytoplankton rapidly clog water filters and may cause a foul taste. Even worse, some species of phytoplankton secrete toxins into the water that may kill other aquatic life and be injurious to human health as well.

Shallow Lakes and Ponds. In lakes and ponds whose water depth is 6 feet (2 m) or less, eutrophication takes a somewhat different course. There, SAV may grow to a height of a meter or more, reaching the surface. Thus, with nutrient enrichment, the SAV is not shaded out but grows abundantly, sprawling over and often totally covering the water surface with dense mats of vegetation that make boating, fishing, or swimming impossible. Any vegetation beneath these mats is shaded out. As the mats of vegetation die and sink to the bottom, they create a BOD that often depletes the water of dissolved oxygen, causing the death of aquatic organisms other than bacteria.

Harmful Algal Blooms. In 1988, a researcher identified a new organism, a kind of microscopic algae called a dinoflagellate, which she named *Pfiesteria piscicida*. Under certain conditions, some members of this genus secrete a toxin that harms fish. *Pfiesteria* species have been implicated in at least 50 major fish kills in North Carolina. These fish kills and other problems caused by large concentrations of *Pfiestreria* are examples of **harmful algal blooms**. *Red tides* are harmful

Figure 20–19 Eutrophication. As nutrients are added from sources of pollution, an oligotrophic system rapidly becomes eutrophic and undesirable.

OLIGOTROPHIC

- Low in nutrients
- Phytoplankton limited

- Water clear
- Light penetrates
- Submerged aquatic vegetation (SAV) thrives

NUTRIENT INPUTS

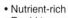

- Nutrient-rich
- Phytoplankton thrive

- Water turbid
- SAV shaded out

EUTROPHIC

- Nutrient-rich
- Rapid turnover of phytoplankton
- Accumulation of detritus of dead algae

- Decomposers feed on detritus
- Depletion of dissolved oxygen
- Fish and shellfish suffocate

algal blooms consisting of high concentrations of various species of toxic dinoflagellates. Red tides occur in coastal waters around the world, often in dead zones.

Scientists believe that nutrient pollution—especially of nitrogen and phosphorus—is the basic culprit in feeding the growth of the algae in estuaries and coastal waters. Recently, five federal agencies joined to sponsor the Ecology and Oceanography of Harmful Algal Blooms, a national research program studying algal blooms in U.S. coastal waters.

Combating Eutrophication

There are two approaches to combating the problem of cultural eutrophication. One is to attack the symptoms—the growth of vegetation, the lack of dissolved oxygen, or both. The other is to get at the root cause—excessive inputs of nutrients and sediments.

Attacking the Symptoms. Attacking the symptoms is appropriate in certain situations where immediate remediation is the goal and costs are not prohibitive. Methods of attacking the symptoms of eutrophication include applying herbicides, aerating, harvesting aquatic weeds, and drawing water down.

Applying Herbicides. Herbicides are often applied to ponds and lakes to control the growth of nuisance plants. To control phytoplankton growth, copper sulfate and di-quat are frequently used. For controlling SAV and emergent vegetation, fluridone, glyphosate, and 2,4-D are employed. However, at the concentrations required to bring the vegetation under control, these compounds are sometimes toxic to fish and aquatic animals. Also, fish are often killed after herbicide is applied, because the rotting vegetation depletes the water of dissolved oxygen. In sum, herbicides provide only cosmetic treatment, and as soon as they wear off, the vegetation grows back rapidly.

Aerating. In the final and most destructive stage of eutrophication, decomposers deplete the water of dissolved oxygen, thus suffocating other aquatic life. *Artificial aeration* of the water can avert this terminal stage. An aeration technique currently gaining in popularity is to lay a network of plastic tubes with microscopic pores on the bottom of the waterway to be treated. High-pressure air pumps force microbubbles from the pores, and the bubbles dissolve directly into the water. The method is proving effective in speeding up the breakdown of accumulated detritus, improving water quality, and enabling the return of more desirable aquatic life. Despite its high cost, the technique is applicable to harbors, marinas, and some water-supply reservoirs, where the demand for better water quality justifies the cost.

Harvesting. In shallow lakes or ponds, where the problem is bottom-rooted vegetation reaching and sprawling over the surface, harvesting the aquatic weeds (see Figure 20–20) may be an expedient way to improve the water's recreational potential and aesthetics. Commercial mechanical harvesters are used, and nearby residents sometimes get together to remove the vegetation by hand.

Figure 20–20 Eutrophication in a shallow pond. In shallow water, sufficient light for photosynthesis continues to reach the submerged aquatic vegetation. The oversupply of nutrients stimulates growth, so that vegetation reaches the surface and forms mats. Here, vegetation has been raked into piles for harvesting.

The harvested vegetation makes good organic fertilizer and mulch. But even harvesting has a limited effect: The vegetation soon grows back because roots are left in the nutrient-rich sediments.

Harvesting can be labor intensive. Because China is so densely populated, its economy and pollution have a significant effect on coastal waters. When Beijing hosted the 2008 Olympics, the sailing competitions were held in the coastal town of Qingdao. Just a few weeks before the competitions, agricultural waste, sewage, warmer-than-usual water, and heavy rains combined to provide the perfect conditions for the growth of green algae, resulting in a record-breaking algal bloom that affected 5,000 square miles (more than 13,000 km^2) of seawater. To clean up the coast, 10,000 Chinese soldiers and volunteers and 1,400 boats were employed to remove more than a million metric tons of algae.

Drawing Water Down. Because many recreational lakes are dammed, another option for shallow-water weed control is to draw the lake down for a period each year. This process kills most of the rooted aquatic plants along the shore, although they grow back in time.

All of these approaches are only temporary fixes for the problem and have to be repeated often, at significant cost, to keep the unwanted plant growth under control.

Getting at the Root Cause. Controlling eutrophication requires long-term strategies for correcting the problem, which ultimately means reducing the inputs of nutrients and sediments—two of the types of pollution discussed at the beginning of this chapter. The first step is to identify the major point and nonpoint sources of nutrients and sediments. Then it is a matter of developing and implementing strategies for correction. Which source or factor is most significant will depend on the human population and the land uses within the particular watershed. Therefore, each watershed must be analyzed as a separate entity, and appropriate measures must be taken to reduce the levels of nutrients and sediments exiting from that watershed. In the sections that follow, we shall discuss major strategies used to control the root causes of eutrophication.

First, consider the concept of limiting factors, in which the lack of only one nutrient can suppress growth (Chapter 3). In natural freshwater systems, phosphorus is the most common limiting factor. In marine systems (like the coastal Gulf of Mexico), the limiting nutrient is most often nitrogen. Both in the environment and in biological systems, phosphorus (P) is present as phosphate (PO_4^{3-}), and nitrogen is present in a variety of compounds, most commonly as nitrate (NO_3^-) or the ammonium ion (NH_4^+).

Ecoregional Nutrient Criteria. Beginning in 2001, the EPA began publishing water-quality nutrient criteria aimed at preventing and reducing the eutrophication that affects so many bodies of water. The agency listed its recommended criteria for *causative* factors—nitrogen and phosphorus—and for *response* factors—chlorophyll a as a measure of both phytoplankton density and water clarity. The EPA divided the country into ecoregions and determined criteria levels for each specific region. Like the other water quality criteria, the nutrient criteria are provided as targets for the states as they address their water-pollution (and especially their eutrophication) problems.

National Pollution Discharge Elimination System (NPDES): Control Strategy for Point Sources. In the past, discharges from sewage-treatment plants were major sources of nutrients entering waterways. The levels of phosphate in effluents from these plants were elevated even more in areas where laundry detergents containing phosphate were used. Phosphate contained in detergents moved through the system and out with the discharge unless the system had a nutrient removal component. In regions where eutrophication had been identified as a problem, a key step toward prevention was to ban the sale of phosphate-based laundry detergents. Because many states established bans on the use of these products, the major laundry detergent manufacturers voluntarily ended their manufacture in 1994. However, these bans do not usually cover dishwashing detergents, some brands of which contain as much as 8.7% phosphate.

Bans on detergents with phosphate and upgrades of sewage-treatment plants have brought about marked improvements in waterways that were heavily damaged by effluents from these plants. In a sense, however, these are the easy measures, as the target for correction (the effluent) is an obvious point source, and methods for correcting the problem are clear cut. The NPDES permitting process is the basic regulatory tool for reducing point-source pollutants. Under the Clean Water Act, "anyone discharging pollutants from any point source into waters of the U.S. [must] obtain an NPDES permit from EPA or an authorized state." More than 140,000 facilities are now regulated by way of NPDES permits. The permit must be drawn up in the context of the total pollutant sources (point and nonpoint) affecting a watershed or body of water.

Total Maximum Daily Load (TMDL): Control Strategy for Nonpoint Sources. Correction becomes more difficult, but no less important, when the source is diffuse, as in the case of farm and urban runoff. Remediation of such nonpoint sources will require thousands—perhaps millions—of individual property owners to adopt new practices regarding land management and the use of fertilizer and other chemicals on their properties. Nevertheless, that is the challenge. Section 303(d) of the Clean Water Act requires states to develop management programs to address nonpoint sources of pollution. Thus, there is a legal mandate to address the issues of agricultural and urban runoff.

The EPA has opted to develop regulations via the TMDL program. The concept is straightforward:

- Identify the pollutants responsible for degrading a body of water.

- Estimate the pollution coming from all sources (point and nonpoint).

- Estimate the ability of the body of water to assimilate the pollutants while retaining water quality.

- Determine the maximum allowable pollution load.

- Within a margin of safety, allocate the allowable level of pollution among the different sources, such that the water quality standards are achieved.

As with most attempts at solving environmental problems, the devil is in the details. The states are responsible for administering TMDL pollution abatement, but the EPA maintains oversight and approval of the results. State administrators have a number of options for managing nonpoint-source pollution, including regulatory programs, technical assistance, financial assistance, education and training, and demonstration projects. By 2015, the states had listed more than 42,000 water bodies as being impaired. More than 7,700 were on the list because of nutrient overload. The EPA has approved more than 68,000 TMDLs submitted by the states to correct the impairments and has substantially closed the gap between the state listings and the approved TMDLs.

Best Management Practices. Reducing or eliminating pollution from nonpoint sources will involve different strategies for different sources. For example, where a significant portion of the watershed is devoted to agricultural activities, the major sources of nutrients and sediments are likely to be (1) erosion and leaching of fertilizer from croplands and (2) runoff of animal wastes from barns and feedlots. All the practices that can be used to minimize such erosion, runoff, and leaching are lumped under a single term, **best management practices**. This includes all the methods of soil conservation available (discussed in Chapter 11). Table 20–2 lists examples of these practices for a variety of nonpoint sources.

Once control measures have been put in place, the polluted body of water must be monitored to determine whether water quality standards are being attained. Several years of data collection may be necessary to detect genuine trends in quality. When water quality standards are reached and sustained, the body of water is removed from the list of impaired waters. To date, 1,125 bodies of water have improved enough to be removed from the list. If standards are not met, the TMDL process must be revisited and new pollution allowances allocated.

Recovery. The good news is that cultural eutrophication and other forms of water pollution can be controlled and often reversed, provided that a total watershed management approach is undertaken. For a small lake fed from a modest-sized watershed, the task may be quite straightforward because major sources of nutrients can be identified and addressed. For large bodies of water, such as the Chesapeake Bay—with a watershed of 63,700 square miles (165,000 km^2) in six states and the District of Columbia—the task is enormous. The Chesapeake Bay has suffered from heavy nutrient pollution, with cultural eutrophication destroying all but a tenth of the 600,000 acres of vital sea grasses that once formed the basis of the bay's rich ecosystem. A dead zone involving up to 40% of the main bay appears for 10 months of every year. The Chesapeake Bay Program, a state-federal partnership begun in 1987, set the ambitious goal of achieving federal clean water standards by 2010. Unfortunately, new development in the watershed contributed nutrients and sediments as fast as the restoration efforts reduced them. Although there have been signs that the dead zone has been decreasing, other water quality indicators have shown little change. In 2010, the EPA designed a longer-term, tributary-based strategy to restore the bay by 2025, with several stages of action to reduce pollution.[2]

Lake Washington. One remarkable success story is Lake Washington, east of Seattle (Figure 20–21). This 34-square-mile lake is at the center of a large metropolitan area. During the 1940s and 1950s, 11 sewage-treatment plants were sending state-of-the-art treated water into the lake at a rate of 20 million gallons per day. At the same time, phosphate-based detergents came into wide use. The lake responded to the massive input of nutrients by developing unpleasant "blooms" of noxious blue-green algae. The water lost its clarity, the desirable fish populations declined, and masses of dead algae accumulated on the shores of the lake.

Table 20–2	Best Management Practices for Reducing Pollution from Nonpoint Sources

Agriculture	Forestry
Animal waste management	Ground cover maintenance
Conservation tillage	Limiting disturbed areas
Contour farming	Log removal techniques
Strip-cropping	Pesticide and herbicide management
Cover crops	
Crop rotation	Proper management of roads
Fertilizer management	Removal of debris
Integrated pest management	Riparian zone management
Range and pasture management	**Mining**
	Underdrains
Terraces	Water diversion
Construction	**Multicategory**
Limiting disturbed areas	Buffer vegetated strips
Nonvegetative soil stabilization	Detention and sedimentation basins
Runoff detention and retention	Devices to encourage infiltration
Urban	Vegetated waterways
Flood storage	Interception and diversion of runoff
Porous pavements	Sediment traps
Runoff detention and retention	Streamside management zones
Street cleaning	Vegetative stabilization and mulching

Source: U.S. Environmental Protection Agency, *Guidance for Water Quality-Based Decisions: The TMDL Process,* EPA440-4-91-001 (Washington, DC: USEPA, 1991).

---

[2]Juliet Eilperin, "EPA Unveils Massive Restoration Plan for Chesapeake Bay," *The Washington Post,* December 30, 2010.

Figure 20–21 Lake Washington. A testimony to the fact that bodies of water can recover from cultural eutrophication, Lake Washington is now thriving because sources of nutrients enriching the lake have been removed.

Citizen concern led to the creation of a system that diverted the treatment-plant effluents into nearby Puget Sound, where tidal flushing would mix them with open-ocean water. The diversion was complete by 1968, and the lake responded quickly. The algal blooms diminished, the water regained its clarity, and by 1975, recovery was complete. Before the diversion, the lake was receiving 220 tons of phosphate per year from all sources. Afterward, the total dropped to 40 tons per year, well within the lake's normal capacity to absorb phosphate by depositing it in deep bottom sediments.

One clear lesson learned was that sewage-treatment effluent must never be allowed to enter a lake unless nutrients are removed as part of the treatment. The major lesson, however, was that eutrophication can be reversed by paying attention to nutrient inputs and addressing the various sources. Lake management, in other words, is a matter of controlling the phosphate loading into a lake from all sources: human and animal sewage, agricultural and yard fertilizers, street runoff, and failing septic systems. Many lakes are now protected by associations of concerned citizens who see themselves as stewards of their lake.

CONCEPT CHECK What strategies can be used to decrease the amount of nonpoint-source pollution entering a river? ☑

20.4 National and International Policy and Water Pollution

Water pollution policy is primarily a national responsibility, although there are also international treaties designed to protect water quality. First we will consider the key U.S. policies relating to water pollution. Then we will consider the key international policies.

U.S. Water Quality Policies. In the United States, the responsibility for overseeing the health of the nation's waters rests with the EPA. However, the EPA can develop regulations only if Congress gives it the authority to do so. Hence, the foundation for public policy must be the laws passed by Congress. Some major legislative milestones in protecting the nation's waters are listed in Table 20–3. The landmark legislation is the Clean Water Act of 1972 (CWA), which gave the EPA jurisdiction over (and for the first time required permits for) all point-source discharges of pollutants; simply put, all discharges into U.S. waters are unlawful unless authorized by a permit. This act and subsequent amendments have provided $85 billion to help cities and towns build treatment plants to meet the federal requirement for secondary treatment of all sewage. As the table shows, the 1987 amendments to the CWA established a revolving loan fund—the Clean Water State Revolving Fund (SRF) program—to replace the direct-grants program. To build treatment facilities, local governments borrow at low interest rates, and as they repay the loans, the funds received are used for more loans.

The fund may also be used to control nonpoint-source pollution. To date, more than $100 billion in SRF loans have been made, funding more than 33,000 projects.

Reauthorization of the Clean Water Act is long overdue. Congress has been hung up on debates over whether regulations should be strengthened or weakened, whether federal regulations in the wetlands permit program intrude on private land-use rights, and how regulatory relief should be provided to industries, states, cities, and individuals required to take actions to comply with the regulations. Its approach for the past two decades has been to reauthorize the provisions of the CWA and its amendments, usually maintaining a status quo in the EPA's annual appropriations for CWA programs.

The EPA has identified nonpoint-source pollution as the nation's number-one water-pollution problem, with the construction of new wastewater-treatment facilities not far behind. Other significant issues—including storm-water discharges, combined and separate sewer overflows, wetland protection, mountaintop mining filling of streams, and animal feeding operations—are also receiving the EPA's attention in the form of new regulations.

Overall, much progress has been made in the 45 years since the enactment of the Clean Water Act. This act is seen as one of the most successful of our nation's environmental laws. The number of people in the United States served by adequate sewage-treatment plants has risen from 85 million to 223 million. Soil erosion has been reduced by 1 billion tons annually, and two-thirds of the nation's waterways are safe for fishing and swimming—double the number from 1972. Many of the nation's most heavily used rivers, lakes, and bays—including the Androscoggin River in Maine, Boston Harbor, the upper Mississippi River, the South Platte River, the upper Arkansas River, Lake Erie, the Illinois River, and the Delaware River—have been cleaned up and restored. Fish now swim in rivers once so polluted that only bacteria and sludge worms could survive. Levels of toxic chemicals in the Great Lakes have been greatly reduced.

International Water Policy. Agreements about lakes and rivers that cross national boundaries, or *transboundary waters,* are critical to the protection of shared resources. The water in many rivers, seas, and oceans arises from watersheds in several countries. Regional treaties are commonly used to protect such bodies of water. For example, more than 143 countries participate in regional seas programs, a part of the UN Regional Seas Programme. Such programs encourage countries to work together to protect specific oceans, seas, and coasts, including the Caribbean, East and South Asian Seas, and southeast Pacific.

The Mediterranean Sea is a good example of a body of water under stress. It is heavily polluted by the many cities, villages, and farms along its coast, while the large amount of international shipping on its waters poses the threat of water-borne diseases and the introduction of exotic species. To protect the Mediterranean, the nations along its coast

Table 20–3 Legislative Milestones in Protecting U.S. Natural Waters and Water Supplies

Date	Name of Act	Key Features
1899	Rivers and Harbors Act	First federal legislation protecting the nation's waters in order to promote commerce.
1948	Water Pollution Control Act	Authorizes the federal government to provide technical assistance and funds to state and local governments to promote efforts to protect water quality.
1972	Clean Water Act	Establishes a comprehensive federal program to achieve the goal of protecting and restoring the physical, chemical, and biological integrity of the nation's waters. Requires permits for any discharge of pollution, strengthens water quality standards, and encourages the use of the best achievable pollution control technology. Provides billions of dollars for construction of sewage-treatment plants.
1972	Marine Protection, Research, and Sanctuaries Act	Prevents unacceptable dumping in oceans.
1974	Safe Drinking Water Act	Authorizes the EPA to regulate the quality and safety of public drinking-water supplies, maintain drinking-water standards for numerous contaminants, and set requirements for the chemical and physical treatment of drinking water.
1977	Clean Water Act amendments	Strengthen controls on toxic pollutants and allow states to assume responsibility for federal programs.
1987	Water Quality Act	Enacts major amendments to the Clean Water Act that (1) create a revolving loan fund to provide ongoing support for the construction of treatment plants, (2) address regional pollution with a watershed approach, and (3) address nonpoint-source pollution, with requirements for states to assess the problem and develop and implement plans for dealing with it. Makes available $400 million in grants to carry out its provisions.
1996	Safe Drinking Water Act amendments	Impose numerous requirements for managing water supplies, establish a revolving loan fund to help municipalities update their water system infrastructure, and provide flexibility for the EPA to base its selection of contaminants on health risk and costs and benefits.
2002	Great Lakes Legacy Act	Enacted amendments to CWA provisions for the Great Lakes and authorized $50 million annually for fiscal years 2004–2008 for the EPA to conduct projects designed to remediate sediment problems in the Great Lakes.

have agreed to control dumping and the run-off of pollution, provide protected areas, and otherwise protect its waters.

There have been several treaties dealing with oil pollution, especially from ships. The *Convention on the Prevention of Marine Pollution by Dumping of Wastes and Other Matter* was implemented in 1975. In 1982, the UN *Convention on the Law of the Sea* increased the amount of control countries had over the waters off their coastlines, in part to give them ways to prevent pollution. The Global Ocean Commission report of 2014 made recommendations for the protection of ocean waters from hazardous waste (Chapter 22) and solid wastes (Chapter 21).

To highlight the importance of water, the United Nations declared 2005–2015 the *International Decade of Action: Water for Life*. Energy was focused on providing both a higher quality and a greater quantity of water to those who lacked it. Actions to improve water quality included efforts to decrease chemical pollution, agricultural runoff, and the contamination of water by sewage; such efforts require coordination between potential polluters and watershed managers.

In conclusion, we need to apply a sense of stewardship to our rivers, lakes, seas, and oceans. The United States and other countries have demonstrated that it is possible to restore waters from a previously polluted condition. Preserving or improving water quality is essential to the well-being of all people, both today and in the future.

CONCEPT CHECK What evidence shows that the Clean Water Act has been highly successful? ✓

REVISITING THE THEMES

SOUND SCIENCE

Solid scientific research and communication were essential in establishing the basic knowledge about pollutants and their impacts on human health and the environment. They continue to be highly important as the EPA carries out its research and regulatory activities for administering the Clean Water Act and the Safe Drinking Water Act. The criteria pollutant standards for natural waters and the maximum contaminant standards for drinking waters must be based on accurate risk assessments as the EPA develops and defends its regulatory activities. Sound science was particularly crucial in the investigation of dead zones and algal blooms and their causes. Experiments such as the Plum Island experiment on marsh eutrophication help us understand the effects of large-scale human interference in ecosystems.

SUSTAINABILITY

Our natural waters represent an enormous bank of ecosystem capital, providing essential goods and services to people everywhere. The sustainability of whole ecosystems is threatened when we fail to handle our wastes in a conscientious manner. The opening story of the dead zone demonstrates how it is possible for nutrient overload to degrade an enormous area of the Gulf of Mexico. Unfortunately, a similar situation exists in the Chesapeake Bay and in many other bodies of water. These are just a couple of the more spectacular examples. The more important battlegrounds are the thousands of smaller lakes, rivers, and coastal estuaries that are also suffering from human impacts. More needs to be done, especially at the state level, to improve our national water quality. Internationally, countries need to cooperate in reducing pollution, especially nutrient pollution and ocean dead zones.

STEWARDSHIP

Stewardship means intentionally caring for people and the natural world. The public-health measures put in place in the past two centuries represent the stewardship of human resources. Also, as we divert public and private resources to cleaning up our waterways, we are acting as good stewards in the best sense of the concept. Lakes and watersheds often benefit from people and associations that have provided leadership in correcting the problems that lead to eutrophication. The Lake Washington story is an excellent example. Stewardship also means caring for our neighbors' needs, and nowhere are those needs more apparent than in the need for clean water and sanitation in so many developing countries. A sense of the global reach of water pollution is difficult to imagine, yet billions of people in the developing countries are held back from achieving their physical and intellectual potential because of the terrible effects of polluted water and lack of sanitation.

REVIEW QUESTIONS

1. What practices and consequences led to passage of the Clean Water Act of 1972?

2. Discuss each of the following categories of water pollutants and the problems they cause: pathogens, organic wastes, chemical pollutants, and sediments.

3. How are water quality standards determined? Distinguish between water quality criteria pollutants and maximum contaminant levels.

4. Name and describe the facility and the process used to remove debris, grit, particulate organic matter, colloidal and dissolved organic matter, and dissolved nutrients from wastewater.

5. Why is secondary treatment also called biological treatment? What is the principle involved? What are the two alternative techniques used?

6. What are the principles involved in removing biological nutrients from waste, and what is accomplished by doing so? Where do nitrogen and phosphate go in the process?

7. Name and describe three methods of treating raw sludge, and give the end product(s) that may be produced from each method.

8. How may sewage from individual homes be handled in the absence of municipal collection systems? What are some problems with these on-site systems?

9. Describe and compare submerged aquatic vegetation and phytoplankton. Where and how does each get nutrients and light?

10. Explain the difference between oligotrophic and eutrophic waters. Describe the sequential process of eutrophication.

11. Distinguish between natural and cultural eutrophication.

12. What is being done to establish nutrient criteria?

13. How does the National Pollution Discharge Elimination System program address point-source pollution by nutrients?

14. Describe the Total Maximum Daily Load program. How does it address nonpoint-source pollution, and what role do water quality criteria play in the program?

15. What are some of the important public-policy issues relating to water quality?

16. What is one international effort to curb water pollution?

THINKING ENVIRONMENTALLY

1. A large number of fish are suddenly found floating dead on a lake. You are called in to investigate the problem. You find an abundance of phytoplankton and no evidence of toxic dumping. Suggest a reason for the fish kill.

2. Nitrogen pollution has damaged the San Diego Creek–Newport Bay ecosystem in California by encouraging the heavy growth of intertidal algae. Report on the TMDL for this system. Consult the EPA Web site: http://water.epa.gov/lawsregs/lawsguidance/cwa/tmdl/nutrients.cfm.

3. Arrange a tour to the sewage-treatment plant that serves your community. Compare it with what is described in this chapter. Is the water being purified or handled in a way that will prevent cultural eutrophication? Are sludges being converted to and used as fertilizer? What improvements, if any, are in order? How can you help promote such improvements?

4. Suppose a new community of several thousand people is going to be built in Arizona (with a warm, dry climate). You are called in as a consultant to design a complete sewage system, including the collection, treatment, and use or disposal of by-products. Write a plan describing the system you recommend, and give a rationale for the choices involved.

MAKING A DIFFERENCE

1. Check out the lakes and ponds in your area, and determine whether cultural eutrophication is occurring. If you find that it is, follow up by consulting a local environmental group, and get involved in combating the problem.

2. Investigate sewage treatment in your community. To what extent is the sewage treated? Where does the effluent go? The sludge? Are there any problems? Are there alternatives that are more ecologically sound? Consult local environmental groups for help.

3. Bring discarded household chemicals to hazardous-waste collection centers; do not pour them down the drain. Pouring chemicals down the drain could upset your septic system or contaminate treatment-plant sludge.

4. Test your soil before using fertilizers. Excess from overfertilizing can leach into groundwater or contaminate rivers or lakes. Also, avoid using fertilizers and pesticides near surface waters.

MasteringEnvironmentalScience®

Students Go to MasteringEnvironmentalScience for assignments, the eText, and the Study Area with animations, practice tests, and activities.

Professors Go to MasteringEnvironmentalScience for automatically graded tutorials and questions that you can assign to your students, plus Instructor Resources.

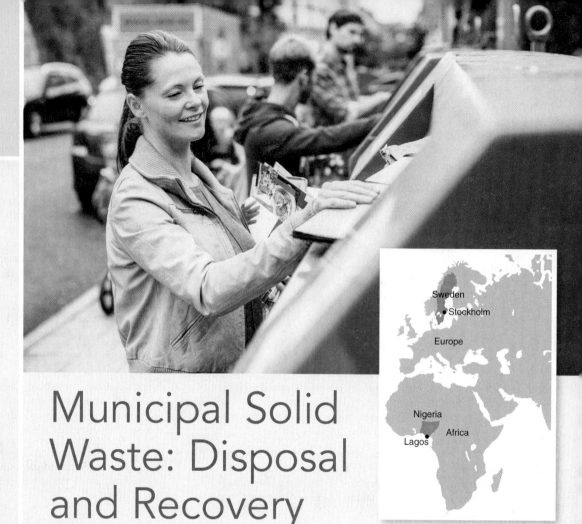

Municipal Solid Waste: Disposal and Recovery

If there were a race for the most sustainable waste management, the country of Sweden would be the frontrunner! Today less than 1% of Sweden's trash ends up in landfills. (In the United States, nearly 70% goes to landfills.) One reason for the Swedes' success is that they actively promote waste reduction when producing and using products; they also recycle, which removes much of the material from their waste stream. But a great deal of their success is the result of 32 power plants that burn trash for fuel. As the huge incinerators burn waste, they heat water, producing steam that turns turbines and generates electricity. The air given off by the incinerators is scrubbed and the ashes are treated to remove metals, gravel, and other components so only a tiny fraction of the ash ends up in landfills. Every year, Sweden's waste-to-energy plants burn two million tons of trash, producing enough electricity for one quarter of a million homes. In addition, the excess heat from the plants is used to heat one million homes (Chapter 14).

Sweden has been so successful with its waste-to-power plants that it imports trash from other countries; as much as 700,000 tons a year. The beauty of the system is that other countries, including the United Kingdom, pay Sweden to take their waste, which Sweden then uses for energy. Sweden may be the world's best example of waste efficiency, but the Swedes envision still better—reducing and recycling even more before materials are disposed of permanently.

Reusing Landfills. Even when materials are used efficiently, some landfills are inevitable, and once a landfill is full, cities must determine how to use the land above it. In the United States, some municipalities are finding that closed landfills can be sustainable-energy goldmines, places where solar arrays, wind turbines, and trapped methane can be used to generate electricity. For example, the town

▲ These young Swedes are recycling; only 1% of the household trash in Sweden ends up in landfills.

of Canton, Massachusetts, worked with Boston-based Southern Sky Renewable Energy to erect an array of 24,000 solar panels on a capped landfill. The 15-acre solar array generates 5.6 megawatts of power, enough to supply 5,600 homes. It is also projected to generate some $16 million in revenues for the town over the next 25 years.

Lagos Forges Forward. In the developing world, the country of Nigeria, Africa's most populous nation, is also facing challenges related to waste management. Its rapidly growing capital, Lagos, a city of 20 million, is known as the "garbage capital of the world." Every day its residents throw away 11,000 tons of waste. At the same time, the city struggles to produce electrical power, sometimes surviving with only four hours of electricity a day. Recently city planners have designed a power plant that would burn the methane produced by the decomposition of the city's waste, providing the city with greatly needed electricity and reducing its greenhouse gas emissions. City planners are also focusing on changing the way waste is recycled. At present, most recycling is done by people who survive by picking through trash, looking for items to use or sell. The new plan would formalize the recycling process, greatly reducing the amount of waste going into landfills and providing much-needed jobs for the city's residents.

As we will see in this chapter, the issues related to solid waste disposal vary around the world. Developed countries have the resources to invest in innovative technologies, such as waste-to-energy power plants and solar arrays on old landfills, while lower-income countries have difficulty affording such solutions. In this chapter we will explore the problems related to waste disposal and the connections between waste disposal and energy, as well as ways to reduce the amount of waste we discard. We will also consider the kinds of local, national, and international policies that are needed for sustainable waste disposal.

21.1 Solid Waste: A Global Problem

From Sweden to Lagos, the disposal of trash is an increasingly critical issue. Over the years, the amount of MSW generated by people around the world has grown steadily, in part because of growing populations, but also because of changing lifestyles and the increasing use of disposable materials and excessive packaging.

The term used to describe all of the materials thrown away from homes and small commercial establishments is **municipal solid waste** (**MSW**). It is also called *trash, refuse,* or *garbage.* MSW does not include hazardous waste and nonhazardous industrial waste, which are covered elsewhere (Chapter 22).

Waste Production in the United States

We produce a lot of trash. In 1960, the average American generated 2.7 pounds (1.2 kg) of MSW per day. In 2012, each of us generated an average of 4.4 pounds (2 kg) of such waste per day. The total MSW that year was 251 million tons (230 million metric tons)—enough waste to fill 95,000 garbage trucks each day.

Composition of Waste. What does all this waste consist of? **Figure 21–1** shows the overall composition of MSW in the United States. However, the proportions of the components of waste vary greatly, depending on whether the source of the trash is commercial or residential, as well as the nature of the neighborhood. The composition of waste also varies throughout the year; during certain seasons, yard wastes, such as grass clippings and raked leaves, add greatly to the solid-waste burden. In many places, little attention is given to what people put in their trash. Even if there are restrictions and prohibitions, many hazardous substances—paint, used motor oil, electronic equipment, and so on—are discarded in trash.

Modern Disposables. Much of the increase in our production of waste stems from our use of disposable products. It is much more common today than in earlier generations to throw away single-use objects such as paper cups and plates. Paper and cardboard also pose a problem as they represent a resource that is wasted when put into a dump or landfill. When they decay, they produce greenhouse gases. Such materials can easily be reused or recycled.

Other difficulties arise from the use of modern materials, especially plastics. Plastics are cheap and easy to make but chemically complex, so they do not decay easily. They can also release toxic chemicals when placed in landfills or when burned. In developing countries, where the open burning of trash is common, burning plastics are a health hazard to those living nearby.

The United States provides an example of the waste produced in one wealthy, developed nation. As we will see next, the composition of waste in developing countries is quite different.

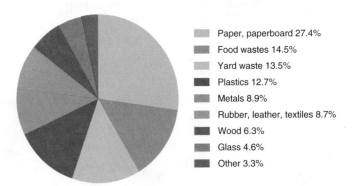

- Paper, paperboard 27.4%
- Food wastes 14.5%
- Yard waste 13.5%
- Plastics 12.7%
- Metals 8.9%
- Rubber, leather, textiles 8.7%
- Wood 6.3%
- Glass 4.6%
- Other 3.3%

Figure 21–1 Composition of U.S. municipal waste. Paper and cardboard are the largest component of solid waste in the United States. Plastics and food and yard wastes are also significant components of the waste stream.

(*Source:* Data from EPA, Office of Solid Waste, Municipal Solid Waste in the United States: 2010 Facts and Figures (Washington, DC: EPA, November, 2013).)

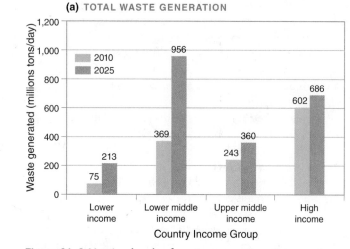

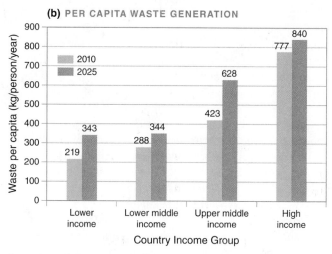

Figure 21–2 Varying levels of waste generation. These graphs compare the amount of waste generated by countries of different income levels in 2010 and 2025. **(a)** The total amount of waste produced by all groups of countries is expected to increase by 2025, although the rate of increase depends upon the economic level of the country. **(b)** The average amount of waste produced per person is greatest in the high-income countries, but is expected to increase in countries of all income levels.

UNDERSTANDING THE DATA

1. In 2010, which group of countries was producing the greatest amount of waste? In 2025, which group of countries will be producing the greatest amount of waste?
2. Between 2010 and 2025, which group of countries will have the greatest percentage increase in total waste production?
3. Between 2010 and 2025, which group of countries will have the greatest increase in the amount of waste produced by each person?

Worldwide Trends

The total amount of waste produced and the components of that waste vary greatly from country to country, as does the ultimate fate of that waste. In 2012, the World Bank released a report on solid waste disposal around the world that highlighted a number of patterns and trends.[1] One major pattern is that wealthy countries produce more trash per person than lower-income countries, as shown in Figure 21–2a. For example, the wealthiest countries of the world, those of the

Organization for Economic Co-operation and Development (OECD), have a total population similar to that of the continent of Africa, home to many of the poorest countries, and yet the OECD countries produce 100 times as much waste as the countries of Africa. As countries become wealthier or more urbanized, their per capita waste production increases. Over time, the people in all groups of nations have been producing more trash, a trend that is expected to continue (Figure 21–2b). Because the lower-middle income countries have rapidly growing populations and economies, that group of countries is expected to produce the largest total volume of trash by 2025.

The types of waste materials also vary by region and income level (Figure 21–3). People in wealthy countries put less

[1]D. Hoornweg and P. Bhada-Tata, *What a Waste: A Global Review of Solid Waste Management* (Washington, D.C.: World Bank, 2012).

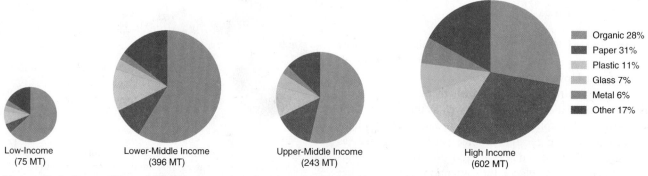

Figure 21–3 Composition of MSW in countries of various income levels. There are significant differences in the kinds of wastes produced by countries of different income levels. Low-income countries produce less solid waste than high-income countries, but a greater percentage of that waste is organic matter. Totals are in metric tons (MT).

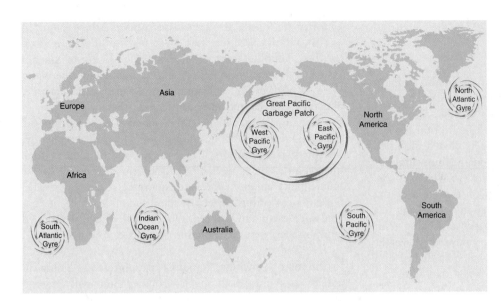

Figure 21–4 Escaped trash. A great deal of uncaptured trash eventually escapes to the ocean, where it may float in currents for years. Areas with a higher-than-normal density of floating trash are called garbage patches or gyres.

organic material, such as yard waste, into the MSW stream. People in wealthy countries discard more paper than those in lower-income countries, but they also recycle more.

The World Bank also found that the fate of collected waste varies by economy. Low- and middle-income countries spend a greater proportion of their waste-management resources on waste collection, while high-income countries spend a greater proportion on waste disposal. Lower-income countries often do not have the resources to invest in modern methods of MSW disposal. As urbanization increases in lower-income countries, the problems associated with waste disposal also increase. Both within nations and between nations, waste is often shipped from higher-income areas to lower-income areas where there are fewer environmental regulations and little protection for those who work with waste.

When Trash Escapes

Trash that is not disposed of properly can cause endless environmental trouble. Often it makes its way through storm sewers and rivers until it reaches beaches or ocean water. Trash in the ocean is now a significant environmental problem (also discussed in Chapters 7 and 20). Even remote patches of the ocean have been found to contain trash.

About 70% of all marine litter sinks onto the seabed, where it damages bottom communities. Swirling currents move the trash that remains floating into endlessly cycling *garbage patches,* or *garbage gyres.* There are six of these gyres in the oceans. The largest, the Pacific garbage gyre, actually comprises two smaller patches (Figure 21–4). One report estimated that the plastic near the surface of the Pacific gyre weighed six times as much as the plankton in that part of the ocean.

The plastic trash in the ocean has a number of features that make it dangerous. Some plastics can leach chemicals into the water or absorb chemicals such as DDT and PCBs. Most plastic objects float, so aquatic animals may become entangled in them (Figure 21–5). Discarded, abandoned, and lost fishing gear alone accounts for about 10% of marine

debris. This gear often continues to trap organisms, a process called "ghost fishing". Scientists estimate that 70% of the endangered North Atlantic right whales have been entangled in fishing gear. Plastics can also break into small pieces; in fact, 90% of the plastic in the ocean is in pieces smaller than a fingernail. Aquatic animals can mistake these plastic pieces for food; unfortunately, the plastic pieces often kill the animals that swallow them.

CONCEPT CHECK How does the kind and amount of MSW produced in high-income countries differ from that produced in low-income countries? How does the disposal of MSW differ between these two groups of countries? ☑

Figure 21–5 A plastic hazard. Solid waste in the ocean often kills animals that eat it or become entangled in it.

21.2 Disposal of Waste in the United States

In the United States, local governments have the primary responsibility for collecting and disposing of MSW. The local jurisdiction may own trucks and employ workers, or it may contract with a private firm to provide the collection service. Traditionally, the cost of waste pickup is passed along to households via taxes. More recently, many municipalities have opted for pay-as-you-throw systems, in which households are charged for waste collection on the basis of the amount of trash they throw away. Some municipalities put all trash collection and disposal in the private sector, with the collectors billing each home by the volume and weight of trash. The MSW that is collected is then disposed of in a variety of ways. It is at the point of disposal that state and federal regulations begin to apply.

Past Practices

During America's early years, many city dwellers simply threw their refuse into alleys or public streets. In coastal cities, wastes were often dumped into the ocean, a common practice in New York Harbor throughout the nineteenth century. By the 1920s, many cities had begun to collect municipal waste and remove it to dumps, largely in response to growing concerns about the spread of disease.

Until the 1960s, most MSW in the United States was disposed of in open dumps, although many homes had a barrel in the backyard for burning paper and wood trash. Older residents of smaller towns and cities can remember gathering at the dump to look through discarded items and socialize. In most dumps, the waste was burned to reduce its volume and lengthen the life span of the dumpsite. Because refuse does not burn well, smoldering dumps produced clouds of smoke that could be seen from miles away, smelled bad, and created a breeding ground for flies and rats.

Some cities began to use incinerators, or *combustion facilities*—huge furnaces in which high temperatures allow the waste to burn more completely than in open dumps. Without controls, however, incinerators were prime sources of air pollution.

During the 1960s and early 1970s, public objections and air pollution laws forced the phase-out of private burning barrels, public open dumps, and many incinerators. Open dumps were then converted to landfills. In 2012, 54% of the MSW in the United States was disposed of in landfills, 35% was recovered for recycling and composting, and the remainder (12%) was burned or incinerated. Over the past 10 years, the landfill and combustion components have been declining, while recycling has shown a steady increase.

Landfills

In a **landfill,** the waste is put on or in the ground and is covered with earth. Because there is no burning and because each day's fill is covered with at least six inches of earth, air pollution and populations of vermin are kept down.

Unfortunately, aside from those concerns and the minimizing of cost, no other factors were given real consideration when the first landfills were opened. Municipal waste managers generally had little understanding of ecology, the water cycle, or what products decomposing wastes would generate, and they had no regulations to guide them. Therefore, in general, any cheap, conveniently located piece of land on the outskirts of town became the site for a landfill. Frequently, this site was a natural gully or ravine, an abandoned stone quarry, a section of wetlands, or a previous dump. Once the municipality acquired the land, dumping commenced without any precautions. After the site was full, it would be covered with earth and ignored. Only recently have abandoned landfills been seen as a valuable open-space resource.

Problems of Landfills. Landfills are subjected to biological and physical factors in the environment and will undergo change over time as a consequence of the operation of those factors on the waste that is deposited. Several of the changes are undesirable because, if they are not dealt with effectively, they present the following problems:

- Leachate generation and groundwater contamination
- Methane production
- Incomplete decomposition
- Settling

Leachate Generation and Groundwater Contamination. The most serious problem by far is groundwater contamination. As water percolates through a landfill, various chemicals in the materials may dissolve in the water and get carried along in a process called *leaching* (Chapter 10). The water with various pollutants in it is called *leachate*. A leachate consists of residues of decomposing organic matter combined with iron, mercury, lead, zinc, and other metals from rusting cans, discarded batteries, and appliances—all generously "spiced" with paints, pesticides, cleaning fluids, newspaper inks, and other chemicals. Poor disposal in landfills can funnel this toxic brew directly into groundwater aquifers.

Leachates from landfills that are near the water table can contaminate water supplies. The flat, wet state of Florida has had particular problems with landfills. No matter where Florida's landfills were located, they were either in wetlands or just a few feet above the water table. As a result, more than 145 former municipal landfill sites in Florida appear on the Superfund list, a national list of heavily polluted sites that jeopardize human health and must be cleaned up (Chapter 22).

Methane Production. Because it is about two-thirds organic material, MSW is subject to natural decomposition. Buried wastes do not have access to oxygen, however, so their decomposition is anaerobic. A major by-product of the process is *biogas*, which is about two-thirds methane and one-third hydrogen and carbon dioxide, a highly flammable mixture (see Chapter 20). Produced deep in a landfill, biogas may seep horizontally through the soil and rock, enter basements, and even cause explosions if it accumulates and is ignited. Homes at distances up to 1,000 feet from landfills

Figure 21–6 Landfill gas project. A well collects biogas from the Smith Creek landfill in St. Clair County, Michigan. There are now more than 635 landfill gas energy systems in operation in the United States.

have been destroyed, and some deaths have occurred as a result of such explosions. Also, gases seeping to the surface kill vegetation by poisoning the roots. Without vegetation, erosion occurs, exposing the unsightly waste.

Yet the production of biogas also presents an opportunity. There is a thriving industry that is exploiting the problem by installing "gas wells" in existing landfills. The wells tap the landfill gas, and the methane is purified and used as fuel. There are now more than 635 such landfill-gas projects in the United States. In 2014, commercial landfill gas in the nation produced 1,600 megawatts of electricity and 100 billion cubic feet of gas for direct-use applications, powering or heating more than 1.7 million homes (Figure 21–6).

The recovery of landfill gas has significant environmental benefits, especially in combating climate change. It directly reduces greenhouse gas emissions because methane is a powerful greenhouse gas and landfills are the largest anthropogenic source of methane emissions in the United States. Methane recovery also reduces the use of coal and oil, which are nonrenewable and highly polluting energy resources.

Incomplete Decomposition. The commonly used plastics in MSW resist natural decomposition because of their molecular structure. Chemically, they are polymers of petroleum-based compounds that microbes are unable to digest. Biodegradable plastic polymers have been developed from sources such as cornstarch, cellulose, lactic acid, and soybean protein as well as petroleum. Two kinds of bioplastics based on corn sugars—polyhydroxyalkanoate and polylactic acid—are in commercial production; a recent estimate from NatureWorks indicates a coming demand for up to 50 billion pounds of bioplastics a year in the next five years. Currently, bioplastics are more expensive than petroleum-based plastics and tend to be used by companies manufacturing natural and certified organic products. They are indeed biodegradable, but some of the corn-based plastics must go to a commercial composting plant to be decomposed.

Studying Trash. A team of "garbologists" from the University of Arizona, led by William Rathje, has been carrying out research on old landfills. The research has shown that even materials formerly assumed to be biodegradable, such as newspapers and wood, often decompose very slowly, if at all, in landfills. In one landfill, 30-year-old newspapers were recovered in a readable state, and layers of telephone directories, practically intact, were found marking each year. Because paper materials are 28% of MSW, this is a serious matter. The reason paper and other organic materials decompose so slowly is the absence of suitable amounts of moisture. The more water percolating through a landfill, the faster the paper materials biodegrade. Unfortunately, the more percolation there is, the greater the amount of toxic leachate produced.

Settling. Waste settles as it compacts and decomposes. Luckily, this eventuality was recognized from the beginning, so buildings are never constructed on landfills. Settling does present a problem in landfills that have been converted to playgrounds and golf courses because of the creation of shallow depressions (and sometimes deep holes) that collect and hold water. This problem can be addressed by continually monitoring the facility and using fill to restore a level surface.

Improving Landfills. Recognizing the foregoing problems, the EPA upgraded siting and construction requirements for new landfills, based on the **Resource Conservation and Recovery Act of 1976 (RCRA)** and subsequent amendments. Under current regulations,

- New landfills are sited on high ground, well above the water table, not in a geologically unstable area, and away from airports (because of bird hazards).

- The floor is contoured so that water will drain into a tile leachate-collection system. The floor and sides are covered with a plastic liner and at least two feet of compacted soil. With such a design, any leachate percolating through the fill will move into the leachate-collection system. Collected leachate can be treated as necessary.

- Layer upon layer of refuse is positioned such that the fill is built up in the shape of a pyramid. Finally, it is capped with at least 18 inches of earthen material and a layer of topsoil and then seeded. The cap and the pyramidal shape help the landfill shed water. In this way, water infiltration into the fill is minimized, and less leachate is formed.

- The entire site is surrounded by a series of groundwater-monitoring wells that are checked periodically, and such checking must go on indefinitely.

These design features are summarized in Figure 21–7, on the following page. Most landfills currently in operation have these improved technologies, which protect both human health and the environment.

The landfill pyramids may well last as long as the Egyptian pyramids. Fortunately, the creative siting and construction of landfills have the potential to address some highly significant future needs, because abandoned landfills can become attractive golf courses, recreational facilities, wildlife preserves, or renewable energy sites.

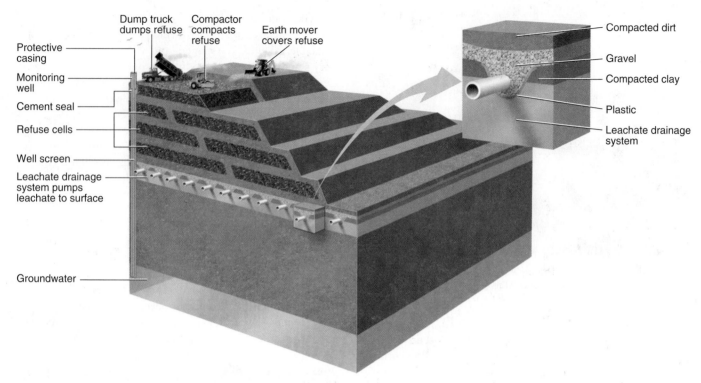

Dump truck
dumps refuse

Compactor
compacts
refuse

Earth mover
covers refuse

Protective
casing

Monitoring
well

Cement seal

Refuse cells

Well screen

Leachate drainage
system pumps
leachate to surface

Groundwater

Compacted dirt

Gravel

Compacted clay

Plastic

Leachate drainage
system

Figure 21–7 Features of a modern landfill. The landfill is sited on a high location, well above the water table. The bottom is sealed with compacted soil and a plastic liner, overlaid by a rock or gravel layer, with pipes to drain the leachate. Refuse is built up in layers as the amount generated each day is covered with soil, so the completed fill has a pyramidal shape that sheds water. The fill is provided with wells for monitoring groundwater.

Siting New Landfills. Between 1988 and 2014, the number of municipal landfills declined from around 8,000 to around 1,900. Because the size of landfills has increased and because recycling is on the rise, the EPA does not believe that landfill capacity is a problem. However, as the agency puts it, "regional dislocations" may be expected. These are due to the one problem associated with landfills that gets more attention than any other: *siting*. It is not that landfills take up enormous amounts of land. One particular landfill occupying 230 acres (Central Landfill) serves the entire state of Rhode Island, and Fresh Kills, which served much of New York City and its environs for many years,

was only 2,200 acres. But as old landfills are closed, it has become increasingly difficult to find land for the new ones needed to take their place.

People in residential communities (where the MSW is generated) invariably reject proposals to site landfills anywhere near where they live, and those who already live close to existing landfills are anxious to close them down. Weary of the odor and heavy truck traffic, residents of Staten Island, New York, pressured the city of New York to shut down the huge Fresh Kills landfill, once the recipient of 27,000 tons of garbage per day (Figure 21–8a). The city fulfilled its promise to phase out the landfill in 2001,

(a)

(b)

Figure 21–8 Fresh Kills landfill before and after. (a) Until 2001, Fresh Kills on Staten Island served as the landfill for the entire city of New York. **(b)** Since its closure, the land has been developed for recreation and other uses.

and Staten Islanders are now enjoying the early stages of the development of Fresh Kills Park, which will eventually include soccer fields, mountain bike and jogging trails, natural areas, and a golf course (Figure 21–8b). Gas from the capped landfill is providing the area with a new energy source. New York City now transports its MSW outside of the city by truck, barge, and rail.

Protests against landfills have been repeated in so many parts of the country (and globe!) that it has given rise to some inventive acronyms: LULU ("locally unwanted land use"), **NIMBY ("not in my backyard")**, NIMTOO ("not in my term of office"), and BANANA (build absolutely nothing anywhere near anything)! You can well imagine how these attitudes apply to the landfill-siting problem.

Outsourcing. The siting problem has some undesirable consequences. First, it drives up the costs of waste disposal. Second, it leads to the inefficient and equally objectionable practice of the long-distance transfer of trash as waste generators look for private landfills whose owners are anxious to receive trash (and cash). Often, this transfer occurs across state and even national lines, leading to resentment and opposition on the part of citizens of the recipient state or nation. Led by Pennsylvania, 11 states import more than 1 million tons of MSW a year (Table 21–1), while 13 states export more than 1 million tons per year, led by New York. Not surprisingly, the heaviest importing states are adjacent to the heaviest exporting states.

One positive result of the siting problem is that it encourages residents to reduce their waste and recycle as much as possible. The siting problem also stimulates the use of combustion for waste disposal.

Combustion: Waste to Energy

Because it has a high organic content, refuse can be burned. Currently, 84 combustion facilities are operating in the United States, burning about 30 million tons of MSW annually—12% of the waste stream. This process is really waste *reduction*, not waste *disposal*, because after incineration the ash must still be disposed of.

Advantages of Combustion. The combustion of MSW has a number of advantages:

- No changes are needed in trash collection procedures or people's behavior. Trash is simply hauled to a combustion facility instead of to a landfill.

- Combustion can reduce the weight of trash by more than 70% and the volume by 90%, thus greatly extending the life of a landfill (which is still required to receive the ash).

- Toxic or hazardous combustion products are concentrated into two streams of ash, which are easier to handle and control than the original MSW. The *fly ash* (captured from the combustion gases by air-pollution control

Table 21–1   Largest Interstate Transfers of MSW, 2005 (most recent date available)

Exporting States and Provinces	Exported MSW (millions of tons)	Importing States	Imported MSW (millions of tons)
New York	7.2	Pennsylvania	7.9
New Jersey	5.8	Virginia	5.7
Illinois	4.4	Michigan	5.4
Ontario, Canada	4.0	Indiana	2.4
Missouri	2.4	Wisconsin	2.1
Maryland	2.0	Illinois	2.1
Massachusetts	2.0	Oregon	1.8
Washington	1.7	Georgia	1.7
Minnesota	1.1	New Jersey	1.7
North Carolina	1.1	Ohio	1.7
Indiana	1.1	South Carolina	1.2
District of Columbia	1.1		
Florida	1.0		
Total, all exports	42.8	Total, all imports	42.2

Source: James E. McCarthy. *Interstate Shipment of Municipal Solid Waste: 2007 Update*, CRS Report for Congress RL34043 (Washington, DC: Congressional Research Service, June 13, 2007).

equipment) contains most of the toxic substances and can be safely put into a landfill. The *bottom ash* (from the bottom of the boiler) can be used as fill in some construction sites and roadbeds. Some combustion facilities process the bottom ash further to recover metals and then convert the remainder into concrete blocks.

- Most combustion facilities are waste-to-energy (WTE) facilities equipped with modern emission-control technology that brings the emissions into compliance with Clean Air Act regulations.

- When burned, unsorted MSW releases about 35% as much energy as coal, pound for pound. WTE facilities produce almost 2,800 megawatts of electricity annually, enough to meet the power needs of more than 2.7 million homes.

- Many of these facilities add resource recovery to their waste processing, in which many materials are separated and recovered before (and sometimes after) combustion.

Drawbacks of Combustion. Combustion has some drawbacks, too:

- Air pollution and offensive odors are two problems that the public associates with combustion facilities. Large reductions in air pollutants have resulted from compliance with stringent air-pollution regulations applied to WTE facilities, making this much less of a problem than it once was.

- Combustion facilities are expensive to build, and their siting has the same problem as that of landfills: No one wants to live near one. Most combustion facilities are located in industrial areas for this reason.

- Combustion ash is often loaded with metals and other hazardous substances and must be disposed of in secure landfills.

- To justify the cost of its operation, the combustion facility must have a continuing supply of MSW. For that reason, the facility enters into long-term agreements with municipalities, and these agreements can lessen the flexibility of the community's solid-waste management options.

- Even if the combustion facility generates electricity, the process wastes both energy and materials unless it is augmented with recycling and recovery.

An Operating Facility. Let us look at the operation of a typical modern WTE facility, which might serve a number of communities or a large metropolitan area. Servicing a population of 1 million or more, the plant receives about 3,000 tons of MSW per day. The waste comes in by rail and truck, and the communities pay disposal fees. Waste processing is efficient because, overall, about 80% of the MSW is burned for energy, 12% is recovered, and 8% is put in landfills. The process, pictured in Figure 21–9, is as follows:

1. Incoming waste is inspected, and obvious recyclable and bulky materials are removed.

2. Waste is then pushed onto conveyers that feed shredders capable of reducing the width of waste particles to six inches or less.

Figure 21–9 Waste-to-energy combustion facility. Schematic flow for the separation of materials and combustion in a typical, modern WTE combustion facility. Numbers refer to steps in the process as described in the text.

3. Strong magnets remove about two-thirds of ferrous metals for recycling before combustion.

4. The waste is then blown into boilers, where light materials burn in suspension and heavier materials burn on a moving grate.

5. Water circulated through the walls of the boilers produces steam, which drives turbines for generating electricity.

6. After the waste is burned, the bottom ash is conveyed to a processing facility, where further separation of metals recovers brass, aluminum, gold, copper, and iron. In some facilities, this process nets $1,000 a day in coins alone!

7. Combustion gases are passed through a lime-based spray dryer/absorber to neutralize sulfur dioxide and other noxious gases. Then the gases go through electrostatic precipitators that remove particles. The resulting air emissions are significantly lower in pollutants than those of an energy-equivalent coal-fired power plant.

8. The fly ash and bottom ash residues are put into landfills.

An appreciation for the impact of such a facility can be gained from looking at the outcome of a year's operation. In one year, 1 million tons of MSW are processed, 40,000 tons of metal are recycled, and 65 megawatts of power are generated—the equivalent of more than 60 million gallons of fuel oil and enough electricity to power 65,000 homes. All this comes from stuff that people have thrown away!

Costs of Municipal Solid-Waste Disposal

The costs of disposing of MSW are escalating, and not just because of the new design features of landfills. More and more, they reflect the expenses of acquiring a site and providing transportation. Landfill disposal fees average about $44 a ton (they are lower than those at WTE facilities because of the high capital costs at WTE facilities). The waste collector must recover this cost, as well as the costs of transporting materials to the site, and generate a profit (if the facility is privately operated). With the closing of the Fresh Kills landfill, the total cost (including sanitation vehicles and workers) of disposing of New York City's MSW rose to $260 a ton. At 12,900 tons per day, this amounts to more than $1 billion a year.

Getting rid of *all* trash is becoming more expensive, and one sad consequence of this increasing expense is illegal dumping. Some towns are charging up to $5 a bag for MSW disposal, and it now costs several dollars to get rid of an automobile tire and $30 or more to dispose of a refrigerator. As a result, tires, refrigerators, yard waste, car parts, and construction waste are appearing all over the landscape. Institutions and apartments that operate with dumpsters often put padlocks on them in order to prevent their unauthorized use by people trying to avoid disposal costs. Many states have established a corps of environmental police to track down midnight dumpers and bring them to justice.

CONCEPT CHECK What are the advantages of disposing of trash in waste-to-energy facilities rather than landfills? ✅

21.3 Better Solutions: Source Reduction and Recycling

Our domestic wastes and their disposal represent a huge stream of material that flows in one direction: from our resource base to disposal sites. The economic and environmental costs of this stream of waste are high, causing us to consider ways to reduce the flow. In general, we can reduce our production of waste by reducing the amount of materials consumed, by reusing materials in the form they were made, and by recycling or composting materials into new products. The best of these strategies is to reduce waste at its source—called *source reduction*. In EPA's waste management hierarchy, source reduction is right at the top (Figure 21–10).

Source Reduction

The EPA defines **source reduction** as "the practice of designing, manufacturing, purchasing, or using materials (such as products and packaging) in ways that reduce the amount or toxicity of trash created." Source reduction accomplishes two goals: It reduces the amount of waste that must be managed, and it conserves resources by preventing the use of virgin materials. It is noteworthy that, after rising rapidly during the latter decades of the 20th century, the amount of waste per capita in the United States peaked at 4.5 pounds in 1990 and has since leveled off.

Source reduction is difficult to measure because it means trying to measure something that no longer exists. The EPA measures source reduction by measuring consumer spending, which reflects the goods and products that ultimately make their way to the trash bin. For example, after 1990, consumer spending continued to grow, but the MSW stream slowed down. If the MSW had grown at the same pace as consumer spending, some 287 million tons would have been generated in 2000 instead of 232 million. Thus, some 55 million tons never made it into the waste stream in 2000, and the EPA considers this to be due to source-reduction activities.

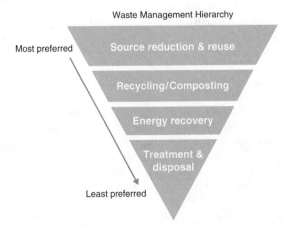

Figure 21–10 EPA waste management hierarchy. This inverted triangle shows the preferable waste management options from most preferred at the top to least at the bottom.

(*Source*: EPA. Non-Hazardous Waste Management Hierarchy. www.epa.gov/epawaste/nonhaz/municipal/wte/nonhaz.htm. March 19, 2012.)

Source reduction can involve a broad range of activities on the part of homeowners, businesses, communities, manufacturers, and institutions. Consider the following examples:

- Reducing the weight of many items has reduced the amount of materials used in manufacturing. Steel cans are 60% lighter than they used to be; disposable diapers contain 50% less paper pulp, due to absorbent-gel technology; and aluminum cans contain only two-thirds as much aluminum as they did 20 years ago.

- The Information Age may be having an impact on the use of paper. Electronic communication, data transfer, and advertising are increasingly performed on personal computers, now present in most homes in the United States. The growth of the Internet as a medium for information transfer has been phenomenal.

- Source reduction includes reuse. Many durable goods are reusable, and indeed, the United States has a long tradition of resale of furniture, appliances, and carpets. Many goods can be donated to charities and put to good use. The popularity of yard sales, flea markets, consignment clothing stores, and other markets for secondhand goods (Figure 21–11) is an encouraging development.

- Online sites such as eBay® and Craigslist link private sellers and buyers. U-Exchange promotes bartering, while free goods can be advertised and acquired on the Freecycle Network™.

- Lengthening a product's life can keep that product out of the waste stream. If products are designed to last longer and to be easier to repair, consumers will learn that they are worth the extra cost.

Figure 21–11 Waste reduction by reuse. Yard sales and flea markets are keeping many materials from becoming MSW after a single use.

The Recycling Solution

Just as natural ecosystems recycle nutrients, we can move in the direction of sustainability by recycling more of the materials in our wastes. Recycling involves reducing the original products to scrap material and remaking them into new products. The process of recycling requires an input of energy and can produce pollutants, which is why it is lower on the EPA's hierarchy than reduction and reuse.

More than 75% of MSW is recyclable material. There are two levels of recycling. In **primary recycling,** the original

---

SUSTAINABILITY

Taking the Waste Out of Take-Out

In today's busy world, we often have to choose between convenience and sustainable actions. For example, take-out meals arrive in pizza boxes or Styrofoam containers, which typically end up in the trash. The *Take Out Without* campaign, started by Canadian designer Lisa Borden, states, "The American population tosses out enough paper bags, plastic cups, forks, and spoons every year to circle the equator 300 times." To cut down on this waste, the campaign recommends bringing reusable containers to stores and restaurants and refusing unneeded items, such as extra straws, napkins, and ketchup.

But sometimes, disposable packaging is necessary. In those cases, it would be helpful to have packaging materials that can be composted. The quest for biodegradable materials to replace plastics is intense. Options include plastic-like materials made from cellulose, starch, and even certain leaves.

Recently researchers have found that *chitosan*, a material made from shrimp shells, can be used to make biodegradable packaging. Chitosan has qualities that are similar to plastic, but unlike plastic, chitosan can decompose quickly in landfills. Chitosan is also antimicrobial, an excellent characteristic for food packaging. Furthermore, it is made from a by-product of shrimp processing, so its use would reduce waste in two ways. Chitosan is not yet ready for scaled up production but researchers are working to get there.

This filament is made of a plastic-like substance produced from the shells of shrimp and crabs.

Source: Take Out Without. http://takeoutwithout.org/behind-the-campaign/; J. Fernandez and D. Ingber. "Manufacturing of Large-Scale Functional Objects Using Biodegradable Chitosan Bioplastic." *Macromolecular Materials and Engineering* (2014). doi:10.1002/mame.201300426.

waste material is made back into the same material—for example, newspapers recycled to make newsprint. In **secondary recycling**, waste materials are made into different products that may or may not be recyclable—for instance, plastic bottles recycled into fleece jackets.

Why Recycle? Recycling is a hands-down winner in terms of energy and resource use and pollution:

- **It saves energy and resources.** One ton of recycled steel cans saves 2,500 pounds of iron ore, 1,000 pounds of coal, and more than 5,400 Btus of energy. One ton of recycled paper saves 17 trees, 6,953 gallons of water, 463 gallons of oil, and 4,000 kilowatt-hours of energy.

- **It decreases pollution.** Making recycled paper requires 64% less energy and generates 74% less air pollution and 35% less water pollution than does using wood from trees. For every ton of waste processed, a thorough recycling program will eliminate 620 pounds of carbon dioxide, 30 pounds of methane, 5 pounds of carbon monoxide, 2.5 pounds of particulate matter, and smaller amounts of other pollutants.

What Gets Recycled? The primary items from MSW that are currently being actively recycled are cans (both aluminum and steel), bottles, plastic containers, newspapers, and yard wastes. Yet there are many alternatives for reprocessing various components of refuse, and people are coming up with new ideas and techniques all the time. A few of the major established techniques, together with their current percentages of recovery by recycling, are as follows:

- Paper and paperboard (70% recovery) can be remade into pulp and reprocessed into recycled paper, cardboard, and other paper products; finely ground and sold as cellulose insulation; or shredded and composted.

- Most glass (34% recovery) that is recycled is crushed, remelted, and made into new containers; a smaller amount is used in fiberglass or "glassphalt" for highway construction.

- Some forms of plastic (13.5% recovery) can be remelted and fabricated into carpet fiber, outdoor wearing apparel, irrigation drainage tiles, building materials (Figure 21–12), and sheet plastic.

- Metals can be remelted and refabricated. Making aluminum (54% recovery) from scrap aluminum saves up to 90% of the energy required to make aluminum from virgin ore. In addition, aluminum ore is imported and represents part of the U.S. trade deficit. National recycling of aluminum saves energy, creates jobs, and reduces the trade deficit.

- Lead-acid batteries have the highest percentage recycling of any U.S. material (96% recovery). Such batteries are almost always made of recycled lead and plastic, representing a closed-loop life-cycle and a sustainability success.

- Old tires (45% recovery) can be remelted or shredded and incorporated into highway asphalt. Not included

Figure 21–12 Deck made from Trex® decking and railing. Trex is made primarily with recycled plastic grocery bags, and plastic film, and waste wood.

(*Source:* www.trex.com.)

in the 45% figure are more than a million tons of tires burned annually in special waste-to-energy facilities.

- Yard wastes (leaves, grass, and plant trimmings—58% recovery) can be composted to produce a humus that can be used to enrich soil.

Recycling is both an environmental and an economic issue. Many people are motivated to recycle because of environmental concern, but the use of recyclable and recycled materials is also driven by economic factors. The Global Recycling Network (www.grn.com) is an Internet information exchange that promotes the trade of recyclables from MSW and the marketing of "ecofriendly" products made from recyclable materials.

Is Recycling Cost Effective? Recycling has its critics, who base their arguments primarily on economics. If the costs of recycling (from pickup to disposal of recyclable components) are compared with the costs of combusting waste or placing it in landfills, recycling frequently comes out second best. Markets for recyclable materials fluctuate wildly, and residents often end up subsidizing the recycling effort. Competition between landfills and combustion facilities often lowers disposal fees, creating an even greater disincentive to recycle. Thus, critics of current recycling practices argue that, unless recycling pays for itself through the sale of the materials recovered, it should not be done.

However, a true environmental assessment should compare the energy costs of recycling with the energy costs of landfill or combustion disposal, as well as comparing the energy costs of making products from recyclable goods with the energy costs of making new products from scratch—called a *life-cycle analysis.* Performed this way, the energy saved by recycling aluminum, for example, is 96%; for newsprint, 45%; for two kinds of plastic, 76–88%; and for glass, 21%.

Figure 21–13 MSW recycling in the United States. MSW recycled from 1960 to 2012, as total waste and percentage recycled.

(*Source:* EPA, Office of Solid Waste and Emergency Response, *Municipal Solid Waste Generation, Recycling, and Disposal in the United States: Facts and Figures for 2012* (Washington, DC: EPA).)

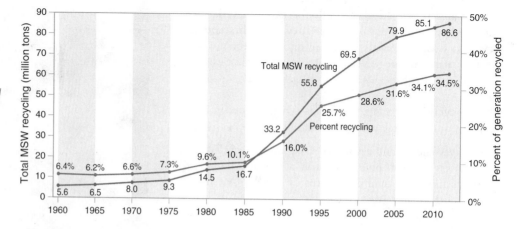

Municipal Recycling

Recycling is probably the most direct way that most people can become involved in environmental issues. If you recycle, you save natural resources from being used and prevent landfills from becoming "landfulls."

Recycling's popularity is well established: Virtually every state has specific recycling goals, met with varying degrees of success. EPA sources report that only 6.4% of MSW was recycled in 1960 versus 34.5% in 2012 (Figure 21–13). There is a great diversity of approaches to recycling in municipalities, from recycling centers requiring miles of driving to curbside recycling with sophisticated separation processes. The most successful programs have the following characteristics:

- There is a strong incentive to recycle where there are PAYT (pay-as-you-throw) charges for general trash and no charge for recyclable goods.

- Recycling is not optional; mandatory regulations are in place, with warnings and sanctions for violations.

- Residential recycling is curbside (Figure 21–14), with free recycling bins distributed to households. (Curbside recycling rose rapidly and then plateaued. Currently, 71% of the U.S. population is served via 9,066 curbside programs.)

- The process employed is single-stream recycling, where all recyclable goods are mixed in a collection truck, allowing residents to put all materials in a single container. Separation of materials for reuse then occurs at a **materials recovery facility (MRF)**.

- Drop-off sites are provided for bulky goods such as sofas, appliances, construction and demolition materials, and yard waste.

- Recycling goals are ambitious, yet clear and feasible. Some percentage of the waste stream is targeted, and progress is followed and communicated.

Nevertheless, municipalities experience very different recycling rates. New York City, for example, has made great progress—from recycling only 5% to recycling 16%—by mandating curbside recycling for all of its 3 million households. New York's commercial recycling is at 40%; reports suggest it could reach as high as 90% if it were better planned. The national prize goes to San Francisco, a city that has achieved a recycling rate of 77%.

Paper Recycling. Newspapers are by far the most important item that is recycled because of their predominance in the waste stream. It is a simple matter to tie up or bag household newspapers, and the amount recovered by recycling is increasing dramatically. Currently, far more newspaper is being recycled (72%) than discarded. Because more than 25% of the trees harvested in the United States are used to make paper, recycling paper saves trees: A 1-meter stack of newspapers equals the amount of pulp from one tree.

The market for used newspapers has fluctuated greatly over the past several decades, rising and then falling. But whatever the price, recycling newspaper is still more sustainable than putting it into a landfill. Today there is a lively international trade in used paper. Forest-poor countries in Europe and Asia purchase wastepaper from the United States and other countries, where there continues to be a surplus of such paper.

Figure 21–14 Curbside recycling. Recyclable materials are picked up in a San Francisco neighborhood, where residents are issued three bins: one for trash, one for recycling, and one for compostable materials. The city now recycles 77% of its waste.

The technology for producing high-quality paper from recyclable stock has improved greatly, to the point where it is virtually impossible to distinguish the recycled product from "virgin" paper. There is often some confusion about what is meant by "recycled paper." The key is the amount of *postconsumer* recycled paper in a given product. In manufacturing processes, much paper is "wasted"; paper that is routinely recovered and rerouted back into processing is called "recycled paper." Thus, the total recycled content of a paper can be 50%, although the actual postconsumer amount recovered by recycling might be only 10% of the total amount of paper.

Glass Recycling and Bottle Laws. The glass in MSW is primarily in the form of containers, most of which held beverages. The average person drinks about a quart of liquid each day. Much of this volume is packaged in single-serving containers that are used once and then thrown away. Nonreturnable glass containers constitute 3.7% of the solid-waste stream in the United States and about 50% of the nonburnable portion; they also make up a large proportion of roadside litter. Broken bottles along roads, beaches, and parklands are responsible for innumerable cuts and other injuries. Both the mining of the materials and the process used to manufacture the beverage containers create pollution. All of these factors produce hidden costs that do not appear on the price tag of the item; however, we pay for them with taxes to clean up litter as well as with our injuries, flat tires, and environmental degradation.

In an attempt to reverse these trends, environmental and consumer groups have promoted **bottle laws** that facilitate the recycling or reuse of beverage containers. Such laws generally call for a deposit on beverage containers, both reusable and throwaway and both glass and plastic. Retailers are required to accept the used containers and pass them along for recycling or reuse. Bottle laws have been proposed in virtually all state legislatures over the past decade. Nevertheless, in every case, the proposals have met with fierce opposition from the beverage and container industries and certain other groups. The reason for their opposition is economic loss to their operations. The container industry contends that bottle laws will result in the loss of jobs and higher beverage costs for the consumer.

As of 2015, 11 states had passed some kind of bottle law. Delaware had such a law and repealed it, and several other states had attempted to pass bottle bills without success. The experience of the states with bottle laws has proved the beverage and bottle industry's claims to be false. In fact, more jobs are gained than lost, costs to the consumer have not risen, a high percentage of bottles are returned, and there is a marked reduction in the can and bottle portion of litter. A final measure of the success of bottle laws is continued public approval.

Repeated attempts have been made to get a national bottle law through Congress—to date, unsuccessfully. Opponents argue that such a law would threaten the continued success of curbside recycling, of which beverage containers represent an important source of revenue. However, as curbside recycling currently reaches only 71% of the U.S. population, a national bottle law would undoubtedly recover a greater proportion of beverage containers. (States with bottle laws report that 75% of containers are returned, a significant proportion from the curbside collections from people who don't bother to redeem the containers.)

In 2012, recycling and bottle laws resulted in the recovery of 34% of glass containers, 55% of aluminum containers, 71% of steel containers, and 31% of plastic containers.

Plastics Recycling. Plastics have a bad reputation in the environmental debate for several reasons. First, plastics have many uses that involve rapid throughput—for example, packaging, bottling, the manufacture of disposable diapers, and the incorporation of plastics into a host of cheap consumer goods. Second, plastics are conspicuous in MSW and litter. Ironically, most trash is disposed of in plastic bags manufactured just for that purpose. Finally, plastics do not decompose in the environment because no microbes (or any other organisms) are able to digest them. (Imagine plastics in landfills delighting archaeologists hundreds of years from now.) Therefore, the possibility of recycling at least some of the plastic in products—containers of liquid, for example—interests the environmental community.

Plastic Bags. Hanging from trees and bushes, blowing along highways, clogging sewer pipes, drifting on ocean currents, plastic bags are everywhere on the planet—everywhere, that is, but the trash or recycling barrel. They cause thousands of deaths of marine mammals and turtles each year. Convenient? Indeed they are, as anyone who has shopped in a supermarket knows. A hundred billion of them are distributed in the United States every year, and close to a trillion are used annually worldwide. Made of low-density polyethylene, plastic bags are difficult to recycle and, like all petroleum-based plastics, are virtually indestructible. China, once a user of 100 billion plastic bags a year, has cracked down on them, banning the production of ultrathin bags—bringing China in line with a number of other countries. In the United States, San Francisco initiated the first citywide ban on plastic grocery bags in 2007; many other cities are instituting similar bans. Instead of using thin plastic bags, customers are encouraged to carry their own cloth bags or to use paper bags or larger, thicker reusable plastic bags that hold five times as much as the standard flimsy bag. Some poverty advocates are concerned that bag laws will make it difficult for the poor, who may not be able to afford reusable bags, but there may be solutions to this problem.

Bottled Water. Throughout the world, the number one "new" drink is bottled water. Health consciousness and suspicions about the safety of tap water have made the plastic water bottle the new symbol of healthy living. More than 10 billion gallons of bottled water were sold in the United States in 2013, making us the world leader in the consumption of bottled water. As a consequence, discarded plastic water bottles now litter our highways and beaches. Only two states have bottle laws that include bottled water and other noncarbonated beverages. People often pay more for bottled water than for gasoline and as much as 10,000 times more

Figure 21–15 Plastic recycling codes. The numbers on plastic and resin products identify the kind of plastic and are used to sort the products for recycling. Unlike glass and metal, plastics degrade somewhat in the recycling process, so they often cannot be recycled into the same material. Some plastics cannot be recycled.

Type of Plastic	Initial Products	Products from Recycled Plastic
1 PETE — Polyethylene Terephalate	Peanut butter jars / Polyester fibers / Water and soda bottles	Beverage bottles / Carpeting / Fleece jackets
2 HDPE — High-Density Polyethylene	Juice bottles / Liquid detergent bottles / Plastic grocery bags	Buckets / Detergent bottles / Outdoor decking
3 PVC — Polyvinyl Chloride	Clear films / Plumbing pipes / Shampoo bottles	Fencing / Garden hoses / Traffic cones
4 LDPE — Low-Density Polyethylene	Food storage containers / Dairy container lids / Dry cleaning bags	Floor tile / Garbage bags / Lawn furniture
5 PP — Polypropylene	Medicine bottles / Yogurt containers / Auto parts	Brooms / Ice scrapers / Shipping pallets
6 PS — Plystyrene	Dairy containers / Vitamin bottles / Styrofoam	Building insulation / Light switch plates / Rulers
7 Other — Other Plastics	Nylon items / Window cleaner bottles / Water coolers	Plastic lumber / Street signs / Pens

than for water from the tap. Only one in six plastic water bottles is recycled, and the rest end up in litter, landfills, or combustion facilities. For these and other reasons, opposition to bottled water is growing; more than 90 colleges and universities and several states and municipalities have either banned or restricted the distribution of bottled water.

Plastics Recycling Codes. If you look on the bottom of a plastic container, you will see a number or some letters inside the little triangle of arrows that represents recyclability (Figure 21–15). The number or the letters are used to differentiate between and sort the many kinds of plastic polymers. The two recyclable plastics in most common use are high-density polyethylene (HDPE, code 2) and polyethylene terephthalate (PETE, code 1). In the recycling process, plastics must be melted down and poured into molds, and some contaminants from the original containers may carry over, making it difficult to reuse some plastics for food containers. While this somewhat restricts uses for recyclable plastic, some interesting uses are appearing, with more in the development stage. PETE, for example, is turned into carpets, jackets, film, strapping, and new PETE bottles; HDPE becomes irrigation drainage tiles, sheet plastic, and, appropriately, recycling bins.

There are, however, some limits to plastic recycling. Certain plastic products, including rugs, lawn furniture,

and polyester clothing, cannot be recycled. Food containers such as water bottles and margarine tubs cannot be made into similar products; instead, they are recycled into less valuable products that may not be recycled. Thus reducing the use of disposable plastics is still the best option for reducing trash.

Recycling Tires. During most of the 20th century, used tires were abandoned in dumps, ravines, and empty lots, where they often trapped water and provided breeding grounds for mosquitoes or caught fires and caused pollution (Figure 21–16). By 1991, there were an estimated 2 billion waste tires in the United States and 242 million scrap tires being produced each year. At that point, 36 states had legislation regulating the disposal of scrap tires and the EPA was promoting longer life for tires, retreading tires, and various ways of recycling tires as well as using them for fuel.

Today tire recycling is a global industry that makes use of the scrap tires from millions of vehicles all over the world. Some building projects reuse whole tires. In other cases, tires are cut, shredded, or ground into "crumb rubber." The crumb material is incorporated into roadways, synthetic turf, playgrounds, building products, flooring, and noise insulation. Nonetheless, more scrap tires are

Figure 21–16 Tire recycling. Tires can be recycled by shredding them into a lightweight, well-draining material that can be used in many construction projects. Here recycled tires are being used to repair a road after a landslide.

Figure 21–17 Composting. In large facilities such as this one in Seattle, Washington, organic materials such as grass and yard clippings are piled into long heaps called windrows, where the action of bacteria causes them to decompose. The windrows are turned periodically, allowing more oxygen to reach the bacteria and speeding up the decomposition process.

being produced than recycled, so tire recycling remains a growth business.

Despite the success of tire recycling, the majority of used tires are now burned rather than recycled. When burned, tires contain more energy per weight than coal; they are often burned in WTE plants or in kilns used to manufacture cement. By the early 2000s, there was a market for up to 80 percent of scrap tires in the United States; about 130 million were burned as fuel.

The Composting Option. An increasingly popular way of treating yard waste and food residuals (currently 13% of MSW) is **composting**. Composting, the natural biological decomposition (rotting) of organic matter in the presence of air, can be carried out in the backyard by homeowners, or it can be promoted by municipalities as a way of dealing with a large fraction of MSW. In addition to diverting wastes away from landfills and combustion facilities, composting provides a product that can be used to amend poor soils, fertilize yards and gardens, and promote higher yields to agricultural crops. In 2012, 58% of yard trimmings were recovered from the waste stream. Small-scale composting can be done in backyards, either by making piles of organic waste or by placing it in bins that are specially made for this use.

Large-scale composting is carried out by recycling facilities and municipalities. Wastes brought into the facility are usually ground up into smaller pieces and then placed outdoors into rows of long piles called *windrows* (Figure 21–17). The windrows are turned regularly to allow oxygen to penetrate the piles, promoting natural biological decomposition. In dry regions, water is added to maintain moisture in the windrows. After several months of this treatment, the compost may be screened for different final uses: finely screened for lawns or more coarsely for garden and landscape beds.

Recycling Electronics. The volume of *e-waste*, or discarded electronic devices such as computers, cell phones, game consoles, and television sets, is growing far more rapidly than our

ability to deal with it. The United States leads the world in e-waste, producing 3.3 million tons a year. One estimate puts the global generation of e-waste at 50 million tons a year. To date, most of this e-waste has been placed in landfills, but an increasing proportion is being recycled.

Because e-waste has numerous toxic components, it is classified as hazardous waste. Heavy metals, including lead, mercury, barium, chromium, arsenic, and copper, are found in the wiring and circuit boards and other components of electronic devices. In a landfill, these substances can leach into groundwater. If e-waste is incinerated, these substances are sent into the air.

Federal law regulates the transport, treatment, and disposal of hazardous wastes, but not from small businesses and households, so the disposal of this e-waste is left up to the states and municipalities. The states that have laws for e-waste are trying to make sure this waste is not incinerated or placed in landfills. There is broad consensus on the need for federal legislation to manage e-waste.

The ideal destination for e-waste is a recycling company that will dismantle the devices, recover the useful materials, and dispose of the rest according to RCRA rules (see Chapter 22). E-waste contains valuable materials such as indium and palladium, which are quite scarce. Precious metals including gold, silver, and copper can also be recovered, but recovery of these materials is labor intensive.

Unfortunately, many recycling companies pack the e-waste into large containers and ship it to China, India, or African countries. An estimated 80% of "recycled" e-waste takes this route, ending up in one of the thousands of small shops in developing countries where laborers pull apart the devices, burn wiring to recover copper, fry circuit boards to free the chips, and dip the chips in buckets of acid to extract gold and silver. Such workers usually have no protection from the hazardous chemicals. Although EU regulations do not allow the transfer of e-waste to poor countries, a black market promotes its smuggling, to the detriment of workers who are too poor to protect themselves.

In contrast, legitimate recycling companies use responsible processes for recycling e-waste. In 2013, the U.S. recycling industry supported a proposed bill to restrict the export of e-waste to countries that are not in the EU or OECD, in order to prevent e-waste dumping. Recently, the nation's largest electronics recycler, Electronics Recyclers International, joined with others to form the e-Stewards Initiative. This initiative certifies e-waste recyclers, assuring consumers and businesses that their e-waste will not be deposited in landfills, incinerated, or exported to developing countries. The regulation of e-waste and the intelligent use of its valuable raw materials demands more attention, but this is a promising start.

Regional Recycling Options

As landfills close down and MSW is transferred out of cities and towns, transfer stations are set up to receive the wastes from collection trucks and transfer them to larger vehicles for long-distance hauling to their final destination. It is an encouraging sign that a number of these transfer stations are being converted into *materials recovery facilities*, referred to in the trade as MRFs, or "murfs." In 2014, there were 1,622 MRFs operating in the United States, handling over 100,000 tons of MSW a day.

More and more, these facilities are designed to receive the collections of single-stream recycling, where all possible recyclable components are commingled. Alternatively, some sorting takes place when waste is collected, either through curbside collection or by town recycling stations (sites to which residents can bring wastes to be recycled). In either case, the waste is trucked to the MRF, where it is moved through the facility by escalators and conveyor belts, tended by workers who inspect and sort the items collected (Figure 21–18). The objective of the process is to prepare materials for the recyclable-goods market. Glass is sorted by color, cleaned, crushed into small pebbles, and then shipped to glass companies, where it replaces the raw materials that go into glass manufacture—sand and soda ash—and saves substantially on energy costs. Cans are sorted, flattened, and sent either to detinning plants or to aluminum-processing facilities. Paper is sorted, baled, and sent to reprocessing mills or exporters. Plastics are sorted, depending on their color and type of polymer, and then sold.

The facility's advantages are its economy of scale and its ability to produce a useful end product for the recyclable-materials market. Some MRFs have gone to high technology, replacing the manual sorting with magnetic pulleys, optical sensors, and air sorters. As the movement grows, it is likely that technology will continue to improve the efficiency of operations and cut costs.

Less common than conventional MRFs are mixed-waste processing facilities, sometimes called "dirty MRFs," which receive unsorted MSW just as if it were going to a landfill or a WTE facility. The waste is loaded on a conveyer and is sorted for recovery of recyclable materials before being landfilled or combusted. Because of extensive contamination of potentially recyclable waste, this method is less desirable.

CONCEPT CHECK Does a law prohibiting single-use bags encourage source reduction or recycling? Which does a bottle bill encourage? ✓

21.4 Public Policy and Waste Management

The management of MSW used to be entirely under the control of local governments. In recent years, however, state and federal agencies have played an increasingly important role in waste management, partly through regulation and partly through encouragement and facilitation. Because a great deal of waste eventually makes its way to the ocean, and because the import of waste from wealthier countries into poor countries raises ethical concerns about dumping, there is increased interest in international coordination about waste.

The U.S. Regulatory Perspective

At the federal level, the following legislation has been passed:

- The first attempt by Congress to address the problem was the Solid Waste Disposal Act of 1965. The legislation gave jurisdiction over solid waste to the Bureau of Solid Waste Management, but the agency's mandate was basically financial and technical rather than regulatory.

- The Resource Recovery Act of 1970 gave jurisdiction over waste management to the newly created EPA and directed attention to recycling programs and other ways of recovering resources in MSW. The act also encouraged the states to develop some kind of waste management program.

- The Resource Conservation and Recovery Act (RCRA) of 1976 used a regulatory (command-and-control) approach to dealing with MSW, giving the EPA power to close local dumps and set regulations for landfills. Combustion

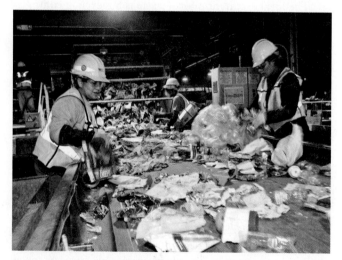

Figure 21–18 Materials recovery facility. Single-stream recycling at a modern MRF in Baltimore. Its purpose is to sort a mixed stream of waste material (cans, bottles, newspapers, etc.) for eventual recycling.

facilities were covered by air-pollution and hazardous-waste regulations, again under the EPA's jurisdiction. RCRA also required the states to develop comprehensive solid-waste management plans.

- The Superfund Act of 1980 (described in Chapter 22) addressed abandoned hazardous-waste sites throughout the country, 41% of which are old landfills.

- The Hazardous and Solid Waste Amendments of 1984 gave the EPA greater responsibility for setting solid-waste criteria for all hazardous-waste facilities. Because even household waste must be assumed to contain some hazardous substances, this meant that the EPA had to determine and monitor all landfill and combustion criteria more closely.

Largely in response to these federal mandates, the states have been putting pressure on local governments to develop *integrated waste management plans*, with goals for recycling, source reduction, and landfill performance.

Integrated Waste Management

It is not necessary to fasten onto just one method of handling wastes, as the example of Sweden demonstrates. Source reduction, waste-to-energy combustion, recycling, materials recovery facilities, landfills, and composting all have roles to play in waste management. Different combinations of these options will work in different regions of the country. A system having several processes in operation is called **integrated waste management.**

Waste Reduction. Today the United States leads the world in per capita waste production and per capita energy consumption—we really are a "throwaway society." How can we turn this situation around? True management of MSW begins in the home (or dorm room). Materialistic lifestyles, affluence, and overconsumption can pack our homes and our trash cans with stuff we simply do not need. Thousands of public storage facilities with interesting names (Stor-U-Self, for example) are packed with stuff that people can't fit into their homes but can't part with.

WasteWise. Several incentives have been put in place on the government level to encourage waste reduction. For example, the EPA sponsors the **WasteWise** program, targeting the reduction of MSW by establishing partnerships with local governments, schools, and organizations, as well as multinational corporations. WasteWise is a voluntary partnership that allows the partners to design their own solid-waste reduction programs. More than 2,700 organizations throughout the country have partnered with the EPA to reduce and recycle waste and, in a new focus of the program, to calculate their reductions of waste-related greenhouse gas emissions.

Pay As You Throw (PAYT). The EPA's PAYT program brings attention to a national trend of unit pricing, or charging households and other customers for the waste they dispose of. Instead of using local taxes to pay for trash collection and disposal, which provides no incentive to reduce waste, communities levy curbside charges for all unsorted

Figure 21–19 Pay-as-you-throw (PAYT) trash pickup. Curbside trash pickup in Hamilton, Massachusetts, features free pickup for single-stream recycling (blue crate) and compost (green bin); bags for other trash are purchased at local stores (blue trash bag) for $2.00.

MSW—perhaps $1–$5 per container (Figure 21–19). PAYT capitalizes on the common sentiment that people should pay their share of disposal costs based on the amount of waste they discard. This practice is found extensively throughout Europe and Asia. A study of PAYT in New England communities[2] found that curbside PAYT communities generated 49% less waste than non-PAYT communities. City managers love it because it saves money and creates jobs. As with the WasteWise program, the EPA's role is that of a facilitator, not a regulator.

Extended Producer Responsibility. Another means of bringing about waste reduction is to establish a program of *extended producer responsibility* (EPR), a concept that involves assigning some responsibility for reducing the environmental impact of a product at each stage of its "life cycle," especially the end. EPR originated in Western Europe. Germany, the Netherlands, and Sweden have well-developed regulations that require companies to take back many items, that encourage the reuse of products, and that promote the manufacture of more durable products. In the United States, EPR is better known as *product stewardship*, a term that implies shared responsibility by producers, consumers, and municipalities. For example, Hewlett-Packard and Xerox make it easy for customers to return spent copier cartridges, and the two companies then recycle the components of the cartridges. Call2Recycle is another good example of this concept, where rechargeable batteries and cell phones are voluntarily taken back by a nonprofit corporation, Rechargeable Battery Recycling Corporation. The program collects up to 7 million pounds of batteries a year.

Again, the EPA is active in providing information to manufacturers and purchasing departments to help them design and buy more environmentally sound products.

---

[2]Environmental Protection Agency, "Pay-As-You-Throw Summer 2010 Bulletin," *Get Smart with Pay-As-You-Throw,* EPA 530-N-09-001, 2010.

STEWARDSHIP
Citizen Power

Do *you* want to help solve the problem of escaped trash? If so, you might consider volunteering for a local cleanup project. Many local and state efforts use volunteers to remove trash from roadways, streams and rivers, lakes, and beaches. Picking up litter is a great example of stewardship and an important activity, especially since even trash that was originally disposed of properly can be swept away by wind and storms.

In Missouri, college students are among the droves of citizens who come out for the annual "No MOre Trash! Bash," an event coordinated by the Missouri Departments of Conservation and Transportation. No MOre Trash! is the state's anti-litter campaign, an effort that uses education and volunteer activities to help protect the environment. In 2014 alone, volunteers collected 127,000 bags of trash.

Another popular cleanup campaign is the Adopt-a-Highway program. In this program, an organization or business keeps a designated section of a roadway free of litter. In exchange, the group is allowed to post its name on the roadway, a form of advertising. Started in the 1980s, the program is now used in 49 states and Puerto Rico, Canada, Japan, Australia, and New Zealand.

Some cleanup efforts are connected to global projects, such as the International Coastal Cleanup, sponsored by the Ocean Conservancy. In 2014, 560,000 volunteers picked up 16 million pounds of trash in 91 countries, part of the world's largest effort to care for the ocean. This cleanup also generates scientific data because volunteers record the number and types of items they find. What have they found? Cigarette butts are, by far, the most common form of coastal trash, but food wrappers, bottles, cans, plastic bags, bottle caps, and straws are also very common.

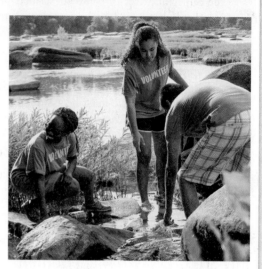

These student volunteers are cleaning up trash along a river.

International Regulation of Waste

As the worldwide volume of waste continues to grow, it is becoming increasingly clear that the problems related to waste management extend beyond national borders. Solutions to these problems require international coordination.

Some international agreements are regional. For example, while many European countries have regulations regarding their own wastes, the European Union coordinates issues related to the production and disposal of MSW among its member countries. A 1999 Landfill Directive established minimum standards for landfills within the member countries. The EU also has several regulations about the transport of hazardous wastes.

Among the issues that have been addressed in international agreements are dumping at sea and the transport of hazardous materials. Because the import of waste from wealthier countries into poor countries raises ethical concerns about dumping, there is increased interest in international coordination of the export of trash to regions that lack environmental regulations.

Ocean Pollution. A great deal of the world's solid waste makes its way to the oceans, much of it ending up in waters that are not under the jurisdiction of any particular nation. The first international agreement on direct dumping into the oceans—the 1972 London Convention on the Prevention of Marine Pollution by Dumping of Wastes and Other Matter—focused on substances harmful to humans or sea life. In 1987, the U.S. Congress passed the Marine Plastic Pollution Research and Control Act (MPPRCA).

About 20% of the MSW in the oceans comes from ocean shipping. The MARPOL Convention (originated in 1973) is the main international convention for the prevention of pollution from ships. Originally it was developed to deal with oil spills, but later other substances, such as sewage, garbage, and air pollution, were added.

Waste Disposal in Less Affluent Countries. As we have seen, the United States and wealthy European countries are addressing many of the problems of solid waste with centralized, capital-intensive solutions such as sanitary landfills and WTE power plants. Yet most poor nations lack the resources to approach MSW disposal in the same way. In addition, solutions that work in developed nations often do not translate well into developing nations, so attempts to fund such solutions frequently fail. For example, between 1974 and 1988, the World Bank gave loans that were used for large trash-compactor trucks. However, such vehicles do not work in crowded, steeply hilly areas or with the dense, highly organic refuse produced in many poorer places. Likewise, incinerators built with western grants have sometimes been abandoned. As a result of these and similar outcomes, many experts believe that our goal should not be to build waste-disposal systems designed for high-income countries in low-income countries. Instead, we need to find solutions that are appropriate to the conditions of the poorer nations.

In developing countries, open dumping and burning are the norm, and the informal collection of refuse is common. Some people make their living by picking up trash or sweeping streets. Others make their homes in dumps, picking over the trash and carrying out a labor-intensive recycling process that often exposes them to unhealthy conditions. Such *waste pickers* usually have low status and are not well treated. However, this informal sector of waste pickers is estimated to employ at least 15 million people, 1% of the urban population of developing countries.[3] For these people, picking valuable components out of trash is an important source of income and a key survival strategy. By reusing materials that would otherwise be wasted, the work they do makes the use of resources more efficient.

Because labor is cheap in developing nations and people need work, researchers argue that waste management in those nations should focus on practices that employ many people. Instead of centralized systems, community waste management makes more sense for these areas. By allowing waste pickers to coordinate themselves, they argue, the pickers will have more rights and better conditions. Pilot projects in Brazil and Egypt suggest that small inputs of money can dramatically improve the conditions of waste pickers. A project in Columbia showed that a sanitary landfill using manual labor to cover waste with earth rather than expensive heavy equipment kept costs down and provided jobs.[4]

In Summary. Solving the problems related to MSW will require us to rethink our approach to the products we use. For those of us in high-income countries, producing less waste needs to be a part of the solution: Purchasing reusable rather than disposable items and recycling can help us make the transition to greater sustainability. Other solutions include using waste to generate energy and safely disposing of the remaining waste. Sweden has already been successful in these approaches, so we know that other developed countries can make this transition, too.

In developing nations, we need to take appropriate community actions, to give rights to those in the waste disposal sector, and to find methods that maintain jobs for those who are already working. Regulations such as those of the EPA and international treaties are necessary to protect the most vulnerable from exploitation, and to keep trash out of the oceans, a commons we all need to protect.

CONCEPT CHECK What are the advantages and disadvantages of pay-as-you-throw rules? ☑

---

[3]M. Medina, "Waste Picker Cooperatives in Developing Countries," Paper at *WIEGO/Cornell/SEWA Conference on Membership-Based Organizations of the Poor*, Ahmedabad, India, January 2005.

[4]M. Medina, *The World's Scavengers: Salvaging for Sustainable Consumption and Production* (Lantham, MD: AltaMira Press, 2007).

REVISITING THE THEMES

SOUND SCIENCE

The role of the EPA and its scientists is crucial. The agency sets the rules for landfill operations, as mandated by Congress, and then takes a proactive role in areas where it has no regulatory authority, encouraging source reduction, recycling, and state MSW management programs. WasteWise and the PAYT program are two excellent examples of initiatives where the EPA plays a facilitating role. The greatest current public-policy needs are a national bottle law, e-waste legislation, and a law dealing with interstate waste transport. Science also plays a role in developing new materials that are less damaging to the environment and in finding ways to reuse and recycle materials.

SUSTAINABILITY

Sustainable natural systems recycle materials, but we create so many products that are difficult to recycle that it becomes necessary to be intentional about source reduction and product responsibility. The interstate transfer of trash wastes energy and allows cities and states to avoid coming to grips with a responsible approach to waste management. The most positive step toward sustainability is intentional source reduction, which both conserves resources and reduces the amount of waste that must be managed. Part of sustainability is finding new ways to reuse wasted resources, to replace environmentally damaging materials, and to combine needs such as the need for energy and need for MSW removal. Living sustainably means wealthier nations will need to produce less waste. Lower-income nations will need solutions that are very different from the modern landfills, compactor trucks, and large incinerators used in developed countries.

STEWARDSHIP

Instead of allowing economic considerations to dictate MSW management choices, many municipalities and organizations are consciously choosing options that reflect stewardship. Such stewardship begins with reducing waste at the source by repairing and reusing items and by selling or giving used items to others. The next step is seeing waste as a resource that can be recycled into other products or burned to produce electricity and heat; stewardship also involves recovering materials from the ashes of items that have been burned. Other steps include ensuring that waste materials are not released into the environment and that only a minimum of the total waste stream ends up in landfills.

Individuals can be stewards of otherwise wasted resources and of their natural environment. Volunteering to clean up litter is an example of stewardship.

REVIEW QUESTIONS

1. List the major components of MSW.

2. Trace the historical development of refuse disposal in the United States. What percentages now go to landfills, combustion, and recycling?

3. What are the major problems with placing waste in landfills? How can these problems be managed?

4. Explain the difficulties accompanying landfill siting and outsourcing. How can these processes be handled responsibly?

5. What are the advantages and disadvantages of WTE combustion?

6. What is the evidence for increasing source reduction, and what are some examples of how it is accomplished?

7. What are the environmental advantages of recycling?

8. Describe the materials that are recycled and how recycling is accomplished.

9. Discuss the attributes of successful recycling programs.

10. Distinguish between MRFs, mixed waste processing, and single-stream recycling.

11. What laws has the federal government adopted to control solid-waste disposal?

12. What is e-waste, and how is it being managed?

13. What is integrated waste management? What decisions would be part of a stewardly and sustainable solid-waste management plan?

14. How are the waste disposal needs of a slum in the developing world different from those in a city in a high-income country?

15. How does the poor disposal of MSW affect the oceans? What existing policies influence the degree of waste in the oceans? What policies do we need to have in the future in order to combat ocean waste?

THINKING ENVIRONMENTALLY

1. Compile a list of all the plastic items you used and threw away this week. Was it more or less than you expected? How can you reduce the number of items on the list?

2. How and where does your school dispose of solid waste? E-waste? Is a recycling program in place? How well does it work?

3. Suppose your town planned to build a combustion facility or landfill near your home. Outline your concerns, and explain your decision to be for or against the site.

4. Does your state have a bottle law? If not, what has prevented such a bill from being adopted?

5. Consider the various methods of managing waste described in the chapter. How many of these methods does your city or town employ?

MAKING A DIFFERENCE

1. Investigate how municipal solid waste is handled in your school or community and what the future plans are for waste management. Promote curbside recycling if it is not happening, and investigate the possibility of adopting a PAYT system.

2. Consider your consumption patterns. Live as simply as possible so as to minimize your impact on the environment.

Buy and use durable products, and minimize your use of disposables.

3. When you shop, bring your own reusable bags for your groceries. When buying small items that you can easily carry, ask that they not be put in a bag.

Chapter 22

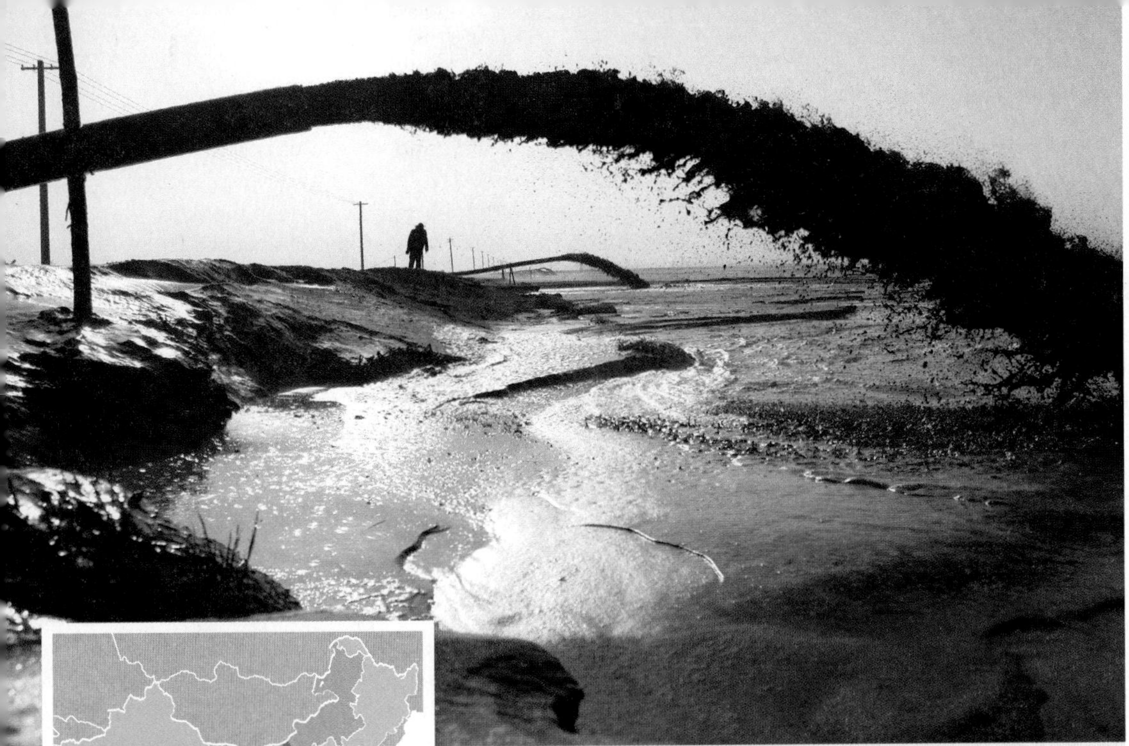

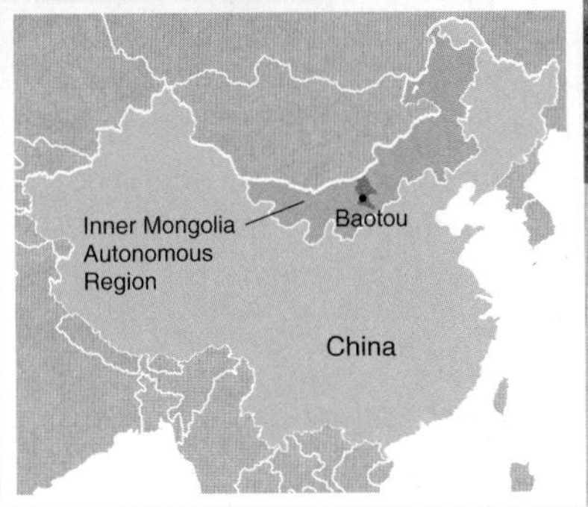

Hazardous Chemicals: Pollution and Prevention

Learning Objectives

22.1 Toxicology and Chemical Hazards: Define toxicology and explain how it applies to many of the chemicals in use in our society; identify the two most toxic chemical groups in use, and assess their involvement in food chains.

22.2 Hazardous Waste Disposal: Describe the three methods employed in land disposal of hazardous wastes; describe and give examples of what happened before land disposal came under regulation.

22.3 Cleaning Up the Mess: Explain how the Superfund program deals with abandoned toxic sites, brownfields, and leaking underground storage tanks.

22.4 Managing Current Toxic Chemicals and Wastes: Review the laws put in place to prevent illegal disposal of toxic wastes, reduce accidents and accidental exposure, and evaluate new chemicals. List several international treaties that regulate hazardous wastes.

22.5 Broader Issues: Discuss the issue of environmental justice, and identify strategies that prevent toxic chemicals from being used.

The Baogang Steel and Rare Earth Complex in Baotou, the largest city in Inner Mongolia, is not a vacationland that you'd ever see in a travel brochure. There is a good-sized lake, but not the kind you might go boating on—there are no sandy beaches and no sparkling waterfalls. Formed by damming a river, the lake serves as a tailings pond for the outfall of mining wastes from multiple industries. The 10-square-kilometer lake has the dubious distinction of being considered one of the most polluted places on Earth.

Rare Earth Production. Some of the refineries in Baotou produce *rare earth minerals,* a group of 17 metals that includes lanthanum, cerium, neodymium, and scandium. Rare earths are used in a wide array of modern products, especially electronics. Cerium, for example, is used in catalytic converters, while neodymium is used to dye the glass used in lasers and to make strong, lightweight magnets. Others are used in the making of computers, smartphones, and even components of wind turbines.

To extract some rare earth minerals, large quantities of rock are crushed and then washed with chemicals such as sulfuric acid, producing enormous amounts of hazardous waste. This waste also contains radioactive elements. When this waste is dumped into tailings ponds, the pollutants may be washed downstream or picked up by wind and spread through the air.

China's Choice. China produces more than 90% of the global rare earths, although these minerals are found worldwide. In part this is because China has

▲ Pipes spew toxic waste from smelting rare earth minerals into a tailings pond near Bautou, Inner Mongolia.

been willing to bear the environmental cost of this industry in order to drive its economy forward. In places such as Baotou, the hazardous nature of producing these minerals is obvious. Visitors and residents of the city can see the stripped earth, smell the acrid tang of chemicals, and watch the black sludge pouring into the lake.

Unfortunately, the hazards involved in producing rare earths are hidden from the millions of people who use cell phones and other electronic products. Yet understanding the life cycle of the goods we use and the hazardous wastes they produce is an essential part of calculating the cost of our actions. To protect ourselves and others, we need to know what substances are dangerous to us and where they are located. We also need to know how to avoid producing them, how to contain them, and how to clean them up.

This chapter focuses on hazardous chemicals. The material in this chapter overlaps with topics in several other chapters—soil pollution (Chapter 11), pesticides (Chapters 12 and 13), environmental health (Chapter 17), air pollution (Chapter 19), and water pollution (Chapter 20). Here we will discuss more specifically the nature of hazardous chemicals and how they are investigated. We will also consider the key U.S. laws and regulations designed to protect human and environmental health.

22.1 Toxicology and Chemical Hazards

Toxicology is the study of the harmful effects of chemicals on human health and the natural environment. A toxicologist might, for example, investigate the potential for *phenolphthalein* (a common chemical laboratory reagent that has also been used in over-the-counter laxatives) to cause health problems. The toxicologist might study the **acute** toxicity effects that could occur upon ingestion of the chemical or upon its contact with the skin, as well as the effects of **chronic** exposure over a period of years, and, finally, the **carcinogenic** (cancer-causing) potential of phenolphthalein. Indeed, such studies have shown that phenolphthalein causes tumors in mice. Because of this finding, the Food and Drug Administration (FDA) canceled all laxative uses of phenolphthalein, which is now listed as "reasonably anticipated to be a human carcinogen."

Data on toxic chemicals are made available to health practitioners and the public via a number of sources, including the National Toxicology Program (NTP) Chemical Repository and the National Institute of Environmental Health Sciences (NIEHS), the latter of which provides an annual updated *Report on Carcinogens*, with a complete list of carcinogenic agents. The Environmental Protection Agency (EPA) also makes available data on toxic chemicals in its Integrated Risk Information System (IRIS). Green Media Toolshed uses these and other data sources in its *Scorecard*, an exhaustive profile of 11,200 chemicals, including information on their manufacture and uses (see http://scorecard.goodguide.com/).

Dose Response and Threshold

In our discussion of risk assessment, we introduced the concepts of dose response and exposure (Chapter 17). When investigating a suspect chemical, a toxicologist would conduct tests on animals, investigate human involvement with the chemical, and present information linking the dose (the level of exposure multiplied by the length of time over which exposure occurs) with the response (some acute or chronic effect or the development of tumors). For example, studies of phenolphthalein toxicity indicate that the LDL_0 (lowest dose at which death occurred in animal testing) was 500 milligrams/kilogram. This is a low toxicity, so concern about the chemical would center on chronic or carcinogenic issues.

Human exposure to a hazard is a vital part of its **risk characterization,** and such exposure can come through the workplace, food, water, or the surrounding environment. For example, the NTP Chemical Repository gives the various uses of phenolphthalein, followed by the precautions to take in handling it and the symptoms of exposure (there are many!). Getting an accurate determination of human exposure is often the most difficult area of risk assessment.

Threshold Level. In the dose-response relationship, there is usually a *threshold*. Organisms are able to deal with certain levels of many substances without suffering ill effects. The level below which no ill effects are observed is called the **threshold level.** Above this level, the effect of a substance depends on both its concentration and the duration of exposure to it. Higher levels may be tolerated if the exposure time is short. Thus, the threshold level is high for short exposures but gets lower as the exposure time increases (Figure 22–1). In other words, it is not the absolute amount, but rather the *dose*, that is important.

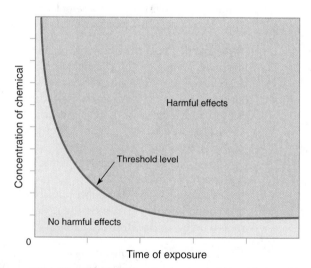

Figure 22–1 The threshold level. The threshold level for harmful effects of toxic pollutants becomes lower as the exposure time increases. It is also different for different chemicals.

Where carcinogens are concerned, the EPA generally takes a zero-dose, zero-response approach. That is, there is no evidence of a discrete threshold level for any carcinogenic chemical. However, the lower the dose, the more likely it is that the response cannot be distinguished from the background level of cancers in a population. In such cases, the risk becomes very low and drops below the level for which regulatory action is needed.

These are just some basics of toxicology. The field is well established and is the most important source of sound scientific information supporting the regulatory efforts of the FDA and the EPA. The NTP was established in 1978 to provide this kind of information and has become the world's leader in assessing chemical toxicity and carcinogenicity. We turn our attention now to the chemicals themselves.

The Nature of Chemical Hazards: HAZMATs

A chemical that presents a certain hazard or risk is known as a **hazardous material (HAZMAT)**. The EPA categorizes substances on the basis of the following hazardous properties:

- *Ignitability:* substances that catch fire readily (e.g., gasoline and alcohol)
- *Corrosivity:* substances that corrode storage tanks and equipment (e.g., acids)
- *Reactivity:* substances that are chemically unstable and that may explode or create toxic fumes when mixed with water (e.g., explosives, elemental phosphorus [not phosphate], and concentrated sulfuric acid)
- *Toxicity:* substances that are injurious to health when they are ingested or inhaled (e.g., chlorine, ammonia, pesticides, and formaldehyde)

Containers in which HAZMATs are stored and vehicles that carry HAZMATs are required to display placards identifying the hazards (flammability, corrosiveness, the potential for poisonous fumes, and so on) presented by the material inside (Figure 22–2). *Radioactive materials* are probably the most hazardous of all and are treated as an entirely separate category. The risks associated with radioactive materials are described in another chapter (Chapter 15).

Sources of Chemicals Entering the Environment

To understand how HAZMATs enter our environment, we need to look at how people in our society live and work. First, the materials making up almost everything we use, from the shampoo and toothpaste we use in the morning to the television we watch in the evening, are products of chemical technology. Our use constitutes only one step in the **total product life cycle,** a term that encompasses all steps, from obtaining raw materials to finally disposing of the product. Implicit in our use of hair spray, for example, is that raw materials were obtained and various chemicals were produced to make both the spray and its container. Chemical wastes and by-products are inevitable in production processes. In addition, consider the risks of accidents or spills occurring in

Figure 22–2 HAZMAT placards. These are examples of some of the placards that are mandatory on trucks and railcars carrying hazardous materials. Numbers in place of the word on the placard or on an additional orange panel identify the specific material. Placards alert workers, police, and firefighters to the kinds of hazards they face in case an accident occurs.

the manufacturing process and in the transportation of raw materials, the finished product, or wastes. Finally, what are the risks of breathing hair spray? What happens to the container that still holds some of the hair spray when you throw it into the trash because the propellant is used up?

Multiply these steps by all the hundreds of thousands of products used by millions of people in homes, factories, and businesses, and you can appreciate the magnitude of the situation. More than 84,000 chemicals are registered for commercial use within the United States. At every stage—from the mining of raw materials through manufacturing, use, and final disposal—various chemical products and by-products enter the environment, with consequences for both human health and the environment.

The use of many products—pesticides, fertilizers, and road salts, for example—directly introduces them into the environment. Alternatively, the intended use may leave a fraction of the material in the environment—the evaporation of solvents from paints and adhesives, for instance. Then there are the product life cycles of materials that are used tangentially to the desired item. Consider lubricants, solvents, cleaning fluids, and cooling fluids, with whatever contaminants they may contain. Likewise, there are the product life cycles of gasoline, coal, and other fuels that are consumed for energy. Figure 22–3, on the following page, illustrates the life cycle of a typical consumer product.

Toxics Release Inventory. The introduction of chemicals into the environment may occur in every sector, from major industrial plants to small shops and individual homes. The EPA tracks the industry sectors that release the

Figure 22–3 Total product life cycle. The life cycle of a product begins when the raw materials are obtained and ends when the used-up product is discarded. At each step—or in transport between steps—wastes, by-products, or the product itself may enter the environment, causing pollution and creating various risks to human and environmental health.

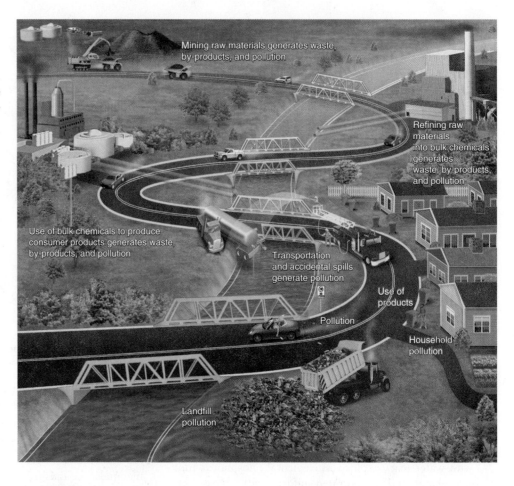

greatest quantities of toxic materials. Information about these releases is reported in the EPA's annual report, the **Toxics Release Inventory (TRI)**.[1] The **Emergency Planning and Community Right-to-Know Act of 1986 (EPCRA)** requires industries to report releases of toxic chemicals to the environment, and the **Pollution Prevention Act of 1990** mandates collection of data on toxic chemicals that are released to the environment, treated on-site, recycled, or burned for energy. The TRI does not cover small businesses, such as dry-cleaning establishments and gas stations, or household hazardous waste. Nor does it cover toxic releases from oil and gas extraction, a limit environmental groups have protested.

The TRI provides an annual record of releases of almost 650 designated chemicals by some 21,600 industrial facilities (Figure 22–4). For 2013, the TRI reveals the following information about toxic chemicals:

- Total production-related toxic wastes: 25.6 billion pounds
- Releases to air: 770 million pounds
- Releases to water: 210 million pounds
- Releases to land disposal sites and underground injection: 2,750 million pounds

- State with the highest releases of toxics: the mining-intensive state of Alaska
- Total environmental releases: 4.14 billion pounds (16% of all production-related toxic wastes; the remainder are recycled or treated on- or off-site)

Add to the last figure the estimated 3 billion pounds of household hazardous waste released each year and an unknown quantity coming from small businesses (facilities do not have to report releases under 500 pounds), and you have some idea of the dimensions of the problem. The good news is that over the more than 30 years since the TRI has been in effect, the quantities of virtually all categories of toxic waste have been going down, and total releases have declined by more than 60%.

The Threat from Toxic Chemicals

All toxic chemicals are, by definition, hazards that pose a risk to humans. Fortunately, a large portion of the chemicals introduced into the environment are gradually broken down and assimilated by natural processes. After that, they pose no long-term human or environmental risk, even though they may be highly toxic in high-level, short-term exposures.

Two major classes of chemicals do not readily degrade in the environment: heavy metals and their compounds and synthetic organics. If sufficiently diluted in air or water, most of these compounds do not pose a hazard, though there are some notable exceptions.

---

[1]U.S. Environmental Protection Agency, *2013 Toxics Release Inventory: National Analysis Overview* (2013), accessed April 13, 2015, http://www.epa.gov/tri.

TOTAL DISPOSAL OR OTHER RELEASES BY INDUSTRY, 2003–2013

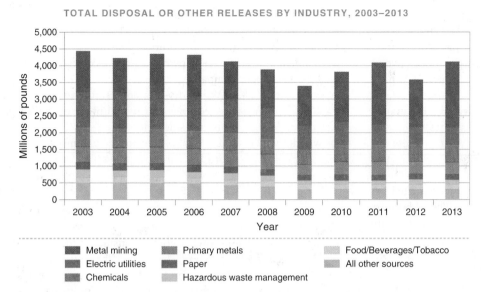

Metal mining   Primary metals   Food/Beverages/Tobacco
Electric utilities   Paper   All other sources
Chemicals   Hazardous waste management

Figure 22–4 Toxics Release Inventory, 2003–2013. There has been a 7% reduction in releases of chemicals since 2003 and a total reduction of more than 60% since 1986. The increase between 2012 and 2013 was largely due to increases in the metal-mining sector. Metal mining, electric utilities, and chemical industries are the industries releasing the most toxics. However, the EPA's monitoring program does not include oil and gas extraction. (*Source:* EPA, 2013 Toxics Release Inventory, April 2015.)

UNDERSTANDING THE DATA

Between 2003 and 2013, the *total* release of toxins in the United States declined by 7%. How did the U.S. *per capita* release of toxins change during that time? Why?

Heavy Metals. The most dangerous heavy metals are lead, mercury, arsenic, cadmium, tin, chromium, zinc, and copper. These metals are used widely in industry, particularly in metalworking or metal-plating shops and in such products as batteries and electronics. They are also used in certain pesticides and medicines. Because compounds of heavy metals can have brilliant colors, they once were heavily used in paint pigments, glazes, inks, and dyes. Millions of children have been treated for lead poisoning because of their exposure to lead-based paint. The paint was banned in the United States in 1978, but many older painted surfaces still hold lead and are a hazard.

Heavy metals are extremely toxic because, as ions or in certain compounds, they are soluble in water and may be readily absorbed into the body, where they tend to combine with and inhibit the functioning of some vital enzymes. Even very small amounts can have severe physiological or neurological consequences. Mercury, in particular, has contaminated rivers, lakes, and streams in the United States because of its release into the air by coal- and oil-fired power plants (mercury is present in fossil fuels in trace quantities). In 2011, using its authority under the Clean Air Act, the EPA issued a new rule requiring a 90% reduction in the mercury released from power plants, bringing long-sought relief to the wildlife and people affected by this toxic pollutant.

Organic Compounds. Petroleum-derived and *synthetic* organic compounds are the chemical basis for all plastics, synthetic fibers, synthetic rubber, modern paint-like coatings, solvents, pesticides, wood preservatives, and hundreds of other products. Because of their chemical structure, many synthetic organics are resistant to biodegradation. This resistance is an important part of what makes many such compounds useful: We wouldn't want fungi and bacteria attacking and rotting the tires on our cars, and paints and wood preservatives function only insofar as they are both nonbiodegradable and toxic to decomposer organisms.

These compounds are toxic because they are often readily absorbed into the body, where they interact with particular enzymes, but their nonbiodegradability prevents them from being broken down or processed further. When a person ingests a sufficiently high dose, the effect may be acute poisoning and death. With low doses over extended periods, however, the effects are insidious and can be mutagenic (mutation causing), carcinogenic, or teratogenic (birth defect causing). Synthetic organic compounds also may cause serious liver and kidney dysfunction, sterility, and numerous other physiological and neurological problems.

Dirty Dozen. A particularly troublesome class of synthetic organics is the **halogenated hydrocarbons,** organic compounds in which one or more of the hydrogen atoms have been replaced by atoms of chlorine, bromine, fluorine, or iodine (**Figure 22–5**). These four elements are classed as *halogens*—hence, the name. Of the halogenated hydrocarbons, the **chlorinated hydrocarbons** (also called organic chlorides) are by far the most common. Organic chlorides

Figure 22–5 A halogenated hydrocarbon. When hydrogen atoms in an organic compound (such as ethylene) are replaced by halogen atoms (chlorine, fluorine, bromine, or iodine), the resulting compound is said to be halogenated. Such compounds are nonbiodegradable and tend to bioaccumulate.

Natural organic compound

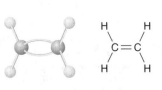

Ethylene

Substitute chlorine for hydrogen

Synthetic halogenated counterpart

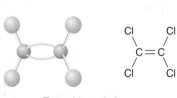

Tetrachloroethylene

are widely used in plastics (polyvinyl chloride), pesticides (DDT, Kepone, and mirex), solvents (carbon tetrachlorophenol and tetrachloroethylene), electrical insulation (polychlorinated biphenyls), and many other products. Twelve **persistent organic pollutants** (called **POPs**, Table 22–1) have been banned or highly restricted globally as a result of the **Stockholm Convention on Persistent Organic Pollutants,** which entered into force in May 2004. Another nine compounds were added to the list in 2009. Most of these are halogenated hydrocarbons. All are toxic to varying extents, and most are known animal carcinogens.

PERC. As an example, let's consider the halogenated hydrocarbon tetrachloroethylene (Figure 22–5). This substance, also called perchloroethylene, or PERC, is colorless and nonflammable and is the major substance in dry-cleaning fluid. It is an effective solvent and is used in all kinds of industrial cleaning operations. It can also be found in shoe polish and other household products. PERC evaporates readily when exposed to air, but in the soil, it can enter groundwater easily because it does not bind to soil particles. Human exposure can occur in the workplace, especially in connection with dry-cleaning operations. It can also occur if people use products containing PERC, such as dry-cleaned garments. PERC enters the body most readily when breathed in with contaminated air. Breathing PERC for short periods can bring on dizziness, fatigue, headaches, and unconsciousness. Over longer periods, PERC can cause liver and kidney damage.

Laboratory studies also show PERC to be carcinogenic to rats and mice, and it is listed in NTP's 2014 *Report on Carcinogens*[2] as "reasonably anticipated to be a human carcinogen." Studies have shown a higher risk of cancer, as well as neurological impairment, in dry-cleaning employees. Approximately 28,000 U.S. dry-cleaning establishments use PERC. The EPA has issued rules that would phase out PERC use in dry-cleaning shops in residential buildings by 2020; critics wonder why it should take so long and why in just a select few shops. The rules for PERC are currently under review.

Effects on Hormones

Hormones are substances that function as messengers in the endocrine system, the group of organs that regulate the growth and functions of vertebrates; insulin, testosterone, and estrogen are hormones. Changes in the concentration of a hormone can have a significant effect on an animal's endocrine system. Disturbances to the endocrine system are not trivial matters—just think of diabetics who lack the hormone insulin and would die if it were not supplied to them. A chemical that affects the function of the endocrine system is called an **endocrine disruptor.**

---

[2]U.S. Department of Health and Human Services, Public Health Service, National Toxicology Program, *Report on Carcinogens*, 12th ed. (Washington, DC: U.S. Department of Health and Human Services, 2011).

Table 22–1 The "Dirty Dozen" Persistent Organic Pollutants

Chemical Substance	Designed Use	Major Concerns
Aldrin	Pesticide to control soil insects and to protect wooden structures from termites	Toxic to humans, probable carcinogen
Chlordane	Broad-spectrum insecticide to protect crops	Biomagnification in food webs, carcinogenic
DDT	Widely used insecticide, malaria control	Biomagnification in food webs, probable carcinogen
Dieldrin	Insecticide to control termites and crop pests	Toxic, biomagnification in food webs, probable carcinogen
Endrin	Insecticide and rodenticide	Toxic, especially in aquatic systems
Heptachlor	General insecticide	Toxic, carcinogenic
Hexachlorobenzene	Fungicide	Toxic, carcinogenic
Mirex	Insecticide to control ants	High toxicity to aquatic animals, carcinogenic
Toxaphene	General insecticide	High toxicity to aquatic animals, carcinogenic
PCBs	Variety of industrial uses, especially in transformers and capacitors	Toxic, teratogenic, carcinogenic
Dioxins	No known use; by-products of incineration and paper bleaching	Toxic, carcinogenic, reproductive system effects
Furans	No known use; by-products of incineration, PCB production	Toxic, especially in aquatic systems

Source: IPCS Assessment Report: L. Ritter et al., *Persistent Organic Pollutants: An Assessment Report on DDT, Aldrin, Dieldrin, Endrin, Chlordane, Heptachlor, Hexachlorobenzene, Mirex, Toxaphene, Polychlorinated Biphenyls, Dioxins, and Furans, International Programme on Chemical Safety,* December 1995.

During the early 1990s, scientists became concerned that synthetic chemicals in food and the environment were acting as endocrine disruptors.[3] At that time, there had been a number of disturbing reports about the abnormal development of wildlife, such as immune disorders in mammals, alligators with reproductive defects, and fish with both ovaries and testes (Chapter 6). Could it be that the release of low levels of synthetic chemicals into the environment was causing these abnormalities? Indeed, the effects of exposure to DDT by fish-eating birds were known to be the outcome of disruption of the birds' reproductive systems, controlled by hormones (Chapter 13).

After pouring through thousands of research articles, scientist Theo Colburn and others were able to show a common thread: Chemicals were disrupting hormone systems. This finding triggered national concern and policy responses. The Food Quality Protection Act of 1996 directed the EPA to develop procedures for testing chemicals for endocrine disruption activity. Accordingly, the EPA established the Endocrine Disruptor Screening Program.

Worldwide, the issue of endocrine disruption remains a serious concern. A 2013 report published by the WHO and UNEP described the current status of the scientific research, noting three disturbing trends: (1) a high and increasing incidence of many endocrine-related disorders in humans; (2) observations of endocrine-related effects in wildlife populations; and (3) laboratory studies linking chemicals with endocrine-disrupting properties to diseases.[4]

Scientific research on endocrine disruptors is ongoing. Unfortunately, we still know little about how dangerous they are (see Sound Science, *Disrupting Hormones with Pollution*).

---

[3]T. Colburn, D. Dumanoski, and J. Peterson Myers, *Our Stolen Future: Are We Threatening Our Fertility, Intelligence, and Survival?—A Scientific Detective Story* (New York: Plume, 1996).

[4]Åke Bergman et al., eds., *State of the Science of Endocrine Disrupting Chemicals 2012* (Geneva: United Nations Environment Programme and the World Health Organization, 2013).

Involvement with Food Chains

Heavy metals and nonbiodegradable synthetic organics are particularly hazardous because they tend to accumulate in organisms. We have discussed bioaccumulation and biomagnification in connection with DDT (Chapter 13).

Minamata. A tragic episode in the 1950s and 60s, known as Minamata disease, revealed the potential for biomagnification of mercury and other heavy metals. The disease is named for a small fishing village in Japan where the episode occurred. In the mid-1950s, cats in Minamata began to show spastic movements, followed by partial paralysis, coma, and death. At first, little attention was paid to it. But when the same symptoms began to occur in people, concern escalated quickly. Additional symptoms, such as mental impairment, insanity, and birth defects, were also observed. Scientists and health experts eventually diagnosed the cause as acute mercury poisoning.

A chemical company near the village was discharging wastes containing mercury into a river that flowed into the bay where the Minamata villagers fished. The mercury was first absorbed by bacteria and then biomagnified as it passed up the food chain through fish to cats or to humans. By the time the situation was brought under control, some 50 persons had died, and 150 had suffered serious bone and nerve damage. Even now, the tragedy lives on in the crippled bodies and disabled minds of some Minamata victims.

Arctic POPs. Much to research scientists' surprise, fish, birds, and mammals all over the Arctic are showing elevated body burdens of a number of persistent organic pollutants—in particular, DDT, toxaphene, chlordane, PCBs, and dioxins. Moreover, because the Inuit people of the far north depend on these animals for food, they, too, are carrying high loads of POPs—some of the highest in the world (Figure 22–6).

The real mystery is how these chemicals—all of which are toxic—get to such remote and pristine environments.

Figure 22–6 An Inuit hazard. This woman is cutting the meat of a seal in Sermilik Fjord, Greenland. Seals and other animals used for food by the Inuit harbor high levels of persistent organic pollutants.

No one is spraying pesticides there, and there are no industries to produce PCBs and dioxins. The key to the presence of such chemicals can be found in several factors that characterize POPs: persistence, bioaccumulation, and the potential for long-range transport. Added to these factors is the unique climate of the Arctic, where the extreme cold promotes a process called *cold condensation*. POPs are volatile chemicals carried in global air patterns to the Arctic, where they condense on the snowpack and are then washed into the water with the spring thaw. Once in the lakes and coastal waters, the chemicals are picked up by plankton and passed up the food chain through bioaccumulation and biomagnification.

A Canadian government study found that three-fourths of Inuit women have PCB levels up to five times above those deemed safe. The potential health effects of the chemical load carried by the Arctic residents range from immune-system disorders to disruption of their hormone systems and, over time, to cancer. The latest impact has been a serious imbalance in births, where twice as many girls as boys are born in many Arctic villages. Scientists from the Arctic Monitoring and Assessment Program, an international body that advises the governments of the eight Arctic nations, measured levels of man-made chemicals in women's blood that were high enough to trigger changes in the sex of unborn children during their first few weeks of gestation. This transboundary pollution by chemicals is exactly why the UN Convention has banned them; hopefully, this action will eventually clear the Arctic air of the POPs.

CONCEPT CHECK How did mercury become concentrated in the bodies of people in Minamata, Japan? How do POPs become concentrated in the bodies of Arctic animals? ☑

22.2 Hazardous Waste Disposal

The Clean Air and Clean Water Acts, passed in the early 1970s, put an end to the widespread disposal of hazardous wastes in the atmosphere and waterways. However, their passage left an enormous loophole. If you can't vent wastes into the atmosphere or flush them into waterways, what do you do with them? Industry turned to *land disposal*, which was essentially unregulated at the time. Indiscriminate air and water disposal became indiscriminate land disposal. Thus, in retrospect, we see that the Clean Air and Clean Water Acts, for all their benefits in improving air and water quality, also succeeded in transferring pollutants from one part of the environment to another.

Methods of Land Disposal

In the early 1970s, there were three primary land disposal methods: surface impoundments, landfills, and deep-well injection. With the conscientious implementation of safeguards, each of these methods has some merit, and each is still heavily used for hazardous-waste disposal. Without adequate regulations or enforcement, however, contamination of groundwater is virtually guaranteed.

Figure 22–7 Surface impoundment. Surface impoundments are often used for preliminary treatment of liquid wastes. Shown here is an impoundment at a paper manufacturer in Alabama.

Surface Impoundments. Surface impoundments are simple excavated depressions (ponds) into which liquid wastes are drained and held (Figure 22–7). They are the least expensive—and, hence, the most widely used—way to dispose of large amounts of water carrying relatively small amounts of chemical wastes. The impoundments may be used to carry out wastewater treatment prior to discharging the wastes. As waste is discharged into the pond, solid wastes settle and accumulate, while water evaporates. If the bottom of the pond is well sealed and if the loss of water through evaporation equals the gain from input, impoundments may receive wastes indefinitely. However, inadequate seals may allow wastes to leach into groundwater, exceptional storms may cause overflows, and volatile materials can evaporate into the atmosphere, adding to air pollution problems. To prevent these problems, hazardous waste surface impoundments are required to be constructed with a double liner system, a leachate collection and removal system, and a leak detection system. The 2014 TRI reports 739 million pounds of toxics released to on-site surface impoundments. (*On-site* refers to disposal by manufacturers or institutions on their own facilities; *off-site* means that the wastes have been transferred to a waste-treatment or disposal facility.)

Landfills. RCRA sets rigid standards for the disposal of hazardous wastes in landfills. When hazardous wastes are in a concentrated liquid or solid form, they are commonly put into drums, collected by a waste management company (for a hefty fee!), and then treated in accordance with their chemical and physical characteristics. *Treatment standards* for wastes are established by the EPA and characterized as **best-demonstrated available technologies (BDATs)**. These standards apply to all solid and liquid hazardous wastes and have the effect of reducing their toxicity and mobility. The BDAT technologies include stabilization and neutralization of sludges, soils, liquids, powders, and slurries; chemical oxidation of organic materials; and various specific techniques applied to metal-bearing wastes that convert them to an insoluble solid material for landfill

SOUND SCIENCE
Disrupting Hormones with Pollution

In recent years, certain widely used chemical compounds have been implicated as potential endocrine disruptors. One of these is bisphenol A (BPA), which is used in plastic containers and the liners of food and beverage cans. Worldwide, about 4.5 billion kilograms (10 billion pounds) of the substance are manufactured annually. Scientists have found that BPA can leach from plastics into liquids in the containers or the environment.

A recent study found BPA in the urine of 93% of people in the United States. Children who were tested showed the highest levels of exposure. Is this BPA causing health problems? It's possible, because BPA is known to act like a reproductive hormone; that is, it simulates the effects of estradiol, one of the most important reproductive hormones in the human body. It does this by binding to estrogen receptors found in many parts of the body. In fact, the compound was first investigated in the 1930s to see if it would work as a synthetic estrogen.

To determine the effects of BPA, over 1,000 tests on animals have been conducted, many of which have shown that doses of BPA can have significant effects on the health of animals, including weight gain, increased aggressiveness, altered sexual behavior, changes in the uterus and ovaries, and even disturbance of glucose metabolism that could lead to diabetes. No experimental studies of the effects of BPA on humans have been conducted because such tests are considered unethical, but about 100 epidemiological studies have been done. Some of these studies show a correlation between BPA exposure and a number of physical problems in humans.

A customer looks over shelves of canned goods, which are likely to be lined with an epoxy resin made with bisphenol A.

However, it is hard to know how dangerous the compound is, and the question has been part of a lively scientific debate. The levels of BPA to which people are exposed may be lower than those used in the animal studies, and other factors may influence the correlation found in the epidemiological studies.

Consumers have pushed to have BPA removed from packaging for food and drinks and other products. In response, manufacturers and chemical industry groups point out that plastic can liners with BPA cut down on corrosion and food poisoning. If we are overzealous, we may ban a useful product. Furthermore, the replacement chemicals may not be any better for human health. The debate over BPA is not over, but the role of science in highlighting its role as an endocrine disruptor has been paramount.

Source: U.S. Department of Health and Human Services. Center for the Evaluation of Risks to Human Reproduction. NTP-CERHR Monograph on the Potential Human Reproductive and Developmental Effects of Bisphenol A. NIH Publication No. 08-5994. September 2008.

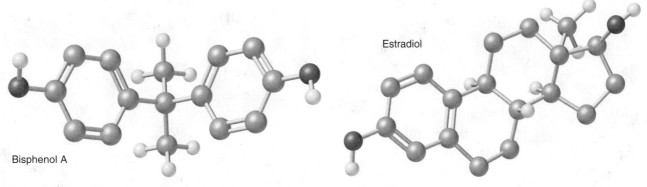

Bisphenol A

Estradiol

These diagrams show the chemical structures of BPA and estradiol. BPA is a known disruptor of estradiol, an important female hormone.

disposal. Some 384 million pounds of hazardous wastes were delivered to on-site landfills in 2013, and an additional 264 million pounds went to off-site landfills. Only 21 landfills in North America are licensed to receive off-site hazardous wastes, and these are owned by just a few waste management companies.

If a landfill is properly lined, supplied with a system to remove leachate, provided with monitoring wells, and appropriately capped, it may be reasonably safe and is referred to as a **secure landfill** (Figure 22–8). The various barriers are subject to damage and deterioration, however, requiring adequate surveillance and monitoring systems to prevent leakage.

Deep-Well Injection. Deep-well injection involves drilling a borehole thousands of feet below groundwater into a porous geological formation. A well consists of concentric pipes and casings that isolate wastes as they are injected and pipes that are sealed at the bottom to prevent wastes from backing up. Over time, the wastes often undergo reactions with naturally occurring material that make them less hazardous. This method is used for various volatile organic compounds, pesticides, fuels, and explosives. Currently, 143 hazardous-waste deep wells are in operation, mostly in the

Gulf Coast region. The EPA's Underground Injection Control Program regulations require that the wells be limited to geologically stable areas so that the hazardous-waste liquids pumped into the well remain isolated indefinitely. The total amount of deep-well injection (225 million pounds in 2013) has declined over the years, but the technique remains useful because it has a greater potential than other methods for keeping toxic wastes from contaminating the hydrologic cycle and the food web.

Mismanagement of Hazardous Wastes

Because early land disposal was not regulated, in numerous instances not even the most rudimentary precautions were taken. In many cases, deep wells injected wastes directly into groundwater, abandoned quarries were used as landfills with no additional precautions being taken, and surface impoundments had no seals or liners whatsoever. Even worse, considerable amounts of waste failed to get to any disposal facility at all.

Midnight Dumping and Orphan Sites. The need for alternative methods to dispose of waste chemicals created an opportunity for a new enterprise: waste disposal. Many reputable businesses entered the field, but—again in the absence of regulations—there were also disreputable operators. As stacks of drums filled with hazardous wastes "mysteriously" appeared in abandoned warehouses, vacant lots, or municipal landfills, it became clear that some operators simply were pocketing the disposal fee and then unloading the wastes in any available location, frequently under cover of darkness—an activity termed **midnight dumping** (Figure 22–9). Authorities trying to locate the individuals responsible would learn that they had gone out of business or were nowhere to be found.

Some companies or individuals simply stored wastes on their own properties and then went out of business, abandoning the property and the wastes. These locations became known as **orphan sites**—hazardous-waste sites without a responsible party to clean them up. As drums containing

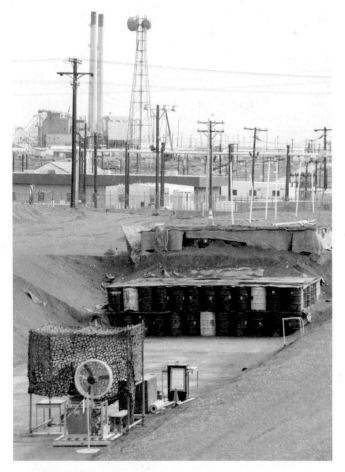

Figure 22–8 Secure landfill. These drums of low-level nuclear waste are being buried in a secure landfill at the Hanford Nuclear Reservation near Richland, Washington.

Figure 22–9 Midnight dumping. Hazardous wastes were often left on remote or unoccupied properties by unscrupulous haulers.

Figure 22–10 **"Valley of the Drums,"** an orphan waste site. Thousands of drums of waste chemicals were unloaded near Louisville, Kentucky, and left to rot, threatening the surrounding environment.

Figure 22–11 **Love Canal.** A bulldozer uncovers one of the tanks holding toxic waste buried beneath the residential area of Love Canal.

hazardous chemicals corroded and leaked, there was great danger of reactive chemicals combining and causing explosions and fires. One of the most infamous abandoned sites is the "Valley of the Drums" in Kentucky (Figure 22–10).

Love Canal, New York. The mounting problem of unregulated land disposal of hazardous wastes was brought vividly to public attention by the events at Love Canal, near Niagara Falls, New York. The area was occupied by a school and a number of houses, all of which were perched on top of a chemical waste dump that had been covered over. The surface of the dump began to collapse, exposing tanks and barrels of chemical wastes (Figure 22–11). Fumes and chemicals began seeping into cellars. Ominously, people began reporting serious health problems, including birth defects and miscarriages. An aroused neighborhood, led by activist Lois Gibbs, began demanding that the state do something about the problem. Following confirmation of the area's contamination by toxic chemicals, President Jimmy Carter signed an emergency declaration in 1978 to relocate hundreds of residents. The state and federal governments closed the school and demolished a number of the homes nearest the site. In all, more than 800 families moved out of the area because of the fear of damage to their health from the toxic chemicals.

Who Owns the Mess? The absence of public policy for dealing with the disposal of hazardous chemicals contributed to the situation at Love Canal. Hooker Chemical and Plastics Company had purchased an abandoned canal near Niagara Falls in 1942 and filled it with an estimated 21,000 tons of hazardous wastes. Hooker covered the canal over with a clay cap and sold it to the school board for a small sum, reportedly after warning the board that there were chemicals buried on the property. Subsequent construction penetrated the clay cap, and rain seeped in and leached chemicals in all directions. Hooker Chemical's parent company, Occidental Petroleum,

eventually spent more than $233 million on the cleanup and subsequent lawsuits.

As both government and independent researchers began surveying the extent of the problem, bad disposal practices were found to be rampant. The World Resources Institute estimated that, in the United States in the early 1980s, there were 75,000 active industrial landfill sites, 180,000 surface impoundments, and 200 other special facilities that were or could be sources of groundwater contamination. In most cases, the contaminated area was relatively small—200 acres (80 hectares) or less—but in total, the problem was immense and affected every state in the country. As studies and tests proceeded, thousands of individual wells and some major municipal wells were closed because they were contaminated with toxic chemicals.

Many cases of private-well contamination caused by careless disposal were discovered only after people experienced "unexplainable" illnesses over prolonged periods. While these incidents did not receive the media attention of Love Canal, they were nevertheless devastating to the people involved. Even more important than the damage already done was the recognition that the problem was ongoing.

The problems concerning toxic chemical wastes can be divided into three areas:

- Cleaning up the "messes" already created, especially where they threaten drinking water supplies
- Regulating the handling and disposal of wastes currently being produced, so as to protect public and environmental health
- Reducing the quantity of hazardous waste produced

These problems are addressed in the following sections.

CONCEPT CHECK   What are the advantages and disadvantages of disposing of hazardous wastes in deep wells? How does that method of disposal compare with landfills? ✓

22.3 Cleaning Up the Mess

A major public health threat from toxic wastes disposed of on land is the contamination of groundwater used for drinking. In dealing with groundwater contamination, the first priority is to ensure that people have safe water. The second is to clean up or isolate the source of pollution so that further contamination does not occur.

Ensuring Safe Drinking Water

To protect the public from the risk of toxic chemicals contaminating drinking water supplies, Congress passed the **Safe Drinking Water Act (SDWA) of 1974.** Under this act, the EPA sets national standards to protect the public health, including allowable levels of specific contaminants. States and public water agencies are required to monitor drinking water to be sure that it meets the standards (see Chapter 20). If any contaminants are found to exceed **maximum contaminant levels (MCLs)**, the water supply must be closed until adequate purification procedures or other alternatives are adopted. The act was amended in 1986, and the EPA now has jurisdiction over groundwater and sets MCLs for 94 contaminants. The contaminants represent a host of chemicals that are known pollutants and are regarded as dangerous to human health (see the EPA Web site for a description of the MCLs and why they are being regulated).

Groundwater Remediation. If dumps, leaking storage tanks, or spills of toxic materials have contaminated groundwater and if such groundwater is threatening water supplies, all is not yet lost. **Groundwater remediation** is a developing and growing technology. Techniques involve drilling wells, pumping out the contaminated groundwater, purifying it, and injecting the purified water back into the ground or discharging it into surface waters. Cleaning up the source of the contamination is mandatory. If, however, the contamination is extensive, remediation may not be possible, and the groundwater must be considered unfit for use as drinking water.

Superfund for Toxic Sites

Probably the most monumental task involved in the management of hazardous wastes is the cleanup of the tens of thousands of toxic sites resulting from the years of mismanaged disposal of toxic materials. Prompted by the public outrage over the Love Canal fiasco, Congress enacted the **Comprehensive Environmental Response, Compensation, and Liability Act of 1980 (CERCLA)**, popularly known as **Superfund.** Through a tax on chemical raw materials, this legislation provided a trust fund for the identification of abandoned chemical waste sites, protection of groundwater near the sites, remediation of groundwater if it has been contaminated, and cleanup of the sites. The Superfund program was greatly expanded by the **Superfund Amendments and Reauthorization Act of 1986 (SARA)**. Superfund, one of the EPA's largest ongoing programs, works as described in the next several subsections.

Setting Priorities. Resources are insufficient to clean up all sites at once. Therefore, a system for setting priorities has been developed:

- Wherever facilities are still operating, pressures are brought to bear on the operators to clean up the sites. In the past, many operators who could not afford to do so simply declared bankruptcy, and the sites joined those already abandoned. As sites are identified—note that many abandoned sites have long since been forgotten—their current and potential threats to groundwater supplies are assessed by taking samples of the waste, determining its characteristics, and testing groundwater around the site for contamination. If it is determined that no immediate threat exists, nothing more may be done.

- If a threat to human health does exist, the most expedient measures are taken immediately to protect the public. These measures may include digging a deep trench around the site, installing a concrete dike, recapping it with impervious layers of plastic and clay to prevent infiltration, and securing the area with a robust fence. Thus, the wastes are isolated, at least for the short term. If the situation has gone past the threat stage and contaminated groundwater is or will be reaching wells, remediation procedures are begun immediately.

- The worst sites (those presenting the most immediate and severe threats) are put on the **National Priorities List (NPL)** and scheduled for total cleanup. A site on this list is reevaluated to determine the most cost-effective method of cleaning it up, and finally, cleanup begins with efforts to identify "potential responsible parties" (PRPs)—namely, industries in which the wastes originated. The industries are "invited" to contribute to the cleanup, and they do so either by contributing financially or by participating in cleanup activities.

Cleanup Technology. If chemical wastes are contained in drums, the drummed wastes can be picked up, treated, and managed in proper fashion. For the "Valley of the Drums" site, the EPA cleanup started with the removal of some 4,200 drums for off-site treatment and disposal. The bigger problem for many sites is the soil (often millions of tons) contaminated by leakage. One procedure is to excavate contaminated soil and run it through an incinerator or kiln assembled on the site to burn off chemicals; synthetic organic chemicals can be broken down by incineration, and heavy metals are converted to insoluble, stable oxides. Another method is to drill a ring of injection wells around the site and a suction well in the center. Water containing a harmless detergent is injected into the injection wells and drawn into the suction well, cleansing the soil along the way. The withdrawn water is treated to remove the pollutants and is then reused for injection.

The "Valley of the Drums" site was situated in a poorly drained shale and limestone deposit. No local use was made of the groundwater. The EPA determined that the most feasible (BDAT) treatment was simply to contain the contaminated soil and groundwater, so the agency installed a clay cap, a perimeter drainage system, monitoring wells, and a security fence. The remedial work was completed in 1989, at a total cost of $2.5 million, and the site was deleted from the NPL in 1996. The third five-year review indicated that the cap was functioning well, the site was still undeveloped, and no human uses were being made of the groundwater.

Bioremediation. Another rapidly developing cleanup technology is *bioremediation.* In many cases, the soil is contaminated with toxic organic compounds that are biodegradable. The problem is that they do not degrade because the soil lacks the necessary organisms, oxygen, or both to aid in biodegradation. In **bioremediation,** oxygen and organisms are injected into contaminated zones. The organisms feed on and eliminate the pollutants (as in secondary sewage treatment) and then die when the pollutants are gone. Considerable research is being done to find and develop microbes that will break down certain kinds of wastes more readily. Bioremediation, which may be used in place of detergents to decontaminate soil, is a rapidly expanding and developing technology that is applicable to leaking storage tanks and spills as well as to waste-disposal sites.

Plant Food? When the soil contaminants are heavy metals and nonbiodegradable organic compounds, phytoremediation has been employed with some success. **Phytoremediation** uses plants to accomplish a number of desirable cleanup steps: stabilizing the soil, preventing further movement of contaminants by erosion, and extracting the contaminants by direct uptake from the soil. After full growth, the plants are removed and treated as toxic-waste products. However, this is a slow process and can be used only at sites where the contaminants and their concentrations are not toxic to plants. Nevertheless, scientists have found sunflowers that will capture uranium, poplar trees that soak up dry-cleaning solvents and mercury, and ferns that thrive on arsenic. Phytoremediation is now used at a level above $100 million a year.

Evaluating Superfund. Of the literally hundreds of thousands of various waste sites, more than 47,000 have been deemed serious enough to be given Superfund status. Over time, the EPA judged that about 33,000 of these sites did not pose a significant public-health or environmental threat. Such sites were accordingly assigned to the category "no further removal action planned." Still, more than 11,300 sites remain on the master list—a figure that, it is assumed, now includes most (if not all) of the sites in the United States that pose a significant risk. Interestingly, some of the worst that have come to light are on military bases, the result of what many characterize as the totally heedless and unconscionable discarding of toxic materials from military operations (13% of the NPL sites are "federal facilities").

Progress. As of 2014, 1,323 sites were still on the NPL (Figure 22–12). Additions are made to the list as sites from the master list are assessed, and subtractions are made when remediation on a site is complete. Since 1980, when CERCLA went into effect, about 1,166 sites have received all the necessary cleanup-related construction, and 387 of these have been deleted from the list (Love Canal was removed in 2004). Cleanup of these sites takes an average of 12 years, at an average cost of $20 million. Total cleanup often takes that long because groundwater remediation is a slow process. Research has shown that cleanup lowers the rate of birth defects in populations living near the site; cleanup also increases home values.

Who Pays? CERCLA is based on the principle that "the polluter pays." More than 70% of the costs to date have been paid by the polluters (the PRPs). Funds collected for a specific site go into *special accounts* until the site cleanup is finished. However, the history of a waste-disposal site may go back 50 years or more, and users may have included everything from schools, hospitals, and small businesses to large corporations, all of which mount legal defenses to

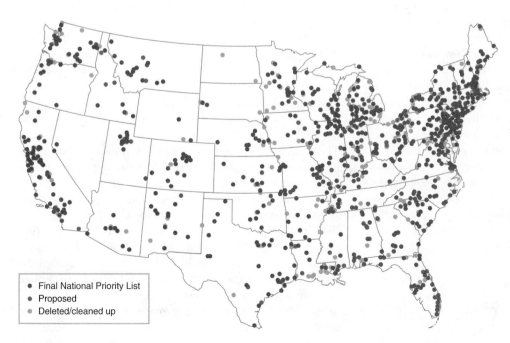

Figure 22–12 Map of Superfund sites in the contiguous United States. Brown dots show sites currently on the National Priorities List, blue dots are sites that have been proposed for the list, and green dots are sites that have been cleaned up and deleted.

(*Source:* Data from U.S. EPA CERCLIS database; map by skew-t, licensed under the Creative Commons Attribution-Share Alike 3.0 Unported http://creativecommons.org/licenses/by-sa/3.0/deed.en.)

- Final National Priority List
- Proposed
- Deleted/cleaned up

disclaim responsibility. When liability is difficult to track down or the responsible parties are unable to pay, the Superfund trust fund kicks in.

Over the years, the EPA has gained a great deal of experience in dealing with Superfund sites. The technology has become quite sophisticated, and many hazardous-waste remediation companies have become established. A great deal of progress has been made; however, the American public still rates contamination by toxic waste at the top of the list of major environmental issues.[5]

Critics. The Superfund program gets its share of criticism. Industries claim that they are unfairly blamed for pollution that reaches back to activities that were legal before the enactment of CERCLA. Many feel that overly stringent standards of cleanup are costing large sums of money without providing any additional benefit to public health. (This problem of finding a suitable balance between costs and benefits is discussed in more depth in Chapter 2.) Congress refused to renew the tax on industry at the end of 1995; the trust fund had a balance of $4 billion at the time. Since that time, the trust fund has been supported by monies in the special accounts and yearly appropriations by Congress. Congress also appropriated $600 million for EPA's Superfund program as a component of the **American Recovery and Reinvestment Act of 2009**, aimed at creating jobs and speeding up the cleanup of NPL sites, but funding cleanup continues to be a problem.

Brownfields. One highly successful recent Superfund development is the brownfields program. **Brownfields** are abandoned, idled, or underused industrial and commercial facilities where expansion or redevelopment is complicated by contamination by hazardous wastes. It has been estimated that environmental hazards not serious enough to be put on the Superfund NPL impair the value of $2 trillion worth of real estate in the United States. The Brownfield Act, which passed in 2002, provides grants for site assessment and remediation work and authorized $250 million/year for the program. Previously authorized under CERCLA, the brownfields program now stands on its own. The legislation limits liability for owners and prospective purchasers of contaminated land, thus clearing the way for more cleanup of the estimated 450,000 sites that would qualify as brownfields. Since its inception, the program has leveraged more than $21 billion in funding for cleanup and redevelopment and more than 100,000 jobs.

Many brownfield sites lie in economically disadvantaged communities, and their rehabilitation contributes jobs and exchanges a functional facility for an unsightly blight on the neighborhood. For example, the city of Chicago recently acquired and cleaned up a former bus barn in west Chicago that was becoming an illegal indoor garbage dump. The Illinois EPA cleared the site for reuse, and it was acquired by Scott Peterson Meats, which erected a $5.2 million smokehouse that employs 100 workers. Frequently, the rehabilitation of brownfield sites provides industries and municipalities with prime land for centrally located facilities that would otherwise have been carved out of suburban or *greenfield* lands (land occupied by natural ecosystems). A further advantage is that the new developments go back on the tax rolls, turning a liability into a community asset that lightens the load of residential taxpayers.

Leaking Underground Storage Tanks (LUSTs). One consequence of our automobile-based society is the millions of underground fuel-storage tanks at service stations and other facilities. Putting such tanks underground greatly diminishes the risk of explosions and fires, but it also hides leaks. Underground storage tanks were traditionally made of bare steel, so they had a life expectancy of about 20 years before they began leaking and contaminating groundwater. By 2011, more than 501,000 failures had been reported, with 6,000 new "releases" in 2011. Indeed, LUSTs are the most common source of groundwater contamination. Without monitoring, small leaks can go undetected until nearby residents begin to smell fuel-tainted water flowing from their faucets.

Regulations regarding underground storage tanks, part of the Resource Conservation and Recovery Act (RCRA), require strict monitoring of fuel supplies, tanks, and piping so that leaks are detected early. When leaks are detected, remediation must begin within 72 hours (Figure 22–13). Rules now also require all such tanks to be upgraded with interior lining and cathodic protection (to retard electrolytic corrosion of the steel), and new tanks must be provided with the same protection if they are steel. Many service stations are turning to fiberglass tanks, which do not corrode. A LUST trust fund, financed by a 0.1-cent-per-gallon tax on motor fuel, pays for federal activities involved with oversight and cleanup. The states are required to have storage-tank programs, and they implement the federal regulations—sometimes adding more stringent requirements. By 2014, nearly 450,000 cleanups were completed, with a backlog of 74,000.

CONCEPT CHECK Why is the Superfund program important? Why is it difficult to fund? ☑

22.4 Managing Current Toxic Chemicals and Wastes

The production of chemical wastes will no doubt continue as long as modern societies persist. It follows that human health and natural ecosystems can be protected only if we have management procedures to handle and dispose of wastes safely. We have already noted that the Clean Water Act and the Clean Air Act limit discharges into water and air, respectively. When problems regarding the disposal of wastes on land became evident in the mid-1970s, Congress passed the Resource Conservation and Recovery Act in order to control land disposal. Any company producing hazardous wastes in the United States today is regulated by these three major environmental acts.

[5]Lydia Saad, Water Issues Worry Americans Most, Global Warming Least, March 28, 2011, http://www.gallup.com/poll/146810/Water-Issues-Worry-Americans-Global-Warming-Least.aspx.

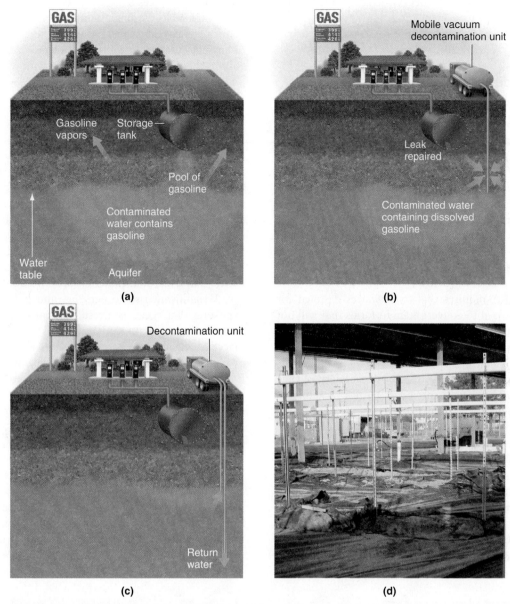

(a)

(b)

(c)

(d)

Figure 22–13 Leaking underground storage tank remediation. (a) Typical subsurface contamination from a leaking fuel tank at a gas station. **(b)** After the leak has been repaired, the vacuum-extraction process causes gasoline and residual hydrocarbons in the soil and on the water table to evaporate and then removes the vapors, preventing further contamination of the groundwater. **(c)** Contaminated groundwater is pumped out, treated, and returned to the ground. **(d)** Soil vapor extraction system removing volatile organic compounds in soil at an abandoned gasoline station.

The Clean Air and Clean Water Acts

The Clean Air Act of 1970, the Clean Water Act of 1972, and their various amendments make up the basic legislation limiting discharges into the air or water (more is said about the Clean Air Act in Chapter 19, and the Clean Water Act in Chapter 20). Under the Clean Water Act, any firm (including facilities such as sewage-treatment plants) discharging more than a certain volume into natural waterways must have a **discharge permit** (specifically, a National Pollution Discharge Elimination System permit). Discharge permits are a means of monitoring who is discharging what. Establishments are required to report all discharges of substances listed in the

TRI. The renewal of the permits is then made contingent on reducing pollutants to meet certain standards within certain periods. Standards are being made continually stricter as technologies for pollution control improve. Some manufacturing firms discharging wastewater into municipal sewer systems are required to pretreat such water to remove any pollutant that cannot be removed by sewage-treatment plants—that is, nonbiodegradable organics and heavy metals.

Nevertheless, even this restriction does not end all water pollution. Certain amounts of wastes are still legally discharged under permits, and although they may account for a low percentage of total emissions, they still add up to large numbers, as we have seen. Moreover, small firms, homes, and farms are exempt from regulation. Further still, a great deal of pollution comes from nonpoint sources such as urban and farm runoff (Chapter 20).

The Resource Conservation and Recovery Act (RCRA)

The 1976 Resource Conservation and Recovery Act (RCRA) and its subsequent amendments are the cornerstone legislation designed to prevent unsafe or illegal disposal of all solid wastes on land (Chapter 21). RCRA has three main features.

First, it requires that all disposal facilities, such as landfills, be sanctioned by *permit*. The permitting process requires that these facilities have all the safety features mentioned in Section 22.2, including monitoring wells. This requirement caused most old facilities to shut down—many subsequently became Superfund sites—and new high-quality landfills with safety measures to be constructed.

Second, RCRA requires that toxic wastes destined for landfills be pretreated to convert them to forms that will not leach. Such treatment now commonly includes biodegradation or incineration in various kinds of facilities, including cement kilns (Figure 22–14). Biodegradation involves the use of systems similar to those used for secondary sewage treatment and perhaps new species of bacteria capable of breaking down synthetic organics (Chapter 20). If treatment is thorough, there may be little or nothing to put in a landfill.

For whatever is still going to disposal facilities, the third major feature of RCRA is the requirement of "cradle-to-grave" tracking of all hazardous wastes. The *generator* (the company that originated the wastes) must fill out a form detailing the exact kinds and amounts of waste generated. Persons transporting the waste—who are also required to be permitted—and those operating the disposal facility must each sign the form, vouching that the amounts of waste transferred

are accurate. Copies of their signed forms go to the EPA. All phases are subject to unannounced EPA inspections. The generator remains responsible for any waste "lost" along the way or any inaccuracies in reporting. This provision of RCRA thus ensures that generators will deal only with responsible parties and curtails midnight dumping.

Reduction of Accidents and Accidental Exposures

In August 2015, a series of massive explosions rocked the port city of Tianjin, China. The explosions were set off by a fire that spread though warehouses containing a variety of hazardous chemicals, including 700 metric tons of toxic sodium cyanide. The disaster killed more than 100 people and injured more than 700. The explosions spread pollution across a wide area, potentially harming thousands more. This tragic event illustrates the importance of regulations to ensure that hazardous materials are stored safely.

A significant risk to personal and public health lies in exposures that occur as a result of leaks, accidents, and the misguided use of hazardous chemicals in the home or workplace. A considerable number of laws bear on reducing the probability of accidents and on minimizing the exposure of both workers and the public if accidents should occur.

Department of Transportation Regulations. Transport is an area that is particularly prone to accidents. As modern society uses increasing amounts and kinds of hazardous materials, the stage is set for accidents to become wide-scale disasters. To reduce this risk, **Department of Transportation regulations (DOT regs)** specify the kinds of containers and methods of packing to be used in the transport of various hazardous materials. Such regulations are intended to reduce the risk of spills, fires, and poisonous fumes that could be generated from mixing certain chemicals in case an accident occurs.

Figure 22–14 **Cement kiln to destroy hazardous wastes.** A cement kiln is a huge rotating "pipe," typically 15 feet in diameter and 230 feet long, mounted on an incline. **(a)** Solid wastes fed in with raw materials are fully incinerated and made to react with cement compounds as they gradually tumble toward the combustion chamber. **(b)** Flammable liquid wastes added with the fuel and air impart fuel value as they are burned. **(c)** Waste dust is trapped and recycled into the kiln.

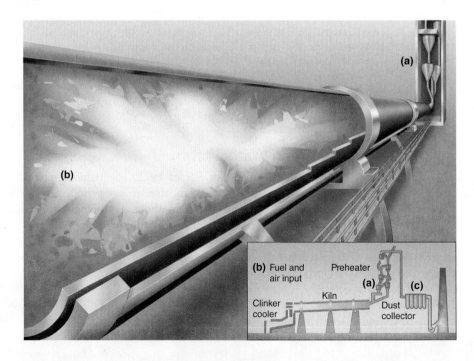

In addition, DOT regs require that every individual container and the outside of a truck or railcar carry a standard placard identifying the hazards (flammability, corrosiveness, the potential for poisonous fumes, and so on) presented by the material inside (Figure 22–2). Such placards enable police and firefighters to identify the potential hazard and respond appropriately in case of an accident. You may also see highway HAZMAT signs restricting truckers with hazardous materials to using particular routes or lanes or to using a highway only during certain hours.

Worker Protection: OSHA and the Worker's Right to Know. In the past, it was not uncommon for industries to require workers to perform jobs that entailed exposure to hazardous materials without informing the workers of the hazards involved. This situation is now addressed by amendments to the **Occupational Safety and Health Act of 1970 (OSHA)**. These amendments make up the **hazard communication standard,** or *worker's right to know*. Basically, the law requires businesses, industries, and laboratories to make available both information regarding hazardous materials and suitable protective equipment. One form in which this information is presented is the **material safety data sheet (MSDS)**, which must accompany more than 600 kinds of chemicals when they are shipped, stored, and handled. These sheets contain information about the reactivity and toxicity of the chemicals, with precautions to follow when one is using the chemical. Notably, however, the responsibility to read the information and exercise proper precautions remains with the worker.

Community Protection and Emergency Preparedness: SARA, Title III. In 1984, an accident at a Union Carbide pesticide plant in Bhopal, India, caused the release of more than 40 tons of methyl isocyanate, an extremely toxic gas (Figure 22–15). An estimated 600,000 people in communities surrounding the plant were exposed to the deadly fumes. Within days, more than 5,000 people died, and more than 10,000 subsequently died from gas-related diseases. At least 50,000 people suffered various degrees of visual impairment, respiratory problems, and other injuries from the exposure, and many more have experienced chronic illnesses related to the disaster. To date, the tragic events in Bhopal have been the world's worst industrial disaster.

An Indian investigation of the accident found that the disaster likely happened because Union Carbide officials had scaled back safety and alarm systems at the plant in order to cut costs. Ironically, most of the deaths and injuries could have been avoided if people had known that methyl isocyanate is highly soluble in water, that a wet towel over the head would have greatly reduced one's exposure, and that showers would have alleviated aftereffects. Unfortunately, neither the people affected nor the medical authorities involved had any idea of what chemical they confronted, much less the way to protect or treat themselves.

After the Bhopal disaster, Congress passed legislation to address the problem of accidents. For expediency, the measure was added as Title III to a bill reauthorizing Superfund: the Superfund Amendments and Reauthorization Act of 1986.

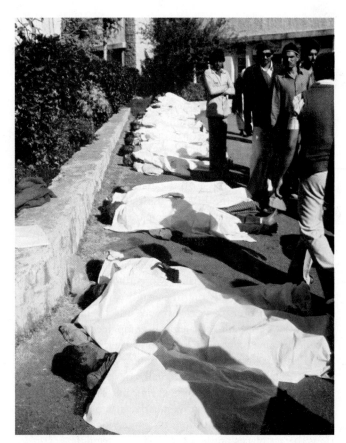

Figure 22–15 Toxic chemical disaster, Bhopal, India. Officials view some of the victims of a deadly gas release at a Union Carbide plant in 1984. At least 10,500 died, and many more suffered injuries.

Title III is better known as the **Emergency Planning and Community Right-to-Know Act (EPCRA)**, which we have already encountered in connection with the TRI.

EPCRA requires companies that handle in excess of five tons of any hazardous material to provide a "complete accounting" of storage sites, feed hoppers, and so on. The information goes to a *local emergency planning committee,* which is also required in every government jurisdiction. The committee is made up of officials representing local fire and police departments, hospitals, and any other groups that might be involved in case of an emergency, as well as the executive officers of the companies in question. The task of the local committees is to develop contingency plans for possible accidents and disasters.

Evaluating New Chemicals

In the past, new synthetic organic compounds were introduced for a specific purpose without any testing of their potential side effects. For example, in the 1960s, a new compound known as TRIS was found to be an effective flame retardant and was widely used in children's sleepwear. It was only later discovered that TRIS was a potent carcinogen. Treated sleepwear was immediately withdrawn from the market. It is not known how many children (if any) developed cancer from TRIS, but this and other such cases revealed a hole in the systems of protection.

Toxic Substances Control Act. Congress responded by passing the **Toxic Substances Control Act of 1976 (TSCA)**, which requires that, before manufacturing a new chemical in bulk, manufacturers submit a "pre-manufacturing notice" to the EPA, in which the potential environmental and human health impacts of the substance are assessed (including those that may derive from the ultimate disposal of the chemical). Depending on the results of the assessment, a manufacturer may be required to test the effects of the product on living things. Following the results of testing, the EPA may restrict a product's uses or keep it off the market altogether. The law, however, also tells the EPA to use the "least burdensome" approach to regulation and to compare the costs and benefits of regulation. TSCA also authorizes the EPA to develop an inventory of the chemicals already in use at the time the legislation was passed and to require testing whenever there are indications of potential risks. More than 80,000 chemical substances are currently on the inventory.

There is serious concern over the effectiveness of TSCA. A huge loophole allows chemical companies to keep their chemical recipes confidential, claiming they are trade secrets. The EPA has required testing for only 200 of the chemicals in commerce and has actually developed regulations for only five of the existing chemicals. The Government Accountability Office (the investigative arm of Congress) has labeled TSCA as a failed program, largely due to the restrictions the act places on the EPA's ability to regulate.

Figure 22–16 summarizes the U.S. laws that apply to hazardous wastes. It is noteworthy that consumer advocate groups have been a major force behind the passage of these laws, as well as the regulations requiring ingredients to be labeled on all products. Citizen action does make a difference!

International Hazardous Waste Regulations

There are a number of international treaties about hazardous waste, all of which are relatively recent. For the most part, these treaties emphasize the importance of informed consent and the precautionary principle, and are designed to protect those who will have the greatest exposure to chemical pollutants. Designing laws to protect those who are the most vulnerable to environmental degradation is part of environmental justice, which we will discuss in the following section.

The Aarhus Convention. The Aarhus Convention on Access to Information, Public Participation in Decision-Making and Access to Justice in Environmental Matters, which went into effect in 2001, is probably the broadest of these agreements (Chapter 2). It establishes the right of the public to information, to a voice in proceedings, and to just treatment when environmental regulations are not kept. Other agreements include the Basel, Rotterdam, and Stockholm Conventions and the REACH treaty.

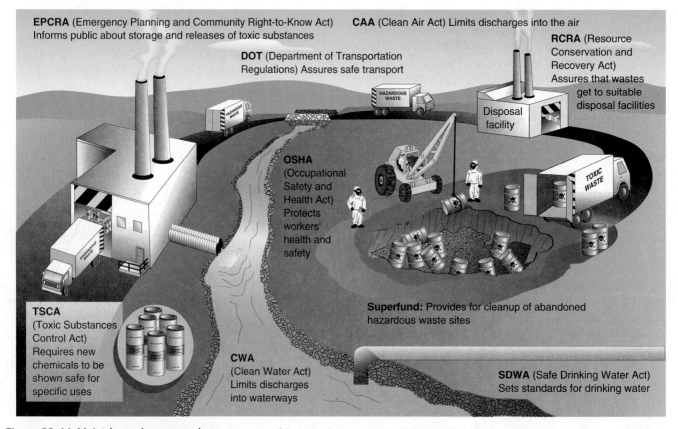

EPCRA (Emergency Planning and Community Right-to-Know Act) Informs public about storage and releases of toxic substances

CAA (Clean Air Act) Limits discharges into the air

DOT (Department of Transportation Regulations) Assures safe transport

RCRA (Resource Conservation and Recovery Act) Assures that wastes get to suitable disposal facilities

Disposal facility

OSHA (Occupational Safety and Health Act) Protects workers' health and safety

TSCA (Toxic Substances Control Act) Requires new chemicals to be shown safe for specific uses

CWA (Clean Water Act) Limits discharges into waterways

Superfund: Provides for cleanup of abandoned hazardous waste sites

SDWA (Safe Drinking Water Act) Sets standards for drinking water

Figure 22–16 Major hazardous-waste laws. Summary of the major laws pertaining to the protection of workers, the public, and the environment from hazardous materials.

Three Important Conventions. The Basel Convention on the Control of Transboundary Movements of Hazardous Wastes and Their Disposal (1989), known as the **Basel Convention,** preceded the Aarhus convention, entering into force in 1992. It regulates the movement of hazardous waste between nations to prevent the transfer of hazardous waste from developed to less developed countries. Its regulations are designed to prevent the types of problems caused by the dumping of e-waste (Chapter 21).

The Stockholm Convention on Persistent Organic Pollutants, or **Stockholm Convention,** came into force in 2004. Countries that signed the treaty agreed to outlaw nine particularly harmful pollutants; to limit the use of DDT to the control of malaria; and to limit the accidental production of pollutants, such as dioxin, by the breakdown of other products.

The **Rotterdam Convention** is an international treaty that went into effect in 2004. It promotes the open exchange of information about hazardous chemicals including pesticides (Chapter 13). One of its important features is a process of *prior informed consent,* in which exporting countries must inform all potential importing countries of actions they have taken to ban or restrict the use of pesticides or other toxic chemicals. Exporting countries must also label toxics clearly (Chapter 13).

European Regulations. In 2007, members of the European Union agreed to the comprehensive **Registration, Evaluation and Authorization of Chemicals Treaty (REACH).** This agreement has resulted in a process of registering and classifying all potentially hazardous chemical substances in use in European countries. Manufacturers must provide a safety report for each chemical that includes human health and environmental hazard assessments, as well as information on how the chemical may be used once it is sold. Many chemicals not currently tested in the United States are included in this regulation, so U.S. companies that export products to Europe must follow these stricter rules.

CONCEPT CHECK What are two key differences between OSHA and the RCRA? ✓

22.5 Broader Issues

Before we leave the topic of hazardous wastes, we still need to address two important broader issues: the problem of environmental justice and the movement toward pollution prevention.

Environmental Justice and Hazardous Wastes

- The largest commercial hazardous-waste landfill in the United States is located in Emelle, Alabama. African Americans make up 90% of Emelle's population. This landfill receives wastes from Superfund sites and every state in the continental United States.

- Residents of Arizona's Navajo Nation reservation are exposed to uranium and radium through airborne dust and contaminated water, a consequence of 520 open uranium mines abandoned by private companies that operated the mines under government contracts.

- One study found that 870,000 U.S. federally subsidized housing units are within a mile of factories that have reported toxic emissions to the EPA.

- Many poor people live in rural communities adjacent to the Louisiana-Mississippi chemical corridor, an 85-mile stretch along the Mississippi River between Baton Rouge and New Orleans with almost 150 petrochemical and other factories.

The issue is *environmental justice,* introduced elsewhere (Chapter 1). The EPA defines **environmental justice** as:

> The fair treatment and meaningful involvement of all people regardless of race, color, national origin, or income with respect to the development, implementation, and enforcement of environmental laws, regulations, and policies. Fair treatment means that no group of people, including racial, ethnic, or socioeconomic group should bear a disproportionate share of the negative environmental consequences resulting from industrial, municipal, and commercial operations or the execution of federal, state, local, and tribal programs and policies.[6]

The U.S. environmental justice movement was an outgrowth of the civil rights movement of the 1960s. One of the prominent events occurred in rural Warren County, North Carolina, in 1982. County residents, poor and largely African American, protested the dumping of toxic PCBs in a government landfill. For six weeks they rallied and marched (Figure 22–17). Although other protests over environmental discrimination preceded it, this was the first environmental protest by people of color to garner nationwide attention,

---

[6]Environmental Protection Agency, *Final Guidance for Incorporating Environmental Justice Concerns in EPA's NEPA Compliance Analyses* (April 1998), available at www.epa.gov.

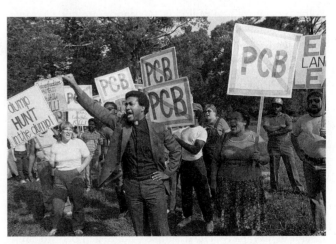

Figure 22–17 Environmental justice protest, 1982, in Warren County, North Carolina. Protesters marched against siting a dump for PCB wastes in a rural, poor, and primarily African American community in one of the early prominent events in the environmental justice movement.

and it triggered further activism. In 1991, a group of prominent civil rights leaders, scientists, and others met at the first National People of Color Environmental Leadership Summit in Washington, DC, an event that prompted government action.

Federal Response. The federal government has taken this problem seriously. In 1994, President Clinton issued Executive Order 12898, focusing federal agency attention on environmental justice. The EPA's response was to establish an Environmental Justice Program, which promotes a number of activities. For example, in connection with the EPA's Brownfields program, communities are putting abandoned properties back into productive use. Many efforts are directed toward addressing justice concerns *before* they become problems. The Obama administration recently coordinated commitments by 16 federal agencies to address environmental justice issues and implement new strategies to combat injustices to minority communities.[7]

International Environmental Justice. One form of trade the developing countries can do without is the international trade in toxic wastes. Some countries or local jurisdictions in search of ready cash have found a source in the form of toxic-waste shipments, which invariably come from more developed countries and are often illegal. This issue appeared earlier in connection with the overseas shipments of electronic wastes (Chapter 21). Close to home, some 20% of spent vehicle batteries (20 million in 2011) are exported from the United States to Mexico, where environmental and safety regulations are more relaxed. As a result, Mexican workers are exposed to levels of lead that far exceed the levels allowed in the United States. The Aarhus, Basel, Stockholm, and Rotterdam conventions, already discussed, are part of the international community's effort to make hazardous waste exposure more equitable.

Pollution Prevention for a Sustainable Society

Pollution control and pollution prevention are not the same. *Pollution control* involves adding a filter or some other device at the "end of the pipe" to prevent pollutants from entering the environment. The captured pollutants still have to be disposed of, and this entails more regulation and control. By contrast, *pollution prevention* involves changing the production process or the materials used so that harmful pollutants won't be produced in the first place. For example, adding a catalytic converter to the exhaust pipe of your car is pollution control, whereas redesigning the engine or switching to a hybrid car so that less pollution is produced is pollution prevention.

Pollution prevention often results in better product or materials management—that is, less waste. Thus, pollution prevention frequently creates a cost savings. For example, Exxon Chemical Company added simple "floating roofs" to tanks

storing its most volatile chemicals, thereby reducing evaporative emissions by 90% and gaining a savings of $200,000 per year. That paid for the cost of the roofs in six months. As another example, United Musical Instruments in Nogales, Arizona, worked with the Pollution Prevention Unit of the Arizona Department of Environmental Quality to install a closed-loop system that reduced water use by 500,000 gallons per year, saving the company $127,000 in its first year alone. The system also reduced hazardous waste by 58% in three years. Both are examples of the *minimization* of pollution.

Green Chemistry. A second approach to pollution avoidance is *substitution*—finding nonhazardous substitutes for hazardous materials. This is known as *green chemistry*. For example, the dry-cleaning industry, which uses large quantities of toxic organic chemicals, is exploring the use of water-based cleaning, or "wet cleaning." Preliminary results from some entrepreneurs indicate that wet cleaning is quite comparable in cost and performance to dry cleaning. There are now more than 150 wet cleaners in North America, and the number is growing.

The American Chemical Society (ACS) and the EPA have been cooperating to promote green chemistry, particularly for undergraduate and graduate chemistry courses. Case studies and laboratory modules have been developed by this cooperation, available on the ACS Web site.

Reuse. A third approach is *reuse*—cleaning up and recycling solvents and lubricants. Some military bases, for example, have been able to distill solvents and reuse them, instead of discarding these solvents in a manner that releases them into the environment. This approach has increased greatly during the past decade, as measured by the TRI data.

All three approaches to preventing pollution—minimization, substitution, and reuse—have been employed increasingly by industries handling hazardous chemicals, and all have contributed to significant reductions in hazardous-waste releases into the environment. (See Figure 22–4.) The public disclosure of TRI data is judged to have played a crucial role in these reductions. The EPA has concluded that the system has been a highly effective environmental program, one of the best ever administered by the EPA.

You, the Consumer. Finally, pollution avoidance can also be applied to the individual consumer. So far as you are able to reduce or avoid the use of products containing harmful chemicals, you are preventing those amounts of chemicals from being released into the environment. You are also reducing the by-products that result from producing those chemicals. The average American home contains as much as 100 pounds of hazardous household waste (HHW)—substances such as paints, stains, varnishes, batteries, pesticides, motor oil, oven cleaners, and more. These materials need to be stored safely, used responsibly, and disposed of properly if they are not to pose a risk of injury or death and a threat to environmental health. (Many communities hold regular HHW collection days; if yours doesn't, you could initiate one!)

[7]Environmental News Service, "Obama Administration Strengthens Environmental Justice Efforts," February 27, 2012, www.ens-newswire.com/ens/feb2012/2012-02-27-091.html.

A growing number of companies are beginning to produce and market **green products,** a term used for products that are more environmentally benign than are their traditional counterparts. For example, Method Products and Clorox-based Green Works are now marketing nontoxic cleaners in Target, Costco, and other large retailers. How fast and to what degree these green products replace traditional products will in large part depend on how we behave as consumers. In other words, your buying power can be an extremely potent force.

In conclusion, there are four ways to address the problems of pollution by hazardous chemicals: (1) pollution prevention, (2) recycling, (3) treatment (breaking down or converting the material to harmless products), and (4) safe disposal. The first three of these methods promote a minimum of waste, in harmony with the principles of sustainability. There is every reason to believe that we can have the benefits of modern chemical technology without destroying the sustainability of our environment by polluting it.

CONCEPT CHECK   How is the story about hazardous waste in Inner Mongolia at the beginning of the chapter an example of an environmental justice issue? ☑

REVISITING THE THEMES

SOUND SCIENCE

Research is under way to evaluate the impacts of bisphenol A and other potential endocrine-disrupting chemicals on humans and wildlife. It took the work of many environmental scientists to uncover the pathways by which some toxic chemicals entered ecological food webs and made their way into some of the most remote regions of Earth, including the Arctic. This work has led to the banning of many persistent organic compounds.

Toxicology is a well-established and crucial branch of science that has helped us to develop a thorough inventory of hazardous chemicals and their health effects. Federal agencies and NGOs have made certain that this information is available to the public. The green chemistry movement is an outcome of the efforts of chemists (and their sponsors) who are determined to design products that are nontoxic yet effective.

SUSTAINABILITY

In our recent past, society was on an unsustainable course in dealing with hazardous chemicals. More and more sites were being contaminated, and we were seeing rising rates of cancer and other health problems. Although the connection between hazardous-waste sites and actual cancer cases has been difficult to prove, the rising tide of hazardous chemicals in the environment was certainly a prime suspect. The legislation and programs discussed in this chapter have turned the situation around, and the move toward pollution prevention and the recycling of hazardous chemicals indicates that we are moving in a sustainable direction in our dealings with toxic chemicals.

STEWARDSHIP

One vital task of stewardship is to protect fellow human beings from unnecessary risks. We were not doing well with this responsibility, and a few wake-up calls led to stewardly action on the part of people such as Lois Gibbs and others who put their lives and reputations on the line to bring some sanity to our permissive dealings with toxic-waste dumps. As a result, our society stopped accepting contamination as the "price of progress." After a great deal of public outcry, Congress responded with legislation to address the many problems. We have made real progress in cleaning up the Superfund sites, the leaking underground storage tanks, and the brownfield sites. Industries can no longer release products without testing their potential for doing harm. The TRI, in particular, has provided the incentive for industry to clean up its act. It is not the time, however, to let down our guard. Injustice to minority groups, pressures by industries to ease up on the regulations, and the attitude that hazardous wastes no longer pose a threat indicate that this issue requires constant surveillance and repeated pressure on Congress and the EPA. New chemicals are constantly appearing, and old ones have not yet been thoroughly researched; the Superfund program is still far from completion.

REVIEW QUESTIONS

1. How do toxicologists investigate hazardous chemicals? How is this information disseminated to health practitioners and the public?

2. What four categories are used to define hazardous chemicals?

3. Define *total product life cycle*, and describe the many stages at which pollutants may enter the environment.

4. What are the two classes of chemicals that pose the most serious long-term toxic risk, and how do they affect food chains?

5. What kind of chemicals are the "dirty dozen" POPs? Why are they on a list?

6. What two laws pertaining to the disposal of hazardous wastes were passed in the early 1970s? Describe how the passage of the laws shifted pollution from one part of the environment to another.

7. Describe three methods of land disposal that were used in the 1970s. How has their use changed over time?

8. What law was passed to cope with the problem of abandoned hazardous-waste sites? What are the main features of the legislation?

9. What is being done about leaking underground storage tanks and brownfields?

10. What law was passed to ensure the safe land disposal of hazardous wastes? What are the main features of the legislation?

11. What laws exist to protect the public against exposures resulting from hazardous-chemical accidents? What are the main features of the legislation?

12. What role does the Toxic Substances Control Act play in the hazardous-waste arena?

13. Why does the EPA have an Environmental Justice Program?

14. Describe the advantages of pollution-prevention efforts.

THINKING ENVIRONMENTALLY

1. Select a chemical product or drug you suspect could be hazardous and investigate it, using the resources made available on the Internet by federal agencies and Scorecard.

2. Before the 1970s, it was not illegal to dispose of hazardous chemicals in unlined pits, and many companies did so. Should they be held responsible today for the contamination those wastes are causing, or should the government (taxpayers) pay for the cleanup? Give a rationale for your position.

3. Use the Toxics Release Inventory on the Internet to investigate the locations and amounts of toxic chemicals released in your state or region.

4. Check out the American Chemical Society Green Chemistry Web site, and find out if your college or university is on the list of schools with green chemistry programs. If not, you could ask your chemistry department to look into it.

MAKING A DIFFERENCE

1. Read labels and become informed about the potentially hazardous materials you use in the home and workplace. Whenever possible, use nontoxic substitutes. If you use toxic substances, make sure that you dispose of them properly.

2. Flammable products, such as gasoline, kerosene, propane gas, and paint thinner, should be stored in approved containers in a garage, shed, or similarly separate, well-ventilated area—never inside the house.

3. To avoid ingesting bisphenol A, eat fresh fruits and vegetables and avoid using bottles that have the recycle codes 3 or 7.

4. Use up your paint—try to buy only the amount of paint that you need, and then finish it. If possible, use only latex-based paint, which can be disposed of in household wastes. Do not put oil-based paints in the trash; save them for a household hazardous-waste collection.

STEWARDSHIP FOR A SUSTAINABLE FUTURE

How will we shape a sustainable future? In previous chapters, we discussed many issues that need to be resolved if we are to achieve a sustainable relationship with the environment. Next we explore one of the most difficult problems faced by every nation today: promoting livable and sustainable communities. With half of humanity now living in cities, urbanization is an irresistible trend, particularly in the developing world, where 95% of urban population growth is occurring. There are advantages to cities, such as the possibility of efficiency in the provision of services. There are also significant disadvantages, including the environmental costs of huge metropolitan areas and the loss of land to urban sprawl.

These are some of the challenges we face: In high-income countries, urban blight plagues many older cities, dooming many residents to second-class citizenship. In low-income countries, migration from the countryside to ever-growing cities results in huge slums and shantytowns. How can the people living under such conditions be brought into the mainstream of a country's economy and achieve sustainable well-being? How should all countries deal with the poverty of the people remaining in rural areas? What will it take to make the energy, water, transportation, and waste management systems of the cities of the future more sustainable?

In Part Six, we explore answers to these questions. We consider examples of cities moving in a sustainable direction. We investigate exciting innovations that connect communities to real-time data, helping them to use resources more efficiently. We also look at some of the work being done by organizations that are grappling with the question of sustainability. Finally, we look at the vital, but difficult, task of adopting lifestyles that embrace the stewardship values capable of shaping a sustainable future.

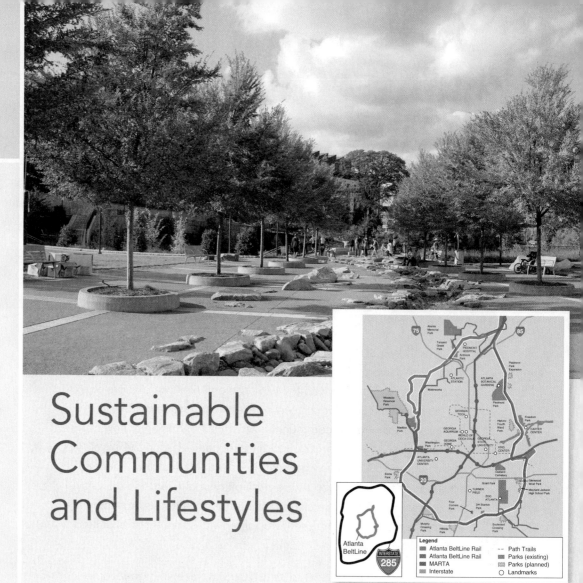

Sustainable Communities and Lifestyles

The word *beltline* conjures up images of a multi-lane highway that runs around a major metropolitan area, facilitating commuting and stimulating a ring of industries, all of which guarantees that not long after such a highway is built, it will become congested. Atlanta, Georgia, has such a belt, Interstate Route 285, a 10-lane highway built to allow through traffic to avoid the city, but now heavily used by commuters who travel between the expanding suburbs and the corporations that have located along the highway. This is the Atlanta Beltway.

The Atlanta BeltLine is something else altogether, an ambitious makeover of an abandoned rail and industrial corridor that loops around the heart of Atlanta, well inside the Beltway (see map above). One writer has called it "the country's most ambitious smart growth project."[1] The project is a remarkable visionary effort that will provide parks, trails, public transportation, brownfields remediation, revitalization of distressed city neighborhoods, and much more.

Connecting Neighborhoods. The "Belt Line" was originally a loop of tracks that provided rail transportation in post–Civil War Atlanta. When trucks and highways came on the scene, most of the rail tracks were abandoned. The new BeltLine concept was the brainchild of Georgia Tech graduate student Ryan Gravel, who, in 1999, proposed a new transit system that would link the inner city

---

[1]Kaid Benfield. "The Atlanta BeltLine: The Country's Most Ambitious Smart Growth Project." *Natural Resources Defense Council Switchboard.* July 26, 2011. http://switchboard.nrdc.org/blogs/kbenfield/the_countrys_most_ambitious_sm.html. February 29, 2012.

▲ The historic Fourth Ward Park, a portion of the Atlanta BeltLine Project in Atlanta, Georgia.

neighborhoods. Gravel's concept quickly attracted grassroots attention, and soon the BeltLine plan included parks and trails, affordable housing, and neighborhood renewal.

Benefits for the City.

The total price tag for the BeltLine is estimated at $2.8 billion, but considering the potential benefits for the city of Atlanta, that seems like a bargain. In concept, it will create 5,600 housing units, 2,000 acres of parks, 33 miles of continuous trails, and a 22-mile light-rail transit system. As a result of the BeltLine, some 40 city neighborhoods will be renewed. The project includes the cleanup and redevelopment of 1,100 acres of contaminated brownfield sites. In addition to thousands of temporary construction jobs, it will eventually provide 30,000 permanent jobs as well as a $20 billion increase in the city's tax base over the next 25 years. These are huge benefits to a city battling the effects of aging and economic struggles.

Many obstacles remain before the Atlanta BeltLine reaches completion, especially obtaining funding. There is no doubt that the BeltLine will be years in the making, but when it is finished, it will represent a monumental improvement in urban Atlanta. With its BeltLine, Atlanta joins a bevy of cities intent on transforming economically depressed neighborhoods into showplaces of urban renewal.

We will begin this chapter by discussing urban and rural communities around the world. Then we will look at trends in U.S. communities, including suburban sprawl and urban reinvigoration. Next we will consider the national and global efforts being undertaken to build sustainable cities and communities in the context of the Sustainable Development Goals. Finally, we will look at some ethical concerns and personal lifestyle issues as we consider the urgent needs of people and the environment in this new millennium.

23.1 Megatrends in Communities

A century ago, less than 5% of the world's people lived in cities; today more than half are city dwellers. And the world's cities are expected to continue growing well into the middle of the 21st century and beyond. By 2050, demographers expect an extra 2.4 billion people will live in cities, which will mean that two-thirds of the world's population will be urban. The general term for this increasing dominance of cities is **urbanization**. Urbanization is a function of natural population growth and migration to cities from rural areas (Chapters 8 and 9).

The pace of urbanization varies greatly from region to region **(Figure 23–1)**. Africa and Asia, the continents with the largest rural populations, are urbanizing the most rapidly; within those continents, Nigeria, India, and China have the highest rates of urbanization. However, the current record for the most rapidly growing city is held by Karachi, Pakistan, which grew by 80% between 2000 and 2010.

This vast urbanization is a *megatrend*, one of the major global changes expected as we move further into the 21st century. Urbanization coincides with a number of other trends we have been following throughout the book. These include positive trends, such as longer life spans, improved health care and education, decreased poverty, and a rising consumer class, as well as negative trends, such as growing populations, increased use of resources, increasing pollution, the loss of biodiversity and ecosystem services, and climate change. In this chapter, we will see how these trends interact with communities—from small villages in rural areas to vast metro regions.

What Is a City?

Before we continue, we need to define some terms that are used to describe various kinds of communities. An **urbanized area** is a densely populated region, dominated by a human-built

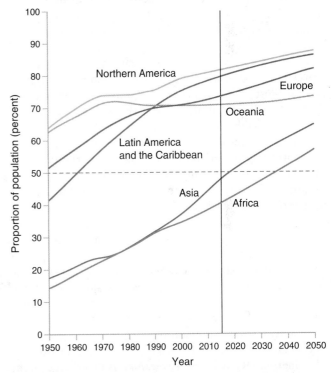

PROPORTION OF POPULATION THAT IS URBAN

Figure 23–1 **Urbanization by region.** Between 1950 and 2014, the proportion of the world population that was urban increased from about 30% to over 50%. By 2050, that number is likely to be 67%. Most of the urban growth in the next decades will occur in Africa and Asia. Oceania, which includes Australia, New Zealand, and the South Pacific Islands, will have the lowest rate of urbanization.

(*Source:* World Urbanization Prospects 2014 Revision. New York: United Nations, 8.)

UNDERSTANDING THE DATA

1. Which area of the world is currently the most urbanized? Which is the least urbanized?

2. In which year (approximately) will 50% of Africans be urban?

Figure 23–2 **Growth of cities of different sizes.** Cities of all sizes are growing. In 2030, most urban people will live in cities with fewer than 500,000 people, but the proportion of people living in larger cities will have increased more rapidly. Megacities can dominate an entire region.

(*Source:* UN World Urbanization Prospects 2014 Highlights. 2014. Geneva: UN, 13.)

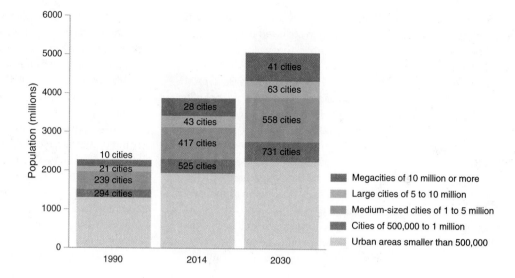

environment, with more than 400 people per square kilometer. Centers with fewer people than urbanized areas may be called *towns, townships,* or *villages,* depending upon their location. Areas with populations below a certain size (in the United States the limit is 2,500 people) are classified as **rural.**

Cities and Suburbs. A **city**, such as Philadelphia or Cleveland, is an urbanized area with a distinct government and, in most places, a mayor. Cities are often surrounded by **suburbs**, areas that are less densely populated than cities but interact with cities more than rural areas do. Suburbs are included in the designation "urban" when demographers are contrasting urban and rural areas. In the United States, suburbs are often wealthy "bedroom communities" for a city, but this is not the case in all countries. **Exurbs** are even further out than suburbs.

A large city may be a **metropolis**, a regional center for a whole area. A city with suburbs and surrounding towns or even smaller cities may be called a metropolitan area or **metropolitan region**. Such an area acts as a single economic zone. That is, people work, shop, and travel throughout the whole. In some places, even large cities have merged so that it is difficult to see where one ends and another begins. A chain of metropolitan regions such as occurs in the eastern United States between Boston and Washington, DC, is called a *megalopolis* or urban agglomeration. The merging of cities and towns around cities and the growth of suburbs are part of the reason why the identities and populations of the largest cities are not the same on all the lists you might read. Researchers have to define the areas being identified as the city: Do they mean the city proper, or the larger metro-region, or even larger combined urban aggregations?

The Growth of Megacities. The very largest cities—those with more than 10 million inhabitants—are called **megacities.** In 1975, Mexico City joined New York and Tokyo as the world's only megacities. By 2014, there were 28 such cities, accounting for 7% of the world's population and 13%

of its urban population.[2] The seven largest megacities are in Asia. The largest is the Pearl River Delta urban area in China, which includes what were once four distinct cities: Dongguan, Foshan, Guangzhou, and Shenzhen. Recently it has surpassed the size of the Tokyo-Yokohama metropolitan region.[3]

Most urban dwellers, however, are not in megacities. About half live in cities of under 500,000. Figure 23–2 shows the increase in cities of different sizes over time.

Advantages of Urbanization

Is urbanization a threat to the environment? It certainly is true that cities have a significant impact on the land and ecosystems around them, consuming large amounts of resources and releasing regionally concentrated levels of pollution. Yet there are a number of environmental and social benefits to urbanization.

Greater Efficiency. If cities provide the basic infrastructure needs—clean water, electricity, mass transit, and the sanitary disposal of solid wastes and sewage—economies of scale can make the per capita resource consumption of city residents far lower than that of people living in suburbs and rural areas. In the United States, for example, the average city-dweller uses only half as much electricity as the average suburbanite. Because city residents drive less than suburbanites, instead walking and using public transportation, they consume less energy and produce less air pollution. It is also more efficient to provide heating to apartment buildings than to single-family homes. Other forms of efficiency available to cities include district heating and waste-to-energy plants, which we have discussed elsewhere (Chapters 14 and 21).

[2]J. Kotkin et al., *The Problem with Megacities* (Orange, CA: Chapman University Press, 2014).

[3]World Bank, *East Asia's Changing Urban Landscape* (Washington, DC: World Bank, 2015).

Figure 23–3 Benefits of urban life. People like these young residents of Melbourne, Australia, enjoy the excitement, opportunities, and diverse culture of city life. Melbourne consistently ranks as a highly livable city because of its amenities, jobs, culture, and safety.

Slower Population Growth. In many parts of the world, urbanization is partly an outcome of rapid population growth. But since urbanization has a natural dampening effect on fertility, the growth of cities should help to slow that trend. A number of factors—economic opportunities for women, access to health services, and less need for children's labor—work to reduce fertility rates in urban populations (see Chapters 8 and 9).

Social and Economic Benefits. Of course, there are social and economic advantages to living in cities, including a high concentration of jobs and cultural opportunities such as museums and theaters, making cities exciting places to live (Figure 23–3). However, realizing these potential benefits requires effective urban planning, a topic we will consider in Section 23.3.

Challenges of Urbanization and Megacities

To provide essential services to their residents, cities require a vast infrastructure. What might be done by individuals in rural areas—getting water from a well, building a pathway, or digging a latrine—is done through an enormous network of water and sewer lines, large highways, and huge landfills. In wealthy and some middle-income countries, cities often coordinate municipal services efficiently and provide millions with employment. But in poorer regions that lack the resources to provide these services, the rapid development of cities can be disastrous. People are moving into these cities at a rate that far exceeds the capacity of the city to assimilate them. For example, in Dhaka, Bangladesh, a megacity of 15 million, two-thirds of the sewage is untreated.

Risks to Public Health. The crowded conditions of many cities have a serious impact on the health of their residents. When large numbers of people live close together without clean water, adequate sanitation, or electricity, diseases can spread rapidly. Mental illness is associated with distressing conditions and lack of exposure to natural landscapes. Air pollution increases with population density, often leading to an increase in respiratory ailments (see Chapter 19). Residents of Mumbai, India, for example, have a life expectancy of 57 years, seven years below the national average.[4]

Environmental Damage. As you might imagine, large cities have an enormous impact on the environment around them. The high concentration of resource use and waste production is very damaging to the surrounding environment, even though the per capita effects of city residents are less than they would be elsewhere. Habitat loss and fragmentation, overuse of groundwater, and the dumping of wastes affect terrestrial ecosystems and the services they provide. In the many large coastal cities, overfishing, heavy pollution, and eutrophication take a toll on oceans and estuaries. One of the chief concerns about large cities, however, is the vast areas of unplanned, crowded development—the slums—that spring up around many of them.

Slums and Shantytowns

Two million residents of Nairobi, the capital of Kenya, live in slums consisting of flimsy shacks made of scrap wood, metal, plastic sheeting, rocks, and mud (Figure 23–4). The slum neighborhoods of Nairobi lie on the outskirts of the city. Burning trash and charcoal cooking fires cloud the air, and the lack of sewers adds to the smell. None of the residents own their homes, because the settlements are built on public land or on private land where landowners exact high rents. Similar neighborhoods can be found around most of the developing-world cities. UN–HABITAT estimates that

---

[4]J. Kotkin, W. Cox, A. Modarres, and A. Renn, *The Problem with Megacities* (Orange, CA: Chapman University Center for Demographics and Policy [Chapman Press], 2014).

Figure 23–4 Shantytown in Nairobi, Kenya. More than half of Nairobi's residents live in such slums, where conditions defy description.

one out of every three people in the developing world cities lives in a slum. Because the buildings in slum areas are usually unauthorized, cities rarely provide them with electricity, water, sanitation, or social services such as schools and hospitals. People must obtain their water and other services from private companies. Crime is rampant, and diseases such as tuberculosis spread readily.

In spite of their living conditions, slum residents are of necessity amazing entrepreneurs, engaging in a widespread informal economy. Residents usually build their own dwellings and erect their own infrastructure. They provide their own food stands, coffee shops, barbershops, and, in some cases, schools for their children. Indeed, slums are often the home of a workforce that is essential to cities—the workers who are willing to take the low-paying jobs that keep the city and its wealthier inhabitants functioning.

The Abandoned Countryside

One outcome of urbanization is the abandonment of rural areas. Over the next 35 years, all of the world's net population growth is expected to occur in urban areas, while rural regions are expected to decline by 300 million. In the past, such changes have led to a disparity in wealth, as well as a breakdown of traditional cultures and institutions.

Fewer Jobs and Aging Workers. Industrial agriculture has increased the efficiency of food production, which means that fewer laborers are needed to perform agricultural tasks. (For example, machines may be used to harvest crops that were once hand-picked.) As a result, there are fewer jobs in rural areas. Because farm work supports a smaller proportion of the population, many people move to cities looking for work. As workers leave rural areas, there are not enough people to support the local businesses, so those jobs dry up, too. This leads to more people leaving, until even the jobs that remain in rural areas may not have people to do them.

As young workers migrate to cities, the remaining rural population becomes increasingly older. In Japan, urban migration has left few rural people to work on farms, care for the elderly, or maintain communities. The majority of rice farming in Japan, for example, is done by people in their 60s and 70s.

Rural Poverty. In much of the world, those who live in rural areas are poorer and have less access to health care and education than those in cities. This difference in wealth is referred to as *spatial inequality*. In fact, the majority of the world's poor are rural. A lack of infrastructure, access to markets, and other problems plague rural areas. Women in rural areas, who do a disproportionate amount of agricultural labor, are particularly disadvantaged.

Rural poverty is worsened by urbanization. A small village or town will eventually have too few jobs and other opportunities to be attractive to the young people who grew up there, and many will feel forced to leave. As they leave, those left behind are even poorer, with fewer opportunities. The effects of rural poverty are pronounced in low-income countries, but can be seen in developed countries as well. Rural poverty is a significant problem in the United States, even though there are social safety nets (Figure 23–5).

Social Changes. The exodus from rural areas is also having a significant impact on rural societies. In China, a massive migration to cities has been triggered by a search for industrial work and encouraged by government authorities who promote urbanization and want to appropriate rural lands for development. Between 2000 and 2010, China's urban population increased by almost 132 million people. In 2000, there were 3.7 million villages in China; by 2010, there were only 2.6 million—a loss of 300 villages a day. This almost geologic shift, the largest human migration in history, has changed the fabric of Chinese society. Because villages were centers of local culture, the loss of traditions is profound. For millennia, villagers passed traditional songs, dances, and techniques of printmaking down from generation to

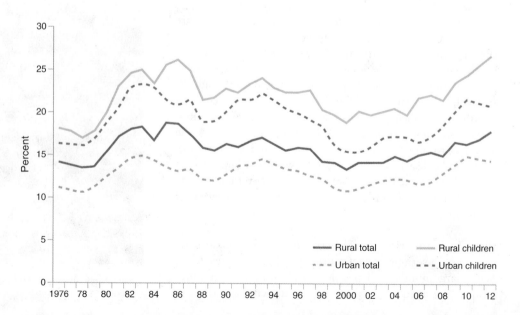

Figure 23–5 Rural and urban poverty in the United States. Poverty is more common in rural areas than in urban areas; the rate of poverty among rural children is greater than 25%. However, there are also significant pockets of urban poverty. This pattern has held true for more than 40 years.

Figure 23–6 Children who are left behind. A volunteer tutors children in China's Jiangxi Province who have been left in the care of others while their parents work in cities. Some 61 million rural Chinese children share the same plight.

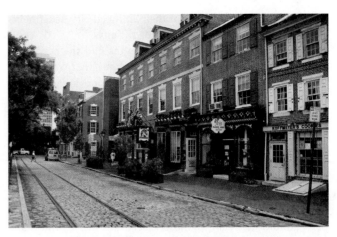

Figure 23–7 A traditional city neighborhood. Before the widespread use of automobiles, cities had an integrated structure. A wide variety of small stores and offices on ground floors, with residences on upper floors, placed everyday needs within walking distance. This scene is from a historic section of Philadelphia.

generation. Today these arts—a part of China's intangible capital—are being lost.

One of the most troubling changes in China is the loss of traditional family structures. When parents travel to cities, regulations about the rights of migrants often cause them to leave their children behind. As a result, today nearly one in five Chinese children is being raised by someone other than his or her parents (Figure 23–6). These almost 61 million "left behind" children are usually being raised by elderly relatives in rural areas. Children who move to cities with their parents also fare poorly. Without legal residence permits, children are not allowed to attend school and are often mistreated.

Any sustainable future will require specific plans to support rural communities as well as urban communities. Meeting the Sustainable Development Goals relating to poverty will also mean a focus on both urban and rural areas.

CONCEPT CHECK Why is the recent large-scale migration to cities a problem for rural areas? ✓

23.2 Trends in U.S. Communities

The pace of urbanization in the United States is far slower than that in the developing world. Even so, an examination of the issues related to urban and suburban growth in this country provides insight into the pressures on modern cities around the world. In this section, we will consider the factors that have driven the movement from city centers to vast sprawling suburbs, leaving behind blighted urban areas. We'll also see how recent efforts have begun to reinvigorate many city centers.

The Growth of Cities

At the beginning of the 20th century, the large majority of Americans lived in rural areas. But as people abandoned less productive farms and mechanized farming reduced the need for farm labor, there was a movement out of rural areas and

into cities. In the South, the unjust conditions of segregation caused many African Americans to leave rural areas, seeking greater economic opportunity in northern cities. An influx of immigrants also increased city populations. At the same time, the growth of industry provided jobs for those moving into cities. By 1940, 56% of the U.S. population was urban or suburban. The majority of people living in cities worked in factories or service jobs. The need for factory workers during World War II only increased the rate of migration to industrial centers.

Until the end of World War II, a relatively small percentage of Americans owned cars. Cities and towns had developed in ways that allowed people to meet their needs by means of the transportation that was available—mainly walking, but also by bicycling. Small groceries, pharmacies, and other stores, as well as professional offices, were integrated with residences. Buildings often had stores at the street level and residences above (Figure 23–7). Schools and parks were scattered throughout neighborhoods. Walking distances were generally short. Forms of public transportation such as electric trolleys, cable cars, and buses were also used. Beyond the city lines lay natural areas and farms, which provided most of the food for the city.

Urban Sprawl

Despite the many advantages of urban life, many people found the pollution, congestion, and noise of industrialized cities to be unpleasant. Many people had the desire—the "American dream"—to live in their own house, on their own piece of land, away from the city.

Cars and Suburbs. At the end of World War II, dramatic social changes occurred. Rationing during the war was replaced by a demand for consumer goods. When the mass production of cars resumed at the end of the war, people flocked to buy them. With private cars, people were no longer restricted to living within walking distance of their workplaces or transit lines. They could move out of their cramped city apartments and into homes of their own outside the city.

At the same time, returning veterans and the accompanying baby boom created a housing demand that was met by developments built on farmlands and natural areas outside the cities. The government aided this trend by providing low-interest mortgages and tax deductions for homeowners, while rent was not tax deductible. These changes meant that making monthly payments on a mortgage for a home in the suburbs was cheaper than paying rent for equivalent or less living space in the city. These economic factors contributed to the development of **urban sprawl,** a far-flung network of low-density residential areas, shopping malls, industrial parks, and other facilities loosely laced together by multilane highways (Figure 23–8).

Highways. The mushrooming development around cities did not proceed according to any plan; rather, it happened wherever developers could acquire land. Local governments in surrounding towns were quickly thrown into the catch-up role of trying to provide schools, sewers, water systems, roads, and other public facilities to accommodate the uncontrolled growth. Local zoning laws kept residential and commercial uses separate. It became more difficult to walk to a grocery store or an office. The influx of commuters created a need for new and larger roads.

In 1956, Congress created a tax on gasoline, producing funds to be used exclusively for building new roads. Ironically, this **Highway Trust Fund** has perpetuated the development of formerly rural areas. A new highway not only alleviates existing congestion, but also encourages development at farther locations, because time, not distance, is the limiting factor for commuters. (When people describe how far they live from work, it is almost always put in

terms of minutes, not distance.) New highways that were intended to reduce congestion actually led to more congestion because they fostered the ability of drivers to commute from more distant locations, leading to the development of more open land and an ever-greater number of commuters. Although the average commuting distance has doubled since 1960, the average commuting time has remained about the same.

Unfortunately, government policy, as an unintended consequence, helped support and promote sprawl. Eventually, shopping malls and other businesses grew up near the residential developments. These commercial centers changed the direction of commuting: Whereas, in the early days of suburban sprawl, the major traffic flow was into and out of cities, it is now between suburban centers.

Breaking the Cycle. Since 1990, there have been some changes that may help break the cycle of urban sprawl. A 1991 act allowed money from the Highway Trust Fund to be used for other modes of transportation, including cycling, walking, and mass transit. The 1998 Transportation Equity Act for the Twenty-First Century and another 2005 law increased funding for mass transit. The economic stimulus package (the American Recovery and Reinvestment Act of 2009) contained $64 billion for roads, bridges, rail, and transit; mass transit has been receiving much of this money.

Measuring Sprawl. A team of researchers from Rutgers University, Cornell University, and Smart Growth America (a nongovernmental organization) worked to measure and evaluate sprawl and its impacts. The research team defined sprawl as "the process in which the spread of development across the landscape far outpaces population growth."[5] The team members created a *sprawl index* based on four measurements: *residential density* (sprawl lowers density), *neighborhood mix* (sprawl separates residences from businesses), *accessibility of the street networks* (sprawl concentrates traffic into large roads with few connections), and *strength of activity centers and downtowns* (sprawl weakens these). They used these numbers to create an index in which lower numbers represent more sprawl. The team analyzed 83 metropolitan areas, representing almost half of the U.S. population. The results ranged from 14 to 178, with the Riverside–San Bernardino, California, metropolitan area the most sprawled and New York City the least (Figure 23–9).

In broad perspective, then, urban sprawl is a process of *exurban migration*—that is, a relocation of residences, shopping areas, and workplaces from their traditional spots in the city to outlying areas. The populations of many eastern and Midwestern U.S. cities, excluding the suburbs, have been declining over the past 60 years (Table 23–1). At the same time, the country's suburban population has increased from 35 million in 1950 to 150 million in 2010, half of the

Figure 23–8 Urban sprawl. Prime land is being sacrificed for development around most U.S. cities and towns. The locations of developments are determined more by the availability of land than by any overall urban planning. Here we see sprawl surrounding the fastest-growing major city in the United States: Las Vegas, Nevada.

[5]Reid Ewing, Rolf Pendall, and Don Chen, *Measuring Sprawl and Its Impact* (Washington, DC: Smart Growth America, 2002). Available at www .smartgrowthamerica.org/research/measuring-sprawl-and-its-impact.

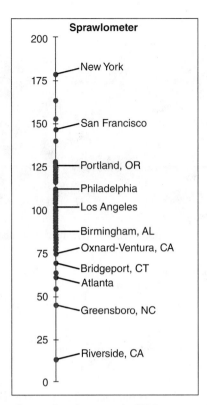

Figure 23–9 Comparing the sprawl of U.S. metropolitan regions. The "sprawlometer" shows rankings of numerous metropolitan areas (each round dot represents a separate area). Low scores mean high sprawl, and vice versa.

(*Source:* Measuring Sprawl and Its Impact, 2002, Smart Growth America.)

U.S. population. Sometimes people from older suburbs move to exurbs, communities that are even farther from cities than suburbs. At the same time, the population of truly rural areas has continued to decline.

Sprawl and Urban Blight

Historically, U.S. cities included people with a wide diversity of economic and ethnic backgrounds. But during the second half of the 20th century, migration to the suburbs and the decline of industrial manufacturing worked together to increase economic and racial segregation, eventually leading to the economic collapse of portions of some cities.

Economic and Racial Segregation. Moving to the suburbs required some degree of affluence—at least the ability to manage a down payment and mortgage on a home and the ability to buy and drive a car. Therefore, exurban migration often excluded the poor, the elderly, and persons with disabilities. White people dominated the mass movement from cities to suburbs, giving the phenomenon the name *white flight*.

The poorest city dwellers were mostly minorities. Some were in poverty because racial discrimination had denied them access to a good education and well-paying jobs, while others had recently immigrated to the United States and were still at the bottom of the social order. Systematic racism also played a role. Discriminatory lending practices by banks and dishonest sales practices by real estate agents (called *redlining*) kept minorities out of the suburbs, even when money was not a factor. Although civil rights laws passed in the 1960s made discriminatory practices illegal, less obvious forms of racism continue to exist, and there remain numerous barriers for residents of economically depressed inner cities to either leave or to improve the economy of the city. Figure 23–10, on the following page, illustrates the outcome of this process.

Decline of Industry. Pittsburgh, Cleveland, Youngstown—these and many other cities of the Midwest are representatives of what has been called the **Rust Belt**. Once known as the

Table 23–1	Decline in Population of American Cities, 1950–2009						
	Population (thousands)						**Percent Change, 1950–2009**
City	**1950**	**1970**	**1990**	**2000**	**2005**	**2009**	
Baltimore, MD	950	905	736	651	640	637	−33
Boston, MA	801	641	574	589	597	645	−19
Buffalo, NY	580	463	328	293	278	270	−53
Cleveland, OH	915	751	505	478	450	431	−53
Detroit, MI	1,850	1,514	1,028	951	921	911	−51
Minneapolis, MN	522	434	368	383	375	385	−26
Philadelphia, PA	2,072	1,949	1,586	1,517	1,460	1,547	−26
St. Louis, MO	857	662	396	348	353	357	−58
Washington, DC	802	757	607	572	582	600	−25
U.S. population (millions)	151	218	249	281	296	307	+103

Source: Data from Population Division, U.S. Bureau of the Census, 2012.

(a)

(b)

Figure 23–10 Segregation by exurban migration. Exurban migration, the driving force behind urban sprawl, has also led to segregation of the population along economic and racial lines. In **(a)** an area of suburbia and **(b)** an area of inner city Baltimore, you see the contrast.

"Foundry of the Nation," these cities and their outskirts had specialized in heavy manufacturing up to the mid-20th century. Then manufacturing migrated to southern states where labor was less costly, globalization of trade gave foreign manufacturers with far lower labor costs access to U.S. markets, and free trade agreements expanded. Finally, jobs were outsourced because of these developments. The result was a wholesale loss of the economic foundations of cities of the Rust Belt during the latter half of the 20th century, and the mass migration of people from these cities to other places. The term *Rust Belt* is a consequence of the all-too-common images of rusting and crumbling buildings and factories in the region (Figure 23–11).

Tax Erosion. Local governments use property taxes to pay for a wide range of public services and infrastructure, including schools, police and fire protection, water and sewer systems, waste collection and disposal, maintenance of local roads and parks, libraries, and welfare services. When large numbers of affluent people leave a city to move to the

suburbs, the city government ends up with a smaller tax base with which to provide these services. Such a reduction in tax revenue is referred to as a **declining tax base,** or **tax erosion,** and it has been a serious handicap for most U.S. cities since the exurban migration started in the late 1940s. Migration to the suburbs has also caused city property values to decline.

The loss of affluent people also increases the proportion of inner-city dwellers who require public assistance of one sort or another, placing a disproportionate burden of welfare obligations on city governments. At the same time, there are fewer affluent people to support traditional inner-city social institutions, such as churches. Combined with the loss of government support, the social fabric of the communities is weakened.

Urban Blight. An eroding tax base forces city governments to cut local services, increase the tax rate, or both. Hence, the property tax on a home in a city is often two to three times greater than the tax on a comparably priced home in the suburbs. Eventually many city properties are abandoned because the owners cannot pay the taxes. Cities "inherit" such abandoned properties, but they are a liability rather than a source of revenue.

Without enough funding, public schools, refuse collection, street repair, libraries, parks, and other services and facilities deteriorate. Increasing taxes and deteriorating services cause yet more people to leave the city, even if they want to stay. Still, it is only the relatively affluent individuals who can afford to move to the suburbs. Thus the whole process of exurban migration, a declining tax base, deteriorating real estate, and worsening schools and other services is perpetuated in a vicious cycle. This downward spiral of conditions is referred to as **urban blight** or **urban decay.**

Loss of Businesses and Jobs. The exurban migration includes more than just individuals and families: The flight of affluent people from the city removes the purchasing power necessary to support stores, professional establishments, and other enterprises. Faced with declining business, merchants

Figure 23–11 Rust Belt image. This abandoned factory is symbolic of many such relics of the heavy manufacturing that once took place in large Midwestern cities.

and practitioners of all kinds are forced to go out of business or to move to new locations in the suburbs. Either way, vacant storefronts are the result, and people remaining in the city lose convenient access to goods, services, and, most importantly, jobs, because each business also represents employment opportunities. Getting to jobs, now located in the suburbs, is impossible for many people, who now have few employment options. As a result, crime rates rise.

Impacts of Sprawl on Public Health and the Environment

In addition to harming the city proper, urban sprawl can have major negative impacts on the quality of life in the surrounding areas. The environmental impacts of urban sprawl are many and serious.

Hazards to Public Health. People living in sprawling suburbs drive more than those who live in more compact communities, who walk more and use mass transit more often. Walking and other moderate physical activities have many health benefits. One study of the health effects of sprawl found a highly significant correlation between the degree of sprawl and the incidence of overweight, obesity, and high blood pressure.

People in high-sprawl areas drive greater distances and experience more traffic fatalities. In Riverside, California, the U.S. city with the greatest degree of sprawl, 18 of every 100,000 people die each year in highway accidents (Table 23–2). In the eight cities that are the lowest on the index, fewer than 8 of 100,000 residents die annually on the highways.

Air Pollution. Shifting to a car-dependent lifestyle has increased our use of petroleum (Chapter 14). From 1950 to 2010, per capita U.S. oil consumption increased 63%. All this car use reduces air quality, making the air less breathable. Vehicles are responsible for an estimated 80% of the air pollution in metropolitan regions. Burning gasoline also releases carbon dioxide; in fact, transportation accounts for 28% of all U.S. greenhouse gas emissions.

Water Resources. The roads and parking lots of sprawling urban and suburban areas increase the amount of impervious surfaces, decreasing the ability of local ecosystems to control floods and recharge groundwater (Chapter 10). Water quality is degraded by the runoff of fertilizer, pesticides, oil, pet droppings, and other pollutants (Chapter 20). These nonpoint-source pollutants are the leading cause of water quality problems, according to the EPA.

Loss of Farmland and Natural Ecosystems. Perhaps the most serous impact of sprawl is the loss of land for agriculture, recreation, and wildlife habitat. New developments are consuming land at an increasing pace. According to the National Resources Inventory, 1.7 million acres a year were developed from 1997 to 2007, compared with 1.4 million acres a year from 1982 to 1992 (Figure 23–12).

In the United States, cropland is declining at a rate of 1.1 million acres per year. Most of the food for cities used to be grown on small family farms surrounding the city. With so many of these farms turned into housing developments, it is estimated that food now travels an average of 1,000 miles (1,500 km) from where it is produced—mostly on huge commercial farms—to where it is eaten.

The fragmentation of wildlife habitat due to urban sprawl has led to marked declines in many species, ranging

Table 23–2	Ten Most Sprawling U.S. Metropolitan Regions	
	Overall Sprawl Index Score	**Rank**
Riverside–San Bernardino, CA	14.2	1
Greensboro–Winston-Salem–High Point, NC	46.8	2
Raleigh–Durham, NC	54.2	3
Atlanta, GA	57.7	4
Greenville–Spartanburg, SC	58.6	5
West Palm Beach–Boca Raton–Delray Beach, FL	67.7	6
Bridgeport–Stamford–Danbury, CT	68.4	7
Knoxville, TN	68.7	8
Oxnard–Ventura, CA	75.1	9
Forth Worth–Arlington, TX	77.2	10

Source: Reid Ewing, Rolf Pendall, and Don Chen. Measuring Sprawl and Its Impact (Washington, DC: Smart Growth America, 2002).

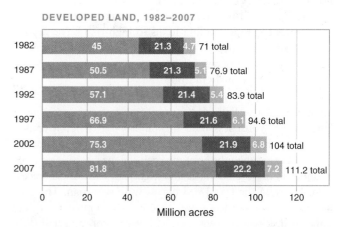

DEVELOPED LAND, 1982–2007

- Large urban and built-up areas
- Rural transportation
- Small built-up areas

Figure 23–12 Conversion of land to developed uses. Between 1982 and 2007, 40 million acres of U.S. farmland and natural areas were converted to other uses, including urban areas and roadways.

(*Source:* U.S. Department of Agriculture. 2007 National Resources Inventory, Natural Resources Conservation Service, Washington, DC, and Center for Survey Statistics and Methodology, Iowa State University, Ames, Iowa. July 2009.)

from birds to amphibians. A 2005 report[6] found that the 35 fastest-growing metropolitan areas are home to nearly one-third of the nation's rarest and most endangered species of plants and animals.

Why Urban Sprawl Persists

With all of these undesirable effects of sprawl, one might wonder why people put up with it. The answer is that people *perceive* that it is better to live in suburbs, so they move there. Quality-of-life issues tend to be decisive for most people, and these issues seem to be heavily weighted in favor of sprawl. In general, sprawl involves lower-density residential living, larger lot sizes, larger single-family homes, better-quality public schools, lower crime rates, better social services, and greater opportunity for participation in local governments. As people move farther away from an urban center, housing costs often decline. People move to the exurbs to live in neighborhoods with lower crime rates, to have access to better schools, or to live in more homogeneous communities.

The result of thousands of people making these choices is to accentuate the concentration of poorer people in inner cities or in aging suburban communities.[7] The environmental costs of sprawl are very real, but the people moving to the sprawling suburbs seldom perceive those costs to be decisive. The costs are more a matter of the common good, and people tend to make choices based on *personal good* rather than the *common good*. The key to controlling sprawl is to provide communities with the quality-of-life benefits that attract people, but without incurring the serious environmental and social costs that are clearly associated with sprawl.

The Housing Bubble and Suburban Blight

The expansion of the suburbs during the second half of the 20th century was remarkable. By the early 2000s, home prices were rising rapidly and many were purchasing homes for their resale value. Banks and mortgage companies began extending loans to high-risk buyers, who were borrowing beyond their limits. But then the housing bubble burst. Property values began to decline in 2006, dropping precipitously in 2008.

When housing prices fell, many people were unable to repay their loans. Fully one-fourth of all homes in the United States were worth less than their mortgaged amount (called underwater mortgages).[8] Loan defaults and foreclosures had a ripple effect through the economy, causing the loss of businesses and jobs in the United States and triggering a worldwide recession. Many of the foreclosed homes were suburban. The economic collapse also left many suburban development projects unfinished, leaving streets lined with vacant lots and half-built

houses. Suburban malls stood empty. This led to an unexpected phenomenon: suburban poverty and *suburban blight.*

Back to the City

In recent years, many college graduates and young professionals have been moving into cities. These new urbanites want the opportunities, convenience, and diversity of experience that cities can provide. Some of these new urban dwellers are looking for a more sustainable lifestyle. Public transit and new approaches to transportation such as car and bicycle sharing make it easier to live without cars. Some city dwellers are involved in a movement called **urban homesteading,** an effort to use skills from earlier eras to reduce their footprint in the city. Fixing up abandoned properties, using alternative sources of energy, and promoting urban agriculture are parts of this approach.

The city of Detroit is one place where new approaches are making a difference. Once a dynamic city with numerous industries that provided well-paying jobs, the city was impoverished by the decades-long flight of the car manufacturers and residents to the suburbs. The 2008 housing collapse was particularly damaging to Detroit. Home values plunged, until almost a third of the city's housing sat vacant. Finally, in 2013, the over-extended city was forced to declare bankruptcy.

Despite its many problems, Detroit is still an important hub for shipping and retains a strong mass transit system. Many residents who could have left have chosen to remain in the communities they love. Investors have been purchasing and rehabilitating homes and office buildings and encouraging the growth of small local businesses. Limited tax revenues make it difficult for the city to maintain its public infrastructure, but some private citizens have stepped up, providing security, repairing streetlights, and planting gardens (Figure 23–13). These residents are a part of the next wave of reinvigoration for a city with a great deal of potential.

CONCEPT CHECK How did the increased use of automobiles in the United States drive urban flight and, by extension, urban blight? ☑

Figure 23–13 Revitalization of Detroit. These young men are planting a neighborhood garden in Detroit, one of America's most economically depressed cities. Such efforts are changing the fortunes of the city.

[6]Reid Ewing et al., *Endangered by Sprawl: How Runaway Development Threatens America's Wildlife* (Washington, DC: National Wildlife Federation, Smart Growth America, and NatureServe, 2005).

[7]Robert W. Burchell et al., *The Costs of Sprawl—2000* (Washington, DC: National Academies Press, 2002).

[8]Michael B. Sauter and Charles B. Stockdale, "American Cities with the Most Underwater Mortgages," *24/7 Wall Street*, December 28, 2011.

23.3 Moving Toward Sustainable Communities

Sustainable Development Goal #11, *Make cities and human settlements inclusive, safe, resilient and sustainable,* is focused on the development of human communities. In order to have sustainable communities and address inequalities in wealth, we must, among other things, rein in urban sprawl, revitalize our cities, and care for rural populations.

Another trend that needs our attention is the overconsumption of resources, which has been increasing. Elsewhere, we have described the negative environmental impacts of that overconsumption. Since 1900, the world population has increased by 12-fold. But during this time, as one researcher noted, "our [global] economy has increased in size by 22-fold, our use of construction minerals has increased 34-fold, ores 27-fold, and fossil fuels (coal, oil and gas) have seen a 12-fold increase. We use 3.6 times as much biomass—crops, residues, and wood—as we did back then."[9] Our challenge today is even harder because our planet is facing or even crossing planetary boundaries (Chapter 1) and experiencing climate change and the loss of biodiversity.

Reining in Sprawl

For cities with large sprawling suburbs, part of the movement toward sustainability will include reining in sprawl. Simply passing laws that bring rural development under control might seem to be the way to stop urban sprawl. Indeed, Japan and a number of European countries have such laws. However, laws cannot be passed without the support of the public. In the United States, the strong sentiment favoring the right of a landowner to develop a property as he or she sees fit has made such restrictions, with few exceptions, politically impossible to pass. Yet there are signs that a sea change is occurring in the attitudes of Americans toward uncontrolled growth.

Smart-Growth Strategies. The recently emerged concept of **smart growth** is inviting communities and metropolitan areas to address sprawl by choosing to develop in more environmentally sustainable ways. While recognizing that communities will continue to grow, advocates of smart growth focus on economic, environmental, and community values that will lead to more livable communities. Table 23–3 presents the basic principles of smart growth. The goal of smart growth is integrated communities, rather than disassociated facilities. By directing growth to particular places, smart growth provides protection for sensitive lands. Early indications in the marketplace—the final determining factor—are that new communities that have followed the principles of smart growth are a great success.

Smart-growth initiatives have been appearing on the ballot in many states and municipalities. The key strategies of these initiatives include setting boundaries on urban sprawl, saving open space, and developing existing urban space. In the 1970s,

Oregon passed a land-use law requiring every city to set a boundary beyond which urban growth is prohibited; beyond that boundary line, expected future growth must be controlled. Some states are enacting programs to acquire crucial remaining open spaces. Maryland's Smart Growth and Neighborhood Conservation initiative includes measures to channel new growth to community sites where key infrastructure is already in place. Abandoned and brownfield sites in metropolitan areas are already well served with roads, electricity, water, and sewerage and can be developed into communities with new homes, workplaces, and shopping areas (Chapter 22).

Creating New Towns. One key to developing sustainable communities is to ensure that zoning laws allow the integration of stores, light industries, professional offices, and higher-density housing. Affordable and attractive housing can be built around industrial parks and shopping malls for the people who work and shop in those places. Safe bicycle and pedestrian access can be provided between residential areas and workplaces.

In the 1960s, several new, self-contained communities were established along these lines, including Reston, Virginia; Columbia, Maryland; and Valencia, California. Today these communities are still highly desirable as business and residential locations. By contrast, almost 100% of Fairfax County, Virginia (next to Washington, DC), is now covered by sprawl-type development. It is estimated that if Fairfax had pursued smart-growth policies, 70% of the county would still be open space, and the county would be far less congested and polluted.

Redesigning Suburbia. The recession and its accompanying impacts on suburbs have created a unique opportunity to redevelop suburbs in more desirable patterns. A recent survey found that most Americans prefer a walkable neighborhood to a large house; this is especially true of Baby Boomers

Table 23–3	Principles of Smart Growth
Communities	Foster distinctive, attractive communities with a strong sense of place.
	Set standards for development that respond to community values.
	Create a range of housing choices for people of all income levels.
	Create walkable neighborhoods.
Development	Encourage the collaboration of citizens, corporations, businesses, and municipal bodies within communities.
	Strengthen and direct development toward existing communities (called *infilling*).
	Take advantage of compact building design instead of conventional land-hogging development.
Transportation	Provide a variety of transportation choices.
Open Space	Preserve open space, farmland, natural beauty, and critical environmental areas.

[9]F. Krausmann et al., "Growth in Global Materials Use, GDP and Population During the 20th Century," *Ecological Economics* 68.10 (2009): 2696–705.

Figure 23–14 Livable cities. Livable cities provide heterogeneity of residences, businesses, and stores and keep layouts on a human dimension so that people can meet over coffee or stroll through an open area. In short, the space is designed for people. Shown here is Quincy Market, Boston.

Figure 23–15 Bicycles in Tokyo. Most people in Tokyo do not own automobiles and instead ride bicycles or walk to stations from which they take fast, inexpensive subways to reach their destinations. Tokyo residents own more than 9 million bicycles.

(now largely empty nesters) and many Millennials (people in their 20s and 30s), who are looking for places where they can socialize.[10] Planners and developers are looking at what can be done to abandoned malls and poorly designed suburban streets and roadways to convert them to more desirable neighborhoods with village centers, greenery, and places to walk.[11] For example, a 100-acre mall in Lakewood, Colorado, has been replaced with 22 blocks of green buildings and public streets, creating a real downtown for the suburban town.

Making Cities Livable

Looking to the future, the only way to sustain the global population is by having livable, resource-efficient cities (Figure 23–14). *Livability* is a general concept based on people's response to the question "Do you like living here, or would you rather live somewhere else?" Crime, pollution, and recreational, cultural, and professional opportunities, as well as many other social and environmental factors, are summed up in the subjective answer to that question. No one wants to, or should be required to, live in the conditions that have come to typify urban blight and urban slums or even to live in new buildings that are densely residential but don't provide neighborhoods or allow mixed use.

Looking at livable cities around the world, we find that common denominators include the following:

- Maintaining a high population density
- Preserving a heterogeneity of residences, businesses, stores, and shops

- Keeping layouts on a human dimension so that people can meet, visit, or conduct business informally over coffee at a sidewalk café or stroll on a promenade through an open area
- Maintaining intangible capital, such as education and employment opportunities (Chapter 2)

Many Japanese cities are highly livable even though they are densely populated. Japan's cities have maintained a heterogeneous urban structure that mixes small shops, professional offices, and residences in such a way that a large portion of the population is able to take trains, walk, or bike rather than drive. Japan has an estimated 72 million bicycles. In Tokyo, millions of people ride bicycles (Figure 23–15), either all the way to work or to subway stations from which they catch fast, efficient trains to their destinations.

Curitiba, Brazil, with a population of 1.6 million, is cited as the most livable city in all of Latin America (see Sustainability, *Curitiba, Brazil—City Planning Meets Growth*). The city has a number of features that make it livable, especially its public transportation, parks, and green spaces.

Coordinated Planning

The key to sustainable urban life is planning. City planners must consider a broad range of topics, including finance and economic development, housing, energy systems, transportation, water supply, solid waste management, and adaptation for climate change. As we saw elsewhere, solving multiple problems at once (planning for co-benefits) allows resources to be used more efficiently.

Green Buildings and Energy Efficiency. Think of rooftops covered in solar panels, and green structures designed to use passive solar. The development of green buildings and the use of alternative energy are central to sustainable cities. Cities use a great deal of energy to heat and cool buildings, most of which are privately owned, so cities often have to partner with the private sector to encourage green buildings.

[10]Belden Russonello and Stewart LLC, "The 2011 Community Preference Survey: What Americans Are Looking for When Deciding Where to Live," March 2011, www.Brspoll.com, Washington, DC.

[11]Ellen Dunham-Jones and June Williamson, *Suburbia: Urban Design Solutions for Redesigning Suburbs.* (Hoboken, NJ: John Wiley & Sons, Inc., 2011); and Galina Tachieva, *Sprawl Repair Manual* (Washington, DC: Island Press, 2010).

SUSTAINABILITY

Curitiba, Brazil—City Planning Meets Growth

Imagine living in a city where there is ample green space, the roads are laid out to improve air quality and minimize congestion, most people recycle, and many people use public transportation. All of this can be found in Curitiba, Brazil's "Green City" in the southern state of Parana. In the 1950s, the city had fewer than 400,000 inhabitants; today it has 2.5 million. What helped the city remain livable while growing so rapidly? The answer is effective urban planning and good design.

The smart growth of Curitiba is due almost entirely to the efforts of Jaime Lerner, an architect who served as the city's mayor for many decades. Lerner guided development that controlled urban sprawl, provided a convenient public transit system, reduced downtown traffic, and preserved historic areas. The city also started a network of free educational programs and has the highest literacy rate of any Brazilian state capital.

Because of its mass transit system, Curitiba's air quality has improved since 1970, even though the city's population has tripled and the city has the highest percentage of car ownership in Brazil. The space saved by not building highways and parking lots has been used for parks and shady walkways. Since 1970, the amount of

Jaime Lerner, former mayor and city planner of Curitiba, Brazil.

green space per inhabitant has increased from 4.5 square feet to 450 square feet.

Some of Curitiba's innovations solve multiple problems at once. For example, public spaces are maintained by grazing goats, and

native vegetation has been planted along riverbanks to lower the risk of floods. Seventy percent of the city's trash is recycled. In shantytowns that cannot be reached by trucks, the removal of wastes is made possible with a "green exchange" program—residents who bring their waste to central areas are given food and bus tickets. As a result of all of these innovations, Curitiba has a Human Development Index value 0.856. In 2010 it was given the Global Sustainable City Award, and in 2014 it was called the "first smart city in the world."

Despite Curitiba's great successes, rapid growth has recently been straining the city's systems. The public transit system is overcrowded and expensive, and people are driving more. The bike paths are underused. Recycling rates have fallen. People have been moving out to the suburbs; unfortunately, the visionary planning in the city did not extend to the now sprawling suburbs. (In the case of Curitiba, the suburbs are the poorer areas, not the city center.) To solve these current problems, the innovative city planning that made Curitiba so livable needs to be upgraded and expanded to include the surrounding regions.

Curitiba, Brazil, is one of the most sustainable cities in the world, with urban parks, walkways, and public transit.

Figure 23–16 A very green building. The Bullitt Center in Seattle, Washington, was built using certified sustainable wood. Among other innovative features, a photovoltaic panel that hangs over the roof and geothermal technology provide the power used by the building. The floor-to-ceiling windows provide natural lighting for workers and can be opened to ventilate the building.

When the Bullitt Center in Seattle, Washington, opened in 2013, it was called the world's greenest commercial building. Designed to last for 250 years, the six-story building captures rainwater, is powered by geothermal energy and solar panels, has its own systems for processing water and sewage, and is carbon- and energy-neutral (Figure 23–16).

The efficient use of energy is another key to sustainable cities. To increase energy efficiency, cities may build power plants that use waste to generate electricity, while also producing heat for homes and businesses (see Chapter 21).

Transportation. The world's most livable and sustainable cities have taken measures to reduce outward sprawl, diminish automobile traffic, and improve access by foot and bicycle in conjunction with mass transit. Such changes require coordination and planning. For example, Geneva, Switzerland, prohibits automobile parking at workplaces in the city's center, forcing commuters to use its excellent public transportation system. Copenhagen bans all on-street parking in its downtown core, and Paris has removed 200,000 parking spaces from its downtown. Among the large cities in developed countries, Amsterdam holds the record for biking: the city's policies are so supportive of biking that fully 38% of the trips within the city are made on bicycles.

Water Use. City planners must also address issues related to the supply and use of water. Many cities face significant problems related to aging water pipes and other parts of their water infrastructure.

Leaking pipes can waste tremendous amounts of water. In Mexico City, for example, the 1,000 kilometers of main pipes and 12,000 kilometers of secondary pipes leak 11 cubic meters of water per second (347 billion liters per year); these leaking pipes waste as much water as some of the world's largest cities use. This loss of water is a significant problem for a city that is experiencing drought. To compensate, the groundwater around the city is being over-pumped; unfortunately, this over-pumping causes the ground to sink, which in turn breaks water pipes. Leaking pipes are also a health hazard: In places where both water and sewage pipes leak, sewage may seep into the water pipes and contaminate drinking water. In order to solve its water problems, Mexico City will need to pump less water, coordinate the more efficient use of water, and replace the broken pipes and infrastructure.

Milwaukee, Wisconsin, has become a global hub for water planning. In 2009, the city joined the UN Global Compact Cities Program in a project that uses technology to manage limited freshwater resources. Milwaukee also houses the Global Water Center, a hub for water education, research, and innovation. Companies at the center are involved in developing products and processes such as permeable paving that allows water filtration, systems to control the runoff of water from green roofs, and the use of algae to process wastewater.

Adapting to Climate Change. One of the critical jobs of city planners is to help cities prepare for future changes, including the conditions that are expected to accompany global climate change. Two-thirds of the world's largest cities are in coastal deltas, making them vulnerable to rising sea levels. In the future, these cities will need to find solutions to the problems caused by higher sea levels and increasingly severe storms.

Rising temperatures will be a particular challenge for cities. Why? The people, buildings, factories, and vehicles in a city give off heat, while the dark surfaces of buildings and roadways absorb solar energy during the day and release it slowly at night, forming a pocket of hot air over the city. As a result, cities are usually 3–8° Celsius hotter than surrounding areas, a phenomenon called the **urban heat island effect** (Figure 23–17). During hot weather, this excess heat can be a serious health hazard. To reduce the heat island effect, cities can set aside parks and green spaces, plant trees, encourage people to grow plants on the roofs of buildings, increase the reflectivity of surfaces by painting them light colors, and use energy sources more efficiently.

Data-Based Decisions. City planners rely on data to make decisions about how to concentrate growth and use resources efficiently. City managers can use big data and cloud computing to build smarter transportation systems, control electrical grids, coordinate waste disposal, and manage emergency responses.

Data-driven technologies are also helping to make cities more sustainable. For example, BigBelly Solar produces solar-powered trash compactors that are used for waste disposal in public areas. These compactors detect their own fullness and signal a need to be emptied, so waste managers do not have to make trips to pick up half-full trashcans

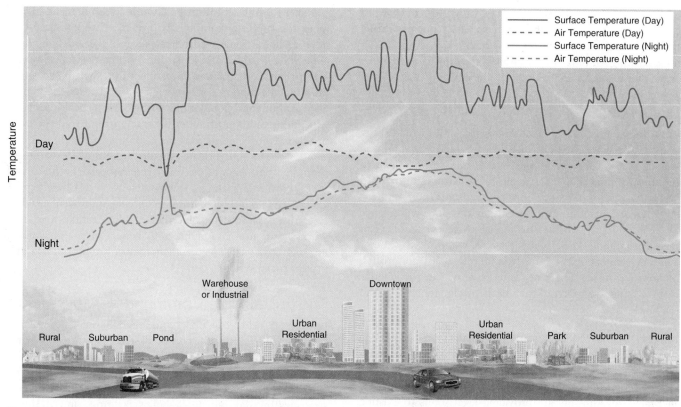

Figure 23–17 **Urban heat island effect.** City air is warmer than that of surrounding rural and suburban areas, as shown in this idealized diagram of day and night temperatures across a landscape. In the city, buildings and human activities release heat, while dark roads and rooftops absorb solar energy and slowly release it into the air. In parks, suburbs, and rural areas, bodies of water and evapotranspiration from plants moderate air temperatures.
(*Source:* EPA. "Urban Heat Island Basics," *Reducing Urban Heat Islands: Compendium of Strategies.* (Washington, DC: 2008)

or to clean up around trashcans that were too full. The city of London, parts of Los Angeles, and Iowa State University (Figure 23–18) all use BigBelly compactors.

Real-time data gathering can also reduce traffic snarls and help solve other problems. However, such tech-heavy approaches may not be appropriate for poorer cities or slum areas. Those places will require solutions that use local labor and provide job opportunities.

Portland, Oregon: A Leader in Sustainability

The 50 most populous cities in the United States have been ranked for sustainability by the organization SustainLane. Criteria for the ranking include transit ridership, street congestion, air and water quality, green building, local food and agriculture, energy and climate change, and green economy. At the top of the ranking was Portland, Oregon, followed by San Francisco, Seattle, Chicago, New York, and Boston.[12]

Why is Portland at the top of this list? Among other things, the city has taken giant steps to curtail automobile use. The first step was to encircle the city with an urban growth boundary, a line outside of which new development

was prohibited. Thus, compact growth, rather than sprawl, was ensured. Second, the city built an efficient light-rail and bus system, which now carries 45% of all commuters to

Figure 23–18 **Solar trash compactors.** An Iowa State University student uses a BigBelly solar trash compactor, which will send real-time data to the owner, signaling the best time to pick up trash.

[12]Green-City Ranking Group SustainLane Explains Its Methodology," *Grist,* May 12, 2008.

downtown jobs. (In most U.S. cities, only 10% to 25% of commuters ride public transit systems.) By reducing traffic, Portland was able to convert a former expressway and a huge parking lot into Tom McCall Waterfront Park, a green space that increases the livability of the city.

Sustainable Communities and Poverty

One look at a list of the most sustainable cities would show you that they are usually in wealthy countries. Making the rest of the world's cities both livable and sustainable will require planning and investment in infrastructure for food systems, sanitation, water, housing, transportation, and energy. According to one estimate, megacities alone will require $10 trillion in investment. It is foolish to believe that we can meet Sustainable Development Goals without investment. That investment may come from foreign aid, nonprofits, or improved trade, but it needs to be provided, particularly if these cities are to avoid the effects of climate change and sea level rise.

Making Slums More Livable. Sustainability is particularly difficult in slum areas, but promoting secure housing and employment are two critical parts to the puzzle. People live in fear of bulldozers coming at any time and leveling the shantytown. In Brazil, governments are providing legal status to existing slums (*favelas*) by granting lawful titles to the land. Peru has undertaken a huge titling program, giving recognition to some 1 million urban land parcels in a four-year period.

The people in slums also need more jobs. At little cost, city governments can employ these people to improve their own neighborhoods and other locations—collecting trash, building sewers, composting organic wastes, and growing food.

Perhaps the greatest need of people living in the informal neighborhoods of city slums is government representation. Their voices are seldom heard because they are among the poorest and most powerless in a society. In some regions, however, community organizations are emerging to deal with the basic issues of the city slums. People are organizing and demanding their rights as citizens. Shack/Slum Dwellers International (SDI), which held its first international conference in 2006, now represents these communities to institutions such as city governments and the World Bank.

As cities devote resources to providing the essential services and infrastructure to the informal slums that surround them, they will only be strengthened. When a city's slums become more livable, the entire city becomes more livable.

Avoiding Gentrification. In the blighted areas of cities in developed countries, it is important that city revitalization efforts do not come at the expense of low-income housing. In many cities, run-down buildings have been rehabilitated and the surrounding land used for the development of professional buildings and high rises. This process of *gentrification* causes real-estate prices to rise, often forcing out life-long, low-income residents and breaking up neighborhoods. Maintaining mixed-income neighborhoods is important in a livable city, and improving blighted areas without harming their residents needs to be part of the plan.

Making Rural Communities Livable. Planning and investment for rural communities will also be important in the next decades. Some researchers suggest that making rural areas more livable will reduce some of the pressures on rapidly growing urban areas. Investment in a sustainable future must include support for better agricultural practices. It must also provide rural populations with access to health services, education, and opportunities for trade, if that trade helps to produce a more just society.

The UN Public Private Alliance for Rural Development promotes partnerships between international agencies and other partners, such as Rotary Clubs and Chambers of Commerce, to promote rural economic improvement. The FAO also has a strategic objective to reduce rural poverty and works with rural communities to better the lot of women, improve crop yields, increase employment opportunities, and protect social systems.

The Sustainable Communities Movement

Revitalizing urban economies and rehabilitating cities requires coordinated efforts on the part of all sectors of society. What is termed a *sustainable communities movement* is taking root in cities around the United States and elsewhere. Some 106 countries have now developed national sustainable-development strategies that lay out public-policy priorities. In the United States, the Department of Housing and Urban Development sponsors the **Sustainable Communities and Development** program, providing grants to foster the development of more-livable urban neighborhoods. Groups such as Sustainable Communities Online and Smart Communities Network provide information, help, and case studies, demonstrating that the movement is gaining ground throughout the nation.

The 1992 UN Conference on Environment and Development (UNCED) created the Commission on Sustainable Development (Chapter 1), which is responsible for monitoring and reporting on the implementation of the agreements that were made at the conference. The UN also has a number of initiatives to help cities become more sustainable, including the Global Compact Cities Program and the Sustainable Cities Program. These programs provide financing and support to help cities figure out how to run vast services efficiently, sometimes with few resources.

The **C40 Cities Climate Leadership Group (C40)** is a network of the world's large city governments. Founded by the former mayor of London and chaired by the mayor of Rio de Janeiro, C40 is committed to addressing climate change. Together, the 75 associated cities account for 25% of the world's GDP and 12% of the world's population. The group provides strategic planning and advice for cities to lower their carbon footprints.

Supporting City Leaders

National and international policies and actions can promote sustainable development. However, cities and larger cooperative regions can improve sustainability even without the support of national governments. Many large cities are larger than small nations. For example, the Tokyo-Yokohama metro-region, with more than 37 million inhabitants, is often listed as the world's largest "city." Tokyo proper has 13 million residents and a budget equivalent to that of the country of Sweden. How cities respond to pressures will depend in part on what citizens tell their leaders, and what directions their leaders take. People who lead large cities, then, can make a huge difference in how sustainable our future is.

CONCEPT CHECK What are three resources that are available to the head of a megacity in a developing nation? ☑

23.4 Sustainable Lifestyles

Sound science, stewardship, and *sustainability*—these are the guiding principles of how we must live on our planet. Sound science is the basis for our understanding of how the world works and how human systems interact with it; stewardship is the actions and programs that manage natural resources and human well-being for the common good; and sustainability is the practical goal towards which our interactions with the natural world should be working. Are we making progress in incorporating these principles into our society?

Human Decisions

Our intention in this book has been to avoid dwelling on the bad news and instead to raise the hope that environmental problems can be addressed successfully. Throughout the text, we have pointed to policies and actions that can help move human affairs in a sustainable direction. To do so requires us to look at causes. As the late Nobel Laureate Henry Kendall said, "Environmental problems at root are human, not scientific or technical."[13] Even though our scientific understanding is incomplete and our grasp of sustainable development is still tentative, we know enough to be able to act in many circumstances. Therefore, it is human decisions that can bring about change, and it is these decisions that define our stewardship relationship with Earth.

Being a good steward of resources is not a simple task. Decision making is affected by our personal values and needs, and competing values and needs exist at every turn. Often good values compete. Making decisions means considering competing needs and values and reaching the best conclusion despite the numbing complexity of demands and circumstances. Although public policies at the national and international levels are absolutely essential elements of our success in turning things around, in the end these policies are the outcome of decisions by very human people. And even if those in power make thoughtful decisions, their policies will fail unless they are supported by the people who are affected by them.

---

[13]Union of Concerned Scientists. *Keeping the Earth: Scientific and Religious Perspectives on the Environment,* videotape transcript (Pittsburgh, PA: New Wrinkle, 1996)

STEWARDSHIP
Living in Tiny Houses

Between 1978 and 2007, the average size of a new single-family home in the United States increased from 1,780 square feet (165 m²) to 2,479 square feet (230.3 m²), even while the average size of American families fell. Yet recently there has been a countermovement, in which many people are choosing to live in houses that occupy less that 1,000 square feet, sometimes even less than 100 square feet. Author Sarah Susanka is credited with starting the movement with her 1997 book, *The Not So Small House,* followed by the founding of the Small House Society in 2002.

People who live in tiny houses are often committed to an alternative lifestyle that is less resource intensive. Tiny houses are far less expensive than typical suburban houses and use far fewer resources. The carbon footprint of a tiny home is smaller than that of a typical single-family home because there is less space to heat and cool and less need for lighting; in addition, fewer resources are used in the construction of tiny homes. Decreasing the size of a house by 50% lowers the greenhouse gas emissions over the life of the house by 36%. Residents of tiny houses have to be very careful about the belongings they keep, which reduces consumption. Sometimes the occupants of a group of tiny houses share facilities, such as a community center or storage area.

Tiny houses, also called *microhousing,* can be used to provide homes for those who are homeless. In 2014, a tiny house community built by and for homeless persons opened in Madison, Wisconsin.

Despite their advantages, tiny houses may not be the best solution in all communities. In some places, zoning regulations make it difficult to build tiny houses. And although tiny houses are more resource efficient than single-family houses, they may use more resources than similarly sized units in apartment buildings, where centralized heating and cooling systems are even more efficient than those of tiny houses.

This couple is moving into their 238 square-foot home in Maryland. The tiny home movement is focused on smaller spaces and simpler living, with houses built to last and green design.

S. Susanka, *The Not So Big House: A Blueprint for the Way We Really Live* (Newtown, CT: Taunton Press, 1998).

Positive Changes. Here are a few examples of significant changes at the level of governments, municipalities, and industries:

- The American Recovery and Reinvestment Act of 2009 (the stimulus package) has invested more than $43 billion in green energy initiatives.

- The EPA has established the Green Power Partnership, a program that encourages organizations to buy green power. The number one purchaser is Intel Corp., buying more than 2.5 billion kilowatt-hours of green power annually.

- Transition Towns (connected by the Transition Network) are more than 400 communities that are working together to respond to the challenges of peak oil and climate change.

- The World Business Council for Sustainable Development is a network of more than 200 global companies promoting greater resource productivity and sustainable product manufacture.

- Mayors from more than 1,000 cities in the United States have signed the U.S. Mayors' Climate Protection Agreement, which pledges the cities to reduce greenhouse gas emissions to 1990 levels or below (in the absence of a federal commitment).

- Australia led the way in eliminating incandescent lightbulbs, an action that has been followed by the European Union nations and other countries. In the United States, the state of California has led the way in the switch to compact fluorescents and LEDs.

- California established the Million Solar Roofs initiative and hit the 250,000 mark (more than 1 gigawatt) at the beginning of 2012. The state's target now is 12 gigawatts of local green energy by 2020.

The Common Good. At the beginning of this book, we defined the common good in the context of public policy: *the improvement of human welfare and the protection of the natural world* (Chapter 2). What compels people to act to promote the common good? We can identify a number of values that can be the basis of stewardship and help us in our decisions involving both public policies and personal lifestyles:

- *Compassion for those less well off*—making compassion a major element of our foreign policy, which would energize our country's efforts in promoting sustainable development in the countries that are most in need. Compassion can motivate young people to spend two years working for the Peace Corps or a hunger relief agency, for example.

- *Concern for justice*—working to make just policies the norm in international economic relations. A concern for justice can move people to protest the placement of hazardous facilities in underprivileged communities.

- *Honesty, or a concern for the truth*—keeping the laws of the land and openly examining issues from different perspectives.

- *Sufficiency*—simplifying lifestyles rather than ever increasing our consumption; using no more than is necessary of Earth's resources.

- *Humility*—instead of demanding our rights, being willing to share and even defer to others.

- *Neighborliness*—being concerned for other members of our communities (even the global community), such that we do not engage in activities that are harmful to them, but instead make positive contributions to their welfare.

To achieve sustainability, we will have to couple values like these with the knowledge that can come from our understanding of how the natural world works and what is happening to it as a result of human activities. For this to work, people must be willing to grapple with scientific evidence and basic concepts and to develop a respect for the consensus that scientists have reached in areas of environmental concern. It has been our objective to convey that consensus to you in this text. Recall also that sustainable solutions have to be economically feasible, socially desirable, and ecologically viable. This applies directly to the political decisions that are shaping our future in so many of our environmental concerns.

The good news is that many people are engaging environmental problems—working with governments and industries, bringing relief to people in need, and making exactly the kinds of decisions that are needed for effective stewardship. Community groups are organizing around sustainable principles, providing models of larger-scale changes that will be essential if we are to achieve a sustainable world. Traditional environmental organizations, as well as people from other sectors of society, are working together on the many issues that we have highlighted in this text, bringing about policy changes as well as changes in personal lifestyles.

Colleges and Universities: Agents of Change. As institutions that train people for a wide variety of future jobs, colleges and universities have unique opportunities to change the world for the better. If they train people to work in more sustainable ways, that impact will be multiplied. Many colleges and universities belong to the Association for the Advancement of Sustainability in Higher Education (AASHE), a network for promoting sustainability initiatives. Colleges often have sustainability initiatives on campus or sustainability coordinators who oversee internships. In addition, many college groups are active in issues related to justice and sustainability. College students who want to make a difference can tap into these opportunities and even be a driving force for positive change (Figure 23–19).

Lifestyle Changes

What are some of the lifestyle changes that are needed, and are they in fact occurring? We should be encouraged and inspired by the literally millions of people in all walks of life who are making outstanding efforts to bring about solutions. Every pathway toward solutions that we have mentioned

Figure 23–19 Sustainability on campus. An environmental science student at Rollins College in Winter Park, Florida, helps install solar panels on the roof of the science building.

represents the work of thousands of dedicated professionals and volunteers, ranging from scientists and engineers to businesspeople, lawyers, and public servants.

Indeed, we are all involved, whether we recognize it or not. Simply by our existence on the planet, every choice we make and action we take—the cars we drive, the products we use, the waste we throw away—has a certain environmental impact and a certain consequence for the future. Therefore, it is not a matter of choosing to have an effect, but a matter of what and how great that effect will be. It is a matter of each of us asking ourselves: Will I be part of the problem or part of the solution? The outcome will depend on how each of us responds to the challenges ahead.

As individuals, there are a number of ways we can participate in building a sustainable future:

- Individual lifestyle choices
- Political involvement
- Membership and participation in nongovernmental environmental organizations
- Volunteer work
- Career choices

Lifestyle choices may involve such things as walking or using a bicycle for short errands; switching to a more fuel-efficient car; recycling paper, cans, and bottles; retrofitting your home with solar energy; starting a backyard garden and composting and recycling food and garden wastes into your soil; choosing low-impact recreation, such as canoeing rather than jet-skiing; and living closer to your workplace. You may even choose to reduce your environmental footprint by living in a tiny house (see Stewardship, *Living in Tiny Houses*).

Political involvement ranges from supporting and voting for particular candidates to expressing your support for particular legislation through letters or phone calls. In effect, you are exercising your citizenship on behalf of the common good.

Membership in nongovernmental environmental organizations can enhance both lifestyle changes and political involvement. As a member of an environmental organization, you will receive, and can help disseminate, information that makes you and others more aware of specific environmental problems and things you can do to help. Also, your

membership and contribution serve to support the lobbying efforts of the organization. A lobbyist representing a million-member organization that can follow up with thousands of phone calls and letters (and, ultimately, votes) can have a powerful impact. In cases where enforcement of the existing law has been the weak link, some organizations, such as the Public Interest Research Groups, the Natural Resources Defense Council, and the Environmental Defense Fund, have been highly influential in bringing polluters or the government to court to see that the law is upheld. Again, this can be done only with the support of members.

Another form of involvement is *joining a volunteer organization*. Groups that depend on volunteer labor carry out many effective actions that care for people and the environment. Political organizations and virtually all NGOs depend on volunteers. Many helping organizations, such as those dedicated to alleviating hunger and to building homes for needy families (Figure 23–20), are adept at mobilizing volunteers to accomplish tasks that often fill in the gaps left behind by inadequate public policies. For example, you might be able to assist at a food bank or soup kitchen to help the homeless.

Finally, you may choose to devote your *career* to implementing solutions to environmental problems. Environmental careers go far beyond the traditional occupations of wildlife ranger and park manager. Many lawyers, journalists, teachers, research scientists, engineers, medical personnel, agricultural extension workers, and others are focusing their talents and training on environmental issues. There are opportunites for careers in pollution control, recycling, waste management, ecological restoration, city planning, environmental monitoring and analysis, nonchemical pest control, the production and marketing of organically grown produce, and the manufacture of solar and wind energy components. Some developers concentrate on rehabilitation and the reversal of urban blight. Some engineers are working on the development of

Figure 23–20 Habitat for Humanity volunteer. By using charitable contributions and volunteer labor, Habitat for Humanity creates quality homes—more than 800,000 built or rehabilitated by 2014 that needy families can afford to buy with no-interest mortgages, enabling the money to be recycled to build more homes. Prospective homebuyers participate in the construction and may gain valuable skills in the process.

pollution-free vehicles to help solve the photochemical smog dilemma of our cities. Indeed, it is difficult to think of a vocation that cannot be directed toward promoting solutions to environmental problems.

Human society is well into a new millennium, and it promises to be an era of rapid changes unprecedented in human history. Will we just survive, or will we thrive and become a sustainable global society? We are engaged in an environmental revolution—a major shift in our worldview and practice from seeing the natural world as resources to be exploited to seeing it as life's supporting structure that needs

our stewardship. You are living in the early stages of this revolution, and we invite you to be one of the many who will make it happen. We close with this quote:

> No one could make a greater mistake than he who did nothing because he could do only a little.
>
> *Edmund Burke, Irish orator, philosopher, and politician (1729–1797).*

CONCEPT CHECK How does the common good relate to the need for sustainability? ☑

REVISITING THE THEMES

SOUND SCIENCE

City planners have a new set of challenges as our society makes the changes that will move us toward greater sustainability. Creating sustainable communities requires the dedicated work of social scientists, natural scientists, engineers, and average citizens as they meet to provide a vision of what they would like their future to look like and then carry out that vision. Many scientific careers are great avenues for finding sustainable solutions; work in the areas of renewable energy, green buildings, water resources, sewage management, transportation, and adaptation to climate change will be particularly important in the coming decades.

SUSTAINABILITY

The story of urban sprawl is a story of poor or nonexistent public policy. Land use is traditionally under local control, and absent any public policy dealing with land use, sprawl happens. In the United States, public policy fed sprawl as the Highway Trust Fund taxed gasoline sales in order to build roads and highways. Urban blight, too, was the outcome of poor or nonexistent public policy. Cities stood by helplessly as wealthier people and businesses moved to the suburbs. In a similar way, cities in the developing world are simply watching the poor flock in.

Crafting effective policy to deal with sprawl, blight, and urban slums means recognizing why people choose to live where they do and changing the way we organize our communities. This will require changes to public policy and effective planning, but changes such as smart growth initiatives are taking root. Reining in sprawl, redesigning suburbia, and building cities that can adapt to climate change are essential moves toward greater sustainability. The most sustainable

cities will use resources efficiently and avoid having such a large resource footprint. As urbanization continues, rural areas will also need investment to prevent the poverty that can result from migration to cities.

Frequently, change can be brought about at the grassroots level, but it helps immensely to have leaders who understand best practices, as was the case in Curitiba, Brazil. City leaders also need adequate resources to carry out these practices. Sustainable solutions are encouraged by a number of philanthropic and international agencies in the United States and around the world. Cities and towns are making an effort to respond to global climate change and limited resources as they engage in community efforts to promote sustainability and encourage sustainable lifestyle changes in their people.

STEWARDSHIP

The stewardship ethic includes a concern for justice. In the United States, the movement from cities into the suburbs has left minorities trapped in the inner cities, held down by racial and economic discrimination. Equally important is the migration of rural people into the cities of the developing world, where they often live in slums and are unjustly denied ordinary social and environmental services. Both of these developments have energized many individuals and numerous NGOs to make common cause with those who are suffering injustice and to help them mobilize resources to bring about change. A similar motivation led Ryan Gravel to propose the Atlanta BeltLine. This is exactly what stewards should be doing. It is a fact that every one of us is given many opportunities to work together to create a truly sustainable society, and stewardship is the ethic that is there to guide our actions.

REVIEW QUESTIONS

1. What is the Atlanta BeltLine, and how does it propose to change urban Atlanta?

2. What is the difference between a metro region and an urban area?

3. What are three advantages of urbanization?

4. What are two things governments can do to help rural areas?

5. How did the structure of U.S. cities begin to change after World War II? What factors were responsible for the change?

6. Why are the terms *car dependent* and *urban sprawl* used to describe our current suburban lifestyle and urban layout?

7. What federal laws and policies tend to support urban sprawl?

8. What is meant by *erosion of a city's tax base*? Why does it follow from exurban migration? What are the results?

9. What are four smart-growth strategies that address urban sprawl?

10. What factors cause the growth of slums in the cities of many developing countries? What do the people who live in shantytowns need most?

11. How can big data be used to make cities more efficient?

12. What are two problems coastal megacities must solve?

13. Why is financial investment in both rural areas and cities so important in a move to sustainability?

14. What is the Sustainable Communities Program?

15. What is the C40 Cities Climate Leadership Group?

16. What are some stewardship virtues, and what role could they play in fostering sustainability?

17. What are five levels at which people can participate in working toward a sustainable future?

THINKING ENVIRONMENTALLY

1. Interview an older person (75 or older) about what his or her city or community was like before 1950 in terms of meeting people's needs for shopping, getting to school or work, and participating in recreation. Was it necessary to have a car?

2. Do a study of your region. What aspects of urban sprawl and urban blight are evident? What environmental and social problems are evident? Are they still going on? Are efforts being made to correct them?

3. Identify nongovernmental organizations in your region that are working to rein in urban sprawl or to preserve or improve a local city. What projects are under way in each of these areas? What roles are local governments playing in the process? How can you become involved?

4. Read about the C40 cities program or another program that supports sustainable cities. Are any of the cities near you part of the group? Learn about one sustainability initiative of the closest C40 city.

MAKING A DIFFERENCE

1. Analyze your environmental impacts as a consumer. What modes of transportation do you use, where do you shop, what foods do you consume or throw away, what appliances do you own, how do you heat or cool your home, and how much water do you use? Consider especially your impact on the four leading consumption-related environmental problems: air pollution, water pollution, changes in natural habitats, and global climate change.

2. Investigate land-use planning and policies in your region. Is urban sprawl or urban blight a problem? Become involved

in zoning issues, and support zoning and policies that will promote, rather than hinder, ecologically sound land use (*smart growth*).

3. Contact local conservation organizations or land trusts, and offer your help in protecting and preserving open land in your region.

4. Find out if there is a Habitat for Humanity project or group in your area; if there is, volunteer some time to support its work.

Appendix A   Answers to Concept Checks and Understanding the Data

Chapter 1

CONCEPT CHECKS

1.1 Today's environmental movement is more focused on global issues and includes climate change, while the focus in 1970 was on obvious pollution such as polluted water. The early 20th century was primarily about over-hunting and the protection of wilderness.

1.2 For the same income level, higher inequality often promotes injustice and civil unrest. People at the bottom are less likely to have their basic needs met. People at the top are more likely to waste.

1.3 You would at least want to know that an experiment was done (presumably many) with replication and control, that the researchers were not biased, that the claim made scientific sense, and that this research had been published in a peer-reviewed journal.

1.4 Stewardship of both the climate and biodiversity are jobs for the global community and require that we protect the atmosphere, species, and ecosystems they rely on.

1.5 Easier: improved communication, improved economies, ease of transportation and financial transactions, consumer interest in environmentally sound products. Harder: graying populations, urbanization, mass migration, the spread of Western culture (in some cases), emerging diseases, the spread of exotic species, pollution, and climate change.

UNDERSTANDING THE DATA
Figure 1-2
1. Economic growth has pulled many people out of poverty. More people have food, education, and water, and mortality rates are lower for younger ages than decades ago in most countries.

2. Countries with the greatest improvement include Saudi Arabia, Oman, China, Indonesia, and Nepal. Zimbabwe, Zambia, and the DR Congo are three poorly performing countries.

3. The collapse of the former Soviet Union in 1991 devastated the Russian economy, pulling down its HDI, which has since rebounded.

Chapter 2

CONCEPT CHECKS

2.1 A centrally planned economy.

2.2 An economist might raise the price of cotton or fish so that the cost of preventing damage to the ecosystem would be included. The down side of this is that the price goes up and poor people may not be able to afford something they need while rich people may be able to afford more than they need and even waste it.

2.3 Wealth inequality leaves the poorest with very little; there is a focus on cash crops for export; households grow maize on poor land; many have a diet with few vegetables and proteins.

2.4 While the Clean Air Act has other provisions that would be a market approach, this particular provision is a good example of a command and control approach.

2.5 **Cost-benefit:** The analysis compares the estimated costs of the proposed action and the main alternatives to the benefits that will be achieved in order to decide whether or not to take the action. Sometimes external costs are not accounted for, which makes the analysis inaccurate. If they are accounted for, a true cost can be determined. **Cost-effectiveness:** Here, the costs of alternative strategies for reaching the goal are analyzed, and the least costly method is adopted.

2.6 Answers may vary: purchase certified products; support policies that eliminate subsidies; use less stuff.

UNDERSTANDING THE DATA
Figure 2-7
1. Loss of natural capital has lowered bio-capacity during that time. This makes the sustainability quadrat smaller.

2. These are countries that are oil producers and use a great deal of petroleum because it is cheap.

Chapter 3

CONCEPT CHECKS

3.1 Organismal: How do emperor penguins survive diving? Population: How do emperor penguins define territories? Community: What predators have the greatest effect on emperor penguins? Answers may vary.

3.2 Nitrogen may have been limiting the growth and, once released from constraint, the grass grew until it was limited by another factor.

3.3 Dissolving salt creates ions, which separate and move around; bonds are weak. Forming water creates bonds that hold the atoms together, although the molecule is slightly ionized.

3.4 Trees usually sequester carbon, meaning they store it in their bodies, because they produce more tree material as a result of photosynthesis than herbivores eat. If the tree dies and is broken down, that stored carbon is released, but it may take a long time.

3.5 Could include burning of fossil fuels, burning forests, and use of artificial fertilizers. Causes water pollution from nitrates, increases crop yields, and causes acid rain.

UNDERSTANDING THE DATA
Figure 3-6
1. It would grow, just slightly more slowly.
2. It would die.

Chapter 4

CONCEPT CHECKS

4.1 Squirrels may be more likely to show logistic growth, to be controlled by predators and food availability, and be in the later stages of logistic growth, although they can also starve. Initially, ragweed is likely to show exponential growth because it is exploiting a new area and it reproduces rapidly. Eventually, competition will likely limit its population.

4.2 It is top-down control because a predator controls numbers of prey. It is density-dependent because the ability of the predator to find prey increases with prey abundance.

4.3 Removal of a predator might cause one of the prey species to take over, kill off other species, overgrow, and even starve. Removal of a competitor could also allow another species to grow without control.

4.4 In general you would expect more rapid change on islands than on the mainland because islands have more small habitats with greater geographic isolation.

4.5 There are many answers. Introducing species to new areas where they lack controls often increases the introduced species' population. An example is the introduction of rodents to marine islands. This also has the effect of decreasing sea bird populations. Lowering food availability will lower populations (such as a loss of habitat for elephants leading to starvation), as will habitat loss.

Sustainability: Elephants in Kruger National Park
1. Number of elephants: 1940: <500, 1960 :about 1500; 1965: about 3000. The exponential growth equation, represented by a J-shaped curve.
2. Nearly 8,000 lived in the park and around 700 were culled.

Chapter 5

5.1 A shrew eating a grasshopper is a second-order consumer, or a first-order carnivore. Earthworms are detritivores that eat dead materials from many trophic levels, but mostly from plants. However, detritivores aren't classified as herbivores, so it is hard to say what trophic classification their predator is.

5.2 Soyburgers. Soyburgers are made from plants. Because they directly produce from the Sun's energy, rather than eating something else and digesting it like cattle do, they represent less of the Sun's energy per burger; that is, it takes less to make them. To raise beef, there is energy lost as the animal eats the plants and digests and uses some. You can grow a lot more soybeans than beef for the same amount of plant productivity.

5.3 While they are not all the same biome, these three areas are all harsh environments that often have dry, well-drained soils. Prickly pear cacti cope well in dry conditions. Coastal areas may have very dry soil even if it rains regularly, if the soil is sandy and drains easily.

5.4 A treeless area that will no longer go back to forest, as the seed banks are gone and the conditions for tree seedlings to survive are not there. Scotland and Iceland are both examples.

5.5 There are many possible answers, such as: involving all stakeholders in decision making, looking at ecosystems on different scales, and properly accounting for the value of ecosystem services.

Figure 5-7
1. You would expect the biomass of the third-order consumers to be about 1/1,000 of the biomass of producers, or 1.2 kg.
2. There are so few organisms at that trophic level and they are so widely dispersed that it would be hard to get enough food.

Figure 5-13
1. Algal beds and reefs do not cover enough area to make as big of a difference as you might expect.
2. We would have very little change in net primary productivity if cultivated land moved to grassland, because they have the same productivity per area. However, because we don't usually fertilize grasslands, that productivity would have less impact on nutrient cycles than cultivated land would.

Chapter 6

6.1 Extrinsic aesthetic value and intrinsic value. That is, panda bears are valuable to you because they are pleasing and they are valuable on their own.

6.2 HIPPO: Habitat loss, Invasive species, Pollution, Population (human) growth, and Overexploitation (such as over-fishing). The passenger pigeon went extinct primarily from overexploitation and habitat loss.

6.3 The Lacey Act is a U.S. federal law, the oldest of the three, and limits interstate commerce in protected species. The ESA is the most comprehensive U.S. law for the protection of species and protects both endangered and threatened species.

6.4 CITES specifically prohibits international trade in endangered species, while the Convention on Biological Diversity is a larger effort and includes issues such as the protection of indigenous rights to genetic resources.

Figure 6-6
1. Birds.
2. Gray, because there are many insect species, and many for which we have no clear estimates.

Chapter 7

7.1 Answers could include: Forests—carbon, regulation of climate, disease and flood, provision of food, drinking water, fuel wood, and recreation. Urban parks—air quality, climate, culture, education, recreation, and water regulation. Both forests and urban parks regulate water in some way, and they are important in recreation. Forests affect air quality. Urban parks may provide food, although that is not listed.

7.2 The common resource is clean air. Because everyone is contributing to the loss of clean air, being the one business that does not do so will make you poorer but won't save the air. Consequently, most people will choose to continue whatever air polluting activities they are doing. Group management of air pollution will solve this.

7.3 Climate change, pollution, and invasive species are three.

7.4 Over-fishing, destruction of coastal habitats, acidification, and pollution are several of them.

7.5 Advantages: You have some type of control and can make a management plan and, presumably, the ecosystem will improve. Disadvantages: People cannot use the resource, and you need active management and monitoring.

Figure 7-5
Increase by 40,000, assuming the fish hovered at carrying capacity and is now well below at 10,000. Fifty thousand is the level of highest productivity.

Chapter 8

8.1 Humans have long life spans, high parental care, and few offspring like K-strategists, but population growth (globally) that looks like that of r-strategists.

8.2 The effect of technologies is not all the same, a fact reflected in ImPACT but not IPAT indices. A solar panel would lower the ImPACT and not affect IPAT, or might increase it.

8.3 The resource use of the 10 people would be highest in the United States, much less in China, and much, much less in Burkina Faso, although in the United States they might not pollute air and water as much.

8.4 The population pyramid becomes more of a column, and the dependency ratio and population momentum increase.

UNDERSTANDING THE DATA

Figure 8-13

1. Tokyo, Delhi, and Shanghai are likely to be largest. Tokyo is also one of the most stable, as is Osaka. New York-Newark will grow but only slightly.

2. Tokyo is likely to shrink slightly, but none of the others are, because of a combination of continuing urbanization, even where population growth is low (Japan) and rapid population growth (other places).

Chapter 9

CONCEPT CHECKS

9.1 China represents direct population control by rules, and Thailand represents direct population control by voluntary efforts as well as the secondary population effects of economic growth. India represents mainly the effects of economic growth, although it has had specific population control efforts as well.

9.2 The Sustainable Development Goals have environmental sustainability as a pillar throughout all of them, rather than only in one MDG (#8).

9.3 Globalization means goods and services travel farther, increasing the energy cost of transportation and its environmental effects. It also means jobs may move to countries where wages are lower. Those jobs may pay less than a living wage even in a poor country. Wealthy consumers may not know the realities of those producing the goods they purchase. Globalization can undermine environmental laws and laws designed to protect indigenous rights. Sometimes globalization leads to dumping cheaper goods and the undermining of weak economies.

UNDERSTANDING THE DATA

Figure 9-8

In 2070, the dependency ratio of South Asia will have risen, and but probably below where East Asia and the Pacific are expected to be in 2050 (60–65%).

Chapter 10

CONCEPT CHECKS

10.1 There is less freshwater and more saltwater as freshwater stored in ice melts into the ocean.

10.2 More water vapor forms at the equator, where solar radiation is more intense and rising air carries the evaporated water higher into the atmosphere. This makes the ground drier.

10.3 Dams slow water movement to the ocean, increase local evaporation, cause earthquakes, and disrupt the movement of fish. However, they provide irrigation water, flood control, and hydroelectric power.

10.4 Individuals can conserve water by controlling water use in the home, not watering outside, using rain collection, and purchasing efficient appliances. The government can conserve water by using water bans, regulating water removal from rivers, regulating large withdrawals such as by industry, stopping infrastructure leaks, and forming a coordinated plan.

UNDERSTANDING THE DATA

Figure 10-18

You can tell that the North African and Upper Ganges aquifers are the most overdrawn in two ways. First, the red color denotes a crisis of overdrawing. Second, the ratio of the aquifer size to the footprint area needed to recharge those aquifers is higher than for the others.

Chapter 11

CONCEPT CHECKS

11.1 The soil would hold a medium amount of water (because of the mix of soil particles and because of the leaf litter), water would infiltrate fairly quickly (because the particles and soil fauna make more air pockets than in silt and clay, but water would infiltrate less quickly than in sand) and evaporate slowly because of shading and leaf litter.

11.2 In a desert, walking off of the path breaks the cryptogamic crust and opens the soil to wind, and possibly water, erosion. In wet mountain regions, it causes compaction of soil, kills vegetation, and allows water erosion.

11.3 No-till farming doesn't break up soil structure, so there is less erosion. While contour plowing breaks up soil, it cuts down on erosion by keeping soil in rows around a hill rather than going downhill.

UNDERSTANDING THE DATA

Figure 11-4

1. Close to 20%, if you look at the corner of sandy clay to the right. The ratio is about 43% sand, 37% clay, and 20% sand.

2. Clay.

Chapter 12

CONCEPT CHECKS

12.1 Fertilizer, high-yield varieties, chemical pesticides, irrigation, and increased mechanization/machinery.

12.2 With GMO technology, genes are taken from unrelated species and put into one crop; other plant breeding techniques relied on the selection of genes already in the organisms, or bred in through crossbreeding with related cultivars.

12.3 They wanted to drive the price of rice up, but it stayed low as others increased production. Tariffs protect the local economy by making the price of imports higher. They may protect jobs, but they make food prices higher. Subsidies of agriculture can make the cost of food lower; they can also undermine the economy and hurt jobs if food is "dumped."

12.4 Answers might include: Effective food aid would be timely, regionally purchased, would not undermine the local economy, would include foods people need, and would be available to the neediest people.

12.5 While efforts to increase efficiency and efforts to produce more food will both be needed, efficiency efforts allow us to feed more people without using more resources such as land or irrigation water.

UNDERSTANDING THE DATA

Sound Science: The Meat Footprint

1. Ratio of N = approximately 176:25, or about 7:1.

2. Pastureland is greater, although it is hard to tell because the graph has a break (the ratio is about 132:15).

3. Yes, pork requires more water although they are close.

4. The difference is too small to be significant.

Chapter 13

CONCEPT CHECKS

13.1 It allows natural predators and competitors to thrive. It is less likely to involve monoculture, and that means crops are less vulnerable to pests.

13.2 Resistance develops when some but not all of a pest population are killed by the pesticides and the pests remaining have a genetically coded higher likelihood of survival with pesticide exposure. Subsequent generations have genetic variations that are increasingly better for surviving the pesticide.

13.3 Answers could include: disrupting pheromone communication, disrupting developmental hormones, and spraying a natural toxin such as the Bt chemical.

13.4 Answers could include at least: less soil pollution, less nitrogen pollution, lower fossil fuel use, less chemical pesticide exposure for humans. Organic farming could increase the problems associated with getting enough yield per acre.

13.5 APHIS, part of the USDA, tracks and prevents pest spread. The EPA monitors and registers pesticides and tries to prevent pollution from pesticides.

UNDERSTANDING THE DATA

Figure 13-9

1. Your fish will likely have 0.17×10^7 parts per billion (or 1,700,000).

2. If you turn that into parts per million, you would get 1,700 ppm, much higher than allowable.

Chapter 14

CONCEPT CHECKS

14.1 More fuels can be used to power steam turbine engines than either gas or water turbines. These steam turbine sources include at least coal, oil, and nuclear fuel. Water turbines are limited in location, and gas turbines require natural gas.

14.2 Because tar sands are expensive to excavate, it is only economically feasible to excavate them if oil prices are high. Thus a drop in oil price will lower tar sands excavation.

14.3 Answers may vary but might include: Benefits: cleaner burning, more efficient than coal and oil, easier to extract than tar sands, shale, and coal. Concerns include: fracking causes a wide array of problems, gas is still a fossil fuel, there are dangers of transport.

14.4 Advantages: Coal is relatively cheap, abundant, and can be shipped as a solid. It is less explosive and prone to spills and fires than are gas and oil. Disadvantages: Coal extraction (especially surface mining) is very destructive, coal ash is hard to dispose of, and coal is polluting (including GHG emissions) when it burns.

14.5 Supply-side policies to reduce dependence on foreign oil focus on increasing our own fossil fuel production. They do not focus on reducing demand or leaving fossil fuels altogether.

UNDERSTANDING THE DATA

Figure 14-8

1. 38.2 quads are produced. Wasted is: 29.73/38.2 = 77.8%. [25.8 quads is not used immediately, and if the electricity that goes to the four use sectors is wasted at the same rate as the energy in each sector is wasted generally, another $4.75(3.98/11.4) + (4.57(3.01/8.59) + (3.26(4.94/24.7) + 0.026(21.3/27.0) = 3.93$ quads]

2. Transportation, because 75% of the energy in transportation is lost as rejected (wasted) heat, a higher percent than for other sectors.

3. The amount of electricity going to transportation, now at 0.026 quads.

Chapter 15

CONCEPT CHECKS

15.1 It takes a long time to build a new reactor and bring it online. Meanwhile, older reactors age out or become uneconomic.

15.2 In general, coal, because of particles and acid-producing ions that are released even under normal operation. Nuclear reactors could spew radioactivity into the air under catastrophic circumstances, although they do not normally do so.

15.3 Answers may vary but could include: Similarities: Both involved evacuation, nuclear reactor meltdown, loss of trust in nuclear power. Differences: Chernobyl harmed more people directly; Fukushima was a result of a large tsunami while Chernobyl originated from failure at the facility; the types of reactors were somewhat different; Fukushima sent more radioactivity into the ocean, Chernobyl to land.

15.4 Obtaining the high temperature and energy requirement to start the fusion reaction and controlling the power of the energy output are both significant hurdles to fusion use.

15.5 Increase: Concern about climate change; desire for energy independence; possibly others. Decrease: Concern about power plant safety; lack of radioactive waste disposal sites; possibly others.

UNDERSTANDING THE DATA

Table 15-1

a. One CT scan.

b. The radon in an average house.

Chapter 16

CONCEPT CHECKS

16.1 No, we still need a zero-carbon future because of the problems of carbon dioxide in the atmosphere.

16.2 A PV panel can be almost any size, but you need a critical level of the Sun's energy focused on a power tower to melt the salt and run a turbine engine. For small-scale installations, PV panels would be appropriate.

16.3 Wind: clean, no burning, but intermittent and hard to store the energy. Biomass: more available and can be stored, but still involves burning. Some biomass fuels are made from food plants.

16.4 Advantages include: clean at point of use, can store well. Disadvantages include: only as green as the primary fuel source, technology not fully developed, hard to refuel.

16.5 Probably not; there are very few places where this is effective and the corrosion and other issues with marine environments make it costly.

16.6 A U.S. federal program that establishes a minimum volume of renewable fuels. It promotes renewables in transportation.

UNDERSTANDING THE DATA

Figure 16-10

1. Probably China.

2. Probably China, because China had increased dramatically in 2012–2013, while Europe's capacity flattened.

Chapter 17

17.1 A hazard is something that can harm you. Risk is the degree to which it is likely to do so. Risk, therefore, is a function of how hazardous something is and how likely you are to be exposed.

17.2 Smoking, working in mines, working with other industrial chemicals, and cooking indoors all expose people to chemical hazards. Living in the tropics, eating food that is not well cooked, exposure to animals or sick people, and drinking contaminated water expose people to biological contaminants.

17.3 Answers might include some of the following: Greater: The hazard is involuntary; it occurs with vulnerable people; it seems unfair; it is unfamiliar or is heavily discussed in the media. Less: The hazard is familiar and does not hurt many people at a time; the media do not cover it; they have chosen the risk.

UNDERSTANDING THE DATA

Figure 17-1
1. The United States.
2. Russia and India are similar, but Russian men die a lot sooner than Russian women, as compared to deaths of men and women in India.
3. The gray circles are all from Africa, are similar to each other, and have lower life expectancies than most other countries.

Sound Science: Water Pollution Drives Malnutrition in India
Bangladesh has low open defecation density, but stunting is higher than India and Haiti so the relationship is not clear. Haiti has lower rates of both factors than India. India is clearly an outlier in density of outdoor defecation and has higher rates of stunting than many countries.

Chapter 18

CONCEPT CHECKS

18.1 Cooling and becoming more saline both make the water more dense, which drives it downward.

18.2 Agriculture, industries, forest fires and other burning are some of the major sources of greenhouse gases. Land use changes, release of aerosols, production, and destruction of ozone are several other factors.

18.3 The amount the climate will still change even if we leveled greenhouse gases today.

18.4 Distrust of science, political divide, lack of willingness to change, the economy, campaign to confuse.

18.5 Answers will vary and could include: Mitigation—lowering emissions through efficiency and conservation, policies such as cap and trade, or renewable energy production. Adaptation—moving back from coasts, storing water, preparing for floods and drought. Geo-engineering—reflecting sunlight more, carbon sequestration.

UNDERSTANDING THE DATA

Figure 18-15
1. CO_2 has the greatest warming effect. Aerosols' effect on cloud albedo have the greatest cooling effect.
2. Tropospheric ozone is a pollutant that harms cells. In the stratosphere, ozone is a necessary part of a layer that protects us from harmful UV radiation.

Chapter 19

CONCEPT CHECKS

19.1 Industrial smog comes directly from fossil fuel burning. Photochemical smog is chemically altered and forms in the presence of sunlight. Atmospheric brown smog is a large-scale phenomenon that is a mix of sources, and may both cool and warm the earth.

19.2 They arise from chemical interactions between pollutants in the presence of sunlight. They are particularly worse in heat and light.

19.3 Less acidity would occur in limestone areas because the bedrock buffers the acid rain.

19.4 The tremendous drop in criteria pollutants and change in pH of water bodies suggest it.

19.5 Tropospheric ozone is a pollutant that kills plants and harms lungs. In the stratosphere, ozone protects from UV rays.

UNDERSTANDING THE DATA

Figure 19-10
1. They increased by 14%.
2. Of those, CO_2 emissions increased the most. Aggregate emissions decreased the most. (Energy consumption per population actually declined because while energy consumption rose, the rise was less than the increase in population over the same time.)

Figure 19-14
Ohio, possibly Pennsylvania and West Virginia.

Figure 19-27
1. Infrared.
2. There is a range: Largest: 100 million times; Smallest: close to 1 million times longer.

Chapter 20

CONCEPT CHECKS

20.1 The loss of oxygen, sediment in the water column, excess nutrients, and the growth of pathogens would all be problems caused by sewage in a waterway. Contamination by other chemicals in sewage would also be one.

20.2 Even when broken down, the nutrients in organic matter can support eutrophication and cause oxygen loss.

20.3 Table 20-2 summarizes quite a number of such strategies, such as protecting water bodies from run-off, limiting disturbed areas, slowing and trapping run-off, and maintaining vegetation.

20.4 Answers may vary: More people have sanitation; erosion is decreased; major rivers are cleaner; and fish have returned to rivers where they were lost.

UNDERSTANDING THE DATA

Figure 20-7
In the decline stage, you will see no fish, but only blackfly larvae and sludge worms. In the recovery phase, the same DO level is associated with sludge worms, filamentous algae, and also fish such as bullheads, carp, and suckers.

Chapter 21

CONCEPT CHECKS

21.1 High-income countries produce more waste, containing more paper but less organic content.

21.2 Retrieving the energy in the trash rather than wasting it.

21.3 A bag bill lowers the use of the bags, thus it is reduction. A bottle bill charges a deposit. It might reduce use, but it more likely increases collection and recycling.

21.4 Advantages: People have an incentive to produce less trash and recycle more. Disadvantages: People have incentive to cheat and dump trash in public areas or waste facilities.

Figure 21-2

1. High-income countries; lower-middle income countries.

2. The lower-middle income countries will have the greatest total group increase, but the lowest income group will have the greatest percent increase.

3. The biggest percentage change in per capita waste is in the low-income group, a 56% increase.

Chapter 22

CONCEPT CHECKS

22.1 Bioaccumulation of methyl mercury in water moved up the food chain. Weather and physics drive pollutants pole-ward, and bioaccumulation moves toxins up the food chain.

22.2 Advantages are it is far away and unlikely to be disturbed, disadvantages are potential ground water pollution, removal of water from use, cost, and causing earthquakes. Disadvantages are that landfills are more likely to leach or be breached but cheaper and less likely to cause earthquakes. Both can contaminate water.

22.3 It is important because otherwise we have no mechanism to clean up huge numbers of polluted sites. It is difficult to fund because paying for pollution is unpleasant since no one wants taxes; some polluting companies have gone bankrupt; and poor people do not have political clout.

22.4 OSHA is focused on workers' rights and the protection of individuals. RCRA is broader and prevents the poor disposal of waste generally.

22.5 The people who live there are not the people making the money off of the products, but they suffer the consequences of mining and steel working. Others purchase cheap products that contain rare earths, although they do not experience the environmental toll.

UNDERSTANDING THE DATA

Figure 22-4

The per capita production of toxics would have been reduced by even more than 7% because population grew during that time.

Chapter 23

CONCEPT CHECKS

23.1 Rural areas become poorer as workers leave and few jobs remain.

23.2 The building of highways allowed people to leave cities and commute from suburbs over long distances.

23.3 IBM's smart city group, the C40 network, and resources from the UN, including the Sustainable Cities Program.

23.4 The common good means consideration of others, including future generations. If we only consider our own good, we often make unsustainable choices.

UNDERSTANDING THE DATA

Figure 23-1

1. North America; Africa.

2. Sometime around the year 2035.

Appendix B Environmental Organizations

This appendix presents a list of selected nonprofit, nongovernmental organizations that are active in environmental matters. Listed are national organizations as well as some smaller, specialized ones. These organizations offer a variety of fact sheets, brochures, newsletters, publications, educational materials, and annual reports, most of which are available on the Internet. Some organizations have internship positions available for those wishing to do work for an environmental group. Web addresses are provided in the list. Most of the Web sites also provide connections via Facebook, Twitter, RSS feeds, Flickr, and YouTube. Also, a useful directory of many more environmental organizations can be accessed at www.earthdirectory.net

American Farmland Trust. Focuses on the preservation of American farmland. 1200 18th St. NW, Suite 800, Washington, DC 20036. www.farmland.org

American Lung Association. Research, education, legislation, lobbying, advocacy: indoor and outdoor air pollution effects, smoking. 1301 Pennsylvania Ave. NW, Suite 800, Washington, DC 20004. www.lung.org

American Rivers, Inc. Mission is "to preserve and restore America's rivers' systems and to foster a river stewardship ethic." Education, litigation, lobbying: wild and scenic rivers, hydropower relicensing. Information on specific rivers available. 1101 14th St. NW, Suite 1400, Washington, DC 20005. www.americanrivers.org

Bread for the World. National advocacy group that lobbies for legislation dealing with hunger. 425 3rd St. SW, Suite 1200, Washington, DC 20024. www.bread.org

Center for Climate and Energy Solutions. The successor to the Pew Center on Global Climate Change, this organization is an independent, nonpartisan, nonprofit organization working to advance strong policy and action to address the twin challenges of energy and climate change. 2101 Wilson Blvd., Suite 550, Arlington, VA 22201. www.c2es.org

Center for Science in the Public Interest. Research and education: alcohol policies, food safety, health, nutrition, organic agriculture. 1220 L St. NW, Suite 300, Washington, DC 20005. www.cspinet.org

Chesapeake Bay Foundation. Research, education, and litigation: environmental defense and management of the Chesapeake Bay and surrounding area. Philip Merrill Environmental Center, 6 Herndon Ave., Annapolis, MD 21403. www.cbf.org

Clean Water Action Project. Lobbying, education, research on water quality. 1010 Vermont Ave. NW, Suite 400 Washington, DC 20005. www.cleanwateraction.org

Climate and Energy. Research, education, and advocacy: affordable, clean and sustainable energy (a function of Public Citizen). 215 Pennsylvania Ave. SE, Washington, DC 20003. www.citizen.org/cmep

Common Cause. Lobbying: government reform, energy reorganization, clean air. 1133 19th Street NW, Washington, DC 20036. www.commoncause.org

Congress Watch. Lobbying: consumer health and safety, pesticides (a function of Public Citizen). 215 Pennsylvania Ave. SE, Washington, DC 20003. www.citizen.org/congress

Conservation International. Education and research to preserve and promote awareness about the world's most endangered biodiversity. 2011 Crystal Dr., Suite 500, Arlington, VA 22202. www.conservation.org

Coral Reef Alliance. Works with the diving community and others to promote coral reef conservation around the world. 351 California St., Suite 650, San Francisco, CA 94104. www.coral.org

Defenders of Wildlife. Research, education, and lobbying: all native animals and plants, especially endangered species. 1130 17th St. NW, Washington, DC 20036. www.defenders.org

Ducks Unlimited, Inc. Hunters working to fulfill the annual life-cycle needs of North American waterfowl by protecting, enhancing, restoring, and managing important wetlands and associated uplands. One Waterfowl Way, Memphis, TN 38120. www.ducks.org

Earthwatch Institute. Environmental research is encouraged by organizing teams to make expeditions to various locations all over the world. Team members contribute to the expenses and spend from several weeks to several months on the site. 114 Western Ave., Boston, MA 02134. www.earthwatch.org

Environmental Defense Fund. Research, litigation, and lobbying: cosmetics safety, drinking water, energy, transportation, pesticides, wildlife, air pollution, cancer prevention, radiation. 257 Park Ave. South, New York, NY 10010. www.edf.org

Environmental Law Institute. Training, educational workshops, and seminars for environmental professionals, lawyers, and judges on institutional and legal issues affecting the environment. 200 L St. NW, Suite 620, Washington, DC 20036. www.eli.org

Forest Stewardship Council. Promotes sustainable forest management and a system of certification of sustainable forest products. 212 Third Ave. North, Suite 504, Minneapolis, MN 55401. www.fscus.org

Freedom from Hunger. Develops programs for the elimination of hunger worldwide. 1644 DaVinci Court, Davis, CA 95618. www.freedomfromhunger.org

Friends of the Earth. Research, lobbying: all aspects of energy development, preservation, restoration, and rational use of the Earth. 1100 15th St. NW, 11th Floor, Washington, DC 20005. www.foe.org

Greenbelt Movement. Environmental organization that empowers communities, especially women, to improve livelihoods by conserving the environment, especially by planting trees in Africa. Adams Arcade, Kilimani Lane off Elgeyo Marakwet Rd. PO Box 67545-00200, Nairobi, Kenya. www.greenbeltmovement.org

Greenpeace,USA, Inc. An international environmental organization dedicated to protecting the planet through nonviolent, direct action and to providing public education, scientific research, and legislative lobbying. 702 H St. NW, Washington, DC 20001. www.greenpeace.org

Growing Power. An urban farm organization begun in Milwaukee, WI, that grows food and provides youth employment and education in sustainable urban farming. 5500 W.

Silver Spring Drive, Milwaukee, Wisconsin 53218. www .growingpower.org/about_us.htm

Habitat for Humanity International. Fosters homebuilding and ownership for the poor. Education on housing issues. 121 Habitat St., Americus, GA 31709-3498. www.habitat.org

Heifer Project International. Works throughout the developing world to bring domestic animals to poor families. 1 World Ave., Little Rock, AR 72203. www.heifer.org

Institute for Local Self-Reliance. Research and education: appropriate technology for community development. 2001 S St. NW, Suite 570, Washington, DC 20009. www.ilsr.org

International Food Policy Research Institute. Mission is to identify and analyze policies for sustainably ending hunger and poverty. 2033 K St. NW, Washington, DC 20006. www.ifpri.org

Land Trust Alliance. Works with more than 1,700 local and regional land trusts to enhance their work. 1660 L St. NW, Suite 1100, Washington, DC 20036. www.landtrustalliance.org

League of Conservation Voters. Political arm of the environmental community. Works to elect candidates to the U.S. House and Senate who will vote to protect the nation's environment, and holds them accountable by publishing the National Environmental Scorecard each year, available from its Web site. 1920 L St. NW, Suite 800, Washington, DC 20036. www.lcv.org

League of Women Voters of the United States. Education and lobbying, general environmental issues. Publications on groundwater, agriculture, and farm policy, with additional topics available. 1730 M St. NW, Suite 1000, Washington, DC 20036. www.lwv.org

Manzanar Project. Provides low-tech solutions to hunger and poverty. Funding for women to plant mangroves along the Eritrean coast, which improves fisheries, food for domestic animals, and family income. PO Box 98, Gloucester, MA 01931. www.themanzanarproject.com

Millennium Ecosystem Assessment (MA). An international work program designed to meet the needs of decision makers and the public for scientific information concerning the consequences of ecosystem change for human well-being and options for responding to those changes. The MA completed its numerous reports in 2005, many of which are referenced in this edition of the text. www.millenniumassessment.org

Mission Blue. Promotes marine protected areas. www.mission-blue.org

National Audubon Society. Research, lobbying, education, litigation, and citizen action: broad-based environmental issues. 225 Varick St., New York, NY 10014. www.audubon.org

National Council for Science and the Environment. Mission is to improve the scientific basis for making decisions about environmental issues. One important service provides access to the Congressional Research Service (briefings on environmental issues and the legislation addressing them). 1101 17th Street NW, Suite 250, Washington, DC 20036. http://ncseonline.org

National Park Foundation. Partners with the National Park Service to encourage education, land acquisition, management of endowments, grant making. 1201 Eye St. NW, Suite 550B, Washington, DC 20005. www.nationalparks.org

National Parks Conservation Association. A private organization with a mission to protect and enhance America's National Park System for present and future generations. 777 6th St. NW, Suite 700, Washington, DC 20001-3723. www.npca.org

National Religious Partnership for the Environment. An association of independent faith groups promoting stewardship of Creation in the context of different faiths. 110 Maryland Ave. NE, Suite 108, Washington, DC 20002. www.nrpe.org

National Wildlife Federation. Research, education, lobbying: general environmental quality, wilderness, and wildlife. 11100 Wildlife Center Dr., Reston, VA 20190. www.nwf.org

Natural Resources Defense Council. Research and litigation: water and air quality, land use, energy, pesticides, toxic waste. 40 W. 20th St., New York, NY 10011. www.nrdc.org

Nature Conservancy. A land conservation agency working around the world to protect ecologically important lands and waters for nature and people. 4525 N. Fairfax Dr., Suite 100, Arlington, VA 22203. www.nature.org

Oxfam America. Funding agency for projects to benefit the "poorest of the poor" in South America, Africa, India, Central America, the Caribbean, and the Philippines. Provides whatever resources are needed. 226 Causeway St., 5th Floor, Boston, MA 02114. www.oxfamamerica.org

The Population Institute. Education, research, and speaking engagements: population control. 107 2nd St. NE, Washington, DC 20002. www.populationinstitute.org

Population Reference Bureau, Inc. Organization engaged in collection and dissemination of objective population information. Excellent publications. 1875 Connecticut Ave. NW, Suite 520, Washington, DC 20009. www.prb.org

Public Employees for Environmental Responsibility. A service organization that helps public employees work as "anonymous activists," enabling them to inform the public and their agencies about environmental abuses. 2000 P St. NW, Suite 240, Washington, DC 20036. www.peer.org

Rachel Carson Council, Inc. Publication and distribution of information on pesticides and toxic substances, educational conferences, and seminars. PO Box 10779, Silver Spring, MD 20914. www.rachelcarsoncouncil.org

Rain Forest Action Network. Information and educational resources: world's rain forests. 221 Pine St., 5th Floor, San Francisco, CA 94104. www.ran.org

Rainforest Alliance. Education, medicinal plants project, timber project to certify "smart wood," and news bureau. 665 Broadway, Suite 500, New York, NY 10012. www.rainforest -alliance.org

Resources for the Future. Research and education think tank: connects economics and conservation of natural resources, environmental quality. Publishes *Resources*. 1616 P St. NW, Washington, DC 20036. www.rff.org

Shack/Slum Dwellers International. Promoting awareness of and support for the urban poor in developing countries; 14 federations and 34 country affiliates, head office in 1st Floor Campground Centre, Cnr Raapenberg and Surrey Rd. Mowbray, Cape Town 7700, S.A. www.sdinet.org

Sierra Club. Education and lobbying: broad-based environmental issues. 85 Second St., 2nd Floor, San Francisco, CA 94105. www.sierraclub.org

Sustainable Goals. The United Nations site providing information on and monitoring progress toward achieving the Sustainable Development Goals. www.un.org/sustainabledevelopment/

Trust for Public Land. Works with citizen groups and government agencies to acquire and preserve open space. 101 Montgomery St., Suite 900, San Francisco, CA 94104. www.tpl.org

Union of Concerned Scientists. Connects citizens and scientists; works through programs and lobbying to inform the public and policy makers on science-related issues. 2 Brattle Square, Cambridge, MA 02238. www.ucsusa.org

U.S. Public Interest Research Group (PIRG). A federation of state PIRGs focusing on research, education, and lobbying: alternative energy, consumer protection, utilities regulation, public interest. 44 Winter St., 4th Floor, Boston, MA 02108. www.pirg.org

Water Environment Federation. Research, education, and lobbying to preserve the global water environment; provides technical education and training for water quality professionals. 601 Wythe St., Alexandria, VA 22314. www.wef.org

Wilderness Society. Research, education, and lobbying: wilderness, public lands; mission is to protect wilderness and inspire Americans to care for our wild places. 1615 M St. NW, Washington, DC 20036. wilderness.org

World Resources Institute. Global think tank that conducts research and publishes reports for educators, policy makers, and organizations on environmental issues. 10 G St. NE, Suite 800, Washington, DC 20002. www.wri.org

World Stove. Non-profit designed to promote small businesses with stoves designed to use local waste fuels, combined with microfinance, sanitation and soil erosion efforts. www.worldstove.com

World Vision. Christian humanitarian organization dedicated to working with children, families, and their communities worldwide to help them reach their full potential by tackling the causes of poverty and injustice. 34834 Weyerhaeuser Way S, Federal Way, WA 98001. www.worldvision.org

Worldwatch Institute. Research and education: energy, food, population, health, women's issues, technology, the environment. Publications: *State of the World, Vital Signs,* Worldwatch papers and reports. 1776 Massachusetts Ave. NW, Washington, DC 20036. www.worldwatch.org

World Wildlife Fund. Preservation of wildlife habitats and protection of endangered species. 1250 24th St. NW, PO Box 97180, Washington, DC 20037. www.worldwildlife.org

Appendix C   Units of Measure

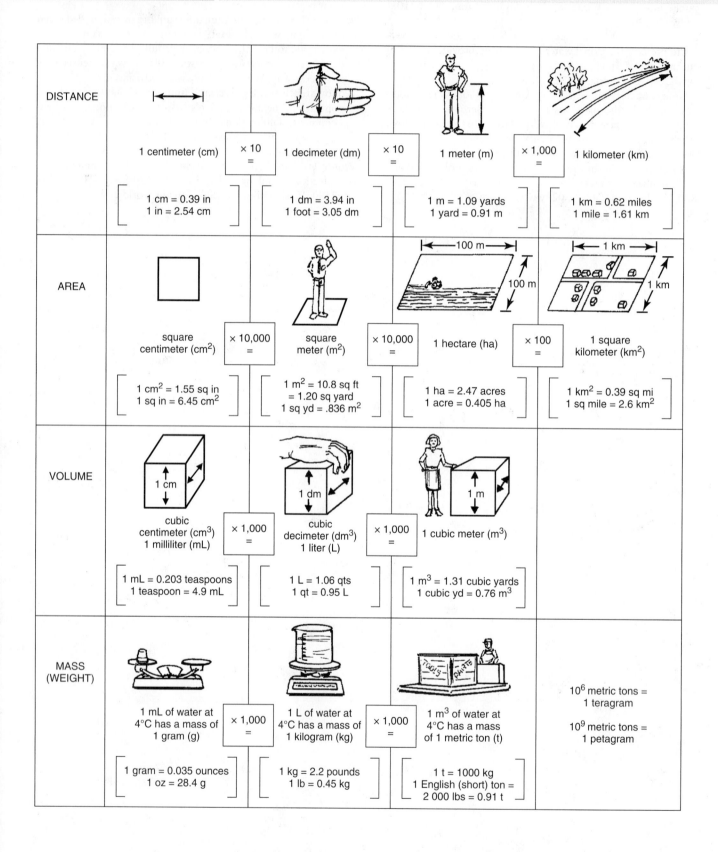

DISTANCE

1 centimeter (cm) × 10 = 1 decimeter (dm) × 10 = 1 meter (m) × 1,000 = 1 kilometer (km)

- 1 cm = 0.39 in
  1 in = 2.54 cm
- 1 dm = 3.94 in
  1 foot = 3.05 dm
- 1 m = 1.09 yards
  1 yard = 0.91 m
- 1 km = 0.62 miles
  1 mile = 1.61 km

AREA

square centimeter (cm^2) × 10,000 = square meter (m^2) × 10,000 = 1 hectare (ha) × 100 = 1 square kilometer (km^2)

- 1 cm^2 = 1.55 sq in
  1 sq in = 6.45 cm^2
- 1 m^2 = 10.8 sq ft
  = 1.20 sq yard
  1 sq yd = .836 m^2
- 1 ha = 2.47 acres
  1 acre = 0.405 ha
- 1 km^2 = 0.39 sq mi
  1 sq mile = 2.6 km^2

VOLUME

cubic centimeter (cm^3)
1 milliliter (mL) × 1,000 = cubic decimeter (dm^3)
1 liter (L) × 1,000 = 1 cubic meter (m^3)

- 1 mL = 0.203 teaspoons
  1 teaspoon = 4.9 mL
- 1 L = 1.06 qts
  1 qt = 0.95 L
- 1 m^3 = 1.31 cubic yards
  1 cubic yd = 0.76 m^3

MASS (WEIGHT)

1 mL of water at 4°C has a mass of 1 gram (g) × 1,000 = 1 L of water at 4°C has a mass of 1 kilogram (kg) × 1,000 = 1 m^3 of water at 4°C has a mass of 1 metric ton (t)

10^6 metric tons = 1 teragram

10^9 metric tons = 1 petagram

- 1 gram = 0.035 ounces
  1 oz = 28.4 g
- 1 kg = 2.2 pounds
  1 lb = 0.45 kg
- 1 t = 1000 kg
  1 English (short) ton = 2 000 lbs = 0.91 t

Energy Units and Equivalents

1 Calorie, food calorie, or kilocalorie – The amount of heat required to raise the temperature of one kilogram of water one degree Celsius (1.8°F)

1 BTU (British Thermal Unit) – The amount of heat required to raise the temperature of one pound of water one degree Fahrenheit

1 joule – a force of one newton applied over a distance of one meter (a newton is the force needed to produce an acceleration of 1 m per sec to a mass of one kg)

1 Calorie = 3.968 BTUs = 4,186 joules

1 BTU = 0.252 Calories = 1,055 joules

1 Therm = 100,000 BTUs

1 Quad = 1 quadrillion BTUs

1 watt = standard of electrical power

1 watt = 1 joule per second

1 watt-hour (wh) = 1 watt for 1 hr. = 3.413 BTUs

1 kilowatt (kw) = 1000 watts

1 kilowatt-hour (kwh) = 1 kilowatt for 1 hr = 3,413 BTUs

1 megawatt (Mw) = 1,000,000 watts

1 megawatt-hour (Mwh) = 1 Mw for 1 hr. = 34.13 therms

1 gigawatt (Gw) = 1,000,000,000 watts or 1,000 megawatts

1 gigawatt-hour (Gwh) = 1 Gw for 1 hr = 34,130 therms

1 horsepower = .7457 kilowatts

1 horsepower-hour = 2545 BTUs

1 cubic foot of natural gas (methane) at atmospheric pressure = 1031 BTUs

1 gallon gasoline = 125,000 BTUs

1 gallon No. 2 fuel oil = 140,000 BTUs

1 short ton coal = 25,000,000 BTUs

1 barrel = 42 gallons

Appendix D Some Basic Chemical Concepts

Atoms, Elements, and Compounds

All matter, whether gas, liquid, or solid; living or nonliving; or organic or inorganic, is composed of fundamental units called **atoms**. Atoms are extremely tiny. If all the world's people, 7 billion of us, were reduced to the size of atoms, there would be room for all of us to dance on the head of a pin. In fact, we would only occupy a tiny fraction (about 1/10,000) of the pin's head. Given the incredibly tiny size of atoms, even the smallest particle that can be seen with the naked eye consists of billions of atoms.

The atoms making up a substance may be all of one kind, or they may be of two or more kinds. If the atoms are all of one kind, the substance is called an **element**. If the atoms are of two or more kinds bonded together, the substance is called a **compound**.

Through countless experiments, chemists have ascertained that there are only 94 distinct kinds of atoms that occur in nature. (An additional 21 have been synthesized.) These natural, and most of the synthetic, elements are listed in Table D–1, together with their chemical symbols. By scanning the table, you can see that a number of familiar substances—such as aluminum, chlorine, carbon, oxygen, and iron—are elements; that is, they are a single, distinct kind of atom. However, most of the substances with which we interact in everyday life—such as water, stone, wood, protein, and sugar—are not on the list. Their absence from the list is indicative that they are not elements; rather, they are compounds, which means that they are actually composed of two or more different kinds of atoms bonded together.

Table D–1 The Elements

Element	Symbol	Atomic Number	Atomic Weight	Element	Symbol	Atomic Number	Atomic Weight
Actinium	Ac	89	(227)	Chlorine	Cl	17	35.5
Aluminum	Al	13	27.0	Chromium	Cr	24	52.0
*Americium	Am	95	(243)	Cobalt	Co	27	58.9
Antimony	Sb	51	121.8	Copper	Cu	29	63.5
Argon	Ar	18	39.9	*Curium	Cm	96	(245)
Arsenic	As	33	74.9	*Dubnium	Db	105	(262)
Astatine	At	85	(210)	Dysprosium	Dy	66	162.5
Barium	Ba	56	137.3	*Einsteinium	Es	99	(254)
*Berkelium	Bk	97	(245)	Erbium	Er	68	167.3
Beryllium	Be	4	9.01	Europium	Eu	63	152.0
Bismuth	Bi	83	209.0	*Fermium	Fm	100	(254)
*Bohrium	Bh	107	(262)	Fluorine	F	9	19.0
Boron	B	5	10.8	Francium	Fr	87	(223)
Bromine	Br	35	79.9	Gadolinium	Gd	64	157.3
Cadmium	Cd	48	112.4	Gallium	Ga	31	69.7
Calcium	Ca	20	40.1	Germanium	Ge	32	72.6
*Californium	Cf	98	(251)	Gold	Au	79	197.0
Carbon	C	6	12.0	Hafnium	Hf	72	178.5
Cerium	Ce	58	140.1	*Hassium	Hs	108	(265)
Cesium	Cs	55	132.9	Helium	He	2	4.00

Table D–1 The Elements (continued)

Element	Symbol	Atomic Number	Atomic Weight	Element	Symbol	Atomic Number	Atomic Weight
Holmium	Ho	67	164.9	Promethium	Pm	61	(145)
Hydrogen	H	1	1.01	Protoactinium	Pa	91	231.0
Indium	In	19	114.8	Radium	Ra	88	226.0
Iodine	I	53	126.9	Radon	Rn	86	(222)
Iridium	Ir	77	192.2	Rhenium	Re	75	186.2
Iron	Fe	26	55.8	Rhodium	Rh	45	102.9
Krypton	Kr	36	83.8	Ruthenium	Ru	44	101.1
Lanthanum	La	57	138.9	*Rutherfordium	Rf	104	(261)
*Lawrencium	Lr	103	(257)	Samarium	Sm	62	150.4
Lead	Pb	82	207.2	Scandium	Sc	21	45.0
Lithium	Li	3	6.94	*Seaborgium	Sg	106	(263)
Lutetium	Lu	71	175.0	Selenium	Se	34	79.0
Magnesium	Mg	12	24.3	Silicon	Si	14	28.1
Manganese	Mn	25	54.9	Silver	Ag	47	107.9
*Meitnerium	Mt	109	(265)	Sodium	Na	11	23.0
*Mendelevium	Md	101	(256)	Strontium	Sr	38	87.6
Mercury	Hg	80	200.6	Sulfur	S	16	32.1
Molybdenum	Mo	42	95.9	Tantalum	Ta	73	180.9
Neodymium	Nd	60	44.2	Technetium	Tc	43	98.9
Neon	Ne	10	20.2	Tellurium	Te	52	127.6
*Neptunium	Np	93	237.0	Terbium	Tb	65	158.9
Nickel	Ni	28	58.7	Thallium	Tl	81	204.4
Niobium	Nb	41	92.9	Thorium	Th	90	232.0
Nitrogen	N	7	14.0	Thulium	Tm	69	168.9
*Nobelium	No	102	(254)	Tin	Sn	50	118.7
Osmium	Os	76	190.2	Titanium	Ti	22	47.9
Oxygen	O	8	16.0	Tungsten	W	74	183.8
Palladium	Pd	46	106.4	Uranium	U	92	238.0
Phosphorus	P	15	31.0	Vanadium	V	23	50.9
Platinum	Pt	78	195.1	Xenon	Xe	54	131.3
Plutonium	Pu	94	(242)	Ytterbium	Yb	70	173.0
Polonium	Po	84	(210)	Yttrium	Y	39	88.9
Potassium	K	19	39.1	Zinc	Zn	30	65.4
Praseodymium	Pr	59	140.9	Zirconium	Zr	40	91.2

*Elements that do not occur in nature.
Parentheses indicate the most stable isotope of the element.

Atoms, Bonds, and Chemical Reactions

In chemical reactions, atoms are neither created nor destroyed, nor is one kind of atom changed into another. What occurs in chemical reactions, whether mild or explosive, is simply a rearrangement of the ways in which the atoms involved are bonded together. An oxygen atom, for example, may be combined and recombined with different atoms to form any number of different compounds, but a given oxygen atom always has been, and always will be, an oxygen atom. The same can be said for all the other kinds of atoms. To understand how atoms may bond and undergo rearrangement to form different compounds, it is necessary to examine several concepts concerning the structure of atoms.

Structure of Atoms

Every atom consists of a central core called the **nucleus** (not to be confused with a cell nucleus). The nucleus of an atom contains one or more **protons** and, except for hydrogen, one or more **neutrons**. Surrounding the nucleus are particles called **electrons**. Each proton has a positive (+) electric charge, and each electron has an equal and opposite negative (−) electric charge. Thus, in any atom, the charge of the protons may be balanced by an equal number of electrons, making the whole atom neutral. Neutrons have no charge.

Atoms of all elements have this same basic structure, consisting of protons, electrons, and neutrons. The distinction between atoms of different elements is in the number of protons. The atoms of each element have a characteristic number of protons that is known as the **atomic number** of the element. (See Table D–1.) The number of electrons characteristic of the atoms of each element also differs, in a manner corresponding to the number of protons. The combined total of protons and neutrons in the nucleus of an element is the **mass number**, or **atomic weight**. The general structure of the atoms of several elements is shown in **Figure D–1**.

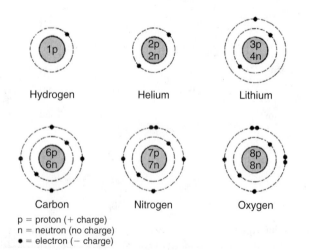

p = proton (+ charge)
n = neutron (no charge)
• = electron (− charge)

Figure D–1 Structure of atoms. All atoms consist of fundamental particles: protons (p), which have a positive electric charge; neutrons (n), which have no charge; and electrons, which have a negative charge.

The number of protons and electrons in an atom of an element (i.e., the atomic number of the element) determines the chemical properties of the element. The number of neutrons may vary. For example, most carbon atoms have six neutrons in addition to the six protons, as indicated in Figure D–1. But some carbon atoms have eight neutrons. Atoms of the same element that have different numbers of neutrons are known as **isotopes** of the element. The total number of protons plus neutrons is used to define different isotopes. For example, the usual isotope of carbon is referred to as carbon-12, while the isotope with eight neutrons is referred to as carbon-14. The chemical reactivities of different isotopes of the same element are identical. (However, certain other properties may differ.) Many isotopes of various elements prove to be radioactive, such as carbon-14. The atomic weights in Table D–1 are expressed as the average of the isotopes of an element as they occur in nature. The weights listed in parentheses are those of the most stable known isotope.

Bonding of Atoms

The chemical properties of an element are defined by the ways in which the element's atoms will react and form bonds with other atoms. By examining how atoms form bonds, we shall see how the number of electrons and protons determines these properties. There are two basic kinds of bonding: (**1**) **covalent bonding** and (**2**) **ionic bonding**.

In both kinds of bonding, it is important to recognize that electrons are not randomly distributed around the atom's nucleus. Rather, there are, in effect, specific spaces in a series of layers, or **orbitals**, around the nucleus. If an orbital is occupied by one or more electrons but is not filled, the atom is unstable; it will then tend to react and form bonds with other atoms to achieve greater stability. A stable state is achieved by having all the spaces in the orbital filled with electrons. It is important, however, to keep the charge neutral (i.e., the total number of electrons should be equal to that of the protons).

Covalent Bonding

These two requirements—filling all the spaces and keeping the charge neutral—may be satisfied by adjacent atoms sharing one or more pairs of electrons, as shown in **Figure D–2**. The sharing of a pair of electrons holds the atoms together in what it is called a **covalent bond**.

Covalent bonding, by satisfying the charge-orbital requirements, leads to discrete units of two or more atoms bonded together. Units of two or more covalently bonded atoms are called **molecules**. A few simple, but important, examples are shown in Figure D–2.

A chemical formula is simply a shorthand description of the number of each kind of atom in a given molecule. The element is given by the chemical symbol, and a subscript following the symbol gives the number of atoms of that element present in the molecule; if there is no subscript, there is only one atom of that chemical. A molecule with two or more different kinds of atoms may be called a compound, but a molecule composed of a single kind of atom—oxygen (O_2), for example—is still defined as an element.

Only a small number of elements—carbon, hydrogen, oxygen, nitrogen, phosphorus, and sulfur—have configurations of electrons that lend themselves readily to the formation of

Figure D–2 Atoms to molecules. Molecules are composed of atoms covalently bonded to each other in stable configurations in which electrons are shared.

covalent bonds. Carbon in particular, with its ability to form four covalent bonds, can produce long, straight, or branched chains or rings (Fig. D–3). Thus, an infinite array of molecules can be formed by using covalently bonded carbon atoms as a "backbone" and filling in the sides with atoms of hydrogen or other elements. Accordingly, it is covalent bonding among atoms of carbon and these few other elements that produces all natural organic molecules, those molecules that make up all the tissues of living things, and also synthetic organic compounds such as plastics.

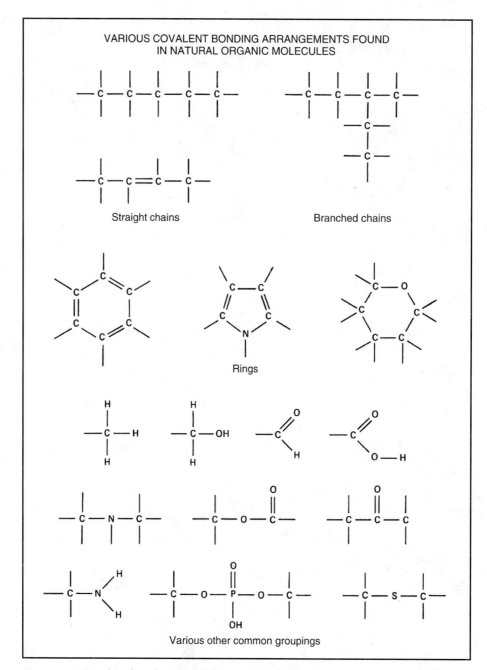

Figure D–3 Covalent bonding and organic molecules. The ability of carbon and a few other elements to readily form covalent bonds leads to an infinite array of complex organic molecules, which constitute all living things. A few major kinds of groupings are shown here. Note that each element forms a characteristic number of bonds: carbon, 4; nitrogen, 3; oxygen, 2; hydrogen, 1; sulfur, 2; and phosphorus, 5. Bonds left hanging (lines without atoms attached) indicate attachments to other atoms or groups of atoms.

Ionic Bonding

Another way in which atoms may achieve a stable electron configuration is to gain additional electrons to complete the filling of an orbital or to lose excess electrons in an incompleted orbital. In general, the maximum number of electrons that can be gained or lost by an atom is three. Therefore, an element's atomic number determines whether one or more electrons will be lost or gained. If an atom's outer orbital is one to three electrons short of being filled, it will always tend to gain additional electrons. If an atom has one to three electrons in excess of its last complete orbital, it will always tend to lose those electrons.

Of course, gaining or losing electrons results in the number of electrons in an atom being greater or less than the number of protons; the atom consequently has an electric charge. The charge will be one negative unit for each electron gained and one positive unit for each electron lost (Fig. D–4). A covalently bonded group of atoms may acquire an electric charge in the same way. An atom or group of atoms that has acquired an electric charge in this way is called an **ion**, positive or negative. Ions are designated by a superscript following the chemical symbol, which denotes the number of positive or negative charges. The absence of superscripts indicates that the atom or molecule is neutral. Some important ions are listed in **Table D–2**.

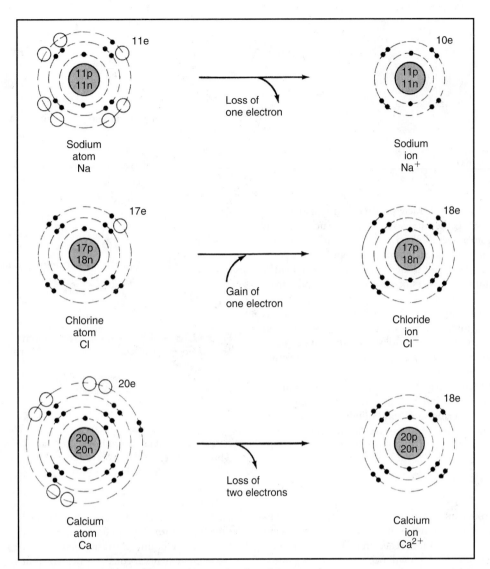

Figure D–4 Formation of ions. Many atoms will tend to gain or lose one or more electrons in order to achieve a state of complete (electron-filled) orbitals. In doing so, these atoms become positively or negatively charged ions, as indicated.

Table D–2   Ions of Particular Importance to Biological Systems

Negative (–) Ions		Positive (+) Ions	
Phosphate	PO_4^{3-}	Potassium	K^+
Sulfate	SO_4^{2-}	Calcium	Ca^{2+}
Nitrate	NO_3^-	Magnesium	Mg^{2+}
Hydroxyl	OH^-	Iron	Fe^{2+}, Fe^{3+}
Chloride	Cl^-	Hydrogen	H^+
Bicarbonate	HCO_3^-	Ammonium	NH_4^+
Carbonate	CO_3^{2-}	Sodium	Na^+

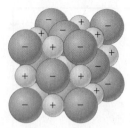

Figure D–5 Crystals. Positive and negative ions bond together by their mutual attraction and form crystals.

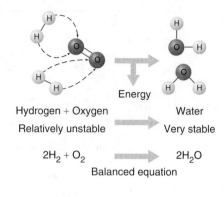

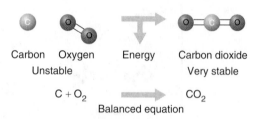

Figure D–6 Stable bonding arrangements. Some bonding arrangements are more stable than others. Chemical reactions will go spontaneously toward more stable arrangements, releasing energy in the process. However, reactions may be driven in the opposite direction with suitable energy inputs.

Because unlike charges attract, positive and negative ions tend to join and pack together in dense clusters in such a way as to neutralize an atom's overall electric charge. This joining together of ions through the attraction of their opposite charges is called **ionic bonding**. The result is the formation of hard, brittle, more or less crystalline substances of which all rocks and minerals are examples (Fig. D–5).

It is significant to note that whereas covalent bonding leads to discrete molecules, ionic bonding does not. Any number and combination of positive and negative ions may enter into an ionic bond to produce crystals of almost any size. The only restriction is that the overall charge of positive ions be balanced by that of negative ions. Thus, substances bonded in an ionic fashion are properly called compounds but not molecules. When chemical formulas are used to describe such compounds, they define the ratio of various elements involved, not specific molecules.

Chemical Reactions

While atoms themselves do not change, the bonds between atoms may be broken and re-formed, with different atoms producing different compounds or molecules. This is essentially what occurs in all chemical reactions. What determines whether a given chemical reaction will occur? Earlier, we noted that atoms form bonds because they achieve a greater stability by doing so. However, some bonding arrangements may provide greater overall stability than others. Consequently, substances with relatively unstable bonding arrangements will tend to react to form one or more different compounds that have more stable bonding arrangements. Common examples are the reaction between hydrogen and oxygen to produce water and the reaction between carbon and oxygen to produce carbon dioxide (Fig. D–6).

Energy is always released when the constituents of a reaction gain greater overall stability, as indicated in the figure. Thus, when energy is released in a chemical reaction, the atoms achieve more stable bonding arrangements. Therefore, it may be said that chemical reactions always tend to go in a direction that releases energy as well as giving greater stability.

However, chemical reactions can be made to go in a reverse direction. With suitable energy inputs and under suitable conditions, stable bonding arrangements may be broken and less stable arrangements formed. As described in Chapter 3 of the text, this is the basis of the photosynthesis that occurs in green plants. Light energy causes the highly stable hydrogen-oxygen bonds of water to split and form less stable carbon-hydrogen bonds, thus creating high-energy organic compounds.

Obviously, much more could be said about the involvement of chemistry in environmental science. We would highly recommend a course (or several courses) in chemistry to the student who wants to gain a deeper understanding of how the subject is involved in environmental concepts and issues.

Glossary

abiotic Pertaining to factors or things that are separate and independent from living things; nonliving.

abortion The termination of a pregnancy by some form of surgical or medicinal intervention.

acid Any compound that releases hydrogen ions when dissolved in water. Also, a water solution that contains a surplus of hydrogen ions.

acid deposition Any form of acid precipitation and also fallout of dry acid particles. (See **acid precipitation**.)

acid neutralizing capacity (ANC) In a water body, the capacity to neutralize acids due to the presence of **buffer** chemicals in solution.

acid precipitation Includes acid rain, acid fog, acid snow, and any other form of precipitation that is more acidic than normal (i.e., less than pH 5.6). Excess acidity is derived from certain air pollutants—namely, sulfur dioxide and oxides of nitrogen.

acquired immunodeficiency syndrome (AIDS) A fatal disease caused by the human immunodeficiency virus (HIV) and transmitted by sexual contact or the use of nonsterile needles (as in drug addiction).

activated-sludge system A system for removing organic wastes from water. The system uses microorganisms and active aeration to decompose such wastes. The system is used most as a means of secondary sewage treatment following the primary settling of materials.

active safety features Those safety features of nuclear reactors that rely on operator-controlled reactions, external power sources, and other features that are capable of failing. (See **passive safety features**.)

activated sludge See activated-sludge system.

acute A response to a disease or chemical pollutant that brings on life-threatening reactions within hours or days.

adaptation An ecological or evolutionary change in structure or function that enables an organism to adjust better to its environment and hence enhances the organism's ability to survive and reproduce.

adiabatic cooling The cooling that occurs when warm air rises and encounters lower atmospheric pressure. **Adiabatic warming** is the opposite process, whereby cool air descends and encounters higher pressure.

Advanced Boiling-Water Reactor (ABWR) A third generation nuclear reactor design for power plants that incorporates many passive safety features. It is the design of choice for most new and planned reactors.

aeration *Soil*: The exchange within the soil of oxygen and carbon dioxide necessary for the respiration of roots. *Water*: The bubbling of air or oxygen through water to increase the amount of oxygen dissolved in the water.

aerosols Microscopic liquid and solid particles originating from land and water surfaces and carried up into the atmosphere.

Agenda 21 A definitive program to promote sustainable development that was produced by the 1992 Earth Summit in Rio de Janeiro, Brazil, and adopted by the UN General Assembly.

age structure Within a population, the different proportions of people who are old, middle aged, young adults, and children.

Aichi Biodiversity Targets International goals for the protection of biodiversity, for the period 2011–2020, developed as a part of the Convention on Biological Diversity.

air pollutants Gases or aerosols in the atmosphere that have harmful effects on people or the environment.

Air Pollution Control Act Federal legislation introduced in 1955 that was the first attempt to bring air pollution under control.

air pollution disaster A short-term situation in industrial cities in which intense industrial smog brings about a significant increase in human mortality.

air toxics A category of air pollutants that includes radioactive materials and other toxic chemicals that are present at low concentrations but are of concern because they often are carcinogenic.

albedo A measure of the reflectivity of the Earth's surface.

alga, pl. algae Any of numerous kinds of photosynthetic plants that live and reproduce while entirely immersed in water. Many species (the planktonic forms) exist as single cells or small groups of cells that float freely in the water. Other species (the seaweeds) may be large and attached.

algal bloom A relatively sudden development of a heavy growth of algae, especially planktonic forms. Algal blooms generally result from additions of nutrients, whose scarcity is normally limiting.

alkaline/alkalinity A *basic* substance; chemically, a substance that absorbs hydrogen ions or releases hydroxyl ions; in reference to natural water, a measure of the base content of the water.

altitude The vertical distance from sea level.

ambient standards Air-quality standards (set by the EPA) determining certain levels of pollutants that should not be exceeded in order to maintain environmental and human health.

amensalism An interaction between species whereby one species is harmed, while the other is unaffected.

American Recovery and Reinvestment Act of 2009 (Recovery Act) Legislation that provided money for a variety of programs, intended to stimulate the economy during the recent recession. Some $90 billion was directed into clean energy programs.

anaerobic Lacking oxygen.

anaerobic digestion The breakdown of organic material by microorganisms in the absence of oxygen. The process results in the release of methane gas as a waste product.

animal testing Procedures used to assess the toxicity of chemical substances, using rats, mice, and guinea pigs as surrogates for humans who might be exposed to the substances.

anthropogenic Referring to pollutants and other forms of impact on natural environments that can be traced to human activities.

aquaculture A propagation or rearing of any aquatic (water) organism in a more or less artificial system.

aquatic succession Process by which a water body such as a lake or pond fills in through a series of vegetative communities.

aquifer An underground layer of porous rock, sand, or other material that allows the movement of water between layers of nonporous rock or clay. Aquifers are frequently tapped for wells.

artificial selection Plant and animal breeders' practice of selecting individuals with the greatest expression of desired traits to be the parents of the next generation.

asthma A chronic disease of the respiratory system in which the air passages tighten and constrict, causing breathing difficulties; acute attacks may be life threatening.

Atlanta BeltLine A redevelopment initiative that employs an old railroad loop around the heart of Atlanta, providing trails and rail transportation that will improve urban Atlanta.

atmosphere The thin layer of gases surrounding Earth. Nitrogen, oxygen, water vapor, and carbon dioxide are major gases, while many minor gases are also present in trace amounts.

atmosphere-ocean general circulation model (AOGCM) A computer-based model for simulating long-term climatic conditions that combines global atmospheric circulation patterns with ocean circulation and cloud-radiation feedback.

atmospheric brown cloud (ABC) A 1- to 3-km-thick blanket of pollution that frequently sits over south and central Asia, made up of black carbon and soot from biomass and fossil fuel burning.

atom The fundamental unit of all elements.

atomic theory Theory that all matter is made up of particles called atoms, represented by more than 100 different elements in nature.

autotroph Any organism that can synthesize all its organic substances from inorganic nutrients, using light or certain inorganic chemicals as a source of energy. Green plants are the principal autotrophs.

Bacillus thuringensis (BT) A type of bacterium that produces a toxin that can kill the larvae of plant-eating insects; genes from this bacterium have been incorporated into plants using genomic techniques.

background radiation Radiation that comes from natural sources apart from any human activity. We are all exposed to such radiation.

bacterium, pl. bacteria Any of numerous kinds of prokaryotic microscopic organisms that exist as simple single cells that multiply by simple division. Along with fungi, bacteria constitute the decomposer component of ecosystems. A few species cause disease.

balance-of-trade deficit A deficit in money flow resulting from purchasing more from other countries than is sold to other countries.

BANANA ("build absolutely nothing anywhere near anything") An acronym expressing a common attitude about siting various facilities.

barrels of oil equivalent (BOE) A measurement that uses the energy content of a barrel of oil to compare different fossil fuel energy amounts.

bar screen In a primary sewage treatment plant, a row of bars about one inch apart that provides initial screening of debris found in incoming sewage.

base Any compound that releases hydroxyl (OH–) ions when dissolved in water. A solution that contains a surplus of hydroxyl ions.

Basel Convention An international agreement that bans most international toxic-waste trade.

bed load The load of coarse sediment, mostly coarse silt and sand, that is gradually moved along the bottom of a riverbed by flowing water rather than being carried in suspension.

benefit-cost analysis A comparison of the benefits of any particular action or project to its costs. (See **cost-benefit ratio**.)

benthic plants Plants that grow under water and are attached to or rooted in the bottom of the body of water. For photosynthesis, these plants depend on light penetrating the water.

benzene An organic chemical present in crude and refined oil products and tobacco smoke; a known human carcinogen.

best-demonstrated available technologies (BDATs) Treatment standards for hazardous wastes that have the effect of reducing their toxicity and mobility.

best management practice Farm management practice that serves best to reduce soil and nutrient runoff and subsequent pollution.

bioaccumulation The accumulation of higher and higher concentrations of potentially toxic chemicals in organisms.

biochemical oxygen demand (BOD) The amount of oxygen that will be absorbed or "demanded" as wastes are being digested or oxidized in both biological and chemical processes. Potential impacts of wastes are commonly measured in terms of the BOD.

biodegradable Able to be consumed and broken down to natural substances, such as carbon dioxide and water, by biological organisms, particularly decomposers. Opposite: **nonbiodegradable.**

biodiesel Diesel fuel made from a mixture of vegetable oil (largely soybean oil) and regular diesel oil.

biodiversity The diversity of living things found in the natural world. The concept usually refers to the different species but also includes ecosystems and the genetic diversity within a given species.

biodiversity hot spots Thirty-four locations around the world that are home to some 60% of the world's known biodiversity.

biofuel Any fuel, but usually liquids, derived from agricultural crops (e.g., ethanol and plant oils).

biogas The mixture of gases—about two-thirds methane, one-third carbon dioxide, and small portions of foul-smelling compounds—resulting from the anaerobic digestion of organic matter. The methane content enables biogas to be used as a fuel.

biogeochemical cycle The repeated pathway of particular nutrients or elements from the environment through one or more organisms and back to the environment. The cycles include the carbon cycle, the nitrogen cycle, the phosphorus cycle, and so on.

biological evolution The theory, attributed to Charles Darwin and Alfred Russell Wallace, that all species now on Earth descended from ancestral species through a process of gradual change brought about by natural selection.

biological nutrient removal (BNR) A process employed in sewage treatment to remove nitrogen and phosphorus from the effluent coming from secondary treatment.

biological wealth The life-sustaining combination of commercial, scientific, and aesthetic values imparted to a region by its biota.

biomagnification Bioaccumulation occurring through several levels of a food chain.

biomass The mass of biological material. Usually the total mass of a particular group or category (e.g., biomass of producers).

biomass energy Energy produced by burning plant-related materials such as firewood.

biomass fuels Fuels in the form of plant-related material (e.g., firewood, grasses, organic farm wastes).

biomass pyramid The structure that is obtained when the respective biomasses of producers, herbivores, and carnivores in an ecosystem are compared. Producers have the largest biomass, followed by herbivores and then carnivores.

biome A group of ecosystems that are related by having a similar type of vegetation governed by similar climatic conditions. Examples include prairies, deciduous forests, arctic tundra, deserts, and tropical rain forests.

biopesticide Naturally occurring substances that control pests; such substances usually clear the EPA registration process quickly. See *Bacillus thuringensis*.

bioremediation The use of microorganisms for the decontamination of soil or groundwater. Usually involves injecting organisms or oxygen into contaminated zones.

biosolid In sewage treatment, biosolids are the relatively stable organic matter that remains after anaerobic digestion has taken place in sludge digesters.

biosphere The overall ecosystem of Earth. The sum total of all the biomes and smaller ecosystems, which ultimately are all interconnected and interdependent through global processes such as the water cycle and the atmospheric cycle.

biota The sum total of all living organisms. The term usually is applied to the setting of natural ecosystems.

biotechnology The use of genetic engineering techniques, enabling researchers to introduce new genes into food crops and domestic animals and thereby produce valuable products and more nutritious crops.

biotic community All the living organisms (plants, animals, and microorganisms) that live in a particular area.

biotic potential Reproductive capacity. The potential of a species for increasing its population and/or distribution. The biotic potential of every species is such that, given optimum conditions, its population will increase. (Contrast **environmental resistance**.)

biotic structure The organization of living organisms in an ecosystem into groups such as producers, consumers, detritus feeders, and decomposers.

birth control Any means, natural or artificial, that may be used to reduce the number of live births.

bisphenol A (BPA) A starter material for many plastics that is suspected of contributing to obesity, diabetes, altered sexual behavior, and cancer because of its estrogenic activity. See **endocrine disruptor**.

Blue Revolution A proposed radical change in managing water supplies intended to make more water available for human use.

blue water In the hydrologic cycle, precipitation and its further movement in infiltration, runoff, surface water, and groundwater; liquid water wherever it occurs.

BOD See **biochemical oxygen demand**.

body mass index A measurement used to calculate overweight and obesity; the ratio of weight in kilograms divided by the square of height in meters.

boiling-water reactor A nuclear reactor that employs boiling water to create steam, which is circulated to turbines that generate electricity; the water is cooled and returned to the reactor tank.

bottle law/bottle bill A law that provides for the recycling or reuse of beverage containers, usually by requiring a returnable deposit upon purchase of the item.

bottom-up regulation Basic control of a natural population that occurs as a result of the scarcity of some resource.

breeder reactor A nuclear reactor that, in the course of producing energy, also converts nonfissionable uranium-238 into fissionable plutonium-239, which can be used as fuel. Hence, a reactor that produces at least as much nuclear fuel as it consumes. Also called a **fast-neutron reactor**.

broad-spectrum pesticides Chemical pesticides that kill a wide range of pests. These pesticides also kill a wide range of nonpest and beneficial species; therefore, they may lead to environmental disturbances and resurgences. The opposite of narrow-spectrum pesticides.

brown economy In contrast to a green or sustainable economy, one that is unsustainable.

brownfields Abandoned, idled, or underused industrial and commercial facilities whose further development is inhibited because of real or perceived chemical contamination.

BTU (British thermal unit) A fundamental unit of energy in the English system. The amount of heat required to raise the temperature of 1 pound of water 1 degree Fahrenheit.

buffer A substance that will maintain a constant pH in a solution by absorbing added hydrogen or hydroxyl ions. An example is the bicarbonate ion, the most important buffer in blood.

Bureau of Land Management (BLM) A federal agency that manages public lands other than national forests and national parks. The agency leases land to ranchers for grazing, among other activities.

bush meat Wild game that is harvested for food, especially in developing countries; often involves poaching and overexploitation.

calorie A fundamental unit of energy. The amount of heat required to raise the temperature of 1 gram of water 1 degree Celsius. All forms of energy can be converted to heat and measured in calories. Calories used in connection with food are kilocalories, or "big" calories, the amount of heat required to raise the temperature of 1 liter of water 1 degree Celsius.

cancer Any of a number of cellular changes that result in uncontrolled cellular growth.

cap-and-trade system A form of market-based environmental policy that sets a maximum level of pollutant (the cap), distributes pollution permits, and allows industries to trade permits to achieve their allowable pollution.

capillary water Water that clings in small pores, cracks, and spaces against the pull of gravity (e.g., water held in a sponge).

capital In classical economic theory, one of the three factors of production (alongside land and labor); more broadly, any form of wealth. See **intangible capital, ecosystem capital** and **natural capital**.

carbon dioxide sinks The oceans and terrestrial environments that serve to absorb some of the carbon dioxide added to the atmosphere by human agency; these are called sinks.

carbon monoxide A highly poisonous gas, the molecules of which consist of a carbon atom with one oxygen attached. Not to be confused with carbon dioxide, a natural gas in the atmosphere.

carbon tax A tax levied on all fossil fuels in proportion to the amount of carbon dioxide that is released as they burn.

carcinogen/carcinogenic Having the property of causing cancer, at least in animals and, by implication, in humans.

carnivore An animal that feeds more or less exclusively on other animals.

carrying capacity The maximum population of a given species that an ecosystem can support without being degraded or destroyed in the long run. Represented symbolically by K.

Cartagena Protocol on Biosafety An international agreement governing trade in genetically modified organisms. The Cartagena Protocol was signed in January 2000.

Carter Doctrine As stated by President Jimmy Carter in 1980, the assertion that the United States would use military force to ensure our access to Persian Gulf oil.

catalyst A substance that promotes a given chemical reaction without itself being consumed or changed by the reaction. Enzymes are catalysts for biological reactions. Also, catalysts are used in some pollution control devices (e.g., the catalytic converter).

catalytic converter The device used by vehicle manufacturers to reduce the amounts of carbon monoxide, hydrocarbons, and nitrogen oxides in a vehicle's exhaust. The converter contains catalysts that change these compounds to less harmful gases as the exhaust passes through.

catch shares See **individual quota system**.

cell The basic unit of life; the smallest unit that still maintains all the attributes of life. Many microscopic organisms consist of a single cell. Large organisms consist of trillions of specialized cells functioning together.

cellular respiration The chemical process that occurs in all living cells whereby organic compounds are broken down to release energy required for life processes. For respiration, higher plants and animals require oxygen, and release carbon dioxide and water as waste products, but certain microorganisms do not require oxygen.

cellulose The organic macromolecule that is the prime constituent of plant cell walls and hence the major molecule in wood, wood products, and cotton.

center-pivot irrigation An irrigation system consisting of a spray arm several hundred meters long, supported by wheels, pivoting around a central well from which water is pumped.

Centers for Disease Control and Prevention (CDC) An agency within the United States Department of Health and Human Services that is the lead federal agency for protecting the health and safety of the public.

centrally planned economy An economic system in which a ruling class makes most of the basic decisions about how the economy will be structured; typical of communist countries.

chain reaction A nuclear reaction wherein each atom that fissions (splits) causes one or more additional atoms to fission.

character displacement An effect of interspecific competition, whereby a physical change occurs in one or both competing species that lessens competition when the two species co-occur.

chemical energy The potential energy that is contained in certain chemicals; most importantly, the energy that is contained in organic compounds such as food and fuels and that may be released through respiration or burning.

chemical treatment The use of pesticides and herbicides to control or eradicate agricultural pests.

chemosynthesis The process whereby some microorganisms utilize the chemical energy contained in certain reduced inorganic chemicals (e.g., hydrogen sulfide) to produce organic matter.

chlorinated hydrocarbons Synthetic organic molecules in which one or more hydrogen atoms have been replaced by chlorine atoms. Chlorinated hydrocarbons are hazardous compounds because they tend to be nonbiodegradable and therefore to bioaccumulate; many have been shown to be carcinogenic. Also called **organochlorides**.

chlorine catalytic cycle In the stratosphere, a cyclical chemical process in which chlorine monoxide breaks down ozone.

chlorofluorocarbons (CFCs) Synthetic organic molecules that contain one or more of both chlorine and fluorine atoms and that are known to cause ozone destruction.

chlorophyll The green pigment in plants responsible for absorbing the light energy required for photosynthesis.

chromosome In a living organism, one of a number of structures on which genes are arranged in a linear fashion.

chronic A reaction to infection or pollutants that leads to a gradual deterioration of health over a period of years.

chronic obstructive pulmonary disease (COPD) A progressive lung disease that makes it hard to breathe, involving three syndromes: asthma, bronchitis, and emphysema.

CITES (Convention on International Trade in Endangered Species) An international treaty conveying some protection to endangered and threatened species by restricting trade in those species or their products.

clay Any particles in soil that are less than 0.004 millimeters in size.

Clean Air Act Amendments of 1970 (CAA) Amended in 1977 and 1990, the act is the foundation of U.S. air pollution control efforts.

Clean Air Interstate Rule (CAIR) EPA rule published in 2005 establishing cap-and-trade programs for sulfur dioxide and nitrogen oxides in 28 eastern states.

Clean Water Act of 1972 (CWA) The cornerstone federal legislation addressing water pollution.

clear-cutting In harvesting timber, the practice of removing an entire stand of trees, leaving an ugly site that takes years to recover.

climate A general description of the average temperature and rainfall conditions of a region over the course of a year.

climate change A change in average long term and large scale climate patterns, from natural and human factors.

Climate Change Science Program (CCSP) See Global Change Research Program (GCRP).

climate commitment The amount climate will continue to change after emissions are stabilized because of greenhouse gases already emitted.

climate sensitivity The response of climate to greenhouse gas concentrations, usually measured as a rise in temperature as a consequence of a rise in greenhouse gas concentration.

climax ecosystem The last stage in ecological succession. An ecosystem in which populations of all organisms are in balance with each other and with existing abiotic factors.

cogeneration Also known as **combined heat and power (CHP)**, the joint production of useful heat and electricity. For example, furnaces may be replaced with gas turbogenerators that produce electricity, while the hot exhaust still serves as a heat source.

combined heat and power (CHP) See **cogeneration**.

combustion The practice of disposing of wastes by incineration in a special facility designed to handle large amounts of waste.

command-and-control strategy The basic strategy behind most public policy having to do with air and water pollution. The strategy involves setting limits on pollutant levels and specifying control technologies that must be used to achieve those limits.

commensalism A relation between two species in which one is benefited and the other is not affected.

common good A recognition of the need for protection of the natural world and improvement of human welfare, especially as applied to public policy.

commons/common-pool resources Resources (usually natural ones) owned by many people in common or, as in the case of the air or the open oceans, owned by no one but open to exploitation.

community In ecological contexts, a community is defined as the group of populations of different species of plants, animals, and microbes living together in one area.

community supported agriculture An arrangement between farms and local citizens where the citizens are shareholders in the farm, supporting the farmers and gaining access to fresh farm produce.

compaction Packing down. *Soil:* Packing and pressing out air spaces present in the soil. Compaction reduces soil aeration and infiltration and thus diminishes the capacity of the soil to support plants. *Trash:* Packing down trash to reduce the space that it requires.

competition In ecology, the interaction between members of the same species or between different species for essential resources or access to those resources.

competitive exclusion principle The concept that when two species compete very directly for resources, one eventually excludes the other from the area.

composting/compost The process of letting organic wastes decompose in the presence of air. A nutrient-rich humus, or compost, is the resulting product.

composting toilet A toilet that does not flush wastes away with water but deposits them in a chamber where they will compost. (See **composting**.)

compound Any substance (gas, liquid, or solid) that is made up of two or more different kinds of atoms bonded together. (Contrast **element**.)

Comprehensive Environmental Response, Compensation, and Liability Act of 1980 (CERCLA) See **Superfund**.

Comprehensive Everglades Restoration Plan (CERP) Multi-billion-dollar plan to restore the Florida Everglades by addressing water flow and storage.

concentrated animal feeding operations (CAFO) Feedlots, areas where animals are confined and fed in high density.

concepts Valid explanations of data from the natural world that often allow predictions but do not reach the level of validity of natural laws.

condensation The collecting of molecules from the vapor state to form the liquid state, as, for example, when water vapor condenses on a cold surface and forms droplets. Opposite: **evaporation**.

condenser A device that turns turbine exhaust steam into water in a power plant.

conditions Abiotic factors (e.g., temperature) that vary in time and space but are not used up by organisms.

coniferous forest biome A biome occupying northern regions or higher mountainous altitudes dominated by coniferous trees.

conservation The management of a resource in such a way as to ensure that it will continue to provide maximum benefit to humans over the long run.

conservation biology The branch of science focused on the protection of populations and species.

conservation reserve An imaginary source of energy that results from policies promoting greater efficiency of energy use, resulting in a reduced energy requirement.

Conservation Reserve Program (CRP) A federal program whereby farmers are paid to place highly erodible cropland in a reserve. A similar program, the **Conservation Stewardship Program**, encourages farmers to conserve soil and water resources on their lands.

consumers In an ecosystem, those organisms that derive their energy from feeding on other organisms or their products.

consumption In ecological use, the process of feeding on organic matter as a source of energy; all heterotrophs are consumers.

consumptive use The harvesting of natural resources in order to provide for people's immediate needs for food, shelter, fuel, and clothing.

consumptive water use The use of water for such things as irrigation, wherein the water does not remain available for potential purification and reuse.

containment building The reinforced concrete building housing a nuclear reactor. Designed to contain an explosion, should one occur.

continental drift See **plate tectonics**.

contour plowing The practice of cultivating land along the contours across, rather than up and down, slopes. In combination with strip cropping, contour farming reduces water erosion.

contraceptive A device or method employed by couples during sexual intercourse to prevent the conception of a child.

contract and converge Plan by which poorer countries can produce more greenhouse gas emissions, richer countries less, and the whole world total must be less, in response to climate change.

control rods A part of the core of a nuclear reactor; the rods of neutron-absorbing material that are inserted or removed as necessary to control the rate of nuclear fissioning.

convection/convection currents The vertical movement of air due to atmospheric heating and cooling, moving generally in a west-to-east pattern.

Convention on Biological Diversity The biodiversity treaty signed by 158 nations at the Earth Summit in Rio de Janeiro in 1992, calling for various actions and cooperative steps between nations to protect the world's biodiversity.

conversion losses Unavoidable energy losses in energy production, usually due to the loss of heat energy.

conveyor system The giant pattern of oceanic currents that moves water masses from the surface to the depths and back again, producing major effects on the climate.

cooling tower A massive tower designed to dissipate waste heat from a power plant (or other industrial facility) into the atmosphere by condensing turbine exhaust steam into water.

Copenhagen Accord The agreement signed at the 2009 Copenhagen meeting of delegates of the Framework Convention on Climate Change, pledging actions to limit global temperature increases to 2°C above preindustrial levels.

coral bleaching A condition, usually brought on by excessively high temperatures in hard corals where the coral animals expel their symbiotic algae and become white in appearance.

corporate average fuel economy (CAFE) Fuel economy standards for vehicles set by the National Highway Traffic Safety Administration.

corrosion In a nuclear power plant, the deterioration of pipes receiving hot, pressurized water in a circulation system that conveys heat from one part of the unit to another.

cosmetic spraying The spraying of pesticides to control pests that damage only the surface appearance of fruits and vegetables.

cost-benefit analysis See **benefit-cost analysis**.

cost-benefit ratio/benefit-cost ratio The value of the benefits to be gained from a project, divided by the costs of the project. If the ratio is greater than unity, the project is economically justified; if the ratio is less than unity, the project is not economically justified.

cost-effectiveness analysis An alternative to benefit-cost analysis, this approach accepts the merits of a goal and attempts to achieve the goal with the least costly alternative.

credit associations Groups of poor people who individually lack the collateral to assure loans but who collectively have the wherewithal to assure each other's loans. Associated with microlending.

criteria maximum concentration (CMC) A water-quality standard; the highest single concentration of a water pollutant beyond which environmental impacts may be expected.

criteria pollutants Certain pollutants whose levels are used as a gauge for the determination of air (or water) quality.

criterion continuous concentration (CCC) A water-quality standard; the highest sustained concentration of a water pollutant beyond which undesirable impacts may be expected.

critical habitat Under the Endangered Species Act, an area provided for a listed species where it can be found or could likely spread as it recovers.

critical number The minimum number of individuals of a given species that is required to maintain a healthy, viable population of the species. If a population falls below its critical number, it will likely become extinct.

Cross State Air Pollution Rule (CSAPR) The ruling by the EPA that replaces the Clean Air Interstate Rule, requiring reductions in NO_x and SO_2 to aid states in the East to achieve ozone and particulate matter reductions.

crude birthrate (CBR) The number of births per 1,000 individuals per year.

crude death rate (CDR) The number of deaths per 1,000 individuals per year.

cryosphere Snow and sea ice in polar regions; it is significant in its contribution to the planetary albedo as it reflects sunlight.

cryptogamic crust Made up of primitive plants, called cryptogams, the crust develops on hardened soils and can stabilize soil, but also prevents water infiltration and seed germination.

crystal A solid form of matter made up of regularly repeating units of atoms, ionic compounds, or molecules.

cultural control A change in the practice of growing, harvesting, storing, handling, or disposing of wastes that reduces their susceptibility or exposure to pests. For example, spraying the house with insecticides to kill flies is a chemical control; putting screens on the windows to keep flies out is a cultural control.

cultural eutrophication The process of natural eutrophication accelerated by human activities. (See **eutrophication**.)

cultural hazards Public health hazards such as smoking, with behavioral or cultural basis.

curie The unit of measurement of radioactivity; one gram of pure radium 226 gives off one curie per second, or 37 billion spontaneous disintegrations into particles and radiation.

DALY (disability-adjusted life year) In assessing the disease burden of different health risks, 1 DALY equals the loss of one healthy year of a person's life (whether through disability or loss of life).

DDT (dichlorodiphenyltrichloroethane) The first and most widely used of the synthetic organic pesticides belonging to the chlorinated hydrocarbon class.

dead zone A region of an aquatic ecosystem (usually a marine coastal area) where oxygen is low or absent from the bottom up well into the water column. Also called a **hypoxic area**.

debt crisis Brought about by the great debt incurred by many less-developed nations, this crisis involves countries building up debts that can never be paid off.

debt relief Steps, including cancellation of debts and other means of reducing poverty, taken by the World Bank and donor nations to alleviate the heavy debt of many poor nations.

declining tax base The loss of tax revenues that occurs when affluent taxpayers and businesses leave an area and property values subsequently decline. Also referred to as an **eroding tax base**.

decommissioning The retirement of nuclear power plants from service after 25 or more years because the effects of radiation will gradually make them inoperable.

decomposers Organisms whose feeding action results in decay or rotting of organic material. The primary decomposers are fungi and bacteria.

deep-well injection A technique used for the disposal of liquid chemical wastes that involves putting them into deep dry wells where they permeate dry strata.

deforestation The process of removing trees and other vegetation covering the soil and converting the forest to another land use, often leading to erosion and loss of soil fertility.

Delaney clause A stipulation in the Federal Food, Drug, and Cosmetic Act of 1938 that prohibited the use of any food additive that is found to induce cancer in humans or animals; this has been replaced by the **negligible risk** standard.

demand-side policies Energy policies that emphasize conserving energy and using more renewable and nuclear energy in the future.

demographic dividend As a country goes through the demographic transition, there is a stage when the dependency ratio is low, giving the country the opportunity to spend more on poverty alleviation and economic growth.

demographic transition The transition of a human population from a condition of a high birthrate and a high death rate to a condition of a low birthrate and a low death rate. A demographic transition may result from economic or social development.

demography/demographer The study of population trends (growth, movement, development, and so on). People who perform such studies and make projections from them.

denitrification The process of reducing oxidized nitrogen compounds present in soil or water to nitrogen gas into the atmosphere, conducted by certain bacteria and now utilized in the treatment of sewage effluents.

density-dependent An attribute of population-balancing factors, such as predation, that increases and decreases in intensity in proportion to population density. May lead to population stability.

density-independent Mortality to a population that occurs regardless of the population density. Tends to be destabilizing to populations.

Department of Transportation regulations (DOT regs) Regulations intended to reduce the risk of spills, fires, and poisonous fumes by specifying the kinds of containers and methods of packing to be used in transporting hazardous materials.

dependency ratio In a human population, the ratio of the nonworking-age population (under 15 and over 65) to the working-age population.

desalination A process that purifies seawater into high-quality drinking water via distillation or microfiltration.

desert biome The pattern of living forms and habitat where rainfall is less than 25 centimeters (10 inches) per year.

desertification The formation and expansion of degraded areas of soil and vegetation cover in arid, semiarid, and seasonally dry areas, caused by climatic variations and human activities. Desertified land is most commonly traced to erosion due to human mismanagement.

desert pavement A covering of stones and coarse sand protecting desert soils from further wind erosion. The covering results from the differential erosion of finer material.

detritivores Organisms (fungi, soil insects, bacteria) that feed on dead or decaying organic matter, important in completing the breakdown of organic matter to inorganic constituents.

detritus The dead organic matter, such as fallen leaves, twigs, and other plant and animal wastes, that exists in any ecosystem.

detritus feeders See **detritivores**.

deuterium (^{2}H) A stable, naturally occurring isotope of hydrogen that contains one neutron in addition to the single proton normally in the nucleus.

developed countries The high-income industrialized countries—the United States, Canada, western European nations, Japan, Korea, Australia, and New Zealand, as well as many middle-income countries such as those in Latin America, China, eastern Europe, and many Arab states.

developing countries Countries in which the gross domestic product is less than $936 per capita. The category includes nations of Africa, India, other countries of southern Asia, and some former Soviet republics.

development A term referring to the continued improvement of human well-being, usually in the developing countries.

Diclofenac An anti-inflammatory drug often administered to animals but toxic to birds. Implicated in the die-off of vultures in the Indian subcontinent.

dioxin A synthetic organic chemical of the chlorinated hydrocarbon class. Dioxin is highly toxic and a widespread environmental pollutant given off in the combustion of many organic compounds.

discharge permit (technically called an **NPDES permit**) A permit that allows a company to legally discharge certain amounts or levels of pollutants into air or water.

discount rate In economics, a rate applied to some future benefit or cost in order to calculate its present real value.

dish-engine system A solar power system consisting of a set of parabolic concentrator dishes that focus sunlight onto a receiver, where the hot fluid is transferred to an engine that generates electricity.

disinfection The killing (as opposed to removal) of microorganisms in water or other media where they might otherwise pose a health threat. For example, chlorine is commonly used to disinfect water supplies.

dissolved oxygen (DO) Oxygen gas molecules (O_2) dissolved in water. Fish and other aquatic organisms depend on dissolved oxygen for respiration. Therefore, the concentration of dissolved oxygen is a measure of water quality.

distillation A process of purifying water or some other liquid by boiling the liquid and recondensing the vapor. Contaminants remain behind in the boiler.

disturbance A natural or human-induced event or process (e.g., a forest fire) that interrupts ecological succession and creates new conditions on a site.

DNA (deoxyribonucleic acid) The natural organic macromolecule that carries the genetic or hereditary information for virtually all organisms.

Doha Development Round Negotiations for trade begun in 2001 with the objective of lowering trade barriers around the world.

donor fatigue The tendency of donor organizations or countries to withhold their aid because of long-term misuse of aid funds in the recipient countries.

dose The mathematical product of the concentration of a hazardous material and the length of exposure to it. The effects of any given material or radiation correspond to the dose received.

dose-response assessment Part of risk assessment, establishing the number of cancers that might develop due to exposure to different doses of a chemical.

drip irrigation A method of supplying irrigation water through tubes that literally drip water onto the soil at the base of each plant.

drought A local or regional lack of precipitation such that the ability to raise crops and water animals is seriously impaired.

drylands Ecosystems characterized by low precipitation (25 to 75 cm per year) and often subject to droughts; some 41% of Earth's land is drylands.

d–t reaction A nuclear fusion reaction employing the hydrogen isotopes deuterium and tritium; the reaction takes extremely high temperatures and is very difficult to contain.

Durban Platform An agreement forged at Durban in 2011 at the 17th Conference of the Parties to the FCCC where delegates agreed to begin negotiations leading to a binding agreement to curb greenhouse gas emissions by 2015.

easement In reference to land protection, an arrangement whereby a landowner gives up development rights into the future but retains ownership of the land.

ecdysone The hormone that promotes molting in insects.

ecological economist An economist who thoroughly integrates ecological and economic concerns; part of a new breed of economist who disagrees with classical economic theory.

ecological footprint A concept for measuring the demand placed on Earth's resources by individuals from different parts of the world, involving calculations of the natural area required to satisfy human needs.

ecological niche See niche (ecological).

ecological pest control Natural control of pest populations through understanding various ecological factors and, so far as possible, utilizing those factors, as opposed to using synthetic chemicals.

ecological risk assessment (ERA) Risk assessment by the EPA designed to protect natural organisms and ecosystems from adverse effects of human activities.

ecological succession The gradual or sometimes rapid change in the species that occupy a given area, with some species invading and becoming more numerous, while others decline in population and disappear.

ecologists Scientists who study ecology.

ecology The study of all processes influencing the distribution and abundance of organisms and the interactions between living things and their environment.

economic exclusion The cutting off of access of certain ethnic or economic groups to jobs, a good education, and other opportunities, and thus preventing them from entering the economic mainstream of society—a condition that prevails in poor areas of cities.

economic production As seen by ecological economists, economic production is the process of converting the natural world to the manufactured world.

economic simplified boiling water reactor A Generation III nuclear power reactor proposed by General Electric; it has many more passive safety feataures than current Advanced Boiling Water Reactors.

economic systems The social and legal arrangements people in societies construct in order to satisfy their needs and wants and improve their well-being.

economic threshold The level of pest damage that, to be reduced further, would require an application of pesticides that is more costly than the economic damage caused by the pests.

economics The social science that deals with the production, distribution, and consumption of goods and services.

economy See **economic systems**.

ecosystem An interactive complex of communities and the abiotic environment affecting them within a particular area. Ecosystems have characteristic forms, such as deserts, grasslands, tundra, deciduous forests, and tropical rain forests.

ecosystem capital The sum of goods and services provided by natural and managed ecosystems, both free of charge and essential to human life and well-being.

ecosystem management The management paradigm, adopted by all federal agencies managing public lands, that involves a long-term stewardship approach to maintaining the lands in their natural state.

ecotone A transitional region between two adjacent ecosystems that contains some of the species and characteristics of each one and also certain species of its own.

ecotourism The enterprises involved in promoting tourism of unusual or interesting ecological sites.

edges Breaks between habitats that may expose sensitive species to predators.

efficiency Producing the greatest amount of output with the lowest amount of input.

electrical power The amount of work done by an electric current over a given time.

electric generator A device that converts mechanical energy into electrical energy by way of wire coils that are rotated in a magnetic field.

electrolysis The use of electrical energy to split water molecules into their constituent hydrogen and oxygen atoms. Hydrogen gas and oxygen gas result.

electrolyzer In a solar receptor that is designed to produce fuel from water using solar energy, a catalyst that uses the electron energy to split water into hydrogen and oxygen.

electrons Fundamental atomic particles that have a negative electrical charge and a very small mass, approximately 9.1×10^{-31} kg. Electrons surround the nuclei of atoms and thus balance the positive charge of protons in the nucleus. A flow of electrons in a wire is identical to an electrical current.

element A substance that is made up of one and only one distinct kind of atom. (Contrast **compound**.)

El Niño/La Niña **Southern Oscillation (ENSO)** A climatic phenomenon characterized by major shifts in atmospheric pressure over the central equatorial Pacific Ocean. An **El Niño** involves the movement of unusually warm surface water into the eastern equatorial Pacific Ocean, and a **La Niña** reverses the pattern. These phenomena result in extensive disruption of weather around the world.

eluviation In soils, the process of leaching many minerals due to the downward movement of water; the E in E horizon of soil profiles stands for eluviation.

embrittlement Becoming brittle. Pertains especially to the reactor vessel of nuclear power plants gradually becoming prone to breakage or snapping as a result of continuous bombardment by radiation.

Emergency Planning and Community Right-to-Know Act (EPCRA) See **SARA**.

emergent vegetation Aquatic plants whose lower parts are underwater but whose upper parts emerge from the water.

endangered species A species whose total population is declining to relatively low levels such that, if the trend continues, the species will likely become extinct.

Endangered Species Act The federal legislation that mandates the protection of species and their habitats that are determined to be in danger of extinction.

endemic species A species that is found only in a particular geographic area or habitat.

endocrine See **hormone**.

endocrine disrupters Any of a class of organic compounds, often pesticides, that are suspected of having the capacity to interfere with hormonal activities in animals.

endocrine system The collection of all of the hormones and glands that produce them in animal bodies; they control metabolism, growth, and reproduction.

energy The capacity to do work. Common forms of energy are light, heat, electricity, motion, and the chemical bond energy inherent in compounds such as sugar, gasoline, and other fuels.

energy carrier Something (e.g., electricity) that transfers energy from a primary energy source to its point of use.

energy conservation and efficiency All of those steps that are taken to avoid or reduce the use of energy, such as raising the Corporate Average Fuel Economy standards for cars and trucks.

energy flow The movement of energy through ecosystems, starting with the capture of solar energy by primary producers and ending with the loss of heat energy.

Energy Independence and Security Act of 2007 Legislation establishing energy policy by emphasizing demand-side policies such as conserving energy and using renewable energy sources.

Energy Policy Act of 2005 Legislation establishing U.S. energy policy for years to come, encouraging greater use of nuclear power, and providing some support for energy conservation and renewable energy.

Energy Star Program The EPA's program focusing on energy conservation in buildings and in appliances. An Energy Star tag signals that the device or building incorporates energy-saving technology.

enhanced geothermal systems (EGS) Systems that tap into the energy in deep, hot rocks by increasing the number of rock fractures and injecting water.

enhanced recovery In the removal of oil from an oil field, a process that increases recovery beyond secondary recovery by injecting carbon dioxide or hydraulic fluids into wells.

enrichment With reference to nuclear power, the separation and concentration of uranium-235 so that, in suitable quantities, it will sustain a chain reaction.

entomologist A scientist who studies insects and their life cycles, physiology, behavior, and economic significance.

entropy A degree of disorder; increasing entropy means increasing disorder.

environment The combination of all things and factors external to the individual or population of organisms in question.

environmental accounting A process of keeping national accounts of economic activity that includes gains and losses of environmental assets.

environmental footprint The area of land or sea required to produce resources for an activity or population; a way of measuring resource use intensity.

environmental health The study of the connections between hazards in the environment and human disease and death.

environmental impact The effect on the natural environment caused by human actions. Includes indirect effects, for example, through pollution, as well as direct effects such as cutting down trees.

environmental impact statement A study of the probable environmental impacts of a development project. The National Environmental Policy Act of 1968 (NEPA) requires such studies prior to proceeding with any project receiving federal funding.

environmental justice The fair treatment and meaningful involvement of all people, regardless of race, color, national origin, or income, with respect to the development, implementation, and enforcement of environmental laws and regulations.

environmental movement The upwelling of public awareness and citizen action that began during the 1960s regarding environmental issues.

environmental public policy All of a society's laws and regulations that deal with that society's interactions with the environment. Its purpose is to promote the common good.

environmental racism Discrimination against people of color whereby hazardous industries are sited in nonwhite neighborhoods and towns.

environmental resistance The totality of factors such as adverse weather conditions, shortages of food or water, predators, and diseases that tend to cut back populations and keep them from growing or spreading. (Contrast **biotic potential**.)

Environmental Revolution In the view of some, a coming change in the adaptation of humans to the rising deterioration of the environment. The Environmental Revolution should bring about sustainable interactions with the environment.

environmental science The multidisciplinary branch of science concerned with environmental issues.

environmental tobacco smoke (ETS) See **secondhand smoke**.

environmentalism An attitude or movement involving concern about pollution, resource depletion, population pressures, loss of biodiversity, and other environmental issues. Usually also implies action to address those concerns.

environmentalist Any person who is concerned about the degradation of the natural world and is willing to act on that concern.

enzyme A protein that promotes the synthesis or breaking of chemical bonds.

EPA (U.S. Environmental Protection Agency) The federal agency responsible for the control of all forms of pollution and other kinds of environmental degradation.

epidemiologic transition In human populations, the pattern of change in mortality from high death rates to low death rates; contributes to the demographic transition.

epidemiology/epidemiological study The study of the causes of disease (e.g., lung cancer) through an examination and comparison of large populations of people living in different locations or following different lifestyles or habits (e.g., smoking versus nonsmoking).

equilibrial species See *K*-strategists.

equilibrium Referring to populations, the condition where births plus immigration are more or less equal to deaths plus emigration over time.

equilibrium climate sensitivity The sustained change in mean global temperature that occurs if atmospheric carbon dioxide is held at double preindustrial values, or 550 parts per million.

equity An ethical principle where people's needs are met in an impartial and just manner.

equity principle An ethical principle claiming that those who produce the most pollution should compensate those who produce less, especially because the less fortunate often suffer the most from, say, global climate change.

EROI (Energy Returned on Investment) The amount of usable energy obtained from a source in comparison to the amount of energy expended to obtain that energy.

erosion The process of soil particles being carried away by wind or water. Erosion moves the smaller particles first and hence degrades the soil to a coarser, sandier, stonier texture.

estimated reserves See **reserves**.

estuary A bay or river system open to the ocean at one end and receiving freshwater at the other. In the estuary, freshwater and saltwater mix, producing brackish water.

ethnobotany The study of the relationships between plants and people.

ethyl alcohol The main product of fermentation of sugars or starches, also called ethanol. It is a 2-carbon compound that can be burned in combustion engines as a substitute for gasoline.

euphotic zone In aquatic systems, the layer or depth of water through which an adequate amount of light penetrates to support photosynthesis.

eutrophic Characterized by nutrient-rich water supporting an abundant growth of algae or other aquatic plants at the surface.

eutrophication The process of becoming eutrophic.

evaporation The process whereby molecules leave the liquid state and enter the vapor or gaseous state, as, for example, when water evaporates to form water vapor. Opposite: **condensation**.

evaporative water loss The loss of moisture from soil by evaporation; it can be lessened by covering the soil with detritus or plant cover.

evapotranspiration The combination of evaporation and transpiration that restores water to the atmosphere.

evolution See **biological evolution**.

e-waste Any of a broad array of electronic devices that are discarded and often are hazardous to handle and difficult to recycle.

executive branch That facet of the government in the United States consisting of the president, his cabinet, and all administrative agencies.

exotic species A species introduced into a geographical area to which it is not native; also, species that are rare and subject to illegal trafficking.

experimentation In the practice of science, the testing of hypotheses by setting up situations where cause and effect are investigated, using careful measurements of conditions and responses.

exponential increase The growth produced when a base population increases by a given percentage (as opposed to a given amount) each year. An exponential increase is characterized by doubling again and again, each doubling occurring in the same period of time. It produces a J-shaped curve.

exposure assessment A part of risk assessment identifying human groups exposed to a chemical and calculating the doses and length of time of their exposure.

extended product responsibility Voluntary or government programs that assign some responsibility for reducing the environmental impact of a product at various stages of its life cycle—especially at disposal.

external cost/externality Any effect of a business process not included in the usual calculations of profit and loss. Pollution of air or water is an example of a *negative* externality—one that imposes a cost on society that is not paid for by the business itself.

extinction The death of all individuals of a particular species. When this occurs, all the genes of that particular line are lost forever.

extreme poverty Poverty that is defined as living on less than $1.25 a day. In 2012, some 1.4 billion were in this condition.

exurban migration The pronounced trend since World War II of relocating homes and businesses from the central city and older suburbs to more outlying suburbs.

exurbs New developments beyond the traditional suburbs, from which most residents still commute to the associated city for work.

facilitation A mechanism of succession where earlier species create conditions that are more favorable to newer occupants and are replaced by them.

family planning Making contraceptives and other health and reproductive services available to couples in order to enable them to achieve a desired family size.

famine A severe shortage of food accompanied by a significant increase in the local or regional death rate.

fast neutron reactors See breeder reactor.

FDA Food Safety Modernization Act A bill passed in 2010 that expanded the authority of the Food and Drug Administration over the food processing industry, in response to repeated outbreaks of food poisoning and pathogens in food.

fecal coliform test A test for the presence of *Escherichia coli*, the bacterium that normally inhabits the gut of humans and other mammals. A positive test result indicates sewage contamination and the potential presence of disease-causing microorganisms.

Federal Agricultural Improvement and Reform (FAIR) Act of 1996 Major legislation removing many subsidies and controls from farming.

Federal Food and Drug Administration (FDA) The federal agency with jurisdiction over foods, drugs, and all products that come in contact with the skin or are ingested.

feedback loops In ecological or environmental systems, feedback loops are processes that can amplify (positive loop) or dampen (negative loop) the behavior of the system. See **positive feedback**.

fermentation A form of respiration carried on by yeast cells and many bacteria in the absence of oxygen. Fermentation involves a partial breakdown of glucose (sugar) that yields energy for the yeast and the release of alcohol as a by-product.

fertility rate See **total fertility rate**.

fertility transition The pattern of change in birthrates in a human society from high rates to low; a major component of the demographic transition.

fertilizer Material applied to plants or soil to supply plant nutrients, most commonly nitrogen, phosphorus, and potassium but possibly others. *Organic* fertilizer is natural organic material, such as manure, that releases nutrients as it breaks down. *Inorganic* fertilizer, also called chemical fertilizer, is a mixture of one or more necessary nutrients in inorganic chemical form.

field scouts Persons trained to survey crop fields and determine whether applications of pesticides or other pest management procedures are actually necessary to avert significant economic loss.

FIFRA (Federal Insecticide, Fungicide, and Rodenticide Act) The key U.S. legislation passed to control pesticides.

fire climax ecosystems Ecosystems that depend on the recurrence of fire to maintain the existing balance.

first-generation pesticides Toxic inorganic chemicals that were the first used to control insects, plant diseases, and other pests. These pesticides included mostly compounds of arsenic and cyanide and various heavy metals, such as mercury and copper.

First Law of Thermodynamics The empirical observation, confirmed innumerable times, that energy is never created or destroyed but may be converted from one form to another (e.g., electricity to light). Also called the **Law of Conservation of Energy**. (See also **Second Law of Thermodynamics**.)

fishery Fish species being exploited, or a limited marine area containing commercially valuable fish.

fish ladder A stepwise series of pools on the side of a dam where the water flows in small falls that fish can negotiate.

fission The splitting of a large atom into two atoms of lighter elements. When large atoms such as uranium or plutonium fission, tremendous amounts of energy are released.

fission products Any and all atoms and subatomic particles resulting from splitting atoms in nuclear reactors or nuclear explosions. Practically all such products are highly radioactive.

fitness The state of an organism whereby it is adapted to survive and reproduce in the environment it inhabits.

flat-plate collector A solar collector that consists of a stationary, flat, black surface oriented perpendicular to the average angle of the sun. Heat absorbed by the surface is removed and transported by air or water (or some other liquid) flowing over or through the surface.

flood irrigation A technique of irrigation in which water is diverted from rivers by means of canals and is flooded through furrows into fields.

food aid Food of various forms that is donated or sold below cost to needy people for humanitarian reasons.

Food and Agriculture Organization of the United Nations (FAO) The UN agency responsible for coordinating international efforts to defeat hunger; also gathers and disseminates information about forests, fisheries, and farming.

food chain The transfer of energy and material through a series of organisms as each one is fed upon by the next.

Food Conservation and Energy Act of 2008 The latest farm bill, maintaining the existing high levels of support and subsidies to farms.

Food Quality Protection Act (FQPA) Passed in 1996, this act amended existing laws governing pesticide use and residues on foods, including the **Delaney clause**.

food security For families, the ability to meet the food needs of everyone in the family, providing freedom from hunger and malnutrition.

food web The combination of all the feeding relationships that exist in an ecosystem.

Forest Stewardship Council An alliance of organizations directed toward the certification of sustainable wood products.

fossil fuels Energy sources—mainly crude oil, coal, and natural gas—that are derived from the prehistoric photosynthetic production of organic matter on Earth.

fracking Refers to hydraulic fracturing, a technique for extracting natural gas and oil involving pumping fluids into a well to open up spaces that release gas or oil to be pumped back up the well.

fragile states Countries that are poor and conflict-prone, often lacking a stable government; an example is present-day Somalia.

fragmentation The division of a landscape into patches of habitat by road construction, agricultural lands, or residential areas.

Framework Convention on Climate Change (FCCC) A result of the 1992 Earth Summit, this international treaty was a start in negotiating agreements on steps to prevent future catastrophic climate change.

Framework Convention on Tobacco Control Adopted in 2003 and put into effect in 2005, this is a global treaty that aims to reduce the spread of smoking with the use of various strategies.

frankenfood A derogatory term used by opponents to genetic modification, applying to any food plant that has been developed with the use of genetic engineering.

free-market economy In its purest form, an economy in which the market itself determines what and how goods will be exchanged. The system is wholly in private hands.

freshwater Water that has a salt content of less than 0.1% (1,000 parts per million).

front The boundary where different air masses meet.

fuel cell A device that produces electrical energy by combining hydrogen and oxygen, used to power vehicles and other electrical machines.

fuel elements (fuel rods) The pellets of uranium or other fissionable material that are placed in tubes, or fuel rods, and that, together with the control rods, form the core of the nuclear reactor.

fuel load Dead trees and dense understories in many forestlands that tend to burn readily if ignited in a forest fire.

fuelwood The use of wood for cooking and heating; the most common energy source for people in developing countries.

fungus, pl. fungi Any of numerous species of molds, mushrooms, bracket fungi, and other forms of nonphotosynthetic plants. Fungi derive energy and nutrients by consuming other organic material. Along with bacteria, they form the decomposer component of ecosystems.

fusion The joining together of two atoms to form a single atom of a heavier element. When light atoms such as hydrogen are fused, tremendous amounts of energy are released.

gasohol A blend of 90% gasoline and 10% ethanol that can be substituted for straight gasoline. Gasohol serves to stretch gasoline supplies and enable gasoline to burn more cleanly.

gene A segment of DNA that codes for one protein, which in turn determines a particular physical, physiological, or behavioral trait.

gene pool The sum total of all the genes that exist among all the individuals of a species.

gene revolution The widespread development of genetically modified crops.

genetically modified (GM) Any organism that has received genes from a different species through the techniques of genetic engineering; also called a transgenic organism.

genetic bank The concept that natural ecosystems with all their species serve as a repository of genes that may be drawn upon to improve domestic plants and animals and to develop new medicines, among other uses.

genetic control The selective breeding of a desired plant or animal to make it resistant to attack by pests. Also, attempting to introduce harmful genes—for example, those that cause sterility—into the pest populations.

genetic engineering The artificial transfer of specific genes from one organism to another.

Genuine Progress Indicator (GPI) An alternative measure of economic progress, the GPI calculates positive and negative economic activities, to reach a more realistic view of sustainable economic activity. It is an alternative to the **gross domestic product**.

geoengineering Large-scale schemes, proposed to reduce global climate change, involving major intrusions into climatic or ecological systems.

Geographic Information Systems A digital technology enabling the location and imaging of landscapes by use of satellite signals. It employs hardware, software, and data to display many forms of geographic information.

geothermal Refers to the naturally hot interior of Earth, where heat is maintained by naturally occurring nuclear reactions.

geothermal energy Useful energy derived from water heated by the naturally hot interior of Earth.

geothermal heat pump (GHP) An energy system whereby heat can be obtained or deposited via pipes buried in the ground and an exchange process that allows heated or cooled air to be circulated via ductwork.

Gini index A way of expressing the inequality of income in a country.

GLASOD A map produced by the Global Assessment of Soil Degradation, prepared in the late 20th century and often cited but based on very suspect data.

Global Assessment of Land Degradation and Improvement (GLADA) A new assessment of land degradation based on satellite remote-sensing techniques.

Global Change Research Program (GCRP) The U.S. government-sponsored program that addresses a number of issues in climate change with research and reports.

global climate change The cumulative effects of rising levels of greenhouse gases on Earth's climate. These effects include global warming, weather changes, and a rising sea level.

Global Climate Change Initiative The U.S. voluntary response to climate change, involving a reduction in **emissions intensity**.

Global Forest Resources Assessment (FRA) A periodic assessment of forest resources conducted by the UN Food and Agricultural Organization (FAO) and made the basis of regular FAO *State of the World's Forest* reports.

Global Gag Rule The U.S. policy initiated by President Reagan and continued by other Republican presidents, whereby U.S. government aid is denied to any agency that provides abortions or abortion counseling.

globalization The accelerating interconnectedness of human activities, ideas, and cultures, especially evident in economic and information exchange.

glucose A simple sugar, the major product of photosynthesis. Glucose serves as the basic building block for cellulose and starches and as the major "fuel" for the release of energy through cell respiration in both plants and animals.

glyphosate (Roundup®) A herbicide that kills most varieties of plants, used to control weeds among crops that have been bioengineered with resistance to the glyphosate.

golden rice New strains of rice produced by genetic engineering that contain carotene (which the body uses to synthesize vitamin A) and iron, intended to compensate for nutritional deficiencies.

goods Products, such as wood and food, that are extracted from natural ecosystems to satisfy human needs.

Grameen Bank A banking network serving the poor with microloans, created by Muhammad Yunus of Bangladesh.

grassland biome Geographical regions that have seasonal rainfall of 10 to 60 inches/yr and are dominated by grass species and large grazing animals.

grassroots Environmental efforts that begin at the level of the populace, often in response to a perceived problem not being addressed by public policy.

gravitational water Water that is not held by capillary action in soil but that percolates downward by the force of gravity.

graying The increasing average age in populations in developed countries and in many developing countries that is occurring because of decreasing birthrates and increasing longevity.

gray water Wastewater, as from sinks and tubs, that does not contain human excrement. Such water can be reused without purification for some purposes.

greenhouse effect An increase in the atmospheric temperature caused by increasing amounts of carbon dioxide and certain other gases that absorb and trap heat, which normally radiates away from Earth.

greenhouse gases Gases in the atmosphere that absorb infrared energy and contribute to the air temperature. These gases are like a heat blanket and are important in insulating Earth's surface. Among the greenhouse gases are carbon dioxide, water vapor, methane, nitrous oxide, chlorofluorocarbons, and other halocarbons.

green products Manufactured products that are more environmentally benign than their traditional counterparts.

Green Revolution The development and introduction of new varieties of (mainly) wheat and rice that have increased yields per acre dramatically in many countries since the 1960s.

green water In the hydrologic cycle, water that is evaporated or transpired and returned as water vapor to the atmosphere; water in vapor form.

grit chamber A part of preliminary treatment in wastewater-treatment plants; a swimming pool–like tank in which the velocity of the water is slowed enough to let sand and other gritty material settle.

gross domestic product (GDP) The total value of all goods and services exchanged in a year within a country.

gross national product (GNP) The value of products and services produced by people of a country annually.

groundwater Water that has accumulated in the ground, completely filling and saturating all pores and spaces in rock or soil. Free to move more or less readily, it is the reservoir for springs and wells and is replenished by infiltration of surface water.

groundwater remediation The repurification of contaminated groundwater by any of a number of techniques.

gully erosion Soil erosion produced by running water and resulting in the formation of gullies.

habitat The specific environment (woods, desert, swamp) in which an organism lives.

habitat destruction Known to be responsible for 36% of known extinctions, the destruction of natural habitats is one of the greatest threats to wild species and ecosystems.

Hadley cell A system of vertical and horizontal air circulation predominating in tropical and subtropical regions and creating major weather patterns.

half-life The length of time it takes for half of an unstable isotope to decay. The length of time is the same regardless of the starting amount. Also refers to the amount of time it takes compounds to break down in the environment.

halogenated hydrocarbon A synthetic organic compound containing one or more atoms of the halogen group, which includes chlorine, fluorine, and bromine.

hard water Water that contains relatively large amounts of calcium or certain other minerals that cause soap to precipitate. (Contrast **soft water.**)

harmful algal blooms Large growths of algae that produce toxins or other negative effects which may kill fish, shellfish, or humans that eat them.

hazard Anything that can cause (1) injury, disease, or death to humans; (2) damage to property; or (3) degradation of the environment. *Cultural* hazards include factors that are often a matter of choice, such as smoking or sunbathing. *Biological* hazards are pathogens and parasites that infect humans. *Physical* hazards are natural disasters like earthquakes and tornadoes. *Chemical* hazards refer to the chemicals in use in different technologies and household products.

hazard communication standard The outcome of amendments to the Occupational Safety and Health Act of 1970, this law guarantees the workers' right to know about hazardous materials they encounter.

hazard identification The process of examining evidence linking a potential hazard to its possible harmful effects; a major element of risk assessment by the EPA.

hazardous material (HAZMAT) Any material having one or more of the following attributes: ignitability, corrosivity, reactivity, and toxicity.

health According to the World Health Organization, "a state of complete physical, social, and mental well-being and not merely the absence of disease or infirmity."

heavy metal Any of the high-atomic-weight metals, such as lead, mercury, cadmium, and zinc. All may be serious pollutants in water or soil because they are toxic in relatively low concentrations and they tend to bioaccumulate.

heavy water Deuterium oxide (D_2O), a form of water incorporating a heavy isotope of hydrogen.

herbicide A chemical used to kill or inhibit the growth of undesired plants.

herbivore An organism such as a rabbit or deer that feeds primarily on green plants or plant products such as seeds or nuts. Such an organism is said to be herbivorous. (Synonym: **primary consumer.**)

herbivory The feeding on plants that occurs in an ecosystem. The total feeding of all plant-eating organisms.

heterotroph Any organism that consumes organic matter as a source of energy. Such an organism is said to be heterotrophic.

high-level wastes In a nuclear reactor, the direct products of fission produce highly radioactive isotopes that make the management of reactor wastes difficult and hazardous.

Highway Trust Fund The monies collected from the gasoline tax and designated for the construction of new highways.

HIPPO An acronym for the major threats to biodiversity: Habitat destruction, Invasive species, Pollution, Population, and Overexploitation.

hormones Natural chemical substances that control the development, physiology, and behavior of an organism. Hormones are produced internally and affect only the individual organism. Hormones are coming into use in pest control. (See also **pheromones.**)

host In feeding relationships, particularly parasitism, refers to the organism that is being fed upon (i.e., the organism that is supporting the feeder).

host-parasite relationship The relationship between a parasite and the organism upon which it feeds.

hot spots See **biodiversity hot spots.**

household hazardous wastes An important component of the hazardous-waste problem in the United States.

Hubbert Peak The concept proposed by M. King Hubbert that peak oil production will follow a bell-shaped curve when half of available oil would have been withdrawn.

human appropriation of net primary productivity (HANPP) A way of measuring human impacts on the environment, this represents the difference between what net primary production would occur without interference by humans and what happens with human actions.

human capital   See **intangible capital.**

Human Poverty Index/Human Development Index Based on life expectancy, literacy, and living standards, these indices are used by the United Nations Development Program to measure progress in alleviating poverty.

human system   The entire system that humans have created for their own support, consisting of agriculture, industry, transportation, communication networks, and so on.

humidity   The amount of water vapor in the air. (See also **relative humidity.**)

humus   A dark brown or black, soft, spongy residue of organic matter that remains after the bulk of dead leaves, wood, or other organic matter has decomposed. Humus does oxidize, but relatively slowly. It is extremely valuable in enhancing the physical and chemical properties of soil.

hunger   A condition wherein the basic food required for meeting nutritional and energy needs is lacking and the individual is unable to lead a normal, healthy life.

hunter-gatherers   Humans surviving by hunting wild game and gathering seeds, nuts, berries, and other edible things from the natural environment.

hybrid   A plant or animal resulting from a cross between two closely related species that do not normally cross.

hybrid electric vehicle   An automobile combining a gasoline motor and a battery-powered electric motor that produces less pollution and gets higher gasoline mileage than do conventional gasoline-powered vehicles.

hybridization   Cross-mating between two more or less closely related species.

hydric soil   Soil that is permanently or seasonally innundated by water, typical of wetland areas.

hydrocarbons   *Chemistry*: Natural or synthetic organic substances that are composed mainly of carbon and hydrogen. Crude oil, fuels from crude oil, coal, animal fats, and vegetable oils are examples. *Pollution*: A wide variety of relatively small carbon-hydrogen molecules resulting from incomplete burning of fuel that are emitted into the atmosphere. (See **volatile organic compounds.**)

hydroelectric dam   A dam and an associated reservoir used to produce electrical power by letting the high-pressure water behind the dam flow through and drive a turbogenerator.

hydroelectric power   Electric power that is produced from hydroelectric dams or, in some cases, natural waterfalls.

hydrogen bonding   A weak attractive force that occurs between a hydrogen atom of one molecule and, usually, an oxygen atom of another molecule. Hydrogen bonding is responsible for holding water molecules together to produce the liquid and solid states.

hydrogen economy   A future economy where renewable energy via hydrogen as a major energy carrier prevails.

hydrogen ions   Hydrogen atoms that have lost their electrons (chemical symbol, H+).

hydrologic cycle   The movement of water from points of evaporation, through the atmosphere, through precipitation, and through or over the ground, returning to points of evaporation.

hydrosphere   The water on Earth, in all of its liquid and solid compartments: oceans, rivers, lakes, ice, and groundwater.

hydroxyl radical   The hydroxyl group (OH), missing the electron. The hydroxyl radical is a natural cleansing agent of the atmosphere. It is highly reactive, readily oxidizes many pollutants upon contact, and thus contributes to their removal from the air.

hygiene hypothesis   The theory that human immune systems need to encounter microbes and even worms at early ages in order to stimulate a regulated anti-inflammatory network; this reduces incidences of allergies and asthma.

hypothesis   A tentative guess concerning the cause of an observed phenomenon that is then subjected to experimentation to test its logical or empirical consequences.

hypoxia   See **dead zone.**

ICPD (International Conference on Population and Development) A UN-sponsored conference held in 1994 in Cairo, Egypt, that produced a program of action that was drafted and later adopted by the United Nations.

immigration   The movement of people from countries of their birth to other countries, usually with the intent to reside there permanently.

ImPACT   A refinement of the IPAT formula that separates the effects of Technology (T in the equation) into two components that incorporate the different effects of consumption of resources.

incremental value   In the measurement of the value of ecosystem services, the incremental value of a service may be calculated when changes in that service occur.

Index of Sustainable Economic Welfare (ISEW) An index of development that takes into account a wide array of economic activities and their effects on human well-being.

indicator organism   An organism, the presence or absence of which indicates certain conditions. For example, the presence of *Escherichia coli* indicates that water is contaminated with fecal wastes and that pathogens may be present.

individual quota system   A system of fishery management wherein a quota is set and individual fishers are given or sold the right to harvest some proportion of the quota. Also called **catch shares.**

industrialized agriculture   The use of fertilizer, irrigation, pesticides, and energy from fossil fuels to produce large quantities of crops and large numbers of livestock with minimal labor for domestic and foreign sale.

Industrial Revolution   During the 19th century, the development of manufacturing processes using fossil fuels and based on applications of scientific knowledge.

industrial smog   The grayish mixture of moisture, soot, and sulfurous compounds that occurs in local areas in which industries are concentrated and coal is the primary energy source.

inequality   In economics, how unevenly wealth is spread.

infant mortality   The number of babies that die before one year of age, per 1,000 babies born.

infiltration   The process in which water soaks into soil as opposed to running off the surface of the soil.

infiltration-runoff ratio   The ratio of the amount of water soaking into the soil to that running off the surface. The ratio is obtained by dividing the first amount by the second.

infrared radiation   Radiation of somewhat longer wavelengths than red light, which comprises the longest wavelengths of the visible spectrum. Such radiation manifests itself as heat.

infrastructure   The sewer and water systems, roadways, bridges, and other facilities that underlie the functioning of a city or a society.

inherently safe reactor   In theory, a nuclear reactor that is designed in such a way that any accident would be automatically corrected, with no radioactivity released.

inorganic   See **inorganic chemical.**

inorganic chemical   In classifying chemical pollutants, some are inorganic, such as the heavy metals (lead, mercury and so forth) and salts.

inorganic compounds/molecules   *Classical definition*: All things such as air, water, minerals, and metals that are neither living organisms nor products uniquely produced by living things. *Chemical definition*: All chemical compounds or molecules that do not contain carbon atoms as an integral part of their molecular structure. (Contrast **organic compounds/molecules.**)

inorganic fertilizer   See **fertilizer.**

instrumental value   The value that living organisms or species have by virtue of their benefit to people; the degree to which they benefit humans. (Contrast **intrinsic value.**)

insurance spraying   The spraying of pesticides when it is not really needed, in the belief that it will ensure against loss due to pests.

intangible capital   One component of the wealth of nations. Comprises the *human capital*, or the population and its attributes; the *social capital*, or the social and political environment that people have created in a society; and the *knowledge assets*, or the human fund of knowledge.

integrated gasification combined cycle (IGCC) plant A coal-fired technology where coal is mixed with water and oxygen and heated under pressure to produce a synthetic gas.

integrated pest management (IPM) A program consisting of two or more methods of pest control carefully integrated and designed to avoid economic loss from pests. The objective of IPM is to minimize the use of environmentally hazardous synthetic chemicals.

Integrated Risk Information System (IRIS) A compendium of risk information published by the EPA providing data on chemical hazards; available on the Internet.

integrated waste management The approach to municipal solid waste that provides for several options for dealing with wastes, including recycling, composting, waste reduction, and landfilling, as well as incineration where unavoidable.

Interdecadal Pacific Oscillation (IPO) One of several ocean-atmosphere cycles, a warm-cool cycle that swings back and forth across the Pacific over periods of several decades.

intergenerational equity Meeting the needs of the present without jeopardizing the ability of future generations to meet their needs in an equitable way.

Intergovernmental Panel on Climate Change (IPCC) The UN-sponsored organization charged with continually assessing the science of global climate change, its potential impacts, and the means of responding to the threat.

International Monetary Fund (IMF) An international organization of nations that facilitates trade and promotes poverty reduction.

International Whaling Commission (IWC) The international body that regulates the harvesting of whales; the IWC placed a ban on all whaling in 1986.

interspecific competition Competition for resources between members of two or more species.

intragenerational equity The attempt to achieve an equitable distribution of assets within a given generation or across existing people groups.

intraspecific competition Competition for resources between members of the same species.

intrinsic value The value that living organisms or species have in their own right; in other words, organisms and species do not have to be useful to have value. (Contrast **instrumental value**.)

invasive species An introduced species that spreads out and often has harmful ecological effects on other species or ecosystems.

inversion See **temperature inversion**.

inverter In a photovoltaic (PV) system, the device that acts as an interface between the solar PV modules and the electric grid or batteries.

ion An atom (or a group of atoms) that has lost or gained one or more electrons and, consequently, has acquired a positive or negative charge. Ions are designated by "+" or "−" superscripts following the chemical symbol for the element(s) involved.

ion-exchange capacity See **nutrient-holding capacity**.

ionic bond The bond formed by the attraction between a positive and a negative ion.

ionizing radiation Radiation that displaces ions in tissues as it penetrates them, leaving behind charged particles (ions).

IPAT formula A conceptual formula relating environmental impact (I) to population (P), affluence (A), and technology (T).

isotope A form of an element in which the atoms have more (or fewer) than the usual number of neutrons. Isotopes of a given element have identical chemical properties but differ in mass (weight) as a result of the superfluity (or deficiency) of neutrons. Many isotopes are unstable and radioactive. (See **radioactive decay, radioactive emissions,** and **radioactive materials**.)

ISTEA (Intermodal Surface Transportation Efficiency Act) Legislation that provides for the funding of alternative transportation (e.g., mass transit or bicycle paths) using money from the Highway Trust Fund. Succeeded by the Transportation Equity Act for the Twenty-first Century (TEA–21) and then the Safe, Accountable, Flexible, Efficient Transportation Equity Act (SAFETEA).

IUCN (International Union for Conservation of Nature) Formerly the **World Conservation Union.** An international organization that maintains a Red List of threatened species worldwide.

J-curve The shape of a growth curve for a population undergoing exponential growth.

jet stream A river of air, generated by Earth's rotation and air-pressure gradients high in the troposphere, that flows eastward at high speeds but meanders considerably.

judicial branch One of the three branches of government set up by the U.S. constitution, involving a system of judges at various levels up to the Supreme Court.

junk science Information presented as valid science but unsupported by peer-reviewed research. Often, politically motivated and biased results are selected to promote a particular point of view.

juvenile hormone The insect hormone that, at sufficient levels, preserves the larval state. Pupation requires diminished levels of juvenile hormone; hence, artificial applications of the hormone may block pupal development.

keystone species A species whose role is essential for the survival of many other species in an ecosystem.

kinetic energy The energy inherent in motion or movement, including molecular movement (heat) and the movement of waves (hence, radiation and therefore light).

knowledge assets See **intangible capital**.

K-strategist A reproductive strategy for a species whereby there is a low reproductive rate but good survival of young due to care and protection by adults. Such species often have populations close to the carrying capacity.

Kyoto Protocol An international agreement among the developed nations to curb greenhouse gas emissions. The Kyoto Protocol was forged in December 1997 and became binding in 2004.

labor In classical economic theory, one of the three elements that constitute the factors of production: the work done by people as they produce goods or services.

Lacey Act Passed in 1900, the first national act that gave protection to wildlife by forbidding interstate commerce in illegally killed animals.

land In classical economic theory, one of the three elements that constitute the factors of production: the natural resources used in the production of goods and services.

land degradation The processes by which land is rendered less capable of supporting plant growth, often through soil erosion.

land ethic First described by Aldo Leopold, this is an ethic that promotes protection of natural areas based on their integrity, stability, and beauty.

landfill A site where municipal, industrial, or chemical wastes are disposed of by burying them in the ground or placing them on the ground and covering them with earth.

landscape A group of interacting ecosystems occupying adjacent geographical areas.

land subsidence The gradual sinking of land. The condition may result from the removal of groundwater or oil, which is frequently instrumental in supporting the overlying rock and soil.

land trust Land that is purchased and held by various private organizations specifically for the purpose of protecting the region's natural environment and the biota that inhabit it.

La Niña See **El Niño/La Niña Southern Oscillation (ENSO)**.

larva, pl. larvae A free-living immature form that occurs in the life cycle of many organisms and that is structurally distinct from the adult. For example, caterpillars are the larval stage of moths and butterflies.

latitude In biomes, latitude is the distance from the equator, which determines the average temperature for a region.

Law of Conservation of Energy See **First Law of Thermodynamics**.

Law of Conservation of Matter The law stating that, in chemical reactions, atoms are neither created nor changed nor destroyed; they are only rearranged.

law of limiting factors The law stating that a system may be limited by the absence or minimum amount (in terms of that needed) of any required factor. Also known as **Liebig's law of minimums**. (See **limiting factor**.)

Law of the Sea A 1994 U.N. treaty that defined national rights and responsibilities with respect to the world's oceans.

leachate The mixture of water and materials that are leaching.

leaching The process in which materials in or on the soil gradually dissolve and are carried by water seeping through the soil. Leaching may eventually remove valuable nutrients from the soil, or it may carry buried wastes into groundwater, thereby contaminating it.

lead One of the heavy metal pollutants, lead was once added to gasoline to improve combustion. It is toxic in very small quantities and has been removed from all but aviation gasoline.

legislative branch One of the three branches of government set up by the U.S. constitution, the legislative branch consists of the House of Representatives and the Senate, all members of which are voted into office by the people of different states.

legumes The group of pod-bearing land plants that is virtually alone in its ability to fix nitrogen; legumes include such common plants as peas, beans, clovers, alfalfa, and locust trees, but no major cereal grains. (See **nitrogen fixation**.)

Liebig's law of minimums See **law of limiting factors**.

life expectancy The number of years a newborn can expect to live under mortality rates current for its country of residence.

life history The progression of changes an organism undergoes in its life from birth to death.

light-water reactor (LWR) A reactor that uses extra-pure ordinary water as a moderator. Nuclear plants in the United States are all LWRs.

limiting factor A factor primarily responsible for determining the growth or reproduction of an organism or a population. The limiting factor in a given environment may be a physical factor such as temperature or light, a chemical factor such as a particular nutrient, or a biological factor such as a competing species.

limits of tolerance The extremes of any factor (e.g., temperature) that an organism or a population can tolerate and still survive and reproduce.

lipids A class of natural organic molecules that includes animal fats, vegetable oils, and phospholipids, the last being an integral part of cellular membranes.

lithosphere The Earth's crust, made up of rocks and minerals.

litter In an ecosystem, the natural cover of dead leaves, twigs, and other dead plant material. This natural litter is subject to rapid decomposition and recycling in the ecosystem, whereas human litter—such as bottles, cans, and plastics—is not.

livability A subjective index of how enjoyable a city is to live in.

loam A soil texture consisting of a mixture of about 40% sand, 40% silt, and 20% clay; loam is usually good soil.

locavore People who try to purchase or grow most of their food within their own region; eating locally.

logistic growth A pattern of growth of a population that results in an S-shaped curve plotted over time, such that the population levels off at the carrying capacity (K).

longevity The maximum life span of individuals of a given species. The known record for humans is 122 years.

loss-of-coolant accident (LOCA) A situation in a nuclear reactor where an accident causes the loss of the water from around the reactor, resulting in overheating and a possible meltdown of the nuclear core.

low-level wastes Radioactive wastes from nuclear power plants and other facilities that employ radioisotopes; they are easier to manage than the high-level wastes.

low-till farming Similar to no-till farming, except that instead of using herbicides, farmers plow just once over residues of a previous crop and plant their new crop.

LULU ("locally unwanted land use") An acronym expressing the difficulty in siting a facility that is necessary but that no one wants in his or her immediate locality.

macromolecules Very large organic molecules, such as proteins and nucleic acids, that constitute the structural and functional parts of cells.

macrosystems ecology A subdiscipline of ecology that looks at phenomena on the scale of continents and how they interact with other scales.

Magnuson-Stevens Fishery Conservation and Management Reauthorization Act. A 2006 reauthorization of the Magnuson Act, an act passed in 1976 that extended the limits of jurisdiction over coastal waters and fisheries of the United States to 200 miles offshore.

malnutrition The lack of essential nutrients such as vitamins, minerals, and amino acids. Malnutrition ranges from mild to severe and life threatening.

marine protected area An area of the coast or open ocean that is closed to commercial fishing and mineral mining; marine reserves.

marker-assisted breeding The science of locating genes with desirable traits using DNA sequencing and then using cross-breeding with standard crop lines, thus avoiding the use of transgenics.

market-based policy An approach to environmental regulation that allows the use of markets to achieve a desired outcome (e.g., a **cap-and-trade system** for reducing emissions of a pollutant).

mass number The number that accompanies the chemical name or symbol of an element or isotope. The mass number represents the number of neutrons and protons in the nucleus of the atom.

material safety data sheets (MSDS) Documents that contain information on the reactivity and toxicity of more than 600 chemicals and that must accompany the shipping, storage, and handling of these chemicals.

materials recovery facility (MRF) A processing plant in which regionalized recycling is carried out. Recyclable municipal solid waste, usually presorted, is prepared in bulk for the recycling market.

matter Any gas, liquid, or solid that occupies space and has mass. (Contrast **energy**.)

maximum achievable control technology (MACT) The best technology available for reducing the output of especially toxic industrial pollutants.

maximum contaminant level (MCL) A drinking water standard; the highest allowable concentration of a pollutant in a drinking water source.

maximum sustainable yield (MSY) The maximum amount of a renewable resource that can be taken year after year without depleting the resource. The maximum sustainable yield is the maximum rate of use or harvest that will be balanced by the regenerative capacity of the system.

MDG Support Team The successor to the Millennium Project, this team works with developing countries to achieve the targets of the Millennium Development Goals (MDGs).

Medical Revolution Medical advances and public sanitation which led to spectacular reductions in mortality, beginning in the late 1800s and extending to the present.

megacities Cities with more than ten million inhabitants.

Megatons to Megawatts Program An exchange program between the United States and Russia whereby weapons-grade uranium from Russia is diluted to produce power-plant uranium that is used in U.S. power plants.

meltdown The event of a nuclear reactor losing its cooling water so that it melts from its own production of heat. (See **loss-of-coolant accident (LOCA)**.)

Mercury and Air Toxics Standards Rules published by the EPA in 2011 that require all coal- and oil-fired power plants to limit their emissions of mercury and other toxic pollutants; industries have four years to comply with the rules.

Meridional Overturning Circulation (MOC) The pattern of ocean currents that involves global seawater movements between the surface and great depths. See **conveyor system**.

mesosphere A layer of atmosphere beyond the stratosphere where only very small quantities of gas are found.

metabolism The sum of all the chemical reactions that occur in an organism.

meteorology The scientific study of the atmosphere, involving both weather and climate.

methane A gas, CH_4. Methane is the primary constituent of natural gas and is also a product of fermentation by microbes. Methane is one of the greenhouse gases.

methyl bromide A soil fumigant to control agricultural pests; its use releases bromine to the atmosphere, and it destroys ozone when it gets up into the stratosphere.

metropolis A large city that acts as a regional economic center.

metropolitan region A city, its suburbs, and surrounding smaller cities and towns that act as a single economy.

microbe Any microscopic organism, although primarily a bacterium, a virus, or a protozoan.

microbial Of or pertaining to microbes.

microclimate The actual conditions experienced by an organism in its particular location. Owing to numerous factors—such as shading, drainage, and sheltering—the microclimate may be quite distinct from the overall climate.

microfiltration A technology for desalination, based on forcing seawater through very small pores in a membrane filter that removes the salt.

microlending The process of providing very small loans (usually $50–$500) to poor people to facilitate their starting a small enterprise and becoming economically self-sufficient.

microorganism See **microbe**.

midnight dumping The illicit dumping of materials, particularly hazardous wastes, frequently under the cover of darkness.

Milankovitch cycle A cycle of major oscillations in the Earth's orbit, taking place over frequencies of thousands of years and known to influence the distribution of solar radiation and therefore global weather patterns.

Millennium Development Goals (MDGs) A comprehensive set of goals aimed at addressing the most important needs of people in the developing countries to improve their well-being, adopted by the United Nations at the UN Millennium Summit in 2000.

Millennium Ecosystem Assessment (MA) A four-year effort by over 1,360 scientists to produce reports on the state of Earth's ecosystems. The reports are aimed at policy makers and the public and are all available on the Internet.

Millennium Project A UN-commissioned project given a mandate to develop an action plan to coordinate the efforts of various agencies and organizations to achieve the Millennium Development Goals. Succeeded by the **MDG Support Team**.

milling A part of the process of producing nuclear fuel from uranium.

Minamata disease A disease named for a fishing village in Japan where an "epidemic" was first observed. Symptoms—which included spastic movements, mental retardation, coma, death, and crippling birth defects in the next generation—were found to be the result of mercury poisoning.

mineral Any hard, brittle, stone-like material that occurs naturally in Earth's crust. All minerals consist of various combinations of positive and negative ions held together by ionic bonds. A pure mineral, or crystal, is one specific combination of elements. Common rocks are composed of mixtures of two or more minerals.

mineralization The process of gradual oxidation of the organic matter (humus) present in soil that leaves just the gritty mineral component of the soil.

mitigation Reducing greenhouse gas emissions by a variety of strategies and policies.

mixture A combination of elements in which there is no chemical bonding between the molecules. For example, air contains (is a mixture of) oxygen, nitrogen, and carbon dioxide.

model In the scientific method, observations may be put together to form a larger picture of how a system works, called a model.

moderator In a nuclear reactor, any material that slows down neutrons from fission reactions so that they are traveling at the right speed to trigger another fission. Water and graphite are two types of moderators.

molecule The smallest unit of two or more atoms forming a compound. A molecule has all the characteristics of the compound of which it is a unit.

monoculture A large area growing only one crop.

monsoon The seasonal airflow created by major differences in cooling and heating between oceans and continents, usually bringing extensive rain.

Montreal Protocol An agreement made in 1987 by a large group of nations to cut back the production of chlorofluorocarbons in order to protect the ozone shield. A 1990 amendment called for the complete phaseout of these chemicals by 2000 in developed nations and by 2010 in less-developed nations.

morbidity The incidence of disease in a population.

mortality The incidence of death in a population.

mountaintop removal mining A form of **strip mining** for coal that takes off whole tops of mountains to get at the seams of coal that are buried as deeply as 1,000 feet below the surface.

municipal solid waste (MSW) The entirety of refuse or trash generated by a residential and business community. Distinct from agricultural and industrial wastes, municipal solid waste is the refuse that a municipality is responsible for collecting and disposing of.

mutagenic Causing mutations.

mutation A random change in one or more genes of an organism. Mutations may occur spontaneously in nature, but their number and degree are vastly increased by exposure to radiation or certain chemicals. Mutations generally result in a physical deformity or metabolic malfunction.

mutualism A close relationship between two organisms from which both derive a benefit.

mycorrhiza, pl. mycorrhizae The mycelia of certain fungi that grow symbiotically with the roots of some plants and provide for additional nutrient uptake.

nanoparticles Extremely small particles with at least one dimension smaller than 100 nanometers (nm).

National Academies of Sciences Top scientists elected by their peers to a group that often speaks to scientific issues.

National Ambient Air Quality Standards (NAAQS) The allowable levels of ambient criteria air pollutants set by EPA regulation.

National Center for Environmental Assessment (NCEA) An agency within the EPA that performs and publishes risk assessments of substances and processes that are widely released into the environment.

National Emission Standards for Hazardous Air Pollutants (NESHAPs) The standards for allowable emissions of certain toxic substances.

National Energy Policy Report The 2001 report by a task force chaired by Vice President Cheney that emphasized dependence on **supply-side policies** to generate the energy needed by the United States into the future.

national forests Public forests and woodlands administered by the National Forest Service for multiple uses, such as logging, mineral exploitation, livestock grazing, and recreation.

national parks Lands and coastal areas of great scenic, ecological, or historical importance administered by the National Park Service, with the dual goals of protecting them and providing public access.

National Pollution Discharge Elimination System (NPDES) An EPA-administered program that addresses point-source water pollution through the issuance of permits that regulate pollution discharge.

National Priorities List (NPL) A list of the chemical waste sites presenting the most immediate and severe threats. Such sites are scheduled for cleanup ahead of other sites.

national wildlife refuges Administered by the U.S. Fish and Wildlife Service, these lands are maintained for the protection and enhancement of wildlife and for the provision of public access.

natural capital Natural assets and the services they perform. One form of the wealth of a nation is its complement of natural capital.

natural chemical control The use of one or more natural chemicals, such as hormones or pheromones, to control a pest.

natural enemies All the predators and parasites that may feed on a given organism. Organisms used to control a specific pest through predation or parasitism.

natural goods The food, fuel, wood, fibers, oils, alcohols, and the like derived from the natural world, on which the world economy and human well-being depend.

natural increase The number of births minus the number of deaths in a given population. The natural increase does not take into account immigration and emigration and is the percent of growth (or decline) of a given population during a year. It is found by subtracting the crude death rate from the crude birth rate and changing the result to a percent.

natural laws Generalizations derived from our observations of matter, energy, and other phenomena. Though not absolute, natural laws have been empirically confirmed to a high degree and are often derivable from higher-level theory.

natural organic compounds The organic compounds that make up living organisms, to be distinguished from synthetic organic compounds, like plastics.

natural resources Features of natural ecosystems and species that are of economic value and that may be exploited. Also, features of particular segments of ecosystems, such as air, water, soil, and minerals.

Natural Resources Conservation Service (NRCS) Formerly the Soil Conservation Service, a nationwide network of offices that provides extension services to farmers and encourages soil and water conservation throughout the United States.

natural sciences Those scientific disciplines that study the natural world, from subatomic particles to the cosmos.

natural selection The evolutionary process whereby the natural factors of environmental resistance tend to eliminate those members of a population that are least well adapted to cope with their environment and thus, in effect, tend to select those best adapted for survival and reproduction.

natural services Functions performed free of charge by natural ecosystems, such as control of runoff and erosion, absorption of nutrients, and assimilation of air pollutants.

neglected tropical diseases (NTDs) Tropical diseases which are underfunded in comparison to their prevalence.

negligible risk standard A new safety standard based on risk analysis, established by the **FQPA** of 1996, that replaces the **Delaney clause** in dealing with pesticide residues on food products.

Neolithic Revolution The development of agriculture begun by human societies around 12,000 years ago, leading to more permanent settlement and population increases.

net metering The practice of allowing individuals on the power grid to subtract the cost of the energy they get from photovoltaics or wind turbines from their electric bills.

neutralism An interaction between two species resulting in a neutral effect on both.

neutron A fundamental atomic particle found in the nuclei of atoms (except hydrogen) and having one unit of atomic mass but no electrical charge.

new forestry Now part of the U.S. Forest Service's management practice, a forestry management strategy that places priority on protecting the ecological health and diversity of forests rather than maximizing the harvest of logs.

niche (ecological) The total of all the relationships that bear on how an organism copes with the biotic and abiotic factors it faces.

NIMBY ("not in my backyard") A common attitude regarding undesirable facilities such as incinerators, nuclear facilities, and hazardous-waste treatment plants, whereby people do everything possible to prevent such facilities from being located near their residences.

NIMTOO ("not in my term of office") An attitude expressing the reluctance of officeholders to make unpopular decisions on matters such as siting waste facilities.

nitric acid A secondary air pollutant and a major component of acid deposition, nitric acid is produced when NO_x in the air reacts with the hydroxyl radical and water.

nitrogen An element important for living organisms, the major component of the atmosphere, atomic number 7.

nitrogen cascade The detrimental complex of ecological effects brought on by reactive nitrogen (Nr) that has been added to natural systems by the burning of fossil fuels and the fertilization of agricultural crops.

nitrogen fixation The process of chemically converting nitrogen gas (N_2) from the air into compounds such as nitrates (NO_3) or ammonia (NH_3) that can be used by plants in building amino acids and other nitrogen-containing organic molecules.

nitrogen oxides A group of nitrogen-oxygen compounds formed when some of the nitrogen gas in the air combines with oxygen during high-temperature combustion. Nitrogen oxides are a major category of air pollutants and, along with hydrocarbons, are a primary factor in the production of ozone and other photochemical oxidants that are the most harmful components of photochemical smog. Nitrogen oxides also contribute to acid precipitation.

nitrous oxide A gas (N_2O). Nitrous oxide, which comes from biomass burning, fossil fuel burning, and the use of chemical fertilizers, is of concern because it is a greenhouse gas in the troposphere and it contributes to ozone destruction in the stratosphere.

nonbiodegradable Not able to be consumed or broken down by biological organisms. Nonbiodegradable substances include plastics, aluminum, and many chemicals used in industry and agriculture. Particularly dangerous are synthetic chemicals that are also toxic and tend to accumulate in organisms. (See **biodegradable** and **bioaccumulation**.)

nonconsumptive water use Uses for water in which the water remains available for further uses.

nongovernmental organization (NGO) Any of a number of private organizations involved in studying or advocating action on different environmental issues.

nonpersistent Said of chemicals that break down readily to harmless compounds, as, for example, natural organic compounds that break down to carbon dioxide and water; applied to pesticides, any pesticide that breaks down within days or a few weeks into nontoxic products.

nonpoint sources Sources of pollution such as the general runoff of sediments, fertilizers, pesticides, and other materials from farms and urban areas, as opposed to specific points of discharge such as factories. Also called **diffuse sources**. (Contrast **point sources**.)

nonreactive nitrogen The main reservoir of nitrogen as nitrogen gas in the atmosphere.

nonrenewable resources Resources, such as ores of various metals, oil, and coal that exist as finite deposits in Earth's crust and that are not replenished by natural processes as they are mined. (Contrast **renewable resources**.)

North Atlantic Deep Water (NADW) The cold, saline water that sinks in the high-latitude North Atlantic and initiates a deep current that spreads the waters throughout the oceans. See **Atlantic Meridional Overturning Circulation**.

North Atlantic Oscillation (NAO) One of several ocean-atmosphere oscillations where atmospheric pressure centers switch back and forth across the Atlantic, with many impacts on winds and storms.

no-till farming The farming practice in which weeds are killed with chemicals (or other means) and seeds are planted and grown without resorting to plowing or cultivation. The practice is highly effective in reducing soil erosion.

NRCS (Natural Resources Conservation Service) Formerly the U.S. Soil Conservation Service. The NRCS provides information to landowners regarding soil and water conservation practices.

nuclear power Electrical power generated by using a nuclear reactor to boil water and produce steam that in turn drives a turbogenerator.

Nuclear Regulatory Commission (NRC) The agency within the U.S. Department of Energy that sets and enforces safety standards for the operation and maintenance of nuclear power plants.

Nuclear Waste Policy Act of 1982 A law that committed the federal government to begin receiving nuclear wastes from commercial power plants by 1998. This has not happened yet, as the nation debates where to take the wastes.

nucleic acids The class of natural organic macromolecules that function in the storage and transfer of genetic information.

nucleus *Biology:* The large body residing in most living cells that contains the genes or hereditary material (DNA). *Physics:* The central core of atoms, which is made up of neutrons and protons. Electrons surround the nucleus.

null hypothesis In a test of a hypothesis, this is a comparison—what one would expect to see if there were no effect of one variable on another.

nutrient *Animal:* Material such as protein, vitamins, and minerals required for growth, maintenance, and repair of the body, and material such as carbohydrates required for energy. *Plant:* An essential element (C, P, N, Ca) in a particular ion or molecule that can be absorbed and used by the plant.

nutrient-holding capacity The capacity of a soil to bind and hold nutrients (fertilizer) against their tendency to be leached from the soil.

nutrient mining   Unreplenished removal of soil nutrients by crops.

observations   Things or phenomena that are perceived through one or more of the basic five senses in their normal state. In addition, to be accepted as factual, observations must be verifiable by others.

Occupational Safety and Health Act of 1970 (OSHA)   The act protecting the rights of workers to know about hazards in their workplace.

ocean acidification   An outcome of the rise in atmospheric carbon dioxide; as the oceans take up more and more of the CO_2, the carbonate ion concentration is reduced, making it more difficult for coral animals to build their calcium carbonate skeletons.

ocean thermal-energy conversion (OTEC)   The concept of harnessing the difference in temperature between surface water heated by the Sun and colder deep water to produce power.

Official Development Assistance (ODA)   Financial assistance provided to developing countries from the developed countries and coordinated by United Nations agencies.

oil field   The underground area in which exploitable oil is found.

oil sand   Sedimentary material containing bitumen, a tar-like hydrocarbon that is subject to exploitation under favorable economic conditions.

oil shale   A natural sedimentary rock that contains kerogen, a material that can be extracted and refined into oil and oil products.

oligotrophic   A body of water that is nutrient poor and hence unable to support much phytoplankton.

omnivore   An animal that feeds on both plant material and other animals.

OPEC (Organization of Petroleum Exporting Countries)   A cartel of oil-producing nations that attempts to control the market price of oil.

optimal population   The population of a harvested biological resource that yields the greatest harvest for exploitation; according to maximum-sustained-yield equations, the optimal population is half the carrying capacity.

optimal range   With respect to any particular factor or combination of factors, the maximum variation that still supports optimal or near-optimal growth of the species in question.

optimum   The condition or amount of any factor or combination of factors that will produce the best result. For example, the amount of heat, light, moisture, nutrients, and so on that will produce the best plant growth.

O&R   The combination of the American Recovery and Reinvestment Act and Obama Administration actions related to energy policy.

organic   May refer to a type of chemical or a type of farming. See organic food/gardening/farming and organic compounds/molecules.

organic chemical   See organic compound/molecules.

organic compounds/molecules   *Classical definition:* All living things and products that are uniquely produced by living things, such as wood, leather, and sugar. *Chemical definition:* All chemical compounds or molecules, natural or synthetic, that contain carbon atoms as an integral part of their molecular structure. The structure of organic compounds is based on bonded carbon atoms with hydrogen atoms attached. (Contrast inorganic compounds.)

organic fertilizer   See fertilizer.

organic food/gardening/farming   Gardening or farming without the use of inorganic fertilizers, synthetic pesticides, or other human-made materials; food raised with organic farming techniques.

Organic Foods Protection Act   A law passed in 1990 that established the National Organic Standards Board, the agency that certifies organic foods.

organic phosphate   Phosphate (PO_4-3) bonded to an organic molecule.

organically grown   Grown without the use of hard chemical pesticides or inorganic fertilizer. Any produce grown according to USDA standards for organic foods is organically grown.

orphan site   The location of a hazardous waste site abandoned by former owners.

OSHA (Occupational Safety and Health Administration)   The federal agency that promulgates regulations to protect workers.

osmosis   The phenomenon whereby water diffuses through a semipermeable membrane toward an area where there is more material in solution (i.e., where there is a relatively lower concentration of water). Osmosis has particular application in the salinization of those soils in which plants are unable to grow because of osmotic water loss.

outbreak   A population explosion of a particular pest. Often caused by an application of pesticides that destroys the pest's natural enemies.

overexploitation   The overharvesting of a species or ecosystem that leads to its decline.

overgrazing   The phenomenon of grazing animals in greater numbers than the land can support in the long term.

overnourishment   The condition of obesity and overweight that results from overeating.

oxidation   A chemical reaction that generally involves a breakdown of some substance through its combining with oxygen. Burning and cellular respiration are examples of oxidation. In both cases, organic matter is combined with oxygen and broken down to carbon dioxide and water.

ozone   A gas, O_3, that is a secondary air pollutant in the lower atmosphere but that is necessary to screen out ultraviolet radiation in the stratosphere. May also be used for disinfecting water.

ozone hole   First discovered over the Antarctic, a region of stratospheric air that is severely depleted of its normal levels of ozone during the Antarctic spring because of CFCs from anthropogenic (human-made) sources.

ozone shield   The layer of ozone gas (O_3) in the stratosphere that screens out harmful ultraviolet radiation from the Sun.

pandemic   A widely spread epidemic of infectious disease.

parasites   Organisms (plant, animal, or microbial) that attach themselves to another organism, the host, and feed on it over a period of time without killing it immediately, although usually doing harm to it. Parasites are commonly divided into ectoparasites—those that attach to the outside of their hosts—and endoparasites—those that live inside their hosts.

parasitoid   Any parasitic insect, usually a wasp, that attacks other insects and can be responsible for bringing pest insects under control.

parent material   Rock material whose weathering and gradual breakdown are the source of the mineral portion of soil.

particulates   See $PM_{2.5}$ and PM_{10} and suspended particulate matter.

parts per million (ppm)   A frequently used expression of concentration. The number of units of one substance present in a million units of another. Equivalent to milligrams per liter.

passive safety features   Those safety features of nuclear facilities that involve processes that are not vulnerable to operator intrusion or electrical power failures. (See active safety features.)

pasteurization   The process of applying enough heat to milk or some other substance to kill pathogens and sufficient other bacteria to extend the shelf life of the product.

pastoralist   One involved in animal husbandry, usually in subsistence agriculture.

pathogen   An organism, usually a microbe, that is capable of causing disease. Such an organism is said to be pathogenic.

pathogen pollution   A situation where human wastes spread diseases to wild species.

payments for ecosystem services (PES)   A market-based system that pays owners who hold properties that provide valuable services.

PAYT (pay as you throw)   A system for collecting municipal wastes that charges residents by the container for nonrecycled wastes.

peat   Compacted plant material that accumulates in wet environments where oxygen is lacking; it may be used for fuel or extracted for use in gardening.

percolation   The process of water seeping downward through cracks and pores in soil or rock.

permafrost   The ground of arctic regions that remains permanently frozen. Defines tundra because only small herbaceous plants can be sustained on the thin layer of soil that thaws each summer.

permanent polyculture   A sustainable agriculture venture where a group of perennial plants of different species are grown together to produce food and fuel.

peroxyacetyl nitrates A group of compounds present in photochemical smog that are extremely toxic to plants and irritating to the eyes, nose, and throat membranes of humans.

persistent organic pollutants (POPs) Any members of a class of organic pollutants that are resistant to biodegradation and that are often toxic; for example, DDT, PCBs, and dioxin are persistent organic pollutants. Such chemicals may remain present in the environment for periods of years.

pest Any organism that is noxious, destructive, or troublesome; usually an agricultural pest.

pesticide A chemical used to kill pests. Pesticides are further categorized according to the pests they are designed to kill—for example, herbicides kill plants, insecticides kill insects, fungicides kill fungi, and so on.

pesticide resistance See **resistance**.

pesticide treadmill The idea that use of chemical pesticides simply creates a vicious cycle of "needing more pesticides" to overcome developing resistance and secondary outbreaks caused by the pesticide applications.

pH The scale used to designate the acidity or basicity (alkalinity) of solutions or soil, expressed as the logarithm of the concentration of hydrogen ions (H^1). pH 7 is neutral; values decreasing from 7 indicate increasing acidity, and values increasing from 7 indicate increasing basicity. Each unit from 7 indicates a tenfold increase over the preceding unit.

pharma crops Crop plants that produce pharmaceutical products with the use of genomics.

pheromones Chemical substances secreted externally by certain members of a species that affect the behavior of other members of the same species. (See also **hormones**.)

phosphorus/phosphate An ion composed of a phosphorus atom with four oxygen atoms attached. Denoted PO_4^{-3}, phosphate is an important plant nutrient. In natural waters, it is frequently the limiting factor; therefore, additions of phosphate to natural water are often responsible for algal blooms.

photochemical oxidants A major category of secondary air pollutants, including ozone, that are formed as a result of interactions between nitrogen oxides and hydrocarbons driven by sunlight.

photochemical smog The brownish haze that frequently forms on otherwise clear, sunny days over large cities with significant amounts of automobile traffic. It results largely from sunlight-driven chemical reactions among nitrogen oxides and hydrocarbons.

photosynthesis The chemical process carried on by green plants through which light energy is used to produce glucose from carbon dioxide and water. Oxygen is released as a by-product.

photovoltaic cell (PV cell) A device that converts light energy into an electrical current.

phthalates A group of chemicals used to soften plastic toys and suspected of being hormone disrupters.

phytoplankton Any of the many species of photosynthetic microorganisms that consist of single cells or small groups of cells that live and grow freely suspended near the surface in bodies of water.

phytoremediation The use of plants to accomplish the cleanup of some hazardous chemical wastes.

planetary albedo The reflection of solar radiation back into space due to cloud cover, snow and ice, contributing to the cooling of the atmosphere.

planetary boundaries Safe limits on a variety of conditions on Earth that, if crossed, could become tipping points and lead to very undesirable changes that would be difficult to reverse. See **tipping point**.

plankton Any and all living things that are found freely suspended in the water and that are carried by currents, as opposed to being able to swim against currents. Plankton includes both plant (phytoplankton) and animal (zooplankton) forms.

plant community The array of plant species, including their numbers, ages, and distribution, that occupy a given area.

plant-incorporated protectants In genetically modified crops, substances that have been bioengineered to incorporate substances that would repel or kill pests.

plate tectonics The theory, proposed by Alfred Wegener, that postulates the movement of sections of the Earth's crust, creating earthquakes, volcanic activity, and the buildup of continental masses over long periods of time, also known as continental drift.

$PM_{2.5}$ and PM_{10} Standard criterion pollutants for suspended particulate matter. PM_{10} refers to particles smaller than 10 micrometers in diameter; $PM_{2.5}$ refers to particles smaller than 2.5 micrometers in diameter. The smaller particles are readily inhaled directly into the lungs.

point sources Specific points of origin of pollutants, such as factory drains or outlets from sewage-treatment plants. (Contrast **nonpoint sources**.)

policy life cycle The typical course of events occurring over time in the recognition of, formulation of an approach to, implementation of a solution to, and control of an environmental problem requiring public policy action.

pollutant A substance that contaminates air, water, or soil.

polluter pays principle An ethical principle whereby those who pollute the environment should pay for the damage their pollution causes.

pollution According to the EPA, "the presence of a substance in the environment that, because of its chemical composition or quantity, prevents the functioning of natural processes and produces undesirable environmental and health effects."

Pollution Prevention Act of 1990 A law that mandates collection of data on toxic chemicals that are released to the environment, treated on site, recycled, or combusted. See **Toxics Release Inventory**.

Pollution tax Taxes on pollutants or goods whose use produces pollutants.

polychlorinated biphenyls (PCBs) A group of widely used industrial chemicals of the chlorinated hydrocarbon class. PCBs have become serious and widespread pollutants because they are resistant to breakdown and are subject to bioaccumulation. They are also carcinogenic.

population A group within a single species whose individuals can and do freely interbreed.

population density The number of individuals per unit of area.

population equilibrium A state of balance between births and deaths in a population.

population explosion The exponential increase observed to occur in a population when conditions are such that a large percentage of the offspring are able to survive and reproduce in turn. A population explosion frequently leads to overexploitation and eventual collapse of the ecosystem.

population growth Positive or negative change in the number of individuals in a population.

population growth rate The change in population numbers divided by the time over which that change occurs.

population momentum A property whereby a rapidly growing human population may be expected to grow for 50–60 years after replacement fertility (2.1 live births per female) is reached. Momentum is sustained because of increasing numbers entering reproductive age.

population profile A bar graph that shows the number of individuals at each age or in each five-year age group, starting with the youngest ages at the bottom of the profile.

population structure See **age structure**.

positive feedback A phenomenon where some process leads to even more intensification of that process; an example is the evaporative rise in water vapor with global warming, adding water, a greenhouse gas, to the atmosphere.

potential energy The ability to do work that is stored in some chemical or physical state. For example, gasoline is a form of potential energy because the ability to do work is stored in the chemical state and is released as the fuel is burned in an engine.

poverty trap In a developing country, the situation where people are so poor that they are unable to earn or work their way out of their poverty and their condition deteriorates.

power grid The combination of central power plants, high-voltage transmission lines, poles, wires, and transformers that makes up an electrical power system.

power tower A solar energy converter that collects solar energy with an array of Sun-tracking mirrors and focuses the energy onto a receiver on a tower in the center of the array; the heat energy drives a conventional turbogenerator.

prairie biome See **grassland biome.**

precautionary principle The principle that says that where there are threats of serious or irreversible damage, the absence of scientific certainty shall not be used as a reason for postponing cost-effective measures to prevent environmental degradation.

precipitation Any form of moisture condensing in the air and depositing on the ground.

predation A conspicuous process in all ecosystems, any situation where one organism feeds on another living organism.

predator An animal that feeds on another living organism, either plant or animal.

predator-prey relationship A feeding relationship existing between two kinds of organisms. The predator is the animal feeding on the prey. Such relationships are frequently instrumental in controlling populations of herbivores.

preservation In protecting natural areas, the objective of preservation is to ensure the continuity of species and ecosystems, regardless of their potential utility.

pressurized-water reactor A nuclear power plant where two loops are employed; one where water is heated in the reactor but does not boil because it is under high pressure, and a second where the pressurized water is circulated through a heat exchanger, boiling other unpressurized water that then produces steam to drive the turbogenerator.

prey An organism that is fed on by a predator.

primary air pollutants Pollutants released directly into the atmosphere, mainly as a result of burning fuels and wastes, as opposed to **secondary air pollutants.**

primary clarifier In a sewage-treatment system, a large tank that allows particulate matter to settle to the bottom before the water moves on to **secondary treatment.**

primary consumer An organism, such as a rabbit or deer, that feeds more or less exclusively on green plants or their products, such as seeds and nuts. (Synonym: **herbivore.**)

primary energy sources Fossil fuels, radioactive material, and solar, wind, water, and other energy sources that exist as natural resources subject to exploitation.

primary producers/primary production Photosynthetic organisms. The activities of these organisms in creating new organic matter in ecosystems.

primary recovery In an oil well or oil field, the oil that can be removed by conventional pumping. (See **secondary recovery.**)

primary recycling A form of recycling where the original waste material is made back into the same material.

primary succession The gradual establishment, through a series of stages, of a climax ecosystem in an area that has not been occupied before (e.g., a rock face).

primary treatment The process that consists of passing the wastewater very slowly through a large tank so that the particulate organic material can settle out. The settled material is **raw sludge,** which is treated separately.

prior informed consent (PIC) A UN-approved procedure whereby countries exporting pesticides must inform importing countries of actions the exporting countries have taken to restrict the use of the pesticides.

private land trust See **land trust.**

produced capital The stock of buildings, machinery, vehicles, and other elements of a country's infrastructure that are essential to the production of economic goods and services. One component of the wealth of nations.

producers In an ecosystem, those organisms (mostly green plants) that use light energy to construct their organic constituents from inorganic compounds.

production In living systems, making organic molecules with simple precursors and energy, usually from the sun.

productive use The exploitation of ecosystem resources for economic gain.

productivity See **primary producers/primary production.**

property taxes Taxes that the local government levies on privately owned properties, proportional to the value of the property. Property taxes are the major source of revenue for local governments.

protein Organic macromolecules made up of amino acids; the major structural component of all animal tissues as well as enzymes in both plants and animals.

proton A fundamental atomic particle with a positive charge, found in the nuclei of atoms. The number of protons present equals the atomic number and is distinct for each element.

protozoon, pl. protozoa Any of a large group of eukaryotic microscopic organisms that consists of a single, relatively large complex cell or, in some cases, a small group of cells. All have some means of movement. *Amoeba* and *Paramecia* are examples.

proved reserves See **reserves.**

proxies These are measurable records, like tree rings, ice cores, and marine sediments, that can provide extensions back in time of climatic factors including temperatures and greenhouse gases.

public good A category of ecosystem services that is not consumed when people use it and cannot be marketed; an example is the air we breathe.

public health Organized efforts to prevent disease and harm by promoting healthy lifestyles though ducation, prevention and policy-making.

Public Trust Doctrine The legal arrangement whereby wildlife resources are considered to be public resources, held in trust for all the people and protected or managed by state or federal agencies.

pumped-storage power plant A hydroelectric plant component that pumps water up to an elevated reservoir when power demand is low and allows it to return to a lower reservoir and generate electricity when demand is higher.

purification For water, purification occurs whenever water is separated from the solutes and particles it contains.

r/r max The intrinsic rate of growth is *r*. In population studies, a variable that represents the rate of growth, a combination of birth and death rates. The maximum rate at which members of a species can produce offspring is *r max.*

radiation sickness The outcome of a high dose of radiation, often leading to death.

radiative forcing Refers to the internal or external factors of the climate system that can influence the climate, usually to promote warming or cooling of the atmosphere as they affect the Earth's energy balance.

radioactive decay The reduction in radioactivity that occurs as an unstable isotope (radioactive substance) gives off radiation, ultimately becoming stable.

radioactive emissions Any of various forms of radiation or particles that may be given off by unstable isotopes. Many such emissions have very high energy and can destroy biological tissues or cause mutations leading to cancer or birth defects.

radioactive materials Substances that are or that contain unstable isotopes and that consequently give off radioactive emissions. (See **isotope** and **radioactive emissions.**)

radioactive wastes Waste materials that contain, are contaminated with, or are radioactive substances. Many materials used in the nuclear industry become radioactive wastes because of such contamination.

radioisotope An isotope of an element that is unstable and that gives off radioactive emissions. (See **isotope** and **radioactive decay.**)

radon A radioactive gas produced by natural processes in Earth and known to seep into buildings. Radon can be a major hazard within homes and is a known carcinogen.

rain shadow The low-rainfall region that exists on the leeward (downwind) side of a mountain range. This rain shadow is the result of the mountain range causing precipitation on the windward side.

range of tolerance The range of conditions within which an organism or population can survive and reproduce—for example, the range from the highest to the lowest temperature that can be tolerated. Within the range of tolerance is the optimum, or best, condition.

rare earth elements About 17 minerals that are important in manufacture of modern electronics.

raw sewage (raw wastewater) The water from a sewer system as it reaches a sewage-treatment plant.

raw sludge The untreated organic matter that is removed from sewage water by letting it settle. Raw sludge consists of organic particles from feces, garbage, paper, and bacteria.

REACH (Registration, Evaluation and Authorization of Chemicals) The European Union's rigorous approach to screening hazardous chemicals for toxicity.

reactive nitrogen All forms of nitrogen in an ecosystem that are usable by organisms, as opposed to the nonreactive nitrogen in the form of nitrogen gas.

reactor vessel A steel-walled vessel that contains a nuclear reactor.

reasonably available control technology (RACT) Applied to the goals of the Clean Air Act, EPA-approved forms of technology that will reduce the output of industrial air pollutants. (See **maximum achievable control technology**.)

recharge area The area over which groundwater will infiltrate and resupply an aquifer.

recruitment The maturation and successful entry of young into an adult breeding population.

recycling The recovery of materials that would otherwise be buried in landfills or be combusted. Primary recycling involves remaking the same material from the waste; in secondary recycling, waste materials are made into different products.

redlining The practice of discriminatory lending and real estate showing designed to keep minorities out of white suburbs.

Red List A list of threatened and endangered species maintained by the International Union for Conservation of Nature (IUCN) and accessible on the Internet.

The Regulatory Right-to-Know Act A 2000 law that requires the White House Office of Management and Budget to provide an annual assessment of the costs and benefits of federal regulations.

relative humidity The percentage of moisture in the air compared with how much the air can hold at the given temperature.

rem An older unit of measurement of the ability of radioactive emissions to penetrate biological tissue. (See **sievert**.)

remittances Money sent back to the country of birth by guest workers and immigrants to help family and friends meet their financial needs.

remote sensing A system of imaging the Earth from satellites, or of accessing data from devices that are distant from the scientist.

renewable fuel standard (RFS) A policy that requires a minimum volume of renewable fuel in gasoline.

renewable portfolio standard A policy that requires utilities to provide a percentage of power from renewable energy sources.

renewable resources Biological resources, such as trees, that may be renewed by reproduction and regrowth. Conservation to prevent overcutting and to protect the environment is still required, however. (Contrast **nonrenewable resources**.)

replacement fertility rate/level The fertility rate (level) that will just sustain a stable population. It is 2.1 for developed countries.

representative concentration pathways (RPCs) Models in the fifth IPCC assessment on climate change that are used to project different possible future outcomes.

reprocessing The recovery of radioactive materials from nuclear reactors and the conversion of those materials into useful nuclear fuel.

reproductive health Health care directed primarily toward the needs of women and infants, strongly emphasized in the 1994 Cairo population conference.

reproductive isolation One of the processes of speciation, involving anything that keeps individuals or subpopulations from interbreeding.

reproductive strategy A life history property of a species that involves a balance between reproduction and death. (See **r-strategist** and **K-strategist**.)

reserves The amount of a mineral resource (including oil, coal, and natural gas) remaining in Earth that can be exploited using current technologies and at current prices. Usually given as proved reserves (those that have been positively identified) and undiscovered reserves (reserves from fields that have not yet been found).

resilience The tendency of ecosystems to recover from disturbances through a number of processes known as resilience mechanisms (e.g., succession after a fire).

resilience mechanisms Processes such as succession that enable an ecosystem to return after disturbance.

resistance The development of more hardy pests through the effects of the repeated use of pesticides in selecting against sensitive individuals in a pest population.

Resource Conservation and Recovery Act of 1976 (RCRA) The cornerstone legislation to control indiscriminate land disposal of hazardous wastes.

resource partitioning The outcome of competition by a group of species where natural selection favors the division of a resource in time or space by specialization of the different species.

resources Biotic and abiotic factors that are consumed by organisms.

respiration See **cell respiration**.

restoration ecology The branch of ecology devoted to restoring degraded and altered ecosystems to their natural state.

resurgence The rapid comeback of a population (especially of pests after a severe die-off, usually caused by pesticides) and the return to even higher levels than before the treatment.

reverse osmosis See **microfiltration**.

Rio+20 A UN conference about sustainable development in 2012.

risk The probability of suffering injury, disease, death, or some other loss as a result of exposure to a hazard.

risk assessment/analysis The process of evaluating the risks associated with a particular hazard before taking some action.

risk characterization The process of determining the level of a risk and its accompanying uncertainties after hazard assessment, dose-response assessment, and exposure assessment have been accomplished.

risk management The task of regulators whose job is to review risk data and make regulatory decisions based on the data. The process often is influenced by considerations of costs and benefits as well as by public perception.

risk multiplier Something that complicates a risk factor, making it worse.

risk perception Nonexperts' intuitive judgments about risks, which often are not in agreement with the level of risk as judged by experts.

roadless rule A moratorium placed on building new roads in the national forests, put in place by the Clinton administration and contested by subsequent administrations.

rotavirus A little-known virus that causes many childhood deaths, responsible for most cases of severe diarrhea.

Rotterdam Convention A mutlilateral treaty about importation of hazardous chemicals.

Roundup® See **glyphosate**.

***r*-strategist** A reproductive strategy of a species that involves producing large numbers of young (a high reproductive rate) and survival of smaller numbers over time. Also called **opportunistic species**.

rural Areas with low population density. In the U.S., aggregations smaller than 2,500 people.

Rust Belt Cities and their outskirts, often in the Midwest, that once specialized in heavy manufacturing and have since suffered a decline, characterized by abandoned factories and crumbling buildings.

Safe, Accountable, Flexible, Efficient Transportation Equity Act of 2005 (SAFETEA) Successor to TEA–21 and ISTEA, this act authorizes funding for highway and transit programs.

Safe Drinking Water Act (SDWA) of 1974 Legislation to protect the public from the risk that toxic chemicals will contaminate drinking water supplies. The act mandates regular testing of municipal water supplies.

safety net In a society, the availability of food and other necessities extended to people who are unable to meet their own needs for a variety of reasons.

salinization The process whereby soil becomes saltier and saltier until, finally, the salt prevents the growth of plants. Salinization is caused by irrigation because salts brought in with the water remain in the soil as the water evaporates.

saltwater intrusion/saltwater encroachment The phenomenon of seawater moving back into aquifers or estuaries. Such intrusion occurs when the normal outflow of freshwater is diverted or removed for use.

salvage logging Established by the Bush administration, this refers to harvesting timber from recent burns in national forestlands; it is controversial because it is suspected of preventing natural regeneration of trees on the burned land.

sand Mineral particles 0.063 to 2.0 mm in diameter.

sanitary sewer A separate drainage system used to receive all the wastewater from sinks, tubs, and toilets.

SARA (Title III) A section of the Superfund Amendments and Reauthorization Act that addresses hazardous waste accidents and promulgates community right-to-know requirements. Also known as the **Emergency Planning and Community Right-to-Know Act of 1986 (EPCRA)**. (See **Toxics Release Inventory (TRI)**.)

scale insect Any of a group of minute insect pests that suck the juices from plant cells.

science A process by which understanding of the natural world is gained; it relies on the scientific method.

scientific method The process of making observations and logically integrating those observations into a model demonstrating how the world works. Often involves forming hypotheses, experimenting, and conducting further testing for confirmation.

secondary air pollutants Air pollutants resulting from reactions of primary air pollutants resident in the atmosphere. Secondary air pollutants include ozone, other reactive organic compounds, and sulfuric and nitric acids. (See **ozone, peroxyacetyl nitrates**, and **photochemical oxidants**.)

secondary clarifier In activated sludge sewage treatment, the tank where organisms and other particles settle out after the aeration tank.

secondary consumer An organism such as a fox or coyote that feeds more or less exclusively on other animals that feed on plants.

secondary energy source An energy source such as electricity, made from a primary energy source such as solar radiation.

secondary pest outbreak The phenomenon of a small and therefore harmless population of a plant-eating insect suddenly exploding to become a serious pest problem. Often caused by the elimination of competitors through the use of pesticides.

secondary recovery In an oil well or oil field, the oil that can be removed by manipulating pressure in the oil reservoir by injecting brine or other substances; more costly than **primary recovery**.

secondary recycling Recycling processes wherein waste materials are made into different products than the starting materials. For example, wastepaper to cardboard.

secondary succession The reestablishment, through a series of stages, of a climax ecosystem in an area from which it was previously cleared.

secondary treatment Also called **biological treatment**. A sewage-treatment process that follows primary treatment. Any of a variety of systems that remove most of the remaining organic matter by enabling organisms to feed on it and oxidize it through their respiration. Trickling filters and activated-sludge systems are the most commonly used secondary-treatment methods.

second-generation pesticides Synthetic organic compounds used to kill insects and other pests, replacing the first-generation pesticides. Started with the use of DDT in the 1940s.

Secondhand smoke Smoke experienced by those breathing air contaminated with tobacco smoke from others; classified as a known human carcinogen.

Second Law of Thermodynamics The empirical observation, confirmed innumerable times, that in every energy conversion (e.g., from electricity to light), some of the energy is converted to heat and some heat always escapes from the system because it always moves toward a cooler place. Therefore, in every energy conversion, a portion of energy is lost, and since, by the **First Law of Thermodynamics**, energy cannot be created, the functioning of any system requires an energy input.

secure landfill A landfill with suitable barriers, leachate drainage, and monitoring systems, such that it is deemed secure against contaminating groundwater with hazardous wastes.

sediment Soil particles—namely, sand, silt, and clay—carried by flowing water. The same material after it has been deposited.

sedimentation The filling in of lakes, reservoirs, stream channels, and so on with soil particles, mainly sand and silt. The soil particles come from erosion. Also called **siltation**.

seep An area where groundwater seeps from the ground. Contrast with a **spring**, which is a single point from which groundwater exits.

selective cutting In forestry, the practice of cutting some mature trees from a stand but leaving enough to maintain normal ecosystem functions and a diverse biota.

selective pressure An environmental factor that causes individuals with certain traits that are not the norm for the population to survive and reproduce more than the rest of the population. Selective pressure is a fundamental mechanism of evolution.

septic system An on-site method of treating sewage that is widely used in suburban and rural areas. Requires a suitable amount of land and porous soil.

services Ecosystem functions that are essential to human life and economic well-being, such as waste breakdown, climate regulation, and erosion control. These can be further categorized as regulating, supporting, cultural, and provisioning services.

sex attractant A natural chemical substance (see **pheromones**) secreted by the female of many insect species that serves to attract males for the function of mating. Sex attractants may be used by humans in traps or to otherwise confuse insect pests in order to control them.

shadow pricing In cost-benefit analysis, a technique used to estimate benefits when normal economic analysis is ineffective. For example, people could be asked how much they might be willing to pay monthly to achieve some improvement in their environment.

sheet erosion The loss of a more or less even layer of soil from the land surface due to the impact of rain and runoff from a rainstorm.

shelterbelts Rows of trees around cultivated fields for the purpose of reducing wind erosion.

shelter-wood cutting In forestry practice, the strategy of cutting the mature trees in groups over a period of years, leaving enough trees to provide seeds and give shelter to growing seedlings.

sievert A unit of measurement of the ability of radioactive emissions to penetrate biological tissues. 1 sievert = 100 rem.

silt Soil particles between the size of sand particles and that of clay particles—namely, particles 0.004 to 0.063 mm in diameter.

siltation See **sedimentation**.

silviculture A term used to denote the practice of forest management.

simplification The human use of habitats that removes natural objects, such as maintaining a forest to produce one kind of tree or removing fallen logs.

sinkhole A large hole in the ground resulting from the collapse of an underground cavern.

slash-and-burn agriculture The practice, commonly found in tropical regions, of cutting and burning vegetation to make room for agriculture. The process destroys soil humus and may lead to rapid degradation of the soil.

sludge cake Treated sewage sludge that has been dewatered to make a moist solid.

sludge digesters Large tanks in which raw sludge (removed from sewage) is treated through anaerobic digestion by bacteria.

slum A concentrated habitation where squalor and disease reign, usually consisting of flimsy shacks and characterized by an almost complete lack of public amenities.

smart grid Control systems for power plants and electricity transmission lines with the ability to monitor problems, react to trouble quickly, and isolate trouble areas.

Smart Growth A movement that addresses urban sprawl by protecting sensitive lands and directing growth to limited areas, creating attractive, sustainable communities.

smog See **industrial smog** and **photochemical smog**.

social capital   See **intangible capital.**

social entrepreneurship   Attempt to draw on business techniques to improve social problems.

social modernization   A process of sustainable development that leads developing countries through the demographic transition by promoting education, family planning, and health improvements rather than simply economic growth.

soft water   Water with little or no calcium, magnesium, or other ions in solution that will cause soap to precipitate. (The soap would otherwise form a curd that makes a "ring" around the bathtub.) (Contrast **hard water.**)

soil   A dynamic terrestrial system involving three components: mineral particles, detritus, and soil organisms feeding on the detritus.

soil classes   Taxonomically arranged major groupings of soil that define the properties of the soils of different regions.

soil degradation   Loss of soil or lessening of its value to plant growth via erosion, salinization or other processes.

soil erosion   The loss of soil caused by particles being carried away by wind or water.

soil fertility   A soil's ability to support plant growth; often refers specifically to the presence of proper amounts of nutrients. The soil's ability to fulfill all the other needs of plants is also involved.

soil horizons   Distinct layers within a soil that convey different properties to the soil and that derive from natural processes of soil formation.

soil order   The highest group in the taxonomy of soils; there are 12 major orders of soil worldwide.

soil profile   A description of the different naturally formed layers, called horizons, within a soil.

soil structure   The composition of soil in terms of particles (sand, silt, and clay) stuck together to form clumps and aggregates, generally with considerable air spaces in between.

soil texture   The relative size of the mineral particles that make up the soil. Generally defined in terms of the soil's sand, silt, and clay content.

solar cells   See **photovoltaic cells.**

solar constant   The radiant energy reaching the top of Earth's atmosphere from the Sun, measured at 1,366 watts per square meter.

solar energy   Energy derived from the Sun; includes direct solar energy (the use of sunlight directly for heating or the production of electricity) and indirect solar energy (the use of wind, which results from the solar heating of the atmosphere, and biological materials such as wood, which result from photosynthesis).

solar-trough collectors   Reflectors, in the shape of a parabolic trough, that reflect sunlight onto a tube of oil at the focal point of the trough. Thus heated, the oil is used to boil water to drive a steam turbine.

solar variability   One proposed mechanism for global warming involving increases in solar output; this explanation has been discredited because of lack of convincing data.

solid waste   The total of materials discarded as "trash" and handled as solids, as opposed to those that are flushed down sewers and handled as liquids.

solution   A mixture of molecules (or ions) of one material in another; most commonly, molecules of air or ions of various minerals in water. For example, seawater contains salt in solution.

sound science   *The basis for our understanding of how the world works and how human systems interact with it.* It stems from scientific work based on peer-reviewed research and is one of the unifying themes of this text.

source reduction   According to the EPA, "the practice of designing, manufacturing, purchasing, or using materials in ways that reduce the amount or toxicity of trash created."

special-interest group   In politics, people organized around a particular issue who lobby for their cause and exert their influence on the political process to bring about changes they favor.

specialization   In evolution, the phenomenon whereby species become increasingly adapted to exploit one particular niche but thereby are less able to exploit other niches.

speciation   The evolutionary process whereby populations of a single species separate and, through being exposed to different forces of natural selection, gradually develop into distinct species.

species   All the organisms (plant, animal, or microbe) of a single kind. The "single kind" is determined by similarity of appearance or by the fact that members do or can mate and produce fertile offspring. Physical, chemical, or behavioral differences block breeding between species.

splash erosion   The compaction of soil that results when rainfall hits bare soil.

springs   Natural exits of groundwater to the surface.

S-shaped curves/S-curve   The curve produced when population growth over time is plotted for a population undergoing logistic growth.

stakeholders   Groups of people who have an interest in the management of a given environmental resource, from those dependent on the resource to those managing or studying the resource.

standing crop biomass   The biomass of primary producers in an ecosystem at any given time.

state capitalism   A hybrid economic system with elements of both central planning and free-market economic systems, seen in China, Russia, and many Gulf oil states.

Statement on Forest Principles   One of the treaties signed at the 1992 Earth Summit wherein the signatory countries agreed to a number of nonbinding principles stressing sustainable management.

steward/stewardship   A steward is one to whom a trust has been given. As one of the unifying themes of this text, stewardship is *the actions and programs that manage natural resources and human well-being for the common good.*

Stockholm Convention on Persistent Organic Pollutants   An international treaty that works toward the banning of most uses of the persistent organic pollutants.

storm drains   Separate drainage systems used for collecting and draining runoff from precipitation in urban areas.

stratosphere   The layer of Earth's atmosphere between 10 and 40 miles above the surface that contains the ozone shield. This layer mixes only slowly; pollutants that enter it may remain for long periods of time. (See **troposphere.**)

strip mining   A mining procedure in which all the earth covering a desired material, such as coal, is stripped away with huge power shovels in order to facilitate removal of the material.

submerged aquatic vegetation (SAV)   Aquatic plants rooted in bottom sediments and growing underwater. Submerged aquatic vegetation depends on the penetration of light through the water for photosynthesis.

subsidy   A payment or other support of produced goods that helps the producers to maintain their production and earn income above what the market would normally provide.

subsistence farmers   Farmers who live on what they grow personally, usually with little other income.

subsistence farming   Farming that meets the food needs of farmers and their families but little more. Subsistence farming involves hand labor and is practiced extensively in the developing world.

subsoil   In a natural situation, the soil beneath topsoil. In contrast to topsoil, subsoil is compacted and has little or no humus or other organic material, living or dead. In many areas, topsoil has been lost or destroyed as a result of erosion or development, and subsoil is at the surface.

suburbs   Areas that are less densely populated than cities, but interact with cities more than rural areas do.

succession   See **ecological succession.**

sulfur dioxide   A major air pollutant and toxic gas formed as a result of burning sulfur. The major sources are burning coal (in coal-burning power plants) that contains some sulfur and refining metal ores (in smelters) that contain sulfur.

sulfuric acid   The major constituent of acid precipitation. Formed when sulfur dioxide emissions react with water vapor in the atmosphere. (See **sulfur dioxide.**)

Superfund   The popular name for the **Comprehensive Environmental Response, Compensation, and Liability Act of 1980 (CERCLA).** This act is the cornerstone legislation that provides the mechanism and funding for the cleanup of potentially dangerous hazardous-waste sites and the protection of groundwater.

Superfund Amendments and Reauthorization Act of 1986 (SARA) Amendments that strengthened the Superfund program for identifying and cleaning up hazardous waste sites.

supply-side policies Those energy policies that emphasize providing greater amounts of energy through more development of oil, coal, and gas resources.

surface impoundments Closed ponds used to collect and hold liquid chemical wastes.

surface water All bodies of water, lakes, rivers, ponds, and so on that are on Earth's surface. Contrast with **groundwater**, which lies below the surface.

survivorship curve A graph depicting the percent of a population surviving at each life stage.

suspended particulate matter (SPM) A category of major air pollutants consisting of solid and liquid particles suspended in the air. (See **PM$_{2.5}$** and **PM$_{10}$**.)

sustainability A property whereby a process can be continued indefinitely without depleting the energy or material resources on which it depends. As one of the unifying themes of the text, *sustainability is the practical goal toward which our interactions with the natural world should be working.*

sustainable agriculture Agriculture that maintains the integrity of soil and water resources such that it can be continued indefinitely.

Sustainable Communities and Development Program in the United States that provides grants to encourage the development of more livable urban neighborhoods.

sustainable development Development that provides people with a better life without sacrificing or depleting resources or causing environmental impacts that will undercut the ability of future generations to meet their needs.

Sustainable Development Goals (SDGs) International goals for world development between 2015–2030, replacing the Millennium Development Goals.

sustainable forest management The management of forests as ecosystems wherein the primary objective is to maintain the biodiversity and function of the ecosystem.

sustainable society A society that functions in a way so as not to deplete the energy or material resources on which it depends. Such a society interacts with the natural world in ways that maintain existing species and ecosystems.

sustainable yield The taking of a biological resource (e.g., fish or forests) in a manner that does not exceed the capacity of the resource to reproduce and replace itself.

sustained yield In forestry, the objective of managing a forest to harvest wood continuously over time without destroying the forest.

symbiosis The intimate living together or association of two kinds of organisms.

synergism The phenomenon whereby two factors acting together have a greater effect than would be indicated by the sum of their effects separately—as, for example, the sometimes fatal mixture of modest doses of certain drugs in combination with modest doses of alcohol.

synergistic effects Where two or more factors interact to affect organisms in a way that makes the impact greater than from the two acting separately.

syngas A combustible gas produced in an integrated gasification combined cycle plant by burning coal; the syngas is then burned in a gas turbine to produce electricity.

synthetic organics Any of a large group of organic compounds that may be synthesized in chemical laboratories and are known by the difficulty with which they degrade in the environment.

tariff Any tax placed on imported goods that raises the price of those goods in the recipient country.

tax erosion The reduction in tax base during urban blight as fewer people have taxable resources.

taxonomy The science of identifying and classifying organisms according to their presumed natural relationships.

tectonic plates The huge slabs of rock in the lithosphere that have moved around and produced major changes in the surface of Earth; they are still moving, creating earthquakes and tsunamis when they do. See **plate tectonics.**

temperate deciduous forest A biome in the temperate zone characterized by deciduous trees, found where annual precipitation is 30–80 inches per year.

temperature inversion The weather phenomenon in which a layer of warm air overlies cooler air near the ground and prevents the rising and dispersion of air pollutants.

tenure Property rights over land and water resources that involve different patterns of use of those resources.

teratogenic Causing birth defects.

terracing The practice of grading sloping farmland into a series of steps and cultivating only the level portions in order to reduce erosion.

territoriality The behavioral characteristic exhibited by many animal species, especially birds and mammalian carnivores, to mark and defend a given territory against other members of the same species. A form of intraspecific competition.

territory The area defended by an animal in territorial behavior.

TESRA (Threatened and Endangered Species Recovery Act) One of a number of acts proposed in Congress in an attempt to weaken the Endangered Species Act.

theory A conceptual formulation that provides a rational explanation or framework for numerous related observations.

thermal pollution The addition of abnormal and undesirable amounts of heat to air or water. Thermal pollution is most significant with respect to the discharging of waste heat from electric generating plants—especially nuclear power plants—into bodies of water.

thermohaline Refers to temperature and salinity, specifically the cold temperature and high salinity of waters that sink in the North Atlantic and initiate the Atlantic Meridional Overturning Circulation.

thermohaline circulation Deep ocean currents, important in climate regulation, driven by differences in salinity and temperature.

thermosphere The highest layer in the atmosphere, between the mesosphere and outer space.

third world A dated term for **developing countries.**

threatened species A species whose population is declining precipitously because of direct or indirect human impacts.

threshold level The maximum degree of exposure to a pollutant, drug, or some other factor that can be tolerated with no ill effect. The threshold level varies, depending on the species, the sensitivity of the individual, the length of exposure, and the presence of other factors that may produce synergistic effects.

tidal barrage A form of renewable energy where barriers are placed across the mouth of a coastal bay and turbines are able to convert tidal water flow to power.

tilth In farming, the ability of a soil to support crop growth.

tipping point A situation in a human-impacted ecosystem where a small action or reaction is the catalyst for a major change in the state of the system.

Tobacco Control Act A law passed in 2009 giving the Food and Drug Administration greater authority to regulate tobacco products.

top-down regulation Basic control of a population (or species) occurs as a result of predation.

topsoil The surface layer of soil, which is rich in humus and other organic material, both living and dead. As a result of the activity of organisms living in the topsoil, it generally has a loose, crumbly structure, as opposed to being a compact mass.

total allowable catch (TAC) In fisheries management, a yearly quota set for the harvest of a species by managers of fisheries.

total fertility rate The average number of children that will be born alive to each woman during her total reproductive years if her fertility is average at each age.

Total Maximum Daily Load (TMDL) program An EPA-administered program to address nonpoint-source water pollution that sets pollution limits according to the ability of a body of water to assimilate different pollutants.

total product life cycle The sum of all steps in the manufacture of a product, from obtaining the raw materials through producing, using, and, finally, disposing of the product. By-products and the pollution resulting from each step are taken into account in a consideration of the total product life cycle.

toxicity The condition of being harmful, deadly, or poisonous.

toxicology The study of the impacts of toxic substances on human health and the pathways by which such substances reach humans.

Toxics Release Inventory (TRI) An annual record of releases of toxic chemicals to the environment and the locations and quantities of toxic chemicals stored at all U.S. sites. Required by the **Emergency Planning and Community Right-to-Know Act of 1986 (EPCRA)**. See **SARA**.

Toxic Substances Control Act of 1976 (TSCA) A law requiring the assessment of the potential hazards of a chemical before the chemical is put on the market.

trace elements Those essential elements, like copper or iron, that are needed in only very small amounts.

trade winds The more or less constant winds blowing in horizontal directions over the surface as part of Hadley cells.

traditional agriculture Farming methods as they were practiced before the advent of modern industrialized agriculture.

TRAFFIC network An acronym for the Trade Records Analysis for Flora and Fauna in International Commerce network, monitoring trade in wildlife and wildlife parts.

tragedy of the commons The overuse or overharvesting and consequent depletion or destruction of a renewable resource that tends to occur when the resource is treated as a commons—that is, when it is open to be used or harvested by any and all with the means to do so.

trait Any physical or behavioral characteristic or talent that an individual is born with.

transgenic organism Any organism with genes introduced from other species via biotechnology to convey new characteristics.

transpiration The loss of water vapor from plants. Water evaporates from cells within the leaves and exits through the stomata.

trapping technique The use of sex attractants to lure male insects into traps.

treadle pump A pump worked like a step exercise machine that pulls water up from groundwater for irrigation, commonly used in developing countries.

treated sludge Solid organic material that has been removed from sewage and treated so that it is nonhazardous.

trickling-filter system A system in which wastewater trickles over rocks or a framework coated with actively feeding microorganisms. The feeding action of the organisms in a well-aerated environment results in the decomposition of organic matter. Used in secondary or biological treatment of sewage.

tritium (^{3}H) An unstable isotope of hydrogen that contains two neutrons in addition to the usual single proton in the nucleus. Tritium does not occur in significant amounts in nature.

trophic level A feeding level defined with respect to the primary source of energy. Green plants are at the first trophic level, primary consumers at the second, secondary consumers at the third, and so on.

trophic structure The major feeding relationships between organisms within ecosystems, organized into trophic levels.

tropical rain forest Biome in the tropic zone where rainfall averages over 95 inches/yr; locations of great biodiversity.

troposphere/tropopause The layer of Earth's atmosphere from the surface to about 10 miles in altitude. The tropopause is the boundary between the troposphere and the stratosphere above. The troposphere is well mixed and is the site and source of our weather as well as the primary recipient of air pollutants. (See also **stratosphere**.)

tsunami A series of ocean waves (often called tidal waves) originating with underwater or coastal earthquakes and moving rapidly across ocean expanses, often resulting in major inundations of coastal areas.

tundra biome In cold latitudes and high altitudes, the pattern of plants and animals able to live where permafrost persists; dominated by low-growing sedges, shrubs, lichens, mosses, and grasses.

turbid Cloudy due to particles present. (Said of water and water purity.)

turbine A rotary engine driven at a very high speed by steam, water, or exhaust gases from combustion, employed in generating electrical energy.

turbogenerator A turbine coupled to and driving an electric generator. Virtually all commercial electricity is produced by such devices.

ultraviolet radiation Radiation similar to light but with slightly shorter wavelengths and with more energy than violet light. Its greater energy causes ultraviolet light to severely burn and otherwise damage biological tissues.

underground storage tanks (USTs) Tanks used to store petroleum products at service stations, now subject to mandated protection against leakage. (See **UST legislation**.)

undernourishment A form of hunger in which the individual lacks adequate food energy, as measured in calories. Starvation is the most severe form of undernourishment.

underweight The world's number one health risk factor, this refers to the effects of undernourishment on children that prevent their normal growth.

undiscovered resources Estimates of oil and natural gas based on geological science but not explored to confirm their presence; the estimates may be way off the mark.

United Nations Development Program (UNDP) A UN agency with offices in 131 countries, with the mission of helping developing countries to attract and use development aid effectively and, in particular, to reach the Millennium Development Goals.

United Nations Environment Program (UNEP) A UN agency with the mission of providing leadership and encouraging partnership in caring for the environment by inspiring, informing, and enabling nations and peoples to improve their quality of life without compromising that of future generations.

unmet need A situation where women who would like to postpone or prevent childbearing are not currently using contraception, often because it is not available to them.

upwelling In oceanic systems, the upward vertical movement of water masses, caused by diverging currents and offshore winds, bringing nutrient-rich water to the surface.

urban blight/decay The general deterioration of structures and facilities such as buildings and roadways, in addition to the decline in quality of services such as education, that has occurred in inner-city areas.

urban food forest A new design for urban landscapes where perennial plants and trees are grown with the intention of allowing people to forage them for food.

urban foragers Those urban dwellers who forage for fruits, berries, mushrooms, and other foods in urban landscapes.

urban heat island A city area usually 3–8° Celsius hotter than surrounding areas because of impervious surface and loss of vegetation.

urban homesteading A movement to use skills from earlier eras to reduce human resource footprints in the city.

urbanization The global movement of people into cities; a megatrend.

urbanized area A densely populated region, dominated by a human-built environment, with more than 400 people per square kilometer.

urban sprawl The widespread expansion of metropolitan areas through building houses and shopping centers farther and farther from urban centers, and lacing them together with more and more major highways, usually without planning.

USDA Organic A food that is certified by the U.S. Department of Agriculture to have been produced without the use of chemical fertilizers, pesticides, or antibiotics.

UST legislation Amendments to the Resources Conservation and Recovery Act of 1976 passed in 1984 to address the mounting problem of leaking underground storage tanks (USTs).

value In ethical considerations, a property attributed to objects, species, or individuals that implies a moral duty to the one assigning the value; intrinsic value and instrumental value are two types of value.

value added The increase in value from raw materials to final production.

vector/vector control An agent, such as an insect or tick, that carries a parasite from one host to another; methods of controlling a parasite or disease by attacking the vector.

virtual water In place of shipping water to water-stressed regions, food may be exported to those regions, thus preventing the direct need for water and replacing it with "virtual" water.

vitamin A specific organic molecule that is required by the body in small amounts but that cannot be made by the body and therefore must be present in the diet.

volatile organic compounds (VOCs) A category of major air pollutants present in the air in the vapor state. The category includes fragments of hydrocarbon fuels from incomplete combustion and evaporated organic compounds such as paint solvents, gasoline, and cleaning solutions. Volatile organic compounds are major factors in the formation of photochemical smog.

vulnerability A measure of the likelihood of exposure to a hazard.

Waste Isolation Pilot Plant (WIPP) A facility built by the U.S. Department of Energy in New Mexico to receive defense-related nuclear wastes.

waste-to-energy combustion A process of combustion of solid wastes that also generates electrical energy.

WasteWise An EPA-sponsored program that targets the reduction of municipal solid waste by a number of partnerships.

water cycle See **hydrologic cycle.**

water-holding capacity The ability of a soil to hold water so that it will be available to plants.

waterlogging The total saturation of soil with water. Waterlogging results in plant roots not being able to get air and dying as a result.

watershed The total land area that drains directly or indirectly into a particular stream or river. The watershed is generally named from the stream or river into which it drains.

water table The upper surface of groundwater, rising and falling with the amount of groundwater.

water vapor Water molecules in the gaseous state.

weather The day-to-day variations in temperature, air pressure, wind, humidity, and precipitation mediated by the atmosphere in a given region.

weathering The gradual breakdown of rock into smaller and smaller particles, caused by natural chemical, physical, and biological factors.

weed Any plant that competes with agricultural crops, forests, lawns, or forage grasses for light and nutrients.

wet cleaning A water-based alternative to dry cleaning that avoids the use of hazardous chemicals.

wetlands Areas that are constantly wet and are flooded at more or less regular intervals—especially marshy areas along coasts that are regularly flooded by tides.

wetland system A biological aquatic system (usually a restored wetland) used to remove nutrients from treated sewage wastewater and return it virtually pure to a river or stream.

Wild and Scenic Rivers Act of 1968 Legislation that protects rivers designated as "wild and scenic," preventing them from being dammed or otherwise harmfully affected.

wilderness Land that is undeveloped and wild; in the U.S., land that is protected by the Wilderness Act.

Wilderness Act of 1964 Federal legislation that provides for the permanent protection of undeveloped and unexploited areas so that natural ecological processes can operate freely in them.

Wildlife Services An agency within the U.S. Department of Agriculture that removes "nuisance" animals for landowners.

wind farms Arrays of numerous, modestly sized wind turbines for the purpose of producing electrical power.

windrows Piles of organic material extended into long rows to facilitate turning and aeration in order to enhance composting.

wind turbines Large "windmills" designed for the purpose of producing electrical power.

workability The relative ease with which a soil can be cultivated.

World Bank A branch of the United Nations that acts as a conduit for handling loans to developing countries.

World Food Program (WFP) A UN administration that receives donations of money and food from donor countries and distributes food aid to needy regions and countries.

World Trade Organization (WTO) A body that sets the rules for international trade agreements, made up of representatives from many countries and often dominated by the rich nations.

xeriscaping Landscaping with drought-resistant plants that need no watering.

yard wastes Grass clippings and other organic wastes from lawn and garden maintenance; major target for composting.

yellowcake So called because of its yellowish color, uranium ore that has been partially purified and is ready for further purification and enrichment.

Younger Dryas event A rapid change in global climate between 10,000 and 12,000 years ago that brought on 1,500 years of colder weather.

Yucca Mountain A remote location in Nevada chosen to be the nation's commercial nuclear waste repository; although well along in development, its future is uncertain.

zones of stress Regions where a species finds conditions tolerable but suboptimal. The species survives but is under stress.

zoonotic Diseases that may be transmitted between animals and humans.

zooplankton Any of a number of animal groups, including protozoa, crustaceans, worms, mollusks, and cnidarians, that live suspended in the water column and that feed on phytoplankton and other zooplankton.

zooxanthellae Photosynthetic algae that live within the tissues of coral species.

Credits

Photo Credits

Figure 15-11: Jack Milton/Portland Press Herald/Getty Images; Figure 15-12: Everett Collection Historical/Alamy; Figure 15-13: STR/AP Images; Figure 15-14: Anton Petrus/Getty Images; Page 376: ITAR-TASS Photo Agency/Alamy; Figure 15-15: Kyodo/Newscom; Figure 15-17: Robert F. Bukaty/File/AP Images

Chapter 16 Page 385: Gyula Gyukli/Fotolia; Figure 16-8: Russ Bishop Stock Connection Worldwide/Newscom; Figure 16-9: National Renewable Energy Laboratory; Figure 16-11: Ethan Miller/Getty Images; Figure 16-12: Maxfx/Shutterstock; Figure 16-13: Newscom; Figure 16-14: Iofoto/Shutterstock; Figure 16-15: Planetpix/Alamy; Figure 16-17: D. Trozzo/Alamy; Figure 16-18: Joerg Boethling/Alamy; Figure 16-19: Ideeone/iStock/Getty Images; Figure 16-20b: Alamy; Figure 16-21: Inga Spence/Alamy

Chapter 17 Page 409: Pascal Guyot/AFP PHOTO/Getty Images; Figure 17-2a: Calvin Larsen/Science Source; Figure 17-2b: Marco Cristofori/Glow Images; Figure 17-2c: TTstudio/Fotolia; Figure 17-2d: Nelson Ching/Bloomberg/Getty Images; Figure 17-4: Donald Weber/VII/Corbis; Figure 17-6a: Aflo Foto Agency/Alamy; Figure 17-6b: Peter Buchacher/Fotolia; Figure 17-6c: Mark Pearson/Alamy; Figure 17-6d: Ryan McGinnis/Alamy; Page 419: Truth Leem/Reuters; Figure 17-10: Scimat/Getty Images; Figure 17-12: Jake Lyell/Alamy; Figure 17-13: Louise Gubb/Corbis; Figure 17-14: David Slater/NHPA/Newscom; Figure 17-16: Dadang Tri/Corbis; Page 427: Bhaskar Krishnamurthy/Robert Harding/Newscom

Chapter 18 Page 435: Shahee Ilyas/Alamy; Page 445: NASA; Figure 18-10a: National Snow and Ice Data Center; Figure 18-10b: National Snow and Ice Data Center; Figure 18-20: Newscom; Figure 18-22: Jenny Matthews/Alamy; Figure 18-23: Michael Urban/AFP/Getty Images

Chapter 19 Page 465: Bettmann/Corbis; Figure 19-4a: Ekash/E+/Getty Images; Figure 19-4b: Michael Ochs Archives/Getty Images; Figure 19-7: Richard T. Wright; Figure 19-10a: Les Cunliffe/Fotolia; Figure 19-10b: Alexandra Gl/Fotolia; Figure 19-10c: Hartphotography/Fotolia; Figure 19-10d: Elenathewise/Fotolia; Figure 19-10e: Stefan Redel/Fotolia; Figure 19-16b (left): Matt Meadows/Photolibrary/Getty Images; Figure 19-16b (right): Matt Meadows/Photolibrary/Getty Images; Figure 19-17: Seth Joel/Photographer's Choice RF/Getty Images; Figure 19-18: U.S. Department of Agriculture; Figure 19-21: Balliolman/Fotolia; Page 485: Vibe Images/Fotolia; Figure 19-22: Anonymous/AP Images

Chapter 20 Page 498: NASA Earth Observatory; Figure 20-1: AP Images; Figure 20-3: Robert Harding Picture Library/SuperStock; Figure 20-5a: Bernard J. Nebel; Figure 20-5b: Bernard J. Nebel; Figure 20-5c: Bernard J. Nebel; Figure 20-5d: Bernard J. Nebel; Figure 20-8: Jerry Mcbride/ZUMA Press/Newscom; Figure 20-9c: Charles R. Belinky/Science Source; Figure 20-10: Bernard J. Nebel; Figure 20-12a: Bernard J. Nebel; Figure 20-12b: Bernard J. Nebel; Figure 20-14: Eric Fowke/Alamy; Figure 20-15a: Bernard J. Nebel; Figure 20-15b: Bernard J. Nebel; Page 515: Photo courtesy of Christopher Neill; Figure 20-20: G. Avila/Alamy; Figure 20-21: Jennifer Weinberg/Alamy

Chapter 21 Page 524: Maskot/Alloy/Corbis; Figure 21-5: Age Fotostock/SuperStock; Figure 21-6: Phil Masturzo/Akron Beacon Journal/AP Images; Figure 21-8a: Greg Williams/Rex/Newscom; Figure 21-8b: Andrew Lichtenstein/Corbis; Figure 21-11: Max Wanger/Crave/Corbis; Page 534: Peter Endig/dpa/picture-alliance/Newscom; Figure 21-12: Image courtesy of Trex Company, Inc., reprinted with permission.; Figure 21-14: Justin Sullivan/Getty Images; Figure 21-16: John Patriquin/Portland Press Herald/Getty Images; Figure 21-17: 145/Patti McConville/Ocean/Corbis; Figure 21-18: Bloomberg/Getty Images; Figure 21-19: Richard T. Wright; Page 542: Ariel Skelley/Getty Images

Chapter 22 Page 545: Weng Huan/ChinaFotoPress/ZUMAPRESS/Newscom; Figure 22-6: Olaf Krüger/imageBROKER/Alamy; Figure 22-7: U.S. Environmental Protection Agency; Page 553: M.Sobreira/Alamy; Figure 22-8: Jackie Johnston/AP Images; Figure 22-9: Nancy J. Pierce/Science Source; Figure 22-10: Van D. Bucher/Science Source; Figure 22-11: Thomas Tolstrup/Getty Images; Figure 22-13d: Image provided courtesy of Tetra Tech EM, Inc.; Figure 22-15: Newscom; Figure 22-17: Greg Gibson/AP Images

Chapter 23 Page 568: Atlanta Beltline; Figure 23-3: CI2/Corbis; Figure 23-6: Hu Guolin/Newscom; Figure 23-13: Jim West/ImageBROKER/Newscom; Figure 23-14: Medioimages/Photodisc/Getty Images; Figure 23-15: Warwick Kent/Photolibrary/Getty Images; Page 581: Alan Weintraub/Architecture/Corbis; Page 581: Marcelo Rudini/Alamy; Figure 23-16: Nic Lehoux/View Pictures/Newscom; Figure 23-18: Ames Tribune/Amy Vinchattle/AP Images; Figure 23-19: Orlando Sentinel/Getty Images; Figure 23-20: Ilene MacDonald/Alamy; Figure 23-4: Marcia Chambers/dbimages/Alamy; Figure 23-7: Esbin-Anderson/The Image Works; Figure 23-8: MaxFX/Shutterstock; Figure 23-10a: Cameron Davidson/Getty Images; Figure 23-10b: Bernard J. Nebel; Figure 23-11: Johnny Stockshooter/Alamy; Page 585: Kenneth K. Lam/MCT/Newscom

Text Credits

Chapter 1 Page 2: Excerpts from *Silent Spring* by Rachel Carson. Copyright (c) 1962 by Rachel L. Carson, renewed 1990 by Roger Christie. Reprinted by permission of Houghton Mifflin Harcourt Publishing Company. All rights reserved.; Page 3: Webster's II New College Dictionary. Boston: Houghton Mifflin Company, 1995; Page 3: Ciara Raudsepp-Hearne et al., Untangling the Environmentalist's Paradox: Why Is Human Well-Being Increasing as Ecosystem Services Degrade? Bioscience 60: 576–589. September 2010.; Page 4, Figure 1.1: Data from United Nations Development Program, Human Development Report 2014: Sustaining Human Progress: Reducing Vulnerabilities and Building Resilience. New York: UNDP. http://hdr.undp.org/en/2014-report. Oct 1, 2014.; Page 5, Figure 1.2: Human Development Report 2010, The Real Wealth of Nations: Pathways to Human Development. Copyright Creative Commons Attribution 3.0 IGO.; Page 6, Table 1.1: Data from Ecosystems and Human Well-Being: Synthesis. Washington, DC: Island Press, 2005.; Page 6, Table 1.2: Source: Adapted from Grimm, N., M. Staudinger, A. Staudt, S. Carter, F. S. Chapin III, P. Kareiva, M. Ruckelhaus, and B. Stein. "Climate-Change Impacts on Ecological Systems: Introduction to a US Assessment." Frontiers in Ecology and the Environment 9, no. 11 (2013): 456–464.; Page 7, Figure 1.3: Dr. Pieter Tans, NOAA/ESRL (www.esrl.noaa.gov/gmd/ccgg/trends/) and Dr. Ralph Keeling, Scripps Institution of Oceanography (scrippsco2.ucsd.edu/).; Page 7, Figure 1.4: Climate at a Glance, National Climate Data Center, National Oceanic & Atmospheric Administration http://www.ncdc.noaa.gov/cag/time-series/global; Page 8: Convention on Biological Diversity. Text of Convention, Article 2. Use of Terms. http://www.cbd.int/convention/articles/?a=cbd-02; Page 9, Figure 1.4: Pearson Education, Inc.; Page 11: United Nations. World Commission on Environment and Development. Report of the World Commission on Environment and Development: Our Common Future. (General Assembly: United Nations, 1987), available from http://www.un-documents.net/our-common-future.pdf; Page 12, Figure 1.7: Pearson Education, Inc.; Page 13: "The Pacific oyster, Crassostrea gigas, shows negative correlation to naturally elevated carbon dioxide levels: Implications for near-term ocean acidification effects". Barton, A, B. Hales, G. G. Waldbusser, C. Langdon, R. A. Feely. (c) 2012 Limnology and Oceanography. Reproduced with permission of John Wiley & Sons, Inc.; Page 14, Figure 1.8: Pearson Education, Inc.; Page 17: Forest Stewardship Council U.S. (FSC-US). Mission and Vision: Protecting Forests for Future Generations. Reprinted with permission.; Page 18: United Nations. Draft Principles On Human Rights And The Environment, E/CN.4/Sub.2/1994/9, Annex I (1994).; Page 20: Institution of Mechanical Engineers, Population: One Planet, Too Many People? 2011. http://www.imeche.org/Libraries/2011_Press_Releases/Population_report.sflb.ashx. September 21, 2011.

Chapter 2 Page 23: By Joowwww [Public domain], via Wikimedia Commons; Page 25, Figure 2.1: Pearson Education, Inc.; Page 25, Figure 2.2: Pearson Education, Inc.; Page 28, Figure 2.4: Pearson Education, Inc.; Page 30, Figure 2.6: Jukka Hoffrén. "Future Trends of Genuine Progress in Finland". Trends and Future of Sustainable Development. Tampere 9-10, 2011. http://www.docstoc.com/docs/84610133/Energy-and-material-flows; Page 31: TEEB. The economics of ecosystems & biodiversity. Mainstreaming the Economics of Nature: A synthesis of the approach, conclusions and recommendations of TEEB. Progress Press, Malta (2010) 39 pp.; Page 31: R. Constanza, et al. "Changes in the global value of ecosystem services". Global Environmental Change 26:152-158. 2014. Denmark; Page 34, Table 2.1: Sources: World Bank, World Development Report, 1992. New York: Oxford University Press, 1992; Millennium Ecosystem Assessment, 2005; UNEP Global Environment Outlook GEO4, 2007.; Page 36, Figure 2.10: Pearson Education, Inc.; Page 37, Figure 2.12: Pearson Education, Inc.; Page 40, Figure 2.14: Pearson Education, Inc.; Page 40, Figure 2.15a-c: Pearson Education, Inc.; Page 41, Figure 2.16: Source: Office of Management and Budget, 2014 Report to Congress on the Benefits and Costs of Federal Regulations and Unfunded Mandates on State, Local and Tribal Entities, June 2014; Page 42, Table 2.2: Pearson Education, Inc.; Page 43: James Skillen, "How Can We Do Justice to Both Public and Private Trusts?" Paper presented at the Global Stewardship Initiative National Conference, Gloucester, MA, October 1996.; Page 44, Table 2.3: Pearson Education, Inc.; Page 45: U.S. Environmental Protection Agency. EPA's Report on the Environment (2003 Draft). U.S. Environmental Protection Agency, Washington, DC, 2008.

Chapter 3 Page 48: Based on Map created bu Dr. Richard Roscie. Photovolcanica.com http://www.photovolcanica.com/PenguinSpecies/Emperor/EmperorPenguinPhotos.htmlImage Courtesu fo Dr. Richard Roscie and Photovolcanica.com; Page 50, Figure 3.1: Pearson Education, Inc.; Page 52: Louise Murray/Alamy; Page 52, Figure 3.4a-b: Pearson Education, Inc.; Page 54, Figure 3.6: Pearson Education, Inc.; Page 56, Figure 3.8: Christopherson, Robert W., Geosystems: An Introduction To Physical Geography 8th Ed., ©2012, p. 13. Reprinted and Electronically reproduced by permission of Pearson Education, Inc., New York, NY.; Page 56, Figure 3.9a-d: Pearson Education, Inc.; Page 57, Figure 3.11: Pearson Education, Inc.; Page 58, Table 3.1: Pearson Education, Inc.; Page 59, Figure 3.12: Pearson Education, Inc.; Page 60, Figure 3.13: Pearson Education, Inc.; Page 61, Figure 3.15: Pearson Education, Inc.; Page 62, Figure 3.16: Pearson Education, Inc.; Page 63, Figure 3.17: Pearson Education, Inc.; Page 65, Figure 3.19: Pearson Education, Inc.; Page 66, Figure 3.20: Pearson Education, Inc.; Page 67, Figure 3.21: Pearson Education, Inc.; Page 69, Table 3.2: Pearson Education, Inc.; Page 70, Figure 3.23: Pearson Education, Inc.; Page 68, Figure 3.22: Pearson Education, Inc.

Chapter 4 Page 73: Based on map of African elephants found [http://www.wildliferanching.com/content/african-elephant-loxodonta-africana]; Page 74, Figure 4.1: Pearson Education, Inc.; Page 75, Figure 4.2: Pearson Education, Inc.; Page 76: Sources: Whyte, I. et al., "Kruger's elephant population: its size and consequences for ecosystem heterogeneity". The Kruger Experience: Ecology And Management Of Savanna Heterogeneity. Ed. H. Biggs. (Washington, D.C.: Island Press, 2003) 339.; Whyte, I. "History of the KNP Elephant population", Elephant Effects on Biodiversity: An Assessment of Current Knowledge and Understanding as a Basis for Elephant Management in SANParks.; Page 77, Figure 4.3: Pearson Education, Inc.; Page 77, Table 4.1: Pearson Education, Inc.; Page 78, Figure 4.4: Pearson Education, Inc.; Page 80, Table 4.2: Pearson Education, Inc.; Page 81, Figure 4.7: Source: John A. Vucetich and Rolf O. Peterson. Ecological Studies of Wolves on Isle Royale: Annual Report 2012–13. 4 March 2013.; Page 84, Figure 4.11: Pearson Education, Inc.; Page 84: G.E. Hutchinson. "Homage to Santa Rosalia or Why Are There So Many Kinds of Animals?" The American Naturalist 93.870 (1959): 145-59.; Page 85, Figure 4.12a: Source: Based on data from Lack, David. Darwin's Finches., Cambridge: Cambridge University Press. 1947.; Page 85, Figure 4.12b: Pearson Education, Inc.; Page 87: Source: From Biology: Life on Earth, 4th ed., by Teresa Audesirk and Gerald Audesirk, copyright © 1994 by Prentice Hall, Inc., reprinted by permission of Pearson Education, Inc., Upper Saddle River, NJ 07458.); Page 88, Figure 4.15: Pearson Education, Inc.; Page 89, Figure 4.17: Pearson Education, Inc.; Page 90, Figure 4.18: Pearson Education, Inc.; Page 91, Figure 4.19: Pearson Education, Inc.; Page 91, Figure 4.19a: Source: For the Artic Foxes from IUCN/SSC Canid Specialist Group [http://www.canids.org/species/view/PREKKS663901]; Page 91, Figure 4.19b: Source: For the Gray Foxes from Robert L. Phillips and Robert H. Schmidt, "Foxes" Prevention and Control of Wildlife Damage. Figure 3 (1994).; Page 93, Figure 4.20 a-b: Christopherson, Robert W.,

Geosystems Introduction To Physical Geography 6th Ed., (c) 2006, p. 346. Reprinted and Electronically reproduced by permission of Pearson Education, Inc., Upper Saddle River, New Jersey.

Chapter 5 Page 101, Figure 5.1: Pearson Education, Inc.; Page 102, Figure 5.2: Pearson Education, Inc.; Page 103, Figure 5.5: Pearson Education, Inc.; Page 103, Figure 5.4: Pearson Education, Inc.; Page 104, Figure 5.6: Pearson Education, Inc.; Page 105, Figure 5.7: Pearson Education, Inc.; Page 107, Figure 5.8: Pearson Education, Inc.; Page 107, Figure 5.9: Christopherson Robert W., Geosystems An Introduction To Physical Geography 8th Ed., ©2012, p. 13. Reprinted and Electronically reproduced by permission of Pearson Education, Inc., New York, NY.; Page 108, Figure 5.10: Source: Audesirk, Gerald; Audesirk, Teresa; Byers, Bruce E., Biology: Life On Earth With Physiology, 9th Ed., ©2011, p. 560. Reprinted and Electronically reproduced by permission of Pearson Education, Inc., New York, NY.; Page 109, Table 5.1: Pearson Education, Inc.; Page 110, Figure 5.11: Pearson Education, Inc.; Page 111, Table 5.2: Pearson Education, Inc.; Page 112, Figure 5.12: Pearson Education, Inc.; Page 113, Figure 5.13: Campbell, Neil.; Reece, Jane B.; Mitchell, Lawrence G.; Taylor, Martha R., *Biology: Concepts And Connections*, 4th Ed., (c) 2003. Reprinted and Electronically reproduced by permission of Pearson Education, Inc., Upper Saddle River, New Jersey.; Page 114, Figure 5.15: Pearson Education, Inc.; Page 116, Figure 5.18: Pearson Education, Inc.; Page 117: Ten Year Review of the National Science Foundation's Long-Term Ecological Research Program. Long Term Ecological Research Network. e National Science Foundation.; Page 119, Figure 5.21: Source: Bazilevich global primary productivity map, NOAA Satellite and Information Service.; Page 121, Figure 5.22: Pearson Education, Inc.; Page 122: Committee of Scientists, Department of Agriculture, Sustaining the People's Lands: Recommendations for Stewardship of the National Forests and Grasslands into the Next Century (1999). Available at www.fs.fed.us/news/science.; Page 122: Millennium Ecosystem Assessment, Living Beyond Our Means: Natural Assets and Human Well-Being (Statement of the Millennium Assessment Board, 2005). Available at www.millenniumassessment.org/en/Products.BoardStatement.aspx

Chapter 6 Page 130, Table 6.1: Pearson Education, Inc.; Page 132: Holmes Rolston III. Environmental Ethics: Duties to and Values in the Natural World (Philadelphia: Temple University Press, 1988), p. 129.; Page 132: Aldo Leopold. "The Land Ethic," in A Sand County Almanac and Sketches Here and There (Oxford, England: Oxford University Press, 1949). Available at http://www.luminary.us/leopold/land_ethic.html.; Page 133, Table 6.2: Source: C. Mora, D.P. Tittensor, S. Adl, A. Simpson, B. Worm. "How Many Species Are There on Earth and in the Ocean?" PLoS Biol (2014) 9(8): e1001127. doi:10.1371/journal.pbio.1001127; Page 134, Figure 6.6: Source: Global Biodiversity Outlook 3; Page 143, Table 6.3: Pearson Education, Inc.; Page 146, Table 6.4: Pearson Education, Inc.; Page 148, Table 6.5: Source: Department of the Interior, U.S. Fish and Wildlife Service, Division of Endangered Species, July 10, 2014; Page 150: Convention on Biological Diversity. Rio de Janeiro. Signed 5 June 1992, ratified December 1993.; Page 152, Figure 6.21: Based on: Robert Kunzig. "Is Focusing on "Hot Spots" the Key to Preserving Biodiversity?." Scientific American: 18, 42 - 49 (2008). Available at http://www.nature.com/scientificamerican/journal/v18/n4s/box/scientificamericanearth0908-42_BX1.html; Page 153: Jefferson, Thomas. Memoir on the Discovery of Certain Bones of a Quadruped of the Clawed King in the Western Parts of Virginia. Transactions of the American Philosophical Society. 1799. Vol. A, pp. 246–260.

Chapter 7 Page 156, Figure 7.1 a-b: Pearson Education, Inc.; Page 157: Source of Data: Millennium Ecosystem Assessment. "Living Beyond Our Means: Natural Assets and Human Well-Being" (Statement of the Millennium Ecosystem Assessment Board, 2005).; Page 161, Figure 7.5: Pearson Education, Inc.; Page 162: M. Weber et al. "Why are rivers Important? Ecosystem Service Definition via Qualitative Research: A Southwestern Pilot Study." Oral Presentation. December 9, 2010. http://www.conference.ifas.ufl.edu/aces10/Presentations/Thursday/C/AM/Yes/0845%20MattWeber.pdf; Page 163, Table 7.2: Pearson Education, Inc.; Page 164, Figure 7.7: Based on map from the FAO Global Forest Resources Assessment 2000; Page 165, Figure 7.9: Food and Agriculture Organization (FAO) of the United Nations. Global Forest Resources Assessment 2010, page xvi.; Page 167, Figure 7.10: Based on Food and Agriculture Organization (FAO) of the United Nations. Global Forest Resources Assessment 2010, page 163.; Page 167: The United Nations Conference on Environment and Development (UNCED) of 1992. Copyright © 1992.; Page 168, Figure 7.12: Benjamin S. Halpern, et al. (2008). "A global map of human impact on marine ecosystems". Science, Vol. 319, No. 5865, pp. 948–952.; Page 169, Figure 7.13: Data from FAO, The State of the World's Fisheries and Aquaculture. 2010.; Page 169, Figure 7.14: Pearson Education, Inc.; Page 171, Figure 7.15: Based on data from NOAA, 2014.; Page 174: Global Ocean Commission. Decline to Recovery: a Rescue Package for the Global Ocean. 2014.; Page 174: Dr. Sylvia Earle. TedPrize Speech, 2009.; Page 175, Figure 7.19: Adapted from Figure 1 of U.S. Government Accountability Office Key Issues "Managing Natural Resources" Report. Available at http://www.gao.gov/key_issues/managing_natural_resources/issue_summary#t=0; Page 176, Figure 7.21: US Forest Service FY 1905–2013 National Summary Cut and Sold Data and Graph.; Page 178: Based on Society for Ecological Restoration, International Science and Policy Working Group, The SER International Primer on Ecological Restoration (2004). Available at www.ser.org./.; Page 180, Figure 7.26: MARS Coral reef restoration, Mars Symbioscience. Image reprinted courtesy of Mars, Inc.; Page 180, Figure 7.27: Lydie Gigerichova/ImageBROKER/Alamy; Page 180: U.N. Secretary-General, Kofi Annan. Millenium Report to the General Assembly, 2000

Chapter 8 Page 184: Pearson Education, Inc.; Page 185, Figure 8.1: Based on Joseph A. McFalls, Jr. "Population: A Lively Introduction," Population Bulletin 46, no. 2 (1991): 4; updated from UN Population Division, medium projection, 2012.; Page 186, Table 8.1: Pearson Education, Inc.; Page 190: Population Summit of the World's Scientific Academies, 1993. National Academy of Sciences, National Academy of Engineering, Institute of Medicine (SEM), p. 7. Available at http://books.nap.edu/openbook.php?record_id=9148&page=7.; Page 190: J. Van Den Bergh and P. Rietveld. "Reconsidering the Limits to World Population: Meta-Analysis and Meta-Prediction," BioScience 54, No. 3 (2004): 195.; Page 191, Figure 8.6: Source: 2014 World Population Data Sheet. http://www.prb.org/Publications/Datasheets/2014/2014-world-population-datasheet/data-sheet.aspx; Page 192, Figure 8.7: "Changes in Country Classifications". The World Bank, 2011. http://data.worldbank.org/news/2010-GNI-income-classifications.; Page 192: "Changes in Country Classifications." The World Bank, 2011. http://data.worldbank.org/news/2010-GNI-income-classifications.; Page 193, Figure 8.8: Total Population poster from Worldmapper website. Copyright © 2006 by SASI Group (University of Sheffield) and Mark Newman (University of Michigan). Reprinted with permission.; Page 194, Figure 8.9: "Population Growth in Different World Regions," from World Population Prospects: 2010 Revision. Copyright © 2010 by United Nations. Reprinted with permission. All rights reserved.; Page 194, Figure 8.10: United Nations, Department of Economic and Social Affairs, Population Division (2013): World Population Prospects: The 2012 Revision. New York http://esa.un.org/wpp/fertility_figures/interactive-maps_tf.htm; Page 195, Table 8.2: Source: Data from 2014 World Population Data Sheet (Washington, DC: Population Reference Bureau, 2014).; Page 199, Figure 8.14b: Data for part (a) from UN Population Division, World Urbanization Prospects: The 2012 Revision (New York: United Nations, 2013.; Page 201, Figure 8.15: Source: Data from Shiro Horiuchi, "World Population Growth Rate," Population Today (June 1993): 7 and (June/July 1996): 1; updated from UN Population Division, World Population Prospects: The 2010 Revision (New York: United Nations, 2010.; Page 202, Figure 8.16a-c: Based on data from U.S. Census Bureau, International Data Base.; Page 203, Figure 8.17: Mark Mather, "The Decline in U.S. Fertility," in 2012 World Population Data Sheet (Washington, DC: Population Reference Bureau, 2012). Reprinted with permission.; Page 203, Figure 8.18: Based on UN Population Division, World Population Prospects: The 2012 Revision (New York: United Nations, 2012); Page 203, Table 8.3: Source: Data from 2014 World Population Data Sheet (Washington, DC: Population Reference Bureau, 2014).; Page 204, Figure 8.19a-d: Source: U.S. Census Bureau, International Data Base.; Page 205, Figure 8.20: U.S. Census Bureau, U.S. Population Projections; Page 206, Figure 8.21a-h: U.S. Census Bureau, International Data Base.; Page 208, Figure 8.22a-b: Population Reference Bureau, "Human Population: Future Growth, Teacher's Guide," (Washington, DC: Population Reference Bureau, 2001). Redrawn with permission.; Page 209, Figure 8.23: Carl Haub and Toshiko Kaneda, 2011 World Population Data Sheet (Washington, DC: Population Reference Bureau, 2011). Reprinted with permission

Chapter 9 Page 211: Pearson Education, Inc.; Page 212, Figure 9.1: Source: Data from 2014 World Population Data Sheet (Washington, D.C.: Population Reference Bureau, 2014.; Page 213, Figure 9.2: World Development Report 2012. Gender Equality and Development. World Bank. Figure 2,Page 9.; Page 214, Figure 9.3: Based on data from the World Data Sheet 2014 and the World Economic Outlook Database, September 2014; Page 215, Figure 9.6: Source: Data from 2014 World Population Data Sheets (Washington, D.C.: Population Reference Bureau, 2014.; Page 216, Figure 9.7: Pearson Education, Inc.; Page 216, Figure 9.8: Population Reference Bureau's World Data Sheet 2011 and International Monetary Fund's World Economic Outlook Database, September 2011.; Page 219: UNFPA State of World Population 2002. People, poverty, and possibilities: making development work for the poor. Ed Thoyaya Ahmed Obaid. UN Population Fund.; Page 219, Table 9.1: Data from various World Population Data Sheets (Washington, D.C.: Population Reference Bureau).; Page 220, Table 9.2: Sources: United Nations, World Bank, Organisation for Economic Co-Operation and Development, Millennium Ecosystem Assessment.; Page 221, Figure 9.12: "The Millennium Development Goals Report." The United Nations. p. 24, 2014.; Page 222, Table 9.3: Based on Open Working Group Proposal for Sustainable Development Goals, by the United Nations, 2014.; Page 223: Making Sustainable Commitments: An Environmental Strategy for the World Bank, in 2002.; Page 223: United Nations Development Programme Mission Statement. UNDP Washington Representation Office. http://www.us.undp.org/WashingtonOffice/Mission.shtml. July 5, 2012.; Page 225, Figure 9.15: Based on information from World Bank DAC2, Center for Global Development 2014.; Page 225, Table 9.4: Source: The Index of Global Philanthropy 2013, Executive Summary. The Center for Global Prosperity. Hudson Institute, 2014 (www.global-prosperity.org).; Page 227, Figure 9.17: Data from World Health Organization, Projections of Mortality and Burden of Disease, 2002–2030 (2008).

Chapter 10 Page 235, Table 10.1: Pearson Education, Inc.; Page 233: Map provided courtesy of NOAA. http://www1.ncdc.noaa.gov/pub/data/cmb/sotc/drought/2014/08/20140902_us; Page 234, Figure 10.1: Pearson Education, Inc.; Page 236, Figure 10.3: Pearson Education, Inc.; Page 237, Figure 10.4: Pearson Education, Inc.; Page 238, Figure 10.5a-c: Pearson Education, Inc.; Page 238, Figure 10.6: Pearson Education, Inc.; Page 239, Figure 10.7: Source: Robert W. Christopherson, Elemental Geosystems, 7th ed., Pearson Education, Upper Saddle River, NJ.; Page 240, Figure 10.8: Tarbuck, Edward J.; Lutgens, Frederick K., *Earth Science*, 8th Ed., ©1997, p. 89. Reprinted and Electronically reproduced by permission of Pearson Education, Inc., New York, NY.; Page 243: Pearson Education, Inc.; Page 244, Figure 10.11: (Source: M. A. Maupin, J. F. Kenny, S. S. Hutson,,J.K. Lovelace, N. L. Barber, and K. S. Linsey, Estimated Use of Water in the United States in 2010 (Circular 1405) (Reston, VA: U.S. Geological Survey, 2014).); Page 244, Table 10.2: M. A. Maupin, J. F. Kenny, S. S. Hutson, J. K. Lovelace, N. L. Barber, and K. S. Linsey, "Estimated use of water in the United States in 2010": U.S. Geological Survey Circular 1405, (2014)56 p.; Page 245, Figure 10.12: Pearson Education, Inc.; Page 246, Figure 10.14a-b: Pearson Education, Inc.; Page 247, Figure 10.15: National Inventory of Dams, United States Army http://geo.usace.army.mil/pgis/f?p=397:5:0::NO; Page 249, Figure 10.17a: Pearson Education, Inc.; Page 250, Figure 10.18: Tom Gleeson, Yoshihide Wada, Marc F. P. Bierkens & Ludovicus P. H. van Beek. "Water Balance of global aquifers revealed by groundwater footprint" Nature 488, pp. 197-200, August 9, 2012 www.nature.co/nature/journal/v488/n7410/full/nature11295.html.; Page 251, Figure 10.20a-b: Pearson Education, Inc.

Chapter 11 Page 260: Pearson Education, Inc.; Page 260: Franklin D. Roosevelt Letter, February 26, 1937, to state governors. Available from http://www.presidency.ucsb.edu/ws/?pid=15373; Page 262, Figure 11.2: Pearson Education, Inc.; Page 263, Figure 11.3: Pearson Education, Inc.; Page 263, Figure 11.4: Pearson Education, Inc.; Page 263,

Table 11.1: Pearson Education, Inc.; Page 264, Figure 11.5: Pearson Education, Inc.; Page 265, Figure 11.6: "Global Soil Regions" by Natural Resources Conservation Service, November 2005. United States Department of Agriculture; Page 266, Figure 11.7: Pearson Education, Inc.; Page 267, Figure 11.8: Pearson Education, Inc.; Page 269, Figure 11.12: Pearson Education, Inc.; Page 270, Figure 11.13: UNEP, International Soil Reference and Information Centre (ISRIC), World Atlas of Desertification, 1997. Phillippe Rekacewicz, UNEP/GRID-Arendal. Reprinted with permission from GRID-Arendal; Page 271, Figure 11.15a: Pearson Education, Inc.; Page 278: Pearson Education, Inc.

Chapter 12 Page 284, Figure 12.3: Source: FAOSTAT.; Page 285, Figure 12.5: Pearson Education, Inc.; Page 289: Adapted from Figure 2 in G. Eshel, A. Shepon, T. Makov, and R. Milo. "Land, irrigation water, greenhouse gas, and reactive nitrogen burdens of meat, eggs, and dairy production in the United States" PNAS (2014) 111 (33) 11996–12001; Page 292, Figure 12.11: Pearson Education, Inc.; Page 293: World Health Organization. "20 Questions on Genetically Modified Foods/Question 8." (2012) Accessed Dec 5, 2014, http://www.who.int/foodsafety/publications/biotech/20questions/en/ June 25, 2012.; Page 295: Cartagena Protocol on Biosafety to the Convention on Biological Diversity. Convention on Biological Diversity (CBD). 15 May 2000.; Page 295: "Rio Declaration on Environment and Development 1992", article 15, upheld in Cartagena Protocol, Articles 1, 10.6 and 11.8:; Page 295, Table 12.1: Pearson Education, Inc.; Page 296, Table 12.2: Source: FAO FAOSTAT, FAO Food Outlook, June 2008.; Page 297, Table 12.3: Source: OECD/FAO Agricultural Outlook 2008–2017.; Page 298: Bread for the World Institute, "Agriculture in the Global Economy Hunger 2003",13th Annual Report on the State of World Hunger, (Washington, D.C: Bread for the World Institute, 2003) p20.; Page 299: Andrew Wardell. "Extreme Hunger in East Africa and the Sahel: Forewarned but Not Forearmed", Poverty MattersBlog, The Guardian. May 9, 2012. http://www.guardian.co.uk/global-development/poverty-matters/2012/may/09/extreme-hunger-east-africa-sahel.; Page 299: UN World Food Conference in 1974; Page 300: World Health Organization, 2012 Obesity and Overweight Factsheet No. 311, Accessed Mar. 6, 2012, http://www.who.int/mediacentre/factsheets/fs311/en/.; Page 300, Figure 12.16: Pearson Education, Inc.; Page 301, Figure 12.18: Adapted from graph "Proportion of children under age five moderately or severely underweight, 1990 and 2012 (Percentage)." The Millennium Development Goals Report 2014. United Nations. New York. p13; Page 301: UN Food and Agriculture Organization, State of Food Insecurity in the World 2002, (Rome: UNFAO, 2003); Page 303, Figure 12.19: Source: http://www.wfp.org/content/food-aid-flows-2012-report; Page 304: Gordon Conway, "The Doubly Green Revolution: Food for All in the 21st Century",(Ithaca, NY: Cornell University Press, 1999); Page 305, Figure 12.20: Pearson Education, Inc.

Chapter 13 Page 313, Figure 13.5: Data from the Environmental Protection Agency, Office of Pesticide Programs.; Page 316, Table 13.1: Source: Private Pesticide Applicator Training Manual, 1993, U.S. Environmental Protection Agency, Region VIII.; Page 318, Figure 13.9: Pearson Education, Inc.; Page 320, Figure 13.10: Pearson Education, Inc.; Page 322, Figure 13.12: Pearson Education, Inc.; Page 323, Figure 13.13: Pearson Education, Inc.; Page 324, Figure 13.15a: Pearson Education, Inc.; Page 327, Figure 13.21: Pearson Education, Inc.; Page 329, Figure 13.23: Pearson Education, Inc.; Page 329, Figure 13.23a: United States Department of Agriculture; Page 331: Federal Food, Drug, and Cosmetic Act (FFDCA) of 1938.

Chapter 14 Page 337, Figure 14.2: Source: Data from BP Statistical Review of World Energy, 2012, p. 41; Page 338, Figure 14.3: Source: Data from Energy Information Administration, U.S. Department of Energy, Monthly Energy Review, Feb. 2015.; Page 339, Figure 14.4: Data from Energy Information Administration, U.S. Department of Energy, Annual Energy Review.; Page 340, Figure 14.5: M.J. Bradley & Associates. (2014). Benchmarking Air Emissions of the 100 Largest Electric Power Producers in the United States. Figure 4, pg. 10. May 2014. Available at http://www.nrdc.org/air/pollution/benchmarking/files/benchmarking-2014.pdf. Reprinted with permission.; Page 341, Figure 14.6: Pearson Education, Inc.; Page 341, Table 14.1: Sources: ISO New England, 2014; U.S. Energy Information Administration,2013; Shift Project Data Portal 2012.; Page 343, Figure 14.8: Source: LLNL 2014 Data is based on DOE/EIA-0035(2014-03), March 2014. Courtesy of Lawrence Livermore National Laboratory and U.S. Department of Energy; Page 345, Figure 14.9: Pearson Education, Inc.; Page 346, Figure 14.10: Source: Data from Energy Information Administration, U.S. Department of Energy, Annual Energy Review 2010, October 2011.; Page 347, Figure 14.11: Adapted from Figure 1.1 "Primary Energy Overview." U.S. Energy Information Administration/ Monthly Energy Review March 2015.; Page 348, Figure 14.12: Source: zFacts.com, using data from U.S. Department of Energy sources; Page 350, Table 14.2: Source: BP Statistical Review of World Energy, 2014; Page 351, Figure 14.14: Pearson Education, Inc.; Page 351, Figure 14.15: Source: U.S. State Department; Page 352, Figure 14.16: Source: U.S. Energy Information Administration; Page 353, Figure 14.17b: © 2015 StudioSayers. All Rights Reserved.; Page 354: Source: Figure 10 in Charles A. S. Hall and John W. Day Jr. "Revisiting the Limits to Growth after Peak Oil." American Scientist 97, May–June 2009, pp. 230–237. Reprinted with permission.; Page 357, Table 14.3: Pearson Education, Inc.; Page 358, Table 14.4: Pearson Education, Inc.

Chapter 15 Page 363: Alfven, Hannes. "Energy and Environment," Bulletin of the Atomic Scientists (May 1972), 6.; Page 364, Figure 15.1: Source: Data from U.S. Department of Energy; Page 365, Figure 15.3: Source: Data from International Atomic Energy Agency; Page 366, Figure 15.4: Pearson Education, Inc.; Page 367, Figure 15.5: Pearson Education, Inc.; Page 368, Figure 15.6: Pearson Education, Inc.; Page 369, Figure 15.8: Pearson Education, Inc.; Page 369, Figure 15.9: Pearson Education, Inc.; Page 370, Figure 15.10: Pearson Education, Inc.; Page 371, Table 15.1: Source: Data from U.S. Nuclear Regulatory Commission; Page 372, Table 15.2: Pearson Education, Inc.; Page 374: Lee H. Hamilton and Brent Scowcroft, chairmen. Draft Report of the Blue Ribbon Commission on America's Nuclear Future. July 2011. www.energy.gov. Nov. 1, 2011.; Page 378, Figure 15.16: George Retseck; Page 378: Nuclear Regulatory Commission. Recommendations for Enhancing Reactor Safety in the 21st Century. Near-Term Task Force. July 12, 2011. www.psr.org. Nov. 3, 2011; Page 379: Union of Concerned Scientists. "Can It Happen Here?" Catalyst, Summer 2011: 7–9. Copyright © 2011.; Page 382, Figure 15.18: Source: Data from International Atomic Energy Agency

Chapter 16 Page 385: Pearson Education, Inc.; Page 386, Figure 16.1: Pearson Education, Inc.; Page 386, Table 16.1: Ren 21, Global Status Report, 2014; Page 387, Figure 16.2: Perez, Richard and Marc Perez. A Fundamental Look at Energy Reserves for the Planet. Figure 1. International Energy Agency Solar Heating and Cooling Programme (SHC) Solar Update.; Page 387: Casey, Zoe. "Fossil fuel subsidies are 'public enemy number one'- IEA Chief." Feb.4, 2013. EWEA. http://www.ewea.org/blog/2013/02/fossil-fuel-subsidies-are-public-enemy-number-one/ Jan 12, 2014.; Page 388, Figure 16.3: Pearson Education, Inc.; Page 389, Figure 16.4: Pearson Education, Inc.; Page 389, Figure 16.5: Pearson Education, Inc.; Page 390, Figure 16.6: Pearson Education, Inc.; Page 390, Figure 16.7: Pearson Education, Inc.; Page 391: National Renewable Energy Laboratory; Page 392, Figure 16.10: Shahan, Zachary. "World Solar Power Capacity Increased 35% In 2013 (Charts)" April 13th, 2014. Available at http://cleantechnica.com/2014/04/13/world-solar-power-capacity-increased-35-2013-charts/. Reprinted courtesy of SolarPower Europe.; Page 396: Dams and Development: A New Framework for Decision-Making. The Report of the World Commission on Dams. Earthscan Publications. 2000.; Page 397, Figure 16.16: Pearson Education, Inc.; Page 402, Figure 16.20a: Pearson Education, Inc.; Page 404, Figure 16.22: Pearson Education, Inc.

Chapter 17 Page 408: Speth, James Gustave. Red Sky at Morning. Yale University Press, New Haven. 2004.; Page 410: World Health Organization Glossary. Available at www.who.int/health.systemsperformance/docs/glossary.htm; Page 411, Figure 17.1: Data is from https://www.cia.gov/library/publications/the-world-factbook/fields/2102.html and https://www.cia.gov/library/publications/the-world-factbook/fields/2119.html.; Page 413, Figure 17.3: Source: Lead Poisoning: Indicators on Children and Youth. Figure 1. Child Trends Data Bank. March 2015. Data from CDC's National Surveillance Data (1997-2013): Number of Children Tested and Confirmed EBLLs by State, Year, and BLL Group, Children < 72 Months Old Available at: http://www.cdc.gov/nceh/lead/data/Website_StateConfirmedByYear_1997_2013_10162014.htm; Page 414, Table 17.1: Data from Institute for Health Metrics and Evaluation (IHME). GBD Database. Seattle, WA: IHME, University of Washington, 2014. Available from http://www.healthdata.org/search-gbd-data?s=global%20mortality. Accessed Jan 30, 2015.; Page 415, Figure 17.5: Pearson Education, Inc.; Page 415, Table 17.2: EM-DAT: The OFDA/CRED International Disaster Database, www.emdat.be, Université Catholique de Louvain, Brussels (Belgium).; Page 417, Figure 17.7: Data from the National Vital Statistics Reports, January 11, 2012; smoking deaths and obesity deaths from the Centers for Disease Control and Journal of the American Medical Association.; Page 417: U.S. Environmental Protection Agency (EPA); Page 418, Figure 17.8: Pearson Education, Inc.; Page 420, Figure 17.9: Reprinted from "Priority Risks and Future Trends," Environment and Health in Developing Countries. World Health Organization, 2014. Available at http://www.who.int/heli/risks/ehindevcoun/en.; Page 420, Figure 17.11: Malaria Atlas Project. Availabe at http://www.map.ox.ac.uk/about-map/open-access. January 16, 2012.; Page 421: World Health Organization; Page 424, Figure 17.15: Pearson Education, Inc.; Page 426: "Hyogo Framework for Action 2005–2015". United Nations Office for Disaster Risk Reduction (UNISDR). Reprinted with the permission of the United Nations; Page 427: Pearson Education, Inc.; Page 427: Source: Data from Worley, Heidi. "Water, Sanitation, Hygiene, and Malnutrition in India." Population Reference Bureau. Available at http://www.prb.org/Publications/Articles/2014/india-sanitation-malnutrition.aspx.; Page 428, Figure 17.17: Knud Juel, Jan Sorensen, and Henrik Bronnum-Hansen. Risk Factors and Public Health in Denmark, Summary Report. National Institute of Public Health and University of Southern Denmark, Copenhagen, 2007.; Page 429, Figure 17.18: EPA, http://epa.gov/risk/risk-characterization.htm. January 19, 2011.; Page 429, Table 17.3: Data from National Vital Statistics Reports 59:4, March 16, 2011, and Health and Safety Executive, UK, September 4, 2012.; Page 430, Figure 17.19: Pearson Education, Inc.; Page 431, Table 17.4: World Health Report, 2009.; Page 431: The World Health Report, "Preventing Risks and Taking Action" World Health Organization. p. 166, 2002.

Chapter 18 Page 435: Pearson Education, Inc.; Page 436, Table 18.1: Pearson Education, Inc.; Page 437, Figure 18.1: Pearson Education, Inc.; Page 438, Figure 18.2: Adapted from: S. Solomon et al., Eds. Climate Change 2007: The Physical Science Basis, Contribution of Working Group I to the Fourth Assessment Report of the Intergovernmental Panel on Climate Change (Cambridge, England: Cambridge University Press, 2007), FAQ 1.3, Figure 1.; Page 440, Figure 18.3: Pearson Education, Inc.; Page 441, Figure 18.4: Adapted from Figure 3.15 from *Our Changing Planet*, 2nd ed., by Fred T. MacKenzie. Copyright © 1998 by Prentice Hall, Inc. Reprinted by permission of Pearson Education, Inc., Upper Saddle River, NJ 07458.; Page 442, Figure 18.5: Data from NOAA Earth System Research Laboratory, 2011.; Page 443, Figure 18.6: Climate Change 2014 Synthesis Report. IPCC Fifth Assessment Report (AR5). Intergovernmental Panel on Climate Change. Topic 1 (a). p 41.; Page 443, Figure 18.7a: Adapted from The Niels Bohr Institute Archives. Reprinted with permission.; Page 443, Figure 18.7b-c: Adapted from Figure 14.26 a & b from *Our Changing Planet*, 2nd ed., by Fred T. MacKenzie. Copyright © 1998 by Prentice Hall, Inc. Reprinted by permission of Pearson Education, Inc., Upper Saddle River, NJ 07458; Page 444, Figure 18.8: Adapted from "The Rise In Global Mean Sea Level (2011_re14)," from CU Sea Level Research Group website. Copyright © 2011 by University of Colorado. Reprinted with permission.; Page 445: Rebecca Lindsey and Robert Simmon, "Correcting Ocean Cooling." NASA Earth Observatory, November 5, 2008.; Page 446, Figure 18.9a: National Snow and Ice Data Center Report, http://nsidc.org/ arcticseaicenews/2011/09/; Page 446, Figure 18.9b: National Snow and Ice Data Center Report.; Page 447: Intergovernmental Panel on Climate Change. Climate Change 2014: Synthesis Report. Contribution of Working Groups I, II and III to the Fifth Assessment. Report of the Intergovernmental Panel on Climate Change (Core Writing Team, Pachauri, R. K., and Reisinger, A. (Eds.)). IPCC, Geneva, Switzerland, 2014.; Page 448, Figure 18.11: CDIAC, Oak Ridge National Laboratory, 2014.; Page 448, Figure 18.12: Data from U.S. Energy Information Administration, December 2014.; Page 448, Figure 18.13: NOAA Earth System Research Laboratory, 2014.; Page 449, Figure 18.14:

U.S. DOE. 2008. Carbon Cycling and Biosequestration: Report from the March 2008 Workshop, DOE/SC-108, U.S. Department of Energy Office of Science, pp. 2–3. http://genomicscience.energy.gov.; Page 449, Table 18.2: IPCC AR5; recent data from CDIAC.; Page 450, Figure 18.15: Based on T. Stocker et al., Eds. "Summary for Policy Makers". Climate Change 2014: The Physical Science Basis, Contribution of Working Group I to the Fourth Assessment Report of the Intergovernmental Panel on Climate Change (Cambridge, England: Cambridge University Press, 2014 pg14.; page 451, Figure 18.16a-b: Mann, MichaelL E.; Kump, Lee R., *Dire Predictions: Understanding Climate Change*, 2nd Ed., ©2016, pp.72–73. Reprinted and Electronically reproduced by permission of Pearson Education, Inc., New York, NY.; Page 453, Figure 18.17: T. Stocker et al., Eds. Climate Change 2014: The Physical Science Basis, Contribution of Working Group I to the Fifth Assessment Report of the Intergovernmental Panel on Climate Change (Cambridge, England: Cambridge University Press, 2014), Figure SPM. 4; Page 455, Figure 18.19: CSSP, Global Climate Change Impacts in the United States: Unified Synthesis Product, 2009.; Page 456, Figure 18.21: Based on Pew Research Center. More Say there is Evidence of Global Warming. 2012, http://www.peoplepress.org/files/legacy-pdf/10-15-12%20 Global%20Warming%20Release.pdf and John Cook et al. "Quantifying the consensus on anthropogenic global warming in the scientific literature," Environmental Research Letters 8 (2) 2014.; Page 458, Table 18.3: Pearson Education, Inc.; Page 459, Table 18.4: Pearson Education, Inc.; Page 460: NAS, Climate Intervention: Carbon Dioxide Removal and Reliable Sequestration (2015) Washington, D.C.:National Academies Press. And NAS,Climate Intervention: Reflecting Sunlight to Cool Earth. (2015) Washington, D.C.:National Academies Press.; Page 461, Table 18.5: Pearson Education, Inc.; Page 462, Figure 18.24: Climate Action Tracker, Dec. 11, 2011. Copyright © 2009 by Ecofys and Climate Analytics.

Chapter 19 Page 466, Figure 19.1: Source: CCSP 1.1. Temperature Trends in the Lower Atmosphere: Steps for Understanding and Reconciling Differences. Thomas R. Karl, Susan J. Hassol, Christopher D. Miller, and William L. Murray, Eds., 2006. A Report by the Climate Change Science Program and the Subcommittee on Global Change Research, Washington, DC.; Page 467, Figure 19.2: Pearson Education, Inc.; Pages 467–477: Charles Dickens. *Hard Times*. Serialized April 1854–12 August 1854; Bradbury & Evans, 1854.; Page 468, Figure 19.3a-b: Pearson Education, Inc.; Page 469, Figure 19.5: Pearson Education, Inc.; Page 470, Table 19.1: Pearson Education, Inc.; Page 471, Figure 19.6: Pearson Education, Inc.; Page 472, Figure 19.8: Source: EPA National Emissions Inventory, 2012. http://www.epa.gov/ttn/chief/trends/index. html; Page 472, Figure 19.9: Source: EPA National Emissions Inventory, 2012. http://www.epa.gov/ttn/chief/trends/index.html.; Page 473, Figure 19.10: Source: EPA Air Trends, 2012. http://epa.gov/airtrends/aqtrends.html; Page 474, Figure 19.11: Pearson Education, Inc.; Page 475, Figure 19.12: Pearson Education, Inc.; Page 475, Table 19.2: Pearson Education, Inc.; Page 476, Figure 19.13: Pearson Education, Inc.; Page 477, Figure 19.14a: National Atmospheric Deposition Program/National Trends Network. Http://nadp.isws.illinois.edu; Page 477, Figure 19.14b: National Atmospheric Deposition Program/National Trends Network. Available at http://nadp.sws.uiuc. edu/maplib/pdf/2012/pH_2012.pdf; Page 477, Figure 19.15: Data from World Health Organization and Yale University Environmental Performance index- Air Pollution Map. 2014; Page 479, Figure 19.16a: Pearson Education, Inc.; Page 481, Figure 19.19: Source: U.S. Department of Agriculture Agricultural Research Service: Effects of Ozone Air Pollution on Plants. 2009. www.ars.usda.gov/Main/docs.htm?docid=12462.; Page 482, Figure 19.20: Pearson Education, Inc.; Page 483, Table 19.3: Source: U.S. EPA, Office of Air Quality Planning and Standards, National Ambient Air Quality Standards. October 2011. www/epa.gov/air/criteria.html.; Page 485: Sources: R. Agrawal et al. Atmospheric Brown Clouds: Regional Assessment Report With Focus on Asia. (2008) UNEP: Nairobi. Kenya. University of Maryland Center for Environmental Science. "Clean Air Act has led to improved water quality in the Chesapeake Bay watershed." ScienceDaily, 6 November 2013. American Chemical Society. "Green plants reduce city street pollution up to eight times more than previously believed." ScienceDaily, 18 July 2012. University of Southampton. "How trees clean the air in London." ScienceDaily, 5 October; Page 486, Figure 19.23: Pearson Education, Inc.; Page 487, Figure 19.24: Source: EPA, Office of Transportation and Air Quality. Light-Duty Automotive Technology, Carbon Dioxide Emissions, and Fuel Economy Trends: 1975 Through 2011. EPA-420-R-12-001a, March 2012. April 30, 2012.; Page 488, Figure 19.25: Source: EPA CSAPR Fact Sheet. www.epa.gov/airtransport/statesmap.html. May 2, 2011.; Page 489, Figure 19.26: Source: Executive Office of the President. National Acid Precipitation Assessment Program Report to Congress 2011: An Integrated Assessment. USGS, 2011. http://ny.water.usgs.gov.; Page 491, Figure 19.27: Pearson Education, Inc.; Page 491, Figure 19.28: Source: Synthesis and Assessment Product 2.4. Report by the U.S. Climate Change Science Program and the Subcommittee on Global Change Research. November 2008.; Page 493, Figure 19.29: Pearson Education, Inc.; Page 494, Figure 19.30: NASA Ozone Hole Watch. http://ozonewatch.gsfc.nasa.gov/. May 15, 2012.; Page 495, Figure 19.31: Source: UN Environment Program, Scientific Assessment of Ozone Depletion: 2006.

Chapter 20 Page 500, Figure 20.2: Pearson Education, Inc.; Page 498: Pearson Education, Inc.; Page 500, Table 20.1: Pearson Education, Inc.; Page 501, Figure 20.4: Reprinted from Progress on Drinking Water and Saitaion, 2014 Update. World Health Organization and UNICEF 2014. Available at http://www.unicef.org/gambia/ Progress_on_drinking_water_and_sanitation_2014_ update.pdf; Page 501, Figure 20.4: Reprinted from Progress on Drinking Water and Saitaion, 2014 Update. World Health Organization and UNICEF 2014. Available at http://www.unicef.org/gambia/Progress_ on_drinking_water_and_sanitation_2014_ update.pdf; Page 502, Figure 20.6: Source: Nancy Rabalais, Louisiana Universities Marine Consortium; R. E. Turner, Louisiana State University, with funding from NOAA, Center for sponsored Coastal Ocean Research and USA EPA Gulf of Mexico Program. Reprinted with permission.; Page 503, Figure 20.7: Pearson Education, Inc.; Page 505, Figure 20.9a-b: Pearson Education, Inc.; Page 508, Figure 20.11: Pearson Education, Inc.; Page 510, Figure 20.13: Pearson

Education, Inc.; Page 512, Figure 20.16: Source: Based on http://www.catawbariver-keeper.org/issues/library-of-documents-on-sewage-issuesand-treatment/septictankdia-gram.jpg/view?searchterm=septic+system; Page 513, Figure 20.17: Ecofilms Australia, "Tania's Compost Toilet Tour" August 6, 2010. Available at http://www.ecofilms.com. au/tanias-compost-toilet-tour/.; Page 514, Figure 20.18: Pearson Education, Inc.; Page 515: Source: L. Deegan et al. "Coastal eutrophication as a driver of salt marsh loss", Nature, 490 (2012):388–392.); Page 516, Figure 20.19: Pearson Education, Inc.; Page 518: Clean Water Act, 1972. Enacted by the 92nd United States Congress; Page 519, Table 20.2: Source: U.S. Environmental Protection Agency, Guidance for Water Quality-Based Decisions: The TMDL Process, EPA440-4-91-001 (Washington, DC:USEPA, 1991).; Page 521, Table 20.3: Pearson Education, Inc.

Chapter 21 Page 524: Pearson Education, Inc.; Page 525, Figure 21.1: Data from EPA, Office of Solid Waste, Municipal Solid Waste in the United States: 2010 Facts andFigures (Washington, DC: EPA, November, 2011).; Page 526, Figure 21.2a-b: Source: Based on data from D. Hoornweg and P. Bhada-Tata. What a Waste: a Global Review of Solid Waste Management. (2012) Washington, D.C.: World Bank pg. 12.; Page 526, Figure 21.3: Source: Based on data from D. Hoornweg and P. Bhada-Tata. What a Waste: a Global Review of Solid Waste Management. (2012) Washington, D.C.: World Bank Figure 9 pg. 20.; Page 527, Figure 21.4: Based on Project Oceanus. The Great Pacific Garbage Patch. Available at: https://projectoceanus.wordpress. com/2014/04/04/the-great-pacific-garbage-patch-in-a-nutshell/.; Page 530, Figure 21.7: Pearson Education, Inc.; Page 531, Table 21.1: James E. McCarthy. Interstate Shipment of Municipal Solid Waste: 2007 Update, CRS Report for Congress RL34043 (Washington, DC: Congressional Research Service, June 13, 2007).; Page 532, Figure 21.9: Pearson Education, Inc.; Page 533, Figure 21.10: EPA. Non-Hazardous Waste Management Heirarchy. www.epa.gov/epawaste/nonhaz/municipal/wte/nonhaz.htm. March 19, 2012.; Page 533: Environmental Protection Agency; Page 534: TakeOut Without. Found at http://takeoutwithout.org/behind-the-campaign/ and J. Fernandez and D. Ingber, "Manufacturing of large-scale functional objects using biodegradable chitosan bioplastic." Macromolecular Materials and Engineering (2014) doi:10.1002/ mame.201300426.; Page 536, Figure 21.13: EPA, Office of Solid Waste and Emergency Response, Municipal Solid Waste Generation, Recycling, and Disposal in the United States: Facts andFigures for 2012 (Washington, DC: EPA).; Page 538, Figure 21.15: Data from the Science Media group, the American Chemical Association, and http://www. tupperware.com.sg/images/quality_assurance_warranty/decoding_chart.jpg

Chapter 22 Page 545: Pearson Education, Inc.; Page 546, Figure 22.1: Pearson Education, Inc.; Page 547, Figure 22.2: Pearson Education, Inc.; Page 548, Figure 22.3: Pearson Education, Inc.; Page 548: U.S. Environmental Protection Agency. 2013 Toxics Release Inventory: National Analysis Overview. April 13, 2015. http://www. epa.gov/tri.; Page 549, Figure 22.4: Source: TRI National Analysis 2013: Introduction. Total Disposal or Other Releases by Industry, 2003-2013. p. 27. United States Environmental Protection Agency. Available at: www2.epa.gov/toxics-release-inven-tory-tri-program/2013-tri-national-analysis; Page 549, Figure 22.5: Pearson Education, Inc.; Page 550: U.S. Department of Health and Human Services, Public Health Service, National Toxicology Program, Report on Carcinogens, 12th ed. (Washington, DC: U.S. Department of Health and Human Services, 2011).; Page 550, Table 22.1: Based on IPCS Assessment Report: L. Ritter, et al., Persistent Organic Pollutants: An Assessment Report on DDT, Aldrin, Dieldrin, Endrin, Chlordane, Heptachlor, Hexachlorobenzene, Mirex, Toxaphene, Polychlorinated Biphenyls, Dioxins, and Furans, International Programme on Chemical Safety, December 1995. Available at http://www.chem.unep.ch/Pops/alts02.html.; Page 553: U.S. Department of Health and Human Services. Center for the Evaluation of Risks to Human Reproduction. NTP—CERHR Monograph on the Potential Human Reproductive and Developmental Effects of Bisphenol A. NIH Publication No. 08—5994. September 2008.; Page 553: Pearson Education, Inc.; Page 553: Pearson Education, Inc.; Page 557, Figure 22.12: Data from U.S. Environmental Protection Agency (EPA) CERCLIS database.; Page 559, Figure 22.13: Pearson Education, Inc.; Page 560, Figure 22.14: Pearson Education, Inc.; Page 562, Figure 22.16: Pearson Education, Inc.; Page 563: Environmental Protection Agency, Final Guidance for Incorporating Environmental Justice Concerns in EPA's NEPA Compliance Analyses (April 1998). Available at www.epa.gov.

Chapter 23 Page 568: Map of Atlanta Beltline courtesy of Atlanta Beltline, Inc. Reprinted with permission.; Page 569, Figure 23.1: Based on World Urbanization Prospects 2014 Revision. New York: United Nations Figure 3 pg. 8. Available at http:// esa.un.org/unpd/wup/Highlights/WUP2014-Highlights.pdf; Page 570, Figure 23.2: World Urbanization Prospects 2014 Revision. New York: United Nations Figure 8 pg. 13. Available at http://esa.un.org/unpd/wup/Highlights/WUP2014-Highlights.pdf; Page 572, Figure 23.5: Source: Tracey Farrigan, Thomas Hertz, and Timothy Parker Rural Poverty & Well-being, USDA, Economic Research Service, May 2015. Data from USDA, Economic Research Serice using Current Population Survey Data.; Page 574: Reid Ewing, Rolf Pendall, and Don Chen, Measuring Sprawl and Its Impact (Washington, DC: Smart Growth America, 2002). Available at www.smartgrowthamerica.org/ research/measuring-sprawl-and-its-impact; Page 575, Figure 23.9: Reid Ewing, Rolf Pendall, and Don Chen. Measuring Sprawl and Its Impact (Washington, DC: Smart Growth America, 2002).; Page 575, Table 23.1: Data from Population Division, U.S. Bureau of the Census, 2012.; Page 577, Figure 23.12: U.S. Department of Agriculture. 2007 National Resources Inventory, Natural Resources Conservation Service, Washington, DC, and Center for Survey Statistics and Methodology, Iowa State University, Ames, Iowa. July 2009.; Page 577, Table 23.2: Reid Ewing, Rolf Pendall, and Don Chen. Measuring Sprawl and Its Impact (Washington, DC: Smart Growth America, 2002).; Page 579: United Nations Sustainable Development Goals; Page 579, Table 23.3: Pearson Education, Inc.; Page 581: Pearson Education, Inc.; Page 583, Figure 23.18: EPA "Chapter 1 Urban Heat Island Basics" in Reducing Urban Heat Islands: Compendium of Strategies. Washington, D.C.:2008.EPA. pg 4.; Page 588: Edmund Burke, Irish orator, philosopher, and politician (1729–1797)

Index